BIOLOGY OF THE HETEROMYIDAE

SPECIAL PUBLICATIONS

This series, published by the American Society of Mammalogists, has been established for papers of monographic scope concerned with some aspect of the biology of mammals.

Correspondence concerning manuscripts to be submitted for publication in the series should be addressed to the Editor for Special Publications, Michael A. Mares (address below).

Copies of Special Publications of the Society may be ordered from the Secretary-Treasurer, H. Duane Smith, 501 Widtsoe Bldg., Dept. Zoology, Brigham Young University, Provo, UT 84602.

COMMITTEE ON SPECIAL PUBLICATIONS

BIOLOGY OF THE HETEROMYIDAE

Edited By

HUGH H. GENOWAYS
University of Nebraska State Museum
212 Morrill Hall
Omaha, Nebraska 68588

JAMES H. BROWN
Department of Biology
University of New Mexico
Albuquerque, New Mexico 87131

SPECIAL PUBLICATION NO. 10
THE AMERICAN SOCIETY OF MAMMALOGISTS
Published 20 August 1993

Library of Congress Catalog Card No. 93-72339

ISBN No. 0-935868-66-6

COMPLETE LIST OF AUTHORS

Troy L. Best
Department of Zoology & Wildlife Sciences
331 Funchess Hall
Auburn University
Auburn, AL 36849-5414

Janet K. Braun
Oklahoma Museum of Natural History
University of Oklahoma
Norman, OK 73019

James H. Brown
Department of Biology
University of New Mexico
Albuquerque, NM 87131

P. Brylski
12 Redwood Tree Lane
Irvine, CA 92715

James N. Derr
Animal Genetics
Animal Science Department
Texas A&M University
College Station, TX 77843

John F. Eisenberg
Florida State Museum
University of Florida
Gainesville, FL 32611

Theodore H. Fleming
Department of Biology
University of Miami
Coral Gables, FL 33124

G. Lawrence Forman
Department of Biology
Rockford College
Rockford, IL 61108

Alan R. French
Department of Biological Sciences
State University of New York
Binghamton, NY 13901

Hugh H. Genoways
University of Nebraska State Museum
Morrill Hall
University of Nebraska
Lincoln, NE 68588-0338

John C. Hafner
Moore Laboratory of Zoology
Department of Biology
Occidental College
Los Angeles, CA 90041

Barbara A. Harney
Department of Ecology and Evolutionary Biology
University of Arizona
Tucson, AZ 85721

Thomas Jones

Douglas M. Lay
Department of Anatomy
University of North Carolina
Chapel Hill, NC 27514

Robert E. Lewis
Department of Entomology
Iowa State University
Ames, IA 50011

Michael A. Mares
Oklahoma Museum of Natural History
University of Oklahoma
Norman, OK 73019

James L. Patton
Museum of Vertebrate Zoology
University of California
Berkeley, CA 94720

Carleton J. Phillips
Department of Biological Sciences
Illinois State University
Normal, IL 61761-6901

Mary V. Price
Department of Biology
University of California
Riverside, CA 92521

O. J. Reichman
Division of Biology
Kansas State University
Manhattan, KS 66506

Duke S. Rogers
Department of Zoology
Brigham Young University
Provo, UT 84602

Victor Sanchez Cordero
Instituto de Biologia
U.N.A.M.
Apartado Postal 70-04510
Mexico, D.F. 04510
MEXICO

David J. Schmidly
Office of the Campus Dean
Texas A&M University at Galveston
Galveston, TX 77553

John H. Wahlert
Department of Vertebrate Paleontology
American Museum of Natural History
Central Park West at 79th Street
New York, NY 10024

John O. Whitaker
Department of Life Sciences
Indiana State University
Terre Haute, IN 47809

Daniel F. Williams
Department of Biological Sciences
California State University–Stanislaus
Turlock, CA 95380

Kenneth T. Wilkins
Department of Biology
Baylor University
Waco, TX 76798

William J. Wrenn
Department of Biology
University of North Dakota
Grand Forks, ND 58202

PREFACE

What is the family Heteromyidae? It is the group of exclusively New World rodents that includes the kangaroo rats, pocket mice, and kangaroo mice of the desert, grasslands, and shrublands of western North America as well as the spiny and pouched rats of the subtropical shrublands and tropical deciduous and evergreen forests of North America, Central America, and northernmost South America. It is a large family, with six genera and 316 species.

Why are these rodents interesting enough to warrant an entire book to review what is known about their biology? The first and most important reason is because heteromyids have served as model organisms for many kinds of biological studies. In their adaptive radiation to colonize diverse temperate and tropical environments, they have acquired specialized morphological, physiological, and behavioral characteristics. In their evolutionary diversification in heterogeneous landscapes, they provide examples of genetic, ecological, and biogeographic processes that have played key roles in speciation and differentiation. In their ecological responses to spatially and temporally varying environments, they offer insights into the mechanisms that control population dynamics and determine community composition. Studies of heteromyid rodents have made seminal contributions to comparative biology, and most of these important discoveries are reviewed in this volume.

A second reason for this review is to synthesize the available information on all aspects of heteromyid biology. Much of what is known about heteromyids was learned by specialists, who elected to study particular species that offered a "good system" for addressing questions in disciplines such as renal physiology, cytogenetics, biomechanics, or community ecology. As important as these studies have been, it is easy to lose sight of the fact that they all describe attributes of living species that make a living in real environments. This book attempts not only to review what is known about heteroymid biology, it also tries to synthesize this information to provide a more integrated view of the lives of these interesting rodents.

A final reason to review the biology of heteromyids is because these special animals deserve a wider audience. Although these rodents have been discovered by many comparative biologists seeking to answer diverse questions, they remain largely unappreciated. Heteromyids belie the stereotypes usually associated with the terms rats and mice. For the most part they are clean, beautiful, and mild-tempered. They live in some of the wildest and most scenic regions of the Western Hemisphere. They have the potential to continue to serve as model systems and to allow researchers to address important new questions. By pointing out what is still unknown, as well as what is known, this book may serve to attract new scientists to the ranks of those who love these rodents for their own sakes as well as for their possibilities for contributing to new discoveries.

The idea to produce this volume began several years ago. The publication in 1968 of an influential volume on *Peromyscus* (American Society of Mammalogists Special Publication No. 2) showed the value of reviewing and synthesizing information on particular taxonomic groups of mammals that have been the subject of much research. With the success of a volume on New World *Microtus* (Special Publication No. 8; 1985), mammalogists who worked on heteromyid rodents began to talk about the desirability of producing a Special Publication devoted to their biology. Making an entire family the subject of this volume obviously represents broader coverage than the previous volumes devoted to single genera of New World rodents. This book is able to call attention to the spectacular diversity of form, function, and environmental relationships that has resulted from the adaptive radiation of the Heteromyidae.

We wish to thank all those individuals who contributed to the production of this volume. Many of these people are acknowledged by the authors in their chapters. In addition to these individuals, we first thank the authors of the chapters for their efforts to prepare broad, careful, authoritative reviews. We appreciate their patience with the delays in publication and their willingness to update their chapters to keep the information current. Second, we thank all those individuals who have improved this volume by preparing constructively critical reviews of early drafts of the chapters. We are especially grateful to R. J. Baker, A. D. Barnosky, S. Berman, M. A. Bowers, J. S. Brown, M. D. Carleton, J. F. Eisenberg, M. D. Engstrom, D. J. Hafner, J. C. Hafner, M. S. Hafner, D. S. Hinds, M. L. Kennedy, B. P. Kotler, R. T. M'Closkey, M. A. Mares, J. E. Martin, J. C. Munger, C. J. Phillips, W. B. Quay, J. A. Randall, O. J. Reichman, V. Sarich, D. A. Schlitter, D. O. Straney, R. M. Sullivan, R. M. Timm, and S. D. Webb, who prepared careful reviews for the editors. Third, we thank David I. Rasmussen, who has enlivened this book with delightful sketches that capture both the attractiveness of heteromyids and interesting features of their biology. Finally, we thank the many members, officers, and editors of the American Society of Mammalogists for their support of this volume and their contributions to its production. We are especially grateful to Michael A. Mares, Editor for Special Publications, and Don E. Wilson and Craig S. Hood, Managing Editors for Mammalian Species and Special Publications, for their work in preparing the final manuscript for the press.

HUGH H. GENOWAYS
Lincoln, Nebraska

JAMES H. BROWN
Albuquerque, New Mexico
January 1993

CONTENTS

THE FOSSIL RECORD

John H. Wahlert

Introduction

This chapter presents a survey of the literature in order to set forth the currently recognized taxa and to identify problems in systematics of the fossil Heteromyidae.

The text is subdivided into topics that illustrate the peculiarities of paleontological data. For each genus the type specimen of the type species is of paramount importance; most of the older types lack detailed stratigraphic data, and there is no guarantee that other specimens, even from the same locality, were near contemporaries. Mandibular fragments containing an incomplete dentition are common remains; upper and lower dentitions found in occlusion are extremely rare. Since type specimens are never complete, diagnosis and description usually include information derived from other specimens found at the same locality; for this reason I have listed the describer's original series of specimens. The temporal and geographic distribution of a genus tells not only when and where it lived, but also points up any discontinuity in its known history. Although the records of taxa in faunal lists are included here, these papers are usually not cited, because they contain no biological information. The bibliographic sources that give access to this literature are noted below.

The revised generic diagnoses are the most problematic part of this paper for two reasons: 1. Describers of new taxa are not consistent in coverage of morphology; 2. Taxonomic boundaries are not sharply drawn because no one has put all the available specimens together in recent years for side by side comparison. I have not redefined Recent taxa, as these are the subjects of other chapters in this volume; Wood (1935*a*) presented thorough descriptions. The list of named species cites the literature in which illustrations of the type specimen and referred material may be seen and gives the figure numbers in those papers. I have not noted republications of original figures. Synonymies of species are indicated in parenthetic reference to original figure captions. Identified skeletal remains are extremely rare and are separately noted; much more material resides in museum collections than has been indicated in print, but identification is uncertain except in the few specimens that include cranial remains. The comments reflect my particular interest in the history of a taxon, its variation, and hypotheses of its relationships. Generic diagnoses of extinct taxa overlap to a degree

Dedicated to the memory of Marvin A. Feldman, who was a fine student.

that, at this point, makes familial definitions unclear.

The literature that is the basis for information about extinct taxa was found primarily through the bibliographies of vertebrate paleontology published by the Geological Society of America and the Society of Vertebrate Paleontology. Three other sources present excellent summaries of the Pleistocene record: Kurtén and Anderson (1980), Lundelius et al. (1983), and Webb (1984). Bibliographic coverage is relatively thorough through 1986, when the original manuscript was completed. I have since added new information that has been a part of my own ongoing research.

Phylogenetic hypotheses and classifications of extinct and living heteromyids are diverse. Wood (1935*a*, fig. 1; 1939) split the Heteromyidae into two groups—the combined Perognathinae and Dipodomyinae versus the Heteromyinae—that extended back to different species of *Heliscomys* in the Oligocene. The subfamilies included the following genera:

Perognathinae:
- *Mookomys*
- *Perognathoides* [=*Cupidinimus*]
- *Perognathus*
- *Microdipodops*

Dipodomyinae:
- *Cupidinimus*
- *Dipodomys*

Heteromyinae:
- *Proheteromys*
- *Peridiomys*
- *Diprionomys*
- *Liomys*
- *Heteromys*

Wood did not assign *Heliscomys* to a subfamily.

Reeder (1957, fig. 91) placed no genus in the ancestral position for his three clades of heteromyids. A classification can be derived from his phylogeny:

Perognathinae:
- *Heliscomys*
- *Mookomys*
- *Apletotomeus*
- *Akmaiomys* [=*Proheteromys*]
- *Trogomys*
- *Perognathus*

Dipodomyinae:
- *Cupidinimus*
- *Perognathoides* [=*Cupidinimus*]
- *Diprionomys*
- *Prodipodomys*
- *Dipodomys*
- *Microdipodops*

Heteromyinae:
- *Proheteromys*
- *Peridiomys*
- *Liomys*
- *Heteromys*

I have omitted Reeder's unpublished taxa.

Lindsay (1972, fig. 38) followed Shotwell's (1967:50) suggestion that the subfamilies of heteromyids and geomyids be united in a single family in recognition of their close relationship. A classification can be derived from his phylogeny:

Entoptychinae

Heteromyinae:
- *Pleurolicus*
- *Proheteromys*
- *Peridiomys*
- *Diprionomys*
- *Liomys*
- *Heteromys*

Perognathinae:
- *Heliscomys*
- *Mookomys*
- *Hitonkala*
- *Perognathus*

Geomyinae

Dipodomyinae:
- *Cupidinimus*
- *Perognathoides* [=*Cupidinimus*]
- *Trogomys*
- *Dipodomys*
- *Microdipodops*

The three classifications share core taxa in each of the subfamilies. Instability arises with *Perognathoides, Diprionomys, Trogomys,* and *Microdipodops.* Despite apparent consistency of core taxa, a basic question has never been asked of most of the extinct genera: Are they indeed heteromyids? The most important heteromyid characters, such as presence of a rostral perforation and of an interorbital distance that is wider than the rostrum, are rarely preserved. Many recorded features of the cheek teeth, such as bilophodont condition with lophs widened by styles, are derived for all of the Geomyoidea (Wahlert, 1985). In many cases the definition of heteromyid seems to be nothing more than: Small rodent with primitive geomyoid crown patten and tooth count I 1/1, P 4/4, M 1-3/1-3.

The relationship of the Heteromyidae to other rodents is firmly established. There is no question today that the Heteromyidae and Geomyidae are each other's closest living relatives. As late as 1945, however, Simpson placed their sister group, the extinct Eomyidae, in the Superfamily Aplodontoidea. Wilson (1949*a*) in a careful analysis of primitive and specialized characters showed that the eomyids are closely related to the living geomyoids. Harris and Wood (1969) proposed that heteromyids were derived from eomyids, perhaps via *Meliakrouniomys.* Hill (1937) found many specializations of the skull, skeleton, and soft anatomy that are shared by geomyoid and muroid rodents. Wilson (1949*a*), Bugge (1985), Luckett (1985), and Wahlert (1985) added characters to this list. The hypothesis that the Eomyoidea and Geomyoidea are members of the Suborder Myomorpha is widely, but not universally, accepted; Fahlbusch (1985), the foremost student of eomyid dental evolution, remains a dissenting voice. Hartenberger (1985) united the Ctenodactyloidea and Geomyoidea, including Eomyidae and Sciuravidae, in a clade that is the sister group to the rest of the Rodentia. The association is based on a similarity in tooth patterns of early forms. To my mind this construction points up the limitation of phylogenetic hypotheses that treat high level taxa on the sole basis of dentition.

Although I follow a cladistic approach to phylogenetic analysis in my own research, I have not altered the language of the many authors cited to fit this mold. In many paleontological works some taxa are based on shared, primitive characters and are admittedly paraphyletic; antiquity is used as a criterion of primitiveness and obviously as a critical factor in determining ancestral status. I have not hesitated to point out that certain genera, as presently constituted, may be ancestral to other taxa. By this I mean that, were the remains of such a genus better known, its species might be partitioned in different clades under other generic names. An ancestral genus is a paraphyletic taxon.

Some papers are so generally important that they merit introductory mention. Wood (1935*a*) presented a thorough study of both extinct and living heteromyid taxa. In addition to adult dentitions and crania, he compared and illustrated deciduous premolars, carpus, and tarsus in those genera in which they are known; he presented comparative tables of measurements that include interparietal widths, dimensions of appendicular elements, and limb proportions. Reeder (1957) compared the fossil remains and is an excellent source of data and ideas; line drawings of cheek teeth in lateral views and of deciduous premolars in crown and lateral views are important, but the scale is not given in the figures or captions, and catalogue numbers of figured specimens are not noted. The many photographs are useless as reproduced. Barnosky (1986*a*) revised the genus *Cupidinimus* and measured crown height of the cheek teeth following the system devised by Rensberger (1971) for the entoptychine geomyids. Shotwell (1967) figured tibiae-fibulae, calcanea, and astragali of several living and extinct taxa. Stehlin and Schaub (1951), the easiest reference for figures of rodent dentitions, is derivative and reproduces illustrations of heteromyid teeth chiefly from Wood.

Abbreviations

Institutions

AMNH—American Museum of Natural History, New York

BS—Biological Survey Collections, Nation-

al Museum of Natural History, Washington, D.C.
CM—Carnegie Museum of Natural History, Pittsburgh
KUMNH—University of Kansas, Museum of Natural History, Lawrence
LACM(CIT)—Los Angeles County Museum (Carnegie Institute of Technology), Los Angeles
MCZ—Museum of Comparative Zoology, Harvard University, Cambridge
SDSM—Museum of Geology, South Dakota School of Mines and Technology, Rapid City
TMM—Texas Memorial Museum, Austin
UCMP—University of California Museum of Paleontology, Berkeley
UF/FGS—University of Florida/Florida Geological Survey, Gainesville
UGV—University of Georgia, Geology Department, Athens
UMMP—University of Michigan Museum of Paleontology, Ann Arbor
USNM—National Museum of Natural History, Washington, D.C.
UWBM—University of Washington, Burke Museum, Seattle

Systematics and the Fossil Record

Except where differences are noted, the following statements could be inserted in the descriptions of any genus below, except *Meliakrouniomys*: Cheek teeth bilophodont; primary cusps and stylar cusps arranged in transverse rows of three cusps each; lophs separated by transverse valley; lophs of uppers widened lingually by stylar cusps, with the usual exceptions of a single cusped protoloph in P4 and a two cusped metaloph in M3; lophids of lowers widened buccally by stylar cuspids, with the usual exceptions of a two cusped metalophid in p4 and hypolophid in m3.

All of the Geomyoidea have bilophodont cheek teeth in which transverse lophs are widened by the addition of styles. Although this character commonly appears in published diagnoses of extinct heteromyid genera, it does not demonstrate that a taxon is, in fact, a heteromyid. Crown height, an important character, is not described in the same way by various authors and is impossible to use here comparatively. Lindsay (1972) composed an excellent generalized description of the cheek teeth of heteromyids from the Barstow Formation.

Suborder Myomorpha Brandt, 1855
Infraorder Geomorpha Thaler, 1966
Superfamily Eomyoidea
Deperet and Douxami, 1902
Family Eomyidae
Deperet and Douxami, 1902
Meliakrouniomys
Harris and Wood, 1969

Type species.—*Meliakrouniomys wilsoni* Harris and Wood, 1969, TMM 40283-80, Chadronian Ash Spring local fauna, Vieja Group, Jeff Davis Co., Texas; partial left mandible with i(broken)-m3.

Temporal and geographic distribution.—OLIGOCENE: Chadronian: Texas, Wyoming.

Revised generic diagnosis.—Eomyid rodent with crown length of p4-m3 6.60 (*M. skinneri*) to 7.15 (*M. wilsoni*) mm. All teeth bilophate, each loph two cusped; minute posterior cingula on m1-2 with slight-largements into hypoconulids. Upper and lower incisors narrow and deep with rounded anterior faces.

Species.—
Meliakrouniomys wilsoni Harris and Wood, 1969, fig. 1.
M. skinneri Emry, 1972, figs. 1, 2.

Comment.—*Meliakrouniomys wilsoni* has a posterior cingulid on p4 but no anterior cingulid; *M. skinneri* exhibits the opposite condition. Anterior cingulids on m1-3 and short posterior cingulids on m1-2 are present in both species.

Harris and Wood (1969:5) said that "the most probable phylogenetic position for

Meliakrouniomys seems to be an eomyid on the way to becoming a heteromyid." The loss of mesoconid and ectolophid (absent in this genus) and reduction of cingulids from this or a more common eomyid crown would be necessary steps toward the heteromyid pattern. The buccal stylids or cingulid seen in heteromyids and geomyids are absent in eomyids including *Meliakrouniomys.* Harris and Wood (1969) and Wood (1974, 1980) referred the genus to the Eomyidae. Emry (1972) found similarities to heteromyids more compelling. Wahlert (1985:14) proposed "transverse lophs of upper and lower cheek teeth widened by styles" as a diagnostic character of the Geomyoidea. In his system *Meliakrouniomys* would be either an eomyid or a branch between the eomyoid and geomyoid clades. Wahlert and Koenigswald (1985) found a complex, 3-layered Schmelzmuster in lower incisors of European and North American eomyids. Geomyoids have the simpler and more primitive enamel type seen in muroids. Knowledge of the incisor enamel microstructure in *Meliakrouniomys* could answer the question of its affinity.

Skeletal remains.—Harris and Wood (1969) described the mandible of the type specimen. Emry (1972) illustrated a partial rostrum. Although Emry found the morphology to be heteromyid-like, Wood (1980) remarked that it is little different from the eomyid *Viejadjidaumo magniscopuli.*

Superfamily Geomyoidea Bonaparte, 1845

Heliscomys Cope, 1873

Type species.—*Heliscomys vetus* Cope, 1873, AMNH 5461, Orellan Cedar Creek Member, White River Formation, northeastern Colorado; fragment of left mandible with i-m2. Original series: AMNH 5462, partial left mandible with i-m1. Note: The type locality cannot be known precisely on the basis of available evidence. Cope was not specific about the locality in his publications. Catalogue cards for the type and paratype specify Cedar Creek, Logan Co., Colorado. The label with the type (not with the paratype as specified by Reeder, 1957: 41) gives Pawnee Creek as the locality. Cope (1884) stated that he traveled east as far as Cedar or Horse Tail Creek on a route parallel to the Chalk Bluffs. He must have crossed Pawnee Creek. In the northernmost part of the state both Pawnee and Cedar Creeks pass from Weld into Logan County. There are not original data that answer the questions of which creek and which county are correct. Galbreath (1953) reported many specimens of *H. vetus* from the Cedar Creek Member of both counties.

Temporal and geographic distribution.—EOCENE: Duchesnean: Saskatchewan. OLIGOCENE: Chadronian: Montana, Wyoming, Saskatchewan; Orellan: Colorado, Montana, Nebraska, South Dakota, Wyoming; Whitneyan: Wyoming. MIOCENE: Arikareean: South Dakota; Hemingfordian: Saskatchewan.

Revised generic diagnosis (modified from Wood, 1980).—Tiny geomyoid rodent with very brachydont cheek teeth; crown length of P 4/4-M 3/3 ranging from 2.8 to 3.1 mm. M 1-2/1-2 roughly square with lingual cingulum continuous or divided and labial cingulid usually divided into two styles; anterior and posterior cingulum/id present, but posterior less prominent. Valleys between cusps slightly shallower than transverse valley between rows so that tranverse lophs form after considerable wear. Hypostylid lacking in p4; hypocone small or absent in M3.

Species.—

Heliscomys vetus Cope, 1873; Cope, 1884, pl. 65, figs. 14–18; Wood, 1931, fig. 2; Wood, 1933 and 1935*a*, fig. 7. Other specimens: Wood, 1935*a*, fig. 6a (*H. senex*); Galbreath, 1962, fig. 1; Sutton and Black, 1975, fig. 28; Setoguchi, 1978, figs. 23a–f; Storer, 1978, fig. 10a; Korth, 1989, fig. 1B–C. Chadronian, Orellan, Whitneyan.

H. gregoryi Wood, 1933, fig. 6. Orellan.

H. hatcheri Wood, 1935*a,* fig. 6b; Wood, 1939, figs. 1–11. Other specimens: Korth, 1989, fig. 1A. Orellan.
H. woodi McGrew, 1941, fig. 17. Arikareean.
H. mcgrewi Korth, 1989, fig. 1D. Orellan.

Comment.—The genus *Heliscomys* is distinct among geomyoid rodents only in its primitiveness, and I place it in the Geomyoidea, *incertae sedis.* Wood (1935*a*:98) described the upper dentition of *Perognathus*: "M1-3 have fundamentally the same pattern as in *Heliscomys gregoryi,* differing chiefly in the greater height of crown and more progressive lophization." The earliest and latest records of its temporal range were reported by Storer (1988) and Skwara (1988), respectively.

The species *Heliscomys vetus* is so variable dentally that both its limits and the validity of other species is uncertain. The upper and lower fourth premolar of *H. vetus* may have one or two anterior cusps. This variation has been recorded in specimens from the lower part of the Orellan Cedar Creek Member of the White River Formation (Galbreath, 1953), and from the Whitneyan Cedar Ridge local fauna (Setoguchi, 1978). Wood (1980) and Wahlert (1983) summarized the evidence that two anterior cusps is the primitive condition in the premolars. Black (1965:45) suggested that the persistence of variation throughout the Oligocene indicates that selection pressure for loss of the protoconid was weak. Setoguchi (1978) described variation in the presence and size of the lingual (upper) and labial (lower) stylar cusps; the tranverse valley in M1 may be blocked lingually by an anteriorly situated entostyle, or it may cleave the lingual cingulum; the cingulum closes the transverse valley in M2. It is unbroken in M3. In lower molars the connection of the labial cingulid to the protostylid and the continuation of the cingulid posterior to the hypostylid vary. There is no doubt that the stylar cusps are a derived feature; whether they arose directly or from a low, continuous cingulum in both the individual upper teeth and the lower teeth has not been demonstrated. Reeder (1957) said that the incisors are narrow in cross section and without ornament except for a somewhat flat anterior face in the lowers. Galbreath (1962) described and figured deciduous upper and lower premolars.

The growing sample size and knowledge of variation in the cheek teeth of *Heliscomys vetus* has led to questions about the validity of other species. Wood (1980:47) said that *H. senex* "almost certainly is a complicated variant of *Heliscomys vetus.*" Sutton and Black (1975) suggested that *H. gregoryi* may be a synonym of *H. vetus.* Korth (1989) observed strong similarity of the upper dentition of *H. hatcheri* and the illustration of the teeth of *H. gregoryi* (type specimen is lost); Korth and Tabrum (in press) present clear evidence that the two species are distinct. Wood (1939) described associated upper and lower dentitions of *H. hatcheri* and confirmed the presence, as in *Proheteromys,* of an accessory cuspule in lower molars at the anterobuccal corner, where the anterior cingulum meets the buccal stylar region; the number of anterior cusps in p4 varies from 1 to 2. Reeder (1957) noted that the diastema of the mandible is longer than in other species. Korth (1989) pointed out that the lingual cingulum in upper molars (M1-2) of *H. hatcheri* consists of two distinct styles, whereas in *H. vetus* it is a continuous feature. *H. woodi* is a distinct species because stylar cusps are weak and m1 is rectangular, clearly longer than wide. *H. mcgrewi* is the largest species of the genus. Korth (1989) suggested including it and *H. woodi* in a lineage, since the size of the premolars is greatly reduced in both.

Galbreath (1962) said that *Heliscomys tenuiceps* belongs in the subfamily Florentiamyinae. Wahlert (1983, 1984) pointed out the merit of this idea and transferred a similar species, *H. schlaikjeri,* to the florentiamyid genus *Kirkomys.* Korth (1989) made

H. tenuiceps Galbreath, 1948, the type species of a new florentiamyid genus *Ecclesimus.*

Cope (1873, 1884) did not assign *Heliscomys* to a higher taxon but felt that the dentition was most like that of genera in the Myomorpha. Zittel (1891–1893) appears to have been the first author to place *Heliscomys* in a taxon Geomyidae together with eomyids, heteromyids, and geomyids. Wood (1939:561) said: "there can be no question but that *Heliscomys* appears to be the most primitive known member not only of the Heteromyidae but also of the Geomyoidea." He proposed (1935*a*, 1937*b*) that the genus could be ancestral to any of the modern heteromyid subfamilies and even to the Geomyidae if the gap between it and the earliest geomyids were not so short. Later discoveries of *Heliscomys* specimens from older Oligocene and late Eocene strata have widened the gap. Wilson (1949*c*) believed that the single anterior cusp, accompanied or not by a weak second cusp, in premolars of *Heliscomys* was a case of reduction that removed the genus from the place of ancestor. Specimens found after his publication showed that the second anterior cusp can be prominent.

Korth, Wahlert, and Emry (in press) created a new family for *Heliscomys* and *Apletotomeus.* Since the dentitions and known skulls show the basic geomyoid synapomorphies, they placed the family as the sister group of the florentiamyids, heteromyids, and geomyids. A new diagnosis of *Heliscomys* includes the important cranial characters. *Heliscomys* retains a combination of primitive cranial features that have been modified in one or more of the other geomyoid rodent groups: posterior maxillary notch not enclosed as a foramen; sphenofrontal foramen present; rostrum imperforate medial to infraorbital foramen; superior angular process of mandible not flared laterally; cheek teeth fully cuspidate. The genus also possesses characters that in combination make it separable from any other geomyoid subfamily: incisive foramina elongated and depressed into diastemal palate; masticatory and buccinator foramina combined with accessory foramen ovale; mental foramen in mandible anterodorsal to tip of masseteric fossa. The authors showed *Heliscomys* as containing two species groups. *H. vetus, H. mcgrewi,* and *H. woodi* have a continuous lingual cingulum in the upper molars. The lingual cingulum is interrupted by the central valley in M1 of *H. gregoryi, H. hatcheri,* and a new species that includes specimens previously described by McGrew (1941, fig. 17B) and Black (1965, fig. 5f–h).

Skeletal remains.—Wood (1939) recorded a partial maxilla of *Heliscomys hatcheri.* Reeder (1957) noted a crushed skull of *H. vetus* that is presumably the basis for his description. Setoguchi (1978) figured a partial skull that he attributed to *H. vetus.* Wood (1939) described the proximal portion of a radius and the left manus in *H. hatcheri.* He concluded that the radius and manus are relatively primitive for heteromyids and that resemblances to *Perognathus* are shared, primitive characters.

Apletotomeus Reeder, 1960*a*

Type species.—*Apletotomeus crassus* Reeder, 1960*a,* UMMP 25893, Orellan Brule Formation, Bill Grimm Ranch, Sioux Co., Nebraska; partial left mandible with i-m3.

Temporal and geographic distribution.—OLIGOCENE: Orellan: Nebraska, South Dakota.

*Revised generic diagnosis (after Reeder, 1960*a*).*—Cheek teeth brachydont, cusps very low. Molars strongly sexcuspidate; slight basal conjunction of laterally adjacent cusps producing weak bilophodonty. p4 with four cusps, nearly quadrate and relatively large. Lower incisor sharply recurved; relatively much broader than those of *Heliscomys* and *Proheteromys*; broad, flat ante-

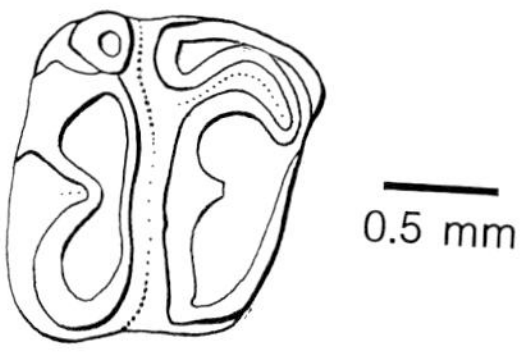

FIG. 1.—*Proheteromys nebraskensis,* MCZ 5051 (holotype). First lower molar, left side, showing Y-shaped median valley. Drawn from a cast. Anterior is to the right.

rior enamel face. Mandible massive; diastema short; masseteric crest shelf-like and terminating abruptly below mental foramen.

Species.—

Apletotomeus crassus Reeder, 1960*a,* figs. 206, 207, 210. Orellan.

Comment.—Black (1965:45) said that *Apletotomeus crassus* may be congeneric with *Heliscomys.* Wood (1980) transferred *A. crassus* to *Proheteromys.* Korth (1989: 35) noted that the complex p4 and massive incisor and mandible distinguish the species from all other Orellan heteromyids. Korth, Wahlert, and Emry (in press) place *Apletotomeus* and *Heliscomys* together in a new family that is the sister taxon of the rest of the Geomyoidea. The sum of crown lengths in lower cheek teeth is 3.65 mm.

Family Heteromyidae Gray, 1868
Proheteromys Wood, 1932

Type species.—*Proheteromys floridanus* Wood, 1932, UF/FGS V-5329, Hemingfordian Hawthorne Formation, The Fullers Earth Company Mine, Midway, Gadsden Co., Florida; fragment of left mandible with p4-m1; specimen lost. Original series: UF/FGS V-5330, M1 right; UF/FGS V-5331, M2 left.

Generic synonyms.—Possibly *Akmaiomys* Reeder, 1960*a.*

Temporal and geographic distribution.—OLIGOCENE: Orellan: Colorado, Nebraska, South Dakota; Whitneyan: Nebraska, Wyoming; Early Arikareean: South Dakota. MIOCENE: Late Arikareean: Florida, South Dakota; Hemingfordian: Florida, New Mexico, South Dakota, Wyoming; Barstovian: California, Colorado.

Revised generic diagnosis.—Small to medium sized heteromyid with brachydont cheek teeth; crown height similar to or higher than *Mookomys.* Crown length of P 4/4-M 3/3 in *Proheteromys floridanus* near 3.0 mm; range in most western species about 4.3 to 4.9 mm; *P. magnus* and probably *P. maximus* larger than 6.0 mm. Molars wider than long, especially upper teeth; primary cusps and stylar cusps subequal in height; cusps less prominent than in *Mookomys.* Transverse valleys much deeper than valleys separating cusps in transverse rows. Anterior and posterior cingula/ids short and variable; most likely to be present on M 1/1. P4 with single anterior cusp and metaloph joining entostyle to form J-shaped loph that is not present in *Heliscomys* and *Mookomys.* "Deep valley (Y-pattern) between external cingulum of lower molars and protoconid" (Wood, 1980:48) (Fig. 1). Hypostylid lacking or small (*P. sulculus*) in p4. Central connection of lophs common in p4 where it produces X-pattern with wear. *P. fedti* exceptional in that buccal stylids of lower molars appear only as a low ridge (Macdonald, 1963, fig. 14).

Species.—

Proheteromys floridanus Wood, 1932, fig. 24. Other specimens: Wood, 1932, figs. 25, 28; Wood, 1947, figs. 1–3; Black, 1963, fig. 4; Olsen, 1964, pl. 69, figs. a–d; Gawne, 1975, figs. 5a–b, 7. Hemingfordian.

P. parvus (Troxell, 1923), figs. 3–5 (*Diplolophus*); Wood, 1931, fig. 1 (*Mookomys*); Wood, 1935*a,* fig. 5. ?Barstovian.

P. magnus Wood, 1932, fig. 26. Other specimens: Wood, 1932, figs. 27, 29; Wilson, 1960, figs. 121–125; Black, 1963, fig. 5; Lindsay, 1974, fig. 8. Hemingfordian.

P. matthewi Wood, 1935*a,* fig. 96a. Arikareean.

P. thorpei Wood, 1935*b,* fig. 1. Arikareean.

P. nebraskensis Wood, 1937*b,* fig. 35. Other specimens: Wood, 1937*b,* fig. 36; Setoguchi, 1978, figs. 22a–d. Whitneyan.

P. sulculus Wilson, 1960, figs. 114a–c. Other specimens: Wilson, 1960, figs. 102–113; Lindsay, 1974, figs. 6, 7. Hemingfordian to Barstovian.

P. incohatus (Reeder, 1960*a*), figs. 208, 209, 211 (*Akmaiomys*). Orellan.

P. maximus James, 1963, fig. 42. Barstovian.

P. fedti Macdonald, J. R., 1963, fig. 14. Arikareean.

P. gremmelsi Macdonald, J. R., 1963, fig. 15. Arikareean.

P. bumpi Macdonald, J. R. 1963, fig. 16. Arikareean.

P. ironcloudi Macdonald, J. R., 1970, fig. 22. Arikareean.

P. cejanus Gawne, 1975, fig. 4. Other specimens: Gawne, 1975, fig. 5a–b. Latest Arikareean or earliest Hemingfordian.

P. sp. Macdonald, L. J., 1972, fig. 11; Gawne, 1975, fig. 8; Martin, J. E., 1976, fig. 5b–c.

Comment.—Wood (1932) defined the genus and included two species, *Proheteromys floridanus* and *P. magnus,* that are the smallest and largest currently recognized. Reeder (1957) pointed out that UF/FGS V-5334, originally considered a paratype of *P. magnus* by Wood, is, in fact, an m1 of *P. floridanus.* The few lower incisors that have been described are narrow in cross section with bowed anterior faces; the upper in *P. cejanus* is narrow and asulcate.

Proheteromys is such a speciose genus that a firm diagnosis is impossible. Wilson (1960) and Black (1963) commented on the highly variable morphology and size of cheek teeth in samples from single localities. Black found the degree of variability, particularly in premolars of *P. floridanus,* to be similar to that in *Heliscomys vetus.* Reeder (1957) had little confidence in the then current systematic definition of the taxon and pointed out features of the mandibles that may be important; his suggested modifications of the genus remain unpublished. Black (1965) referred Reeder's species *Akmaiomys incohatus* to *Proheteromys.* Heteromyid specimens recorded from the Oligocene of Florida (Patton, 1969) may be related to *Proheteromys* and should certainly contribute to understanding of heteromyid evolution.

The phylogenetic position of *Proheteromys* is disputed. Wood (1935*a,* 1980) asserted that the genus is a primitive member of the Heteromyinae and clearly separated from the perognathine *Mookomys.* However, taxa and specimens such as *Mookomys bodei, Proheteromys sulculus,* and *Mookomys* cf. *M. altifluminis* (Lindsay, 1974, fig. 5k) combine supposed diagnostic features of the two genera. Wilson (1960) thought that the Heteromyinae and Perognathinae were not clearly separable in the early Miocene.

The crown pattern of *Proheteromys* is similar to that in members of the Florentiamyidae (Wahlert, 1983). Setoguchi (1978) considered *Heliscomys schlaikjeri* to be conspecific with *Proheteromys nebraskensis,* and he described similarity of the latter species to *Heliscomys tenuiceps.* Both species have been transferred to the Florentimyidae. It is possible that part of the known sample of *Proheteromys* is the remains of florentiamyid taxa. The H-pattern formed by wear in the molars of *P. nebraskensis* (Wood, 1937*b*) does not occur in florentiamyids. However, sorting of genera is not possible at present, because the Florentiamyidae are best distinguished by cranial features.

Skeletal remains.—Reeder (1957) commented on mandibular structure and figured mandibles of *P. crassus* and *P. incohatus* (1960*a,* figs. 206–209). Gawne (1975) described the partial skull of *P. cejanus* and noted the presence of a large rostral perforation that is a defining character of the modern Heteromyidae.

Hitonkala Macdonald, J. R., 1963

Type species.—*Hitonkala andersontau* Macdonald, J. R., 1963, SDSM 56120, early Arikareean Sharps Formation, locality SDSM 5354, north of Porcupine, Shannon Co., South Dakota; skull with dP 4/4, P

4/4-M 3/3. Original series: SDSM 56141, partial skull with dP 4/4-M 3/3.

Temporal and geographic distribution.—OLIGOCENE: Early Arikareean: South Dakota.

Revised generic diagnosis.—Medium sized heteromyid with brachydont cheek teeth; range of crown length of P 4/4-M 3/3 about 4.3 to 5.0 mm. Tooth crown patterns as in *Proheteromys* except: upper molars with strong anterior cingula; lower molars with variable development of anterior and posterior cingulids; p4 having six principal cusps.

Species.—

Hitonkala andersontau Macdonald, J. R., 1963, fig. 17A, B. Other specimens: Macdonald, J. R., 1963, fig. 17C, D. Early Arikareean.

Comment.—J. R. Macdonald neither commented on nor illustrated tooth crown height in this genus. The asulcate upper incisors are laterally compressed and have a rounded anterior face; the lower incisors have a similar shape. Macdonald (1963) described deciduous upper and lower premolars in detail. The six cusped p4 (fig. 17b) is an unerupted tooth, and I wonder if that will be its final form. The other defining character of the genus, considerable development of cingula, is probably within the range of variation in *Proheteromys.*

Macdonald (1963) saw a striking similarity in lower dentitions of *Hitonkala* and the type of *Peridiomys rusticus* but admitted that the upper molars are not so close in morphology. He proposed that *Hitonkala* may bridge the gap between *Proheteromys* and *Peridiomys.*

Skeletal remains.—Macdonald (1963, 1970) noted cranial remains. He observed (1963:187): "Palate wide, cheek teeth mounted on raised and expanded base; palatines extending posteriorly to form tubular posterior choanae; bullae moderately large, not visible dorsally." These characters are probably not diagnostic for the genus; they are more widespread among heteromyids.

Harrymys Munthe, 1988

Type species.—*Harrymys irvini* Munthe, 1988, UCMP 122004, late Hemingfordian Split Rock local fauna, Arikaree Formation, Fremont Co., Wyoming; skull and mandibles with complete dentition. Original series: See Munthe, 1988:68.

Temporal and geographic distribution.—MIOCENE: Hemingfordian: Montana, Wyoming.

Revised generic diagnosis.—Large heteromyid with rostral perforation but retaining primitive geomyoid features: complete accessory foramen ovale; temporal foramen present; flange over orbits absent; superior angular process of mandible not flared; incisors without ornament. Derived characters of skull: elongated incisive foramina, 45 to 46 percent of diastemal length; auditory bullae, consisting of thin bone, enlarged ventrally with anteromedial processes that meet in midline; mastoid chambers enlarged posteriorly and dorsally; squamosal emarginate posteriorly; parietals not reaching occipital plane. Cheek teeth moderately protohyposdont and rooted. Upper cheek teeth wearing to U pattern open buccally; M3 early enclosing central valley as basin. Retention of ectolophid in lower cheek teeth forming H pattern with wear; proximity of protostylid and hypostylid producing subsequent R pattern; metalophid embayed posteriorly by tip of lingual flexid; hypolophid V shaped with anterior point joining ectolophid and with posterior embayment between hypoconid and entoconid that may be occupied by small cusp.

Species.—

Harrymys irvini Munthe, 1988, figs. 15–17, 19; other specimens: Munthe, 1988, fig. 18. Hemingfordian.

Harrymys woodi (Black, 1961), fig. 5B (*Dikkomys woodi*). Hemingfordian.

Comment.—Munthe (1988) named *Harrymys* as a new genus of florentiamyid. It has neither the cranial nor the dental hallmarks of that family. Its large rostral perforation is a prominent heteromyid character. The retained primitive features, together with highly specialized characteristics of the auditory region have led me (in manuscript) to distinguish the taxon as a new family that is an early branch of the Heteromyidae. A complete description with new illustrations is in progress; I have merely restated the chief points made there. Other extinct heteromyids may share the same mixture of primitive and derived cranial features, but adequate comparative material for each genus is not known. The crown length of the cheek tooth row in *H. irvini* is about 8.0 mm; incisors lack ornament.

Subfamily Heteromyinae Gray, 1868
Heteromys Desmarest, 1817

Type species.—*Mus anomalus* Thompson, 1815, probably BMNH, Trinidad.

Comment.—I have found no published fossil record of the genus.

Liomys Merriam, 1902

Type species.—*Heteromys alleni* Coues, 1881 (now *Liomys irroratus alleni*), MCZ 5889, Rio Verde, San Luis Potosi.

Temporal and geographic distribution (excluding Recent).—Rancholabrean to sub-Recent: Nuevo Leon, Tamaulipas.

Comment.—A summary of the fossil record of *Liomys irroratus* is given by Dowler and Genoways (1978). Hibbard (1972) transferred the species *L. centralis* to the genus *Prodipodomys.*

Subfamily Perognathinae
Coues, 1875
Mookomys Wood, 1931

Type species.—*Mookomys altifluminis* Wood, 1931, AMNH 21360, Barstovian Deep River Beds, 7 miles south of Logan, Gallatin Co., Montana; partial mandibles with left i-m2 and right i; skull fragment with left I; partial skeleton of young individual.

Temporal and geographic distribution.—MIOCENE: Arikareean: Nebraska, ?California, ?New Mexico, ?Texas; Hemingfordian: California, Colorado; Barstovian: California, Montana, Saskatchewan.

Revised generic diagnosis.—Small heteromyid with brachylophodont cheek teeth; similar to *Proheteromys* in many features; crown length of p4-m3 in *Mookomys altifluminis* approximately 3.90 mm (sum of individual tooth lengths from Black, 1961). M 1-3/1-3 wider than long, higher crowned than in *Heliscomys*; transverse valleys deeper than valleys separating cusps in transverse rows; anterior cingulum short, connecting only protoconid and protostylid (more extensive in *M. bodei*), posterior lacking; anterior and posterior cingulids weak or absent in m1 and m2. M 1-2/1-2 and anterior half of M3 with primary cusps and stylar cusps of nearly equal height. Cusps uniting to form lophs at earlier stage of wear than in *Heliscomys*; with extreme wear lophids of lower cheek teeth in some specimens uniting at center of tooth to give H-pattern, p4 with two strong anterior cusps and hypostylid.

Species.—

Mookomys altifluminis Wood, 1931, fig. 4; Wood, 1935*a*, fig 4. Other specimens: Black, 1961, fig. 5a; Lindsay, 1974, fig. 5. Hemingfordian to Barstovian.

M. formicarum Wood, 1935*a,* fig. 8 (emendation of spelling, based on feminine, Latin noun, was suggested by Wood, personal communication). Other specimens: Wilson, 1960, figs. 116–118; Lindsay, 1972, fig. 20. Wood identified the type specimen, CM 10177, as a right m1; Lindsay (1972, p. 45) said that Wood's illustration indicates that it is a left M1; Wood (personal communication) maintains his original identification. Hemingfordian to Barstovian.

M. bodei Wilson, 1949*b,* pl. 1, figs. 5, 5a, 6, 6a–b. Arikareean.
M. subtilis Lindsay, 1972, fig. 21. Barstovian.
M. sp.: Stevens, et al., 1969, figs. 6g–h (?*M.*); Gawne, 1975, fig. 8 (?*M.*); Storer, 1975, fig. 68; Whistler, 1984, figs. 9–12.

Comment.—In the type species, *Mookomys altifluminis,* the anterior face of the upper incisor is grooved as in *Perognathus* (Reeder, 1957). Wood (1935*a*) pointed out that the cheek teeth of *M. formicarum* are lower crowned and the separation of individual cusps is greater than in *M. altifluminis.* He proposed that the species fills the gap between *Heliscomys gregoryi* and *M. altifluminis.* Wood (1935*a*:89) stated that "the genus could easily be ancestral to the recent *Perognathus,* as well as to other Perognathines." Although he said that the H-pattern is absent in lower molars, Black (1961) described and figured such a pattern in extremely worn specimens. Reeder (1957: 113–114) argued that the height of crown combined with primitive cusp pattern made such union of lophids unlikely; he concluded that *Mookomys altifluminis* "probably cannot be considered to be ancestral to any known later heteromyid." He transferred the species *M. formicarum* to the genus *Perognathus. M. subtilis* is a small species in which the protostylids and hypostylids are posterior to the transverse lophids with which they are associated. *M. bodei* is larger than the other species; lophids in p4 unite early in wear to form an X-pattern as in *Perognathus*; the upper incisor, however, is not sulcate. For these reasons Wilson (1949*b*) said that the taxonomic position of *M. bodei* is uncertain. Korth et al. (1990) transferred one of Wilson's specimens, identified as *Mookomys* sp., cf. *M. formicorum,* to a new genus *Stratimus.* The known lower incisors of *Mookomys* are narrow with bowed anterior faces in cross section. Lindsay figured deciduous upper (1974, fig. 5d, h) and lower (1972, fig. 21) premolars.

Skeletal remains.—Wood (1931, 1935*a*) described skeletal elements associated with the type of *Mookomys altifluminis.* These include a skull fragment, partial humerus, distally fused tibia and fibula, astragalus, and calcaneum. Comparison of these remains to *Perognathus* and *Paramys* led him to conclude that *Mookomys* is close to *Perognathus* in most characters; differences were similar to the morphology of *Paramys* and were presumed to be primitive characters. Reeder (1957) recorded the morphology of the premaxillary fragment associated with the type species.

Trogomys Reeder, 1960*b*

Type species.—*Trogomys rupinimenthae* Reeder, 1960*b,* LACM(CIT) 5184, late Arikareean or early Hemingfordian Tick Canyon Formation, near narrows of Vasquez Canyon, approximately 10 miles by road NE of Saugus, Los Angeles Co., California; fragmentary rostrum and palate with right and left I-M3. Original series: numerous specimens, see Reeder (1960*b*).

Temporal and geographic distribution.—Miocene: ?late Arikareean or early Hemingfordian: California.

Revised generic diagnosis.—Small heteromyid with brachydont, rooted cheek teeth; crown length of P 4/4-M 3/3 ranging approximately from 3.5 to 4.0 mm. Molars wider than long. M 1-2/1-2 and anterior half of M 3/3 with primary cusps and stylar cusps subequal in height; stylar cusps joined to lophs and lophids with moderate wear. Transverse valleys much deeper than valleys separating cusps in transverse rows. Posterior cingula/ids small or absent; anterior cingula/ids long and low on M 1-2/1-2. P4 with single, strongly slanted anterior cusp; metaloph joining entostyle. Deep valley (Y-pattern) between protostylid and protoconid of lower molars. Hypostylid lacking in p4. First junction of lophids in m1-2 between protoconid and hypoconid forming H-pattern.

Species.—
Trogomys rupinimenthae Reeder, 1960*b,* pl. 39, fig. 1a, and pl. 40. Other specimens:

Reeder, 1960*b*, pl. 39, fig. 1b, and pl. 41; Whistler, 1984, figs. 13–16.

Comment.—The type specimen of the type species is a fragmentary rostrum and palate with complete upper dentition. Most of the particular diagnostic characters of the genus concern the lower dentition, known from the original series of specimens. *Trogomys* differs from *Proheteromys* chiefly in its small size, though it may be larger than *P. floridanus,* and in the prominence and slant of the anterior cusp in P4; the stylar cusps in the upper molars appear better formed in comparison with those of *P. sulculus* (Wilson, 1960, fig. 108) and *P. cejanus* (Gawne, 1975, fig. 4). I1 is asculate; the wear facet is grooved by the narrow lower incisor. Deciduous teeth are not known.

Reeder (1960*b*) suggested that *Perognathus, Mookomys,* and *Trogomys* be united as the subfamily Perognathinae. The genera share moderate lophodonty in low crowned molars and small size. *Trogomys* lacks a sulcus in the upper incisors; this and possession of strong cingula may be primitive characters. Lindsay (1972) stated that *Trogomys* was probably ancestral to *Cupidinimus* in the dipodomyine lineage, but he gave no reason for this view.

Skeletal remains.—Reeder (1960*b*) described morphology of the mandible, rostrum, and palate as strongly perognathine in character. Reeder (1960*b*) stated that the premaxillary–maxillary suture seems to cross the incisive foramina at or near their posterior borders. His figure (pl. 40), however, shows the suture approaching the middle of the foramen on the left side, a derived feature of the dipodomyines.

Stratimus
Korth, Bailey, and Hunt, 1990

Type species.—*Stratimus strobeli* Korth, Bailey, and Hunt, 1990, USNM 26688, early Hemingfordian Runningwater Formation, Dawes Co., Nebraska; left mandible with p4-m2.

Temporal and geographic distribution.—Miocene: Hemingfordian: Colorado, Nebraska, Saskatchewan.

Generic diagnosis.—"Small perognathine; p4 with central anteroposteriorly directed loph originating from anterolingual corner of hypoconid and anterobuccal corner of entoconid; small lophules originating from the hypoconid and entoconid on lower molars join anteriorly to form V-shaped hypolophid; upper molars with short anterior cingulum; protocone of P4 circular without accessory cuspules" (Korth et al., 1990:27).

Species.—

Stratimus strobeli Korth, Bailey, and Hunt, 1990, fig. 1A, C, D. Other specimens: Wilson, 1960, fig. 119 (*Mookomys* sp., cf. *M. formicorum*); Skwara, 1988, pl. 23, fig. 1, pl. 24, fig. 4, pl. 25, fig. 1, pl. 26, fig. 4 (*Proheteromys* sp. indeterminate species). Hemingfordian.

Comment.—Korth et al. (1990) remarked that *Stratimus* differs from all contemporary heteromyids in possession of the anteriorly pointing, V-shaped hypolophid in lower molars, and the central, anteroposterior lophule of p4. A similar hypolophid occurs in the supposed geomyoid *Lignimus* Storer, 1970, and a similar lophule is known in *Cupidinimus saskatchewanensis* Storer, 1970, and *Proheteromys* sp., cf. *P. magnus* (Wilson, 1960, fig. 121). The authors do not explain how *Stratimus* is related to other perognathines. i1 is narrow with a small, flattened area on the anterior surface.

Cheek teeth are brachydont. The lower incisor is narrow and tapers posteriorly; it has a small flattened area on the anterior surface. The mandible is stated to be similar to that of *Proheteromys*; the ventral ridge of the masseteric fossa is prominent and terminates in an elongated knob that is anterior and ventral to p4. The mental foramen is just anterior to this knob.

Korth et al. (1990) placed in *Stratimus* a specimen from Colorado referred by Wilson (1960, fig. 119) to *Mookomys.* The authors also noted similarity to *Stratimus strobeli* of some of the isolated heteromyid teeth from the Hemingfordian of Saskatchewan that Skwara (1988, pls. 23–26) identified as "*Proheteromys* sp., indeterminate species." I have taken the liberty of indicating specific illustrations on these plates that may be examples of *Stratimus.*

Perognathus Wied, 1839
(including *Chaetodipus* Merriam, 1889)

Type species.—*Perognathus fasciatus* Wied, 1839; neotype BS 168,599 (Williams and Genoways, 1979), Buford, Williams Co., North Dakota; skin and skull.

Temporal and geographic distribution (excluding Recent).—MIOCENE: Hemingfordian: Oregon, Wyoming; Barstovian: California, Nebraska; Clarendonian: California, Kansas, Nebraska, Oregon, South Dakota, Wyoming; Hemphillian: Arizona, Nebraska, Oregon. PLIOCENE: Hemphillian: Arizona, Kansas, Chihuahua; Blancan: Arizona, California, Idaho, Kansas, Oklahoma, Texas. PLEISTOCENE: Irvingtonian: Arizona, Kansas, Texas; Rancholabrean: Arizona, California, Colorado, Idaho, Kansas, Missouri, Nevada, New Mexico, Oklahoma, Texas, Wyoming, Washington, Tamaulipas.

Species (known as fossils).—

Perognathus brevidens Korth, 1987, fig. 1–2; Korth, 1979*b,* fig. 2 (*P. furlongi*). Other specimens: Korth, 1979*b,* fig. 1 (*P. furlongi*); Korth, 1987, fig. 1-1, 1-3, 1-4. Barstovian.

P. californicus. Rancholabrean to Recent.

P. carpenteri Dalquest, 1978, fig. 6. Blancan.

P. coquorum Wood, 1935*a,* figs 28–29. Clarendonian to Hemphillian (Miocene).

P. dunklei Hibbard, 1939, fig. 2. Other specimens: Hibbard, 1939, fig. 3. Hemphillian (Pliocene).

P. furlongi Gazin, 1930, pl. 3, fig. 5–6; Wood, 1935*a,* fig. 27. Other specimens: Wood, 1937*a,* figs. 6–7; James, 1963, figs. 43–44; Lindsay, 1972, figs. 23–24. Barstovian to Clarendonian.

P. gidleyi Hibbard, 1941*b,* fig. 9. Other specimens: Hibbard, 1942, pl. 1, figs. 2, 5. Blancan to Irvingtonian.

P. henryredfieldi Jacobs, 1977, fig. 1c. Hemphillian (Pliocene).

P. hispidus. Other specimens: Hibbard, 1955, fig. 4a; Hibbard and Taylor, 1960, fig. 11a; Schultz, 1965, fig. 3b. Irvingtonian to Recent.

P. huastecensis Dalquest and Roth, 1970, fig. 2. Rancholabrean.

P. inornatus. Rancholabrean to Recent.

P. intermedius. Rancholabrean to Recent.

P. madei Zakrzewski, 1969, fig. 5b. Blancan.

P. mclaughlini Hibbard, 1949, fig. 1. Other specimens: Hibbard, 1949, fig. 2h; Jacobs, 1977, fig. 1a–b. Hemphillian (Pliocene) to Blancan.

P. minutus James, 1963, fig. 45. Other specimens: Lindsay, 1972, fig. 22. Barstovian to Clarendonian.

P. parvus. Rancholabrean to Recent.

P. pearlettensis Hibbard, 1941*a,* pl. 1, fig. 9. Other specimens: Hibbard, 1941*a,* pl. 1, fig. 12, pl. 2, figs. 10, 14; Hibbard, 1950, figs. 8b–e; Hibbard, 1956, fig. 10. Blancan to Irvingtonian.

P. rexroadensis Hibbard, 1950, fig. 9. Other specimens: Hibbard, 1950, figs. 8a, f, 10a–e. Blancan.

P. stevei Martin, 1984, fig. 9a. Other specimens: Martin, 1984, figs. 8b, 9b–f, 10a–c. Hemphillian (Miocene).

P. trojectioansrum Korth, 1979*b,* fig. 3b–c. Other specimens: Korth, 1979*b,* figs. 3a, 7. Barstovian.

P. sp. Hibbard, 1942, pl. 1, fig. 9; Klingener, 1968, fig. 2c; Hibbard and Taylor, 1960, fig. 11b; Schultz, 1969, fig. 4f; Shotwell, 1967, fig. 7; Akersten, 1972, fig. 11c; Voorhies, 1974, figs. 2, 3; Martin, 1984, figs. 8c, 10d–f; Lindsay and Jacobs, 1985, fig. 5a, pl. 1h.

Comment.—I have not separated *Perognathus* from *Chaetodipus* because the distinction has not been made in the paleontological literature and cannot be made yet on the basis of dentition alone. The genus *Perognathus* is ancient. The earliest specimens, not identified to species, are of Hemingfordian age. One was listed by Munthe (1979, 1988) in the Split Rock local fauna of Wyoming. Another was described by James (1963) from the upper part of the John Day Formation of Oregon; it is similar to *P. furlongi* but has asulcate upper incisors. Fifteen extinct species have been described from later mammal ages. The possibility of discovering the interrelationships of these species from the literature is nil. Comparisons are to be found in Reeder (1957) and Martin (1984).

Martin (1984) measured teeth in eleven of the extinct species and compared their size, degree of hypsodonty, and crown morphology. He found that all Barstovian and Clarendonian specimens have much more brachydont dentitions than a new Hemphillian species, *Perognathus stevei.* Crown height in this species is similar to that of Hemphillian and later taxa and somewhat lower than in *P. maldei* and the living *P. parvus.* His comparison revealed dental variation, especially in the fourth premolars. The P4 in *P. furlongi* has a bulbous crown and in some specimens a small accessory cuspule on the protoloph buccal to the protocone. In both *P. furlongi* and *P. parvus* the protocone is slanted so that its root projects far anteriorly. *P. stevei* has the modern crown configuration of a single cusped protoloph, but its root is directed downward and forms an abbreviated anterior slope profile. In the living *P. parvus* the profile of some specimens is also abbreviated. The p4 is variable in *P. stevei*: an anteroconid is present in some specimens between the protoconid and metaconid as it is in some other extinct species; anteroposterior connection of lophids can be buccal, medial, or in both positions; the hypolophid is extended by a small buccal cusp in two specimens. James (1963, fig. 43a) illustrated variation in P4 of *P. furlongi.* The right and left P4 in a single palate differ markedly in union of protoloph and metaloph (the specimen number appears to be incorrect in the figure, when compared with that in the text). Wood (1935*a*:96) stated in the diagnosis of *Perognathus* that the lower molars generally lack an H-pattern, which is well developed in *Microdipodops* and in his Dipodomyinae. The presence of the H-pattern in some specimens of *Perognathus* is clearly illustrated by Martin (1984, fig. 8). No deciduous teeth have been described from fossils.

Korth (1979*b*) suggested that *Perognathus saskatchewanensis* is probably referable to *Cupidinimus*; Martin (1984) left the species in *Perognathus,* but Whistler (1984) and Barnosky (1986*a*) transferred it. Martin (1984) created a new heteromyine genus, *Oregonomys,* and placed three other species, *Perognathus sargenti* Shotwell, 1956, *P. magnus* Zakrzewski, 1969, and *Diprionomys agrarius* Wood, 1935*a,* in it.

Skeletal remains.—Gazin (1930, pl. 3, figs. 5, 6) figured the rostrum and palate of *Perognathus furlongi.* These structures are broad; the diastema is described as rather long. Martin (1984) described differences in the anterior root of the zygomatic process among several species. Dice (1925) noted 24 skulls of *P. c. californicus* from Rancho La Brea. Wood (1937*a*) figured cervical vertebrae of *P. furlongi* and said that they are typically perognathine. Voorhies (1974) described and illustrated Clarendonian age burrows of *Perognathus.*

Subfamily Dipodomyinae Gervais, 1853

Cupidinimus Wood, 1935A

Type species.—*Cupidinimus nebraskensis* Wood, 1935*a,* CM 10193, late Barstovian Crookston Bridge Member (Skinner et al., 1968, p. 405), Valentine Formation, about 3 mi S of Valentine, Cherry Co., Nebraska; partial skull with left P4 and M3, left mandible with i-m3, partial skeleton. Original series: CM 10170, partial left maxilla with P4-M3; CM 10175, partial right maxilla with M1; CM 10173, partial right mandible with I, dp4-m1; CM 10171, partial left ulna; CM 10172, partial left radius; CM 10174, two fused caudal vertebrae.

Generic synonyms.—*Perognathus* Wood, 1936

Temporal and geographic distribution.—MIOCENE: Hemingfordian: California; Barstovian: California, Montana, Nebraska, Utah, Saskatchewan; Clarendonian: California, Nevada, Oregon, South Dakota; Hemphillian: Arizona.

Revised generic diagnosis.—Small heteromyid with moderately hypsodont but rooted cheek teeth. Alveolar length of P 4/4-M 3/3 about 3.4 to 5.0 mm; crown length about 3.0 to 4.6 mm. Cheek teeth strongly bilophodont. p4 nearly quadrate with 4 chief cusps; anteroposterior junction of metalophid and hypolophid in middle of tooth apparent with wear. Lophs of M1-2 fusing first at lingual side; labial fold in side of tooth retained until late in wear. Lophs of m1-2 fusing first between bases of protoconid and hypoconid at middle of tooth to produce H-pattern. Enamel lakes within lophids rare and transient. Slight dentine tracts (chevrons) present on lingual sides of lower molars.

Species.—

Cupidinimus nebraskensis Wood, 1935*a*, figs. 35–36, 38–39, 41–64; Korth, 1979*b*, fig. 4. Other specimens: Wood, 1935*a*, figs. 37, 40; Klingener, 1968, fig. 2b; Korth, 1979*b*, figs. 5–7; Barnosky, 1986*a*, figs. 3d, 4a, and i, 5j. Barstovian.

C. quartus (Hall, 1930), figs. 5–6 (*Diprionomys*); Barnosky, 1986*a*, fig. 5l. Other specimens: Hall, 1930, figs. 7–8 (*D.*). Clarendonian.

C. tertius (Hall, 1930), figs. 2–4 (*Diprionomys*). Clarendonian.

C. halli (Wood, 1936), fig. 2 (*Perognathoides*); Barnosky, 1986*a*, figs. 3h, 5k. Other specimens: Wood, 1936, figs. 2–9 (*P. halli* and *Peridiomys kellogi*); Lindsay, 1972, figs. 27–28 (*Perognathoides*). Barstovian.

C. cuyamensis (Wood, 1937*a*), figs. 2–5 (*Perognathoides*). Clarendonian.

C. madisonensis (Dorr, 1956), pl. 17, figs., 6–7 (*Perognathoides*). Barstovian.

C. kleinfelderi (Storer, 1970), fig. 4. (*Perognathoides*); Storer, 1975, fig. 72d (*P.*). Other specimens: Storer, 1975, figs. 70, 72e (*P.*). Barstovian.

C. saskatchewanensis (Storer, 1970), fig. 3. (*Perognathus*); Storer, 1975, figs. 69a, 72b (*P.*). Other specimens: Storer, 1975, figs. 69b–e, 72c (*P.*). Barstovian.

C. eurekensis (Lindsay, 1972), fig. 29 (*Perognathoides*). Barstovian.

C. bidahochiensis (Baskin, 1979), figs. 5, 6 (*Perognathoides*). Other specimens: Baskin, 1979, fig. 7 (*P.*). Hemphillian.

C. boronensis Whistler, 1984, figs. 20–39. Hemingfordian.

C. whitlocki Barnosky, 1986*a*, fig. 4f, j. Other specimens: Barnosky, 1986*a*, figs. 3a, e, i and l, 4b, d, and m, 5a, d and g. Barstovian.

C. avawatzensis Barnosky, 1986*a*, figs. 4g, k; Wilson, 1939, pl. 1, fig. 5 (M1 only) (*Perognathoides* cf. *tertius*). Other specimens: Wilson, 1939, pl. 1, figs. 1–5, 7–9, 12 (*P.* cf. *tertius*); Barnosky, 1986*a*, figs. 3b, f, j, and m, 4c, e, and n, 5b, e, and h. Clarendonian.

C. lindsayi Barnosky, 1986*a*, figs. 4h, l; Lindsay, 1972, figs. 26c, f (*C. nebraskensis*). Other specimens: Lindsay, 1972, figs. 25, 26 (*C. nebraskensis*); Barnosky, 1986*a*, figs. 3c, g, k, and n, 4o, 5c, f, and i. Barstovian.

C. sp.: Shotwell and Russell, 1963, fig. 41b; Green, 1971, fig. 2; Barnosky, 1986*b*, pl. 4, figs. a-1.

Comment.—The type specimen of the genotypic species is remarkable in its completeness and fragility. Wood (1935*a*:121) pointed out that "wear affects the upper molars progressively from rear to front; that is, M3 is the first worn to a circle, while M1 is the last." In fig. 39 he showed a median connection of the metalophid and hypolophid in both p4 and m3, whereas the lophids are separate in m1 and m2 (Fig. 2). Wear produces either a U- or H-pattern depending on the connection that forms first; the one connection is quickly followed by the other. Lindsay (1972, p. 50) noted that the tooth crowns are straight walled, and a slight basal inflation occurs only in the upper premolar.

The upper incisor is asulcate; the lower is much deeper than wide and has a rounded anterior face. Deciduous premolars have been described and figured by Wood (1935*a*), Wilson (1939), Lindsay (1972), and Whistler (1984).

Korth (1979*b*) described a large sample of *Cupidinimus nebraskensis* from the Crookston Bridge Member of the Valentine Formation (not at the locality of the holotype). He observed variation in the upper and lower premolars in which accessory cuspules are present in a notable percent of the sample. He pointed out that the cheek teeth of *C. nebraskensis* are not low crowned (Wood, 1935*a*:118: "teeth medium to low crowned") but are equal or nearly equal in crown height to that of species of *Perognathoides.* Korth synonymized *Cupidinimus* and *Perognathoides* and suggested that *Cupidinimus* be selected as the generic name. *Perognathoides* has only page priority, and the dentition of the type specimen is so worn as to be generically indeterminate. Lindsay (1972) had noted the great similarity of *Perognathoides halli* and *Cupidinimus nebraskensis* and suggested that they might be congeneric. Whistler (1984) and Barnosky (1986*a*) followed Korth in the synonymy. Korth referred *Perognathus saskatchewanensis* Storer, 1970, to the genus; he pointed out special features of Lindsay's Barstovian sample of *C. nebraskensis* that make these specimens a new species of the genus, and Barnosky (1986*a*) created the species *C. lindsayi* to receive them. As mentioned above, *Cupidinimus magnus* was formally transferred to the geomyid genus *Pliosaccomys* by Shotwell (1967). A still larger specimen attributed to *Cupidinimus* (Chaffee, 1936) might belong there also.

Barnosky (1986*a*) named new species of *Cupidinimus* and carefully reexamined the existing taxa. He identified two major clusters of species that are divided by the Rocky Mountains. He found evidence of parallel evolution chiefly in characters associated with increased crown height of the cheek teeth; these trends are in step with geologic age. Barnosky presented a clear phylogenetic hypothesis of *Cupidinimus* but did not place the genus in any heteromyid subfamily.

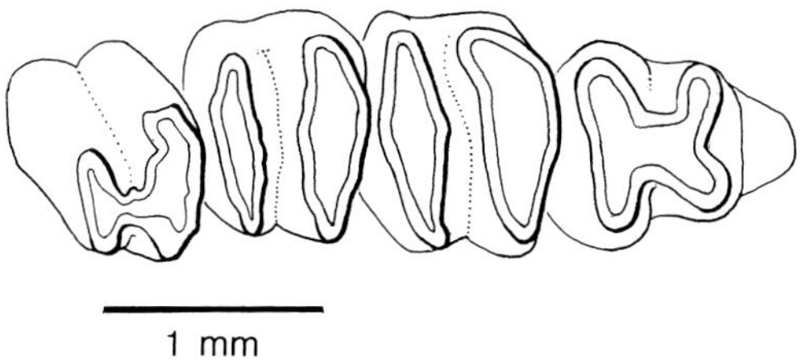

FIG. 2.—*Cupidinimus nebraskensis* CM 10193 (holotype). Lower left cheek teeth, redrawn from Wood, 1935a, fig. 39. Anterior is to the right.

Wood (1935*a*) selected the partial skull of *Diprionomys quartus* as the type of a new genus *Perognathoides.* Lindsay's (1972) sample of teeth of *Perognathoides halli* showed a range of variability that included the special characteristics of *P. madisonensis, P. kelloggi,* and possibly *P. cuyamensis,* and he synonymized these species. Barnosky (1986*a*) again recognized the species *C. madisonensis* and *C. cuyamensis.*

Hall (1930:304) stated: "A careful study of the fossil and Recent forms leaves one with the impression that *Diprionomys quartus* [now *Cupidinimus quartus*] is, at least, an approximation of the ancestral type of *Liomys* or, and, *Dipodomys.*" "*Diprionomys tertius* and probably *D. parvus* have closer affinities with *Perognathus* than with *Dipodomys.*" Wood (1935*a*) found the cheek tooth crown pattern to be perognathine but said that *Perognathoides* cannot have been ancestral to *Perognathus* because its cheek teeth are higher crowned, and it occurs too late. Other differences from perognathines are the presence of accessory cusps on P4 in many specimens and the asulcate upper incisors in *Perognathoides.* Reeder (1957) proposed that *Diprionomys* and *Perognathoides* were closely related and included them in a branching system of heteromyid phylogeny with *Cupidinimus, Microdipodops,* and *Dipodomys*; he placed *Perognathus* in a different lineage. Lindsay (1972) and Baskin (1979) agreed with this general arrangement and said that *Microdipodops* may be derived from *Perognathoides*.

Wood (1935*a*:144) stated that detailed comparison of cranial and postcranial remains showed that resemblances of *Cupidinimus* to *Perognathus* are chiefly shared, primitive characters; unique features of *Cupidinimus* appear to be stages in specialization toward the morphology of *Dipodomys*. Reeder (1957) related *Cupidinimus* to *Microdipodops*; Lindsay (1972) showed the genus as an early branch of the Dipodomyinae. Whistler (1984) placed *Cupidinimus* in the Dipodomyinae. He pointed out the importance of *C. boronensis*, which is the earliest known species of the genus. He stated (1984:18) that "*C. boronensis* could easily be derived from any perognathine by a significant increase in crown height and a reduction of cingula." He felt that *Trogomys* is the most likely candidate for ancestor among fossil perognathines despite the short time and morphological differences separating it from *C. boronensis*.

Skeletal remains.—Wood (1935*a*) described the mandible, cranium, and postcranial elements of *Cupidinimus nebraskensis*. Reeder (1957) included only the cranium and mandible in his redescription; Korth (1979*b*) commented on ankle structure. Skeletal comparisons of limb bones and feet led Wood (1935*a*) to conclude that *Cupidinimus* was sub-ricochetal in locomotion, although there is no fusion of the cervical vertebrae. Korth (1979*b*) took issue with Wood's statement that the calcaneal-navicular contact was established in the foot. Korth (1979*a*) noted other postcranial elements of the genus in a taphonomic study. Wood's illustration of the skull (1935*a*, fig. 41) suggests to me that a rostral perforation was probably present. Wood found the skull shape similar to that of *Perognathus*, which he used as a model in reconstruction. Reeder (1957:365) believed that Wood underestimated the amount of bullar expansion. He contradicted Wood on bullar structure and said that the bulla was not filled with cancellous bone but "is formed of an exceedingly thin layer of hard, brittle bone, such as is seen in *Dipodomys* and *Microdipodops*".

Mandibular and rostro-palatine morphology of *Perognathoides* was described by Hall (1930) and Reeder (1957). The infraorbital canal is depressed into the rostrum at its anterior end, but neither author mentions the presence or absence of a rostral perforation medial to its anterior end. The parapterygoid fossae, a characteristic of the Geomyoidea, are partially preserved. Wood (1937*a*) tentatively associated calcanea and possibly limb bone fragments with the genus.

Prodipodomys Hibbard, 1939

Type species.—*Prodipodomys kansensis* (Hibbard, 1937), KUMNH VP 3945, late Hemphillian Ogallala Formation, Edison Quarry, Sherman Co., Kansas; partial left mandible with i, p4, and molar aveoli.

Temporal and geographic distribution.—MIOCENE: Barstovian: California. PLIOCENE: Hemphillian: Arizona, Kansas, Chihuahua; Blancan: Arizona, Idaho, Kansas, Nebraska, Texas. PLEISTOCENE: Irvingtonian: Kansas.

Revised generic diagnosis.—Moderately large heteromyid with high crowned cheek teeth that are pillar-like but rooted; alveolar length of p4-m3 ranging approximately from 5.0 to 5.5 mm, and crown length from about 4.0 to 4.7 mm. Cheek teeth with cusps distinct only at earliest stages of wear; crown height and root length approximately the same in adults. P4 with separate anteroloph and 3 roots; p4 wearing to X-pattern, having 2 roots. m1 and m2 subequal with H-pattern developing prior to simpler ovoid shape; 2 rooted. m3 much smaller than anterior teeth and having 2 fused roots.

Species.—

Prodipodomys kansensis (Hibbard, 1937), fig. 3 (*Dipodomys kansensis*); Hibbard, 1962, fig. 1a. Other specimens: Hibbard,

1939, fig. 4; Jacobs, 1977, fig. 1d, e. Hemphillian (Pliocene).

P. minor (Gidley, 1922), pl. 34, fig. 16 (*Dipodomys minor*). Other specimens: Wood, 1935*a,* fig. 73 (*D. minor*). Blancan.

P. centralis (Hibbard, 1941*b*), fig. 8 (*Liomys centralis*). Other specimens: Hibbard, 1954, figs. 3a–e (*P. rexroadensis*); Hibbard, 1972, figs. 35a–g; Hager, 1974, fig. 4f; Dalquest, 1978, fig. 7. Blancan.

P. tiheni (Hibbard, 1943), fig. 9 (*Etadonomys tiheni*). Other specimens: Hibbard, 1943, figs. 5–8; Hibbard, 1962, fig. 1b (*E. tiheni*); Zakrzewski, 1981, figs. 1, 2a, 2d. Blancan.

P. mascallensis Downs, 1956, fig. 7. Other specimens: Shotwell, 1967, fig. 15. Barstovian.

P. idahoensis Hibbard, 1962, fig. 1c. Other specimens: Zakrzewski, 1969, fig. 5a; Harrison, 1978, fig. 4; Shotwell, 1967, fig. 25o (*Dipodomys* sp.); Lindsay and Jacobs, 1985, fig. 5b–h, pl. 1g, i–l. Hemphillian (Pliocene) to Blancan.

P. griggsorum Zakrzewski, 1970, figs. 1a, b. Other specimens: Zakrzewski, 1970, fig. 1c; Hibbard, 1953*b,* fig. 4d (*P.* sp.). Blancan.

P. sp.: Hibbard, 1953*b,* fig. 4b; Hibbard, 1956, fig. lli; Akersten, 1972, fig. 11b; Zakrzewski, 1970, fig. 1d.

Comment.—Hibbard (1939) described the genus *Prodipodomys* based on the species *P. kansensis* that he had assigned previously to *Dipodomys.* He distinguished the genus from *Cupidinimus* by the presence of a *Dipodomys*-like foramen between m3 and the base of the coronoid process; the presence of roots on the cheek teeth clearly divides the new genus from *Dipodomys.* Gazin (1942) noted the conformity of Gidley's (1922) species *Dipodomys minor* with the new genus and transferred it. Hibbard (1972) recognized that the species *Liomys centralis* (Hibbard, 1941*b*) was merely a young specimen of *Prodipodomys,* and he synonymized with it the species *P. rexroadensis.* Eshelman (1975) commented that he was unable to find characters based on isolated teeth that would separate *Etadonomys tiheni* from *Prodipodomys.* Zakrzewski (1981) formally synonymized the two. Harrison (1978), however, suggested that *Etadonomys* may be a primitive line divergent from *Prodipodomys.* Reeder (1957) transferred *P. mascallensis* to the genus *Peridiomys*; Shotwell (1967) found specimens less worn than the type; these were also higher crowned and fit better in *Prodipodomys.*

Upper dentitions are very rare. The upper incisors of *Prodipodomys centralis* are sulcate, and the tube of the nasal bones extends well beyond their anteriormost curvature. The lower incisor appears rather flat with an open curvature. I have found no comment on its form. Zakrzewski (1969, fig. 5d) described and illustrated a dP4 of *P. idahoensis.*

Zakrzewski (1969) examined the dentine tracts in 13 Recent species of *Dipodomys* and found that they are clearly visible in all but young specimens. Dentine tracts are absent or only slightly developed in *Prodipodomys*; they have been figured in *P. idahoensis* (Harrison, 1978) and *P. tiheni* (Zakrzewski, 1981). The slightly worn cheek teeth are heteromyine in appearance; this suggests that formation of transitory enamel lakes within the metalophid of m1 may occur very early in wear. Zakrzewski (1981) discussed the occurrence of a fossa at the site of the mandibular foramen and found it to be a variable feature in both *Prodipodomys* and *Dipodomys.*

Hibbard (1939) clearly stated the similarities of *Prodipodomys* to both *Cupidinimus* and the living genus *Dipodomys.* That these taxa are part of a natural clade has not been questioned by subsequent authors. Zakrzewski (1981) proposed a chronocline for specimens from Kansas that begins with *Prodipodomys kansensis* from Edson Quarry and continues through *P. griggsorum,* then *P. centralis,* and finally to the most hypsodont member of the series, *P. tiheni* from the Rexroad local fauna. Hibbard (1962) said that the absence or slight development of dentine tracts, lesser degree of hypsodonty, and presence of larger roots may remove this last species at least from close relationship to *Dipodomys.* Voorhies (1975) presented further evidence of the improbability of direct relationship between *Prodipodomys* and *Dipodomys* in his discussion of *Eodipodomys* (see section on that genus). Dalquest and Carpenter (1986) doubted that any known species of *Prodipodomys* was ancestral to the living genus.

Skeletal remains.—The similarity of mandibular morphology of *Prodipodomys* and *Dipodomys* has been described by many authors including Hibbard (1939, 1954, 1962, 1972), Gazin (1942), and Zakrzewski (1969, 1981). Maxillary fragments are rare, and only one partial skull has been described (Hibbard, 1972). Shotwell (1967, fig. 25o) figured a calcaneum that may belong to *Prodipodomys idahoensis.*

Eodipodomys Voorhies, 1975

Type species.—*Eodipodomys celtiservator* Voorhies, 1975, UGV 109, Clarendonian Ash Hollow Formation, Antelope County, Nebraska; fragmented skull with rostrum, anterior portion of palate, right I-M1 and left I-P4, partial occiput with bulla, mandibles with i-m3, right and left humeri and femora, left calcaneum and metatarsals I, IV, and V, other fragments.

Temporal and geographic distribution.—MIOCENE: Clarendonian: Nebraska.

Generic diagnosis.—Large heteromyid; occlusal length of lower cheek teeth 6.55 mm, alveolar length 6.95 mm; "hypsodont but rooted teeth, hypertrophied auditory bullae, forelimbs shortened relative to hind limbs, dentine tracts present on all cheek-teeth, incisors slender and ungrooved, mandibular fossa in shallow pit" (Voorhies, 1975:164).

Species.—

Eodipodomys celtiservator Voorhies, 1975, figs. 3–5.

Comment.—I suspect that mandibular foramen, not fossa, was intended in the passage quoted above. The single specimen of the genus was found fossilized in its burrow. Voorhies pointed out its strong similarities to *Dipodomys* in dentition, bullar enlargement, mandibular morphology, and limb proportions. Roots are better developed on the cheek teeth, and the anterior loph of p4 is semicircular rather than indented to the base as in *Dipodomys.* Voorhies reported the presence of a large perforation in the rostrum, and flaring angular process and weak coronoid process in the mandible; these are distinctive heteromyid characteristics. Enlarged bullae that compress the squamosal and parietal bones of the skull roof, dentine tracts on the lingual and buccal sides of the cheek teeth, and elongation of the hind limbs are dipodomyine characters. The femur has some features that are present in *Cupidinimus.*

The specimen is of great phylogenetic interest because of its early age and degree of morphological specialization. Dental similarity to *Dipodomys* suggested to Voorhies that the genus is intermediate between *Cupidinimus* and *Dipodomys.* He proposed that the more primitive genus *Prodipodomys* is an offshoot from the main lineage and led to no modern form. Zakrzewski (1981) took strong exception to this view and was not convinced that *Eodipodomys* is a dipodomyine. Voorhies pointed out that *Cupidinimus* could be ancestral to *Eodipodomys* because of its primitiveness and greater biostratigraphic age; the amount of time between known earliest occurrences, however, would require a period of very rapid evolution. The asulcate upper incisors of *Eodiopodomys* are primitive. A sulcus is present in incisors of *Prodipodomys* (Hibbard, 1972), and *Dipodomys* and *Microdipodops* (Wood, 1935*a*).

Shotwell (1967) proposed a close, early relationship between the Dipodomyinae and the Geomyinae based on dental similarity of *Prodipodomys* and *Pliosaccomys.* Lindsay (1972) concurred with this view. Voorhies (1975) compared *Eodipodomys* and the early geomyoid *Parapliosaccomys.* He found the latter to differ in possession of derived geomyine characters: flat faced incisors; narrow and strongly ribbed palate; wide, arched rostrum; and bilobed P4.

Skeletal remains.—Important features of the type specimen have been mentioned above.

Dipodomys Gray, 1841

Type species.—*Dipodomys phillipsii* Gray, 1841, BMNH, Recent, Mineral del Monte (formerly Real del Monte), Hidalgo State, Mexico; skin.

Temporal and geographic range (excluding Recent).—MIOCENE: Barstovian: California; Clarendonian: California, Nevada; Hemphillian: Oregon. PLIOCENE: Blancan: California, Kansas. PLEISTOCENE: Irvingtonian: Arizona; Rancholabrean: Arizona, California, Colorado, Idaho, Kansas, Nevada, New Mexico, Texas, Washington.

Species (known as fossils).—

Dipodomys agilis. Rancholabrean to Recent.

D. gidleyi Wood, 1935*a*, fig. 74. Irvingtonian.

D. hibbardi Zakrzewski, 1981, fig. 2b. Other specimens: Zakrzewski, 1981, fig. 2e. Blancan.

D. ingens. Rancholabrean to Recent.

D. merriami. Rancholabrean to Recent.

D. ordii. Other specimens: Hibbard, 1955, figs. 6h, k; Hibbard and Taylor, 1960, fig. 11f. Rancholabrean to Recent.

D. pattersoni Dalquest and Carpenter, 1986. Irvingtonian.

D. spectabilis. Rancholabrean to Recent.

D. sp.: Wilson, 1939, figs. 10, 10a (Dipodomyine(?), n. gen. and sp.); Shotwell, 1967, fig. 25i (*Pliosaccomys* sp.), p; Miller, W. E., 1971, fig. 9.

Comment.—Paleontologists have paid attention to the particular characteristics of the cheek teeth in *Dipodomys* and made no more than brief mention of other remains. Hibbard (1953*a*:265) revised the traditional diagnosis of the genus with the emendation, "the fourth premolar rooted or generally rooted in adult specimens and rooted in old adult specimens. As far as known, the first to third molars are 'evergrowing'." Zakrzewski's (1981:82) observations suggested "that all species of *Dipodomys* have closed roots at some post-adult stage of life. The exact stage at which closure occurs varies among species and most likely among individuals in any particular species." He found that dentine tract height is related to the reduction of roots. In the species *D. hibbardi* root development and dentine tract height are intermediate between *Prodipodomys* and the dentally primitive living species *Dipodomys compactus.* Zakrzewski (1969) discussed the degree to which dentine tracts are developed on the different teeth. Dalquest and Carpenter (1986) described variability and wear in teeth of living species and discussed the characteristics of *Prodipodomys.*

The Pleistocene distribution of *Dipodomys* is easily available in Kurtén and Anderson (1980), who briefly described each species. The abundant late Pleistocene faunas were summarized by Lundelius et al. (1983). The earliest Pleistocene and pre-Pleistocene record of *Dipodomys* is of especial interest; here, certainty of specific identification declines, and the relationship of living and extinct taxa should be sought. Lindsay (1978) noted the presence of the genus in the early Irvingtonian near Benson, Arizona. Golz et al. (1977) recorded cf. *Dipodomys* in California from strata of Blancan age. Shotwell (1967) described several specimens as *Dipodomys* sp. from Hemphillian localities in Oregon. The earliest specimens attributed to the genus were identified by Reeder (1957) from the Clarendonian of Nevada and California (including the "Dipodomyine(?), n. gen. and sp." of Wilson, 1939), and from the Barstovian of California. Gazin (1942) transferred *D. minor* to *Prodipodomys.* No deciduous teeth have been recorded for fossil *Dipodomys.*

Skeletal remains.—Most of the literature deals with the simple identification of the modern genus in fossil faunas. Hibbard

(1955, fig. 6h) figured a baculum attributed to *Dipodomys ordii.* Dice (1925) recorded the collection of 44 partial skulls of *D. a. agilis* from Rancho a Brea. Hibbard and Taylor (1960, fig. 11f) and Miller, W. E., (1971, fig. 9) illustrated mandibles. It is likely that skeletal remains are known but are undescribed. Wood (1935*a*) remarked on limb bones associated with the type specimen of *D. gidleyi.* Shotwell (1967, figs. 25i, p) described and figured calcanea that may belong to the genus; he suggested that the specimen in fig. 25o may represent *Prodipodomys.*

Microdipodops Merriam, 1891

Type species.—Microdipodops megacephalus Merriam, 1891, USNM 24417/31823, Halleck, Elko Co., Nevada; skin and skeleton.

Temporal and geographic distribution (excluding Recent).—PLEISTOCENE: Rancolabrean: Nevada.

Species (known as fossils).—
Microdipodops megacephalus

Comment.—S. J. Miller (1979:287) reported the presence of *Microdipodops* in the "red silt deposits" of Smith Creek Cave; the introduction to this section notes only pink silt deposits, which are presumably the same. Mead et al. (1982) reviewed and updated Pleistocene and Holocene faunas from the region and listed *Microdipodops,* as reported by Miller, in a reddish-brown silt that they dated at greater than or equal to 12,000 years before the present.

Schizodontomys Rensberger, 1973*a*

Type species.—Schizodontomys greeni Rensberger, 1973*a,* UCMP 39435, early Hemingfordian Haystack Valley Member of John Day Formation, Picture Gorge 1, Grant Co., Oregon; partial left mandible with i-m1.

Temporal and geographic distribution.—MIOCENE: ?late Arikareean: South Dakota; Hemingfordian: Nebraska, Oregon, South Dakota, Wyoming.

Revised generic diagnosis.—Large heteromyids; cheek teeth moderately hypsodont, strongly bilophodont. "Metalophid of p4 with two cusps; lingual cusp larger, oval or D-shaped; accessory cusps other than small anteroconid absent" (Rensberger, 1973*a*: 60). Skull with rostral perforation; auditory bullae and mastoid chambers enlarged; bullae meeting anteriorly in midline.

Species.—

Schizodontomys greeni Rensberger, 1973*a,* figs. 27, 28, 29b, pls. 6a, 14b. Hemingfordian.

S. harkseni (Macdonald, 1970), fig. 20 (*Grangerimus harkseni*). Other specimens: Rensberger, 1973a, figs. 29a, 30–32, pls. 5b–d, 6b, 10e, 13f, 14e, 15a–b, 17b; Munthe, 1981, figs. 1–6. Hemingfordian.

S. sulcidens Rensberger, 1973*a,* figs. 29c, 35, 36, pls. 6c, 15c–d, 17d. ?late Arikareean.

S. amnicolus Korth, Bailey, and Hunt, 1990, figs. 2–3. Hemingfordian.

Comment.—Schizodontomys was described by Rensberger (1973*a*) as a pleurolicine rodent; the subfamily is usually included with the Entoptychinae in the Family Geomyidae (Wahlert and Souza, 1988). Wahlert (1985) cited Munthe's (1981) description of the skull as suggesting heteromyid and, within that family, dipodomyine affinity. Korth, Bailey, and Hunt (1990) disputed this placement and found cranial and dental characteristics that support heteromyine relationship. I am not fully convinced of either position. The specimens of *Schizodontomys* have yet to be placed side by side for complete comparative description. The cusp pattern of p4 of *S. amnicolus* does not fit the original diagnosis of the genus. Crown length of the cheek teeth is 5.7–6.1 mm; incisors lack ornament.

Skeletal remains.—Munthe (1981) described and figured a nearly complete skeleton of *S. harkseni*. She found that the species was not fossorial and was at least quadrupedally saltatorial.

Diprionomys Kellogg, 1910

Type species.—*Diprionomys parvus* Kellogg, 1910, UCMP 12566, Hemphillian Thousand Creek Formation, 5¾ mi SW Hot Spring and ½ mi N Thousand Creek (Hall, 1930), Humboldt Co., Nevada; partial right mandible with i-m1.

Temporal and geographic distribution.—MIOCENE: Barstovian: Nebraska, Saskatchewan; Clarendonian: Nebraska, Nevada; Hemphillian: Nevada, Oregon.

Revised generic diagnosis.—Moderately large to large heteromyids with hypsodont cheek teeth; alveolar length of p4-m3 in *Diprionomys agrarius* 6.0 mm (individual tooth lengths close to those of *D. parvus* and larger than *D. minimus*). Cusps distinct on lophs at earliest stages of wear; cingula not apparent in slightly worn teeth. p4 consisting of two subequal lophids that unite at ends with wear and surround an enamel lake; width of hypolophid suggesting presence of hypostylid. Molars lacking H-pattern; first union of lophids between protostylid and hypostylid, union occurring first in m1 (*D. agrarius*); no trace of enamel lakes within lophids.

Species.—

Diprionomys parvus Kellogg, 1910, figs. 17a, b; Hall, 1930, figs. 9–10. Other specimens: Clark, Dawson, and Wood, 1964, fig. 11; Shotwell, 1967, figs. 10, 11, 24j, 25n, 26h. Clarendonian to Hemphillian.

D. minimus (Kellogg, 1910), fig. 15 (*Entoptychus minimus*); Wood, 1936, fig. 10 (differs significantly from Kellogg's illustration). Hemphillian.

D. agrarius Wood, 1935*a*, figs. 102–104, 105b, 106–128. Other specimens: Klingener, 1968, figs. 2a, d; Storer, 1975, fig. 73; Korth, 1979*b*, figs. 8–9; Barnosky, 1986*b*, pl. 2, figs. d–i. Barstovian to Clarendonian.

Comment.—Kellogg (1910) originally named two species of *Diprionomys*, and Hall (1930) described two more. Hall said that the four species, though similar, included a range of variation that made framing a generic definition difficult. Since then *D. magnus* was transferred to *Cupidinimus* (Wood, 1935*a*); Wilson (1936, p. 25) pointed out the strong resemblance of *D. magnus* to the extinct geomyid *Pliosaccomys*; Reeder (1957) offered corroborative information, and Shotwell (1967) transferred the species to this genus of geomyid. Wood (1935*a*) created the genus *Perognathoides* to receive Hall's additional species of *Diprionomys*. Wood described *D. agrarius* from an excellent specimen, and he transferred Kellogg's species *Entoptychus minimus* to the genus in 1936.

There is little argument over the combination of *Diprionomys parvus* and *D. minimus* in a genus; the type specimens of both are incomplete, worn lower dentitions. The placement of *D. agrarius*, however, is disputed; the type includes a complete lower dentition that is but lightly worn. Reeder (1957) placed the species in a new genus (unpublished), because the cheek teeth arc clearly lower crowned than those of *D. parvus*. Korth (1979*b*) suggested that *D. agrarius* be left in the genus *Diprionomys* until more detailed study is made. Martin (1984) transferred the species to a new genus, *Oregonomys*, but I have not followed him. Korth (1979*b*:303) suggested that specimens of *D.* cf. *parvus*, identified by Clark et al. (1964), "may represent a distinct species assignable to a new genus which would also include '*D.*' *agrarius*." Barnosky (1986*b*:31) agreed with this possibility, but his statement, "I include *D. parvus* in *Diprionomys*," is misleading, since *D. parvus* is the type species of the genus.

Kellogg (1910:433) stated in the generic and specific definitions of *Diprionomys parvus* that "the coronoid process rises abruptly from behind M3, with no depression for the dental foramen." Wood (1935*a*) included this character in his generic diagnosis. Such a depression is lacking in living heteromyids except some specimens of *Dipodomys* (Wahlert, 1985:11). Wood's figure (1935*a*, fig. 103) of *D. agrarius* shows a broad depression.

The lower incisor in *Diprionomys parvus* is narrow and appears to have a curved antrior surface (Hall, 1930, fig. 9); Reeder (1957:288), however, said that the anterior face is nearly flat, and he reported a similar condition in *D. minimus*. The upper incisor in *D. agrarius* bears a broad lateral sulcus. The deciduous p4 of *D. agrarius* was described by Klingener (1968).

Wood (1935*a*:177) placed *Diprionomys* in the Heteromyinae as a distinct branch. He said that "the tendency to develop lakes on the lophs, by the double union of the protoconid and protostylid, in the lower molars, which is so characterstic of *Heteromys* and is already developed in *Proheteromys* and *Peridiomys*, is apparently entirely absent in *Diprionomys*." Reeder (1957) placed the genus as a relative of *Microdipodops*; he stated that it is closely related to *Perognathoides* in the state of hypsodonty and variable but usually tricuspidate metalophid in p4.

Skeletal remains.—Maxillary morphology was described by Kellogg (1910), Hall (1930), Wood (1935*a*), and Reeder (1957); cranial and postcranial remains by Wood (1935*a*) and Reeder (1957). The type specimen of *Diprionomys agrarius* includes a mandible, partial skull, and nearly complete skeleton. Wood (1935*a*) observed that the occipital bone is highly compressed transversely, and the contact with the auditory region is clearly displayed on its lateral margin. He concluded that auditory inflation was considerable and may well have reached the dorsal surface of the skull. Study of the postcranial remains and comparison of the revised intermembral index with those of living heteromyids suggested that *Diprionomys* had a sub-ricochetal mode of locomotion (Wood, 1935*a*). Shotwell (1967) described and figured the distal end of a tibiafibula and a calcaneum and astragalus that he assigned to the species *D. parvus*.

Peridiomys Matthew, 1924

Type species.—*Peridiomys rusticus* Matthew, 1924, AMNH 18894, early Barstovian Olcott Formation, Far Surface Quarry (Skinner, Skinner, and Gooris, 1977), Sioux Co., Nebraska; partial right mandible with p4-m2.

Temporal and geographic distribution.—MIOCENE: Hemingfordian: Wyoming; Barstovian: Nebraska, Oregon, Saskatchewan.

Revised generic diagnosis.—Moderately large to large heteromyids with brachylophodont cheek teeth; relative crown height similar to *Proheteromys*. Range of crown length of p4-m3 approximately 5.2–7.0 mm. Cheek teeth like those of *Proheteromys* except some stylar cusps equal in height to major cusps; only portion of anterior cingulum that connects protoconid and protostylid present; P4 lacking distinct central basin but possessing hypostylid; H-pattern appearing progressively from posterior to anterior; isolation of small labial lakes in lower molars of some specimens.

Species.—

- *Peridiomys rusticus* Matthew, 1924, fig. 9 (a thoroughly inaccurate representation); Wood, 1935*a*, figs. 97, 98. Barstovian.
- *P. oregonensis* (Gazin, 1932), figs. 1–4 (*Diprionomys? oregonensis*); Wood, 1935*a*, figs. 99, 100. Other specimens: Shotwell, 1967, fig. 9. Barstovian.
- *P. borealis* Storer, 1970, fig. 5; Storer, 1975, fig. 72f. Other specimens: Storer, 1975, figs. 71b–e, 72g–h. Barstovian.
- *P.* sp.: Barnosky, 1986*b*, pl. 4, figs. m–p. Barstovian.

Comment.—Reeder (1957:194) stated that "the term *Peridiomys* as here used can be considered only as a broad phylad of related genera." He synonymized the species *P. kelloggi* with *Perognathoides halli* and was left with an assemblage that was similar in both tooth morphology and possession of a short, strongly concave diastema.

Reeder (1957:193) described the upper incisors as asulcate and the lower incisors as having a "cross-section narrow to broadly triangulate with anterior surface round or flattened." The crown length of cheek teeth in *Peridiomys rusticus* is the high end of the range given; that in *P. oregonensis,* the low end. No deciduous premolars have been described.

Matthew (1924) found similarities of *Peridiomys rusticus* to both *Perognathus* and *Heteromys.* Wood (1935*a*) placed the genus in the Heteromyinae and said that, except for the strong H-patterns, the genus could be ancestral to all later heteromyines. Reeder placed the specimens that Downs (1956) identified as *Peridiomys* cf. *oregonensis* in a genus, still unpublished, allied closely with *Diprionomys.* A Hemingfordian record of *Peridiomys* sp. is merely a listing of the genus in the Split Rock local fauna by Munthe (1979, 1988).

Skeletal remains.—Maxillary morphology was described by Downs (1956) and Reeder (1957). Gazin's sample of *Peridiomys oregonensis* included four partial skulls; the specimens are similar in size to *Heteromys,* and the long, broad cranium suggests that the absent bullae were not inflated (Gazin, 1932). Reeder (1957) reexamined these specimens and described cranial remains of a still unpublished new species. In it he suspected that the bullae may have been somewhat enlarged compared to those of *Heteromys,* because the basisphenoid is relatively narrower. Neither author noted the presence or absence of a rostral perforation. Shotwell (1967) described and figured an isolated calcaneum of *Peridiomys oregonensis*; he did not give a reason for making this specific identification.

Oregonomys Martin, 1984

Type species.—Oregonomys pebblespringsensis Martin, 1984, UWBM 57116, Hemphillian Dalles Formation, Arlington, Gilliam Co., Oregon; left mandible (angle broken) with i-m3. Original series: numerous specimens, see Martin (1984:107).

Temporal and geographic distribution.—Miocene: Hemphillian: Oregon. Pliocene: Blancan: Idaho.

Generic diagnosis (after Martin, 1984).—Moderately large heteromyid with brachyodont cheek teeth; crown length of P 4/4-M 3/3 ranging approximately from 4.3 to 5.5 mm. Cheek teeth strongly bilophodont; cusps obliterated early in wear. Protoloph of P4 with buccal cuspule. Anterolophids connecting protoconid and protostylid in p4-m3. Upper incisors grooved. Auditory region inflated.

Species.—

Oregonomys pebblespringsensis Martin, 1984, figs. 14b, 15a. Other specimens: Martin, 1984, figs. 12, 13, 14c, d, e. Hemphillian.

O. magnus (Zakrzewski, 1969), fig. 5c (*Perognathus magnus*). Other specimens: Martin, 1984, fig. 14g. Blancan.

O. sargenti (Shotwell, 1956), figs. 5j, 6a (*Perognathus sargenti*); Shotwell, 1967, fig. 8 (*P. sargenti*); Martin, 1984, fig. 14f. Other specimens: Martin, 1984, fig. 15b, c. Hemphillian.

Comment.—Martin (1984) described the similarity of *Diprionomys* to *Oregonomys* and transferred the species with closest resemblance, *D. agrarius,* to the new genus. I have not followed him, because the evidence appears equivocal. Inclusion of the species *O. sargenti* and *O. magnus,* formerly *Perognathus,* in the genus was based on additional fossil material. All of these species have a complex metalophid in p4 that consists of three or more cusps and cuspules. The character has been noted also in the much earlier genus *Hitonkala* (Macdonald, 1963).

Martin's (1984) description of *Oregonomys pebblespringsensis* is especially thorough. Enough specimens were collected from the Ordnance locality that he could statistically compare the dimensions of unworn, moderately worn, and very worn teeth; the differences in sum are considerable. The large sample revealed that the first point of union between lophs and lophids, especially in the fourth premolars, is variable and not a diagnostic character: in P4 lophs connect medially or lingually; in p4 lophid connection is normally medial, sometimes buccal; in lower molars it is normally buccal with a tendency for medial union. Martin (1984, fig. 13e) figured a dP4.

Martin (1984:120) proposed a phylogenetic tree springing from the species *Oregonomys agrarius* (which I have retained in *Diprionomys*) in which the complexity of p4 increases over time. He placed *Oregonomys* in the Heteromyinae and said that it "appears to have paralleled *Dipodomys*, which has more hypsodont cheek teeth and occupied a similar environmental position."

Skeletal remains.—Martin (1984) described mandibular and cranial morphology. The mandible of *Oregonomys pebblespringsensis* is more primitive than that of *Dipodomys*. The palate is primitively broad and the tooth rows parallel. The posterior part of a skull and a portion of an inflated mastoid of *Oregonomys* revealed an expanded auditory region similar in size and character to that in *Dipodomys*. The mastoids in both genera have thin, non-cancellous walls, a derived character (Wahlert, 1985), unlike those in *Perognathus* and other living geomyoids. The number of supporting struts is greater in *Oregonomys* than in *Dipodomys*. The nature of the mastoid suggests to me that *Oregonomys* may be an early branch of the Dipodomyinae. Partial ossification of the stapedial canal is similar to *Perognathus* and unlike the derived, complete ossification in *Dipodomys*. Neither of these important cranial remains has teeth in association. Martin's remarks about the skeleton are based on the specimen of *Diprionomys agrarius*.

Conclusion

The rodent family Heteromyidae is purported to have arisen from within the Eomyidae, to have a fossil record that extends back to the early Oligocene, and to contain at least three subfamilies—Perognathinae, Dipodomyinae, and Heteromyinae—that were distinct from one another in early Miocene time (Wood, 1935*a*; Lindsay, 1972). Four problems cast doubt on this picture: (1) Eomyid and heteromyid teeth, despite similarities, have important morphological and functional differences; (2) the Oligocene genus *Heliscomys*, often considered to be a primitive heteromyid, may not belong to the clade of living heteromyids; (3) heteromyid subfamilies were not sharply divided in the Miocene, and certain genera have been assigned to one or another subfamily by different authors; (4) in the last few decades new genera and species have been named, but with few exceptions no thorough comparison of specimens has been done. A diagram of temporal ranges of genera described in this chapter is given in Figure 3.

The extinct Eomyidae, known from North America and Europe, have long been recognized as possible ancestors (Matthew, 1910, fig. 19; Wilson, 1949*c*; Harris and Wood, 1969) or as a sister taxon (Wahlert, 1978, 1985) of the Geomyoidea. The primitive eomyid dental crown pattern (Fig. 4) may resemble that from which the bilophodont heteromyid morphology (Fig. 5) was derived. A variety of early Oligocene dentitions of North American eomyids were illustrated by Black (1965) and Wood (1974). Eomyid cheek teeth are strongly lophate, but the crown has two distinct levels. In upper teeth the buccal paracone and metacone stand high and shear against the high lingual metaconid and entoconid. The lingual protocone, hypocone, and connecting

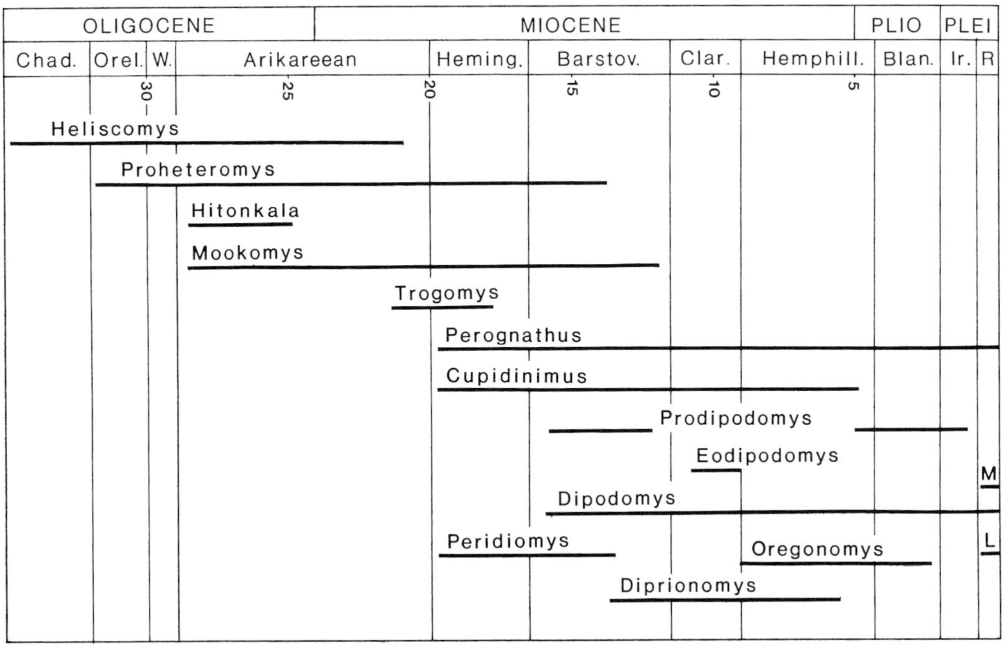

FIG. 3.—Temporal ranges of heteromyid genera. Abbreviations: Genera—L, *Liomys*; M, *Microdipodops*. Epochs—Plio, Pliocene; Plei, Pleistocene. North American Land Mammal Ages—Barstov., Barstovian; Blan., Blancan; Chad., Chadronian; Clar., Clarendonian; Heming., Hemingfordian; Hemphill., Hemphillian; Ir., Irvingtonian; Orel., Orellan; R, Rancholabrean; W., Whitneyan. Numbers indicate millions of years before present. See text for ranges of *Apletotomeus, Harrymys, Schizodontomys* and *Stratimus* and for complete range of *Heliscomys*.

endoloph of the upper teeth wear to a low, flat surface that occludes with a similar low, buccal surface in lower teeth that consists of protoconid, hypoconid, and connecting ectolophid. Long, low anterior and posterior cingula/ids complete the crown except at the anterior end of the premolar and posterior end of the third molar. In the eomyid *Ritteneria* the teeth, which are strongly bilophodont and flat crowned, are convergent on the heteromyid design but appear to lack stylar cusps (Stehlin and Schaub, 1951, figs. 190, 506).

Rensberger (1973*b*) described chewing in eomyids as having had a strong medial component seen also in the paramyids, which are more primitive, but lacking in geomyoids. The geomyoids are characterized by new stylar cusps, lingual in upper teeth and buccal in lower. These limit occlusal motion to an anteroposterior direction. Other scenarios for the origin of the geomyoids (Wilson, 1949*c*; Wood, 1980) are concerned only with crown pattern. It is easy to conceive of a change from the eomyoid to geomyoid pattern with gradual loss of anterior and posterior cingula, disappearance of endoloph and ectolophid, and addition of styles that widen the teeth. The marked difference in tooth function, however, suggests that the story is not so simple. Recently Wahlert and Koenigswald (1985) showed that the incisor enamel structure of Oligocene and Miocene eomyids is highly derived relative to that of geomyoids; if the same unusual structure was also present in earlier eomyids, then the eomyids could not have been ancestors of geomyoids.

Harris and Wood (1969:5) proposed that *Meliakrouniomys* "seems to be an eomyid on the way to becoming a heteromyid." The bilophodont lower teeth have short, narrow

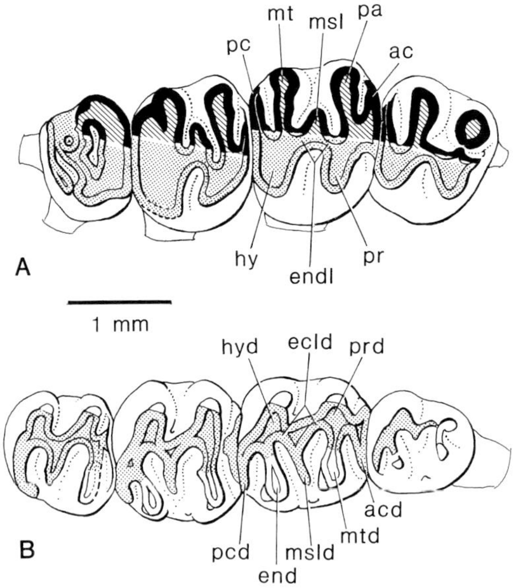

FIG. 4.—Primitive eomyid crown pattern seen in *Adjidaumo minutus.* 4A. Upper cheek teeth, AMNH 5390, left side reversed. 4B. Lower cheek teeth, AMNH 5363, right side reversed. Cedar Creek, Logan Co., Colorado; Orellan. Anterior is to the right. Blackened wear surfaces tilt toward bottom of page, white toward top; stippled area is flat. Abbreviations: **ac**, anterior cingulum; **acd**, anterior cingulid; **ecld**, ectolophid; **end**, entoconid; **endl**, endoloph; **hy**, hypocone; **hyd**, hypoconid; **hys**, hypostyle; **hysd**, hypostylid; **msl**, mesostyle; **msld**, mesostylid; **mt**, metacone; **mtd**, metaconid; **pa**, paracone; **pc**, posterior cingulum; **pcd**, posterior cingulid; **pr**, protocone; **prd**, protoconid; **prs**, protostyle; **prsd**, protostylid.

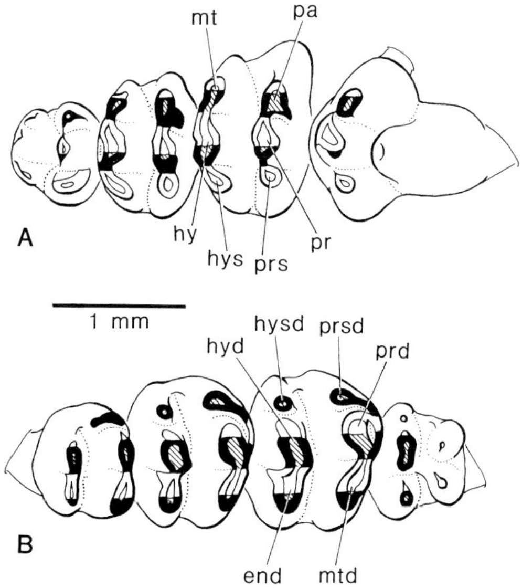

FIG. 5.—Heteromyid crown pattern in *Perognathus parvus,* AMNH 33504. Ironside, Malheur Co., Oregon; Recent. 5A. Upper right cheek teeth. 5B. Lower left cheek teeth. Blackened wear surfaces tilt toward bottom of page, white toward top. Anterior is to the right. See Fig. 4 for abbreviations.

anterior and posterior cingulids and lack the ectolophid that is typical of eomyids. As drawn in their fig. 1, however, the buccal cusps appear to be worn to a low, flat plane. The buccal stylids of geomyoids are lacking. The genus is certainly not a heteromyid. The late Eocene genus *Griphomys* Wilson, 1940, has also been suggested (Wilson, 1949*c*) as a structural stage in the origin of the geomyoid crown pattern; its crown pattern is already bilophodont. Wahlert (1976) described the lower dentition of *Jimomys* in which the difference in height between lingual and buccal parts of the teeth is slight and decreases as the teeth are evened by wear. The absence of lingual facets suggests that neither upper nor lower teeth were widened by styles. *Jimomys* may be a geomyoid more primitive than any heteromyid or geomyid.

Most authors subscribe to the hypothesis that eomyoids and geomyoids are descended from sciuravids. Transverse lophs and rectangular cheek tooth shape are the chief features of dental similarity. Dawson (1968) figured several different members of the Sciuravidae from the middle Eocene.

Wood (1935*a*:82–83) did not assign *Heliscomys* to a heteromyid subfamily: "Genus incertae sedis, perhaps ancestral to all three subfamilies, with characters common to all. Perhaps nearest to the Perognathinae." He emphasized dental primitiveness of the genus and stated that "*Heliscomys,* as far as its dental development is concerned, is structurally ancestral to the geomyids." Wood (1939) described new specimens and questioned his earlier view that the 3-cusped lower premolar and single cusped anterior loph of the upper premolar were primitive. He suggested that these conditions could also point to *Heliscomys* as a slightly degenerate descendant of the common ancestor of heteromyids, or as a case of evolutionary reversal. Reeder (1957, fig. 91) agreed that the genus was already specialized in extreme reduction of p4, and he showed it as a side branch in his phylogeny of heteromyids. Rensberger (1973*b*, fig. 5) suggested that the origin of *Heliscomys* involved reduction in size and complexity of premolars as well as in body size; he showed it as an ancestor of all geomyoids.

The cheek teeth in *Heliscomys* are widened by styles that can be a continuous lingual ridge in upper molars and are distinct buccal cusps in lower teeth. The crown surface is relatively flat but not lophate and thus differs from both eomyids and heteromyids. The eomyid cheek teeth have a highly lophodont bi-level crown. The heteromyids have lophodont teeth that wear quickly to a surface of low relief. Korth, Wahlert, and Emry (in press) proposed that cranial and dental features of *Heliscomys* are so primitive that the genus cannot be included in any of the three geomyoid families. They have created a new geomyoid family for it and *Apletotomeus*. Here, I designate *Heliscomys* and *Apletotomeus* as Geomyoidea *incertae sedis* to avoid confusion about creation of the new family.

Proheteromys, which appeared slightly later than the earliest *Heliscomys*, has a lophate crown morphology that is better suited as an ancestral pattern for heteromyids. One species, *P. cejanus*, is known to have a rostral perforation. The premolars of *Proheteromys* are not as simple as those of *Heliscomys* and may be a better starting point for the origin of later morphologies. A common feature in some species is the appearance late in wear of an anteroposterior connection between the transverse lophs of the lower molars (Wood, 1937*b*, fig. 35); the feature is lacking in premolars of *Proheteromys*. The connection could be the remnant of an ectolophid homologous to that in eomyids. The X-pattern of p4 (which looks like an H to me) and the H-patterns of molars that are mentioned in the descriptions of many taxa would be such remnants; they do not, therefore, indicate relationships. Careful attention should be paid to the nature of the anteroposterior connection of transverse lophids. It can occur through retention of the ectolophid, a primitive feature, or simply by basal broadening of cusps that lead to union with considerable wear. Inclusion in the metalophid of a shallow valley between the protoconid and protostylid is a typical feature of geomyoid molars that persists in many heteromyids. The genus *Proheteromys* has a remarkably wide geographic range, and in the Miocene is known from both Florida and California.

Harrymys, known from nearly complete cranial and dental remains, presents a mixture of primitive and specialized features. I currently favor placement of the genus in its own heteromyid subfamily; I intend to publish a redescription of the specimen.

Although phylogenies of the Heteromyidae show subfamilies extending back in time to the latest Oligocene, another body of evidence belies this view. The evidence is simply the difficulty that individual authors have faced in assigning Miocene genera to subfamilies. Barnosky (1986*a*:46) stated the problem in the introduction to his work on the heteromyid genus *Cupidinimus*: "Unfortunately, the species are known mainly by fossil cheek teeth whose specialized geomyoid cusp pattern (Fig. 2) changes very little throughout the range of the genus." The interesting point is that the cusp pattern is geomyoid and not even as specific as heteromyid.

The Dipodomyinae is a subfamily that can be traced well into the Miocene. *Prodipodomys* and *Dipodomys* share hypertrophied bullae and high crowned cheek teeth with dentine tracts; *Eodipodomys* differs in having asulcate upper incisors. *Cupidinimus* has been perceived traditionally as the ancestor of the Dipodomyinae. Enamel chevrons on the sides of the moderately hypsodont teeth may be the beginnings of dentine tracts. Reeder (1957) said the mastoid bulla is expanded and the bullar wall is thin bone as in *Dipodomys* and *Microdipodops*. Barnosky (1986*a*:55) was not so bold and stated merely that "the cheek teeth in *Cupidinimus* are characteristic of the perognathine pattern (Fig. 2) described in detail by Wood (1935) and Lindsay (1972)." Here, Barnosky has referred to the same figure as above and has made the association of the perognathine pattern, itself, with the basic geomyoid pattern.

Evidence of bullar enlargement has been noted in five other Miocene taxa: *Harrymys, Schizodontomys, Diprionomys,* which has a compressed occiput; *Peridiomys,* a narrow basisphenoid; and *Oregonomys,* an inflated mastoid of the thin walled, dipodomyine kind. It is possible that some of these genera are associated with the dipodomyine clade.

The living genus *Perognathus* has a fossil record as ancient as any dipodomyine genus. It, too, has enlarged auditory bullae, but these have a thick wall of trabecular bone. Similar bone in bullae of living heteromyines and geomyids suggests that the texture is primitive for geomyoids (Wahlert, 1985). The low crowned cheek teeth of *Perognathus* retain a crown pattern that is primitive to geomyoids. The dental characters that have been used to unite *Perognathus, Heliscomys, Mookomys,* and *Trogomys* in the subfamily Perognathinae are shared, primitive ones that do not define a clade. Dental similarity of *Trogomys* and *Proheteromys* again suggests a central role of the latter in heteromyid evolution. Wilson (1960) pointed out the similarity of *Mookomys* and *Proheteromys* in shared, primitive features. *Perognathus* is probably allied to the Dipodomyinae (Wahlert, 1985), but common ancestry must have been in the early Miocene or late Oligocene. I place *Mookomys* and *Trogomys* in the Perognathinae in accord with the traditional view; however, I would be just as content to leave these genera in the Heteromyidae *incertae sedis*. It is interesting that most of the Miocene heteromyids, known from the United States and Canada, can be associated with genera that live today in the same area.

The Recent genera, *Heteromys* and *Liomys* have a negligible fossil history, possibly because the later Cenozoic record is so poorly known within their present range. Only *Liomys* has a geographic range that extends a short distance into the United States. In contrast to the perognathines and dipodomyines these genera, the Heteromyinae, have a primitive looking braincase and auditory region. It is possible that *Proheteromys* is also ancestral to this subfamily.

A classification that contains my stated views follows:

Suborder Myomorpha
  Infraorder Geomorpha
  Superfamily Eomyoidea
    Family Eomyidae
      †*Meliakrouniomys* (other genera not treated)
  Superfamily Geomyoidea
    incertae sedis
      †*Heliscomys*
      †*Apletotomeus*
    Family Heteromyidae
    incertae sedis
      †*Proheteromys*
      †*Hitonkala*
      †*Harrymys*
    Subfamily Heteromyinae
      Heteromys
      Liomys
    Subfamily Perognathinae
      †*Mookomys*
      †*Trogomys*
      †*Stratimus*
      Perognathus
    Subfamily Dipodomyinae
      †*Cupidinimus*
      †Prodipodomys
      †*Eodipodomys*
      Dipodomys
      Microdipodops
      †*Schizodontomys*
      †*Diprionomys*

†*Peridiomys*
†*Oregonomys*
Geomyoid families—Florentiamyidae, Entoptychidae, and Geomyidae—not treated.

Extinct heteromyid genera and species need reexamination and new diagnoses. The basic conservatism of tooth crown pattern and variability of the premolars in both presence of accessory cusps and anteroposterior union of lophs and lophids make the dentition a problematic source of phylogenetic data. Genera that are known to include such variation are *Heliscomys* (now removed from the Heteromyidae), *Proheteromys, Cupidinimus, Diprionomys, Oregonomys, Dipodomys,* and *Perognathus.* When few specimens were known, differences in premolar morphology seemed important in defining genera and species. As sample size grew, the definitions became invalid, but corrections, though noted, were not used in broad generic revisions. The positions and sequence of union of lophs may relate to the suppression of primitive morphology and have little relevance to the characterization of taxa. A general trend toward increased crown height appears to be the chief influence in modification of crown pattern, and it is likely that similar modifications of crown pattern arose in parallel in most subfamilial clades. I have refrained from presenting a cladogram of the Heteromyidae, because placement of many taxa is uncertain, and the characterization of most nodes is, as yet, not possible.

Shotwell's (1967) and Lindsay's (1972) view that the Geomyinae merit no higher rank than the heteromyine subfamilies is symptomatic of the poor characterization of extinct geomyoid taxa. The Recent Geomyidae and Heteormyidae are clearly separate groups that are defined by specific, derived cranial characters (Wahlert, 1985; Wahlert and Souza, 1988). The low level of cranial differentiation among living geomyines, however, suggests fairly late origin of the group. If the Heteromyidae includes *Heliscomys* and *Proheteromys,* then the Heteromyidae is much more ancient than the Geomyidae. This view yields two possible conclusions: (1) the early history of the Geomyidae is unknown, or (2) the Geomyidae are descended from some extinct heteromyid.

I have excluded *Heliscomys* and *Apletotomeus* from the Heteromyidae and proposed *Proheteromys* as a stem heteromyid or geomyid. The geomyids and heteromyids could then share common ancestry prior to the Arikareean. The oldest geomyids, the entoptychines, are of early Miocene age. Confusion exists in the constitution of *Diprionomys.* The Hemphillian *Diprionomys magnus* has been considered a species of *Cupidinimus,* as similar to the geomyid *Pliosaccomys,* and even as a geomyid. Although the species is of late Miocene age, it illustrates dental similarity of the two families. I suspect that the early history of the Geomyidae is indeed poorly known. Wahlert and Souza (1988) proposed that the early center of evolution of geomyid and heteromyid rodents was in Mexico and Central America.

Several taxa have at one time or another been placed in the Heteromyidae. The Whitneyan and Arikareean Florentiamyidae, including *Florentiamys, Sanctimus, Kirkomys,* and *Ecclesimus* have a dentition with geomyoid crown pattern. The premolars are more complex than those of geomyids and heteromyids and may be more primitive. On the basis of derived cranial features, Wahlert (1983, 1984) and Korth (1989) placed the genera in a separate family within the Geomyoidea.

Certain trends in heteromyid evolution since the Oligocene are clear. These are various combinations of greater lophodonty and crown height, bullar enlargement, and elongation and change in proportions in the hind limbs toward increased jumping ability. Only the first of these can be fully documented in the specimens at hand. Adequate information about the auditory region and mode of locomotion will only be acquired through a continuation of collection of fossils and good fortune in locating such complete materials.

Acknowledgments

I appreciate the bibliographic assistance of Drs. Karl F. Koopman, Malcolm C. McKenna, and Richard H. Tedford and Ms. Marie A. Lawrence of the American Museum of Natural History, and Ida B. Lowe of the Baruch College Library. Dr. Koopman was an invaluable resource for solving systematic problems and Dr. Tedford for stratigraphic ones. The Departments of Vertebrate Paleontology and Mammalogy provided research space. I am especially grateful to Prof. Albert E. Wood and Dr. Anthony D. Barnosky, who meticulously reviewed the manuscript. Wood's method of phylogenetic analysis is very different from my cladistic one, and I was unable to incorporate his most significant disagreements. An ongoing dialogue and collaboration with Dr. William W. Korth has kept me alert to the evolving concepts of heteromyid phylogeny. The Committee on Released Time and Prof. Martin Stevens, Dean, School of Liberal Arts and Sciences, Baruch College, City University of New York, granted me release time from teaching in support of this project. Ms. Lauren Wilson typed the bibliography. Mr. Chester Tarka and Ms. Lorraine Meeker advised me on the illustrations.

Literature Cited

AKERSTEN, W. A. 1972. Red Light local fauna (Blancan) of the Love Formation, southeastern Hudspeth County, Texas. Texas Memorial Museum Bulletin, 20:1–53.

BARNOSKY, A. D. 1986*a*. New species of the Miocene rodent *Cupidinimus* (Heteromyidae) and some evolutionary relationships within the genus. Journal of Vertebrate Paleontology, 6:46–64.

———. 1986*b*. Arikareean, Hemingfordian, and Barstovian mammals from the Colter Formation, Jackson Hole, Teton County, Wyoming. Bulletin of Carnegie Museum on Natural History, 26:1–69.

BASKIN, J. A. 1979. Small mammals of the Hemphillian Age White Cone local fauna, northeastern Arizona. Journal of Paleontology, 53:695–708.

BLACK, C. C. 1961. Rodents and lagomorphs form the Miocene Fort Logan and Deep River Formations of Montana. Postilla, 48:1–20.

———. 1963. Miocene rodents from the Thomas Farm local fauna, Florida. Bulletin of the Museum of Comparative Zoology, 128:483–501.

———. 1965. Fossil mammals from Montana. Pt. 2. Rodents from the early Oligocene Pipestone Springs local fauna. Annals of Carnegie Museum, 38:1–48.

BONAPARTE, C. L. J. L. 1845. Catalogo metodico dei mammiferi europei. Giacomo Pirola, Milan, 36 pp.

BRANDT, J. F. 1855. Beiträge zur nähern Kentniss der Säugethiere Russlands. Mémoires de l'Académie Impériale des Sciences de St. Pétersbourg, 7:1–365, pls. 1–3, and pls. 1–11.

BUGGE, J. 1985. Systematic value of the carotid arterial pattern in rodents. Pp. 355–379, *in* Evolutionary relationships among rodents, a multidisciplinary analysis (W. P. Luckett and J.-L. Hartenberger, eds.). Plenum Press, New York, 721 pp.

CHAFFEE, R. G. 1936. A new heteromyid skull of the genus *Cupidinimus*. Journal of Mammalogy, 17:419–420.

CLARK, J. B., M. R. DAWSON, AND A. E. WOOD. 1964. Fossil mammals from the Lower Pliocene of Fish Lake Valley, Nevada. Bulletin of the Museum of Comparative Zoology, 131:27–63.

COPE, E. D. 1873. Synopsis of new Vertebrata from the Tertiary of Colorado, obtained during the summer of 1873. Washington, Government Printing Office, pp. 1–19.

———. 1884. The Vertebrata of the Tertiary formations of the West. Book I. Report of the United States Geological Survey of the Territories, F. V. Hayden, United States Geologist-in-charge. Vol. 3. Washington, pp. 1–1009, pls. i–lxxva (work issued in 1885).

COUES, E. 1875. A critical review of the North American Saccomyidae. Proceedings of the Academy of Natural Sciences of Philadelphia, 27:272–327.

———. 1881. Pp. 187–189, *in* List of mammals collected by Dr. Edward Palmer in northeastern Mexico, with field notes by the collector (J. A. Allen, author). Bulletin of the Museum of Comparative Zoology, 8:183–189.

DALQUEST, W. W. 1978. Early Blancan mammals of the Beck Ranch local fauna of Texas. Journal of Mammalogy, 59:269–298.

DALQUEST, W. W., AND E. ROTH. 1970. Late Pleistocene mammals from a cave in Tamaulipas, Mexico. The Southwestern Naturalist, 15:217–230.

DALQUEST, W. W., AND R. M. CARPENTER. 1986. Dental characters of some fossil and Recent kangaroo rats, with description of a new species of Pleistocene *Dipodomys*. The Texas Journal of Science, 38:251–263.

DAWSON, M. R. 1968. Middle Eocene rodents (Mammalia) from northeastern Utah. Annals of Carnegie Museum, 39:327–370.

DEPERET, C., AND H. DOUXAMI. 1902. Les vertébrés oligocène de Pyrimont-Challonges (Savoie). Mémoires de la Société Paléontologique Suisse, 29:1–90, pls. 1–6.

DESMAREST, A. G. 1817. Neuvième espèce—le hamster anomal. Pp. 180–181. Hétéromys. *In* Nouveau dictionaire d'histoire naturelle, 14:627 pp. Deterville, Paris.

DICE, L. R. 1925. Rodents and lagomorphs of the Rancho La Brea deposits. Carnegie Institution of Washington Publication, 349:119–130.

DORR, J. A. 1956. Anceney local mammal fauna, latest Miocene, Madison Valley Formation, Montana. Journal of Paleontology, 30:62–74.

DOWLER, R. C., AND H. H. GENOWAYS. 1978. *Liomys irroratus.* Mammalian Species, 82:1–6.

DOWNS, T. 1956. The Mascall fauna from the Miocene of Oregon. University of California Publications in Geological Sciences, 31:199–354.

EMRY, R. J. 1972. A new heteromyid rodent from the early Oligocene of Natrona County, Wyoming. Proceedings of the Biological Society of Washington, 85:179–190.

ESHELMAN, R. E. 1975. Geology and paleontology of the early Pleistocene (late Blancan) White Rock fauna from north-central Kansas. Claude W. Hibbard Memorial Volume 4, University of Michigan Papers on Paleontology, 13:1–60.

FAHLBUSCH, V. 1985. Origin and evolutionary relationships among geomyoids. Pp. 617–629, *in* Evolutionary relationships among rodents, a multidisciplinary analysis (W. P. Luckett and J.-L. Hartenberger, eds.). Plenum Press, New York, 721 pp.

GALBREATH, E. C. 1948. A new species of heteromyid rodent from the Middle Oligocene of northeast Colorado with remarks on the skull. University of Kansas Publications, Museum of Natural History, 1:285–300.

———. 1953. A contribution to the Tertiary geology and paleontology of northeastern Colorado. University of Kansas Paleontological Contributions, Vertebrata, 4:1–120.

———. 1961. The skull of *Heliscomys,* an Oligocene heteromyid rodent. Transactions of the Kansas Academy of Science, 64:225–230.

———. 1962. The deciduous premolars of the Oligocene heteromyid rodent *Helyiscomys* [sic]. Transactions of the Kansas Academy of Science, 65:263–265.

GAWNE, C. E. 1975. Rodents from the Zia Sand Miocene of New Mexico. American Museum Novitates, 2586:1–25.

GAZIN, C. L. 1930. A Tertiary vertebrate fauna from the upper Cuyama drainage basin, California. Carnegie Institution of Washington Publication, 404: 55–76, pls. 1–4.

———. 1932. A Miocene mammalian fauna from southeastern Oregon. Carnegie Institution of Washington Publication, 418:37–86, pls. 1–6.

———. 1942. The late Cenozoic vertebrate faunas from the San Pedro Valley, Ariz. Proceedings of the United States National Museum, 92:475–518.

GERVAIS, P. 1853. Description ostéologique de l'*Anomalurus,* et remarques sur la classification naturelle des Rongeurs. Annales des Sciences Naturelles, Paris, Troisième Série, Zoologie, 20:238–246.

GIDLEY, J. W. 1922. Preliminary report on fossil vertebrates of the San Pedro Valley, Arizona, with descriptions of new species of Rodentia and Lagomorpha. United States Geological Survey, Professional Paper, 131-E:119–131, pls. 34–35.

GOLZ, D. J., G. T. JEFFERSON, AND M. P. KENNEDY. 1977. Late Pliocene vertebrate fossils from the Elsinore fault zone, California. Journal of Paleontology, 51:864–866.

GRAY, J. E. 1841. A new genus of Mexican glirine Mammalia. The Annals and Magazine of Natural History, series 1, 7:521–522.

———. 1868. Synopsis of the species of Saccomyinae or pouched mice in the collection of the British Museum. Proceedings of the Zoological Society of London, 1868:199–206.

GREEN, M. 1971. Additions to the Mission vertebrate fauna, Lower Pliocene of South Dakota. Journal of Paleontology, 45:486–490.

HAGER, M. W. 1974. Late Pliocene and Pleistocene history of the Donnelly Ranch vertebrate site, southeastern Colorado. University of Wyoming, Contributions to Geology, Special Paper, 2:1–62.

HALL, E. R. 1930. Rodents and lagomorphs form the later Tertiary of Fish Lake Valley, Nevada. University of California Publications, Bulletin of the Department of Geological Sciences, 19:295–312.

HARRIS, J. M., AND A. E. WOOD. 1969. A new genus of eomyid rodent form the Oligocene Ash Spring local fauna of Trans-Pecos Texas. Pearce-Sellards Series, Texas Memorial Museum, 14:1–7.

HARRISON, J. A. 1978. Mammals of the Wolf Ranch local fauna, Pliocene of the San Pedro Valley, Arizona. Occasional Papers of the Museum of Natural History, the University of Kansas, 73:1–18.

HARTENBERGER, J.-L. 1985. The Order Rodentia: major questions on their evolutionary origin, relationships and suprafamilial systematics. Pp. 1–33, *in* Evolutionary relationships among rodents, a multidisciplinary analysis (W. P. Luckett and J.-L. Hartenberger, eds.). Plenum Press, New York, 721 pp.

HIBBARD, C. W. 1937. Additional fauna of Edson Quarry of the Middle Pliocene of Kansas. The American Midland Naturalist, 18:460–464.

———. 1939. Notes on additional fauna of Edson Quarry of the Middle Pliocene of Kansas. Transactions of Kansas Academy of Science, 42:457–462.

———. 1941*a.* The Borchers fauna, a new Pleistocene interglacial fauna from Meade County, Kansas. State Geological Survey of Kansas, Bulletin, 38:197–220.

———. 1941*b.* New mammals from the Rexroad fauna, Upper Pliocene of Kansas. The American Midland Naturalist, 26:337–368.

———. 1942. Pleistocene mammals from Kansas. State Geological Survey of Kansas, Bulletin, 41:261–269.

———. 1943. *Etadonomys,* a new Pleistocene heteromyid rodent, and notes on other Kansas mammals. Transactions of Kansas Academy of Science, 46:185–191.

———. 1949. Pliocene Saw Rock Canyon fauna in Kansas. Contributions from the Museum of Paleontology, University of Michigan, 7:91–105.

———. 1950. Mammals of the Rexroad Formation from Fox Canyon, Meade County, Kansas. Contributions from the Museum of Paleontology, University of Michigan, 8:113–192.

———. 1953*a.* A change in the definition of the generic characters on the genus *Dipodomys* Gray. Journal of Mammalogy, 34:264–265.

———. 1953*b*. The Saw Rock Canyon fauna and its stratigraphic significance. Papers of the Michigan Academy of Science, Arts and Letters, 38:387–411.

———. 1954. Second contribution to the Rexroad fauna. Transactions of Kansas Academy of Science, 57:221–237.

———. 1955. The Jinglebob Interglacial (Sangamon?) fauna from Kansas and its climatic significance. Contributions from the Museum of Paleontology, University of Michigan, 12:179–228.

———. 1956. Vertebrate fossils from the Meade Formation of southwestern Kansas. Papers of the Michigan Academy of Science, Arts, and Letters, 41:145–203, pls. 1–2.

———. 1962. Two new rodents from the early Pleistocene of Idaho. Journal of Mammalogy, 43:482–485.

———. 1972. Class Mammalia. Pp. 77–130, *in* Early Pleistocene pre-glacial and glacial rocks and faunas of north-central Nebraska (M. F. Skinner, C. W. Hibbard, et al.). Bulletin of the American Museum of Natural History, 148:1–148.

HIBBARD, C. W., AND D. W. TAYLOR. 1960. Two late Pleistocene faunas from southwestern Kansas. Contributions from the Museum of Paleontology, the University of Michigan, 16:1–223.

HILL, J. E. 1937. Morphology of the pocket gopher mammalian genus *Thomomys*. University of California Publications in Zoology, 42:81–172.

JACOBS, L. L. 1977. Rodents of the Hemphillian Redington local fauna, San Pedro Valley, Arizona. Journal of Paleontology, 51:505–519.

JAMES, G. T. 1963. Paleontology and nonmarine stratigraphy of the Cuyama Valley Badlands, California. Part I. Geology, faunal interpretations, and systematic descriptions of Chiroptera, Insectivora, and Rodentia. University of California Publications in Geological Sciences, 45:i–iv, 1–171.

KELLOGG, L. 1910. Rodent fauna of the late Tertiary beds at Virgin Valley and Thousand Creek, Nevada. University of California Publications, Bulletin of the Department of Geology, 5:421–437.

KLINGENER, D. 1968. Rodents of the Mio-Pliocene Norden Bridge local fauna, Nebraska. The American Midland Naturalist, 80:65–74.

KORTH, W. W. 1979*a*. Taphonomy of microvertebrate fossil assemblages. Annals of Carnegie Museum, 48:235–285.

———. 1979*b*. Geomyoid rodents from the Valentine Formation of Knox County, Nebraska. Annals of Carnegie Museum, 48:287–310.

———. 1987. New rodents (Mammalia) from the late Barstovian (Miocene) Valentine Formation, Nebraska. Journal of Paleontology, 61:1058–1064.

———. 1989. Geomyoid rodents (Mammalia) from the Orellan (middle Oligocene) of Nebraska. Pp. 31–46, *in* Papers on fossil rodents in honor of Albert Elmer Wood (C. C. Black and M. R. Dawson, eds.). Science Series, Natural History Museum of Los Angeles County, 33:1–192.

KORTH, W. W., B. E. BAILEY, AND R. M. HUNT, JR. 1990. Geomyoid rodents from the early Hemingfordian (Miocene) of Nebraska. Annals of Carnegie Museum, 59:25–47.

KORTH, W. W., AND A. R. TABRUM. In press. Fossil mammals of the Orellan Cook Ranch local fauna, southwest Montana. Part 1: Eomyid and heteromyid rodents. Annals of Carnegie Museum.

KORTH, W. W., J. H. WAHLERT, AND R. J. EMRY. In press. A new species of the Oligocene rodent *Heliscomys* and recognition of a new family of the Geomyoidea (Mammalia). Journal of Vertebrate Paleontology.

KURTÉN, B., AND E. ANDERSON. 1980. Pleistocene mammals of North America. Columbia University Press, New York, 442 pp.

LINDSAY, E. H. 1972. Small mammal fossils from the Barstow Formation, California. University of California Publications in Geological Sciences, 93:1–104.

———. 1974. The Hemingfordian mammal fauna of the Vedder locality, Branch Canyon Formation, Santa Barbara County, California. Part II: Rodentia (Eomyidae and Heteromyidae). PaleoBios, 16:1–20.

———. 1978. Late Cenozoic vertebrate faunas, southeastern Arizona. Pp. 269–275, *in* Land of Cochise; southeastern Arizona (J. F. Callender et al., eds.). New Mexico Geological Society Annual Field Conference Guidebook, 29:372 pp.

LINDSAY, E. H., AND L. L. JACOBS. 1985. Pliocene small mammal fossils from Chihuahua, Mexico. Paleontologia Mexicana, 51:1–53.

LUCKETT, W. P. 1985. Superordinal and intraordinal affinities of rodents: developmental evidence from the dentition and placentation. Pp. 227–276, *in* Evolutionary relationships among rodents, a multidisciplinary analysis (W. P. Luckett and J.-L. Hartenberger, eds.). Plenum Press, New York, 721 pp.

LUNDELIUS, E. L., JR., R. W. GRAHAM, E. ANDERSON, J. GUILDAY, J. A. HOLMAN, D. W. STEADMAN, AND S. D. WEBB. 1983. Terrestrial vertebrate faunas. Pp. 311–353, *in* Late-Quaternary environments of the United States. Vol. 1. The late Pleistocene (S. C. Porter, ed.). University of Minnesota Press, 480 pp.

MACDONALD, J. R. 1963. The Miocene faunas from the Wounded Knee area of western South Dakota. Bulletin of the American Museum of Natural History, 125:139–238, 2 maps.

———. 1970. Review of the Miocene Wounded Knee faunas of southwestern South Dakota. Bulletin of the Los Angeles County Museum of Natural History Science, 8:1–82.

MACDONALD, L. J. 1972. Monroe Creek (early Miocene) microfossils from the Wounded Knee area, South Dakota. South Dakota Geological Survey Report of Investigations, 105:1–43.

MCGREW, P. O. 1941. Heteromyids from the Miocene and Lower Oligocene. Geological Series of Field Museum of Natural History, 8:55–57.

MARTIN, J. E. 1976. Small mammals from the Miocene Batesland Formation of South Dakota. Contributions to Geology, the University of Wyoming, 14:69–98.

———. 1984. A survey of Tertiary species of *Perognathus* (Perognathinae) and a description of a new genus of Heteromyinae. Pp. 90–112, *in* Papers in vertebrate paleontology honoring Robert Warren Wilson (R. M. Mengel, ed.). Special Publications, Carnegie Museum of Natural History, 9:1–186.

MATTHEW, W. D. 1910. On the osteology and relationships of *Paramys*, and the affinities of the Is-

chyromyidae. Bulletin of the American Museum of Natural History, 28:43–71.
———. 1924. Third contribution to the Snake Creek fauna. Bulletin of the American Museum of Natural History, 50:59–210.
MEAD, J. I., R. S. THOMPSON, AND T. R. VAN DEVENDER. 1982. Late Wisconsinan and Holocene fauna from Smith Creek Canyon, Snake Range, Nevada. Transactions of the San Diego Society of Natural History, 20:1–26.
MERRIAM, C. H. 1889. Revision of the North American pocket mice. North American Fauna, 1:1–36, pls. 1–4.
———. 1891. Description of a new genus and species of dwarf kangaroo rat from Nevada (*Microdipodops megacephalus*). North American Fauna, 5:115–117.
———. 1902. Twenty new pocket mice (*Heteromys* and *Liomys*) from Mexico. Proceedings of the Biological Society of Washington, 15:41–50.
MILLER, S. J. 1979. The archaeological fauna of four sites in Smith Creek Canyon. Pp. 271–329, *in* The archaeology of Smith Creek Canyon, eastern Nevada (D. R. Tuohy and D. L. Rendall, eds.). Nevada State Museum Anthropological Papers, 17:1–394.
MILLER, W. E. 1971. Pleistocene vertebrates of the Los Angeles Basin and vicinity (exclusive of Rancho La Brea). Bulletin of the Los Angeles County Museum of Natural History Science, 10:1–124.
MUNTHE, J. 1979. Summary of Miocene vertebrate fossils of the Granite Mountains Basin, central Wyoming. Contributions to Geology, the University of Wyoming, 18:33–46.
———. 1988. Miocene mammals of the Split Rock area, Granite Mountains Basin, central Wyoming. University of California Publications, Geological Sciences, 126:i–vii, 1–136, pls. 1–3.
MUNTHE, K. 1981. Skeletal morphology and function of the Miocene rodent *Schizodontomys harkseni*. PaleoBios, 35:1–33.
OLSEN, S. J. 1964. The stratigraphic importance of a Lower Miocene vertebrate fauna from north Florida. Journal of Paleontology, 38:477–482.
PATTON, T. H. 1969. An Oligocene land vertebrate fauna from Florida. Journal of Paleontology, 43:543–546.
REEDER, W. G. 1957. A review of Tertiary rodents of the Family Heteromyidae. University Microfilms International, Ann Arbor, 618 pp.
———. 1960*a*. Two new rodent genera from the Oligocene White River Formation (family Heteromyidae). Fieldiana Geology, 10:511–524.
———. 1960*b*. A new rodent genus (family Heteromyidae) from the Tick Canyon Formation of California. Bulletin, Southern California Academy of Sciences, 59:121–132.
RENSBERGER, J. M. 1971. Entoptychine pocket gophers (Mammalia, Geomyoidea) of the early Miocene John Day Formation, Oregon. University of California Publications in Geological Sciences, 90: 1–209.
———. 1973*a*. Pleurolicine rodents (Geomyoidea) of the John Day Formation, Oregon, and their relationships to taxa from the early and middle Miocene, South Dakota. University of California Publications in Geological Sciences, 102:1–95, pls. 1–17.
———. 1973*b*. *Sanctimus* (Mammalia, Rodentia) and the phyletic relationships of the large Arikareean geomyoids. Journal of Paleontology, 47:835–853.
SCHULTZ, G. E. 1965. Pleistocene vertebrates from the Butler Spring local fauna, Meade County, Kansas. Papers of the Michigan Academy of Science, Arts, and Letters, 50:235–265.
———. 1969. Geology and paleontology of a late Pleistocene basin in southwest Kansas. Geological Society of America, Special Papers, 105:1–85.
SETOGUCHI, T. 1978. Paleontology and geology of the Badwater Creek area, central Wyoming. Part 16. The Cedar Ridge local fauna (late Oligocene). Bulletin of Carnegie Museum of Natural History, 9:1–61.
SHOTWELL, J. A. 1956. Hemphillian mammalian assemblage from northeastern Oregon. Bulletin of the Geological Society of America, 67:717–738.
———. 1967. Late Tertiary geomyoid rodents of Oregon. Bulletin of the Museum of Natural History, University of Oregon, 9:1–51.
SHOTWELL, J. A., AND D. E. RUSSELL. 1963. Mammalian fauna of the Upper Juntura Formation, the Black Butte local fauna. Pp. 42–69, *in* The Juntura Basin: studies in earth history and paleoecology (J. A. Shotwell, ed.). Transactions of the American Philosophical Society, n.s., 53:1–77.
SIMPSON, G. G. 1945. The principles of classification and a classification of mammals. Bulletin of the American Museum of Natural History, 85:1–350.
SKINNER, M. F., S. M. SKINNER, AND R. J. GOORIS. 1968. Cenozoic rocks and faunas of Turtle Butte, south-central South Dakota. Bulletin of the American Museum of Natural History, 138:379–436.
———. 1977. Stratigraphy and biostratigraphy of late Cenozoic deposits in central Sioux County, western Nebraska. Bulletin of the American Museum of Natural History, 158:263–370.
SKWARA, T. 1988. Mammals of the Topham local fauna: Early Miocene (Hemingfordian), Cypress Hills Formation, Saskatchewan. Saskatchewan Parks, Recreation and Culture, Museum of Natural History, Natural History Contributions, 9:1–169.
STEHLIN, H. G., AND S. SCHAUB. 1951. Die Trigonodontie der simplicidentaten Nager. Schweizerische palaeontologische Abhandlungen, 67:1–385.
STEVENS, M. S., J. B. STEVENS, AND M. R. DAWSON. 1969. New early Miocene formation and vertebrate local fauna, Big Bend National Park, Brewster County, Texas. The Pearce–Sellards Series, Texas Memorial Museum, 15:1–53.
STORER, J. E. 1970. New rodents and lagomorphs from the Upper Miocene Wood Mountain Formation of southern Saskatchewan. Canadian Journal of Earth Sciences, 7:1125–1129.
———. 1975. Tertiary mammals of Saskatchewan. Part III: the Miocene fauna. Life Sciences Contribution, Royal Ontario Museum, 103:1–134.
———. 1978. Rodents of the Calf Creek local fauna (Cypress Hills Formation, Oligocene, Chadronian), Saskatchewan. Saskatchewan Culture and Youth, Museum of Natural History, Natural History Contributions, 1:1–54.
———. 1988. The rodents of the Lac Pelletier lower fauna, late Eocene (Duchesnean) of Saskatchewan. Journal of Vertebrate Paleontology, 8:84–101.
SUTTON, J. F., AND C. C. BLACK. 1975. Paleontology of the earliest Oligocene deposits in Jackson Hole,

Wyoming. Part I. Rodents exclusive of the family Eomyidae. Annals of Carnegie Museum, 45:299–315.

Thaler, L. 1966. Les Rongeurs fossiles du Bas-Languedoc dans leurs rapports avec l'histoire des faunes et la stratigraphie du Tertiare d'Europe. Mémoires du Muséum National d'Histoire Naturelle, Sér. C., Sciences de la Terre, 17:1–295, pls. 1–27.

Thompson, J. V. 1815. Description of a new species of the genus *Mus,* belonging to the section of pouched rats. Transactions of the Linnean Society of London, 11:161–163, pl. 10.

Troxell, E. L. 1923. *Diplolophus,* a new genus of rodents. American Journal of Science, 5:157–159.

Voorhies, M. R. 1974. Fossil pocket mouse burrows in Nebraska. The American Midland Naturalist, 91: 492–498.

———. 1975. A new genus and species of fossil kangaroo rat and its burrow. Journal of Mammalogy, 56:160–176.

Wahlert, J. H. 1976. *Jimomys labaughi,* a new geomyoid rodent from the early Barstovian of North America. American Museum Novitates, 2591:1–6.

———. 1978. Cranial foramina and relationships of the Eomyoidea (Rodentia, Geomorpha). Skull and upper teeth of *Kansasimys.* American Museum Novitates, 2645:1–16.

———. 1983. Relationships of the Florentiamyidae (Rodentia, Geomyoidea) based on cranial and dental morphology. American Museum Novitates, 2769:1–23.

———. 1984. *Kirkomys,* a new florentiamyid (Rodentia, Geomyoidea) form the Whitneyan of Sioux County, Nebraska. American Museum Novitates, 2793:1–8.

———. 1985. Skull morphology and relationships of geomyoid rodents. American Museum Novitates, 2812:1–20.

Wahlert, J. H., and R. A. Souza. 1988. Skull morphology of *Gregorymys* and relationships of the Entoptychinae (Rodentia, Geomyidae). American Museum Novitates, 2922:1–13.

Wahlert, J. H., and W. v. Koenigswald. 1985. Specialized enamel in incisors of eomyid rodents. American Museum Novitates, 2832:1–12.

Webb, S. D. 1984. Ten million years of mammal extinctions in North America. Pp. 189–210, *in* Quaternary extinctions: a prehistoric revolution (P. S. Martin and R. G. Klein, eds.). University of Arizona Press, Tucson, 892 pp.

Whistler, D. B. 1984. An early Hemingfordian (early Miocene) fossil vertebrate fauna from Boron, western Mojave Desert, California. Contributions in Science, Natural History Museum of Los Angeles County, 355:1–36.

Wied, Maximilian, Prince of. 1839. Über einige Nager mit äusseren Backentaschen aus dem westlichen Nord-America. I. Über ein paar neue Gattungen der Nagethiere mit äusseren Backentaschen. Nova Acta Physico-Medica, Academiae Caesareae Leopoldino-Carolinae, Naturae Curiosorum, 19:365–374, pl. 34.

Williams, D. F., and H. H. Genoways. 1979. A systematic review of the olive-backed pocket mouse, *Perognathus fasciatus* (Rodentia, Heteromyidae). Annals of Carnegie Museum, 48:73–102.

Wilson, R. W. 1936. A Pliocene rodent fauna from Smiths Valley, Nevada. Carnegie Institution of Washington Publication, 473:15–34, pls. 1–2.

———. 1939. Rodents and lagomorphs of the late Tertiary Avawatz fauna, California. Carnegie Institution of Washington Publication, 514:31–38.

———. 1940. Two new Eocene rodents from California. Carnegie Institution of Washington Publication, 514:85–95, pls. 1–2.

———. 1949*a*. On some White River fossil rodents. Carnegie Institution of Washington Publication, 584: 27–50.

———. 1949*b*. Rodents and lagomorphs of the Upper Sespe. Carnegie Institution of Washington Publication, 584:51–65.

———. 1949*c*. Early Tertiary rodents of North America. Carnegie Institution of Washington Publication, 584:67–164.

———. 1960. Early Miocene rodents and insectivores from northeastern Colorado. The University of Kansas, Paleontological Contributions, Vertebrata, 7:1–92.

Wood, A. E. 1931. Phylogeny of the heteromyid rodents. American Museum Novitates, 501:1–19.

———. 1932. New heteromyid rodents from the Miocene of Florida. Florida State Geological Survey Bulletin, 10:43–51.

———. 1933. A new heteromyid rodent from the Oligocene of Montana. Journal of Mammalogy, 14: 134–141.

———. 1935*a*. Evolution and relationships of the heteromyid rodents with new forms from the Tertiary of western North America. Annals of Carnegie Museum, 24:73–262.

———. 1935*b*. Two new rodents from the John Day Miocene. American Journal of Science, 30:368–372.

———. 1936. Fossil heteromyid rodents in the collection of the University of California. American Journal of Science, 32:112–119.

———. 1937*a*. Additional material from the Tertiary of the Cuyama Basin of California. American Journal of Science, 33:29–43.

———. 1937*b*. Rodentia. Part II. Pp. 155–269, *in* The mammalian fauna of the White River Oligocene (W. B. Scott, G. L. Jepsen, and A. E. Wood, eds.). Transactions of the American Philosophical Society, n.s., 28:155–269.

———. 1939. Additional specimens of the heteromyid rodent *Heliscomys* from the Oligocene of Nebraska. American Journal of Science, 237:550–561.

———. 1947. Miocene rodents from Florida. Bulletin of the Museum of Comparative Zoology, 99: 489–494, pl. 1.

———. 1974. Early Tertiary vertebrate faunas, Vieja Group, Trans-Pecos Texas: Rodentia. Texas Memorial Museum Bulletin, 21:1–112.

———. 1980. The Oligocene rodents of North America. Transactions of the American Philosophical Society, 70(5):1–68.

Zakrzewski, R. J. 1969. The rodents from the Hagerman local fauna, Upper Pliocene of Idaho. Contributions from the Museum of Paleontology, the University of Michigan, 23:1–36.

———. 1970. Notes on kangaroo rats from the Pliocene of southwestern Kansas, with the description of a new species. Journal of Paleontology, 44:474–477.

———. 1981. Kangaroo rats from the Borchers local fauna, Blancan, Meade County, Kansas. Transactions of the Kansas Academy of Science, 84:78–88.

ZITTEL, K. A. V. 1891–1893. Handbuch Der Paleontologie. I Abt. Palaeozoologie. IV Band. Vertebrata (Mammalia). R. Oldenbourg, Munchen, 799 pp.

TAXONOMY

Daniel F. Williams, Hugh H. Genoways, and Janet K. Braun

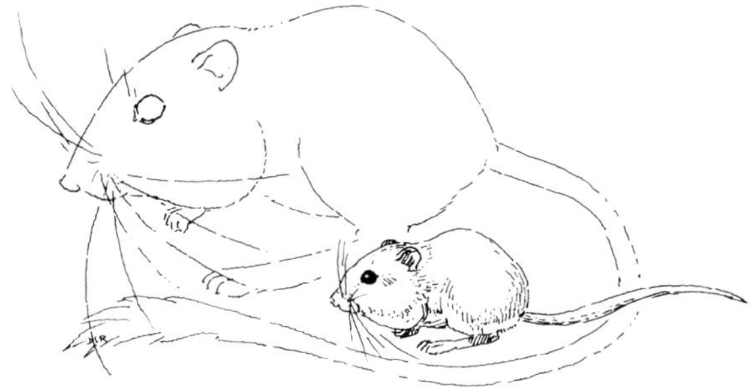

Introduction

Herein the taxonomy of the family is reviewed and diagnoses of Recent species and accounts of all currently recognized Recent species and subspecies are provided. The objective is to present a current taxonomy for the family rather than a systematic review. Unfortunately, there have been no comprehensive reviews for most genera for 80 years or more and the current taxonomy of many species clearly is unsatisfactory, as is the understanding of the relationships between most of the described fossil and Recent taxa.

Arrangement of subfamilies, genera, species, and subspecies is alphabetical rather than phylogenetic. Recognition of species and subspecies, and taxonomic treatment of taxa at the generic level and above, were generally based on the most recent systematic reviews and published analyses of arrangements. For cases where there was disagreement in the literature on taxonomy, opinion derived from studies of distribution and variation was favored over unsubstantiated belief.

Measurements of holotypes (except weight in g) are given in mm. Measurements are as listed in the original description, except where conversion to mm was necessary. Where different terms were used in the literature for the same measurement, we used a single term for consistency except in cases where we were unsure how an author measured a trait. If no measurements were given for holotypes (more than half of the species and subspecies) or where the measurements in the original description differed from our own, either the published measurements in a subsequent revision, with appropriate citation, or our measurements were used. For a few taxa, measurements were made by others at our request. These are noted in the acknowledgments. Museums where holotypes are stored are identified by acronyms preceding the catalog numbers; acronyms are those listed by Yates et al. (1987) for North American collections or by convention for those few outside North America. Type localities are as given in the original description except all numbers ≥1,000 have commas inserted for clarity, accent marks are used in spelling locality names where appropriate, and ft and m have no period. Other necessary amendments to the type localities are enclosed in brackets.

Taxonomic History of the Family Heteromyidae

The name Saccomyna Gray, 1843 was the first applied to members of the family, based on the genus *Saccomys* Fr. Cuvier, 1823, with type species *S. anthophilus.* The origin of the type specimen was uncertain and its type locality, given as North America, was thought to be one of the West Indian islands (Baird, 1858). Because Cuvier's name could not be applied to any of the known species, it was later considered a nomen dubium and *Saccomys* was relegated to the synonymy of *Heteromys* Desmarest, 1817 (Peters, 1874, as quoted in translation in Coues, 1877:487). Coues (1877), based on remarks by Peters (as quoted in Coues, 1877:487), reached this conclusion but believed it unnecessary to discard the synonym *Saccomys* as the basis of the family name. Waterhouse (1848) used the name Saccomyina to contain all of the North American rodents with external cheek pouches, encompassing the pocket gophers and heteromyids. The family name, Dipodomyna Gervais, 1853, which applied only to the currently defined heteromyids, was proposed next. Baird (1858) amended Gray's name to Saccomyidae, and used the subfamily names Geomyinae for the pocket gophers and Saccomyinae for the heteromyids. Later, Gray (1868) used the name Saccomyinae for the known species of heteromyids and the subdivisions (tribes), Dipodomyina (*Dipodomys*) and Heteromyina (*Abromys, Cricetodipus, Dasynotus, Heteromys, Perognathus,* and *Saccomys—Abromys* and *Cricetodipus* are synonyms of *Perognathus,* and *Dasynotus* and *Saccomys* are synonyms of *Heteromys*). Gray (1868:199) used Baird's (1858) subfamily spelling but wrote only of "the family" of pouched mice, from which he excluded the pocket gophers. Gray (1868) treated *Saccomys anthophilus* as a valid species, not mentioning its possible synonymy with *Heteromys* (Hafner and Hafner, 1983:4, incorrectly implied that Gray recognized that *Saccomys* was a synonym of *Heteromys*; see Peters above). Gill (1872) first formally used the name Saccomyidae as equivalent to the subfamily Saccomyinae of Baird (1858) and Gray (1868), and combined the Saccomyidae and Geomyidae in the superfamily Saccomyoidea. Alston (1876) first used the name Heteromyinae to contain the species included in the family Saccomyidae of Gill (1872). Alston (1876) discarded the family and subfamily names of earlier authors because *Saccomys* was a synonym of *Heteromys* and he believed it necessary to derive the name from the latter. Alston (1876) combined the Geomyinae and Heteromyinae in the family Geomyidae. Coues (1877) reviewed the prior taxonomic history of the heteromyids and included other family or subfamily names that had been used for various heteromyids. Allen and Chapman (1893) first used the name Heteromyidae to constitute the family as currently defined. Weber (1904) first used the superfamily name Geomyoidea for the Geomyidae and Heteromyidae. Wood (1936) described a new subfamily of Heteromyidae, Florentiamyinae, based on *Florentiamys loomisi*; Rensberger (1973) transferred Florentiamyinae to the Geomyidae, stating that it shared lineage with Entoptychinae and Pleurolicinae.

Recognition of one or two families for the geomyids and heteromyids and their grouping as a superfamily have generated persistent disagreements for nearly 150 years. Most recently, Shotwell (1967) and Lindsay (1972) classified geomyids and heteromyids in a single family, while others (e.g., Hafner, 1982; Hafner and Hafner, 1983; Wahlert, 1985) classified them as separate families in the superfamily Geomyoidea. Although the two groups are recognizably distinct biochemically and structurally, recognition of one or two families is arbitrary; thus, there probably will continue to be disagreements about their classification at the family level. Herein, the taxonomic arrangement of Heteromyidae above the family level follows the classification presented by Carleton (1984).

Classification and Species Accounts

Class Mammalia Linnaeus, 1758
Subclass Theria Parker and Haswell, 1897
Infraclass Eutheria Gill, 1872

Order Rodentia Bowdich, 1821
Suborder Sciurognathi Tullberg, 1899
Infraorder Myomorpha Brandt, 1855
Superfamily Geomyoidea Weber, 1904

Family Heteromyidae Allen and Chapman, 1893

Diagnosis.—Size small to medium-small for order, total length varying from about 100 to 370 mm and mass ranging from about 5 to 170 g; tail shorter to much longer than length of head and body; body forms quadrupedal, subricochetal (some fossil forms), and ricochetal; locomotion scansorial to ricochetal; hind limbs considerably larger than forelimbs; hind feet longer than typical murids and cricetids; manus with four, clawed digits; pes with four or five clawed digits; claws of manus elongate and slender; fore- and hind feet slender; soles of hind feet naked to clothed with dense covering of fur; invagination of skin on face forms fur-lined cheek pouch opening anteriorly near the mouth; hairs of two or three general types, long overhairs with or without a flattened cross-section and concave trough on the dorsal surface, and underfur of shorter, often slightly curved hairs; hairs with imbricate, flattened cuticular scales; scales of hairs in single layer; medulla cells of hairs variable, but always compound; dental formula i 1/1, c 0/0; p 1/1, m 3/3, total 20 teeth; anterior face of upper incisor smooth or with weak or strong groove; cheek teeth brachydont to hypsodont and definitive to evergrowing; usually six cusps per molar, three on each loph; enamel of molars rarely divided into two plates, never reduced to one; skull light and thin; width across skull at mastoid bullae often greater and never much less than width across maxillary arches; mastoids inflated and bullous and forming part of lateral and (in all but Heteromyinae) dorsal cranial surface; occipital small and limited in area, but extending onto dorsal surface of cranium; interorbital region wider than rostrum; tympanic inflated or highly inflated, and vesicular; palate nearly horizontal, little if any below level of zygomatic arches; nasals extend far beyond incisors; zygomatic arches slender, with greatly reduced jugal (malar) and nearly contacting the tympanic; frontal trapezoid and broad; parietal broad and quadrate to pentagonal or triangular; interparietal broad to exceedingly tiny or obsolete; squamosal mostly or entirely confined to orbit; infraorbital canal long and opening at a large perforation anteriorly on rostrum and protected from muscle pressure by countersinking in vacuity; mastoid foramen minute or absent; sphenopterygoid foramen occupied only by a large vein or by a large vein and part of the internal pterygoid muscle; vascular canal in basisphenoid vestigial; jugular (posterior lacerate) foramen situated between tympanic bulla and basioccipital and large and slit-like; coronoid process small, sloping, and below level of mandibular condyle; mandibles small and widely diverging posteriorly; baculum with bulbous base tapering to a thin, upturned tip or, exceptionally, an ornate, trifid tip; male reproductive tract typically with the full complement of accessory glands common for muroid rodents, or without ventromedial prostate or preputial or both glands (*Heteromys* in so far as is known is atypical in having only three of the seven glands); origin of temporal muscle restricted to lateral part of roof of skull; M. platysma cericale inserts into pouch retractor muscle at fusion zone; M. cervico-auricularis divided into three slips; M. buccinatorius pars intermaxillaris anterior on palate; M. adductor brevis double; M. semitendinosus with one head; and pedal flexor muscles reduced (Arata, 1964; Burt, 1936; Ellerman, 1940; Hafner and Hafner, 1983; Hill, 1935; Homan and Genoways, 1978; Merriam, 1889; Ryan, 1986, 1989; Wahlert, 1985; Wood, 1935).

Remarks.—The diagnosis is based mostly on Recent species. The Heteromyidae is divided into three subfamilies: Dipodomyinae, Heteromyinae, and Perognathinae. The earliest known heteromyid dates from the Middle Oligocene genus *Heliscomys* Cope,

1873, which was not assigned to any of the recognized subfamilies (Wood, 1935), but was later included in the subfamily Perognathinae (Lindsay, 1972). Rensberger (1973), however, depicted *Heliscomys* as the common ancestor of geomyids and heteromyids. For a comprehensive review of fossil forms see the chapter, Fossil Record, in this volume by J. H. Wahlert.

Key to the Subfamilies

1. Body form ricochetal; soles of hind feet densely covered with short hairs; mastoid bullae enormously enlarged, forming half or more of the dorsal surface of the skull posterior to the orbits; mastoid bulla with hollow interior .... Dipodomyinae

1′. Body form scansorial; soles of hind feet naked or partly covered with sparse, short hairs; mastoid bullae moderately to strongly inflated but not forming more than about half of the dorsal surface of the skull posterior to the orbits; bullae with spongy network of bony trabeculae in interior 2

2. Adult pelage coarse with abundant stiff, flattened spine-like hairs on dorsal and lateral surfaces of body; juvenile pelage and adult pelage of a few taxa of fine hairs and soft, flattened and anteriorly-grooved bristles; mastoid bullae moderately inflated; mastoid bullae never projecting to postero-dorsal plane of occiput (dorsal to the external auditory meatus the lateral wall of the skull is composed entirely of the squamosal); ventral plane of tympanic bulla higher than occlusal plane of cheek teeth .... Heteromyinae

2′. Pelage silky or with a sparse to moderate mix of coarse, spine-like hairs and silky hairs on dorsal and lateral surfaces of body; mastoid bullae strongly inflated; mastoid bullae projecting to or beyond posterior plane of occiput; except for a thin spine of the squamosal extending over the external auditory meatus, lateral wall of the skull dorsal to the meatus is composed of mastoid bulla; ventral plane of tympanic bulla lower than plane of occlusal surface of cheek teeth .... Perognathinae

Subfamily Dipodomyinae Coues, 1875

Diagnosis.—Size varying from small to large for family; total length from 130 to 370 mm; weights from 10 to about 170 g; body form ricochetal in Recent species, subricochetal in some fossil forms; forelimbs slender; hind limbs enormously developed compared to forelimbs; length of tail from slightly greater than head and body to 1.65 times head and body length; soles of hind feet with dense covering of hair; hind feet long; manus with four, clawed digits; pes with four or five, clawed digits; hair long, smooth, slightly curved, and oval in cross section; no obvious distinction between overhair and underfur; upper incisors strongly grooved; molars hypsodont in Recent forms, rooted, and tuberculate; cheek teeth nearly evergrowing or evergrowing in advanced forms; cusps short relative to elongated base of tooth ("alveolar part" of Wood, 1935:110); enamel pattern of cheek teeth not persisting with age; lophs of P4 unite first at or near center of tooth; protoloph of P4 usually single-cusped; lophs of upper molars unite first at the lingual margin and progress to buccal margins; lophs of p4 unite at center of tooth, presenting an X-pattern; lophs of lower molars unite first at buccal margin and next at the lingual margin, always forming an H-pattern; foramen and pit between m3 and the base of the coronoid process present or absent; palate broad, extending to or behind level of M3; center of palate between cheek teeth with or without ridges; squamosal perforated by auditory bulla; incisive foramen small; auditory bullae enormously expanded; mastoid bullae appearing on dorsal surface of skull; inflated interior of bullae mostly hollow, without spongy network of trabeculae; middle ear with stapedius muscle; auditory bullae extend posteriorly well beyond plane of occiput; median ventral foramina in the "central" of caudal vertebrae present or absent; scapula prolonged posteriorly; obturator foramen triangular; articulation of trapezium with scapholunar;

articulation of pes bones variable: calcaneonavicular, calcaneo-cuneiform, or astragalus-cuboid; dorsal surface of ectocuneiform hour-glass-shaped; baculum of moderate length relative to body size, slender, with a swollen base, and an upturned tip; tip of baculum simple; phallus moderate in length, slender, and with external spines and urethral lappets; M. bucco-naso-labialis with anterior origin from dorsal margin of the bulge created by the incisor root; M. cleidomastoideus absent; extensores breves muscles absent; M. adductor minimum superficial at origin; M. abductor hallucis absent; lumbricales muscles reduced or absent (Burt, 1936; Hafner, 1978; Hafner and Hafner, 1983; Hall, 1941; Nader, 1966; Ryan, 1989; Webster and Webster, 1975; Wood, 1935).

Remarks.—The above diagnosis is based largely on Recent species. The Dipodomyinae includes the kangaroo mice, *Microdipodops,* which share many structural features with the Perognathinae. Hafner and Hafner (1983) considered sharing to be coincidental retention of ancestral traits. The union of *Dipodomys* and *Microdipodops* in the subfamily Dipodomyinae, first advocated by Reeder (1956) and sustained by Hafner and Hafner (1983), Lindsay (1972), Ryan (1989), Wahlert (1985), and Webster and Webster (1975), is based on structural similarities of muscular, skeletal, phallic, and auditory systems, and patterns of protein electrophoresis. Development of ricochetal locomotion in the two living genera of the subfamily may be the result of parallel evolution (Hafner and Hafner, 1983; Wood, 1935).

Key to the Genera

1. Size relatively small, total length from about 130 to 180 mm; tail relatively short, averaging only slightly longer than length of head and body, wider in middle than at base, and without a crest of long hairs along its distal one-third or a prominent tuft at its tip; digit 1 of hind foot not reduced to vestige nor absent; auditory bullae enormously inflated, tympanic extending below level of occlusal surface of upper cheek teeth and anteriorly to beyond the mandibular fossa ... *Microdipodops*

1′. Size relatively large, total length from about 205 to 370 mm; tail relatively long, averaging from about 1.05 to 1.6 or more of head and body length, not wider in middle than at base, and with a prominent distal crest and terminal tuft of long hairs; digit 1 of hind foot reduced to vestige located about half-way up the foot or absent externally; auditory bullae greatly inflated, but tympanic rarely extending below level of occlusal surface of cheek teeth and never extending anteriorly beyond mandibular fossa ... *Dipodomys*

Genus *Dipodomys*

1841. *Dipodomys* Gray, Ann. Mag. Nat. Hist., ser. 1, 7:521, August.

Type species.—*Dipodomys phillipsii* Gray, 1841, Ann. Mag. Nat. Hist., ser. 1, 7:521, August.

Diagnosis.—Body form ricochetal, with greatly enlarged hind limbs, short forelimbs, short neck with compressed and partly fused cervical vertebrae, and with tail usually longer than length of head and body; digit 1 of hind feet vestigial or absent externally; underparts, foreleg (with individual exceptions), forefoot, dorsal surface of hind foot, upper lip, spot above eye, and spot behind ear white; white stripe across flank, extending to base of tail; base of tail white all around; white stripes extending along each side of tail from base to near tip; dorsal side of tail dark (not white); ventral side of tail usually dark (except most individuals of *D. deserti*); hairs long and narrow for family; hairs smooth and soft or silky; no distinct underfur layer to pelage; hairs oval in cross section; base of hairs often wide compared to shafts; tail with crest of longer hairs distally, terminating in a tuft; upper parts some shade of buff or brown; large, dermal sebaceous gland on back between shoulders; cheek teeth hypsodont, crown height in adults generally exceeding length of root; molars evergrowing or nearly evergrowing, single-rooted in advanced species, the root tubular and continuous with the crown; molars of primitive species with more than one root, but roots generally fused; p4 two-rooted and P4 three-rooted in primitive species; in advanced species, upper and lower premolars single-rooted; anterocone of P4 lost early by wear; in advanced species, cheek teeth with enamel limited to anterior and posterior plates; p4 never with more than 5 cusps, the fifth, when present, appearing in center of metalophid; third upper and lower molars small, with oval enamel pattern; auditory bullae huge; ventral surface of tympanic bullae rarely reaching level of occlusal surface of cheek teeth, never extending noticeably below that level; frontal without ethmoid foramen; zygomatic process of maxilla expanded in anteroposterior plane; center of palate between premolars ridged; pterygoid fossae double; masticatory and buccinator foramina usually united; median ventral foramina in caudal vertebrae; notch in transverse processes of caudal vertebrae gently curved; calcaneo-navicular or calcaneo-cuneiform articulation of foot; phallus with spines and urethral lappets, and without dorsal groove; tip of phallus strongly upturned; baculum moderate in length relative to body size and with strongly upturned tip and swollen base; preputial gland absent (Burt, 1936; Hall, 1981; Hafner and Hafner, 1983; Homan and Genoways, 1978; Wahlert, 1985; Wood, 1935; Zakrzewski, 1981).

Remarks.—The genus *Dipodomys* includes as synonyms: *Perodipus* Fitzinger, 1867 (type species *D. agilis* Gambel, 1848), a name that was subsequently applied to the species of *Dipodomys* with five toes on the hind feet; *Dipodops* Merriam, 1890; and *Macrocolus* Wagner, 1846 (Coues, 1875; Hall, 1981; Jones and Genoways, 1975). Some species of *Dipodomys* have also been listed as species of *Cricetodipus* Peale, 1848. Grinnell (1919) synonymized *Perodipus* with *Dipodomys* and provided the most recent review of the genus.

Key to the Species

1. Four, clawed toes on hind foot (minute vestigial, fifth toe [digit 1 or hallux] rarely present, but without a claw) 2

1′. Five, clawed toes on hind foot (the hallux vestigial and located about halfway up the inner side of the foot) 10

2. Tip of tail with conspicuous, white tuft of hairs; tuft usually measures about 20 mm or greater in length, but ranges from 6 to about 40 mm or more 5

2′. Tip of tail without a white tuft of hairs .. 3

3. Distal one-third of tail, including tuft, black; found only in central México .. .. *Dipodomys phillipsii*

3′. Distal one-third of tail not all black, rather a mix of light- and dark-colored hairs; distributed widely in western North America, including central México 4

4. Length of hind foot usually less than 36 mm; length of nasals usually less than 12.4 mm; occurs only in central California north of the Tehachapi Mountains and west of the Sierra Nevada .. .. *Dipodomys nitratoides*

4′. Length of hind foot usually 36 mm or greater; length of nasals greater than 12.6, averaging greater than 13.0 mm; not occurring in central California, north of the Tehachapi Mountains and west of the Sierra Nevada .... *Dipodomys merriami*

5. Width across auditory bullae less than 25 mm; length of hind foot usually less than 46 mm; greatest length of skull usually less than 41 mm 8

5′. Width across bullae greater than 25.5 mm; length of hind foot usually greater than 46 mm; greatest length of skull usually greater than 41 mm 6

6. White tail tuft not bordered proximally by band of blackish hairs; underside of tail usually not darker than sides of tail; interparietal usually not visible in dorsal view .... *Dipodomys deserti*

6′. White tail tuft bordered proximally by band of blackish hairs; underside of tail

dark, contrasting with light sides of tail; interparietal conspicuous from dorsal view 7
7. White tail tuft small, length from about 6 to 20 mm; width across maxillary arches less than 23.5 mm .... *Dipodomys nelsoni*
7′. White tail tuft large, length from about 25 to 40 mm; width across maxillary arches greater than 25 mm .... *Dipodomys spectabilis*
8. Length of hind foot usually between about 42 and 49 mm; not occurring in central México 9
8′. Length of hind foot usually between about 34 and 42 mm; found only in central México ...... *Dipodomys phillipsii*
9. Found only in north-central California and south-central Oregon .... *Dipodomys californicus*
9′. Occurs only in north-central Texas and southwestern Oklahoma .... *Dipodomys elator*
10. Lower incisors with flattened anterior faces (chisel-like) .... *Dipodomys microps*
10′. Lower incisors with rounded anterior faces (awl-like) 11
11. Length of head and body normally greater than 130 mm and ratio of lengths of tail to head and body less than 1.35; width of skull across bullae greater than 27 mm; breadth across maxillary arches greater than 26 mm .. *Dipodomys ingens*
11′. Length of head and body less than 130 mm; width of skull across bullae less than 26.2 mm; breadth across maxillary arches less than 25.5 mm 12
12. Ratio of lengths of tail to head and body less than 1.30; not occurring in western California or Baja California 13
12′. Ratio of lengths of tail to head and body more than 1.30; may or may not occur in western California or Baja California 14
13. Width of skull across bullae less than 22.2 mm; ratio of lengths of tail to head and body less than 1.08 .... *Dipodomys compactus*
13′. Width of skull across bullae greater than 22.2, ratio of lengths of tail to head and body greater than 1.13 ... *Dipodomys ordii*
14. Ratio of breadth across maxillary arches to greatest length of skull greater than 0.548 18
14′. Ratio of breadth across maxillary arches to greatest length of skull less than 0.548 15
15. Numerous white hairs measuring 2.0 to 2.5 mm on inner surface of ear pinna; crown length of ear generally more than 16 mm 16
15′. No noticeable long, white hairs on inner surface of ear pinna; crown length of ear generally less than 16 mm 17
16. Crown length of ear generally greater than 16.7 mm .. *Dipodomys elephantinus*
16′. Crown length of ear generally less than 16.7 mm .... *Dipodomys venustus*
17. Length of hind foot greater than 43 mm; greatest length of skull greater than 39.4 mm .... *Dipodomys agilis*
17′. Length of hind foot less than 42 mm; greatest length of skull less than 39.3 mm .... *Dipodomys simulans*
18. Ratio of lengths of tail to head and body normally greater than 1.42; not in Baja California 19
18′. Ratio of lengths of tail to head and body normally less than 1.42; occurs only in Baja California ..... *Dipodomys gravipes*
19. Occurs only in western Nevada and southeastern California east of the coastal mountain ranges .... *Dipodomys panamintinus*
19′. Not in Nevada and southeastern California east of the coastal mountain ranges 20
20. Found only in central California north of the Transverse ranges .... *Dipodomys heermanni*
20′. Found only in southern California south and west of the San Bernardino, San Jacinto, and Santa Rosa ranges .... *Dipodomys stephensi*

Species Accounts

Dipodomys agilis

Diagnosis.—A medium-sized kangaroo rat with large ears, five toes on the hind feet, and a skull narrow across the maxillary arches; breadth of skull across maxillary arches less than 54.8% of greatest skull length; head and body length generally averages greater than 119 mm in males and 116 in females; greatest length of skull averages greater than 39.5 mm in females; and breadth of maxillary arches averages greater than 21 mm in females (male are larger).

Comparisons.—*D. agilis* is most similar to *D. simulans, D. venustus, D. elephantinus,* and *D. heermanni.* From *D. heermanni, D. agilis* can be distinguished by the narrower breadth across the maxillary arches, smaller hind feet, and larger ears. From *D. elephantinus* and *D. venustus, D. agilis* can be distinguished by its smaller size, shorter ears, and narrower maxillary arch (breadth at midpoint less than 5.2 rather than greater than 5.2 mm). From *D. simulans, agilis* differs in being larger with wider maxillary arches and longer ears; length of hind foot averages greater than 43 in *agilis* and less than 42 mm in *simulans*; greatest length of skull averages greater than 39.4 in *agilis* and less than 39.3 mm in *simulans*; size differences are greater when comparisons are made between animals of the same sex (Best, 1978, 1983*a*; Grinnell, 1922; Hall, 1981; Sullivan and Best, in press).

Distribution.—*Dipodomys agilis* occurs primarily in open woodland and chaparral communities from the Temblor Mountains and Transverse ranges (Tehachapi Mountains on the east and Santa Ynez Mountains on the west) in west-central California, southward in the San Gabriel and San Bernardino ranges to the lower hills of the Los Angeles Basin (Sullivan and Best, in press; unpubl. data).

Remarks.—Grinnell (1922) and subsequent authorities have remarked that *D. venustus* is similar to *D. agilis,* and may prove not to warrant specific recognition; if this is true, *D. elephantinus* may also prove to be conspecific with *D. agilis.* Stock (1974), however, found identical karyotypes for *D. simulans, D. venustus,* and *D. elephantinus,* which differed from *D. agilis perplexus* by having 2n = 60 versus 62 and FN = 116 versus 110. Best et al. (1986) reported on karyologic, morphologic, and genic investigations of the *agilis* group, finding that the northern, 62-chromosome forms (*agilis, perplexus,* and *fuscus*) were consistently distinguishable from the southern group with 60 chromosomes (*simulans* and others). Best et al. (1986) provided analyses of genic differences between the northern and southern karyotypic forms; and Sullivan and Best (in press) determined current taxonomic arrangements from bacular, morphometric, and previous genic and karyotypic studies.

Dipodomys agilis agilis Gambel, 1848

1848. *Dipodomys agilis* Gambel, Proc. Acad. Nat. Sci. Philadelphia, 4:77.

1853. *D*[*ipodomys*]. *wagneri* Le Conte, Proc. Acad. Nat. Sci. Philadelphia, 6:224, January.

Holotype.—Holotype lost, sex unknown, from Los Angeles, Los Angeles Co., California; obtained by William Gambel (Grinnell, 1922).

Measurements of holotype.—Total length, 267; length of tail, 165.

Distribution.—Coastal plains and Pacific slope of mountains from Cuyama Valley, Santa Barbara County, southward to Orange County, California. In the Los Angeles Basin, generally occupies chaparral communities above 500 m in elevation.

Remarks.—Taxonomy follows the arrangement suggested by Best et al. (1986) and Sullivan and Best (in press).

Dipodomys agilis perplexus (Merriam, 1907)

1907. *Perodipus perplexus* Merriam, Proc. Biol. Soc. Washington, 20:79, 22 July.

1943. *Dipodomys agilis fuscus* Boulware, Univ. California Publ. Zool., 46:393, 16 September.

Holotype.—Adult male, skin and skull, USNM 29261/41328, from Walker Basin, 3,400 ft, Kern Co., California; obtained on 15 July 1891 by Vernon Bailey.

Measurements of holotype.—Total length, 320; length of tail, 195; length of hind foot,

46; greatest length of skull, 41.75; width across bullae, 25.55; breadth across maxillary arches, 22.55; nasal length, 15.55; interorbital breadth, 12.10.

Distribution.—Found in semiarid communities in the mountains bordering the southern end of the San Joaquin Valley and the western edge of the Mojave Desert; from the Temblor Range, Kern and San Luis Obispo counties and southern Sierra Nevada and Tehachapi Mountains on the northeast to the desert slopes of the San Bernardino Mountains on the southeast, westward to the vicinity of Mt. Pinos, Ventura County, and coastal chaparral communities, Santa Barbara Co., California. Generally occupies chaparral communities in areas above 500 m in elevation.

Remarks.—Grinnell (1921) first treated *perplexus* as a subspecies of *D. agilis.* Best (1983*b*) remarked that he could find no distinct separation of specimens of *perplexus* and *fuscus,* based upon multivariate analyses using 5 skeletal and 14 cranial characters.

Dipodomys californicus

Diagnosis.—A medium-sized, four-toed (with exceptions) kangaroo rat with a broad face, dark color, and with a tail with broad, dark dorsal and ventral stripes and a white tuft at its tip; length of head and body averages from about 112 to 117 mm; skull with widely spreading maxillary arches, small bullae, thick incisors, relatively long and broad (heavy) rostrum, and broad supraoccipital and interparietals, the length and breadth of the latter about 3.5 and 3.0, respectively; average and extreme measurements for 10 adult and subadult specimens were given by Grinnell (1922).

Comparisons.—*Dipodomys californicus* is readily distinguished from all other species of kangaroo rats occupying the same or adjacent geographic areas by the combination of four toes on the hind foot, medium size, dark color, and small auditory bullae. Overall, *D. californicus* is perhaps most similar in size, structure, and color pattern to *D. elator,* a species of north central Texas and adjacent areas in Oklahoma. The interorbital region of *D. elator* is broader, the bullae larger, and the rostrum shorter than in *D. californicus.*

Distribution.—California kangaroo rats primarily occur in chaparral and other shrub communities from central California, north of the San Pablo-Suisun bays/Sacramento-San Joaquin rivers estuary system, northward in the coastal mountains and Sacramento Valley to south-central Oregon in the Klamath and Rogue river watersheds, and eastward on the Modoc Plateau of California to the Surprise Valley on the California-Nevada border, and thence southward on the eastern slopes of the Sierra Nevada to about the northern border of Sierra County, California.

Remarks.—Grinnell (1922) treated *D. californicus* as a subspecies of *D. heermanni,* primarily because he found a few individuals with rudimentary fifth toes on the hind feet, and because of what he termed "*roundabout* intergradation—from *berkeleyensis* through *tularensis, heermanni,* and *eximius,* to *californicus.*" He apparently failed to consider, however, *D. heermanni dixoni,* the population geographically intermediate to *heermanni* and *tularensis.* He remarked that the latter two taxa were alike in external appearance (but failed to note the distinctive white tip of the tail of *eximius*), but noted great differences in the skulls (see above diagnosis). Stock (1974) found widely divergent karyotypes for *heermanni* and *californicus* (also Fashing, 1973), and noted the great similarity of the karyotypes of *D. merriami* and *D. nitratoides* to that of *D. californicus.* Patton et al. (1976) compared allele frequencies at 18 loci among populations of *D. californicus, D. heermanni, D. panamintinus, D. merriami,* and *D. nitratoides,* and concluded that the differences between the four- and five-toed populations of *D. heermanni* were greater than either population was to *D. panamintinus,* and that the electrophoretic data clearly supported karyotypic evidence that *D. californicus* was not conspecific with *D. heermanni.* Hall (1981) did not follow Patton et al. (1976) in recognizing *D. californicus* as a species separate from *D. heermanni.*

Dipodomys californicus californicus Merriam, 1890

1890. *Dipodomys californicus* Merriam, N. Amer. Fauna, 4:49, 8 October.
1899. *Dipodomys californicus pallidulus* Bangs, Proc. New England Zool. Club, 1:65, 31 July.
1916. *Dipodomys californicus trinitatis* L. Kellogg, Univ. California Publ. Zool., 12:366, 27 January.
1925. *Dipodomys heermanni gabrielsoni* Goldman, Proc. Biol. Soc. Washington, 38:33, 12 March.

Holotype.—Adult male, skin and skull, USNM 16618/23544, from Ukiah, Mendocino Co., California; obtained on 4 May 1889 by Theodore S. Palmer.

Measurements of holotype.—Total length, 302; length of tail, 183; length of hind foot, 43; greatest length of skull, 39.60; width across bullae, 24.75; breadth across maxillary arches, 22.70; nasal length, 14.55; interorbital breadth, 12.90.

Distribution.—Occurs from north of the Sacramento-San Joaquin estuary system in the coastal mountains and on the western edge of the Sacramento Valley, northward to southern Oregon, and eastward on the Modoc Plateau to the border between California and Nevada.

Remarks.—Grinnell and Linsdale (1929) regarded *gabrielsoni* as inseparable from *californicus.*

Dipodomys californicus eximius Grinnell, 1919

1919. *Dipodomys californicus eximius* Grinnell, Proc. Biol. Soc. Washington, 32:205, 31 December.

Holotype.—Adult male, skin and skull, MVZ 18347, from Marysville Buttes, 300 ft, 3 mi NW Sutter, Sutter Co., California; obtained by F. H. Holden on 5 April 1912.

Measurements of holotype.—Total length, 303; length of tail, 192; length of hind foot, 42; length of ear (crown), 13; greatest length of skull, 36.7; width across bullae, 22.0; length of nasals, 13.9; breadth across maxillary arches, 21.9; width of maxillary arch at middle, 5.3; width of rostrum near end, 4.3.

Distribution.—Found only in the vicinity of Marysville Buttes, Sutter Co., California.

Remarks.—This population shares with *D. c. saxitilis* the characteristic of being substantially smaller in all dimensions than individuals of *D. c. californicus*; *eximius* differs from *saxitilis* primarily by slightly lighter color, differences that perhaps do not warrant recognition as separate subspecies.

Dipodomys californicus saxitilis Grinnell and Linsdale, 1929

1929. *Dipodomys heermanni saxitilis* Grinnell and Linsdale, Univ. California Publ. Zool., 30: 453, 15 June.

Holotype.—Adult female, skin and skull, MVZ 34963, from Mesa near Dale's, on N side Paine's Creek, 700 ft, Tehama Co., California; obtained on 27 December 1924 by Joseph Grinnell.

Measurements of holotype.—Total length, 300; length of tail, 183; length of hind foot, 41; length of ear (crown), 12; weight, 57.5 g; greatest length of skull, 37.3; width across bullae, 22.85; breadth across maxillary arches, 21.4; nasal length, 14.9; interorbital breadth, 12.0.

Distribution.—Foothills on the eastern edge of the Sacramento Valley, from about the South Fork of the American River, northward to the south bank of Battle Creek, Tehama County, California.

Remarks.—Measurement of 14 adult males from the type locality were listed by Grinnell and Linsdale (1929).

Dipodomys compactus

Diagnosis.—A medium-sized, five-toed kangaroo rat with a remarkably short tail; the tail has a relatively small crest; pelage is relatively short and coarse; color of dorsal parts has an orangeish hue; skull with relatively broad interorbital region, small bullae, and rounded or rectangular-shaped interparietal. The ratio of lengths of tail to head and body averages from about 1.02 to 1.08, values from lower than other species of *Dipodomys.* The skull is narrower across the bullae than all other species, averaging 21.7 mm in males and 22.1 in females (Schmidly and Hendricks, 1976).

Comparisons.—*Dipodomys compactus* is most likely to be confused only with *D. ordii,* the only other medium-sized, five-toed kangaroo rat in the southeastern portion of the geographic range of *Dipodomys. D. compactus compactus* has a shorter tail which is less bushy and with a shorter dorsal crest; the ventral pencil of the tail is lighter colored and does not extend to the tip. The pelage of *D. compactus* is shorter and coarser than that of *D. ordii.* The auditory bullae of *D. ordii* are more inflated, giving the skull a triangular outline. In *D. ordii,* the supraoccipital is narrower and the interparietal is narrower, more pointed posteriorly, and more triangular in outline than the rectangular to roundish shape of the interparietal of *D. compactus.*

Distribution.—Southeastern Texas, including Padre and Mustang islands, and the mainland eastward from Zapata Co. and southward from Bexar and Gonzales counties to the barrier islands of Tamaulipas, México.

Remarks.—The characteristics listed by Schmidly and Hendricks (1976) as indicating specific separation of *D. compactus* and *D. ordii* were considered by Hall (1981) to indicate "hardly more than subspecific grade;" therefore, Hall provisionally retained *compactus, largus, parvabullatus,* and *sennetti* as subspecies of *D. ordii.* Baumgardner and Schmidly (1981) provided additional analyses demonstrating specific separation of *D. compactus* and *D. ordii* and their sympatric occurrence in portions of the geographic range of *compactus.* Schmidly and Hendricks (1976) listed average measurements for 12 males and 13 females. Setzer (1949) gave means and ranges of measurements for each sex of each subspecies.

Dipodomys compactus compactus True, 1889

1889. *Dipodomys compactus* True, Proc. U.S. Natl. Mus., 11:160, 5 January.

1942. *Dipodomys ordii parvabullatus* Hall, Univ. Kansas Publ., Mus. Nat. Hist., 5:38, 1 October.

1951. *Dipodomys ordii largus* Hall, Univ. Kansas Publ., Mus. Nat. Hist., 5:40, 1 October.

Holotype.—None designated, but Poole and Schantz (1942) assumed that it was an adult female, skin only (skull lost), USNM 19665/35227, from Padre Island, Cameron Co., Texas; obtained on 3 April 1888 by C. K. Worthen.

Measurements of holotype.—Total length, 209.5; length of tail, 114.5; length of hind foot, 31; length of ear (crown), 6.5.

Distribution.—Occurs on Mustang and Padre Islands of southeastern Texas and on the barrier islands of Tamaulipas, México.

Remarks.—Baumgardner and Schmidly (1981), using both univariate and multivariate statistical routines, found no distinct separation of the Mustang Island population from that of Padre Island or the island populations from Tamaulipas, and therefore treated *parvabullatus* and *largus* as junior synonyms of *compactus.*

Dipodomys compactus sennetti (J. A. Allen, 1891)

1891. *Dipodops sennetti* J. A. Allen, Bull. Amer. Mus. Nat. Hist., 3:226, 29 April.

Holotype.—Adult male, skin and skull,

AMNH 3478/2733, from near Brownsville, Cameron Co., Texas (amended to "Santa Rosa, near Sarita, Kenedy Co., Texas" by Baumgardner [1989:103]); obtained on 9 March 1888 by J. M. Priour.

Measurements of holotype.—[Approximated by Allen from the dried skin] Total length, 210; length of tail, 110; length of hind foot, 35; length of ear (crown), 9; greatest length of skull, 38.1; width across bullae, 23.7; length of nasals, 14.0; breadth across maxillary arches, 19.4; interorbital breadth, 13.1; length of rostrum, 20.1; length of maxillary toothrow, 5.0.

Distribution.—Found in the eastern two-thirds of the mainland of southern Texas, east of Zapata and Willacy counties and south of Atascosa, Bexar, and Gonzales counties (Baumgardner, 1989).

Remarks.—Davis (1942) treated *sennetti* as a subspecies of *D. ordii*; Schmidly and Hendricks (1976) assigned it to *D. compactus*. According to Baumgardner and Schmidly (1981), topotypes of *sennetti* are the most atypical of the mainland samples of *D. compactus,* showing more similarity to the samples of *D. c. compactus* of the barrier islands than do other samples from the mainland.

Dipodomys deserti

Diagnosis.—A large kangaroo rat with relatively large, four-toed hind feet and a tail with ventral coloration the same as or only slightly darker than the lateral light stripes, and with no dark band proximal to the white, distal tuft; length of head and body averaging from about 134 to 155 mm (Hoffmeister, 1986; Nader, 1978); skull with greatly inflated auditory bullae; interparietal usually absent in dorsal view in adult specimens; supraoccipital so compressed as to be barely visible dorsally (1 mm or less). External and skull measurements were given by Hoffmeister (1986) and Nader (1978).

Comparisons.—*Dipodomys deserti* is readily distinguished from all other species of kangaroo rats by its large size, absence of a vestigial fifth toe on the hind feet, and the usual absence of a darker ventral stripe on the tail. Overall, *D. deserti* is most similar in size and proportions to *Dipodomys spectabilis,* but lacks a dark-colored band of hairs bordering the white tail tuft, generally has larger hind feet, and a more inflated skull with a narrower interorbital region and narrower breadth across the maxillary arches (Hall, 1981; Nader, 1978).

Distribution.—Desert kangaroo rats are found almost exclusively on loose sandy soils in association with *Larrea* or *Prosopis* in the low, hot deserts of Nevada, extreme southwestern Utah, western Arizona, southeastern California, northwestern Sonora, and northeastern Baja California (Nader, 1978).

Remarks.—Nader (1978) reviewed individual, secondary sexual, and geographic variation in external and skull dimensions and provided the most recent systematic review of the species.

Dipodomys deserti aquilus Nader, 1965

1965. *Dipodomys deserti aquilus* Nader, Proc. Biol. Soc. Washington, 78:52, 21 July.

Holotype.—Adult male, skin and skull, MVZ 126411, from 1.5 mi NW High Rock Ranch, about 12 mi SE Wendel, 4,080 ft, Lassen Co., California; obtained 21 July 1960 by Charles S. Thaeler.

Measurements of holotype.—Total length, 321; length of tail, 191; length of hind foot, 54; length of ear, 18; weight, 108.9 g; width across bullae, 30.00; breadth across maxillary arches, 23.50; nasal length, 11.85; interorbital breadth, 15.40; length of maxillary toothrow, 5.40; depth of cranium, 14.40.

Distribution.—Sandy deserts of east-central Lassen Co., California and adjacent northwestern Nevada in Humboldt and Washoe counties (Nader, 1978).

Remarks.—*D. d. aquilus* is characterized by its small size in comparison to other subspecies.

Dipodomys deserti arizonae Huey, 1955

1955. *Dipodomys deserti arizonae* Huey, Trans. San Diego Soc. Nat. Hist., 12:99, 10 February.

Holotype.—Adult male, skin and skull, SDSNH 12532, from 3 mi SE Picacho, Pinal Co., Arizona; obtained on 14 May 1937 by Laurence M. Huey.

Measurements of holotype.—Total length, 345; length of tail, 193; length of hind foot, 52; length of ear, 14; greatest length of skull, 45.3; width across bullae, 30.8; breadth across maxillary arches, 23.7; length of nasals, 16.2; width of maxillary arch at middle, 6.5.

Distribution.—Generally limited to sandy soils in south central Arizona in Maricopa and Pinal counties (Nader, 1978).

Remarks.—According to Nader (1978), W. Goodpaster collected desert kangaroo rats living in gravelly soil, the only recorded observations of this species occupying substrates other than sand. Measurements for 13 males and 10 females were given by Nader (1978).

Dipodomys deserti deserti Stephens, 1887

1887. *Dipodomys deserti* Stephens, Amer. Nat., 21:42, January.
1903. *Dipodomys deserti helleri* Elliot, Field Columbian Mus., Publ. 87, Zool. Ser., 3:249, 7 January.

Holotype.—Juvenile female, skin and skull, USNM 15629/22522, from Mojave River [3 or 4 miles from, and opposite Hesperia], San Bernardino Co., California; obtained on 29 June 1886 by Frank Stephens.

Measurements of holotype.—Total length, 328; length of tail, 196; length of hind foot, 48.3; greatest length of skull, 44.2; width across bullae, 29.5; length of nasals, 16.0; breadth across maxillary arches, 21; interorbital breadth, 13.1; length of rostrum, 22.2; length of maxillary toothrow, 6.0; depth of cranium, 13.3.

Distribution.—Found on loose, sandy soils in west-central and southern Nevada, extreme southeastern Utah, western Arizona, northwestern Sonora, northeastern Baja California, and east-central and southeastern California.

Remarks.—Measurements for 8 males and 6 females were given by Nader (1978), and Hall (1946) gave measurements for 10 males and 6 females.

Dipodomys deserti sonoriensis Goldman, 1923

1923. *Dipodomys deserti sonoriensis* Goldman, Proc. Biol. Soc. Washington, 36:139, 1 May.

Holotype.—Adult male, skin and skull, USNM 242306, from La Libertad Ranch, 30 mi E Sierra Seri, Sonora, México; obtained on 3 January 1922 by Charles Sheldon.

Measurements of holotype.—[No external measurements recorded] Greatest length of skull, 46.6; width across bullae, 31.5; length of nasals, 16.9; breadth across maxillary arches, 24.7; interorbital breadth, 14.6; length of rostrum, 22.2; length of maxillary toothrow, 6.0; depth of cranium, 13.3.

Distribution.—Found on the coastal plain of west central Sonora.

Remarks.—*D. d. sonoriensis* is the largest subspecies and the darkest colored, being "nearly blackish" (Nader, 1978). Measurements for seven males and eight females were presented by Nader (1978).

Dipodomys elator

Diagnosis.—A medium-large sized kangaroo rat with four toes on the hind foot and a prominent white tuft at the tip of the tail; skull with relatively widely positioned auditory bullae, large interparietals and wide supraoccipital; rostrum wide, interorbital region narrow, and orbits large; incisors relatively thick and stout (Carter et al., 1985; Merriam, 1894*a*).

Comparisons.—*Dipodomys elator* is superficially similar to *D. spectabilis,* but is smaller. It resembles *D. phillipsii ornatus,* except for the white, rather than black-tipped tail (Dalquest and Collier, 1964). It is also similar in size, proportions, and coloration to *D. californicus* (see the account of the latter species for distinguishing characters).

Distribution.—The Texas kangaroo rat is known from a small area in north-central Texas and one locality in adjacent Oklahoma in association with mesquite (*Prosopis*) and short grasses (Carter et al., 1985).

Remarks.—*Dipodomys elator* is considered to be related to *D. phillipsii,* although its specific relationships are not clear (Jannett, 1976).

Dipodomys elator Merriam, 1894

1894. *Dipodomys elator* Merriam, Proc. Biol. Soc. Washington, 9:109, 21 June.

Holotype.—Adult male, skin and skull, USNM 64802, from Henrietta, Clay Co., Texas; obtained on 13 April 1894 by J. Alden Loring.

Measurements of holotype.—Total length, 292; length of tail, 173; length of hind foot, 47; length of ear (dry, from notch), 14; greatest length of skull, 40.0; width across bullae, 24.3; length of nasals, 14.5; breadth across maxillary arches, 23.2; interorbital breadth, 12.7; length of rostrum, 21.2; length of maxillary toothrow, 5.4; depth of cranium, 12.9.

Remarks.—Ranges of external measurements and means of cranial measurements for 15 adults were given by Carter et al. (1985). Best (1987) provided summary statistics of 19 characters by sex for 206 adults in three geographic samples, finding significant sexual dimorphism and geographic variation.

Dipodomys elephantinus

Diagnosis.—A moderately large-sized, large-eared, five-toed kangaroo rat with a relatively long tail, dark coloration, and a narrow face; auditory bullae large, supraoccipital and interparietal very narrow, and incisors relatively robust; the length of the ear pinna exceeds that of all other species of *Dipodomys* (Grinnell, 1922).

Comparisons.—*Dipodomys elephantinus* can be distinguished from all other kangaroo rats except *D. venustus* by a combination of five toes on the hind feet, ear (crown length) in excess of 16.7 mm, a relatively long, heavily crested tail averaging 155% of length of head and body, large body size, and relatively narrowly spread maxillary arches; from typical individuals of *D. venustus, D. elephantinus* differs in being lighter in color and having larger ears. Characteristics of coloration and size listed by Grinnell (1922) as distinguishing *elephantinus* and *venustus* are not diagnostic for populations of the *venustus* group in the Diablo Range of western Stanislaus and southeastern San Benito counties (D. F. Williams, unpubl. data).

Distribution.—The big-eared kangaroo rat is known from chaparral communities in the southern Gabilan Range in San Benito and Monterey counties, California (Best, 1986).

Remarks.—*Dipodomys elephantinus* is closely similar in structure, genetics, and ecology to *D. venustus*; Best (1986) noted that they possibly were conspecific.

Dipodomys elephantinus (Grinnell, 1919)

1919. *Perodipus elephantinus* Grinnell, Univ. California Publ. Zool., 21:43, 29 March.

Holotype.—Adult male, skin and skull, MVZ 28511, from 1 mi N Cook P. O., 1,300 ft, Bear Valley, San Benito Co., California; obtained on 9 July 1918 by Halsted G. White.

Measurements of holotype.—Total length, 333; length of tail, 200; length of hind foot, 48; length of ear (crown), 18; weight, 90.7

g; greatest length of skull, 43.2; width across bullae, 26.2; breadth across maxillary arches, 22.95; nasal length, 15.5; interorbital breadth, 12.15.

Remarks.—Grinnell (1922) gave measurements for 10 adults.

Dipodomys gravipes

Diagnosis.—The San Quintín kangaroo rat has five toes on the hind feet, is medium-large in size with a small ear pinna, and a thick tail of medium length; the skull is relatively wide, the width across the maxillary arches averaging more than 54.8% of the greatest length of skull; the maxillary root of the zygomatic arch has a sharp postero-external angle (Best and Lackey, 1985).

Comparisons.—*Dipodomys gravipes* is larger than the sympatric species, *D. merriami* and *D. simulans.* It can be distinguished from the former by having five toes on the hind feet; in comparison to *D. simulans, D. gravipes* has shorter ears, longer hind feet, larger body size, and a tail that is thicker, paler, and less sharply bicolored (Best and Lackey, 1985); the head is wider, which is apparent externally as well as being reflected in measures of skull width. *D. gravipes* is most similar to the allopatric species, *D. stephensi,* from which it differs in having larger hind feet (length greater than 44 versus less than 44 mm), larger body size (weights averaged about 79 to 85 versus 67 g), relatively smaller auditory bullae that are less globular in shape, and less constriction of the interbullar region dorsally (Best and Lackey, 1985; Bleich, 1977; Grinnell, 1922).

Distribution.—*Dipodomys gravipes* is known from an area in Baja California bordering the Pacific Ocean, from near San Telmo on the north to just south of El Rosario on the south. Its range extends from the seashore to approximately 20 km inland (Best and Lackey, 1985).

Remarks.—Best and Lackey (1985) reviewed the relationships of *D. gravipes* to other species of *Dipodomys.*

Dipodomys gravipes Huey, 1925

1925. *Dipodomys gravipes* Huey, Proc. Biol. Soc. Washington, 38:83, 26 May.

Holotype.—Adult male, skin and skull, SDSNH 4703, from 2 mi W Santo Domingo Mission, Baja California Sur, México, lat. 30°45′N, long. 115°58′West; obtained on 28 February 1925 by Laurence M. Huey.

Measurements of holotype.—Total length, 310; length of tail, 180; length of hind foot, 44; length of ear, 10; greatest length of skull, 41.8; width across bullae, 26.5; length of nasals, 14.7; breadth across maxillary arches, 24.0; width of maxillary arch at middle, 5.9.

Remarks.—Best (1983*a*) presented external and cranial measurements for *D. gravipes.*

Dipodomys heermanni

Diagnosis.—A medium large-sized kangaroo rat with five toes on the moderately-sized hind feet, a relatively broad face, and moderate-sized ear pinnae and auditory bullae; the width across the maxillary arches averages more than 54.8% of the greatest length of skull and the maxillary root of the zygomatic arch has a sharp postero-external angle.

Comparisons.—*Dipodomys heermanni* is similar in size and general appearance to contiguous populations of *D. agilis* and *D. venustus,* and the allopatric species *D. stephensi* and *D. gravipes.* From *D. agilis, venustus,* and *elephantinus, D. heermanni* differs in having smaller ear pinnae, lighter coloration, a less heavily crested tail, and a broader face that is apparent externally and in greater width across the maxillary arches. From *D. gravipes, heermanni* differs in having shorter hind feet. The auditory bullae of *D. stephensi* are more globular in shape (i.e., shorter in length and deeper and wider) than those of *heermanni,* and the length of tail of *heermanni* averages greater than 147% of the length of head and body versus 145%

in *D. stephensi* (Grinnell, 1922). There is no known physical characteristic that will distinguish all *D. heermanni* from *D. panamintinus,* although they are not known to be sympatric. From the sympatric species, *D. nitratoides, heermanni* differs in being much larger (average weights about 70 to 80 versus 35 to 40 g) and having five rather than four toes on the hind feet. From the sympatric *D. ingens, heermanni* differs in being smaller (*ingens* averages greater than 110 g in weight), having shorter hind feet (45 or less versus greater than 45 mm), and having a relatively longer tail (the length of tail of *ingens* averages about 128% of head and body length).

Distribution.—Found in west central California south of the Sacramento Valley; known from south of the American River on the east and the Suisun Bay on the west, southward through the San Joaquin Valley below about 3,000 ft; and extending through the interior valleys and mountains to the coast at Morro Bay and from Point Sal to Point Conception in Santa Barbara County (Grinnell, 1922; Hall, 1981). Within this area, Heermann kangaroo rats are found on a variety of substrates in a diversity of plant communities.

Remarks.—Grinnell (1922) and Hall (1981) included the four-toed populations of *D. californicus* in *D. heermanni* (see account of *D. californicus* for details). External and skull measurements and details of coloration were given by Grinnell (1922).

Dipodomys heermanni arenae Boulware, 1943

1943. *Dipodomys heermanni arenae* Boulware, Univ. California Publ. Zool., 46:392, 16 September.

Holotype.—Adult male, skin and skull, MVZ 84840, from C. A. Davis Ranch, 2 mi NNW Lompoc, 400 ft, Santa Barbara Co., California; obtained on 16 February by Richard M. Bond.

Measurements of holotype.—Total length, 290; length of tail, 171; length of hind foot, 40; length of ear, 16; greatest length of skull, 39.2; width across bullae, 23.9; breadth across maxillary arches, 22.3; nasal length, 15.2; interorbital breadth, 13.05.

Distribution.—Open sandy soils along the Pacific slope of the coastal ranges in San Louis Obispo and Santa Barbara counties, California, from about Oceano on the north to the Santa Inez River on the south (Boulware, 1943).

Remarks.—Boulware (1943) gave external and skull measurements for five males and seven females.

Dipodomys heermanni berkeleyensis Grinnell, 1919

1919. *Dipodomys berkeleyensis* Grinnell, Proc. Biol. Soc. Washington, 32:204, 31 December.

Holotype.—Adult male, skin and skull, MVZ 28729, from head of Dwight Way, Berkeley, Alameda Co., California; obtained on 6 October 1918 by Joseph Grinnell and D. D. McLean.

Measurements of holotype.—Total length, 301; length of tail, 180; length of hind foot, 41; length of ear (crown), 12; greatest length of skull, 39.5; width across bullae, 24.1; length of nasals, 14.5; breadth across maxillary arches, 23.0; width of maxillary arch at middle, 5.0; greatest width of rostrum near end, 4.5; weight, 77 g.

Distribution.—Known from the hills and valleys east of San Francisco Bay, in Alameda and Contra Costa counties, California (Grinnell, 1922).

Remarks.—Heermann kangaroo rats from adjacent areas in Santa Clara and Stanislaus counties, in the Diablo Range, are probably of this subspecies. Grinnell (1922) listed measurements for two specimens.

Dipodomys heermanni dixoni (Grinnell, 1919)

1919. *Perodipus dixoni* Grinnell, Univ. California Publ. Zool., 21:45, 29 March.

Holotype.—Adult male, skin and skull, MVZ 26805, from Delhi, near Merced River, Merced Co., California; obtained on 23 March 1917 by Joseph Dixon.

Measurements of holotype.—Total length, 280; length of tail, 165; length of hind foot, 40; length of ear (crown), 14; weight, 72.5 g; greatest length of skull, 37.5; width across bullae, 23.5; breadth across maxillary arches, 21.65; nasal length, 13.55; interorbital breadth, 12.45.

Distribution.—Known from grassland and savanna communities on the eastern margin of the San Joaquin Valley, California, in Merced and Stanislaus counties (Grinnell, 1922).

Remarks.—Grinnell (1922) gave measurements for 10 adult and subadult specimens. This is the smallest sized subspecies of *D. heermanni* (Grinnell, 1922).

Dipodomys heermanni goldmani (Merriam, 1904)

1904. *Perodipus goldmani* Merriam, Proc. Biol. Soc. Washington, 17:143, 14 July.

Holotype.—Adult male, skin and skull, USNM 118924, from Salinas, mouth of Salinas Valley, Monterey Co., California; obtained on 4 September 1902 by Luther J. Goldman.

Measurements of holotype.—Total length, 312; length of tail, 185; length of hind foot, 45.4; greatest length of skull, 40.45; width across bullae, 25.35; breadth across maxillary arches, 20.80; nasal length, 14.55; interorbital breadth, 12.55.

Distribution.—Ranges from near sea level to about 1,300 ft, from the sea coast at Monterey Bay northeastward to San Jose, Santa Clara Co., and southward in the Salinas Valley to the vicinity of Soledad, Monterey Co. (Grinnell, 1922).

Remarks.—External and cranial measurements for 10 adult males and females were given by Grinnell (1922).

Dipodomys heermanni heermanni Le Conte, 1853

1853. *D[ipodomys]. heermanni* Le Conte, Proc. Acad. Nat. Sci. Philadelphia, 6:224.

Holotype.—Juvenile [sex unknown], partial skin and skull, ANSP 229, from Sierra Nevada, California [probably along the Calaveras River, Upper Sonoran life zone, Calaveras Co.; Grinnell, 1922].

Measurements of holotype.—Length of hind foot, 37.5; length of ear (crown), 10; greatest length of skull, 32.3; width across bullae, 20.0; breadth across maxillary arches, 17 (measured on intact side and doubled); nasal length, 10.2; interorbital breadth, 10.8 (external measurements from mounted skin by Grinnell, 1922).

Distribution.—Found in grassland and savanna communities between about 500 and 3,200 feet in the northeastern San Joaquin Valley, California, from Amador Co. on the north to Mariposa Co. on the south.

Remarks.—Grinnell (1922) augmented the scanty diagnosis of Le Conte (1853) and discussed features of the type specimen and the probable type locality.

Dipodomys heermanni jolonensis Grinnell, 1919

1919. *Dipodomys jolonensis* Grinnell, Proc. Biol. Soc. Washington, 32:203, 31 December.

Holotype.—Adult male, skin and skull, MVZ 29087, from valley floor, 1 mi SW Jolon, Monterey Co., California; obtained on 18 October 1918, by Joseph Dixon.

Measurements of holotype.—Total length, 310; length of tail, 185; length of hind foot, 44; length of ear (crown), 15; weight, 82.9 g; greatest length of skull, 43.0; width across bullae, 27.0; length of nasals, 16.3; breadth across maxillary arches, 22.9; width of maxillary arch at middle, 5.7; greatest width of rostrum near end, 4.6.

Distribution.—Chiefly occupies Lower Sonoran associations in west central California in the Salinas Valley, from near Peachtree and San Lucas, Monterey Co., southward at least to Creston, San Luis Obispo Co.

Remarks.—External and cranial measurements for 10 adults were given by Grinnell (1922).

Dipodomys heermanni morroensis (Merriam, 1907)

1907. *Perodipus morroensis* Merriam, Proc. Biol. Soc. Washington, 20:78, 22 July.

Holotype.—Adult female, skin and skull, USNM 31626/43499, from Morro [south side Morro Bay, about 4 mi S of town; Grinnell, 1922], San Luis Obispo Co., California; obtained on 11 November 1891, by Edward W. Nelson.

Measurements of holotype.—Total length, 300; length of tail, 182; length of hind foot, 45; greatest length of skull, 40.00; width across bullae, 23.60; breadth across maxillary arches, 21.45; nasal length, 14.80; interorbital breadth, 13.10.

Distribution.—Limited to coastal plant communities on sandy soils along the shore of Morro Bay, San Louis Obispo Co., California.

Remarks.—Boulware (1943) first treated *morroensis* as a subspecies of *D. heermanni.* Grinnell (1922) gave external and cranial measurements for 10 adults.

Dipodomys heermanni swarthi (Grinnell, 1919)

1919. *Perodipus swarthi* Grinnell, Univ. California Publ. Zool., 21:44, 29 March.

Holotype.—Adult male, skin and skull, MVZ 14440, from 7 mi SE Simmler, Carrizo Plain, San Luis Obispo Co., California; obtained on 26 May 1911, by H. S. Swarth and W. L. Chandler.

Measurements of holotype.—Total length, 313; length of tail, 187; length of hind foot, 45; length of ear (crown), 11; greatest length of skull, 42.35; width across bullae, 25.95; breadth across maxillary arches, 23.55; nasal length, 15.00; interorbital breadth, 13.45.

Distribution.—Occurs in *Bromus, Atriplex,* and *Ephedra* associations along the southwestern margin of the San Joaquin Valley, Kern Co., westward to the Carrizo Plain, San Louis Obispo Co., and the upper Cuyama Valley, Santa Barbara Co., California.

Remarks.—Measurements of six adults were listed by Grinnell (1922).

Dipodomys heermanni tularensis (Merriam, 1904)

1904. *Perodipus agilis tularensis* Merriam, Proc. Biol. Soc. Washington, 17:143, 14 July.

Holotype.—Adult female, skin and skull, USNM 127158, from Alila [Earlimart], Tulare Co., California; obtained on 23 June 1903 by Luther J. Goldman.

Measurements of holotype.—Total length, 308; length of tail, 182; length of hind foot, 41; length of skull, 40.10; width across bullae, 24.80; breadth across maxillary arches, 20.65; nasal length, 14.75; interorbital breadth, 10.70.

Distribution.—Grassland and savanna associations on the floor of the San Joaquin Valley, California, from the vicinity of Tracy, San Joaquin Co., on the north to the foothills of the Tehachapi and Temblor ranges, Kern Co., on the south and southwest, respectively.

Remarks.—Specimens from the Temblor Mountains, 3,000 ft, 12 mi W McKittrick, were assigned to *tularensis* by Grinnell (1922), but are surrounded geographically by specimens Grinnell (1922) assigned to *D. h. swarthi.*

Dipodomys ingens

Diagnosis.—*Dipodomys ingens* is the largest kangaroo rat with five toes on the hind feet, and has a relatively short tail and short ears; other diagnostic traits are the wide maxillary arches (width at mid-length averaging 5.9 mm).

Comparisons.—Giant kangaroo rats can be readily distinguished from the other large-sized kangaroo rats, *D. deserti, D. nelsoni,* and *D. spectabilis,* by having five, rather than four, toes on each hind foot and by its greater average weight and relatively shorter tail (about 128% of head and body length, range about 112 to 135%, D. F. Williams, unpubl. data). From the sympatric kangaroo rats, *D. heermanni* and *D. nitratoides, D. ingens* can be distinguished by having a hind foot longer than 47 mm (rarely to 46 mm in immature individuals), whereas the length of the hind foot in *D. heermanni* does not exceed 46 mm, and is usually less than 45 mm. The huge hind feet will distinguish even juvenile giant kangaroo rats from sympatric species. The hind foot of *D. nitratoides* has only four toes and its length does not exceed 39 mm (Grinnell, 1922). The skull of *D. ingens* is large and broad, averaging 27 mm or more across the maxillary arches, a width much larger than that of all other kangaroo rats except *D. spectabilis,* which averages from 25.8 to 26.9 mm (Grinnell, 1922; Nader, 1978). The average greatest length of the skull of *D. ingens,* 45.9 (Grinnell, 1932), is equaled only by *D. deserti* and *D. spectabilis* (Nader, 1978).

Distribution.—*Dipodomys ingens* occupies a narrow band of gently sloping ground along the western edge of the San Joaquin Valley, California. Records of occurrence extend from the base of the Tehachapi Mountains on the south to a point about 16 km S Los Banos, Merced Co. Inhabited areas west of the San Joaquin Valley include the Carrizo and Elkhorn plains west of the Temblor Mountains, and the upper Cuyama Valley, adjacent to and nearly contiguous with the Carrizo Plain, and scattered colonies in the Ciervo, Kettleman, Panoche, and Tumey hills, and the Panoche Valley (Grinnell, 1922; Hall, 1981; Williams, in press; Williams and Kilburn, 1991).

Remarks.—Measurements for eight subadult and adult specimens were given by Grinnell (1922).

Dipodomys ingens (Merriam, 1904)

1904. *Perodipus ingens* Merriam, Proc. Biol. Soc. Washington, 17:141, 14 July.

Holotype.—Adult male, skin and skull, USNM 128805, from Painted Rock, 20 [= 25.5] mi SE Simmler, Carrizo Plain, San Luis Obispo Co., California; obtained on 6 August 1903 by Luther J. Goldman.

Measurements of holotype.—Total length, 360; length of tail, 191; length of hind foot, 54; greatest length of skull, 48; width across bullae, 30.5; length of nasals, 18; breadth across maxillary arches, 23.5.

Remarks.—*Dipodomys ingens* appears to be most closely related to the sympatric species, *D. heermanni.*

Dipodomys merriami

Diagnosis.—A small-sized kangaroo rat (head and body length averages less than about 105 mm and greatest length of skull averages less than about 37 mm in all populations; Lidicker, 1960) with four toes on the relatively slender hind feet, and with a tail terminating in a crest and tuft of dusky or blackish-brown hairs; mastoid bullae usually relatively more inflated and rostrum narrower than in most other species; rostrum not decidedly shortened and face not noticeably narrower than most other species.

Comparisons.—*Dipodomys merriami* is most similar in size and appearance to the other small-sized, four-toed species, *D. nitratoides*; all other four-toed species are considerably larger and most have a prominent tuft of white hairs at the tip of the tail. *D. merriami* differs from *D. nitratoides* in generally being larger in size, and in having a longer and wider rostrum (nasal length usually averages greater than 13.1 in *D.*

merriami and less than 12.3 mm in *D. nitratoides*; Hoffmann, 1975; Lidicker, 1960).

Distribution.—*Dipodomys merriami* is widely distributed in desert and arid grassland associations in southwestern North America, extending from northwestern Nevada southward through southeastern California, the length of the Baja California Peninsula, and along the mainland of México through Sonora to northern Sinaloa; and ranging eastward across southern Nevada to extreme southeastern Utah, western and southern Arizona, central and southern New Mexico to western Texas; and thence southward through the plateau regions of México to Aguascalientes and Zacatecas (Hall, 1981).

Remarks.—Hall (1981) treated *Dipodomys margaritae* as a species distinct from *D. merriami,* although Lidicker (1960) considered it to be a subspecies of *D. merriami*; we concur with Lidicker. Lidicker (1960) provided measurements for each of the subspecies of *D. merriami.*

Dipodomys merriami ambiguus Merriam, 1890

1890. *Dipodomys ambiguus* Merriam, N. Amer. Fauna, 4:42, 8 October.

Holotype.—Adult male, skin and skull, USNM 18147/25045, from El Paso, El Paso Co., Texas; obtained by Vernon Bailey on 13 December 1889.

Measurements of holotype.—Total length, 233; length of tail, 133; length of hind foot, 37; length of ear (crown), 7, (notch, dry), 12; greatest length of skull, 36.7; width across bullae, 23.0; length of nasals, 13.6; breadth across maxillary arches, 20.8; interorbital breadth, 13.5; length of nasals, 13.6; length of rostrum, 19.7; length of maxillary toothrow, 4.3; depth of cranium, 10.6.

Distribution.—Occurs in the northern parts of the central Mexican deserts, from western New Mexico and western Texas southward to southern and eastern Chihuahua, northeastern Durango, southern Coahuila and central Nuevo León (Lidicker, 1960).

Remarks.—Lidicker (1960) gave measurements and comparisons for *D. m. ambiguus*; Anderson (1972) listed means and ranges of measurements for a sample of 37 from Chihuahua.

Dipodomys merriami annulus Huey, 1951

1951. *Dipodomys merriami annulus* Huey, Trans. San Diego Soc. Nat. Hist., 11:224, 30 April.

Holotype.—Adult female, skin and skull, SDSNH 15522, from El Barril, Gulf of California, lat. 28°20'N, long. 112°50'W, Baja California, México; obtained on 24 March 1928 by Laurence M. Huey.

Measurements of holotype.—Total length, 251; length of tail, 150; length of hind foot, 35; length of ear, 10; greatest length of skull, 37.9; width across bullae, 23.8; length of nasals, 14.1; breadth across maxillary arches, 21.0; width of maxillary arch at middle, 5.5.

Distribution.—Found along the coastal plain of the Gulf of California from near Bahía Los Angeles on the north to Bahía Santa Teresa on the south (Huey, 1951).

Remarks.—Huey (1951) and Lidicker (1960) gave measurements for samples of *D. m. annulus.*

Dipodomys merriami arenivagus Elliot, 1904

1904. *Dipodomys m[erriami]. arenivagus* Elliot, Field Columbian Mus., Publ. 87, Zool. Ser., 3:249, 7 January.

Holotype.—Adult skin and skull, FMNH 10733, from San Felipe, Baja California, México; obtained on 1 April 1902 by Edmund Heller.

Measurements of holotype.—Total length, 225; length of tail, 134; length of hind foot, 36; length of ear (notch), 15; greatest length of skull, 34; width across bullae, 22; length of nasals, 12; length of maxillary toothrow, 3; greatest width of rostrum, 5; palatal length, 11.

Distribution.—Arid communities east of the Sierra San Pedro Mártir and Sierra Juárez, from just south of the U.S. border and west of the Colorado River Delta, southward to San Felipe, all in Baja California, México (Huey, 1951).

Remarks.—Huey (1951) provided means and ranges of measurements for five males and five females. Elliot's paper was listed as being published in December, 1903, but was actually published on 7 January 1904.

Dipodomys merriami atronasus Merriam, 1894

1894. *Dipodomys merriami atronasus* Merriam, Proc. Biol. Soc. Washington, 9:113, 21 June.

Holotype.—Adult male, skin and skull, USNM 50276, from Hacienda La Parada [about 25 mi NW San Luis Potosí], San Luis Potosí, México; obtained by Edward W. Nelson on 20 August 1892.

Measurements of holotype.—Total length, 250; length of tail, 162; length of hind foot, 40; length of nasals, 13.6; breadth across maxillary arches, 21.6; length of maxillary toothrow, 4.8.

Distribution.—Occupies the southern portions of the central Mexican desert, from central Durango and southeastern Coahuila, southward to southeastern Aguascalientes, southwestern San Luis Potosí, and extreme southwestern Tamaulipas (Lidicker, 1960).

Remarks.—Dalquest (1953) gave means for measurements of two samples from San Luis Potosí; Lidicker (1960) listed average measurements for specimens of *D. m. atronasus.*

Dipodomys merriami brunensis Huey, 1951

1951. *Dipodomys merriami brunensis* Huey, Trans. San Diego Soc. Nat. Hist., 11:225, 30 April.

Holotype.—Adult male, skin and skull, SDSNH 6904, from Llano de San Bruno, Baja California, México; obtained by Laurence M. Huey on 24 March 1928.

Measurements of holotype.—Total length, 252; length of tail, 152; length of hind foot, 37; length of ear (crown), 10; greatest length of skull, 37.6; length of nasals, 13.8; breadth across maxillary arches, 19.8; width of maxillary arch at middle, 4.8.

Distribution.—Costal areas along the Gulf of California from near El Valle de Yaqui, about 12 km NW Santa Rosalía, southward to the southern end of Bahía Concepción, Baja California (Huey, 1951).

Remarks.—Huey (1951) gave means and ranges of measurements for five males and three females; Lidicker (1960) listed average measurements for an unspecified sample size.

Dipodomys merriami collinus Lidicker, 1960

1960. *Dipodomys merriami collinus* Lidicker, Univ. California Publ. Zool., 67:194, 4 August.

Holotype.—Adult female, skin and skull, MVZ 123455, from 3.25 mi S, 2.25 mi E Scissors Crossing, Earthquake Valley, San Diego Co., California; obtained on 10 September 1958 by William Z. Lidicker, Jr.

Measurements of holotype.—Total length, 252; length of tail, 148; length of hind foot, 39; length of ear, 14; weight, 39.3 g; greatest length of skull, 35.8; width across bullae, 22.55; breadth across maxillary arches, 18.65; nasal length, 12.9; interorbital breadth, 13.0.

Distribution.—Occupies areas in San Felipe, Earthquake, and Mason (La Puerta) valleys in eastern San Diego Co., and

Aguanga Valley in southern Riverside Co., California (Lidicker, 1960).

Remarks.—Lidicker (1960) listed average measurements for an unspecified number of individuals of this subspecies.

Dipodomys merriami frenatus Bole, 1936

1936. *Dipodomys merriami frenatus* Bole, Sci. Publs., Cleveland Mus. Nat. Hist., 5:1, 17 January.

Holotype.—Adult female, skin and skull, UMMZ 121258 from Toquerville, 3,200 ft, Washington Co., Utah; obtained on 14 May 1934 by F. J. Tobin.

Measurements of holotype.—Total length, 207; length of tail, 113; length of hind foot, 35; occipitonasal length of skull, 33.5; width across bullae, 21.9; length of nasals, 12.7; breadth across maxillary arches, 19.5; width of maxillary arch at middle, 5.2; width of nasals in front of incisors, 2.7; length of maxillary toothrow, 4.1.

Distribution.—Occupies arid communities from the Virgin River Valley in Washington Co., Utah, westward to the edge of the Beaver Dam Mountains, north as far as Veyo, east to Springdale, and southward to the Kanab Plateau in northwestern Arizona (Lidicker, 1960).

Remarks.—Durrant and Setzer (1945) considered *frenatus* to be a synonym of *D. m. vulcani,* but Lidicker (1960) found that the two subspecies were recognizably distinct. Hoffmeister (1986) considered it to be most like, and possibly a synonym of, *D. m. merriami.* Hoffmeister (1986) listed statistics for measurements of 12 males and 11 females from Arizona. The holotype was originally in the Cleveland Museum of Natural History (CMNH 6771).

Dipodomys merriami insularis Merriam, 1907

1907. *Dipodomys insularis* Merriam, Proc. Biol. Soc. Washington, 20:77, 22 July.

Holotype.—Adult female, skin and skull, USNM 79053, from San José Island, Gulf of California, Baja California Sur, México; obtained on 6 August 1895 by J. E. McLellan.

Measurements of holotype.—Total length, 243; length of tail, 143; length of hind foot, 39; greatest length of skull, 36.0; width across bullae, 22.7; length of nasals, 13.2; breadth across maxillary arches, 21.3; interorbital breadth, 13.0; length of rostrum, 20.1; length of maxillary toothrow, 4.4; depth of cranium, 10.5.

Remarks.—Lidicker (1960) included samples of *insularis* in his analysis of geographic variation in *D. merriami* because of the possibility that it might prove to be a subspecies of the latter species. He retained specific rank for *D. insularis,* however, because of its greater structural divergence from the norm among samples of all subspecies of *D. merriami.* Best and Janecek (1992) found that *insularis* was significantly different in several morphological traits from samples of *D. merriami* from the mainland, but they classified it as a subspecies of *D. merriami* based on allozymic similarities. Lidicker (1960) found considerable sexual dimorphism, with males being larger than females; he provided external and cranial measurements for samples of males and females.

Dipodomys merriami margaritae Merriam, 1907

1907. *Dipodomys margaritae* Merriam, Proc. Biol. Soc. Washington, 20:76, 22 July.

Holotype.—Young adult male, skin and skull, USNM 146058, from [Santa] Margarita Island, Baja California Sur, México; obtained on 1 December 1905 by Edward W. Nelson and Edward A. Goldman.

Measurements of holotype.—Total length, 234; length of tail, 144; length of hind foot, 38; greatest length of skull, 34.6; width across bullae, 21.7; length of nasals, 12.2; breadth across maxillary arches, 19.5; length of rostrum, 19.0; length of maxillary toothrow, 4.7; depth of cranium, 11.1.

Distribution.—Occurs only on Santa Margarita Island off the Pacific Coast, in Baja California Sur.

Remarks.—Lidicker (1960), in his analysis of geographic variation in *D. merriami*, concluded that *margaritae* did not warrant specific status and assigned it as a subspecies of *D. merriami*. Huey (1964) concluded that "specific instead of subspecific rank seems preferable." Hall (1981) listed *margaritae* as a species without reference to Huey (1964) or Lidicker (1960). We believe Lidicker's (1960) conclusions are based on the most comprehensive investigations, and concur with his assignment of *margaritae* as a subspecies of *D. merriami*. Huey (1951) and Lidicker (1960) listed measurements for samples of this subspecies.

Dipodomys merriami mayensis
Goldman, 1928

1928. *Dipodomys merriami mayensis*, Proc. Biol. Soc. Washington, 41:141, 15 October.

Holotype.—Adult male, skin and skull, USNM 96437, from Alamos, Sonora, México; obtained on 19 December 1898 by Edward A. Goldman.

Measurements of holotype.—Total length, 240; length of tail, 138; length of hind foot, 37; greatest length of skull, 35.8; occipitonasal length of skull, 34.3; width across bullae, 22.8; length of nasals, 13.1; breadth across maxillary arches, 20.5; interorbital breadth, 12.4; length of rostrum, 19.5; length of maxillary toothrow, 4.1; depth of cranium, 10.9; least width of supraoccipital, 1.6.

Distribution.—Occupies open shrublands of the coastal plains from the Río Yaqui, southern Sonora, southward to extreme northern Sinaloa to the south side of the Río del Fuerte (Lidicker, 1960).

Remarks.—Lidicker (1960) gave average measurements for an unspecified number of specimens.

Dipodomys merriami melanurus
Merriam, 1893

1893. *Dipodomys merriami melanurus* Merriam, Proc. California Acad. Sci., Ser. 2, 3:345, 5 June.

1951. *Dipodomys merriami llanoensis* Huey, Trans. San Diego Soc. Nat. Hist., 11:226, 30 April.

Holotype.—Adult male, CAS 539, from San José del Cabo, Baja California Sur, México; obtained by Walter E. Bryant on 19 March 1892.

Measurements of holotype.—Total length, 239; length of tail, 144; length of hind foot, 35.

Distribution.—From near San Jorge on the north end of the Magdalena Plain, southward through the Cape region of Baja California Sur (Lidicker, 1960).

Remarks.—The holotype was destroyed in the San Francisco earthquake and fire of 1906. After the fire, a new catalog was started; thus, CAS 539 now represents a different specimen. Lidicker (1960) synonymized *llanoensis* with *melanurus*, because the former exhibited intermediate characteristics between typical *melanurus* to the south and *platycephalus* to the north. Huey (1964) retained *llanoensis* "as a means of cataloging the geographic variation shown by the specimens." Hall (1981) cited Huey (1964) as his reason for recognizing *llanoensis*. We concur with Lidicker (1960).

Dipodomys merriami merriami
Mearns, 1890

1890. *Dipodomys merriami* Mearns, Bull. Amer. Mus. Nat. Hist., 2:290, 21 February.

1894. *Dipodomys simiolus* Rhoads, Proc. Acad. Nat. Sci. Philadelphia, 45:410, 27 January.

1894. *Dipodomys similis* Rhoads, Proc. Acad. Nat. Sci. Philadelphia, 45:411, 27 January.

1894. *Dipodomys merriami nevadensis* Merriam, Proc. Biol. Soc. Washington, 9:111, 21 June.

1894. *Dipodomys merriami nitratus* Merriam,

Proc. Biol. Soc. Washington, 9:112, 21 June.
1904. *Dipodomys merriami mortivallis* Elliot, Field Columbian Mus., Publ. 87, Zool. Ser., 3:250, 7 January.
1907. *Dipodomys merriami kernensis* Merriam, Proc. Biol. Soc. Washington, 20:77, 22 July.
1937. *Dipodomys merriami regillus* Goldman, Proc. Biol. Soc. Washington, 50:75, 22 June.

Holotype.—Adult male, skin and skull, AMNH 2394, from New River, between Phoenix and Prescott, Maricopa Co., Arizona; obtained by Edgar A. Mearns on 16 May 1885.

Measurements of holotype.—Total length, 259; length of tail, 149; length of hind foot, 36; length of ear (crown), 10, (meatus), 13.

Distribution.—Extends from northwestern Nevada, southward through the deserts of southeastern California, extreme southeastern Utah, western and southern Arizona as far east as the Peloncillo Mountains, and to the Río Yaqui in southern Sonora (Lidicker, 1960).

Remarks.—Durrant and Setzer (1945) gave measurements for 6 males and 2 females from Utah; Hall (1946) listed measurements for 10 males and 10 females from Nevada; and Hoffmeister (1986) gave statistics for measurements of two samples (18 males, 13 females; 6 males, 10 females) from Arizona. Mearns (1890) referred to the skull in his description of the type specimen; but provided no measurements. Goodwin (1953) stated that the skull was not catalogued into the collections at the American Museum of Natural History, and was presumed to be lost. Goodwin (1953) listed the total length of the holotype as 259 mm but Mearns (1890) listed it as 281 mm, the difference apparently being that Mearns measured to the end of the hairs on the tail.

Dipodomys merriami mitchelli Mearns, 1897

1897. *Dipodomys mitchelli* Mearns, Proc. U.S. Natl. Mus., 19:719, 30 July.

Holotype.—Adult female, skin and skull, USNM 63188, from Tiburón Island, Gulf of California, Sonora, México; obtained by J. W. Mitchell on 23 December 1895.

Measurements of holotype.—No external measurements recorded; greatest length of skull, 36.1; width across bullae, 22.4; length of nasals, 12.9; breadth across maxillary arches, 19.8; interorbital breadth, 12.3; length of rostrum, 19.0; length of maxillary toothrow, 4.8; depth of cranium, 10.2.

Distribution.—Occurs only on Tiburón Island, Gulf of California, Sonora (Hall, 1981).

Remarks.—Lidicker (1960) listed average measurements for a sample of unspecified size.

Dipodomys merriami olivaceus Swarth, 1929

1929. *Dipodomys merriami olivaceus* Swarth, Proc. California Acad. Sci., ser. 4, 18:356, 26 April.

Holotype.—Adult male, skin and skull, CAS 6235, from Fairbank, Cochise Co., Arizona; obtained by Sam Davidson on 28 October 1928.

Measurements of holotype.—Total length, 243; length of tail, 141; length of hind foot, 37; length of ear, 12; greatest length of skull, 36.45; width across bullae, 22.80; length of nasals, 14.40; breadth across maxillary arches, 18.90; length of maxillary toothrow, 4.65; depth of cranium, 12.15; width of maxillary arch at middle, 5.0; greatest width of rostrum near end, 3.2.

Distribution.—Desert associations from southeastern Arizona in Cochise and Santa Cruz counties and western New Mexico east of the Peloncillo Mountains, eastward to eastern Luna and western Doña Ana counties, and southward around the northern end of the Sierra Madre Occidental into northeastern Sonora on the west and central Chihuahua on the east (Hall, 1981).

Remarks.—Anderson (1972) listed measurements for a sample of 19 from Chihuahua; Hoffmeister (1986) listed statistics for measurements of 15 males and 11 females from Arizona; and Swarth (1929) listed means and ranges of measurements for two samples of *olivaceus* from Arizona.

Dipodomys merriami parvus Rhoads, 1894

1894. *Dipodomys parvus* Rhoads, Amer. Nat., 28:70, January.

Holotype.—Adult female, skin and skull, ANSP 8213, from San Bernardino [Reche Canyon, 4 mi SE Colton], San Bernardino Co., California; obtained on 12 June 1892 by R. B. Herron.

Measurements of holotype.—Total length, 248; length of tail, 154; length of hind foot, 35; length of ear (dry, from crown), 10; basilar length of skull, 21; width across bullae, 22.5; length of nasals, 13; interorbital breadth, 13; crown length of mandibular toothrow, 3.6.

Distribution.—Found in the San Bernardino and San Jacinto valleys, in San Bernardino and Riverside counties, California.

Remarks.—Grinnell (1922) amended the type locality and provided measurements of 10 adult and subadult specimens. Lidicker (1960) noted that *parvus* differed from typical samples of *merriami* in being much smaller, and remarked that it was one of the most highly differentiated subspecies of *D. merriami.* He believed that it had nearly achieved species rank.

Dipodomys merriami platycephalus Merriam, 1907

1907. *Dipodomys platycephalus* Merriam, Proc. Biol. Soc. Washington, 20:76, 22 July.
1927. *Dipodomys merriami semipallidus* Huey, Trans. San Diego Soc. Nat. Hist., 5:66, 6 July.

Holotype.—Adult male, skin and skull, USNM 139882, from Calmallí, Baja California, México; obtained on 1 October 1905 by Edward W. Nelson and Edward A. Goldman.

Measurements of holotype.—Total length, 238; length of tail, 145; length of hind foot, 38; greatest length of skull, 35.6; occipitonasal length, 34; width across bullae, 23.6; length of nasals, 12.5; breadth across maxillary arches, 20.9; interorbital breadth, 13.7; length of rostrum, 18.7; length of maxillary toothrow, 4.5; depth of cranium, 11.0.

Distribution.—Found along the Pacific slope from the southern end of the Sierra San Pedro Mártir and San Fernando, southward to about lat. 26°15′N, including the entire Viscaíno Desert, but not extending to the Gulf coastal regions (Lidicker, 1960).

Remarks.—Huey (1951) listed means and ranges for males and females; Lidicker (1960) listed ranges of means for unspecified samples. Lidicker (1960) synonymized *semipallidus* with *platycephalus* because he found them to be intergrades between *quintinensis* and *platycephalus.* Huey (1951, 1964) distinguished *semipallidus* entirely on the basis of slightly darker coloration, which Lidicker (1960) did not consider worthy of subspecific recognition. Hall (1981) retained *semipallidus* for reasons not stated. We concur with Lidicker (1960) that the slight and inconsistent differences in color attributed to *semipallidus* by Huey (1951, 1964) do not merit recognition as a separate subspecies.

Dipodomys merriami quintinensis Huey, 1951

1951. *Dipodomys merriami quintinensis* Huey, Trans. San Diego Soc. Nat. Hist., 11:222, 30 April.

Holotype.—Adult male, skin and skull, SDSNH 4205, from 5 mi E San Quintín, Baja California México; obtained on 9 April 1923 by Laurence M. Huey.

Measurements of holotype.—Total length, 240; length of tail, 148; length of hind foot, 39; length of ear (crown), 10; greatest length of skull, 36.4; width across bullae, 22.6;

length of nasals, 13.0; breadth across maxillary arches, 20.1; width of maxillary arch at middle, 5.1.

Distribution.—Occurs on the San Quintín Plain along the Pacific coast, from near Santo Domingo, southward to the region of El Rosario, Baja California.

Remarks.—Huey (1951) gave means and ranges of measurements for five males and five females.

Dipodomys merriami trinidadensis
Huey, 1951

1951. *Dipodomys merriami trinidadensis* Huey, Trans. San Diego Soc. Nat. Hist., 11:220, 30 April.

Holotype.—Adult male, skin and skull, SDSNH 11531, from Aguajito Spring, El Valle de la Trinidad, Baja California, México; obtained by Laurence M. Huey on 13 March 1936.

Measurements of holotype.—Total length, 243; length of tail, 144; length of hind foot, 38; length of ear (crown), 10; greatest length of skull, 35.5; width across bullae, 22.1; length of nasals, 13.0; breadth across maxillary arches, 19.5; width of maxillary arch at middle, 5.0.

Distribution.—Lidicker (1960) outlined a discontinuous range for *trinidadensis*: the Jacumba Valley and Mountain Springs region of southern San Diego Co., California, and nearby Baja California; and from El Valle de la San Rafael and El Valle de la Trinidad along the western border of the Sierra Juárez, Baja California, México.

Remarks.—Lidicker (1960) discussed the possibility that *trinidadensis* was polyphyletic, and that the diagnostic characters exhibited by the disjunct populations were examples of convergence. A more likely hypothesis, according to Lidicker (1960), is that the populations are relicts of a formerly more widely-distributed population. The northern populations of this taxon were classified as *D. m. simiolus* by Grinnell (1922). Huey (1951) listed means and ranges of measurements for samples of males and females from Baja California.

Dipodomys merriami vulcani
Benson, 1934

1934. *Dipodomys merriami vulcani* Benson, Proc. Biol. Soc. Washington, 47:181, 2 October.

Holotype.—Adult male, skin and skull, MVZ 56002, from lower end Toroweap Valley, about 0.5 mi E Vulcan's Throne, Mohave Co., Arizona; obtained by Annie M. Alexander on 11 November 1932.

Measurements of holotype.—Total length, 241; length of tail, 138; length of hind foot, 39; length of ear (crown), 10; weight, 39.5 g; greatest length of skull, 35.3; width across bullae, 23.45; breadth across maxillary arches, 19.6; nasal length, 12.85; interorbital breadth, 12.9.

Distribution.—Known only from Toroweap Valley, Mohave Co., Arizona.

Remarks.—Durrant and Setzer (1945) listed measurements for six males and three females; Hoffmeister (1986) listed statistics for measurements of five males and six females.

Dipodomys microps

Diagnosis.—A medium-sized kangaroo rat (head and body length averages from about 108 to 120 mm) with a relatively short tail (averaging about 135% of head and body length), with five toes on the hind feet, a narrow face, reflected in the breadth across the maxillary arches, and with lower incisors that are flattened on their anterior faces and chisel-like.

Comparisons.—Other species of *Dipodomys* that are most similar in size and general appearance to *D. microps,* and which occupy the same geographic areas, are *D. ordii* and *D. merriami.* Other, larger-sized, sympatric species are *D. panamintinus* and *D. deserti. D. agilis* occupies a contiguous area along the southwestern margin of the geographic range of *D. microps,* and may be marginally sympatric. From *D. merriami* and *D. deserti, D. microps* differs in having five, rather than four toes on the hind feet. From *D. agilis* and *D. panamintinus, D. microps* differs in being decidedly smaller and having chisel-like rather than awl-like lower incisors. From *D. ordii, D. microps* differs in having chisel-like, rather than awl-like lower incisors, and a narrower face, reflected in a shorter breadth across the maxillary arches in sympatric populations (Grinnell, 1922; Hall and Dale, 1939; Setzer, 1949).

Distribution.—Occurs throughout most of the Great Basin desert of western North America, from southeastern Oregon and adjacent Owyhee Co., Idaho, southward through Nevada and the Mojave Desert region of southeastern California, eastward in northwestern Utah to the Great Salt Lake. Disjunct populations occur in extreme southwestern Utah and adjacent northwestern Arizona, and in the vicinity of Joshua Tree National Monument in Riverside and San Bernardino counties, California (Csuti, 1979; Hall and Dale, 1939).

Remarks.—Hall and Dale (1939) provided the only systematic review of *Dipodomys microps*; they listed measurements for each of the subspecies then known (two were described subsequent to their revision). Durrant (1952) and Hall (1946) gave measurements for subspecies found in Utah and Nevada, respectively. Csuti (1979) reviewed patterns of variation in structure, plasma and tissue proteins, physiology, and ecology, but provided no taxonomic review of the subspecies.

Dipodomys microps alfredi Goldman, 1937

1937. *Dipodomys microps alfredi* Goldman, Proc. Biol. Soc. Washington, 50:221, 28 December.

Holotype.—Adult female, skin and skull, USNM 262846, from Gunnison Island, 4,300 ft, Great Salt Lake, Box Elder Co., Utah; obtained on 1 June 1937 by Alfred M. Bailey and Robert J. Niedrach.

Measurements of holotype.—Total length, 286; length of tail, 170; length of hind foot, 44; greatest length of skull, 38.5; width across bullae, 25.0; length of nasals, 12.8; breadth across maxillary arches, 20.6; interorbital breadth, 12.5; length of rostrum, 19.2; length of maxillary toothrow, 4.3; depth of cranium, 11.5.

Distribution.—Confined to Gunnison Island, Great Salt Lake, Box Elder Co., Utah (Hall, 1981).

Remarks.—Hall and Dale (1939) listed measurements for three males and three females and noted that members of this subspecies were the largest of the species.

Dipodomys microps aquilonius Willett, 1935

1935. *Dipodomys microps aquilonius* Willett, J. Mamm., 16:63, 14 February.

Holotype.—Adult female, skin and skull, LACM 3935, from 3 mi E Eagleville, Modoc Co., California; obtained by George Willett on 4 June 1934.

Measurements of holotype.—Total length, 272; length of tail, 156; length of hind foot, 41; greatest length of skull, 36.1; width across bullae, 23.9; width of maxillary arch at middle, 3.2.

Distribution.—Lower portions of the Great Basin Desert in northeastern California and adjacent Nevada, from the Surprise Valley on the northwest to the south end of Pyramid Lake on the southeast.

Remarks.—Hall and Dale (1939) gave measurements for four males and two females.

Dipodomys microps bonnevillei Goldman, 1937

1937. *Dipodomys microps bonnevillei* Goldman, Proc. Biol. Soc. Washington, 50:222, 28 December.

Holotype.—Adult female, skin and skull, USNM 31894/43755, from Kelton, about 4,300 ft, Box Elder Co., Utah; obtained on 7 November 1891 by Vernon Bailey.

Measurements of holotype.—Total length, 260; length of tail, 156; length of hind foot, 41; greatest length of skull, 36.0; width across bullae, 23.7; length of nasals, 11.9; breadth across maxillary arches, 18.9; interorbital breadth, 11.2; length of rostrum, 17.8; length of maxillary toothrow, 4.6; depth of cranium, 11.1.

Distribution.—Great Basin Desert of northeastern Nevada and northwestern Utah, corresponding closely with the former outline of the Pleistocene Lake Bonneville.

Remarks.—Hall and Dale (1939) listed means and ranges of measurements for 8 males and 10 females; Durrant and Setzer (1945) gave measurements for 17 males and 15 females.

Dipodomys microps celsus Goldman, 1924

1924. *Dipodomys microps celsus* Goldman, J. Washington Acad. Sci., 14:372, 19 September.
1942. *Dipodomys microps woodburyi* Hardy, Proc. Biol. Soc. Washington, 55:89, 25 June.

Holotype.—Adult male, skin and skull, USNM 243101, from 6 mi N Wolf Hole, 3,500 ft, Mohave County, Arizona; obtained on 16 October 1922 by Edward A. Goldman.

Measurements of holotype.—Total length, 284; length of tail, 170; length of hind foot, 44; greatest length of skull, 37.9; width across bullae, 25.3; length of nasals, 13.0; breadth across maxillary arches, 19.9; interorbital breadth, 11.4; length of rostrum, 18.5; length of maxillary toothrow, 5.1; depth of cranium, 11.5.

Distribution.—Occupies an apparently disjunct range in extreme southwestern Utah in the Virgin River Valley and in adjacent northwestern Arizona north of the Colorado River and east to Kanab Creek (Hall and Dale, 1939).

Remarks.—Hall and Dale (1939) listed the type specimen as USNM 243093, but the number of the designated type in the U.S. National Museum is number 243101; measurements are for the latter numbered specimen. Stock (1970) regarded *woodburyi* as indistinguishable from *celsus.*

Dipodomys microps centralis Hall and Dale, 1939

1939. *Dipodomys microps centralis* Hall and Dale, Occas. Papers Mus. Zool., Louisiana State Univ., 4:52, 10 November.

Holotype.—Adult male, skin and skull, MVZ 70817, from 4 mi SE Romano, Diamond Valley, Eureka Co., Nevada; obtained by William B. Richardson on 3 June 1936.

Measurements of holotype.—Total length, 282; length of tail, 164; length of hind foot, 43; length of ear, 13; weight, 72.8 g; greatest length of skull, 37.65; width across bullae, 23.55; breadth across maxillary arches, 19.65; nasal length, 12.6; interorbital breadth, 12.2.

Distribution.—Great Basin Desert of central Nevada, from the Humboldt River Valley southward to Pahute Mesa; from northeastern Pershing Co. on the northwest, to Spring Valley on the northeast.

Remarks.—Measurements for three males and three females were given by Hall and Dale (1939).

Dipodomys microps idahoensis Hall and Dale, 1939

1939. *Dipodomys microps idahoensis* Hall and Dale, Occas. Papers Mus. Zool., Louisiana State Univ., 4:53, 10 November.

Holotype.—Adult male, skin and skull, MVZ 67568, from 5 mi SE Murphy, Owyhee Co., Idaho; obtained on 26 May 1935 by Howard Twining.

Measurements of holotype.—Total length,

201; length of tail, 155; length of hind foot, 43; length of ear, 13; greatest length of skull, 37.15; width across bullae, 24.05; breadth across maxillary arches, 20.75; nasal length, 12.85; interorbital breadth, 12.55.

Distribution.—Known only from south of the Snake River in the Snake River Valley of northwestern Owyhee Co., Idaho.

Remarks.—Measurements for three males and one female were listed by Hall and Dale (1939).

Dipodomys microps leucotis Goldman, 1931

1931. *Dipodomys microps leucotis* Goldman, Proc. Biol. Soc. Washington, 44:135, 17 October.

Holotype.—Adult male, skin and skull, USNM 250036, from 6 mi W Colorado River Bridge, Houserock Valley, N [= W] side Marble Canyon, Colorado River, 3,700 ft, Coconino Co., Arizona; obtained on 8 June 1931 by Edward A. Goldman.

Measurements of holotype.—Total length, 281; length of tail, 170; length of hind foot, 43.5; greatest length of skull, 35.5; occipitonasal length, 34.6; width across bullae, 23.7; length of nasals, 12.0; breadth across maxillary arches, 19.6; interorbital breadth, 11.7; length of rostrum, 18.4; length of maxillary toothrow, 4.9; depth of cranium, 11.1.

Distribution.—Known only from a narrow strip between the Vermilion Cliffs and the brink of Marble Canyon of the Colorado River, in northwestern Arizona (Goldman, 1931).

Remarks.—Hall and Dale (1939) listed measurements for five male and five female topotypes, and Hoffmeister (1986) gave measurements for one male and three females.

Dipodomys microps levipes (Merriam, 1904)

1904. *Perodipus microps levipes* Merriam, Proc. Biol. Soc. Washington, 17:145, 14 July.

Holotype.—Adult male, skin and skull, USNM 27176/34575, from Perognathus Flat, Emigrant Gap, 5,200 ft, Panamint Mountains, Inyo Co., California; obtained by Vernon Bailey on 16 April 1891.

Measurements of holotype.—Total length, 288; length of tail, 156; length of hind foot, 43; greatest length of skull, 38.20; width across bullae, 24.25; breadth across maxillary arches, 20.55; nasal length, 13.50; interorbital breadth, 12.30.

Distribution.—Limited to an area in Inyo Co., in the Panamint Valley, California from the Panamint Mountains westward to the vicinity of Darwin.

Remarks.—Grinnell (1922) listed means and ranges of measurements for five individuals of each sex. Hall and Dale (1939) restricted the application of the name *levipes* to the population in the Panamint Valley; the name had been applied widely in previous publications.

Dipodomys microps microps (Merriam, 1904)

1904. *Perodipus microps* Merriam, Proc. Biol. Soc. Washington, 17:145, 14 July.

Holotype.—Adult male, skin and skull, USNM 25288/32701, from Lone Pine, Owens Valley, Inyo Co., California; obtained by Edward W. Nelson on 22 December 1890.

Measurements of holotype.—Total length, 282; length of tail, 165; length of hind foot, 41; greatest length of skull, 36.05; width across bullae, 22.50; breadth across maxillary arches, 18.35; nasal length, 12.10; interorbital breadth, 11.75.

Distribution.—Limited to the western Mojave Desert, California, in the Owens River watershed, from near the Nevada border N of Benton Station, southward to near Victorville.

Remarks.—Measurements were given by Csuti (1979), Grinnell (1922), and Hall and Dale (1939).

Dipodomys microps occidentalis Hall and Dale, 1939

1939. *Dipodomys microps occidentalis* Hall and Dale, Occas. Papers Mus. Zool., Louisiana State Univ., 4:56, 10 November.

Holotype.—Adult female, skin and skull, MVZ 64119, from 3 mi S Schurz, 4,100 ft, Mineral Co., Nevada; obtained by E. Raymond Hall on 8 July 1934.

Measurements of holotype.—Total length, 273; length of tail, 160; length of hind foot, 41; length of ear, 13; weight, 55.8 g; greatest length of skull, 35.1; width across bullae, 23.25; breadth across maxillary arches, 18.3; nasal length, 12.0; interorbital breadth, 11.95.

Distribution.—Great Basin Desert in western and southern Nevada, and eastern California in Mono and Inyo counties; south from Humboldt Co., Nevada to Death Valley, Inyo Co., California and Las Vegas, Clark Co., Nevada; the eastern limits in Nevada are Smiths Creek Valley, Lander Co., Mud Lake, Nye Co., and west of Panaca in Desert Valley, Lincoln Co. (Hall and Dale, 1939). A disjunct population occurs in the vicinity of Joshua Tree National Monument in Riverside and San Bernardino counties, California (Csuti, 1979).

Remarks.—Hall (1946) and Grinnell (1922) gave means and ranges of measurements for samples from Nevada and California, respectively.

Dipodomys microps preblei (Goldman, 1921)

1921. *Perodipus microps preblei* Goldman, J. Mamm., 2:233, 29 November.

Holotype.—Adult female, skin and skull USNM 79340, from Narrows, Malheur Lake, Harney Co., Oregon; obtained on 23 July 1896 by E. A. Preble.

Measurements of holotype.—Total length, 262; length of tail, 154; length of hind foot, 41; greatest length of skull, 33.5; width across bullae, 22.4; length of nasals, 11.6; breadth across maxillary arches, 19.2; length of maxillary toothrow, 4.2; least width of supraoccipital, 2.5.

Distribution.—Great Basin Desert of southeastern Oregon, and northwestern Nevada; westernmost locality is Summer Lake, Lake Co. Oregon, and the easternmost is near the Idaho-Nevada-Oregon border in the Owyhee River Valley, Humboldt Co., Nevada.

Remarks.—Hall and Dale (1939) gave measurements for three males and one female.

Dipodomys microps russeolus Goldman, 1939

1939. *Dipodomys microps russeolus* Goldman, J. Mamm., 20:353, 14 August.

Holotype.—Adult male, skin and skull, USNM 263895, from Dolphin Island, Great Salt Lake, 4,250 ft, Box Elder Co., Utah; obtained by William H. Marshall on 5 June 1938.

Measurements of holotype.—Total length, 248; length of tail, 136; length of hind foot, 41; greatest length of skull, 37.5; width across bullae, 24.0; length of nasals, 12.1; breadth across maxillary arches, 19.2; interorbital breadth, 11.4; length of rostrum, 18.5; length of maxillary toothrow, 5.0; depth of cranium, 11.0.

Distribution.—Limited to Dolphin Island, Great Salt Lake, Box Elder Co., Utah.

Remarks.—According to Durrant (1952), *russeolus* is known only from the type specimen.

Dipodomys microps subtenuis Goldman, 1939

1939. *Dipodomys microps subtenuis* Goldman, J. Mamm., 20:354, 14 August.

Holotype.—Adult female, skin and skull, USNM 263917, from Carrington Island, Great Salt Lake, 4,250 ft, Tooele Co., Utah;

obtained by William H. Marshall on 30 June 1938.

Measurements of holotype.—Total length, 267; length of tail, 157; length of hind foot, 40; length of ear (crown), 10; greatest length of skull, 37.6; width across bullae, 23.7; length of nasals, 12.6; breadth across maxillary arches, 19.7; length of rostrum, 19.4; length of maxillary toothrow, 4.7; depth of cranium, 11.5.

Distribution.—Known from Badger, Carrington, and Stansbury islands, Great Salt Lake, Tooele Co., and south on the mainland through Rush Valley to Cedar Valley west of Utah Lake, Utah Co., Utah (Durrant and Setzer, 1945).

Remarks.—Durrant and Setzer (1945) remarked that no specimens of *D. microps* had ever been taken on the eastern or southern shore of Great Salt Lake, adjacent to the islands supporting populations of this species—of hundreds of specimens trapped on the shore of the mainland, all were *D. ordii.*

Dipodomys nelsoni

Diagnosis.—A relatively large, four-toed kangaroo rat with a white tuft of hair at the tip of the tail; length of head and body of adults ranges from about 118 to 134 mm and length of hind foot ranges from about 46 to 52 mm; white tail tuft relatively small, measuring from about 6 to 20 mm in length; face of moderate width, not narrow nor broad; rostrum relatively short and auditory bullae relatively large.

Comparisons.—*Dipodomys nelsoni* is most similar in size and appearance to *D. spectabilis,* from which it can be distinguished by its average smaller size, shorter white tail tuft (the tuft of *spectabilis* is typically 25 mm or longer), and lesser breadth across the maxillary arches (breadth in *D. nelsoni* is less than about 24.5 and in *D. spectabilis* is greater than 25 mm in all but *D. s. intermedius*). It can be distinguished from all other four-toed kangaroo rats by either its larger size or darker coloration with a conspicuous black border to the terminal white tuft of hairs on the tail, and a dark underside to the tail (see account of *D. californicus, D. deserti,* and *D. elator* for other diagnostic information).

Distribution.—Open desert communities of north-central México, from southeastern Chihuahua and western Coahuila southward through northeastern Durango and southern Coahuila to northern Zacatecas, northern San Luis Potosí, and southern Nuevo León (Nader, 1978).

Remarks.—Controversy about the specific status of *D. nelsoni* is not satisfactorily resolved. Nader (1978), in reviewing the distribution, variation, and systematics of *D. deserti, D. spectabilis,* and *D. nelsoni,* found evidence of intergradation between *D. s. zygomaticus* and *D. nelsoni* in a narrow zone of contact in south central Chihuahua and north central Durango. He characterized the intergradation as secondary, the zone narrow, and the character gradients between taxa as steep. Nader (1978) concluded that *nelsoni* was best treated as a subspecies of *D. spectabilis.* Anderson (1972), working with less material found no overlap in a combination of characters, maxillary breadth plotted against total length of skull, between the two taxa. Hall (1981) stated that specimens unavailable to Anderson (1972) were "less easily identified to taxon, by means of the graph," but even so, concluded that they were not intergrades, and did not concur with Nader (1978). Matson (1980) interpreted Nader's (1978) analyses as providing "no direct evidence of intergradation between *D. nelsoni* and *D. spectabilis zygomaticus.*" We believe, however, that the indirect evidence is strong. Matson (1980) analyzed structural relationships between samples of the *spectabilis* group from areas near the boundaries of the ranges of *nelsoni* and *D. s. cratodon,* which ranges to the south of *nelsoni.* Discriminant function analyses clearly separated reference samples of the two taxa and unequivocally placed "unknowns" from areas between the ranges of the two taxa with one or the other of the reference groups. He interpreted these analyses as demonstrating the *D. nelsoni* and *D. spectabilis cratodon* were separate species. Questions that remain to be resolved concern the relationship of *cratodon* to *D. spectabilis* (see remarks in the account of *D. spectabilis*), and the significance of the narrow zone of intergradation in structural characteristics between *D. s. zygomaticus* and *D. nelsoni*. Considering all of the evidence available, we believe that it points to specific status for *D. nelsoni*. Additional samples from possible zones of contact and genic analyses are needed to help resolve remaining issues, however.

Dipodomys nelsoni Merriam, 1907

1907. *Dipodomys nelsoni* Merriam, Proc. Biol. Soc. Washington, 20:75, 22 July.

Holotype.—Adult male, skin and skull, USNM 79439, from La Ventura, Coahuila, México; obtained on 2 July 1896 by Edward W. Nelson and Edward A. Goldman.

Measurements of holotype.—Total length, 330; length of tail, 204; length of hind foot, 50; greatest length of skull, 42.7; width across bullae, 27.0; length of nasals, 14.3; breadth across maxillary arches, 22.8; interorbital breadth, 14.1; length of rostrum, 21.9; length of maxillary toothrow, 5.8; depth of cranium, 12.5.

Remarks.—Means and ranges of measurements for 24 specimens from Chihuahua were given by Anderson (1972). Dalquest (1953) gave cranial measurements of 3 specimens, and Nader listed measurements for 11 males and 3 females.

Dipodomys nitratoides

Diagnosis.—A small-sized, four-toed kangaroo rat with a tail terminating in a small tuft of buffy and blackish-brown hairs, and with a face that is not noticeably narrow; length of head and body averaging from about 88 to 100 mm; length of hind foot generally less than 36 mm; skull with relatively short and narrow rostrum with relatively straight sides; nasal length averaging less than 12.3 mm, and rostral breadth averaging 3.1 mm or less.

Comparisons.—*Dipodomys nitratoides* is most similar in size and appearance to the other small-sized, four-toed species, *D. merriami* and *D. insularis*; all other four-toed species are considerably larger and most have a prominent tuft of white hairs at the tip of the tail. See the account of *D. insularis* for characters distinguishing that species. *D. merriami* differs from *D. nitratoides* in generally being larger in size, and in having a longer and wider rostrum (nasal length usually averages greater than 13.1 in *D. merriami*; Hoffmann, 1975; Lidicker, 1960).

Diagnosis.—*Dipodomys nitratoides* occupies arid, alkaline grassland and desert shrub associations in the San Joaquin Valley and contiguous areas to the west, in west-central California. Its range extends from the valley floor in Merced Co., south of the Merced and San Joaquin rivers to the southern edge of the valley; it also occurs in the Panoche Valley of eastern San Benito Co., the Carrizo Plain, San Luis Obispo Co., and the upper Cuyama Valley in San Luis Obispo and Santa Barbara counties.

Remarks.—Lidicker (1960) did not include taxa assigned to *D. nitratoides* in his systematic review of *D. merriami*, although he acknowledged the two taxa were closely related and possibly conspecific. Hoffmann (1975) found that diagnostic characteristics for *D. nitratoides* clearly separated specimens of that species from *D. merriami*, except for a sample from the Cuyama Valley, California, geographically and ecologically contiguous to populations of *D. n. brevinasus*. These specimens exhibited larger size and longer, more robust rostra, characteristics attributed to *D. merriami*. Populations of similar sized *D. merriami*, however, such as *D. m. parvus*, are not geographically adjacent to populations of *D. nitratoides*. Measures of genetic similarity (Johnson and

Selander, 1971; Patton et al., 1976), and chromosomal data (Hoffmann, 1975; Stock, 1974) provide strong evidence that *D. nitratoides* is related to *D. merriami*, but is specifically distinct.

Dipodomys nitratoides brevinasus Grinnell, 1920

1920. *Dipodomys merriami brevinasus* Grinnell, J. Mamm., 1:179, 24 August.

Holotype.—Adult male, skin and skull, MVZ 28634, from Hayes Station, near bench mark 503, 19 mi SW Mendota, Fresno Co., California; obtained on 30 June 1918 by Richard Hunt.

Measurements of holotype.—Total length, 252; length of tail, 145; length of hind foot, 36; length of ear (crown), 11; weight, 43.9 g; greatest length of skull, 35.0; width across bullae, 23.2; length of nasals, 12.4; breadth across maxillary arches, 19.5; width of maxillary arch at middle, 4.7; greatest width of rostrum near end, 3.1; width of rostrum close to base, 5.2.

Distribution.—Annual grassland and desert shrub associations (principally *Atriplex* spp.) along the western margin of the San Joaquin Valley, California, from Merced Co. on the north to the mouth of San Emigdio Creek, Kern Co., on the south; then extending eastward in a narrow band above the valley floor to the east edge of the valley and northward, east of Bakersfield to approximately Poso Creek; also occurs in the Panoche Valley in eastern San Benito Co., on the Carrizo Plain, San Luis Obispo Co., and in the upper Cuyama Valley in San Luis Obispo and Santa Barbara counties.

Remarks.—Specimens from iodine bush (*Allenrolfea*) shrublands on the floor of the San Joaquin Valley, in Merced Co., exhibit characteristics intermediate to *brevinasus* and *exilis*, but overall, seem closest to the latter subspecies (unpubl. data). Hoffmann (1975) provided means and ranges of measurements for samples of *brevinasus*.

Dipodomys nitratoides exilis Merriam, 1894

1894. *Dipodomys merriami exilis* Merriam, Proc. Biol. Soc. Washington, 9:113, 21 June.

Holotype.—Young adult male, skin and skull, USNM 34843/43823, from Fresno, San Joaquin Valley, Fresno Co., California; obtained on 23 September 1891 by Vernon Bailey.

Measurements of holotype.—Total length, 241; length of tail, 143; length of hind foot, 33; greatest length of skull, 32.5; width across bullae, 20.9; length of nasals, 10.8; breadth across maxillary arches, 17.8; interorbital breadth, 11.9; length of rostrum, 16.8; length of maxillary toothrow, 3.8; depth of cranium, 10.3.

Distribution.—Occupies alkaline grassland and shrub communities on the floor of the San Joaquin Valley, from approximately the Merced River, Merced Co., on the north, to Kings River on the south and Fresno Slough on the west; the area in southern Fresno and northern Kings counties, north of historic Tulare Lake, was probably an area of intergradation between typical populations of *exilis* and *nitratoides*.

Remarks.—*Dipodomys nitratoides exilis* is the smallest of the kangaroo rats. Its current range is restricted by irrigated cultivation of its habitat to a few small, isolated areas of no more than a few hundred acres. Hoffmann (1975) gave measurements for samples of *exilis*.

Dipodomys nitratoides nitratoides Merriam, 1894

1894. *Dipodomys merriami nitratoides* Merriam, Proc. Biol. Soc. Washington, 9:112, 21 June.

Holotype.—Adult male, skin and skull, USNM 54674, from Tipton, San Joaquin Co., Tulare Co., California; obtained by Clark P. Streator on 25 June 1893.

Measurements of holotype.—Total length,

246; length of tail, 148; length of hind foot, 36; greatest length of skull, 34.5; length of nasals, 11.8; breadth across maxillary arches, 18.6; interorbital breadth, 11.6; length of rostrum, 17.6; length of maxillary toothrow, 4.4; depth of cranium, 10.6.

Distribution.—Occupies the San Joaquin Valley floor, in alkaline shrub and annual grassland communities, from approximately the northern boundaries of Kings County, east of the Kettleman and Lost Hills, southward to the edge of the Tehachapi Mountains, in Kern County. Limits on the southeast are approximately Bakersfield, and on the southwest, the old shore line of Lake Buena Vista.

Remarks.—Extant populations are small and isolated, being scattered in a few sites throughout the former range of the subspecies. Hoffmann (1975) listed measurements for samples of *nitratoides.*

Dipodomys ordii

Diagnosis.—A small- to medium-sized kangaroo rat with five toes on the hind feet and relatively short tail and ears; the maxillary arch is relatively slender; and the rostrum is comparatively short. The ratio of lengths of tail to head and body normally averages between about 1.16 and 1.26 and the width of the skull across the bullae ranges between about 22.3 and 26.3 (Grinnell, 1922; Setzer, 1949).

Comparisons.—*D. ordii* exhibits great variation in size over its extensive geographic range, but it is sympatric with few other five-toed kangaroo rats. *D. ordii* can be distinguished from *D. compactus* by its longer tail and wider skull (see account of the latter species); from *D. microps, ordii* differs in having a shorter tail and awl-like lower incisors; from *panamintinus, ordii* differs in being smaller with a relatively shorter tail and, within their area of sympatry, a shorter hind foot (length less than 44 mm; Hall, 1981). All other species sympatric with *D. ordii* have four toes on the hind feet. All other allopatric, five-toed species, except *D. ingens,* have significantly greater ratios of lengths of tail to head and body; the hind foot of *ingens* measures 48 mm or more, whereas the hind foot of *ordii* measures 43 mm or less.

Distribution.—*Dipodomys ordii* occupies an extensive range in the interior grasslands and deserts of western North America; extending from southeastern Washington southward through the Great Basin region, and from southwestern Saskatchewan and southeastern Alberta southward through the western Great Plains region to western Texas and Coahuila, southern Texas nearly to the Gulf coast, and Tamaulipas, Nueva León, and Coahuila; its range also includes the intermountain basins and plateaus of Arizona, Colorado, New Mexico, and Utah, and extensive areas of the tablelands of central México; the southernmost portion of its range is in eastern Zacatecas, Aguascalientes, northeastern Jalisco, San Luis Potosí, Hidalgo, and Guanajuato (Baumgardner and Schmidly, 1981; Hall, 1981; Setzer, 1949).

Remarks.—Setzer (1949) reviewed the species and provided documentation of distributions and means and ranges of measurements for subspecies. Schmidly and Hendricks (1976) and Baumgardner and Schmidly (1981) reported on distribution, structure, and systematics of populations of *D. ordii* and *D. compactus* in the southeastern portion of the range of *D. ordii.* Hall (1981) treated *compactus* as conspecific with *D. ordii* (see account of *compactus*). Kennedy and Schnell (1978) reported on multivariate analyses of geographic variation and sexual dimorphism of samples of *D. ordii* from throughout its geographic range.

Dipodomys ordii celeripes Durrant and Hall, 1939

1939. *Dipodomys ordii celeripes* Durrant and Hall, Mammalia, 3:10, March.

Holotype.—Adult male, skin and skull, UU 1956, from Trout Creek, 4,600 ft, Juab

Co., Utah; obtained by Stephen D. Durrant on 5 May 1937.

Measurements of holotype.—Total length, 225; length of tail, 126; length of hind foot, 41; length of ear, 13; greatest length of skull, 36.5; width across bullae, 22.95; breadth across maxillary arches, 19.70; nasal length, 13.30; interorbital breadth, 11.75.

Distribution.—Great Basin Desert of northeastern Nevada, and northwestern Utah.

Remarks.—Setzer (1949) gave means and ranges of measurements for four males and two females; Hall (1946) gave measurements for six males and nine females.

Dipodomys ordii chapmani Mearns, 1890

1890. *Dipodomys chapmani* Mearns, Bull. Amer. Mus. Nat. Hist., 2:291, 21 February.

Syntypes.—Adult male and female, skins and skulls, AMNH 2400/10560 and 2398/10561, respectively, from Fort Verde, Yavapai Co., Arizona; obtained on 26 January 1887 and 1 October 1885, respectively, by Edgar A. Mearns.

Measurements of holotype.—[2400, 2398] Total length, 256, 280; length of tail, 137, 148; length of hind foot, 36, 38; length of ear (crown), 11, 11, (notch), 13, 13; greatest length of skull, 36.7, 37.1; width across bullae, 23.2, 22.7; length of nasals, 12.3, 12.7; breadth across maxillary arches, 19.7, 18.7; interorbital breadth, 12.4, 12.7; length of rostrum, 18.9, 19.2; length of maxillary toothrow, 4.6, 4.7.

Distribution.—Found in west-central Arizona from near Kingman and Seligman south to the Bradshaw Mountains and southeastward to the Verde Valley (Hoffmeister, 1986).

Remarks.—Setzer (1949) noted that populations of this taxon were relatively isolated from those of other subspecies. Hoffmeister (1986) gave statistics for 16 males and 16 females. Setzer (1949) listed means and ranges for six males and four females.

Dipodomys ordii cinderensis Hardy, 1944

1944. *Dipodomys ordii cinderensis* Hardy, Proc. Biol. Soc. Washington, 57:53, 31 October.

Holotype.—Adult male, skin and skull, UU 4611, from immediately N of the northern of two large cinder cones, about 4,000 ft, Diamond Valley, 10 mi N Saint George, Washington Co., Utah; obtained by Ross Hardy on 13 February 1944.

Measurements of holotype.—Total length, 232; length of tail, 124; length of hind foot, 38; length of ear (notch), 14; condylobasal length, 29.9; occipitonasal length, 34.4; width across bullae, 23.2; length of nasals, 13.9; breadth across maxillary arches, 19.5; interorbital breadth, 12.1; length of maxillary toothrow, 5.0; greatest length of bulla as measured parallel with auditory opening, 15.0; ventral symphysis of bulla to ventral edge of auditory opening, 13.0; length of diastema, 7.9; length of left auditory opening, 3.7; width of supraoccipital at dorsal crest between bullae, 2.6.

Distribution.—Occupies sandy soils in sagebrush communities in southwestern Utah in northern Washington Co. and most of Iron County, Utah.

Remarks.—Hardy (1944) listed measurements for seven males and two females; Setzer (1949) gave measurements for two males and two females.

Dipodomys ordii cineraceus Goldman, 1939

1939. *Dipodomys ordii cineraceus* Goldman, J. Mamm., 20:352, 14 August.

Holotype.—Adult male, skin and skull, USNM 263890, from Dolphin Island, Great Salt Lake, 4,250 ft, Box Elder Co., Utah; obtained on 4 June 1938 by William H. Marshall.

Measurements of holotype.—Total length, 231; length of tail, 130; length of hind foot, 37; occipitonasal length, 35.7; width across bullae, 23.8; length of nasals, 13.8; breadth across maxillary arches, 20.1; length of maxillary toothrow, 5.1; width of nasals (in front of incisors), 3.5; least width of supraoccipital, 2.9; width of cutting edge of upper incisor, 1.8.

Distribution.—Limited to Dolphin Island, Great Salt Lake, Box Elder Co., Utah.

Remarks.—Goldman (1939) gave measurements for three male and one female topotypes.

Dipodomys ordii columbianus (Merriam, 1894)

1894. *Perodipus ordii columbianus* Merriam, Proc. Biol. Soc. Washington, 9:115, 21 June.

Holotype.—Adult female, skin and skull, USNM 24181/31594, from Umatilla, at mouth of Umatilla River, plains of Columbia, Umatilla Co., Oregon; obtained by Clark P. Streator on 18 October 1890.

Measurements of holotype.—Total length, 254; length of tail, 148; length of hind foot, 40; length of ear (anterior base, dry), 13; greatest length of skull, 37.30; width across bullae, 23.30; breadth across maxillary arches, 20.30; nasal length, 13.65; interorbital breadth, 12.60.

Distribution.—Found on the Columbia Plateau and parts of the Great Basin from southeastern Washington, eastern Oregon, and southwestern Idaho southward to northeastern California and parts of northern Nevada (Hall, 1981).

Remarks.—Hall (1946) listed means and ranges of measurements for six males and six females from Nevada. Setzer (1949) gave measurements for three male and two female topotypes.

Dipodomys ordii cupidineus Goldman, 1924

1924. *Dipodomys ordii cupidineus* Goldman, J. Washington Acad. Sci., 14:372, 19 September.

Holotype.—Adult male, skin and skull, USNM 243093, from Kanab Wash, at southern boundary of Kaibab Indian Reservation, Mohave Co., Arizona; obtained by Edward A. Goldman on 12 October 1922.

Measurements of holotype.—Total length, 257; length of tail, 150; length of hind foot, 41; greatest length of skull, 38.3; occipitonasal length, 36.4; width across bullae, 24.9; length of nasals, 13.3; breath across maxillary arches, 21.1; interorbital breadth, 11.9; length of rostrum, 19.1; length of maxillary toothrow, 4.8; depth of cranium, 11.6; least width of supraoccipital, 2.5.

Distribution.—Known from an area north of the Grand Canyon of the Colorado in northwestern Arizona and southwestern Utah.

Remarks.—Measurements were given by Hoffmeister (1986) for 10 males and 6 females from Arizona.

Dipodomys ordii durranti Setzer, 1952

1949. *Dipodomys ordii fuscus* Setzer, Univ. Kansas Publ. Mus. Nat. Hist., 1:555, 27 December.

1952. *Dipodomys ordii durranti* Setzer, J. Washington Acad. Sci., 42:391, 17 December. Renaming of *fuscus* Setzer, 1949.

Holotype.—Adult male, skin and skull, USNM 93886, from Juamave [Jaumave], Tamaulipas, México; obtained on 3 June 1898 by Edward W. Nelson and Edward A. Goldman.

Measurements of holotype.—Total length, 241; length of tail, 146; length of hind foot, 38; greatest length of skull, 37.5; width across bullae, 23.8; length of nasals, 12.3; breadth across maxillary arches, 19.6; interorbital breadth, 12.3; length of rostrum, 18.8; length of maxillary toothrow, 4.3; depth of cranium, 11.4.

Distribution.—Occurs on the northern half of the Mesa Central, from southern Coahuila, Nuevo León, and adjacent Tamaulipas to northeastern Zacatecas and northern San Luis Potosí (Baumgardner and Schmidly, 1981; Matson and Baker, 1986).

Remarks.—The name *fuscus*, first given to this taxon by Setzer (1949), was preoccupied by *D. agilis fuscus* Boulware, 1943. Baumgardner and Schmidly (1981) provided the most recent review of *D. ordii durranti*. Measurements were given by Setzer (1949).

Dipodomys ordii evexus Goldman, 1933

1933. *Dipodomys ordii evexus* Goldman, J. Washington Acad. Sci., 23:468, 15 October.

Holotype.—Adult male, skin and skull, USNM 150990, from Salida, 7,000 ft, Chaffee County, Colorado; obtained by Merrit Cary on 10 November 1907.

Measurements of holotype.—Total length, 266; length of tail, 149; length of hind foot, 42; greatest length of skull, 38.9; occipitonasal length, 37.3; width across bullae, 23.9; length of nasals, 13.6; width of maxillary arch at middle, 20.9; interorbital breadth, 12.7; length of rostrum, 20.0; length of maxillary toothrow, 5.0; depth of cranium, 11.7; width of nasals (in front of incisors), 3.8; least width of supraoccipital, 3.0.

Distribution.—Valley of the upper Arkansas River of south-central Colorado, above the Royal Gorge.

Remarks.—Setzer (1949) listed measurements of seven specimens.

Dipodomys ordii extractus Setzer, 1949

1949. *Dipodomys ordii extractus* Setzer, Univ. Kansas Publ., Mus. Nat. Hist., 1:534, 27 December.

Holotype.—Adult male, skin and skull, MVZ 76562, from 1 mi E Samalayuca, 4,500 ft, Chihuahua, México; obtained on 15 May 1937 by William B. Richardson.

Measurements of holotype.—Total length, 244; length of tail, 132; length of hind foot, 39; length of ear, 13; weight, 49.2 g; greatest length of skull, 38.6; width across bullae, 24.35; breadth across maxillary arches, 20.35; nasal length, 14.85; interorbital breadth, 12.85.

Distribution.—Known only from the sand dunes of Samalayuca in northern Chihuahua, occupying an area approximately 48 km in diameter (Anderson, 1972).

Remarks.—Anderson (1972) listed means and ranges of measurements for nine specimens. Anderson (1972) could find no significant differences, other than paler color, to distinguish *extractus* from nearby populations of *D. ordii ordii*.

Dipodomys ordii fetosus Durrant and Hall, 1939

1939. *Dipodomys ordii fetosus* Durrant and Hall, Mammalia, 3:14, March.

Holotype.—Adult female, skin and skull, MVZ 48451, from 2 mi N of Panaca, 4,800 ft, Lincoln Co., Nevada; obtained by Ward C. Russell on 24 June 1931.

Measurements of holotype.—Total length, 225; length of tail, 125; length of hind foot, 41; length of ear, 13; greatest length of skull, 36.65; width across bullae, 23.85; breadth across maxillary arches, 20.25; nasal length, 13.3; interorbital breadth, 11.9.

Distribution.—Found in southeastern Nevada, in Lincoln Co., and southwestern Utah in western Beaver and Millard counties.

Remarks.—Setzer (1949) gave measurements for four males and four females; Hall (1946) listed measurements for six males and seven females.

Dipodomys ordii fremonti Durrant and Setzer, 1945

1945. *Dipodomys ordii fremonti* Durrant and Setzer, Bull. Univ. Utah, 35(26):21, 30 June.

Holotype.—Adult female, skin and skull, CM 15661, from Torrey, 7,000 ft, Wayne Co., Utah; obtained on 19 July 1938 by W. F. and F. H. Wood.

Measurements of holotype.—Total length,

246; length of tail, 132; length of hind foot, 38; condylobasal length, 23.1; occipitonasal length, 30.3; length of nasals, 13.4; interorbital breadth, 12.1; length of bulla, 14.7; width of bulla, 9.6; depth of bulla, 12.1.

Distribution.—Known only from the vicinity of the type locality in the upper reaches of the Fremont River watershed in west-central Wayne Co., Utah (Setzer, 1949).

Remarks.—Setzer (1949) listed measurements for two males and two females.

Dipodomys ordii inaquosus Hall, 1941

1941. *Dipodomys ordii inaquosus* Hall, Proc. Biol. Soc. Washington, 54:58, 20 May.

Holotype.—Adult male, skin and skull, MVZ 73580, from 11 mi E, 1 mi N Jungo, 4,200 ft, Humboldt Co., Nevada; obtained by Ward C. Russell on 26 July 1936.

Measurements of holotype.—Total length, 261; length of tail, 142; length of hind foot, 41; length of ear, 12; weight, 49 g; greatest length of skull, 37.55; width across bullae, 24.15; breadth across maxillary arches, 20.4; nasal length, 14.3; interorbital breadth, 11.45.

Distribution.—Found in north-central Nevada in southeastern Humboldt and northern Lander counties (Hall, 1946).

Remarks.—Hall (1946) gave measurements for six males and two females.

Dipodomys ordii longipes (Merriam, 1890)

1890. *Dipodops longipes* Merriam, 1890, N. Amer. Fauna, 3:72, 11 September.
1933. *Dipodomys ordii cleomophila* Goldman, J. Washington Acad. Sci., 23:469, 15 October.

Holotype.—Young adult male, skin and skull, USNM 17703/24639, from foot of Echo Cliffs, Painted Desert, Coconino Co., Arizona; obtained by C. Hart Merriam on 22 September 1889.

Measurements of holotype.—Total length, 275; length of tail, 165; length of hind foot, 42; length of ear (crown, dry), 8; greatest length of skull, 39.8; width across bullae, 25.5; length of nasals, 14.0; breadth across maxillary arches, 19.8; interorbital breadth, 12.9; length of rostrum, 19.8; length of maxillary toothrow, 4.4.

Distribution.—Ranges through the Painted Desert region of southeastern Utah, extreme southwestern Colorado, northwestern New Mexico, and northeastern Arizona.

Remarks.—Setzer (1949) placed *cleomophila* as a synonym of *longipes.* Hoffmeister (1986) gave measurements for eight males and nine females from Arizona; Durrant (1952) gave measurements for two males and five females from Utah.

Dipodomys ordii luteolus (Goldman, 1917)

1917. *Perodipus ordii luteolus* Goldman, Proc. Biol. Soc. Washington, 30:112, 23 May.

Holotype.—Adult male, skin and skull, USNM 160408, from Casper, Natrona Co., Wyoming; obtained by Merritt Cary on 2 September 1903.

Measurements of holotype.—Total length, 272; length of tail, 154; length of hind foot, 42; occipitonasal length, 38.1; width across bullae, 24.3; breadth across maxillary arches, 21.2; length of maxillary toothrow, 4.8; least width of supraoccipital, 2.8.

Distribution.—Occupies the northern Great Plains from western South Dakota to southern Nebraska, northeastern Colorado, and southeastern Wyoming (Hall, 1981; Setzer, 1949).

Remarks.—Armstrong (1972), Jones (1964), Long (1965), and Setzer (1949) listed measurements.

Dipodomys ordii marshalli Goldman, 1937

1937. *Dipodomys ordii marshalli* Goldman, Proc. Biol. Soc. Washington, 50:223, 28 December.

Holotype.—Adult female, skin and skull, USNM 262655, from Bird Island, about 4,300 ft, Great Salt Lake, Tooele Co., Utah; obtained by William H. Marshall on 22 June 1937.

Measurements of holotype.—Total length, 273; length of tail, 123; length of hind foot, 36; length of ear, 14; greatest length of skull, 35.5; width across bullae, 22.8; length of nasals, 13.6; breadth across maxillary arches, 20.0; interorbital breadth, 11.0; length of maxillary toothrow, 14.3; depth of cranium, 10.6; occipitonasal length, 35.0; width of nasals (in front of incisors), 3.4; least width of supraoccipital, 2.6.

Distribution.—Found on Badger, Bird, Carrington, and Stansbury islands, Great Salt Lake, and along the western edge of Great Salt Lake, north to Kelton, Box Elder Co., Utah; and around the southern and southeastern shores of the lake to the mouth of the Jordan River, Davis and Salt Lake counties, Utah.

Remarks.—The skull of the holotype was damaged sometime after Goldman's description and after Poole and Schantz (1942) remarked on the condition of the skull. Setzer (1949) and Durrant (1952) gave measurements for specimens of *marshalli.*

Dipodomys ordii medius Setzer, 1949

1949. *Dipodomys ordii medius* Setzer, 1949, Univ. Kansas Publ., Mus. Nat. Hist., 1:519, 27 December.

Holotype.—Adult male, skin and skull, USNM 118526, from Santa Rosa, Guadalupe Co., New Mexico; obtained by Jason H. Grant on 5 October 1902.

Measurements of holotype.—Total length, 256; length of tail, 146; length of hind foot, 38; greatest length of skull, 38.7; width across bullae, 24.9; length of nasals, 13.7; breadth across maxillary arches, 21.0; length of maxillary toothrow, 4.8; depth of cranium, 12.0.

Distribution.—Found on the High Plains of southeastern and north-central New Mexico and western Texas (Hall, 1981).

Remarks.—Setzer (1949) listed measurements for six male and two female topotypes.

Dipodomys ordii monoensis (Grinnell, 1919)

1919. *Perodipus monoensis* Grinnell, Univ. California Publ. Zool., 21:46, 29 March.

Holotype.—Adult female, skin and skull, MVZ 27002, from Pellisier Ranch, 5,600 ft, 5 mi N Benton Station, Mono Co., California; obtained by Joseph Dixon on 21 September 1917.

Measurements of holotype.—Total length, 245; length of tail, 124; length of hind foot, 38; length of ear, 12; weight, 48.1 g; greatest length of skull, 37.15; width across bullae, 23.40; breadth across maxillary arches, 19.90; nasal length, 13.55; interorbital breadth, 11.50.

Distribution.—Found in the Mono Basin of Mono, Co., California and a wider area of west-central Nevada, from Pyramid Lake and the Humboldt Sink on the north, southward to Beatty, Nye Co., and, with the exception of the Reese River Valley, eastward over Nye Co. to the Quinn Canyon Mountains (Hall, 1946, 1981).

Remarks.—Hall (1946) gave measurements for 10 males and 3 females from Nevada, and Grinnell (1922) gave measurements for 5 males and 5 females from California.

Dipodomys ordii montanus Baird, 1855

1855. *Dipodomys montanus* Baird, Proc. Acad. Nat. Sci. Philadelphia, 7:334, April.

Holotype.—Adult [sex unknown], skin and skull, ANSP 490/1631, from Fort Massachusetts [now Fort Garland], Costilla Co., Colorado; obtained by F. Kreuzfeldt in 1853.

Measurements of holotype.—Total length, 247.7; length of tail, 133.4; length of hind foot, 38.1; greatest length of skull, 36.8; length of nasals, 12.8; breadth across maxillary arches, 19.5; interorbital breadth, 12.5; length of rostrum, 19.5; length of maxillary toothrow, 4.7.

Distribution.—Found in the San Luis Valley of south-central Colorado and adjacent north-central New Mexico.

Remarks.—External measurements and length of skull were converted from inches to mm from values published by Baird (1858); measurements were not recorded with the holotype and the skull is damaged such that a measure of greatest length could not be taken by us. Armstrong (1972) provided statistics for 10 males and 6 females and Setzer (1949) published means and ranges of measurements for 11 males and 11 females.

Dipodomys ordii nexilis Goldman, 1933

1933. *Dipodomys ordii nexilis* Goldman, J. Washington Acad. Sci., 23:470, 15 October.

Holotype.—Adult male, skin and skull, USNM 149938, from 5 mi W Naturita, Montrose Co., Colorado; obtained by Merritt Cary on 20 July 1907.

Measurements of holotype.—Total length, 268; length of tail, 147; length of hind foot, 45; greatest length of skull, 40.9; width across bullae, 25.9; length of nasals, 14.0; breadth across maxillary arches, 21.2; interorbital breadth, 13.1; length of rostrum, 19.6; length of maxillary toothrow, 5.0; depth of cranium, 12.1; occipitonasal length, 39.0; width of nasals (in front of incisors), 4.3; least width of supraoccipital, 2.3.

Distribution.—Found in southeastern Utah, between the Colorado and San Juan rivers, and in adjacent southwestern Colorado (Hall, 1981).

Remarks.—Armstrong (1972) listed measurements of 10 males and 4 females from Colorado; Goldman (1933) listed measurements for 3 adult topotypes.

Dipodomys ordii obscurus (J. A. Allen, 1903)

1903. *Perodipus obscurus* J. A. Allen, Bull. Amer. Mus. Nat. Hist., 19:603, 12 November.

1939. *Dipodomys ordii attenuatus* Bryant, Occas. Papers Mus. Zool., Louisiana State Univ., 5:65, 10 November.

1949. *Dipodomys ordii idoneus* Setzer, Univ. Kansas Publ., Mus. Nat. Hist., 1:546, 27 December.

Holotype.—Adult male, skin and skull, AMNH 20957, from Río Sestín, northwestern Durango, México; obtained by J. H. Batty on 13 April 1903.

Measurements of holotype.—Total length, 232; length of tail, 130; length of hind foot, 35; length of ear (notch), 12.7; greatest length of skull, 35.6; width across bullae, 23.0; length of nasals, 12.8; breadth across maxillary arches, 19.7; interorbital breadth, 12.4; length of rostrum, 19.0; length of maxillary toothrow, 4.4.

Distribution.—Occurs on the northern portion of the Mexican Plateau north of northern Zacatecas and southern Coahuila (Mesa del Norte), and the adjoining regions of the Big Bend Basin and Rio Grande Plain of Texas (Baumgardner and Schmidly, 1981; Bryant, 1939; Matson and Baker, 1986).

Remarks.—Baumgardner and Schmidly (1981) synonymized *attenuatus* and *idoneus* with *obscurus,* to which they also referred the northern populations previously assigned to *durranti.* Baumgardner and Schmidly (1981) gave means and ranges of measurements for samples of *obscurus.*

Dipodomys ordii oklahomae Trowbridge and Whitaker, 1940

1940. *Dipodomys oklahomae* Trowbridge and Whitaker, J. Mamm., 21:343, 13 August.

Holotype.—Young adult female, skin and skull, USNM 265454, from north bank of South Canadian River, 2.25 mi S Norman, Cleveland Co., Oklahoma; obtained by H. L. Whitaker on 16 March 1934.

Measurements of holotype.—Total length, 243; length of tail, 128; length of hind foot, 41; greatest length of skull, 38.4; occipitonasal length, 36.5; width across bullae, 23.8; length of nasals, 13.6; breadth across maxillary arches, 21.6; interorbital breadth, 13.1; length of rostrum, 19.6; length of maxillary toothrow, 4.8; depth of cranium, 12.1; width of nasals (in front of incisors), 3.9; least width of supraoccipital, 3.3; width of palate between first molars, 3.4; length of interparietal, 4.0; width of interparietals, 3.1.

Distribution.—Known only from the South Canadian River Valley of central Oklahoma from Canadian Co. eastward into Cleveland Co.

Remarks.—Kennedy et al. (1980) analyzed intraspecific variation in anatomical structure of *oklahomae* and other populations of *D. ordii* from Oklahoma, and provided lists of measurements.

Dipodomys ordii ordii Woodhouse, 1853

1853. *D[ipodomys]. ordii* Woodhouse, Proc. Acad. Nat. Sci. Philadelphia, 6:224.

Holotype.—None designated, species characterized from specimens obtained by Dr. Woodhouse at El Paso, El Paso Co., Texas. Neotype—Adult female, skin and skull, USNM 18135/25033, from El Paso, El Paso Co., Texas; obtained on 11 December 1889 by Vernon Bailey (designated by Merriam, 1890:45, as the "Duplicate type"; see remarks below).

Measurements of neotype.—Total length, 240; length of tail, 134; length of hind foot, 38; length of ear (crown), 7, (anterior base, dry) 12; greatest length of skull, 38.60; width across bullae, —; breadth across maxillary arches, 20.70; nasal length, 14.00; interorbital breadth, 12.40.

Distribution.—Ranges through desert communities of southeastern Arizona, southern New Mexico, Trans-Pecos Texas, northern Chihuahua, and northeastern Sonora (Hall, 1981).

Remarks.—Although Merriam (1890) used the term duplicate type rather than neotype, we believe his designation met the essential criteria of the International Code of Zoological Nomenclature for designating a neotype, and so regard his duplicate type. Anderson (1972) provided measurements for 14 specimens from Chihuahua, Hoffmeister (1986) gave statistics for measurements of 16 males and 10 females from Arizona, and Setzer (1949) listed measurements for 5 male and 6 female topotypes.

Dipodomys ordii pallidus Durrant and Setzer, 1945

1945. *Dipodomys ordii pallidus* Durrant and Setzer, Bull. Univ. Utah, 35(26):24, 30 June.

Holotype.—Adult male, skin and skull, UU 3526, from Old Lincoln Highway, 18 mi SW Orr's Ranch, Skull valley, 4,400 ft, Tooele Co., Utah; obtained by Stephen D. Durrant on 6 June 1940.

Measurements of holotype.—Total length, 251; length of tail, 141; length of hind foot, 45; occipitonasal length, 30.7; condylobasal length, 23.5; length of nasals, 13.0; length of bulla, 15.1; width of bulla, 10.0; depth of bulla, 12.2.

Distribution.—Occupies the lower-lying valleys of west-central Utah in Juab, Tooele, and Millard counties (Durrant, 1952).

Remarks.—Durrant listed means and ranges of measurements for seven males and two females.

Dipodomys ordii palmeri (J. A. Allen, 1891)

1891. *Dipodops ordii palmeri* J. A. Allen, Bull. Amer. Mus. Nat. Hist., 3:276, 30 June.

Syntypes.—Two adult males, skins and skulls, MCZ 5886, 5887, from San Luis Potosí, San Luis Potosí, México on 1 May and 1 September 1878 by Edward Palmer.

Measurements of syntypes.—[5886, 5887; external measurements approximated by Allen (1891) from skins and reported as single values] Total length, 249; length of tail, 141; length of hind foot, 35; length of ear (notch, dry), 11.4; greatest length of skull, 36.0, 36.3; width across bullae, —, 23.0; length of nasals, 12.5, 13.3; breadth across maxillary arches, 19.1, 19.6; interorbital breadth, 12.8, 12.9; length of rostrum, 17.5, 18.0; length of maxillary toothrow, 4.2, 4.3.

Distribution.—Occupies the southern portion of the Central Plateau of México, from northern San Luis Potosí and central Zacatecas, south to Hidalgo (Hall, 1981; Matson and Baker, 1986).

Remarks.—Baumgardner and Schmidly (1981) questioned the subspecific distinctness of *palmeri* and *durranti,* based on multivariate analysis of anatomical structure. Previous authors have listed the paper by Allen (1881), as the first description of *D. o. palmeri*; Allen (1891) first named and described *palmeri,* noting that he had earlier (Allen, 1881) referred specimens of this subspecies to *Dipodomys phillipsii.* Dalquest (1953) gave means for measurements of five males and nine females from San Luis Potosí.

Dipodomys ordii panguitchensis Hardy, 1942

1942. *Dipodomys ordii panguitchensis* Hardy, Proc. Biol. Soc. Washington, 55:90, 25 June.

Holotype.—Adult male, skin and skull, UU 4375, from 1 mi S Panguitch, 6,666 ft, Garfield Co., Utah; obtained on 31 August 1940 by Ross Hardy.

Measurements of holotype.—Total length, 257; length of tail, 145; length of hind foot, 41; length of ear (notch), 14; basal length of skull, 25.2; width across bullae, 23.0; length of nasals, 13.4; breadth across maxillary arches, 20.0; interorbital breadth, 12.3; length of maxillary toothrow, 4.7.

Distribution.—Known only from the type locality in south-central Utah.

Remarks.—Hardy (1940) listed measurements for two male and two female holotypes.

Dipodomys ordii priscus Hoffmeister, 1942

1942. *Dipodomys ordii priscus* Hoffmeister, Proc. Biol. Soc. Washington, 55:167, 31 December.

Holotype.—Adult female, skin and skull, MVZ 89119, from Kinney Ranch, 21 mi S Bittercreek, 7,100 ft, Sweetwater Co., Wyoming; obtained by Donald T. Tappe on 16 September 1939.

Measurements of holotype.—Total length, 267; length of tail, 145; length of hind foot, 41; length of ear, 13; weight, 64.4 g; greatest length of skull, 39.5; width across bullae, 24.45; breadth across maxillary arches, 20.7; nasal length, 14.3; interorbital breadth, 13.1.

Distribution.—Located in the arid intermountain basins of southwestern Wyoming, northwestern Colorado, and extreme northeastern Utah.

Remarks.—Armstrong (1972) gave measurements for one male and four females. Setzer (1949) provided means and ranges of measurements of 11 specimens.

Dipodomys ordii pullus Anderson, 1972

1972. *Dipodomys ordii pullus* Anderson, Bull. Amer. Mus. Nat. Hist., 148:317, 8 September.

Holotype.—Adult female, skin and skull, KU 73733, from El Rosario, 6,700 ft, Chihuahua, México; obtained on 28 July 1957 by Sydney Anderson.

Measurements of holotype.—Total length, 254; length of tail, 147; length of hind foot, 40.5; length of ear, 15; weight, 57.3 g; greatest length of skull, 36.95; width across bullae, 24.10; breadth across maxillary arches, 21.70; nasal length, 13.20; interorbital breadth, 13.40.

Distribution.—Found in west-central

Chihuahua, including the high valleys of the Río Papigochic, Laguna de los Méxicanos, and Laguna de Bustillos (Anderson, 1972).

Remarks.—Anderson (1972) listed means and ranges of measurements for 51 specimens.

Dipodomys ordii richardsoni (J. A. Allen, 1891)

1891. *Dipodops richardsoni* J. A. Allen, Bull. Amer. Mus. Nat. Hist., 3:277, 30 June.

Holotype.—Adult male, skin and skull, AMNH 3025/2345, from Beaver River [confluence of Cienquilla and Currumpaw creeks, Sec. 32, T2N, R2E], Cimarron Co., Oklahoma; obtained by Jenness Richardson and John Rowley, Jr., on 26 October 1889.

Measurements of holotype.—Greatest length of skull, 40.4; width across bullae, 25.0; length of nasals, 14.2; breadth across maxillary arches, 21.6; interorbital breadth, 11.9; length of rostrum, 19.2; length of maxillary toothrow, 5.2.

Distribution.—Occupies the west-central Great Plains, from southwestern Nebraska, to western Oklahoma and the Panhandle region of west Texas, and westward in southeastern Colorado and northeastern New Mexico (Hall, 1981).

Remarks.—Armstrong (1972), Jones (1964), Kennedy et al. (1980), and Setzer (1949) gave measurements for samples of *richardsoni.* Glass (1971) provided information to amend the type locality and the year that the holotype was collected.

Dipodomys ordii sanrafaeli Durrant and Setzer, 1945

1945. *Dipodomys ordii sanrafaeli* Durrant and Setzer, Bull. Univ. Utah, 35(26):26, 30 June.

Holotype.—Adult female, skin and skull, UU 4612, from 1.5 mi N Price, 5,567 ft, Carbon Co., Utah; obtained on 5 June 1940 by Ross Hardy and H. Higgins.

Measurements of holotype.—Total length, 249; length of tail, 138; length of hind foot, 42; length of ear, 16; greatest length of skull, 37.80; width across bullae, 24.80; breadth across maxillary arches, 20.20; nasal length, 13.10; interorbital breadth, 12.75.

Distribution.—Occurs in the high desert of east-central Utah eastward into Colorado along the Colorado River Valley to Grand Junction, Mesa Co. (Hall, 1981).

Remarks.—Measurements were given by Armstrong (1972), Durrant and Setzer (1945), and Setzer (1949).

Dipodomys ordii terrosus Hoffmeister, 1942

1942. *Dipodomys ordii terrosus* Hoffmeister, Proc. Biol. Soc. Washington, 55:165, 31 December.

Holotype.—Adult male, skin and skull, MVZ 93477, from Yellowstone River, 5 mi W Forsyth, 2,750 ft, Rosebud Co., Montana; obtained on 2 June 1940 by J. R. Alcorn.

Measurements of holotype.—Total length, 266; length of tail, 143; length of hind foot, 43; weight, 71.8 g; occipitobasal length, 28.4; width across bullae, 25.8; length of nasals, 15.5; breadth across maxillary arches, 22.8; interorbital breadth, 13.2.

Distribution.—Occupies the northwestern Great Plains, from southern Alberta and Saskatchewan, southward through eastern Montana, extreme southwestern North Dakota, and northwestern South Dakota to northern Wyoming (Hall, 1981).

Remarks.—Long (1965) provided measurements for 3 males and 10 females; Setzer (1949) gave measurements of 5 adults.

Dipodomys ordii uintensis Durrant and Setzer, 1945

1945. *Dipodomys ordii uintensis* Durrant and Setzer, Bull. Univ. Utah, 35(26):27, 14 July.

Holotype.—Adult male, skin and skull,

CM 11634, from Red Creek, 6,700 ft, 2 mi N Fruitland, Duchesne Co., Utah; obtained by J. K. and M. T. Doutt on 15 August 1936.

Measurements of holotype.—Total length, 253; length of tail, 140; length of hind foot, 40; greatest length of skull, 37.9; width across bullae, 23.3; length of nasals, 13.5; breadth across maxillary arches, 20.3; interorbital breadth, 12.5; width of rostrum, 3.9; basilar length, 23.9.

Distribution.—Occupies sandy soils of the Uintah Basin in northeastern Utah.

Remarks.—Durrant (1952) gave measurements for three adult males.

Dipodomys ordii utahensis (Merriam, 1904)

1904. *Perodipus montanus utahensis* Merriam, Proc. Biol. Soc. Washington, 17:143, 30 June.

Holotype.—Adult male, skin and skull, USNM 55115, from Ogden, Weber Co., Utah; obtained on 15 July 1893 by Vernon Bailey.

Measurements of holotype.—Total length, 260; length of tail, 150; length of hind foot, 41; greatest length of skull, 37.50; width across bullae, 23.40; breadth across maxillary arches, 20.15; nasal length, 13.75; interorbital breadth, 11.75.

Distribution.—From near the Idaho border southward in Utah along the eastern margin of the Pleistocene Lake Bonneville basin to northern Sevier Co.; ranges to the west side of the Onaqui Mountains, Tooele Co., on the west (Durrant, 1952).

Remarks.—Durrant (1952) gave average measurements of five specimens.

Dipodomys panamintinus

Diagnosis.—A relatively large, five-toed kangaroo rat with short ear pinnae, a broad face, stout maxillary arch, and relatively small auditory bullae; length of head and body averages from about 123 to 128 mm; width across the maxillary arches averages from about 23.2 to 23.6 or greater; width of maxillary arch, at middle, averages from about 5.4 to 5.7 (Grinnell, 1922).

Comparisons.—*Dipodomys panamintinus* is most similar in size and proportions to some populations of *D. heermanni,* which occupies an allopatric range in central California; *panamintinus* differs in having much less black in the markings on the face, shorter ears, and having auditory bullae that are about half the volume of those of *D. heermanni.* From the sympatric five-toed species, *D. panamintinus* can be distinguished by its considerably larger size, and proportionately longer tail than *D. ordii,* and its awl-shaped lower incisors in comparison to the chisel-shaped incisors of *D. microps.* Overall, *D. panamintinus* is nearly indistinguishable externally from the allopatric species, *D. stephensi.* It differs from *stephensi* in having smaller auditory bullae, longer nasals, wider maxillary arches, and in other proportions of the skull (Grinnell, 1922).

Distribution.—Occurs in desert shrub and arid woodland associations along the boundary between California and Nevada, from the Virginia Mountains, Washoe Co., Nevada and the Honey Lake Valley, Lassen Co., California, southward through the Mojave Desert to northeastern Los Angeles Co. and Hesperia, San Bernardino Co., California; it also occupies the slopes of the desert mountain ranges in this region. A disjunct population is found in extreme southern Nevada and adjacent California in *Yucca* associations of the Providence Mountain region (Hall, 1981).

Remarks.—Grinnell (1922) and Hall (1946) gave measurements for samples of *D. panamintinus* from California and Nevada, respectively. Populations of *D. panamintinus* and *D. heermanni* are similar in chromosome structure (Stock, 1974), and exhibit a high degree of genic similarity (Patton et al., 1976), although Schnell et al. (1978) determined that *D. panamintinus* was phenetically an outlier in terms of any well-defined species group of *Dipodomys.*

Dipodomys panamintinus argusensis Huey, 1945

1945. *Dipodomys mohavensis argusensis* Huey, Trans. San Diego Soc. Nat. Hist., 10:131, 9 March.

Holotype.—Adult male, skin and skull, SDSNH 9552, from Junction Ranch, 5,725 ft, Argus Mountains, Inyo Co., California; obtained by Samuel G. Harter on 13 August 1931.

Measurements of holotype.—Total length, 297; length of tail, 172; length of hind foot, 44; length of ear, 13; greatest length of skull, 40.2; width across bullae, 24.5; length of nasals, 15.6; breadth across maxillary arches, 24.0; width of maxillary arch at middle, 5.5; greatest width of rostrum near end, 4.5.

Distribution.—Known only from the vicinity of the type locality in the Argus Mountains, Inyo Co., California.

Remarks.—Miller and Kellogg (1955) first treated *argusensis* as a subspecies of *D. panamintinus.*

Dipodomys panamintinus caudatus Hall, 1946

1946. *Dipodomys panamintinus caudatus* Hall, Mammals of Nevada, p. 409, 1 July.

Holotype.—Adult female, skin and skull, MVZ 80028, from 6 mi S Granite Well, 3,800 ft, Providence Mountains, San Bernardino Co., California; obtained on 18 December 1937 by F. Wallace Taber.

Measurements of holotype.—Total length, 299; length of tail, 180; length of hind foot, 43; length of ear, 16; weight, 68.8 g; greatest length of skull, 40.2; width across bullae, 24.95; breadth across maxillary arches, 22.55; nasal length, 15.0; interorbital breadth, 13.05.

Distribution.—Chiefly found in *Yucca* associations of the Providence Mountains region of San Bernardino Co., California and adjacent Clark Co., Nevada.

Remarks.—Hall (1946) listed means and ranges for 3 males and 3 females from Nevada and 10 adults of each sex from California.

Dipodomys panamintinus leucogenys (Grinnell, 1919)

1919. *Perodipus leucogenys* Grinnell, Univ. California Publ. Zool., 21:46, 29 March.

Holotype.—Adult male, skin and skull, MVZ 26933, from Pellisier Ranch, 5,600 ft, 5 mi N Benton Station, Mono Co., California; obtained by Joseph Dixon on 20 September 1917.

Measurements of holotype.—Total length, 310; length of tail, 176; length of hind foot, 48; length of ear (crown), 13; weight, 85.5 g; greatest length of skull, 41.1; width across bullae, 24.3; breadth across maxillary arches, 23.85; nasal length, 16.0; interorbital breadth, 13.05.

Distribution.—Northeastern California and western Nevada from Lassen Co., California and Washoe Co., Nevada, southward along the state boundary to the Excelsior Mountains, Nevada, then southward in California through the upper Owens Valley to the vicinity of Independence, Inyo Co.

Remarks.—Grinnell (1922) gave average measurements for 5 adults of each sex from California and Hall (1946) listed means and ranges for 9 males and 10 females from Nevada.

Dipodomys panamintinus mohavensis (Grinnell, 1918)

1918. *Perodipus mohavensis* Grinnell, Univ. California Publ. Zool., 17:428, 25 April.

Holotype.—Adult male, skin and skull, MVZ 26835, from 0.5 mi E railway station of Warren (about 5 mi N Mohave [Mojave]), 3,275 ft, Kern Co., California; obtained on 27 March 1917 by Joseph Grinnell.

Measurements of holotype.—Total length, 305; length of tail, 178; length of hind foot, 44, length of ear (crown), 12; weight, 88 g;

greatest length of skull, 40.7; width across bullae, 25.0; length of nasals, 15.8; width of maxillary arch at middle, 5.2.

Distribution.—Mojave Desert from near Lone Pine in the Owens Valley, southward into the Antelope Valley in extreme northeastern Los Angeles Co., and to near Hesperia, San Bernardino Co., California.

Remarks.—Grinnell (1922) listed means and ranges of measurements for five adults of each sex.

Dipodomys panamintinus panamintinus (Merriam, 1894)

1894. *Perodipus panamintinus* Merriam, Proc. Biol. Soc. Washington, 9:114, 21 June.

Holotype.—Adult male, skin and skull, USNM 28566/40670, from head of Willow Creek [about 6,200 ft, about 3 mi NE Jackass Spring], Panamint Mountains, Inyo Co., California; obtained on 12 May 1891 by Edward W. Nelson.

Measurements of holotype.—Total length, 293; length of tail, 179; length of hind foot, 43; length of ear from anterior base (dry skin), 15; greatest length of skull, 39.75; width across bullae, 24.50; breadth across maxillary arches, 22.80; nasal length, 15.55; interorbital breadth, 13.00.

Distribution.—Occupies a known area in the Panamint Mountains measuring about 10 by 13 km in size, in the Upper Sonoran and arid Transition life zones, in the vicinity of Jackass Spring, Inyo Co., California (Grinnell, 1922).

Remarks.—Measurements for five adults of each sex were listed by Grinnell (1922); Merriam (1894*a*) listed average external measurements for 16 specimens from the type locality.

Dipodomys phillipsii

Diagnosis.—A medium-sized kangaroo rat with four toes on the hind feet, a relatively long tail with the dark dorsal and ventral stripes uniting in the distal third, and usually a white, terminal tuft of hairs on the tail; auditory bullae relatively small, cranium flattened, maxillary region broad, and postrostral region nearly quadrate; maxillary plate projecting posteriorly to level of second or third molar; rostrum relatively narrow and straight-sided; incisors slender in comparison to other species of *Dipodomys* (Jones and Genoways, 1975).

Comparisons.—*Dipodomys phillipsii* can be distinguished from the other small- to medium-sized, four-toed kangaroo rats by the combination of a tail that has an all blackish distal third, with or without a white terminal tuft; a hind foot less than 42 mm in length; and a relatively narrow skull, measuring less than 25.0 mm across the bullae. It is most similar in size and appearance to *D. merriami, D. californicus* and *D. elator*; both of the latter are larger, have broader rostra and heavier incisors, and are widely separated geographically. Other four-toed kangaroo rats with white tail tufts are *D. deserti, D. nelsoni,* and *D. spectabilis,* all of which are considerably larger than *D. phillipsii,* with larger hind feet and broader skulls. From *D. phillipsii, D. merriami* differs most notably in coloration of the tail, lacking an all-blackish distal one third, and lacking a prominent, all white tip.

Distribution.—Primarily occupies sandy soils on the Mexican Plateau and adjacent areas, from central Durango to northern Oaxaca (Jones and Genoways, 1975).

Remarks.—*Macrocolus halticus* Wagner, 1845, Abhandl. K. Baier. Akad., 22:319, is a synonym of *Dipodomys phillipsii,* but its type locality, other than México, is unknown; thus it is not listed in the synonymy of any of the subspecies of *phillipsii.* Genoways and Jones (1971) provided a systematic review of *D. phillipsii* and gave means and ranges of measurements for samples from throughout the geographic range of the species. Jones and Genoways (1975) provided a review of the biology of the species.

Dipodomys phillipsii oaxacae
Hooper, 1947

1947. *Dipodomys phillipsii oaxacae* Hooper, J. Mamm., 28:48, 17 February.

Holotype.—Adult male, skin and skull, UMMZ 88652, from Teotitlán, 950 m, Oaxaca, México; obtained on 15 September 1944 by Helmuth O. Wagner.

Measurements of holotype.—Total length, 255; length of tail, 160; length of hind foot, 37; length of ear (notch, dry), 10.5; greatest length of skull, 34.5; length of nasals, 12.9; breadth across maxillary arches, 19.6; width of maxillary arch at middle, 4.8; interorbital breadth, 12.5; width of nasals near end, 3.2; greatest length of interparietal, 3.2; greatest breadth of interparietals, 2.2.

Distribution.—Known from the type locality and a single locality in southern Puebla (Genoways and Jones, 1971).

Remarks.—This subspecies is distinguished by small size and pale coloration. Genoways and Jones (1971) listed measurements.

Dipodomys phillipsii ornatus
Merriam, 1894

1894. *Dipodomys ornatus* Merriam, Proc. Biol. Soc. Washington, 9:110, 21 June.

Holotype.—Adult female, skin and skull, USNM 57990, from Berriozábal, Zacatecas, México; obtained by Edward A. Goldman on 29 December 1893.

Measurements of holotype.—Total length, 274; length of tail, 167; length of hind foot, 39; greatest length of skull, 36.5; width across bullae, 22.6; length of nasals, 13.0; breadth across maxillary arches, 21.4; interorbital breadth, 13.6; length of rostrum, 20.9; length of maxillary toothrow, 4.8; depth of cranium, 11.0.

Distribution.—Ranges from central Durango southeastward to Querétaro on the Mexican Plateau (Genoways and Jones, 1971).

Remarks.—Genoways and Jones (1971) gave means and ranges of measurements for nine samples of *ornatus.*

Dipodomys phillipsii perotensis
Merriam, 1894

1894. *Dipodomys perotensis* Merriam, Proc. Biol. Soc. Washington, 9:111, 21 June.

Holotype.—Adult female, skin and skull, USNM 54285, from Perote, Vera Cruz, México; obtained by Edward W. Nelson on 21 May 1893.

Measurements of holotype.—Total length, 265; length of tail, 162; length of hind foot, 40; length of ear (anterior base, dry), 14; greatest length of skull, 36.3; width across bullae, 22.8; length of nasals, 12.9; length of rostrum, 19.9; length of maxillary toothrow, 4.9; depth of cranium, 11.2.

Distribution.—Known only from around Tlaxcala, west-central Veracruz, and from eastern Puebla (Genoways and Jones, 1971).

Remarks.—Genoways and Jones (1971) gave means and ranges of measurements for three samples of *perotensis.*

Dipodomys phillipsii phillipsii
Gray, 1841

1841. *Dipodomys phillipii* [*sic*] Gray, Ann. Mag. Nat. Hist., ser. 1, 7:522, August.

Holotype.—Adult male, skin and skull, BM(NH) 45.1580, from near Real del Monte, Hidalgo, México; obtained by John Phillips.

Measurements of holotype.—Total length, 292; length of tail, 165; length of hind foot, 38.

Distribution.—Limited to the Valle de México and adjacent areas in Hidalgo and the Distrito Federal, México (Genoways and Jones, 1971).

Remarks.—All of the skull except the upper and lower incsiors of the holotype are missing. See Genoways and Jones (1971) for a discussion of controversy in the literature about the type locality. Coues (1875) discussed the spelling of the specific epithet. Genoways and Jones (1971) listed means and ranges of measurements for a sample of *phillipsii.*

Dipodomys simulans

Diagnosis.—A medium-sized kangaroo rat with five toes on the hind feet, moderately long ear pinnae, relatively narrow breadth across the maxillary arches, and 60 chromosomes. Length of head and body averages about 115 mm and length of hind foot averages less than 42 mm; greatest length of skull averages less than 39.5 mm and maxillary breadth averages less than 21.0 mm.

Comparisons.—*Dipodomys simulans* is most similar to *D. agilis,* and must be distinguished from sympatric individuals of *D. gravipes,* and *D. stephensi.* From *D. gravipes, simulans* can be distinguished by the narrower breadth across the maxillary arches, smaller hind feet, and larger ears. See account of *agilis* for comparison with that species; from *stephensi, simulans* differs in being smaller with a narrower skull across the maxillary arches and across the bullae; the bullae of *stephensi* are more globose in shape.

Distribution.—Occupies coastal chaparral and grassland communities from the Los Angeles Basin and San Jacinto mountains of southern California, southward in Baja California to the vicinity of Bahía Almejas, Pacific Coast, Baja California Sur.

Remarks.—Best et al. (1986) and Sullivan and Best (in press) discussed the relationships of *D. agilis* and *D. simulans,* and provided methods of distinguishing the two species. Huey (1951) treated *peninsularis* as a species, and Best (1978) listed it as a subspecies of *D. agilis.* Huey (1962) considered *D. antiquarius,* from Sierra Borja, Baja California, to be closest to *D. stephensi,* but Lackey (1967) could find no significant differences between samples of *antiquarius* and *D. peninsularis.* Stock (1974) postulated that *antiquarius* was a subspecies of *D. agilis* (now *simulans*).

Dipodomys simulans peninsularis (Merriam, 1907)

1907. *Perodipus simulans peninsularis* Merriam, Proc. Biol. Soc. Washington, 20:79, 22 July.
1951. *Dipodomys peninsularis pedionomus* Huey, Trans. San Diego Soc. Nat. Hist., 11: 247, 30 April.
1951. *Dipodomys peninsularis eremoecus* Huey, Trans. San Diego Soc. Nat. Hist., 11:248, 30 April.
1951. *Dipodomys peninsularis australis* Huey, Trans. San Diego Soc. Nat. Hist., 11:249, 30 April.
1962. *Dipodomys antiquarius* Huey, Trans. San Diego Soc. Nat. Hist., 12:477, 30 August.

Holotype.—Young adult male, USNM 139872, from Santo Domingo [Landing], 28°51′N lat., 114°W long., Baja California, México; obtained on 27 September 1905 by Edward W. Nelson and Edward A. Goldman.

Measurements of holotype.—Total length, 312; length of tail, 203; length of hind foot, 45; greatest length of skull, [40.80]; width across bullae, 26.75; breadth across maxillary arches, 20.95; nasal length, [13.60—tips broken]; interorbital breadth, 11.85.

Distribution.—Occurs on the central and southern portions of the Baja California Peninsula, from the vicinity of San Fernando Mission, Baja California southward to the vicinity of Magdalena and Almejas bays, near lat. 24°30′N.

Remarks.—Best (1978) reported on systematic investigations of *D. s. peninsularis* and its related subspecies in Baja California, determining that *D. peninsularis* was not specifically distinct from *D. simulans* and that *D. antiquaris* was indistinguishable from *D. s. pedionomus.* Best (1983*b*) showed a close similarity in structure between *Dipodomys s. peninsularis* and *D. s. australis,* whose geographic ranges are contiguous. The current arrangement of subspecies and their synonyms follows Sullivan and Best (in press).

Dipodomys simulans simulans (Merriam, 1904)

1904. *Perodipus streatori simulans* Merriam, Proc. Biol. Soc. Washington, 17:144, 14 July.
1904. *Perodipus cabezonae* Merriam, Proc. Biol. Soc. Washington, 17:144, 14 July.
1925. *Dipodomys agilis latimaxillaris* Huey, Proc. Biol. Soc. Washington, 38:84, 26 May.
1927. *Dipodomys agilis martirensis* Huey, Trans. San Diego Soc. Nat. Hist., 5:7, 20 February.
1951. *Dipodomys agilis plectilis* Huey, Trans. San Diego Soc. Nat. Hist., 11:240, 30 April.
1951. *Dipodomys paralius* Huey, Trans. San Diego Soc. Nat. Hist., 11:241, 30 April.

Holotype.—Adult female, skin and skull, USNM 33105/45103, from Dulzura, San Diego Co., California; obtained on 24 November 1891 by C. H. Marsh.

Measurements of holotype.—Total length, 280; length of tail, 165; length of hind foot, 40; greatest length of skull, 38.00; width across bullae, 24.40; breadth across maxillary arches, 20.15; nasal length, 13.30; interorbital breadth, 12.65.

Distribution.—On the north found in Pacific coastal chaparral and grassland communities generally below about 500 m in elevation in Los Angeles and San Bernardino counties; also found on both the Pacific and desert slopes of coastal mountain ranges from San Gorgonio Pass, Riverside Co., California, southward. Range extends through the coastal basins and mountains of southern California southward on the slopes of the Sierra Juárez and Sierra San Pedro Mártir and the Pacific coastal plains of Baja California to the vicinity of Santa Catarina and Santa Catarina Landing.

Remarks.—Grinnell (1922) noted only slight and inconstant differences in specimens of *D. a. agilis* and *simulans*; the major difference was the greater inflation of the bullae of *simulans*. After obtaining specimens from several additional localities in Baja California, Huey (1951) determined that specimens he referred to *D. agilis latimaxillaris* did not differ substantially from specimens from some populations of *simulans*. Morphometric analyses by Best (1981, 1983*b*) showed little differentiation among some samples of *martirensis, plectilis,* and *simulans*. Best (1978) considered that *D. paralius* was indistinguishable from *plectilis*.

Dipodomys spectabilis

Diagnosis.—One of the largest species of kangaroo rats, with four toes on the hind feet and with a tail terminating in a large, white tuft of hairs (tuft exceeds about 25 mm in length, typically about 40 mm) bordered proximally by a band of black hairs; lateral white stripes of tail present only on about the proximal half; underside of tail, proximal to white tuft, dark colored; head and body length averaging from about 128 to 150 mm; the auditory bullae relatively large and inflated; skull relatively wide across the maxillary arches (breadth averaging from about 25.0 to 26.3).

Comparisons.—*Dipodomys spectabilis* is most similar in size and appearance to *D. nelsoni*; see the account of the latter species for diagnostic characters. It is also similar in size to the four-toed species, *D. deserti,* from which *D. spectabilis* can be distinguished by the prominent black border to the white tail tuft, the dark underside to its tail, generally smaller hind feet (usually averaging from 47 to 52 in *spectabilis* and from 52 to 53.5 in *deserti*), and a less inflated skull with a wider interorbital region and wider breadth across the maxillary arches (Nader, 1978). See the accounts of *D. californicus* and *D. elator* for other comparisons.

Distribution.—Banner-tailed kangaroo rats occur in desert associations from south-central Arizona and northern Sonora, eastward to the Trans-Pecos region of Texas; they also extend from extreme north-central

Arizona and the Four Corners Region of the San Juan River Basin, northeastern Arizona, southward through lower-lying portions of New Mexico; their range continues from northwestern Chihuahua, southward and eastward through central Chihuahua to northeastern Durango. A disjunct population occurs south of the range of *D. nelsoni* in east-central Zacatecas, northeastern Aguascalientes, and western San Luis Potosí.

Remarks.—*Dipodomys nelsoni* is treated as a separate species herein despite evidence of limited hybridization in a narrow zone of contact with *D. spectabilis* in southern Chihuahua (Nader, 1978). The relationships between *spectabilis* and *nelsoni*, and the relationship of the disjunct population, *cratodon*, in central México to other populations of *D. spectabilis* and to *D. nelsoni* require further study.

Dipodomys spectabilis baileyi Goldman, 1923

1923. *Dipodomys spectabilis baileyi* Goldman, Proc. Biol. Soc. Washington, 36:140, 1 May.
1933. *Dipodomys spectabilis clarencei* Goldman, J. Washington Acad. Sci., 23:467, 15 October.

Holotype.—Adult male, skin and skull, USNM 97185, from 40 mi W Roswell, Chaves Co. [Lincoln Co.], New Mexico; obtained by Vernon Bailey on 13 June 1899.

Measurements of holotype.—Total length, 385; length of tail, 283; length of hind foot, 58; greatest length of skull, 47.4; width across bullae, 30.7; length of nasals, 15.5; breadth across maxillary arches, 26.8; interorbital breadth, 15.6; length of rostrum, 24.8; length of maxillary toothrow, 6.3; depth of cranium, 14.5; least width of supraoccipital, 1.7.

Distribution.—Found from northeastern Arizona (where it no longer occurs; Hoffmeister, 1986) through western, central, and southeastern New Mexico, and the Trans-Pecos area of western Texas (Nader, 1978).

Remarks.—According to Nader (1978), the type locality is in Lincoln Co., not Chaves Co. as listed by Goldman. Nader (1978) found no consistent differences between samples of populations of *clarencei* and *baileyi*, and regarded *clarencei* as indistinguishable from *baileyi*. Hall (1981) did not cite Nader's (1978) revision and did not explain his reasoning for retention of *clarencei* as a subspecies. We concur with Nader (1978). Nader (1978) provided ranges and means of measurements for *D. s. baileyi*. Hoffmeister (1986) reported on specimens from near Rainbow Lodge, Coconino Co., Arizona, approximately 112 km west and 80 km north of the populations in northwestern Arizona studied by Nader (1978). He tentatively assigned these specimens to *baileyi*.

Dipodomys spectabilis cratodon Merriam, 1907

1907. *Dipodomys spectabilis cratodon* Merriam, Proc. Biol. Soc. Washington, 20:75, 22 July.

Holotype.—Adult male, skin and skull, USNM 78953, from Chicalote, Aguascalientes, México; obtained on 2 July 1896 by Edward W. Nelson and Edward A. Goldman.

Measurements of holotype.—Total length, 342; length of tail, 217; length of hind foot, 54; greatest length of skull, 46.5; width across bullae, 29.7; length of nasals, 15.6; breadth across maxillary arches, 27.5; interorbital breadth, 17.0; length of rostrum, 25.2; length of maxillary toothrow, 6.6; depth of cranium, 14.2.

Distribution.—Occupies an area disjunct from other populations of *D. spectabilis* in east-central Zacatecas, northeastern Aguascalientes, and western San Luis Potosí.

Remarks.—See the account of *D. nelsoni* and the species diagnosis above concerning the relationships of *cratodon* to other *spectabilis* group kangaroo rats. Dalquest (1953) and Nader (1978) listed measurements.

Dipodomys spectabilis intermedius
Nader, 1965

1965. *Dipodomys spectabilis intermedius* Nader, Proc. Biol. Soc. Washington, 78:50, 21 July.

Holotype.—Adult female, skin and skull, MVZ 82782, from 16.7 mi SW Bámori, about 1,900 ft, Sonora, México; obtained by Seth B. Benson on 25 April 1938.

Measurements of holotype.—Total length, 314; length of tail, 180; length of hind foot, 47; length of ear, 17; weight, 105.7 g; greatest length of skull, 41.9; width across bullae, 27.1; breadth across maxillary arches, 24.8; nasal length, 15.0; interorbital breadth, 15.65.

Distribution.—West central Sonora, México, from Querobabi on the north, southward to Carbo and westward to a locality about 17 mi southwest of Bámori (Nader, 1978).

Remarks.—Nader (1965) listed means and ranges of measurements for seven specimens from the type locality. This race is the smallest in size of the subspecies of *D. spectabilis.*

Dipodomys spectabilis perblandus
Goldman, 1933

1933. *Dipodomys spectabilis perblandus* Goldman, J. Washington Acad. Sci., 23:466, 15 October.

Holotype.—Adult female, skin and skull, USNM 17748/24689, from Calabasas, about 3,500 ft, Santa Cruz Co., Arizona; obtained by Vernon Bailey on 27 October 1889.

Measurements of holotype.—Total length, 315; length of tail, 184; length of hind foot, 48; greatest length of skull, 42.9; width across bullae, 27.7; length of nasals, 14.4; breadth across maxillary arches, 25.9; interorbital breadth, 15.2; length of rostrum, 22.7; length of maxillary toothrow, 5.5; depth of cranium, 13.1; width of nasals in front of incisors, 4.2; least width of supraoccipital, 2.3.

Distribution.—Found in south central Arizona south of the Gila River, west of the Santa Catalina and Santa Rita ranges and east from the Ajo Valley, southward into north central Sonora.

Remarks.—Nader (1978) listed means and ranges of measurements. Populations of this subspecies are among the smallest of the species.

Dipodomys spectabilis spectabilis
Merriam, 1890

1890. *Dipodomys spectabilis* Merriam, N. Amer. Fauna, 4:46, 8 October.

Holotype.—Young adult male, skin and skull, USNM 17886/24823, from Dos Cabezos [= Dos Cabezas], Cochise Co., Arizona; obtained by Vernon Bailey on 22 November 1889.

Measurements of holotype.—Total length, 350; length of tail, 211; length of hind foot, 52; length of ear (crown), 10, (anterior base, dry skin), 16; greatest length of skull, 45.6; width across bullae, 29.5; length of nasals, 15.9; interorbital breadth, 15.8; length of rostrum, 23.9; length of maxillary toothrow, 5.8; depth of cranium, 13.3.

Distribution.—Desert associations in southeastern Arizona east of the Galiuro and Santa Rita ranges, northeastern Sonora, southwestern New Mexico, and northern and central Chihuahua.

Remarks.—Nader (1978) and Anderson (1972) gave means and ranges of measurements for samples of *D. spectabilis spectabilis.*

Dipodomys spectabilis zygomaticus
Goldman, 1923

1923. *Dipodomys spectabilis zygomaticus* Goldman, Proc. Biol. Soc. Washington, 36:140, 1 May.

Holotype.—Young adult male, skin and skull, USNM 96432, from Parral [= Hidalgo del Parral], southern Chihuahua,

México; obtained by Edward A. Goldman on 17 September 1898.

Measurements of holotype.—Total length, 340; length of tail, 197; length of hind foot, 54; greatest length of skull, 46.1; width across bullae, 30.6; length of nasals, 16.2; breadth across maxillary arches, 27.1; interorbital breadth, 16.4; length of rostrum, 24.7; length of maxillary toothrow, 6.5; depth of cranium, 13.5.

Distribution.—Occupies a range in south central Chihuahua and north central Durango (Nader, 1978).

Remarks.—Anderson (1972) and Nader (1978) gave means and ranges for measurements of samples of *D. s. zygomaticus.*

Dipodomys stephensi

Diagnosis.—A medium-sized kangaroo rat with five toes on the hind feet, a wide face, reflected in broad maxillary arches, and relatively large and inflated auditory bullae with a unique, "globose" shape (Grinnell, 1922); length of head and body averages about 119 to 121 mm; width across the maxillary arches averages about 22.5 mm; greatest breadth of skull, across the bullae, averages about 24.7 to 25.1 mm or more (Lackey, 1967).

Comparisons.—*Dipodomys stephensi* is similar to the allopatric species *D. heermanni* and *D. panamintinus.* From *panamintinus, stephensi* is distinguished by its greater breadth across the auditory bullae (average 24.7 mm or greater in *stephensi,* 24.2 or less in *panamintinus*); the ratio of interparietal length to width is 0.3 in *stephensi* and 0.6 in *panamintinus* (see account of *D. panamintinus* and Lackey, 1967, for other diagnostic characters). Lackey (1967) stated that *D. stephensi* was more similar structurally to the southern subspecies of *D. heermanni* than to other species; compared to *D. heermanni tularensis, D. stephensi* has a shorter basioccipital width and greater jugal curvature. The face of *D. stephensi* is wider than the sympatric five-toed species, *D. simulans* (distance across maxillary arches averages about 20.8 to 20.9 in *simulans*; Lackey, 1967).

Distribution.—Confined to open, arid grassland associations in a relatively small area in southern California, in the San Jacinto Valley and adjacent lowlands of western Riverside, southwestern San Bernardino, and northwestern San Diego counties (Bleich, 1977; Lackey, 1967); an apparently disjunct population occurs in north-central San Diego Co., in the vicinity of Warner Springs (O'Farrell et al., 1986).

Remarks.—Bleich (1977) reviewed papers discussing the relationships of *D. stephensi* to other species of *Dipodomys*; although most authors agree that *D. stephensi* is related to the other broad-faced kangaroo rats of the *heermanni* group, its interspecific relationships are unresolved.

Dipodomys stephensi (Merriam, 1907)

1907. *Perodipus stephensi* Merriam, Proc. Biol. Soc. Washington, 20:78, 22 July.
1962. *Dipodomys cascus* Huey, Trans. San Diego Soc. Nat. Hist., 12:479, 30 August.

Holotype.—Adult male, skin and skull, USNM 186503, from San Jacinto Valley [a little W of Winchester, toward Menifee], Riverside Co., California; obtained on 27 November 1885 by Frank Stephens.

Measurements of holotype.—Greatest length of skull, [39.95]; width across bullae, 25.45; nasal length, [14.45]; interorbital breadth, 12.30.

Remarks.—No external measurements were taken for the holotype, and its maxillary arches and tips of nasals are missing. According to Lackey (1967), the population named *cascus* is similar to, but recognizably distinct from the more northern population, *stephensi*; Lackey did not recommend subspecific recognition, however. The two populations apparently are currently allopatric. Lackey (1967) provided measurements and comparisons to other species.

Dipodomys venustus

Diagnosis.—A moderately large-sized, large-eared, five-toed kangaroo rat with a relatively long tail, large feet, and dark coloration; length of the ear pinna exceeded only by *D. elephantinus*; face narrow and spread of maxillary arches usually less than 25.5 mm; auditory bullae large and back of skull wide; supraoccipital and interparietal very narrow; incisors relatively robust; nasals typically not flaring at end (Grinnell, 1922).

Comparisons.—*Dipodomys venustus* can be distinguished from all other kangaroo rats except *D. elephantinus* by having five toes on the hind feet, long ears (length from crown in excess of 15.5 mm), a relatively long, heavily crested tail averaging about 155% of length of head and body, and large body size; from typical individuals of *D. venustus*, *D. elephantinus* differs in being lighter in color, having longer ears (length of ear from crown in *venustus* averages less than about 16 mm), and more flaring nasals (see account of *elephantinus* for additional remarks). *D. venustus* differs from *D. agilis* in being slightly larger in most dimensions and in having longer ears; length of hind foot in *D. venustus* averages 45 mm or more and less than 45 mm in *D. agilis*.

Distribution.—*Dipodomys venustus* occupies chaparral communities on loose soils in the coastal ranges of west-central California. Its range extends from the Santa Cruz Mountains and hills near the south end of San Francisco Bay in Santa Cruz and Santa Clara counties, eastward in the Diablo Range in western Stanislaus Co., thence southward along the outer coastal ranges to the Santa Lucia Range, San Luis Obispo Co., and in the inner coastal ranges (Diablo Range) to south of the boundary between San Benito and Fresno counties, in western Fresno Co. (Grinnell, 1922; D. F. Williams, unpubl. data).

Remarks.—Some authorities have remarked on the close similarity between *D. venustus* and *D. elephantinus* and opined that they are conspecific (see account of *D. elephantinus*). *D. venustus* also is similar to the northern populations of *D. agilis* and may be conspecific (Stock, 1974). Means and ranges of measurements were given by Grinnell (1922).

Dipodomys venustus santiluciae Grinnell, 1919

1919. *Dipodomys santiluciae* Grinnell, Proc. Biol. Soc. Washington, 32:204, 31 December.

Holotype.—Adult male, skin and skull, MVZ 29023, from 1 mi SW Jolon, Monterey Co., California; obtained by Joseph Grinnell on 21 October 1918.

Measurements of holotype.—Total length, 315; length of tail, 119; length of hind foot, 46; length of ear (crown), 16; weight, 82 g; greatest length of skull, 42.5; width across bullae, 25.8; length of nasals, 15.5; breadth across maxillary arches, 23.0; width of maxillary arch at middle, 5.3; width of rostrum near end, 4.3.

Distribution.—Chaparral associations of the coastal mountains, from the south end of Monterey Bay, Monterey Co., southward to the Santa Lucia Mountains east of Morro Bay, San Luis Obispo Co., California. The Salinas Valley marks the eastern boundary of the range.

Remarks.—Grinnell (1922) provided means and ranges of measurements for two males and four females.

Dipodomys venustus venustus (Merriam, 1904)

1904. *Perodipus venustus* Merriam, Proc. Biol. Soc. Washington, 17:142, 14 July.

Holotype.—Adult male, skin and skull, USNM 51852, from Santa Cruz, Santa Cruz Co., California; obtained by G. B. Badger on 12 March 1893.

Measurements of holotype.—Total length, 339; length of tail, 211; length of hind foot,

46; greatest length of skull, 41.60; width across bullae, 24.75; breadth across maxillary arches, 22.00; nasal length, 15.35; interorbital breadth, 12.10.

Distribution.—Apparently disjunct populations occupy the Santa Cruz Mountains and adjacent area west of the Santa Clara Valley, the Diablo Range in Santa Clara, Stanislaus, Merced, San Benito, and Fresno counties, and the northern end of the Gabilan Range in Monterey and San Benito counties (Grinnell, 1922; Hall, 1981; unpubl. data).

Remarks.—Samples of populations from the Diablo Range are virtually indistinguishable from *D. elephantinus.* Grinnell (1922) gave ranges and means of measurements for six males and two females from the westernmost populations.

Genus *Microdipodops*

1891. *Microdipodops* Merriam, N. Amer. Fauna, 5:115, 30 July.

Type species.—*Microdipodops megacephalus* Merriam, 1891, N. Amer. Fauna, 5:116, 30 July.

Diagnosis.—Size small, total length 130–180 mm and mass 10–17 g; body form ricochetal; tail thickest near its midlength and only slightly longer than length of head and body; tail with short hairs, not crested or tufted at end; tail with fat deposit centered at proximal one-third to one-half of tail; body hairs relatively long, lax, and silky; no large dermal sebaceous gland on back; soles of hind feet densely covered with long hair; pes with five well-developed digits; manus longer and more slender than other members of family; protoloph of P4 with one cusp; lophs of P4 unite first at or near center of tooth; p4 with 5 or 6 cusps; molars with H-pattern; cusp patterns worn away quickly leaving occlusal surface of cheek teeth as an island of dentine surrounded by enamel; cheek teeth extremely high-crowned and with more than one root except M3/m3; roots of cheek teeth develop after adulthood; upper incisors grooved; knob of incisor pulp cavity on mandible at lower edge of ascending ramus; incipient postorbital process; center of palate between premolars not ridged; one pterygoid fossa apparent (anterior fossa nearly invisible); zygomatic process of maxilla not much expanded; masseter fossa separated from infraorbital foramen by slight crest; posterior palantine foramen located in palantine; auditory bullae more inflated than in other genera of the family—bullae meeting in symphysis across ventral face of basisphenoid; tympanic bulla extends anterior to mandibular fossa (glenoid); ventral surface of tympanic bulla extending below occlusal surface of cheek teeth; obturator foramen of pelvis subtriangular and much compressed dorso-ventrally; astragalar-cuboid contact in foot and no contact between navicular and calcaneum; trapezium-scapholunar articulation; cervical vertebrae mostly unfused; fewer caudal vertebrae than other members of family, but individual vertebrae elongated; median ventral foramen in caudal vertebrae; tibia and fibula fused for about 60% of their length; temporalis muscle reduced; M. tensor fasciae latae insertion includes medial thigh; pallus with spines, urethral lappets, and a dorsal groove; tip of phallus not upturned; ventromedial prostrate gland absent; baculum with large, bulbous basal part, tapering into a moderately upward-curving shaft; spermatozoa large with long, roughly triangular head with rounded vertices, and tail of medium length (Burt, 1960; Hafner and Hafner, 1983; Hall, 1941; Hatt, 1932; Homan and Genoways, 1978; Quay, 1965; Ryan, 1989; Wahlert, 1985; Wood, 1935).

Remarks.—The subfamilial relationships of *Microdipodops* have been controversial since its description by Merriam (1891). Wood (1935), in a major review of heteromyid evolution, placed the kangaroo mice in the subfamily Perognathinae, allying them with the silky pocket mice, *Perognathus.* Another possible arrangement, in which kangaroo mice represent a distinct evolutionary line, was raised when Setzer (1949) declined to place them in any recognized subfamily. Reeder (1956) first placed them in the subfamily Dipodomyinae based on dental characters (see remarks in the diagnosis for Dipodomyinae for additional information).

Key to the Species

1. Length of hind foot usually 25 mm or less; anterior palatine foramina wide posteriorly and tapering to a sharp point anteriorly; premaxillae terminating at, or extending less than 1 mm posterior to the posterior end of nasals; upper parts dark brownish or buffy with overwash of blackish; dorsal surface of tail blackish distally and tipped in black .... *Microdipodops megacephalus*

1′. Length of hind foot usually 25 mm or more; anterior palatine foramina parallel-sided; premaxillae terminating posteriorly more than 1–2 mm beyond posterior end of nasals; upper parts pale, near Light Pinkish Cinnamon (Ridgway, 1912); top of tail distally approximately same color as near base .... *Microdipodops pallidus*

Species Accounts

Microdipodops megacephalus

Diagnosis.—Anterior palatine foramina wide posteriorly and tapering to a sharp point anteriorly; nasals terminating posteriorly at, or almost at, the same level as the premaxillae; skull with less inflated auditory bullae, thus a proportionally narrower skull; upper incisors relatively curved; hind foot length averages about 23–35 mm; post-auricular spot buffy; upper parts dark, brownish or buffy, washed with blackish; hairs of underparts grayish for most of length, but either subterminally white with buffy tips or lacking buffy and tipped with white (a few populations have hairs white throughout their length); tail above tipped with black; diploid number 40.

Comparisons.—Many of the characters given in the key and diagnosis above are variable geographically. Individuals of the two species from close geographic proximity should be compared to confirm identification. In as far as is known, the diploid chromosome numbers are diagnostic (40 and 42; Hafner et al., 1979). The dorsal tip of the tail of *M. megacephalus* is generally black, whereas in *M. pallidus* this area is the same color as the back. The shape of the anterior palatine foramina and the point of the posterior termination of the nasals appear to be the most useful cranial characters for separating the species, but there is some geographic variation in these traits in *M. pallidus* (Hall, 1941).

Distribution.—Found on a variety of substrates in the Great Basin region including Nevada, southeastern Oregon, extreme southwestern Idaho, west-central Utah, and extreme northeastern and east-central (Mono Co.) California.

Remarks.—The two species of kangaroo mice occur sympatrically, but areas of sympatry are not large. Hall (1941) reported sympatry at 12 localities. At one of these—Penoyer Valley, Nevada—he reported three hybrid individuals. Later examination of these and 89 additional specimens from the same area found only one individual that was structurally intermediate between the species (Hafner et al., 1979). Karyotypes and electrophoretic patterns, however, showed no evidence of hybridization.

Microdipodops megacephalus albiventer Hall and Durrant, 1937

1937. *Microdipodops pallidus albiventer* Hall and Durrant, J. Mamm., 18:357, 14 August.

Holotype.—Adult male, skin and skull, MVZ 52803, from Desert Valley, 5,300 ft, 21 mi W Panaca, Lincoln Co., Nevada; obtained on 30 May 1932 by Ward C. Russell.

Measurements of holotype.—Total length, 150; length of tail, 80; length of hind foot, 24; length of ear, 9.5; weight, 11.6 g; greatest length of skull, 28.30; width across bullae, 18.85; length of nasals, 9.80; breadth across maxillary arches, 11.90; interorbital breadth, 6.50; length of maxillary toothrow, 3.50; depth of cranium, 8.00.

Distribution.—Restricted to Desert Valley in central Lincoln Co., Nevada.

Remarks.—Hall and Durrant (1937) described *albiventer* as a subspecies of *M. pallidus* because of its light coloration. They gave measurements for 16 adult topotypes. Hall (1941) classified it as a subspecies of *M. megacephalus* based upon cranial characters. *M. m. albiventer,* as currently defined, is restricted to Desert Valley, Nevada, although specimens from Coal Valley to the west assigned to *sabulonis* approach them in coloration.

Microdipodops megacephalus ambiguus Hall, 1941

1941. *Microdipodops megacephalus ambiguus* Hall, Field Mus. Nat. Hist., Zool. Ser., 27:252, 8 December.

Holotype.—Adult male, skin and skull, MVZ 73840, from 1.25 mi N Sulphur, 4,050 ft, Humboldt Co., Nevada; obtained on 25 July 1936 by E. Raymond Hall.

Measurements of holotype.—Total length, 162; length of tail, 89; length of hind foot, 25; length of ear, 9.8; weight, 13.2 g; greatest length of skull, 29.00; width across bullae, 19.10; length of nasals, 10.35; breadth across maxillary arches, 11.30; interorbital breadth, 6.55; length of maxillary toothrow, 3.45; depth of cranium, 8.10.

Distribution.—Smoke Creek and Black Rock deserts and the lower part of Humboldt River Valley in Nevada and extreme eastern Lassen Co., California.

Remarks.—According to Hall (1941), specimens of *ambiguus* are widely variable, particularly in external proportions and coloration. This taxon is in contact with four other subspecies. Hall (1941) found evidence of integradation between *ambiguus* and each of these four other subspecies. Hall (1941) listed means and ranges of measurements for 17 adult topotypes.

Microdipodops megacephalus atrirelictus Hafner, 1985

1985. *Microdipodops megacephalus atrirelictus* Hafner, Proc. Biol. Soc. Washington, 98:3, 20 March.

Holotype.—Adult female, skin, skull, and skeleton, MVZ 160039, from 11 mi S, 44.2 mi W Riddle, 5,000 ft, Owyhee Co., Idaho; obtained on 8 October 1978 by John C. Hafner.

Measurements of holotype.—Total length, 169; length of tail, 93; length of hind foot, 26; length of ear, 12; weight, 12.5 g; greatest length of skull, 29.27; width across bullae, 19.16; basal length of skull, 19.06; bullar length, 14.60; length of nasals, 10.35; breadth across maxillary arches, 11.62; interorbital breadth, 6.69; greatest length of incisive foramina, 2.19; length of maxillary toothrow, 3.45; depth of cranium, 7.95.

Distribution.—Known only from southwestern Owyhee Co., Idaho.

Remarks.—This is among the most highly differentiated subspecies of *Microdipodops.* It generally is larger than other subspecies of *M. megacephalus,* and its dorsal pelage is nearly black. These kangaroo mice are isolated from other members of the genus by over 100 km of ground unsuitable as habitat (Hafner, 1985). Hafner (1985) listed statistics for four individuals.

Microdipodops megacephalus californicus Merriam, 1901

1901. *Microdipodops californicus* Merriam, Proc. Biol. Soc. Washington, 14:128, 19 July.

Holotype.—Young adult male, skin and skull, USNM 101227, from Sierra Valley, near Vinton, Plumas Co., California; obtained on 7 August 1900 by Walter K. Fisher.

Measurements of holotype.—Total length,

158; length of tail, 91; length of hind foot, 25; greatest length of skull, 26.3; width across bullae, 17.7; length of nasals, 8.3; breadth across maxillary arches, 11.3; interorbital breadth, 6.2; length of maxillary toothrow, 3.4; depth of cranium, 7.8; length of rostrum 10.4.

Distribution.—Occurs in the intermontane valleys of eastern Plumas Co., California, and southern Washoe and Ormsby counties, Nevada.

Remarks.—Hall (1941) found evidence of intergradation between *californicus* and *ambiguus* at several points of contact between their ranges, and therefore relegated *californicus* to subspecific status. Hall (1941) provided means and ranges of measurements for 5 adults and Hafner (1985) listed statistics for measurements of 12 individuals.

Microdipodops megacephalus leucotis Hall and Durrant, 1941

1941. *Microdipodops megacephalus leucotis* Hall and Durrant, The Murrelet, 22:6, 30 April.

Holotype.—Adult female, skin and skull, UU 3525 from 18 mi SW Orr's Ranch, 4,400 ft, Tooele Co., Utah; obtained on 6 June 1940 by Stephen D. Durrant.

Measurements of holotype.—Total length, 142; length of tail, 75; length of hind foot, 24; basal length (to anterior face of incisor), 17.5; width across bullae, 19.2; length of nasals, 9.3; breadth across maxillary arches, 11.3; interorbital breadth, 6.1.

Distribution.—Restricted to the sand dunes along the valley floors of the Bonneville Basin in south-central Tooele Co., Utah.

Remarks.—This taxon is one of the palest members of the genus, being paler than many populations of *M. pallidus*. However, most cranial and external features ally *leucotis* with *M. megacephalus* (Hall, 1941). Later, Hall (1981) remarked that *leucotis* might be a distinct species, but Hafner and Hafner (1983) found that genic and karyotypic evidence did not support specific status.

Microdipodops megacephalus medius Hall, 1941

1941. *Microdipodops megacephalus medius* Hall, Field Mus. Nat. Hist., Zool. Ser., 27:256, 8 December.

Holotype.—Adult female, skin and skull, MVZ 73890, from 3 mi S Vernon, 4,250 ft, Pershing Co., Nevada; obtained on 28 July 1936 by E. Raymond Hall.

Measurements of holotype.—Total length, 165; length of tail, 89; length of hind foot, 25; length of ear, 10.5; greatest length of skull, 28.55; width across bullae, 18.65; length of nasals, 10.10; breadth across maxillary arches, 11.35; interorbital breadth, 6.25; length of maxillary toothrow, 3.55; depth of cranium, 7.85.

Distribution.—Southwestern Pershing Co., Nevada.

Remarks.—The only other subspecies with which *medius* is in contact is *ambiguus*. From *ambiguus, medius* is distinguished by its darker coloration and narrower skull resulting from less inflated auditory bullae. Hall (1941) listed means and ranges of measurements for 18 adult topotypes, and Hafner (1985) listed statistics for 12 individuals.

Microdipodops megacephalus megacephalus Merriam, 1891

1891. *Microdipodops megacephalus* Merriam, N. Amer. Fauna, 5:116, 30 July.

Holotype.—Adult male, skin and skull, USNM 24417/31823, from Halleck, Elko Co., Nevada; obtained 23 October 1890 by Vernon Bailey.

Measurements of holotype.—Total length, 157; length of tail, 83; length of hind foot, 24; greatest length of skull, 28.4; width across bullae, 19.3; length of nasals, 9.4; breadth across maxillary arches, 11.9; interorbital breadth, 6.7; length of maxillary toothrow, 3.4; depth of cranium, 7.9; length of rostrum, 10.6.

Distribution.—Central and northeastern

Nevada, from northern Elko Co. to northern Nye and Lincoln counties, and from western Lander Co. eastward to near the Utah border.

Remarks.—Specimens from Eureka Co., Nevada, represent possible intergrades between *megacephalus* and *nexus.* There is a broad zone of integradation with *M. m. sabulonis* in south-central Nevada. Hall (1941) gave means and ranges of measurements for 10 adults, and Hafner (1985) listed statistics for 12 individuals.

Microdipodops megacephalus nasutus Hall, 1941

1941. *Microdipodops megacephalus nasutus* Hall, Field Mus. Nat. Hist., Zool. Ser., 27:251, 8 December.

Holotype.—Adult female, skin and skull, MVZ 40439, from Fletcher, 6,098 ft, Mineral Co., Nevada; obtained on 22 July 1928 by Louise Kellogg.

Measurements of holotype.—Total length, 158; length of tail, 88; length of hind foot, 25; greatest length of skull, 28.00; width across bullae, 18.25; length of nasals, 10.10; breadth across maxillary arches, 11.8; interorbital breadth, 6.65; length of maxillary toothrow, 3.40; depth of cranium, 7.80.

Distribution.—Known only from the type locality.

Remarks.—From *polionotus,* the only subspecies in contact with *nasutus, nasutus* differs in having a longer body and shorter tail, longer hind foot, and skull that is broader across the auditory bullae. Hall (1941) listed means and ranges of measurements for 4 adults.

Microdipodops megacephalus nexus Hall, 1941

1941. *Microdipodops megacephalus nexus* Hall, Field Mus. Nat. Hist., Zool. Ser., 27:257, 8 December.

Holotype.—Adult male, skin and skull, MVZ 70917, from 3 mi S Izenhood, Lander Co., Nevada; obtained on 22 May 1936 by Ward C. Russell.

Measurements of holotype.—Total length, 167; length of tail, 99; length of hind foot, 25; length of ear, 10; greatest length of skull, 29.25; width across bullae, 19.25; length of nasals, 10.40; breadth across maxillary arches, 11.55; interorbital breadth, 6.60; length of maxillary toothrow, 3.30; depth of cranium, 7.85.

Distribution.—Found in Humboldt and Lander counties, Nevada.

Remarks.—The subspecies *nexus* is separated from *ambiguus* to the west by a low range of mountains and there is a 75-mi gap to the eastward with no dark kangaroo mice. In color, *nexus* is intermediate between the western, gray mice and the blackish mice to the east. Hall (1941) gave means and ranges of measurements for 20 adult topotypes and Hafner (1985) listed statistics for 12 individuals.

Microdipodops megacephalus oregonus Merriam, 1901

1901. *Microdipodops megacephalus oregonus* Merriam, Proc. Biol. Soc. Washington, 14:127, 19 July.

Holotype.—Young adult male, skin and skull, USNM 80128, from Wild Horse Creek, 4 mi NW Alvord Lake, Harney Co., Oregon; obtained on 18 August 1896 by Clark P. Streator.

Measurements of holotype.—Total length, 153; length of tail, 88; length of hind foot, 24; greatest length of skull, 27.8; width across bullae, 18.1; length of nasals, 9.6; interorbital breadth, 6.6; length of maxillary toothrow, 3.5; depth of cranium, 7.8; length of rostrum, 11.0.

Distribution.—Extends from southeastern Oregon to Modoc and Lassen counties, in northeastern California, and Washoe and Humboldt counties in northwestern Nevada.

Remarks.—Hall (1941) noted integradation between *ambiguus* and *oregonus* in individuals from Smoke Creek, Washoe Co.,

Nevada, and the vicinity of Denio, Oregon. From *M. m. ambiguus, oregonus* differs in having grayer upper parts and a narrower cranium. Hall (1941) listed means and ranges of measurements for 13 adults and Hafner (1985) listed statistics for 11 individuals.

Microdipodops megacephalus paululus Hall and Durrant, 1941

1941. *Microdipodops megacephalus paululus* Hall and Durrant, The Murrelet, 22:5, 30 April.

Holotype.—Subadult male, skin and skull, MVZ 74660, from Pine Valley, 0.5 mi E headquarters building of the Desert Range Experiment Station, U.S. Forest Service, T25S, R17W, Sec. 33, Salt Lake B. M., Millar Co., Utah; obtained on 17 July 1936 by Stephen D. Durrant.

Measurements of holotype.—Total length, 147; length of tail, 76; length of hind foot, 24; length of ear, 11; greatest length of skull, 27.65; width across bullae, 18.30; length of nasals, 9.40; breadth across maxillary arches, 11.45; interorbital breadth, 6.30; length of maxillary toothrow, 3.25; depth of cranium, 7.70.

Distribution.—Occurs in west-central Utah in the Pine, White, and Snake valleys.

Remarks.—Hall (1941) only provisionally assigned specimens from Snake and White valleys to this subspecies because only two young specimens and one adult were available from these areas, respectively. Hall (1981), however, maintained this arrangement. Hall (1941) gave means and ranges of measurements for 9 adult topotypes.

Microdipodops megacephalus polionotus Grinnell, 1914

1914. *Microdipodops polionotus* Grinnell, Univ. California Publ. Zool., 12:302, 15 April.

Holotype.—Adult male, skin and skull, MVZ 17031, from McKeever's Ranch, 2 mi S Benton Station, 5,200 ft, Mono Co., California; obtained on 10 July 1912 by Charles D. Holliger.

Measurements of holotype.—Total length, 145; length of tail, 80; length of hind foot, 24; length of ear, 9; width across bullae, 18.35; breadth across maxillary arches, 12.10; interorbital breadth, 7.00; length of maxillary toothrow, 3.45.

Distribution.—Found in the Mono Lake Basin and head of Owens Valley, Mono Co., California.

Remarks.—The coloration of the upper parts of individuals of *polionotus* are quite variable (Hall, 1941). This taxon differs from other subspecies occurring along the California-Nevada border in having less black on the dorsal surface of the distal portion of the tail, although some black is present. Hall (1941) listed means and ranges of measurements for 20 adults.

Microdipodops megacephalus sabulonis Hall, 1941

1941. *Microdipodops megacephalus sabulonis* Hall, Proc. Biol. Soc. Washington, 54:59, 20 May.

Holotype.—Adult male, skin and skull, MVZ 49381, from 5 mi SE Kawich P. O., 5,400 ft, Kawich Valley, Nye Co., Nevada; obtained on 27 September 1931 by Robert T. Orr.

Measurements of holotype.—Total length, 155; length of tail, 83; length of hind foot, 25; length of ear, 10; weight, 11.3 g; greatest length of skull, 27.90; width across bullae, 19.05; length of nasals, 9.85; breadth across maxillary arches, 11.35; interorbital breadth, 6.40; length of maxillary toothrow, 3.15; depth of cranium, 7.80.

Distribution.—Occurs in south-central Nevada.

Remarks.—The auditory bullae of specimens of *sabulonis* are proportionately much inflated so that their greatest width exceeds the basal length. Hall (1941) gave means and ranges of measurements for 9 adult topotypes. Hafner et al. (1979) listed means ± *SD* for 14 specimens from Penoyer Valley.

Microdipodops pallidus

Diagnosis.—Anterior palatine foramina parallel-sided; premaxillae extending more than 1–2 mm posterior to the posterior border of the nasals; auditory bullae greatly inflated; upper incisors relatively straight; length of hind foot averages 25–27 mm; postauricular spot pure white; upper parts near Light Pinkish Cinnamon (Ridgway, 1912) with a light overwash of buffy or blackish; tail about same color on dorsal side as the dorsal body parts; tail lacks black tip; hairs of underparts white to their bases; diploid chromosome number 42.

Comparisons.—See account for *M. megacephalus.*

Distribution.—Found almost exclusively on fine sands supporting some vegetation in the Great Basin region of west-central Nevada, extreme eastern Mono Co., California, and as a disjunct population in Deep Spring Valley, Inyo Co., California.

Remarks.—Pale kangaroo mice are much more restricted to fine, loose sands. Relationships with *M. megacephalus* are discussed in the account of that species.

Microdipodops pallidus ammophilus Hall, 1941

1941. *Microdipodops pallidus ammophilus* Hall, Field Mus. Nat. Hist., Zool. Ser., 27:273, 8 December.

Holotype.—Adult female, skin and skull, MVZ 58208, from Railroad Valley, Able Spring, 12.5 mi S Locks Ranch, 5,000 ft, Nye Co., Nevada; obtained on 29 July 1933 by E. Raymond Hall.

Measurements of holotype.—Total length, 162; length of tail, 90; length of hind foot, 26.2; length of ear, 11.0; weight, 14.6 g; greatest length of skull, 28.55; width across bullae, 19.00; length of nasals, 9.75; breadth across maxillary arches, 11.80; interorbital breadth, 6.90; length of maxillary toothrow, 3.50; depth of cranium, 8.30.

Distribution.—Occurs in Railroad Valley, Nye Co., Nevada.

Remarks.—This subspecies has the pale dorsal ground color of the species, but this is overlaid with a frosting of black. Specimens from east of New Revielle show slight reddish tint of *M. p. ruficollaris.* Hall (1941) listed means and ranges of measurements for 12 adult topotypes.

Microdipodops pallidus pallidus Merriam, 1901

1901. *Microdipodops pallidus* Merriam, Proc. Biol. Soc. Washington, 14:127, 19 July.

1926. *Microdipodops megacephalus lucidus* Goldman, Proc. Biol. Soc. Washington, 39:127, 27 December.

1927. *Microdipodops megacephalus dickeyi* Goldman, Proc. Biol. Soc. Washington, 40:115, 26 September.

Holotype.—Adult female, skin and skull, USNM 93520, from Mountain Well [fide Vernon Bailey in litt.—Hall, 1941], Churchill Co., Nevada; obtained on 11 May 1898 by Harry C. Oberholser.

Measurements of holotype.—Total length, 171; length of tail, 102; length of hind foot, 25.5; greatest length of skull, 29.4; width across bullae, 19.5; length of nasals, 9.0; breadth across maxillary arches, 12.2; interorbital breadth, 6.7; length of maxillary toothrow, 3.7; depth of cranium, 8.1; length of rostrum 11.6.

Distribution.—Found in the lower, western part of Nevada, Pershing Co., through Fish Lake Valley to Oasis, and in the Deep Spring Valley, Inyo Co., California.

Remarks.—Integradation between *ruficollaris* and *pallidus* is suggested by specimens from west of Millers Wells, Esmeralda Co., Nevada. The specimens from Deep Spring Valley are isolated from other populations of the species, but are very similar to specimens from Fish Lake Valley, Nevada. Hall (1941) listed means and ranges of measurements for 20 adult topotypes; Hafner (1985) gave statistics for measurements of 14 individuals.

Microdipodops pallidus purus Hall, 1941

1941. *Microdipodops pallidus purus* Hall, Field Mus. Nat. Hist., Zool. Ser., 27:273, 8 December.

Holotype.—Adult male, skin and skull, MVZ 52753, from 14.5 mi S Groom Baldy, Lincoln Co., Nevada; obtained on 1 June 1932 by H. Robert Poultney.

Measurements of holotype.—Total length, 160; length of tail, 88; length of hind foot, 26; length of ear, 10; weight, 12.9 g; greatest length of skull, 28.85; width across bullae, 19.85; length of nasals, 10.00; breadth across maxillary arches, 12.45; interorbital breadth, 6.75; length of maxillary toothrow, 3.50; depth of cranium, 7.90.

Distribution.—Occurs in the Emigrant and Desert valleys, Lincoln Co., Nevada.

Remarks.—Specimens from the type locality are the palest of the species. This taxon also has a wide skull at the maxillary arches and a narrow interorbital region. The range of *purus* may not be disjunct from that of *ruficollaris*, which occurs close by to the west. Hall (1941) listed means and ranges of measurements for 20 adult topotypes.

Microdipodops pallidus restrictus Hafner, 1985

1985. *Microdipodops pallidus restrictus* Hafner, Proc. Biol. Soc. Washington, 98:6, 20 March.

Holotype.—Adult male, skin, skull, and partial skeleton, MVZ 159970, from 8.9 mi S, 1.2 mi E Mina, 4,400 ft, Mineral Co., Nevada; obtained on 2 August 1979 by John C. Hafner.

Measurements of holotype.—Total length, 158; length of tail, 89; length of hind foot, 25.5; length of ear, 12; weight, 9.5 g; greatest length of skull, 28.01; width across bullae, 19.53; length of nasals, 10.14; breadth across maxillary arches, 12.49; interorbital breadth, 6.59; greatest length of incisive foramina, 2.23; length of maxillary toothrow, 3.35; depth of cranium, 8.15.

Distribution.—Known only from the type locality, which is located at the southern end of Rhodes Salt Marsh in Soda Spring Valley.

Remarks.—Although *restrictus* is located in the western part of the range of the species, it is similar structurally to ruficollaris to the east. Yet, *restrictus* differs karyotypically from the eastern subspecies and is similar to the taxa to the west (Hafner, 1985). Hafner (1985) gave statistics for measurements of seven individuals.

Microdipodops pallidus ruficollaris Hall, 1941

1941. *Microdipodos* [misspelling of *Microdipodops*] *pallidus ruficollaris*. Hall, Proc. Biol. Soc. Washington, 54:60, 20 May.

Holotype.—Adult female, skin and skull, MVZ 49254, from 5 mi SE Kawich P. O., 5,400 ft, Kawich Valley, Nye Co., Nevada; obtained on 25 September 1931 by Robert T. Orr.

Measurements of holotype.—Total length, 160; length of tail, 90; length of hind foot, 25; length of ear, 9; weight, 12.0 g; greatest length of skull, 28.35; width across bullae, 19.35; length of nasals, 9.85; breadth across maxillary arches, 12.15; interorbital breadth, 7.00; length of maxillary toothrow, 3.45; depth of cranium, 7.90.

Distribution.—Ranges from the valleys of western Nye Co. eastward to western Lincoln Co., Nevada.

Remarks.—This taxon is characterized by reddish upper parts and a broad cinnamon-colored collar. Hall (1941) found little variation within this subspecies except for the population in the Penoyer Valley. Hall (1941) gave means and ranges of measurements for 10 adult topotypes. Hafner et al.

(1979) gave means ± *SD* for 35 individuals from the Penoyer Valley; and Hafner (1985) gave statistics for measurements of 8 individuals.

Subfamily Heteromyinae Coues, 1875

Diagnosis.—Size medium to large, from 180 to 360 mm in total length; body form quadrupedal and generally mouse- or rat-like; tail not tufted or crested; locomotion scansorial with slight tendency to subricochetal; hind limbs larger than forelimbs; pes with five clawed digits; pes soles naked or partly clothed with a scant covering of short hairs; pelage stiff or hispid; body hairs of three types—straight, relatively long and widened overhairs, wide, troughed overhairs, and a thin underfur of slightly wavy hairs; cuticular scales in trough of hairs form series of longitudinal ridges; anterior face of upper incisor either smooth or with shallow groove; molars progressively hypsodont, rooted, and tuberculate; enamel cusp pattern of cheek teeth lost to wear early in life; lophs of P4 unite first at lingual then labial side; protoloph of P4 with more than one cusp; lophs of upper molars unite first at lingual then labial side, surrounding a central basin in a majority of spccics (most prominent in M1); lophs of p4 unite first at lingual then labial side; stylids on any point of p4, developing progressively in geologic time; lophs of lower molars united first at labial then lingual side; no foramina and usually no pit between m3 and the base of the coronoid process; palate extending beyond level of M3; center of palate between cheek teeth not ridged; squamosal in broad contact with parietal on dorsal surface of skull; squamosal not perforated by auditory bulla; anterior zygomatic root not greatly enlarged on joining lachrymal; ossification of orbital walls complete; incisive foramen small; interorbital foramen present in orbitosphenoid bone ventral to optic foramen; masticatory and buccinator foramina united; stapedial and sphenofrontal foramina absent; no median ventral foramina in the "central" of caudal vertebrae; two pterygoid fossae present; auditory bullae moderately expanded; inflated interior of auditory bullae filled with spongy trabeculae; mastoid bulla not appearing on dorsal surface of skull and not projecting to posterodorsal plane of occiput; lateral wall of skull dorsal to external auditory meatus composed entirely of squamosal; ventral plane of tympanic bulla higher than occlusal plane of cheek teeth; no stapedius muscle in middle ear; no fusion or broadening of cervical vertebrae; scapula not prolonged posteriorly; obturator foramen not triangular; no articulation of trapezoid and scapholunar; astragalus articulating with cuboid; dorsal surface of ectocuneiform not hourglass-shaped; phallus without external spines and with 2–3 lobed urethral lappets; phallus long relative to bacular length; baculum with a swollen base and an upturned tip; male accessory reproductive glands variable—preputial gland absent, ventromedial prostate present or absent, anterior and dorsal prostates present or absent, ampullary present or absent; vesicular glands elongate, hook-shaped, and translucent; nasolabialis profundus pars maxillaris muscles originate from infraorbital foramen; M. temporalis not reduced, origin far lateral, narrow, and reaching back of squamosal; M. cleidomastoideus present; extensores breves muscles present; M. abductor hallucis present; lumbricales muscles not reduced or absent (Burt, 1936; Hafner and Hafner, 1983; Genoways, 1973; Homan and Genoways, 1978; Merriam, 1889; Osgood, 1900; Ryan, 1989; Wahlert, 1985; Webster and Webster, 1975; Wood, 1935).

Remarks.—Heteromyines are the most murinelike members of the family, and are, in most ways, more similar structurally to the ancestral heteromyids and geomyoids than other living heteromyids. Of the two genera, *Liomys* shows more development of subricochetal tendencies than *Heteromys.* Genetic studies by Rogers (1990) suggest that *Heteromys* and *Liomys* should be combined into one genus, perhaps containing three to five, or more, subgenera. One group would include species classified here as *Liomys, H. australis* and an undescribed species of *Heteromys.*

Key to the Genera

1. Cheek teeth with medium-high crowns; lower premolar with 2 lophids; upper and lower premolars wider than last upper and lower molars; length of P4/greatest length of skull averages greater than 0.046; posterior portion of soles of hind feet with sparse covering of hairs .... *Liomys*

1′. Cheek teeth with high crowns; lower premolar with 3 or 4 lophids; upper and lower premolars wider than last upper and lower molars; length of P4/greatest length of skull averages less than 0.043; posterior portion of soles of hind feet usually naked (except *H. gaumeri*) .... *Heteromys*

Genus *Heteromys*

1817. *Heteromys* Desmarest, Nouv. Dict. Hist. Nat., 14:181, May.

Type species.—Mus anomalus Thompson, 1815, Trans. Linn. Soc. London, 11: 161, pl. 10.

Diagnosis.—Cheek teeth high crowned, with most complicated pattern in family; anterior cingulum in lower molars and a posterior cingulum in the upper molars nearly as high as remainder of crown giving them three lophs before wear; upper incisors asulcate; three or four lophids on lower premolar; skull elongate; auditory region uninflated; pelage hispid, consisting of stiff spines mingled with slender soft hairs; spines are flattened and anteriorly grooved; tail usually longer than head and body length; soles of hind feet naked, except in *H. gaumeri*; second digit of hind foot with weakly-developed spoon-like claw; interpterygoid fossa V-shaped anteriorly; adapted for scampering.

Remarks.—The genus *Heteromys* was last reviewed by Goldman (1911). The research of Rogers and Schmidly (1982), Engstrom et al. (1987), and Rogers (1989, 1990) has contributed much to elucidate the relationships of species within this genus. Herein, two subgenera are recognized, but their validity and relationships are questionable (see below). The subgenus *Heteromys* contains five species, whereas *Xylomys* is monotypic. Additionally, Rogers (1990) showed the existence in Costa Rica of an undescribed species related to *H. anomalus*; biochemically, these two taxa are more closely linked with *Liomys* than with other species of *Heteromys.* The following key was adapted from Schmidt et al. (1989). We know couplet 4 to be unreliable, but have not found key characters to distinguish all populations of these species.

Key to the Subgenera and Species

1. Adult pelage harsh but bristles soft 2

1′. Adult pelage with numerous stiff bristles or spines .... Subgenus *Heteromys* 3

2. Ears edged with white; premaxillary bones terminating posterior to nasals; known only from Costa Rica .... *Heteromys (Heteromys) oresterus*

2′. Ears without white edges; premaxillary and nasals bones terminating at same level; known only from southeastern Chiapas, México and southwestern Guatemala .... *Heteromys (Xylomys) nelsoni*

3. Soles of hind feet hairy from posterior tubercle to heel; orange buff lateral line broad and conspicuous .... *Heteromys gaumeri*

3′. Soles of hind feet naked posteriorly; lateral line absent or when present never broad and conspicuous 4

4. Inner side of forearm clouded with dusky 5

4′. Inner side of forearm white .... *Heteromys desmarestianus*

5. Tail usually less than 140; slender hairs on dorsum clear grayish .... *Heteromys australis*

5′. Tail more than 140; slender hairs on dorsum dull ochraceous buffy .... *Heteromys anomalus*

Subgenus *Heteromys*

1817. *Heteromys* Desmarest, Nouv. Dict. Hist. Nat., 14:181, May.

Type species.—*Mus anomalus* Thompson, 1815, Trans. Linn. Soc. London, 11: 161, pl. 10.

Diagnosis.—Pelage with numerous stiff bristles or spines; inner sides of forelimbs white or slightly dusky; braincase more or less flat; parietals not extending laterally to the mastoids; posterior molars narrower than premolars; molars not complex but possessing accessory enamel islands in posterior loops of upper molars and anterior loops of lower molars; small posterior cingulum, which is closely compressed with metaloph, present on last upper molar.

Remarks.—Members of the subgenus are widespread, covering the entire known distribution of the genus. This subgenus is currently recognized as containing five species, although Rogers (1990) stated that the current taxonomy is incongruent with biochemical and chromosomal data. Goldman (1911) and Hall (1981) divided this subgenus into the *anomalus* species group containing *anomalus* and *australis,* and the *desmarestianus* species group which consisted of *desmarestianus, goldmani,* and *gaumeri.* Removal of *gaumeri* from the *desmarestianus* species group was advocated by Engstrom et al. (1987) based on morphology and karyology. Rogers (1990) advocated splitting thc *anomalus* species group and including *oresterus* within the *desmarestianus* species group of the subgenus *Heteromys.*

Species Accounts

Heteromys anomalus

Diagnosis.—Size large for species of genus; adult head and body length generally between 130 and 140 mm; tail longer than head and body length; tail dark above, lighter below, distinctly bicolored; dorsum dark mouse gray to grayish brown with a sprinkling of ochraceous hairs; venter white; lateral line absent; inner forelimbs clouded with dusky (except in *H. anomalus jesupi*); ears large; skull large; narrow interorbital region; bullae small; nasals broad; 2n = 60, FN = 68.

Comparisons.—The range of this species may contact that of *H. australis* and *H. desmarestianus.* From both species, *anomalus* may be distinguished by its larger ears, narrower interorbital region, and broader nasals. *H. anomalus* is larger than *australis* in head and body size. Compared with *H. desmarestianus, anomalus* has dusky, instead of white, inner forelimbs.

Distribution.—Extreme eastern Panamá, western and northern Colombia, northern Venezuela, Trinidad, Tobago, and Margarita Island.

Remarks.—Best (1992) gave external and cranial measurements for this species. Rogers (1990) indicated that this species occurred in eastern Panamá, near a locality where *H. australis* was collected, but at a slightly higher elevation. This specimen, originally identified as *H. australis,* was genically and morphologically most similar to *H. anomalus* from eastern Venezuela (Rogers, 1990). This species also was found by Rogers (1990) to be the sister taxon to several taxa including *Liomys, H. australis,* and an undescribed species of *Heteromys* from Costa Rica. Engstrom et al. (1987) reported no karyotypic variation for 20 specimens from Venezuela; based on distribution these specimens would be referred to *H. a. anomalus.* Based on an examination by H. Genoways, one specimen from Palomino, Colombia and two unregistered specimens from Bogotá, Colombia in the British Museum (Natural History), are grayer and appear to have more spines than three additional specimens from near Bogotá that are tentatively assigned *H. australis.* The former specimens are assigned to *H. anomalus,* although they might be *desmarestianus.*

Heteromys anomalus anomalus (Thompson, 1815)

1815. *Mus anomalus* Thompson, Trans. Linn. Soc. London, 11:161, pl. 10.

1827. *Heteromys thompsonii* Lesson, Manuel de Mammalogie, p. 264.

1868. *Perognathus bicolor* Gray, Proc. Zool. Soc. London, p. 202, May.

1868. *Heteromys melanoleucus* Gray, Proc. Zool. Soc. London, p. 203, May.

Holotype.—Adult of unknown sex, poorly preserved skin and skull, BM(NH) unregistered, from Trinidad.

Measurements of holotype.—Interorbital breadth, 8.9; length of nasals, 15.7; length of rostrum, 16.8; length of maxillary toothrow, 5.6.

Distribution.—Northern and eastern Venezuela and the islands of Trinidad and Margarita.

Remarks.—We believe the holotype of *Mus anomalus* and the cotypes of *Perognathus bicolor* and *Heteromys melanoleucus* are the same taxon; the senior synonym is *Heteromys anomalus.* The lectotype (BM[NH] 47.2.1.7) and lectoparatype (BM[NH] 47.2.1.5) of *P. bicolor* are both juveniles with silky juvenile pelage and deciduous premolars. The lectotype (BM[NH] 47.2.1.4) and lectoparatype (BM[NH] 47.2.1.6) are both subadults with juvenile pelage except on the head, neck, and rump and permanent premolars. Genoways's examination of the holotype indicates that Gray apparently described juveniles from this series as *P. bicolor* and subadult and adult material as *H. melanoleucus.* Gray described *Liomys irroratus* and *L. albolimbatus* on a similar basis (Genoways, 1973: 111–112). These four specimens were all collected by Dyson; Gray described them as coming from Honduras but the specimen tags and the Museum register (Alston, 1879–1882) indicate that they are from Venezuela. A note on the tag states "described and stand painted as from Honduras." External measurements for this subspecies (including *melanoleucus*) were given by Allen (1899). Goldman (1911) gave measurements for five adult specimens from Trinidad.

Heteromys anomalus brachialis
Osgood, 1912

1912. *Heteromys anomalus brachialis* Osgood, Field Mus. Nat. Hist., Zool. Ser., 10 (5):54, 10 January.

Holotype.—Adult female, skin and skull, FMNH 18623, from El Panorama, Rio Aurare, eastern shore of Lake Maracaibo, Venezuela; obtained on 19 January 1911 by Wilfred H. Osgood and Stanley G. Jewett.

Measurements of holotype.—Total length, 284; length of tail, 150; length of hind foot, 34; greatest length of skull, 35.2; zygomatic breadth, 16.1; interorbital breadth, 7.3; length of nasals, 14.2; width of braincase, 14.8; length of maxillary toothrow, 5.6 (Osgood, 1912).

Distribution.—Confined to the area east of Lake Maracaibo.

Remarks.—Osgood (1912) described *H. a. brachialis* primarily because its upper parts are paler than in other members of the species and because the front legs are nearly white. The skull of *brachialis* was noted as being practically the same as those of other populations of *H. anomalus.* The type and five other specimens were collected under mayas and thorn shrubs that formed hedgelike borders of the trails leading into El Panorama (Osgood, 1912). Specimens from the region west of Lake Maracaibo (Handley, 1976) may be assignable to this subspecies.

Heteromys anomalus hershkovitzi
Hernández-Camacho, 1956

1956. *Heteromys* (*Heteromys*) *anomalus hershkovitzi* Hernández-Camacho, Lozania, 10:3, 26 April.

Holotype.—Adult female, Instituto Carlos Finlay, Bogotá no. 2701, but now deposited in the Instituto de Ciencias Naturales, Bogotá, from (according to Cabrera, 1961:512) "volcanes, cerca de la cabecera del corregimiento de Córdoba, municipio de Caparrapi, departmento de Cundinamarca," 250 m; obtained 18 February 1944.

Measurements of holotype.—No measurements of the holotype are available.

Distribution.—Known with certainty only from the type locality. Cabrera (1961) indicated that this subspecies is distributed in the Magdalena River Valley of the Andean region of Colombia at least on the forested western slopes of the Eastern Cordillera.

Heteromys anomalus jesupi J. A. Allen, 1899

1899. *Heteromys jesupi* J. A. Allen, Bull. Amer. Mus. Nat. Hist., 12:201, 20 December.

Holotype.—Adult female, AMNH 15347, from below Minca, 1,000 ft, Santa Marta District, Colombia; obtained on 30 July 1899 by Herbert H. Smith.

Measurements of holotype.—Total length, 330; length of tail, 163; length of hind foot, 33; greatest length of skull, 36.2; zygomatic breadth, 16.4; interorbital constriction, 8.4; mastoid breadth, 5.3; length of nasals, 15.0; length of rostrum, 15.7; length of maxillary toothrow, 5.5.

Distribution.—Northern Colombia, the mountain slopes of the Sierra Nevada de Santa Marta, sea level to 610 m.

Remarks.—External measurements for females of this taxon are greater than those for females of either *H. a. anomalus* or *H. a. brachialis* (Allen, 1899). Measurements for males are not reported. Osgood (1912) used the trinomial *H. a. jesupi* in comparing this taxon to *H. a. brachialis.*

Heteromys australis Thomas, 1901

Diagnosis.—Medium-sized species for genus; head and body length less than 130 mm; tail equal to or slightly longer than head and body length; tail brownish above, lighter below, but not distinctly bicolored; dorsum gray to blackish slate, grizzled with gray; venter white; inner sides of forelimbs dusky; lateral line absent; ears without a white edging; skull short, broad; braincase inflated; auditory bullae small; zygoma anteriorly spreading.

Comparisons.—This species can be distinguished from *H. anomalus* and *H. desmarestianus* by its smaller size and shorter tail (less than 140 mm); both of the above have a tail length greater than 140 mm. In *H. australis* the inner side of the forearm is dusky, while the inner side of the forearm in *H. desmarestianus* is white. The dorsal pelage of *H. desmarestianus* and *H. anomalus* is usually sprinkled with ochraceous hairs; in *H. australis* these hairs are gray. The braincase of *H. australis* is inflated; it is not inflated in *H. anomalus* and *H. desmarestianus.*

Distribution.—*H. australis* is distributed from eastern Panamá into northwestern Colombia southward to northwestern Ecuador. It is also distributed eastward in Colombia to near Bogotá.

Remarks.—The karyotype for this species has not been reported. External and cranial measurements were given by Best (1992). Rogers (1990) found that, genically, *H. australis* was not closely related to *H. anomalus,* but instead was more related to an undescribed species from Costa Rica. All three taxa were more closely related to *Liomys* than to other species of *Heteromys* (Rogers, 1990). Goldman (1911), later followed by Hall (1981), placed *australis* and *anomalus* in the *anomalus* species group. Although the distribution of *H. australis* contacts that of *H. desmarestianus crassirostris* in eastern Panamá, *H. australis* is found at lower elevations. *H. australis* may come into contact with *H. anomalus* in Panamá and north-central Colombia.

Heteromys australis australis Thomas, 1901

1901. *Heteromys australis* Thomas, Ann. Mag. Nat. Hist., ser. 7, 7:194, February.

1912. *Heteromys lomitensis* J. A. Allen, Bull. Amer. Mus. Nat. Hist., 31:77, 19 April.

Holotype.—Female, skin and skull, BM(NH) 1.3.19.23, from St. Javier, 60 ft, northwestern Ecuador; obtained on 23 June 1900 by G. Flemming and R. Miketta.

Measurements of holotype.—Total length, 272; length of tail, 137; length of hind foot, 30; length of ear, 15; greatest length of skull, 34.5; zygomatic breadth, 16.0; interorbital breadth, 9.2; mastoid breadth, 14.0; length of nasals, 13.8; length of rostrum, 15.4; length of maxillary toothrow, 5.3.

Distribution.—This subspecies is distributed from northwestern Colombia southward into northwestern Ecuador and eastward in Colombia to near Bogotá. Elevations range from 20 m to 2,750 m. The localities in Colombia are intercordilleran.

Remarks.—Four specimens at the British Museum (Natural History) from localities near Bogotá, Colombia may represent *australis.* Allen (1916) placed *lomitensis* in synonymy with *australis,* noting that the differences in pelage and coloration upon which the description of *lomitensis* were based were probably accidental or of a seasonal nature. Goldman (1911) gave measurements for specimens from the type locality.

Heteromys australis conscius Goldman, 1913

1913. *Heteromys australis conscius* Goldman, Smithsonian Misc. Coll., 60(22):8, 28 February.

Holotype.—Adult male, skin and skull, USNM 178699, from Cana, 2,000 ft; mountains of eastern Panamá; obtained on 8 March 1912 by Edward A. Goldman.

Measurements of holotype.—Total length, 260; length of tail, 133; length of hind foot, 32; greatest length of skull, 34.3; zygomatic breadth, 16.0; interorbital breadth, 8.0; mastoid breadth, 13.8; length of nasals, 13.7; length of rostrum, 15.2; length of maxillary toothrow, 5.5; depth of cranium, 8.7.

Distribution.—Known from extreme eastern Panamá and perhaps into northwestern Colombia at elevations ranging from 150 m to 800 m.

Remarks.—External measurements for two adult topotypes were given by Goldman (1913). Specimens have been collected under logs in the forest at low elevation slopes of the Pirre Range (Goldman, 1920).

Heteromys australis pacificus Pearson, 1939

1939. *Heteromys australis pacificus* Pearson, Notulae Naturae, Acad. Nat. Sci. Philadelphia, 6:4, 8 June.

Holotype.—Adult male, skin and skull, ANSP 19499, from Amagal, 1,000 ft, S of Guayabo Bay, Darién, Panamá; obtained on 10 June 1938 by Oliver P. Pearson.

Measurements of holotype.—Total length, 275; length of tail, 144; length of hind foot, 31; greatest length of skull, 35.0; zygomatic breadth, 16.8; interorbital breadth, 9.0; length of nasals, 14.4; length of maxillary toothrow, 5.4 (Pearson, 1939).

Distribution.—Known only from the area near the type locality at elevations ranging from 300 m to 610 m (Handley, 1966). Pearson (1939) noted that no specimens were collected near sea level.

Remarks.—According to Pearson (1939: 5), the nine specimens from the type locality were captured at elevations ranging from 300 m to 610 m in a "habitat . . . characterized by an enormous amount of rainfall and a luxurious growth of many palms, constituting an infinitely richer vegetation than found near Cana, the type locality of *conscius.*" Pearson (1939) also gave averages for external and cranial measurements for three adult female topotypes.

Heteromys desmarestianus

Diagnosis.—Large-sized species for genus; head and body length generally between 130 and 140 mm; tail longer than head and body length; ears without white edging; lateral line occasionally present, but never pronounced; dorsum gray to slaty black with sprinkling of ochraceous hairs; venter white; inner sides of forelimbs white; skull medium to large size; bullae small; interparietal variable; zygoma narrow or only slightly spreading anteriorly; base of baculum one-third total bacular length; baculum tapers to the slightly upturned and slightly laterally compressed tip; 2n = 60, FN = 67–86.

Comparisons.—This species may be distinguished from *H. nelsoni* and *H. oresterus* by smaller body size, smaller ears, and by a greater development of the posterior portion of the toothrow. In addition, specimens of *H. nelsoni* and *H. oresterus* have soft, not stiff, bristles. Compared with *H. gaumeri, desmarestianus* is larger and lacks the furred soles of the hind feet; *H. australis* is smaller and the ochraceous hairs on the dorsum are lacking. The ears of *H. anomalus* are much larger than those of *desmarestianus.*

Distribution.—This species is distributed southward from the southern Mexican province of Veracruz, including the southern Yucatán Peninsula, Belize, Guatemala, Honduras, El Salvador, Costa Rica, Panamá, and into northwestern Colombia.

Remarks.—This is the most widely distributed species of the genus. Rogers (1989) described standard karyotypes and used C-banding to evaluate the chromosomal variation within this species. Seven chromosomal forms are known for this species, each of which exhibits no interpopulational variation (Rogers, 1989). Mascarello and Rogers (1988) described G-banded chromosomes. Comparisons of the external, cranial, and bacular morphology of this species with *Liomys* were made by Genoways (1973). Rogers and Schmidly (1982) placed *lepturus* and *longicaudatus* in synonymy with *H. d. desmarestianus*; *H. temporalis* was recognized as a subspecies of *desmarestianus.* Goodwin (1969) treated *nigricaudatus* as another synonym of *H. d. desmarestianus.* Rogers (1990) found that *H. goldmani* was conspecific with *H. desmarestianus,* but did not remark on its subspecific status. Herein we recognize it as a subspecies. Best (1992) gave external and cranial measurements.

Heteromys desmarestianus chiriquensis Enders, 1938

1938. *Heteromys desmarestianus chiriquensis* Enders, Proc. Acad. Nat. Sci. Philadelphia, 90: 141, 20 September.

Holotype.—Adult male, skin and skull, ANSP 17835, from Cerro Pando, a hill 4,000 ft in elevation situated between the Río Chiriquí Viejo and its tributary, the Río Colorado, about 10 mi from the Post Office of El Volcán, Chiriquí, Panamá; obtained on 20 August 1935 by Robert K. Enders.

Measurements of holotype.—Total length, 302; length of tail, 154; length of hind foot, 35; length of ear, 19; greatest length of skull, 37.6; zygomatic breadth, 17.2; interorbital breadth, 10.1; length of nasals, 17.0 (Enders, 1938).

Distribution.—Southeastern Costa Rica and adjacent Chiriquí Province in west-central Panamá at elevations from 1,150 m to 1,600 m.

Heteromys desmarestianus crassirostris Goldman, 1912

1912. *Heteromys crassirostris* Goldman, Smithsonian Misc. Coll., 60(2):10, 20 September.

Holotype.—Adult male, skin and skull, USNM 179016, from Mount Pirri, 5,000 ft, near head of Río Limon, Panamá; obtained on 26 April 1912 by Edward A. Goldman.

Measurements of holotype.—Total length, 267; length of tail, 132; length of hind foot, 32.5; greatest length of skull, 33.0; zygomatic breadth, 15.8; interorbital constriction, 9.4; mastoid breadth, 13.5; length of nasals, 13.6; length of rostrum, 15.1; length of maxillary toothrow, 4.8; depth of cranium, 8.9.

Distribution.—Densely forested upper slopes of extreme eastern Panamá into northwestern Colombia at elevations from 1,350 m to 1,600 m.

Remarks.—This is the only subspecies of *desmarestianus* with a distribution that extends into South America. Goldman (1920) placed this taxon as a subspecies of *desmarestianus.* External measurements for eight adult topotypes and cranial measurements for five adults were given by Goldman (1912).

Heteromys desmarestianus desmarestianus Gray, 1868

1868. *Heteromys desmarestianus* Gray, Proc. Zool. Soc. London, p. 204, May.
1868. *Heteromys longicaudatus* Gray, Proc. Zool. Soc. London, p. 204, May.

1902. *Heteromys griseus* Merriam, Proc. Biol. Soc. Washington, 15:42, 5 March.
1902. *Heteromys goldmani lepturus* Merriam, Proc. Biol. Soc. Washington, 15:42, 5 March.
1928. *Heteromys desmarestianus psakastus* Dickey, Proc. Biol. Soc. Washington, 41:10, 1 February.
1956. *Heteromys nigricaudatus* Goodwin, Amer. Mus. Novit., 1791:4, 28 September.

Holotype.—Adult of unknown sex, skin and broken skull, BM(NH) 43.6.13.1 from Coban, Guatemala; obtained on an unknown date by an unknown collector.

Measurements of holotype.—Length of hind foot, 35; length of ear, 15; length of nasals 15.6; length of rostrum, 16.7; length of maxillary toothrow, 5.3.

Distribution.—Humid forested mountain slopes and coastal plains of southern México, Guatemala, Belize, and El Salvador.

Remarks.—The characteristics attributed to *H. longicaudatus* are within the variation of several samples of *H. d. desmarestianus*; the latter name was chosen by Rogers and Schmidly (1982) because both names were published simultaneously by Gray (1868). Also see remarks in the species account above.

Heteromys desmarestianus fuscatus
J. A. Allen, 1908

1908. *Heteromys fuscatus* J. A. Allen, Bull. Amer. Mus. Nat. Hist., 24:652, 13 October.

Holotype.—Adult male, skin and skull, AMNH 28451, from Tuma, Nicaragua; obtained on 1 December 1907 by William B. Richardson.

Measurements of holotype.—Total length, 300; length of tail, 150; length of hind foot, 30; greatest length of skull, 36.4; zygomatic breadth, 16.9; interorbital constriction, 10.0; mastoid breadth, 14.7; length of nasals, 14.8; length of rostrum, 16.8; length of maxillary toothrow, 5.3.

Distribution.—Southern, western, and northeastern Honduras and central Nicaragua.

Remarks.—Goldman (1920) recognized this taxon as a subspecies of *desmarestianus*. Measurements for three specimens are given in Allen (1908). Distribution in northeastern Honduras was established by Benshoof et al. (1984). At this site, specimens were collected in dry areas on the tops of hills in second growth and mature forest at elevations of 50 m and 100 m. These authors also gave external measurements for three males and two females.

Heteromys desmarestianus goldmani
Merriam, 1902

1902. *Heteromys goldmani* Merriam, Proc. Biol. Soc. Washington, 15:41, 5 March.

Holotype.—Adult male, skin and skull, USNM 77576, from Chicharras, Chiapas; obtained on 7 February 1896 by Edward W. Nelson and Edward A. Goldman.

Measurements of holotype.—Total length, 347; length of tail, 199; length of hind foot, 40; greatest length of skull, 39.0; zygomatic breadth, 17.4; interorbital breadth, 9.7; mastoid breadth, 16.0; length of nasals, 15.8; length of rostrum, 17.4; length of maxillary toothrow, 5.6; depth of braincase, 10.0.

Distribution.—Restricted to the heavily forested Pacific slope of the Sierra Madre in extreme southern Chiapas and adjacent Guatemala between 45 m and 1,860 m.

Remarks.—Considerable morphologic geographic variation among three samples from Chiapas, México was found by Rogers and Schmidly (1982). Based on morphology, Rogers and Schmidly (1982) recognized this taxon as a distinct species and included *H. goldmani* in the *H. desmarestianus* species group. Later, Rogers (1990) found that, based on genetic data, *H. goldmani* was indistinguishable from nearby populations of *H. desmarestianus* and proposed that *H. goldmani* be synonymized with *H. desmarestianus.* External and cranial measurements for this species were given by Goldman (1911), Rogers and Schmidly (1982), and Best (1992). Goldman (1951) described collecting localities at two sites in Chiapas.

Heteromys desmarestianus panamensis Goldman, 1912

1912. *Heteromys panamensis* Goldman, Smithsonian Misc. Coll., 56(36):9, 19 February.

Holotype.—Adult male, skin and skull, USNM 171107, from Cerro Azul, 2,800 ft, near headwaters of Chagres River, Panamá; obtained on 23 March 1911 by Edward A. Goldman.

Measurements of holotype.—Total length, 283; length of tail, 148; length of hind foot, 35; greatest length of skull, 34.4; zygomatic breadth, 16.9; interorbital constriction, 9.3; mastoid breadth, 14.8; length of nasals, 13.9; length of rostrum, 14.9; length of maxillary toothrow, 5.2; depth of cranium, 8.9.

Distribution.—North-central Panamá at elevations between 610 m and 910 m.

Remarks.—Goldman (1920) placed this taxon as a subspecies of *desmarestianus.*

Heteromys desmarestianus planifrons Goldman, 1937

1937. *Heteromys desmarestianus planifrons* Goldman, J. Washington Acad. Sci., 27 (10): 418, 15 October.

Holotype.—Adult female, skin and skull, USNM 250348, from San Geronimo, Pirris, Costa Rica; obtained 12 April 1931 by C. F. Underwood.

Measurements of holotype.—Total length, 303; length of tail, 169; length of hind foot, 35; length of ear, 18; greatest length of skull, 37.8; interorbital breadth, 9.8; mastoid breadth, 15.7; length of nasals, 16.2; length of rostrum, 17.8; length of maxillary toothrow, 5.6; depth of cranium, 9.3.

Distribution.—Known only from the lowlands of western Costa Rica.

Heteromys desmarestianus repens Bangs, 1902

1902. *Heteromys repens* Bangs, Bull. Mus. Comp. Zool., Harvard Univ., 39:45, April.

Holotype.—Adult female, skin and skull, MCZ 10356, from Boquete, southern slope Volcán de Chiriquí, 4,000 ft, Panamá; obtained on 8 April 1901 by W. W. Brown, Jr.

Measurements of holotype.—Total length, 282; length of tail, 150; length of hind foot, 33; length of ear, 15; greatest length of skull, 34.9; interorbital breadth, 9.4; mastoid breadth, 13.6; length of nasals, 14.8; length of rostrum, 15.6; length of maxillary toothrow, 4.9.

Distribution.—Extreme southwestern Panamá.

Remarks.—The type locality is at approximately 1,200 m. Goldman (1911) gave external and cranial measurements for this taxon. Later, Goldman (1920) recognized *repens* as a subspecies of *desmarestianus.*

Heteromys desmarestianus subaffinis Goldman, 1937

1937. *Heteromys desmarestianus subaffinis* Goldman, J. Washington Acad. Sci., 27(10): 420, 15 October.

Holotype.—Adult male, skin and skull, USNM 12904/38591, from Angostura, about 1,980 ft, southern side Río Reventazon, opposite Turrialba, Costa Rica; obtained in May 1876 by José C. Zeledon.

Measurements of holotype.—Zygomatic breadth, 17.2; interorbital breadth, 10.0; mastoid breadth, 14.5; length of maxillary toothrow, 5.2.

Distribution.—Lowlands of northeastern Costa Rica.

Remarks.—Goldman (1911) had earlier referred the specimens on which this subspecies is based to *repens.* This subspecies appears to prefer relatively low elevations; the type locality is approximately 600 m. The range of this subspecies is bordered by that of *H. d. underwoodi*; *underwoodi* is smaller and is found at higher elevations.

Heteromys desmarestianus temporalis Goldman, 1911

1911. *Heteromys temporalis* Goldman, N. Amer. Fauna, 34:26, 7 September.

Holotype.—Adult female, skin and skull, USNM 63719, from Motzorongo, Veracruz; obtained on 3 March 1894 by Edward W. Nelson and Edward A. Goldman.

Measurements of holotype.—Total length, 320; length of tail, 180; length of hind foot, 37; greatest length of skull, 39.2; zygomatic breadth, 17.8; interorbital breadth, 10.3; mastoid breadth, 15.5; length of nasals, 15.7; length of rostrum, 17.7; length of maxillary toothrow, 5.3; depth of cranium, 9.6.

Distribution.—Forested eastern basal slopes of mountains in central Veracruz between 210 m and 460 m. Goldman (1951) described the type locality.

Remarks.—This is the northernmost extension of the genus. Rogers and Schmidly (1982) recognized *temporalis* as a subspecies of *desmarestianus* and presented external and cranial measurements.

Heteromys desmarestianus underwoodi Goodwin, 1943

1943. *Heteromys desmarestianus underwoodi* Goodwin, Amer. Mus. Novit., 1227:1, 22 April.

Holotype.—Adult female, skin and skull, AMNH 131729, from Escazú, 7 mi SW San José, 3,000 ft, Province of San José, Costa Rica; obtained on 28 April 1938 by C. F. Underwood.

Measurements of holotype.—Total length, 277; length of tail, 148; length of hind foot, 30; length of ear, 15; greatest length of skull, 34.5; zygomatic breadth, 15.6; interorbital breadth, 9.7; mastoid breadth, 14.3; length of nasals, 14.0; length of rostrum, 16.0; length of maxillary toothrow, 5.2.

Distribution.—Central Costa Rica at elevations ranging from 1,350 m to 1,700 m.

Remarks.—The range of this subspecies is located between that of *H. d. planifrons* and *H. d. subaffinis. H. d. underwoodi* is found at higher elevations and is smaller than either of the two preceding subspecies. Goodwin (1943) presented external and cranial measurements for the type and one male and three females from Los Higuerones, Escazú. He (Goodwin, 1943:2) noted that the type specimen was collected at an altitude between 5,000 and 5,500 ft "in humid virgin country that is interspersed with patches of maize as well as clearings for cattle."

Heteromys desmarestianus zonalis Goldman, 1912

1912. *Heteromys zonalis* Goldman, Smithsonian Misc. Coll., 56(36):9, 19 February.

Holotype.—Adult female, skin and skull, USNM 170976, from Río Indio, near Gatún, Canal Zone, Panamá; obtained on 15 February 1911 by Edward A. Goldman.

Measurements of holotype.—Total length, 263; length of tail, 134; length of hind foot, 35; greatest length of skull, 36.3; zygomatic breadth, 16.8; interorbital breadth, 9.6; mastoid breadth, 14.5; length of nasals, 13.6; length of nostrum, 15.7; length of maxillary toothrow, 5.2; depth of cranium, 9.5.

Distribution.—Widely distributed in Panamá from the northeastern part of the country through central Panamá and into eastern Panamá.

Remarks.—Goldman (1920) recognized this taxon as a subspecies of *desmarestianus.*

Heteromys gaumeri

Diagnosis.—Size medium for genus; head and body length less than 130 mm; tail longer than head and body, well-haired, grayish brown above, whitish below, and with terminal tuft of hair; ears dusky, edged with dull white; dorsal pelage dark- to medium-gray with orange-buff hairs; broad, ochraceous lateral line, extending from the cheeks to the base of the tail; venter and feet white; soles of feet haired posterior to the last plantar tubercle; skull medium-sized; large auditory bullae; lower premolar with three lophs; upper and lower molars with three lophs, enamel island formed between metaloph (id) and cingulum disappearing with wear; baculum with broad base approximately one-third the total bacular length; baculum with narrow oval shaft making up the remaining two-thirds of the length ending in a slightly upturned tip; 2n = 56, FN = 76.

Comparisons.—This species can be distinguished from all other species of *Heteromys* by the presence of hair on the posterior portion of the sole of the hind foot (hair is absent in other species) and a distinct, broad, ochraceous lateral line. Only the range of *H. desmarestianus* may overlap that of *H. gaumeri.* Compared with *H. desmarestianus, H. gaumeri* is smaller in both external and cranial measurements; has larger auditory bullae; has a broader ochraceous lateral line; a well-haired tail with a terminal tuft; and a baculum with a narrow shaft.

Distribution.—Endemic to the Yucatán Peninsula at elevations from sea level to 100 m. The range includes: northern Belize; El Petén, Guatemala; eastern Tabasco, México; and the Mexican states of Campeche, Quintana Roo, and Yucatán. This species prefers tropical deciduous and subdeciduous-subperennial tropical rain forests and thorn scrub forest (Engstrom et al., 1987; Schmidt et al., 1989). Goldman (1951) described the localities for specimens collected in the Mexican states of Campeche, Quintana Roo, and Yucatán.

Remarks.—*H. gaumeri* is monotypic. Based on morphologic, karyologic, and genic information, Engstrom et al. (1987) and Rogers (1990) recommended that *H. gaumeri* be removed from the *desmarestianus* species group and suggested that this taxon might warrant separate subgeneric recognition. Extensive morphologic comparisons of this species with species of *Liomys* were made by Genoways (1973).

Heteromys gaumeri
Allen and Chapman, 1897

1897. *Heteromys gaumeri* Allen and Chapman, Bull. Amer. Mus. Nat. Hist., 9:9, 23 February.

Holotype.—Adult male, skin and skull, AMNH 12028 and 10461, from Chichén-Itzá, Yucatán, México; obtained on 17 March 1896 by Frank M. Chapman.

Measurements of holotype.—Total length, 292; length of tail, 162; length of hind foot, 32; length of ear, 14.5; interorbital constriction, 8.9; length of nasals, 15.6; length of rostrum, 16.4; length of maxillary toothrow, 4.8.

Remarks.—A review of this species was given by Engstrom et al. (1987) and Schmidt et al. (1989). External and cranial measurements were presented by Best (1992), Engstrom et al. (1987), Goldman (1911), and Schmidt et al. (1989). This species exhibits little geographic variation, and populations are chromosomally monomorphic (Engstrom et al., 1987). Rogers (1990) found little genic variation.

Heteromys oresterus

Diagnosis.—Size very large for genus; head and body greater than 140 mm; tail length greater than or equal to length of head and body; dorsum blackish gray, grizzled with ochraceous or buffy hairs; sides more buffy than the dorsum; venter and feet white; forelimbs more buffy than dorsum; hind limbs above and ankles dark gray; ears blackish with white edgings; tail black above and white below except for black base and white tip; premaxillae terminating posterior to nasals; palate narrow; 2n = 60, FN = 78.

Comparison.—See account of *H. nelsoni* for a comparison with that species. Only the range of three subspecies of *H. desmarestianus* (*planifrons, subaffinis,* and *underwoodi*) are near the distribution of this species. From *H. desmarestianus, H. oresterus* may be identified by its larger body size; the soft, instead of stiff, spines or bristles of the pelage; and the overall greater dimensions of the skull.

Distribution.—Known only from the provinces of Cartago and San José in central Costa Rica at elevations from 1,800 m to 2,650 m. It appears to be restricted to cloud forest in the western portion of the Talamanca Range (Rogers and Rogers, 1992*a*).

Remarks.—This is a monotypic species. Mascarello and Rogers (1988) and Rogers (1989) described the chromosomal variation. Rogers (1990) recommended that *H. oresterus* be removed from the subgenus *Xylomys* and instead be included in the *desmarestianus* species group (*desmarestianus* and *oresterus*) within the subgenus *Heteromys*. External and cranial measurement for this taxon were given by Harris (1932), Best (1992), and Rogers and Rogers (1992*a*).

Heteromys oresterus Harris, 1932

1932. *Heteromys oresterus* Harris, Occas. Papers Mus. Zool., Univ. Michigan, 248:4, 4 August.

Holotype.—Adult male, skin and skull, UMMZ 64027, from El Copey de Dota, Cordillera de Talamanca, 6,000 ft, Costa Rica; obtained on 25 May 1931 by Austin Smith.

Measurements of holotype.—Total length, 357; length of tail, 174; length of hind foot, 40; length of ear, 16; greatest length of skull, 39.6; zygomatic breadth, 16.8; interorbital breadth, 9.6; length of nasals, 16.2; breadth of braincase, 15.7; length of maxillary toothrow, 5.2 (Harris, 1932).

Remarks.—This species is restricted to cloud forest habitats (Rogers and Rogers, 1992*a*). The type locality, which is about 25 miles south of Cartago, is on "the Pacific watershed on the boundary between the subtropical and temperate zones" (Harris, 1932:1). Rogers and Rogers (1992*a*) reviewed the biology of this species.

Subgenus *Xylomys*

1902. *Xylomys* Merriam, Proc. Biol. Soc. Washington, 15:43, 5 March.

Type species.—*Heteromys nelsoni* Merriam, 1902, Proc. Biol. Soc. Washington, 15:43, 5 March.

Diagnosis.—Size large for genus; posterior borders of nasals equal to posterior margin of premaxillae; braincase arched, not flat; palate narrow; rostrum not tapering anteriorly; posterior molars equal to or broader than premolars; more intricate enamel folds in posterior upper molars; posterior cingulum on last upper molar as well developed as either of the other two lophs, joined at center with metaloph; parietals extend laterally along lambdoidal crest, nearly reaching the mastoids; $2n = 42$, $FN = 72$.

Remarks.—*Xylomys* currently is considered to be monotypic. Rogers (1989, 1990), based on chromosomal and genic data, found that the two traditional species (*H. oresterus, H. nelsoni*) of *Xylomys* do not form a natural group. He recommended that *H. oresterus* be included within the *desmarestianus* species group; *H. nelsoni* may be allied to this group, although the affinities are not clear. Thus, the subgeneric ranking of *Xylomys* is doubtful.

Species Account

Heteromys nelsoni

Diagnosis.—Size largest of the genus; external and cranial measurements the largest for the genus; head and body length greater than 140 mm; tail longer than length of head and body; dorsum mouse gray, darker along the mid-dorsum; lateral line absent; inner sides of fore- and hind limbs dusky; venter white; ears large, blackish, without white edge; tail dusky above, whitish beneath except tip which is dark all around; pelage harsh, but bristles soft (also see subgeneric diagnosis).

Comparisons.—*Heteromys nelsoni* differs from *H. oresterus* in overall greater size; the ears of *nelsoni* lack the white edge found in *oresterus*; the nasals and premaxillae end at the same level in *nelsoni* rather than having the premaxillae longer than the nasals in *oresterus*; the tail of *nelsoni* is dark-tipped, whereas it is white-tipped in *oresterus*. Only *H. d. goldmani* occurs near the range of *H.*

nelsoni; *H. nelsoni* occurs at higher elevations, is larger in overall size, has soft, instead of stiff, bristles, and has wider posterior molars.

Distribution.—Known only from southern Chiapas and western Guatemala. Specimens have been collected in cloud forest at elevations ranging from 2,500 m to 2,800 m, near Cerro Mozotol in southeastern Chiapas, México, and from Volcán Tajumulco in western Guatemala (Rogers and Rogers, 1992*b*). The habitat of the type locality was described by Goldman (1951).

Remarks.—This is a monotypic species known only from four localities (Rogers and Rogers, 1992*b*). The habitat of the type locality was characterized as humid, heavily forested Pacific slopes of the Sierra Madre (Goldman, 1951).

Heteromys nelsoni Merriam, 1902

1902. *Heteromys nelsoni* Merriam, Proc. Biol. Soc. Washington, 15:43, 5 March.

Holotype.—Adult male, skin and skull, USNM 77920, from Pinabete, 8,200 ft, Chiapas; obtained on 11 February 1896 by Edward W. Nelson and Edward A. Goldman.

Measurements of holotype.—Total length, 358; length of tail, 195; length of hind foot, 43.5; greatest length of skull, 40.6; interorbital breadth, 9.3; mastoid breadth, 15.8; length of nasals, 15.8; length of maxillary toothrow, 6.0; depth of braincase, 10.1.

Remarks.—Goldman (1911) suggested that this species represented a survivor of an ancestral group with more complicated dentition. External and cranial measurements for this taxon were given by Best (1992) and Rogers and Rogers (1992*b*).

Genus *Liomys*

1902. *Liomys* Merriam, Proc. Biol. Soc. Washington, 14:44, 5 March.

Type species.—*Heteromys alleni* Coues, 1881, *in* Allen, Bull. Mus. Comp. Zool., 8:187, March.

Diagnosis.—Cheek teeth with medium-high crowns; upper incisors asulcate; only two lophids on lower premolar; accessory enamel island on molars present only for short period (visible only in unworn molars); entostyle closely united to hypocone so that Y-shape of median valley of upper premolar is poorly formed; auditory region uninflated; pelage hispid, consisting of stiff spines mingled with slender soft hairs; soles of hind feet sparsely haired; interpterygoid fossa U-shaped anteriorly; claw of second digit on hind foot spoon-like; adapted for scampering.

Remarks.—Although *Liomys* is a common inhabitant of México and Central America, Gray (1868) did not describe the first representative of the genus until 1868. The first species were described in the genus *Heteromys* until Merriam (1902) described the genus *Liomys* just after the turn of the 20th century. Goldman (1911) reviewed both *Liomys* and *Heteromys*. Genoways (1973) was the last author to review the genus *Liomys* and his arrangement is generally followed in these accounts. Systematic relationships among *Liomys* species, based on genic data, were investigated by Rogers (1990).

Key to the Species

1. Five planter tubercles; pterygoid bones with broad wings; shaft of baculum oval to tip; glans penis long (more than 75%) in comparison with baculum; FN of chromosomes 60; upper parts greyish brown, lateral stripe pale pinkish to buffy; occurring on Mexican Plateau and in adjacent areas of northern and central México, south as far as south-central Oaxaca . . . . *Liomys irroratus*

1′. Usually six plantar tubercles; pterygoid bones with narrow wings; shaft of baculum either flattened dorsoventrally or flattened dorsoventrally and compressed laterally at some point; glans penis short (less than 75%) in comparison with baculum; FN of chromosomes 48 or 56; upper parts either reddish brown or chocolate brown, or somewhat paler, with either an ochraceous lateral stripe or lateral stripe absent; not occurring on the Mexican Plateau 2

2. Upper parts reddish brown with an ochraceous lateral stripe; interorbital region broad in comparison with greatest length of skull; distal end of the shaft of the baculum with a laterally compressed ventral keel and just posterior to this region, the shaft is flattened dorsoventrally; tip of glans penis long when compared with its total length; FN of chromosomes 48; hairs on back not curled upward and not visible above spines; occurring along the Pacific slope of western México as far south as the vicinity of Tonalá, Chiapas, in the central valley of Chiapas, and in the southern half of Veracruz 3

2′. Upper parts chocolate brown to somewhat paler, lateral stripe absent; interorbital region narrow in comparison with greatest length of skull; shaft of baculum dorsoventrally flattened just posterior to slightly upturned tip, no laterally compressed ventral keel present; tip of glans penis short when compared with its total length; FN of chromosomes 56; hairs on back curled upward and visible above the spines; occurring in Central America and into México along the Pacific coast as far as the vicinity of Reforma, Oaxaca 4

3. In southeastern Jalisco, size small (greatest length of skull, 28.9 to 32.0 mm; specimens approaching *L. spectabilis* in size occur in Guerrero and Oaxaca but these are still slightly smaller and have proportionally deeper braincases); in Jalisco hind foot rarely more than 30; laterally compressed ventral keel on baculum short, 0.85 to 1.25 mm; FN of chromosomes 66; occurring along Pacific coast of western México, in the central valley of Chiapas, and in southern Veracruz... .... *Liomys pictus*

3′. In southeastern Jalisco, size large (greatest length of skull, 33.0 to 35.3 mm); hind foot rarely less than 30 mm; laterally compressed ventral keel on baculum relatively long, 1.30 mm; FN of chromosomes 64; occurring only in southeastern Jalisco .... *Liomys spectabilis*

4. Size small (greatest length of skull averaging less than 33.5 mm); FN of chromosomes 86; occurring from southern Oaxaca to central Costa Rica .... *Liomys salvini*

4′. Size large (greatest length of skull averaging over 34.5 mm); FN of chromosomes 84 or less; occurring only in central Panamá .... *Liomys adspersus*

Species Accounts

Liomys adspersus

Diagnosis.—External and cranial measurements large; premolars similar in structure to those of *L. salvini*; baculum with large rounded base, shaft oval to a point just posterior to the slightly upturned tip where it is dorsoventrally flattened; glans penis medium-sized in comparison with length of baculum, tip of glans short, glans highly sculptured and with deeply incised ventral folds; 2n = 56; FN probably 84; wings of pterygoids narrow; six plantar tubercles; upper parts usually chocolate brown (some paler individuals may show grayish tones); no lateral stripe; hairs on back curled upward so as to be conspicuous above spines.

Comparisons.—Specimens of *Liomys adspersus* can easily be distinguished from specimens of *L. salvini* by their much larger external and cranial measurements. In both species 2n = 56, but FN = 86 in *salvini* and FN = 84 in *adspersus.*

Distribution.—Central Panamá principally on the Pacific versant.

Remarks.—*Liomys adspersus* is monotypic. This species is not sympatric with any other member of the genus; the distribution of *Liomys salvini* is geographically nearest, with the ranges of the two species separated by a distance of approximately 300 km. These two taxa have been shown to be closely related morphologically, karyotypically (Genoways, 1973), and genically (Rogers, 1990).

Liomys adspersus (Peters, 1874)

1874. *Heteromys adspersus* Peters, Monatsb. preuss. Akad. Wiss., Berlin, p. 357, May.

Holotype.—Young adult male, mounted skin with skull, unknown number in Berlin Museum, from Panamá. Type locality restricted to City of Panamá by Goldman (1920).

Measurements of holotype.—Total length, 240; length of tail, 95; length of hind foot, 30.

Remarks.—Genoways (1973) found little variation among populations of this species.

Liomys irroratus

Diagnosis.—A medium-sized spiny pocket mouse with the cranium relatively broad in comparison with length; protoloph of upper permanent premolar composed of three discernible cusps; metaloph of upper premolar with three cusps (hypocone largest, metacone only slightly smaller than hypocone); entostyle distinct but not widely separated from hypocone; re-entrant angle on labial side of lower premolar not united with median valley; baculum simple, with large rounded base, oval-shaped shaft, and slightly upturned tip (extreme tip of baculum may be slightly laterally compressed); glans penis relatively long in comparison with length of baculum; 2n = 60, FN = 62; wings of pterygoids relatively broad; five plantar tubercles; upper parts grayish brown; lateral stripe, which is usually present, generally pale pink to buff.

Comparisons.—Specimens of *Liomys irroratus* are most likely to be confused with *L. pictus* or *L. spectabilis.* In areas of potential sympatry, *L. irroratus* is larger than *L. pictus,* but smaller than *L. spectabilis.* There are five plantar tubercles in *L. irroratus,* whereas the other species have six (except one population of *pictus*). The karyotype of *L. irroratus* is composed of 60 chromosomes as opposed to 48 in *L. pictus* and *L. spectabilis.* The baculum of *irroratus* has an ovoid shaft to the slightly upturned tip, whereas the baculum of *pictus* and *spectabilis* has a terminal ventral keel that is laterally compressed and a section of the shaft that has been dorsoventrally flattened posterior to the keel. The wings of the pterygoid bones are broad rather than narrow in *irroratus.* The upper parts of *irroratus* are grayish brown with a pale pinkish to buffy lateral stripe as opposed to upper parts reddish brown with an ochraceous lateral stripe in *pictus* and *spectabilis.*

Distribution.—The range of *Liomys irroratus* includes the Mexican Plateau and adjacent areas from southern Chihuahua in the northwest and southern Texas in the northeast to south-central Oaxaca in the south. It generally occurs to the east of the Sierra Madre Occidental, but also is found in various parts of the Transverse Volcanic Belt.

Remarks.—Hooper and Handley (1948) presented a synopsis of subspecies of *irroratus* and analyzed the trends in geographic variation within the species. Prior to the study by Genoways (1973), three species (*bulleri, guerrerensis,* and *irroratus*) were recognized in the *irroratus* group by Goldman (1911). However, Genoways (1973) reduced *bulleri* and *guerrerensis* to subspecific status while recognizing seven subspecies in the remaining species *irroratus.* The relationship of *irroratus* with the other species of *Liomys* has been little clarified by morphologic, karyotypic, or genic data (Genoways, 1973; Rogers, 1990). *Liomys irroratus* and *L. pictus* occur sympatrically in a zone from central Jalisco southward through Michoacán to the region of the Balsas Basin and Sierra Madre del Sur of Guerrero and the Sierra Madre of Oaxaca. *Liomys irroratus* is potentially sympatric with *L. spectabilis* in southeastern Jalisco, but the two species have not been reported from the same locality. The biology of this species was reviewed by Dowler and Genoways (1978).

Liomys irroratus alleni (Coues, 1881)

1881. *Heteromys alleni* Coues, *in* Allen, Bull. Mus. Comp. Zool., 8:187, March.

1902. *Liomys canus* Merriam, Proc. Biol. Soc. Washington, 15:44, 5 March.

1947. *Liomys irroratus pullus* Hooper, J. Mamm., 28:47, 17 February.

1948. *Liomys irroratus acutus* Hall and Villa-

R., Univ. Kansas Publ., Mus. Nat. Hist., 1:253, 26 July.

Holotype.—Probably an adult, sex unknown, skull in skin, MCZ 5889, from Hacienda Angostura, Río Verde, San Luis Potosí; obtained on 26 February 1878 by Edward Palmer.

Measurements of holotype.—No measurements are available.

Distribution.—Extensive geographic area on the Mexican Plateau north of the Transverse Volcanic Belt; also occurring in Sierra Madre Oriental of Tamaulipas and Nuevo León.

Remarks.—Populations of *L. i. alleni* are characterized by large size and a relatively high percentage of individuals with the interparietal bone divided in half and the posterior termination of nasals truncate. *L. i. alleni* intergrades with *L. i. jaliscensis* in an area from near Lago de Chapala in Jalisco and Michoacán northward through Jalisco into extreme southern Zacatecas.

Liomys irroratus bulleri (Thomas, 1893)

1893. *Heteromys bulleri* Thomas, Ann. Mag. Nat. Hist., ser. 6, 11:330, April.

Holotype.—Adult female, skin in alcohol with skull removed, BM(NH) 93.3.6.39, from La Laguna, Sierra de Juanacatlán, Jalisco; obtained in December 1892 by A. C. Buller.

Measurements of holotype.—Greatest length of skull, 34.1; zygomatic breadth, 16.7; interorbital constriction, 8.6; mastoid breadth, 15.5; length of nasals, 13.4; length of rostrum, 15.3; length of maxillary toothrow, 5.9; depth of braincase, 10.0.

Distribution.—Known only from the mountains of west-central Jalisco.

Remarks.—Until the work of Genoways (1973), *bulleri* was considered to be a monotypic species, but was placed as a subspecies of *irroratus* in this study. *L. i. bulleri* is a large-sized subspecies characterized by a small, subtriangular interparietal bone. *L. i. bulleri* is potentially in contact with *L. i. jaliscensis,* although no direct intergrades have been found.

Liomys irroratus guerrerensis Goldman, 1911

1911. *Liomys guerrerensis* Goldman, N. Amer. Fauna, 34:62, 7 September.

Holotype.—Subadult female, skin and skull, USNM 127523, from Omilteme, Guerrero; obtained 17 May 1903 by Edward W. Nelson and Edward A. Goldman.

Measurements of holotype.—Total length, 255; length of tail, 127; length of hind foot, 34; greatest length of skull, 33.3; interorbital constriction, 8.6; mastoid breadth, 15.8; length of nasals, 12.5; length of rostrum, 14.9; length of maxillary toothrow, 9.2.

Distribution.—Pacific slope of the Sierra Madre del Sur of Guerrero.

Remarks.—Until the study of Genoways (1973), *guerrerensis* was considered to be a monotypic species of the *irroratus*-group. Genoways (1973) found specimens from Chilpancingo, Guerrero, that were intermediate between *guerrerensis* and *torridus. L. i. guerrerensis* can be distinguished from all other subspecies of *irroratus* by larger size and darker upper parts.

Liomys irroratus irroratus (Gray, 1868)

1868. *Heteromys irroratus* Gray, Proc. Zool. Soc. London, p. 205, May.
1868. *Heteromys albolimbatus* Gray, Proc. Zool. Soc. London, p. 205, May.
1956. *Liomys irroratus yautepecus* Goodwin, Amer. Mus. Novit., 1757:7, 8 March.

Holotype.—Subadult of unknown sex, skin and skull, BM(NH) 59.7.10.2, from Oaxaca; obtained on an unknown date by A. Sallé. Type locality restricted to Oaxaca, Oaxaca, by Genoways (1973).

Measurements of holotype.—Interobital constriction, 9.1; length of nasals, 12.5; length of rostrum, 15.2; length of maxillary toothrow, 6.3; mastoid breadth, 15.2.

Distribution.—Confined to central and south-central Oaxaca.

Remarks.—This large-sized subspecies was the first member of the genus to be described (Gray, 1868). Besides large size, this subspecies has a large percentage of individuals with the posterior margin of the nasals emarginate.

Liomys irroratus jaliscensis (J. A. Allen, 1906)

1906. *Heteromys jaliscensis* J. A. Allen, Bull. Amer. Mus. Nat. Hist., 22:251, 25 July.

Holotype.—Adult male, skin and skull, AMNH 26325, from Las Canoas, 7,000 ft, Jalisco; obtained on 6 August 1905 by J. H. Batty.

Measurements of holotype.—Greatest length of skull, 31.6; interorbital constriction, 8.0; mastoid breadth, 14.1; length of nasals, 11.6; length of rostrum, 13.4; length of maxillary toothrow, 5.2; depth of braincase, 9.2.

Distribution.—Known from southern, central northern Jalisco and extreme southern Zacatecas and Nayarit.

Remarks.—*L. i. jaliscensis* is a medium-sized subspecies that is located geographically between two larger-sized subspecies, *alleni* and *bulleri.*

Liomys irroratus texensis Merriam, 1902

1902. *Liomys texensis* Merriam, Proc. Biol. Soc. Washington, 15:44, 5 March.
1911. *Liomys irroratus pretiosus* Goldman, N. Amer. Fauna, 34:58, 7 September.

Holotype.—Adult female, skin and skull, USNM 58670, from Brownsville, Cameron Co., Texas; obtained on 19 February 1894 by J. Alden Loring.

Measurements of holotype.—Total length, 231; length of tail, 114; length of hind foot, 30; greatest length of skull, 31.0; zygomatic breadth, 14.5; interorbital constriction, 8.0; mastoid breadth, 14.1; length of nasals, 12.8; length of maxillary toothrow, 5.2; depth of braincase, 8.6.

Distribution.—Gulf coastal lowlands from southern Texas to northern Puebla and central Veracruz.

Remarks.—This small- to medium-sized subspecies intergrades with *L. i. alleni* for a considerable distance along the eastern foothills of the Sierra Madre Oriental and the eastern edge of the Mexican Plateau. Some geographic variation within the subspecies *texensis* was noted by Genoways (1973). Specimens from Texas and northern Tamaulipas averaged larger than other samples. The smallest specimens of *texensis,* on the average, were from the vicinity of Ebano, San Luis Potosí.

Liomys irroratus torridus Merriam, 1902

1902. *Liomys torridus* Merriam, Proc. Biol. Soc. Washington, 15:45, 5 March.
1902. *Liomys torridus minor* Merriam, Proc. Biol. Soc. Washington, 15:45, 5 March.
1903. *Heteromys exiguus* Elliot, Field Columb. Mus., Zool. Ser., 3:146, 20 March.

Holotype.—Adult female, skin and skull, USNM 69645, from Cuicatlán, Oaxaca; obtained on 14 October 1894 by Edward W. Nelson and Edward A. Goldman.

Measurements of holotype.—Total length, 242; length of tail, 134; length of hind foot, 28; greatest length of skull, 30.3; zygomatic breadth, 14.2; interorbital constriction, 7.9; mastoid breadth, 14.0; length of nasals, 11.6; length of rostrum, 13.6; length of maxillary toothrow, 4.5; depth of braincase, 8.9.

Distribution.—South of the Transverse Volcanic Belt in the Mexican states of Puebla, Morelos, Guerrero, and northern Oaxaca.

Remarks.—The subspecies *torridus* is

composed of individuals of medium to small size for the species. The subspecies *torridus* appears to intergrade with the subspecies *irroratus* in central Guerrero and with the subspecies *alleni* along the southern slope of the Transverse Volcanic Belt in Morelos and Puebla, and in the lower country of eastern Puebla.

Liomys pictus

Diagnosis.—External and cranial measurements medium to small for the genus, although some populations (*annectens*) are relatively large in size; cranium relatively narrow in comparison with length; protoloph of upper premolar generally appearing to be composed of a single cusp; three cusps of metaloph connected by loph so as not to form discrete cones; hypocone largest cusp on metaloph; entostyle always connected to hypocone by loph; re-entrant angle on labial margin of lower premolar not reaching median valley; baculum long and with a small rounded base, distal end of shaft with ventral keel that is laterally compressed and the shaft dorsoventrally compressed posterior to terminal keel; 2n = 48; FN = 66; wings of pterygoids narrow; six plantar tubercles on most specimens, although some individuals of *L. p. planitarensis* have only five; upper parts reddish brown; lateral stripe generally ochraceous, but may be rather pale.

Comparisons.—*Liomys pictus* is generally smaller than *L. spectabilis* and this is especially true in sympatric populations. The hind foot of *pictus* is rarely more than 30 mm, but rarely less than 30 mm in *spectabilis*. In *pictus*, the FN = 66, whereas in *spectabilis* FN = 64.

Distribution.—West coast and adjacent slopes of Sierra Madre Occidental and Sierra Madre del Sur from near Nogales, Sonora, to Tonalá, Chiapas; central valley of Chiapas into adjacent Guatemala; across the Isthmus of Tehuantepec into southern Veracruz on the east coast of México.

Remarks.—Genoways (1973) reduced *L. annectens* to a subspecies of *L. pictus*. The species described by Goodwin (1956), *L. pinetorum*, was placed as a junior synonym of *L. p. pictus*. Genic data indicate that this species is paraphyletic (Morales and Engstrom, 1989; Rogers, 1990). The closest relative of *L. pictus* in the genus is *L. spectabilis* (Genoways, 1973; Rogers, 1990). These two species occur sympatrically in southeastern Jalisco at three localities. *L. pictus* is also sympatric with *L. salvini* in coastal Oaxaca and Chiapas and with *L. irroratus* in a narrow band from southern Jalisco through Michoacán and Guerrero to Oaxaca. This is the only member of the genus that has been reported as occurring sympatrically with more than one other species of the genus. The biology of this species was reviewed by McGhee and Genoways (1978).

Liomys pictus annectens (Merriam, 1902)

1902. *Heteromys annectens* Merriam, Proc. Biol. Soc. Washington, 15:43, 5 March.

Holotype.—Adult male, skin and skull, USNM 71510, from Pluma Hidalgo, Oaxaca; obtained on 18 March 1895 by Edward W. Nelson and Edward A. Goldman.

Measurements of holotype.—Total length, 300; length of tail, 165; length of hind foot, 33; greatest length of skull, 34.3; zygomatic breadth; 16.1; interorbital constriction, 7.9; mastoid breadth, 14.7; length of nasals, 14.7; length of maxillary toothrow, 5.3; depth of braincase, 8.9.

Distribution.—High elevations in the Sierra Madre del Sur of Guerrero and Oaxaca.

Remarks.—Until the study by Genoways (1973), this taxon was considered a species distinct from *L. pictus*. However, there is clear evidence for intergradation between the taxa in the vicinity of Candelaria, Oaxaca, along the lower slope of the Sierra Madre del Sur. *L. p. annectens* is distinguished from other subspecies by larger size and darker color dorsally with a darker orange lateral stripe.

Liomys pictus hispidus (J. A. Allen, 1897)

1897. *Heteromys hispidus* J. A. Allen, Bull. Amer. Mus. Hist., 9:56, 15 March.
1902. *Liomys sonorana* Merriam, Proc. Biol. Soc. Washington, 15:47, 5 March.
1906. *Heteromys pictus escuinapae* J. A. Allen, Bull. Amer. Mus. Nat. Hist., 22:221, 25 July.

Holotype.—Subadult female, skin and skull, AMNH 8333/6667, from Rancho El Colomo, Compostela, Nayarit; obtained on 11 February 1893 by A. C. Buller.

Measurements of holotype.—Greatest length of skull, 30.5; interorbital constriction, 7.1; mastoid breadth, 13.9; length of nasals, 12:3; length of rostrum, 13.4; length of maxillary toothrow, 5.0.

Distribution.—Northwestern México from near Nogales, Sonora, southward in coastal Sinaloa and northern Nayarit and then inland as far as central Jalisco.

Remarks.—The subspecific name *hispidus* applies to populations along the northwestern coast of México that vary in a clinal fashion from medium-sized individuals in Sonora and northern Sinaloa to relatively small individuals in northern Nayarit. In the vicinity of San Blas, Nayarit, *hispidus* intergrades with *pictus* from coastal areas to the south. Specimens from the vicinity of Tepic, Nayarit, and southeastward through Nayarit to just south of Guadalajara, Jalisco, are intermediate in size between *pictus* to the west and *hispidus* to the north, but possess cranial characters that ally them with *hispidus.*

Liomys pictus pictus (Thomas, 1893)

1893. *Heteromys pictus* Thomas, Ann. Mag. Nat. Hist., ser. 6, 12:233, September.
1902. *Liomys pictus rostratus* Merriam, Proc. Biol. Soc. Washington, 15:46, 5 March.
1902. *Liomys pictus isthmius* Merriam, Proc. Biol. Soc. Washington, 15:46, 5 March.
1902. *Liomys veraecrucis* Merriam, Proc. Biol. Soc. Washington, 15:47, 5 March.
1902. *Liomys obscurus* Merriam, Proc. Biol. Soc. Washington, 15:48, 5 March.
1902. *Liomys phaeura* Merriam, Proc. Biol. Soc. Washington, 15:48, 5 March.
1902. *Liomys orbitalis* Merriam, Proc. Biol. Soc. Washington, 15:48, 5 March.
1903. *Heteromys paralius* Elliot, Field Columb. Mus., Zool. Ser., 3:233, 3 September.
1956. *Liomys pinetorum* Goodwin, Amer. Mus. Novit., 1791:2, 28 September.

Holotype.—Adult female, skin and skull, BM(NH) 93.8.12.2, from San Sebastian, 4,300 ft, Jalisco; obtained on 9 May 1893 by A. C. Buller.

Measurements of holotype.—Zygomatic breadth, 15.1; interorbital constriction, 7.6; mastoid breadth, 13.7; length of maxillary toothrow, 5.2; depth of braincase, 8.5.

Distribution.—Coastal western México from Santiago, Nayarit, southward through Jalisco, Colima, Michoacán, Guerrero, Oaxaca to Chiapas; central valley of Chiapas and into and adjacent Guatemala; across the Isthmus of Tehuantepec into southern Veracruz.

Remarks.—*L. p. pictus* is characterized by a medium size, undivided interparietal bone, and a premaxillary bone terminating posterior to the nasal bones. Goodwin (1956) described *L. pinetorum* based upon a single specimen from near Tonalá, Chiapas. Genoways (1973) showed that all characters of this specimen fell within the range of variation of *L. p. pictus* and placed *pinetorum* as a junior synonym of *pictus.* Genoways (1973) described the intergradation of *pictus* and *plantinarensis* in Colima. The extent of morphological differentiation between *pictus* and *plantinarensis* in Colima and Jalisco was assessed by Morales and Engstrom (1989). They (Morales and Engstrom, 1989: 2) suggested that, based on morphological variation, "genetic exchange between the two subspecies is limited and that introgression is minimal." In this area of contact, *L. p. pictus* has an interparietal length greater than 4 mm, a hindfoot length greater than 28 mm, and is found in more mesic areas.

Liomys pictus plantinarensis Merriam, 1902

1902. *Liomys plantinarensis* Merriam, Proc. Biol. Soc. Washington, 15:46, 5 March.
1904. *Liomys parviceps* Goldman, Proc. Biol. Soc. Washington, 17:82, 21 March.

Holotype.—Adult female, skin and skull, USNM 45630/33595, from Platanar, Jalisco; obtained on 4 April 1892 by Edward W. Nelson and Edward A. Goldman.

Measurements of holotype.—Total length, 202; length of tail, 102; length of hind foot, 26; greatest length of skull, 29.4; interorbital constriction, 7.4; mastoid breadth, 13.0; length of nasals, 11.8; length of maxillary toothrow, 5.0.

Distribution.—Interior drainage basins and river systems of southeastern Jalisco, eastern Colima, central Michoacán, and northern Guerrero.

Remarks.—Members of the subspecies *L. p. plantinarensis* are characterized by small size, high percentage of individuals with the posterior border of the interparietal notched, and a low percentage of individuals with the termination of the nasal bones truncate. The number of plantar tubercles is usually constant in species of *Liomys,* but the development of the sixth tubercle is highly variable in northern populations of *L. p. plantinarensis.* Morales and Engstrom (1989) found this taxon to be morphologically distinct from *L. p. pictus* and a monophyletic lineage within the *pictus* species-group. They suggested that specific status may be warranted. Similar results were found by Rogers (1990) using genic data. In the area of contact, *L. p. plantinarensis* is smaller and inhabits drier areas, whereas *L. p. pictus* is larger and is found in more mesic areas (Morales and Engstrom, 1989).

Liomys salvini

Diagnosis.—External and cranial measurements small for the genus, tail being especially short; protoloph of permanent upper premolar appears to be composed of one cusp, metaloph composed of three and sometimes four cusps, metacone of metaloph sometimes larger than hypocone, entostyle distinctly separated from other cusps of metaloph; re-entrant angle on labial margin of lower premolar connected with median valley; baculum with large rounded base, shaft oval to point just posterior to the slightly upturned tip where it is dorsoventrally flattened; 2n = 56; FN = 86; wings of pterygoids narrow; six planter tubercles; upper parts grayish brown to deep chocolate brown; no lateral stripe; hairs on back curled so as to be conspicuous above spines.

Comparisons.—*Liomys salvini* can be distinguished from *L. pictus* by its smaller overall size. The entostyle on the upper premolar is more distinctly separated from other cusps in *salvini* than *pictus* and the reentrant angle on the labial margin of the lower premolar reaches the median valley in specimens of *salvini* but not *pictus.* The baculum of *salvini* lacks the laterally compressed distal keel found in *pictus.* The 2n of *L. salvini* is 56, whereas that of *L. pictus* is 48. The upper parts of *salvini* are chocolate brown or paler and lack the lateral stripe whereas in *pictus* the upper parts are reddish brown with an ochraceous lateral stripe. The hairs on the back of specimens of *salvini* are curled upward so that they are visible above the spines, but not in specimens of *pictus.*

Distribution.—Occurs along the Pacific coast and adjacent mountain slopes from Reforma, Oaxaca, southward through Chiapas, Guatemala, Honduras, El Salvador, Nicaragua to Monte Rey, Costa Rica; also in the Caribbean drainage of central and eastern Guatemala.

Remarks.—Before the revision by Genoways (1973), five species were recognized in the *crispus*-group. Genoways (1973) recognized two species—*salvini* and *adspersus*—in a group that was renamed the *salvini*-group. He reduced *crispus* to a subspecies of *L. salvini* and placed *L. anthonyi* and *L. heterothrix* as a junior synonym of *L. s. salvini*. Goodwin (1946) had earlier placed *L. vulcani* as a subspecies of *L. salvini,* which Genoways continued to follow. Morphologic, genic, and karyotypic data indicate that *L. salvini* is most closely related to *L. adspersus* (Genoways, 1973; Rogers, 1990). *L. salvini* occurs sympatrically only with *L. pictus* among the four other species in the genus. Carter and Genoways (1978) reviewed the biology of this species.

Liomys salvini crispus Merriam, 1902

1902. *Liomys crispus* Merriam, Proc. Biol. Soc. Washington, 15:49, 5 March.
1902. *Liomys crispus setosus* Merriam, Proc. Biol. Soc. Washington, 15:49, 5 March.

Holotype.—Adult male, skin and skull, USNM 75105, from Tonalá, Chiapas; obtained on 7 August 1895 by Edward W. Nelson and Edward A. Goldman.

Measurements of holotype.—Total length, 210; length of tail, 99; length of hind foot, 27.5; greatest length of skull, 31.8; interorbital constriction, 6.4; mastoid breadth, 13.8; length of nasals, 12.2; length of rostrum, 13.5; length of maxillary toothrow, 5.0; depth of braincase, 9.0.

Distribution.—Coastal areas of southeastern Oaxaca, Chiapas, and southwestern Guatemala from Reforma, Oaxaca, in the north to vicinity of Mazatenago, Guatemala, in the south.

Remarks.—The taxon *crispus* was considered to be a distinct species by Goldman (1911) and subsequent authors until the work of Genoways (1973). He considered it to be a subspecies of *salvini* characterized by short total length and length of tail, high percentage of individuals with the posterior margin of the interparietal bone deeply notched, and no individuals with the interparietals divided. The subspecies *crispus* and *salvini* appear to come into contact along the slopes of the Guatemalan highlands on the southern coast of Guatemala.

Liomys salvini salvini (Thomas, 1893)

1893. *Heteromys salvini* Thomas, Ann. Mag. Nat. Hist., ser. 6, 11:331, April.
1893. *Heteromys salvini nigrescens* Thomas, Ann. Mag. Nat. Hist., ser. 6, 12:234, September.
1902. *Liomys heterothrix* Merriam, Proc. Biol. Soc. Washington, 15:50, 5 March.
1932. *Liomys anthonyi* Goodwin, Amer. Mus. Novit., 528:2, 23 May.
1938. *Liomys salvini aterrimus* Goodwin, Amer. Mus. Novit., 987:4, 13 May.

Holotype.—Adult male, skin and skull, BM(NH) 75.2.27.35, from Dueñas, Sacatépequez, Guatemala; obtained on 31 July 1873 by Osbert Salvin.

Measurements of holotype.—Greatest length of skull, 32.8; zygomatic breadth, 15.0; interorbital constriction, 7.3; mastoid breadth, 14.1; length of nasals, 12.5; length of rostrum, 14.2; length of maxillary toothrow, 5.4; depth of braincase, 9.3.

Distribution.—Guatemala (along the southern coast, in the highlands around Guatemala City, and in the valleys of the Río Negro and Río Motagua as far as San Pedro Sula, Honduras), most of El Salvador, south-central Honduras, north-central and central Nicaragua, Isla de Ometepe in Lake Nicaragua, extreme southwestern Nicaragua, and western and central Costa Rica.

Remarks.—*L. s. salvini* is characterized by large external and cranial measurements, at least some individuals in each population with the interparietal bone divided, and a low percentage of individuals having the posterior margin of the interparietal deeply notched. Genoways (1973) reduced two recognized species (*anthonyi* Goodwin, 1932, and *heterothrix* Merriam, 1902) and two recognized subspecies (*nigrescens* Thomas, 1893, *aterrimus* Goodwin, 1938) to junior synonyms of *salvini.*

Liomys salvini vulcani
(J. A. Allen, 1908)

1908. *Heteromys vulcani* J. A. Allen, Bull. Amer. Nat. Hist., 24:652, 13 October.

Holotype.—Adult female, skin and skull, AMNH 28315, from Volcán de Chinandega, about 4,000 ft, Chinandega, Nicaragua; obtained on 7 May 1909 by William B. Richardson.

Measurements of holotype.—Total length, 220; length of tail, 110; length of hind foot, 25; interorbital constriction, 6.8; mastoid breadth, 13.5; length of nasals, 11.1; length of rostrum, 12.5; length of maxillary toothrow, 4.9.

Distribution.—Confined to western Nicaragua on the volcanoes that make up the Cordillera los Marrabios and the lowland to the west of them, west of Lake Managua, on the Meseta de los Pueblos west of Lake Nicaragua, and on Isla de Zapatera.

Remarks.—*L. s. vulcani,* which was recognized as a distinct species until the work of Goodwin (1946), is characterized by generally smaller size, especially cranially. Intergradation between *vulcani* and *salvini* was detected in specimens from 8 km N Las Maderas, Nicaragua, and the taxa also appear to be in contact between La Celera and San Pedro near the southern edge of the Meseta de los Pueblos of Nicaragua.

Liomys spectabilis

Diagnosis.—External and cranial measurements large; premolar resembles that of *L. pictus*; baculum long with a small rounded base; distal end of shaft with ventral keel that is laterally compressed; shaft dorsoventrally compressed posterior to terminal keel; 2n = 48; FN = 64; wings of pterygoids narrow; six plantar tubercles; upper parts reddish brown; lateral stripe ochraceous.

Comparisons.—See accounts for *L. irroratus* and *L. pictus.*

Distribution.—Confined to southeastern Jalisco.

Remarks.—*L. spectabilis* is most closely related to *L. pictus* as indicated by morphologic, genic, and karyotypic data (Genoways, 1973; Rogers, 1990). These two taxa occur sympatrically at three locations in southeastern Jalisco. It is potentially sympatric with *L. irroratus,* but the two have not been collected together. *L. spectabilis* is monotypic.

Liomys spectabilis Genoways, 1971

1971. *Liomys spectabilis* Genoways, Occas. Papers Mus. Nat. Hist., Univ. Kansas, 5:1, 18 June.

Holotype.—Adult male, skin and skull, KU 96051, from 2.2 mi NE Contla, 3,850 ft, Jalisco; obtained on 20 September 1963 by Percy L. Clifton.

Measurements of holotype.—Total length, 280; length of tail, 142; length of hind foot, 31; length of ear, 17; greatest length of skull, 35.1; zygomatic breadth, 16.3; interorbital constriction, 8.2; mastoid breadth, 15.1; length of nasals, 14.0; length of rostrum, 16.0; length of maxillary toothrow, 5.0; depth of braincase, 8.6.

Remarks.—Because only 21 specimens of this species are known, very little information is available on variation in the species (Genoways, 1973).

Subfamily Perognathinae Coues, 1875

Diagnosis.—Size small to medium, from 100 to 230 mm in total length and 5 to 60 g in mass; body form quadrupedal and generally mouselike; locomotion scansorial to subricochetal; hind limbs considerably larger than forelimbs; hind feet and legs longer than typical murids and cricetids; pes with five clawed digits; pes soles naked or partly clothed with a scant covering of short hairs; body hairs of two types—straight, relatively long overhairs and a thin underfur of short, often slightly curved hairs; upper incisors strongly grooved; molars progressively brachydont, rooted, and tuberculate; enamel pattern of cheek teeth persisting into late life of individual, but lost by wear in older individuals; lophs of P4 unite first at or near center of tooth; protoloph of P4 usually single-cusped; lophs of upper molars unite first at the lingual margin and progress to buccal margins; those of p4 unite at center of tooth, presenting an X-pattern; lophs of lower molars united first at buccal margin and progressively at center of tooth, sometimes forming an H-pattern; no foramina and usually no pit between m3 and the base of the coronoid process; palate broad, extending beyond level of M3; center of palate between cheek teeth not ridged; squamosal in broad contact with parietal on dorsal surface of skull; squamosal not perforated by auditory bulla; anterior zygomatic root not greatly enlarged on joining lachrymal; large orbital nonossification usually including ethmoid foramen; incisive foramen small; one prominent pterygoid fossa (the anterior fossa nearly invisible); auditory bullae greatly expanded, mastoid bullae appearing on dorsal surface of skull; inflated interior of mastoid and tympanic bullae filled with spongy trabeculae; no stapedius muscle in middle ear; no fusion of cervical vertebrae; in so far as is known, no median ventral foramina in the "central" of caudal vertebrae; scapula not prolonged posteriorly; obturator foramen not triangular; no articulation of trapezoid and scapholunar; astragalus articulating with cuboid; cuboid with long process between calcaneum and navicular, meeting astragalus; dorsal surface of ectocuneiform hour-glass-shaped; phallus with gently upcurved tip, and with or without external spines and urethral lappets; baculum relatively long and slender with a swollen base and an upturned tip; tip of baculum usually simple, but one species with an ornate, trifid tip; nasolabialis profundus pars maxillaris muscles of dual origin, from lateral zygoma and infraorbital foramen; M. cleidomastoideus present; extensores breves muscles present; M. abductor hallucis present; lumbricales muscles not reduced or absent (Burt, 1936; Hafner and Hafner, 1983; Homan and Genoways, 1978; Merriam, 1889; Osgood, 1900; Ryan, 1989; Wahlert, 1985; Webster and Webster, 1975; Wood, 1935).

Remarks.—This diagnosis was developed primarily from information about Recent species. As defined above, the subfamily Perognathinae includes the genera *Perognathus* and *Chaetodipus,* but not *Microdipodops* (see remarks in the account of Dipodomyinae). Wood (1935:88) amended the name, Perognathidinae, given to the subfamily by Coues (1875).

Key to the Genera

1. Sole of hind foot naked; pelage relatively coarse, often with stiff, spine-like bristles on rump; stiff, coarse hairs usually project across anterior margin of ear pinna; antitragus of ear pinna lobed; mastoid bulla usually not projecting posteriorly beyond plane of occiput; postero-medial border of mastoid bulla usually projecting as a distinct indentation into the supraoccipital; interparietal width equal to or greater than interorbital breadth .... *Chaetodipus*

1′. Posterior one-third to one-half of sole of hind foot with sparse covering of short hairs; pelage relatively fine and soft, never with stiff, spine-like bristles on rump; no long, stiff, coarse hairs projecting across anterior margin of ear pinna; antitragus of ear pinna usually not lobed; mastoid bulla projecting posteriorly beyond the plane of the occiput; postero-medial border of mastoid bulla not projecting as a distinct indentation into the supraoccipital; interparietal width nearly always less than interoribital breadth .... *Perognathus*

Genus *Chaetodipus*

1889. *Chaetodipus* Merriam, N. Amer. Fauna, 1:5, 25 October.

Type species.—*Perognathus* [*Chaetodipus*] *spinatus* Merriam, 1889, N. Amer. Fauna, 1:5, 25 October.

Diagnosis.—Size small to medium, total length from about 150 to 230 mm, weight from about 15 to 60 g; body form quadrupedal and scansorial with slight tendency to ricochetal; tail usually relatively long, length averaging greater than length of head and body (except *C. hispidus*); tail with distinct distal, dorsal crest and terminal tuft of hairs or without crest and terminal tuft; tail some shade of brown or buffy above, whitish below; sole of hind foot naked; ear pinna short, rounded, and with a lobed antitragus; coarse, long hairs project from in front of anterior margin of ear pinna to partly cover opening; usually a small, whitish spot present below the external auditory meatus; dorsal surfaces generally some shade of buffy, brownish, or grayish; dorsal parts usually tinged with black; usually a clear, buffy lateral stripe without blackish tinge; undersides usually whitish; hairs relatively short to medium in length and straight; texture of hair relatively harsh; hairs flattened in cross section; hairs have dorsal trough; cuticular scales of troughed hairs extending laterally across trough, not forming longitudinal ridges; stiff, spine-like hairs present or absent in dorsal pelage; mastoid bullae not extending beyond the plane of the occiput (except *C. baileyi* and *C. formosus,* whose mastoid bullae extend slightly beyond the occiput); tympanic bullae relatively widely spaced anteriorly on the ventral surface of the skull; supraoccipital with strong lateral indentations of mastoid bullae; interparietals not compressed and wider than interorbital breadth; phallus relatively long; soft tissue of phallus extends about one-half the length of the baculum; phallus lacking urethral lappets; rim of terminal crater of phallus forms vent-like urethral opening; baculum relatively long, with only slightly enlarged proximal end, and slender, strongly upturned distal end, or shaft thick and straight with trifid tip; vesicular glands of male short, round or bulb-shaped, and yellow- to gray-colored (pinkish and granular in fresh specimens); head of sperm approximating a triangle, with acute vertices; enamel patterns of cheek teeth less persistent with age than *Perognathus*; tendon at origin of M. rectus femoris Y-shaped (Hall, 1981; Hafner and Hafner, 1983; Homan and Genoways, 1978; Ryan, 1989; Wood, 1935).

Remarks.—Merriam (1889) included *formosus* in the subgenus *Perognathus,* an arrangement that endured until Patton et al. (1981) demonstrated conclusively that it should be classified as a *Chaetodipus*; Hafner and Hafner (1983) formally elevated *Chaetodipus* to generic rank. In the following key, presence or absence of spines (actually stiff bristles) are not infallible characters, as juvenile and molting animals often exhibit no obvious spines, individuals of some species normally having spines are found without spines, and individuals of species normally lacking spines are found with spines.

Key to the Subgenera and Species

1. Tail without distal crest and tuft of long hairs at tip; length of tail usually shorter than length of head and body .... *Chaetodipus* (*Burtognathus*) *hispidus*

1′. Tail with distal crest and long tuft of hairs at tip; length of tail averaging greater than length of head and body . ..... *Chaetodipus* (*Chaetodipus*) 2

2. Rump with mix of long, stiff, whitish or blackish, bristly hairs or spines and normal hairs 9

2′. Rump without stiff bristles or spines in pelage 3

3. Posterior end of mastoid bullae projecting slightly beyond the posterior plane of the occipitals 4

3′. Posterior end of mastoid bullae anterior to or about level with posterior plane of occipitals 5

4. Size moderate, length of head and body generally less than 90 mm; length of hind foot usually less than 26 mm; width of interparietals averaging 5.9 mm or less, rarely ranging to 6.5 .... *Chaetodipus formosus*

4′. Size large, length of head and body generally 90 mm or more; length of hind foot usually 26 mm or greater; width of interparietals averaging 6.1 mm or greater .... *Chaetodipus baileyi*

5. Ear pinna large and rounded, length from notch usually 10 mm or greater .... *Chaetodipus artus*
5′. Ear pinna short to moderate size, length from notch usually less than 10 mm .. 6
6. Tail thinly haired and with small crest; occurring along the coastal lowlands of western México, from southern Sonora to northern Nayarit; in southern part of range, broad lateral stripe of buffy or yellowish contrasting with brownish-appearing dorsal parts; in northern Sinaloa and southern Sonora, lateral stripe narrow or faint ...... *Chaetodipus pernix*
6′. Tail with moderate to large crest; not occurring south of southern Sonora along the Pacific coast of mainland México; lateral stripe on sides between fore- and hind limbs faint or absent 7
7. Dorsal color dull gray with buffy wash (no blackish overwash); buffy lateral stripe narrow, but distinct; occurring only on the Mexican Plateau in San Luis Potosí and possibly eastern Zacatecas .... *Chaetodipus lineatus*
7′. Dorsal color some shade of brownish or yellowish-gray with faint to strong overwash or admixture of blackish; ranging widely, including northern and western San Luis Potosí 8
8. Occurring only on the Baja California peninsula and its islands; tail faintly annulated in fresh specimens and only slightly longer than length of head and body; interoribital breadth averaging greater than 25.8% of occipitonasal length, except for some insular populations .... *Chaetodipus arenarius*
8′. In Baja California, only occurring on the Gulf plains in the northeastern corner, from San Felipe north; elsewhere ranging widely in the creosote bush deserts of the U.S. and México; tail considerably longer than length of head and body and annulated in life; interorbital breadth averaging less than 25% of occipitonasal length in most populations, including those in Baja California .... *Chaetodipus penicillatus*
9. Ear pinna long, usually measuring more than 9 mm 13
9′. Ear pinna short, usually measuring less than 9 mm 10
10. Occurring only west of the Colorado River in southern Nevada, California, or the Baja California peninsula 11
10′. Not occurring west of the Colorado River in southern Nevada, California, or the Baja California peninsula 12
11. Broad lateral stripe of buffy (yellowish with strong orangish tone) contrasting with brownish-appearing dorsal parts, shading to more blackish on the rump; skull relatively arched dorsally .... *Chaetodipus fallax*
11′. Lateral stripe narrow and pale yellowish-gray, or obsolete; upper parts brownish to pale buffy-yellow and not shading to more blackish on rump; skull relatively flattened dorsally .... *Chaetodipus spinatus*
12. Length of head and body averages about 77 to 80 mm; tail heavily crested and pelage of upper parts relatively coarse or harsh; spines on rump usually moderate in number and easily detected; in western Texas and Chihuahua, dorsal wing of premaxilla extending distinctly beyond posterior tip of nasal .... *Chaetodipus nelsoni*
12′. Length of head and body averaging from about 70 to 75 mm; tail moderately created and pelage of upper parts not markedly coarse; spines on rump often sparse and difficult to detect; in western Texas and Chihuahua, dorsal wing of premaxilla terminating only slightly posterior to tip of nasal .. *Chaetodipus intermedius*
13. Occurring only west of the Colorado River in California and Baja California 14
13′. Occurring only on the Pacific slopes and plains of the coastal mainland of México 15
14. A few, weakly developed bristles on rump; length of ear usually less than 10 mm; occurs only in the Cape region of Baja California Sur .... *Chaetodipus arenarius*
14′. Strong, whitish spines on rump; length of ear usually 10 mm or greater; occurs in California and the central highlands of northern Baja California .... *Chaetodipus californicus*
15. Dorsal wing of premaxilla extending distinctly beyond posterior tip of nasal; length of mastoid bullae averaging less than 6.0 mm .... *Chaetodipus artus*

15′. Dorsal wing of premaxilla terminating about even with posterior end of nasal; length of mastoid bullae averaging greater than 6.5 mm .... *Chaetodipus goldmani*

Subgenus *Burtognathus* Hoffmeister, 1986

1986. *Burtognathus* Hoffmeister, Mammals of Arizona, Univ. Arizona Press, p. 247, 31 July.

Type Species.—*Perognathus hispidus* Baird, 1858, Mammals, *in* Repts. Expl. Surv. . . . , 8(1):421, 14 July.

Diagnosis.—Body large for genus and tail relatively short; length of tail slightly shorter or about the same as length of head and body, both average about 100–115 mm; mass averages from about 35 to 60 g; no crest or long tuft at the end of the tail; pelage harsh, with flattened, troughed hairs, crescent-shaped in cross section; trough of hairs deep; hairs not differentiated into spines or bristles on rump; frontal bone of skull with strong supraorbital bead; squamosal portion of zygomatic arch usually free of or only loosely attached to auditory bulla; supraoccipital does not indent into mastoid bulla; mastoid bulla small; baculum relatively long (14–18 mm) and straight with a unique, trilobed tip (two ventral and one dorsal lobes); diploid karyotype of 34 biarmed chromosomes.

Comparisons.—*Burtognathus* differs from other genera and subgenera of pocket mice in the unique, three-lobed tip of its baculum (the bacula of others taper to simple tips). From the subgenus *Chaetodipus, Burtognathus* differs further in having a supraorbital bead on the frontal; squamosal portion of zygomatic arch free of meatal part of bulla (or at best a loose attachment); smaller bullae; and absence of an indentation into mastoid bulla by the supraoccipital. *Burtognathus* can be distinguished from *Perognathus* by a variety of characters, including larger size, hispid pelage of flattened, troughed hairs, less inflated mastoid and tympanic bullae, presence of supraorbital bead on frontal; and other diagnostic characters of the genus *Chaetodipus.*

Remarks.—*Burtognathus* is monotypic. The unique characters of *Burtognathus* suggest an early divergence and long, independent evolution from species of *Chaetodipus* (Burt, 1936; Hafner and Hafner, 1983; Merriam, 1889; Patton et al., 1981). *Burtognathus* was named by Hoffmeister (1986) in recognition of William H. Burt, for his pioneering work, "Bacula of North American Mammals" (Burt, 1960), in which he described the baculum of *C. hispidus.* Although the name *Burtognathus* literally means "Burt jaw" or "Burt face," we believe Dr. Burt, who coined the name "Bailey's pocket-faced fornicator" (*Perognathus baileyi fornicatus*), would see both the humor and honor in his patronym.

Species Account

Chaetodipus hispidus

Diagnosis.—See diagnosis of subgenus.

Comparisons.—*Chaetodipus hispidus* is distinguished from all other species of *Chaetodipus* by its large body size and short, noncrested tail; length of head and body generally ranges from about 95 to 115 mm. In body size, *C. hispidus* is similar only to *C. baileyi*; *hispidus* differs in being somewhat larger, and having longer, broader interparietals, smaller mastoid bullae, beaded supraorbital ledges, and a short, noncrested tail.

Distribution.—*Chaetodipus hispidus* occupies the Great Plains from south-central North Dakota southward, east of the Rocky Mountains and generally west of the Missouri River and Ozark Plateau through central New Mexico, virtually all of Texas, eastern and central Chihuahua, extreme northeastern Durango, northeastern Coahuila, northern Nuevo León, and south-central Tamaulipas; on the east, extending to west-central Louisiana; on the west, extending across southern New Mexico to southeastern Arizona and probably extreme northeastern Sonora; on the south, an apparently disjunct population occupies the central plateau of México, ranging from southeastern Coahuila and probably southwestern Nuevo León on the northeast, and east-central Durango on the northwest, southward through eastern San Luis Potosí, Guanajuato, to México on the east, and through most of Zacatecas and eastern Jalisco on the west (Hall, 1981).

Remarks.—Glass (1947) reviewed the species exclusive of the Mexican subspecies, *zacatecae,* and gave average measurements and diagnoses for subspecies.

Chaetodipus hispidus hispidus (Baird, 1858)

1858. *Perognathus hispidus* Baird, Mammals, *in* Repts. Expl. Surv. . . . , 8(1):421, 14 July.

Holotype.—Adult female, skin and skull, USNM 577/1696, from Charco Escondido, Tamaulipas, México; obtained in 1853 by D. N. Couch.

Measurements of holotype.—Length of head and body, 79; length of tail (terminal portion broken off), 72; length of hind foot, 21.5; length of ear (crown, dry), 4.5; occipitonasal length (approximated by Merriam from broken skull), 27.5; width across mastoid bullae, 14.5; length of interparietal, 3.8; width of interparietals, 7.0; least intermastoid distance, 8.0.

Distribution.—Ranges from near the Oklahoma boundary in north-central and eastern Texas and from west-central Louisiana southward to northeastern Coahuila and south-central Tamaulipas.

Remarks.—Baker (1956) listed means and ranges for measurements of four males and four females from Coahuila; Osgood (1900) gave average measurements for three specimens.

Chaetodipus hispidus paradoxus (Merriam, 1889)

1889. *Perognathus paradoxus* Merriam, N. Amer. Fauna, 1:24, 25 October.

1894. *Perognathus latirostris* Rhoads, Amer. Nat., 28:185, February.

1894. *Perognathus conditi* J. A. Allen, Bull. Amer. Mus. Nat. Hist., 6:318, 7 November.

Holotype.—Adult female, skin and skull, USNM 940/1544, from Trego Co., Kansas; obtained on 17 October 1884 by A. B. Baker.

Measurements of holotype.—Total length, 205; length of tail, 105; length of hind foot, 26; length of ear (crown, dry), 5.5; occipitonasal length, 32.0; basilar length, 26.6; basilar length of Hensel, 23.7; width across mastoid bullae, 15.7; length of interparietal, 4.5; width of interparietals, 8.0; least intermastoid distance, 9.0.

Distribution.—Occupies the northern and western Great Plains and Chihuahuan Desert area, ranging from south-central North Dakota southward through western and central South Dakota and Nebraska, western Kansas, Oklahoma, and Texas; on the west, from extreme southeastern Montana, eastern Wyoming and Colorado, central and southern New Mexico, southeastern Arizona, and west-central Chihuahua to extreme north-central Durango (Hall, 1981). Also known from a disjunct area near Camp Verde, Yavapai Co., Arizona.

Remarks.—Measurements were given by Anderson (1972), Armstrong (1972), Cockrum (1952), Hoffmeister (1986), Jones (1964), and Long (1965). Hoffmeister (1986) compared a sample of 15 specimens from Arizona with a sample of 14 *paradoxus* from the Great Plains, as well as with other samples. Because he found that the Arizona sample was distinguishable from the Great Plains sample, he believed that *conditi* was a valid subspecies.

Chaetodipus hispidus spilotus (Merriam, 1889)

1889. *Perognathus paradoxus spilotus* Merriam, N. Amer. Fauna, 1:25, 25 October.

1904. *Perognathus hispidus maximus* Elliot, Field Columbian Mus., Publ. 87, Zool. Ser., 3:253.

Holotype.—Adult female, skin only, USNM 186514, from Gainesville, Cook [= Cooke] Co., Texas; obtained on 8 October 1886 by G. H. Ragsdale.

Measurements of holotype.—Total length, 196; length of tail, 95; length of hind foot, 23; length of ear (crown, dry), 5.

Distribution.—Ranges from southeastern Nebraska southward through east-central Kansas and eastern and central Oklahoma to extreme north-central Texas along the Oklahoma border.

Remarks.—Merriam (1889), while not designating it as a cotype, characterized cranial features of *spilotus* using USNM 23096, a young adult female from the type locality, obtained on 24 September 1888 by G. H. Ragsdale; its measurements, listed by Merriam (1889), are: occipitonasal length, 31.0; basilar length, 25.2; basilar length of Hensel, 23.0; width across mastoid bullae, 14.5; length of interparietals, 4.2; width of interparietals, 7.3; least intermastoid distance, 8.6. Cockrum (1952) gave measurements for two males and three females from Kansas.

Chaetodipus hispidus zacatecae (Osgood, 1900)

1900. *Perognathus hispidus zacatecae* Osgood, N. Amer. Fauna, 18:45, 20 September.

Holotype.—Young adult female, skin and skull, USNM 91877, from Valparaiso, Zacatecas, México; obtained by Edward A. Goldman on 16 December 1897.

Measurements of holotype.—Total length, 211; length of tail, 105; length of hind foot, 27.5; occipitonasal length, 30.2; basilar length of Hensel, 22.5; interorbital breadth, 7.0; width across mastoid bullae, 15.0; length of interparietal, 4.0; width of interparietals, 8.0; length of nasals, 12.0.

Distribution.—Found from southeastern Coahuila and probably southwestern Nuevo León on the northeast, and east-central Durango on the northwest, southward through eastern San Luis Potosí, Guanajuato, and México; on the west, through most of Zacatecas and eastern Jalisco.

Remarks.—Baker (1956) listed measurements of one specimen from Coahuila, and Genoways and Jones (1973) gave measurements for an adult male from Zacatecas.

Subgenus *Chaetodipus*

1889. *Chaetodipus* Merriam, N. Amer. Fauna, 1:5, 25 October.

Type species.—*Perognathus* [*Chaetodipus*] *spinatus* Merriam, 1889, N. Amer. Fauna, 1:5, 25 October.

Diagnosis.—Characters as for genus except baculum tapers to a simple, non-lobed tip; no supraorbital bead or ridge on frontal; squamosal portion of zygomatic arch firmly attached to meatal part of auditory bulla; supraoccipital forms indentation into mastoid bulla; diploid number of chromosomes in so far as is known 36 to 56.

Comparisons.—See accounts of *Burtognathus* and *Perognathus* for comparison with those taxa.

Remarks.—The subgenus *Chaetodipus* as applied here, differs from the subgenus of Merriam (1889) as it was applied by him and subsequent authorities except Hoffmeister (1986).

Species Accounts

Chaetodipus arenarius

Diagnosis.—A small species of *Chaetodipus,* with a tail that is longer than length of head and body, small ears, and with relatively soft pelage usually lacking stiff bristles or spines; a clear, buffy lateral stripe is usually absent; length of head and body ranges from about 66 to 80 mm; length of tail ranges from about 70 to 103 mm; length of ear ranges from about 7 to 10 mm.

Comparisons.—*Chaetodipus arenarius* is most similar in general appearance to the other small, soft-haired species, *C. penicillatus* and *C. pernix*; *arenarius* differs from *penicillatus* in being smaller, having a less annulated appearing tail in life, and broader skull; *pernix* differs from *arenarius* in having a well developed lateral stripe of clear buff. From other species of *Chaetodipus* occurring on the baja California Peninsula (*C. baileyi, californicus, formosus, fallax,* and *spinatus*), *C. arenarius* can be distinguished by its smaller size; it further differs from *californicus, fallax,* and *spinatus* by lacking stiff bristles or spines in the pelage of the rump and flanks, although weakly developed bristles or spines may be found rarely; of the sympatric soft-haired species, the hind foot measures less than 25 mm in *arenarius* and greater than 25 mm in *baileyi*; length of ear generally measures greater than 9 mm in *formosus* and less than 9 in *arenarius*; the mastoid bullae of *formosus* project slightly beyond the plane of the occipitals at the back of the skull, whereas those of *arenarius* do not extend to the posterior plane of the occipitals.

Distribution.—Generally limited to sandy soils on the Baja California Peninsula, from near De Mara's Well, Laguna Salada, near the international boundary in northeastern Baja California, and San Quintín on the Pacific coast on the northwest, southward to near the southern cape in Baja California Sur; not known from the eastern, gulf slope from about El Barril, Baja California south to near La Paz, Baja California Sur, nor from coastal areas south of La Paz (Hall, 1981).

Remarks.—We have treated *Chaetodipus dalquesti* (Roth, 1976) as a synonym of *C. arenarius arenarius* based on studies by M. A. Bogan (in litt.), although Bogan's investigations left subspecific status of *dalquesti* unresolved. Presence or absence of stiff bristles or spines on the rump is not constant within other species of *Chaetodipus,* such as *C. intermedius* and *C. penicillatus* (Hoffmeister and Lee, 1967), and alone does not indicate specific status for *dalquesti.* Hafner and Hafner (1983) found no differences between the karyotypes of *dalquesti* and *C. arenarius.*

Chaetodipus arenarius albescens (Huey, 1926)

1926. *Perognathus arenarius albescens* Huey, Proc. Biol. Soc. Washington, 39:67, 30 July.

Holotype.—Adult male, skin and skull, SDSNH 5103, from San Felipe, Baja California, México; obtained by Laurence M. Huey on 23 March 1926.

Measurements of holotype.—Total length, 170; length of tail, 95; length of hind foot, 22; length of ear (crown), 5; weight, 15 g; condylobasal length, 23.2; interorbital breadth, 6.0; length of maxillary toothrow, 3.0; width across mastoid bullae, 12.3; length of nasals, 8.8.

Distribution.—Known from sandy soils along the Gulf of California coast in the vicinity of the type locality in northeastern Baja California.

Remarks.—Huey (1926) listed measurements for 10 adults.

Chaetodipus arenarius albulus (Nelson and Goldman, 1923)

1923. *Perognathus penicillatus albulus* Nelson and Goldman, Proc. Biol. Soc. Washington, 36:159, 1 May.

Holotype.—Adult male, skin and skull, USNM 146864, from Magdalena Island, Baja California Sur, México; obtained by Edward W. Nelson and Edward A. Goldman on 3 December 1906.

Measurements of holotype.—Total length, 155; length of tail, 83; length of hind foot, 22; greatest length of skull, 22.8; interorbital breadth, 6.1; length of maxillary toothrow, 3.4; width across mastoid bullae, 12.0; length of interparietal, 3.4; width of interparietals, 7.0; length of nasals, 8.7; zygomatic width, 11.2.

Distribution.—Found only on Magdalena Island, Baja California Sur.

Remarks.—Nelson and Goldman (1923) listed means and ranges of measurements of nine adults.

Chaetodipus arenarius ambiguus (Nelson and Goldman, 1929)

1929. *Perognathus arenarius ambiguus* Nelson and Goldman, Proc. Biol. Soc. Washington, 42:108, 25 March.

Holotype.—Young adult male, skin and skull, USNM 140011, from Yubay, 30 mi SE Calamahué [= Calamajué], 2,000 ft, Baja California; obtained on 18 September 1905 by Edward W. Nelson and Edward A. Goldman.

Measurements of holotype.—Total length, 150; length of tail, 88; length of hind foot, 22; greatest length of skull, 22.6; interorbital breadth, 6.2; length of maxillary toothrow, 3.2; width across mastoid bullae, 12.0; length of interparietal, 2.8; width of interparietals, 5.2; length of nasals, 8.3; width of nasals, 2.3.

Distribution.—Occupies sandy soils in the central Baja California Peninsula, from near Chapala southward to the middle of the Vizcaíno Desert in Baja California Sur (Huey, 1964).

Remarks.—Nelson and Goldman (1929) listed average and extremes of measurements of five adults.

Chaetodipus arenarius ammophilus (Osgood, 1907)

1907. *Perognathus penicillatus ammophilus* Osgood, Proc. Biol. Soc. Washington, 20:20, 23 February.

Holotype.—Adult male, skin and skull, USNM 146859, from Santa Margarita Island, Baja California Sur, México; obtained on 20 November 1905 by Edward W. Nelson and Edward A. Goldman.

Measurements of holotype.—Greatest length of skull, 26.1; basilar length, 18.0; interorbital breadth, 6.6; length of maxillary toothrow, 3.8; width across mastoid bullae, 13.2; length of interparietal, 3.8; width of interparietals, 7.3; length of nasals, 9.4; zygomatic width, 13.1; length of diastema, 6.4.

Distribution.—Known only from Santa Margarita Island, Baja California Sur, off the west coast of the Baja California Peninsula.

Remarks.—Osgood (1907) gave averages and extremes of measurements of nine topotypes.

Chaetodipus arenarius arenarius (Merriam, 1894)

1894. *Perognathus arenarius* Merriam, Proc. Calif. Acad. Sci., ser. 2, 4:461, 25 September.

1976. *Perognathus dalquesti* Roth, J. Mamm., 57:562, 27 August.

Holotype.—Adult female, skin and skull, CAS 99, from San Jorge, near Comondú, Baja California Sur, México; obtained by Walter E. Bryant on 17 March 1889.

Measurements of holotype.—[External measurements from dry skin] Total length, 136; length of tail, 70; length of hind foot, 20; length of ear (anterior base), 7; greatest length of skull, 22.5; occipitonasal length, 23.0; basilar length, 15.3; interorbital breadth, 6.2; width across mastoid bullae, 12.0; length of interparietal, 3.5; width of interparietals, 6.4; length of nasals, 8.8; zygomatic breadth anteriorly, 11.0.

Distribution.—Found on sandy soils from near the cape region northward to the southern part of the Vizcaíno Desert on the Pacific coast of the Baja California peninsula, in Baja California Sur.

Remarks.—Roth (1976:564) characterized *dalquesti* as being larger, with a longer, darker colored tail than *arenarius,* and having "distinct rump spines," similar to *C. intermedius,* but "not as stout." Hafner and Hafner (1983) stated that the karyotype of *dalquesti* was identical to that of *C. arenarius.* Banks (1964) listed means and ranges of measurements of 8 males and 12 females from Todos Santos. Total length of skull, listed by Merriam (1894*c*) differs from occipitonasal length reported by Osgood (1900).

Chaetodipus arenarius helleri (Elliot, 1903)

1903. *Perognathus helleri* Elliot, Field Columbian Mus., Publ. 74, Zool. Ser., 3:166, 7 May.

Holotype.—Adult male, skin and skull, FMNH 10355, from San Quentin [= San Quintín], Baja California, México; obtained by Edmund Heller on 26 July 1902.

Measurements of holotype.—Total length, 159; length of tail, 83; length of hind foot, 20.5; length of ear, 8; occipitonasal length, 23.0; basilar length of Hensel, 14.0; interorbital breadth, 6.0; length of maxillary toothrow, 3.0; width across mastoid bullae, 11.5; length of nasals, 7.5; width of rostrum, 4.0; zygomatic breadth, 11.5; greatest width of brain case, 10.5; palatal arch to alveoli of incisors, 8.5.

Distribution.—Known from sandy soils on the San Quintín Plain on the Pacific coast of northern Baja California.

Remarks.—Elliot (1903) listed extremes for 15 specimens.

Chaetodipus arenarius mexicalis (Huey, 1939)

1939. *Perognathus arenarius mexicalis* Huey, Trans. San Diego Soc. Nat. Hist., 9:57, 31 August.

Holotype.—Adult female, skin and skull, SDSNH 12127 from Los Muertos Canyon fan at Gaskill's Tank, near Laguna Salada, lat. 32°27′N, long. 115°53′W, Baja California, México; obtained on 23 November 1936 by Laurence M. Huey.

Measurements of holotype.—Total length, 176; length of tail, 102; length of hind foot, 23; length of ear (crown), 5; greatest length of skull 23.7; interorbital breadth, 6.1; length of maxillary toothrow, 3.2; width across mastoid bullae, 12.0; length of nasals, 8.5.

Distribution.—Known from sandy areas on the western side of Laguna Salada in northeastern Baja California.

Remarks.—According to Huey (1939*b*), this subspecies may range northward as far as a few miles into the state of California, although the species has not been taken in the United States.

Chaetodipus arenarius paralios (Huey, 1964)

1964. *Perognathus arenarius paralios* Huey, Trans. San Diego Soc. Nat. Hist., 13:113, 15 January.

Holotype.—Adult male, skin and skull, SDSNH 15542, from [El] Barril, lat. 28°20′N, on the Gulf of California, Baja California, México; obtained by Laurence M. Huey on 23 March 1947.

Measurements of holotype.—Total length, 150; length of tail, 81; length of hind foot, 22; length of ear (crown), 5; greatest length of skull, 24.0; interorbital breadth, 6.3; length of maxillary toothrow, 2.9; width across mastoid bullae, 13.0; length of nasals, 8.9.

Distribution.—Occurs on sandy soils on the Gulf of California slope from near Bahía Los Angeles southward to near the type locality at El Barril, Baja California.

Remarks.—No other measurements are available for this subspecies.

Chaetodipus arenarius sabulosus (Huey, 1964)

1964. *Perognathus arenarius sabulosus* Huey, Trans. San Diego Soc. Nat. Hist., 13:114, 15 January.

Holotype.—Adult male, skin and skull, SDSNH 5300, from mainland on S side Scammon's Lagoon, Baja California Sur, México; obtained by Laurence M. Huey on 22 May 1926.

Measurements of holotype.—Total length, 165; length of tail, 91; length of hind foot, 21; length of ear (crown), 5; greatest length of skull, 23.5; interorbital breadth, 6.1; length of maxillary toothrow, 2.9; width across mastoid bullae, 12.3; length of nasals, 8.8.

Distribution.—Found on sandy ground from Bahía Santa Rosalía, Baja California, southward through the western Vizcaíno Desert to the vicinity of Laguna San Ignacio.

Remarks.—There are no published measurements other than those of the holotype.

Chaetodipus arenarius siccus (Osgood, 1907)

1907. *Perognathus penicillatus siccus* Osgood, Proc. Biol. Soc. Washington, 20:20, 23 February.

Holotype.—Adult male, skin and skull, USNM 146890, from Ceralbo [= Cerralvo] Island, Baja California Sur, México; obtained by Edward W. Nelson and Edward A. Goldman on 13 February 1906.

Measurements of holotype.—Greatest length of skull, 25.9; basilar length, 17.7; interorbital breadth, 6.6; length of maxillary toothrow, 4.0; width across mastoid bullae, 13.8; length of interparietal, 3.6; width of interparietals, 7.2; length of nasals, 8.9; zygomatic width, 12.6; length of diastema, 6.0.

Distribution.—Known only from Cerralvo Island in the Gulf of California, Baja California Sur.

Remarks.—Osgood (1907) listed means and extremes of external measurement of 10 topotypes. Banks (1964) gave means and ranges of measurements for 18 males and 17 females.

Chaetodipus arenarius sublucidus (Nelson and Goldman, 1929)

1929. *Perognathus arenarius sublucidus* Nelson and Goldman, Proc. Biol. Soc. Washington, 42:109, 25 March.

Holotype.—Adult male, skin and skull, USNM 146896, from La Paz, Baja California Sur, México on 16 February 1906 by Edward W. Nelson and Edward A. Goldman.

Measurements of holotype.—Total length, 167; length of tail, 95; length of hind foot, 22; greatest length of skull, 24.8; interorbital breadth, 6.3; length of maxillary toothrow, 3.2; width across mastoid bullae, 12.6; length of interparietal, 3.7; width of interparietals, 6.4; length of nasals, 9.4; width of nasals, 2.3.

Distribution.—Known from sandy areas in the desert basin in the vicinity of La Paz, on the Gulf coast of Baja California Sur.

Remarks.—Nelson and Goldman (1929) listed means and extremes for external measurements of 10 adult topotypes. Banks (1964) listed means and ranges of measurements for 13 specimens of each sex.

Chaetodipus artus

Diagnosis.—A species of *Chaetodipus* of medium-large size, with weakly developed spines present or absent on the rump, relatively large, rounded ears, and a tail of about average length for the genus; skull with broad, nearly elliptically-shaped interparietals and small mastoid bullae; baculum relatively long with a strongly curved tip.

Comparisons.—*Chaetodipus artus* shares parts of its geographic range with *C. goldmani,* to which it is most similar in appearance, and with *C. baileyi* and *C. pernix.* No single character of skin or skull will separate all *artus* from all *goldmani* (Anderson, 1964), but *C. artus* typically differs from *C. goldmani* in being slightly smaller; having a less hairy tail; fewer and more weakly-developed rump spines; darker color, including the ear pinnae; shorter, more strongly curved baculum; smaller, narrower skull; smaller mastoid bullae; and in having the ascending process of the premaxillae extending posteriorly beyond the end of the nasals a distance equal to or greater than the least breadth of one nasal bone (in *goldmani* the premaxillae barely extends posteriorly beyond the end of the nasal). *C.*

pernix is considerably smaller than *artus* and lacks any spines on the rump; interorbital breadth of *pernix* is less than 5.8 mm, whereas it is greater than 5.8 in *artus* (Hall, 1981). Total length of adult *C. baileyi* generally exceeds 200 mm, whereas *artus* rarely measures more than 200 mm; length of hind foot is 26 mm or more in *baileyi* and rarely exceeds 25.5 mm in *artus* (Anderson, 1964; Hall, 1981); the mastoid bullae of *baileyi* are relatively large; and *baileyi* always lacks spines in the pelage.

Distribution.—Found along the Pacific coastal slope of western México, from the barrancas of southwestern Chihuahua and northwestern Durango westward to southern Sonora and southward to northern Nayarit (Hall, 1981).

Remarks.—Refer to Anderson (1964) for characteristics distinguishing *C. artus* and *C. goldmani.* Anderson (1972) listed means and ranges of measurements for 13 specimens from Chihuahua.

Chaetodipus artus (Osgood, 1900)

1900. *Perognathus artus* Osgood, N. Amer. Fauna, 18:55, 20 September.

Holotype.—Adult female, skin and skull, USNM 96298, from Batopilas, Chihuahua, México; obtained by Edward A. Goldman on 6 October 1898.

Measurements of holotype.—Total length, 197; length of tail, 109; length of hind foot, 24; greatest length of skull, 26.30; width across bullae, 12.55; breadth across maxillary arches, 11.95; nasal length, 10.10; interorbital breadth, 5.95.

Remarks.—Osgood listed means for measurements of three specimens. Anderson (1964) gave means and ranges of measurements for 14 samples of mixed ages.

Chaetodipus baileyi

Diagnosis.—A large species of *Chaetodipus* with a strongly crested tail of moderate length, and soft, spineless pelage; skull large and robust with narrow interparietals and relatively large mastoid bullae that project slightly posteriad beyond the plane of the occipitals; width of interparietals about equal to or less than interorbital breadth; dorsal and lateral color grayish washed with yellowish.

Comparisons.—*Chaetodipus baileyi* is exceeded in length of head and body only by some *C. hispidus,* from which it can be distinguished by its grayish color and its long, crested tail (the tail of *hispidus* is not crested and is shorter than length of head and body); it can be distinguished from all other *Chaetodipus* species lacking spines in the pelage by its larger size (average length of head and body greater than 90 mm), longer hind foot (generally exceeding 26 mm), larger mastoid bullae, and grayish rather than brownish color of upper parts (discounting the yellowish tinge present on both the grayish- and brownish-colored species).

Distribution.—*Chaetodipus baileyi* occupies rocky hillsides and brushy, desert associations from extreme southwestern New Mexico westward across southern Arizona to extreme southeastern California, and southward throughout most of the Baja California peninsula, and along the lowlands of Sonora to northwestern Sinaloa (Hall, 1981).

Remarks.—Patton et al. (1981) reviewed the patterns of biochemical and karyotypic variation in *C. baileyi,* noting that the populations west of the Colorado River in California and the Baja California peninsula, and east of the Colorado River in Arizona, New Mexico and mainland México formed two relatively distinct units. Both geographic units showed relatively high uniformity in biochemical and karyotypic traits (Patton, 1977) within their units and reduced levels of similarity between units. They noted that the biochemical differentiation was generally unrelated to the current intraspecific taxonomy based on structural features. The species has never had a comprehensive systematic review.

Chaetodipus baileyi baileyi (Merriam, 1894)

1894. *Perognathus baileyi* Merriam, Proc. Acad. Nat. Sci. Philadelphia, 46:262, 27 September.
1928. *Perognathus baileyi domensis* Goldman, Proc. Biol. Soc. Washington, 41:204, 18 December.

Holotype.—Adult female, skin and skull, USNM 17838/24775, from Magdalena, Sonora, México; obtained by Vernon Bailey on 3 November 1889.

Measurements of holotype.—Total length, 210; length of tail, 122; length of hind foot, 27; length of ear (anterior base, dry), 11; greatest length of skull, 29.5; basal length, 24.0; basilar length of Hensel, 20.5; length of nasals, 11.5; greatest zygomatic breadth (posteriorly), 15.5.

Distribution.—Ranges from the Colorado river in southwestern Arizona, the southern edge of the Mogollon Plateau in central Arizona, and southwestern New Mexico southward in the foothills of the Sierra Madre Occidental and the coastal plain of Sonora to northern Sinaloa (Hall, 1981; Hoffmeister, 1986).

Remarks.—Hoffmeister (1986) listed statistics for measurements of 13 samples from Arizona. Merriam (1894*b*) listed average external measurements for eight topotypes. Hoffmeister (1986) considered *domensis* to be indistinguishable from *baileyi*; Goldman (1928) noted that *domensis* mainly differed from *baileyi* in its more palely-colored upper parts.

Chaetodipus baileyi extimus (Nelson and Goldman, 1930)

1930. *Perognathus baileyi extimus* Nelson and Goldman, J. Washington Acad. Sci., 20:223, 19 June.

Holotype.—Adult female, skin and skull, USNM 146672, from Tres Pachitas, 700 ft, 36 mi S La Paz, Baja California Sur, México; obtained by Edward W. Nelson and Edward A. Goldman on 25 December 1905.

Measurements of holotype.—Total length, 198; length of tail, 107; length of hind foot, 25; greatest length of skull, 29.0; interorbital breadth, 6.9; length of maxillary toothrow, 4.5; width across mastoid bullae, 14.3; length of interparietal, 3.4; width of interparietals, 6.1; length of nasals, 10.2; width of nasals, 2.9; zygomatic breadth, 15.5.

Distribution.—Found on the Pacific slope of the cape region of Baja California Sur, from near San José de Comondú on the north to near Todos Santos on the south.

Remarks.—Nelson and Goldman (1930) listed measurements for three adults; Burt (1932) gave ranges and means for measurements of six specimens.

Chaetodipus baileyi fornicatus (Burt, 1932)

1932. *Perognathus baileyi fornicatus* Burt, Trans. San Diego Soc. Nat. Hist., 7:164, 31 October.

Holotype.—Adult male, skin and skull, CIT 50289, from Monserrate Island, lat. 25°38′N, long. 111°02′W, Gulf of California, Baja California Sur, México; obtained by William H. Burt on 16 December 1931.

Measurements of holotype.—Total length, 202; length of tail, 108; length of hind foot, 26; length of ear (crown), 7, (notch, dry), 8.9; greatest length of skull, 29.1; basal length, 24.6; interorbital breadth, 6.9; length of maxillary toothrow, 4.2; length of mastoid bulla, 8.8; width across mastoid bullae, 14.3; length of interparietal, 3.8; width of interparietals, 6.6; length of nasals, 11.1.

Distribution.—Known only from Monserrate Island, Gulf of California, Baja California Sur.

Remarks.—Burt (1932) gave means and ranges for measurements of 10 specimens. The holotype is now housed in the UCLA collections.

Chaetodipus baileyi hueyi (Nelson and Goldman, 1929)

1929. *Perognathus baileyi hueyi* Nelson and Goldman, Proc. Biol. Soc. Washington, 42:106, 25 March.

Holotype.—Young adult female, skin and skull, SDSNH 5220, from San Felipe, northeastern Baja California, México; obtained by Laurence M. Huey on 17 April 1926.

Measurements of holotype.—Total length, 196; length of tail, 106; length of hind foot, 24; greatest length of skull, 27.5; interorbital breadth, 6.4; length of maxillary toothrow, 3.7; width across mastoid bullae, 14.3; length of interparietal, 4.4; width of interparietals, 5.7; length of nasals, 10.5; width of nasals, 2.5; zygomatic breadth, 14.5.

Distribution.—Found on the desert slopes of the coastal mountains of southern San Diego Co., California eastward to the Colorado River and southward in Baja California to the vicinity of San Felipe on the Gulf of California.

Remarks.—No other measurements are available.

Chaetodipus baileyi insularis (Townsend, 1912)

1912. *Perognathus baileyi insularis* Townsend, Bull. Amer. Mus. Nat. Hist., 31:122, 14 June.

Holotype.—Male, skin and skull, USNM 198410, from Tiburón Island, Gulf of California, Sonora, México; obtained on 13 April 1911 by H. E. Anthony.

Measurements of holotype.—Total length, 212; length of tail, 118; length of hind foot, 27; greatest length of skull, 28.50; width across bullae, 15.10; breadth across maxillary arches, 13.15; nasal length, 11.25; interorbital breadth, 6.55.

Distribution.—Known only from Tiburón Island, Gulf of California, Sonora, México.

Remarks.—Townsend (1912) listed average external measurements of three specimens; Burt (1932) listed means and extremes of measurements of eight specimens.

Chaetodipus baileyi mesidios (Huey, 1964)

1964. *Perognathus baileyi mesidios* Huey, Trans. San Diego Soc. Nat. Hist., 13:112, 15 January.

Holotype.—Adult male, skin and skull, SDSNH 14470, from San Borja Mission, near lat. 28°45′N, Baja California, México; obtained by Laurence M. Huey on 13 October 1941.

Measurements of holotype.—Total length, 212; length of tail, 121; length of hind foot, 25; length of ear (crown), 6; greatest length of skull, 29.7; interorbital breadth, 7.0; length of maxillary toothrow, 4.1; width across mastoid bullae, 15.6; length of nasals, 7.0.

Distribution.—Ranges over the middle portion of the Baja California peninsula from Onyx, near lat. 30°N, in Baja California, southward to Bahía Concepción, Baja California Sur.

Remarks.—There are no published measurements other than for the holotype.

Chaetodipus baileyi rudinoris (Elliot, 1903)

1903. *Perognathus baileyi rudinoris* Elliot, Field Columbian Mus., Publ. 74, Zool. Ser., 3:167, 7 May.

1903. *Perognathus knekus* Elliot, Field Columbian Mus., Publ. 74, Zool. Ser., 3:169, 7 May.

Holotype.—Adult male, skin and skull, FMNH 10329, from San Quentin [= San Quintín], Baja California, México; obtained by Edmund Heller, on 24 July 1902.

Measurements of holotype.—Total length, 232; length of tail, 128; length of hind foot, 27; length of ear, 11.5; occipitonasal length, 31.0; basal length of Hensel, 22.0; interorbital breadth, 6.5; length of maxillary toothrow, 4.0; width across mastoid bullae, 15.0; width of interparietals, 6.5; length of nasals, 10.0; width of nasals (anterior), 3; width of rostrum, 4.0; zygomatic breadth, 16.0; distance from palatal arch to alveolus of incisor, 12.0; greatest width of basioccipital between bullae, 5.0.

Distribution.—Found on the coastal plain around San Quintín and the slopes of the southern foothills of the Sierra San Pedro Mártir, southward to the region of El Mármol and San Fernando Mission, in Baja California.

Remarks.—Except for the holotypes of *rudinoris* and *knekus,* no measurements have been published.

Chaetodipus californicus

Diagnosis.—A *Chaetodipus* species of medium-large size, with a relatively long, strongly crested tail, numerous stiff spines on the rump and flanks, and a relatively long ear pinna; mastoid bulla relatively small and interparietal broad; ratio of lengths of tail to head and body ranges from about 1.15 to 1.55; length of ear usually greater than 9 mm (range 9-14), and width of interparietals usually greater than 8.1 mm.

Comparisons.—*Chaetodipus californicus* is most similar in general appearance to other medium-sized species of *Chaetodipus* with spines on the rump; it can be distinguished from *C. fallax* by its longer ears (the length of the ear of *fallax* usually is less than 9 mm), slightly larger size, relatively longer tail, and broader interparietal; from *C. spinatus, C. californicus* differs in having fewer spines, a conspicuous stripe of buffy-colored fur on the sides (lateral stripe), a relatively longer tail, longer ear (length 8.5 mm or less in *spinatus*), and wider interparietals; *californicus* is larger, has a relatively longer tail, larger ears, and more spines on the rump than the allopatric species, *C. intermedius* and *C. nelsoni*; *californicus* differs from the allopatric species, *C. goldmani,* in having wider parietals, a longer tail and, in typical individuals, more spines on the rump; from the allopatric species, *C. artus, californicus* is distinguished by its longer tail and greater number of spines.

Distribution.—Occurs mostly in chaparral and less often in annual grassland and desert shrub communities of coastal southwestern North America, ranging from south of San Francisco Bay southward in the coastal ranges west of the Central Valley to the Sierra San Pedro Mártir, Baja California, México; in California, also ranges eastward in the Transverse Range to the southwestern slope of the Sierra Nevada and thence northward to Placer Co. (Hall, 1981).

Remarks.—*Chaetodipus californicus* has never been reviewed systematically; little information is available on structural and biochemical variation, and the published records of occurrence suggest errors in taxonomic assignment to subspecies.

Chaetodipus californicus bensoni (von Bloeker, 1938)

1938. *Perognathus californicus bensoni* von Bloeker, Proc. Biol. Soc. Washington, 51:197, 23 December.

Holotype.—Subadult male, skin and skull, MVZ 81579, from Stonewall Creek, 1,300 ft, 6.3 mi NE Soledad, Monterey Co., California; obtained on 16 June 1937 by Jack C. von Bloeker, Jr.

Measurements of holotype.—Total length, 197; length of tail, 107; length of hind foot, 29; length of ear, 12; greatest length of skull, 26.7; interorbital breadth, 6.5; length of mastoid bulla, 7.9; width across mastoid bullae, 13.85; length of interparietal, 4.0; width of interparietals, 7.6; length of nasals, 10.3; width of rostrum, 4.7.

Distribution.—Occupies chaparral and arid shrub communities in the Diablo Range of west-central California, from eastern Stanislaus Co., on the north, southward to the Temblor Range in Kern and San Luis Obispo counties; ranging on the west to the Salinas Valley, Monterey Co.

Remarks.—Means and ranges of measurements of 10 adults and 5 subadults were listed by von Bloeker (1938).

Chaetodipus californicus bernardinus (Benson, 1930)

1930. *Perognathus californicus bernardinus* Benson, Univ. California Publ. Zool., 32:449, 6 September.

Holotype.—Adult male, skin and skull, MVZ 44094, from 2 mi E Strawberry Peak,

5,750 ft, San Bernardino Mountains, San Bernardino Co., California; obtained on 19 September 1920 by Laurence M. Huey and Donald R. Dickey.

Measurements of holotype.—Total length, 227; length of tail, 131; length of hind foot, 28; length of ear, 12; greatest length of skull, 28.65; interorbital breadth, 7.15; length of mastoid bulla, 8.3; width across mastoid bullae, 13.3; length of interparietal, 3.95; width of interparietals, 7.5; length of nasals, 11.35; width of rostrum, 4.8.

Distribution.—Known from the slopes of the San Gabriel and San Bernardino Mountains in Los Angeles and San Bernardino counties, California.

Remarks.—Benson (1930) listed means and ranges for measurements of eight specimens.

Chaetodipus californicus californicus (Merriam, 1889)

1889. *Perognathus californicus* Merriam, N. Amer. Fauna, 1:26, 25 October.
1889. *Perognathus armatus* Merriam, N. Amer. Fauna, 1:27, 25 October.

Holotype.—Adult female, skin and skull, USNM 186506, from Berkeley, Alameda Co., California; obtained on 8 November 1888 by T. S. Palmer and Charles A. Keeler.

Measurements of holotype.—Total length, 186; length of hind foot, 24; length of ear (crown), 6; occipitonasal length, 26.7; basilar length, 20.5; basilar length of Hensel, 18.1; width across mastoid bullae, 13.4; length of interparietal, 4.3; width of interparietals, 8.5; least intermastoid breadth, 8.7.

Distribution.—Occurs in the hills around San Francisco Bay, California, from the San Francisco peninsula on the northwest and Mt. Diablo on the northeast, southward to Portola, San Mateo Co., and Gilroy, Santa Clara Co.

Remarks.—Merriam (1889) listed measurements for two specimens (holotype of *armatus* included) in addition to the holotype; Osgood (1900) gave measurements for five adults.

Chaetodipus californicus dispar (Osgood, 1900)

1900. *Perognathus californicus dispar* Osgood, N. Amer. Fauna, 18:58, 20 September.

Holotype.—Adult male, skin and skull, USNM 32116/43928, from Carpenteria, Santa Barbara Co., California; obtained by Edward W. Nelson on 19 December 1891.

Measurements of holotype.—Total length, 218; length of tail, 120; length of hind foot, 27; greatest length of skull, 28.40; width across bullae, 13.95; breadth across maxillary arches, 12.70; nasal length, 11.80; interorbital breadth, 6.50.

Distribution.—Occupies coastal chaparral communities in California, from San Luis Obispo, San Luis Obispo Co., California, southward to Santa Monica, Los Angeles Co.

Remarks.—The population ranging along the western Sierra Nevada foothills between Placer and Fresno counties (Hall, 1981) is excluded here from *dispar* on geographic grounds. Benson (1930) gave statistics for measurements of 15 specimens.

Chaetodipus californicus femoralis (J. A. Allen, 1891)

1891. *Perognathus* (*Chaetodipus*) *femoralis* J. A. Allen, Bull. Amer. Mus. Nat. Hist., 3:281, 30 June.

Holotype.—Adult male, skin and skull, AMNH 3386/2659, from Dulzura, San Diego Co., California; obtained on 12 February 1891 by C. H. Marsh.

Measurements of holotype.—Total length, 241.3; length of tail, 152.4; length of hind foot (dry), 27.5; length of ear, 9.5; greatest length of skull, 27.4; interorbital breadth, 6.7; length of maxillary toothrow, 3.8; width across mastoid bullae, 13.6; length of nasals,

10.0; zygomatic breadth, 14.0; length of rostrum, 11.5.

Distribution.—Occupies chaparral communities on the Pacific slope in San Diego Co., California, ranging southward in the central mountain range (Hanson Laguna Mountains) of northern Baja California to the northwestern slopes of the Sierra San Pedro Mártir.

Remarks.—Benson (1930) listed measurements of 20 specimens.

Chaetodipus californicus marinensis (von Bloeker, 1938)

1938. *Perognathus californicus marinensis* von Bloeker, Proc. Biol. Soc. Washington, 51:199, 23 December.

Holotype.—Adult male, skin and skull, MVZ 81550, from Indian Harbor, 50 ft, 1.5 mi S Marina, Monterey Co., California; obtained by J. C. von Bloeker, Jr., on 27 May 1937.

Measurements of holotype.—Total length, 203; length of tail, 114; length of hind foot, 26; length of ear, 14; greatest length of skull, 28.1; interorbital breadth, 6.6; length of mastoid bulla, 7.9; width across mastoid bullae, 13.5; length of interparietal, 3.8; width of interparietals, 8.1; length of nasals, 11.3; width of rostrum, 5.15.

Distribution.—Occupies coastal chaparral communities in California, from the north end of the Gabilan Range and Monterey Bay, southward to Morro Bay.

Remarks.—Means and ranges of measurements for 12 adult and subadult specimens were listed by von Bloeker (1938).

Chaetodipus californicus mesopolius (Elliot, 1903)

1903. *Perognathus femoralis mesopolius* Elliot, Field Columbian Mus., Publ. 74, Zool. Ser., 3:168, 7 May.

Holotype.—Adult female, skin and skull, FMNH 10374, from Penon [= Piñon], 5,000 ft, San Pedro Mártir Mountains, Baja California, México; obtained by Edmund Heller on 6 October 1902.

Measurements of holotype.—Total length, 232; length of tail, 136; length of hind foot, 27; length of ear, 14; occipitonasal length, 27.0; basilar length of Hensel, 18.0; interorbital breadth, 6.0; width across mastoid bullae, 13.0; length of nasals, 10.0; width of nasals (anterior), 3.0; posterior width of nasals, 2.0; width of rostrum, 5; zygomatic breadth, 13.0; distance from palatal arch to alveolus of incisor, 10.0.

Distribution.—Occurs in the Sierra San Pedro Mártir, Baja California.

Remarks.—No other published measurements are available for this subspecies. The description of color suggests an animal in juvenile pelage; characterization of the subspecies was primarily on the basis of its grayer color compared to *femoralis.*

Chaetodipus californicus ochrus (Osgood, 1904)

1904. *Perognathus californicus ochrus* Osgood, Proc. Biol. Soc. Washington, 17:128, 9 June.

Holotype.—Young adult female, skin and skull, USNM 130348, from Santiago Springs, 16 mi SW McKittrick, Kern Co., California; obtained on 30 July 1903 by Luther J. Goldman.

Measurements of holotype.—Total length, 196; length of tail, 117; length of hind foot, 25; greatest length of skull, 27.05; width across bullae, 13.50; breadth across maxillary arches, 12.45; nasal length, 10.15; interorbital breadth, 6.15.

Distribution.—Found in desert shrub, chaparral, annual grassland, oak woodland, and pine communities of the Transverse ranges and southeastern slopes of the southern Sierra Nevada, California; ranging from the southern Temblor Range and Carrizo Plain, San Luis Obispo Co., south and eastward to eastern Tulare Co. on the desert slope of the Tehachapi Mountains and Sierra Nevada, and northward on the western slope of the Sierra Nevada to Placer Co.

Remarks.—Osgood (1904) listed statistics for external measurements of 10 young adult topotypes. There are no published measurements for skulls of *ochrus*. The subspecies was characterized entirely on the basis of color.

Chaetodipus fallax

Diagnosis.—A medium-sized species of *Chaetodipus* with relatively coarse pelage, a moderate number of spines on the rump, a clear, buffy lateral stripe, a relatively short ear pinna, and a long, crested tail; mastoid bullae relatively large; length of head and body ranging from about 80 to 90 mm; length of tail ranging from about 88 to 120 mm; width of interparietals usually less than 8.0 mm.

Comparisons.—*Chaetodipus fallax* is most similar to *C. californicus* in general appearance (see account of the latter species for distinguishing characteristics), and also resembles other species of *Chaetodipus* with spines on the rump. From *C. spinatus*, *fallax* differs in having fewer spines (spines on the flanks are less numerous and conspicuous), and a more conspicuous lateral stripe of clear buff. Other species of *Chaetodipus* with spines in the pelage (*artus*, *goldmani*, *intermedius*, and *nelsoni*) have geographic ranges allopatric to *fallax*, and generally have fewer and more weakly-developed spines; the ear pinnae of *artus* and *goldmani* are longer (generally greater than 9 mm); typical specimens of *fallax* are larger than typical specimens of *intermedius* and *nelsoni*.

Distribution.—Found mainly on the Pacific slopes and the basins of southwestern California and Baja California, but also occupies the desert slopes of the southern California mountain chain; ranges from Oro Grande and Twenty-nine Palms, San Bernardino Co., in the Mojave Desert, southward and westward to the Sierra Vizcaíno in extreme northwestern Baja California Sur; also occurs on Cedros Island off the Pacific coast of Baja California.

Remarks.—Huey (1960*a*) provided brief discussions of the subspecies in Baja California. *Chaetodipus anthonyi* is treated as a subspecies of *C. fallax* herein, based primarily on the work of C. L. Blount (in litt. and pers. comm.).

Chaetodipus fallax anthonyi (Osgood, 1900)

1900. *Perognathus anthonyi* Osgood, N. Amer. Fauna, 18:56, 20 September.

Holotype.—Adult female, skin and skull, USNM 81058, from South Bay, Cerros [= Cedros] Island, Baja California, México; obtained on 29 July 1896 by A. W. Anthony.

Measurements of holotype.—Total length, 168; length of tail, 92; length of hind foot, 23.5; occipitonasal length, 25.4; basilar length of Hensel, 17.4; interorbital breadth, 6.0; width across mastoid bullae, 12.9; length of interparietal, 2.6; width of interparietals, 5.8; length of nasals, 10.2.

Distribution.—Known only from Isla Cedros, Pacific Ocean, Baja California.

Remarks.—*C. fallax anthonyi* is characterized as being smaller than typical *fallax*, but is similar in size to the subspecies, *C. f. inopinus*, on the mainland opposite Cedros Island.

Chaetodipus fallax fallax (Merriam, 1889)

1889. *Perognathus fallax* Merriam, N. Amer. Fauna, 1:19, 25 October.

Holotype.—Adult male, skin and skull, USNM 15889/22684, from San Bernardino [Reche Canyon, 3 mi SE Colton, 1,250 ft], San Bernardino Co., California; obtained by Frank Stephens on 21 April 1887.

Measurements of holotype.—Total length, 183; length of tail, 104; length of hind foot (dry), 24.0; length of ear (crown, dry), 6; occipitonasal length, 26.7; basilar length, 20.8; basilar length of Hensel, 18.7; width across mastoid bullae, 13.8; length of interparietal, 3.7; width of interparietals, 8.1; least intermastoid distance, 8.4.

Distribution.—Occupies the basins and slopes on the Pacific side of the mountains of southern California, from the base of the San Bernardino Mountains, Los Angeles Co., southward to Ensanada, Baja California along the coast, and inland to the northern end of the Sierra San Pedro Mártir near Cerro San Matías, Baja California.

Remarks.—Osgood (1900) gave average external measurements for six individuals and average cranial measurements for three individuals.

Chaetodipus fallax inopinus (Nelson and Goldman, 1929)

1929. *Perognathus fallax inopinus* Nelson and Goldman, Proc. Biol. Soc. Washington, 42:110, 25 March.

Holotype.—Adult male, skin and skull, USNM 81059, from Turtle Bay [= Bahía Tortugas; also known as Bahía San Bartolomé], Baja California Sur, México; obtained by A. W. Anthony on 1 August 1896.

Measurements of holotype.—Total length, 180; length of tail, 104; length of hind foot (dry), 23; greatest length of skull, 25.5; interorbital breadth, 6.2; length of maxillary toothrow, 3.5; width across mastoid bullae, 13.2; length of interparietal, 4.3; width of interparietals, 6.6; length of nasals (median line), 9.3; width of nasals, 2.5.

Distribution.—Found along the Pacific coast of the Baja California Peninsula, from near Punta San Carlos and Desembarcadero Santa Catarina, Baja California, southward to Bahía Tortugas, Baja California Sur.

Remarks.—Other than those of the holotype, there are no published measurements for *inopinus.*

Chaetodipus fallax majusculus (Huey, 1960)

1960. *Perognathus fallax majusculus* Huey, Trans. San Diego Soc. Nat. Hist., 12:418, 1 February.

Holotype.—Adult male, skin and skull, SDSNH 15952, from San Quintín, Baja California, México; obtained by Laurence M. Huey on 29 June 1947.

Measurements of holotype.—Total length, 191; length of tail, 105; length of hind foot, 24; length of ear, 6; greatest length of skull, 26.7; interorbital breadth, 6.9; length of maxillary toothrow, 3.6; width across mastoid bullae, 14.7; length of nasals, 9.8.

Distribution.—Found primarily on the coastal mesas of Baja California, from south of Santo Tómas southward to El Rosario.

Remarks.—No measurements have been published for specimens other than the holotype.

Chaetodipus fallax pallidus (Mearns, 1901)

1901. *Perognathus fallax pallidus* Mearns, Proc. Biol. Soc. Washington, 14:135, 9 August.

Holotype.—Adult female, skin and skull, USNM 61007, from Mountain Spring, halfway up E slope of Coast Range Mountains, on the Mexican Boundary Line, in San Diego Co., California; obtained on 16 May 1894 by Edgar A. Mearns.

Measurements of holotype.—Total length, 188, length of tail, 98; length of hind foot, 23.7; length of ear, 7; greatest length of skull, 27.20; width across bullae, 13.85; breadth across maxillary arches, 12.70; nasal length, 10.40; interorbital breadth, 6.65.

Distribution.—Southern margin of the Mojave Desert, California, along the northern slopes of the San Bernardino Mountains and the western edge of the Colorado Desert, south to the Mexican boundary.

Remarks.—Mearns (1901) listed means and ranges of external measurements for six adult females, and characterized *pallidus* as being exactly like *C. fallax fallax* in size and cranial characters, the only difference being its lighter gray color.

Chaetodipus fallax xerotrophicus (Huey, 1960)

1960. *Perognathus fallax xerotrophicus* Huey, Trans. San Diego Soc. Nat. Hist., 12:419. 1 February.

Holotype.—Adult male, skin and skull, SDSNH 8310, from 2 mi NE Chapala, Baja California, México; obtained on 15 October 1930 by Laurence M. Huey.

Measurements of holotype.—Total length, 210; length of tail, 127; length of hind foot, 24; length of ear, 6; greatest length of skull, 26.8; interorbital breadth, 6.2; length of maxillary toothrow, 3.7; width across mastoid bullae, 14.0; length of nasals, 10.4.

Distribution.—Found at inland sites in Baja California, from the central Sierra San Pedro Mártir south to near Rosarito.

Remarks.—No measurements, other than those of the holotype, are available.

Chaetodipus formosus

Diagnosis.—A medium-sized species of *Chaetodipus,* with soft, spineless pelage, relatively large ear pinnae and long tail, large mastoid bullae and narrow interparietals; mastoid bullae projecting slightly posteriad beyond the plane of the occipitals; tympanic bullae nearly meeting over the basisphenoid anteriorly; length of ear is usually greater than 9 mm; width of interparietals usually averages less than 5.9 mm and rarely ranges to about 6.5 mm.

Comparisons.—*Chaetodipus formosus* is generally similar in appearance to most other species of *Chaetodipus* without spines. Refer to the account of *arenarius* for comparison with that species. From *C. penicillatus, formosus* can be distinguished by its finer and longer pelage, more heavily haired and crested tail, longer ear, narrower interparietals, and larger mastoid bullae; the mastoid bullae of *penicillatus* do not extend posteriad to the plane of the occipitals. From *C. baileyi, formosus* can be distinguished by its smaller size, narrower interparietals, and shorter hind foot (length of hind foot of *formosus* averages less than 25 mm and rarely ranges to 26, the hind foot of *baileyi* measures 26 mm or greater and averages 27 or more). Other species of *Chaetodipus* with spineless pelage (some *artus, hispidus, lineatus, pernix*) have allopatric geographic ranges to *formosus* and differ in size and appearance (see accounts of those species for details).

Distribution.—Occupies desert slopes of the middle and southern Great Basin, Mojave and Colorado deserts, and the Gulf coastal slope of the Baja California peninsula. Ranges on the northwest from the Honey Lake Valley in extreme eastern Lassen Co., California, southward through western Nevada, and on the northeast, from near the western and southern shores of the Great Salt Lake, Utah, southward through western Utah and extreme eastern Nevada to south-central Utah and northwestern Arizona, north of the Grand Canyon of the Colorado River; thence across the southern one-third of Nevada and southward through southeastern California and the eastern slope of the Baja California peninsula to near Santa Rosalía, in northeastern Baja California Sur.

Remarks.—Refer to Hafner and Hafner (1983) and Patton et al. (1981) for detailed discussions of the systematic position of *C. formosus.*

Chaetodipus formosus cinerascens (Nelson and Goldman, 1929)

1929. *Perognathus formosus cinerascens* Nelson and Goldman, Proc. Biol. Soc. Washington, 42:105, 25 March.

Holotype.—Young adult female, skin and skull, MVZ 37685, from San Felipe, northeastern Baja California, México; obtained by Chester C. Lamb on 10 April 1926.

Measurements of holotype.—-Total length, 155; length of tail, 75; length of hind foot, 22.8; weight, 15.8 g; greatest length of skull, 24.3; interorbital breadth, 6.1; length of maxillary toothrow, 3.4; width across mastoid bullae, 13.6; length of interparietal, 3.5; width of interparietals, 5.0; length of nasals, 9.3; width of nasals, 2.2.

Distribution.—Known from an area along the Gulf slope of central Baja California, from San Felipe southward to near El Mármol.

Remarks.—No additional measurements for *cinerascens* are available.

Chaetodipus formosus formosus (Merriam, 1889)

1889. *Perognathus formosus* Merriam, N. Amer. Fauna, 1:17, 25 October.
1956. *Perognathus formosus domisaxensis* Cockrum, J. Washington Acad. Sci., 46:131, 7 May.
1956. *Perognathus formosus melanocaudus* Cockrum, J. Washington Acad. Sci., 46:132, 7 May.

Holotype.—Adult male, skin and skull, USNM 186508, from St. George, Washington Co., Utah; obtained by Vernon Bailey on 9 January 1889.

Measurements of holotype.—Total length, 195; length of tail, 111; length of hind foot, 26; length of ear (crown, dry), 6; occipitonasal length, 27.5; basilar length, 21.1; basilar length of Hensel, 19.2; width across mastoid bullae, 14.6; length of interparietal, 4.0; width of interparietals, 5.8; least intermastoid distance, 6.0.

Distribution.—Occupies south-central and southwestern Utah and northwestern Arizona; in Utah, found near St. George eastward to the Colorado River; in Arizona, ranges north and west of the Colorado River.

Remarks.—Cockrum (1956) listed measurements for 6 adults under the two subspecific synonyms listed above; Durrant (1952) gave statistics for measurements of 7 adult female topotypes; and Hoffmeister (1986) gave statistics for measurements of 4 samples ranging from 15 to 29 individuals each. Hoffmeister (1986) considered *domisaxensis* and *melanocaudus* to be synonyms of *formosus.*

Chaetodipus formosus incolatus (Hall, 1941)

1941. *Perognathus formosus incolatus* Hall, Proc. Biol. Soc. Washington, 54:56, 20 May.

Holotype.—Adult female, skin and skull, MVZ 78555, from 2 mi W Smith Creek Cave, 6,300 ft, Mt. Moriah, White Pine Co., Nevada; obtained on 18 June 1937 by Lee W. Arnold.

Measurements of holotype.—Total length, 190; length of tail, 103; length of hind foot, 23; length of ear, 12; weight, 21.5 g; greatest length of skull, 22.30; width across bullae, 13.70; breadth across maxillary arches, 12.35; nasal length, 11.00; interorbital breadth, 7.05.

Distribution—Ranges over the basin of the Pleistocene Lake Bonneville in northwestern Utah and northeastern Nevada.

Remarks.—Hall (1946) gave means and ranges of measurements for four females from the type locality.

Chaetodipus formosus infolatus (Huey, 1954)

1954. *Perognathus formosus infolatus* Huey, Trans. San Diego Soc. Nat. Hist., 12:1, 1 March.

Holotype.—Adult male, skin and skull, SDSNH 15664, from 7 mi W San Francisquito Bay, lat. 28°30′N, Gulf of California, Baja California Sur, México; obtained by Laurence M. Huey on 3 April 1947.

Measurements of holotype.—Total length, 187; length of tail, 104; length of hind foot, 25; length of ear (crown), 6; occipitonasal length, 27.5; interorbital breadth, 6.6; length of maxillary toothrow, 3.8; width across mastoid bullae, 14.6; length of nasals, 10.2.

Distribution.—Occupies the southern portion of the range of the species, from near El Mármol, Baja California, southward to near Santa Rosalía, northeastern Baja California Sur.

Remarks.—No other measurements are available for *infolatus.*

Chaetodipus formosus melanurus (Hall, 1941)

1941. *Perognathus formosus melanurus* Hall, Proc. Biol. Soc. Washington, 54:57, 20 May.

Holotype.—Adult female, skin and skull, MVZ 73442, from 6 mi E California boundary, 4,000 ft, lat. 40°28′N, Washoe Co., Nevada; obtained by E. Raymond Hall on 7 July 1936.

Measurements of holotype.—Total length, 199; length of tail, 108; length of hind foot, 25.3; length of ear, 11; weight, 21.2 g; greatest length of skull, 27.20; width across bullae, 14.40; breadth across maxillary arches, 14.40; nasal length, 11.00; interorbital breadth, 7.00; length of maxillary toothrow, 3.95; depth of cranium, 3.55.

Distribution.—Extreme northeastern California in the eastern Honey Lake Valley, Lassen Co., and northwestern Nevada in the Smoke Creek desert and adjacent areas; in Nevada, from southwestern Humboldt Co. southward to the Truckee River near Toulon, Pershing Co., and westward to near Reno, Washoe Co.

Remarks.—Hall (1946) listed means and ranges for measurements of 10 males and 11 females.

Chaetodipus formosus mesembrinus (Elliot, 1904)

1904. *Perognathus mesembrinus* Elliot, Field Columbian Mus., Publ. 87, Zool. Ser., 3:251, 7 January.

Holotype.—Adult male, skin and skull, FMNH 11784, from Palm Springs, Riverside Co., California; obtained by Edmund Heller on 24 February 1903.

Measurements of holotype.—Total length, 195; length of tail, 114; length of hind foot, 23; length of ear, 11; total length of skull, 21; basilar length of Hensel, 18; interorbital breadth, 7.0; length of maxillary toothrow, 4.0; length of mastoid bulla, 9.0; width across mastoid bullae, 14.0; length of nasals, 9.4; zygomatic breadth, 13.0; greatest parietal width, 10.5; palatal length, 10.0; length of mandible, 12.5; length of mandibular toothrow, 3.5.

Distribution.—Occupies the Colorado Desert region of southeastern California and northeastern Baja California, ranging from near the type locality on the northwest and near Needles, San Bernardino Co., on the northeast, southward on the Gulf slope of Baja California to the east-central Sierra San Pedro Mártir.

Remarks.—No other published measurements for *mesembrinus* are known. Length of skull given by Elliot (1904) suggests either an error in measuring or recording, or a juvenile specimen.

Chaetodipus formosus mohavensis (Huey, 1938)

1938. *Perognathus formosus mohavensis* Huey, Trans. San Diego Soc. Nat. Hist., 9:35, 21 November.

Holotype.—Adult male, skin and skull, SDSNH 11317, from Bonanza King Mine, Providence Mountains, San Bernardino Co., California; obtained on 4 April 1935 by Laurence M. Huey.

Measurements of holotype.—Total length, 181, length of tail, 100; length of hind foot, 24; length of ear (crown), 8; occipitonasal length, 25.8; interorbital breadth, 7.0; length of maxillary toothrow, 3.7; width across mastoid bullae, 13.8; length of nasals, 9.5.

Distribution.—Occupies the Mojave Desert area of southeastern California, and western Nevada from the Walker River southward and eastward to the extreme southwestern corner of Utah.

Remarks.—Hall (1946) listed means and ranges for measurements of four males and seven females from Nevada; Durrant (1952) listed measurements for one male and two females from Utah. Specimens from northwestern Arizona, assigned to this subspecies by Hall (1981) were considered to be *C. f. formosus* by Hoffmeister (1986).

Chaetodipus goldmani

Diagnosis.—A medium-large species of *Chaetodipus* with coarse pelage with few short, stout spines on the rump, long, rounded ear pinna, a moderately long and strongly crested tail, and a conspicuous, buffy-colored lateral stripe; baculum comparatively long and tip weakly curved.

Comparisons.—*Chaetodipus goldmani* is most likely to be confused with *C. artus,* with which it shares a portion of its geographic range (see account of *artus* for comparison). *C. goldmani* is sympatric with no other species of *Chaetodipus* with spines in the pelage, although its range approaches that of *C. intermedius*; *intermedius* is smaller, with a relatively longer tail, smaller hind foot, and shorter ear pinnae (less than 9 mm). From the potentially sympatric species without spines in the pelage, *C. goldmani* differs as follows: size smaller than *C. baileyi,* with relatively smaller mastoid bullae, and wider interparietals; size larger than *C. pernix,* hind foot significantly longer (length averages greater than 24 mm in *goldmani*), and tail much more strongly crested; size larger, pelage coarser, and lateral stripe more conspicuous than in *C. penicillatus.*

Distribution.—Occupies thornscrub and short-tree forest associations in the Río Yaqui drainage of extreme northeastern Sonora, México, southward through the coastal plains of Sonora to northern Sinaloa; on the east, ranges in the barrancas of the Pacific slope of the Sierra Madre Occidental into southwest-central Chihuahua (Patton, 1969*a*; Hall, 1981).

Remarks.—Patton (1969*a*) documented the distribution of six allopatric chromosomal races of *C. goldmani*; the species is currently monotypic. Straney and Patton (1980) documented geographic variation in structure.

Chaetodipus goldmani (Osgood, 1900)

1900. *Perognathus goldmani* Osgood, N. Amer. Fauna, 18:54, 20 September.

Holotype.—Adult female, skin and skull, USNM 96673, from Sinaloa, Sinaloa, México; obtained by Edward A. Goldman on 15 February 1899.

Measurements of holotype.—Total length, 202; length of tail, 108; length of hind foot, 28; length of ear (anterior base), 11; greatest length of skull, 28.80; width across bullae, 14.10; breadth across maxillary arches, 13.00; nasal length, 10.70; interorbital breadth, 6.25.

Remarks.—Osgood (1900) listed means of skull measurements for three specimens; Anderson (1964, 1972) and Straney and Patton (1980) provided statistics for measurements of several samples.

Chaetodipus intermedius

Diagnosis.—A smaller than average to medium-sized species of *Chaetodipus,* with relatively weak spines usually present on the rump, a long, crested tail, relatively short ear pinna, conspicuous, buffy lateral stripe, wide interparietals, and narrow skull, measured at the anterior point of the zygomatic arches; head and body length varies from about 70 to 77 mm; width of interparietals ranges from about 7.2 to 8.0 mm; anterior zygomatic breadth ranges from about 10.5 to 11.8 mm (Anderson, 1972; Weckerly and Best, 1985).

Comparisons.—*Chaetodipus intermedius* is most often confused with *C. penicillatus,* which is similar in size and proportions, and with which it is broadly sympatric. *C. intermedius* differs from *penicillatus* in having rump spines present (they may be rarely absent individually and seemingly so when animals are molting); a shorter hind foot (length usually averages 22 mm or less in *intermedius* and over 22 in *penicillatus*); narrower rostrum; narrower breadth across the anterior portion of the zygomatic arches (Anderson, 1972); and anterior extension of the supraoccipital, between the mastoid and interparietal, is shorter and comes to an acute angle anteriorly (in *penicillatus* it is longer, broader, and squared off at the anterior end; Hoffmeister and Lee, 1967). *C. intermedius* is probably most closely related to *C. nelsoni, C. artus,* and *C. goldmani,* the latter two being larger than and having allopatric geographic ranges to *intermedius*; from *nelsoni, intermedius* differs in being slightly smaller, with slightly finer, softer pelage with fewer rump spines, and in having shorter interparietals and a narrower rostrum. Other species of *Chaetodipus* found within the geographic range of *intermedius* are *C. baileyi, C. formosus* (possible marginal sympatry), and *C. hispidus*; all three are larger and lack spines in the pelage.

Distribution.—*Chaetodipus intermedius* is broadly distributed in rocky habitats in the southwestern deserts of North America, ranging from south and east of the Colorado River in extreme south-central Utah through most of western and southern Arizona to northern Sonora, central and southern New Mexico, Trans-Pecos Texas, and northern and central Chihuahua (Hall, 1981).

Remarks.—*Chaetodipus intermedius* generally is limited to rocky areas and several black colored subspecies, confined to relatively small, isolated lava flows, have been described. The species has never been reviewed systematically.

Chaetodipus intermedius ater (Dice, 1929)

1929. *Perognathus intermedius ater* Dice, Occas. Papers Mus. Zool., Univ. Michigan, 203: 2, 19 June.

Holotype.—Adult male, skin, skull, and body skeleton, UMMZ 58474, from Malpais Spring, 15 mi W Three Rivers, Otero Co., New Mexico; obtained on 17 July 1927 by Lee R. Dice.

Measurements of holotype.—Total length, 167; length of tail, 86; length of hind foot, 20; length of ear (notch), 8.0; weight, 14.3 g; occipitonasal length, 21.4; interorbital breadth, 6.3; width across mastoid bullae, 12.3; length of nasals, 9.6.

Distribution.—Known from the Carrizozo lava beds in western Lincoln and Otero counties, New Mexico (Hall, 1981).

Remarks.—Weckerly and Best (1985) provided statistics for measurements of 16 males and 14 females.

Chaetodipus intermedius beardi Weckerly, Gennaro, and Best, 1988

1988. *Chaetodipus intermedius beardi* Weckerly, Gennaro, and Best, Southwestern Nat., 33:100, 30 March.

Holotype.—Adult female, skin and skull, ENMU 8919, from 26 mi N, 15.5 mi E Engle, Socorro Co., New Mexico; obtained by Floyd W. Weckerly on 23 August 1982.

Measurements of holotype.—Total length, 157; length of tail, 91; length of hind foot, 19; length of ear, 7.7; greatest length of skull, 23.5; width across mastoid bullae, 12.8; nasal length, 8.2; interorbital breadth, 6.2; greatest depth of cranium, 8.7; rostral width, 3.6; nasal width, 2.1; interparietal length, 3.7; interparietal width, 7.1; length of maxillary toothrow, 3.6; length of mandible, 10.6.

Distribution.—Occurs only on the Pedro Armendariz lava field in Sierra and Socorro counties, New Mexico.

Remarks.—Weckerly et al. (1988) provided measurements for four adult topotypes in addition to the holotype, and means and ranges of measurements for samples of 68 adult males and 82 adult females.

Chaetodipus intermedius crinitus (Benson, 1934)

1934. *Perognathus intermedius crinitus* Benson, Proc. Biol. Soc. Washington, 47:199, 2 October.

Holotype.—Adult male, skin and skull, MVZ 55883, from 2.6 mi W Wupatki Ruins, Coconino Co., Arizona; obtained by Louise Kellogg on 8 October 1932.

Measurements of holotype.—Total length, 177; length of tail, 101; length of hind foot, 23.5; length of ear, 5; weight, 13.5 g; greatest length of skull, 25.3; interorbital breadth, 6.35; length of mastoid bulla, 7.5; width across mastoid bullae, 12.9; length of interparietal, 3.35; width of interparietals, 7.25; length of nasals, 9.35; width of rostrum, 4.45.

Distribution.—Occurs in north-central and northwestern Arizona and south-central Utah; ranging the south side of Navajo Mountain, south and east of the Colorado River and north of the Mogollon Rim; marginal localities on the east are Aztec Creek, river mile 68.5, Glen Canyon, San Juan Co., Utah; Moa Ave and Walnut Tank (10 mi N Angell Augusta), Coconino Co., Arizona (Hall, 1981).

Remarks.—Benson (1934) listed averages and ranges of measurements for 15 males; Hoffmeister (1986) gave statistics for measurements of samples of 9 and 13 individuals.

Chaetodipus intermedius intermedius (Merriam, 1889)

1889. *Perognathus intermedius* Merriam, N. Amer. Fauna, 1:18, 25 October.
1889. *Perognathus obscurus* Merriam, N. Amer. Fauna, 1:20, 25 October.
1933. *Perognathus intermedius nigrimontis* Blossom, Occas. Papers Mus. Zool., Univ. Michigan, 265:1, 21 June.

Holotype.—Adult male, skin and skull, USNM 186509, from Mud Spring, Mohave Co., Arizona; obtained by Vernon Bailey on 26 February 1889.

Measurements of holotype.—Total length, 183; length of tail, 106; length of hind foot, 21; length of ear (crown, dry), 4.5; occipitonasal length, 24.4; basilar length, 19.0; basilar length of Hensel, 16.6; width across mastoid bullae, 13.5; length of interparietal, 3.0; width of interparietals, 7.8; least intermastoid distance, 8.0.

Distribution.—Ranges from the Colorado River in western Arizona southward and eastward across Arizona south of the Mogollon Rim (but not including the lower Colorado River Valley or the western Gila River Valley) to southern and central New Mexico, Trans-Pecos Texas, and northern and central Chihuahua; in New Mexico, ranges northward in the Rio Grande lowlands to near Albuquerque; in Texas, ranges east to about western Winkler and Ward counties; in Chihuahua, ranges southward to near Ojinaga and westward to near Casas Grandes and 5 mi W San Francisco (Hall, 1981).

Remarks.—Hoffmeister (1986) noted that the type locality for *nigrimontis,* a dark-colored but not black population, was in a zone of intergradation between *intermedius* and *phasma.* Hoffmeister (1986) gave statistics for measurements of four samples from Arizona; Weckerly and Best (1985) gave statistics for measurements of two samples (68 males, 82 females; 11 males, 16 females) from New Mexico; Anderson (1972) gave statistics for external measurements of three samples and cranial measurements for one sample from Chihuahua; Genoways et al. (1977) listed measurements for two males and three females from Texas.

Chaetodipus intermedius lithophilus (Huey, 1937)

1937. *Perognathus intermedius lithophilus* Huey, Trans. San Diego Soc. Nat. Hist., 8:355, 15 June.

Holotype.—Adult male, skin and skull, SDSNH 11211, from Porto [Puerto] Libertad, summit of rocky hill 1.5 mi NNW fresh water spring on beach, Sonora, México; obtained by Laurence M. Huey on 5 February 1935.

Measurements of holotype.—Total length, 166; length of tail, 91; length of hind foot, 19; length of ear (crown), 5; occipitonasal length, 23.5; interorbital breadth, 6.2; length of maxillary toothrow, 3.4; width across mastoid bullae, 12.7; length of nasals, 9.3.

Distribution.—Known only from the type locality in northwestern Sonora.

Remarks.—Huey (1937) gave means and ranges of measurements for eight specimens, including the holotype.

Chaetodipus intermedius minimus (Burt, 1932)

1932. *Perognathus penicillatus minimus* Burt, Trans. San Diego Soc. Nat. Hist., 7:164, 31 October.

Holotype.—Adult male, skin and skull, CIT 50424, from Turner's Island, lat. 28°43′N, long. 112°19′W, Gulf of California, Sonora, México; obtained on 31 December 1931 by William H. Burt.

Measurements of holotype.—Total length, 162; length of tail, 97; length of hind foot, 20; length of ear (crown), 5, (notch, dry), 6.2; greatest length of skull, 23.9; basal length, 19.5; interorbital breadth, 6.2; length of maxillary toothrow, 3.5; length of mastoid bulla, 7.1; width across mastoid bullae, 12.5; length of interparietal, 3.6; width of interparietals, 7.1; length of nasals, 9.2.

Distribution.—Known only from the type locality.

Remarks.—Hoffmeister and Lee (1967), in their review of *Chaetodipus penicillatus,* decided that *minimus* was a subspecies of *C. intermedius*; in so far as is recorded, *minimus* is represented only by the holotype. The holotype is now in the UCLA collections.

Chaetodipus intermedius phasma (Goldman, 1918)

1918. *Perognathus intermedius phasma* Goldman, Proc. Biol. Soc. Washington, 31:22, 16 May.

1933. *Perognathus intermedius pinacate* Blossom, Occas. Papers Mus. Zool., Univ. Michigan, 273:4, 31 October.

Holotype.—Adult male, skin and skull, USNM 203003, from Tinajas Atlas, 1,400 ft, Gila Mountains, Yuma Co., Arizona; obtained by Edward A. Goldman on 23 November 1913.

Measurements of holotype.—Total length, 165; length of tail, 97; length of hind foot, 20.5; greatest length of skull, 23.0; interorbital breadth, 5.8; length of maxillary toothrow, 3.4; width across mastoid bullae, 12.3; length of interparietal, 2.6; width of interparietals, 6.1; length of nasals, 8.5.

Distribution.—Occurs in westernmost Arizona along the Colorado River, from near Hoover Dam southward, extending eastward in the Gila River Valley to Phoenix and southward to Organ Pipe National Monument; also known from a small area in northwestern Sonora and southwestern Arizona encompassed by the Pinacate Mountain lava flows and in Sonora in the Tinajas and Tule ranges (Hall, 1981; Hoffmeister, 1986).

Remarks.—Hoffmeister (1986) found that specimens of *pinacate* did not differ from other samples of *phasma* except for darker color. Hoffmeister (1986) gave statistics for four samples of *phasma* from Arizona.

Chaetodipus intermedius rupestris (Benson, 1932)

1932. *Perognathus intermedius rupestris* Benson, Univ. California Publ. Zool., 38:337, 14 April.

Holotype.—Young adult male, skin and skull, MVZ 50595, from that part of lava beds nearest Kenzin, Doña Ana Co., New Mexico; obtained on 24 October 1931 by Annie M. Alexander.

Measurements of holotype.—Total length, 169; length of tail, 92; length of hind foot, 20.5; length of ear (crown), 4; weight, 12.9 g; occipitonasal length, 23.9; frontonasal length, 16.0; interorbital breadth, 5.8; length of mastoid bulla, 7.8; width across mastoid bullae, 12.4; distance between stylomastoid foramina, 9.7.

Distribution.—Known only from the Afton Lava Flow, Doña Ana Co., south-central New Mexico.

Remarks.—Benson (1932) gave measurements of 2 paratypes and Weckerly and Best (1985) listed statistics for measurements of 50 males and 55 females.

Chaetodipus intermedius umbrosus (Benson, 1934)

1934. *Perognathus intermedius umbrosus* Benson, Proc. Biol. Soc. Washington, 47:200, 2 October 1934.

Holotype.—Adult male, skin and skull, MVZ 55964, from Camp Verde, Yavapai Co., Arizona; obtained on 3 October 1932 by Louise Kellogg.

Measurements of holotype.—Total length, 173; length of tail, 99; length of hind foot, 23; length of ear (crown), 5; weight, 16.7 g; greatest length of skull, 25.5; interorbital breadth, 6.55; length of mastoid bulla, 7.95; width across mastoid bullae, 13.3; length of interparietal, 3.6; width of interparietals, 7.65; length of nasals, 9.6; width of rostrum, 4.35.

Distribution.—Occurs in central Arizona, from the Verde Valley, Yavapai Co., on the southeast, northwestward to near Drake, in extreme south-central Coconino Co. (Hoffmeister, 1986).

Remarks.—Hoffmeister (1986) restricted the distribution of *umbrosus,* assigning specimens from the southern and eastern portion of the range previously attributed to *umbrosus* (Hall, 1981) to *intermedius* instead. Hoffmeister (1986) gave statistics for a sample of 50 individuals from near Camp Verde; Benson (1934) gave means and extremes of measurements of 7 adult males.

Chaetodipus lineatus

Diagnosis.—A medium-sized species of *Chaetodipus* with a long, crested tail; color is dull-gray dorsally, consisting of a mix of gray and blackish and lined with buffy, with a narrow, buffy lateral stripe; no spines on the rump; skull essentially the same as *C. nelsoni*; length of head and body averages around 76 to 79 mm; length of tail averaging from about 95 to 98 mm.

Comparisons.—*Chaetodipus lineatus* shares its geographic range with *C. nelsoni* and *C. penicillatus,* both of which have been taken in the same traplines with *lineatus* (Dalquest, 1951). From *C. nelsoni,* it is distinguished by its dull gray rather than dark brownish color and the lack of spines in the pelage; from *C. penicillatus, lineatus* differs in being grayer without the strong, black overwash of sympatric *penicillatus,* and in its larger size and larger and broader skull (Dalquest, 1951). *C. lineatus* may also be found with *C. hispidus,* which is much larger with a relatively short, non-crested tail.

Distribution.—Known only from a small area in southwestern San Luis Potosí and extreme southeastern Zacatecas (Dalquest, 1951; Hall, 1981; Matson and Baker, 1986).

Remarks—The status of *C. lineatus* has been questioned by several mammalogists, primarily because of the inability of others to capture specimens clearly referable to this species and because of its great similarity to *C. nelsoni,* except for its color and lack of rump spines. Variability in presence or absence of rump spines has been noted for a number of species of *Chaetodipus,* including those that typically have spines (e.g., *C. artus, C. intermedius,* molting and/or juveniles of several species), and species normally without spines (e.g., *C. penicillatus,* Hoffmeister and Lee, 1967). Furthermore, there is no definite distinction between the normal, coarse, troughed overhairs, weak bristles, and stiffer spines of all *Chaetodipus* spp. (Homan and Genoways, 1978). Three species of similar-sized *Chaetodipus* sharing the same area would be unusual. Specimens of *C. lineatus* may be spineless individuals of *C. nelsoni.* That they are also dull gray rather than the normal dark brownish suggests that a simple mutation or age is responsible for the differences. However, this matter only can be resolved by additional study.

Chaetodipus lineatus (Dalquest, 1951)

1951. *Perognathus lineatus* Dalquest, J. Washington Acad. Sci., 41:362, 14 November.

Holotype.—Adult male, skin and skull, LSUMZ 5253, from 1 km S Arriaga, San

Luis Potosí, México; obtained on 21 September 1950 by Walter W. Dalquest.

Measurements of holotype.—Total length, 175; length of tail, 98; length of hind foot, 23; length of ear, 8; greatest length of skull, 25.20; width across bullae, 14.00; breadth across maxillary arches, 11.90; nasal length, 8.80; interorbital breadth, 6.55.

Remarks.—Dalquest (1951) listed means for measurements of eight males and seven females.

Chaetodipus nelsoni

Diagnosis.—A medium-sized species of *Chaetodipus* with a long, crested tail, relatively coarse hair with spines on the rump, and short ear pinnae (8 mm or less); skull moderately narrow across anterior portion of zygomatic arches; interparietals relatively long; length of head and body averages between about 76 and 80 mm; length of tail averages about 98 to 110 mm; length of interparietal averages 3.7 mm (range 3.4 to 3.9); width of interparietals ranges from about 6.7 to 8.1 mm (mean 7.5); anterior zygomatic breadth averages 11.9 mm (range 11.0 to 12.9).

Comparisons.—*Chaetodipus nelsoni* is most likely to be confused with *C. penicillatus,* which is similar in size and proportions and with which it is broadly sympatric. *C. nelsoni* differs from *penicillatus* in usually having rump spines, narrower rostrum, and narrower breadth across the anterior portion of the zygomatic arches (Anderson, 1972). In *nelsoni,* the anterior extension of the supraoccipital, between the mastoid and interparietal, is shorter and comes to an acute angle anteriorly (in *penicillatus* it is longer, broader and squared off at the anterior end). *C. nelsoni* typically has more spines and a broader interorbital region and longer interparietals than *artus* (Anderson, 1972). *C. goldmani* is larger with relatively smaller mastoid bullae. See account of *intermedius* for comparison with that species. The only other potentially sympatric species is *C. hispidus,* a much larger animal with a relatively short, noncrested tail and without spines on the rump.

Distribution.—Ranges from southeastern New Mexico, southward through south-central Chihuahua, the eastern one-half of Durango (except for the northeastern corner), extreme northeastern Jalisco, Zacatecas, and extreme east-central Jalisco; on the east, ranges to south-central Nuevo León, southwestern Tamaulipas, and central San Luis Potosí (Hall, 1981).

Remarks.—The relationship between *C. nelsoni* and *C. lineatus* appears to be close; *lineatus* may not be a valid species, but rather a name applied to variant specimens of *nelsoni* (see account of *C. lineatus*). Aside from *C. lineatus, C. nelsoni* appears to be most closely related to *C. artus, C. goldmani,* and *C. intermedius,* the former two having allopatric geographic ranges to *nelsoni.*

Chaetodipus nelsoni canescens (Merriam, 1894)

1894. *Perognathus* (*Chaetodipus*) *intermedius canescens* Merriam, Proc. Acad. Nat. Sci. Philadelphia, 46:267, 27 September.
1938. *Perognathus collis* Blair, Occas. Papers Mus. Zool., Univ. Michigan, 381:1, 20 June.
1938. *Perognathus collis popei* Blair, Occas. Papers Mus. Zool., Univ. Michigan, 381:3, 20 June.

Holotype.—Young adult male, skin and skull, USNM 51016, from Jaral, Coahuila, México; obtained by Clark P. Streator on 14 January 1893.

Measurements of holotype.—Total length, 193; length of tail, 117; length of hind foot, 22; length of ear (anterior base, dry), 8; occipitonasal length, 25.0; basilar length of Hensel, 17.5; interorbital breadth, 6.1; width across mastoid bullae, 13.5; length of interparietal, 3.7; width of interparietals, 7.2; length of nasals, 9.3.

Distribution.—Ranges from southeastern New Mexico, near the boundary with Texas, southward through eastern Chihuahua to northeastern Durango and southern Coahuila; probably also occurs in extreme west-central Nuevo León; easternmost localities in Texas are Sheffield, Pecos Co., and Comstock, Val Verde Co. (Hall, 1981).

Remarks.—Anderson (1972) gave statistics for measurements of 14 specimens from Chihuahua, and Baker (1956) listed measurements (means, ranges) for 5 males and 8 females from Coahuila.

Chaetodipus nelsoni nelsoni (Merriam, 1894)

1894. *Perognathus (Chaetodipus) nelsoni* Merriam, Proc. Acad. Nat. Sci. Philadelphia, 46: 266, 27 September.

Holotype.—Old adult female, skin and skull, USNM 50214, from Hacienda La Parada, San Luis Potosí, México; obtained by Edward W. Nelson on 19 August 1892.

Measurements of holotype.—Total length, 190; length of tail, 105; length of hind foot, 24; length of ear (anterior base, dry), 8; greatest length of skull, 26.05; width across bullae, 13.00; breadth across maxillary arches, 12.30; nasal length, 10.05; interorbital breadth, 6.75.

Distribution—Ranges from south-central Chihuahua southward through the eastern one-half of Durango (except for the northeastern corner), extreme northeastern Jalisco, Zacatecas (except montane forests in western part), and extreme east-central Jalisco; on the east, ranges to south-central Nuevo León, southwestern Tamaulipas, and central San Luis Potosí (Hall, 1981; Matson and Baker, 1986).

Remarks.—Alvarez (1963) listed means and ranges of measurements for nine specimens from Tamaulipas. Baker (1956) gave means and ranges of measurements for seven males and two females from Coahuila. Dalquest (1953) listed means of measurements for 10 males and 10 females from San Luis Potosí. Genoways and Jones (1973) gave statistics for seven adults from Zacatecas, and Osgood (1900) listed means for measurements of three specimens from the type locality.

Chaetodipus penicillatus

Diagnosis.—A medium-sized species of *Chaetodipus,* with a relatively long, crested, and sparsely-haired tail, short ear pinnae, no spines on the rump, and a buffy lateral stripe that is comparatively faint or absent; proximal two-thirds of tail relatively sparsely haired and with annulated pattern; anterior extension of the supraoccipital between interparietal and mastoid relatively broad, straplike, and squared at the anterior end; mastoid bulla relatively small, but interparietals not comparatively broader; average length of head and body ranges from about 77 to 87 mm; average length of tail ranges from about 88 to 101 mm; length of ear averages 8 mm or less.

Comparisons.—*Chaetodipus penicillatus* is similar in general appearance to other small and medium-sized species of *Chaetodipus* without spines on the rump, and to *C. intermedius* and *C. nelsoni,* both of which normally have spines (accounts of the latter two species have comparisons). See accounts of *C. arenarius, C. baileyi,* and *C. formosus* for methods of distinguishing those species from *C. penicillatus. C. penicillatus* also is generally similar to *C. pernix*; *penicillatus* differs in being larger, lacking a conspicuous, buffy lateral stripe, and having a larger crest of hairs on the tail, interorbital breadth averaging 5.9 mm or greater (versus less than 5.8 in *pernix*), and longer mastoid bullae usually measuring more than 7.5 mm rather than less (Anderson, 1972; Hall, 1981; Hoffmeister and Lee, 1967).

Distribution.—Nearly always associated with creosote bush (*Larrea* spp.) communities in the deserts of southwestern North America, ranging from southeastern California and northwestern Baja California, eastward and southward through southern Nevada to extreme southwestern Utah, central and southern Arizona south of the Mogollon Rim, all of Sonora except the southern and southeastern one-fifth, southern New Mexico, the Trans-Pecos region of western Texas, and the desert portions of the Mexican Plateau, including the northern and eastern halves of Chihuahua, all but the northeastern segment of Coahuila, eastern Durango, northeastern Zacatecas, the northern half of San Luis Potosí, and the southern extreme of Nuevo León; may also occur in the extreme southwestern portion of Tamaulipas (Hall, 1981).

Remarks.—Patton (1969*b*) and Patton et al. (1981) found three chromosomal and biochemical races of *C. penicillatus* which corresponded to the groups, based on geographic variation of structure, documented by Hoffmeister and Lee (1967). These "genetic" races did not correspond precisely to Hoffmeister and Lee's taxonomic assignments, however. The locations of the three major races are: Sonora and eastern Arizona (*pricei* and the eastern segment of *penicillatus*); Chihuahuan Desert (*eremicus* and *atrodorsalis*); Mojave and Colorado desert areas of California, Baja California, and western Arizona (*angustirostris,* western Arizona segment of *penicillatus*). *C. p. stephensi* and *sobrinus* of the northern Mojave Desert, California and southern Nevada and Utah, respectively, have not been studied electrophoretically or karyotypically. *C. p. seri* of Tiburón Island has the same karyotype as *pricei* from the mainland of Sonora (Patton, 1969*b*). Individuals of the first two genetic races have been taken at one locality near the boundary of the Sonoran and Chihuahuan deserts, on the Deming Plain of southwestern New Mexico, without evidence of hybridization. Some evidence of hybridization of the first and third genetic races has been found, however (Patton et al., 1981). The significance of these findings to the specific status of the three races has not yet been determined. Hoffmeister and Lee (1967) reviewed *C. penicillatus* systematically and provided statistics for measurements and diagnoses for each subspecies. Although generally considered to lack spines on the rump, Hoffmeister and Lee (1967) noted that a few individuals from scattered localities in Arizona and many individuals in a population from the Graham Mountains, Arizona, had spines.

Chaetodipus penicillatus angustirostris (Osgood, 1900)

1900. *Perognathus penicillatus angustirostris* Osgood, N. Amer. Fauna, 18:47, 20 September.

Holotype.—Adult male, skin and skull, USNM 73881, from Carriso [= Carrizo] Creek, Colorado Desert, Imperial Co., California; obtained by A. W. Anthony on 31 March 1895.

Measurements of holotype.—Total length, 191; length of tail, 105; length of hind foot (dry), 24.4; greatest length of skull, 27.10; width across bullae, 13.30; breadth across maxillary arches, 12.90; nasal length, 11.10; interorbital breadth, 6.40.

Distribution.—Found in southeastern California, from near Peck's Butte, Los Angeles Co., southward to the region around San Felipe, Baja California on the Gulf slope, and eastward to the Colorado River at Fort Yuma, Imperial Co., California.

Remarks.—Length of foot (dry) listed on the tag of the holotype was 23.9 mm. Based on research of Patton (1969*b*) and Patton et al. (1981), the western Arizona population of *C. penicillatus penicillatus* probably should be allied with *angustirostris* rather than with populations of *penicillatus* from central Arizona.

Chaetodipus penicillatus atrodorsalis (Dalquest, 1951)

1951. *Perognathus penicillatus atrodorsalis* Dalquest, J. Washington Acad. Sci., 41:362, 14 November.

Holotype.—Adult male, skin and skull, LSUMZ 5226, from 7 km W Presa de Guadalupe, San Luis Potosí, México; obtained by Walter W. Dalquest on 12 October 1950.

Measurements of holotype.—Total length, 169; length of tail, 96; length of hind foot, 22; length of ear, 8; greatest length of skull, 24.90; width across bullae, 12.80; breadth across maxillary arches, 12.05; nasal length, 9.90; interorbital breadth, 5.90.

Distribution.—Known from northern and eastern San Luis Potosí and southern Nuevo León (Hall, 1981).

Remarks.—Dalquest (1951) listed means for 9 adult males and 12 adult females from the vicinity of the type locality; Dalquest (1953) listed means for 8 males from a second locality in San Luis Potosí.

Chaetodipus penicillatus eremicus (Mearns, 1898)

1898. *Perognathus* (*Chaetodipus*) *eremicus* Mearns, Bull. Amer. Mus. Nat. Hist., 10:300, 31 August.

Holotype.—Adult female, skin and skull, USNM 21052/36094, from Fort Hancock, El Paso [now in Hudspeth] Co., Texas; obtained by Edgar A. Mearns on 27 June 1893.

Measurements of holotype.—Total length, 163; length of tail, 83; length of hind foot, 22.1; basilar length of Hensel, 17.5; interorbital breadth, 6.4; width across mastoid bullae, 12.6; length of interparietal, 3.0; width of interparietals, 7.0; length of nasals, 9.8.

Distribution.—Ranges from southern New Mexico, southward through Trans-Pecos Texas (Winkler Co. on the northwest and Val Verde Co. on the southeast) and the desert regions of Chihuahua, Coahuila, Durango, Zacatecas, and the western corner of San Luis Potosí (Hall, 1981).

Remarks.—Anderson (1972) listed statistics for measurements of 18 specimens from Chihuahua, Genoways et al. (1977) gave measurements for 4 specimens from Texas, and Baker (1956) gave means and ranges of measurements for 8 adult males and 5 adult females from Coahuila.

Chaetodipus penicillatus penicillatus (Woodhouse, 1852)

1852. *Perognathus penecillatus* [*sic*] Woodhouse, Proc. Acad. Nat. Sci. Philadelphia, 6:200, December.

Holotype.—Type specimen not specified by Woodhouse, but later regarded as adult male, mounted skin, skull, and baculum, USNM 2676/37437 (Merriam, 1889; Osgood, 1900), from San Francisco Mountains, New Mexico (later thought to be northeast side of San Francisco Mountains, Coconino Co., Arizona); type locality subsequently fixed as: 1 mi SW Parker, Yuma Co., Arizona (Hoffmeister and Lee, 1967).

Measurements of holotype.—Total length, 204; length of tail, 115; length of hind foot, 25.5; occipitonasal length, 27.55; interorbital breadth, 6.6; length of maxillary toothrow, 3.9; length of mastoid bulla, 8.5; width across mastoid bullae, 13.7; length of nasals, 10.9.

Distribution.—Known from west-central Arizona, from the Colorado River Valley in the vicinity of Toprock, Mohave Co., south to near Yuma, Yuma Co., and eastward south of the Mogollon Rim to San Carlos Reservoir, Gila County; the southern boundary generally parallels the course of the Gila River, although populations assigned to this subspecies are found at some localities several miles south of the river (Hall, 1981; Hoffmeister, 1986; Hoffmeister and Lee, 1967).

Remarks.—The karyotypic and biochemical information reported by Patton (1969*b*) and Patton et al. (1981) suggests that western Arizona populations of *C. penicillatus penicillatus* are allied with populations of *angustirostris* from California and Baja California.

Chaetodipus penicillatus pricei (J. A. Allen, 1894)

1894. *Perognathus pricei* J. A. Allen, Bull. Amer. Mus. Nat. Hist., 6:318, 7 November.

Holotype.—Subadult male, skin and skull, AMNH 8359/6685, from Oposura [= Moctezuma], Sonora, México; obtained by B. C. Condit on 31 May 1894.

Measurements of holotype.—Total length, 157; length of tail, 90; length of hind foot, 23; length of ear, 7.5; greatest length of skull, 23.0; basilar length, 18; interorbital breadth, 5.5; length of maxillary toothrow, 3.5; width across mastoid bullae, 11.5; length of interparietal, 4.0; width of interparietals, 8.0; length of nasals, 7.7; zygomatic breadth, 11.5; length of rostrum, 9.6.

Distribution.—Found across southern Arizona, generally south of the Gila River, southward to the vicinity of the Río Mayo, near Navajoa, Sonora, and eastward to southwestern New Mexico near Deming (Hall, 1981; Patton et al., 1981).

Remarks.—The boundary between *pricei* and *eremicus* in southwestern New Mexico is not established in detail; Patton et al. (1981) reported capturing specimens of both subspecies, based on chromosomes and tissue allele patterns, from the same traplines near Deming, Luna Co., New Mexico.

Chaetodipus penicillatus seri (Nelson, 1912)

1912. *Perognathus penicillatus goldmani* Townsend, Bull. Amer. Mus. Nat. Hist., 31:122, 14 June.

1912. *Perognathus penicillatus seri* Nelson, Proc. Biol. Soc. Washington, 25:116, 29 June. Renaming of *goldmani* Townsend, 1912.

Holotype.—Male, skin and skull, USNM 198411, from Tiburón Island, Gulf of California, Sonora, México; obtained by H. E. Anthony on 13 April 1911.

Measurements of holotype.—Total length, 171; length of tail, 90; length of hind foot, 23.0; greatest length of skull, 24.80; width across bullae, 11.90; breadth across maxillary arches, 11.25; nasal length, 9.60; interorbital breadth, 6.10.

Distribution.—Restricted to Tiburón Island, Gulf of California, Sonora.

Remarks.—Hoffmeister and Lee (1967) listed the type specimen as AMNH 31845; it is as given above. Burt (1932) gave means and ranges of measurements of 18 specimens.

Chaetodipus penicillatus sobrinus (Goldman, 1939)

1939. *Perognathus penicillatus serosus* Goldman, Proc. Biol. Soc. Washington, 52:34, 11 March.

1939. *Perognathus penicillatus sobrinus* Goldman, J. Mamm., 20:257, 15 May. Renaming of *serosus* Goldman, 1939.

Holotype.—Adult male, skin and skull, USNM 27598/39697; from sand flat along Virgin River, 7 mi above Bunkerville, Clark Co., Nevada [considered to be Mohave Co., Arizona by Hardy, J. Mamm., 30:435, 17 November 1949]; obtained on 9 May 1891 by Vernon Bailey.

Measurements of holotype.—Total length, 202; length of tail, 102; length of hind foot, 26; occipitonasal length, 27.6; interorbital breadth, 6.4; length of maxillary toothrow, 4.0; width across mastoid bullae, 13.5; length of interparietal, 3.2; width of interparietals, 7.5; length of nasals, 11.1; zygomatic breadth, 14.2.

Distribution.—Ranges from the extreme southwestern corner of Utah and extreme northwestern corner of Arizona to near Davis, Mohave Co., southward and westward through southern Nevada (Hall, 1981).

Remarks.—The state in which the type locality is found has been disputed (Hoffmeister and Lee, 1967). Hall (1946) gave means and ranges of measurements of 15 males and 7 females from Nevada, listed as *sobrinus,* and 10 males and 10 females listed as *penicillatus.* Hoffmeister (1986) gave statistics for six males and six females from Arizona.

Chaetodipus penicillatus stephensi (Merriam, 1894)

1894. *Perognathus* (*Chaetodipus*) *stephensi* Merriam, Proc. Acad. Nat. Sci. Philadelphia, 46:267, 27 September.

Holotype.—Adult male, skin and skull, USNM 27774/39873, from Mesquite Valley, NW arm Death Valley, −13 ft, Inyo Co., California; obtained by Frank Stephens on 6 April 1891.

Measurements of holotype.—Total length, 177; length of tail, 96; length of hind foot, 21; length of ear (anterior base, dry), 7.5; occipitonasal length, 22.7; basilar length of Hensel, 16.0; interorbital breadth, 6.0; width across mastoid bullae, 12.0; length of interparietal, 3.0; width of interparietals, 6.7; length of nasals, 9.0.

Distribution.—Known only from the northern end of Death Valley, Inyo Co., California (Hoffmeister and Lee, 1967).

Remarks.—Hoffmeister and Lee (1967) noted that the specimens of *stephensi* available to them were too few and too variable in size to characterize adequately the subspecies or determine its relationships to *angustirostris* and *penicillatus.*

Chaetodipus pernix

Diagnosis.—A smaller than average species of *Chaetodipus,* with spineless pelage, a long, small-crested tail and a conspicuous, buffy lateral stripe; relatively small mastoid bullae, wide interparietals, and wide rostrum; anterior extensions of supraoccipital narrow and ending in a more or less acute point between the mastoid bulla and the interparietal; length of head and body ranges from about 68 to 78 mm; length of tail from about 94 to 97 mm.

Comparisons.—*Chaetodipus pernix* is most likely to be confused with *C. penicillatus,* with which it shares the northern portion of its range; see account of *penicillatus* for comparison. *C. pernix* is also similar in appearance to the allopatric species, *C. arenarius* (see account of that species for comparison). *C. pernix* differs from the sympatric species, *C. goldmani,* by being much smaller, having a smaller crest on the tail, and lacking spines on the rump. From the sympatric species, *C. baileyi, C. pernix* differs in being much smaller (see also the account of *baileyi*). Refer to the account of *C. artus* for characters distinguishing that potentially sympatric species.

Distribution.—Occurs on the coastal plains of western México, from central Sonora southward to northern Nayarit (Hall, 1981).

Remarks.—Patton et al. (1981) noted extensive differences in the karyotypes of *C. pernix pernix* and *C. pernix rostratus* (2n = 36 or 38 and 52, respectively) and a corresponding biochemical difference of lesser magnitude. Little data are available on structural variation.

Chaetodipus pernix pernix (J. A. Allen, 1898)

1898. *Perognathus pernix* J. A. Allen, Bull. Amer. Mus. Nat. Hist., 10:149, 12 April.

Holotype.—Adult male, skin and skull, BM(NH) 98.3.2.126, from Rosario, Sinaloa, México; obtained by P. O. Simmons on 22 February 1897 [field number of collector, 139].

Measurements of holotype.—Total length, 165; length of tail, 90; length of hind foot, 22; length of ear, 9; greatest length of skull, 25.0; basal length, 19.6; interorbital breadth, 6.0; width across mastoid bullae, 12.0; length of nasals, 8.5.

Distribution.—Ranges from north-central Sinaloa, near Pericos, southward to northern Nayarit.

Remarks.—Osgood (1900) gave average measurements for three specimens from the type locality.

Chaetodipus pernix rostratus (Osgood, 1900)

1900. *Perognathus pernix rostratus* Osgood, N. Amer. Fauna, 18:51, 20 September.

Holotype.—Young adult male, skin and skull, USNM 95818, from Camoa, Río

Mayo, Sonora, México; obtained by Edward A. Goldman on 28 October 1898.

Measurements of holotype.—Total length, 162; length of tail, 94; length of hind foot, 23.5; greatest length of skull, 23.30; width across bullae, 11.40; breadth across maxillary arches, 11.15; nasal length, 8.75; interorbital breadth, 5.45.

Distribution.—Ranges from near Tecoripa, in south-central Sonora, southward to near Guamuchil, north-central Sinaloa (Hall, 1981).

Remarks.—Osgood (1900) listed average measurements of three specimens from the type locality.

Chaetodipus spinatus

Diagnosis.—A medium- to medium-large sized species of *Chaetodipus,* with a long, crested tail, small ear pinnae, numerous spines on the rump and often on the flanks, and without, or nearly without, a lateral stripe of buffy color; skull comparatively flat dorsally; skull relatively narrow; mastoid bullae small; interparietals moderate in width; length of head and body ranging from about 75 to 95 mm; length of ear from about 5 to 8.5 mm; width of interparietals ranges from about 7.0 to 8.4 mm; length of mastoid bulla ranges from about 6.4 to 8.2 mm.

Comparisons.—*Chaetodipus spinatus* is generally similar in appearance to *C. californicus* and *C. fallax,* both of which may be marginally sympatric with *spinatus*; and to *C. intermedius,* which is found on the opposite side (eastern) of the Colorado River from *spinatus,* along the Arizona-California boundary. *C. spinatus* differs from these species in lacking or having a weakly developed, buffy, lateral stripe (see accounts of *californicus* and *fallax*); from *intermedius, spinatus* also differs in having more numerous and more strongly developed rump spines and usually flank spines as well, and in its larger size; other potentially sympatric species of *Chaetodipus* all lack spines in the pelage (*arenarius, baileyi, formosus,* and *penicillatus*).

Distribution.—Generally found on rocky, desert slopes and other rocky areas in the Colorado Desert of southeastern California and the Baja California peninsula; ranges from the extreme southern tip of Nevada along the Colorado River, and from near Palm Springs, Riverside Co., California, southward in Baja California to Cabo San Lucas, Baja California Sur; on the Baja California peninsula, not recorded from the cooler Pacific coastal plains north of the Sierra de la Laguna in the cape region, but found on Magdalena and Margarita islands off of the Pacific coast of Baja California Sur (Hall, 1981; Huey, 1964).

Remarks.—Populations of this species are found on most of the islands on the Baja California peninsula side of the Gulf of California, as well as two islands on the Pacific side of the peninsula. With one exception, each of these populations is considered to be a different subspecies, resulting in the relatively large number of subspecies for *C. spinatus.* The species has never been reviewed systematically.

Chaetodipus spinatus broccus (Huey, 1960)

1960. *Perognathus spinatus broccus* Huey, Trans. San Diego Soc. Nat. Hist., 12:410, 1 February.

Holotype.—Adult male, skin and skull, SDSNH 6891, from San Ignacio, lat. 27°17'N, Baja California Sur, México; obtained by Laurence M. Huey on 18 March 1928.

Measurements of holotype.—Total length, 204; length of tail, 118; length of hind foot, 22; length of ear (crown), 5; greatest length of skull, 25.6; interorbital breadth, 6.6; length of maxillary toothrow, 3.4; width across mastoid bullae, 13.1; length of nasals, 9.8.

Distribution.—Ranges over the northern three-fourths of the peninsula in Baja California Sur, generally on the slopes of the Sierra de la Giganta.

Remarks.—Huey (1960*b*) gave means of

measurements of five males from the type locality.

Chaetodipus spinatus bryanti (Merriam, 1894)

1894. *Perognathus bryanti* Merriam, Proc. California Acad. Sci., ser. 2, 4:458, 25 September.

Holotype.—Adult male, skin and skull, CAS 551/594, from San José Island, Gulf of California, Baja California Sur; obtained by Walter E. Bryant on 2 May 1892.

Measurements of holotype.—Total length, 216; length of tail, 127; length of hind foot, 25; length of ear, 11; greatest length of skull, 26.65; width across bullae, 12.90; breadth across maxillary arches, 12.30; nasal length, 10.65; interorbital breadth, 6.7; length of maxillary toothrow, 3.55; depth of cranium, 3.50.

Distribution.—Known only from Isla San José, Baja California Sur.

Remarks.—Benson (1930) gave means and ranges of measurements of 7 adults, Burt (1932) listed means and ranges of measurements for 23 specimens, and Osgood (1900) gave means for 3 individuals, including the holotype.

Chaetodipus spinatus evermanni (Nelson and Goldman, 1929)

1929. *Perognathus evermanni* Nelson and Goldman, Proc. Biol. Soc. Washington, 42:111, 25 March.

Holotype.—Adult male, skin and skull, CAS 3937, from Mejía Island, near north end of Angel de la Guardia [= Guarda] Island, Gulf of California, Baja California; obtained by V. W. Owen on 28 June 1921.

Measurements of holotype.—[External measurements from dry skin] Total length, 156; length of tail, 80; length of hind foot, 20.7; greatest length of skull, 24.2; interorbital breadth, 5.6; length of maxillary toothrow, 3.4; width across mastoid bullae, 11.7; length of interparietal, 3.0; width of interparietals, 6.5; length of nasals, 9.6; width of nasals, 2.5.

Distribution.—Found only in Isla Mejía, Gulf of California, Baja California.

Remarks.—Burt (1932) and Banks (1967) gave means and ranges of measurements for 4 and 27 specimens, respectively.

Chaetodipus spinatus guardiae (Burt, 1932)

1932. *Perognathus spinatus guardiae* Burt, Trans. San Diego Soc. Nat. Hist., 7:165, 31 October.

Holotype.—Adult male, skin and skull, CIT 50495, from Puerto Refugio, north end Angel de la Guardia [= Guarda] Island, 30 ft, Gulf of California, Baja California, México; obtained on 7 January 1932 by William H. Burt.

Measurements of holotype.—Total length, 164; length of tail, 89; length of hind foot, 22; length of ear (crown), 5, (notch, dry), 6.7; greatest length of skull, 24.1; basal length, 19.6; interorbital breadth, 6.1; length of maxillary toothrow, 4.0; length of mastoid bulla, 6.9; width across mastoid bullae, 12.1; length of interparietal, 3.0; width of interparietals, 7.0; length of nasals, 9.7.

Distribution.—Occurs only on Isla Angel de la Guarda, Gulf of California, Baja California.

Remarks.—Burt (1932) listed means and extremes of measurements of 36 specimens; Banks (1967) gave means and ranges of measurements of 12 specimens. The holotype is now in the UCLA collections.

Chaetodipus spinatus lambi (Benson, 1930)

1930. *Perognathus spinatus lambi* Benson, Univ. California Publ. Zool., 32:452, 6 September.

Holotype.—Young adult female, skin and skull, MVZ 42938, from San Gabriel, Espíritu Santo Island, Gulf of California, Baja California Sur, México; obtained by Chester C. Lamb on 9 January 1929.

Measurements of holotype.—Total length, 175; length of tail, 105; length of hind foot,

23; length of ear (crown), 6; greatest length of skull, 24.85; interorbital breadth, 6.1; length of mastoid bulla, 6.85; width across mastoid bullae, 11.85; length of interparietal, 3.6; width of interparietals, 7.25; length of nasals, 9.85; width of rostrum, 3.9.

Distribution.—Known only from Isla Espíritu Santo, Gulf of California, Baja California Sur.

Remarks.—Benson (1930) and Burt (1932) listed means and ranges of measurements for five and eight adults, respectively.

Chaetodipus spinatus latijugularis (Burt, 1932)

1932. *Perognathus spinatus latijugularis* Burt, Trans. San Diego Soc. Nat. Hist., 7:168, 31 October.

Holotype.—Adult male, skin and skull, Donald R. Dickey collection (UCLA) 18020, from San Francisco Island, lat. 24°50'N, long. 110°34'W, Gulf of California, Baja California Sur, México; obtained on 19 March 1930 by H. H. Sheldon.

Measurements of holotype.—Total length, 188; length of tail, 110; length of hind foot, 22; length of ear (notch, fresh), 9, (dry), 7.1; greatest length of skull, 25.7; basal length, 21.2; interorbital breadth, 6.7; length of maxillary toothrow, 4.0; length of mastoid bulla, 7.8; width across mastoid bullae, 12.7; length of interparietal, 3.7; width of interparietals, 7.5; length of nasals, 9.3.

Distribution.—Known only from Isla San Francisco, Gulf of California, Baja California Sur.

Remarks.—Burt (1932) gave means and ranges for measurements of nine specimens.

Chaetodipus spinatus lorenzi (Banks, 1967)

1967. *Perognathus spinatus lorenzi* Banks, Proc. Biol. Soc. Washington, 80:101, 28 July.

Holotype.—Adult male, skin and skull, SDSNH 19901, from South San Lorenzo Island, lat. 28°36'N, long. 112°51'W, Gulf of California, Baja California, México; obtained by Richard C. Banks on 22 October 1964.

Measurements of holotype.—Total length, 169; length of tail, 93; length of hind foot, 20; length of ear, 8; weight, 13.4 g; greatest length of skull, 23.9; interorbital breadth, 6.1; length of maxillary toothrow, 3.5; width across mastoid bullae, 12.2; length of nasals, 9.4; depth of skull, 7.7.

Distribution.—Occurs on South and North San Lorenzo islands, Gulf of California, Baja California.

Remarks.—Banks (1967) gave means and ranges of measurements for 20 specimens. South San Lorenzo Island is listed on recent Mexican maps as Isla San Lorenzo; North San Lorenzo Island (28°42'N, 112°57'W) is also known as Isla Las Animas.

Chaetodipus spinatus magdalenae (Osgood, 1907)

1907. *Perognathus spinatus magdalenae* Osgood, Proc. Biol. Soc. Washington, 20:21, 23 February.

Holotype.—Adult female, skin and skull, USNM 146102, from Magdalena Island, Pacific Ocean, Baja California Sur, México; obtained by Edward W. Nelson and Edward A. Goldman on 25 November 1905.

Measurements of holotype.—Greatest length of skull, 26.4; basilar length, 17.9; interorbital breadth, 6.9; length of maxillary toothrow, 4.3; width across mastoid bullae, 12.6; length of interparietal, 3.6; width of interparietals, 8.1; length of nasals, 10.5; zygomatic width, 12.8; length of diastema, 6.0.

Distribution.—Known only from Isla Magdalena, Pacific Ocean, Baja California Sur.

Remarks.—Osgood (1907) listed external measurements (means, extremes) for 10 topotypes.

Chaetodipus spinatus macrosensis (Burt, 1932)

1932. *Perognathus spinatus macrosensis* Burt, Trans. San Diego Soc. Nat. Hist., 7:166, 31 October.

Holotype.—Adult male, skin and skull, CIT 50604, from San Marcos Island, lat. 27°13′N, long. 112°05′W, Gulf of California, Baja California Sur, México; obtained on 18 January 1932 by William H. Burt.

Measurements of holotype.—Total length, 174; length of tail, 102; length of hind foot, 22; length of ear (crown), 5, (notch, dry), 7.8; greatest length of skull, 25.4; basal length, 20.9; interorbital breadth, 6.2; length of maxillary toothrow, 3.6; length of mastoid bulla, 7.5; width across mastoid bullae, 12.5; length of interparietal, 3.6; width of interparietals, 7.1; length of nasals, 9.8.

Distribution.—Known only from Isla San Marcos, Gulf of California, Baja California Sur.

Remarks.—Burt (1932) gave means and ranges of measurements of 17 topotypes. The holotype is now in the UCLA collections.

Chaetodipus spinatus margaritae (Merriam, 1894)

1894. *Perognathus margaritae* Merriam, Proc. California Acad. Sci., ser. 2, 4:459, 25 September.

Holotype.—Adult female, skin and skull, CAS 90, from Santa Margarita Island, Pacific Ocean, Baja California Sur, México; obtained on 2 March 1889 by Walter E. Bryant.

Measurements of holotype.—[External measurements from dry skin] Total length, 170; length of tail, 102; length of hind foot, 22.5; length of ear (anterior base), 8.5; occipitonasal length, 25.9; basilar length of Hensel, 18.0; interorbital breadth, 6.5; width across mastoid bullae, 12.0; length of interparietal, 3.7; width of interparietals, 8.0; length of nasals, 10.3.

Distribution.—Known only from Isla Santa Margarita, Pacific Ocean, Baja California Sur.

Remarks.—Merriam (1894*c*) gave external measurements of two specimens (measured fresh); no other data on measurements are available for *margaritae.*

Chaetodipus spinatus occultus (Nelson, 1912)

1912. *Perognathus spinatus nelsoni* Townsend, Bull. Amer. Mus. Nat. Hist., 31:122, 14 June.

1912. *Perognathus spinatus occultus* Nelson, Proc. Biol. Soc. Washington, 25:116, 29 June. Renaming of *nelsoni* Townsend, 1912.

Holotype.—Male, skin and skull, USNM 198409, from Carmen Island, Gulf of California, Baja California Sur, México; obtained by H. E. Anthony on 3 April 1911.

Measurements of holotype.—Total length, 170; length of tail, 88; length of hind foot, 24.5; greatest length of skull, 26.05; width across bullae, 12.35; breadth across maxillary arches, 12.20; nasal length, 10.50; interorbital breadth, 6.45.

Distribution.—Known only from Isla del Carmen, Gulf of California, Baja California Sur.

Remarks.—Townsend (1912) gave average external measurements of two males and one female; Burt (1932) listed means and ranges of measurements for five specimens.

Chaetodipus spinatus oribates (Huey, 1960)

1960. *Perognathus spinatus oribates* Huey, Trans. San Diego Soc. Nat. Hist., 12:409, 1 February.

Holotype.—Adult male, skin and skull, SDSNH 18742, from San Fernando Mission, lat. 30° N, Baja California, México; obtained by Laurence M. Huey on 27 February 1958.

Measurements of holotype.—Total length, 192; length of tail, 112; length of hind foot, 22; length of ear (crown), 5; greatest length of skull, 24.9; interorbital breadth, 6.5; length of maxillary toothrow, 3.5; width across mastoid bullae, 12.8; length of nasals, 9.1.

Distribution.—Occupies rocky terrain on the lower slopes of the central mountain mass of the northern Baja California peninsula; found in the Sierra San Pedro Mártir and the Sierra San Miguel, southward to the vicinity of La Ramona, northeast of Santa Catarina, all in Baja California.

Remarks.—Huey (1960*b*) listed averages for measurements of five specimens.

Chaetodipus spinatus peninsulae (Merriam, 1894)

1894. *Perognathus spinatus peninsulae* Merriam, Proc. California Acad. Sci., ser. 2, 4:460, 25 September.

Holotype.—Young adult female, skin and skull, CAS 274, from San José del Cabo, Baja California Sur, México; obtained on 11 September 1891 by Walter E. Bryant.

Measurements of holotype.—Total length, 198; length of tail, 107; length of hind foot, 23; length of ear (anterior base, dry), 9; greatest length of skull, 26.60; width across bullae, 12.80; breadth across maxillary arches, 13.25; nasal length, 10.05; interorbital breadth, 6.60; length of maxillary toothrow, 3.50; depth of cranium, 3.55.

Distribution.—Occupies the Baja California peninsula in the cape region from approximately the level of Bahía de la Paz to the southern end, on both coasts (Hall, 1981).

Remarks.—Benson (1930) listed means and ranges of measurements of 7 specimens, Huey (1930, 1960*b*) listed means of 12 and 5 specimens, respectively, from different localities, and Osgood (1900) gave means for 4 specimens.

Chaetodipus spinatus prietae (Huey, 1930)

1930. *Perognathus spinatus prietae* Huey, Trans. San Diego Soc. Nat. Hist., 6:232, 24 December.

Holotype.—Adult male, skin and skull, SDSNH 8450, from 25 mi N Punta Prieta, lat. 29°24′N, long. 114°24′W, Baja California, México; obtained by Laurence M. Huey on 26 October 1930.

Measurements of holotype.—Total length, 194; length of tail, 112; length of hind foot, 21; length of ear (crown), 5; greatest length of skull, 25.2; interorbital breadth, 5.9; length of maxillary toothrow, 3.3; width across mastoid bullae, 12.3; length of nasals, 9.5.

Distribution.—Ranges over rocky areas of the central Baja California peninsula, from near San Augustín, Baja California, south to near Santa Gertrudis Mission, inland, and El Barril on the Gulf coast.

Remarks.—Banks (1967) listed means and ranges of measurements for 22 specimens and Huey (1930) gave means of measurements of 5 topotypes.

Chaetodipus spinatus pullus (Burt, 1932)

1932. *Perognathus spinatus pullus* Burt, Trans. San Diego Soc. Nat. Hist., 7:166, 31 October.

Holotype.—Adult male, skin and skull, CIT 50324, from Coronados Island, lat. 26°06′N, long. 111°18′W, Gulf of California, Baja California Sur, México; obtained by William H. Burt on 20 December 1931.

Measurements of holotype.—Total length, 192; length of tail, 111; length of hind foot, 23; length of ear (crown), 6 (notch, dry), 8.2; greatest length of skull, 25.8; basal length, 21.3; interorbital breadth, 6.4; length of maxillary toothrow, 3.6; length of mastoid bulla, 7.3; width across mastoid bullae, 12.6; length of interparietal, 4.1; width of interparietals, 7.8; length of nasals, 10.0.

Distribution.—Found only on Isla Coronados, Gulf of California, Baja California Sur.

Remarks.—Burt (1932) listed means and ranges of measurements for seven specimens. The holotype is now in the UCLA collections.

Chaetodipus spinatus rufescens (Huey, 1930)

1930. *Perognathus spinatus rufescens* Huey, Trans. San Diego Soc. Nat. Hist., 6:231, 24 December.

Holotype.—Adult male, skin and skull, SDSNH 7446, from mouth of Palm Canyon, Borego Valley, San Diego Co., California; obtained by Laurence M. Huey on 10 November 1929.

Measurements of holotype.—Total length, 194; length of tail, 112; length of hind foot, 21; length of ear (crown), 5; greatest length of skull, 25.2; interorbital breadth, 5.9; length of maxillary toothrow, 3.3; width across mastoid bullae, 12.3; length of nasals, 9.5.

Distribution.—Rocky, desert slopes of the southern California coastal ranges, from near Palm Springs, Riverside Co., south at least to the Mexican Boundary (Hall, 1981).

Remarks.—Huey (1930, 1960*b*) listed means for measurements of eight specimens from San Diego Co., California.

Chaetodipus spinatus serosus (Burt, 1932)

1932. *Perognathus spinatus serosus* Burt, Trans. San Diego Soc. Nat. Hist., 7:167, 31 October.

Holotype.—Adult male, skin and skull, CIT 50307, from Danzante Island [= Isla Danzante Primero], lat. 25°47′N, long. 111°11′W, Gulf of California, Baja California Sur, México, on 17 December 1931 by William H. Burt.

Measurements of holotype.—Total length, 187; length of tail, 104; length of hind foot, 22; length of ear (crown), 5, (notch, dry), 7; greatest length of skull, 25.7; basal length, 21.4; interorbital breadth, 6.7; length of maxillary toothrow, 3.6; length of mastoid bulla, 7.2; width across mastoid bullae, 12.3; length of interparietal, 3.6; width of interparietals, 7.6; length of nasals, 10.1.

Distribution.—Occurs only on Isla Danzante Primero, Gulf of California, Baja California Sur.

Remarks.—Burt (1932) provided means and extremes of measurements of seven specimens. The holotype is now in the UCLA collections.

Chaetodipus spinatus spinatus (Merriam, 1889)

1889. *Perognathus spinatus* Merriam, N. Amer. Fauna, 1:21, 25 October.

Holotype.—Adult male, skin and skull, USNM 186516, from lower Colorado River, 25 mi below [S] the Needles, San Bernardino Co., California; obtained by Vernon Bailey on 23 March 1889.

Measurements of holotype.—Total length, 179; length of tail, 104; length of hind foot, 21; length of ear (crown, dry), 3.5; occipitonasal length, 23.7; basilar length, 18.8; basilar length of Hensel, 16.5; width across mastoid bullae, 12.2; length of interparietal, 3.5; width of interparietals, 7.3; least intermastoid distance, 8.0.

Distribution.—Ranges along the Colorado River, from Granite Springs in extreme southern Clark Co., Nevada, south to near Fort Yuma, Imperial Co., California, and southward on rocky slopes in northeastern Baja California to the vicinity of San Felipe on the Gulf of California.

Remarks.—Huey (1930, 1960*b*) and Osgood (1900) listed means for five and four specimens, respectively.

Genus *Perognathus*

1839. *Perognathus* Wied, Nova Acta Phys.-Med. Acad. Caesar. Leop.-Carol., 19(1):368.

Type Species.—*Perognathus fasciatus* Wied, 1839, Nova Acta Phys.-Med. Acad. Caesar. Leop.-Carol., 19(1):368.

Diagnosis.—Size small, total length from

about 100 to 200 mm, weight from about 5 to 30 g; body form quadrupedal and scansorial; hind limbs longer than forelimbs; tail relatively short, length usually averaging less to slightly more than length of head and body; tail without prominent distal, dorsal crest or terminal tuft of hairs, although several species have a slight crest and terminal tuft (pencil); tail some shade of brown or buffy above, whitish below; proximal one-fourth to one-half of sole of hind foot with sparse covering of short hairs; ear pinna short and rounded, and without a lobed antitragus except in *P. parvus* and *P. alticola*; antero-lateral edge of ear pinna without covering of long, coarse hairs; generally a contrasting, light buffy area on the head around ear pinna (post-auricular patch), and a small, whitish spot below the external auditory meatus; dorsal surfaces generally some shade of buffy or brownish, usually tinged with black; usually a clear, buffy lateral stripe without blackish tinge; undersides usually whitish; in general, hairs relatively short, soft, and oval to flattened in cross section; hairs lack dorsal trough except, in so far as is known, *P. amplus*; longer, stiff, spine-like hairs never present in pelage of dorsal and lateral surfaces; mastoid bullae usually extending beyond the plane of the occiput; tympanic bullae nearly meeting anteriorly on the ventral surface of the skull; exoccipital without strong lateral indentations of mastoid bullae; interparietals compressed and narrower than interorbital breadth; phallus relatively short in length; soft tissue of phallus extends about two-thirds of the length of the baculum; phallus lacks external spines but has urethral lappets; baculum relatively short in length, with swollen, bulbous proximal end and slender, slightly upturned distal end; vesicular glands of male reproductive system elongated and tube-like with hooked end and translucent; head of spermatozoa approximating a triangle, with rounded vertices; tendon at origin of M. rectus femoris fan-shaped (Hafner and Hafner, 1983; Hall, 1981; Homan and Genoways, 1978; Ryan, 1989; Wood, 1935).

Remarks.—This diagnosis is based on Recent species. The genus *Perognathus* includes as synonyms *Cricetodipus* Peale, 1848 (type species *C. parvus*), a name that also was subsequently applied, for a time, to some species of *Dipodomys*; *Abromys* Gray, 1868 (type species *A. lordi*, a synonym of *P. parvus*), and *Otognosis* Coues, 1875 (type species *O. longimembris*). Considerable taxonomic confusion arose as a result of S. F. Baird, E. Coues, J. E. Gray, T. R. Peale, and others misapplying the names *Perognathus fasciatus, Perognathus hispidus, Abromys lordi,* and *Cricetodipus parvus.* Merriam (1889) and Osgood (1918) were largely responsible for clarifying the nomenclature and stabilizing the taxonomy of pocket mice. As conceived herein, the genus *Perognathus* does not include the hispid-haired species of *Chaetodipus* nor *Perognathus formosus* Merriam, which has been shown to be a species of *Chaetodipus* (Homan and Genoways, 1978; Patton et al., 1981).

Most of the structural characteristics that are of value in distinguishing species are related to size, proportions, and color, all of which exhibit considerable individual and geographic variation, making identification keys based on skins and skulls cumbersome.

Key to the Species

1. Antitragus of ear pinna lobed 2

1′. Antitragus of ear pinna not lobed 3

2. Inner surface of ear pinna with white or yellowish hairs; distal one-third of tail with blackish hairs dorsally; mastoid bulla forms small, sharp indentation in exoccipital; occurring only in southern California in the Transverse ranges and the San Bernardino Mountains .... *Perognathus alticola*

2′. Hairs of inner surface of ear pinna buffy rather than white or yellowish; distal onc-third of tail with mix of sooty brown and black hairs dorsally; mastoid bulla with little or no indentation into exoccipital; occurring on the Columbia Plateau, in the Great Basin, and the eastern slopes of the Sierra Nevada .... *Perognathus parvus*

3. Length of tail averages less than length of head and body (a few individuals have tails longer than length of head and body); not occurring west of eastern Utah or west-central Arizona, but found in Sonora along the Gulf coast 4
3′. Length of tail averages greater than length of head and body (a few individuals have tails shorter than head and body); not occurring east of central Arizona and south-central Utah 9
4. Interparietal length less than 2.9 mm; length of tail less than 66 mm (mean less than 60 mm); interorbital breadth averages 4.7 mm or less and maximum less than 5.0; width of interparietals less than 4.2 mm, averaging 3.6 or less 5
4′. Interparietal length greater than or equal to 2.9 mm; length of tail averages greater than 60 mm; interorbital breadth 4.5 mm or more, mean 4.8 or greater; width of interparietals 3.15 mm or more, mean 4.5 or greater 6
5. Length of tail averages 56 mm or more; interorbital breadth averages 4.5 or more; width of interparietals averages 3.3 mm or more; pelage sleek, not noticeably lax; dorsal color yellowish tinged with blackish from black-tipped hairs; slight contrast between the darker mid-dorsal and lighter dorso-lateral color; post-auricular light spot relatively small .... *Perognathus merriami*
5′. Length of tail averages 55 mm or less; interorbital breadth averages 4.5 mm or less; width of interparietals averages 3.3 or less; pelage lax, not sleek in appearance; dorsal color buffy-yellow with a pinkish hue and a tinge of blackish from black-tipped hairs; dark mid-dorsal area contrasting markedly with lighter dorso-lateral color; post-auricular light spot relatively large .... *Perognathus flavus*
6. Occurring on the Great Plains, or in the northern Chihuahuan Desert region in southern Arizona and New Mexico, western Texas, and northern Chihuahua.. 7
6′. Occurring in the intermountain plateaus and basins of south-central and southwestern Wyoming, eastern Utah, southern and western Colorado, eastern Arizona, and New Mexico 8
7. Dorsal yellowish color with an olive-yellow (olivaceous) tone; length of interparietal generally less than 3.0 mm; length of head and body averaging greater than 68 mm; not occurring south and east of western Nebraska on the Great Plains and southeastern Colorado along the eastern front of the Rocky Mountains .... *Perognathus fasciatus*
7′. Dorsal yellowish color with yellowish-orange tone; length of interparietal generally greater than 3.0 mm; length of head and body averaging less than 66 mm; not occurring north and west of a line from approximately the east base of the Rocky Mountains in northern Colorado, north and east through south-central North Dakota .... *Perognathus flavescens*
8. Dorsal yellowish color with an olive-yellow (olivaceous) tone; least interbullar distance (on dorsal surface of skull) averaging 4.3 mm or more; width of first upper molar 1.16 mm of less (mean 1.10); not occurring south of the Uintah Basin of Utah and Colorado .... *Perognathus fasciatus*
8′. Dorsal yellowish color with yellowish-orange tone; least interbullar distance averaging less than 4.0 mm, generally not exceeding 4.3; width of first upper molar 1.13 mm or more (mean 1.22); not occurring north of the Uintah Basin of Colorado and Utah .... *Perognathus flavescens*
9. Not occurring in California 10
9′. Occurring in California 11
10. Greatest length of skull generally less than 23.0 mm; frontonasal length averaging less than about 15 mm; length of hind foot 20 mm or less (mean 19 or less) .... *Perognathus longimembris*
10′. Greatest length of skull generally greater than 23.0 mm; frontonasal length averaging greater than 15.1 mm; length of hind foot 19 mm or more (mean greater than 20) .... *Perognathus amplus*
11. Occurring in central California in the San Joaquin and Sacramento valleys, north of the Tehachapi Mountains ... . *Perognathus inornatus*
11′. Occurring in southern California from the Tehachapi Mountains southward 12
12. Occipitonasal length of adults (perma-

nent upper premolar with moderate to heavy wear) generally greater than 22.5 mm .... *Perognathus inornatus*

12′. Occipitonasal length of adults (see above) generally less than 22.1 mm .. .. *Perognathus longimembris*

Species Accounts

Perognathus alticola

Diagnosis.—Size medium-large for genus, total length from about 130 to 180; greatest length of skull from about 23 to 26 mm; size small compared to most subspecies of *P. parvus*; length of tail equal to or slightly longer than length of head and body; tail slightly crested for distal one-third; upper side of tail shading to blackish toward tip; antitragus of ear pinna lobed; inner side of ear pinna with white or yellowish hairs; auditory bullae relatively small; posterior border of mastoid bulla about even with posterior extent of supraoccipital; mastoid bulla forms small indentation in border of exoccipital on back of skull; interparietal compressed compared to most subspecies of *P. parvus*; ascending branches of supraoccipital relatively broad; interorbital region broad; phallus and baculum, in so far as is known, relatively short compared to *P. parvus*, and relatively short, compared to head and body length, for genus (Burt, 1936, 1960; Osgood, 1900; Rhoads, 1894; Sulentich, 1983).

Comparisons.—*Perognathus alticola* is most similar to *P. parvus*, from which it can be distinguished only by its smaller average size, having white or yellowish hairs on the inner surface of the pinna, having a tail with more blackish hairs on the dorsal surface of its distal one-third, and by having a more pronounced indentation of the mastoid bulla into the border of the exoccipital. *P. alticola* can be distinguished from all other species of *Perognathus* by its lobed antitragus and more prominent crest on the distal portion of the tail.

Distribution.—Occupies arid shrub and forest communities in south-central California, in the Transverse Ranges of Kern and Los Angeles counties, and the San Bernardino Mountains, San Bernardino Co.

Remarks.—*Perognathus alticola* is closely related to, and perhaps only subspecifically distinct from, *P. parvus*. The members of the *parvus* species group exhibit relatively great diversity in chromosome structure (Williams, 1978*a*) and biochemical variation (Sulentich, 1983). Sulentich (1983) considered *P. a. alticola* to be specifically distinct from *P. a. inexpectatus*, primarily on the basis of the size and shape of the interparietals. Greater variation, however, is seen in size and proportions of the interparietals of *P. parvus* (for example, those of *P. p. bullatus* are extremely compressed compared to those of adjacent populations of *P. p. clarus* in eastern Utah; D. Williams, unpubl. data), and variation in size and shape of the interparietals in *P. flavescens* and *P. fasciatus* was shown to be strongly related to degree of bullar inflation and significantly correlated with degree of environmental aridity (Williams, 1978*b*; Williams and Genoways, 1979).

Perognathus alticola alticola Rhoads, 1894

1894. *Perognathus alticolus* Rhoads, Proc. Acad. Nat. Sci. Philadelphia, 45:412, 27 January.

Holotype.—Adult male, skin and skull, ANSP 1615, from Squirrel Inn, near Little Bear Valley, 5,500 ft, San Bernardino Mountains, San Bernardino Co., California; obtained on 22 September 1893 by R. B. Herron.

Measurements of holotype.—Total length, 157; length of tail, 77; length of hind foot, 20; length of ear (crown), 5 (external measurements from dry skin); basilar length of skull, 16.0; width across bullae, 12.5; interorbital breadth, 6.0; length of nasals, 8.6; length of mandible, 10.5; height of coronoid process, 4.6.

Distribution.—Known only from arid ponderosa pine communities in the vicinity of Little Bear Valley and Strawberry Peak, San Bernardino Mountains, San Bernardino Co., California.

Remarks.—The specific epithet, *alticolus,* constructed by Rhoads (1894), apparently referred to the mountainous home of this species (*altus* = high; *cola* = inhabiting). His alteration of the word *cola* to conform in gender to *Perognathus* was in error, however. *Cola* is not based on the root, *col*; the latter is a Latin prefix to be used before r and meaning "with" or "together." Osgood (1900) amended the spelling without comment, although Rhoads' spelling is still used by some.

Perognathus alticola inexpectatus Huey, 1926

1926. *Perognathus alticola inexpectatus* Huey, Proc. Biol. Soc. Washington, 39:121, 27 December.

Holotype.—Adult male, skin and skull, SDSNH 5724, from 14 mi W Lebec, 6,000 ft, Kern Co., California; obtained by George G. Cantwell on 28 August 1926.

Measurements of holotype.—Total length, 176; length of tail, 97; length of hind foot, 22; greatest length of skull, 25.9; width across bullae, 13.8; breadth across maxillary arches, 13.0; length of nasals, 10.0; interorbital breadth, 6.2; length of maxillary toothrow, 3.4.

Distribution.—Occupies arid shrub-steppe communities in the Tehachapi Mountains of south-central California, from the vicinity of Tehachapi Pass, Kern Co., on the northeast to the vicinity of Mt. Pinos, Ventura Co. on the northwest, and Elizabeth and Quail lakes, Los Angeles Co., on the south (unpubl. data).

Remarks.—Sulentich (1983) found relatively great biochemical differences between *P. a. inexpectatus* and *P. parvus olivaceus* and *P. p. xanthonotus* (Rogers' similarity coefficients of 0.404 and 0.402, respectively). This, together with the more inflated bullae and compressed interparietals, and the shorter phallus and baculum of *P. alticola inexpectatus,* led him to conclude that *inexpectatus* was specifically distinct from both *P. parvus* and *P. alticola.* He did not have data on biochemical variation or phallic structure of *P. alticola alticola,* however. Although we concur that individuals of *inexpectatus* are distinct from *P. parvus* samples from adjacent geographic regions, we think it best to retain the current taxonomic arrangement until information on other populations of *P. parvus* and *P. alticola* is available. Sulentich (1983) provided means and ranges of measurements for 20 specimens.

Perognathus amplus

Diagnosis.—A medium-sized species of *Perognathus* with a relatively long tail, large hind feet, inflated mastoid bullae, and compressed interparietals; head and body length normally ranges between about 71 and 88 mm; length of tail generally ranges between about 72 and 88, and the ratios of length of tail to head and body range between about 0.88 and 1.33; length of hind foot generally measures 20 mm or more; width of skull varies from about 12.8 to 15.2 mm and width of interparietal from 3.2 to 4.45 mm.

Comparisons.—*Perognathus amplus* is most similar to *P. inornatus* and *P. longimembris.* No single character of skin or skull is known that will distinguish all individuals of *P. amplus* from these other species; *amplus* averages larger than *inornatus* in most dimensions, has a relatively longer and more penciled tail and an interorbital breadth in excess of 5.1 mm (average between about 5.3 and 5.6), whereas interorbital breadth in *inornatus* averages between about 4.75 and 4.9, and individuals rarely exceed 5.1 mm. *P. amplus* differs from *P. longimembris* in larger average size and in having a wider upper premolar; greatest length of skull generally exceeds 23.2 mm in *amplus* and less than 23.1 in *logimembris*; length of hind foot usually measures 20 mm or more in *amplus* and 19.5 or less in *logimembris*. *P. amplus* is also similar in size and general appearance to some subspecies of *P. flavescens*; the tail of *amplus* is longer relative to head and body length and is more penciled, the interparietal is narrower, and the length of hind foot is greater (20 mm or more versus 20 or less).

Distribution.—Occurs as apparently disjunct populations in north-central and west-central and southwestern Arizona, and adjacent areas in northwestern Sonora (Hall, 1981).

Remarks.—The relationships among the taxa of the *longimembris* species group (the nominate forms, *amplus, inornatus,* and *longimembris*) are poorly known and a new taxonomy probably will be required when these relationships are clarified. Hoffmeister (1986) discussed ways to distinguish *amplus* from *logimembris* in Arizona.

Perognathus amplus amplus
Osgood, 1900

1900. *Perognathus amplus* Osgood, N. Amer. Fauna, 18:32, 20 September.
1932. *Perognathus amplus rotundus* Goldman, J. Washington Acad. Sci., 22:387, 19 July.
1933. *Perognathus amplus jacksoni* Goldman, J. Washington Acad. Sci., 23:465, 15 October.

Holotype.—Adult male, skin and skull, USNM 34626/46711, from Fort Verde, Yavapai Co., Arizona; obtained by J. Alden Loring on 26 June 1892.

Measurements of holotype.—Total length, 155; length of tail, 80; length of hind foot, 20; greatest length of skull, 25.05; occipitonasal length, 23.65; interorbital breadth, 5.20; length of maxillary toothrow, 3.55; width across maxillary toothrows, 5.20; length of mastoid bulla, 9.85; width across mastoid bullae, 13.90; length of interparietal, 2.70; width of interparietals, 3.20; length of nasals, 9.40; width of nasals, 2.00; width of rostrum, 3.95.

Distribution.—Occupies central and southwestern Arizona, from Fort Verde, Yavapai Co., southeastward to San Carlos Lake, Gila Co., and westward to near Signal, Mojave Co., thence southward through central Yuma Co. into coastal Sonora to the vicinity of Puerto Libertad.

Remarks.—Hoffmeister (1986) placed *rotundus* and *jacksoni* in the synonymy of *amplus,* and provided statistics for measurements of 190 specimens in 6 samples. Benson (1933) gave measurements of one topotype of *amplus*; Goldman (1933) gave measurements of four adults from Yavapai Co.; and Goldman (1932*a*) listed average measurements for five adults from Yuma Co.

Perognathus amplus cineris
Benson, 1933

1933. *Perognathus amplus cineris* Benson, Proc. Biol. Soc. Washington, 46:109, 27 April.
1933. *Perognathus amplus ammodytes* Benson, Proc. Biol. Soc. Washington, 46:110, 27 April.

Holotype.—Adult male, skin and skull, MVZ 55771, from near the Wupatki Ruins, Wupatki National Monument (about 27 mi NE Flagstaff), Coconino Co., Arizona; obtained by Annie M. Alexander on 12 October 1932.

Measurements of holotype.—Total length, 197; length of tail, 107; length of hind foot, 20; length of ear (crown), 4; weight, 10.3 g; greatest length of skull, 23.45; interorbital breadth, 5.8; length of mastoid bulla, 8.85; width across mastoid bullae, 12.85; length of interparietal, 2.3; width of interparietals, 3.8; length of nasals, 8.7; width of rostrum, 4.0.

Distribution.—Found south of the Colorado River, from near Navajo Spring, Echo Cliffs, Coconino Co., on the north, southward along the Little Colorado River Valley to the vicinity of Wupatki National Monument, Coconino Co., Arizona.

Remarks.—Hoffmeister (1986) considered *ammodytes* to be indistinguishable from *cineris* except for the darker color of populations in the vicinity of the type locality. Hoffmeister (1986) listed statistics for measurements of samples of 20 and 32; Benson (1933) gave measurements for 7 specimens from the vicinity of Wupakti National Monument.

Perognathus amplus pergracilis Goldman, 1932

1932. *Perognathus amplus pergracilis* Goldman, J. Washington Acad. Sci., 22:387, 19 July.

Holotype.—Adult male, skin and skull, USNM 227528, from Hackberry, 3,500 ft, Mohave Co., Arizona; obtained on 14 September 1917 by Edward A. Goldman.

Measurements of holotype.—Total length, 143; length of tail, 80; length of hind foot, 21; greatest length of skull, 22.60; occipitonasal length, 22.00; interorbital breadth, 5.35; length of maxillary toothrow, 3.50; width across maxillary toothrows, 4.25; length of mastoid bulla, 7.85; width across mastoid bullae, 12.40; length of interparietal, 2.90; width of interparietals, 3.80; length of nasals, 8.65; width of nasals, 1.95; width of rostrum, 2.65.

Distribution.—Occurs in desert associations in northwestern Arizona, from south of the Grand Canyon, Mohave Co., southward to near Wikieup, Mohave Co.

Remarks.—Hoffmeister (1986) gave measurements and associated statistics for samples of 17, 49, and 38; Goldman (1932*a*) listed measurements for 5 adult topotypes, and Benson (1933) gave statistics for measurements of 18 specimens.

Perognathus amplus taylori Goldman, 1932

1932. *Perognathus amplus taylori* Goldman, J. Washington Acad. Sci., 22:488, 19 October.

Holotype.—Adult female, skin and skull, USNM 250533, from Santa Rita Range Reserve (near Northeast Station), 35 mi S Tucson, about 4,000 ft, Pima Co., Arizona; obtained on 3 August 1930 by Walter P. Taylor.

Measurements of holotype.—Total length, 155; length of tail, 84; length of hind foot, 20; greatest length of skull, 24.35; occipitonasal length, 23.35; interorbital breadth, 5.30; length of maxillary toothrow, 3.30; width across maxillary toothrows, 4.20; length of mastoid bulla, 9.00; width across mastoid bullae, 13.35; length of interparietal, 3.15; width of interparietals, 3.35; length of nasals, 9.20; width of nasals, 2.25; width of rostrum, 3.55.

Distribution.—Desert areas of south-central Arizona, south of the Gila River in Pinal Co. (and probably southeastern Maricopa Co.), southward east of Organ Pipe Cactus National Monument to 13 mi W of Caborca, Sonora (Hoffmeister, 1986).

Remarks.—Goldman (1932*b*) gave averages and ranges of measurements for 10 adult topotypes; Hoffmeister (1986) gave statistics for measurements of samples of 14 and 26 individuals.

Perognathus fasciatus

Diagnosis.—A medium-sized species of *Perognathus,* with a relatively short, nonpenciled tail and olivaceous (olive-yellow) tone to the dorsal color; skull with relatively small auditory bullae and wide interorbital region. Width of lower premolar relatively wider than other species of *Perognathus,* ranging from 0.52 to 0.77 mm (mean, 0.66). Length of head and body averages between about 68.9 and 70.2 mm; length of tail averages between about 59.3 and 64.3 mm; interorbital breadth ranges between 4.5 and 5.6 mm with means ranging from about 4.9 to 5.1; length of mastoid bulla averages from about 7.6 to 8.5 mm (Williams and Genoways, 1979).

Comparisons.—*Perognathus fasciatus* is most similar to some populations of *Perognathus flavescens,* from which it can be distinguished by its unique, olivaceous dorsal color. Comparisons of cranial characters must be made by subspecies: *P. fasciatus callistus* differs from *P. flavescens caryi* by smaller size, shorter interparietal, narrower rostrum, and wider interbullar region; *P. fasciatus fasciatus* differs from Great Plains populations of *P. flavescens* by being larger, with a narrower interorbital breadth, larger auditory bullae, shorter interparietals, and larger molariform teeth. *P. fasciatus* is significantly larger with less inflated auditory bullae and wider interbullar region than *P. flavus* of the Great Plains area. From *P. parvus, P. fasciatus* is distinguished by its significantly smaller size (greatest length of skull less than 25 versus greater than 25.5 mm) and non-penciled tail. No other species of *Perognathus* occurs within or near the geographic range of *P. fasciatus* (Williams and Genoways, 1979).

Distribution.—Occupies grasslands of the northern Great Plains from southeastern Alberta, southern Saskatchewan, and southwestern Manitoba southward through northwestern Nebraska and east-central Colorado along the east flank of the Rocky Mountains to Huerfano Co. On the west, ranging to central Montana, western Wyoming, and the Uintah Basin of northeastern Utah and northwestern Colorado (Williams and Genoways, 1979).

Remarks.—Williams (1978*a*), redefined the *fasciatus* species group of Osgood (1900) to exclude *P. flavus*; thus it consists of *P. fasciatus* and *P. flavescens.* Williams and Genoways (1979) gave means and ranges of measurements for four samples of *P. fasciatus* and provided a comprehensive, systematic review of the species.

Perognathus fasciatus callistus Osgood, 1900

1900. *Perognathus callistus* Osgood, N. Amer. Fauna, 18:28, 20 September.

Holotype.—Adult male, skin and skull, USNM 88245, from Kinney Ranch [about 22 mi S Bitter Creek], Sweetwater Co., Wyoming; obtained on 14 May 1897 by J. Alden Loring.

Measurements of holotype.—Total length, 135; length of tail, 63; length of hind foot, 18; greatest length of skull, 22.80; occipitonasal length, 22.65; interorbital breadth, 5.05; length of maxillary toothrow, 3.10; width across maxillary toothrows, 4.45; length of mastoid bulla, 8.55; width across mastoid bullae, 13.10; length of interparietal, 2.70; width of interparietals, 4.65; length of nasals, 8.10; width of nasals, 2.35; width of rostrum, 3.55.

Distribution.—Occupies desert and steppe grassland associations in the Uintah, Bridger, and Great Divide basins and contiguous areas of Colorado, Utah, and Wyoming (Williams and Genoways, 1979).

Remarks.—Most populations included in *P. fasciatus litus* prior to the review by Williams and Genoways (1979) were regarded by them as inseparable from *P. f. callistus*; however *litus* is a synonym of *P. fasciatus fasciatus.*

Perognathus fasciatus fasciatus Wied, 1839

1839. *Perognathus fasciatus* Wied, Nova Acta Phys.-Med., Acad. Caesar. Leop.-Carol., 19: 369.

1893. *Perognathus infraluteus* Thomas, Ann. Mag. Nat. Hist., ser. 6, 11:406, May.

1911. *Perognathus fasciatus litus* Cary, Proc. Biol. Soc. Washington, 24:61, 22 March.

1940. *Perognathus flavescens olivaceogriseus* Swenk, Missouri Valley Fauna, 3:6, 5 June.

Neotype.—Adult male, skin and skull, USNM 168599, from Buford, Williams Co., North Dakota; obtained on 6 May 1910 by H. E. Anthony.

Measurements of neotype.—Total length, 140; length of tail, 66; length of hind foot, 18; greatest length of skull, 23.35; occipitonasal length, 23.35; interorbital breadth, 5.30; length of maxillary toothrow, 3.20; width across maxillary toothrows, 4.45; length of mastoid bulla, 7.70; width across mastoid bullae, 11.90; length of interparietal, 2.65; width of interparietals, 4.90; length of nasals, 8.75; width of nasals, 2.15; width of rostrum, 4.00.

Distribution.—Occupies light, sandy soils in the northern Great Plains, from near Saskatoon, Saskatchewan, southward to near La Veta, Huerfano Co., Colorado, and from near the North Dakota–Minnesota boundary westward through Wyoming, central Montana, and southeastern Alberta (Williams and Genoways, 1979).

Remarks.—There is no information on the disposition of the holotype of Wied (1839). Williams and Genoways (1979) presumed that it was lost and therefore designated a neotype from near the type locality (upper Missouri River near junction with the Yellowstone River, near Buford, Williams Co., North Dakota). They chose a specimen different from the "duplicate type" of Merriam (1889) because the specimen chosen by Merriam was not from the original type locality and because the right, lower premolar exhibited a unique anomaly.

Perognathus flavescens

Diagnosis.—A small- to medium-sized species of *Perognathus* with a tail typically shorter than length of head and body (ratio of lengths of tail to head and body averages from about 0.86 to 0.97), relatively broad interparietals, and wide interorbital region. Dorsal yellowish color lacks an olivaceous tone. Length of head and body ranges from about 60 to 73 mm; length of tail ranges from 56 to 73 mm; interorbital breadth ranges from about 4.5 to 5.6 mm; interparietal width varies from 3.15 to 5.4 mm (Williams, 1978*b*; Williams and Genoways, 1979).

Comparisons.—*Perognathus flavescens* is most similar to *Perognathus fasciatus* (see account of *fasciatus* for distinguishing characteristics). Great Plains populations exhibit similarity in external size and color to some *P. merriami merriami* and some populations of *P. m. gilvus*; from *P. merriami* and *P. flavus, P. flavescens* can be distinguished by its wider interparietals (width averages greater than 4.0 mm in Great Plains populations of *flavescens* versus less than 3.6 in *P. merriami* and *P. flavus*); *P. flavus* generally has a darker color dorsally due to numerous black-tipped guard hairs, and its post-auricular patch appears larger and contrasts more strongly with surrounding areas than in *P. flavescens*; *P. flavescens* generally averages larger, with a longer tail and hind foot than *P. merriami* and *P. flavus*; differences in size are most pronounced in sympatric populations from the intermountain basins of southwestern North America. *P. flavescens* is marginally sympatric with *P. parvus* in eastern Utah; *flavescens* is smaller (length of hind foot and greatest skull length less than 21 and 25.5 mm in *flavescens,* respectively, versus greater than 21 and 25.5 mm in *parvus*) and lacks the small crest on the distal, dorsal one-third of the tail (pencil) exhibited by *parvus*. *P. flavescens* is marginally sympatric with *P. amplus* in north-central Arizona and possibly sympatric with *P. longimembris* in north-central Arizona and south-central Utah; *flavescens* has a relatively shorter and less penciled tail than *amplus* or *longimembris* (ratio of lengths of tail to head and body averages 0.95 or less for *flavescens* versus greater than 1.10 for *amplus* and *longimembris*); *amplus* is generally larger, with a length of hind foot 20 mm or greater versus 20 or less in *flavescens*; interparietal width averages wider in *flavescens* than *amplus* from nearby populations (3.9 mm or greater versus about 3.5). From *longimembris,* nearby populations of *flavescens* differ in larger average body size, reflected in most external and cranial measurements (Williams, 1978*b*; Williams and Genoways, 1979; D. Williams, unpubl. data).

Distribution.—Occupies sandy soils in grassland and desert communities of the Great Plains and intermountain basins of west-central North America, ranging from southeastern North Dakota south and east nearly to the Mississippi River in southern Minnesota; on the east extending from Minnesota southwesterly through the extreme northwestern corner of Missouri, central Kansas and central Oklahoma; on the west, extending southwesterly from southeast North Dakota to extreme southeastern Wyoming and then southward through eastern Colorado and New Mexico to western Texas; also occupies the intermountain basins and plateaus, from the Uintah Basin of northeastern Utah and northwestern Colorado, southward to northeastern Arizona and most of central and western New Mexico, extending westward in the northern Chihuahuan Desert to near Willcox, Arizona, and southward to the vicinity of Casas Grandes, Chihuahua (Williams, 1978*b*).

Remarks.—*Perognathus flavescens* exhibits extreme geographic variation in color, size, bullar inflation, constriction of the post-cranial region between the mastoid bullae, and relative size of the rostrum and nasal bones, all of which are strongly correlated with environmental temperature and moisture (Williams, 1978*b*). This great variation makes it difficult to characterize and distinguish this species from other *Perognathus*. Williams (1978*b*) provided a systematic review of the intermountain populations, formerly designated as *P. apache*, and gave means and ranges of measurements for 17 samples; Williams and Genoways (1979) provided measurements for an additional sample from the northern Great Plains. Reed and Choate (1986) reviewed geographic variation in the Great Plains populations and gave statistics for measurements of six samples, representing the four subspecies occupying the Great Plains. Hoffmeister (1986) remarked that the evidence developed by Williams (1978*b*) did not unequivocally prove that *apache* and *flavescens* were conspecific, and chose to treat *apache* as a distinct species. However, no evidence reviewed suggests recognition of two species.

Perognathus flavescens apache Merriam, 1889

1889. *Perognathus apache* Merriam, N. Amer. Fauna, 1:14, 25 October.
1918. *Perognathus apache cleomophila* Goldman, Proc. Biol. Soc. Washington, 31:23, 16 May.

Holotype.—Adult male, skin and skull, USNM 186504, from near Keam's Canyon, Navajo Co., Arizona; obtained by Jere Sullivan on 22 May 1888.

Measurements of holotype.—Total length, 140; length of tail, 68; length of hind foot, 18.5; greatest length of skull, 24.05; occipitonasal length, 23.95; interorbital breadth, 5.10; length of maxillary toothrow, 3.20; width across maxillary toothrows, 4.60; length of mastoid bulla, 8.45; length of interparietal, 2.50; width of interparietals, 3.75; length of nasals, 8.70; width of nasals, 2.45; width of rostrum, 4.65.

Distribution.—Found on loose sandy soils in arid grassland and open woodland communities in northeastern Arizona, north and east of the Mogollon rim, west of the Chuska Mountains, and east of the Coconino Plateau, northward into southeastern Utah east of the Colorado River and south of the San Juan River (Williams, 1978*b*).

Remarks.—Williams (1978*b*) treated *cleomophila* as a synonym of *apache*. Merriam (1889) listed the type locality as in Apache Co.

Perognathus flavescens caryi Goldman, 1918

1918. *Perognathus apache caryi* Goldman, Proc. Biol. Soc. Washington, 31:24, 16 May.

Holotype.—Adult male, skin and skull, USNM 148206, from 8 mi W Rifle, Garfield Co., Colorado; obtained by Merritt Cary on 4 October 1906.

Measurements of holotype.—Total length, 154; length of tail, 73; length of hind foot, 21; greatest length of skull, 25.15; occipitonasal length, 25.15; interorbital breadth, 5.60; length of maxillary toothrow, 3.50; width across maxillary toothrows, 4.65; length of mastoid bulla, 9.00; width across mastoid bullae, 13.40; length of interparietal, 3.35; width of interparietals, 4.15; length of nasals, 9.30; width of nasals, 2.30; width of rostrum, 4.35.

Distribution.—Typically found on sandy substrates in semiarid grassland and pinyon-juniper associations, from the Uintah Basin of Utah and Colorado, southward through western Colorado and eastern Utah north of the San Juan River, thence south and eastward in northwestern New Mexico and southwestern Colorado from the San Juan Basin to the Rio Grande Valley near Vale Verde, Socorro Co.; on the east to the upper Pecos River Valley, San Miguel Co., New Mexico (Williams, 1978*b*).

Remarks.—Williams provided means and ranges of measurements for samples of *P. flavescens caryi.*

Perognathus flavescens cockrumi Hall, 1954

1954. *Perognathus flavescens cockrumi* Hall, Univ. Kansas Publ., Mus. Nat. Hist., 7:589, 15 November.

Holotype.—Adult female, skin and skull, KU 13045, from 4.5 mi NE Danville, Harper Co., Kansas; obtained on 1 December 1939 by Sam Tihen.

Measurements of holotype.—Total length, 114; length of tail, 51; length of hind foot, 17; condylobasal length, 18.5; interorbital breadth, 4.7; length of maxillary toothrow, 3.1; length of mastoid bulla, 6.8; width across mastoid bullae, 10.5; width of interparietals, 4.3.

Distribution.—Known from sandy soils from central Kansas southward to west-central Oklahoma (Hall, 1981).

Remarks.—Hall (1954) characterized the holotype as subadult, but by the classification scheme of Williams (1978*b*), it would be adult (age class 4). This small, dark-colored subspecies is easily confused with *P. flavus* on the basis of external features. According to Reed and Choate (1986) *cockrumi* only differs from samples of *flavescens* from Kansas in darker color.

Perognathus flavescens copei Rhoads, 1894

1894. *Perognathus copei* Rhoads, Proc. Acad. Nat. Sci. Philadelphia, for 1893, 46:404, 27 January.

Holotype.—Adult male, skin and skull, ANSP 1612, from Staked Plains, near Mobeetie, Wheeler Co., Texas; obtained by E. D. Cope on 26 August 1893.

Measurements of holotype.—Total length, 114; length of tail, 49; length of hind foot, 15; greatest length of skull, 19.5; interorbital breadth, 4.5; width across mastoid bullae, 10.0; length of nasals, 7.0; length of mandible, 9.7; height of coronoid process from angle, 4.2; ratio of length to breadth of interparietal, 0.62.

Distribution.—Sandy desert and arid grassland associations of western Oklahoma, western Texas, and eastern New Mexico.

Remarks.—The structural differences between *P. f. copei* and *P. f. flavescens* are relatively slight and the variation appears to be largely clinal, a situation that may not warrant recognition of *copei* as a distinct subspecies (Williams, 1978*b*; D. Williams, unpubl. data).

Perognathus flavescens flavescens Merriam, 1889

1889. *Perognathus fasciatus flavescens* Merriam, N. Amer. Fauna, 1:11, 25 October.

Holotype.—Adult male, skin and skull, USNM 186507, from Kennedy, Cherry Co., Nebraska; obtained on 11 June 1888 by Vernon Bailey.

Measurements of holotype.—Total length,

136; length of tail, 73; length of hind foot, 17; length of ear (crown, dry), 4; greatest length of skull, 22.10; occipitonasal length, 22.10; interorbital breadth, 5.25; length of maxillary toothrow, 3.00; width across maxillary toothrows, 4.45; length of mastoid bulla, 7.25; width across mastoid bullae, 11.85; length of interparietal, 3.10; width of interparietals, 5.20; length of nasals, 7.90; width of nasals, 2.20; width of rostrum, 3.75.

Distribution.—Occupies sandy soils of the west-central Great Plains, from southwestern North Dakota through southwestern Kansas; on the west to extreme southeastern Wyoming and southward east of the Rocky Mountain front at least to Pueblo Co., Colorado.

Remarks.—Armstrong (1972), Jones (1964), Reed and Choate (1986), Williams (1978*b*), and Williams and Genoways (1979) listed means and ranges of measurements for samples of *P. f. flavescens.*

Perognathus flavescens melanotis Osgood, 1900

1900. *Perognathus apache melanotis* Osgood, N. Amer. Fauna, 18:27, 20 September.
1929. *Perognathus gypsi* Dice, Occas. Papers Mus. Zool., Univ. Michigan, 203:1, 19 June.

Holotype.—Adult female, skin and skull, USNM 97416, from Casas Grandes, Chihuahua, México; obtained on 21 May 1899 by Edward A. Goldman.

Measurements of holotype.—Total length, 133; length of tail, 65; length of hind foot, 19.5; greatest length of skull, 22.20; occipitonasal length, 22.20; interorbital breadth, 5.10; length of maxillary toothrow, 2.85; width across maxillary toothrows, 4.20; length of mastoid bulla, 7.60; width across mastoid bullae, 11.75; length of interparietal, 2.60; width of interparietals, 3.95; length of nasals, 8.25; width of nasals, 2.25; width of rostrum, 3.65.

Distribution.—Sandy areas in desert and arid grassland associations from Gran Quivira and the San Augustine Plains of central New Mexico, southward to the Samalayucca Sands and Casas Grandes, Chihuahua; on the southwest, extending to Willcox Playa, Arizona, and on the southeast, at least to El Paso Co., Texas (Williams, 1978*b*).

Remarks.—Populations of *melanotis* from the White Sands of the Tularosa Basin are whitish dorsally and ventrally; color, however, is extremely variable geographically, depending upon the color of the local substrate and indirectly on the amount of annual precipitation (Williams, 1978*b*).

Perognathus flavescens perniger Osgood, 1904

1904. *Perognathus flavescens perniger* Osgood, Proc. Biol. Soc. Washington, 17:127, 9 June.

Holotype.—Subadult female, skin and skull, USNM 57725, from Vermillion, Clay Co., South Dakota; obtained on 22 August 1889 by G. S. Agersborg.

Measurements of holotype.—Total length, 140; length of tail, 68; length of hind foot, 17 (dry); greatest length of skull, 21.05; occipitonasal length, 21.05; interorbital breadth, 4.90; length of maxillary toothrow, 3.10; width across maxillary toothrows, 4.15; length of mastoid bulla, 6.40; width across mastoid bullae, 10.70; length of interparietal, 2.55; width of interparietals, 4.70; length of nasals, 7.50; width of nasals, 2.00; width of rostrum, 3.70.

Distribution.—Occupies sandy soils in grassland associations in the northeastern sector of the species range, from southeastern North Dakota and southern Minnesota southward to northeastern Iowa, and westward and southward to central South Dakota and eastern Nebraska (Hall, 1981).

Remarks.—Typical specimens of *perniger* are larger than typical specimens of *cockrumi,* but both share the characteristic of dark color, which is associated with the high precipitation in the eastern sector of the geographic range of the species. Jones (1964) gave external and cranial measurements for a sample of *perniger* from Nebraska; Reed and Choate (1986) gave statistics for measurements for a grouped sample of 20.

Perognathus flavescens relictus Goldman, 1938

1938. *Perognathus apache relictus* Goldman, J. Mamm., 19:495, 14 November.

Holotype.—Adult male, skin and skull, USNM 150768, from Medano Springs Ranch, 7,600 ft, 15 mi NE Mosca, Alamosa Co., Colorado; obtained on 2 November 1907 by Merritt Cary.

Measurements of holotype.—Total length, 137; length of tail, 68; length of hind foot, 19.0; greatest length of skull, 22.70; occipitonasal length, 22.70; interorbital breadth, 5.45; length of maxillary toothrow, 3.15; width across maxillary toothrows, 4.20; length of mastoid bulla, 8.00; width across mastoid bullae, 12.40; length of interparietal, 3.10; width of interparietals, 3.75; length of nasals, 7.90; width of nasals, 2.35; width of rostrum, 3.60.

Distribution.—In so far as is known, occupies sandy substrates in the San Luis Valley, south-central Colorado (Williams, 1978*b*).

Remarks.—Williams (1987*b*) gave means and ranges of measurements for 17 adults, and Armstrong (1972) listed measurements for 6 adult males.

Perognathus flavus

Diagnosis.—The smallest species of *Perognathus,* with the shortest tail and smallest ratio of lengths of tail to head and body (ratio averaging less than 0.9); dorsal color is heavily overlain with blackish-tipped hairs in most subspecies, contrasting sharply with the clear buffy post-auricular patch and a narrow line without black tinge on the side adjacent to the white underparts; the posterior region of the cranium has relatively large auditory bullae and is relatively constricted, with short, narrow interparietals; length of head and body averages about 60 mm or less; length of tail averages less than 57 mm and rarely exceeds 60; interparietal length averages from about 2.3 to 2.5 mm and rarely exceeds 2.9; width of interparietals averages between about 3.0 and 3.3 mm and rarely exceeds 3.9.

Comparisons.—*P. flavus* is most similar to *P. merriami*; they may occasionally hybridize at a few localities. Generally, *flavus* differs from *merriami* in having: a shorter tail; longer, softer, and laxer pelage; darker, more contrasting mid-dorsal color with a pinkish rather than yellowish or yellowish-orange hue; larger post-auricular spots; relatively shorter tail; larger, more inflated auditory bullae; narrower interorbital and interparietal widths; and wider P4. No single set of characters is known that will distinguish all *P. flavus* from all *P. merriami.* Comparisons of individuals from areas of sympatry show *P. f. flavus* and *P. m. gilvus,* respectively, to average: bullar length/occipitonasal length, 0.388, 0.359; tail length/total length, 0.452, 0.488; distance across mastoid bullae, 11.93, 11.44 mm; width of P4, 0.90, 0.82 mm. See accounts of *P. fasciatus* and *P. flavescens* for comparisons with those sympatric species. *P. flavus* is also sympatric with *P. amplus* and *P. longimembris*; it differs from those species in having an absolutely and relatively shorter, non-penciled tail; from *amplus, P. flavus* also differs in being significantly smaller (length of hind foot rarely exceeds 18 mm versus rarely less than 19); from *longimembris, flavus* differs in smaller average size, smaller hind feet, narrower interorbital width, and significantly wider upper molars. *P. flavus* is significantly smaller and has a relatively shorter, non-penciled tail than the allopatric species *P. alticola, P. inornatus,* and *P. parvus.*

Distribution.—*Perognathus flavus* occupies a broad range in the west-central and southweatern Great Plains, intermountain basins, and central plateau of México; a disjunct population is found along the coastal plain of the Gulf of California in Sonora. Its geographic range extends from eastern Wyoming and western Nebraska, southward through western Texas and western Coahuila on the Mexican Plateau to eastern Jalisco, Morelos, Puebla, and tiny area of east-central Veracruz; on the west, it ranges across New Mexico and southern Colorado to southeastern Utah and west-central Arizona both north and south of the Mogollon Mountains (Hall, 1981).

Remarks.—Wilson (1973) treated *P. merriami merriami* and *P. m. gilvus* as subspecies of *P. flavus.* Anderson (1972) and Lee and Engstrom (1991) presented structural and genic evidence that *flavus* and *merriami* are not conspecific. A few individuals of *P. f. flavus* and *P. m. gilvus* apparently hybridize at a few localities in southeastern New Mexico, complicating identification. Williams (1971, 1978*a*) placed *P. flavus* with *P. merriami* in the *flavus* species group, separate from the *fasciatus* group to which *flavus* and *merriami* had been assigned by Osgood (1900) and subsequent researchers.

Perognathus flavus bimaculatus Merriam, 1889

1889. *Perognathus flavus bimaculatus* Merriam, N. Amer. Fauna, 1:12, 25 October.

Holotype.—Adult male, skin and skull, USNM 8455/23789, from Fort Whipple, Yavapai Co., Arizona; obtained on 21 May 1865 by Elliott Coues.

Measurements of holotype.—Total length, 117; length of tail, 40; length of hind foot, 17; length of ear (crown, dry), 4.

Distribution.—Occupies the plateau region of west-central Arizona, from the south rim of the Grand Canyon of the Colorado River in central Coconino Co., southward around the west rim of the Mogollon Plateau to near Prescott, Yavapai Co., and westward to the Aubrey Valley in western Coconino Co. (Hall, 1981).

Remarks.—The skull of the holotype is so fragmented that it precludes taking standard measurements. Hoffmeister (1986) gave statistics for measurements of 10 specimens; Osgood (1900) gave average external measurements for 10 individuals and cranial measurements for 3 individuals.

Perognathus flavus bunkeri Cockrum, 1951

1951. *Perognathus flavus bunkeri* Cockrum, Univ. Kansas Publ., Mus. Nat. Hist., 5:205, 15 December.

Holotype.—Adult female, skin and skull, KU 11716, from Conrad Farm, 1 mi E Coolidge, Hamilton Co., Kansas; obtained by F. Parks and Claude W. Hibbard on 1 July 1936.

Measurements of holotype.—Total length, 104; length of tail, 44; length of hind foot, 16; length of ear, 7; occipitonasal length, 20.6; frontonasal length, 14.0; interorbital breadth, 4.5; length of maxillary toothrow, 3.3; length of mastoid bulla, 8.0; width across mastoid bullae, 12.1; width of interparietals, 3.2.

Distribution.—Occurs in the west-central Great Plains from eastern Colorado though western Kansas, and south to western Oklahoma.

Remarks.—Armstrong (1972) listed measurements for seven individuals from Colorado.

Perognathus flavus flavus Baird, 1855

1855. *Perognatus* [*sic*] *flavus* Baird, Proc. Acad. Nat. Sci. Philadelphia, 7:332.

Holotype.—Sex and age unknown, skin, USNM 148/1130, from El Paso, El Paso Co., Texas; obtained by J. H. Clark in 1851 (see remarks below).

Measurements of holotype.—Total length, 102.1; length of tail, 50.8; length of hind foot, 15.7.

Distribution.—Ranges from the extreme

northern Panhandle of Texas westward through most of northern New Mexico, excluding the San Juan Basin, and southward in central New Mexico through extreme western Texas and northern and eastern Chihuahua to north-central Durango; also extends westward across southern New Mexico to southeastern Arizona, south of the Gila River and east of the Santa Cruz River, and probably into northeastern Sonora (Hall, 1981).

Remarks.—Some measurements given by Baird (1858), in a more complete description of the species, differ from the original description. According to Merriam (1889) the holotype was lost; he designated USNM 4328/5047 (adult male) from Mason, Texas as a duplicate type, but noted that Mason was about 400 miles east of El Paso, and remarked that "it would not be strange if future collections show the El Paso animal to be different from the one here described." The duplicate type is from the easternmost area of intergradation between *P. merriami gilvus* and *P. m. merriami,* and clearly does not represent *P. f. flavus.* Anderson (1972) gave means and ranges of measurements for 9 specimens from Chihuahua, Hoffmeister (1986) listed statistics for measurements of 14 specimens from Arizona; Williams (1971) gave statistics for measurements for 30 adults from Bernalillo Co., New Mexico; and Wilson (1973) gave means ± *SD* for 37 specimens.

Perognathus flavus fuliginosus Merriam, 1890

1890. *Perognathus fuliginosus* Merriam, N. Amer. Fauna, 3:74, 11 September.

Holotype.—Immature male, skin and skull, USNM 17708/24644, from cedar belt, northeast of San Francisco Mountain, 7,000 ft, Coconino Co., Arizona; obtained by Vernon Bailey on 4 October 1889.

Measurements of holotype.—Total length, 116; length of tail, 58; length of hind foot, 18; length of ear (crown, dry), 4; nasal length, 7.70; interorbital breadth, 4.65.

Distribution.—Occupies arid grassland and woodland associations in the volcanic region around the San Francisco Mountains and Mogollon Plateau near Flagstaff, northward west of the Little Colorado River to near the south rim of the Grand Canyon of the Colorado River, all in Coconino Co., Arizona.

Remarks.—The skull of the holotype is broken, preventing the measuring of several traits. Hoffmeister (1986) gave statistics for a sample of 20, and Osgood (1900) listed average measurements for 3 individuals.

Perognathus flavus fuscus Anderson, 1972

1972. *Perognathus flavus fuscus* Anderson, Bull. Amer. Mus. Nat. Hist., 148:304, 8 September.

Holotype.—Adult male, skin and skull, KU 81168, from 2 mi W Minaca, 6,900 ft, Chihuahua; obtained by Charles A. Long on 2 July 1959.

Measurements of holotype.—Total length, 103; length of tail, 45; length of hind foot, 16; length of ear, 6; weight 7.4 g; greatest length of skull, 20.40; width across bullae, 11.70; breadth across maxillary arches, 10.30; nasal length, 7.60; interorbital breadth, 4.40.

Distribution.—Known from the upper valley of the Río Papigochic and the watershed of the Laguna Bustillos above 6,000 ft elevation, west-central Chihuahua, México (Anderson, 1972).

Remarks.—Anderson (1972) gave means and ranges of measurements for five individuals.

Perognathus flavus goodpasteri Hoffmeister, 1956

1956. *Perognathus flavus goodpasteri* Hoffmeister, Proc. Biol. Soc. Washington, 69:55, 21 May.

Holotype.—Adult male, skin and skull, UIMNH 6312, from 2.75 mi NW Springerville, Apache Co., Arizona; obtained by Charles A. McLaughlin on 16 August 1953.

Measurements of holotype.—Total length, 107; length of tail, 54; length of hind foot, 17; length of ear (notch), 7; occipitonasal length, 20.5; basilar length, 14.4; interorbital breadth, 4.4; length of maxillary toothrow, 3.3; width across mastoid bullae, 12.2; length of interparietal, 3.1; width of interparietals, 3.1; length of nasals, 7.2; width of nasals, 2.0.

Distribution.—Occupies grassland associations in east-central Arizona, ranging from near Springerville through Snowflake to south of Holbrook on the northeast Mogollon Plateau, and also occurs on the south side of the Mogollon Plateau near Nash Creek Reservoir, Navajo Co. (Hoffmeister, 1986).

Remarks.—Hoffmeister (1986) gave statistics for a sample of 10 individuals.

Perognathus flavus hopiensis Goldman, 1932

1932. *Perognathus flavus hopiensis* Goldman, Proc. Biol. Soc. Washington, 45:89, 21 June.

Holotype.—Adult female, skin and skull, USNM 248014, from Oraibi, 6,000 ft, Hopi Indian Reservation, Navajo Co., Arizona; obtained by George G. Cantwell on 5 June 1927.

Measurements of holotype.—Total length, 115; length of tail, 50; length of hind foot, 15; occipitonasal length, 19.7; interorbital breadth, 4.4; length of maxillary toothrow, 2.9; width across mastoid bullae, 12.0; length of interparietal, 2.5; width of interparietals, 3.2; length of nasals, 7.0; width of nasals, 2.2.

Distribution.—Occurs from southeastern Utah, mostly south of the San Juan River, and extreme southwestern Colorado in Montezuma Co., southward through northeastern Arizona; in Arizona occurs mostly east of the Little Colorado River and south and east of the Colorado River, but occurs west of the Little Colorado River near Wupatki National Monument (Hoffmeister, 1986).

Remarks.—Hoffmeister (1986) gave statistics for measurements for a sample of 23 individuals.

Perognathus flavus medius Baker, 1954

1954. *Perognathus flavus medius* Baker, Univ. Kansas Publ., Mus. Nat. Hist., 7:343, 15 February.

Holotype.—Adult female, skin and skull, KU 48583, from 1 mi S, 6 mi E Rinóon de Romos, 6,550 ft, Aguascalientes, México; obtained on 14 July 1952 by Rollin H. Baker.

Measurements of holotype.—Total length, 111; length of tail, 52; length of hind foot, 17; occipitonasal length, 20.7; frontonasal length, 13.9; interorbital breadth, 4.5; length of maxillary toothrow, 3.0; length of mastoid bulla, 7.8; width of interparietals, 2.9.

Distribution.—Occurs on the plateau of north-central México, from southeastern Durango and southeastern Coahuila southward through San Luis Potosí to Aguascalientes, extreme northern and eastern Jalisco, and Guanajuato.

Remarks.—Baker (1954) listed statistics for measurements of 17 adults.

Perognathus flavus mexicanus Merriam, 1894

1894. *Perognathus flavus mexicanus* Merriam, Proc. Acad. Nat. Sci. Philadelphia, 46:265, 27 September.

Holotype.—Young adult male, skin and skull, USNM 50714, from Tlalpam, Valley of México, Distrito Federal, México; obtained by Edward W. Nelson on 4 December 1892.

Measurements of holotype.—Total length, 118; length of tail, 55; length of hind foot, 17.5; length of ear (from anterior base, dry), 6; greatest length of skull, 21.00; width across bullae, 11.90; breadth across maxillary arches, 10.20; nasal length, 6.90; interorbital breadth, 4.60.

Distribution.—Occupies south-central México from Querétaro and Hidalgo southward to Morelos, Puebla, and west-central Veracruz (Baker, 1954).

Remarks.—Baker (1954) gave means and extremes of measurements for 27 specimens.

Perognathus flavus pallescens Baker, 1954

1954. *Perognathus flavus pallescens* Baker, Univ. Kansas Publ., Mus. Nat. Hist., 7:345, 15 February.

Holotype.—Adult male, skin and skull, KU 40298, from 1 mi SW San Pedro de las Colonias, 3,700 ft, Coahuila, México, on 9 February 1951 by J. R. Alcorn.

Measurements of holotype.—Total length, 107; length of tail, 52; length of hind foot, 16; greatest length of skull, 20.5; frontonasal length, 13.7; interorbital breadth, 4.5; length of maxillary toothrow, 3.0; length of mastoid bulla, 8.0; width across mastoid bullae, 11.9; width of interparietals, 2.8.

Distribution.—Occupies portions of the Chihuahuan Desert in central and western Coahuila, southeastern Chihuahua, northeastern Durango, and extreme northern Zacatecas (Hall, 1981; Matson and Baker, 1986).

Remarks.—Baker (1954) gave measurements for four specimens from Coahuila.

Perognathus flavus parviceps Baker, 1954

1954. *Perognathus flavus parviceps* Baker, Univ. Kansas Publ., Mus. Nat. Hist., 7:344, 15 February.

Holotype.—Adult female, skin and skull, KU 38402, from 4 mi W, 2 mi S Guadalajara, 5,100 ft, Jalisco, México; obtained by J. R. Alcorn on 15 June 1950.

Measurements of holotype.—Total length, 113; length of tail, 52; length of hind foot, 17; occipitonasal length, 20.9; frontonasal length, 13.9; interorbital breadth, 4.6; length of maxillary toothrow, 3.2; length of mastoid bulla, 7.9; width across mastoid bullae, 12.2; width of interparietals, 3.3.

Distribution.—Known from the plateau region of central and eastern Jalisco and extreme southern Zacatecas (Baker, 1954; Matson and Baker, 1986).

Remarks.—Baker (1954) listed means and ranges of measurements for 15 specimens.

Perognathus flavus piperi Goldman, 1917

1917. *Perognathus flavus piperi* Goldman, Proc. Biol. Soc. Washington, 30:148, 27 July.

Holotype.—Adult male, skin and skull, USNM 168650, from 23 mi SW Newcastle, Weston Co., Wyoming; obtained on 25 May 1910 by S. E. Piper.

Measurements of holotype.—Total length, 113; length of tail, 51; length of hind foot, 17; greatest length of skull, 22.0; interorbital breadth, 4.6; length of maxillary toothrow, 3.5; width across mastoid bullae, 12.4; length of interparietal, 1.6; width of interparietals, 2.7; length of nasals, 8.1.

Distribution.—Known from the northern Great Plains in eastern Wyoming and western Nebraska (Hall, 1981).

Remarks.—Jones (1964) gave measurements for four specimens from Nebraska.

Perognathus flavus sanluisi Hill, 1952

1942. *Perognathus flavus sanluisi* Hill, Amer. Mus. Novitates, 1212:1, 7 December.

Holotype.—Adult female, skin and skull, AMNH 137669, from 9 mi E of Center,

7,580 ft, Saguache Co. (20 mi NNW of Alamosa, Alamosa Co.), Colorado; obtained by Peter E. Crowe on 18 August 1940.

Measurements of holotype.—Total length, 109; length of tail, 46; length of hind foot, 15; length of ear (notch), 6; greatest length of skull, 20.0; basilar length, 14.1; interorbital breadth, 4.5; length of maxillary toothrow, 3.1; width across mastoid bullae, 11.7; length of interparietal, 2.6; width of interparietals, 3.1; zygomatic breadth, 9.7; length of nasals, 7.5; length of rostrum, 8.2.

Distribution.—Occurs only in the San Luis Valley of south-central Colorado.

Remarks.—Armstrong (1972) gave statistics for measurements of seven males and four females.

Perognathus flavus sonoriensis Nelson and Goldman, 1934

1934. *Perognathus flavus sonoriensis* Nelson and Goldman, J. Washington Acad. Sci., 24:267, 15 June.

Holotype.—Adult female, skin and skull, USNM 250885, from Costa Rica Ranch, lower Río Sonora, Sonora, México; obtained on 13 December 1932 by Vernon Bailey and Frederic Winthrop.

Measurements of holotype.—Length of hind foot, 15; occipitonasal length, 19.3; interorbital breadth, 4.2; length of maxillary toothrow, 2.8; width across mastoid bullae, 11.7; length of interparietal, 3.0; width of interparietals, 3.2; length of nasals, 6.5; width of nasals, 2.0; zygomatic breadth (posteriorly), 10.3.

Distribution.—Known only from the coastal plains of central Sonora (Hall, 1981).

Remarks.—No other published information is known.

Perognathus inornatus

Diagnosis.—A medium-sized species of *Perognathus,* with a slightly penciled tail averaging longer than length of head and body; interorbital breadth relatively narrow compared to other *longimembris*-group species; width of interparietal highly variable, generally broader than other *longimembris*-group species; dorsal profile of skull relatively flat; length of head and body averages from about 70 to 75 mm; ratio of lengths of tail to head and body averages from about 1.02 to 1.07; interorbital breadth ranges from about 4.75 to 5.7 mm (means from about 4.9 to 5.1); width of interparietal ranges from about 3.3 to 4.7 mm, averaging from about 3.6 to 3.9.

Comparisons.—*Perognathus inornatus* is most similar to *P. amplus* and *P. longimembris* (see account of *amplus* for methods of distinguishing that species). The similarities between *P. inornatus* and *longimembris* are great, but generally *P. inornatus* averages slightly larger with a relatively shorter and less penciled tail; the interorbital region of *inornatus* is relatively narrower, but there is broad overlap in absolute measurements. *P. inornatus* is potentially sympatric with *P. alticola,* and possibly also with *P. parvus*; from these two species, *inornatus* differs in having a less-penciled tail, smaller average size (hind foot generally less than 21 versus 21 mm or more); relatively larger mastoid bullae and greater mastoid breadth, and relatively narrower interparietals. From the allopatric species, *P. fasciatus, P. flavescens, P. flavus,* and *P. merriami, inornatus* differs in having a slightly penciled tail averaging longer than length of head and body (the tails of the other species lack pencils and average shorter in length than length of head and body).

Distribution.—Occupies arid, annual grassland, savanna, and desert shrub associations of west-central California, ranging from the upper Sacramento Valley, Tehama Co., southward through the San Joaquin and Salinas Valleys and contiguous areas to the Mojave Desert, Los Angeles, Kern, and extreme western San Bernardino counties; also occupies the Tehachapi mountains and the foothills of the western Sierra Nevada below about 2,000 ft (Hall, 1981; D. Williams, unpubl. data).

Remarks.—The literature of *P. inornatus* shows great confusion about the taxonomy of *longimembris* and *inornatus* (e.g., Osgood, 1918), and the systematic relationships among populations of these two species. These relationships are not adequately clarified; at least two and possibly three species probably are currently included under the name *P. inornatus.* With one exception, we retain the current taxonomy for *longimembris* group pocket mice from central California, although our unpublished information suggests a different taxonomy will be required when studies are completed.

Perognathus inornatus inornatus
Merriam, 1889

1889. *Perognathus inornatus* Merriam, N. Amer. Fauna, 1:15, 25 October.

Holotype.—Adult male, skin (in alcohol) and skull, USNM 13394/23790 from Fresno, Fresno Co., California; obtained by Gustav Eisen (no date; cataloged 6 February 1882).

Measurements of holotype.—[Externals taken by Merriam from alcoholic specimen before skinning out skull, except ear from notch] Total length, 137; length of tail, 71; length of hind foot, 18.5; length of ear (crown), 4, notch 6.6; greatest length of skull, 22.80; occipitonasal length, 22.35; interorbital breadth, 4.60; length of maxillary toothrow, 3.10; width across maxillary toothrows, 4.05; length of mastoid bulla, 8.70; width across mastoid bullae, 12.35; length of interparietal, 2.80; width of interparietals, 3.55; length of nasals, 8.20; width of nasals, 2.05; width of rostrum, 3.40.

Distribution.—Ranges through the Sacramento Valley, from Tehama Co. on the north to the coastal mountains in eastern Lake Co. on the west, southward along the eastern side and floor of the San Joaquin Valley to near its southern end. The specific status of pocket mice of the *inornatus* complex from the western side of the San Joaquin Valley floor and the Tehachapi Mountains remains to be determined.

Remarks.—Pocket mice of this complex from the eastern side of the San Joaquin Valley in Fresno, Madera, and Stanislaus and the floor of the valley in Fresno, Kings, and Kern counties have 50 chromosomes and differ structurally from those of the western edge of the San Joaquin Valley and areas to the west in Kern, San Luis Obispo, Fresno, San Benito, Merced, San Joaquin, and Alameda counties. Individuals of the latter populations have 56 chromosomes and are generally larger, with relatively longer tails. A single individual from Lake Co. had 60 chromosomes; no others from the Sacramento Valley region have been studied cytologically (Williams, 1978*a*; D. Williams, unpubl. data). Largely on the basis of the consistent chromosomal difference, populations in the foothills of the western edge of the San Joaquin Valley and nearby mountains are assigned to *P. inornatus neglectus* rather than *inornatus* in deference to Hall (1981).

Perognathus inornatus neglectus
Taylor, 1912

1912. *Perognathus longimembris neglectus* Taylor, Univ. California Publ. Zool., 10:155, 21 May.

Holotype.—Adult male, skin and skull, MVZ 14526, from McKittrick, 1,111 ft, Kern Co., California; obtained on 18 May 1911 by H. S. Swarth and W. L. Chandler.

Measurements of holotype.—Total length, 157; length of tail, 77; length of hind foot, 22; length of ear, 6; greatest length of skull, 25.75; interorbital breadth, 5.75; length of mastoid bulla, 8.85; width across mastoid bullae, 13.85; length of interparietal, 2.80; width of interparietals, 3.65; length of nasals, 9.60; width of rostrum, 3.85.

Distribution.—Occurs in the hills and piedmont slopes along the western edge of the San Joaquin Valley from near Suisun

Bay, Contra Costa Co., southward to the Mojave Desert, western San Bernardino and northeastern Los Angeles counties; and westward in the Panoche Valley, San Benito Co., the Carrizo Plain, San Luis Obispo Co., and the upper Cuyama Valley in Santa Barbara and San Luis Obispo counties.

Remarks.—See remarks above under *P. inornatus inornatus* and below under *P. inornatus psammophilus.* We consider individuals from the Carrizo Plain to be indistinguishable from those near the type locality on the western edge of the San Joaquin Valley. This arrangement differs from that of von Bloeker (1937). The taxonomic assignment of populations from the extreme western floor of the San Joaquin Valley and the Salinas River Valley is undecided. Specimens from the Mojave Desert are assigned to *neglectus* on the basis of similar structure. Taylor (1912) gave means of measurements for an unknown number of individuals of each sex.

Perognathus inornatus psammophilus von Bloeker, 1937

1937. *Perognathus longimembris psammophilus* von Bloeker, Proc. Biol. Soc. Washington, 50:153, 10 September.
1937. *Perognathus inornatus sillimani* von Bloeker, Proc. Biol. Soc. Washington, 50:154, 10 September.

Holotype.—Subadult female, skin and skull, MVZ 74681, from west side of Arroyo Seco Wash, 150 ft, 4 mi S Soledad, Monterey Co., California; obtained on 5 June 1936 by Jack C. von Bloeker, Jr.

Measurements of holotype.—Total length, 130; length of tail, 62; length of hind foot, 18; length of ear, 7; greatest length of skull, 21.10; interorbital breadth, 4.85; length of mastoid bulla, 7.35; width across mastoid bullae, 11.55; length of interparietal, 2.55; width of interparietals, 4.50; length of nasals, 7.50; width of rostrum, 3.50.

Distribution.—Occupies the Salinas Valley, from near Soledad, southward at least to Hog Canyon, Monterey Co.

Remarks.—The only differences we can find between the type specimens and assigned specimens of *P. longimembris psammophilus* and *P. inornatus sillimani* of von Bloeker (1937) are due to age. Nearly all of von Bloeker's (1937) specimens of *psammophilus* are juveniles and subadults, and nearly all of his specimens of *sillimani* are adults; the type series of the latter are old adults (age class 5 of Williams, 1978*b*). On the basis of page precedence, *sillimani* is a junior synonym of *psammophilus. P. inornatus psammophilus* is most similar to *P. inornatus neglectus*; we have assigned specimens from the Carrizo Plain that von Bloeker (1937) treated as *sillimani* and *psammophilus* to *P. inornatus neglectus,* and consider it likely that *psammophilus* will prove to be a synonym of *neglectus.*

Perognathus longimembris

Diagnosis.—A small species of *Perognathus* with a relatively long, slightly penciled tail, medium-sized interparietals, and narrow upper premolar. Ratio of length of tail to head and body nearly always exceeds 1.03, ranging to about 1.40; length of hind foot ranges from about 15 to 20 mm (means between about 18 and 19); occipitonasal length rarely exceeds 22.0 mm (population means always less than 22.0); width of upper premolar ranges from about 0.75 to 0.95 mm.

Comparisons.—*Perognathus longimembris* is most similar to *P. amplus* and *P. inornatus* (see accounts of those species for comparisons). *P. longimembris* is similar in size to *P. flavescens,* but differs in having a tail longer than length of head and body and with a slight pencil; from *P. alticola* and *P. parvus, P. longimembris* differs in being significantly smaller in nearly all dimensions; length of hind foot rarely exceeds 20 mm in *longimembris* whereas the hind feet of *alticola* and *parvus* usually exceed 20 mm. From *P. merriami, P. longimembris* differs in larger size with a relatively and absolutely longer tail; length of tail in *longimembris* typically exceeds 66 mm whereas in *merriami,* the tail rarely measures as long as 66 mm. See account of *P. flavus* for comparison with that species.

Distribution.—Occupies desert, shrub-steppe, and open, arid woodland communities of the Great Basin, southward from southeastern Oregon, extreme southwestern Idaho, and western Utah; the Mojave Desert and Tehachapi Mountains, southern California coastal basins, Colorado Desert, and adjacent areas along the Colorado River in south-central Utah, Arizona, and California; also extends into northern Baja California and around the northern perimeter of the Gulf of California in Sonora; and occurs as apparently disjunct populations in Pinal and Maricopa counties, south-central Arizona, and around Bahía Kino in west-central Sonora (Hall, 1981).

Remarks.—Distribution records for *P. longimembris* from west-central California in the Sacramento Valley, most or all of the San Joaquin Valley, Salinas Valley, and on the Carrizo Plain are based on misidentified specimens of *P. inornatus.* The northern distributional limit of *P. longimembris* in central California has not been determined, but appears to coincide approximately with the Transverse ranges, from which populations are known. The identity of small pocket mice on the northern slopes of the Transverse ranges in the San Joaquin Valley is uncertain (unpubl. data).

Perognathus longimembris aestivus Huey, 1928

1928. *Perognathus longimembris aestivus* Huey, Trans. San Diego Soc. Nat. Hist., 5:87, 18 January.

Holotype.—Adult male, skin and skull, SDSNH 6110, from Sangre de Cristo, Valle San Rafael on western base of Sierra Juárez, lat. 31° 52′ N, long. 116° 06′ W, Baja California, México; obtained on 10 June 1927 by Laurence M. Huey.

Measurements of holotype.—Total length, 135; length of tail, 71; length of hind foot, 17; length of ear, 5; weight, 8.7 g; greatest length of skull, 22.0; interorbital breadth, 5.4; length of maxillary toothrow, 2.8; width across mastoid bullae, 13.2; length of nasals, 8.0.

Distribution.—Known from the western base of the Sierra Juárez, from near the type locality on the north, south to El Valle de la Trinidad (Hall, 1981).

Remarks.—Like other populations of *P. longimembris* from the southwestern part of its range, the mastoid bullae are greatly inflated; whether it is subspecifically distinct from adjacent subspecies is equivocal, however. Huey (1939*a*) listed measurements for five males and three females.

Perognathus longimembris arizonensis Goldman, 1931

1931. *Perognathus longimembris arizonensis* Goldman, Proc. Biol. Soc. Washington, 44:134, 17 October.

1935. *Perognathus longimembris arcus* Benson, Univ. California Publ. Zool., 40:451, 31 December.

1939. *Perognathus longimembris virginis* Huey, Trans. San Diego Soc. Nat. Hist., 9:55, 31 August.

Holotype.—Adult female, skin and skull, USNM 250032, from 10 mi S Jacobs Pools, Houserock Valley, north side of Marble Canyon of Colorado River, 4,000 ft, Coconino Co., Arizona; obtained on 17 June 1931 by Edward A. Goldman.

Measurements of holotype.—Total length, 137; length of tail, 79; length of hind foot, 18.5; greatest length of skull, 22.05; occipitonasal length, 21.35; interorbital breadth, 4.95; length of maxillary toothrow, 3.00; width across maxillary toothrows, 4.05; length of mastoid bulla, 8.50; width across mastoid bullae, 12.15; length of interparietal, 3.15; width of interparietals, 3.10;

length of nasals, 8.00; width of nasals, 2.10; width of rostrum, 3.35; zygomatic breadth, 10.9.

Distribution.—Occurs in extreme southwestern and south-central Utah and northwestern and north-central Arizona generally north of the Colorado River, and in southeastern Nevada (Hall, 1981; Hoffmeister, 1986). Known from south (east) of the Colorado River only from an area near Rainbow Bridge, San Juan Co., Utah, and the vicinity of Page, Coconino Co., Arizona (Hall, 1981; Hoffmeister, 1986).

Remarks.—Hoffmeister (1986) considered *virginis* not to be subspecifically distinct from *arizonensis,* and he found the two specimens known of *arcus* to provide little justification for recognition as a subspecies. Goldman (1931) listed mean external measurements for four topotypes. Hall (1946) gave means and ranges of measurements for 10 individuals of each sex from Nevada; Hoffmeister (1986) gave statistics for 201 specimens in four samples from Arizona.

Perognathus longimembris bangsi Mearns, 1898

1898. *Perognathus longimembris bangsi* Mearns, Bull. Amer. Mus. Nat. Hist., 10:300, 31 August.

1900. *Perognathus panamintinus arenicola* Stephens, Proc. Biol. Soc. Washington, 13:151, 13 June.

Holotype.—Adult female, skin and skull, AMNH 5304, from Palm Springs, Colorado Desert, Riverside Co., California; obtained on 13 April 1896 by E. C. Thurber.

Measurements of holotype.—Total length, 138; length of tail, 80; length of hind foot, 19; greatest length of skull, 20.6; interorbital breadth, 4.8; length of maxillary toothrow, 3.4; width across mastoid bullae, 11.4; length of nasals, 7.7; length of rostrum, 9.5; zygomatic breadth, 9.8.

Distribution.—Occurs in the Colorado Desert region west of the Imperial Valley, in southern California, from north of Palm Springs, Riverside Co., southward to near the Mexican boundary in western Imperial Co. (unpubl. data).

Remarks.—Huey (1939*a*) listed measurements for five males and five females.

Perognathus longimembris bombycinus Osgood, 1907

1907. *Perognathus bombycinus* Osgood, Proc. Biol. Soc. Washington, 20:19, 23 February.

Holotype.—Adult male, skin and skull, USNM 136123, from Yuma, Yuma Co., Arizona; obtained on 18 March 1905 by Edward A. Goldman.

Measurements of holotype.—Total length, 134; length of tail, 79; length of hind foot, 18.5; greatest length of skull, 21.55; occipitonasal length, 20.40; basilar length, 14.5; interorbital breadth, 4.50; length of maxillary toothrow, 2.95; width across maxillary toothrows, 3.40; length of mastoid bulla, 8.60; width across mastoid bullae, 11.7; length of interparietal, 2.5; width of interparietals, 2.6; length of nasals, 7.3; width of nasals, 1.80; width of rostrum, 2.90; zygomatic width, 9.8; diastema, 5.1.

Distribution.—Found along the lower Colorado River Valley in southeastern California and southwestern Arizona from near Parker Dam on the north, and around the northern end of the Gulf of California in northwestern Sonora and northeastern Baja California. Isolated populations extend eastward in southwestern Arizona to near Phoenix and Casa Grande. In Baja California, extends to near San Felipe, and in Sonora, to near the Pinacate Lava Flows (Hall, 1981).

Remarks.—Huey (1939*a*) listed measurements of three males and four females.

Perognathus longimembris brevinasus Osgood, 1900

1900. *Perognathus panamintinus brevinasus* Osgood, N. Amer. Fauna, 18:30, 20 September.

Holotype.—Adult female, skin and skull, USNM 186515, from San Bernardino, San Bernardino Co., California; obtained on 2 May 1885 by Frank Stephens.

Measurements of holotype.—Total length, 124; length of tail, 66; length of hind foot (dry), 17.4; greatest length of skull, 20.65; occipitonasal length, 20.15; interorbital breadth, 5.05; length of maxillary toothrow, 3.00; width across maxillary toothrows, 3.85; length of mastoid bulla, 8.00; width across mastoid bullae, 11.80; length of interparietal, 3.10; width of interparietals, 4.20; length of nasals, 7.20; width of nasals, 1.90; width of rostrum, 3.35.

Distribution.—Occupies the arid coastal basins of southern California in grassland and coastal sage associations, from approximately Burbank and San Fernando, Los Angeles, Co., on the northwest, to San Bernardino, San Bernardino Co., and Cabazon, Hemet, and Aguanga, Riverside Co., on the east; the southern limit is near the northern boundary of San Diego Co. (Huey, 1939*a*).

Remarks.—Osgood (1900) gave average measurements for three topotypes; Huey (1939*a*) gave measurements of five individuals of each sex.

Perognathus longimembris gulosus Hall, 1941

1941. *Perognathus longimembris gulosus* Hall, Proc. Biol. Soc. Washington, 54:55, 20 May.

Holotype.—Adult female, skin and skull, MVZ 78764, from near [0.25 mi S] Smith Creek Cave, 5,800 ft, Mount Moriah, White Pine Co., Nevada; obtained by Lee W. Arnold on 4 June 1937.

Measurements of holotype.—Total length, 132; length of tail, 72; length of hind foot, 17; length of ear, 7; weight, 8.9 g; occipitonasal length, 21.1; frontonasal length, 14.2; interorbital breadth, 5.3; length of maxillary toothrow, 3.0; length of mastoid bulla, 8.2; width across mastoid bullae, 12.5.

Distribution.—Occurs in desert associations along the western margin of the Pleistocene Lake Bonneville in eastern Nevada and western Utah, from near Kelton, Box Elder Co., Utah, southward to near Pruss Lake, Millard Co., Utah (Hall, 1981).

Remarks.—Durrant (1952) gave measurements for six males and three females from Utah; Hall (1946) listed measurements for five females and three males from Nevada.

Perognathus longimembris internationalis Huey, 1939

1939. *Perognathus longimembris internationalis* Huey, Trans. San Diego Soc. Nat. Hist., 9:47, 31 August.

Holotype.—Adult male, skin and skull, SDSNH 11972, from Baja California side of the International Boundary at Jacumba, San Diego Co., California; obtained by Laurence M. Huey on 24 April 1936.

Measurements of holotype.—Total length, 141; length of tail, 78; length of hind foot, 19; length of ear (crown), 5; greatest length of skull, 21.9; interorbital breadth, 5.2; length of maxillary toothrow, 3.1; width across mastoid bullae, 12.6; length of nasals, 7.5.

Distribution.—Known from La Puerta and San Felipe valleys of San Diego Co., California and adjacent area in Baja California (Hall, 1981).

Remarks.—Huey (1939*a*) gave measurements for five adults of each sex. This subspecies seems to represent a form structurally intermediate between the coastal basin subspecies, *brevinasus* and *pacificus,* and the inland desert forms, *bangsi* and *aestivus,* a situation to be expected on the basis of its intermediate geographic position. Subspecific recognition is equivocal.

Perognathus longimembris kinoensis Huey, 1935

1935. *Perognathus longimembris kinoensis* Huey, Trans. San Diego Soc. Nat. Hist., 8:73, 24 August.

Holotype.—Adult male, skin and skull, SDSNH 11300, from Bahía Kino (northern end of the sand dune peninsula that borders the bay and forms the northern arm of the estuary), Sonora, México; obtained by Laurence M. Huey on 26 February 1935.

Measurements of holotype.—Total length, 135; length of tail, 80; length of hind foot, 17; length of ear (crown), 4; greatest length of skull, 20.7; interorbital breadth, 4.6; length of maxillary toothrow, 2.6; width across mastoid bullae, 11.4; length of nasals, 7.2.

Distribution.—Known only from the vicinity of the type locality on the Gulf Coast of central Sonora, México.

Remarks.—Like other *Perognathus* living in relatively humid environments, the bullae are less inflated and the posterior cranial region less constricted than populations from arid environments. This subspecies appears to be little differentiated from *bombycinus* from farther north.

Perognathus longimembris longimembris (Coues, 1875)

1875. *O*[*tognosis*]. *longimembris* Coues, Proc. Acad. Nat. Sci. Philadelphia, 27:305, 31 August.

1904. *Perognathus elibatus* Elliot, Field Columbian Mus., Publ. 87, Zool. Ser., 3:252, 7 January.

1904. *Perognathus pericalles* Elliot, Field Columbian Mus., Publ. 87, Zool. Ser., 3:252, 7 January.

Holotype.—Adult female, skin (in alcohol) and skull, USNM 9856/37356, from Fort Tejon, Tehachapi Mountains, Kern Co., California; obtained by John Xantus sometime in 1857 or 1858.

Measurements of holotype.—[Total length determined from length of head and body given by Coues (1875) plus length of tail taken from specimen in alcohol; other external measurements made from specimen in alcohol.] Total length, 121; length of tail, 63; length of hind foot, 16.1; length of ear (notch), 5.7; greatest length of skull, 20.55; occipitonasal length, 20.30; interorbital breadth, 4.90; length of maxillary toothrow, 3.00; width across maxillary toothrows, 3.90; length of mastoid bulla, 7.50; width across mastoid bullae, 11.45; length of interparietal, 2.85; width of interparietals, 4.35; length of nasals, 7.10; width of nasals, 1.75; width of rostrum, 3.65.

Distribution.—Occurs in the Mojave Desert region and Transverse ranges of southeastern California, from near Independence in the Owens Valley on the northeast, to near Mount Pinos on the southwest, and the Providence Mountains on the southeast.

Remarks.—Specimens from the north of the Transverse ranges in west-central California, assigned to this species (Hall, 1981), are *P. inornatus*. The northern distributional limits of *P. longimembris longimembris* are not known. The type locality is in a position where either or both *P. longimembris* and *P. inornatus neglectus* may occur, but no pocket mice have been taken from there in this century. The holotype may not have come from Ft. Tejon, but rather from somewhere in the surrounding country, perhaps the Mojave Desert side of the Transverse ranges (Osgood, 1918). Bole (1937) gave measurements of 17 specimens.

Perognathus longimembris nevadensis Merriam, 1894

1894. *Perognathus nevadensis* Merriam, Proc. Acad. Nat. Sci. Philadelphia, 46:264, 27 September.

Holotype.—Adult male, skin and skull, USNM 54828, from Halleck, East Humboldt Valley, Elko Co., Nevada; obtained by Vernon Bailey on 4 July 1893.

Measurements of holotype.—Total length, 127; length of tail, 72; length of hind foot, 19; length of ear (anterior base, dry), 7; greatest length of skull, 21.40; occipitonasal length, 20.80; interorbital breadth, 5.05; length of maxillary toothrow, 2.90; width across maxillary toothrows, 4.15; length of mastoid bulla, 7.90; width across mastoid bullae, 12.05; length of interparietal, 2.25; width of interparietals, 4.10; length of nasals, 8.05; width of nasals, 1.75; width of rostrum, 3.20.

Distribution.—Occupies Great Basin desert associations from the southeastern corner of Oregon, southwestward to the Surprise Valley, Modoc Co., California, and southward in northern and Central Nevada; in Nevada, south to Smiths Creek Valley and eastward to Halleck and near Eureka (Hall, 1946, 1981).

Remarks.—Hall (1946) gave statistics for measurements of 10 males and 5 females.

Perognathus longimembris pacificus Mearns, 1898

1898. *Perognathus pacificus* Mearns, Bull. Amer. Mus. Nat. Hist., 10:299, 31 August.
1932. *Perognathus longimembris cantwelli* von Bloeker, Proc. Biol. Soc. Washington, 45:128, 9 September.

Holotype.—Adult female, skin and skull, USNM 61022, from the shore of the Pacific Ocean at Mexican boundary monument No. 258, San Diego Co., California; obtained on 12 July 1894 by Edgar A. Mearns.

Measurements of holotype.—Total length, 113; length of tail, 53; length of hind foot, 15.5; length of ear, 5; greatest length of skull, 19.20; occipitonasal length, 19.00; interorbital breadth, 4.65; length of maxillary toothrow, 2.85; width across maxillary toothrows, 3.75; length of mastoid bulla, 7.15; width across mastoid bullae, 10.85; length of interparietal, 2.30; width of interparietals, 3.70; length of nasals, 7.05; width of nasals, 1.85; width of rostrum, 3.10.

Distribution.—Occurs on the coastal plains of southern California, from near El Segundo, Los Angeles Co., southward to near Tiajuana, Baja California, México.

Remarks.—Huey (1939*a*) listed measurements for 10 males and 6 females.

Perognathus longimembris panamintinus Merriam, 1894

1894. *Perognathus longimembris panamintinus* Merriam, Proc. Acad. Nat. Sci. Philadelphia, 46:265, 27 September.

Holotype.—Adult male, skin and skull, USNM 27767/39866, from Perognathus Flat, 5,200 ft, Panamint Mountains, Inyo Co., California; obtained by Vernon Bailey on 16 April 1891.

Measurements of holotype.—Total length, 152; length of tail, 83; length of hind foot, 20; greatest length of skull, 22.80; occipitonasal length, 22.10; interorbital breadth, 5.45; length of maxillary toothrow, 3.10; width across maxillary toothrows, 4.25; length of mastoid bulla, 8.15; width across mastoid bullae, 12.35; length of interparietal, 3.00; width of interparietals, 4.05; length of nasals, 8.55; width of nasals, 2.00; width of rostrum, 3.55.

Distribution.—Occupies Great Basin Desert associations in western Nevada and southeastern California, ranging from Quinn River Crossing, Humboldt Co., Nevada south into Clark Co., Nevada and westward in Inyo Co., California to the Panamint Mountains (Hall, 1981).

Remarks.—Hall (1946) gave statistics for measurements of 10 males and 7 females from Nevada. Bole (1937) listed measurements of 20 specimens from California.

Perognathus longimembris pimensis Huey, 1937

1937. *Perognathus longimembris pimensis* Huey, Trans. San Diego Soc. Nat. Hist., 8:355, 15 June.

Holotype.—Adult male, skin and skull, SDSNH 12579, from 11 mi W Casa Grande, Pinal Co., Arizona; obtained by Laurence M. Huey on 22 May 1937.

Measurements of holotype.—Total length, 144; length of tail, 83; length of hind foot,

18; length of ear (crown), 4; greatest length of skull, 21.2; interorbital breadth, 5.0; length of maxillary toothrow, 2.7; width across mastoid bullae, 12.2; length of nasals, 7.4.

Distribution.—South-central Arizona, from near Marinette, Maricopa Co., on the north to near Casa Grande, Pinal Co., on the southeast and Gila Bend, Pinal Co., on the southwest.

Remarks.—This population is apparently isolated from others of *P. longimembris.*

Perognathus longimembris salinensis Bole, 1937

1937. *Perognathus longimembris salinensis* Bole, Sci. Publs., Cleveland Mus. Nat. Hist., 5(2):3, 4 December.

Holotype.—Adult male, skin and skull, UMMZ 121257, from 1 mi N Salt Camp, 1,060 ft, west edge salt lake, Saline Valley, Inyo Co., California; obtained by P. N. Moulthrop on 29 March 1934.

Measurements of holotype.—Total length, 130; length of tail, 74; length of hind foot, 17.5; occipitonasal length, 19.5; basilar length, 13.9; interorbital breadth, 4.9; length of maxillary toothrow, 3.1; width across mastoid bullae, 11.6; length of interparietal, 2.0; width of interparietals, 3.6; length of nasals, 7.5; width of rostrum, 2.1.

Distribution.—Known only from the Saline Valley, Inyo Co., California.

Remarks.—The holotype was originally CMNH 6242 (Cleveland Museum of Natural History). Bole (1937) listed measurements for 20 specimens.

Perognathus longimembris tularensis Richardson, 1937

1937. *Perognathus longimembris tularensis* Richardson, J. Mamm., 18:510, 22 November.

Holotype.—Adult male, skin and skull, MVZ 74668, from 1 mi W Kennedy Meadows, 6,000 ft, S Fork Kern River, Tulare Co., California; obtained on 14 August 1936 by William B. Richardson.

Measurements of holotype.—Total length, 127; length of tail, 68; length of hind foot, 19; length of ear (notch), 7; occipitonasal length, 21.0; frontonasal length, 14.6; interorbital breadth, 5.2; length of mastoid bulla, 8.2; width across mastoid bullae, 11.7; distance between stylomastoid foramina, 10.5.

Distribution.—Known only from the vicinity of the type locality in the upper valley of the Kern River in eastern Tulare Co., California (Hall, 1981).

Remarks.—Richardson (1937) gave means and extremes for measurements of five subadult and adult specimens. Aside from slightly darker color, this subspecies differs from *P. longimembris longimembris* only in negligibly smaller size of the mastoid bullae and related structure, differences that may reflect the immaturity of most of the type series.

Perognathus longimembris venustus Huey, 1930

1930. *Perognathus longimembris venustus* Huey, Trans. San Diego Soc. Nat. Hist., 6:233, 24 December.

Holotype.—Adult female, skin and skull, SDSNH 8196, from San Agustín, lat. 30°N, long. 115°W, Baja California, México; obtained on 4 October 1930 by Laurence M. Huey.

Measurements of holotype.—Total length, 130; length of tail, 78; length of hind foot, 19; length of ear (crown), 5; greatest length of skull, 21.8; interorbital breadth, 5.0; length of maxillary toothrow, 3.3; width across mastoid bullae, 12.3; length of nasals, 7.9.

Distribution.—Known only from the type

locality in north-central Baja California (Hall, 1981).

Remarks.—Huey (1939*a*) gave measurements for three specimens. We see no significant differences in the measurements listed by Huey (1939*a*) for *venustus* and *aestivus,* leaving only the "much darker color" of the former as apparently diagnostic, a situation that may not warrant taxonomic separation.

Perognathus merriami

Diagnosis.—A small pocket mouse, with length of head and body seldom exceeding 62 mm, and with a relatively short tail; ratio of length of tail to head and body averaging 0.82–0.95; dorsal color yellowish or yellowish-orange with a slight blackish tinge and little contrast between the mid-dorsal and dorso-lateral areas; buffy post-auricular spot relatively small; auditory bullae moderately inflated; rostrum proportionately broad and short.

Comparisons.—See account of *P. flavus* for comparison with that species. Size smaller, tail shorter, and interparietals not as wide as *P. flavescens. P. merriami* is sympatric only with the preceding two species of *Perognathus.* See accounts of the other species for additional comparisons.

Remarks.—*P. merriami* was considered a species distinct from *P. flavus* until Wilson (1973), based on morphology, treated them as conspecific. The distribution of *P. m. gilvus* is geographically intermediate to the main range of *P. flavus* and *P. m. merriami,* and is in many ways structurally intermediate to *P. m. merriami* and *P. f. flavus,* giving the impression that the two taxa broadly hybridize. However, genic analysis shows distinct genotypes for *P. m. gilvus* and *P. f. flavus* from several localities of sympatry, and only a few probable hybrids from one locality in southeastern New Mexico (Lee and Engstrom, 1991). On this basis, Lee and Engstrom (1991) recommended the two nominate forms be recognized as distinct species. There is no known character or small set of characters that will reliably distinguish all *P. m. gilvus* from all *P. f. flavus.*

Perognathus merriami gilvus Osgood, 1900

1900. *Perognathus merriami gilvus* Osgood, N. Amer. Fauna, 18:22, 20 September.

Holotype.—Adult male, skin and skull, USNM 35939/48273, from Eddy, Eddy Co., New Mexico; obtained by B. H. Dutcher on 18 September 1892.

Measurements of holotype.—Total length, 118; length of tail, 58; length of hind foot, 16.5; greatest length of skull, 21.00; width across bullae, 11.75; breadth across maxillary arches, 10.00; nasal length, 7.25; interorbital breadth, 4.65.

Distribution.—Occurs in western Texas and southeastern New Mexico, southward through eastern Chihuahua into extreme northern and western Coahuila, México (Hall, 1981).

Remarks.—Anderson (1972) listed ways of distinguishing *P. m. gilvus* and *P. f. flavus* and included a ratio diagram for this purpose, although he had measurements for only two *P. m. g.* from Chihuahua. Wilson (1973) gave statistics for measurements of 42 specimens from throughout the geographic range, and Baker (1956) gave measurements for 4 individuals from western Coahuila.

Perognathus merriami merriami J. A. Allen, 1892

1892. *Perognathus merriami* J. A. Allen, Bull. Amer. Mus. Nat. Hist., 4:45, 25 March.
1896. *Perognathus mearnsi* J. A. Allen, Bull. Amer. Mus. Nat. Hist., 8:237, 21 November.

Holotype.—Adult male, skin and skull, AMNH 4145/3177, from Brownsville, Cameron Co., Texas; obtained by F. B. Armstrong on 10 August 1891.

Measurements of holotype.—Total length, 124; length of tail, 58; length of hind foot,

18; greatest length of skull, 20.3; interorbital breadth, 4.6; length of maxillary toothrow, 2.9; width across mastoid bullae, 11.2; length of nasals, 7.2; length of rostrum, 7.7.

Distribution.—Ranges from central Texas near the Oklahoma boundary, southward to eastern Coahuila, north and central Nuevo León, and central Tamaulipas.

Remarks.—Wilson (1973) gave means and ranges of measurements for 57 specimens from throughout the range of the subspecies; Baker (1956) listed measurements for 10 adults from eastern Coahuila.

Perognathus parvus

Diagnosis.—The largest species of *Perognathus,* with a slightly crested and tufted tail (pencil) that averages longer than length of head and body; distal, dorsal one-third of tail with a mix of buffy, brownish, and blackish hairs; antitragus of ear pinna lobed; hairs of inner surface of pinna not white or yellowish; interparietal relatively short; mastoid breadth relatively narrow and posterior cranial region relatively unconstricted by the mastoid bullae; mastoid bulla forms slight indentation in lateral border of exoccipital on back of skull; interorbital region broad; baculum relative to head and body length long for species group, but about average for genus (Burt, 1960; Osgood, 1900; Sulentich, 1983).

Comparisons.—*Perognathus parvus* is most similar to *P. alticola*; see account of the latter species for comparisons. *P. parvus* can be distinguished from all other species of *Perognathus* by its larger size, lobed antitragus, and more prominent crest on the distal portion of the tail.

Distribution.—Occupies steppe and open, arid shrub and woodland communities in the Columbia River Basin, Great Basin, and contiguous areas in western North America; ranges from sagebrush communities in the Okanogan Valley of south-central British Columbia, southward through eastern Washington, Oregon, and California; eastward across the Snake River Plains of Idaho into southwestern Montana, southeastern Wyoming, and northeastern Utah; also ranges through all but the southern tip of Nevada and eastward through the western two-thirds of Utah, and southward into northwestern Arizona north of the Grand Canyon of the Colorado River (Hall, 1981).

Remarks.—*Perognathus parvus* and the closely related *P. alticola* comprise the *parvus* species group; herein we treat *Perognathus xanthonotus* as a subspecies of *P. parvus.* The subspecies and species of the *parvus* group are poorly known; a systematic review of the group is called for.

Perognathus parvus bullatus Durrant and Lee, 1956

1956. *Perognathus parvus bullatus* Durrant and Lee, Proc. Biol. Soc. Washington, 69:183, 31 December.

Holotype.—Adult male, skin and skull, UU 8771, from Ekker's Ranch, Robbers Roost, 25 mi (airline) E of Hanksville, 6,000 ft, Wayne Co., Utah; obtained by John Bushman on 18 May 1951.

Measurements of holotype.—Total length, 160; length of tail, 85; length of hind foot, 22; length of ear, 8; greatest length of skull, 25.40; width across bullae, 14.00; breadth across maxillary arches, 11.80; length of nasals, 9.15; interorbital breadth, 6.10.

Distribution.—Known from southeastern Utah in an area bounded by the San Rafael (Huntington Creek tributary), Green, Colorado, and Fremont rivers, in Emery and Wayne counties (Hall, 1981).

Remarks.—Durrant and Lee (1956) gave statistics for measurements of 19 specimens.

Perognathus parvus clarus Goldman, 1917

1917. *Perognathus parvus clarus* Goldman, Proc. Biol. Soc. Washington, 30:147, 27 July.

Holotype.—Adult male, skin and skull, USNM 178939, from Cumberland, Lincoln

Co., Wyoming; obtained by Stanley G. Jewett on 18 May 1912.

Measurements of holotype.—Total length, 181; length of tail, 84; length of hind foot, 22; greatest length of skull, 25.35; occipitonasal length, 25.25; interorbital breadth, 5.75; length of maxillary toothrow, 3.70; width across maxillary toothrows, 4.95; length of mastoid bulla, 8.35; width across mastoid bullae, 13.15; length of interparietal, 3.45; width of interparietals, 5.70; length of nasals, 10.00; width of nasals, 2.45; width of rostrum, 4.30.

Distribution.—Occupies the northeastern sector of the species' range, in southeastern Idaho, southwestern Montana, southwestern Wyoming, and northeastern Utah (Hall, 1981).

Remarks.—Williams (1978*a*) considered specimens from the Uintah Basin of northeastern Utah to belong to this subspecies rather than *P. p. olivaceous*; in either case there is little to distinguish the two subspecies except color, which varies considerably over the range of the two subspecies. Long (1965) listed means and ranges of measurements for eight males from the type locality and one male and one female from 21 mi W of Rock Springs, Wyoming.

Perognathus parvus columbianus Merriam, 1894

1894. *Perognathus columbianus* Merriam, Proc. Acad. Nat. Sci. Philadelphia, 46:263, 27 September.

Holotype.—Adult male, skin and skull, USNM 27351/39450, from Pasco, Plains of the Columbia, east side of Columbia River near mouth of Snake River, Franklin Co., Washington; obtained by Clark P. Streator on 9 May 1891.

Measurements of holotype.—Total length, 187; length of tail, 96; length of hind foot, 23; greatest length of skull, 27.60; occipitonasal length, 27.60; interorbital breadth, 6.00; length of maxillary toothrow, 4.10; width across maxillary toothrows, 5.15; length of mastoid bulla, 9.40; width across mastoid bullae, 14.45; length of interparietal, 3.30; width of interparietals, 4.80; length of nasals, 11.25; width of nasals, 2.60; width of rostrum, 4.30.

Distribution.—Occurs on the Columbia Plateau of central Washington, generally north and east of the Columbia River and north and west of the Snake River on the south, and south of the Columbia River on the north; recorded from west of the Columbia River only from near Wenatchee, Chelan Co., Washington (Hall, 1981).

Remarks.—Merriam (1894*b*) indicated that the holotype was a young adult; its cheek teeth show considerable wear (age class 5 of Williams, 1978*a*), indicating that it is old, not young. Osgood (1900) gave average external measurements for 5 adult topotypes, Broadbooks (1954) listed statistics for 52 specimens by age and sex, and Dalquest (1948) listed means for external measurements of 21 males and 9 females.

Perognathus parvus idahoensis Goldman, 1922

1922. *Perognathus parvus idahoensis* Goldman, Proc. Biol. Soc. Washington, 35:105, 17 October.

Holotype.—Adult male, skin and skull, USNM 236394, from Echo Crater, 20 mi SW Arco, Blaine [Butte] Co., Idaho; obtained by Luther J. Goldman on 14 June 1921.

Measurements of holotype.—Total length, 189; length of tail, 102; length of hind foot, 24; greatest length of skull, 27.40; occipitonasal length, 27.40; interorbital breadth, 5.90; length of maxillary toothrow, 3.85; width across maxillary toothrows, 5.15; length of mastoid bulla, 9.00; width across mastoid bullae, 14.25; length of interparietal, 3.60; width of interparietals, 5,50; length of nasals, 10.55; width of nasals, 2.75; width of rostrum, 4.35.

Distribution.—Occurs on lava soils in south-central Idaho, on a portion of the

Snake River Plain and slopes of volcanos in the vicinity of Craters of the Moon National Monument (Hall, 1981).

Remarks.—The type locality, 20 mi SW Arco, is in Butte Co., rather than Blaine Co.

Perognathus parvus laingi
Anderson, 1932

1932. *Perognathus laingi* Anderson, Bull. Natl. Mus. Canada, 70:100, 24 November.

Holotype.—Adult male, skin and skull, NMC 9200, from Anarchist Mountain, near Osoyoos-Bridesville summit, about 8 mi E of Osoyoos Lake, about 3,500 ft, lat. 49° 08′ N, long. 119° 32′ W, British Columbia; obtained by Hamilton Laing on 29 August 1928.

Measurements of holotype.—Total length, 191; length of tail, 102; length of hind foot, 25; occipitonasal length, 27.0; basal length of Hensel, 20.0; interorbital breadth, 5.0; width across mastoid bullae, 14.5; length of interparietal, 4.0; width of interparietals, 5.0; length of nasals, 10.0; length of incisive foramina, 2.0.

Distribution.—Occupies the upper Okanogan Basin of south-central British Columbia, from Ashcroft on the northwest to near Osoyoos on the southeast (Hall, 1981).

Remarks.—Anderson (1932) gave measurements for three specimens.

Perognathus parvus lordi
(Gray, 1868)

1868. *Abromys lordi* Gray, Proc. Zool. Soc. London, p. 202, May.

Holotype.—Age and sex unknown, BM(NH) 62.12.12.19, obtained from British Columbia by J. K. Lord.

Measurements of holotype.—[External measurements and greatest length of skull from Merriam, 1889] Total length, 155; length of tail, 81; length of hind foot, 21.8; length of ear (crown), 6.5; greatest length of skull, 25.5; width across bullae, 13.9; nasal length, 9.4; interorbital breadth, 6.0.

Distribution.—Ranges from near the international boundary in south-central British Columbia, southward in north-central and eastern Washington to west-central Idaho in Nez Perce Co., and south eastern Washington in Asotin Co.; generally ranges north of the Columbia River in north-central Washington and east of the Columbia Basin in eastern Washington (Hall, 1981).

Remarks.—The holotype was originally mounted; external measurements provided by Merriam (1889) were estimated from the dry skin. The occipital and maxillary regions of the skull of the holotype are broken. Anderson (1932) listed statistics for measurements of 24 males and 12 females. Broadbooks (1954) listed statistics for measurements by age and sex for 29 specimens from British Columbia and Washington, and Dalquest (1948) gave means for external measurements of 29 males and 10 females. Rhoads (1894) supplemented the extremely scanty description of Gray, and provided measurements for a "duplicate type."

Perognathus parvus mollipilosus
Coues, 1875

1875. *P[erognathus]. mollipilosus* Coues, Proc. Acad. Nat. Sci. Philadelphia, 27:296, 31 August.

Holotype.—Female, USNM 7251, from Fort Crook [about 2 mi NE of Burgettville], Shasta Co., California; obtained by J. Feilner.

Measurements of holotype.—[Measurements from specimen in alcohol, before skinning] Total length, 145; length of tail, 81.3; length of hind foot, 20.3; length of ear, 7.6 (see remarks).

Distribution.—Occupies the Modoc Plateau and adjacent areas of northeastern California, and south-central Oregon in the Klamath Basin; ranges from Crater Lake National Park, Oregon on the north, to Edgewood, Siskiyou Co., California on the west, and Susanville, Lassen Co., California on the southeast (Hall, 1981).

Remarks.—Coues (1875) listed the catalog number for the holotype but did not indicate the museum. Merriam (1889) remarked that the specimen should be in the U.S. National Museum but it could not be found and the number was wrong. Robert Fisher (in litt.) noted that USNM 7251 was a *Scalopus aquaticus* that also was missing. Osgood (1900) gave average measurements for three topotypes.

Perognathus parvus olivaceus Merriam, 1889

1889. *Perognathus olivaceus* Merriam, N. Amer. Fauna, 1:15, 25 October.
1889. *Perognathus olivaceus amoenus* Merriam, N. Amer. Fauna, 1:16, 25 October.
1900. *Perognathus parvus magruderensis* Osgood, N. Amer. Fauna, 18:38, 20 September.
1939. *Perognathus parvus plerus* Goldman, J. Mamm., 20:352, 14 August.

Holotype.—Adult male, skin and skull, USNM 186511, from Kelton, near N end Great Salt Lake, Box Elder Co., Utah; obtained by Vernon Bailey on 24 October 1888.

Measurements of holotype.—Total length, 184; length of tail, 101; length of hind foot, 23; greatest length of skull, 26.70; occipitonasal length, 26.70; interorbital breadth, 6.10; length of maxillary toothrow, 3.70; width across maxillary toothrows, 4.55; length of mastoid bulla, 8.90; width across mastoid bullae, 13.30; length of interparietal, 2.95; width of interparietals, 5.15; length of nasals, 10.65; width of nasals, 2.75; width of rostrum, 4.40.

Distribution.—Occupies the Great Basin, its range extending from extreme south-central Idaho southward through all but the extreme northwestern corner and extreme southern portion of Nevada; into east-central California from near Lake Tahoe, southward to Inyo Co.; through western Utah west of the central mountain axis, except, according to Durrant (1952) and Hall (1981) for populations in Emery, Duchesne, and Uintah counties, east of the Wasatch Range.

Remarks.—Durrant (1952) gave statistics for measurements of 5 males and 4 females from central Utah, and Hall (1946) provided measurements for 10 individuals of each sex from Nevada. Williams (1978*a*) assigned specimens from the Uintah Basin, Duchesne Co., Utah, to *P. p. clarus.*

Perognathus parvus parvus (Peale, 1848)

1848. *Cricetodipus parvus* Peale, Mammalia and ornithology, *in* U.S. Expl. Exped. . . . , 8:53.
1858. *Perognathus monticola* Baird, Mammals, *in* Repts. Expl. Surv. . . . , 8(1):422, 14 July.

Holotype.—Juvenile female [age assumed from the description], Exploration Expeditions Collections, USNM [no information recorded concerning holotype and assumed to be lost or destroyed—originally preserved in alcohol]; assumed to be from The Dalles, Wasco Co., Oregon.

Measurements of holotype.—Total length, 106; length of tail, 58; length of hind foot, 20.5.

Distribution.—Ranges from southeastern Washington, south of the Snake River, east of the Columbia River, and west of the Blue Mountains, through eastern Oregon to the extreme northwestern corner of Nevada, and in southwestern Idaho along the lowlands of the Snake and Salmon rivers; in Oregon, ranges from the eastern edge of the Cascade Mountains southward east of the Winter and Warner ranges (Hall, 1981).

Remarks.—The original description obviously was of a very young animal. Its juvenile characteristics led to considerable taxonomic confusion during the 1800's between *Perognathus* and *Dipodomys.* Rhoads (1894) and Osgood (1900) were primarily responsible for determining the identity of *Cricetodipus parvus* of Peale. Hall (1946) listed measurements of 4 specimens from

Nevada, Anderson (1932) listed means and ranges of measurements for 6 specimens from Oregon, and Dalquest (1948) gave means of external measurements of 31 males and 9 females from Washington.

Perognathus parvus trumbullensis Benson, 1937

1937. *Perognathus parvus trumbullensis* Benson, Proc. Biol. Soc. Washington, 50:181–182, 28 October.

Holotype.—Young adult male, skin and skull, MVZ 60929, from Nixon Spring, 6,250 ft, Mount Trumbull, Mohave Co., Arizona; obtained on 26 May 1933 by Annie M. Alexander.

Measurements of holotype.—Total length, 174; length of tail, 90; length of hind foot, 23; length of ear, 9; greatest length of skull, 26.85; interorbital breadth, 5.85; length of mastoid bulla, 9.25; width across mastoid bullae, 13.90; length of interparietal, 3.45; width of interparietals, 4.80; length of nasals, 10.45; width of rostrum, 4.00.

Distribution.—Occupies the high plateau country of northwestern Arizona, north of the Colorado River in Mohave and Coconino counties, and adjacent areas of south-central and southwestern Utah in Washington, Kane, and Garfield counties (Hall, 1981).

Remarks.—Benson (1937) gave means and ranges of measurements for 20 adult males; Hoffmeister (1986) listed statistics for measurements of a sample of 18 individuals from Arizona.

Perognathus parvus xanthonotus Grinnell, 1912

1912. *Perognathus xanthonotus* Grinnell, Proc. Biol. Soc. Washington, 25:128, 31 July.

Holotype.—Adult male, skin and skull, MVZ 16154, from Freeman Canyon, 4,900 ft, east slope Walker Pass, Kern Co., California; obtained by H. A. Carr on 27 June 1911.

Measurements of holotype.—Total length, 170; length of tail, 85; length of hind foot, 22.5; greatest length of skull, 25.40; interorbital breadth, 6.00; length of mastoid bulla, 8.10; width across mastoid bullae, 12.95; length of nasals, 9.10; width of rostrum, 4.55.

Distribution.—Occupies the eastern desert slope of the northern Tehachapi Mountains, Kern Co., California; known range extends from Indian Wells Canyon on the north to Horse Canyon on the south (unpubl. data).

Remarks.—We concur with Sulentich (1983) that available evidence does not support recognizing *xanthonotus* as a species separate from *P. parvus.* The ranges of *P. parvus olivaceous* and *P. parvus xanthonotus* are probably continuous along the eastern front of the Sierra Nevada. Sulentich (1983) listed statistics for measurements of 20 specimens.

Perognathus parvus yakimensis Broadbooks, 1954

1954. *Perognathus parvus yakimensis* Broadbooks, J. Mamm., 35:96, 10 February.

Holotype.—Young adult male, skin, skull, and baculum, UMMZ 95978, from Rocky Flat (or Rocky Prairie), 16 mi NW Naches, 3,800 ft, Yakima Co., Washington; obtained by Harold E. Broadbooks on 27 August 1947.

Measurements of holotype.—Total length, 174; length of tail, 91; length of hind foot, 24; length of ear, 8; weight, 17.2 g; greatest length of skull, 25.9; width across bullae, 13.1; breadth across maxillary arches, 11.9; nasal length, 11.2; interorbital breadth, 5.8.

Distribution.—Occurs in sagebrush and open, arid ponderosa pine forests in south-central Washington, west of the Columbia River and east of the Cascade Mountains; ranges from the Kittitas Valley, Kittitas Co., on the north, southward to the Columbia river, thence westward to the vicinity of

Dallesport, Klickitat Co. (Broadbooks, 1954).

Remarks.—Broadbooks listed statistics for measurements of 49 specimens by age and sex. Hooper (1977) indicated that the type locality was 17 mi NW Naches.

Acknowledgments

Troy L. Best, Kerry S. Kilburn and Duke S. Rogers reviewed portions of the manuscript and provided helpful suggestions for its improvement. Troy also provided information on his "in press" manuscripts that allowed us to keep abreast of some taxonomic changes. Susan D. Williams provided invaluable editorial assistance by proofreading and verifying bibliographic and nomenclatural data and measurements of type specimens. Donald F. Hoffmeister kept us appraised of the progress of his book, "Mammals of Arizona," in which he proposed several taxonomic changes. The following curators and collection managers provided information on, or took measurements of, listed holotypes: Sydney Anderson, American Museum of Natural History; Ted Daeschler, The Academy of Natural Sciences, Philadelphia—*Dipodomys h. heermanni*; Robert D. Fisher, National Museum of Natural History—*D. heermanni tularensis, D. o. ordii* (neotype), *D. o. utahensis, Perognathus flavus mexicanus*; Mark S. Hafner, Louisiana State University, Museum of Zoology—*P.* [*Chaetodipus*] *penicillatus atrodorsalis, P.* [*C.*] *lineatus*; Thor Holmes, University of Kansas, Museum of Natural History—*D. ordii pullus, P. flavus fuscus*; Paula Jenkins, British Museum (Natural History)—*Abromys* [Perognathus] *lordi*; Philip Myers, University of Michigan, Museum of Zoology—*P. parvus yakimensis*; Eric A. Richard, University of Utah, Utah Museum of Natural History—*D. o. celeripes, D. o. sanrafaeli, P. parvus bullatus*; Robert M. Timm, University of Kansas, Museum of Natural History. We extend our appreciation to these persons and the curators and curatorial staffs who allowed us access to their holotype collections and provided so generously of their time and hospitality.

Literature Cited

ALLEN, J. A. 1881. List of mammals collected by Dr. Edward Palmer in northeastern Mexico, with field-notes by the collector. Bulletin of the Museum of Comparative Zoology, 8:183–189.

———. 1891. Notes on new and little-known North American mammals, based on recent additions to the collection of mammals in the American Museum of Natural History. Bulletin of the American Museum of Natural History, 3:263–310.

———. 1899. New rodents from Colombia and Venezuela. Bulletin of the American Museum of Natural History, 12:195–218.

———. 1908. Mammals from Nicaragua. Bulletin of the American Museum of Natural History, 24:647–670.

———. 1916. List of mammals collected in Colombia by the American Museum of Natural History Expeditions, 1910-1915. Bulletin of the American Museum of Natural History, 35:191–238.

ALLEN, J. A., AND F. M. CHAPMAN. 1893. On a collection of mammals from the Island of Trinidad, with descriptions of new species. Bulletin of the American Museum of Natural History, 5:203–234.

ALSTON, E. R. 1876. On the classification of the order Glires. Proceedings of the Zoological Society of London, 1876:61–98.

———. 1879–1882. Biologia Centrali-Americana. Mammalia. Published for the Editors by R. H. Porter, London, 220 pp.

ALVAREZ, T. 1963. The Recent mammals of Tamaulipas, Mexico. University of Kansas Publications, Museum of Natural History, 14:363–473.

ANDERSON, R. M. 1932. Five new mammals from British Columbia. National Museum of Canada Bulletin, 70:99–119.

ANDERSON, S. 1964. The systematic status of *Perognathus artus* and *Perognathus goldmani* (Rodentia). American Museum Novitates, 2184:1–27.

———. 1972. Mammals of Chihuahua taxonomy and distribution. Bulletin of the American Museum of Natural History, 148:149–410.

ARATA, A. A. 1964. The anatomy and taxonomic significance of the male accessory reproductive glands of muroid rodents. Bulletin of the Florida State Museum, 9:1–42.

ARMSTRONG, D. M. 1972. Distribution of mammals in Colorado. Monograph, University of Kansas Publications, Museum of Natural History, 3:1–415.

BAIRD, S. F. 1858. Mammals, *in* Explorations and surveys, to ascertain the most practical and economical route for a railroad from the Mississippi River to the Pacific Ocean. United States Department of the Interior, Washington, D.C., 8(1):1–757.

BAKER, R. H. 1954. The silky pocket mouse (Perognathus flavus) of Mexico. University of Kansas Publications, Museum of Natural History, 7:339–347.

———. 1956. Mammals of Coahuila, Mexico. University of Kansas Publications, Museum of Natural History, 9:125–335.

BANKS, R. C. 1964. The mammals of Cerralvo Island, Baja California. Transactions of the San Diego Society of Natural History, 13:397–404.

———. 1967. A new insular subspecies of spiny pocket mouse (Mammalia; Rodentia). Proceedings of the Biological Society of Washington, 80:101–104.

BAUMGARDNER, G. D. 1989. Refinement of the type locality and distribution of *Dipodomys compactus sennetti* (J. A. Allen, 1891). The Texas Journal of Science, 41:103–104.

BAUMGARDNER, G. D., AND D. J. SCHMIDLY. 1981. Systematics of the southern races of two species of kangaroo rats (*Dipodomys compactus* and *D. ordii*). Occasional Papers, The Museum, Texas Tech University, 73:1–27.

BENSHOOF, L., T. L. YATES, AND J. W. FROEHLICH. 1984. Noteworthy records of mammals from eastern Honduras. Southwestern Naturalist, 29:511–514.

BENSON, S. B. 1930. Two new pocket mice, genus Perognathus, from the Californias. University of California Publications in Zoology, 32:449–454.

———. 1932. Three new rodents from lava beds of southern New Mexico. University of California Publications in Zoology, 38:335–344.

———. 1933. Descriptions of two races of Perognathus amplus from Arizona. Proceedings of the Biological Society of Washington, 46:109–112.

———. 1934. Descriptions of two races of Perognathus intermedius from Arizona. Proceedings of the Biological Society of Washington, 47:199–202.

———. 1937. Description of a race of Perognathus parvus from Arizona. Proceedings of the Biological Society of Washington, 50:181–182.

BEST, T. L. 1978. Variation in kangaroo rats (genus *Dipodomys*) of the *heermanni* group in Baja California, Mexico. Journal of Mammalogy, 59:160–175.

———. 1981. Bacular variation in kangaroo rats (genus *Dipodomys*) of the *heermanni* group in Baja California, Mexico. Southwestern Naturalist, 25:529–534.

———. 1983*a*. Morphologic variation in the San Quintin kangaroo rat (*Dipodomys gravipes* Huey 1925). The American Midland Naturalist, 109:409–413.

———. 1983*b*. Intraspecific variation in the agile kangaroo rat (*Dipodomys agilis*). Journal of Mammalogy, 64:426–436.

———. 1986. Dipodomys elephantinus. Mammalian Species, 255:1–4.

———. 1987. Sexual dimorphism and morphometric variation in the Texas kangaroo rat (*Dipodomys elator* Merriam 1894). Southwestern Naturalist, 32:53–59.

———. 1993. Patterns of morphologic and morphometric variation in heteromyid rodents. Pp. 197–235, *in* Biology of the Heteromyidae (H. H. Genoways and J. H. Brown, eds.). Special Publication, American Society of Mammalogists, 10:1–719.

BEST, T. L., AND L. L. JANECEK. 1992. Allozymic and morphologic variation among *Dipodomys insularis, Dipodomys nitratoides,* and two populations of *Dipodomys merriami* (Rodentia: Heteromyidae). The Southwestern Naturalist, 37:1–8.

BEST, T. L., AND J. A. LACKEY. 1985. Dipodomys gravipes. Mammalian Species, 236:1–4.

BEST, T. L., R. M. SULLIVAN, J. A. COOK, AND T. L. YATES. 1986. Chromosomal, genic, and morphologic variation in the agile kangaroo rat, *Dipodomys agilis* (Rodentia: Heteromyidae). Systematic Zoology, 35:311–324.

BLEICH, V. C. 1977. Dipodomys stephensi. Mammalian Species, 73:1–3.

BOLE, B. P., JR. 1937. A new pocket mouse from eastern California. Science Publications of the Cleveland Museum of Natural History, 5(2):3–10.

BOULWARE, J. T. 1943. Two new subspecies of kangaroo rats (genus *Dipodomys*) from southern California. University of California Publications in Zoology, 46:391–396.

BROADBOOKS, H. E. 1954. A new subspecies of the pocket mouse (*Perognathus parvus*) from Washington. Journal of Mammalogy, 35:95–99.

BRYANT, M. D. 1939. A new kangaroo rat of the Dipodomys ordii group from the Big Bend region of Texas. Occasional Papers of the Museum of Zoology, Louisiana State University, 5:65–66.

BURT, W. H. 1932. Descriptions of heretofore unknown mammals from islands in the Gulf of California, Mexico. Transactions of the San Diego Society of Natural History, 7:163–182.

———. 1936. A study of the baculum in the genera Perognathus and Dipodomys. Journal of Mammalogy, 17:145–156.

———. 1960. Bacula of North American mammals. Miscellaneous Publications of the Museum of Zoology, University of Michigan, 113:1–76.

CABRERA, A. 1961. Catálogo de los mamíferos de América del Sur. Revista del Museo Argentino de Ciencias Naturales "Bernardino Rivadavia", 4:309–732.

CARLETON, M. D. 1984. Introduction to rodents. Pp. 255–265, *in* Orders and families of Recent mammals of the world (S. Anderson and J. K. Jones, Jr., eds.). John Wiley and Sons, New York, 686 pp.

CARTER, C. H., AND H. H. GENOWAYS. 1978. Liomys salvini. Mammalian Species, 84:1–5.

CARTER, D. C., W. D. WEBSTER, J. K. JONES, JR., C. JONES, AND R. D. SUTTKUS. 1985. Dipodomys elator. Mammalian Species, 232:1–3.

COCKRUM, E. L. 1952. Mammals of Kansas. University of Kansas Publications, Museum of Natural History, 7:1–303.

———. 1956. Two new long-tailed pocket mice (*Perognathus formosus*) from Arizona. Journal of the Washington Academy of Sciences, 46:131–132.

COUES, E. 1875. A critical review of the North American Saccomyidae. Proceedings of the Academy of Natural Sciences of Philadelphia, 27:272–327.

———. 1877. Saccomyidae. Pp. 483–542, *in* Report of the United States Geological Survey of the Territories (F. V. Hayden, ed.). United States Geological Survey, Miscellaneous Publications, 2:1–1091.

CSUTI, B. 1979. Patterns of adaptation and variation in the Great Basin kangaroo rat (*Dipodomys microps*). University of California Publications in Zoology, 111:1–71.

DALQUEST, W. W. 1948. Mammals of Washington. University of Kansas Publications, Museum of Natural History, 2:1–444.

———. 1951. Six new mammals from the state of San Luis Potosi, Mexico. Journal of the Washington Academy of Sciences, 41:361–364.

———. 1953. Mammals of the Mexican state of San Luis Potosi. Louisiana State University Studies, Biological Sciences Series, 1:1–229.

DALQUEST, W. W., AND G. COLLIER. 1964. Notes on *Dipodomys elator,* a rare kangaroo rat. Southwestern Naturalist, 9:146–150.

DAVIS, W. B. 1942. The systematic status of four kangaroo rats. Journal of Mammalogy, 23:328–333.

DOWLER, R. C., AND H. H. GENOWAYS. 1978. Liomys irroratus. Mammalian Species, 82:1–6.

DURRANT, S. D. 1952. Mammals of Utah, taxonomy and distribution. University of Kansas Publications, Museum of Natural History, 6:1–549.

DURRANT, S. D., AND M. R. LEE. 1956. A new pocket mouse from southeastern Utah. Proceedings of the Biological Society of Washington, 69:183–186.

DURRANT, S. D., AND H. W. SETZER. 1945. The distribution and taxonomy of kangaroo rats (genus *Dipodomys*) of Utah. Bulletin of the University of Utah, 35(26):1–38.

ELLERMAN, J. R. 1940. The families and genera of living rodents. British Museum of Natural History, London, 1:1–689.

ELLIOT, D. G. 1903. Descriptions of apparently new species and subspecies of mammals from California, Oregon, the Kenai Peninsula, Alaska, and Lower California, Mexico. Field Columbian Museum, Publication 74, Zoology Series, 3:153–173.

———. 1904. Descriptions of twenty-seven apparently new species and subspecies of mammals. Field Columbian Museum, Publication 87, Zoology Series, 3:239–261.

ENDERS, R. K. 1928. Description of Heteromys desmarestianus chiriquensis. Proceedings of the Academy of Natural Sciences of Philadelphia, 90:141–142.

ENGSTROM, M. D., H. H. GENOWAYS, AND P. K. TUCKER. 1987. Morphological variation, karyology, and systematic relationships of *Heteromys gaumeri* (Rodentia: Heteromyidae). Pp. 289–303, *in* Studies in Neotropical mammalogy: essays in honor of Philip Hershkovitz (B. D. Patterson and R. M. Timm, eds.). Fieldiana: Zoology, 39:1–506.

FASHING, N. J. 1973. Implications of karyotypic variation in the kangaroo rat, *Dipodomys heermanni*. Journal of Mammalogy, 54:1018–1020.

GENOWAYS, H. H. 1973. Systematics and evolutionary relationships of spiny pocket mice, genus Liomys. Special Publications, The Museum, Texas Tech University, 5:1–368.

GENOWAYS, H. H., AND J. K. JONES, JR. 1971. Systematics of southern banner-tailed kangaroo rats of the *Dipodomys phillipsii* group. Journal of Mammalogy, 52:265–287.

———. 1973. Notes on some mammals from Jalisco, Mexico. Occasional Papers, The Museum, Texas Tech University, 9:1–22.

GENOWAYS, H. H., R. J. BAKER, AND J. E. CORNELY. 1977. Mammals of the Guadalupe Mountains National Park, Texas. Pp. 271–332, *in* Biological investigations in the Guadalupe Mountains National Park, Texas (H. H. Genoways and R. J. Baker, eds.). Proceedings and Transactions Series, National Park Service, 4:1–442.

GILL, T. H. 1872. Arrangement of the families of mammals. With analytical tables. Prepared for the Smithsonian Institution. Smithsonian Miscellaneous Collections, 230:1–98.

GLASS, B. P. 1947. Geographic variation in Perognathus hispidus. Journal of Mammalogy, 28:174–179.

———. 1971. The type locality of *Dipodomys ordii richardsoni*. Southwestern Naturalist, 15:497–499.

GOLDMAN, E. A. 1911. Revision of the spiny pocket mice (genus Heteromys and Liomys). North American Fauna, 34:1–70.

———. 1912. New mammals from eastern Panama. Smithsonian Miscellaneous Collections, 60:1–18.

———. 1913. Description of new mammals from Panama and Mexico. Smithsonian Miscellaneous Collections, 60:1–20.

———. 1920. Mammals of Panama. Smithsonian Miscellaneous Collections, 69:1–309.

———. 1928. Three new rodents from western Arizona. Proceedings of the Biological Society of Washington, 41:203–206.

———. 1931. Three new rodents from Arizona and New Mexico. Proceedings of the Biological Society of Washington, 44:133–136.

———. 1932*a*. Two new pocket mice from Arizona. Journal of the Washington Academy of Sciences, 22: 386–388.

———. 1932*b*. A new pocket mouse from southern Arizona. Journal of the Washington Academy of Sciences, 22:488–489.

———. 1933. New mammals from Arizona, New Mexico, and Colorado. Journal of the Washington Academy of Sciences, 23:463–473.

———. 1939. Nine new mammals from islands in Great Salt Lake, Utah. Journal of Mammalogy, 20: 351–357.

———. 1951. Biological investigations in Mexico. Smithsonian Miscellaneous Collections, 115:1-476.

GOODWIN, G. G. 1943. Four new rodents from Costa Rica. American Museum Novitates, 1227:1–4.

———. 1946. Mammals of Costa Rica. Bulletin of the American Museum of Natural History, 87:271–474.

———. 1953. Catalogue of type specimens of Recent mammals in the American Museum of Natural History. Bulletin of the American Museum of Natural History, 102:211–411.

———. 1956. Seven new mammals from Mexico. American Museum Novitates 1791:1–10.

———. 1969. Mammals from the state of Oaxaca, Mexico in the American Museum of Natural History. Bulletin of the American Museum of Natural History, 141:1–269.

GRAY, J. E. 1868. Synopsis of the species of Saccomyinae, or pouched mice, in the collection of the British Museum. Proceedings of the Zoological Society of London, 1868:199–206.

GRINNELL, J. 1919. Four new kangaroo rats from west-central California. Proceedings of the Biological Society of Washington, 32:203–205.

———. 1921. Revised list of the species in *Dipodomys*. Journal of Mammalogy, 2:94–97.

———. 1922. A geographical study of the kangaroo rats of California. University of California Publications in Zoology, 24:1–124.

———. 1932. Habitat relations of the giant kangaroo rat. Journal of Mammalogy, 13:305–320.

GRINNELL, J., AND J. M. LINSDALE. 1929. A new kangaroo rat from the upper Sacramento Valley, California. University of California Publications in Zoology, 30:453–459.

HAFNER, D. J., J. C. HAFNER, AND M. S. HAFNER. 1979. Systematic status of kangaroo mice, genus *Microdipodops*: morphometric, chromosomal, and protein analyses. Journal of Mammalogy, 60:1–10.

HAFNER, J. C. 1978. Evolutionary relationships of kangaroo mice, genus *Microdipodops*. Journal of Mammalogy, 59:354–366.

———. 1985. New kangaroo mice, genus *Microdipodops* (Rodentia: Heteromyidae), from Idaho and Nevada. Proceedings of the Biological Society of Washington, 98:1–9.

HAFNER, J. C., AND M. S. HAFNER. 1983. Evolutionary relationships of heteromyid rodents. Great Basin Naturalist Memoirs, 7:3–29.

HAFNER, M. S. 1982. A biochemical investigation of geomyoid systematics (Mammalia: Rodentia). Zeitschrift für Zool. Systematik und Evolutionsforschung, 20:118–130.

HALL, E. R. 1941. Revision of the rodent genus Microdipodops. Field Museum of Natural History, Zoology Series, 27:233–277.

———. 1946. Mammals of Nevada. University of California Press, Berkeley, 710 pp.

———. 1954. A new subspecies of pocket mouse from Kansas. University of Kansas Publications, Museum of Natural History, 7:587–590.

———. 1981. The mammals of North America. Second edition. John Wiley and Sons, New York, 1:1–600 + *90*.

HALL, E. R., AND F. H. DALE. 1939. Geographic races of the kangaroo rat, Dipodomys microps. Occasional Papers of the Museum of Zoology, Louisiana State University, 4:47–63.

HALL, E. R., AND S. D. DURRANT. 1937. A new kangaroo mouse (Microdipodops) of Utah and Nevada. Journal of Mammalogy, 18:357–359.

HANDLEY, C. O., JR. 1966. Checklist of the mammals of Panama. Pp. 753–795, *in* Ectoparasites of Panama (R. L. Wenzel and V. J. Tipton, eds.). Field Museum of Natural History, Chicago, 861 pp.

———. 1976. Mammals of the Smithsonian Venezuelan Project. Brigham Young University Science Bulletin, 20:1–89.

HARDY, R. 1940. Three new rodents from southern Utah. Proceedings of the Biological Society of Washington, 55:87–92.

———. 1944. A new kangaroo-rat from southwestern Utah. Proceedings of the Biological Society of Washington, 57:53–56.

HARRIS, W. P., JR. 1932. Four new mammals from Costa Rica. Occasional Papers of the Museum of Zoology, University of Michigan, 248:1–6.

HATT, R. T. 1932. The vertebral columns of ricochetal rodents. Bulletin of the American Museum of Natural History, 63:599–738.

HILL, J. E. 1935. The cranial foramina in rodents. Journal of Mammalogy, 16:121–129.

HOFFMANN, W. M. 1975. Geographic variation and taxonomy of *Dipodomys nitratoides* from the California San Joaquin Valley. M.A. thesis, California State University, Fresno, 75 pp.

HOFFMEISTER, D. F. 1986. Mammals of Arizona. University of Arizona Press, Tucson, 602 pp.

HOFFMEISTER, D. F., AND M. R. LEE. 1967. Revision of the pocket mice, *Perognathus penicillatus*. Journal of Mammalogy, 48:361–380.

HOMAN, J. A., AND H. H. GENOWAYS. 1978. An analysis of hair structure and its phylogenetic implications among heteromyid rodents. Journal of Mammalogy, 59:740–760.

HOOPER, E. T. 1977. Type specimens of Recent mammals in the Museum of Zoology, the University of Michigan. Occasional Papers of the Museum of Zoology, University of Michigan, 680:1–21.

HOOPER, E. T., AND C. O. HANDLEY, JR. 1948. Character gradients in the spiny pocket mouse, *Liomys irroratus*. Occasional Papers of the Museum of Zoology, University of Michigan, 514:1–34.

HUEY, L. M. 1926. The description of a new subspecies of *Perognathus* from Lower California with a short discussion of the taxonomic position of other peninsular members of this genus. Proceedings of the Biological Society of Washington, 39:67–70.

———. 1930. Two new pocket mice of the spinatus group and one of the longimembris group. Transactions of the San Diego Society of Natural History, 6:231–234.

———. 1937. Descriptions of new mammals from Arizona and Sonora, Mexico. Transactions of the San Diego Society of Natural History, 8:349–359.

———. 1939*a*. The silky pocket mice of southern California and northern Lower California, Mexico, with the description of a new race. Transactions of the San Diego Society of Natural History, 9:47–54.

———. 1939*b*. A new Perognathus arenarius from near the U.S.-Mexican boundary in Lower California. Transactions of the San Diego Society of Natural History, 8:57–58.

———. 1951. The kangaroo rats (Dipodomys) of Baja California, Mexico. Transactions of the San Diego Society of Natural History, 11:205–256.

———. 1960*a*. Comments on the pocket mouse, Perognathus fallax, with descriptions of two new races from Baja California, Mexico. Transactions of the San Diego Society of Natural History, 12:415–419.

———. 1960*b*. Two new races of Perognathus spinatus from Baja California, Mexico. Transactions of the San Diego Society of Natural History, 12:409–412.

———. 1962. Two new species of broad-faced, five-toed kangaroo rats (genus Dipodomys). Transactions of the San Diego Society of Natural History, 12:477–480.

———. 1964. The mammals of Baja California, Mexico. Transactions of the San Diego Society of Natural History, 13:85–168.

JANNETT, F. J., JR. 1976. Bacula of *Dipodomys ordii compactus* and *Dipodomys elator*. Journal of Mammalogy, 57:382–387.

JOHNSON, W. E., AND R. K. SELANDER. 1971. Protein variation and systematics in kangaroo rats (genus *Dipodomys*). Systematic Zoology, 20:377–405.

JONES, J. K., JR. 1964. Distribution and taxonomy of mammals of Nebraska. University of Kansas Publications, Museum of Natural History, 16:1–356.

JONES, J. K., JR., AND H. H. GENOWAYS. 1975. Dipodomys phillipsii. Mammalian Species, 51:1–3.

KENNEDY, M. L., M. L. BECK, AND T. L. BEST. 1980. Intraspecific morphologic variation in Ord's kangaroo rat, *Dipodomys ordii*, from Oklahoma. Journal of Mammalogy, 61:311–319.

KENNEDY, M. L., AND G. D. SCHNELL. 1978. Geographic variation and sexual dimorphism in Ord's kangaroo rat, *Dipodomys ordii*. Journal of Mammalogy, 59:45–59.

LACKEY, J. A. 1967. Biosystematics of the *heermanni* group kangaroo rats in southern California. Transactions of the San Diego Society of Natural History, 14:313–344.

Le Conte, J. L. 1853. [Remarks on the genus Dipodomys and Perognathus.]. Proceedings of the Academy of Natural Sciences of Philadelphia, 6:224–225.

Lee, T. E., Jr., and M. D. Engstrom. 1991. Genetic variation in the silky pocket mouse (*Perognathus flavus*) in Texas and New Mexico. Journal of Mammalogy, 72:273–285.

Lidicker, W. Z., Jr. 1960. An analysis of intraspecific variation in the kangaroo rat Dipodomys merriami. University of California Publications in Zoology, 67: 125–218.

Lindsay, E. H. 1972. Small mammal fossils from the Barstow Formation, California. University of California Publications in Geological Sciences, 93:1–104.

Long, C. A. 1965. The mammals of Wyoming. University of Kansas Publications, Museum of Natural History, 14:493–758.

Mascarello, J. T., and D. S. Rogers. 1988. Banded chromosomes of *Liomys salvini, Heteromys oresterus,* and *H. desmarestianus.* Journal of Mammalogy, 69:126–130.

Matson, J. O. 1980. The status of the banner-tailed kangaroo rats, genus *Dipodomys,* from central Mexico. Journal of Mammalogy, 61:563–566.

Matson, J. O., and R. H. Baker. 1986. Mammals of Zacatecas. Special Publications, The Museum, Texas Tech University, 24:1–88.

McGhee, M. E., and H. H. Genoways. 1978. Liomys pictus. Mammalian Species, 83:1–5.

Mearns, E. A. 1890. Description of supposed new species and subspecies of mammals, from Arizona. Bulletin of the American Museum of Natural History, 2:277–307.

———. 1901. A new pocket mouse from southern California. Proceedings of the Biological Society of Washington, 14:135–136.

Merriam, C. H. 1889. Revision of the North American pocket mice. North American Fauna, 1:1–29.

———. 1890. Descriptions of three new kangaroo rats, with remarks on the identity of Dipodomys ordii of Woodhouse. North American Fauna, 4:41–49.

———. 1891. Description of a new genus and species of dwarf kangaroo rat from Nevada (Microdipodops megacephalus). North American Fauna, 5:115–117.

———. 1894*a*. Preliminary descriptions of eleven new kangaroo rats of the genera Dipodomys and Perodipus. Proceedings of the Biological Society of Washington, 9:109–116.

———. 1894*b*. Descriptions of eight new pocket mice (genus Perognathus). Proceedings of the Academy of Natural Sciences of Philadelphia, 46:262–268.

———. 1894*c*. Descriptions of four new pocket mice from Lower California, collected by Walter E. Bryant. Proceedings of the California Academy of Sciences, Series 2, 4:457–462.

———. 1902. Twenty new pocket mice (Heteromys and Liomys) from Mexico. Proceedings of the Biological Society of Washington, 15:41–50.

Miller, G. S., Jr., and R. Kellogg. 1955. List of North American Recent mammals. Bulletin of the United States National Museum, 205:1–954.

Morales, J. C., and M. D. Engstrom. 1989. Morphological variation in the painted spiny pocket mouse, *Liomys pictus* (Family Heteromyidae), from Colima and southern Jalisco, México. Royal Ontario Museum, Life Sciences Occasional Paper, 38:1–16.

Nader, I. A. 1965. Two new subspecies of kangaroo rats, Genus *Dipodomys.* Proceedings of the Biological Society of Washington, 78:49–54.

———. 1966. Roots of teeth as a generic character in the kangaroo rats *Dipodomys.* Bulletin of the Biological Research Centre, University of Baghdad, 2:62–68.

———. 1978. Kangaroo rats: intraspecific variation in *Dipodomys spectabilis* Merriam and *Dipodomys deserti* Stephens. Illinois Biological Monographs, 49: 1–116.

Nelson, E. W., and E. A. Goldman. 1923. A new pocket mouse from Lower California. Proceedings of the Biological Society of Washington, 36:159–160.

———. 1929. Six new pocket mice from Lower California and notes on the status of several described species. Proceedings of the Biological Society of Washington, 42:103–112.

———. 1930. A new pocket mouse from southern Lower California. Journal of the Washington Academy of Sciences, 20:223–224.

O'Farrell, M. J., S. M. Juarez, and C. E. Uptain. 1986. An addition to the known range of Stephen's kangaroo rat, *Dipodomys stephensi,* in San Diego Co., California. California Fish and Game, 72:187–189.

Osgood, W. H. 1900. Revision of the pocket mice of the genus Perognathus. North American Fauna, 18:1–65.

———. 1904. Two new pocket mice of the genus *Perognathus.* Proceedings of the Biological Society of Washington, 17:127–128.

———. 1907. Four new pocket mice. Proceedings of the Biological Society of Washington, 20:19–22.

———. 1912. Mammals from western Venezuela and eastern Colombia. Field Museum of Natural History, Zoology Series, 10:31–67.

———. 1918. The status of Perognathus longimembris (Coues). Proceedings of the Biological Society of Washington, 31:93–100.

Patton, J. L. 1969*a*. Chromosome evolution in the pocket mouse, *Perognathus goldmani* Osgood. Evolution, 23:645–662.

———. 1969*b*. Karyotypic variation in the pocket mouse, *Perognathus penicillatus* Woodhouse (Rodentia-Heteromyidae). Caryologia, 22:351–358.

———. 1977. B-chromosome systems in the pocket mouse, *Perognathus baileyi*: meiosis and C-band studies. Chromosoma, 60:1–14.

Patton, J. L., H. MacArthur, and S. Y. Yang. 1976. Systematic relationships of the four-toed populations of *Dipodomys heermanni.* Journal of Mammalogy, 57:159–163.

Patton, J. L., S. W. Sherwood, and S. Y. Yang. 1981. Biochemical systematics of chaetodipine pocket mice, genus *Perognathus.* Journal of Mammalogy, 62:477–492.

Pearson, O. P. 1939. Three new small mammals from eastern Panama. Notulae Naturae, 6:1–5.

Poole, A. J., and V. S. Schantz. 1942. Catalog of

the type specimens of mammals in the United States National Museum, including the biological survey collection. Bulletin of the United States National Museum, 178:1–705.

Quay, W. B. 1965. Integumentary modifications of North American desert rodents. Pp. 59–74, *in* Biology of the skin and hair growth (A. G. Lyne and B. F. Short, eds.). American Elsevier Publishing Company, New York, 806 pp.

Reed, K. M., and J. R. Choate. 1986. Geographic variation in the plains pocket mouse (*Perognathus flavescens*) on the Great Plains. The Texas Journal of Science, 38:227–240.

Reeder, W. G. 1956. A review of Tertiary rodents of the family Heteromyidae. Ph.D. dissertation, University of Michigan, Ann Arbor, 618 pp.

Rensberger, J. M. 1973. *Sanctimus* (Mammalia, Rodentia) and the phyletic relationships of the large Arikareean geomyoids. Journal of Paleontology, 47: 835–853.

Rhoads, S. N. 1894. Descriptions of new species of North American mammals with remarks on species of the genus Perognathus. Proceedings of the Academy of Natural Sciences of Philadelphia [for 1893], 45:404–412.

Richardson, W. B. 1937. Description of a race of Perognathus longimembris from Tulare County, California. Journal of Mammalogy, 18:510–511.

Ridgeway, R. 1912. Color standards and color nomenclature. Published by the Author, Washington, D.C., 44 pp. + LIII plates.

Rogers, D. S. 1989. Evolutionary implications of chromosomal variation among spiny mice, genus *Heteromys* (Order Rodentia). Southwestern Naturalist, 34:85–100.

———. 1990. Genic evolution, historical biogeography, and systematic relationships among spiny mice (subfamily Heteromyinae). Journal of Mammalogy, 71:668–685.

Rogers, D. S., and J. E. Rogers. 1992*a*. Heteromys oresterus. Mammalian Species, 397:1–2.

———. 1992*b*. Heteromys nelsoni. Mammalian Species, 396:1–2.

Rogers, D. S., and D. J. Schmidly. 1982. Systematics of spiny pocket mice (genus *Heteromys*) of the *desmarestianus* species group from México and northern Central America. Journal of Mammalogy, 63:375–386.

Roth, E. L. 1976. A new species of pocket mouse (*Perognathus*: Heteromyidae) from the Cape region of Baja California Sur, Mexico. Journal of Mammalogy, 57:562–566.

Ryan, J. M. 1986. Comparative morphology and evolution of cheek pouches in rodents. Journal of Morphology, 190:27–41.

———. 1989. Comparative myology and phylogenetic systematics of the Heteromyidae (Mammalia, Rodentia). Miscellaneous Publications, Museum of Zoology, University of Michigan, 176:1–103.

Sanborn, C. C. 1947. Catalogue of type specimens of mammals in the Chicago Natural History Museum. Fieldiana-Zoology, Chicago Museum of Natural History, 32:209–293.

Schmidly, D. J., and F. S. Hendricks. 1976. Systematics of the southern races of Ord's kangaroo rat, *Dipodomys ordii*. Bulletin of the Southern California Academy of Sciences, 75:225–237.

Schmidt, C. A., M. D. Engstrom, and H. H. Genoways. 1989. Heteromys gaumeri. Mammalian Species, 345:1–4.

Schnell, G. D., T. L. Best, and M. L. Kennedy. 1978. Interespecific variation in kangaroo rats (*Dipodomys*): degree of concordance with genic variation. Systematic Zoology, 27:34–48.

Setzer, H. W. 1949. Subspeciation in the kangaroo rat, Dipodomys ordii. University of Kansas Publications, Museum of Natural History, 1:473–573.

Shotwell, J. A. 1967. Later Tertiary geomyoid rodents of Oregon. University of Oregon Museum of Natural History Bulletin, 9:1–51.

Straney, D. O., and J. L. Patton. 1980. Phylogenetic and environmental determinants of geographic variation of the pocket mouse *Perognathus goldmani*. Evolution, 34:888–903.

Stock, A. D. 1970. Notes on mammals from southwestern Utah. Journal of Mammalogy, 51:429–433.

———. 1974. Chromosome evolution in the genus *Dipodomys* and its taxonomic and phylogenetic implications. Journal of Mammalogy, 55:505–526.

Sulentich, J. M. 1983. The systematics and evolution of the *Perognathus parvus* species group in southern California (Rodentia: Heteromyidae). M.S. thesis, California State University, Long Beach, 85 pp.

Sullivan, R. M., and T. L. Best. In press. Systematics and biogeography of the agile kangaroo rat (*Dipodomys agilis*): rediscovery of a species (*D. simulans*). Systematic Zoology.

Swarth, H. S. 1929. The faunal areas of southern Arizona: a study in animal distribution. Proceedings of the California Academy of Sciences, Series 4, 18: 267–383.

Taylor, W. P. 1912. A new *Perognathus* from the San Joaquin Valley, California. University of California Publications in Zoology, 10:155–166.

Townsend, C. H. 1912. Mammals collected by the 'Albatross' expedition in Lower California in 1911, with descriptions of new species. Bulletin of the American Museum of Natural History, 31:117–130.

von Bloeker, J. C., Jr. 1937. Four new rodents from Monterey County, California. Proceedings of the Biological Society of Washington, 50:153–158.

———. 1938. Two new pocket mice, genus *Perognathus*, from west-central California. Proceedings of the Biological Society of Washington, 51:197–200.

Wahlert, J. H. 1985. Skull morphology and relationships of geomyoid rodents. American Museum Novitates, 2812:1–20.

Waterhouse, G. R. 1848. A natural history of mammalia. Vol. 2, containing the order Rodentia or gnawing animals. Hippolyte Bailliere, London, 500 pp.

Weber, M. 1904. Die Säugetiere. Einfühing in die anatomie und systematik der Recenten und fossilen Mammalia. Jena, Gustav Fischer, 866 pp.

Webster, D. B., and M. Webster. 1975. Auditory systems of Heteromyidae: functional morphology and evolution of the middle ear. Journal of Morphology, 146:343–376.

Weckerly, F. W., and T. L. Best. 1985. Morphologic variation among rock pocket mice (*Chaetodipus intermedius*) from New Mexico lava fields. Southwestern Naturalist, 30:491–501.

Weckerly, F. W., A. L. Gennaro, and T. L. Best. 1988. Description of a new rock pocket mouse, *Chaetodipus intermedius,* from New Mexico. Southwestern Naturalist, 33:100–102.

Wied, Maximilian, Prince of. 1839. Über einige Nager mit äusseren Backentaschen aus dem westlichen Nord-America. I. Über ein paar neue Gattungen der Nagethiere mit äusseren Backentaschen. Nova Acta Physico-Medica, Academiae Caesareae Leopoldino-Carolinae, 19:367–374.

Williams, D. F. 1971. The systematics and evolution of the *Perognathus fasciatus* group of pocket mice. Ph.D. dissertation, University of New Mexico, Albuquerque, 176 pp.

———. 1978*a*. Karyological affinities of the species groups of silky pocket mice (Rodentia, Heteromyidae). Journal of Mammalogy, 59:599–612.

———. 1978*b*. Systematics and ecogeographic variation of the Apache pocket mouse (Rodentia: Heteromyidae). Bulletin of Carnegie Museum of Natural History, 10:1–57.

———. In press. Geographic distribution and population status of the giant kangaroo rat, *Dipodomys ingens* (Rodentia, Heteromyidae). *In* Endangered and sensitive species of the San Joaquin Valley, California (D. F. Williams, S. Byrne, and T. A. Rado, eds.). The California Energy Commission, Sacramento.

Williams, D. F., and H. H. Genoways. 1979. A systematic review of the olive-backed pocket mouse, *Perognathus fasciatus* (Rodentia, Heteromyidae). Annals of Carnegie Museum, 48:73–102.

Williams, D. F., and K. S. Kilburn. 1991. Dipodomys ingens. Mammalian Species, 377:1–7.

Wilson, D. E. 1973. The systematic status of *Perognathus merriami.* Proceedings of the Biological Society of Washington, 86:175–192.

Wood, A. E. 1935. Evolution and relationships of the heteromyid rodents with new forms from the Tertiary of western North America. Annals of Carnegie Museum, 24:73–262.

———. 1936. A new subfamily of heteromyid rodents from the Miocene of western United States. American Journal of Science, series 5, 31:41–49.

Yates, T. L., R. W. Barber, and D. M. Armstrong. 1987. Survey of North American collections of Recent mammals. Journal of Mammalogy, 68(2), Suppl.: 1–76.

Zakrzewski, R. J. 1981. Kangaroo rats from the Borchers Local Fauna, Blancan, Meade County, Kansas. Transactions of the Kansas Academy of Science, 84:78–88.

PATTERNS OF MORPHOLOGIC AND MORPHOMETRIC VARIATION IN HETEROMYID RODENTS

TROY L. BEST

Introduction

Studies of morphologic variation in heteromyids historically have centered on delineating taxonomic relationships and rarely contained quantitative information beyond the cursory presentation of measurements of type specimens (e.g., Dice, 1929; Grinnell, 1919; Merriam, 1894, 1902, 1904, 1907; Stephens, 1887). Descriptions such as "mastoid bullae more fully inflated" (Goldman, 1923:139), "mandible small for size of skull" (Merriam, 1894:110), or "outline of the skull is more nearly that of an equilateral triangle" (Huey, 1951:241) were used to describe morphologic differences among taxa. Later, descriptions of morphology became oriented toward use of standard statistical descriptions, that is, mean, range, and standard deviation (e.g., Hooper and Handley, 1948; Huey, 1951; Setzer, 1949). As the need for more detailed taxonomic assessments became apparent, partially because of the larger numbers of specimens from a larger number of collecting localities, so did the need for analyses with greater discriminating abilities. In the past 20 years there has been a trend toward quantifying morphologic variation using a variety of univariate and multivariate statistical techniques. In addition to their use as tools in taxonomic studies, these techniques have provided the basis for detailed studies of patterns of morphologic variation within and among species.

Although assessments of morphologic variation in heteromyids once centered on taxonomic implications, other aspects of these rodents' lifestyles have been addressed by studying morphologic traits. For example, information on morphologic variation within this family has led to assessments of predator avoidance and detection (e.g., Dice and Blossom, 1937; Kotler, 1985; Webster, 1962; Webster and Webster, 1971), movement and locomotion (e.g., Bartholomew and Cary, 1954; Bartholomew and Caswell, 1951; Biewener and Blickhan, 1988; Biewener et al., 1988; Hatt, 1932; Howell, 1933, 1944; Williamson and Frederick, 1977), functional anatomy and behavior (e.g., Dressler, 1979; Forman and Phillips, 1988; Hafner and Hafner, 1984; Kenagy and Trombulak, 1986; Nikolai and

Bramble, 1983; Pfaffenberger et al., 1985; Reichman, 1983; Ryan, 1986, 1989; Rylander, 1981; Thompson, 1985; Tibbitts and King, 1975; Van De Graaff, 1973; Vimtrup and Schmidt-Nielsen, 1952), water balance and physiology (e.g., Lawler and Geluso, 1986; MacMillen, 1983; Schmidt-Nielsen and Schmidt-Nielsen, 1951, 1952), seed-husking abilities (Rosenzweig and Sterner, 1970), resource partitioning and community structure (e.g., Bowers and Brown, 1982; Munger et al., 1983; Price, 1983, 1984; Price and Brown, 1983; Price and Heinz, 1984), integumentary modifications (e.g., Quay, 1965; Westerhaus, 1983), fossil history (e.g., Dalquest and Carpenter, 1986; Reeder, 1956; Voorhies, 1975; Wood, 1935), effects of long-term environmental changes (Roth, 1976*a*), life-history variables (Jones, 1985), and burrow structure (Best, 1982; Best et al., 1988). Morphologic studies also have addressed pelage and coloration, bacula, geographic variation, and environmental-morphologic relationships, which are discussed below.

Heteromyids occupy a range that extends over western North America and into northern South America. The 57 species within this family occupy a wide variety of habitats and are morphologically diverse (Fig. 1). Gray (1868), Coues (1875, 1877), and Elliot (1901) provided early summaries of what was known about this family. However, Wood (1931) was the first to review the fossil history and phylogeny, and he provided an interpretation of the evolutionary relationships of various taxa. Later, he presented a detailed review of evolutionary relationships among the extinct and extant forms of heteromyids (Wood, 1935). In the interim, Hatt (1932) and Howell (1933) presented interpretations of evolutionary relationships of heteromyids. Reeder (1956), employing mainly dental characters, performed an extensive review of fossil and Recent heteromyids. Hafner (1978) examined evolutionary relationships of *Microdipodops* using phenetic characteristics of four genera of heteromyids. Hafner (1982) and Hafner and Hafner (1983) compared heteromyids and geomyids to ascertain phylogenetic relationships among these families, and Wahlert (1985) presented an interpretation of relationships among extinct and extant forms of Geomyoidea. Wahlert (1993) reviewed the fossil record of heteromyids and suggested a classification of Recent and fossil genera.

To assess patterns of morphologic variation within heteromyids, a variety of morphologic traits have been studied including color (e.g., Benson, 1933; Dice and Blossom, 1937; Lidicker, 1960*a*), hair structure (Homan and Genoways, 1978), cheek pouch capacity (Morton et al., 1980), internal anatomy (Setzer, 1949), bacula (e.g., Best, 1981*a*; Best and Schnell, 1974; Burt, 1936, 1960; Genoways, 1973; Hoffmeister, 1986; Lidicker, 1960*b*), spermatozoa (Genoways, 1973; Hafner and Hafner, 1983), skeletons (Best, 1978; Schnell et al., 1978; Shaver, 1973), interparietal bones (e.g., Beer, 1965; Thompson, 1969), middle and inner ear structure (e.g., Webster and Webster, 1975, 1980), and crania (e.g., Baumgardner, 1989; Best, 1983*a*, 1983*b*, 1987; Best and Janecek, 1992; Engstrom et al., 1987; Grinnell, 1922; Hoffmeister, 1986; Kennedy and Schnell, 1978; Lidicker, 1960*a*; Morales and Engstrom, 1989; Rogers and Schmidly, 1982; Setzer, 1949; Smith, 1986). These studies have centered on variation within populations (e.g., Desha, 1967; Schitoskey, 1968; Schmidly, 1971; Webster and Jones, 1985), species (Best, 1987; Best and Janecek, 1992; Best et al., 1986; Dale, 1939; Engstrom et al., 1987; Hall and Dale, 1939; Hartman, 1980; Hoffmeister and Lee, 1967; Kennedy and Schnell, 1978; Lidicker, 1960*a*; Weckerly and Best, 1985; Williams, 1978; Williams and Genoways, 1979), or genera (e.g., Baumgardner, 1989; Genoways, 1973; Hafner, 1981; Hall, 1941; Schnell et al., 1978).

The purposes of this chapter are to provide a summary of studies related to pelage and coloration, bacula, geographic variation, and environmental-morphologic relationships, and to present new data regard-

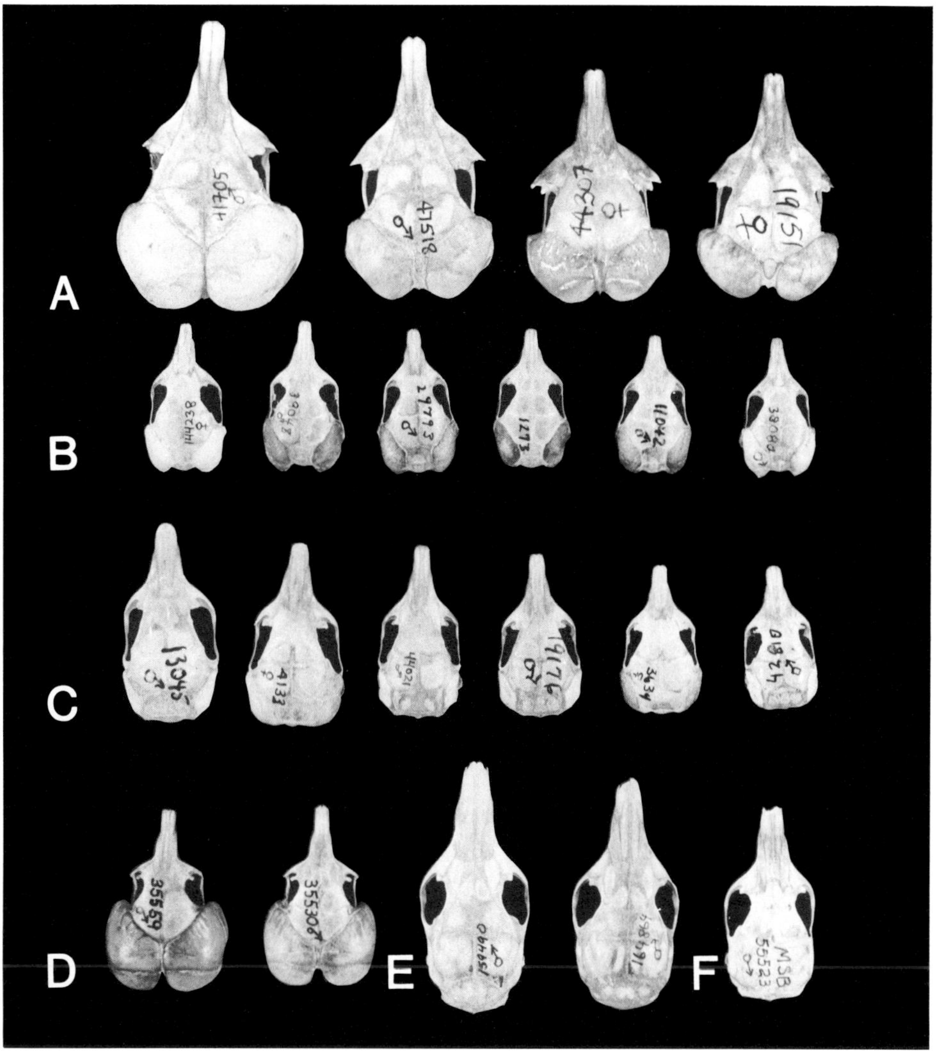

FIG. 1.—Crania of 21 species of the six genera of heteromyid rodents: A) *Dipodomys deserti* (Museum of Southwestern Biology 41705), *D. agilis* (MSB 47518), *D. insularis* (MSB 44307), *D. phillipsii* (MSB 19151); B) *Perognathus inornatus* (Museum of Vertebrate Zoology 144238), *P. amplus* (MSB 38048), *P. fasciatus* (MSB 29793), *P. flavus* (MSB 1273), *P. flavescens* (MSB 11042), *P. longimembris* (MSB 38080); C) *Chaetodipus hispidus* (MSB 13045), *C. baileyi* (MSB 4133), *C. spinatus* (MSB 44021), *C. nelsoni* (MSB 19176), *C. intermedius* (MSB 3634), *C. arenarius* (MSB 42810); D) *Microdipodops pallidus* (MSB 35559), *M. megacephalus* (MSB 35530); E) *Heteromys nelsoni* (MVZ 154490), *H. oresterus* (MVZ 164864); F) *Liomys pictus* (MSB 55523).

ing sexual dimorphism in size and phenetic patterns of morphologic variation among the 57 species of heteromyids. No attempt has been made to cover all aspects of these topics or of non-geographic variation, although authors have addressed these topics further (e.g., age variation is assessed by Anderson, 1964; Best and Schnell, 1974; Engstrom et al., 1987; Hoffmeister and Lee, 1967; Nader, 1978; Reeder, 1953; Rogers and Schmidly, 1982; Schitoskey, 1968; Webster and Jones, 1985).

TABLE 1.—*Secondary sexual dimorphism*[a] *in size of 19 external and skeletal characters of 20 species of kangaroo rats (*Dipodomys*). Minimally significant sexual dimorphism was assumed where* P ≤ *0.05 (one asterisk) and* P ≤ *0.01 (two asterisks). Measurements are mean values (mm).*

		Character							
Dipodomys	Sex	1	2	3	4	5	6	7	8
agilis	♂♂	288.8**	116.1**	172.8**	42.1**	15.3**	22.0**	39.4**	20.9**
	♀♀	283.9	114.0	169.9	41.6	15.0	21.8	39.0	20.7
californicus	♂♂	307.3	119.5	187.9**	44.1	15.8	22.4	38.9	22.6
	♀♀	304.8	120.5	184.3	43.9	16.0	22.3	38.7	22.6
compactus	♂♂	234.4	112.5	121.9	37.7*	12.5	21.3	37.1	19.8
	♀♀	231.9	110.6	121.3	36.8	12.4	21.2	36.9	19.9
deserti	♂♂	342.4**	141.2**	201.1**	53.7**	15.9	24.7**	45.8**	23.9**
	♀♀	330.7	135.5	195.2	52.6	15.5	24.0	44.7	23.1
elator	♂♂	306.2	124.3	182.0	45.7	13.4	23.7*	40.4*	23.7**
	♀♀	303.0	124.0	178.9	45.4	13.7	23.5	40.2	23.4
elephantinus	♂♂	326.2	129.0	196.7	46.7	19.2	24.2	42.6	22.8
	♀♀	322.5	128.9	192.6	46.8	19.5	24.3	42.5	22.8
gravipes	♂♂	306.8**	130.6**	176.1	44.8**	13.3	23.0	41.6*	23.6
	♀♀	300.0	127.1	173.2	44.1	13.5	22.9	40.6	23.4
heermanni	♂♂	300.4**	121.6**	178.8**	43.3**	15.0*	22.6**	40.2**	22.4**
	♀♀	295.1	119.4	175.6	42.6	14.6	22.4	39.7	22.2
ingens	♂♂	333.2	147.6*	185.7	50.2*	15.9	25.6*	45.2*	26.9
	♀♀	328.9	144.5	184.4	49.2	15.5	25.3	44.5	26.6
insularis	♂♂	258.2	108.2**	150.0	40.1	13.0	20.8	36.4	20.7
	♀♀	243.9	97.3	146.6	38.4	13.5	20.7	36.0	20.9
merriami	♂♂	245.7**	100.6**	145.7**	37.9**	12.6	19.6**	36.0**	19.6**
	♀♀	241.2	99.2	142.7	37.4	15.9	19.4	35.5	19.3
microps	♂♂	273.0*	113.5	159.6*	42.0**	12.9	22.2	39.0	19.3*
	♀♀	268.4	111.8	156.6	41.3	12.7	20.8	36.4	19.1
nelsoni	♂♂	318.9**	128.3	190.6**	66.6	23.1	24.6*	42.6**	23.1*
	♀♀	311.8	127.1	184.7	62.0	20.9	24.4	42.1	22.8
nitratoides	♂♂	240.0**	97.1	140.7**	35.3**	12.1	18.8**	34.4**	18.8**
	♀♀	235.1	98.1	137.0	34.9	11.9	18.6	34.0	18.7
ordii	♂♂	242.6	114.2	128.5**	39.1**	12.5	22.1**	39.4**	21.3*
	♀♀	241.5	114.0	127.3	38.6	12.4	21.9	39.1	21.3
panamintinus	♂♂	292.4**	120.2	172.3**	44.5**	14.0	22.7**	39.9**	23.1**
	♀♀	287.8	121.1	169.6	43.8	13.9	22.5	39.4	22.8
phillipsii	♂♂	276.0*	105.0	171.0*	41.4**	14.7	21.4	36.9*	21.5
	♀♀	271.4	104.2	167.2	40.7	14.6	21.2	36.5	21.5
spectabilis	♂♂	342.2**	142.3	199.2*	52.1	15.8	26.5**	45.7**	26.5**
	♀♀	338.0	142.0	195.9	51.8	15.7	26.2	45.2	26.2
stephensi	♂♂	284.2	115.7	168.4	41.7	13.8	22.1	39.1	22.6
	♀♀	281.8	115.8	166.0	41.3	13.8	22.1	39.0	22.6
venustus	♂♂	318.2*	128.7	192.9	45.8*	18.6	23.6**	41.6**	22.1**
	♀♀	313.5	122.9	190.7	45.1	18.2	23.3	41.0	21.7

[a] Sample sizes for *Dipodomys* used in these analyses are: *agilis* (1,741 adult males, 1,425 adult females), *californicus* (191, 150), *compactus* (48, 29), *deserti* (254, 204), *elator* (120, 86), *elephantinus* (38, 32), *gravipes* (56, 54), *heermanni* (474, 366), *ingens* (55, 47), *insularis* (9, 16), *merriami* (433, 397), *microps* (156, 174), *nelsoni* (112, 87), *nitratoides* (276, 200), *ordii* (691, 662), *panamintinus* (467, 385), *phillipsii* (93, 77), *spectabilis* (296, 232), *stephensi* (81, 70), *venustus* (65, 73).

TABLE 1.—*Continued.*

Character										
9	10	11	12	13	14	15	16	17	18	19
10.6**	14.2**	7.4**	4.9	3.6	4.9**	5.7**	13.2**	24.5**	19.1**	3.8**
10.5	14.1	7.3	4.9	3.6	4.9	5.6	13.2	24.3	19.0	3.7
10.8	15.0	7.4	5.1*	4.3	5.5	5.7	12.9	23.6	20.2	4.2*
10.7	14.9	7.4	5.1	4.3	5.5	5.7	12.9	23.5	20.1	4.1
10.4	14.4	7.0	5.0	3.8	4.9	5.1	11.9	22.2	17.8	3.8
10.3	14.3	6.9	5.0	3.9	4.9	5.0	11.8	22.1	17.6	3.8
12.4**	17.1**	8.3**	6.0	4.3**	4.8**	6.6**	15.2**	30.6**	21.7**	4.2**
12.2	16.6	8.1	5.9	4.2	4.6	6.5	14.9	29.9	21.2	4.1
11.6	16.0	7.9	5.6*	4.4*	6.4**	5.7**	13.4	24.8**	19.8**	4.3
11.6	15.9	7.8	5.5	4.3	6.3	5.5	13.4	24.5	19.6	4.2
11.0	15.7	7.9	5.4	3.9*	5.2	6.1*	14.0	25.9	20.5	4.5
11.0	15.6	7.8	5.5	4.1	5.2	5.8	14.0	25.9	19.9	4.6
10.9	14.8*	7.9	5.3	4.5	6.1	6.2*	13.7	26.0	21.2	4.0*
10.8	14.6	8.0	5.3	4.5	6.1	6.1	13.6	25.7	20.9	3.9
11.1**	14.8**	7.6*	5.3	4.2	5.2*	5.9**	13.5**	24.9**	20.4**	4.0**
11.0	14.6	7.6	5.3	4.1	5.2	5.8	13.3	24.6	20.1	3.9
12.3	16.7	8.5	5.7	5.0	6.4	6.8*	15.1*	29.2*	24.4	4.7*
12.3	16.6	8.5	5.7	4.8	6.3	6.6	14.9	28.7	24.0	4.6
11.1	13.4	7.4*	4.9**	3.5	5.7	4.7	11.6	22.8	17.8	3.6
11.2	13.7	7.2	4.6	3.6	5.7	4.7	11.5	22.7	17.6	3.5
11.1*	13.3**	7.0*	4.6	3.2*	5.2**	4.9**	11.9**	22.9**	17.0**	3.2**
11.0	13.1	7.0	4.6	3.2	5.1	4.8	11.8	22.6	16.8	3.1
10.0**	13.0**	6.8	4.8	3.7	3.7	5.6*	12.6*	23.5	18.3*	3.6*
9.8	12.7	6.7	4.7	3.7	3.7	5.5	12.5	23.3	18.1	3.5
11.9*	15.1*	7.7	5.8	4.3	5.2*	6.3**	14.5*	27.2**	21.9	4.1**
11.7	15.0	7.6	5.8	4.2	5.1	6.2	14.4	27.0	21.8	4.0
10.7	12.3*	6.8	4.5*	3.1	4.7*	5.2**	11.7**	22.4**	16.1	3.1**
10.6	12.2	6.8	4.4	3.1	4.7	5.1	11.5	22.1	16.0	3.0
13.2	14.6**	7.7	5.4	3.9	5.0*	5.7**	13.1*	24.7**	18.6*	3.9
13.1	14.5	7.7	5.5	3.9	5.0	5.7	13.0	24.6	18.5	3.9
11.5	15.5**	7.8**	5.3	4.2*	5.4	5.7**	13.1**	24.3**	20.6**	4.0**
11.5	15.2	7.7	5.2	4.2	5.3	5.6	12.9	24.0	20.3	3.9
11.7	13.7	7.4	5.2*	3.7	5.9	5.2	12.4	23.0	17.9	3.6
11.8	13.6	7.4	5.1	3.7	6.0	5.1	12.4	22.9	17.9	3.6
13.1	16.6	8.5	6.3*	5.0	5.7**	6.7**	15.3**	29.3**	24.9**	4.7**
13.1	16.5	8.5	6.2	5.0	5.6	6.7	15.2	29.0	24.7	4.5
11.1	14.3	7.5	5.2	4.1	5.8	5.5	13.5*	25.1	20.4	4.0
11.1	14.4	7.5	5.1	4.1	5.8	5.5	13.4	25.0	20.3	4.0
11.0	15.6	7.7	5.3	3.8	5.3*	6.0	13.7**	25.1**	20.6**	4.2
11.0	15.4	7.6	5.3	3.7	5.2	6.0	13.5	24.7	19.9	4.2

Methods

To assess patterns of morphologic variation and secondary sexual dimorphism in size, >19,500 specimens of the 57 species of heteromyid rodents were examined. Only the 12,563 adult specimens were used in statistical analyses. Examination of patterns of sexual dimorphism in size among taxa included assessment of 20 adult males and 20 adult females of each species. The only species with smaller samples were: *Dipodomys insularis* (9 males, 16 females); *D. margaritae* (3, 1); *Chaetodipus lineatus* (16, 10); *Heteromys goldmani* (14, 20); *H. nelsoni* (3, 6); *H. oresterus* (10, 9); and *Liomys spectabilis* (8, 12). For more detailed analyses of interspecific variation in sexual dimorphism, larger numbers of *Dipodomys* were examined (Table 1). Five external and 14 cranial measurements were analyzed (for description of characters, see Best, 1978). External characters were recorded to the nearest mm (from specimen tags) and cranial measurements were made to the nearest 0.1 mm using dial calipers. *Dipodomys* were aged according to the criteria of Best and Schnell (1974), and other genera were considered to be adult if the first premolar exhibited wear.

Character heterogeneity between sexes of each species was tested using a one-way analysis of variance, and mean values of each character for males and females of each species were used in multivariate procedures. Characters were standardized and correlation and distance matrices were calculated (Sneath and Sokal, 1973). Clusters of species and characters were obtained with the unweighted pair-group method using arithmetic averages (UPGMA). Principal components were calculated from a correlation matrix among characters, and projections of species were plotted on the first two components. A shortest minimally-connected network was computed from the matrix of distances between taxa. To elucidate correlations among characters, dendrograms were constructed from correlation matrices of the 19 standardized characters for males and for females of the genera *Dipodomys, Perognathus, Chaetodipus, Heteromys, Liomys,* and for the 57 species of Heteromyidae. Multivariate assessment was not conducted within *Microdipodops* since this genus is represented by only two species. However, *Microdipodops* is included in analyses of the family. Analyses were performed using an IBM mainframe computer and the programs UNIVAR (D. M. Power, unpublished) and NT-SYS (Rohlf et al., 1974).

Results

Character Correlations

For *Dipodomys,* most characters were highly correlated ($r > 0.88$). For both sexes, interorbital width, length of ear, and width of maxillary arch were not highly correlated with other characters ($r < 0.56$). Character correlations for *Perognathus* were slightly less than those for *Dipodomys.* Although most characters were highly correlated ($r > 0.83$), length of lacrimal, greatest width of cranium, and length of ear were the least correlated with other characters for either sex ($r < 0.75$). Females had lower correlations among characters than males. For *Chaetodipus,* characters grouped into two main clusters. In males, except for width of maxillary arch, which was not correlated highly with any other character ($r = 0.38$), one cluster contained the four external measurements obtained from specimen tags and the other contained the remaining characters. In females, maxillary arch width grouped with the four external characters and was well separated ($r = 0.64$) from the other cluster. For both sexes, the four external characters were not as highly correlated as were the other characters.

In *Heteromys,* all characters except interorbital width, basioccipital length, and lacrimal length were highly correlated ($r > 0.77$). For *Liomys,* characters were likewise highly correlated ($r > 0.80$). Characters for *Heteromys* and *Liomys* were the most highly correlated character sets, and those for *Perognathus* and *Chaetodipus* were the least correlated.

Character correlations within each sex for the 57 species of Heteromyidae were similar; both sexes had two large clusters. One contained total length, length of tail, length of body, nasal width, basal length of cranium, nasal length, alveolar length, basioccipital length, and length of ear. These characters were separated from those in the other cluster at a correlation of 0.78 for males and 0.79 for females. The most highly correlated pairs of characters in both sexes were total length with length of tail and basal length of cranium with nasal length.

Sexual Dimorphism

Means of characters and results of analyses of sexual dimorphism using large samples of 20 species of *Dipodomys* are presented in Table 1. For most characters males were larger than females, including those showing statistically significant differences between sexes. The sample of *D. margaritae* was inadequate for statistical analyses. Some species exhibited a large number of dimorphic characters (*D. agilis, D. deserti, D. heermanni, D. merriami, D. nelsoni, D. nitratoides, D. ordii, D. panamintinus,* and *D. spectabilis*), while others exhibited almost no difference between sexes (*D. californicus, D. compactus, D. elephantinus, D. insularis,* and *D. stephensi*). Species such as *D. elator, D. gravipes, D. ingens, D. microps, D. phillipsii,* and *D. venustus* exhibited significant sexual dimorphism in size in few characters. It was not surprising that so few characters showed dimorphism in *D. insularis* and *D. compactus* due to the small samples, but the nearly complete lack of differences in *D. californicus, D. stephensi,* and *D. elephantinus* was not expected. Qualitative examination of habitat differences, body size, or number of specimens used in analyses did not reveal any relationships with the number of sexually dimorphic characters.

When sexual dimorphism was assessed using only 20 males and 20 females of each species of *Dipodomys* (Table 2), results were different from those obtained with large samples. Little sexual dimorphism was evident using smaller samples. *D. deserti* showed sexual dimorphism in nine characters, *D. ingens* and *D. venustus* in five, and *D. spectabilis* in six. In the remaining species four or fewer characters exhibited differences between sexes. While there was little or no effect of sample size on detection of sexual dimorphism among the larger samples in Table 1, reduction to only 20 males and 20 females greatly affected detection of sexually dimorphic characters. However, mean values for each measurement remained similar for the two data sets.

Although detection of sexual differences appears to be sample-size dependent, it is of interest to see what patterns appeared among the other species and genera. In *Perognathus,* more than one-half of the characters for *P. alticola* and *P. parvus* were sexually dimorphic, as were seven characters for *P. amplus* and three for *P. inornatus* (Table 2). The remaining species had one or no sexually dimorphic characters. For *Chaetodipus,* four or more characters were dimorphic in *C. arenarius, C. artus, C. baileyi, C. goldmani, C. intermedius, C. lineatus, C. pernix,* and *C. spinatus.* No sexual differences were found in *C. formosus* or *C. hispidus.* The most sexually dimorphic species were *C. artus, C. goldmani, C. intermedius,* and *C. pernix.*

No sexual differences were detected in *Microdipodops.* For *Heteromys,* only *H. australis* and *H. nelsoni* were sexually dimorphic in four or more characters. One character was dimorphic in *H. desmarestianus* and two in *H. oresterus. L. irroratus* was dimorphic in four characters, *L. spec-*

TABLE 2.—*Secondary sexual dimorphism in size of 19 external and skeletal characters of 57 species of heteromyid rodents. Minimally significant sexual dimorphism was assumed where* P ≤ *0.05 (one asterisk) and* P ≤ *0.01 (two asterisks). Measurements are mean values (mm).*

Species	Sex	Character 1	2	3	4	5	6	7
Dipodomys								
1. *agilis*	♂♂	284.7	117.2	167.5	42.2	16.9*	21.7	39.3
	♀♀	285.1	116.1	169.0	42.2	17.6	21.7	38.3
2. *californicus*	♂♂	308.0	118.8	189.2	44.2	15.8	22.5	39.1
	♀♀	302.9	118.6	184.3	43.5	15.7	22.0	38.2
3. *compactus*	♂♂	233.6	113.6	120.0	38.1**	12.7	21.4	37.3
	♀♀	230.6	108.8	121.8	36.3	12.5	21.1	36.8
4. *deserti*	♂♂	334.9	141.8	193.1	54.2**	16.5	24.7*	45.8**
	♀♀	327.3	135.6	191.7	52.2	15.6	24.0	44.3
5. *elator*	♂♂	308.8	123.1	185.7	46.2	12.9	23.7	40.4
	♀♀	305.5	121.9	183.6	45.2	13.5	23.6	40.4
6. *elephantinus*	♂♂	327.2	129.7	197.5	46.8	19.4	24.4	42.8
	♀♀	320.7	130.3	190.4	46.7	19.5	24.3	42.5
7. *gravipes*	♂♂	299.1	127.4	171.7	44.5	13.1	22.7	41.1
	♀♀	298.0	126.0	172.0	44.0	13.4	22.7	40.7
8. *heermanni*	♂♂	298.1	119.1	179.0	43.1	14.2	22.4	39.6
	♀♀	291.8	119.5	172.3	42.6	14.5	22.4	39.5
9. *ingens*	♂♂	334.4	148.9	185.5	50.3	16.3	25.8*	45.6*
	♀♀	327.7	144.8	182.9	49.3	15.2	25.4	44.8
10. *insularis*	♂♂	258.2	108.2**	150.0	40.1	13.0	20.8	36.4
	♀♀	243.9	97.3	146.6	38.4	13.5	20.7	36.0
11. *margaritae*	♂♂	238.7	91.3	147.3	38.0	13.0	19.8	35.0
	♀♀	247.0	97.0	150.0	39.0	—	19.6	34.6
12. *merriami*	♂♂	240.8	97.7	143.1	37.5	11.2	19.4	35.8
	♀♀	244.5	101.2	143.3	38.3	11.7	19.6	35.9
13. *microps*	♂♂	276.7	116.4	160.4	42.6*	13.9	20.9	37.0*
	♀♀	270.9	115.6	155.3	41.3	13.1	20.6	36.0
14. *nelsoni*	♂♂	326.3	133.3**	193.0	46.3	16.8*	24.6	42.6
	♀♀	320.7	128.4	192.3	46.2	15.9	24.5	42.4
15. *nitratoides*	♂♂	240.6	101.4	139.2	35.2	12.4	19.0	34.7
	♀♀	241.7	102.2	139.5	35.2	11.8	18.9	34.6
16. *ordii*	♂♂	249.7	110.7	139.1	40.2	12.5	21.3	38.1
	♀♀	247.3	111.7	135.5	40.2	12.9	21.5	38.3
17. *panamintinus*	♂♂	291.3	119.1	172.2	43.9	14.0	22.4	39.4
	♀♀	286.5	119.5	167.1	44.0	14.4	22.6	39.4
18. *phillipsii*	♂♂	279.4	110.0	169.4	40.6	14.8	21.3	37.0**
	♀♀	276.4	109.3	167.1	39.9	14.8	21.0	36.3
19. *spectabilis*	♂♂	345.1	140.6	204.5	53.3	14.9*	27.0**	46.3**
	♀♀	339.1	141.9	197.2	49.7	16.2	26.0	45.1
20. *stephensi*	♂♂	284.4	116.3	168.1	41.4	14.2	22.2	38.9
	♀♀	282.6	116.6	166.0	41.3	13.9	22.2	39.0
21. *venustus*	♂♂	314.0	118.7	195.2	45.6	18.9	23.7*	41.6
	♀♀	316.2	123.0	193.3	45.4	19.2	23.3	41.0
Perognathus								
22. *alticola*	♂♂	163.6**	77.6**	86.0**	21.9**	5.9	15.3**	24.9**
	♀♀	149.5	72.5	77.1	20.7	5.6	14.6	23.8

TABLE 2.—*Continued.*

Character											
8	9	10	11	12	13	14	15	16	17	18	19
21.0	10.4	14.1	7.4	4.9	3.7	4.9*	5.5	13.2	24.6	18.9	3.6
21.1	10.5	14.0	7.5	5.1	3.7	5.1	5.5	13.2	24.5	18.9	3.6
23.0**	10.8	15.3	7.4	5.0	4.2	5.6	5.7	12.8	23.7	20.4	4.2
22.0	10.6	14.8	7.3	5.0	4.2	5.4	5.5	12.8	23.2	20.0	4.0
20.0	10.3	14.4	6.9	5.0	3.9	5.0	5.2	11.9	22.2	17.6	3.8
19.9	10.3	14.3	6.9	5.0	3.9	4.9	5.0	11.8	22.1	17.6	3.8
24.0**	12.2	17.1**	8.2	5.9	4.3	4.9*	6.5	15.2*	30.8**	21.6*	4.2
22.9	12.1	16.3	8.1	5.8	4.1	4.6	6.4	14.8	29.6	21.0	4.0
23.7	11.5	16.0	7.8	5.5	4.3	6.4	5.7	13.3	24.7	19.7	4.3**
23.4	11.5	15.7	7.8	5.4	4.2	6.3	5.6	13.3	24.5	19.5	4.1
22.9	11.0	15.8	8.0*	5.4	3.9	5.3	6.1	14.1	26.1	21.1	4.6
22.8	11.0	15.6	7.8	5.5	4.1	5.2	6.0	14.0	25.9	21.2	4.6
23.3	10.8	14.7	7.9	5.4	4.5	6.1	6.1	13.6	25.6	21.1	4.0
23.3	11.0	14.6	7.9	5.3	4.5	6.0	6.1	13.5	25.6	20.7	4.0
22.0*	10.9	14.6	7.6*	5.2	4.0	5.2	5.8	13.3	24.4	20.0	3.9
22.5	10.9	14.5	7.5	5.1	4.1	5.2	5.7	13.3	24.5	20.0	3.9
26.9	12.4	17.0	8.5	5.7	4.8	6.4	6.7	15.1*	29.3*	24.4	4.8**
26.5	12.2	16.9	8.5	5.7	4.8	6.3	6.7	14.9	28.7	24.0	4.6
20.7	11.1	13.4	7.4*	4.9**	3.5	5.7	4.7	11.6	22.8	17.8	3.6
20.9	11.2	13.7	7.2	4.6	3.6	5.7	4.7	11.5	22.7	17.6	3.5
19.4	11.2	13.2	7.0	4.9	3.3	5.3	4.6	11.4	21.9	17.6	3.3
19.3	11.0	12.9	7.0	4.8	3.2	4.9	4.4	11.7	21.5	17.5	3.0
19.7	11.3	13.2	7.0	4.6	3.2*	5.2	5.0	11.9	22.9	17.1*	3.2
19.4	11.3	13.5	7.0	4.5	3.0	5.2	4.9	11.8	22.8	16.7	3.1
19.3	10.0	12.9	6.7	4.6	3.5	3.8**	5.6	12.7*	23.6	18.2	3.5
19.1	9.8	12.5	6.7	4.7	3.5	3.5	5.5	12.5	23.3	17.9	3.5
23.2	12.0	15.2	7.8	5.8	4.2	5.1	6.4	14.4	27.1	21.6	4.1
23.2	12.1	15.1	7.7	5.7	4.1	5.2	6.3	14.5	27.2	21.8	4.0
19.0	10.9	12.4	6.9	4.4	3.0	4.8	5.1	11.7	22.5	16.1	3.1
18.8	10.7	12.3	6.9	4.4	3.1	4.7	5.1	11.7	22.4	16.2	3.0
20.8	10.5	14.1	7.3	4.9	3.7	4.6	5.5	12.7	23.9	18.0	3.8
20.7	10.9	14.1	7.4	5.1	3.8	4.7	5.5	12.7	24.0	18.0	3.8
22.9	11.4	15.2	7.7	5.2	4.0	5.4	5.5	12.9	23.9	20.2	3.8
22.7	11.4	15.1	7.7	5.2	4.1	5.3	5.6	13.0	24.0	20.2	3.8
21.0	11.2	13.8	7.5	5.1	3.7	5.9	5.0	12.4	22.9	17.8	3.6
20.9	11.2	13.7	7.5	4.9	3.6	5.9	5.0	12.4	22.7	17.5	3.5
26.6	12.9	16.9**	8.6	6.4	5.0	5.7	6.9	15.6	29.6*	25.1*	4.8
25.8	12.8	16.3	8.5	6.3	4.9	5.5	6.7	15.3	28.9	24.5	4.6
22.4	11.0	14.2	7.6	5.1	4.0	5.7	5.4	13.5	25.0	20.4	4.1
22.5	10.9	14.3	7.5	5.2	4.1	5.9	5.4	13.4	25.0	20.3	4.0
22.3**	11.1	15.4	7.7	5.3	3.7	5.3**	6.0	13.8	25.2*	20.4**	4.2
21.5	11.0	15.4	7.6	5.3	3.6	5.0	6.0	13.6	24.7	19.8	4.1
12.0**	6.1**	10.0**	4.7*	3.8*	1.6	1.2	3.9*	8.2**	12.9**	12.4**	2.6*
11.5	5.8	9.2	4.6	3.7	1.7	1.2	3.7	8.0	12.4	11.8	2.5

TABLE 2.—*Continued.*

Species	Sex	Character 1	2	3	4	5	6	7
23. *amplus*	♂♂	153.9	70.8	83.1*	20.7	7.2	14.8**	24.7**
	♀♀	149.1	70.1	79.1	20.3	6.7	14.3	23.9
24. *fasciatus*	♂♂	136.8	71.2	65.7	17.4	7.2	13.9	23.0
	♀♀	137.1	72.1	65.0	17.5	6.9	13.9	22.9
25. *flavescens*	♂♂	135.1	72.1	63.0	18.8	6.6	13.8	22.8
	♀♀	134.8	70.4	64.4	19.1	6.7	14.0	23.0
26. *flavus*	♂♂	113.2	59.0	54.2*	16.5	6.2	12.7	20.8
	♀♀	111.6	60.4	51.3	16.7	6.2	12.5	20.5
27. *inornatus*	♂♂	148.8	72.6	76.2	19.6	7.8	14.5	23.6
	♀♀	146.6	72.1	74.5	19.4	7.4	14.2	23.0
28. *longimembris*	♂♂	134.1	58.5	72.5	18.8	7.0	13.0	22.1
	♀♀	137.4	56.9	74.2	19.0	6.9	13.1	22.0
29. *parvus*	♂♂	178.7**	83.6**	95.1**	23.8**	9.5*	16.2**	26.3**
	♀♀	163.9	78.3	85.6	22.3	8.3	15.4	25.2
Chaetodipus								
30. *arenarius*	♂♂	155.8	69.5	86.3	21.3	7.9	13.9	22.9*
	♀♀	152.3	67.3	85.1	20.9	7.8	13.8	22.4
31. *artus*	♂♂	190.5**	92.1**	98.5	23.5*	10.8	16.7**	26.5**
	♀♀	180.0	85.9	94.1	22.9	10.9	16.3	25.7
32. *baileyi*	♂♂	209.9	94.5	115.4*	25.4*	10.1*	18.1	28.9
	♀♀	218.1	92.5	125.6	28.8	11.3	18.0	28.6
33. *californicus*	♂♂	218.3	88.5	125.2	26.5	13.1	17.1	27.9
	♀♀	213.8	85.3	124.1	26.3	13.7	17.0	27.7
34. *fallax*	♂♂	190.9	84.6	106.3	24.0	9.7	16.3	26.6
	♀♀	186.6	82.0	104.6	23.3	9.7	16.3	26.4
35. *formosus*	♂♂	193.7	82.7	106.7	25.1	11.6	16.4	27.0
	♀♀	189.5	79.0	106.4	24.8	11.5	16.4	27.0
36. *goldmani*	♂♂	197.9	81.4	107.5	24.2	11.3	16.7**	27.0*
	♀♀	189.4	83.7	105.8	23.4	10.4	16.1	26.0
37. *hispidus*	♂♂	205.1	101.5	97.9	24.6	11.3	19.6	30.7
	♀♀	201.2	106.9	94.3	23.7	11.4	19.8	30.7
38. *intermedius*	♂♂	166.6	74.0	92.6	21.5**	7.5	15.1*	24.6**
	♀♀	167.1	74.1	93.0	20.9	7.5	14.9	24.1
39. *lineatus*	♂♂	168.9	74.4	94.5	23.3	7.4	15.5**	25.0
	♀♀	163.2	73.0	90.2	22.3	7.5	14.8	24.3
40. *nelsoni*	♂♂	179.2	80.7	98.6	21.2	8.0	15.9	25.5
	♀♀	176.4	78.4	98.1	21.0	7.8	15.7	25.2
41. *penicillatus*	♂♂	167.7	76.6	91.1	21.6**	8.2	15.1	24.2
	♀♀	167.3	75.8	91.5	23.1	8.5	15.0	24.3
42. *pernix*	♂♂	162.1*	75.2	83.0	21.6	9.9*	15.3**	24.4**
	♀♀	152.5	68.9	80.1	21.3	9.1	14.6	23.2
43. *spinatus*	♂♂	198.7	85.0	113.7	23.8	10.1	16.1*	26.3**
	♀♀	198.1	82.6	115.4	23.7	10.2	15.6	25.5
Microdipodops								
44. *megacephalus*	♂♂	149.2	65.0	84.2	24.9	10.1	14.6	28.3
	♀♀	148.9	66.8	82.1	24.6	10.0	14.5	28.0
45. *pallidus*	♂♂	154.8	65.8	89.0	25.6	10.3	14.8	28.9
	♀♀	154.4	66.3	88.1	25.7	10.3	14.7	28.7

TABLE 2.—*Continued.*

Character											
8	9	10	11	12	13	14	15	16	17	18	19
11.6**	5.6	9.5**	4.3	3.5	1.8	1.0	3.9	8.1**	13.8**	12.4	2.4
11.2	5.5	9.1	4.3	3.4	1.6	1.0	3.8	8.0	13.4	12.1	2.4
10.7	5.3	8.5	4.5	3.3	1.4	1.0	3.9	8.2	12.9	12.3	2.3
10.6	5.2	8.6	4.4	3.3	1.4	1.0	3.9	8.1	12.9	12.3	2.4
10.5	5.4	8.4	4.3	3.4	1.6	0.9	3.9	7.9	12.6	11.6	2.4
10.5	5.3	8.5	4.4	3.4	1.6	0.9	3.9	7.9	12.8	11.9	2.4
9.9	4.6	8.0	4.0	3.2	1.4	0.9	3.4	7.3	12.0	10.9	2.1
9.8	4.6	7.8	4.0	3.2	1.5	0.9	3.4	7.3	11.9	10.8	2.2
11.7**	5.2	8.9**	4.2	3.4	1.7**	1.1	3.8	7.7	13.1	12.1	2.4
11.3	5.2	8.5	4.2	3.4	1.4	1.1	3.7	7.8	13.0	11.8	2.4
10.5	5.4	8.4	4.2	3.1	1.6*	1.0	3.5	7.6	12.7	11.3	2.3
10.4	5.5	8.4	4.3	3.2	1.3	1.1	3.5	7.5	12.4	11.2	2.3
12.6**	6.1**	10.6**	4.9	3.8	1.8	1.2	4.1	8.6	13.5	13.0	2.7
11.9	5.8	9.8	4.8	3.8	1.8	1.2	4.0	8.5	13.3	12.7	2.5
11.0*	6.1	8.8	4.3	3.3	1.6	1.3*	3.3*	7.6	12.1	11.5**	2.5*
10.5	5.9	8.7	4.3	3.3	1.5	1.2	3.1	7.5	11.8	11.1	2.4
13.0**	6.5*	10.0**	4.6	3.9	1.7	1.3	4.0**	8.6	13.4**	13.3*	3.1
12.5	6.3	9.6	4.6	3.8	1.6	1.2	3.7	8.5	12.9	12.9	3.1
13.4	6.9	11.7	4.9	4.3	2.2	1.8	4.4*	9.7	15.2	15.0	3.0
13.5	6.8	11.5	5.0	4.3	2.2	1.8	4.2	9.5	15.3	15.0	2.9
13.0	6.8	11.0	4.9	4.1	1.8	1.6	4.1	9.0*	14.1	14.3	3.2
13.0	6.8	11.1	4.9	4.1	1.9	1.6	4.1	8.8	13.9	14.3	3.1
12.4	6.4	10.3	4.9	4.0*	1.9	1.6	4.0	8.9	13.8	13.4	2.7
12.3	6.5	10.4	4.9	4.1	2.0	1.6	3.9	8.7	13.9	13.5	2.7
12.5	6.7	10.8	4.9	4.0	1.9	1.6	4.0	9.0	14.3	13.7	2.7
12.5	6.7	10.8	4.9	4.0	2.0	1.5	4.0	9.0	14.3	13.6	2.7
12.5**	6.2*	10.6*	4.6*	3.9*	2.0	1.4	4.3	8.6	13.6*	13.0**	3.3
11.8	6.1	10.0	4.5	3.7	1.8	1.4	4.1	8.5	13.2	12.6	3.1
14.9	7.4	12.3	5.4	4.6	2.0	1.2	5.0	10.2	15.7	15.5	3.5
15.1	7.3	12.2	5.3	4.7	2.1	1.3	5.0	10.3	15.6	15.6	3.5
11.3	6.3*	9.5	4.6	3.7	1.7	1.3	3.8**	8.2**	13.0	12.3*	2.7**
11.2	6.2	9.4	4.5	3.7	1.8	1.3	3.6	8.1	12.9	12.1	2.6
11.7**	6.2	9.7*	4.6	3.7	1.7	1.3	3.8	8.3	13.0	12.9*	2.7
11.2	6.2	9.3	4.6	3.7	1.7	1.3	3.7	8.2	12.7	12.4	2.6
11.8*	6.4	10.0	4.6	3.7	1.7	1.3	3.8	8.3	13.5*	13.1	2.8
11.5	6.4	9.7	4.6	3.7	1.8	1.4	3.9	8.3	13.2	12.8	2.7
11.1	5.8	9.1	4.4	3.5	1.7	1.3	3.9	8.0	12.5	12.3	2.9
11.2	5.9	9.2	4.4	3.6	1.7	1.4	3.9	7.9	12.5	12.2	2.8
11.9**	5.8*	9.4**	4.4	3.6	1.6	1.2	3.8*	8.0**	12.4**	12.4**	2.9*
11.4	5.6	8.6	4.3	3.5	1.6	1.2	3.6	7.7	11.9	11.8	2.7
11.9	6.6	10.4	4.4	3.8	1.7	1.4	4.0*	8.4	13.3	12.9	2.8**
11.7	6.5	10.0	4.4	3.7	1.7	1.3	3.8	8.3	13.1	12.7	2.6
11.5	6.6	10.4	5.3	3.5	1.4	1.4	4.1	9.7	18.6	11.2	2.3
11.3	6.4	10.1	5.2	3.5	1.4	1.4	4.0	9.6	18.7	11.2	2.3
12.1	6.8	10.1	5.3	3.5	1.3	1.5	4.0	10.1	19.9	11.5	2.3
12.0	6.8	10.0	5.4	3.5	1.3	1.5	4.1	10.0	19.9	11.6	2.3

TABLE 2.—*Continued.*

Species	Sex	Character 1	2	3	4	5	6	7
Heteromys								
46. *anomalus*	♂♂	289.6	133.8	155.8	34.1	19.2	22.9	36.1
	♀♀	279.1	128.9	150.2	33.9	19.2	22.3	35.4
47. *australis*	♂♂	267.5	127.5*	140.0	33.3	16.8*	21.5*	35.8*
	♀♀	256.7	120.3	136.4	33.0	15.9	21.1	35.1
48. *desmarestianus*	♂♂	281.1*	133.1	148.0	34.1	16.0	21.8	35.4
	♀♀	271.6	129.7	141.9	34.2	15.9	21.9	35.3
49. *gaumeri*	♂♂	276.6	125.3	151.3	34.1	15.9	21.4	34.7
	♀♀	269.3	123.4	145.9	33.2	16.3	21.5	34.7
50. *goldmani*	♂♂	336.8	148.5	188.2	37.1	19.1	24.3	38.9
	♀♀	335.1	143.6	190.9	36.0	17.9	23.6	38.0
51. *nelsoni*	♂♂	358.0*	161.5	196.5	43.3	22.0**	25.5	41.0*
	♀♀	336.4	150.6	185.8	41.5	20.3	24.8	39.3
52. *oresterus*	♂♂	334.0**	159.4*	174.6	40.4	18.3	24.2	39.2
	♀♀	309.9	141.1	168.7	40.3	18.8	24.2	38.7
Liomys								
53. *adspersus*	♂♂	253.5	126.9	126.6	30.8	15.5	22.9	35.4
	♀♀	248.8	123.4	125.4	30.3	15.2	22.7	35.0
54. *irroratus*	♂♂	254.8**	125.4**	129.4	31.9	—	21.9	33.7
	♀♀	246.1	118.6	127.6	30.9	—	21.6	33.2
55. *pictus*	♂♂	239.5	110.3	129.2	28.8	—	19.8	31.4
	♀♀	232.0	110.5	121.5	28.4	—	19.7	31.0
56. *salvini*	♂♂	226.7	114.2*	112.4	26.8	14.4	20.0	31.6
	♀♀	216.5	107.5	109.0	26.0	13.7	19.6	30.9
57. *spectabilis*	♂♂	267.6*	109.4	142.6**	30.6	16.4	21.4	33.7
	♀♀	244.8	101.5	123.0	30.7	16.7	21.0	33.5

tabilis in two, and *L. salvini* in one. Most of the dimorphic characters in *Liomys* were non-cranial measurements.

Patterns of Variation

To illustrate morphologic diversity among the six genera, crania of 21 species are shown in Fig. 1. Phenograms, constructed from correlation and distance matrices, for each sex of five genera and for the family Heteromyidae are presented in Fig. 2 through 5. For *Dipodomys,* each of the correlation phenograms can be divided into two primary clusters at a correlation of about −0.25 (Figs. 2A and 2C). In males, the upper cluster contains *D. agilis, D. elephantinus, D. venustus, D. deserti, D. microps, D. nelsoni, D. californicus,* and *D. heermanni,* and the remaining species compose the second cluster. Similarly for females, *D. agilis, D. elephantinus, D. venustus, D. microps, D. deserti,* and *D. nelsoni* are in the upper cluster, but *D. spectabilis* also is present, and *D. californicus* and *D. heermanni* are in the lower cluster. The most highly correlated pairs of species for both sexes are *D. elephantinus* with *D. venustus* and *D. insularis* with *D. phillipsii.*

Distance phenograms for *Dipodomys* may be interpreted as containing two clusters representing variation in size among species (Figs. 2B and 2D). For males, *D. deserti, D. ingens,* and *D. spectabilis* compose one cluster and the smaller species the other. Similarly for females, *D. ingens* and *D. spectabilis* make up one cluster and the remaining species make up the other. If the

TABLE 2.—*Continued.*

Character											
8	9	10	11	12	13	14	15	16	17	18	19
16.1	8.4	15.1	5.5	5.5	2.0	1.0	5.2	9.6	15.0	—	4.9
16.0	8.4	14.6	5.5	5.5	1.7	1.1	5.2	9.7	14.8	—	4.8
16.1*	9.1	14.5	5.4	5.3	1.9	1.0	5.5**	9.6	14.8	16.0	4.5*
15.8	9.0	14.3	5.4	5.4	1.9	1.0	5.2	9.5	14.7	15.9	4.3
16.2	9.4	14.5	5.4	5.4	1.5	1.0	4.9	9.5	15.1	—	4.4
16.2	9.3	14.3	5.4	5.5	1.6	1.0	5.0	9.6	15.0	—	4.3
15.7	8.6	14.0	5.3	5.3	2.0	1.0	5.1	10.1	15.6	15.3	4.0
15.8	8.5	14.3	5.3	5.3	1.8	1.0	5.0	10.0	15.4	15.4	4.1
17.6	9.9	16.4	5.9	5.8	2.0	1.3	5.5	10.5	16.2	17.4	4.8
17.1	9.7	15.8	5.9	5.8	2.0	1.2	5.4	10.4	16.1	17.1	4.7
18.2	9.5	16.1	6.5*	6.4	1.7	1.2	6.1	11.1	16.8	18.0	5.0
17.7	9.6	15.7	6.2	6.1	1.8	1.3	6.0	11.0	16.6	17.7	5.1
17.2	9.4	15.9	5.9	5.8	2.0	1.2	5.4	10.5	16.3	16.9	4.8
17.1	9.2	15.9	5.8	5.6	1.9	1.3	5.2	10.6	16.1	16.9	4.8
16.5	7.7	14.7	6.1	5.5	1.9	1.2	5.3	9.9	15.3	16.3	4.0
16.4	7.8	14.3	6.1	5.5	1.9	1.3	5.2	9.9	15.2	16.4	4.0
15.8	8.2	13.8*	6.2	5.5	1.5	1.2	4.9	9.9*	15.8	—	4.0
15.7	8.2	13.2	6.3	5.5	1.4	1.2	4.8	9.7	15.6	—	3.9
14.8	7.6	13.1	5.5	5.1	1.6	1.1	4.7	9.2	14.4	—	3.7
14.6	7.5	12.9	5.5	4.9	1.5	1.0	4.5	9.2	14.3	—	3.7
14.3	6.7	12.8	5.4	5.0	1.8	0.9	4.8	9.3	14.3	14.5	3.5
14.1	6.7	12.4	5.5	5.0	1.7	0.9	4.7	9.2	14.1	14.5	3.5
15.0	8.1	14.2	6.2	5.6	2.2	1.1	5.1	9.4	15.3	15.5	3.9
15.0	8.2	13.8	6.1	5.6	2.0	1.0	5.1	9.4	15.0	15.6	3.9

phenograms are considered to have three clusters, the general size separation for both sexes is still evident. The larger species are in the lower clusters for both sexes, the intermediate-sized species are in the upper cluster, and the generally smaller species are in the middle cluster.

Correlation and distance phenograms for both sexes of *Perognathus* are presented in Fig. 3. Because the correlations among species are low, it seems appropriate not to separate them into clusters (Figs. 3A and 3C). The species having the highest correlation for both sexes are *P. flavus* with *P. parvus*. The distance phenograms can be divided into two clusters (Figs. 3B and 3D). The phenogram for males has *P. alticola* and *P. parvus* in the same cluster separated from the other species (Fig. 3B). For females, *P. parvus* and *P. flavus* each form single-member clusters (Fig. 3D). The most similar species pairs for both sexes are *P. amplus* with *P. inornatus* and *P. fasciatus* with *P. flavescens*.

Also shown in Fig. 3 are correlation and distance phenograms for male and female *Chaetodipus*. In the correlation phenograms, there are two primary clusters (Figs. 3E and 3G). For males, the lower cluster contains *C. artus, C. pernix, C. hispidus, C. goldmani,* and *C. penicillatus*. For females, except for the inclusion of *C. hispidus* in the upper cluster, the same species occur in the lower cluster. The most highly correlated pairs of species for males were *C. baileyi* with *C. fallax* and *C. goldmani* with *C. penicillatus*; for females, they were *C. intermedius* with *C. nelsoni* and *C. goldmani* with

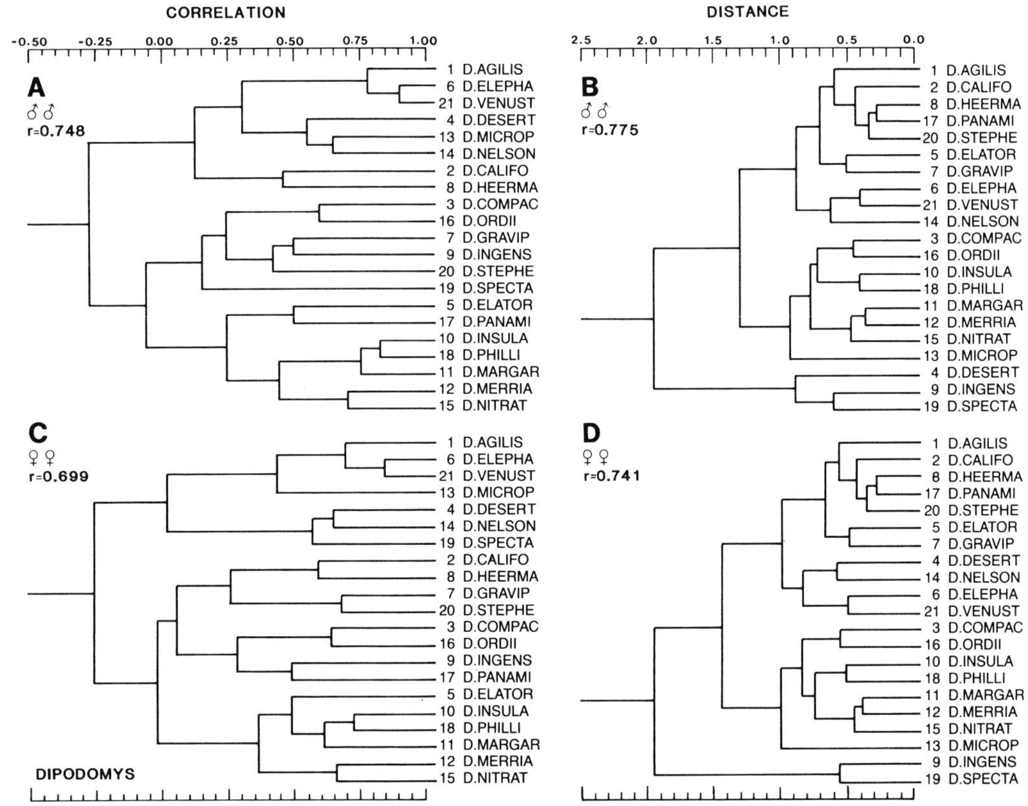

FIG. 2.—Phenograms constructed from correlation and distance matrices for the 21 species of male (A and B, respectively) and female (C and D, respectively) *Dipodomys*. Clusters were obtained using the UPGMA. Accuracy of the diagrams in depicting the relationships increases from left to right. Numerical identifications are the same as in Table 2. The cophenetic correlation coefficients (*r*) are indicated.

C. penicillatus. Distance phenograms for both sexes separated *C. hispidus,* the largest species of this genus, from other taxa (Figs. 3F and 3H). Remaining species are divided into two clusters for each sex. For males, *C. arenarius, C. intermedius, C. lineatus, C. nelsoni, C. penicillatus,* and *C. pernix* are separated from the other species, and for females, *C. baileyi, C. californicus, C. fallax,* and *C. formosus* form a separate cluster. This second cluster appears to be related to differences in body size because smaller species tend to group in the upper cluster for each sex.

Correlation and distance phenograms for both sexes of *Heteromys* are presented in Fig. 4. The correlations among species for both sexes are low (Figs. 4A and 4C). Thus, separation into clusters is not appropriate. The most highly correlated species pair for males is *H. desmarestianus* with *H. goldmani* and for females *H. desmarestianus* with *H. nelsoni.* Distance phenograms for both sexes are similar (Figs. 4B and 4D). One cluster contains *H. anomalus, H. australis, H. gaumeri,* and *H. desmarestianus,* and the second contains the larger-sized taxa, *H. goldmani, H. oresterus,* and *H. nelsoni.* The most similar pairs of species in the distance phenograms for males and females are *H. goldmani* with *H. oresterus* and *H. australis* with *H. gaumeri,* respectively.

For *Liomys,* correlation phenograms show less correlation among species than within *Heteromys* (Figs. 4E and 4G). The occurrence of *L. adspersus* with *L. salvini* is the

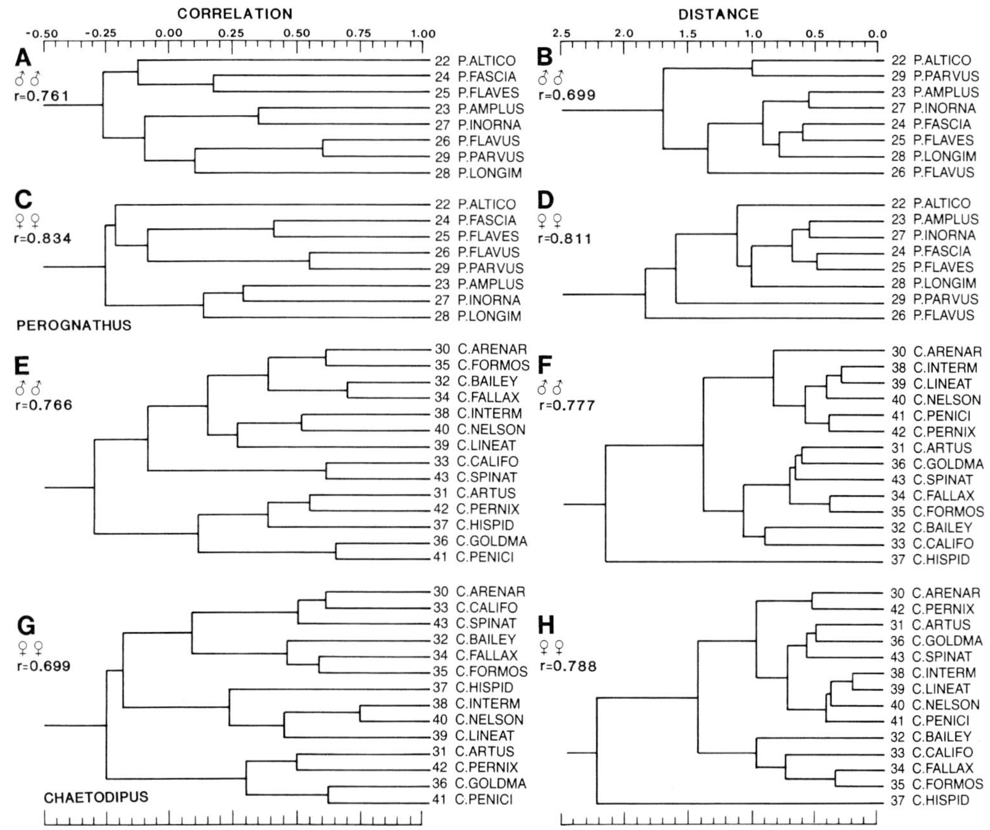

FIG. 3.—Phenograms constructed from correlation and distance matrices for the eight species of male (A and B) and female (C and D) *Perognathus,* and for the 14 species of male (E and F) and female (G and H) *Chaetodipus.* The format is the same as in Fig. 2.

only consistant pairing of species among correlation phenograms for males and females. Distance phenograms are similar between sexes (Figs. 4F and 4H); both have two clusters separated at a distance of about 1.6. For both sexes, *L. adspersus, L. irroratus,* and *L. spectabilis* form one cluster and *L. pictus* and *L. salvini* form the other. *Liomys pictus* and *L. salvini* are the smallest taxa in most measurements (Table 2).

Correlation and distance phenograms for both sexes of the 57 species of Heteromyidae are presented in Fig. 5. Each correlation phenogram can be divided into three clusters at a correlation of about zero (Figs. 5A and 5C). In males, the lower cluster contains *C. hispidus* and all the species of *Heteromys* and *Liomys,* the middle cluster contains *P. parvus, C. baileyi, C. artus, C. goldmani, C. penicillatus, C. pernix, C. californicus, C. spinatus, C. fallax,* and *C. formosus,* and the upper cluster contains *Dipodomys, Microdipodops,* and the remaining species of *Perognathus* and *Chaetodipus.* The most highly correlated species are *M. megacephalus* with *M. pallidus, D. merriami* with *D. nitratoides,* most of the *Heteromys* with each other, *L. adspersus* with *L. salvini,* and *L. irroratus* with *L. pictus.* For females, the same species are in the lower cluster as in the male phenogram, the middle cluster does not have *P. parvus,* and the upper is the same with the addition of *P. parvus.* The most highly correlated species are the same for females as for males.

Distance phenograms also show three major clusters for both sexes that generally

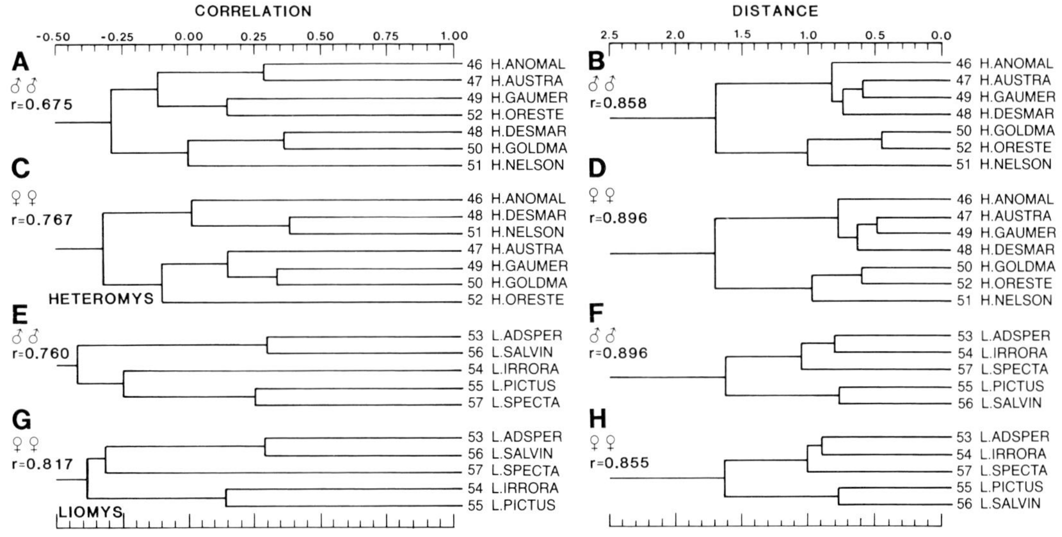

FIG. 4.—Phenograms constructed from correlation and distance matrices for the seven species of male (A and B) and female (C and D) *Heteromys,* and for the five species of male (E and F) and female (G and H) *Liomys.* The format is the same as in Fig. 2.

appear to represent size variation among species (Figs. 5B and 5D). For both sexes, *Dipodomys* are in the upper cluster, *Liomys, Heteromys,* and *C. hispidus* are in the middle cluster, and the remaining *Chaetodipus, Perognathus,* and both species of *Microdipodops* are in the lower cluster. For males, *Microdipodops* are not as close to *Perognathus* and *Chaetodipus* as in the female phenogram. Also, there are some differences in the arrangement of species in each of the primary clusters when male and female phenograms are examined. For example, within the upper cluster for males, *Dipodomys* generally is separated into small, medium, and large body size; for females, it is separated into medium-large, large, and small body size.

Loadings of characters on the first two principal component axes are presented in Table 3, and two-dimensional projections are depicted in Figs. 6 and 7. For *Dipodomys,* character correlations with principal component I for both males and females are >0.73 for all characters except length of ear and maxillary arch width. For both sexes, species 6 (*D. elephantinus*), 14 (*D. nelsoni*), 4 (*D. deserti*), 9 (*D. ingens*), and 19 (*D. spectabilis*) have the highest loadings along component I (Figs. 6A and 6B). This component accounts for about 80% of the phenetic variation (Table 3) and has separated larger species to the right side of the figures. In analyses of morphologic characters, this component may be taken to represent overall variation in size. On principal component II, which accounts for about 7% of the phenetic variation (Table 3), species 13 (*D. microps*) and 5 (*D. elator*) are widely separated for males and for females (Figs. 6A and 6B). The second component has its highest loading on length of ear and maxillary arch width for both sexes (Table 3).

Both sexes of *Perognathus* have loadings >0.70 on principal component I for all characters except length of ear (Table 3). For both sexes, species 26 (*P. flavus*) has the lowest loading along this component and species 29 (*P. parvus*) the highest (Figs. 6C and 6D). This component accounts for 82% of the phenetic variation in males and 77% in females (Table 3). Larger species are to the right side of Figs. 6C and 6D. On the second component, which accounts for 6

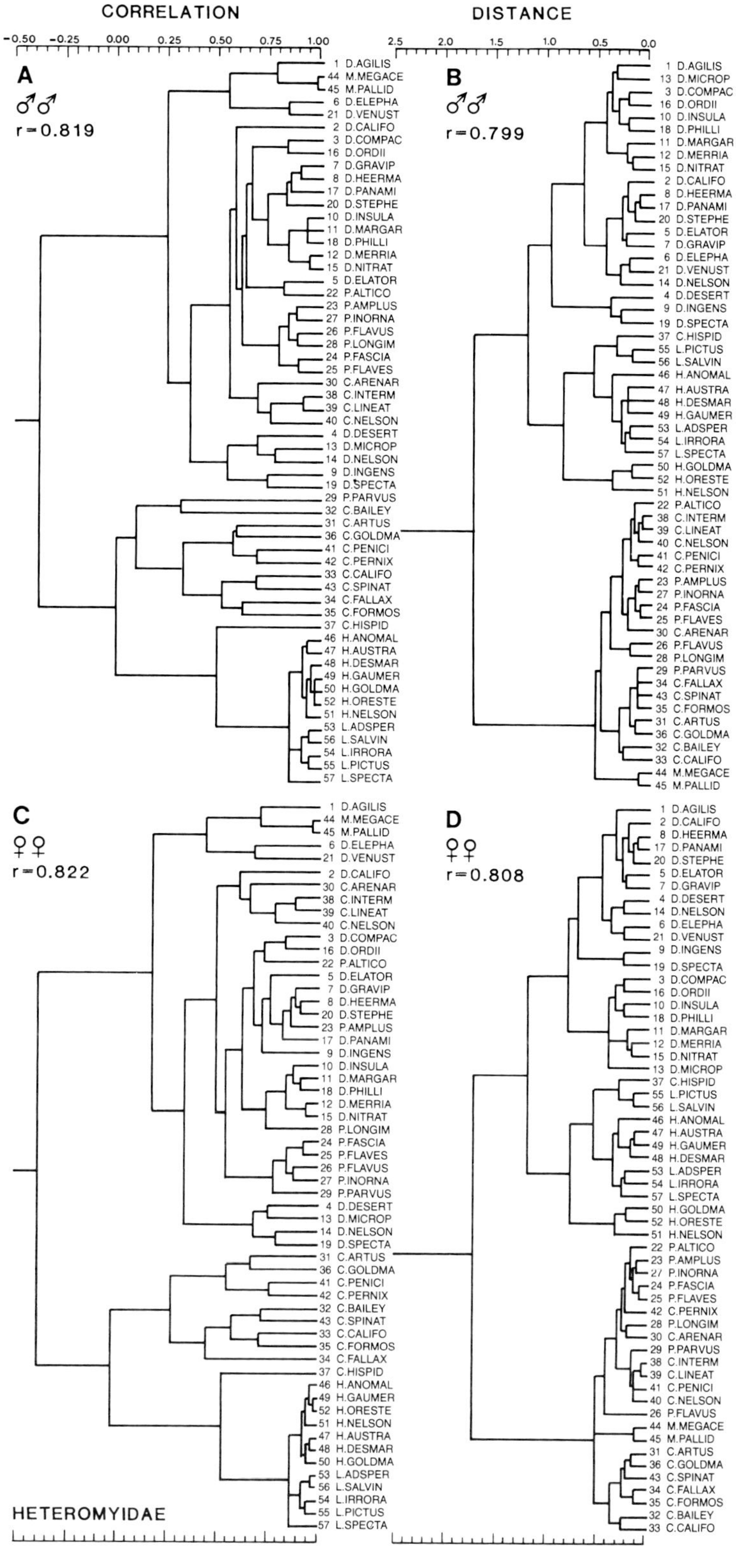

FIG. 5.—Phenograms constructed from correlation and distance matrices for the 57 species of male (A and B) and female (C and D) Heteromyidae. The format is the same as in Fig. 2.

TABLE 3.—*Character loadings*[a] *on the first two principal components of phenetic variation among 19 characters for five genera and the family Heteromyidae.*

Character	Char. no.	Sex	*Dipodomys* I	*Dipodomys* II	*Perognathus* I	*Perognathus* II	*Chaetodipus* I	*Chaetodipus* II	*Heteromys* I	*Heteromys* II	*Liomys* I	*Liomys* II	Heteromyidae I	Heteromyidae II
External														
Total length	1	♂♂	0.952	−0.157	0.994	−0.038	0.910	−0.299	0.986	−0.031	0.871	−0.436	0.958	−0.250
		♀♀	0.942	−0.201	0.958	−0.075	0.900	−0.345	0.961	−0.134	0.968	−0.013	0.965	−0.211
Length of body	2	♂♂	0.945	−0.109	0.914	0.167	0.930	0.154	0.955	−0.005	0.574	0.816	0.885	−0.433
		♀♀	0.962	−0.120	0.853	0.306	0.923	0.266	0.979	0.011	0.515	0.815	0.913	−0.376
Length of tail	3	♂♂	0.860	−0.172	0.945	−0.145	0.682	−0.646	0.964	−0.049	0.635	−0.680	0.965	−0.109
		♀♀	0.855	−0.233	0.866	−0.229	0.666	−0.687	0.933	−0.201	0.871	0.104	0.961	−0.088
Length of hind foot	4	♂♂	0.959	−0.119	0.956	−0.060	0.851	−0.395	0.937	0.021	0.939	−0.028	0.986	0.091
		♀♀	0.942	−0.183	0.914	−0.269	0.777	−0.532	0.922	0.086	0.928	−0.103	0.984	0.088
Length of ear	5	♂♂	0.613	−0.562	0.618	−0.544	0.732	−0.176	0.879	−0.190	0.802	−0.777	0.832	−0.462
		♀♀	0.601	−0.526	0.511	0.526	0.774	−0.220	0.816	0.476	0.743	−0.803	0.844	−0.443
Cranium														
Basal length	6	♂♂	0.984	−0.033	0.986	0.006	0.973	0.204	0.986	−0.104	0.935	0.280	0.962	−0.254
		♀♀	0.985	−0.003	0.988	0.055	0.975	0.197	0.979	0.043	0.946	0.232	0.963	−0.250
Greatest length	7	♂♂	0.984	−0.102	0.996	−0.037	0.996	0.070	0.993	−0.036	0.932	0.194	0.994	−0.043
		♀♀	0.984	−0.043	0.992	0.052	0.994	0.095	0.983	−0.077	0.961	0.011	0.993	−0.041
Maxillary arch spread	8	♂♂	0.961	0.235	0.962	−0.112	0.946	0.269	0.987	0.084	0.878	0.433	0.970	0.222
		♀♀	0.943	0.298	0.944	−0.138	0.953	0.243	0.988	−0.063	0.919	0.355	0.972	0.212
Interorbital width	9	♂♂	0.739	0.348	0.920	0.175	0.897	0.013	0.674	0.502	0.821	−0.301	0.960	0.182
		♀♀	0.732	0.345	0.859	−0.369	0.926	0.054	0.727	−0.349	0.843	−0.181	0.964	0.169
Nasal length	10	♂♂	0.945	0.057	0.974	0.043	0.985	0.014	0.917	−0.136	0.942	−0.011	0.960	−0.238
		♀♀	0.932	0.128	0.978	−0.132	0.986	−0.003	0.942	−0.223	0.912	−0.105	0.962	−0.228
Intermaxillary width	11	♂♂	0.949	0.190	0.889	0.355	0.913	0.134	0.971	0.077	0.946	−0.210	0.947	0.278
		♀♀	0.951	0.196	0.918	−0.126	0.916	0.154	0.990	0.050	0.918	−0.100	0.951	0.266
Alveolar length	12	♂♂	0.939	0.096	0.910	0.301	0.974	0.126	0.964	0.066	0.968	−0.182	0.914	−0.361
		♀♀	0.947	−0.017	0.898	−0.246	0.960	0.173	0.940	0.214	0.889	−0.305	0.906	−0.380
Lacrimal length	13	♂♂	0.898	0.261	0.752	−0.491	0.873	−0.136	0.017	−0.983	0.293	−0.667	0.851	0.489
		♀♀	0.879	0.263	0.703	−0.120	0.906	−0.139	0.499	−0.662	0.341	−0.773	0.850	0.489
Maxillary arch width	14	♂♂	0.435	0.781	0.824	0.104	0.437	−0.828	0.938	−0.103	0.909	0.247	0.779	0.601
		♀♀	0.395	0.784	0.803	−0.412	0.676	−0.652	0.963	0.059	0.914	0.384	0.786	0.591
Basioccipital length	15	♂♂	0.931	−0.212	0.874	0.019	0.913	0.291	0.831	−0.139	0.828	0.014	0.970	−0.069
		♀♀	0.928	−0.174	0.790	0.508	0.895	0.298	0.812	0.242	0.829	−0.243	0.967	−0.072

TABLE 3.—*Continued.*

Character	Char. no.	Sex	*Dipodomys* I	*Dipodomys* II	*Perognathus* I	*Perognathus* II	*Chaetodipus* I	*Chaetodipus* II	*Heteromys* I	*Heteromys* II	*Liomys* I	*Liomys* II	Heteromyidae I	Heteromyidae II
Greatest depth	16	♂♂	0.964	−0.166	0.903	0.166	0.975	0.092	0.909	−0.036	0.797	0.594	0.942	0.300
		♀♀	0.971	−0.136	0.946	0.219	0.975	0.173	0.945	−0.107	0.899	0.380	0.949	0.281
Greatest width	17	♂♂	0.926	−0.122	0.804	−0.412	0.966	0.023	0.911	−0.004	0.935	0.025	0.876	0.448
		♀♀	0.924	−0.075	0.752	0.479	0.972	0.014	0.917	−0.239	0.936	0.090	0.880	0.439
Zygomatic width	18	♂♂	0.967	0.054	0.934	−0.008	0.977	0.041	0.990	0.364	0.994	0.149	0.986	0.111
		♀♀	0.963	0.047	0.893	0.409	0.990	0.037	0.997	0.143	1.000	0.016	0.988	0.088
Nasal width	19	♂♂	0.922	0.030	0.940	0.232	0.733	0.417	0.820	−0.197	0.988	0.117	0.877	−0.441
		♀♀	0.907	−0.011	0.918	−0.136	0.760	0.415	0.856	0.469	0.983	−0.080	0.861	−0.476
Total[b]		♂♂	80.79	7.38	81.77	5.88	78.79	9.71	81.28	7.94	73.66	17.31	86.27	10.52
		♀♀	79.81	7.50	76.57	8.74	80.41	9.96	82.87	7.01	76.53	13.79	86.71	10.02

[a] Correlations of species mean values of individual characters with the component axes.
[b] Percent of total phenetic variance explained.

and 9% of the phenetic variation for males and females, respectively, the highest character loadings are for length of ear and greatest width of cranium for both sexes, plus lacrimal length for males and basioccipital length for females. Species 27 (*P. inornatus*), 23 (*P. amplus*), and 22 (*P. alticola*) are the most widely separated along this component for males (Fig. 6C), and species 28 (*P. longimembris*) and 24 (*P. fasciatus*) are the most divergent for females (Fig. 6D).

For *Chaetodipus,* loadings on principal component I are >0.70 for all characters, except length of tail and maxillary arch width for both sexes (Table 3). For both sexes the smallest taxon, species 30 (*C. arenarius*), has the lowest loading on this component and species 35 (*C. formosus*), 33 (*C. californicus*), 32 (*C. baileyi*), and 37 (*C. hispidus*) have the highest loadings (Figs. 6E and 6F). The largest species is *C. hispidus* (species 37). This component accounts for about 80% of the variation for both sexes. The second principal component, which represents about 10% of the variation in both sexes, has loadings $> \pm 0.50$ for length of tail and maxillary arch width for both sexes (Table 3). In addition, females have a high loading on length of hind foot for this component. Species 37 (*C. hispidus*) is the most widely separated taxon along this component for both sexes (Figs. 6E and 6F).

Heteromys has high loadings for all characters on principal component I, except lacrimal length (Table 3). About 82% of the variance is explained by this component. For both sexes, species 47 (*H. australis*) and 49 (*H. gaumeri*), which are the smallest taxa in this genus, have lowest loadings on this component. Species 52 (*H. oresterus*), 51 (*H. nelsoni*), and 50 (*H. goldmani*) have highest loadings (Figs. 7A and 7B). The second principal component, representing about 7% of the variation, has loadings $> \pm 0.50$ for lacrimal length for both sexes, and interorbital width for males (Table 3). Species 48 (*H. desmarestianus*) for males (Fig. 7A) and species 46 (*H. anomalus*) and 50 (*H. goldmani*) for females had the most divergent scores on this component.

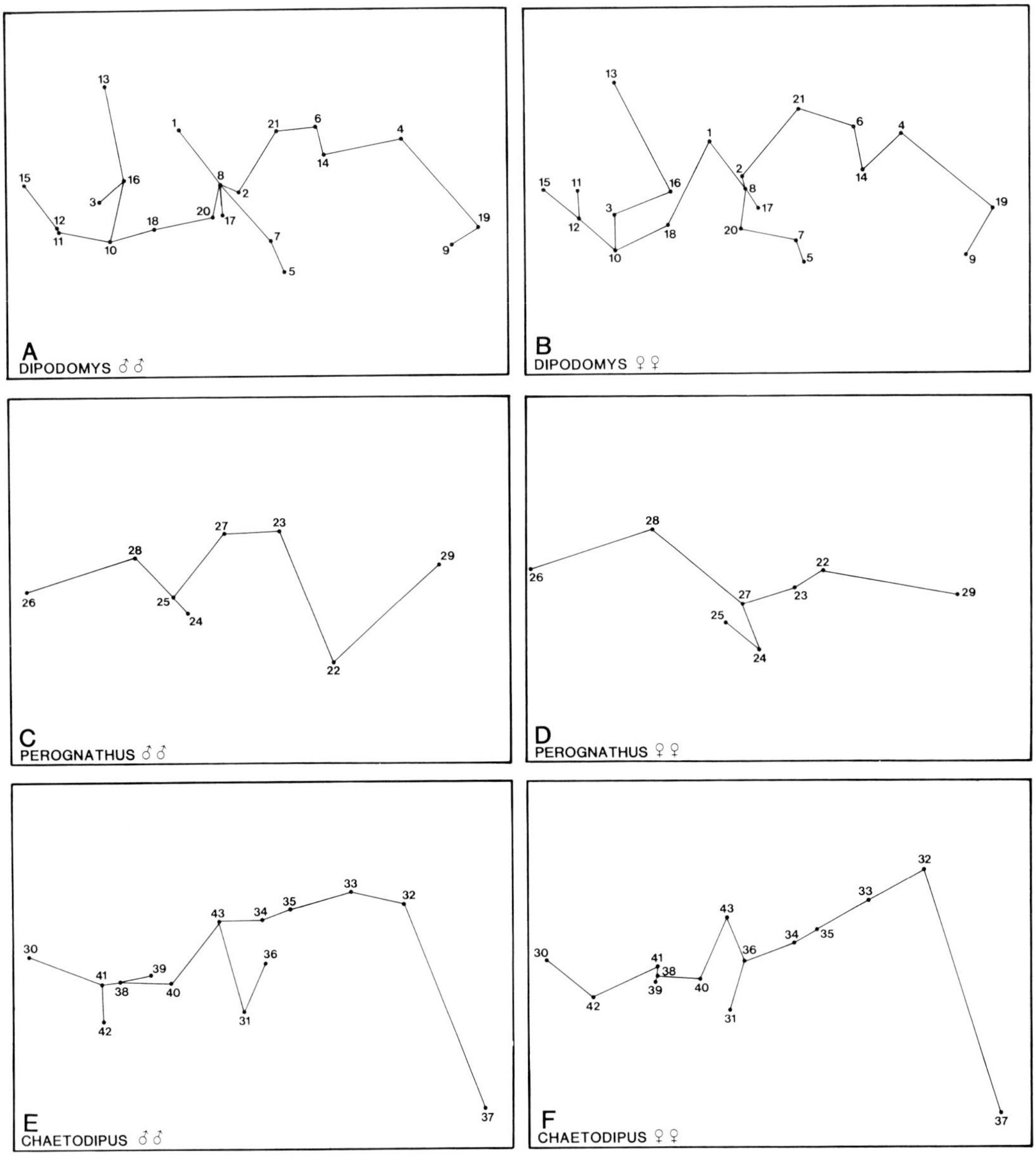

FIG. 6.—Projections of species onto the first two principal component axes of variation in the matrix of correlations of 19 morphologic characters for male (A) and female (B) *Dipodomys,* male (C) and female (D) *Perognathus,* and male (E) and female (F) *Chaetodipus.* The range of values along principal component I (horizonal axis) is from −1.5 to 2.0 (left to right), and for principal component II (vertical axis) from 1.0 to −1.0 (front to back). The shortest simply-connected networks, derived from the matrix of distance coefficients for the same character, are drawn between the species. Numbers correspond to the species listed in Table 2.

In *Liomys,* all characters except length of body and lacrimal length for both sexes and length of tail for males had loadings >0.74 on principal component I (Table 3), which accounted for about 75% of the variation. Species 56 (*L. salvini*) and 55 (*L. pictus*) had the least loadings on this component. These two species are smaller than the others. The

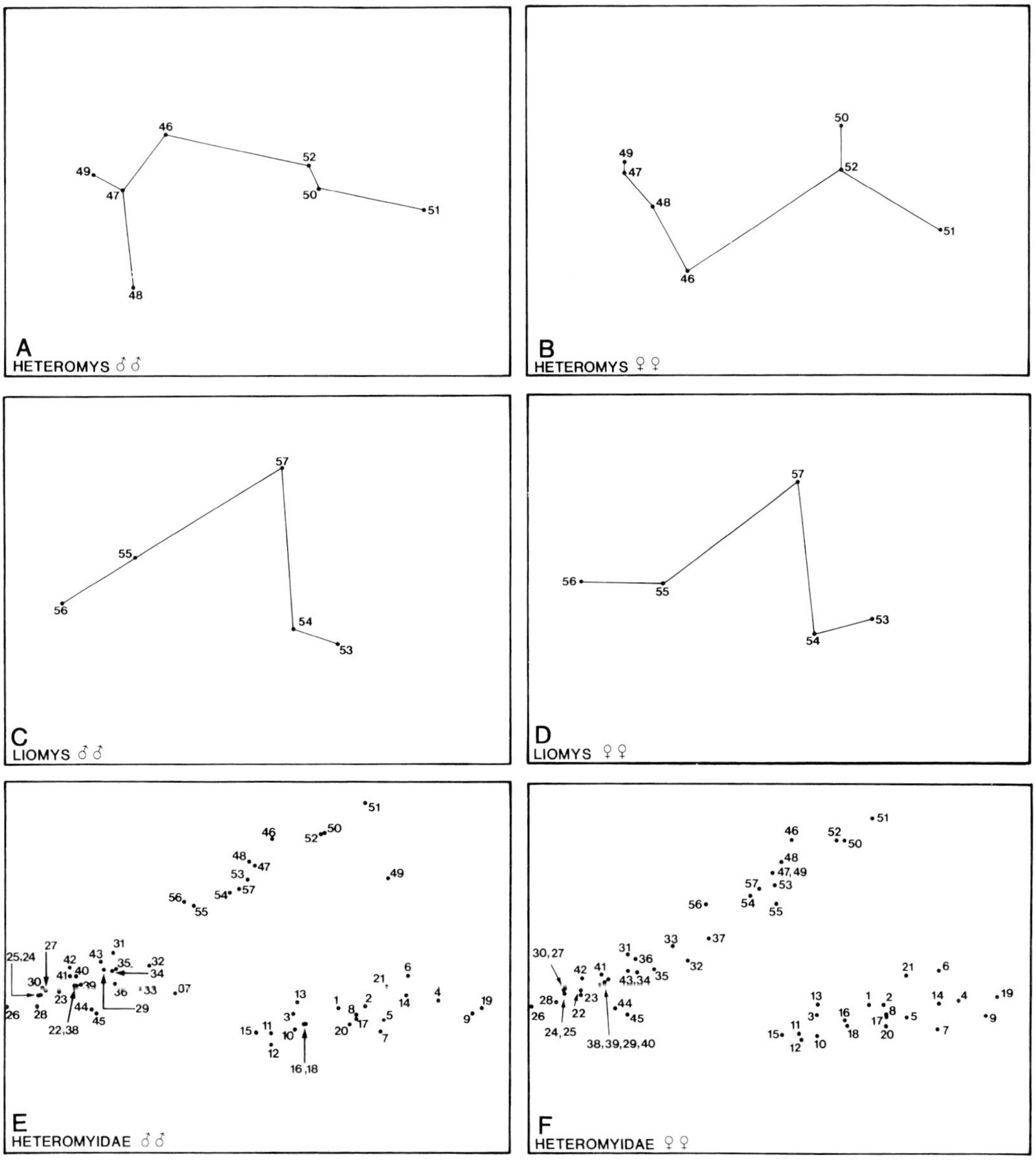

FIG. 7.—Projections of species onto the first two principal component axes of variation in the matrix of correlations for male (A) and female (B) *Heteromys,* male (C) and female (D) *Liomys,* and for 57 species of male (E) and female (F) Heteromyidae. The format is the same as in Fig. 6, except that the shortest simply-connected network for the Heteromyidae plots are listed below to simplify Figs. 7E and 7F. For the Heteromyidae (E and F) the networks are as follows: males, 1-8, 8-17, 8-20, 8-2, 8-7, 7-5, 1-18, 18-10, 10-11, 11-12, 12-15, 10-16, 16-3, 16-13, 2-21, 21-6, 6-14, 14-4, 4-9, 9-19, 13-49, 49-48, 49-47, 49-54, 54-53, 53-57, 47-46, 54-55, 55-56, 56-37, 37-32, 32-33, 33-35, 35-34, 34-29, 34-43, 29-40, 40-39, 39-38, 38-41, 41-42, 39-22, 38-23, 23-27, 27-24, 24-25, 34-31, 31-36, 25-28, 27-30, 28-26, 29-44, 44-45, 46-52, 52-50, and 50-51; females, 1-2, 2-8, 8-17, 8-20, 8-7, 17-5, 1-18, 18-10, 10-12, 12-15, 12-11, 10-3, 3-16, 2-21, 21-6, 6-14, 14-4, 16-13, 4-9, 9-19, 13-53, 53-54, 53-47, 47-49, 49-48, 54-57, 48-46, 54-55, 55-56, 56-37, 46-52, 52-50, 50-51, 37-32, 32-33, 33-35, 35-34, 34-43, 43-36, 36-31, 43-40, 40-39, 39-38, 39-41, 40-29, 39-22, 22-23, 23-27, 27-25, 25-24, 41-42, 27-28, 28-30, 28-26, 29-44, and 44-45.

second component represented 17% of the variation for males and 14% for females. This is about twice the amount of variability represented on principal component II by any of the other genera. Characters with loadings $> \pm 0.50$ on this component were length of body, length of ear, and lacrimal length for both sexes, and length of tail and greatest depth of cranium for males (Table 3). Species 57 (*L. spectabilis*) had the greatest divergence on this component for both sexes (Figs. 7C and 7D).

When the entire family Heteromyidae was examined, there were high loadings on all characters relative to principal component I (Table 3). This component accounted for about 86% of the variation. As in the analyses by genus, this component may be taken to represent overall size in both sexes because it accounts for most of the covariation among characters. For both sexes, species 24 (*P. fasciatus*), 25 (*P. flavescens*), 26 (*P. flavus*), 27 (*P. inornatus*), 28 (*P. longimembris*), and 30 (*C. arenarius*) had the lowest loadings along this component (Figs. 7E and 7F). These are among the smallest heteromyids (Table 2). Highest loadings along component I (Table 3) were for species 4 (*D. deserti*), 9 (*D. ingens*), and 19 (*D. spectabilis*). These are among the largest species (Table 2). Along principal component II, 10% of the variance is explained. The only character with a loading >0.50 was maxillary arch width for both sexes (Table 3). Length of ear, lacrimal length, greatest width of cranium, and nasal width had loadings that approached ± 0.50 for both sexes. This component tends to separate *Dipodomys* from most other taxa for both sexes (Figs. 7E and 7F). In addition, along component II *Heteromys* and *Liomys* are well separated from *Dipodomys, Perognathus,* and *Chaetodipus.* There is some separation of *Microdipodops* (species 44 and 45) from the *Perognathus–Chaetodipus* grouping along principal component II for both males and females.

Discussion

Sexual Dimorphism

Previous studies of sexual dimorphism in *Dipodomys* have included data for *D. agilis* (Best, 1978, 1983*a*), *D. californicus* (Dale, 1939; Dunmire, 1955); *D. compactus* (Baumgardner and Schmidly, 1981; Schmidly and Hendricks, 1976), *D. deserti* (Hall, 1946; Nader, 1978), *D. elator* (Best, 1987; Webster and Jones, 1985), *D. gravipes* (Best, 1978, 1983*b*), *D. merriami* (Hall, 1946; Lidicker, 1960*a*), *D. microps* (Csuti, 1979; Hall and Dale, 1939), *D. ordii* (Baumgardner and Schmidly, 1981; Desha, 1967; Hall, 1946; Kennedy and Schnell, 1978; Schmidly, 1971; Schmidly and Hendricks, 1976; Setzer, 1949), *D. panamintinus* (Hall, 1946), *D. phillipsii* (Genoways and Jones, 1971), and *D. spectabilis* (Nader, 1978). Results presented here generally agree with these earlier studies, but there are differences that should be pointed out. These differences may reflect the effect of sample size, selection of specimens, or the characters examined. Hall (1946) found slight secondary sexual variation in *D. merriami* from Nevada, but Lidicker (1960*a*) observed considerable secondary sexual variation in specimens of *D. merriami*; 17 of 19 characters examined herein show statistically significant differences (14 of which are highly significantly different between sexes). Genoways and Jones (1971) found that males and females of *D. phillipsii* were significantly different in total length and length of tail, but not in other measurements. In the present analyses, five of 19 characters of this species showed sexual dimorphism. Nader (1978) noted that no significant differences existed between sexes of *D. spectabilis,* but the present analyses clearly show it to be among the most sexually dimorphic species of *Dipodomys.* For *D. deserti,* Nader (1978) examined 18 characters and found length of tail, basal length of cranium, greatest length of cranium, breadth of maxillary arches, least interorbital breadth, greatest breadth of bullae, rostral depth, breadth of exoccipitals, and mandibular length to be significantly larger for males. Herein, 17 of 19 characters were different between sexes.

Best (1983*a*) pointed out that because of the small absolute differences between the means of measurements for male and female *D. agilis,* a large sample probably was necessary to detect the dimorphism. Later, Best (1987) found little difference in the number of sexually dimorphic characters when sample sizes of *D. elator* were reduced to make uniform comparisons among the three populations studied. However, the effect of sample size on the detection of sexually dimorphic characters of *Dipodomys* is considerable. When the number of individuals was reduced to only 20 males and 20 females of each species, the number of sexually dimorphic characters was reduced dramatically. Since the samples used by Best (1987) were 29 males and 20 females, it appears that samples should contain >20 individuals per sex to most accurately determine sexual differences.

The effect of having more than one species in a sample also has been shown to influence the degree of sexual differences that were detected in *Dipodomys.* When Best (1978) evaluated sexual dimorphism in the *heermanni* group of kangaroo rats he found only five of 19 characters exhibited significant secondary sexual dimorphism. This was likely the result of including specimens of both *D. gravipes* and *D. agilis* in the same analyses, which would increase the variance within sexes. The small number of dimorphic characters also could have been due to the relatively small sample examined. When he analyzed *D. gravipes* separately (Best, 1983*b*), there were seven characters that exhibited significant secondary sexual dimorphism, and when he analyzed *D. agilis* separately (Best, 1983*a*), 17 of 19 characters were significantly dimorphic.

Sexual dimorphism also has been examined in *Perognathus.* Hall (1946) did not find significant sexual differences in *P. longimembris,* but noted that males were larger than females for *P. parvus.* Baker (1954) did not detect sexual dimorphism in his examination of *P. flavus.* For *P. flavescens,* Williams (1978) and Reed and Choate (1986) found a few significant differences between sexes, but thought they could be attributed to sampling errors, and Williams and Genoways (1979) found females from one of their geographic groups had significantly greater lengths of interparietals than males, but no other sexual differences were detected. The difference in several characters between sexes of *P. alticola, P. amplus,* and *P. parvus* shown here, coupled with data on other species, indicates a large variation in sexual dimorphism among species in this genus.

Hall (1946) found no significant sexual differences in *C. formosus,* nor did Glass (1947) for *C. hispidus.* No differences were found in this study as well. Differences between sexes were found in *C. artus* and *C. goldmani* in the present study, and by Anderson (1964) who grouped sexes for analyses. Hall (1946) and Hoffmeister and Lee (1967) found males usually averaged larger than females of *C. penicillatus,* but only length of hind foot differed significantly herein. Wilkins and Schmidly (1979) found significant differences in external and cranial dimensions between males and females of *C. intermedius, C. nelsoni,* and *C. penicillatus.* Straney and Patton (1980) found that sexual differences within localities were of about the same magnitude as were differences between localities within races of *C. goldmani.* They considered sexes separately in analyzing geographic trends, and combined sexes to examine relationships between morphologic, environmental, and lineage trends. Weckerly and Best (1985) found varying degrees of differences between sexes of *C. intermedius* in southern New Mexico.

Hall (1941, 1946) reported a lack of significant sexual dimorphism in both species of *Microdipodops*, and Schitoskey (1968) found one of 14 measurements (zygomatic breadth) differed significantly between sexes of *M. megacephalus*. There were no sexual differences shown in the present analyses.

Sexual differences in *H. gaumeri* were addressed by Engstrom et al. (1987). Using age classes described by Genoways (1973), they compared sexes and found one character differed between sexes in their youngest age class and seven of 14 measurements differed in age class III. However, they concluded that only a minor component of their total variance was attributable to sex. Only *H. australis* and *H. nelsoni* were sexually dimorphic in four or more characters in the present study.

Genoways (1973) determined that males were larger than females in approximately one-half of the 13 measurements tested for *L. adspersus, L. irroratus, L. pictus*, and *L. salvini*. For most of the other means, where no significant differences were found, males averaged larger than females. Only three species of *Liomys* examined herein had dimorphic characteristics, and none to the degree observed by Genoways (1973).

Geographic variation in sexual dimorphism is a factor that could significantly affect the detection of sexual differences in heteromyids. The selection of specimens for analyses of sexual dimorphism appears to be important. Except for *Dipodomys*, which were selected more or less randomly from large data matrices, those animals examined during the present study were measured as encountered, and because of the limited numbers of specimens available for study no attempt was made to insure that specimens were from the same geographic area. Had specimens been selected from one locality for each species, it is likely that a different number of sexually dimorphic traits would have been found among the 57 species.

Schmidly (1971), in his study of *D. ordii*, indicated that sexual dimorphism varied geographically and suggested that the variability may result from genetic and hormonal sex differences or may, to some extent, be due to nongenetic modification of the phenotype caused by local environmental conditions. In his study, females were larger than males in northern Texas, whereas the reverse was true in the southern samples. Baumgardner and Schmidly (1981) noted that sexual differences varied among samples of *D. ordii*. Best (1987) found significant geographic variation in sexual dimorphism across the range of *D. elator* in northcentral Texas.

Geographic variation in sexual dimorphism has been found in *C. intermedius* from three lava fields in New Mexico by Weckerly and Best (1985). They speculated that the varying degree of sexual differences may be related to age of the lava fields or selective pressure of climate, vegetative cover, food, physiology, reproduction, competition, etc. The three samples of *C. penicillatus* examined by Hoffmeister and Lee (1967) and those of *P. flavescens* analyzed by Williams (1978) also exhibited differences in sexual dimorphism among localities. It seems clear that an accurate assessment of sexual dimorphism within a species cannot be achieved without an examination of differences at the population level.

Patterns of Variation

Phenetic analyses have been used to assess relationships among and within species of *Dipodomys* (Baumgardner, 1989; Baumgardner and Schmidly, 1981; Best, 1978, 1983*a*, 1983*b*, 1987; Best and Janecek, 1992; Best and Schnell, 1974; Best et al., 1986; Brownlee, 1973; Genoways and Jones, 1971; Kennedy and Schnell, 1978; Schmidly and Hendricks, 1976; Schnell et al., 1978), *Perognathus* (Williams, 1978; Williams and Genoways, 1979), *Chaetodipus* (Caire, 1976; Straney and Patton, 1980), *Microdipodops* (Hafner, 1978, 1981), *Liomys* (Genoways, 1973), and *Heteromys* (Engstrom et al., 1987; Genoways, 1973; Rogers and Schmidly, 1982). Other than the comparisons of *Liomys* with *Heteromys* (Genoways, 1973), only Hafner (1978) has used phenetic analyses of morphologic traits to elucidate relationships among genera of heteromyids. Separation of individuals or populations of closely related species of *Dipodomys* (Best, 1978, 1981*a*) and *Microdipodops* (Hafner, 1981) also have been accomplished using phenetic analyses.

Analyses presented in this chapter are the first to investigate morphologic variation among 57 species of Heteromyidae. Analyses of species within the genus *Dipodomys* produced dendrograms that generally grouped taxa on the basis of size. A comparison with the dendrograms in Schnell et al. (1978) indicates several similarities with the present classification. Schnell et al. (1978) grouped sexes and used 41 characters (including all characters examined here, except length of body). The large species, *D. ingens, D. spectabilis,* and *D. deserti* were well separated from the others in their distance phenogram, as well as in the distance phenogram for males presented in this chapter. For females, only *D. ingens* and *D. spectabilis* were well separated, and *D. deserti* was placed into a cluster adjacent to another large species, *D. nelsoni.* Other similarities between this study and Schnell et al. (1978) were the close associations of *D. compactus* with *D. ordii, D. agilis* with *D. heermanni, D. stephensi, D. panamintinus,* and *D. californicus, D. elator* with *D. gravipes, D. insularis* with *D. phillipsii,* and *D. merriami* with *D. nitratoides.* When Schnell et al. (1978) divided their mean measurements by unstandardized principal component I, the phenetic relationships changed considerably. However, their resulting distance phenogram did not have clusters that were any better defined than before the effect of size had been reduced. An examination of that phenogram indicates there is still a tendency for species to group by size.

Grinnell (1921, 1922) grouped *Dipodomys* into species groups that indicated their closest relatives. Burt (1936), Davis (1942), Setzer (1949), and Lidicker (1960*a*) revised these species groupings, and subsequently other authors have referred to them in assessing patterns of variation in karyotypes (Stock, 1974), proteins (Hamilton et al., 1987; Johnson and Selander, 1971), bacula (Best and Schnell, 1974), and skeletal morphology (Schnell et al., 1978). Species in Lidicker's (1960*a*) *heermanni* group are clustered relatively closely in the distance phenograms and the plots of the first two principal components presented here. The occurrence of *D. elator* within the *heermanni* group is interesting since there is a general lack of knowledge relative to its affinities with other *Dipodomys.* In the present analyses, *D. ingens,* considered a member of the *heermanni* group, is morphologically most similar to *D. spectabilis.* Thus, the two largest species have been grouped together; usually near *D. deserti.* The occurrence of *D. ordii* (with *D. compactus*) and *D. microps* somewhat separately from other species, and the presence of *D. merriami, D. nitratoides,* and *D. margaritae* in the same group, likewise, are similar to Lidicker's groupings. The close affinity of *D. insularis* and *D. phillipsii* shown here is not shown in previous groupings, nor is the placement of *D. ingens* with *D. spectabilis,* nor is the placement of *D. nelsoni* in the *heermanni* group. The large and morphologically similar species *D. nelsoni* and *D. spectabilis* were expected to group more closely. Analyses by Baker (1956), Anderson (1972), Matson (1980), and others (see Nader, 1978) separated these species, although Nader (1978) considered them conspecific. The analyses here indicate they are more similar to other species than to each other.

Anderson (1972) used ratio diagrams for comparing morphologic differences among *P. merriami, P. apache,* and *P. flavus,* and Wilson (1973) used discriminant function analyses to compare *P. merriami* and *P. flavus.* Williams (1978) assessed patterns of phenetic variation in *P. flavescens,* and Williams and Genoways (1979) similarly studied *P. fasciatus.* Hafner (1978) and Hafner (1982) used *Perognathus* in their assessments of relationships among the geomyoid rodents. However, the present study is the first to assess morphologic variation for the eight species of this genus simultaneously. Differences between results of analyses of males and females are in part due to sexual differences within species. However, the primary separation of the species is on the basis of size. The smallest species, *P. flavus,* as well as the largest, *P. parvus,* generally are separated from the others. The morphologically most similar species (*P. amplus* with *P. inornatus* and *P. fasciatus* with *P. flavescens*) group together in the distance phenograms and the principal component plot for males, but are not as clearly aligned on the principal component plot for females. Morphologically, this genus exhibits considerable variation in overall body size, although no species represents a large divergence from the others.

Studies by Hafner (1978) and Hafner and Hafner (1983) included *Chaetodipus* in their assessments of geomyoid phylogenies. The most striking attribute of the analyses of *Chaetodipus* herein was the large difference in body size of *C. hispidus* from the other species; a difference clearly shown in the ratio diagram of Anderson (1972). An examination of bacular characters, such as overall size and configuration of the bacular tip, also shows a similar separation of *C. hispidus* from the other taxa (see Burt, 1960). Within the remaining 13 species, the greatest degrees of consistency in analyses are the close affinities of *C. intermedius* with *C. lineatus* and *C. fallax* with *C. formosus.* The coupling of *C. artus* and *C. goldmani* reflects the similarity studied by Hall and Ogilvie (1960), Anderson (1964), and Patton (1967). Other groupings are more variable, although larger body size generally differentiates *C. baileyi* and *C. californicus* from the other taxa.

Caire (1976) examined phenetic relationships among species of *Chaetodipus,* and concluded that his numerical analyses supported previous species groupings and demonstrated new groupings that were worthy of future consideration. When the results obtained by Caire (1976) are compared with those presented herein, some differences are apparent. In Caire's analysis, *C. hispidus* couples with *C. baileyi,* and *C. formosus, C. californicus, C. pernix, C. artus,* and *C. goldmani* are each well separated from the remaining species. In the analyses presented herein, *C. hispidus* is well separated from the other taxa, and those that remain form two clusters (the clusters differ between sexes). Since the same statistical treatment was used, it is suspected that differences in characters, number of characters, sample sizes, and treatments of sexual differences are responsible for the dissimilarity between the studies. Data on sample size, characters, and sexual differences were not presented by Caire (1976).

Genoways (1973) used *Heteromys* in phenetic comparisons with *Liomys.* Rogers and Schmidly (1982) examined phenetic variation of *Heteromys* in the *desmarestianus* species group, and concluded that this group is represented in northern Middle America by two species, *H. desmarestianus* and *H. goldmani.* Herein, the genus formed two distinct groups in the assessment of patterns of morphologic variation. The larger species, *H. goldmani, H. oresterus,* and *H. nelsoni,* formed a group well separated from the remaining four; a phenetic classification at variance with Rogers and Schmidly (1982).

The only phenetic assessment of relationships among the *Liomys* has been that of Genoways (1973). He found *L. pictus* and *L. spectabilis* were the most similar species pairs; they were most closely associated with *L. irroratus,* and *L. salvini* and *L. adspersus* formed a more distantly related pair of species. In the present analyses, *L. pictus* and *L. salvini* were the most similar species and they grouped separately from *L. adspersus, L. irroratus,* and *L. spectabilis.* The presence of *Heteromys* in Genoways' distance analysis could have affected the placement of the *Liomys* taxa. The inclusion of additional species in phenetic analyses has been shown to affect the relationships resulting from cluster analyses in heteromyids (Best, 1978, 1981*a*). Considering the differences in body size between *L. spectabilis* and *L. pictus* (Genoways, 1971), separation of these species was expected in the phenetic analyses herein. Along the second principal component axis, *L. spectabilis* is quite divergent from the other taxa.

Hafner (1978) examined the relationship of *Microdipodops* to *Perognathus, Chaetodipus,* and *Dipodomys,* and found *Microdipodops* to be phenetically most similar to *Perognathus.* Phenetic analyses in this chapter also show a close relationship between *Microdipodops* and the *Perognathus-Chaetodipus* cluster. Most recently, Hafner and Hafner (1983) summarized the evolutionary relationships of heteromyid rodents. Using a variety of data they concluded that extant heteromyids comprise three main lineages, *Chaetodipus-Perognathus, Dipodomys-Microdipodops,* and *Liomys-Heteromys.* Results presented here differ somewhat from their evolutionary scheme. *Dipodomys* is phenetically most similar to *Liomys* and *Heteromys* (including *C. hispidus*), *Perognathus* and *Chaetodipus* are intermixed, and *Microdipodops* shows closest affinities to the *Perognathus-Chaetodipus* cluster. As mentioned previously, the phenetic analyses are heavily biased by size, which likely is not a good indicator of evolutionary relationships. Overall, *Dipodomys* shows more intrageneric variation, for both males and females, than do the four other genera examined.

Pelage and Coloration

Homan and Genoways (1978) analyzed hair structure and its phylogenetic implications among heteromyids. They used both light and scanning electron microscopy and investigated variables such as length and width of hair, imbricate scale pattern, and medullary characteristics. Although the hair of individual species could be characterized with detailed study, they did not believe that hair structure would be of value in evolutionary studies of this group below the generic level. They found that the overhair of heteromyids falls into two morphologic types, that is, hair that is round to oval in outline and hair that has a trough along the dorsal surface. Hair of the first type is found in most perognathines and members of the genera *Dipodomys* and *Microdipodops.* Troughed hairs were found in chaetodipines, *Liomys, Heteromys, P. amplus,* and *P. formosus.*

Odd pelage colors and characteristics have received some attention (e.g., Blair, 1940; Howell, 1923; von Bloeker, 1930), but coloration of desert species has received much attention because of the variability in color among populations occupying habitats with unusually colored substrates (e.g., Benson, 1933; Dice and Blossom, 1937; Sumner, 1921; Sumner and Swarth, 1924). Cloudsley-Thompson (1979) reviewed accounts of the colors of desert animals, including what is known of color variation in heteromyids, and discussed adaptive functions of colors in a wide variety of taxa.

In local areas, such as lava fields or areas with pale-colored substrates, dark or pale-colored populations often are found. Merriam (1890) mentioned *P. flavus fuliginosus* from an Arizona lava field that differed strikingly in its dark coloration from populations in the neighboring desert where the soil was pale in color. Other accounts of unusually colored populations of heteromyids include those of Dice (1929, 1930, 1940), Blossom (1931, 1933), Benson (1932), Bradt (1932), Blair (1943), Baker (1960), Findley (1967), Koschmann (1972), and Elder (1977).

There are many taxonomic descriptions in addition to those cited below that contain information on coloration and pelage characteristics of heteromyids. Osgood (1900) described color and pelage of *Perognathus* and *Chaetodipus.* Color, molt, and pelage descriptions for some *Perognathus, Chaetodipus, Microdipodops,* and *Dipodomys* were presented by Hall (1946). Goldman (1911) provided descriptions of color and molt in *Liomys* and *Heteromys.* Variation in coloration among populations of *P. fasciatus* (Williams and Genoways, 1979), *P. flavescens* (Williams, 1978), and *P. flavus* (Wilson, 1973) have been described. Speth

(1969) described color variation and patterns of molt in *P. parvus*. Color variation in a southern New Mexico population of *C. intermedius* was examined by Elder (1977). Anderson (1964) illustrated and described pelage characters that differentiated *C. artus* from *C. goldmani*. Coloration was a significant factor in Hall's (1941, 1946) discussions of variation in *Microdipodops*. Hafner et al. (1979) separated *M. megacephalus* from *M. pallidus* collected in Penoyer Valley, Nevada, on the basis of three qualitative pelage characters, which closely agreed with identification based upon their 14 mensural characters. Schitoskey (1968) was not able to discern variations in coat color in a Nevada population of *M. megacephalus*. Genoways (1971) used mean reflectance values to demonstrate differences in coloration of *L. spectabilis* and *L. pictus*, and Hooper and Handley (1948) described variation in color of *L. irroratus*.

Grinnell (1922) described variation in coloration and pelage of *Dipodomys* in California, Setzer (1949) examined variation in color of *D. ordii*, and Blair (1949) and Lidicker (1960*a*) assessed variation in color among populations of *D. merriami*. For *D. phillipsii*, Genoways and Jones (1971) described pelage, found color was variable geographically, and noted that color varied to a greater degree among animals from a single locality than did external or cranial measurements. Among their 14 geographic samples, reflectance of red ranged from 5.5 to 13.1 in coefficient of variation. Baker (1960) described specimens of *D. phillipsii* from the Guadiana lava field, Durango, as being darker in dorsal coloration than typical examples of the species. His is the only account that documents darker-colored *Dipodomys* associated with lava fields. Nader (1978) described color variation in *D. spectabilis* and *D. deserti*.

Specimens of *D. agilis* from the northern portion of the range are noticably darker (northern Pacific coast and mountains of Baja California). Paler-colored populations occur southward in Baja California, and populations are very pale in the vicinity of San Francisquito Bay. Darker forms are in areas with darker soils or areas that have a relatively dense cover of vegetation. Pale-colored *Dipodomys* occur in other areas as well (e.g., *D. ordii oklahomae* in Oklahoma and *D. compactus* from Padre Island, Texas, are among the palest-colored *Dipodomys*). Nader (1978) found the palest-colored *D. deserti* in the hottest and driest area within the range of the species (Death Valley, California), and noted that high alkalinity of the soil affects color. However, the original color is restored after molting.

The difference in substrate coloration and degree of isolation required for unusually colored populations to evolve also has resulted in a great deal of morphologic variation, even among geographically close lava fields (Weckerly and Best, 1985; Weckerly et al., 1988). Weckerly and Best (1985) found several morphologic characters that differed statistically among populations from three lava fields in southern New Mexico. Although differences in coloration among their populations were detectable, the degree of morphologic variation was not expected because of the closeness of the populations.

Geographic variation in color was assessed throughout the range of *D. merriami* by Lidicker (1960*a*). He found an extremely broad spectrum of fur color in the dorsal pelage, reflecting variation in both the dusky bases of the bicolored hairs and their reddish tips. In addition to the complex variation in color of the dorsum, he found that *D. merriami* possessed a number of other pelage characters that proved to be useful in examining geographic variation in coloration. Along with a discussion of coloration of *D. merriami*, he included a figure depicting geographic variation in color of the arietiform markings over the species' range.

Huey (1951) pointed out that the darkest race of *D. merriami* in Baja California lives in the relatively cool and humid San Quintin region, whereas directly eastward on the torrid, arid desert east of the Sierra San Pedro Martir, the most pallid race occurs. Hooper and Handley (1948) concluded that pelage coloration in *L. irroratus* may be associated with humidity or, more probably, with color of the substrate. Grinnell (1922) suggested that color intensity in *Dipodomys* was correlated with cloudiness. Lidicker (1960*a*) pointed out that it seems more likely that *Dipodomys* are affected directly by vegetation types and soil colors, which in turn may be related to cloudiness. Color of pelage seems closely related to soil color and moisture.

Bacular Variation

Burt (1936) presented the first comparative data on bacula of the Heteromyidae. He found variation in bacula of adult *Perognathus, Chaetodipus,* and *Dipodomys* was not great within given races and that there was considerable age variation. Bacula of young individuals were smaller with less bulbous basal ends than were those of adults. Burt (1960) later summarized his bacular observations of heteromyids by pointing out that, with the exception of *C. hispidus,* bacula in this family fall into a general pattern. They are simple rods, usually with expanded basal ends, and with tapering shafts that vary in dimensions and amounts of curvature. Burt (1960) included illustrations of many additional heteromyids in his later paper. Kelly (1969) provided data on bacula of *Dipodomys, Perognathus, Chaetodipus, Microdipodops, Heteromys,* and *Liomys,* Hoffmeister (1986) illustrated bacula of *Perognathus* and *Chaetodipus,* and Patterson and Thaeler (1982) assessed patterns of variation in bacula of *Dipodomys, Perognathus,* and *Chaetodipus.*

Additional descriptions of the bacula of *Dipodomys* have been presented (Best, 1981*a*; Blair, 1954; Boulware, 1943; Csuti, 1979; Desha, 1967; Genoways and Jones, 1971; Lackey, 1967; Lidicker, 1960*b*). Best and Schnell (1974) provided data on most species of *Dipodomys* and provided an estimate of relationships within the genus using phenetic analyses of bacular characters. Subsequently, Jannett (1976) provided additional data on *D. compactus* and *D. elator* and pointed out the presence of variation in the tip of bacula and suggested that it would be an informative character in future analyses. Best (1981*a*) studied intraspecific, interspecific, and geographic variation in bacula of Baja California *Dipodomys.*

Following comparison of phenograms for *Dipodomys* bacula, with and without body size considered, Best and Schnell (1974) concluded there was apparently no definite trend in the relationship between bacular size and body size. However, Best (1981*a*) later discovered that if *D. deserti* and *D. nitratoides* were omitted from correlation analyses, there was a significant relationship between body size and bacular length.

Burt (1960) illustrated and described bacula of *Perognathus.* He found that *Perognathus* have a short baculum with a relatively large, bulbous basal end that tapers rapidly into the slender shaft, which turns up at a nearly right angle distally, and terminates in a point.

Burt (1960) also illustrated and described bacula of *C. arenarius, C. baileyi, C. californicus, C. formosus, C. goldmani, C. hispidus, C. intermedius, C. nelsoni, C. penicillatus, C. pernix,* and *C. spinatus.* He described the baculum of *Chaetodipus* (except *C. baileyi* and *C. hispidus*) as relatively longer and more slender than *Perognathus.* The basal portion is slightly bulbous and the distal end is upturned. As viewed from the side, the baculum is roughly sigmoid in outline. In *C. formosus* and *C. baileyi,* the bone is extremely slender, only slightly enlarged at the basal end, the shaft has a gentle curve upward, and the point does not bend abruptly. The baculum of *C. hispidus* differs from others in the genus, as well as in the family Heteromyidae, in having a three-lobed distal end instead of a terminating point. Anderson (1964) provided detailed illustrations of variation in bacula of *C. artus* and *C. goldmani,* and noted that every baculum studied could be identified to species. When Roth (1976*b*) described *C. dalquesti,* he included illustrations of bacula of *C. arenarius, C. dalquesti,* and *C. penicillatus.*

The baculum of *M. pallidus* was described by Burt (1960) as definitely intermediate between those of *Perognathus* and *Dipodomys.* The large basal part tapers into a shaft that curves moderately upward. Schitoskey (1968) described differences among bacula of *M. megacephalus,* and found no variations in shape of adult specimens.

Burt (1960) described bacula of three species of *Liomys* (*L. irroratus, L. pictus,* and *L. salvini* under the name *L. crispus*) and figured the structure of two (*L. irroratus* and *L. pictus*). Genoways (1973) illustrated and assessed bacular variation in the five recognized species of *Liomys.* He found the bacula of *L. salvini* and *L. adspersus* were similar, as were those of *L. pictus* and *L. spectabilis,* and that the baculum of *L. irroratus* was most similar to *H. desmarestianus* and *H. gaumeri.*

Genoways (1973) described and illustrated the baculum of *H. desmarestianus* and *H. gaumeri.* Rogers and Schmidly (1982) used bacular characteristics to evaluate taxonomic relationships of *H. desmarestianus* and *H. goldmani* to other *Heteromys.* They found considerable variation in bacula and concluded only two species were represented in their samples.

Geographic variation has been shown in bacular characters of heteromyids. Anderson (1964) figured individual and geographic variation in bacula among populations of *C. artus* and *C. goldmani.* Csuti (1979) found geographic variation among samples of bacula from *D. microps* and concluded that bacular characters may be of considerable systematic value. Best (1981*a*) demonstrated significant geographic variation among bacula of the *heermanni* group of kangaroo rats in Baja California (*D. agilis* and *D. gravipes*). Patterson and Thaeler (1982) found no evidence of geographic patterns in bacular lengths of *Dipodomys.* Rogers and Schmidly (1982) elucidated interspecific and geographic variation in bacula of *Heteromys,* and used their findings to reach taxonomic conclusions.

Geographic Variation

Before and after the review of geographic variation by Gould and Johnston (1972), numerous articles have addressed the subject for heteromyids. Many studies of geographic variation in heteromyid rodents that were conducted to assess taxonomic relationships have not been included below because of space limitations and because this chapter deals with morphologic variation (not taxonomic relationships). Under the topics of sexual dimorphism, pelage and coloration, and bacular variation elsewhere in this chapter, comments on geographic variation of those traits have been included.

As the number of specimens from more collecting localities increased near the turn of the century, it became imperative that students of taxonomy evaluate the degree of geographic variation present within species. Once specimens from intermediate localities were shown to have intermediate characters (often between different species), it became clear that geographic variation was an important aspect of examining taxonomic relationships among populations. The early studies and taxonomic reviews of heteromyids by Merriam (1889), Osgood (1900), Goldman (1911), Grinnell (1922), and others, were influenced by examination of geographic variation. Taxa often were named on the basis of a few specimens from an isolated collecting locality. Because of the availability of more specimens from more localities, the investigators that followed have spent even more effort elucidating patterns of geographic variation within or among species (e.g., Best, 1983*a*; Best et al., 1986; Genoways, 1973; Hafner, 1981; Hall, 1941; Hooper and Handley, 1948; Kennedy and Schnell, 1978; Kennedy et al., 1980; Lidicker, 1960*a*; Rogers and Schmidly, 1982; Schmidly, 1971; Setzer, 1949; Williams, 1978).

Within the genus *Dipodomys,* several studies have focused upon geographic variation in morphology within species. The most frequently studied taxon has been *D. ordii* (Anderson, 1972; Baumgardner and Schmidly, 1981; Grisham, 1967; Hall, 1946; Hartman, 1980; Kennedy and Schnell, 1978; Kennedy et al., 1980; Schmidly, 1971; Schmidly and Hendricks, 1976; Setzer, 1949; Shaver, 1973). Other taxa also have been examined, including *D. agilis* (Best, 1978, 1983*a*; Best et al., 1986; Lackey, 1967), *D. californicus* (Dale, 1939), *D. compactus* (Baumgardner and Schmidly, 1981; Shaver, 1973), *D. deserti* (Hall, 1946; Nader, 1978), *D. elator* (Best, 1987), *D. gravipes* (Best, 1978, 1983*b*), *D. merriami* (Best and Janecek, 1992; Hall, 1946; Lidicker, 1960*a*), *D. microps* (Csuti, 1979; Hall, 1946; Hall and Dale, 1939; Lester, 1973), *D. nelsoni* (Nader, 1978), *D. panamintinus* (Hall, 1946), *D. phillipsii* (Genoways and Jones, 1971), *D. spectabilis* (Nader, 1978), and *D. stephensi* (Lackey, 1967). Variation in several species was examined by Grinnell (1922), Villa-R. (1941), Durrant and Setzer (1945), and Huey (1951).

Techniques for studying geographic variation have stressed statistical analyses of mensural data, but analyses of nonmetric characters also have been used. Hartman (1980) examined geographic variation in 18 nonmetric cranial traits for 11 populations of *D. ordii* from Oklahoma, Texas, and New Mexico. He examined frequencies of traits for sex, size, and side dependencies, and found nonmetric traits were clearly of value in studying geographic variation.

Coupling of geographic variation in morphology with other data has allowed a clearer understanding of taxonomic and evolutionary relationships. An example is the separation of *D. heermanni* from *D. californicus,* which was accomplished by use of genetic and morphologic attributes (Patton et al., 1976), and the comparison of results obtained from genetic analyses with those derived from morphologic analyses (Best and Janecek, 1992; Best et al., 1986; Johnson and Selander, 1971; Stock, 1974).

Within *Perognathus,* geographic variation has been examined for relatively few species. Those species that have been examined in detail include *P. flavus* (Baker, 1954; Wilson, 1973), *P. fasciatus* (Williams and Genoways, 1979), and *P. flavescens* (Reed and Choate, 1986; Williams, 1978). Villa-R. (1941) presented information on several species in Baja California and northern Mexico.

Geographic variation in *C. hispidus* was discussed by Glass (1947). Other species of *Chaetodipus* that have been examined are *C. artus* (Anderson, 1964; Hall and Ogilvie, 1960), *C. goldmani* (Anderson, 1964; Hall and Ogilvie, 1960; Patton, 1969), *C. intermedius* (Weckerly and Best, 1985; Weckerly et al., 1988), and *C. penicillatus* (Hoffmeister and Lee, 1967). Villa-R. (1941) presented data for several species of *Chaetodipus.* The study of geographic variation in morphology of *C. penicillatus* by Hoffmeister and Lee (1967) disclosed several significant aspects of intraspecific differentiation. Some seemingly prominent geographic barriers have not been important in differentiation of *C. penicillatus,* whereas others have. They observed some marked morphologic divergence in areas with no obvious geographic barrier, although no large-scale geographic trends were revealed.

Patton (1969) examined geographic variation in *C. goldmani* and ascertained that phenotypic variation (in terms of shifts in direction or magnitude of clines) did not correspond in any significant way to his chromosomal data. He attributed the lack of correspondence to different levels of organization of gross morphology and chromosomes, and pointed out that the expression of the phenotype is more strongly dependent on ecologic factors than on gross chromosomal arrangements. Weckerly and Best (1985) examined morphologic variation among *C. intermedius* from three lava fields in southern New Mexico and found a large amount of variability. Their study has shown that coloration is not the only variable affected by isolation on lava fields. They found significant geographic variation occurred in 15 of 16 characters.

Geographic variation in morphologic and color traits of *Microdipodops* was examined by Hall (1941, 1946). Hafner (1981) has studied patterns of evolutionary concordance among morphometric, colorimetric, karyologic, electromorphic, and climatic data sets within *M. megacephalus* and *M. pallidus.*

Using univariate and multivariate statistical techniques, the degree of geographic variation within each of the species of *Liomys* was assessed by Genoways (1973). Previously, Hooper and Handley (1948) examined geographic variation in *L. irroratus* and presented taxonomic conclusions.

Rogers and Schmidly (1982) examined geographic variation among *H. desmarestianus* and *H. goldmani,* and their analyses formed the basis for synonomizing *H. longicaudatus, H. lepturus,* and *H. temporalis* with *H. desmarestianus.* Engstrom et al. (1987) evaluated geographic variation in *H. gaumeri.* They found only six of 14 characters were significantly heterogeneous among grouped localities. They concluded that patterns and level of intralocality variation appeared similar to other heteromyines, but geographic variation in *H. gaumeri* was relatively conservative. They postulated that the relative lack of interlocality variance in *H. gaumeri* might be attributable to a restricted geographic distribution, to relative environmental homogeneity on the Yucatan Peninsula, or to a lack of genetic divergence among populations.

Environmental-morphologic Relationships

As eluded to in the previous section, many studies of morphologic variation in heteromyids have examined geographic variation among populations, yet few studies have assessed the relationship of environmental variation with morphologic traits. Roth (1976*a*) quantified and compared morphologic features for *Perognathus, Chaetodipus,* and *Dipodomys* with various degrees of desertification found in Baja California. He found that where there was an increase in openness of habitat to 20–30%, there was a greater development of morphologic specialization of the auditory bullae, hind feet, and tail in the heteromyid population. Further increase in openness (to 80% or more) did not necessarily result in additional specialization. He concluded that desertification was not the source of morphologic adaptations in the heteromyids he studied, but pre-adaptation in a more mesic environment was a significant factor.

Relationships between ecogeographic and morphologic variation in *D. agilis* in Baja California were examined by Best (1981*b*). He analyzed variation in temperature and precipitation and determined those data were correlated with morphologic parameters. His morphologic principal component I (size) was significantly correlated with latitude and longitude for both sexes. The morphologic principal component II of females (nasal width, length of ulna, and length of hind foot) was correlated with July mean temperature and January mean precipitation.

In studies of *D. ordii,* Kennedy and Schnell (1978) and Kennedy et al. (1980) suggested that small body size might be selected for when there is only a limited amount of desirable space available. The reasoning was that small size could reduce the amount of food and space needed by each individual, thereby enhancing survival of small individuals and lowering the probability of extinction in local populations.

For *Perognathus,* Williams (1978) found a strong north-south size cline in *P. flavescens.* Latitude showed 23 significant positive correlations with the morphometric traits; climatic severity index, growing season, and mean July minimum temperature were not significantly correlated with any of the morphometric traits. Mean annual temperature was negatively correlated with body size, indicating that larger animals were in cooler regions. Thus, latitude was more highly correlated with size than the temperature variables.

Straney and Patton (1980) examined geographic variation in 15 external and cranial characters of *C. goldmani,* and compared their findings to ecogeographic predictor variables including temperature, precipitation, and isophane (a measure of growing season length). They found several morphologic characters to be significantly related to isophane, one to be related to precipitation, and two related to temperature variables.

The morphologic differences between *M. megacephalus* and *M. pallidus* in size and shape of the angular processes, pterygoids, and incisive foramina suggested a means of ecologic separation related to the food resource base (Hafner et al., 1979). These authors reasoned that since these characters are related to, or are direct components of, the masticatory apparatus, it appeared that the functional significance of the differences might be explained by differing food habits between these two species. Hafner (1981) compared results of his assessment of morphologic variation and environmental parameters and found the environment was a good predictor of pelage color patterns, but not morphometric variation in *Microdipodops.*

Geographic variation in size observed in *L. irroratus* by Hooper and Handley (1948) appeared to be correlated with altitude and latitude, and factors associated therewith. Small size was characteristic of low elevations and low latitudes, and largeness was correlated with high altitudes and latitudes.

Since heteromyids vary geographically and there are statistical associations between that variation and some environmental characters, further investigations of these relationships seem warranted. One of many environmental–morphologic relationships of heteromyids yet to be examined in detail is that of the degree of morphologic variation among years. If genetically similar animals such as domesticated livestock can be treated in different ways to get them to grow larger (e.g., a steer in a feedlot will develop larger bones and overall size than one placed onto the open range), then rodents should be expected to differ under varying environmental conditions. Perhaps in favorable years, when food and resources are abundant, developing rodents are larger than those growing up in years with limited resources. This might be difficult to test in wild populations, but laboratory manipulations of resources available to developing heteromyids could help clarify the degree of difference and the factors controlling annual variation in morphologic traits.

Summary

Patterns of morphologic variation were assessed in 57 species of heteromyid rodents using 19 external and cranial measurements for 12,563 adult specimens. Results of this study, coupled with those of previous workers, indicate varying degrees of secondary sexual differences in size among species, and geographic variation in sexual dimorphism of *Chaetodipus* and *Dipodomys.* Phenetic relationships are similar to previous phenetic and phyletic analyses; however, affinities of genera and placement of *C. hispidus* with *Heteromys* and *Liomys* differ from previous findings. Heteromyids exhibit pronounced geographic variation in color attributable to substrate coloration, moisture, or other environmental factors. Bacula are similar among genera, differing primarily in length, diameter, curvature, and size of base, and they vary geographically in some species of *Chaetodipus, Dipodomys,* and *Heteromys.* Geographic variation in morphology is evident across the range of most species, and significant differences are evident over short distances in *Chaetodipus* and *Dipodomys.* Relationships between morphologic and environmental variation indicate a tendency for body size to be associated with latitude and some temperature and precipitation characters.

Acknowledgments

I thank the curators and museum personnel at the following collections for assistance with loans of specimens and for allowing me to examine specimens in their care: American Museum of Natural History, New York; Arizona State University, Tempe; Brigham Young University, L. L. Bean Museum, Provo; California Academy of Sciences, San Francisco; Carnegie Museum, Pittsburgh; Eastern New Mexico University, Portales; Field Museum of Natural History, Chicago; Harvard University, Museum of Comparative Zoology, Cambridge; Los Angeles County Museum, Los Angeles; Louisiana State University, Baton Rouge; Michigan State University, The Museum, East Lansing; Midwestern State University, Wichita Falls; Oregon State University, Corvallis; San Diego Natural History Museum, San Diego; Texas A&M University, Cooperative Wildlife Collection, College Station; Texas Tech University, The Museum, Lubbock; United States National Museum of Natural History, Washington; University of Arizona, Tucson; University of California, Museum of Vertebrate Zoology, Berkeley; University of California at Los Angeles, Donald R. Dickey Collection, Los Angeles; University of Illinois, Champagne-Urbana; University of Kansas, Museum of Natural History, Lawrence; University of Michigan, Museum of Zoology, Ann Arbor; University of New Mexico, Museum of Southwestern Biology, Albuquerque; University of Oklahoma, Oklahoma Museum of Natural History, Norman; University of Puget Sound, Tacoma; University of Texas, Natural History Laboratory, Austin.

W. R. Barber, P. A. Chappell, and L. L. Janecek assisted with loans at The University of New Mexico, Museum of Southwestern Biology. B. Dennis, H. T. Haagenstad, C. R. Lambert, and R. M. Sullivan assisted in the preparation of the figures. F. H. Best helped verify data, R. L. Garrison assisted in data entry and verification, and T. E. Garrison gathered literature, entered and verified data, and helped with other aspects of the study. D. E. Kidd provided encouragement and assistance throughout the preparation of this chapter. R. C. Calabro, H. H. Genoways, J. C. Hafner, R. M. Lee, and R. M. Sullivan provided helpful suggestions in their reviews of an early draft. I am especially grateful to S. A. Cole who spent many days during the summer of 1985 helping record data from thousands of heteromyids. This is a journal article of the Alabama Agriculture Experiment Station.

Literature Cited

ANDERSON, S. 1964. The systematic status of *Perognathus artus* and *Perognathus goldmani* (Rodentia). American Museum Novitates, 2184:1–27.

———. 1972. Mammals of Chihuahua: taxonomy and distribution. Bulletin of the American Museum of Natural History, 148:149–410.

BAKER, R. H. 1954. The silky pocket mouse (Perognathus flavus) of Mexico. University of Kansas Publications, Museum of Natural History, 7:339–347.

———. 1956. Mammals of Coahuila, México. University of Kansas Publications, Museum of Natural History, 9:125–335.

———. 1960. Mammals of the Guadiana lava field, Durango, Mexico. Publications of The Museum, Michigan State University, Biological Series, 1:305–327.

BARTHOLOMEW, G. A., JR., AND G. R. CARY. 1954. Locomotion in pocket mice. Journal of Mammalogy, 35:386–392.

BARTHOLOMEW, G. A., JR., AND H. H. CASWELL, JR. 1951. Locomotion in kangaroo rats and its adaptive significance. Journal of Mammalogy, 32:155–169.

BAUMGARDNER, G. D. 1989. Morphologic variation in kangaroo rats (genus *Dipodomys*): I. Nongeographic variation and character relationships; II. Geographic patterns of size and correlation with selected abiotic variables; III. patterns of interspecific variation. Ph.D. dissert., Memphis State University, Memphis, Tennessee, 469 pp.

BAUMGARDNER, G. D., AND D. J. SCHMIDLY. 1981. Systematics of the southern races of two species of kangaroo rats (Dipodomys compactus and D. ordii). Occasional Papers, The Museum, Texas Tech University, 73:1–27.

BEER, J. R. 1965. The interparietal in kangaroo rats. The Southwestern Naturalist, 10:145–150.

BENSON, S. B. 1932. Three new rodents from lava beds of southern New Mexico. University of California Publications in Zoology, 38:335–345.

———. 1933. Concealing coloration among some desert rodents of the southwestern United States. University of California Publications in Zoology, 40: 1–70.

BEST, T. L. 1978. Variation in kangaroo rats (genus *Dipodomys*) of the *heermanni* group in Baja California, Mexico. Journal of Mammalogy, 59:160–175.

———. 1981*a*. Bacular variation in kangaroo rats (genus *Dipodomys*) of the *heermanni* group in Baja California, Mexico. The Southwestern Naturalist, 25: 529–534.

———. 1981*b*. Relationships between ecogeographic and morphologic variation of the agile kangaroo rat (*Dipodomys agilis*) in Baja California, Mexico. Bulletin of the Southern California Academy of Sciences, 80:60–69.

———. 1982. Relationships of the burrows of Baja California kangaroo rats to ecogeographic and morphologic variation. Journal of Mammalogy, 63:532–536.

———. 1983*a*. Intraspecific variation in the agile kangaroo rat (*Dipodomys agilis*). Journal of Mammalogy, 64:426–436.

———. 1983*b*. Morphologic variation in the San Quintin kangaroo rat (Dipodomys gravipes Huey 1925). The American Midland Naturalist, 109:409–413.

———. 1987. Sexual dimorphism and morphometric variation in the Texas kangaroo rat (*Dipodomys elator* Merriam 1894). The Southwestern Naturalist, 32: 53–59.

BEST, T. L., AND L. L. JANECEK. 1992. Allozymic and morphologic variation among *Dipodomys insularis, Dipodomys nitratoides,* and two populations of *Dipodomys merriami* (Rodentia: Heteromyidae). The Southwestern Naturalist, 37:1–8.

BEST, T. L., AND G. D. SCHNELL. 1974. Bacular variation in kangaroo rats (genus Dipodomys). The American Midland Naturalist, 91:257–270.

BEST, T. L., C. INTRESS, AND K. D. SHULL. 1988. Mound structure in three taxa of Mexican kangaroo rats (Dipodomys spectabilis cratodon D. s. zygomaticus and D. nelsoni). The American Midland Naturalist, 119:216–220.

BEST, T. L., R. M. SULLIVAN, J. A. COOK, AND T. L. YATES. 1986. Chromosomal, genic, and morphologic variation in the agile kangaroo rat, *Dipodomys agilis* (Rodentia: Heteromyidae). Systematic Zoology, 35:311–324.

BIEWENER, A. A., AND R. BLICKHAN. 1988. Kangaroo rat locomotion: design for elastic energy storage or acceleration? Journal of Experimental Biology, 140: 243–255.

BIEWENER, A. A., R. BLICKHAN, A. K. PERRY, N. C. HEGLUND, AND C. R. TAYLOR. 1988. Muscle forces during locomotion in kangaroo rats: force platform and tendon buckle measurements compared. Journal of Experimental Biology, 137:191–205.

BLAIR, W. F. 1940. Two cases of abnormal coloration in mammals. Journal of Mammalogy, 21:461–462.

———. 1943. Ecological distribution of mammals in the Tularosa Basin, New Mexico. Contributions from the Laboratory of Vertebrate Biology, University of Michigan, 20:1–24.

———. 1949. Shade of pelage color in two populations of kangaroo rats and remarks on the status of *Dipodomys merriami ambiguus* Merriam. Journal of Mammalogy, 30:388–390.

———. 1954. Mammals of the mesquite plains biotic district in Texas and Oklahoma, and speciation in the central grasslands. The Texas Journal of Science, 6:235–264.

BLOSSOM, P. M. 1931. Relation between color of desert rodents and of the soil. Carnegie Institution of Washington Year Book, 30:266.

———. 1933. Description of a new rock pocket-mouse and a new desert-mouse from southern Arizona. Occasional Papers of the Museum of Zoology, University of Michigan, 265:1–4.

BOULWARE, J. T. 1943. Two new subspecies of kangaroo rats (genus Dipodomys) from southern California. University of California Publications in Zoology, 46:391–396.

BOWERS, M. A., AND J. H. BROWN. 1982. Body size and coexistence in desert rodents: chance or community structure? Ecology, 63:391–400.

BRADT, G. W. 1932. The mammals of the malpais, an area of black lava rock in the Tularosa Basin, New Mexico. Journal of Mammalogy, 13:321–328.

BROWNLEE, A. S. 1973. Differentiation of nine species of *Dipodomys* (Rodentia: Heteromyidae): a numerical taxonomy study based on morphology. Ph.D. dissert., University of Mississippi, University, 81 pp.

BURT, W. H. 1936. A study of the baculum in the genera Perognathus and Dipodomys. Journal of Mammalogy, 17:145–156.

———. 1960. Bacula of North American mammals. Miscellaneous Publications of the Museum of Zoology, University of Michigan, 113:1–76.

CAIRE, W. 1976. Phenetic relationships of pocket mice in the subgenus *Chaetodipus* (Rodentia: Heteromyidae). Journal of Mammalogy, 57:375–378.

CLOUDSLEY-THOMPSON, J. L. 1979. Adaptive functions of the colours of desert animals. Journal of Arid Environments, 2:95–104.

COUES, E. 1875. A critical review of the North American Saccomyidae. Proceedings of the Academy of Natural Sciences of Philadelphia, 27:272–327.

———. 1877. Saccomyidae. Pp. 481–542, *in* Monographs of North American Rodentia (E. Coues and J. A. Allen). Report of the United States Geological Survey of the Territories, 11:1–1091.

CSUTI, B. A. 1979. Patterns of adaptation and variation in the Great Basin kangaroo rat (*Dipodomys microps*). University of California Publications in Zoology, 111:1–69.

DALE, F. H. 1939. Variability and environmental responses of the kangaroo rat, Dipodomys heermanni saxatilis. The American Midland Naturalist, 22:703–731.

DALQUEST, W. W., AND R. M. CARPENTER. 1986. Dental characters of some fossil and Recent kangaroo rats, with descriptions of a new species of Pleistocene *Dipodomys.* The Texas Journal of Science, 38:251–263.

DAVIS, W. B. 1942. The systematic status of four kangaroo rats. Journal of Mammalogy, 23:328–333.

DESHA, P. G. 1967. Variation in a population of kangaroo rats, *Dipodomys ordii medius* (*Rodentia*: *Heteromyidae*) from the High Plains of Texas. The Southwestern Naturalist, 12:275–289.

DICE, L. R. 1929. Description of two new pocket mice and a new woodrat from New Mexico. Occasional Papers of the Museum of Zoology, University of Michigan, 203:1–4.

———. 1930. Mammal distribution in the Alamogordo region, New Mexico. Occasional Papers of the Museum of Zoology, University of Michigan, 213: 1–32.

———. 1940. The Tularosa malpais. The Scientific Monthly, 50:419–424.

DICE, L. R., AND P. M. BLOSSOM. 1937. Studies of mammalian ecology in southwestern North America with special attention to the colors of desert mammals. Carnegie Institution of Washington Publication, 485:1–129.

DRESSLER, J. B. 1979. An anatomical study of the brains of *Dipodomys* (Mammalia: Rodentia: *Heteromyidae*). Anatomischer Anzeiger, 145:359–368.

DUNMIRE, W. W. 1955. Sex dimorphism in the pelvis of rodents. Journal of Mammalogy, 36:356–361.

DURRANT, S. D., AND H. W. SETZER. 1945. The distribution and taxonomy of kangaroo rats (genus *Dipodomys*) of Utah. Bulletin of the University of Utah, 35:1–39.

ELDER, F. F. B. 1977. The ecological distribution of the rock pocket mouse *Perognathus intermedius* Merriam, in the Afton lava flows of southern New Mexico. Studies in Natural Science, The Natural Science Research Institute, Eastern New Mexico University, Portales, 2(3):1–23.

ELLIOT, D. G. 1901. A synopsis of the mammals of North America and the adjacent seas. Publication of the Field Columbian Museum, Zoölogical Series, 2:1–522.

ENGSTROM, M. D., H. H. GENOWAYS, AND P. K. TUCKER. 1987. Morphological variation, karyology, and systematic relationships of *Heteromys gaumeri* (Rodentia: Heteromyidae). Pp. 289–303, *in* Studies of Neotropical mammalogy: essays in honor of Philip Hershkovitz (B. P. Patterson and R. M. Timm, eds.). Fieldiana: Zoology (new series), 39:1–506.

FINDLEY, J. S. 1967. A black population of the Goldman pocket mouse. The Southwestern Naturalist, 12:191–192.

FORMAN, G. L., AND C. J. PHILLIPS. 1988. Histological variation in the proximal colon of heteromyid and cricetid rodents. Journal of Mammalogy, 69:144–149.

GENOWAYS, H. H. 1971. A new species of spiny pocket mouse (genus *Liomys*) from Jalisco, Mexico. Occasional Papers of the Museum of Natural History, The University of Kansas, 5:1–7.

———. 1973. Systematics and evolutionary relationships of spiny pocket mice, genus Liomys. Special Publications, The Museum, Texas Tech. University, 5:1–368.

GENOWAYS, H. H., AND J. K. JONES, JR. 1971. Systematics of southern banner-tailed kangaroo rats of the *Dipodomys phillipsii* group. Journal of Mammalogy, 52:265–287.

GLASS, B. P. 1947. Geographic variation in Perognathus hispidus. Journal of Mammalogy, 28:174–179.

GOLDMAN, E. A. 1911. Revision of the spiny pocket mice (genera *Heteromys* and *Liomys*). North American Fauna, 34:1–70.

———. 1923. Three new kangaroo rats of the genus Dipodomys. Proceedings of the Biological Society of Washington, 36:139–142.

GOULD, S. J., AND R. F. JOHNSTON. 1972. Geographic variation. Annual Review of Ecology and Systematics, 3:457–498.

GRAY, J. E. 1868. Synopsis of the species of *Saccomyinae*, or pouched mice, in the collection of the British Museum. Proceedings of the Zoological Society of London, 1868:199–206.

GRINNELL, J. 1919. Four new kangaroo rats from west-central California. Proceedings of the Biological Society of Washington, 32:203–205.

———. 1921. Revised list of the species in the genus Dipodomys. Journal of Mammalogy, 2:94–97.

———. 1922. A geographical study of the kangaroo rats of California. University of California Publications in Zoology, 24:1–125.

GRISHAM, K. B. 1967. Geographic variation in the Ord's kangaroo rat in the upper Rio Grande Valley. M.S. thesis, University of New Mexico, Albuquerque, 37 pp.

HAFNER, D. J., J. C. HAFNER, AND M. S. HAFNER. 1979. Systematic status of kangaroo mice, genus *Microdipodops*: morphometric, chromosomal, and protein analyses. Journal of Mammalogy, 60:1–10.

HAFNER, J. C. 1978. Evolutionary relationships of kangaroo mice, genus *Microdipodops*. Journal of Mammalogy, 59:354–366.

———. 1981. Evolution, systematics, and historical biogeography of kangaroo mice, genus *Microdipodops*. Ph.D. dissert., University of California, Berkeley, 269 pp.

HAFNER, J. C., AND M. S. HAFNER. 1983. Evolutionary relationships of heteromyid rodents. Great Basin Naturalist Memoirs, 7:3–29.

HAFNER, M. S. 1982. A biochemical investigation of geomyoid systematics (Mammalia: Rodentia). Zeitschrift fuer Zoologische Systematik und Evolutionsforschung, 20:118–130.

HAFNER, M. S., AND J. C. HAFNER. 1984. Brain size, adaptation and heterochrony in geomyoid rodents. Evolution, 38:1088–1098.

HALL, E. R. 1941. Revision of the rodent genus Microdipodops. Field Museum of Natural History, Zoology Series, 27:233–277.

———. 1946. Mammals of Nevada. University of California Press, Berkeley, 710 pp.

HALL, E. R., AND F. H. DALE. 1939. Geographic races of the kangaroo rat, Dipodomys microps. Occasional Papers of the Museum of Zoology, Louisiana State University, 4:47–63.

HALL, E. R., AND M. B. OGILVIE. 1960. Conspecificity of two pocket mice, Perognathus goldmani and P. artus. University of Kansas Publications, Museum of Natural History, 9:513–518.

HAMILTON, M. J., R. K. CHESSER, AND T. L. BEST. 1987. Genetic variation in the Texas kangaroo rat, *Dipodomys elator* Merriam. Journal of Mammalogy, 68:775–781.

HARTMAN, S. E. 1980. Geographic variation analysis of *Dipodomys ordii* using nonmetric cranial traits. Journal of Mammalogy, 61:436–448.

HATT, R. T. 1932. The vertebral columns of ricochetal rodents. Bulletin of the American Museum of Natural History, 63:599–738.

HOFFMEISTER, D. F. 1986. Mammals of Arizona. University of Arizona Press and Arizona Game and Fish Department [Tucson], 602 pp.

HOFFMEISTER, D. F., AND M. R. LEE. 1967. Revision of the pocket mice, *Perognathus penicillatus*. Journal of Mammalogy, 48:361–380.

HOMAN, J. A., AND H. H. GENOWAYS. 1978. An analysis of hair structure and its phylogenetic implications among heteromyid rodents. Journal of Mammalogy, 59:740–760.

HOOPER, E. T., AND C. O. HANDLEY. 1948. Character gradients in the spiny pocket mouse, *Liomys irroratus*. Occasional Papers of the Museum of Zoology, University of Michigan, 514:1–34.

HOWELL, A. B. 1923. Abnormal hairy growths upon the tails of the Heteromyidae. Journal of Mammalogy, 4:56–58.

———. 1933. The saltatorial rodent Dipodomys: the functional and comparative anatomy of its muscular

and osseous systems. Proceedings of the American Academy of Arts and Sciences, 67:377–536.
———. 1944. Speed in animals: their specialization for running and leaping. University of Chicago Press, Chicago, 270 pp.
HUEY, L. M. 1951. The kangaroo rats (Dipodomys) of Baja California, Mexico. Transactions of the San Diego Society of Natural History, 11:205–256.
JANNETT, F. J., JR. 1976. Bacula of *Dipodomys ordii compactus* and *Dipodomys elator*. Journal of Mammalogy, 57:382–387.
JOHNSON, W. E., AND R. K. SELANDER. 1971. Protein variation and systematics in kangaroo rats (genus *Dipodomys*). Systematic Zoology, 20:377–405.
JONES, W. T. 1985. Body size and life-history variables in heteromyids. Journal of Mammalogy, 66: 128–132.
KELLY, T. S. 1969. The comparative morphology of the male phallus in the genus *Dipodomys*. M.S. thesis, San Fernando Valley State College, Northridge, California, 133 pp.
KENAGY, G. J., AND S. C. TROMBULAK. 1986. Size and function of mammalian testes in relation to body size. Journal of Mammalogy, 67:1–22.
KENNEDY, M. L., AND G. D. SCHNELL. 1978. Geographic variation and sexual dimorphism in Ord's kangaroo rat, *Dipodomys ordii*. Journal of Mammalogy. 59:45–59.
KENNEDY, M. L., M. L. BECK, AND T. L. BEST. 1980. Intraspecific morphologic variation in Ord's kangaroo rat, *Dipodomys ordii*, from Oklahoma. Journal of Mammalogy, 61:311–319.
KOSCHMANN, J. R. 1972. Melanism in rodents of the Afton lava flows, Dona Ana County, New Mexico. M.S. thesis, University of Texas at El Paso, El Paso, 49 pp.
KOTLER, B. P. 1985. Owl predation on desert rodents which differ in morphology and behavior. Journal of Mammalogy, 66:824–828.
LACKEY, J. A. 1967. Biosystematics of *heermanni* group kangaroo rats in southern California. Transactions of the San Diego Society of Natural History, 14:313–344.
LAWLER, R. M., AND K. N. GELUSO. 1986. Renal structure and body size in heteromyid rodents. Journal of Mammalogy, 67:367–372.
LESTER, L. A. 1973. An analysis of intraspecific variation in the chisel-toothed kangaroo rat, *Dipodomys microps*. M.A. thesis, California State University, Long Beach, 134 pp.
LIDICKER, W. Z., JR. 1960*a*. An analysis of intraspecific variation in the kangaroo rat Dipodomys merriami. University of California Publications in Zoology, 67:125–218.
———. 1960*b*. The baculum of *Dipodomys ornatus* and its implication for superspecific groupings of kangaroo rats. Journal of Mammalogy, 41:495–499.
MACMILLEN, R. E. 1983. Adaptive physiology of heteromyid rodents. Great Basin Naturalist Memoirs, 7:65–76.
MATSON, J. O. 1980. The status of banner-tailed kangaroo rats, genus *Dipodomys*, from central Mexico. Journal of Mammalogy, 61:563–566.
MERRIAM, C. H. 1889. Preliminary revision of the North American pocket mice (genera Perognathus et Cricetodipus auct.) with descriptions of new species and subspecies and a key to the known forms. North American Fauna, 1:1–36.
———. 1890. Results of a biological survey of the San Francisco Mountain region and desert of the Little Colorado, Arizona. North American Fauna, 3:43–86.
———. 1894. Preliminary descriptions of eleven new kangaroo rats of the genera Dipodomys and Perodipus. Proceedings of the Biological Society of Washington, 9:109–115.
———. 1902. Twenty new pocket mice (*Heteromys* and *Liomys*) from Mexico. Proceedings of the Biological Society of Washington, 15:41–50.
———. 1904. New and little known kangaroo rats of the genus *Perodipus*. Proceedings of the Biological Society of Washington, 17:139–145.
———. 1907. Descriptions of ten new kangaroo rats. Proceedings of the Biological Society of Washington, 20:75–80.
MORALES, J. C., AND M. D. ENGSTROM. 1989. Morphological variation in the painted spiny pocket mouse, *Liomys pictus* (family Heteromyidae), from Colima and southern Jalisco, Mexico. Royal Ontario Museum, Life Sciences Occasional Paper, 38:1–16.
MORTON, S. R., D. S. HINDS, AND R. E. MACMILLEN. 1980. Cheek pouch capacity in heteromyid rodents. Oecologia (Berlin), 46:143–146.
MUNGER, J. C., M. A. BOWERS, AND W. T. JONES. 1983. Desert rodent populations: factors affecting abundance, distribution, and genetic structure. Great Basin Naturalist Memoirs, 7:91–116.
NADER, I. A. 1978. Kangaroo rats: intraspecific variation in *Dipodomys spectabilis* Merriam and *Dipodomys deserti* Stephens. Illinois Biological Monographs, 49:1–116.
NIKOLAI, J. C., AND D. M. BRAMBLE. 1983. Morphological structure and function in desert heteromyid rodents. Great Basin Naturalist Memoirs, 7:44–64.
OSGOOD, W. H. 1900. Revision of the pocket mice of the genus Perognathus. North American Fauna, 18:1–73.
PATTERSON, B. D., AND C. S. THAELER, JR. 1982. The mammalian baculum: hypotheses in the nature of bacular variability. Journal of Mammalogy, 63:1–15.
PATTON, J. L. 1967. Chromosome studies of certain pocket mice, genus *Perognathus* (Rodentia: Heteromyidae). Journal of Mammalogy, 48:27–37.
———. 1969. Chromosome evolution in the pocket mouse, *Perognathus goldmani* Osgood. Evolution, 23:645–662.
PATTON, J. L., H. MACARTHUR, AND S. Y. YANG. 1976. Systematic relationships of four-toed populations of *Dipodomys heermanni*. Journal of Mammalogy, 57: 159–163.
PFAFFENBERGER, G. S., F. W. WECKERLY, AND T. L. BEST. 1985. Male pseudohermaphroditism in a population of kangaroo rats, *Dipodomys ordii*. The Southwestern Naturalist, 31:124–136.
PRICE, M. V. 1983. Ecological consequences of body size: a model for patch choice in desert rodents. Oecologia (Berlin), 59:384–392.
———. 1984. Microhabitat use in rodent commu-

nities: predator avoidance or foraging economics? Netherlands Journal of Zoology, 34:63–80.
Price, M. V., and J. H. Brown. 1983. Patterns of morphology and resource use in North American desert rodent communities. Great Basin Naturalist Memoirs, 7:117–134.
Price, M. V., and K. M. Heinz. 1984. Effects of body size, seed density, and soil characteristics on rates of seed harvest by heteromyid rodents. Oecologia (Berlin), 61:420–425.
Quay, W. B. 1965. Integumentary modifications of North American desert rodents. Pp. 59–74, *in* Biology of the skin and hair growth (A. G. Lyne and B. F. Short, eds.). American Elsevier, New York, 806 pp.
Reed, K. M., and J. R. Choate. 1986. Geographic variation in the plains pocket mouse (*Perognathus flavescens*) on the Great Plains. The Texas Journal of Science, 38:227–240.
Reeder, W. G. 1953. Age variation in enamel patterns in the spiny pocket mouse, *Liomys pictus sonoranus*. Journal of Mammalogy, 34:59–64.
———. 1956. A review of Tertiary rodents of the family Heteromyidae. Ph.D. dissert., University of Michigan, Ann Arbor, 618 pp.
Reichman, O. J. 1983. Behavior of desert heteromyids. Great Basin Naturalist Memoirs, 7:77–90.
Rogers, D. S., and D. J. Schmidly. 1982. Systematics of spiny pocket mice (genus *Heteromys*) of the *desmarestianus* species group from Mexico and northern Central America. Journal of Mammalogy, 63:375–386.
Rohlf, F. J., J. Kishpaugh, and D. Kirk. 1974. Numerical taxonomy system of multivariate statistical programs (NT-SYS). State University of New York, Stony Brook, 98 pp.
Rosenzweig, M. L., and P. W. Sterner. 1970. Population ecology of desert rodent communities: body size and seed-husking as bases for heteromyid coexistence. Ecology, 51:217–224.
Roth, E. L. 1976*a*. Evolutionary significance of adaptations in heteromyid rodents in Baja California, Mexico. Ph.D. dissert., University of Arizona, Tucson, 81 pp.
———. 1976*b*. A new species of pocket mouse (*Perognathus*: Heteromyidae) from the Cape Region of Baja California, Mexico. Journal of Mammalogy, 57: 562–566.
Ryan, J. M. 1986. Comparative morphology and evolution of cheek pouches in rodents. Journal of Morphology, 190:27–41.
———. 1989. Comparative myology and phylogenetic systematics of the Heteromyidae (Mammalia, Rodentia). Miscellaneous Publications of the Museum of Zoology, University of Michigan, 176:1–103.
Rylander, M. K. 1981. Brain volume and cell density in two kangaroo rats, *Dipodomys merriami* and *D. ordii*. The Texas Journal of Science, 33:39–41.
Schitoskey, F., Jr. 1968. Notes on morphological variation in the dark kangaroo mouse. The Southwestern Naturalist, 13:243–248.
Schmidly, D. J. 1971. Population variation in *Dipodomys ordii* from western Texas. Journal of Mammalogy, 52:108–120.
Schmidly, D. J., and F. S. Hendricks. 1976. Systematics of the southern races of Ord's kangaroo rat, *Dipodomys ordii*. Bulletin of the Southern California Academy of Sciences, 75:225–237.
Schmidt-Nielsen, B., and K. Schmidt-Nielsen. 1951. A complete account of the water metabolism in kangaroo rats and an experimental verification. Journal of Cellular and Comparative Physiology, 38: 165–181.
Schmidt-Nielsen, K., and B. Schmidt-Nielsen. 1952. Water metabolism of desert mammals. Physiological Reviews, 32:135–166.
Schnell, G. D., T. L. Best, and M. L. Kennedy. 1978. Interspecific morphologic variation in kangaroo rats (*Dipodomys*): degree of concordance with genic variation. Systematic Zoology, 27:34–48.
Setzer, H. W. 1949. Subspeciation in the kangaroo rat, Dipodomys ordii. University of Kansas Publications, Museum of Natural History, 1:473–573.
Shaver, W. M. 1973. Skeletal morphology as an index of variation among selected sub-species of Ord's kangaroo rat *Dipodomys ordii* (Rodentia: Heteromyidae). M.S. thesis, University of Mississippi, University, 48 pp.
Smith, E. H. 1986. Morphological and karyotypic variation of the Gulf Coast kangaroo rat, *Dipodomys compactus* (Rodentia: Heteromyidae). M.S. thesis, Corpus Christi State University, Corpus Christi, Texas, 46 pp.
Sneath, P. H. A., and R. R. Sokal. 1973. Numerical taxonomy: the principles and practices of numerical classification. W. H. Freeman and Co., San Francisco, 573 pp.
Speth, R. L. 1969. Patterns and sequences of molts in the Great Basin pocket mouse, *Perognathus parvus*. Journal of Mammalogy, 50:284–290.
Stephens, F. 1887. Description of a new species of Dipodomys, with some account of its habits. The American Naturalist, 21:42–49.
Stock, A. D. 1974. Chromosome evolution in the genus *Dipodomys* and its taxonomic and phylogenetic implications. Journal of Mammalogy, 55:505–526.
Straney, D. O., and J. L. Patton. 1980. Phylogenetic and environmental determinants of geographic variation of the pocket mouse *Perognathus goldmani* Osgood. Evolution, 34:888–903.
Sumner, F. B. 1921. Desert and lava-dwelling mice, and the problem of protective coloration in mammals. Journal of Mammalogy, 2:75–86.
Sumner, F. B., and H. S. Swarth. 1924. The supposed effects of the color tone of the background upon the coat color of mammals. Journal of Mammalogy, 5:81–113.
Thompson, J. N., Jr. 1969. Variations in the interparietal of the kangaroo rat, Dipodomys ordii. The American Midland Naturalist, 82:625–627.
Thompson, S. D. 1985. Bipedal hopping and seed-dispersion selection by heteromyid rodents: the role of locomotion energetics. Ecology, 66:220–229.
Tibbitts, F. D., and B. F. King. 1975. The fine structure of the trophospongial layer of the kangaroo rat placenta. Anatomical Record, 183:567–578.
Van De Graaff, K. M. 1973. Comparative developmental osteology in three species of desert rodents,

Peromyscus eremicus, Perognathus intermedius, and *Dipodomys merriami.* Journal of Mammalogy, 54: 729–741.

VILLA-R., B. 1941. Nota acerca de algunas especies de roedores de los generos Dipodomys, Perognathus y Peromyscus. Anales del Instituto de Biologia, Universidad Nacional Autonoma de Mexico, 12:335–399.

VIMTRUP, B., AND B. SCHMIDT-NIELSEN. 1952. The histology of the kidney of kangaroo rats. Anatomical Record, 114:515–528.

VON BLOEKER, J. C., JR. 1930. An albino kangaroo rat. Journal of Mammalogy, 11:237.

VOORHIES, M. R. 1975. A new genus and species of fossil kangaroo rat and its burrow. Journal of Mammalogy, 56:160–176.

WAHLERT, J. H. 1985. Skull morphology and relationships of geomyoid rodents. American Museum Novitates, 2812:1–20.

———. 1993. The fossil record. Pp. 1–37, *in* Biology of the Heteromyidae (H. H. Genoways and J. H. Brown, eds.). Special Publication, The American Society of Mammalogists, 10:1–719.

WEBSTER, D. B. 1962. A function of the enlarged middle-ear cavities of the kangaroo rat, Dipodomys. Physiological Zoology, 35:248–255.

WEBSTER, D. B., AND M. WEBSTER. 1971. Adaptive value of hearing and vision in kangaroo rat predator avoidance. Brain, Behavior, and Evolution, 4:310–322.

———. 1975. Auditory systems of Heteromyidae: functional morphology and evolution of the middle ear. Journal of Morphology, 146:343–376.

———. 1980. Morphological adaptations of the ear in the rodent family Heteromyidae. The American Zoologist, 20:247–254.

WEBSTER, W. D., AND J. K. JONES, JR. 1985. Nongeographic variation, reproduction, and demography in the Texas kangaroo rat, *Dipodomys elator* (Rodentia: Heteromyidae). The Texas Journal of Science, 37:51–61.

WECKERLY, F. W., AND T. L. BEST. 1985. Morphological variation among rock pocket mice (*Chaetodipus intermedius*) from New Mexico lava fields. The Southwestern Naturalist, 30:491–501.

WECKERLY, F. W., A. L. GENNARO, AND T. L. BEST. 1988. Description of a new rock pocket mouse, *Chaetodipus intermedius,* from New Mexico. The Southwestern Naturalist, 33:100–102.

WESTERHAUS, M. D. 1983. A histological comparison of the dorsal and generalized holocrine skin glands in the kangaroo rat, *Dipodomys ordii.* Ohio Journal of Science, 83:253–255.

WILKINS, K. T., AND D. J. SCHMIDLY. 1979. Identification and distribution of three species of pocket mice (genus *Perognathus*) in Trans-Pecos Texas. The Southwestern Naturalist, 24:17–32.

WILLIAMS, D. F. 1978. Systematics and ecogeographic variation of the Apache pocket mouse (Rodentia: Heteromyidae). Bulletin Carnegie Museum of Natural History, 10:1–57.

WILLIAMS, D. F., AND H. H. GENOWAYS. 1979. A systematic review of the olive-backed pocket mouse, *Perognathus fasciatus* (Rodentia, Heteromyidae). Annals of Carnegie Museum, 48:73–102.

WILLIAMSON, R. G., JR., AND E. C. FREDERICK. 1977. A functional analysis of ankle extension in the ricochetal rodent (Dipodomys merriami). Zentralblatt fuer Veterinaermedizin. Reihe C. Anatomia, Histologia, Embryologia, 6:157–166.

WILSON, D. E. 1973. The systematic status of *Perognathus merriami* Allen. Proceedings of the Biological Society of Washington, 86:175–192.

WOOD, A. E. 1931. Phylogeny of the heteromyid rodents. American Museum Novitates, 501:1–19.

———. 1935. Evolution and relationships of the heteromyid rodents with new forms from the Tertiary of western North America. Annals of Carnegie Museum, 24:73–262.

CYTOGENETICS

JAMES L. PATTON AND DUKE S. ROGERS

Dipodomys tracks

Introduction

The chromosomal complement of heteromyid rodents, either diploid number estimates or karyotypic descriptions, has played a continuing role in understanding cytological variability in mammals from the earliest days of such research (e.g., Cross, 1931; Makino, 1953; Matthey, 1952, 1956). These rodents, particularly the genera *Chaetodipus* and *Dipodomys,* represent some of the first examples of mammals utilized in cytosystematic approaches to such diverse evolutionary topics as the analysis of patterns of geographic variation, modes of speciation, and estimates of phylogenetic relatedness. These studies quickly followed the introduction, in the late 1950s, of the now standard colchicine and hypotonic salt technique for the characterization of mitotic metaphase chromosomes.

This review will be uneven because the available data base is uneven. All genera and virtually every species of living heteromyid have been characterized for their basic chromosomal complement, diploid number and standard, non-differentially stained karyotype (the alpha and beta levels of karyology defined by White [1978]). However, the amount and depth of data for each species varies extensively. For example, the karyotypes of only a few taxa have been dissected by the longitudinal banding techniques used to identify homologous elements across taxa, unambiguously determine mechanisms of chromosomal change, and identify gross underlying DNA components. Even fewer studies have dealt with molecular aspects of heteromyid chromosomes, such as the characterization and localization of satellite DNA sequences, nucleolar organizers, and so forth. Ironically, while heteromyid karyology represented part of the initial burst of scientific excitement in mammalian chromosome systems, this excitement seemed to wane early. Few attempts have been made within the past decade to employ the modern techniques of molecular cytogenetics to unravel further the cytological problems identified by earlier workers.

Our task here is to provide a synopsis of what is known about the chromosomal complements of heteromyid rodents, but in so doing to identify what we still need to know before firm conclusions can be drawn. At the same time, emphasis is placed on the role played by heteromyid taxa in fundamental issues of mammalian cytogenetics.

***Chromosomal Variation in Kangaroo Rats,* Dipodomys**

Numerical Variation, Modes of Karyotypic Change, and Phyletic Implications

Descriptions of karyotypic variation within *Dipodomys* are available from several authors (Csuti, 1971, 1979; Dingman et al., 1971; Fashing, 1973; Futcher, 1974; Jackson and Hunsaker, 1971; Stock, 1974). Stock (1974) provided the only comprehensive survey for the genus, including interpretations of phyletic relationships based on chromosomal characteristics. A summary of available data for the genus is presented in Table 1.

The diploid number for species of *Dipodomys* ranges from 52 to 74, with three major, and non-overlapping, species clusters: 2n = 52–54, 60–64, and 70–74. The number of autosomal arms (NA) ranges from 70 to 144, and all species exhibit a simple XX/XY sex chromosome system. Longitudinal banding techniques have not been employed in any systematic study of cytological variation in the genus, although examples of G- and C-banded karyotypes are presented in Stock (1974), Mascarello et al. (1974), Bostock et al. (1972), and Bostock and Christie (1974, 1975). Hence, interpretations of patterns and pathways of chromosomal evolution in the genus, as well as phyletic implications derived from the chromosomal data, are based solely on simple non-differentially stained karyomorphology. With this serious limitation in mind, a summary of the major conclusions of Stock (1974) is as follows:

(1) The primitive karyotype is assumed to be 2n = 72/NA = 70; one in which all autosomes are essentially uni-armed. This assumption is based largely on the fact that species with a high 2n (70 to 74) karyotype occur within all but one of the major systematic groups of kangaroo rats recognized by morphology (Grinnell, 1922; Lidicker, 1960; Setzer, 1949). This assumption represents the common equals primitive criterion for establishing character polarity.

(2) Three principal modes of chromosomal change have taken place in various kangaroo rat lineages, each converting uni-armed chromosomes to bi-armed ones: fusion, pericentric inversion, and heterochromatin addition. These modes have acted differentially in generating the karyotypic characteristics of each of the three diploid number classes from the 2n = 72/NA = 70 ancestral condition: (a) independent fusion plus subsequent pericentric inversion and/or heterochromatic addition to generate the 2n = 60–64/NA = 94–116 and 2n = 52–54/NA = 100–104 groups, respectively; and (b) pericentric inversion and/or heterochromatin addition to give the 2n = 70–74/NA = 134–144 group.

(3) Five groups of kangaroo rats are recognized, based on these presumed changes and on the presumptive morphological relationships of the taxa (Fig. 1).

These views are at best untested hypotheses, at least until additional data (including banding studies, DNA characterization, and DNA sequence position) become available. Clearly, however, a simple tri-partite direction of chromosomal change does not characterize the genus. For example, under Stock's hypothesis of a 2n = 72/NA = 70 ancestral condition, a low 2n/NA has been independently achieved in two separate lineages (*D. californicus* of the *heermanni*-group, 2n = 52/NA = 100, and *D. merriami* and *D. nitratoides* of the *merriami*-group, 2n = 52–54/NA = 100–104; Fashing, 1973; Stock, 1974; independence corroborated by allozyme analysis, Patton et al., 1976). Moreover, not all karyotypic variation recorded for the genus can be attributed solely to the mechanisms offered by Stock (1974). For example, *D. ordii monoensis* (2n = 72/NA = 140) has both a lower diploid and autosome arm number than does *D. compactus* (2n = 74/NA = 144), yet it has a higher total amount of nuclear DNA (C-value; Table 2). The nature of this difference cannot be accounted for by any structural

TABLE 1.—*Summary of chromosomal characteristics for taxa of kangaroo rats, genus* Dipodomys. *For simplicity, and to minimize subjectivity, only the bi-armed and uni-armed autosomal classes are used, with the uni-armed class equivalent to the definition of 'acrocentric' provided by Patton (1967a). Species are listed alphabetically rather than in any particular species-group arrangement; karyotypic variants recorded for each species are indicated by subspecies. 2n = diploid number; NA = number of autosomal arms; SM = meta-submetacentric; ST = subtelocentric; A = acrocentric.*

Taxon	2n	NA	Autosome pairs bi-	Autosome pairs uni-	X	Y	Reference[a]
agilis agilis	62	116	28	2	SM	A	1
a. peninsularis	60	116	29	0	SM	A	6
a. perplexus	62	116	28	2	SM	A	1
a. perplexus	62	110	25	5	SM	A	6
a. plectilis	60	116	29	0	SM	A	6
a. simulans	60	116	29	0	SM	A	5
californicus	52	96	23	2	SM	SM	4
compactus	74	144	36	0	SM	A-ST	6
deserti	64	110	24	7	SM	A	5
deserti	64	108	25	8	SM	A-ST	6
elator	72	82	6	29	SM	A	6
elephantinus	60	116	29	0	SM	A	6
gravipes[b]	70	71	1.5	32.5	SM	A	6
heermanni	64	94	16	15	SM	A	6
heermanni	64	90	14	17	SM	SM	1
ingens	64	98	18	13	SM	A	6
merriami	52	100	25	5	SM	A	5, 6
microps	60	116	29	0	SM	A	1, 2, 6
microps	60	90	16	13	SM	A	2
nelsoni	72	134	32	3	SM	A	6
nitratoides	54	104	25	0	SM	A	5, 6
ordii	72	140	35	0	SM	A-ST	6
phillippsii ornatus	72	138	34	1	SM	A	6
p. panamintinus	64	96	17	14	SM	A	3,6
p. caudatus	64	94	16	15	SM	SM	3
p. mohavensis	64	98	18	13	SM	SM	5
p. mohavensis	64	96	17	14	SM	SM	1
s. spectabilis	72	70	0	35	SM	A	6
s. perblandus	72	78	4	31	SM	A	6
s. perblandus	72	80	5	30	SM	SM	5
s. baileyi	72	94	12	23	SM	A	6
stephensi	70	86	10	24	SM	A	5, 6
venustus	60	116	29	0	SM	A	6

[a] 1 = Csuti, 1971; 2 = Csuti, 1979; 3 = Dingman et al., 1971; 4 = Fashing, 1973; 5 = Jackson and Hunsaker, 1971; 6 = Stock, 1974.

[b] Figure 2E from Stock (1974) indicates that *gravipes* is polymorphic in centromere placement for a single pair of autosomal elements.

rearrangement, or by simple whole-arm addition of heterochromatin.

Within-Species Variation

Karyotypic differentiation has been reported within seven species of kangaroo rats (Table 1). With the exception of apparent polymorphism within *D. gravipes,* other examples comprise fixed variation among subspecific or intra-subspecific geographic units (e.g., *D. agilis, D. deserti, D. heermanni, D. microps, D. panamintinus,* and *D. spectabilis*).

Two points must be kept in mind when viewing data on geographic variation. First, in some cases variation rests on interpre-

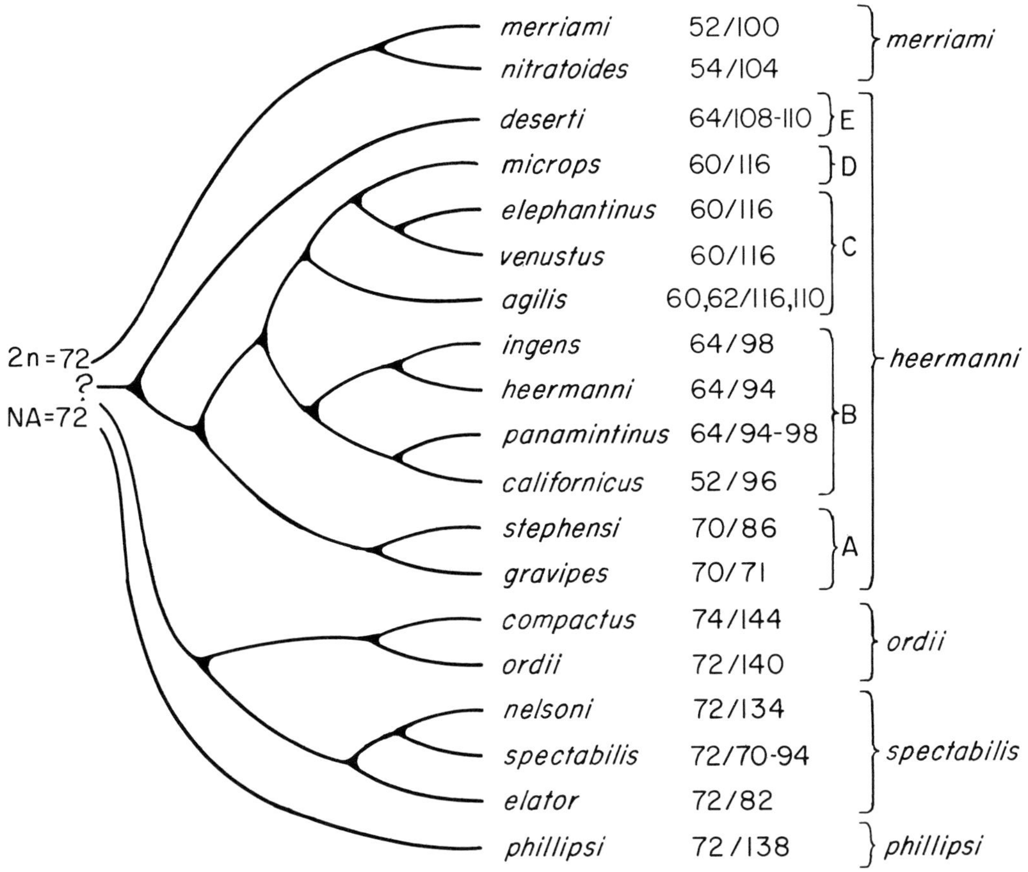

FIG. 1.—Suggested phylogenetic relationships among extant species of kangaroo rats, *Dipodomys*, based on chromosomal data. Path of chromosomal change and identified species groups follow Stock (1974). The diploid number and number of autosomal arms is given for each species.

tational differences among authors regarding individual chromosome morphology (e.g., *D. deserti, D. heermanni*, and *D. panamintinus*). Such variation may be real, but it is just as likely to be an artifact of preparation or interpretation. Second, without benefit of longitudinal banding studies, or other characterization of the chromosomes, identification of the mechanisms of karyotypic change in any of these examples is not possible.

Nevertheless, clear-cut examples of karyotypic polytypy do characterize *D. agilis, D. microps*, and *D. spectabilis* (Table 1). It is not clear as yet, however, if this variability reflects chromosomally differentiated, cryptic species, as was the case for *D. californicus* relative to *D. heermanni* (see Fashing, 1973; Patton et al., 1976). Similarly, it is not known if the geographic distribution of variants corresponds to morphologically defined subspecies; no thorough sampling program for karyotypic variation has been undertaken for any species of kangaroo rat. Karyotypic variation may, or may not, be concordant with morphological patterns (see section on *Chaetodipus* for examples).

Molecular Cytogenetic Characteristics of Dipodomys

Variation in C-value.—The nuclear DNA volume (2C-value) of 11 species of kangaroo rats has been established by flow-cy-

TABLE 2.—*Nuclear DNA compositions in kangaroo rats,* Dipodomys. *2C-value is the total amount of DNA per diploid cell (pg/nucleus). The DNA fractions are the components isolated in buoyant density gradients; data are given as a proportion of the 2C-value. Data from Hatch et al. (1976:Table 1 and Fig. 2c).*

		DNA fractions[a]			
				Satellite DNA	
Taxon	2C-value	P 1.689	I 1.702	MS 1.707	HS 1.713
agilis	9.5	47.4	19.0	16.8	16.8
californicus	8.5	50.6	20.0	11.8	17.7
compactus	9.7	35.1	22.7	12.4	29.9
deserti	—	52.5	24.9	19.1	3.5
elator	—	12.8	57.0	22.0	8.2
elephantinus	—	36.0	34.0	13.0	17.0
gravipes	—	66.0	24.0	7.5	2.5
heermanni	6.9	53.6	20.3	17.0	8.7
ingens	—	37.0	35.0	17.0	11.0
merriami	8.1	42.0	32.1	19.8	6.1
microps	8.0	43.8	18.7	12.5	25.0
nelsoni	—	49.0	20.0	14.0	17.0
nitratoides	—	52.0	28.0	20.0	0.0
ordii	10.9	32.1	18.4	22.9	26.6
phillippsii ornatus	—	68.0	10.0	4.0	18.0
panamintinus	7.9	51.9	21.5	12.7	13.9
s. spectabilis baileyi	9.8	79.6	10.2	8.2	2.0
	9.6	39.6	34.4	20.8	5.2
stephensi	—	77.0	9.0	13.0	11.0
venustus	—	24.0	45.0	19.0	12.0

[a] P = principal band; I = intermediate density band; MS = medium satellite fraction; HS = heavy satellite fraction.

tometric methods (Hatch et al., 1976; Table 2). The 2C-value ranges from 6.9 to 10.9 picograms (pg)/nucleus among species, with an average of 8.9 ± 1.1 pg/nucleus. Highest DNA amounts are found within the 2n = 72–74 group of species (*D. compactus, D. ordii, D. spectabilis*), regardless of arm number; those species with low (2n = 50–54) or intermediate diploid numbers (2n = 60–64) have smaller amounts. Total DNA content is positively correlated with the total amount of satellite DNA of both the MS and HS density classes (Hatch et al., 1976; see below).

Satellite DNAs.—Neutral CsCl bouyant densities of principal and satellite DNA fractions have been examined for each of the 19 species of kangaroo rat (Hatch and Mazrimas, 1974; Hatch et al., 1976; Mazrimas and Hatch, 1972). The same four fractions are recognized in all species, although the proportional amount present differs among them (Hatch et al., 1976; Table 2): principal, 1.698 gm/ml; intermediate density, 1.702 gm/ml; MS satellite, 1.707 gm/ml; and two heavy satellites (HS-a and HS-b), 1.713 gm/ml. The principal fraction is the dominant component of the genome in each species (Mazrimas and Hatch, 1972; see John and Miklos, 1979, for correction of data in Hatch et al., 1976). Both heavy density satellites have been sequenced (HS-a, Fry and Salser, 1977; HS-b, Fry et al., 1973). The chromosomal replication pattern for all three satellites has been examined in three species (Bostock et al., 1972, 1976). General relationships among satellite DNAs, C-value, and karyotype drawn by these workers include the following:

(1) Species with a reduced 2n show, on average, a proportionally lower C-value, the difference resulting primarily in the satellite

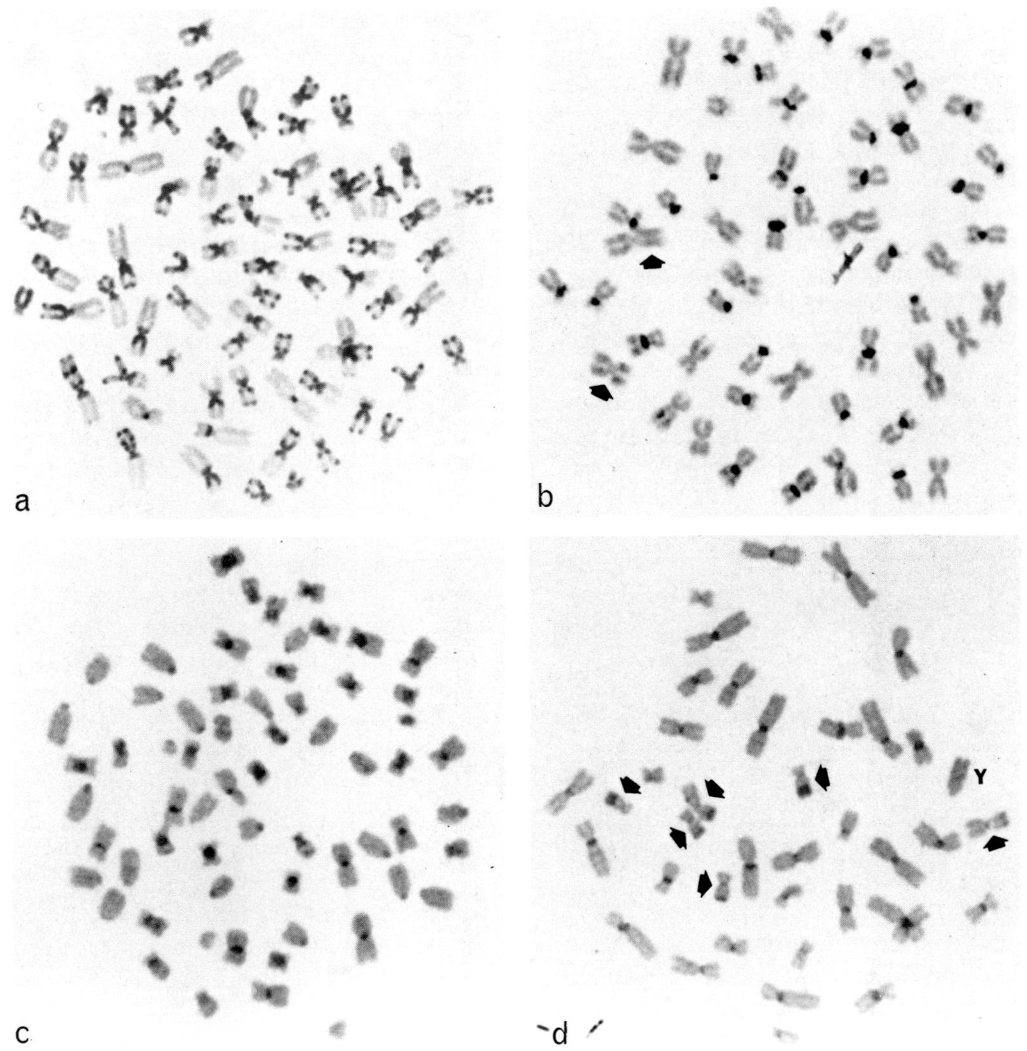

FIG. 2.—C-band mitotic metaphase figures of three species of kangaroo rats, *Dipodomys,* and one kangaroo mouse, *Microdipodops.* (a) *D. ordii columbianus* (MVZ 171001)—note large pericentromeric and telomeric blocks of heterochromatin. (b) *D. merriami merriami* (MVZ 170591)—note large interstitial C-bands of different staining intensity from centromeric heterochromatin on a few chromosomes (arrows). (c) *D. panamintinus mohavensis* (MVZ 170988)—note C-bands restricted to centromeric and pericentromeric regions. (d) *M. megacephalus californicus,* 2n = 40-a karyotypic race (MVZ 171005)—note telomeric blocks of heterochromatin on three pairs of small bi-armed autosomes (arrows); the heterochromatic Y-chromosome is indicated.

fractions. However, *D. californicus* (2n = 52) has a higher 2C-value (8.5 pg) than does its relative, *D. heermanni* (2n = 64, 2C-value = 6.9 pg); this difference is associated with a 250% increase in the HS satellite fraction in *D. californicus.* Similarly, *D. ordii* has a higher 2C-value than does *D. compactus,* although it has a lower 2n (72 versus 74). These two species differ in the respective amount of the MS satellite fraction present (Table 2).

(2) Increases in the number of autosomal arms are accompanied by a trend toward increasing the HS satellite fraction. This is

particularly true within species where karyotypic variants differ only by arm number, not diploid number (as in *D. s. spectabilis* versus *D. s. baileyi*; Table 2).

(3) Hatch and his co-workers (Hatch and Carrano, 1978; Hatch et al., 1973, 1976) have suggested that these satellite DNAs have had a direct role in the process of differential fixation of chromosome rearrangements in the karyotypic evolution of the genus, and thus in the process of speciation. Without banding studies to corrobrate the mode of chromosomal change within and among species, this conclusion is an untested hypothesis at best (see review by John and Miklos, 1979). Indeed, the differences among the karyotypic variants of *D. spectabilis,* for example, probably result from heterochromatin additions/deletions rather than structural rearrangements, something that C-band analyses would reveal.

C-bands and chromosomal localization of satellite DNAs.—The relationship between satellite DNAs and their localization relative to heterochromatic and euchromatic chromosomal segments has been examined in three species, *D. ordii* (Bostock et al., 1972; Prescott et al., 1973; Schreck et al., 1977), *D. merriami* (Bostock and Christie, 1974) and *D. panamintinus* (Bostock and Christie, 1975). The C-band patterns are complex and distinctly different among these three taxa. In all species, most chromosomes band positively at the centromere. For *D. ordii,* most biarmed autosomes exhibit large blocks of pericentromeric and one arm with telomeric heterochromatin (Fig. 2a); in *D. merriami,* several autosomal pairs have interstitial C-bands which differ in staining quality from centromeric bands (Fig. 2b); whereas in *D. panamintinus,* C-bands are limited to centromeric and pericentromeric regions (Fig. 2c).

c-RNA transcripts of the HS-b satellite bind to the centromeres of all but three chromosome pairs in *D. ordii,* while transcripts of a mixture of HS-a and MS satellites bind to the centromeres of the three pairs lacking HS-b specificity, and to the whole-arm heterochromatic regions. Both *D. merriami* and *D. panamintinus,* however, lack the HS-b fraction; in these two species the C-band positive centromeres as well as non-centromeric regions presumably show specificity to either, or both, HS-a and MS satellites.

Satellite DNA conservation within Dipodomys.—Although the three major satellite DNAs of kangaroo rats are not simply related to one another, the single-strand densities for a given satellite are nearly identical among species examined (Mazrimas and Hatch, 1977). Thus, at the level of bouyant density characterization, there is apparently a conserved satellite system throughout the genus. Because a similar HS-a satellite, with the same common repeated nucleotide sequence, also occurs in pocket gophers (*Thomomys bottae*), antelope ground squirrels (*Ammospermophilus leucurus*), and even guinea pigs (*Cavia porcellus*), this conserved sequence might extend throughout the Rodentia (Mazrimas, 1976; Mazrimas and Hatch, 1977). This conservation forms the basis of the hypothesis that species of rodents share a common library of satellite sequences, differences among them resulting only from differential amplification of particular variants of a basic sequence (Fry and Salser, 1977; Salser et al., 1976).

Chromosomal Variation in Kangaroo Mice, Microdipodops

Standard karyotypes have been described for both species of kangaroo mice (Hafner, 1981; Hafner et al., 1979; O'Farrell and Blaustein, 1974*a,* 1974*b*). Hafner (1981) placed the known chromosomal variants in phylogenetic context within the genus *Microdipodops,* and Hafner and Hafner (1983) discussed the karyological relationships of kangaroo mice to other heteromyid genera.

The two living species differ in diploid number, although there is geographic variation in the number of arms within each taxon (see Hafner, 1981; Table 3). *Micro-*

TABLE 3.—*Summary of chromosomal characteristics for taxa of kangaroo mice, genus* Microdipodops. *Data are arranged as in Table 1. Chromosomal variants within each species are indicated by letter, following Hafner (1981).*

			Autosome pairs				
Taxon	2n	NA	bi-	uni-	X	Y	Reference[a]
megacephalus-a	40	74	36	2	SM	A	1, 2
-b	40	76	38	0	SM	A	2
pallidus-a	42	70	15	5	SM	A	2
-b	42	80	20[b]	0	SM	A	1, 2
-c	42	80	20[b]	0	SM	A	2

[a] 1 = Hafner et al. (1979); 2 = Hafner (1981).
[b] Race 42-b and 42-c differ in the number of subtelocentric autosomes present in their respective complements.

dipodops pallidus has 42 chromosomes, and the number of autosomal arms varies from 70 to 80 among populations; *M. megacephalus* populations have 2n = 40, and the NA varies from 74 to 76. Both species exhibit a normal XX/XY sex chromosome system; the X is bi-armed and the Y uni-armed (see Hafner, 1981, for correction of sex chromosome morphology given by O'Farrell and Blaustein, 1974*a*, 1974*b*).

Hafner (1981) described three geographic chromosomal variants for *M. pallidus* and two for *M. megacephalus* (Table 3). In both cases, intraspecific karyotypic differences are attributed to heterochromatin additions to the smallest pairs of autosomes, although C-band patterns are available only for *M. megacephalus* race b. Large terminal blocks of heterochromatin are present on three autosomal pairs of the 2n = 40-b karyotype; all remaining C-bands are centromeric in placement (see Fig. 2d). The 2n = 42-a karyotype of *M. pallidus* and the 2n = 40-a karyotype of *M. megacephalus* were considered primitive for the respective species, based on geographic extent and habitat and elevational range. Hafner (1981) suggested that the 40-b race of *M. megacephalus* was independently derived twice, once each from western and eastern segments of the 40-a karyotypic range. This hypothesis is supported by protein similarity data. The primitive karyotypes of the two species are considered to differ by a single Robertsonian rearrangement, most likely a fission of a pair of small bi-armed autosomes in the 40-a *M. pallidus* complement (Hafner, 1981).

The possibility of natural hybridization between the two species of *Microdipodops* was investigated by Hafner et al. (1979), using cytological as well as morphological and genic analyses. Evidence for current hybridization was not found.

Chromosomal Variation in Pocket Mice, Chaetodipus

Numerical Variation, Modes of Karyotypic Change, and Phyletic Implications

Descriptions of standard, non-differentially stained karyotypes are available for all species in the genus (with the exception of *C. lineatus*; Table 4). These descriptions include reasonably detailed analyses of geographic variation in several species (see below). To date, however, no published report has attempted to use all available data either to estimate the major modes of chromosomal change within the genus or to suggest phylogenetic relationships among member species based on shared chromosomal characteristics.

The range in diploid number for *Chaetodipus* species is rather extensive (from 34 in *C. hispidus* to 56 in one race of *C. goldmani*), although the number of autosomal arms is similar among them (48–66, 52–58

TABLE 4.—*Chromosomal characteristics of pocket mice, genus* Chaetodipus. *Data are arranged as in Table 1. Chromosomal variants within species are indicated by letter unless they are known to coincide with recognized subspecies.*

Taxon	2n	NA	Autosome pairs bi-	Autosome pairs uni-	X	Y	Reference[a]
arenarius[b]	42	64	12	8	SM	A	5
artus	54	54	1	25	SM	A	1, 3, 5
bailey-a[c]	46	64	10	12	SM	A	1, 5, 6
-b	46	66	11	11	SM	A	6
californicus	44	54	6	15	SM	A	5
fallax	44	56	7	14	SM	A	5, 9
formosus	36	52	9	8	SM	A	2, 5, 8, 10
goldmani-a	52	54	2	23	A	A	1, 3
-b	56	56	1	26	SM	A	3
-c	52	54	2	23	SM	A	3
-d	50	54	3	21	A	A	3
-e	54	54	1	25	SM	A	3
-f	52	54	2	23	SM	A	3
hispidus	34	64	16	0	SM	A	1, 5
intermedius	46	58	7	15	SM	A	1, 5
nelsoni-a	46	58	7	15	SM	A	5
-b	48	58	6	17	SM	A	5
penicillatus-a	46	48	2	20	SM	A	1, 4, 5
-b	46	56	6	16	SM	A	4, 5
-c	46	54	5	17	SM	A	4, 5
pernix pernix-a	38	56	10	8	SM	A	1, 5
-b	36	56	12	6	SM	A	5
rostratus	52	56	2	24	SM	A	5, 7
spinatus	44	54	6	15	SM	A	5

[a] 1 = Patton, 1967*a*; 2 = Patton, 1967*b*; 3 = Patton, 1969*a*; 4 = Patton, 1969*b*; 5 = Patton, 1970; 6 = Patton, 1972; 7 = Patton and Soule, 1967; 8 = Towner, 1965; 9 = Cross, 1931; 10 = Williams, 1978*a*.

[b] Includes *C. dalquesti* Roth, which is karyotypically identical to *C. arenarius* (J. C. Hafner and M. S. Hafner, unpubl. data).

[c] *C. baileyi* exhibits extensive intrapopulation variation in diploid number due to the presence of supernumerary (B) chromosomes (see Patton, 1972, 1977; Sherwood, 1983); data are given here only for the A-chromosome set.

excluding three outliers; Table 4). All species have a simple XX/XY sex chromosome system, and both X and Y chromosomes are similar in size and morphology in all species (excepting two races of *C. goldmani,* which have acrocentric X chromosomes; Patton, 1969*a*).

The following hypotheses of phyletic relationships and mechanisms of chromosomal change within *Chaetodipus* are based solely on chromosomal measurements and arm-ratios of the standard karyotypes, supplemented by G-band analyses of *C. baileyi, C. penicillatus, C. californicus, C. fallax,* and *C. formosus.*

(1) The X-chromosome is homologous across all species in the genus. This is based on non-significant differences in both total length (5.7–5.9% of haploid genome length, including the acrocentric X-chromosomes of two races of *C. goldmani*) and arm ratios (1.70 to 1.78).

(2) The karyotypes of *C. hispidus* and *C. formosus* are so highly modified relative to other taxa that there are few, if any, potential homologies between elements of their respective autosomal complements and those of any other member of the genus. The karyotypic uniqueness of *C. formosus* is supported by a CMA-banding comparison between it and *C. baileyi* (Sherwood, 1983).

TABLE 5.—*Suggested homologies among bi-armed autosomes among species of pocket mice, genus* Chaetodipus *based on comparative length and arm-ratio data from standard karyotypes. Chromosomes are numbered in order of descending size. The number of bi-armed autosomes unique to each species is also indicated.*

Taxon	Chromosome														Number of unique bi-arms
	1	2	3	4	5	6	7	8	9	10	11	12	13	14	
californicus	X		X		X	X		X	X						0
spinatus	X		X		X	X		X	X						0
penicillatus-A		X	X		[a]	[a]		[a]							0
-B		X	X		X	X		X							1
-C		X	X		X	X		X							0
arenarius		X	X		X	X		X				X			7
pernix-A		X	X		X	X	X	X		X	X	X	X		0
-B		X	X		X	X	X	X		X	X	X	X		1
rostratus							X	X							0
intermedius			X	X	X	X		X							2
nelsoni-A			X	X	X	X		X						X	1
-B			X		X	X		X						X	2
fallax			X		X	X		X							2
baileyi					X	X		X							7
artus															1
goldmani															5
formosus															9
hispidus															16

[a] Present as uni-armed elements.

(3) The remaining species, and their racial variants, belong to one of three diploid number groups: 50–56, with mostly uni-armed complements; 42–48, with from 5 to 8 large bi-armed autosomes; and 36–38, with 10–16 bi-armed elements of varying size (Table 4).

(4) Table 5 provides the suggested homologies of the autosomal bi-armed elements among species, excepting *C. hispidus* and *C. formosus.* A cladogram based on the inferred homologies is presented in Figure 3. Assumptions upon which this cladogram is constructed include (a) monophyly of all known chromosomal races at the species level, (b) chromosomal evolution toward decreasing diploid number, and (c) homoplasy (convergences and reversals) minimized. The first assumption requires, for example, that the highly divergent karyotypes of *C. pernix* achieved their respective characteristics subsequent to the development of that species from a common ancestor. The second assumption is used simply as a matter of convenience; there are no data to support either reduction or increase in number as the primary direction of change in the genus.

(5) The cladogram (Fig. 3) and genic (=allozyme) analyses (Patton et al., 1981) do not corroborate the traditional species-groups of *Chaetodipus* recognized by Osgood (1900) and subsequent workers. However, the chromosomal and genic views of relationships are not terribly concordant with each other. The validity of the chromosomal cladogram must await confirmation by banding and other cytogenetic studies.

(6) Based on this analysis, the mode of chromosomal change appears to be balanced between Robertsonian events (fu-

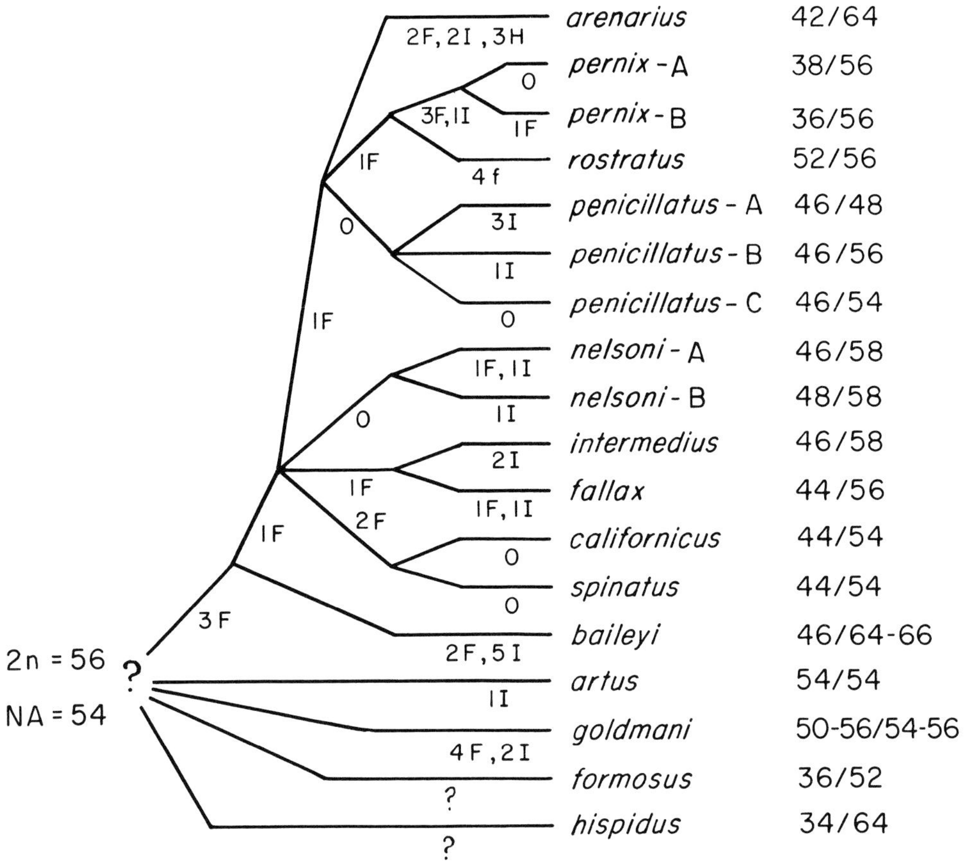

FIG. 3.—Suggested phylogenetic relationships among extant species of pocket mice, *Chaetodipus*, based on chromosomal data. The number and type of chromosomal rearrangement occurring along each branch are indicated (F = fusions, f = fissions, I = pericentric inversions, H = heterochromatin additions). Data are from non-differentially stained and C-banded karyotypes; homologies are based on proportional length and arm ratio.

sion/fission; 27 independent occurrences) and non-Robertsonian ones (inversions/translocations, including whole-arm heterochromatin shifts; 23 occurrences; see Fig. 3). This tabulation excludes consideration of *C. hispidus* and *C. formosus.*

(7) Heterochromatin variation, with the exception of *C. baileyi* (see below), appears to be minimal among species of *Chaetodipus.* C-bands are restricted to the centromeric regions in *C. goldmani* (Fig. 4a), *C. intermedius, C. penicillatus*-race A, *C. spinatus,* and *C. californicus.* In *C. arenarius,* C-bands are present both at the centromeres and as whole arms of three small pairs of bi-armed autosomes. The centromeres of *C. formosus* are C-band positive, as are the short arms of 6 of the 8 large subtelocentric autosomes; staining quality of the latter, however, is quite different from that of centromeric C-bands, suggesting differences in underlying DNA sequences.

Within-Species Variation

Local polymorphism for supernumerary (B) chromosomes has been described for *C. baileyi* (see below); all other examples of within-species variation are limited to polytypic geographic races. Five species are known to be comprised of such races (see Table 4). Of these, *C. goldmani* has been most thoroughly examined (Patton, 1969*a*).

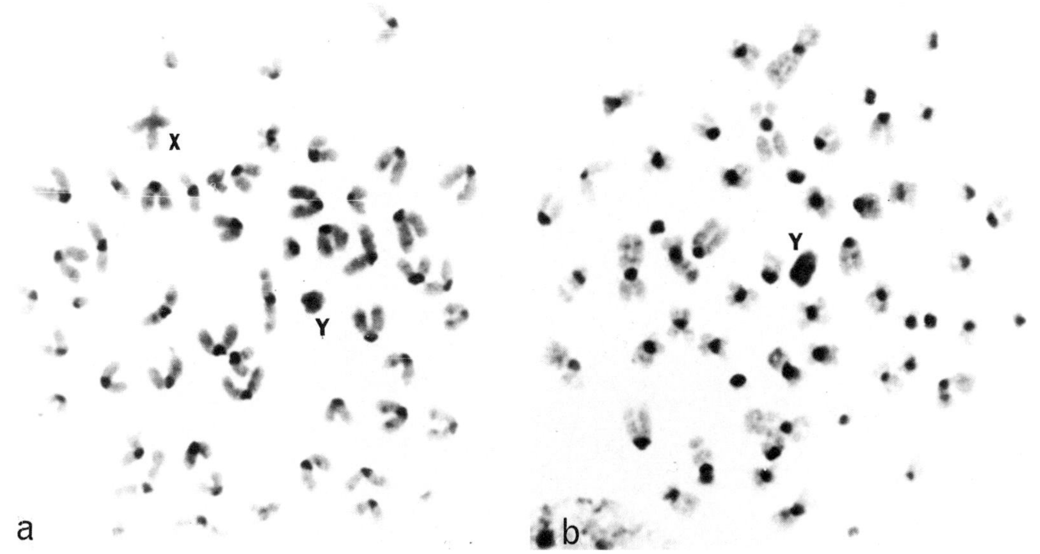

FIG. 4.—C-band mitotic metaphase plates of (a) *C. goldmani* [chromosome race-e, MVZ 148974] and (b) *P. amplus* [chromosome race IV, MVZ 149951]. Note the restriction of heterochromatin to centromeric regions on all autosomes; the Y-chromosome of each species is also heterochromatic.

In this example, chromosomal differentiation has proceeded without marked morphological shifts, although the geographic pattern of morphology mirrors the phyletic history of the races (Straney and Patton, 1980). Chromosomal evolution has also not been accompanied by demonstrable genetic differentiation, as measured by allozyme divergence (Patton et al., 1981). However, in the cases of *C. baileyi, C. penicillatus,* and *C. pernix,* chromosomal geographic patterns do generally map onto recognized subspecific groupings (Patton, 1969*b*); these races are also significantly differentiated at the genic level, relative to within-race population similarity (Patton et al., 1981). Indeed, the chromosomal races of *C. penicillatus* likely represent cryptic species; at least individuals of races A and B have been collected together without evidence of hybridization (Patton et al., 1981). The karyotypic divergence between *C. p. pernix* and *C. p. rostratus* might also indicate the presence of cryptic species. With available data, the two chromosome races of *C. nelsoni* do not appear to map onto recognized subspecific units (Patton, 1970).

The factors involved in geographic differentiation of the karyotype within species are not at all understood. Patton (1969*a*) invoked parapatric divergence with rivers forming local barriers for racial development in *C. goldmani.* The pattern of morphological variation within as opposed to among races is consistent with this hypothesis (see Straney and Patton, 1980). Certainly, the chance fixation of rearrangements in allopatry cannot be the sole driving force in all cases, as none of the populations/subspecies of species occurring on islands in the Gulf of California exhibit chromosomal differentiation from their respective mainland forms. This applies to samples of *C. baileyi* from Tiburon, *C. intermedius* from Tiburon and Datil, *C. penicillatus* from Tiburon, and *C. spinatus* from San Lorenzo (Patton, 1969*b,* 1970, 1972).

Interspecific Hybridization

Natural hybridization between two species, *C. pernix rostratus* and *C. penicillatus pricei* (chromosome race-A), has been reported from a single locality in southern Sonora, based on karyotypic intermediacy

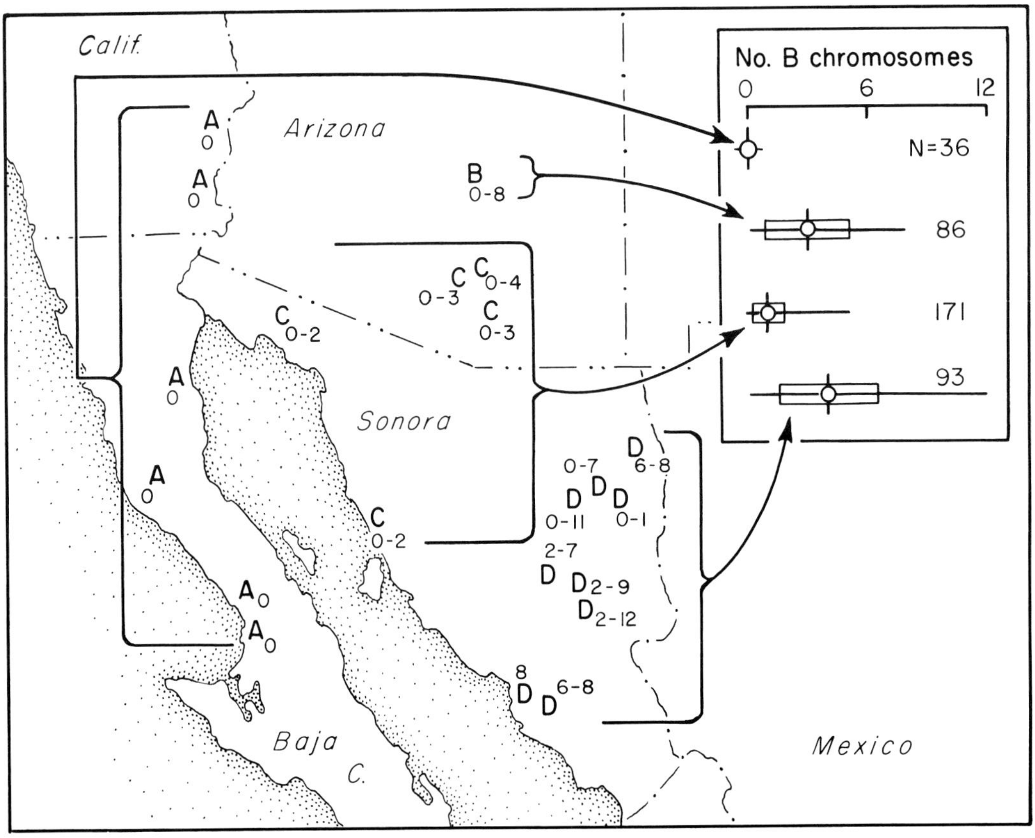

FIG. 5.—Mean and range of B-chromosome number in population samples of *C. baileyi*. Geographic samples are: A—California and Baja California; B—vicinity of Superior, Pinal Co., Arizona; C—vicinity of Tucson, Pima Co., Arizona; D—eastern and southern Sonora.

(Patton and Soule, 1967). However, only two F1 hybrids have been observed within a total sample of more than 200 individuals collected over a seven year period. This suggests that hybridization is infrequent, at best, and that the species remain genetically isolated, a view supported by the extensive allozyme differentiation among populations of the two species from the locality of known hybridization (Patton et al., 1981).

Molecular Cytogenetic Characteristics

Molecular characterization of the genome for *Chaetodipus* species has been limited to buoyant density gradient analysis of nuclear DNA in *C. formosus, C. spinatus,* and *C. baileyi* (Sherwood, 1983; see below). All three species show a main band at 1.700 gm/ml, which represents an overall G-C content of 43%. In addition, *C. spinatus* has a small heavy satellite in neutral CsCl gradients, and both *C. formosus* and *C. baileyi* exhibit an identical light satellite in Ag-Cs_2SO_4 gradients. In both species, this light satellite represents a significant portion of the entire genome, an average of 28% in *C. baileyi* and approximately 20% in *C. formosus*. Sherwood (1983) suggested that this satellite is AT-rich relative to mainband DNA. It is not known if the satellites for both *C. baileyi* and *C. formosus* represent homologous sequences, although the one for

C. baileyi corresponds to a 180 bp repeat located in chromosomal heterochromatin (see below). These two species do not share obvious chromosomal relationships (Fig. 3). Hence, the density gradients suggest that the light satellite may be conserved over a relatively wide phylogenetic range (see Patton et al., 1981). If true, this would represent conservation equivalent to that seen for satellite DNAs among kangaroo rat species (see above). The similarity between these satellites and those of kangaroo rats has not been examined.

Supernumerary Chromosomes

Supernumerary (or "B") chromosomes represent a form of karyotypic variability that may be both quantitative, in terms of diploid number and genome size, and qualitative, in terms of genomic components as well as meiotic, developmental, or other effects. These systems are not common among mammals in general (see reviews, Jones and Rees, 1982; Volobujev, 1980, 1982), but 18 species of rodents are known to exhibit this type of cytological variation. Perhaps the most thoroughly studied example of a mammalian supernumerary system is that of *C. baileyi* (Patton, 1972, 1977; Sherwood, 1983).

Numerical and geographic variability.—B-chromosomes within *C. baileyi* are restricted to those populations east of the Colorado River in Arizona and Sonora, Mexico. With limited sampling in southeastern California and Baja California, diploid numbers in these regions are uniform at 2n = 46, the basic number for the species (Patton, 1967*a,* 1972). However, numerical variation within and among populations of *C. baileyi* in Arizona and Sonora is extreme, with significant differences among populations in mean and range of Bs (Patton, 1972; Sherwood, 1983; Fig. 5). In general, populations in the northern and western parts of the range in Arizona are characterized by a smaller diploid number (2n = 46 to 50), whereas those from central to southern Sonora have both a higher range and mean (2n = 46 to 58; Patton 1972; Fig. 5).

Three types of B-chromosomes have been identified based on overall size and C-band staining qualities (Patton, 1977). Type I elements are small, usually bi-armed, and totally heterochromatic (Figs. 6a and 6b); they are found in all populations of *C. baileyi* that are known to exhibit B-variation. Type IIs chromosomes are large, nearly equivalent in size to the dominant bi-armed components of the A-chromosome set, and are totally heterochromatic (Fig. 6b). This type of chromosome has been found in a few populations in central Sonora (Patton, 1977) and in eastern Arizona (Sherwood, 1983). Finally, Type IIa elements are large, bi-armed chromosomes with centromeric and peri-centromeric blocks of heterochromatin but with large euchromatic terminal portions (Fig. 6c). Sherwood (1983) has noted that the distinction between each of these "classes" can be arbitrary. He also showed that the DNA comprising these different elements has the same characteristics (see below). It should be noted that the geographic extent of both Type IIa and IIs elements is poorly understood since C-band analysis, the only basis for the distinction between the classes, has been limited to only a few populations in Arizona and northeastern Sonora.

Meiotic behavior.—The A-chromosome set exhibits 23 bivalents at meiosis I, with the X and Y chromosomes pairing end-to-end (Fig. 6d and 6c). The larger, bi-armed autosomes usually show chiasmata on both arms while smaller, uni-armed elements exhibit a single chiasma in diplotene (Patton, 1977). Both Type I and IIa B-chromosomes are usually seen as univalents at meiosis I (Fig. 6e), although multivalent associations of up to tetravalents are observed (Fig. 6f). B-chromosomes have not been observed to pair with any member of the A-chromosome set. Pairing data suggest considerable

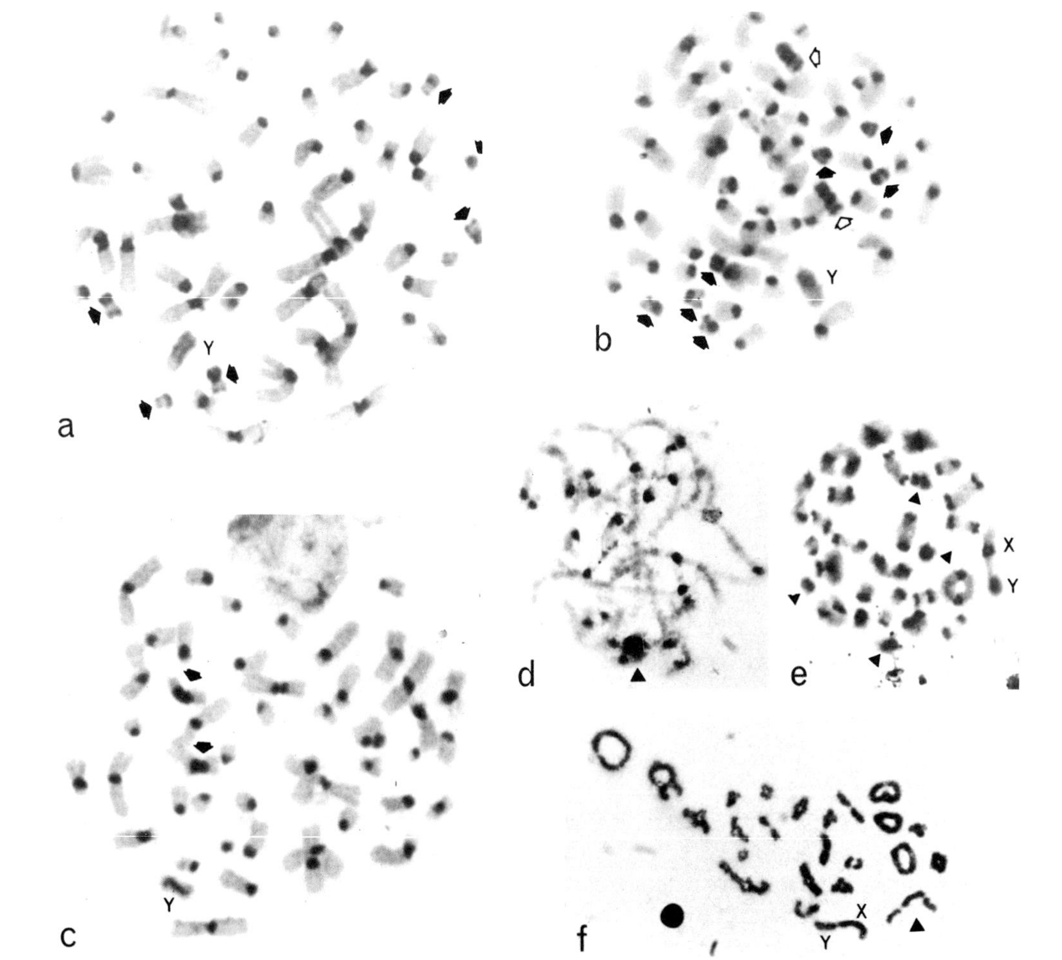

FIG. 6.—Chromosomal characteristics of B-chromosome variation in *C. baileyi.* (a) C-banded metaphase; Type I B-chromosomes are indicated by arrows. (b) C-banded metaphase; solid arrows identify Type I Bs, open arrows indicate Type IIs Bs. (c) C-banded metaphase; Type IIA Bs are indicated by arrows. (d) C-banded pachytene meiotic figure; note restriction of heterochromatin to centromeric regions of the A-chromosome set. The sex-vesicle is identified by the dart. (e) C-banded diakinesis figure. The X and Y bivalent is identified. Note the unpaired condition of the 4 B-chromosomes present (darts). (f) Diplotene figure, with X-Y pair indicated. Note 3 B-chromosomes pairing as a trivalent (dart).

homology among B-chromosome types, an hypothesis supported by their molecular characteristics (see below, and Sherwood, 1983).

Segregation of Type IIa B-chromosomes into secondary spermatocytes is not random as the number of Bs present in metaphase II figures is higher than expected, with the degree of meiotic distortion as high as 20% (Patton, 1977). Meiosis of females has not been examined, so it is unknown if this type of accumulation mechanism, often a general feature of B-systems (see Jones and Rees, 1982), characterizes both sexes of *C. baileyi.* However, unless females preferentially lose B-chromosomes in egg formation, offsetting the increase in B number per sperm, an increase in the number of B-chromosomes per individual in offspring relative to their parents is expected. Without some balancing mechanism, populations would be expected to show a steady increase in diploid number with each generation (see below).

B-chromosomes are associated with a very slight but statistically significant increase in chiasma frequency in the A-chromosome complement (Patton, 1977). The biological importance of this effect is not known.

Seasonal variation and demographic correlates.—S. W. Sherwood (1983, in litt.) examined characteristics of the B-chromosome system in populations of *C. baileyi* in eastern Arizona, where the number of Bs per individual ranges from 0 to 9, but where the sampled populations show significantly different frequencies of B-chromosomes. General observations for these populations include: (1) Males and females do not differ in mean number of Bs per individual, and the number per individual is randomly distributed about the population mean. (2) Populations exhibit temporal stability in B number, at least over a 10 year sampling period. (3) An annual cycle of B frequency characterized one studied population; young animals entering the population tended to have slightly higher numbers of Bs than breeding adults. These data suggest that Bs are maintained in a dynamic equilibrium within populations, with natural selection involved in maintaining the seasonal cyclicity observed, thus offsetting the meiotic tendency to increase B number. The factors responsible for the observed level of population differentiation in mean B frequency are totally unknown.

Molecular characterization and origin.—The presence of B-chromosomes adds significantly to total DNA levels of individuals within populations of *C. baileyi.* Using flow-cytometry, Sherwood (1983) showed that the difference in average cellular DNA content between 0B and 4B individuals was 20% (6.25–7.43 pg/diploid nucleus, respectively).

Sherwood (1983) characterized genomic DNA by both density gradient centrifugation and restriction endonuclease digestion. The former technique demonstrated a large satellite comprising about 28 percent of the genome. It was cryptic in neutral CsCl but banded as a light satellite in $AgCs_2SO_4$ gradients. Digestion of genomic DNA with the restriction endonucleases TaqI, AluI, and HphI revealed the presence of an abundant tandemly repeated sequence, 180 base pairs in length. A *c*RNA probe made to the TaqI fragment binded exclusively to the C-band positive (=heterochromatic) regions of both the A and B chromosomes. The satellite and the TaqI fragment were present in the genomes of individuals with and without B-chromosomes, as well as in populations from west of the Colorado River in California which lack Bs altogether. Clearly, therefore, B-chromosomes are not qualitatively different in their underlying DNA characteristics from the A-chromosome set. They very likely evolved simply as centric fragments or from non-disjuction of members of the A-chromosome set, followed by differential amplification of the basic repeated sequence.

Chromosomal Variation in Pocket Mice, Perognathus

Numerical Variation and Chromosomal Groups

Descriptions of standard, non-differentially stained karyotypes of all species of silky pocket mice, genus *Perognathus,* were provided primarily by Patton (1967*b*) and Williams (1978*a*, 1978*b*). Williams (1978*a*) interpreted differences among species in an evolutionary framework for the genus. Both authors included *formosus* within their perognathine groups, a species now placed in the genus *Chaetodipus* (see Patton et al., 1981; Hafner and Hafner, 1983).

Diploid numbers for all nine species fall into four classes, 44, 50, 54, and 56, with extensive arm number variation within both the 54 and 56 groups (Table 6). The sex chromosomes are of the standard XX/XY

TABLE 6.—*Chromosomal characteristics of pocket mice, genus* Perognathus. *Data are arranged as in Table 1. Designation of geographic chromosomal variants within taxa follows Williams (1978*a*).*

Taxon	2n	NA	Autosome pairs bi-	Autosome pairs uni-	X	Y	References[a]
alticola	54	74	11	15	ST	A	3
amplus I	56	84	15	12	SM	M	2, 3
II	56	86	16	11	SM	M	3
III	56	88	17	10	SM	M	3
IV	56	90	18	9	SM	M	1, 2, 3
V	56	92	19	8	SM	M	3
VI	56	94	20	7	SM	M	1, 2, 3
apache I	44	48	3	18	ST	ST	3, 4
II	44	50	4	17	SM	SM	3, 4
fasciatus	44	48	3	18	A	A	3, 4
flavescens	44	48	3	18	ST	ST	3, 4
flavus	50	86	19	5	SM	M	1, 2, 3
inornatus I	50	88	20	4	SM	M	3
II	56	86	16	11	SM	?	3
III	56	88	17	10	SM	M	3
longimembris I	56	86	16	11	SM	M	2, 3
II	56	88	17	10	SM	M	1, 2, 3
III	56	88	17	10	SM	M	1, 2, 3
parvus I	54	70	8	18	ST	A	1, 3
II	54	74	11	15	ST	A	2, 3
III	54	76	12	14	ST	A	3
IV	54	104	26	0	SM	SM	3
xanthonotus	54	76	12	14	ST	A	3

[a] 1 = Patton, 1967*b*; 2 = Patton, 1970; 3 = Williams, 1978*a*; 4 = Williams, 1978*b*.

system, with both elements typically biarmed.

With a single exception, each of the diploid number classes delineates a group of species recognized by traditional taxonomy (Osgood, 1900). The exception is the 2n = 50 group of *P. flavus* and one population of *P. inornatus*. These two karyotypes are quite similar, a feature that suggests close relationship (Williams, 1978*a*). Since other populations of *P. inornatus* have 2n = 56, as do other members of the *longimembris*-group, either the 2n = 50 karyotype is primitive to the two groups, or it has been independently derived in each.

Modes of Chromosomal Change

Williams (1978*a*) argued that chromosomal evolution within *Perognathus* proceeded primarily by non-Robertsonian events. While Robertsonian rearrangements must be responsible, at least in part, for differences among the diploid number groups, inversions could account for all observed arm number variation within each. C-band data are available only for *P. amplus* VI and *P. inornatus* I (Fig. 4b). In both cases, heterochromatin is restricted to the centromeric regions of all autosomes; there are no whole-arm or interstitial C-bands. Therefore, arm number variation within the 2n = 56 karyotypes, at least, probably results from actual structural rearrangements, most likely pericentric inversions.

Within-Species Variation

Chromosome race formation is known for five species in the genus, and is particularly extensive in both *P. amplus* and *P. parvus* (Table 6). No detailed geographic analysis

TABLE 7.—*Chromosomal characteristics of spiny pocket mice, genus* Heteromys. *Data are arranged as in Table 1. Designation of geographic variants within* H. desmarestianus *follows Rogers (1986, 1989).*

Taxon	2n	NA	Autosome pairs bi-	Autosome pairs uni-	X	Y	Reference[a]
anomalus	60	68	5	24	SM	ST	2
desmarestianus A	60	86	14	15	SM	SM	1
B	60	82	12	17	SM	ST	1
C	60	72	7	22	SM	ST	1
D	60	67[b]	4	24	?	?	1
E	60	68	5	24	?	?	1
F	60	86	14	15	ST	ST-A	1
G	60	80	11	18	SM	ST	1
	60	82	12	17	SM	?	3
gaumeri	56	76	11	16	SM	ST	2
goldmani	60	76	9	20	SM	ST	1
nelsoni	42	72	16	4	SM	ST	1
oresterus	60	80	10	19	SM	A	1
Species[c]	60	90	16	13	SM	A	1

[a] 1 = Rogers, 1986; 2 = Engstrom et al., 1987; 3 = Genoways, 1973.
[b] This karyotypic form possesses a heteromorphic pair of autosomes.
[c] This taxon represents an undescribed species from Costa Rica.

of karyotypic variation is available for any species, however, so it is not clear at present if the known variation is reflective of, or independent of, morphological trends within species. It is also not clear if the recorded variation for taxa such as *P. inornatus,* which involves both 2n and NA change with geography, signals cryptic species.

Chromosomal Variation in Spiny Pocket Mice, Heteromys

Numerical Variation and Chromosome Groups

Standard karyotypes have been described for all species of *Heteromys* with the exception of *H. australis* (Burton et al., 1987; Engstrom et al., 1987; Genoways, 1973; Mascarello and Rogers, 1988; Rogers, 1986, 1989). Diploid numbers fall into three classes, 42, 56, and 60, with the majority of species possessing the 2n = 60 karyotype (Table 7). The sex chromosomes are of the standard XX/XY system, and the X is larger than the Y. Both elements are typically biarmed. A series of karyotypic variants has been described within *H. desmarestianus* (Engstrom et al., 1987; Rogers, 1986, 1989; Table 7), with the number of autosomal arms varying from 66 to 86. Three other species possessing the 2n = 60 karyotype (*H. anomalus, H. goldmani,* and *H. oresterus*) fall within the range of NA reported for *desmarestianus.*

In general, the diploid number classes are not concordant with traditional systematic groupings of the genus (Mascarello and Rogers, 1988; Rogers, 1986, 1989). For example, Hall (1981) places *H. nelsoni* with *H. oresterus* in the subgenus *Xylomys,* but karyotypically, *oresterus* most closely resembles members within the *desmarestianus* species group of the subgenus *Heteromys.*

Modes of Chromosome Change

Genoways (1973) suggested that the primitive karyotype for the subfamily Heteromyinae was similar to that of *Liomys irroratus* (2n = 60, low NA). However, this hypothesis has been tested with longitudinal banding data, specifically G-bands, only minimally (see Mascarello and Rogers, 1988).

Regardless of the primitive karyotypic condition for *Heteromys,* the presence of extensive 2n and NA variation requires that both Robertsonian and non-Robertsonian events have occurred in the evolution of the genus. C-band data are available for *H. desmarestianus, H. oresterus, H. nelsoni,* and *H.* sp. (Rogers, 1986, 1989). Within *H. nelsoni,* heterochromatin is restricted to the centromeres of certain autosomal pairs. *Heteromys desmarestianus* and *H. oresterus* possess both centromeric and interstitial bands of heterochromatin located in the pericentromeric region of some autosomes. In addition, certain elements possess telomeric knobs of heterochromatin located on the short arms of biarmed chromosomes. However, there are no whole-arm segments of heterochromatin. An undescribed species of *Heteromys* from Costa Rica (termed *H. desmarestianus* #2 by Mascarello and Rogers, 1988) has centromeric, interstitial, and telomeric C-band positive chromatin on certain autosomal elements in addition to two autosomal pairs that are C-band positive over their entire length (Mascarello and Rogers, 1988; Rogers, 1986). Thus, NA variation within the 2n = 60 class of *Heteromys* species is due probably to euchromatic structural transpositions, such as pericentric inversions or reciprocal translocations.

Within-Species Variation

Nine karyotypic variants, based on extensive arm number variation, have been documented for *H. desmarestianus* (Burton et al., 1987; Engsrom et al., 1987; Genoways, 1973; Rogers, 1986, 1989; Table 7). Most of these are known only from one or two localities, thus no detailed geographic picture can be drawn. Nevertheless, these chromosome races appear to form discrete geographic subdivisions within the range of *desmarestianus* and have not been found to occur sympatrically at any location. Based on available C-banding data (see above), these races appear to differ by structural rearrangements that could, theoretically, result in meiotic imbalances when heterozygous. It remains to be established, however, if the recorded cytological variation within *desmarestianus* signals more than one biological unit.

Chromosomal Variation in Spiny Pocket Mice, Liomys

Genoways (1973) described non-differentially stained karyotypes for all species included within *Liomys,* while Patton (1965) and Beck and Kennedy (1977) examined karyotypes of *L. pictus.* Three classes of diploid numbers have been recorded (2n = 48, 56, and 60; Table 8). These diploid number classes parallel the major generic subdivisions based on traditional systematic treatments (Genoways, 1973), and phyletic relationships suggested by allozyme data (Rogers, 1990). The sex chromosomes are usually bi-armed and are the typical XX/XY mammalian system, with the X substantially larger than the Y.

As is the case for *Heteromys,* both Robertsonian and non-Robertsonian events are required to derive the various karyotypes within *Liomys.* Both a 2n = 60 (low NA) and a 2n = 56 (high NA) karyotype have been proposed as primitive for heteromyines (see above). Differentially stained material is limited to a description of the C-banding pattern representative of *L. salvini* (Mascarello and Rogers, 1988; Rogers, 1986). In this species, heterochromatin is restricted to the centromeric regions of a few autosomal pairs.

Known intraspecific chromosomal variation is limited to the description of two karyotypic forms of *L. pictus* that differ both in autosomal arm number and in the centromeric position of the Y-chromosome

TABLE 8.—*Chromosomal characteristics of spiny pocket mice, genus* Liomys. *Data are arranged as in Table 1. Designation of geographic chromosomal variants in* L. pictus *follows Rogers (1986).*

Taxon	2n	NA	Autosome pairs bi-	Autosome pairs uni-	X	Y	Reference[a]
adspersus	56	84	15	12	SM	M	1
irroratus	60	62	2	27	SM	ST	1, 2
	58	60	?	?	?	?	3
pictus A	48	62	8	15	M	A	4
B	48	66	10	13	M	M	1, 5, 6
salvini	56	86	16	11	SM	M	1, 7, 8
spectabilis	48	64	9	14	M	M	1

[a] 1 = Genoways, 1973; 2 = Dowler and Genoways, 1978; 3 = Makino, 1953; 4 = Patton, 1965; 5 = Beck and Kennedy, 1977; 6 = McGhee and Genoways, 1978; 7 = Carter and Genoways, 1978; 8 = Rogers, 1986.

(Table 8). The existence of these races parallels the within-species variation in allozymes recorded by Rogers (1990). Taken together, these data suggest that *L. pictus* may be a composite of two or more biological species.

Conclusions and Prospectus

In this concluding section, we identify those areas of future research on heteromyid cytogenetic systems which are likely to further an understanding of both the components of genome evolution within the family as well as across mammalian systems in general.

Systematic relationships among currently recognized species within most heteromyid genera, based on chromosomal characters, are not always concordant with traditional morphologically based species groups. This is particularly true for *Chaetodipus* and *Heteromys,* and to a lesser extent for *Dipodomys.* Whether or not this lack of concordance reflects real independence between evolution at these levels remains to be established. Certainly, there is statistical independence between chromosomal, morphologic, and genic (=allozyme) divergence patterns for kangaroo rat species (Cothran and Smith, 1983; Schnell and Selander, 1981) but not for chaetodipine pocket mice (Cothran and Smith, 1983). This points out the continued need for more refined karyotypic analyses that can assess phyletic relationships, particularly G- and other longitudinal banding studies. Despite more than 20 years of cytogenetic research on heteromyids, the major modes and mechanisms of karyotypic change remain to be firmly established for most groups.

Chromosomal race formation is common among many species of heteromyid rodents, within all genera. In only very few cases has the geographic extent of these races been fully described (*C. goldmani* [Patton, 1969*a*] is a notable exception), and the possibility that such race formation represents cryptic species remains to be investigated for most situations. The recognition of concordant cytological and genic divergence among populations formerly allocated to the kangaroo rat species *D. heermanni* in California clearly supports the possibility of cryptic chromosomal species within other taxa (Fashing, 1973; Patton et al., 1976). Thorough analyses of geographic variation patterns, coupled with an understanding of the underlying cytological mechanisms responsible for the observed variants, are needed for systematic resolution. Simple descriptions of such variation, however, will unlikely lead one closer to an understanding of the role that chromosomal change plays in evolutionary divergence (see Patton and Sherwood, 1983). The latter may only be achieved by the combination of careful analyses of the meiotic, and other, effects of chromosomal rearrangements placed in the context of the genetic demography of populations.

Finally, known cytogenetic systems of heteromyid rodents represent a gold mine of potential molecular cytogenetic problems. While many molecular genomic components of kangaroo rats have been documented, these analyses were done prior to the advent of modern recombinant DNA technology. The high variability in C-band amount and chromosomal position, the wide range in C-value reported for taxa (particularly *Dipodomys*), the expressed variability in satellite DNA fractions, and the likelihood that the latter may be conserved over a broad taxonomic range (e.g., Fry and Salser, 1977) provide enormous potential for both the molecular dissection of heteromyid karyotypes and for understanding of major components of genome evolution in general. Certainly, no real understanding of either the systematic relationships within genera or major modes of chromosomal change can be achieved without detailed consideration of the molecular genomic characterization of heteromyid cytogenetic systems.

Acknowledgments

We are indebted to Drs. Mark S. Hafner, John C. Hafner, and Steven W. Sherwood for supplying data summarized in this review, and to Drs. Mark D. Engstrom and David J. Hafner for critical editorial review. Ms. Patty Garvey-Darda prepared the photographs provided in figures 2 and 4, and Gene M. Christman drew figures 1, 3, and 5. Aspects of this research have been supported by the Museum of Vertebrate Zoology, University of California, Berkeley, and the National Science Foundation.

Literature Cited

BECK, M. L., AND M. L. KENNEDY. 1977. Chromosome study of *Liomys pictus*. Journal of the Tennessee Academy of Science, 52:109–111.

BOSTOCK, C. J., AND S. CHRISTIE. 1974. Chromosome banding and DNA replication studies on a cell line of *Dipodomys merriami*. Chromosoma, 48:73–87.

———. 1975. Chromosomes of a cell line of *Dipodomys panamintinus* (Kangaroo rat). A banding and autoradiographic study. Chromosoma, 51:25–34.

BOSTOCK, C. J., D. M. PRESCOTT, AND F. T. HATCH. 1972. Timing of replication of the satellite and main band DNA in cells of the kangaroo rat *Dipodomys ordii*. Experimental Cell Research, 74:487–495.

BOSTOCK, C. J., ET AL. 1976. S phase patterns of replication of different satellite DNA in three species of *Dipodomys* (Kangaroo rat). Journal of Molecular Biology, 108:417–433.

BURTON, D. W., J. W. BICKHAM, H. H. GENOWAYS, AND T. J. MCCARTHY. 1987. Karyotypic analysis of five rodents and a marsupial from Belize, Central America. Annals of the Carnegie Museum, 56, Art. 4:103–112.

CARTER, C. H., AND H. H. GENOWAYS. 1978. *Liomys salvini*. Mammalian Species, 84:1–5.

COTHRAN, E. G., AND M. H. SMITH. 1983. Chromosomal and genic divergence in mammals. Systematic Zoology, 32:360–368.

CROSS, J. C. 1931. A comparative study of the chromosomes of rodents. Journal of Morphology, 52: 373–402.

CSUTI, B. A. 1971. Karyotypes of kangaroo rats from southern California. Journal of Mammalogy, 52:202–206.

———. 1979. Patterns of adaptation and variation in the Great Basin kangaroo rat (*Dipodomys microps*). University of California Publications in Zoology, 111:1–69.

DINGMAN, R. E., R. J. DAVIS, AND D. HUNSAKER, II. 1971. Karyotypes of the Panamint kangaroo rat (*Dipodomys panamintinus* (Merriam)). Experientia, 27 (12):1491–1492.

DOWLER, R. C., AND H. H. GENOWAYS. 1978. *Liomys irroratus*. Mammalian Species, 82:1–6.

ENGSTROM, M. D., H. H. GENOWAYS, AND P. K. TUCKER. 1987. Morphological variation, karyology, and systematic relationships of *Heteromys gaumeri* (Rodentia: Heteromyidae). Pp. 296–303, *in* Studies in Neotropical mammalogy: essays in honor of Philip Hershkovitz (B. D. Patterson and R. M. Timm, eds.). Fieldiana: Zoology, (New Series), 39:1–506.

FASHING, N. J. 1973. Implications of karyotypic variation in the kangaroo rat, *Dipodomys heermanni*. Journal of Mammalogy, 54:1018–1020.

FRY, K., AND W. SALSER. 1977. Nucleotide sequences of HS-a satellite DNA from the kangaroo rat *Dipodomys ordii* and characterization of similar sequences in other rodents. Cell, 12:1069–1084.

FRY, K., ET AL. 1973. Nucleotide sequence of HS-b satellite DNA from the kangaroo rat *Dipodomys ordii*. Proceedings of the National Academy of Sciences (USA), 70(9):2642–2646.

FUTCHER, A. D. 1974. Biosystematics of the heermanni subgroup A of the genus *Dipodomys* (Rodentia: Heteromyidae). Ph.D. dissert., Loma Linda University, 74 pp.

GENOWAYS, H. H. 1973. Systematics and evolutionary relationships of the spiny pocket mice, genus *Liomys*. Special Publications of The Museum, Texas Tech University, 5:1–368.

GRINNELL, J. 1922. A geographical study of kangaroo rats in California. University of California Publications in Zoology, 24:1–124.

HAFNER, D. J., J. C. HAFNER, AND M. S. HAFNER. 1979. Systematic status of kangaroo mice, genus *Microdipodops*: morphometric, chromosomal, and protein analyses. Journal of Mammalogy, 60:1–10.

HAFNER, J. C. 1981. Evolution, systematics, and historical zoogeography of kangaroo mice, genus *Microdipodops*. Ph.D. dissert., University of California, Berkeley, 269 pp.

HAFNER, J. C., AND M. S. HAFNER. 1983. Evolutionary relationships of heteromyid rodents. Great Basin Naturalist Memoirs, No. 7, pp. 3–29.

HALL, E. R. 1981. The mammals of North America. John Wiley and Sons, New York, vol. 1.

HATCH, F. T., AND A. V. CARRANO. 1978. Satellite DNA, heterochromatin and rearrangement of the genome. Mutation Research, 53:196–197.

HATCH, F. T., AND J. A. MAZRIMAS. 1974. Fractionation and characterization of satellite DNAs of the kangaroo rat (*Dipodomys ordii*). Nucleic Acids Research, 1:559–575.

HATCH, F. T., A. J. BODNER, J. A. MAZRIMAS, AND D. H. MOORE, II. 1976. Satellite DNA and cytogenetic evolution. DNA quantity, satellite DNA and karyotypic variations in kangaroo rats (genus *Dipodomys*). Chromosoma, 58:155–168.

HATCH, F. T., ET AL. 1973. Properties and chromosomal location of satellite DNA in *Dipodomys* kangaroo rat species. Genetics, 74:s111–s112.

JACKSON, L., AND D. HUNSAKER, II. 1971. Chromosomes in the genus *Dipodomys*. Mammalian Chromosome Newsletter, 12(2):50–51.

JOHN, B., AND G. L. G. MIKLOS. 1979. Functional aspects of satellite DNA and heterochromatin. International Review of Cytology, 58:1–113.

JONES, R. N., AND H. REES. 1982. B chromosomes. Academic Press, New York and London.

LIDICKER, W. Z., JR. 1960. An analysis of intraspecific variation in the kangaroo rat *Dipodomys merriami*. University of California Publications in Zoology, 67: 125–218.

MAKINO, S. 1953. Chromosome numbers of some American rodents. Science, 118:630.

MASCARELLO, J. T., AND D. S. ROGERS. 1988. Banded chromosomes of *Liomys salvini, Heteromys oresterus*, and *H. desmarestianus*. Journal of Mammalogy, 69:126–130.

MASCARELLO, J. T., A. D. STOCK, AND S. PATHAK. 1974. Conservatism in the arrangement of genetic material in rodents. Journal of Mammalogy, 55:695–704.

MATTHEY, R. 1952. Chromosome de Muridae (II). Experientia, 8:389–390.

———. 1956. Nouveau apports a la cytologie comparee des rongeurs. Chromosoma, 7:670–692.

MAZRIMAS, J. A. 1976. Similarity of satellite DNAs from four rodent families: Caviidae, Geomyidae, Heteromyidae, and Sciuridae. Federation Proceedings, 35(7):1676.

MAZRIMAS, J. A., AND F. T. HATCH. 1972. A possible relationship between satellite DNA and the evolution of kangaroo rat species (genus *Dipodomys*). Nature, New Biology, 240:102–105.

———. 1977. Similarity of satellite DNA properties in the Order Rodentia. Nucleic Acids Research, 4:3215–3227.

MCGHEE, H. E., AND H. H. GENOWAYS. 1978. *Liomys pictus*. Mammalian Species, 83:1–5.

O'FARRELL, M. J., AND A. R. BLAUSTEIN. 1974*a*. *Microdipodops megacephalus*. Mammalian Species, 46: 1–3.

———. 1974*b*. *Microdipodops pallidus*. Mammalian Species, 47:1–2.

OSGOOD, W. F. 1900. Revision of the pocket mice of the genus *Perognathus*. North American Fauna, 18:1–63.

PATTON, J. L. 1965. Cytotaxonomy of the pocket mice, genus *Perognathus* (Rodentia: Heteromyidae). M.S. thesis, University of Arizona, Tucson.

———. 1967*a*. Chromosomal studies of certain pocket mice, genus *Perognathus* (Rodentia: Heteromyidae). Journal of Mammalogy, 48:27–37.

———. 1967*b*. Chromosomes and evolutionary trends in the pocket mouse subgenus *Perognathus* (Rodentia: Heteromyidae). Southwestern Naturalist, 12:429–438.

———. 1969*a*. Chromosome evolution in the pocket mouse, *Perognathus goldmani* Osgood. Evolution, 23:645–662.

———. 1969*b*. Karyotypic variation in the pocket mouse, *Perognathus penicillatus* Woodhouse (Rodentia-Heteromyidae). Caryologia, 22:351–358.

———. 1970. Karyotypes of five species of pocket mice, genus *Perognathus* (Rodentia: Heteromyidae), and a summary of chromosome data for the genus. Mammalian Chromosome Newsletter, 11:3–8.

———. 1972. A complex system of chromosomal variation in the pocket mouse, *Perognathus baileyi* Merriam. Chromosoma, 36:241–255.

———. 1977. B-chromosome systems in the pocket mouse *Perognathus baileyi*: meiosis and C-band studies. Chromosoma, 60:1–14.

PATTON, J. L., AND S. W. SHERWOOD. 1983. Chromosome evolution and speciation in rodents. Annual Review of Ecology and Systematics, 14:139–158.

PATTON, J. L., AND O. H. SOULE. 1967. Natural hybridization in pocket mice, genus *Perognathus*. Mammalian Chromosome Newsletter, 8:263–264.

PATTON, J. L., H. MACARTHUR, AND S. Y. YANG. 1976. Systematic relationships of the four-toed populations of *Dipodomys heermanni*. Journal of Mammalogy, 57:159–163.

PATTON, J. L., S. W. SHERWOOD, AND S. Y. YANG. 1981. Biochemical systematics of chaetodipine pocket mice, genus *Perognathus*. Journal of Mammalogy, 62:477–492.

PRESCOTT, D. M., C. J. BOSTOCK, F. T. HATCH, AND J. A. MAZRIMAS. 1973. Location of satellite DNA in the chromosomes of the kangaroo rat *Dipodomys ordii*. Chromosoma, 42:205–213.

ROGERS, D. S. 1986. Evolutionary relationships within the subfamily Heteromyinae (genera *Heteromys* and *Liomys*). Ph.D. dissert., University of California, Berkeley, 247 pp.

———. 1989. Evolutionary implications of chromosomal variation among spiny pocket mice, genus *Heteromys* (Order Rodentia). Southwestern Naturalist, 34:85–100.

———. 1990. Genic evolution, historical biogeography, and systematic relationships among spiny pocket mice (Subfamily Heteromyinae). Journal of Mammalogy, In press.

SALSER, W., ET AL. 1976. Investigation of the organization of mammalian chromosomes at the DNA sequence level. Federation Proceedings, 35:23–35.

SCHNELL, G. D., AND R. K. SELANDER. 1981. Environmental and morphological correlates of genetic variation in mammals. Pp. 60–99, *in* Mammalian population genetics (M. H. Smith and J. J. Joule, eds.). University of Georgia Press, Athens.

SCHRECK, R. R., B. F. ERLANGER, AND O. J. MILLER. 1977. Binding of anti-nucleoside antibodies reveals different classes of DNA in the kangaroo rat, *Dipodomys ordii.* Experimental Cell Research, 108:403–412.

SETZER, H. W. 1949. Subspeciation in the kangaroo rat *Dipodomys ordii.* University of Kansas Publications, Museum of Natural History, 1:473–573.

SHERWOOD, S. W. 1983. Supernumerary chromosomes in *Perognathus baileyi* (Rodentia: Heteromyidae): molecular and population cytogenetics of a mammalian B-chromosome system. Ph.D. dissert., University of California, Berkeley, 194 pp.

STOCK, A. D. 1974. Chromosome evolution in the genus *Dipodomys* and its taxonomic and phylogenetic implications. Journal of Mammalogy, 55:505–526.

STRANEY, D. O., AND J. L. PATTON. 1980. Phylogenetic and environmental determinants of geographic variation in the pocket mouse *Perognathus goldmani* Osgood. Evolution, 34:888–903.

TOWNER, J. W. 1965. The effect of radio-active fallout at the Nevada Test Site on the chromosomes of the pocket mouse. Health Physics, 11:1569–1571.

VOLOBUJEV, V. T. 1980. The B-chromosome systems of mammals. Genetica, 53:333–337.

———. 1982. B-chromosome systems of mammals. Caryologia, 34:1–22.

WHITE, M. J. D. 1978. Modes of speciation. W. H. Freeman and Co., San Francisco, 455 pp.

WILLIAMS, D. F. 1978*a*. Karyological affinities of the species groups of silky pocket mice (Rodentia, Heteromyidae). Journal of Mammalogy, 59:599–612.

———. 1978*b*. Systematics and ecogeographic variation of the Apache pocket mouse (Rodentia: Heteromyidae). Bulletin of the Carnegie Museum of Natural History, No. 10, 57 pp.

BIOCHEMICAL GENETICS

JAMES L. PATTON AND DUKE S. ROGERS

Introduction

General Content and Perspective

This review will focus on the degrees of genetic variability and differentiation at the population and species level in heteromyid rodents. Aspects of phyletic relatedness among species within genera will be considered as far as the published literature permits, but the general questions of the definition of major clades and their relationships within the family, and within the Geomyoidea, based on biochemical data, will not be addressed here. These are topics that other review chapters can deal with more effectively and more comprehensively.

The title of this chapter is misleading, however, as it suggests that a more comprehensive knowledge of the biochemical genetics of heteromyid rodents is known, and will be summarized here, than is in fact the case. Indeed, and with very few exceptions (e.g., Hafner, 1982; Thomas et al., 1990), the only biochemical methodology that has been applied to aspects of heteromyid biology is that of protein electrophoresis (isozyme and allozyme analyses). Moreover, the scope of the questions asked in the published studies has been both narrow and descriptive in content, centering only on (1) amounts of within and among population genic variation as measured by this approach and (2) suggestions of phyletic relationship among taxa within and among the recognized genera in the family. Most studies are relatively old, at least in the context of the time frame over which protein electrophoresis has been applied to these issues. More recent developments, including both empirical and theoretical advances, are therefore generally lacking from the existing heteromyid literature on this topic. Clearly, much work remains to be accomplished, and the recent developments in DNA sequencing and fingerprinting offer considerable excitement for future studies. Hence, the data summarized here should be considered only as an introduction both to our understanding of the levels of biochemical diversity within heteromyid taxa as well as to the types of questions to which these data can be applied. Thus, while this review necessarily will be short, it will, we hope, point out major gaps in our current knowledge and indicate where future research likely will yield significant advances.

The Focus of Biochemical Systematic Studies in Heteromyid Rodents

Studies of biochemical diversity in mammalian taxa have generally centered on one or more of three broad aspects: First, there has been much interest in measuring levels of within population genetic variability, primarily by recording the numbers of segregating alleles at genetic loci and by estimating the levels of individual heterozygosity for these loci. These data, in turn, have been used to support specific hypotheses explaining patterns and degrees of observed variability, such as relating that variability to niche characteristics, temporal patterns and degrees of population bottlenecks, or other aspects of historical versus selective perspectives on the nature of the maintenance of genetic variation in natural populations (Lewontin, 1974; Nevo, 1978; Nevo et al., 1984; Selander, 1976; Soulé, 1976). Secondly, protein variation has been used to estimate relatedness among taxa, from the population to higher taxonomic levels, with an emphasis on understanding both cladogenic and anagenic patterns and processes (e.g., Avise, 1976; Buth, 1984; Straney, 1981). Finally, allozymes have been used to assess aspects of population structure, or what can generally be referred to as genetic demography, with a focus on issues as diverse as speciation models, mating systems, dispersal patterns, and so forth (Selander and Whittam, 1983; Gaines, 1985). Studies of biochemical variability within heteromyids have been used almost entirely to address the first two points.

In the following sections, we summarize available data on allozyme variability and differentiation for heteromyid taxa. In doing so, we point out areas where we consider there are unanswered questions, to which genetic perspectives can and should be applied with increased interest. Data are available only for the genera *Dipodomys* (Beck et al., 1981; Best et al., 1986; Johnson and Selander, 1971; Patton et al., 1976), *Microdipodops* (Hafner et al., 1979; Hafner, 1981); *Chaetodipus* (Patton et al., 1981), *Heteromys* (Rogers, 1990), and *Liomys* (Rogers, 1990); to date, no data for *Perognathus* have been published.

Electromorphic Variation Within Populations

General Levels

Measures of genic variability within species for each genus of heteromyid for which data are available are given in Tables 1 through 4. Several features of these data, however, need to be emphasized prior to initiating any discussion of the observed levels and their potential meaning. For one, heterozygosity estimates of the kind produced from electromorphic analyses have quite large inter- and intra-locus theoretical sampling errors (Nei and Roychourdhuri, 1974); these errors are usually neglected in summary statistics, such as those presented here. Secondly, most studies try to maximize the sample size of individuals within a population, or number of populations, rather than the number of loci examined, yet it is the latter which may be most critical for estimating true heterozygosity (Gorman and Renzi, 1979; Nei, 1978). Finally, the large intra-locus sampling variance is due, in part, to quite disparate evolutionary rates among proteins (Nei, 1975; Sarich, 1977). With few exceptions (Elliott et al., 1989; Hafner, 1981; Hamilton et al. 1987), most studies on heteromyids have neglected esterases or general serum proteins, a group of proteins characterized by high levels of variation within populations because of higher than average rates of evolutionary change (Sarich, 1977). Hence, any comparisons among the various heteromyid species and genera, or between these and other groups of mammals, should take into account differences in the numbers as well as the specific protein loci examined by particular studies.

TABLE 1.—*Genic variation in pocket mice,* Chaetopidus *(data from Patton et al., 1981).*

Taxon	N_P*	N_I*	N_L*	Mean alleles/ locus	Mean proportion of loci	
					P*	H* (range)
arenarius	1	18	26	1.115	0.115	0.041
artus	2	12	26	1.400	0.192	0.072 (0.058–0.087)
baileyi	21	289	26	2.808	0.209	0.049 (0.015–0.154)
californicus	2	21	26	1.192	0.115	0.096 (0.077–0.115)
fallax	5	16	27	1.346	0.105	0.059 (0.039–0.074)
formosus	7	49	26	1.308	0.079	0.022 (0.000–0.038)
goldmani	19	256	26	1.418	0.276	0.027 (0.000–0.045)
hispidus	1	2	26	1.000	0.000	0.000
intermedius	6	47	24	1.792	0.160	0.054 (0.037–0.063)
nelsoni	2	35	28	1.250	0.143	0.022 (0.018–0.026)
penicillatus	22	179	26	1.603	0.167	0.047 (0.008–0.077)
pernix	6	40	26	1.404	0.250	0.084 (0.067–0.103)
spinatus	2	3	28	1.107	0.071	0.045 (0.018–0.071)
Unweighted mean:				1.442	0.145	0.048

* N_P = number of sampled populations; N_I = total number of individuals; N_L = number of loci; P = proportion of loci polymorphic (0.95 criterion) per population; H = proportion of loci heterozygous per individual (direct count method).

TABLE 2.—*Genetic variation in kangaroo rats,* Dipodomys. *Data from Johnson and Selander (1971), Patton et al. (1976), Beck et al. (1981), Best et al. (1986), Hamilton et al. (1987), and Elliott et al. (1989). See Table 1 for explanation of abbreviations.*

Taxon	N_P	N_I	N_L	Mean alleles/ locus	Mean proportion of loci	
					P	H (range)
agilis	1	14	18	1.11	0.11	0.040
agilis	12	179	16	1.31	0.24	0.011 (0.000–0.030)
californicus	3	41	22	1.09	0.02	0.006 (0.000–0.019)
compactus	3	34	18	1.03	0.03	0.023
deserti	4	22	18	1.06	0.06	0.010
elator	1	23	18	1.06	0.00	0.002
elator	3	21	29	1.24	—	—
gravipes	1	2	16	1.00	0.00	0.000
heermanni	3	49	18	1.20	0.17	0.042
heermanni	3	30	22	1.23	0.08	0.024 (0.000–0.037)
merriami	11	251	18	1.14	0.16	0.051
merriami	1	6	22	1.14	0.14	0.061
merriami	1	10	16	1.00	0.06	0.000
microps	7	103	18	1.03	0.03	0.007
nitratoides	1	7	18	1.22	0.22	0.040
nitratoides	1	14	22	1.23	0.14	0.065
ordii	12	405	18	1.03	0.03	0.008
ordii	10	220	14	1.71	0.21	0.024 (0.008–0.051)
panamintinus	2	10	18	1.00	0.00	0.000
panamintinus	1	8	22	1.05	0.05	0.006
spectabilis	2	48	18	1.06	0.06	0.008
spectabilis	2	~150	~27	1.21	0.21	0.051
Unweighted mean:				1.143	0.088	0.0228

TABLE 3.—*Genetic variation in kangaroo mice,* Microdipodops. *Data from Hafner et al. (1979) and Hafner (1981). See Table 1 for explanation of abbreviations.*

Taxon	N_P	N_I	N_L	Mean alleles/ locus	Mean proportion of loci P	H (range)
megacephalus	1	20	22	1.272	0.217	0.064
megacephalus	1		?*	1.448	0.344	0.029 (0.028–0.031)
pallidus	1	20	22	1.272	0.217	0.064
pallidus	1		?*	1.375	0.344	0.005
Unweighted mean:				1.317	0.281	0.041

* Hafner (1981) used polyacrylamide gel electrophoresis to separate non-specific plasma proteins and liver esterases. The exact number of loci comparable to data from starch gel electrophoresis is unknown.

Variability Levels Within and Among Species

With the above caveats in mind, the following generalizations can be made with regard to genetic variability within and among populations, species, or genera of heteromyid rodents. Species within genera generally exhibit rather uniform average levels of genic variability, as measured by either the number of alleles per locus, loci polymorphic per population, or loci heterozygous per individual. These levels are also somewhat low in comparison to those reported for other mammals (see reviews by Nevo, 1978; Powell, 1975; Selander and Kaufman, 1973). The average rodent species, for example, is heterozygous at between 4 and 5 percent of its allozyme loci (Nevo, 1978; Selander and Kaufman, 1973). Among heteromyids, only *Chaetodipus* and *Microdipodops* are characterized by levels of this magnitude (Tables 1 and 3). *Dipodomys* has approximately one-half this average level (Table 2), and both *Heteromys* and *Liomys* have about one-fourth this level (Table 4).

Nevertheless, extensive variation in heterozygosity does characterize many species of both *Chaetodipus* and *Dipodomys*. Indeed, the observed ranges for *C. baileyi* (0.015–0.154; Patton et al., 1981), for *D. ordii* (0.006–0.051; Beck et al., 1981), and for *D. spectabilis* (0.006–0.068; Johnson and Selander, 1971; Elliott et al., 1989), for example, span the range across all species in each genus. Such infra-specific variation must be considered in any discussion of variability patterns among species. Moreover, following from the caveats given above, such variability must be viewed with caution when building broadscale conclusions regarding pattern and process related to heterozygosity values. In the case of the first two examples listed above, the measures are based on the same set of loci and nearly equivalent sample sizes. However, for *D. spectabilis,* the variance in heterozygosity reported comes from separate studies, one (Johnson and Selander, 1971) that included mostly rather conservative intermediary metabolism enzymes, while the second (Elliott et al., 1989) included data from a number of typically more highly variable esterase and general protein loci. For this species, therefore, it is unclear as to how much of the reported variance in individual heterozygosity results from real geographic variation or only from the choice of proteins from which data were collected.

As yet, there are no convincing arguments to support any particular interpretation for either the range of variability levels observed across heteromyid genera or for the expressed low levels in some taxa. For example, Johnson and Selander (1971) reported a positive correlation between heterozygosity and habitat tolerance among species of kangaroo rats. Whether such a correlation in fact exists remains to be es-

TABLE 4.—*Genetic variation in spiny pocket mice,* Liomys *and* Heteromys, *subfamily Heteromyinae (data from Rogers, 1986). See Table 1 for explanation of abbreviations.*

Taxon	N_P	N_I	N_L	Mean alleles/ locus	Mean proportion of loci	
					P	H (range)
Liomys						
adespersus	1	1	30	1.000	0.033	0.003
irroratus	2	8	30	1.050	0.050	0.011 (0.000–0.022)
pictus	3	12	30	1.067	0.050	0.018 (0.022–0.033)
salvini	2	10	30	1.100	0.084	0.016 (0.000–0.033)
spectabilis	1	1	30	1.000	0.000	0.000
Unweighted mean:				1.043	0.043	0.016
Heteromys						
anomalus	3	10	30	1.130	0.055	0.010 (0.000–0.020)
australis	1	2	30	1.000	0.033	0.033
desmarestianus	15	74	30	1.050	0.044	0.010 (0.000–0.033)
gaumeri	4	25	30	1.020	0.025	0.002 (0.000–0.007)
nelsoni	1	6	30	1.000	0.033	0.006
oresterus	1	6	30	1.000	0.033	0.000
sp. A	2	9	30	1.050	0.033	0.007 (0.000–0.014)
Unweighted mean:				1.031	0.036	0.013

tablished (see, for example, Beck et al., 1981). Even so, it is not possible to relate it in a causal sense to any one of the many competing general hypotheses proposed to explain patterns of intrapopulational heterozygosity (Selander and Kaufman, 1973; Soulé, 1976, Valentine, 1976), as Johnson and Selander (1971) correctly noted. For example, habitat tolerance may have both a direct and positive relationship to effective population size and, thereby, to the probability of historical and/or continued population bottlenecks, as well as to a measure of habitat variability experienced by a population and species, which might be taken to infer general niche dimensions. One such attempt was the selective based niche-width variation hypothesis (Nevo et al., 1984; Soulé, 1976). Consequently, the low levels of variability that characterize *Heteromys,* for example, can be explained for the moment just as easily by historical processes, including phylogenetic inertia, as by any other hypothesis. This general conundrum is, unfortunately, not peculiar to heteromyid rodents, but is a general problem of population genetics (see Lewontin, 1974, for review).

Beyond simple correlations of the type developed by Johnson and Selander (1971), however, no investigation has explored the variation patterns exhibited by heteromyids in an explicit way to attempt a test of competing explanations (such as has been done for birds by Barrowclough et al., 1985).

Genetic Demographic Studies

The use of genetic markers as a tool to investigate aspects of the demography of populations of heteromyid rodents has been extremely limited. Only in the case of *D. merriami* (Johnson and Selander, 1971) has such an attempt been made, and even this one is only at the most simple, descriptive level. In their analysis, Johnson and Selander examined the spatial relationships among individual genotypes within a population of *D. merriami* over a 10-acre area near Kramer, California. They concluded, without statistical corroboration, that spa-

TABLE 5.—*Levels of within-species electromorphic distance (Rogers' D) for selected species of pocket mice,* Chaetodipus. *Data from Patton et al. (1981).*

Taxon	N_P*	D-value ± *SD*
baileyi	21	0.088 ± 0.019
fallax	5	0.002 ± 0.003
formosus	7	0.012 ± 0.003
goldmani	19	
Within cytotypes		0.036 ± 0.023
Among 5 cytotypes		0.056 ± 0.028
intermedius	6	0.025 ± 0.004
nelsoni		
Between 2 cytotypes		0.131
penicillatus	21	
Within cytotypes		0.036 ± 0.012
Among 3 cytotypes		0.184 ± 0.017
pernix	6	
Within cytotypes		0.025 ± 0.007
Between 2 cytotypes		0.079 ± 0.007

* N_P = number of sampled populations.

tial clustering of genotypes was evident at two loci, and suggested that local structure, including the possibility of inbreeding, may characterize local kangaroo rat populations. Such structure may thus be a contributing factor to the observed variance in among-population differentiation (see below).

Only two recent studies (Elliott et al., 1989 for *D. spectabilis* and Hamilton et al., 1987 for *D. elator*) have examined the level of inbreeding in local populations through the use of Wright's (1965) *F*-statistics, specifically calculation of F_{is}, a measure of nonrandom breeding within subpopulations. In the first case, there was no evidence for inbreeding, as there was no consistency of significant and positive F_{is} values across loci. In the second, however, these values were all positive and high, indicating the heterozygote deficiency expected from high inbreeding. However, the small sample sizes and the possibility of pooling across subpopulations (Wahlund effect; Wahlund, 1928) provide other, currently equally viable, explanations.

Considering the variation in mating systems, dispersal patterns, and other attributes contributing to demography in heteromyid species (e.g., Jones, this volume), genetic data may be of considerable use in examining the validity of various behavioral ecological models, as well as the efficacy of evolutionary paradigms associated with speciation mode. This avenue of investigation is wide open, but it is likely that techniques more refined than protein electrophoresis will need to be used, considering the general low levels of variability in standard allozymes exhibited by many heteromyid taxa, especially kangaroo rats and mice and spiny pocket mice. Both DNA fingerprinting and denaturing gradient gel electrophoresis are two such techniques now readily available.

Levels of Genetic Distance Among Taxa

Genic data based on allozyme surveys, and summarized in the form of genetic similarity/distance measures (Nei, 1972; Rogers, 1972), are available for all taxa of heteromyids with the exception of *Perognathus* species (Tables 5–7). As with estimates of genic heterozygosity, genetic distance is sensitive to evolutionary rate differences among proteins (Sarich, 1977) and to the sample sizes of both number of loci and number of individuals examined (Gorman and Renzi, 1979; Nei, 1978). For example, the latter authors have shown that the variance of the genetic distance becomes small and constant only when 50 loci or more are studied, if individual sample sizes are low. While there is a wide variance in the number of individuals and populations examined for any given species of heteromyid, in no case does the number of loci approach this theoretical optimum. For general conclusions regarding the overall degree of differentiation at the population or species level, this is probably not a substantive handicap. However, the topological reliability of dendrograms generated from genic distance data is a function of the number of loci examined (Nei et al., 1983). Thus, caution must be used in interpreting genetic distance data available for most heteromyids, particularly those summarized below regarding phylogenetic conclusions.

Differentiation Within Species

Species of all heteromyid rodents examined to date are characterized generally by high degrees of genetic similarity among their sampled populations, with the genetic distance (Rogers, 1972; *D*) less than 0.09 for most species of *Chaetodipus* (Patton et al., 1981; Table 5) and *Dipodomys* (Best et al., 1986; Hamilton et al., 1987; Johnson and Selander, 1971; Table 6). These levels are equivalent to general degrees of local population differentiation observed for most mammals (Avise, 1976; Selander and Johnson, 1973). However, there is a notable increase in genetic distance among segments of the geographic range of several species of *Chaetodipus, Dipodomys, Heteromys,* and *Liomys,* an increase that is coincidental with fixed chromosomal differentiation. For example, while within-cytotype distance averages only 0.036 (range 0.023–0.056) for *Chaetodipus penicillatus,* the among-cytotype distance value is significantly higher, averaging 0.186 (range 0.104–0.251; Patton et al., 1981). A similar pattern of appreciably higher among-cytotype divergence values relative to within-cytotype ones characterizes most species of heteromyids that exhibit chromosome race formation, including *penicillatus, baileyi, nelsoni,* and *pernix* of the genus *Chaetodipus, Dipodomys agilis, Heteromys desmarestianus,* and *Liomys pictus* (Tables 5–7). While the exact level of divergence is different among these species, this pattern suggests that some of these chromosome forms may represent separate biological species. Available data support such an hypothesis, at least for the cytotypes of *C. penicillatus* (Patton et al., 1981), but the detailed contact zone analyses necessary to test this possibility for any of the cases cited above are lacking to date.

TABLE 6.—*Levels of within-species electromorphic distance among selected species of kangaroo rats,* Dipodomys. *Data from Johnson and Selander (1971), unless otherwise noted. See Table 5 for explanation of abbreviations.*

Taxon	N_P	D-value (range)
agilis	12[a]	0.130 (0.21–0.04)
Within cytotypes		0.116 (0.21–0.04)
Between cytotypes		0.151 (0.21–0.10)
ordii	12	0.037 (0.00–0.07)
	10[b]	0.070 (0.03–0.13)
merriami	11	0.046 (0.01–0.08)
microps	7	0.006 (0.00–0.02)
deserti	4	0.003 (0.00–0.01)
compactus	3	0.013 (0.00–0.02)
heermanni	3	0.025 (0.02–0.03)
panamintinus	2	0.059
spectabilis	2	0.084

[a] Data from Best et al. (1986); [b] data from Beck et al. (1981).

Certainly, a large number of heteromyid species are characterized by inter-population karyotypic differentiation (see Patton and Rogers, this volume); each of these should be examined in detail for allozyme divergence if general relationships between these genetic variables are to be sought.

Phylogenetic Hypotheses Based on Genetic Distances

Estimates of relatedness among species of the various heteromyid rodent genera based on genetic distances have provided valuable insights into three components of higher classification in these groups. At one level, genetic data have uncovered apparent cryptic species; at another, they have been used to explore patterns of phylogenesis; and, finally, at a third level, these data have challenged the traditional taxonomy of the groups in question. Selected examples of each of these contributions are given here.

Recognition of cryptic species.—The concordance between genetic distance values and chromosome race formation in some currently recognized heteromyid species,

TABLE 7.—*Levels of within-species electromorphic distance among species of the heteromyine genera* Heteromys *and* Liomys. *Data from Rogers (1986). See Table 5 for explanation of abbreviations.*

Taxon	N_P	D-value (range)
Heteromys		
desmarestianus[a]	14	0.164 (0.005–0.280)
species A	2	0.045
guameri	4	0.025 (0.000–0.030)
anomalus	2	0.114
Liomys		
salvini	2	0.203
irroratus	2	0.128
pictus[a]	3	0.293 (0.170–0.364)

[a] These two species are characterized by extensive chromosome race formation; 8 races have been described for *H. desmarestianus* and 2 for *L. pictus* (see Rogers, 1986; Patton and Rogers, this volume).

and their combined relationship to interpretation of biological species status, has been mentioned above. There are at least four other cases, however, where the degree of genetic divergence observed among populations of a single presumptive species supports the hypothesis that two separate biological units are in fact involved. This applies to *Dipodomys compactus* (Johnson and Selander, 1971; Schmidly and Hendricks, 1976) and *D. californicus* (Patton et al., 1976), both of which were elevated to specific status primarily by virtue of the degree of their genetic divergence from *D. ordii* and *D. heermanni*, respectively. Similarly, Rogers (1990) recognized an undescribed species of *Heteromys* from Costa Rica, based on the large genetic distances of two representative samples to all other species and populations of the genus that he examined.

Patterns of phylogenesis.—The temporal pattern of species formation based on genetic differentiation has only been examined for pocket mice of the genera *Chaetodipus* (Patton et al., 1981), *Liomys*, and *Heteromys* (Rogers, 1990). In these cases, phyletic events leading to the extant species are concentrated early in the radiation of each genus. This conclusion is based on either Fitch-Margoliash (1967) or Wagner (Farris, 1972) distance analyses, both of which apportion the amount of accumulated genetic change to internal versus terminal branches in a resulting tree. For the Wagner analysis, for example, Patton et al. (1981) found that while the distances between internal nodes averaged only 0.033 (range 0.01–0.10; Rogers' distance), differentiation along the terminal branches was five times higher, averaging 0.151 (range 0.05–0.29). This in turn suggests that cladogenic events leading to the extant species within the genus were both relatively old and nearly simultaneous, a critical realization for the eventual understanding of patterns of adaptive radiation in the group.

Systematic conclusions derived from electromorphic data.—It is an often recognized fact that phylogenetic estimates based on electromorphic analyses may not be particularly concordant with those suggested from other data sources (Straney, 1981). The available data for heteromyid rodents fully support this observation. One of the earliest attempts to examine the relationship between classifications suggested by morphological and biochemical data was that for *Dipodomys*, initially by Johnson and Selander (1971) and subsequently by Schnell et al. (1978). The latter study clearly shows that phyletic relationships based on electromorphic distance data are discordant with those derived from morphological data. Moreover, the resulting allozyme classification is the most divergent when comparisons are made between the schemes suggested by all previous authors.

This general picture of discordance between estimates of species relationships based on traditional morphological and biochemical data also characterizes the grouping of species in both *Chaetodipus* and *Heteromys*. For the former, the species groups of Merriam (1899) and Osgood (1900) are not substantiated by the electromorphic data (Patton et al., 1981). Similarly, the subgenus *Xylomys* of *Heteromys* is clearly not the monophyletic unit suggested by morphology when the genetic distance between its member species relative to each other and to those of the nominate subgenus are considered (Rogers, 1986, 1990).

There are several different explanations for this set of observations, none of which is mutually exclusive and none of which has been examined in the case of heteromyid rodents. For one, current views of relationships among species for each heteromyid genus are based on a phenetic assessment of a variety of morphological characters. Until these are examined in a cladistic fashion, something yet to be accomplished for any heteromyid group, the various suggested classificatory schemes based on morphology are in themselves open to question. Secondly, the published electromorphic data have been treated largely phenetically and solely as distance matrix summaries. Mickevich and Johnson (1976) and Farris (1981, 1983), among many other recent authors, have emphasized the need to construct trees from character states, rather than distance matrices, an approach that is fully amenable to, but historically rarely used for, electromorphic data (see general review by Buth, 1984). Since this has not been done for any heteromyid taxon, except for the family and subfamily levels (Hafner, 1982), it may be just as premature to conclude phylogenetic relationships within genera from the published electromorphic data as it is from morphology.

However, even if data are reanalyzed by other procedures, it is still likely that discordance will exist between estimates of relatedness based on biochemical and morphological characters. For example, there is a clear indication that evolution at the level of structural proteins, or other molecular variables, is generally uncoupled from that at the phenotypic level of gross morphology (Schnell and Selander, 1981; Straney, 1981; Wilson, 1976). While biochemical divergence may proceed relatively uniformly along all branches subsequent to a cladogenic event, morphological divergence need not, and indeed usually does not (Gould, 1980; Wilson et al., 1977). Hence, many authors argue that allozyme differentiation will mirror more closely phyletic events than will morphological characters. Whether or not this holds for the various heteromyid genera reviewed here remains to be determined by future studies, which will involve, at a minimum, more sophisticated biochemical techniques, appropriate methods of data analysis and phylogenetic reconstruction, and a thorough cladistic analysis of morphological characters to complement the biochemical data.

Acknowledgments

We are indebted to Mark S. Hafner, John C. Hafner, Margaret F. Smith, and Donald O. Straney for discussions on numerous aspects of heteromyid biology and genetic variation over the past decade. Particular appreciation is extended to Suh Y. Yang, Monica Frelow, and Juliana H. Feder for invaluable assistance in the laboratory. Biochemical studies on heteromyines were supported by National Science Foundation grant DEB 82-02270 to DSR. This review was prepared, in part, with financial help from NSF grant BSR 84-09175 to JLP, and the Museum of Vertebrate Zoology.

Literature Cited

AVISE, J. 1976. Genetic differentiation during speciation. Pp. 106–122, *in* Molecular evolution (F.J. Ayala, ed.). Sinauer Assoc., Inc., Sunderland, Massachusetts.

BARROWCLOUGH, G. F., N. K. JOHNSON, AND R. M. ZINK. 1985. On the nature of genic variation in birds. Pp. 135–154, *in* Current ornithology, vol. 2 (R. F. Johnston, ed.). Plenum Publ. Corp., New York.

BECK, M. L., M. L. KENNEDY, AND C. J. BIGGERS. 1981. Genic variation in Ord's kangaroo rat *Dipodomys ordii* in Oklahoma. American Midland Naturalist, 105:77–83.

BEST, T. L., R. M. SULLIVAN, J. A. COOK, AND T. L. YATES. 1986. Chromosomal, genic, and morphologic variation in the agile kangaroo rat, *Dipodomys agilis* (Rodentia: Heteromyidae). Systematic Zoology, 35:311–324.

BUTH, D. G. 1984. The application of electromorphic data in systematic studies. Annual Review of Ecology and Systematics, 15:501–522.

ELLIOTT, L. F., P. M. WASER, G. F. MCCRACKEN, N.

E. Link, and M. K. Gustin. 1989. Genetic variation in two populations of *Dipodomys spectabilis*. Journal of Mammalogy, 70:852–855.

Farris, J. S. 1972. Estimating phylogenetic trees from distance matrices. American Naturalist, 106:645–668.

———. 1981. Distance data in phylogenetic analysis. Pp. 3–23, *in* Advances in cladistics, vol. 1, Proceedings of 1st Meeting of the Willi Hennig Society (V. A. Funk and D. R. Brooks, eds.). New York Bot. Garden, New York.

———. 1983. The logical basis of phylogenetic analysis. Pp. 7–36, *in* Advances in cladistics, vol. 2, Proceedings of 2nd Meeting of the Willi Hennig Society (N. I. Platnik and V. A. Fund, eds.). Columbia University Press, New York.

Fitch, W. M., and E. Margoliash. 1967. Construction of phylogenetic trees. Science, 155:279–284.

Gaines, M. S. 1985. Genetics. Pp. 845–883, *in* Biology of New World *Microtus* (R. H. Tamarin, ed.). American Society of Mammalogists, Special Publications No. 8, Allen Press, Lawrence, Kansas.

Gorman, G. C., and J. Renzi, Jr. 1979. Genetic distance and heterozygosity estimates in electrophoretic studies: effects of sample size. Copeia, 1979: 242–249.

Gould, S. J. 1980. Is a new and general theory of evolution emerging? Paleobiology, 6:119–130.

Hafner, D. J., J. C. Hafner, and M. S. Hafner. 1979. Systematic status of kangaroo mice, genus *Microdipodops*: morphometric, chromosomal, and protein analyses. Journal of Mammalogy, 60:1–10.

Hafner, J. C. 1981. Evolution, systematics, and historical zoogeography of kangaroo mice, genus *Microdipodops*. Ph.D. dissert., University of California, Berkeley, 269 pp.

Hafner, M. S. 1982. A biochemical investigation of geomyoid systematics (Mammalia: Rodentia). Zeitschrift für Zoologishe Systematik and Evolutionsforschung, 20:118–130.

Hamilton, M. J., R. K. Chesser, and T. L. Best. 1987. Genetic variation in the Texas kangaroo rat, *Dipodomys elator* Merriam. Journal of Mammalogy, 68:775–781.

Johnson, W. E., and R. K. Selander. 1971. Protein variation and systematics in kangaroo rats (genus *Dipodomys*). Systematic Zoology, 20:377–405.

Lewontin, R. C. 1974. The genetic basis of evolutionary change. Columbia Univ. Press, New York, 346 pp.

Merriam, C. H. 1899. Revision of North American pocket mice. North American Fauna, 1:1–29.

Mickevich, M. F., and M. S. Johnson. 1976. Congruence between morphological and allozyme data in evolutionary inference and character evolution. Systematic Zoology, 25:260–270.

Nei, M. 1972. Genetic distance between populations. American Naturalist, 106:283–292.

———. 1975. Molecular population genetics and evolution. North-Holland, Amsterdam, 288 pp.

———. 1978. Estimation of average heterozygosity and genetic distance from a small number of individuals. Genetics, 89:583–590.

Nei, M., and A. K. Roychourdhuri. 1974. Sampling variance of heterozygosity and genetic distance. Genetics, 76:379–390.

Nei, M., F. Tajima, and Y. Tateno. 1983. Accuracy of estimated phylogenetic trees from molecular data. Journal of Molecular Evolution, 19:153–170.

Nevo, E. 1978. Genetic variation in natural populations: patterns and theory. Theoretical Population Biology, 13:121–177.

Nevo, E., A. Beiles, and R. Ben-Shlomo. 1984. The evolutionary significance of genetic diversity: ecological, demographic and life history correlates. Pp. 13–213, *in* Evolutionary dynamics of genetic diversity (G. S. Mani, ed.). Lecture notes in biomathematics, vol. 53. Springer-Verlag, Berlin-Heidelberg-New York.

Osgood, W. F. 1900. Revision of the pocket mice of the genus *Perognathus*. North American Fauna, 18:1–63.

Patton, J. L., H. MacArthur, and S. Y. Yang. 1976. Systematic relationships of the four-toed populations of *Dipodomys heermanni*. Journal of Mammalogy, 57:159–163.

Patton, J. L., S. W. Sherwood, and S. Y. Yang. 1981. Biochemical systematics of chaetodipine pocket mice, genus *Perognathus*. Journal of Mammalogy, 62:477–492.

Powell, J. R. 1975. Protein variation in natural populations of animals. Pp. 79–119, *in* Evolutionary biology, vol. 8 (Th. Dobzhansky, M. K. Hecht, and W. C. Steere, eds.). Plenum Press, New York.

Rogers, D. S. 1986. Evolutionary relationships within the subfamily Heteromyinae (genera *Heteromys* and *Liomys*). Ph.D. dissert., University of California, Berkeley, 247 pp.

———. 1990. Genic evolution, historical biogeography, and systematic relationships among spiny pocket mice (Subfamily Heteromyinae). Journal of Mammalogy, in press.

Rogers, J. S. 1972. Measures of genetic similarity and genetic distance. University of Texas Publication No. 7213:145–153.

Sarich, V. M. 1977. Rates, sample sizes, and the neutrality hypothesis for electrophoresis in evolutionary studies. Nature, 265:24–28.

Schmidly, D. J., and F. S. Hendricks. 1976. Systematics of the southern races of *Dipodomys ordii*. Bulletin of the Southern California Academy of Sciences, 75:225–237.

Schnell, G. D., and R. K. Selander. 1981. Environmental and morphological correlates of genetic variation in mammals. Pp. 60–99, *in* Mammalian population genetics (M. H. Smith and J. Joule, eds.). University of Georgia Press, Athens.

Schnell, G. D., T. L. Best, and M. L. Kennedy. 1978. Interspecific morphologic variation in kangaroo rats (*Dipodomys*): degree of concordance with genic variation. Systematic Zoology, 27:34–48.

Selander, R. K. 1976. Genic variation in natural populations. Pp. 21–45, *in* Molecular evolution (F. J. Ayala, ed.). Sinauer Assoc., Inc., Sunderland, Massachusetts.

Selander, R. K., and W. E. Johnson. 1973. Genetic variation among vertebrate species. Annual Review of Ecology and Systematics, 4:75–91.

Selander, R. K., and D. W. Kaufman. 1973. Genetic variability and strategies of adaptation in animals. Proceedings of the National Academy of Sciences, USA, 70:1875–1877.

Selander, R. K., and T. S. Whittam. 1983. Protein polymorphism and the genetic structure of populations. Pp. 89–114, *in* Evolution of genes and proteins (M. Nei and R. K. Koehn, eds.). Sinauer Assoc., Inc., Sunderland, Massachusetts.

Soulé, M. 1976. Allozyme variation: its determinants in time and space. Pp. 60–77, *in* Molecular evolution (F. J. Ayala, ed.). Sinauer Assoc., Inc., Sunderland, Massachusetts.

Straney, D. O. 1981. The stream of heredity: genetics in the study of phylogeny. Pp. 100–138, *in* Mammalian population genetics (M. H. Smith and J. Joule, eds.). University of Georgia Press, Athens.

Thomas, W. K., S. Pääbo, F. X. Villablanca, and A. C. Wilson. 1990. Spatial and temporal continuity of kangaroo rat populations shown by sequencing mitochondrial DNA from museum specimens. Journal of Molecular Evolution, 31:101–112.

Valentine, J. W. 1976. Genetic strategies of adaptation. Pp. 78–94, *in* Molecular evolution (F. J. Ayala, ed.). Sinauer Assoc., Inc., Sunderland, Massachusetts.

Wahlund, S. 1928. Zusammensetzung von Population and Korrelationserscheinungen vom Standpunkt der Verebungslehre aus Betrachtet. Hereditas (Lund), 11:65–106.

Wilson, A. C. 1976. Gene regulation in evolution. Pp. 225–234, *in* Molecular evolution (F. J. Ayala, ed.). Sinauer Assoc., Inc., Sunderland, Massachusetts.

Wilson, A. C., S. S. Carlson, and T. J. White. 1977. Biochemical evolution. Annual Review of Biochemistry, 46:573–639.

Wright, S. 1965. The interpretation of population structure by F-statistics with special regard to systems of mating. Evolution, 19:395–420.

ANATOMY OF THE HETEROMYID EAR

DOUGLAS M. LAY

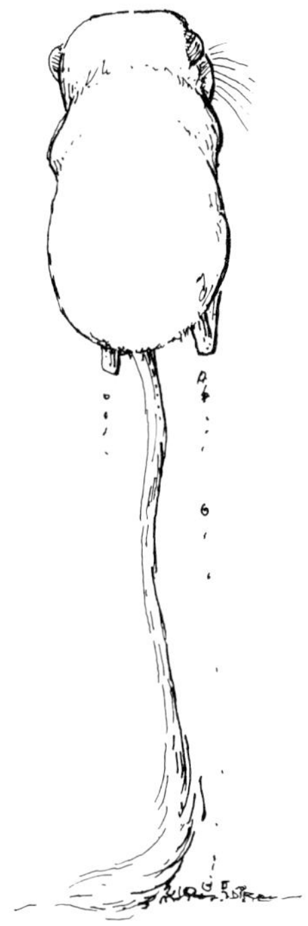

Introduction

Middle and inner ear anatomy exhibit striking differences in detail and scale among the heteromyid genera. The relative hypertrophy of middle ear structures is greatest in *Dipodomys* and *Microdipodops,* least in *Liomys* and *Heteromys,* and intermediate in *Perognathus* and *Chaetodipus.* Patterns of anatomical variation in the cochlea parallel those of the middle ear.

Howell (1932) provided the first anatomical descriptions of the middle ear of *Dipodomys.* A series of papers by Webster and coworkers (Webster, 1961, 1962; Webster and Webster, 1971, 1975, 1977) have produced a significant body of data describing middle and inner ear anatomy and physiology of most heteromyid genera and have elucidated the probable functional significance of the highly specialized ears of *Dipodomys* and *Microdipodops.* This work is summarized in a review by Webster and Webster (1984). Many facets of Webster's work are utilized in the present review and the reader should consult the original papers for details omitted here (see also Hafner, this volume).

Anatomy of the Heteromyid Middle Ear

The middle ear, located within the temporal bone, is formed by components which function in combination to transmit the energy of airborne sound waves to transducers in the fluid-filled inner ear. The middle ear, thus, consists of air-filled cavities and a conducting system, the tympanum (ear drum), malleus, incus and stapes, which receive and transmit airborne sound energy across the middle ear to the inner ear.

The middle ear functions to transform acoustical energy in air to acoustical energy in the cochlear fluids. Since acoustic impedance in air is much lower than in fluid, the middle ear functions to match these impedances with minimum loss. Webster and Webster (1975) and Durrant and Lovrinic

(1984) present excellent discussions of this topic. All components of the middle ear are involved in impedance matching. Pressure at the small stapedial footplate is increased in two ways. The force impinging on the large area of the tympanum (=tympanic membrane or eardrum) is concentrated on the smaller area of the stapedial footplate and is increased essentially as the ratio of the two areas. Because the manubrium of the malleus is longer than the long process of the incus, the pressure is increased as the ratio of these two lever arms. The total pressure increase is the product of these two factors and is called the transformer ratio. The air-filled middle ear cavities allow the tympanum to vibrate freely (Durrant and Lovrinic, 1984; Wever and Lawrence, 1954).

The walls of the middle ear cavity are formed primarily by periotic and ectotympanic elements which fuse to form the auditory bulla. Each bulla contains three compartments: the hypotympanum, the epitympanum and the mastoid or antrum (Fig. 1). Webster and Webster (1975) use the term antrum in reference to the mastoid air cell(s) and have clarified the earlier terminology of Howell (1932).

The hypotympanic cavity lies below the level of the horizontal semicircular canal. Its medial wall contributes to the formation of the braincase and supports the cochlea which extends into thc hypotympanic cavity. The tympanum forms most of the lateral wall. The hypotympanic cavity communicates with the mastoid air cells posteriorly and the epitympanic cavity superiorly (Figs. 1–3).

The epitympanic cavity lies above the level of the horizontal semicircular canal and its walls articulate with the parietal, interparietal and occipital bones. The head of the malleus and body of the incus project into this cavity (Figs. 1–3, 5B).

The mastoid lies posterior to the hypo and epitympanic cavities and abuts the occipital bone superiorly and medially (Figs. 1, 2). The bony auditory canal passes medially, inferiorly and anteriorly from the external auditory meatus to end at the tym-

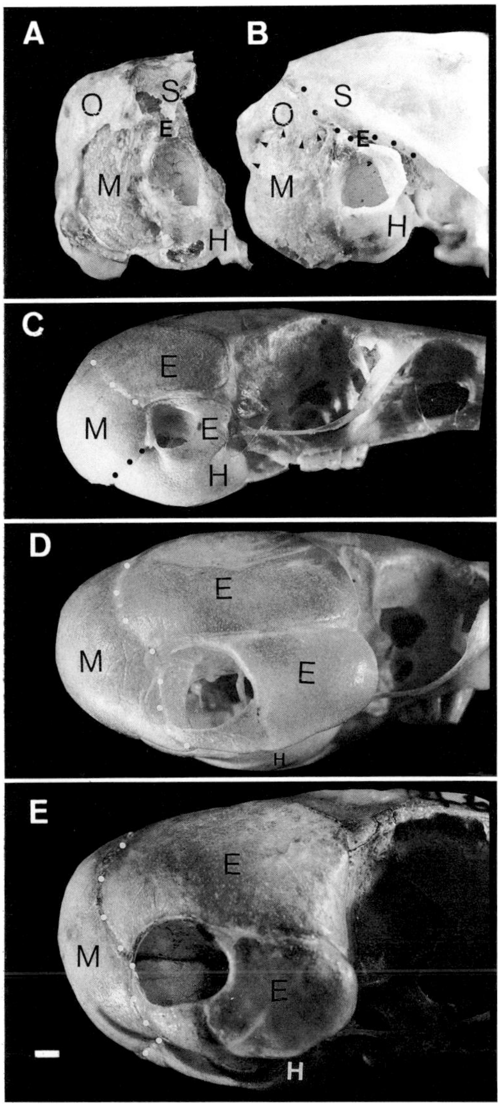

FIG. 1.—Lateral view of auditory bullae. A. *Heteromys desmarestianus* USNM 319464. B. *Liomys irroratus* USNM 41943. C. *Perognathus longimembris* DML 4715. D. *Microdipodops pallidus* DML 4779. E. *Dipodomys merriami* DML 4602. O = occipital, S = squamosal, E = epitympanum, H = hypotympanum, M = mastoid, black dots in B = posterior border of squamosal covering the epitympanum at E. Black arrowheads = mastoid-occipital suture, white and black dots in C, D and E outline the boundaries of the mastoid cell, white bar lower left = 1 mm. Note the large expansion of the epitympanum into the anterior wall of the auditory canal in D and E. Note the broken squamosal in A and its relationship to the epitympanum.

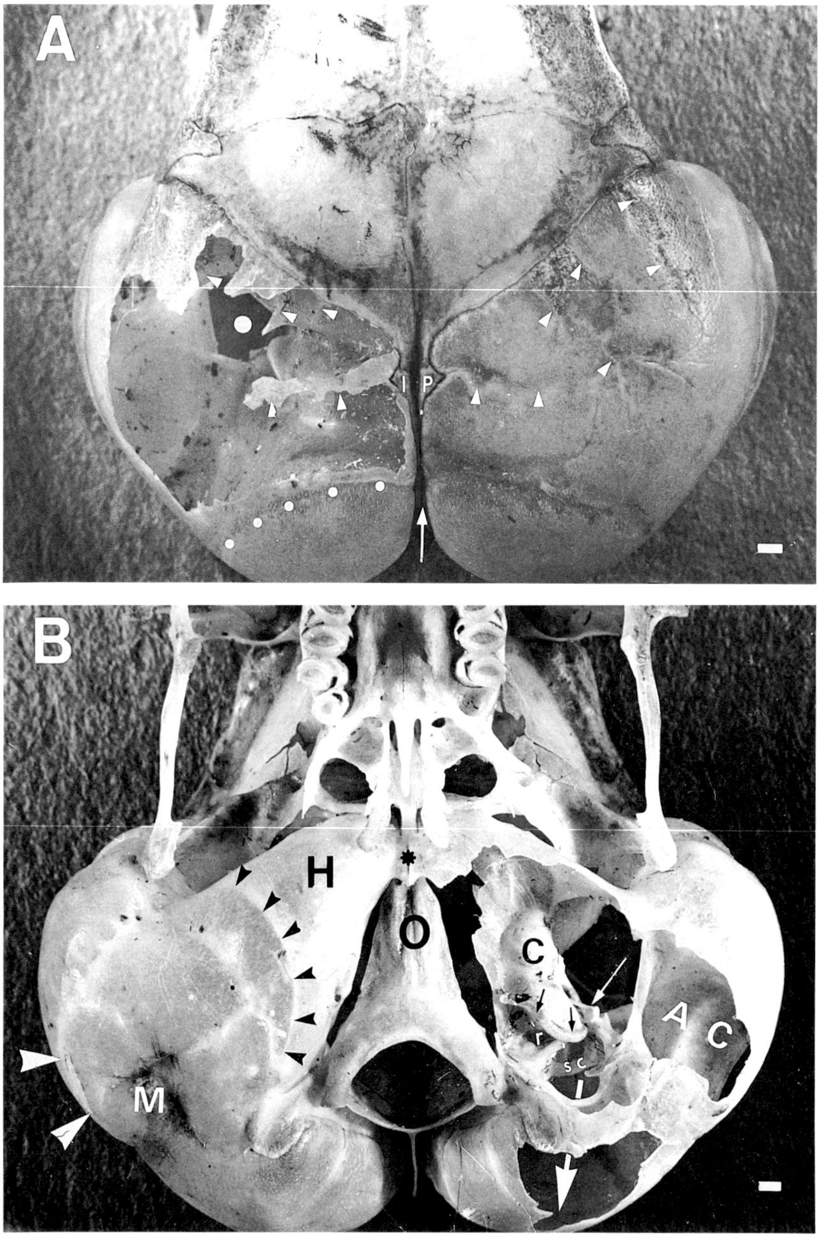

FIG. 2.—Middle ear morphology of *Dipodomys deserti* DML 4637. A. Dorsal view. Arrowheads = buttresses between brain case and roof of epitympanum (left) and surface markings of buttresses on roof of epitympanum (right), large dot = opening between epitympanum and hypotympanum, small dots = partition between epitympanum and mastoid, arrow = occipital, IP = interparietal. B. Ventral view. AC = auditory canal opened ventrally, C = cochlea, H = hypotympanum, M = origin of sternocleidomastoid muscle, O = occipital, r = round window of cochlea, sc = horizontal semicircular canal, small white arrow = head of stapes (note the anterior crus medial to the arrow tip passes into oval window), large white arrow = mastoid cavity (head) and communication between mastoid and hypotympanum (tail), large white arrowheads = anterior and posterior ends of external auditory meatus, small black arrows = canal of stapedial artery on basal turn of the cochlea, black arrowheads = tympanic annulus (internally), * = juncture of right and left hypotympanic chambers ventral to sphenoid, white bar lower right = 1 mm.

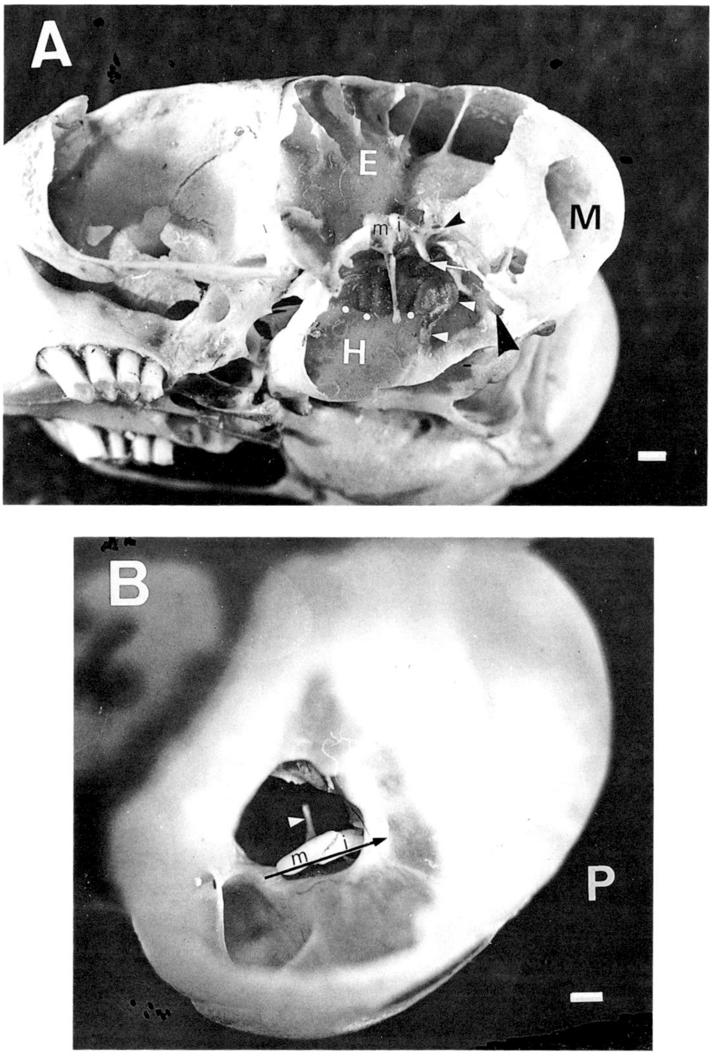

FIG. 3A.—Middle ear cavity of *Dipodomys ordii* DML 4755 as seen ventrolaterally. E = epitympanum, H = hypotympanum, M = mastoid, m = malleus, i = incus, small white dots = ventral margin of cochlea, white arrow = articulation between long processes of incus and stapes, small white arrowheads = bony canal for stapedial artery, large black arrowhead = site of communication between mastoid and hypotympanic chambers, small black arrowhead = short process of incus. B. The epitympanum of *D. deserti* DML 4637. P = posterior, m = malleus, i = incus, white arrowhead = manubrium of malleus, black arrow = alignment of axis of rotation of malleus and incus, white bars lower right = 1 mm. Note the broad communication between the epitympanum and hypotympanum that surrounds the malleus and incus.

panum. It lies above the hypotympanic cavity, below the epitympanic cavity and anterior to the mastoid. The medial end of the auditory canal is formed by the circular tympanic annulus to which the tympanum attaches. The diameter of the annulus is usually considerably greater than that of the auditory canal. The tympanic annulus is oriented so that the inferior border lies medial to the superior border. Thus, the eardrum lies at an angle of about 45° to the perpendicular of the skull (Figs. 1, 2B, 4B).

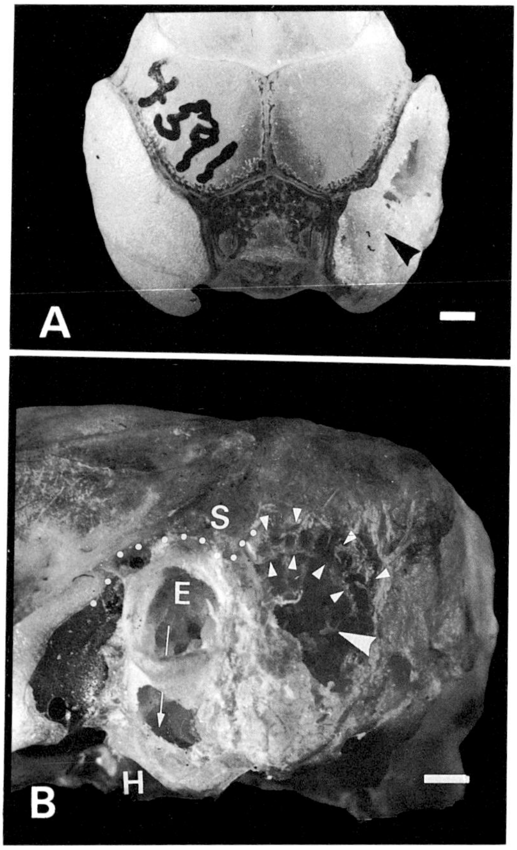

FIG. 4A.—The epitympanum of *P. longimembris* DML 4591. Black arrowhead = layer of trabecular air cells with cavity anteriorly. B. The middle ear of *Heteromys desmarestianus* USNM 319464. E = epitympanum (note the air cells below E where the superior wall of the auditory canal has been removed), H = hypotympanum opened ventrally, S = squamosal, white dots = suture between squamosal and temporal, white arrow = auditory canal (the arrowhead lies near the medial end of the canal) large white arrowhead = mastoid cavity and trabecular air cells, small white arrowheads (two rows) = double layered outer wall of mastoid, white bar lower right = 1 mm.

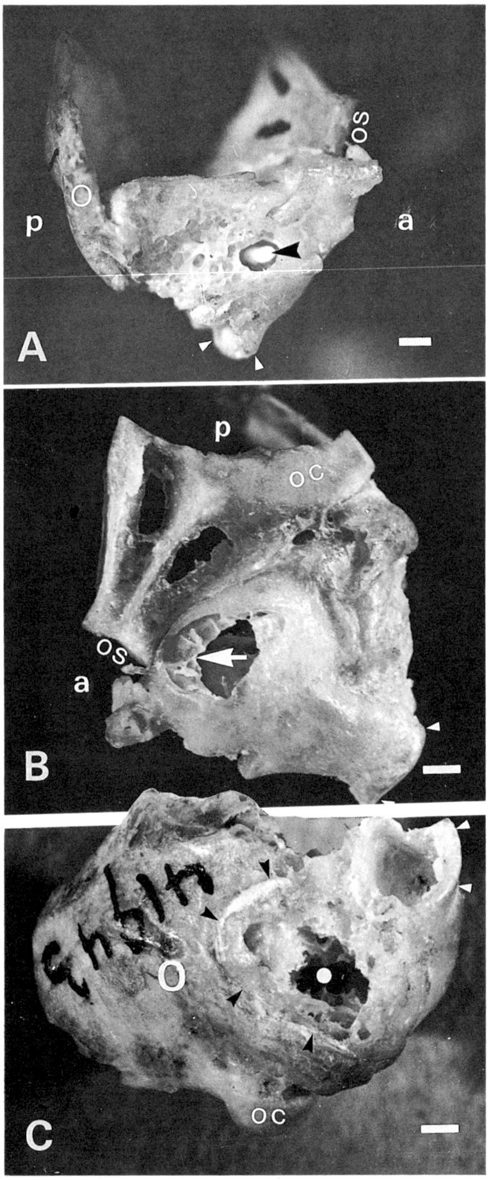

FIG. 5.—Middle ear dissection of *Liomys irroratus* USNM 41943. A. Dorsal view of epitympanum with heads of malleus and incus in cavity and air cells posteriorly. B. Ventral view of hypotympanum showing layer of trabecular bone. C. Posterior view of mastoid. O = occipital, os = articular site of occipital and sphenoid, oc = occipital condyle, a = anterior, p = posterior, large white arrow = trabecular layer of hypotympanum, small white arrowheads = lip of external auditory meatus, large black arrowhead = head of malleus in epitympanum, small black arrowheads = occipital-mastoid suture, large white dot = mastoid cavity and air cells, white bar in lower right = 1 mm.

The malleus has a head, neck, manubrium, anterior process and muscular process. The incus consists of a body, short process and long process. The joint between the malleus and incus is multiplanar and effectively locks these bones so that they move in phase together (Fig. 3).

The stirrup shaped stapes consists of a head, anterior and posterior crura and a base (or footplate). The oval shaped footplate is loosely sealed into the fenestra ovalis (oval window) of the cochlea (Figs. 2B, 3A).

The eardrum is interposed between the air of the outside environment via the auditory canal and the air contained in the middle ear. The tympanic membrane takes the shape of a flattened cone and is attached peripherally to the tympanic annulus. The centrally located peak of the cone, termed the umbo, lies at the tip of the manubrium. The entire manubrium is embedded in the middle layer of the tympanum.

The tympanic membrane is trilaminar consisting of external and internal layers of simple squamous epithelium separated by a middle fibrous layer and is divided into two parts. The largest portion attaches to the manubrium of the malleus and is called the pars tensa, whereas a small portion located superior to the manubrium is much thicker and is referred to as the pars flaccida. It is generally accepted that the effective vibratory surface of the eardrum consists of approximately ⅔ of its total area (Wever and Lawrence, 1954).

The malleus and incus are aligned on an antero-posterior axis about which these two elements rotate in and out through the anterior process and head of the malleus and the body and short process of the incus (Fig. 3). Thus, the incursions and excursions of the tympanum are followed in synchrony by the malleus and incus. The long process of the incus through its terminal lenticular facet is joined to the head of the stapes completing the ossicular chain across the middle ear (Fig. 3A).

Comparative Anatomy of the Middle Ear

Analysis of middle ear volume by Webster and Webster (1975) led them to suggest "two major divergent lines of evolution within the family Heteromyidae." These included *Microdipodops* and *Dipodomys* in one group and *Liomys* and *Perognathus* (including *Chaetodipus*) in the other. I feel that *Heteromys* should also be included in the latter and the descriptions will follow this dichotomy. Because the available data refer to total middle ear volume (Webster and Webster, 1975), references to the relative volume of the three respective middle ear chambers are my subjective opinions unless indicated otherwise.

The Hypotympanic Cavity

The relative volume of this chamber appears to be smallest in *Heteromys* and *Liomys*, and slightly larger in *Perognathus* and *Chaetodipus*, though it is not voluminous in any of these taxa if compared to *Dipodomys*. The outer wall is formed by compact bone, the internal surface of which is lined by a distinctive layer of trabeculated bone in all four genera (Fig. 1).

The following distinctions were noted. In *Heteromys*, the trabecular layer is thin, waferlike and separated from the outer shell by a narrow space. In *Liomys*, the trabecular bone extends further internally and the space between it and the compact outer wall is subdivided by numerous bony air cells which communicate with the hypotympanic cavity proper (Fig. 5B). In *Perognathus*, the outer wall appears thinner and the entire surface is lined by tiny bony air cells that communicate with the hypotympanic cavity. Histologically, the trabecular bone and the compact bone are separated by a layer of connective tissue and blood vessels (Fig. 6).

The anterior ends of the hypotympanic chambers contact the basisphenoid in all

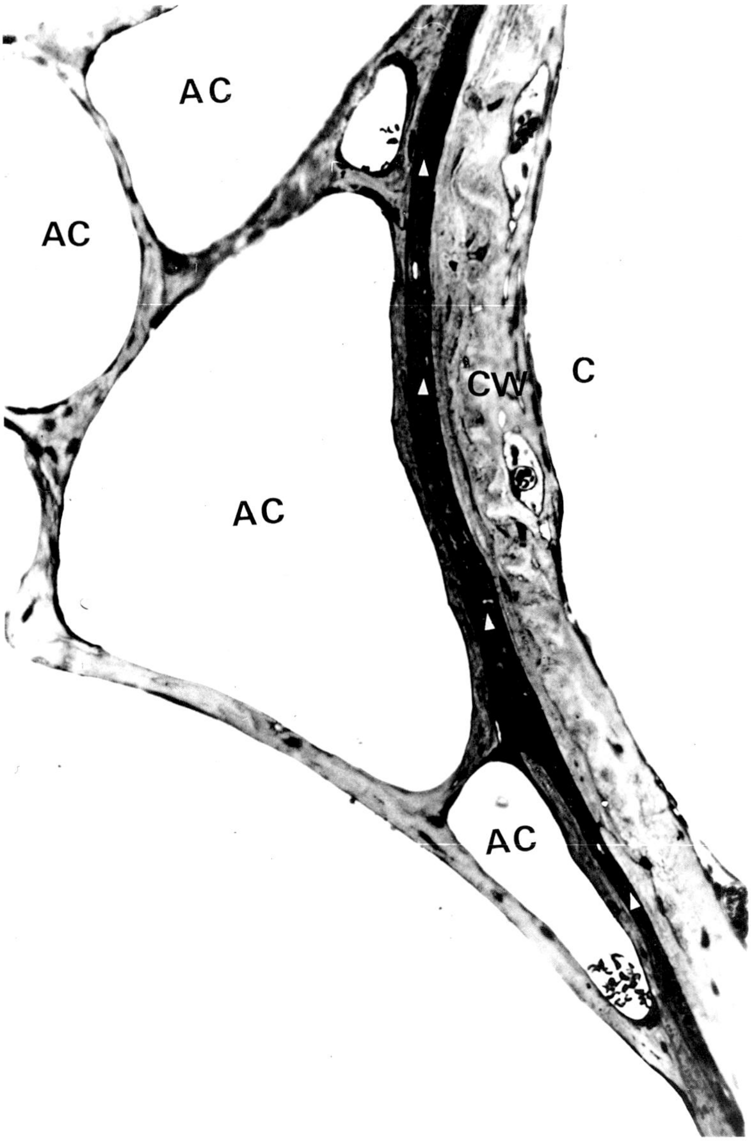

FIG. 6.—Section of trabecular air cells in hypotympanum of *Perognathus longimembris*. AC = air cells, CW = bony capsule of cochlea, C = cochlear chamber, white arrowheads = connective tissue layer between cochlea and air cells.

genera, but are widely separated by the broad basisphenoid in *Heteromys*. The basisphenoid of *Liomys* is noticeably narrower than in *Heteromys*, yet the anterior ends remain clearly separated. In *Perognathus*, the anterior ends come into close proximity and the basisphenoid is greatly narrowed. Concomitant with these differences the volume of this chamber expands laterally, ventrally and antero-medially (Figs. 1A, 1B, 1C). Although the hypotympanic cavity of *Microdipodops* and *Dipodomys* resembles that of *Perognathus* in overall shape, it differs strikingly in two features not found in *Perognathus*, *Liomys* and *Heteromys*. No trabecular bone or bony air cells are found in the cavity and the anteriormost portions of both sides extend to the midline ventral to the basisphenoid to abut one another (Fig. 2B). Moreover, the volume of this cavity is larger in all dimensions relative to the condition seen in *Perognathus*.

The Epitympanic Chamber

The capsule of the tiny epitympanic cavity in *Heteromys* and *Liomys* is lined by relatively large trabecular bony air cells which appear to communicate with the cavity via a few openings. This chamber is covered laterally by the posterior portion of the squamosal, but may be entered by perforating the superior wall of the auditory canal (Figs. 1A, 1B, 4B, 5A).

The epitympanic chamber of *Perognathus* is much larger than in *Heteromys* or *Liomys* and forms a prominent surface feature of the skull (Figs. 1C, 4A). Concomitant with this enlargement is a major reduction of the posterior portion of the squamosal and a narrowing of the interparietal and the superior portion of the occipital. The cavity has a relatively small lumen that is lined peripherally by a thick layer of delicate bony air cells (Fig. 4A). These cells communicate with the main cavity by numerous openings.

In contrast, the epitympanic chambers of *Microdipodops* and *Dipodomys* are enormous. Expansion has occurred in all dimensions and the other chamber contains no trabecular bone. One portion has expanded to occupy space anterior to the auditory canal (Figs. 1D, 1E). This expansion is also seen in *Perognathus* on a smaller scale but it is filled with trabecular air cells (Figs. 1C, 4A). Dorsally, the interparietal and occipital are narrower than in *Perognathus* and in extreme cases, exemplified by *M. pallidus* and *D. deserti,* are reduced to a thin strap (Fig. 2A). The epitympanic chambers of *Dipodomys* and *Microdipodops* enlarge to form the roof of the skull and both sides are in virtual contact midsagittally. Vertical buttresses pass between the roof and braincase in these two genera (Figs. 2A, 3A). The epitympanic chamber communicates with the hypotympanic via a wide opening medial to the tympanum (Figs. 2, 3B).

The Mastoid Air Cells

The mastoid region of *Heteromys* and *Liomys* is relatively the smallest found in the family. These two genera differ from all others in two ways. The outer mastoid wall is relatively thick, consisting of inner and outer tables of compact bone which are separated by a layer of trabecular bone (Fig. 4B). Internally, the entire mastoid is filled with a meshwork of bony walled air cells that are considerably larger in volume than those of *Perognathus.* These cells communicate with both the epi- and hypotympanic chambers by means of a limited number of openings (Figs. 4B, 5C).

The shape and position of the mastoid is similar in *Perognathus, Chaetodipus, Microdipodops* and *Dipodomys.* A thin bony partition completely separates the mastoid and epitympanic chambers in the last two genera (Fig. 2A) but is incomplete in the first two. This partition, visible on the external surface of the bulla as a line passing from the occipital to approximately the middle of the posterior side of the external auditory meatus, extends from the dorsal aspect of the bulla to the crista parotica, ventrally (Figs. 1C, 1D, 1E, 2A, 7). The mastoid chamber communicates with the hypotympanic compartment via an opening inferior to the horizontal semicircular canal and medial to the distal aspect of the facial canal. The mastoid chambers of *Microdipodops* and *Dipodomys* are devoid of trabecular air cells while all internal surfaces of the mastoid are covered with a layer of fine bony air cells in *Perognathus* and *Chaetodipus* (Fig. 4A).

Bullar Volume

Webster (1961) determined the following percentages of total middle ear volume for *Dipodomys*: hypotympanum, 18%; epitypanum, 49%; mastoid, 33%. Data regarding the relative contribution of each chamber to middle ear volume are not available for other genera. Webster and Webster (1975) measured total middle ear volume for eight species of *Dipodomys,* 14 species of *Perognathus* and two species for both *Microdipodops* and *Liomys.* Mean volumes in cm^3 were as follows: *Liomys,* 0.03; *Perognathus,* 0.06; *Dipodomys,* 0.68; *Microdipodops,* 0.31. These values were converted to relative volumes by dividing the cube root of middle ear volume by nasooccipital length. The mean relative volumes are: *Liomys,* 0.10; *Perognathus,* 0.18; *Dipodomys,* 0.21; *Microdipodops,* 0.31. I estimate the total and relative middle ear volume for *Heteromys* to be less than that reported for *Liomys.* Variation in relative volume for the eight *Dipodomys* species ranged from 0.23 to 0.29 and that for 14 *Perognathus* species was even greater, 0.13–0.24.

The Middle Ear Transformer

Tympanic membrane area increases in the sequence from *Liomys* to *Perognathus* to *Microdipodops* to *Dipodomys.* This trend is partially due to size differences among taxa. The functional significance of tympanum size may be obtained from the ratio of area of stapes footplate/two-thirds of the area (effective size) of the tympanum. Webster and Webster (1975) have determined these values for the same taxa enumerated above and there are only minor differences across the same four genera. They note that a favorable transformer ratio is achieved by reduction of stapedial footplate area in *Perognathus,* the opposite of that which is observed in *Dipodomys* and *Microdipodops.* It is unclear how this was determined and this point needs verification.

The lever ratio of the length of the long process of incus/length of the manubrium of the malleus displays a clear and functionally significant trend among the same taxa. Mean values for genera show increasing mechanical advantage in the sequence: *Liomys* (0.57); *Perognathus* (0.42); *Dipodomys* (0.30); *Microdipodops* (0.28) (Webster and Webster, 1975).

Webster and Webster (1975), used these values to compute impedance transform ratios according to the formula of Dallos (1973). The ranges of the values obtained were: *Liomys,* 0.019–0.021; *Perognathus,* 0.006–0.011; *Dipodomys,* 0.005–0.008; *Microdipodops,* 0.004–0.005. These calculations suggest that the middle ear is least efficient in impedance matching in *Liomys* and most efficient in *Dipodomys* and *Microdipodops.*

Mass is more important in impeding the transfer of acoustical energy at higher frequencies, while stiffness is the greater impediment at lower frequencies (Dallos, 1973; Webster and Webster, 1975).

The mass of the middle ear transformer is contributed by the tympanum and ossicles. The mass is small simply because of the small size of these elements. It is further reduced in *Microdipodops* and *Dipodomys* by a thinning of the stapedial footplate and portions of the malleus, and in *Perognathus* by a thinning of the stapedial crura (Webster and Webster, 1975).

The anatomical specializations associated with a reduction of stiffness, i.e., low frequency impedence, are profound. The middle layer of the tympanic membrane is notably thin in *Dipodomys* (Webster, 1961) and, presumably, *Microdipodops.* I am unaware of data on the tympani of *Liomys, Heteromys* or *Perognathus.* Only two fine ligaments attach to the malleus and incus, where most mammals have two additional ligaments attaching to the malleus (Kobayashi, 1955; Webster, 1961).

Two muscles, the tensor tympani and stapedius, attach to the malleus and stapes, respectively, and on contraction damp the oscillations of these bones (Wever and Lawrence, 1954). The stapedius muscle is absent in *Perognathus* and *Liomys* (Webster and Webster, 1975) and *Heteromys* has not been studied. Presumably, the loss of the stapedius reduces the stiffness of the system. The annular ligament of the stapedial footplate is extremely thin in *Liomys, Perognathus,*

Dipodomys and *Microdipodops* (Webster and Webster, 1975). All these features suggest stiffness reduction.

The volume of air within the middle ear plays an important role in determining stiffness. If tympanic membrane area remains constant, then the smaller the middle ear volume the greater the stiffness. Increases in tympanic membrane area are highly correlated with increases in total bullar volume in *Dipodomys* and *Microdipodops* (Webster and Webster, 1975) and are also requisite for increases in the incudo-manubrial lever ratio.

The Cochlea

The cochlea bulges into the medial side of the hypotympanic cavity and houses the acoustic receptors of the inner ear. These receptors are coiled spirally and the external appearance of the cochlea resembles a snail shell (Fig. 2B). Pye (1965) reported 3.5 to 4 complete turns of the cochlear spiral in *Heteromys, Liomys, Microdipodops* and *Dipodomys*. Her figure shows 3.5 turns for *Heteromys*. According to Webster and Webster (1977) *Dipodomys, Microdipodops* and *Liomys* have 3.5 turns and *Perognathus* 3.0 turns.

The cochlea is oriented so that the broad basal turn is posterior and the narrower apex is anterior. The fenestra ovalis, which receives the stapes footplate, lies on the superior aspect of the basal turn and faces laterally (Figs. 2B, 3A). A bony canal which begins on the ventral side of the medial wall of the hypotympanum arches across the basal turn, passes through the stapedial crura, continues onto the dorsal cochlear surface and ultimately enters the brain case. This canal transmits the stapedial artery (Figs. 2B, 3A). Because of the relative incompressibility of fluid, a relief valve, the secondary tympanic or round window membrane, is present in the cochlea. This membrane functions to relieve pressure produced by inward excursions of the stapes. The round window (fenestra rotundum) is located on the posterior aspect of the basal turn posterior to the stapedial artery or its vestiges and faces posteriorly and inferiorly (Fig. 2B).

The internal structure of the cochlea is complex; an excellent description is presented by Durrant and Lovrinic (1984). The basic morphology of the heteromyid cochlea is typically eutherian.

The auditory transducers are located within the Organ of Corti which consists of two groups of sensory cells, the inner and outer hair cells, a variety of supporting cells and the entire structure rests on a basilar membrane (Fig. 7). This structure spirals around the cochlea from base to apex (Fig. 8). Large spaces filled with perilymph (scala tympani and scala vestibuli) lie on either side of the basilar membrane, but a separate compartment, the scala media or cochlear duct filled with endolymph, encompasses the organ of Corti. The cochlear duct is separated from the scala vestibuli by the vestibular of Reissner's membrane and from the scala tympani by the basilar membrane (Fig. 7). The unit composed by the cochlear duct, Reissner's membrane and the basilar membrane, including the Organ of Corti, is referred to as the cochlear partition. A bony core, the modiolus, forms the central axis of the cochlea (Fig. 8). A fine double ridge of bone, the osseous spiral lamina, spirals around the modiolus (Fig. 7). The basilar membrane is centrally suspended between the osseous spiral lamina and the spiral ligament of the peripheral cochlear wall (Fig. 7). A tectorial membrane attaches into the spiral limbus on the vestibular side of the osseous spiral lamina and at its opposite end has fine connections to some of the stereocilia of the hair cells (Fig. 7).

The basilar membrane is narrowest in the basal turn and widens progressively along its spiral to the apex and is characterized by a decreasing gradient of stiffness from base to apex (Fig. 11). Sound reception is related to these changes in width and stiffness and is tonotopically organized along the basilar membrane. This membrane consists of two portions, a zona tecta located between the

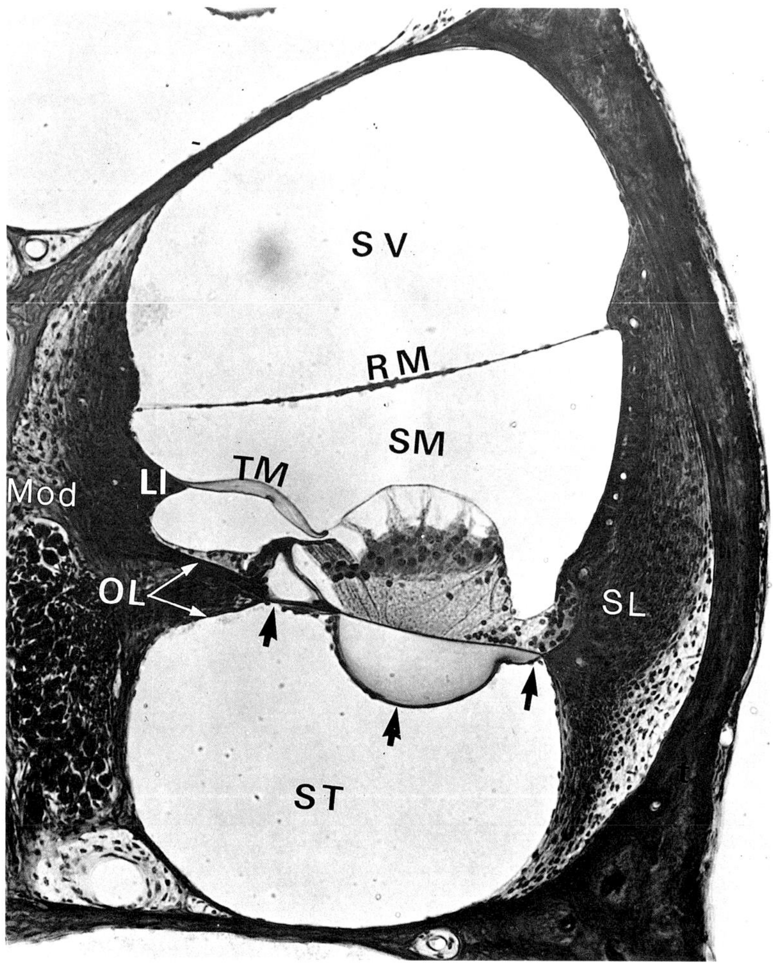

FIG. 7.—Enlargement of the third half turn of the cochlea of *D. merriami.* LI = spiral limbus, Mod = modiolus, OL with white arrows = osseous spiral lamina, SL = spiral ligament, RM = Reisner's membrane, TM = tectorial membrane, SM = scala media (endolymph), ST = scala tympani (perilymph), SV = scala vestibuli (perilymph), black arrows = basilar membrane.

osseous spiral lamina and the outer pillar cell and a zona pectinata that occupies the space between the outer pillar cell and the spiral ligament (Fig. 9).

Comparative Anatomy of the Cochlea and the Organ of Corti

Cochlea anatomy of all heteromyid genera but *Heteromys* has been described by Webster and Webster (1977). Pye (1965) presented some data on the inner ear of *Heteromys.*

The following cochlear features display unusual variation among the Heteromyidae: structure of the basilar membrane; the supporting cells of the Organ of Corti; the scala typani.

The zona pectinata of the basilar membrane encloses a large hyaline mass in *Microdipodops, Perognathus* (Fig. 10), and *Dipodomys* (Fig. 7), and a much smaller quantity of hyaline material in *Liomys* (Webster and Webster, 1977) and *Heteromys* (Pye, 1965). The shape and quantity of this mass varies along the basilar membrane and is difficult to quantify (Fig. 11). Vari-

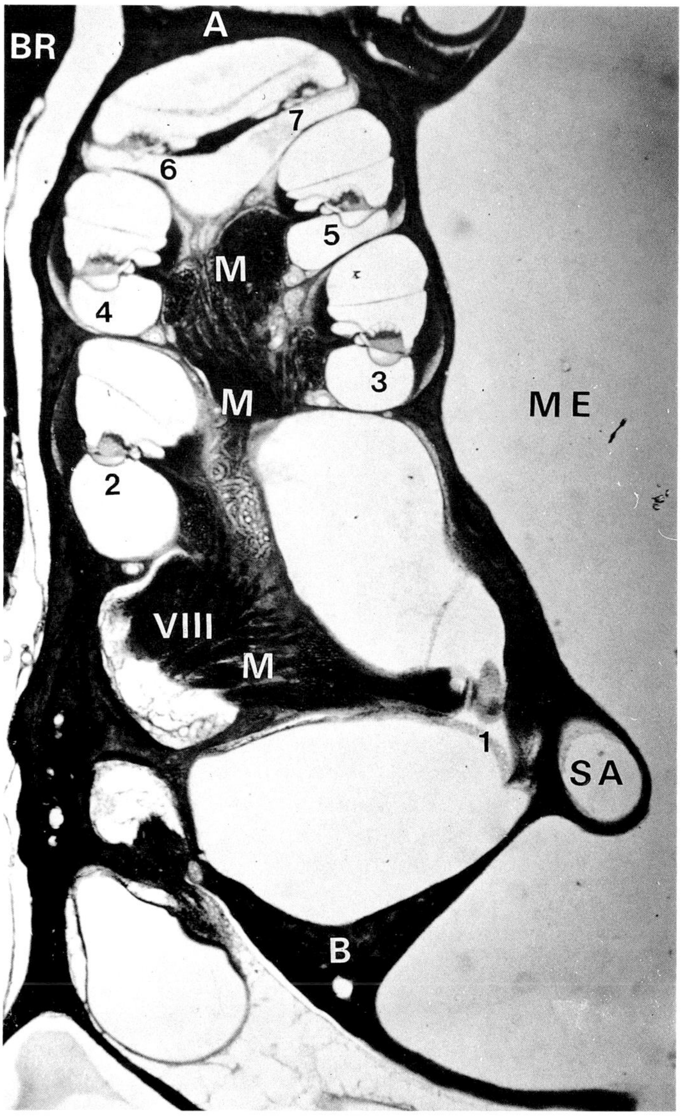

FIG. 8.—A nearly mid modiolar section of the cochlea of *D. merriami.* B = base, A = apex, ME = hypotympanum of middle ear, BR = brain, M = modiolus, SA = stapedial artery, VIII = acoustic division of cranial nerve VIII, 1 through 7 = cochlear half turns from base to apex.

ation in the distribution of the hyaline mass along the basilar membrane has been treated by measuring its thickness at the point of greatest extension below the basilar membrane. While not entirely satisfactory this method does convey an idea of changes in basilar membrane thickness. Data on all genera but *Heteromys* reveal a similar pattern in the distribution of hyaline material within the zona pectinata (Webster and Webster, 1977). The thickness of the pars pectinata is least in the basal turn of all genera, increases rapidly through the first turn to reach a maximum in the second turn, and decreases beyond this point until at the apex it is about the same thickness as in the basal turn. Relative thickness is greatest in *Microdipodops* and least in *Liomys*; *Dipod-*

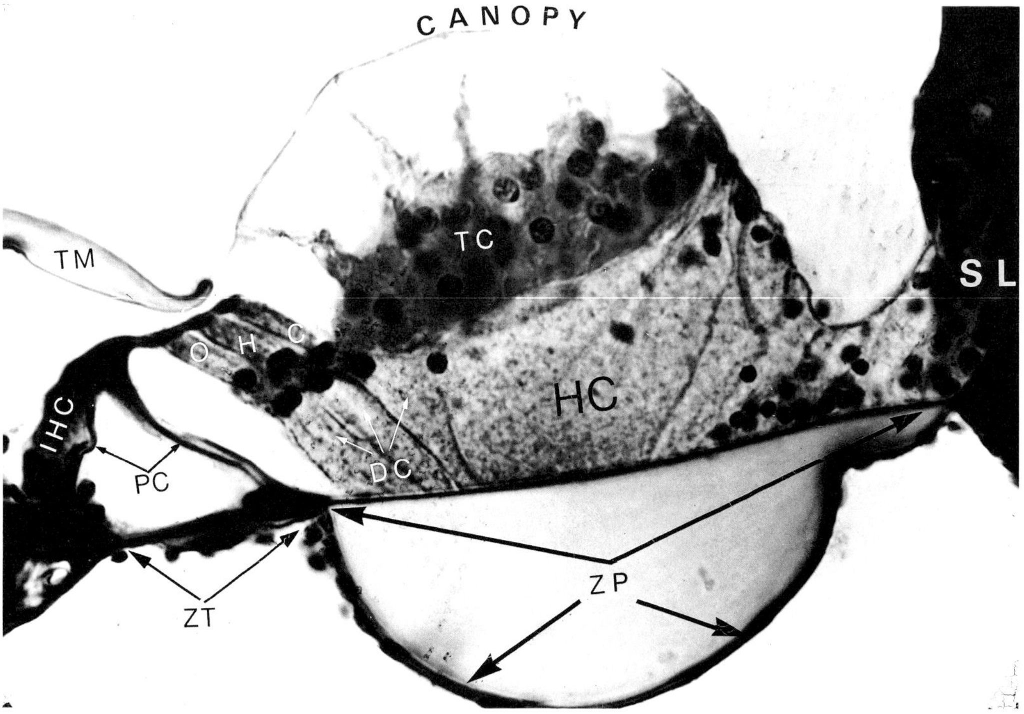

FIG. 9.—The Organ of Corti in the third half turn of *D. merriami.* DC and three white arrows = Deiter's cells (three rows), HC = Henson's cells, IHC = inner hair cell, OHC = outer hair cells (three rows), PC = pillar cells (inner and outer), SL = spiral ligament, TC and CANOPY = tectal cells and tectal canopy, TM = tectorial membrane, ZP and arrows = zona pectinata of basilar membrane containing hyaline mass, ZT = zona tecta of basilar membrane.

omys, Perognathus, and *Chaetodipus* are each similar and intermediate to the extremes. The variation in quantity and distribution of hyaline material is striking from base to apex and it seems reasonable to predict that these changes affect the vibratory characteristics of the basilar membrane (Fig. 11). Pye's (1965) observations on *Heteromys* suggest that the distribution of hyaline material in the zona pectinata is similar to that observed for *Liomys.*

The basilar membrane is narrowest (±100 μm) at the beginning of the first half turn of the cochlea in all genera (Webster and Webster, 1977). In *Dipodomys* and *Microdipodops,* the width doubles in the first half turn and increases less rapidly thereafter to a maximum of 241 and 254 μm, respectively, until it declines slightly in the apical turn. Width increases dramatically in *Perognathus* and *Chaetodipus* to reach a maximum of about 150 μm at the end of the first turn beyond which it decreases slightly to the apex. *Liomys* shows a slight and uniform increase from 110 μm basally to a maximum of 160 μm at the end of the second turn and decreases slightly to the apex (Webster and Webster, 1977). In comparison, the basilar membrane of the bat, *Pteronotus parnellii,* varies in width from ca. 40 μm at the base to ca. 112 μm at the apex. Relative width shows the same pattern among the heteromyid genera as described above for relative thickness.

As a group the several cell types which lie on the radial side of the outer hair cells and Deiter's cells are referred to as border, sustentacular, or supporting cells (Fig. 9).

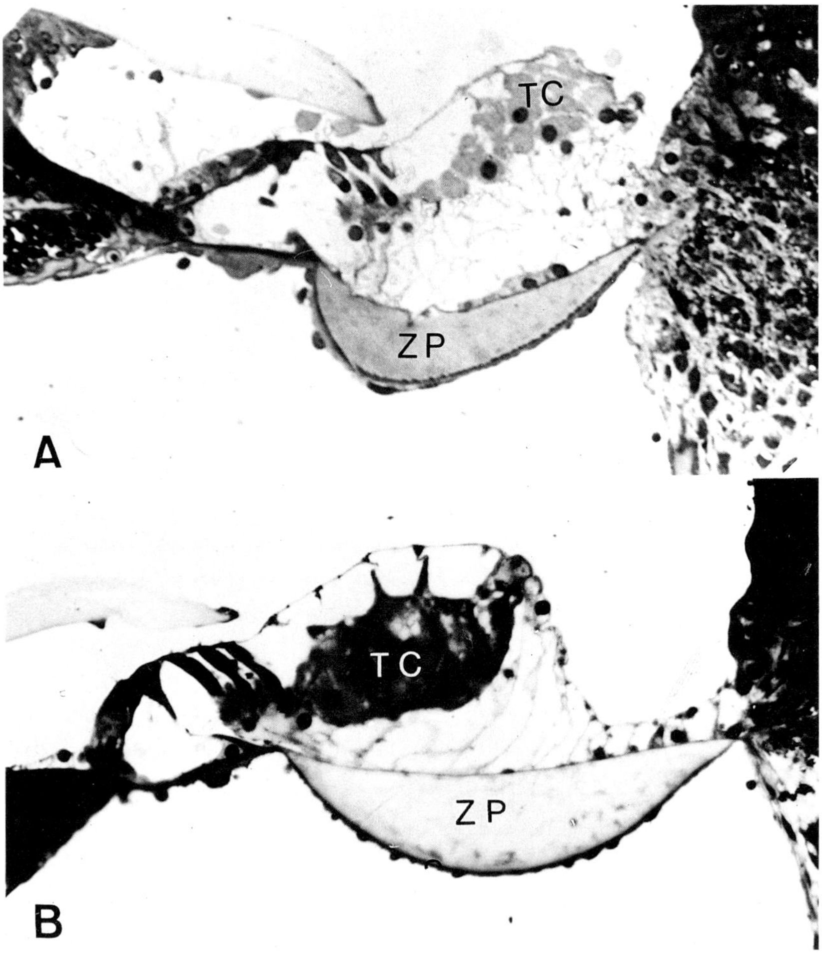

FIG. 10.—Organ of Corti in the second cochlear turn. A. *Perognathus longimembris.* B. *Microdipodops pallidus.* TC = tectal cells, ZP = zona pectinata of basilar membrane.

The degree of development and morphology of one type of these border cells is unique to *Microdipodops, Dipodomys, Perognathus,* and *Chaetognathus* among all mammal species studied to date. Webster (1961) and Webster and Webster (1975, 1977) contend that these cells are homologous with the Henson's cells as described for all other mammal species studied. Lay (1972) questioned this identification and noted similarities to Deiter's cells. The identification of these cells was seemingly resolved by Henson et al. (1983). They presented a variety of data which distinguish a previously undescribed population of cells in a number of mammalian taxa and named these the tectal cells. To bring the terminology applied to the supporting cells of *Dipodomys, Microdipodops,* and *Perognathus* by Webster into compliance with that used for other species, Henson et al. (1983) suggest ". . . that the canopy [Henson's cells of Webster] . . . represents an elaboration of the tectal cells and the cells on which they rest perhaps represent what we have called tunnel floor cells rather than cells of Claudius. Thus, the first row of Henson's cells would lie along the outer margin of the canopy." In the sec-

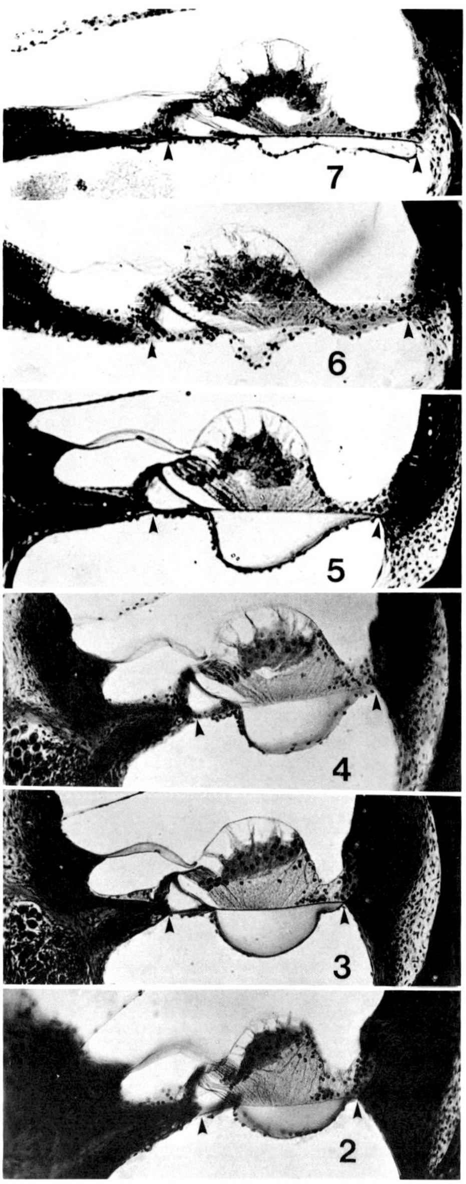

FIG. 11.—Morphological change in the Organ of Corti of *D. merriami* from the second half turn to the apex. Sections numbered 3, 5 and 7 are printed in reverse to facilitate comparisons. Numbers 2 through 7 are identical with the same numbers in Fig. 8. Black arrowheads = attachment of basilar membrane to osseous spiral lamina and spiral ligament. Note the progressive increase in width from 2 to 7.

tions I have studied it is clear that the tectal cells largely rest atop Henson's cells, which have apical processes directed toward the spiral ligament (Fig. 9). It is not possible to comment on the tunnel floor cells until thin sections are prepared.

The elaboration of the tectal cells upon the Henson's cells effectively increases the height of the Organ of Corti above the basilar membrane. Height increases three to four fold from the first turn to the second or third turns then declines toward the apex (Webster and Webster, 1977). Relative height is greatest and sustained over the greatest length on the basilar membrane in *Microdipodops. Perognathus, Chaetodipus* and *Dipodomys* are similar to one another and somewhat less developed than *Microdipodops. Liomys* and *Heteromys* lack a tectal canopy and the height of the Organ of Corti produced by Henson's cells, not tectal cells, changes little but reaches a maximum in the second turn thereafter declining toward the apex. The development of a tectal canopy results in a major increase in Organ of Corti height in the second and third turns relative to the conservative condition of *Liomys* and *Heteromys.*

The cross sectional area of the scala tympani exhibits a striking localized constriction in *Chaetodipus, Liomys,* and *Heteromys.* The cross sectional areas of both perilymphatic chambers are approximately equal at any point along the basilar membrane in most species of mammals that have been studied and such is the case for *Microdipodops, Dipodomys* and *Perognathus* (Webster and Webster, 1977). Nevertheless, the scala tympani changes shape dramatically between the fourth and fifth half turns (Fig. 8) in *Dipodomys* and at the beginning of the third cochlear turn in both *Liomys* and *Chaetodipus* the volume of the scala tympani is greatly reduced relative to that of the scala vestibuli (Webster and Webster, 1977). Pye's (1965) figures of the cochlea of *Heteromys* show a dramatic reduction of the volume of the scala tympani at the apical end of the cochlea as in *Liomys.*

The tectorial membrane is long, thin and delicate in *Dipodomys* and *Microdipodops* (Figs. 10B, 11) and slightly more heavily constructed in *Perognathus* (Fig. 10A) and that of *Liomys* and *Heteromys* has not been described. This membrane is short, thick and massive in the basal turn of the bat, *Pteronotus parnellii* (O. W. Henson, pers. comm.). Thus, the tectorial membrane may also exhibit specialization for low frequency reception and deserves study.

Significance of Cochlear Specializations

The properties of the basilar membrane and the cochlear partition are of the greatest importance in respect to the ultimate utilization of sound energy. As the stapes pushes inward a pressure gradient develops across the cochlear partition. Since the cochlear fluids are incompressible the basilar membrane deflects toward the scala tympani and the displaced fluids therein produce an outward displacement of the round window. As the stapes moves outward, the opposite occurs. When the stapes vibrates in the oval window the entire cochlear portion is set into an up and down motion following the alternating pressure gradients across it. In this way most of the vibratory energy introduced by the stapes is coupled to the basilar membrane.

Once energy has been coupled to the basilar membrane, the displacement of the membrane for any given frequency varies in amplitude along its length. The amplitude of displacement gradually increases until a specific distance is reached where displacement is maximum and beyond which there is a rapid decline in amplitude. This pattern of displacement is called a travelling wave. Travelling waves always progress from the base towards the apex. Due to variation in stiffness and mass from base to apex different frequencies optimally activate different but specific regions of the basilar membrane. Low frequency sounds are received toward the apex and high frequencies toward the base (cf. Durrant and Lovrinic, 1984, for an excellent account of cochlear mechanics).

The cochlear anatomical specializations of *Microdipodops, Dipodomys,* and *Perognathus* show maximal development in the second and third turns, regions that should be associated with low frequency hearing. Auditory sensitivity may be determined by recording the bioelectrical responses of the inner ear to sound stimuli. The most widely used test records the cochlear microphonic potential, the voltage of which is proportional to the intensity of the sound stimulus and the frequency of which mimics the frequency of the stimulus. Webster and Webster (1984) determined mean auditory sensitivity by recording cochlear microphonic responses over the frequency range of 0.05 to 100 kHz for all heteromyid genera. All genera exhibited a broad area of sensitivity up to 30 kHz. The greatest sensitivities in cochlear microphonic potentials, especially at low frequencies, were displayed in *Dipodomys* (75–3,000 Hz) and *Microdipodops* (200–2,000 Hz). These two genera are about 20 dB (decibels) or 100 times more sensistive at frequencies lower than 3,000 Hz than *Perognathus* and about 40 dB or 10,000 times more sensitive than *Liomys* and *Heteromys.*

No direct evidence exists to indicate that the specializations in the second and third turns of the Organ of Corti of *Dipodomys, Microdipodops,* and *Perognathus* function to increase low frequency sensitivity. Basilar membrane thickenings similar to those of heteromyids have been described in *Rhinolophus* (Bruns, 1976), *Pteronotus* (Henson, 1978), *Tursiops* (Wever et al., 1971) and several genera and species of Gerbillinae (Lay, 1972). These basilar membrane specializations occur in the basal cochlear turn of the bats and dolphin each of which utilize ultrasound for echolocation whereas they occur in the apical turns in gerbils that exhibit high sensitivity to low frequency sound. The location of specialized features along the basilar membrane correlates reasonably well with the physiological performance of the ear in these species. Because the areas of specialization in the Organ of Corti correlate with regions associated with unusual hearing abilities, most investigators have assumed that they may function to increase sensitivity of the Organ of Corti to particular frequencies. Direct evidence from frequency mapping of the basilar membrane is required in order to document these anatomical and physiological observations.

Although the means by which the inner ear specializations function to achieve the increases in sensitivity observed for *Dipodomys* and *Microdipodops* remain obscure, Lay (1972) and Webster and Webster (1977) have considered some possible mechanisms.

The abrupt constriction of the scala tympani, found in *Chaetodipus, Liomys,* and *Heteromys,* has not been observed in any other mammal species. Functional explanation of these observations requires further research.

The Adaptive Significance of Specialized Ears

Webster (1962) studied the behavioral responses of *Dipodomys* to natural predators confined in total darkness. He observed that kangaroo rats avoided the attack strikes by owls and sidewinder rattlesnakes (*Crotalus cerastes*) at the last instant by a sudden vertical jump. The owls used in this experiment were identified as screech owls (*Otus asio*) in 1962 and as barn owls (*Tyto alba*) in Webster and Webster (1984). These avian and reptilian predators possess sensory specializations which allow them to locate prey in total darkness (cf. Knudsen and Konishi, 1978, 1979; Knudsen et al., 1979; Webster and Webster, 1984). When *Dipodomys* with middle ear volumes reduced by 75% bilaterally were placed in the enclosures no effort to evade the strikes by either type of predator was observed and all of the experimentally impaired specimens were captured. Tape recordings revealed that as the owls braked flight for the strike, their wings produced low intensity sound that contained frequencies up to 1,200 Hz. Bursts of weak sound produced by the rattlesnake strikes contained frequencies up to 2,000 Hz. Reduction of bullar volume resulted in diminished auditory sensitivity of the cochlear microphonic response over all frequencies tested, particularly lower frequencies (1,000–3,000 Hz). Further, the reduction in sensitivity was in proportion to the amount of volume reduction. From these results Webster (1962) concluded that the enlarged middle ear cavities of *Dipodomys* function to reduce damping of the tympano-ossicular system sufficiently to increase low frequency sensitivity and that this increased sensitivity enables them to detect predators. A later study (Webster and Webster, 1971) determined that vision allowed *Dipodomys* to evade strikes if sufficient light was available, even when middle ear volume was reduced surgically. In a field test designed to clarify these observations, saturation trapping of an area 190 m × 365 m yielded 27 *Dipodomys merriami.* These were divided into three groups (normal; control-sham operated, experimental-middle ear volume reduced) and released. Recapture trapping was performed for one month and no differences at the 0.05 confidence level were noted among the three groups for activity and movement nor in retrapability. Only 14 animals were present at the end of the study in the following proportions of the original total: normal—67%; control—67%; operated—22%. The 78% of the operated group disappeared during the period of greatest darkness. These results suggest that vision and hearing may be used together or separately depending on conditions to avoid predation and that hearing is especially vital under conditions of darkness.

Evolution of the Heteromyid Ear

Many features of ear anatomy are common to all heteromyid genera. In the middle ear all possess three chambers, similar morphology of the malleus and incus, and the presence of trabecular bone as adults or juveniles. Middle ear volume exhibits striking increases from generalized forms as *Liomys* and *Heteromys* through *Perognathus/Chaetodipus* to specialized taxa as *Dipodomys* and *Microdipodops.* A stapedius muscle is absent in *Liomys* and *Perognathus* while the stapedial artery has atrophied in *Liomys* and *Heteromys.*

For the inner ear, significant widening of the basilar membrane, a tectal cell canopy in the Organ of Corti and a well developed hyaline mass characterize *Perognathus, Dipodomys* and *Microdipodops. Liomys* and *Heteromys* lack the tectal cell canopy and exhibit only a minimal inclusion of hyaline material in the apical end of the zona pectinata. Basilar membrane width varies conservatively in *Liomys* and presumably in *Heteromys,* but widens rapidly and markedly in all other genera.

A generalized ancestral form with ear morphology similar to that of modern *Liomys* or *Heteromys* but possessing both a stapedius muscle and a stapedial artery is postulated. This ancestral group gave rise to more advanced forms of which the modern genera segregate into three groups: 1) *Liomys* and *Heteromys* share the following characters: small generalized middle ear with retention of trabecular bone in the adult; generalized inner ear; lack of a stapedial artery; stapedius muscle absent (note: this is not verified for *Heteromys*). 2) *Perognathus* and *Chaetodipus* lack a stapedius muscle, have a stapedial artery, possess a specialized inner ear and exhibit a greatly enlarged middle ear relative to that of *Liomys* and *Heteromys*; and all three middle ear chambers are lined by a thick layer of tiny trabecular air cells; 3) *Dipodomys* and *Microdipodops* possess a stapedius muscle and stapedial artery, highly specialized inner ear and greatly expanded middle ear which exhibits trabecular bone only in juvenile stages (Webster and Webster, 1975). The absence of a stapedius muscle links *Perognathus* and *Chaetodipus* with *Liomys* and *Heteromys* but the persistence of a stapedial artery and the major differences in middle and inner ear anatomy fail to support close relationship. A separate line that retained both stapedius muscle and stapedial artery gave rise to *Dipodomys* and *Microdipodops.* The morphology of the auditory bulla and the adjacent cranial elements of *Perognathus* and *Chaetodipus* adumbrate the morphology of these structures in *Dipodomys* and *Microdipodops.* I prefer to consider the extant heteromyids as representing three separate groups evolved from distant common ancestors.

Selective pressures leading to the evolution of the auditory specializations of *Perognathus, Dipodomys* and *Microdipodops* were probably similar to those postulated by Lay (1972) for the Gerbillinae. The presence of trabecular air cells in all middle ear chambers of all genera suggests a common pattern of development. These air cells persist in the adults of *Heteromys, Liomys* and *Perognathus,* but are lost during the development of the ear in *Dipodomys* and presumably *Microdipodops.* These differences may have resulted by heterochronous development. Selection acting on the mechanisms which control the timing of differentiation has been postulated to result in major morphological change and novelties during phyletic evolution (Alberch and Alberch, 1981). The morphological differences between the *Perognathus/Chaetodipus* and *Dipodomys/Microdipodops* groups are not extreme, the principal difference being the presence of a layer of trabeculated air cells lining the middle ear cavities in the former. One of the forms of peramorphosis, as defined by Alberch et al. (1979), may have functioned in the evolution of the heteromyid ear. Small perturbations in growth rate may alter the relative adult proportions of an organ or surrounding tissue types in major ways. Because the course of development is directed to a large extent by tissue interactions, distortion of tissue juxtapositions can result in quantitatively different organ structure and a complete restructuring of local architecture (Alberch et al., 1979). Webster's (1975) study of postnatal ear development in *Dipodomys merriami* reveals that the anatomy of middle ear cavities of 14 day postpartum specimens are very similar to those of adult *Perognathus*. It seems possible that the adult condition of *Dipodomys* is the result of an incremental change in the time of onset of development. Alberch et al. (1979) refer to such changes as pre- and post-displacement and note that the morphological consequences of such can be profound, resulting in qualitative differences, including loss of bony elements.

Size reduction in the sphenoid, squamosal, parietal, interparietal and occipital bones has occurred in *Perognathus/Chaetodipus* and is carried to a greater extreme in *Dipodomys/Microdipodops* relative to these elements in *Liomys* or *Heteromys*. The degree of bone reduction is inversely proportional to the volume increase of the middle ear. The functional significance of the trabeculated air cell lining of the middle ear in *Perognathus* and *Chaetodipus* remains unknown (Webster and Webster, 1975). Is it possible that this particular feature merely reflects a developmental stage and is neutrally adaptive in these genera? A comparative study of development and developmental physiology may reveal important clues for interpreting the evolution of the heteromyid ear (Hafner, 1983).

Information regarding predation on heteromyid rodents is scanty. Snakes and carnivorous mammals certainly consume these rodents, but I am unaware of any long term study documenting the extent of predation relative to prey population densities. Reports of owl feeding habits in arid regions exist (e.g., Marks and Marks, 1981; Sonnenberg and Powers, 1976; etc.). Only two studies, however, have attempted to correlate prey captured by owls with prey populations available to owls. Marti's (1974) study of owl feeding habits in north central Colorado reveals that *Dipodomys* and *Perognathus* are eaten by Great-Horned Owls (*Bubo virginianus*), Long-eared Owls (*Asio otus*), Burrowing Owls (*Speotyto cunicularia*) and Barn Owls (*Tyto alba*). In this three year study, 9,491 individual mammals were identified and only 234 or 1% were *Dipodomys ordi,* and these represented approximately 8% of the total biomass consumed annually. Many other rodent species occurred in the study area apparently in larger numbers than *Dipodomys*. Thus, one conclusion that may be drawn from this work is that owls do capture *Dipodomys,* and that it is not possible to assess the adaptive significance of auditory specializations from the data available. Kotler (1985) studied Long-eared Owl predation on a community of desert rodents near Tonopah, Nevada. Rodent populations were censused and owl pellets collected simultaneously at seven intervals over two years. Kotler concluded that owls did not capture species in the proportions in which his census technique indicated they occurred in the environment and that owls specialized on species lacking bipedal locomotion and inflated auditory bullae (*Perognathus longimembris, Peromyscus maniculatus* and *Reithodontomys megalotis*) and had low selectivity for the specialized kangaroo rats and kangaroo mice. Conversion of Kotler's skull counts from Table 1 to biomass and then comparing owl consumption of quadrupedal and bipedal forms to the census data leads to a different interpretation. Mean weights for a sample of 12 wild caught live specimens from Fishlake Valley, Nevada, about 30 miles W of Tonopah are: *Peromyscus maniculatus,* 17 g; *Reithrodontomys megalotus,* 9.5 g; *Perognathus longimembris,* 7 g; *Dipodomys merriami,* 38 g; *Dipodomys ordi,* 48 g; *Di-*

podomys microps, 58 g (Lay, unpublished data). Mean weights of the quadrupedal group of 11 g and the bipedal group of 48 g were used to estimate biomass from the skull data. Mean weight of bipedal species was 57 g when *Microdipodops pallidus* and *D. deserti* were included, but these two species represented a small proportion of the bipedal population on dunes in Fishlake Valley. More accurate biomass comparisons should be made, but the census numbers were not provided for individual species. Quadrupedal forms averaged 12% of biomass consumed by owls and 29% of the total population over the entire sampling period. The ratio of quadruped biomass to total biomass was strikingly lower than the census estimates of the proportions of live quadrupedal species in the population in all but one sampling period. Kotler's study only demonstrates that owls are quite successful in capturing heteromyids with specialized ears and other desert rodents and reiterates some of the difficulties associated with field studies designed to estimate differential predation. No study has yet demonstrated unequivocably that the specialized ears of *Dipodomys* and *Microdipodops* confer a selective advantage in open desert situations.

Literature Cited

Alberch, P., and J. Alberch. 1981. Heterochronic mechanisms of morphological diversification and evolutionary change in the neotropical salamander, *Bolitoglossa occidentalis*. Journal of Morphology, 167: 249–264.

Alberch, P., S. J. Gould, G. F. Oster, and D. B. Wake. 1979. Size and shape in ontogeny and phylogeny. Paleobiology, 5:296–317.

Bruns, V. 1976. Peripheral auditory tuning for fine frequency analysis by the CF-FM bat, *Rhinolophus ferrumequinum*. I. Mechanical specializations of the cochlea. Journal of Comparative Physiology, 106: 77–86.

Dallos, P. 1973. The auditory periphery. Academic Press Inc., New York, XII + 548 pp.

Durrant, J. D., and J. H. Lovrinic. 1984. Bases of hearing science. Williams and Wilkins, Baltimore, XXIV + 276 pp.

Hafner, J. C. 1993. Macroevolutionary diversification in heteromyid rodents: heterochrony and adaptation in phylogeny. Pp. 291–318, *in* Biology of the family Heteromyidae (H. H. Genoways and J. H. Brown, eds.). Special Publication, American Society of Mammalogists, 10:1–719.

Henson, M. M. 1978. The basilar membrane of the bat, *Pteronotus p. parnellii*. American Journal of Anatomy, 153:143–158.

Henson, M. M., D. B. Jenkins, and O. W. Henson, Jr. 1983. Sustentacular cells of the organ of Corti—the tectal cells of the outer tunnel. Hearing Research, 10:153–166.

Howell, A. B. 1932. The saltatorial rodent *Dipodomys*: the functional and comparative anatomy of its muscular and osseous systems. Proceedings of the American Academy of Arts and Science, 67:377–536.

Knudsen, E. I., and M. Konishi. 1978. A neural map of auditory space in the owl. Science, 200:795–797.

———. 1979. Mechanisms of sound localization in the Barn Owl (*Tyto alba*). Journal of Comparative Physiology, 133:13–21.

Knudsen, E. I., G. G. Blasdel, and M. Konishi. 1979. Sound localization by the Barn Owl (*Tyto alba*) measured with the search coil technique. Journal of Comparative Physiology, 133:1–11.

Kobayashi, M. 1955. The articulations of the auditory ossicles, with a note suggesting their function. Hiroshima Journal of Medical Science, 4:319–349.

Kotler, B. P. 1985. Owl predation on desert rodents which differ in morphology and behavior. Journal of Mammalogy, 66:824–828.

Lay, D. M. 1972. The anatomy, physiology, functional significance and evolution of specialized hearing organs of gerbilline rodents. Journal of Morphology, 138:41–120.

Marks, J. S., and V. A. Marks. 1981. Comparative food habits of the screech owl and the long-eared owl in southwestern Idaho. Murrelet, 62:80–82.

Marti, C. D. 1974. Feeding ecology of four sympatric owls. Condor, 76:45–61.

Pye, A. 1965. The auditory apparatus of the Heteromyidae. Journal of Anatomy, London, 99:161–174.

Sonnenberg, E. L., and L. R. Powers. 1976. Notes on the food habits of long-eared owls (*Asio otus*) in southwestern Idaho. Murrelet, 57:63–64.

Webster, D. B. 1961. The ear apparatus of the kangaroo rat, *Dipodomys*. American Journal of Anatomy, 108:123–148.

———. 1962. A function of the enlarged middle-ear cavities of the kangaroo rat, *Dipodomys*. Physiological Zoology, 35:248–255.

———. 1975. Auditory systems of Heteromyidae: postnatal development of the ear of *Dipodomys merriami*. Journal of Morphology, 146:377–393.

Webster, D. B., and M. Webster. 1971. Adaptive value of hearing and vision in kangaroo rat predator avoidance. Brain, Behaviour and Evolution, 4:310–322.

———. 1975. Auditory systems of Heteromyidae: functional morphology and evolution of the middle ear. Journal of Morphology, 146:343–376.

———. 1977. Auditory systems of Heteromyidae:

cochlear diversity. Journal of Morphology, 152:153–170.

———. 1984. The specialized auditory system of kangaroo rats. Contributions to Sensory Physiology, 8:161–196.

Wever, E. G., and M. Lawrence. 1954. Physiological acoustics. Princeton University Press, Princeton, 454 pp.

Wever, E. G., J. G. McCormick, J. Palin, and S. H. Ridgway. 1971. Cochlea of the dolphin, *Tursiops truncatus*: the basilar membrane. Proceedings of the National Academy of Sciences (U.S.A.), 68:2708–2711.

MACROEVOLUTIONARY DIVERSIFICATION IN HETEROMYID RODENTS: HETEROCHRONY AND ADAPTATION IN PHYLOGENY

JOHN C. HAFNER

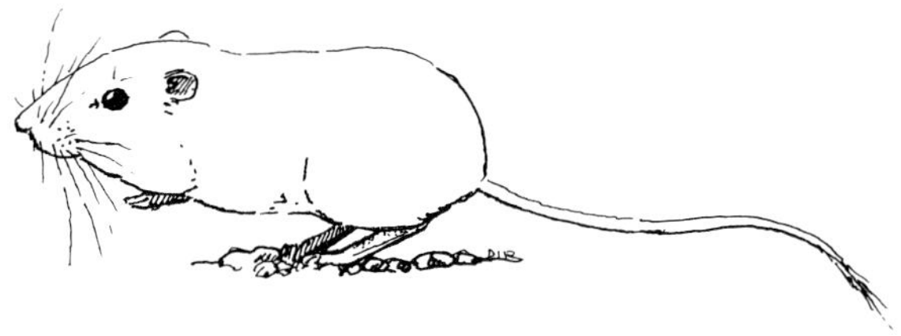

Introduction

Macroevolution, although variously defined and redefined (e.g., Bock, 1979; Dobzhansky, 1937; Goldschmidt, 1940; Mayr, 1963; Rensch, 1959; Simpson, 1944, 1953; Stanley, 1979), is probably best considered to be a phenomenological description of the large-scale morphological change that usually accompanies transpecific evolution (cladogenesis at the species level and higher taxonomic categories). Importantly, this large-scale morphological differentiation includes the eventual production of a macroevolutionary novelty: a unique morphology or *Bauplan* that, when considered in retrospect, seemingly allowed the bearer (a new taxon) to enter a new adaptive zone (sensu Simpson, 1944). One of the most exciting tasks in evolutionary biology is that of providing a convincing explanation, attendant with testable propositions, for the macroevolutionary novelty. To provide a thorough explanation of macroevolution in a taxonomic group, one must carefully consider three separate aspects of the problem: 1) phylogenetic relationships among the taxa; 2) adaptive significance of novel morphologies; and, 3) the causal mechanism(s) responsible for morphological diversification.

The intent of this contribution is to provide a preliminary assessment of macroevolution in the rodent family Heteromyidae (pocket mice, kangaroo rats, and their allies). This family is a morphologically and ecologically diverse group whose extant members display an array of adaptive types from scansorial, mesic-adapted genera to bipedal, xeric-adapted genera. Although differences at the generic level are generally considered to be at the lower end of the scale of macroevolutionary divergence (e.g., Bock, 1979), the high degree of structural divergence among the heteromyid genera provides an unusual opportunity to investigate major evolutionary change.

The evolutionary and taxonomic history of the Heteromyidae is intimately associated with that of the Geomyidae (pocket gophers; all extant members are fossorial). Together, these two families form a cohesive superfamily, the Geomyoidea, whose members are united by the presence of externally opening, fur-lined cheek pockets (among other features). The superfamily Geomyoidea is autochthonous in continental North America and is an old, monophyletic lineage that is distantly related to other rodent groups (Wood, 1935; Hafner, 1982). The geomyoids experienced major phyletic diversification in the Oligocene to Pliocene coincident with the climatic trend towards increasing coolness and aridity (Flint, 1971) and the development of the Madro-Tertiary Geoflora (Axelrod, 1950, 1958, 1976). Due to marked similarities in heteromyid and geomyid biogeographic histories and their close phyletic association (for review, see Hafner, 1982; Hafner and Hafner, 1983), relevant geomyid information will be presented in this review to facilitate a more thorough understanding of macroevolutionary divergence within the Heteromyidae.

Evolutionary Relationships of Heteromyid Rodents

During the past one-half century a tremendous volume of literature pertaining to heteromyid evolution has accumulated. The most recent statement of the evolutionary relationships of the Heteromyidae was provided by Hafner and Hafner (1983); that study integrated the classic morphological treatises of the 1930s (e.g., Hatt, 1932; Howell, 1932; Wood, 1935) with the more recent systematic treatments (e.g., *Chaetodipus*: Patton, 1967*a*; Patton et al., 1981; *Perognathus*: Patton, 1967*b*; Williams, 1978; *Dipodomys*: Johnson and Selander, 1971; *Microdipodops*: Hafner, 1978, 1981; Hafner et al., 1979; *Liomys*: Genoways, 1973; *Heteromys*: Rogers and Schmidly, 1982). The following account is a current synopsis of the patterns of supraspecific relationships within the extant Heteromyidae. I use as my point of departure Hafner and Hafner (1983) and included references.

Hafner's (1982) molecular study indicated that the divergence between the heteromyids and geomyids occurred in the early Eocene (approximately 50 my before present); this estimate was based on an average of 90 immunological distance (*ID*) units measured between the families and molecular clock calculations available at that time. As Hafner noted, his time estimate for the heteromyid-geomyid split predates the earliest geomyoid fossil (*Heliscomys*; early Oligocene) by 10 to 15 my. I have re-evaluated Hafner's (1982) immunological data using Sarich's (1985) revised calibration for the molecular clock. This new calibration rectifies some recently discovered miscalibrations and resets the molecular clock for albumin immunological studies (for discussion see Sarich, 1985:429–433). Accordingly, the time, t, of separation between two lineages may be estimated by the relation $t = k(100 - QP)$, where k is the new calibration factor (0.67–0.71) relating the difference in albumin cross-reaction to time (in millions of years) and QP is the quantitative precipitin value for micro-complement fixation data ($QP = 100 - [ID/2]$). This corrected time calibration places the initial geomyoid radiation in the early Oligocene (approximately 30–32 mybp); hence, the molecular data are in accord with the paleontological evidence (see also Wahlert, this volume).

Figure 1 summarizes current views on heteromyid relationships. The extant heteromyids comprise three principal lineages (including six genera) that diverged in the Oligocene: 1) subfamily Heteromyinae (*Liomys* and *Heteromys*); 2) subfamily Perognathinae (*Perognathus* and *Chaetodipus*); and, 3) subfamily Dipodomyinae (*Dipodomys* and *Microdipodops*). The Heteromyinae (spiny pocket mice) form the most distinct and internally cohesive lineage within the family and have experienced a long evolutionary history independent of the other subfamilies (Fig. 1). The spiny

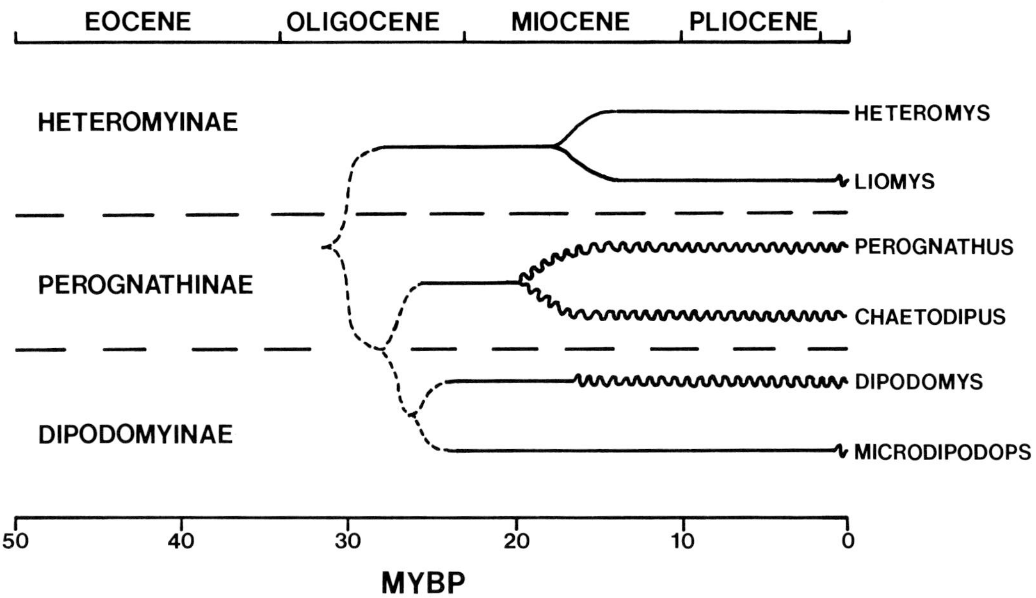

FIG. 1.—Phylogeny of the extant Heteromyidae indicating supraspecific relationships. Solid lines indicate probable affinities, sinuous lines relate the fossil record (the paleontological literature does not distinguish between *Perognathus* and *Chaetodipus*), and the dashed lines signal areas in need of further investigation. The time scale is based on fossil evidence (Lindsay, 1972; Wood, 1935) and immunological and allozymic data (Hafner, 1982; see text for molecular clock calibration).

pocket mice show generalized (mouse-like) rodent morphology and, unlike the other heteromyids, show a marked ecological propensity for tropical to subtropical environs. The Perognathinae (pocket mice) are also morphologically conservative in body plan, yet these pocket mice inhabit a broad spectrum of arid environments (e.g., sandy deserts, arid grasslands, chaparral and thornscrub forests). The Dipodomyinae includes the morphologically aberrant kangaroo mice, *Microdipodops,* and kangaroo rats, *Dipodomys.* The kangaroo mice are narrowly adapted to xeric, sandy habitats, whereas the kangaroo rats show a broad tolerance to generally arid environments (e.g., arid grasslands, chaparral, and desert habitats).

The subfamilial affinity of *Microdipodops* has plagued heteromyid systematists virtually since its discovery and description a century ago (for review, see Hafner and Hafner, 1983). Over the years, the taxonomic placement of kangaroo mice has vacillated between the Perognathinae and the Dipodomyinae, but it is now recognized that kangaroo mice may represent an independent lineage with no close relatives in the extant fauna (Hafner, 1978; Hafner and Hafner, 1983). However, recent biochemical evidence (Hafner, 1982; Hafner and Hafner, 1983) indicates that *Microdipodops* may be genetically slightly more closely related to *Dipodomys* than to extant perognathines. More detailed biochemical analyses are needed to confirm or refute the hypothesized alignment of kangaroo mice with kangaroo rats, but for now *Microdipodops* is placed provisionally in the Dipodomyinae (Fig. 1). It is important to recognize, however, that kangaroo mice, although sharing with kangaroo rats obvious superficial traits (e.g., large head and long hind feet), are not merely scaled-down versions of kangaroo rats; kangaroo mice are physiologically, ecologically and morphologically quite different from kangaroo rats.

The family Heteromyidae, including morphologically disparate scansorial and ricochetal forms, has undergone flamboyant morphological diversification by rodent standards; *Microdipodops* and *Dipodomys* clearly represent evolutionary novelties (but see Mares, this volume). In marked contrast to the heteromyids, members of the closely related family Geomyidae (pocket gophers) are remarkably conservative morphologically. With the exception of size differences, the extant pocket gophers are nearly uniform in morphology. Presumably, stringent selective constraints associated with the fossorial habitus restricted the realm of possible morphologies in the Geomyidae. Nevertheless, pocket gopher morphology is also a novel (derived) body plan when compared with the generalized (mouse-like) rodent condition. If macroevolution is considered as large-scale morphological change, then it seems that macroevolutionary diversification in the superfamily has resulted in several distinct evolutionary novelties among the extant taxa: kangaroo rats, kangaroo mice and pocket gophers.

Adaptation and the Evolution of Novel Features

The tremendous breadth of morphological differentiation seen in the Geomyoidea provides an exceptional opportunity for studies in evolutionary morphology. Although this superfamily is geographically restricted when compared with most other major rodent groups, geomyoids inhabit both desert and tropical environments and show remarkable modifications attendant with fossoriality (pocket gophers) as well as scansorial (pocket mice) and ricochetal (kangaroo rats and kangaroo mice) habits. It is indeed a challenge to explain the evolution of these extreme and conspicuous morphological modifications.

The remarkable morphological features characteristic of the ricochetors (*Dipodomys* and *Microdipodops*) are a popular case in point. Conventional explanations for the morphological novelties shown by these forms focus on the adaptive aspects (advantages) of the functional design (for example, see review by Eisenberg, 1975). Inherent in these explanations is the assumption that random mutation produces sufficient variation in form such that natural selection will continually shape the morphological features into a better adapted form. For kangaroo mice and kangaroo rats, one's attention is drawn immediately to the enormous head (due, in part, to inflation of the auditory bullae) and huge hind feet, as well as the large eyes and long tail. There are many opinions as to the function of each of these features (see Table 1), and virtually all of these adaptive explanations focus on what is termed the "anti-predator morphology" of *Microdipodops* and *Dipodomys* (e.g., Kotler, 1985). These explanations demonstrate "the enormous power of the principle [of natural selection] as a weapon of explanation" (Waddington, 1975: 41). Morphological evolution in the Heteromyidae is usually explained as a result of long-term, directional selection (orthoselection): natural selection favors certain adaptations present in ancestral species and these adaptations are accentuated in descendant species in response to the same selective pressures. For example, Hall (1946: 406) remarked on "evolution towards perfecting rapid locomotion by use of the hind limbs in *Dipodomys* . . .", while Grinnell (1922:23) wrote, "The reduction of the toes [in *Dipodomys*] is, then, a sort of orthogenetic tendency inherent in the group as a whole, but it is no less, in the writer's view, an adaptational process. . . ." The question of adaptation is, of course, central to our understanding of morphological evolution in the Geomyoidea. However, previous attempts to address this topic have relied on an oversimplified accounting of the mechanism of evolution. Evolution involves more than simply natural selection acting on a "genetic system" and, as Waddington (1975: 58) observes, we must consider at least two other crucial components: the "exploitive system" and the "epigenetic system." A

TABLE 1.—*Adaptive explanations for certain conspicuous morphological features of kangaroo mice (*Microdipodops*) and kangaroo rats (*Dipodomys*).*

Morphological feature	Adaptive function	Source
Enlarged auditory bullae	1) delicate balance in ricochetal locomotion	Seton (1928); Hatt (1932); Setzer (1949)
	2) highly specialized acoustic sense	Howell (1932); Setzer (1949)
	3) sounding boards to monitor vibrations through the ground	Howell (1932)
	4) low-frequency hearing sensitivity for predator avoidance	Webster (1962); Webster and Webster (1971, 1975, 1980)
Large eyes	1) good nocturnal or crepuscular vision	Hall and Linsdale (1929)
Long tail	1) counterbalancing organ	Hatt (1932); Howell (1932); Hall (1941); Bartholomew and Caswell (1951)
	2) support prop or "third leg" for animal at rest	Hatt (1932); Howell (1932); Bartholomew and Caswell (1951)
	3) mid-air rudder for trajectory control	Hatt (1932); Howell (1932); Bartholomew and Caswell (1951)
	4) protection: misdirection of enemy's attack to terminal tuft (*Dipodomys* only)	Howell (1932); Hatt (1932); Mares (1983)
Long hind feet	1) "sand paddles" for locomotion on sandy soils	Hall and Linsdale (1929); Hall (1941)
	2) richochetal locomotion for rapid dodging and quick escape from predators	Howell (1932); Hall (1946); Bartholomew and Caswell (1951); Kotler (1985)
	3) bipedal saltation for an energetically efficient mode of locomotion and successful exploitation of a hyperdispersed resource base	Howell (1932); Reichman and Oberstein (1977); Price (1978)

more thorough understanding of adaptation is possible when one takes this more eclectic view of the evolutionary process.

Functional Significance of Novel Features

There are two types of explanations in biology: functional and causal explanations. Functional explanations explain morphology in terms of its purpose to the animal and disregard prior states, whereas causal explanations focus on prior morphological states and attempt to predict future states from earlier ones. It is important to observe that functional and causal explanations are not directly competitive and, in fact, address different sets of questions. Functional explanations focus on present use of a feature, whereas causal explanations are concerned solely with the evolution of the feature (regardless of its present use). Both kinds of explanations are important, but it seems that problems arise when functional explanations are used to make unwarranted extrapolations as to the evolution of a feature. Until recently, explanations for the extreme morphological variation found among the heteromyids were all of the functional (or adaptationist) type (see Table 1). These functional explanations, unfortunately, often infer evolution and, therefore, obfuscate the other important set of questions that needs to be addressed (see also Brookfield, 1982). Clearly, some of these oft-cited explanations may pertain to actual "aptations" (either adaptive or exaptive characteristics; Gould and Vrba, 1982), but functional explanations are epistemologically unsatisfactory hypotheses for the evolution of a novel feature. Below I address in detail several cases that are relevant to this discussion of adaptation and the evolution of novel features.

Large auditory bullae.—Evidence indicating that the enlarged bullae in *Microdipodops* and *Dipodomys* function to facilitate low-frequency hearing and represent an adaptation to avoid predatory strikes by owls and snakes (Webster, 1962; Webster and Webster, 1971, 1972, 1975, 1980) has been received with virtually universal acceptance. Not surprisingly, the enlarged bullae of kangaroo rats are now cited as a textbook example of vertebrate adaptation (e.g., Gunderson, 1976:331–332; Stebbins, 1983: 84; Vaughan 1978:467–468; Willson, 1984: 158). Nonetheless, this hypothesized antipredator adaptation deserves critical evaluation (see Lay, this volume). Webster (1962) originally determined that in *Dipodomys* hypertrophy of the middle-ear cavity is associated with unusually sensitive cochlear microphonic responses between 1,000 and 3,000 Hz and that experimental reduction of middle-ear volume dramatically reduces this sensitivity. Webster (1962) further showed that the predatory strikes of owls and rattlesnakes have pre-strike sounds that contain these same (1,000 to 3,000 Hz) frequencies. The conclusion drawn from these studies, coupled with predator-prey experiments (Webster, 1962; Webster and Webster, 1971), is that the enlarged middle ear of the kangaroo rat facilitates low-frequency reception, which is particularly adaptive for nocturnal animals in open areas and, therefore, plays an adaptive role in predator avoidance. Importantly, it is not generally recognized that the heightened reception of key frequencies within the 1,000 to 3,000 Hz range originally reported by Webster (1962) was later "regarded as an artifact of the method of analysis" by Webster and Webster (1972:50). Subsequent data by Webster and Webster (1975, 1980) indicate that the auditory sensitivity curves for various species of heteromyids (including *Perognathus, Microdipodops, Liomys, Heteromys,* and *Dipodomys*) do not exhibit a pronounced peak between 1,000 to 3,000 Hz, but show rather flat sensitivity curves from low (≈100 Hz) to high (≈30,000 Hz) frequencies. Further, although *Microdipodops* and *Dipodomys* are said to have more sensitive cochlear microphonics than do other heteromyids with smaller middle-ear cavities (Webster and Webster, 1980), no statistical tests were performed to document that significant differences in sensitivity actually exist. From the available data (Webster and Webster, 1980:252) it is clear that all genera of heteromyids are actually more sensitive between 300 to 1,000 Hz than in the "predatory range" of 1,000 to 3,000 Hz; hence, the validity of this predator-avoidance hypothesis is questionable. Most significantly, and contrary to the predictions of their model, Webster and Webster (1971: 314) have demonstrated that kangaroo rats with experimentally reduced middle-ear volume were still able to avoid the predatory strikes of the rattlesnake (see below). Moreover, if enlarged middle ear cavities do indeed function to facilitate low-frequency hearing for predator avoidance in open desert environments, it is not clear why there is no demonstrated relationship between auditory bullar size and environment within the family (Grinnell, 1922; Setzer, 1949; Webster and Webster, 1975).

Large eyes.—Kangaroo rats and kangaroo mice are nocturnal rodents that have large eyes. If enlargement of the eyes is an adaptation for enhanced nocturnal vision, then it is necessary to document that the large eyes of the kangaroo rats and kangaroo mice facilitate more keen vision than do the smaller eyes of other rodent genera with which they are sympatric. So far, no such tests have been undertaken. One must be cautious in assuming that the large eyes represent a special adaptation for enhanced nocturnal vision; actually, the antelope ground squirrel (*Ammospermophilus*), commonly found sympatric with the heteromyid ricochetors, also has large eyes, yet is a diurnal rodent.

Elongated tail.—Kangaroo mice and kangaroo rats have long tails relative to body length (for comparative rodent data, see Hatt, 1932). The tails of *Microdipodops* and *Dipodomys* undoubtedly function to some degree in maintaining balance during locomotion; this locomotory function of the tail (and a thermoregulatory function as well) applies to rodents in general (for review, see Thorington, 1966). But if the long tails of *Microdipodops* and *Dipodomys* represent adaptations for the richochetal mode of locomotion (Bartholomew and Caswell, 1951; Hall, 1941; Hatt, 1932; Howell, 1932), one would expect that tail length and hind foot length would be correlated across ricochetal taxa. Importantly, Grinnell (1922) and Setzer (1949) have determined that such a relationship does not exist among heteromyid species. Also, if natural selection has acted to perfect the ricochetal mode of locomotion, it is unclear why kangaroo rats hurl through the air "in a more or less lopsided fashion" (Howell, 1932:386), and often land off-balance "owing apparently to clumsy use of the tail" (Howell, 1944:40). In a rigorous examination of tail function, Bartholomew and Caswell (1951) cut off the tail of a kangaroo rat to test the importance of the tail in the maintenance of equilibrium during bipedal locomotion. Their experiment determined that "removal of the tail had no apparent effects despite frequently heard statements to the contrary" (Bartholomew and Caswell, 1951:165). Coupled with this experimental result is Howell's (1923) and my personal observations that short-tailed kangaroo rats (those whose tails have been shortened presumedly by accidental injury) frequently occur in nature. It also seems incongruent that the tail of *Microdipodops* is "relatively less flexible" than that of pocket mice (Hatt, 1932:646) if it is adapted to serve as a counterpoise and a mid-air rudder during saltation. Actually, the unusual tail of kangaroo mice, being thickened in the mid region, is a site of fat deposition and storage and probably serves important physiological needs of the animal. In sum, the arguments that the elongated tail was shaped by natural selection to function specifically in ricochetal locomotion seem unconvincing.

Long hind feet.—Certain heteromyids have long hind feet and are able to move rapidly in open environments. However, the assumption that the long hind foot of kangaroo rats and kangaroo mice is an adaptation that has been finely tuned by natural selection to facilitate ricochetal locomotion in open habitats remains unchallenged. The argument would be stronger if all bipedal forms inhabited open environments. However, Grinnell (1922) and Setzer (1949) concluded that there is no relationship between saltatorial specialization and habitat type in species of *Dipodomys.* Indeed, kangaroo rats inhabit both xeric grasslands and coastal (often fog-shrouded) chaparral hillsides, as well as sparsely vegetated sand dunes; thus, the explanation that the ricochetal mode of locomotion is a finely tuned adaptation specifically for life in open environments appears incorrect. Further, recent studies have shown, contrary to conventional opinion, that bipedal saltation does not impart kangaroo rats an energetic advantage over quadrupedal locomotion (MacMillen, 1983; Thompson et al., 1980) and energy saving by elastic storage (in tendons and muscle fibers) appears to be unimportant in animals of this size (Biewener et al., 1981). With respect to the explanation that bipedal saltation is a special adaptation for predator avoidance, I point out that both the bipedal kangaroo rats and the quadrupedal pocket mice rely on the same escape response (long, erratic leaping) to avoid predation (Bartholomew and Cary, 1954; Price and Brown, 1983; personal observations).

Questioning adaptationist explanations.—The ad hoc adaptationist explanations designed to account for the existence of the unusual morphological features discussed above may themselves be impediments to a clear understanding of morphological evolution in the Heteromyidae. Surprisingly, however, only a few workers have questioned these functional explanations. Wood (1935:143) remarked, "There is a strong relation among rodents between a highly inflated auditory region and ricochetal locomotion, though the reason for this correlation is obscure." Also, Bartholomew and Cary (1954:391), in observing that the food habits, food-gathering style, habitat preferences, and escape responses were similar in both kangaroo rats and pocket mice, posed the question, "Why should one be bipedal and the other quadrupedal?" Pye (1965:169), in an analysis of the auditory apparatus of the Heteromyidae, observes, "The possible functional significance of these modified cochlear structures has been discussed by Webster, but more ecological and physiological work needs to be carried out before any firm conclusions can be reached." More recently, Thompson et al. (1980:224) asked, "If there is no energetic advantage for small animals to hop, why do they do so?" These questions and many others prompt critical reconsideration of these functional explanations. Functional (or adaptationist) explanations are important in biology, but one must keep in mind that the theory of natural selection does not necessarily legitimize ad hoc functional explanations for each smallest component of an atomized organism (for discussion see Brookfield, 1982; Mayr, 1983). Future workers should evaluate critically these long-accepted explanations as well as consider other explanations for the evolution of these features.

Predator-prey Studies: an Evaluation

Most of the functional explanations that have been proffered (Table 1) argue that the novel morphology of ricochetal heteromyids is, in one way or another, adaptive in predator avoidance. If we are to understand the present utility of these novel morphologies, then it is crucial that these adaptive hypotheses be evaluated and tested thoroughly. Several important studies have been conducted that were designed to test the view that kangaroo mice and kangaroo rats have particularly adaptive "anti-predator morphology" (Kotler, 1985, Webster, 1962; Webster and Webster, 1971). A critical review of these predator-prey studies follows.

Predator-prey experiments with kangaroo rats.—Webster's (1962) experimental study of Screech Owl (*Otus asio*) and Sidewinder (*Crotalus cerastes*) predation on the kangaroo rat (*Dipodomys merriami*) was the first investigation of the functional significance of the "anti-predator morphology." Webster's (1962) studies focused on the function of the inflated auditory bullae. Having first reported that normal kangaroo rats have unusually sensitive cochlear microphonics between the range of 1,000 to 3,000 Hz (a finding later dismissed as erroneous; see above), Webster compared the ability of kangaroo rats with experimentally reduced middle-ear volume (and reduced microphonic response) with normal kangaroo rats in avoiding the predatory strikes of owls and rattlesnakes. Eight kangaroo rats (four unoperated and four with reduced middle-ear volumes) were tested for their ability to avoid predation by screech owls: two of four normal individuals avoided the owls' attacks, but none of the animals with reduced middle-ear volume was able to avoid capture. Six kangaroo rats (three normal and three with reduced middle-ear volumes) were used in the Sidewinder experiments: all three normal kangaroo rats avoided strikes and all three operated individuals were struck and killed by the rattlesnake. Webster (1962) further reported that sounds of an attacking owl and rattlesnake contained frequencies of 1,200 Hz and 2,000 Hz, respectively. Webster (1962) concluded from these experiments that the hypertrophied bullae confer to the kangaroo rat a special auditory sensitivity at low frequencies that enables the detection of predators.

In a more extensive study, Webster and Webster (1971) carried out observation sessions to test kangaroo rat (*D. merriami*) avoidance of the predatory strikes of the Sidewinder. The predator-prey sessions involved kangaroo rats in six different physical conditions: 1) normal (unoperated); 2) operated (plasticene placed in bullae to reduce middle-ear volume); 3) control (plasticene placed between skin and bullae); 4) blinded; 5) control and blinded; and, 6) operated and blinded. These predator avoidance experiments demonstrated that kangaroo rats with experimentally reduced middle-ear volumes were, in fact, *able* to avoid the strikes of the Sidewinder, contradicting Webster's (1962) earlier findings. Actually, kangaroo rats in five of the six test states (involving 18 forty-minute observation sessions) were routinely able to avoid the predatory strikes of the Sidewinder; only those kangaroo rats that were both surgically blinded and had their middle-ear volumes reduced were occasionally struck and killed (three of eight such encounters resulted in the rattlesnake striking and killing the kangaroo rat).

Webster and Webster (1971) concluded from their studies that natural selection should favor individuals with good low-frequency hearing (for predator avoidance) and that the enlarged bullae evolved in response to those selection pressures (see also Fleischer, 1978; Lay, 1972). However, the predator-prey experiments conducted by Webster (1962) and Webster and Webster (1971) do not support the idea that the hypertrophied bullae endow the possessor with an enhanced ability to avoid predation. It is now known that kangaroo rats (and kangaroo mice) have a broad range of hearing sensitivity (Webster and Webster, 1980) but do not exhibit the peaks of sensitivity in the "predatory frequencies" of 1,000 to 3,000 Hz (see discussion above). Also, a recent study of hearing in other small mammals (Heffner and Heffner, 1985) contradicts the idea that small mammals require enlarged auditory bullae to hear low frequencies. It seems that the predator-prey experiments conducted by Webster (1962) and Webster and Webster (1971) are inconclusive and more experiments are needed to gain a more thorough understanding of the functional significance of the enlarged auditory bullae.

Natural selection experiments.—Webster and Webster (1971) live-trapped 27 kangaroo rats (*D. merriami*) for an experiment to determine whether animals with reduced middle-ear volume showed reduced survivorship in the wild compared with those individuals with normal, unoperated bullae. The kangaroo rats were divided into three groups of nine: 1) normal, unoperated; 2) operated (plasticene placed in bullae to reduce middle-ear volume); and, 3) control (plasticene weights placed above middle-ear cavities). The kangaroo rats were released on the study site (approximately 7 ha) and survivorship was estimated by mark-and-recapture censusing methods involving 136 live traps set out approximately once very four nights. Individuals were presumed dead if they were not recaptured during four consecutive nights of trapping at the end of the study (approximately one month after the initiation of the experiment). At the conclusion of the study, 14 kangaroo rats were recaptured: two of the operated kangaroo rats and six individuals each from the normal and control groups. Webster and Webster (1971) concluded from these data that fewer operated animals were retrapped because these kangaroo rats, having reduced middle-ear volume, were more vulnerable to predation by the Sidewinder (the only kangaroo rat predator observed on the study site) than were those animals without middle-ear reduction.

The results of Webster and Webster's (1971) natural selection study are provocative but, given the design of the study and the nature of the animals themselves, there are important considerations that should be addressed before meaningful conclusions can be drawn. For example, a *G*-test for goodness of fit performed on Webster and Webster's (1971) survivorship data reveals that the observed frequencies of recaptured individuals do not depart significantly from random expectation ($G = 2.64$; $P \gg 0.05$). Thus, Webster and Webster (1971) were not justified in concluding that the apparent disappearance of the operated animals during the study resulted from their increased vulnerability to predation relative to normal animals. Their conclusion, in addition to being statistically unwarranted, seems to be inconsistent with their own predator-prey experiments that demonstrated that kangaroo rats with reduced middle-ear volume and normal kangaroo rats were equally able to avoid the strikes of the Sidewinder (also in Webster and Webster, 1971). Webster and Webster's (1971) experiments do *not* demonstrate that the enlarged middle-ear cavities in kangaroo rats, as opposed to the reduced condition in many other rodents, represent a special adaptation for predator avoidance.

Predator selectivity studies.—Kotler (1985) stated that *Microdipodops* and *Dipodomys* exhibit conspicuous "anti-predator morphology" and he pointed out that the genera possess many morphological characteristics (e.g., hyperinflated auditory bullae and elongated hind legs) that are beneficial in avoiding predation. He reasoned that bipedal heteromyids, by virtue of their "anti-predator morphology," should suffer lower rates of predation than coexisting species that were quadrupedal and lacked such adaptive morphology. To assess this, Kotler (1985) studied predation by long-eared owls (*Asio otus*) on desert rodents in a sand-dune community. Densities of rodent species were determined by census trapping and those values were compared with the proportional occurrence of rodent species in regurgitated pellets found beneath a nearby roost of long-eared owls. Prey selectivity indices were calculated (proportion of a rodent species in pellets divided by the proportional density of that rodent) and selectivity values for "bipedal species" (*Dipodomys* and *Microdipodops*) were compared with "quadrupedal species" (*Perognathus, Peromyscus, Reithrodontomys*) over six sample time periods. Kotler (1985) concluded that the owls did not capture rodents in the proportion in which they occurred on the sand-dune site but, instead, selectively preyed upon the quadrupedal rodents because of their lack of anti-predator adaptations. Although the results of his study are consistent with the premise that bipedal heteromyids possess "anti-predator morphology," the conclusions are suspect because of problems in the experimental design and analysis of data.

The experimental design rests on a tenuous assumption that the desert rodents can be partitioned into two locomotory categories: bipedal and quadrupedal rodents. While it is fashionable to categorize *Dipodomys* and *Microdipodops* as bipedal rodents (essentially everyone does this as a kind of shorthand for describing their curious, kangaroo-like morphology), in the present study that seeks to explore whether bipeds are superior to quadrupeds in predator avoidance this categorization cannot pass without careful scrutiny. Despite the fact that nearly everyone refers to kangaroo mice as being bipedal, Eisenberg (1963:29) states that, "Analysis of several films and observation of *Microdipodops* for long periods of time reveal that a quadrupedal ricochet is the predominant mode of locomotion". Also, most workers have seemingly overlooked Seth Benson's observation of a pocket mouse running bipedally on a lengthy (15-foot) dash (see Hatt, 1932:629–630). In view of what is known about the locomotion of kangaroo rats, kangaroo mice and pocket mice, it is incorrect for Kotler (1985) to lump kangaroo rats and kangaroo mice together as "bipeds" and compare them to other "quadrupeds" in this predator selectivity study.

The experimental design of the study assumes that the owls hunted for rodents in the same area (or at least in the same habitats) as that censused by live trapping; this is a basic assumption upon which the accuracy of the selectivity indices are dependent. Given that there is no assurance that the owls restricted their hunting activities to the semi-stabilized dune area, it is incorrect to report selectivity values that are based on the comparison of the proportion of rodents in owl pellets with the density of rodents on the dunes. Although Kotler was careful to raise this caveat (Kotler, 1985: 826), the fact that several rodent species that never occurred on the sand dunes were recovered in the owl pellets strongly suggests that the owls hunted in a variety of habitats. Although Kotler eliminated the non-dune species from his analysis, it is, nevertheless, impossible to disentangle this sampling bias from the selectivity indices.

In analyzing the selectivity information, Kotler (1985) compared the pooled data for bipedal species with pooled data for quadrupedal species. Pooling the species based on locomotory morphology may have obscured important information on the selectivity of long-eared owls for individual prey species. While the categorical division of the rodents into bipedal and quadrupedal species was designed to represent those species that do and do not possess the "anti-predator morphology," this subdivision reflects, more realistically, body size differences between the two categories. The so-termed bipedal group includes four species of kangaroo rats (ranging in body size from 41 g to 100 g) and one species of kangaroo mouse (13 g), whereas the quadrupedal group includes one pocket mouse (8 g) and two cricetids (13 g and 26.5 g; body mass data from Burt and Grossenheider, 1952; Jones, 1985). Thus, the typical animal in the bipedal group was considerably larger than the typical quadrupedal animal. Body size is an important criterion of prey vulnerability (see Craighead and Craighead, 1969) and, in a separate study, Kotler (1984) noted that large size helps reduce the vulnerability of kangaroo rats to predators. It seems that Long-eared Owl prey selectivity in Kotler's (1985) study may simply reflect this owl's preference for small-sized prey and has little or nothing to do with the morphological conformation of the prey species. Hence, there is no control for the owl's predatory behavior, which may well be biased toward one functional prey type for reasons unrelated to the prey themselves. Inasmuch as both bipedal and quadrupedal heteromyids exhibit the same predator escape response (Bartholomew and Cary, 1954), prey size may be of paramount importance to the hunting owl; the morphological conformation of the prey may be a minor consideration. Such an interpretation is supported by other predator-prey studies that show that a different species of owl, the Great Horned Owl (*Bubo virginianus*), preys selectively on larger desert rodents (Longland, 1983; Longland and Jenkins, 1987).

Conclusions on predator-prey studies.— The kind of research that seeks to demonstrate the adaptedness of individuals and their features, termed the "adaptationist program" (Gould and Lewontin, 1979; Mayr, 1983), has proven itself to be of fundamental importance in many areas of biology. The adaptationist program, doubtless, can also be of great heuristic value in the study of the Heteromyidae. This is because there is much need for a clearer understanding of the functional significance of the novel morphologies seen in the heteromyid species.

Experimental predator-prey studies would seem to offer the most promising avenue for future research designed to investigate the adaptedness of individuals and their features. Such studies, by the nature of their design, are able to eliminate the confounding variables that often plague natural predation and selection studies. Future studies should be more comparative in nature, involving nonricochetal heteromyids as experimental controls against which the predator avoidance success of species with the hypothesized "anti-predator morphology" is compared. It is important to approach future studies with the purpose of attempting to refute the "anti-predator morphology" hypothesis rather than to embark on a study of adaptation with the premise that the hypothesis is already known to be correct (for philosophical perspective on scientific research, see Wenner and Wells, 1990). It is in this manner that rigorous studies can be designed and executed to provide unambiguous answers to tedious questions pertaining to the functional significance of morphological features.

Heterochrony and Macroevolution in Geomyoid Rodents

Among all geomyoid rodents, the bizarre morphology of the ricochetal kangaroo mice and kangaroo rats has attracted the greatest amount of attention from biologists and these forms are considered by many to be the epitome of desert specialization (e.g., Mares, 1983). The evolution of the enormous head, huge hind feet, large eyes, and long tail in kangaroo mice and kangaroo rats is explained conventionally in terms of the adaptive aspects of their morphology (see discussion above and Table 1). Despite how these features may function in their present context, the very different question remains: how did they evolve? Because a feature functions in a certain way today, we might assume that it originated for that purpose. However, as emphasized by Gould and Vrba (1982:13), "current utility [of a feature] carries no automatic implication about [its] historical origin."

The proper alternative to a functional explanation is a causal explanation for the evolution of a feature. Causal explanations decouple evolution from present use and thereby avoid the inherent ad-hoc nature of functional explanations that have been misappropriated to infer evolution. The causal explanation that seems to account for macroevolution in the Geomyoidea focuses on the mechanisms of heterochrony (mutations that effect changes in developmental programs). Heterochrony provides a general, unifying explanation to account for the evolution of the wide variety of morphological novelties evident in this group (Hafner and Hafner, 1983, 1988; Hafner and Hafner, 1984).

Several years ago I observed that many of the obvious morphological features of kangaroo rats and kangaroo mice, most notably the large head and eyes and the long hind feet, were traits commonly attributable to a paedomorph. This initial observation, coupled with the subsequent discovery that young pocket gophers (approximately two weeks of age) look remarkably like mature pocket mice, prompted further investigation and culminated in the hypothesis that evolutionary epigenetics might resolve the riddles of morphological transformation in geomyoid rodents (Hafner and Hafner, 1983, 1988; Hafner and Hafner, 1984). Below, I rely on phenomenological descriptions to demonstrate how regulatory changes in ontogeny may, in affecting the timing of gene action and rates of morphogenesis and growth, lead to morphological phyletic evolution (for review see Alberch, 1980; Alberch et al., 1979; Gould, 1977; Løvtrup, 1981*a,* 1981*b*; Rachootin and Thomson, 1981).

The view that morphological evolution is the result of regulatory shifts in development has been championed by many workers, most notably by Goldschmidt (1940), Waddington (1957, 1962), and Gould (1977). Waddington (1957, 1962) and, more recently, Alberch (1980), observed that morphologies do not appear in nature in a random or continuous manner, but that there is a repetition of several distinctive morphotypes. Such is the case with geomyoid rodents. A perusal of both fossil and extant forms reveals that virtually any geomyoid rodent fits into one of three general morphological categories (*Baupläne*): generalized mice, kangaroo-like rats and mice, and the fossorial, pocket gopher morphotype (Fig. 2). As pointed out by Alberch (1980), epigenetic interactions may reduce the scope of potential novelties and impose a sense of order in morphological transformations through phylogeny; as a consequence, we see developmental constraints effecting phyletic parallelism. The Geomyoidea is rife with phyletic parallelism (Hafner and Hafner, 1983; Wood, 1935) and, indeed, the bipedal pocket gopher ancestor, *Schizodontomys,* from the Miocene is an excellent case in point.

The paedomorphs.—Microdipodops and *Dipodomys* share a variety of features that

FIG. 2.—Silhouettes of the three principal morphotypes found in the extant Geomyoidea. The body form of the pocket mouse (*Perognathus, Chaetodipus, Liomys, Heteromys*; lower) represents the generalized rodent condition. The fossorial pocket gopher (*Thomomys, Geomys, Pappogeomys, Orthogeomys, Zygogeomys*; right) is a robust, large-bodied form. The kangaroo-like form (*Dipodomys, Microdipodops*; left) is highly specialized with a large head, long hind feet, small fore feet, and long tail (the tail of *Microdipodops* [not shown] is thick at the midsection and lacks the terminal tuft). See text for discussion.

are characteristic of the juvenile state, or paedomorph (Hafner and Hafner, 1983, 1988; Hafner and Hafner, 1984). The obvious traits include the large head, large eyes, and long hind feet; these are among the endearing qualities of kangaroo mice and kangaroo rats and, doubtless, have contributed to the popularity of the heteromyids. As Gould (1977:350) observes, "Our concept of 'cute' is strongly determined by the common traits of babyhood: relatively large eyes, short face, smooth features, bulbous cranium." The manifestation of this character complex (plus other features) in the adults of *Microdipodops* and *Dipodomys* argues strongly for paedomorphosis.

The large heads of *Microdipodops* and *Dipodomys* have attracted much comment from biologists. At first glance, extreme inflation for the auditory bullae appears singularly responsible for the large crania of these ricochetal rodents. While it is unclear how and why the bullae have become so greatly hypertrophied, bullar expansion alone does not seem to account for the large crania of these forms. As noted by Hafner and Hafner (1988), precaudal vertebral length (a measure of body size excluding tail) and condylonasal length of skull (a measure of skull size excluding bullae) are allometrically related and follow the equation *skull length* = *k* (*precaudal vertebral length*)$^{\alpha}$, where k is the allometric coefficient and α is the allometric exponent. The allometric exponent (regression coefficient) on a double logarithmic plot of skull length on precaudal vertebral length across the Rodentia is estimated to be 0.65 (Fig. 3; $r = 0.97$, $P \ll 0.01$). Hence, animals with shorter precaudal vertebral lengths (heteromyids in general) have, necessarily, proportionately larger heads. It is this allometric relationship (Fig. 3) that explains, in part, the large heads of kangaroo mice and kangaroo

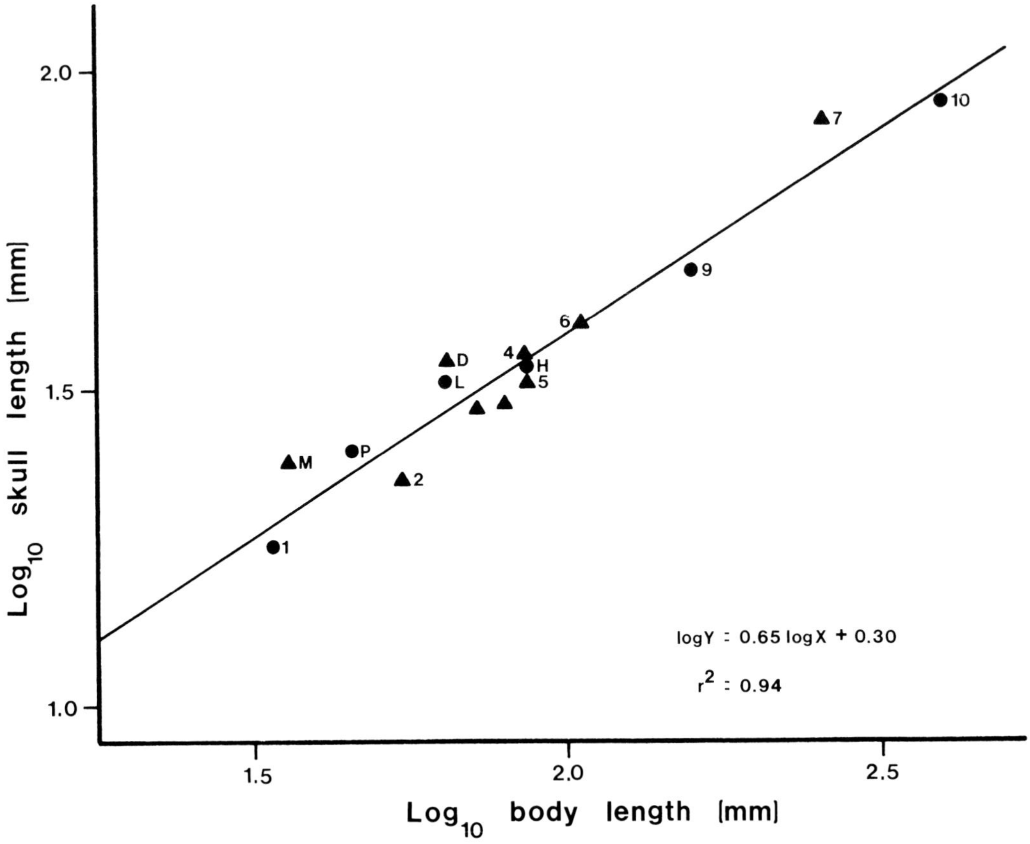

FIG. 3.—Relationship between skull (condylonasal) length and body (precaudal vertebral) length for 15 genera of ricochetal and nonricochetal rodents (data from Hatt, 1932:722). Ricochetors are indicated with triangles and nonricochetors by dots. Genera are as follows: M, *Microdipodops*; P, *Perognathus*; D, *Dipodomys*; L, *Liomys*; H, *Heteromys*; 1, *Sicista*; 2, *Zapus*; 3, *Scirtopoda*; 4, *Jaculus*; 5, *Dipus*; 6, *Allactaga*; 7, *Pedetes*; 8, *Notomys*; 9, *Rattus*; 10, *Paramys*.

rats. Note, however, that the regression of skull length against precaudal vertebral length (Fig. 3) reveals that *Microdipodops* and *Dipodomys* have skulls that are 135% and 125% larger, respectively, than predicted from precaudal vertebral length alone. Apparently, functional constraints associated with locomotion do not explain this deviation. Inspection of Fig. 3 shows that there is no relationship between the ricochetal habitus and proportionate head size in rodents; that is, some bipedal rodents have heads that are larger than predicted (e.g., *Jaculus* and *Pedetes*), whereas other bipedal rodents have heads that are smaller than predicted (e.g., *Scirtopoda* and *Notomys*).

The large heads of kangaroo mice and kangaroo rats may also be due to their relatively large brains (see Table 2; Hafner and Hafner, 1984). Enlargement of the brain seems to be a common result of paedomorphosis (Gould, 1977). The relatively large brains of kangaroo mice and kangaroo rats may result from time hypermorphosis, as is well established for human encephalization (for discussion see Shea, 1988). Time hypermorphosis here results in a relatively enlarged brain by allowing for a protracted fetal period of high relative growth of the brain; as a consequence, time hypermorphosis yields a high brain/body ratio that is characteristically paedomorphic.

Kangaroo rats and kangaroo mice share several other paedomorphic features. Both

TABLE 2.—*Selected morphological and life-history characteristics of heteromyids.*

Genus	Mean adult body mass (g)[a]	Mean brain *EQ* value[b]	Mean rear foot in percentage head and body[c]	Mean tail in percentage head and body[c]	Mean number of caudal vertebrae[c]	Mean litter size[a]	Mean gestation time (days)[d]
Perognathus	10.6	0.96	28	107	—	4.2	25.1
Chaetodipus	22.6	0.92	28	121	26[e]	4.3	27.0
Liomys	43.8[f]	0.94	25[e]	118[e]	26[e]	4.0[g]	26.5
Heteromys	74.4[h]	0.96[i]	27[e]	128[e]	27.5	3.0[j]	27[i]
Microdipodops	12.5	1.21	36	130	24	3.9	—
Dipodomys	76.1	1.26	37	143	29.9	3.1	29.4

[a] Means for genera (except *Liomys* and *Heteromys*) calculated from species data in Jones (1985) and included references.

[b] Means for intra-geomyoid encephalization quotients (*EQ* values) calculated from species data in Hafner and Hafner (1984). *EQ* values less than 1.00 denote relatively small brains, whereas values greater than 1.00 describe relatively large brains.

[c] Hatt (1932).

[d] Means for genera calculated from species data in Jones (1985) and included references.

[e] Value based on one specimen.

[f] Mean calculated from species data in Burt and Grossenheider (1952), MacMillen and Hinds (1983) and Jones (1985). When a range was given instead of a mean, the average of the range limits was used. Also, several estimates available for the same species were averaged.

[g] Mean calculated from data for species in Eisenberg (1963) and Fleming (1977); several estimates for the same species were averaged.

[h] Mean calculated from data on *H. desmarestianus* in MacMillen and Hinds (1983) and Jones (1985).

[i] Data from a single species.

[j] Mean calculated from data for species in Eisenberg (1963) and Fleming (1977). When a range was given instead of a mean for a species, the average on the range limits was used. Also, several estimates available for the same species were averaged.

Microdipodops and *Dipodomys* retain the stapedial canal and artery (Howell, 1932). The stapedial artery, derived from the second aortic arch, is likely present in all mammals at some stage of development, yet in many mammals it commonly disappears at an early ontogenetic stage. Importantly, Howell (1932:503–504) observes that, among the geomyoids, only *Microdipodops* and *Dipodomys* have a complete stapedial canal and artery, whereas species of *Perognathus, Liomys* and *Thomomys* show various degrees of degeneration of the artery and canal. Lastly, kangaroo mice and kangaroo rats, in comparison with other geomyoids, have very light, delicate skeletons whose osseous elements are smooth and paper thin, and show a low degree of fusion. Unfused and/or definitive sutures are commonplace in the skulls of *Microdipodops* and *Dipodomys* and this reduced level of ossification further describes a paedomorphic condition.

Kangaroo mice and kangaroo rats do, however, differ by a vast number of fundamental morphological features (e.g., see Hafner, 1978; Hatt, 1932; Howell, 1932; Wood, 1935), despite the many shared paedomorphic characters. Indeed, the entire natural history of *Microdipodops* is unlike that of *Dipodomys*; the genera differ markedly in body size, life-history strategy, locomotion, ecology, physiology, and geographic distribution. Paedomorphosis is a gross morphological expression (a shape phenomenon) that appears to be shared between these genera as a result of phyletic parallelism. As hypothesized by Hafner and Hafner (1983, 1988), two separate heterochronic processes may be involved in the juvenilization of kangaroo mice and kangaroo rats.

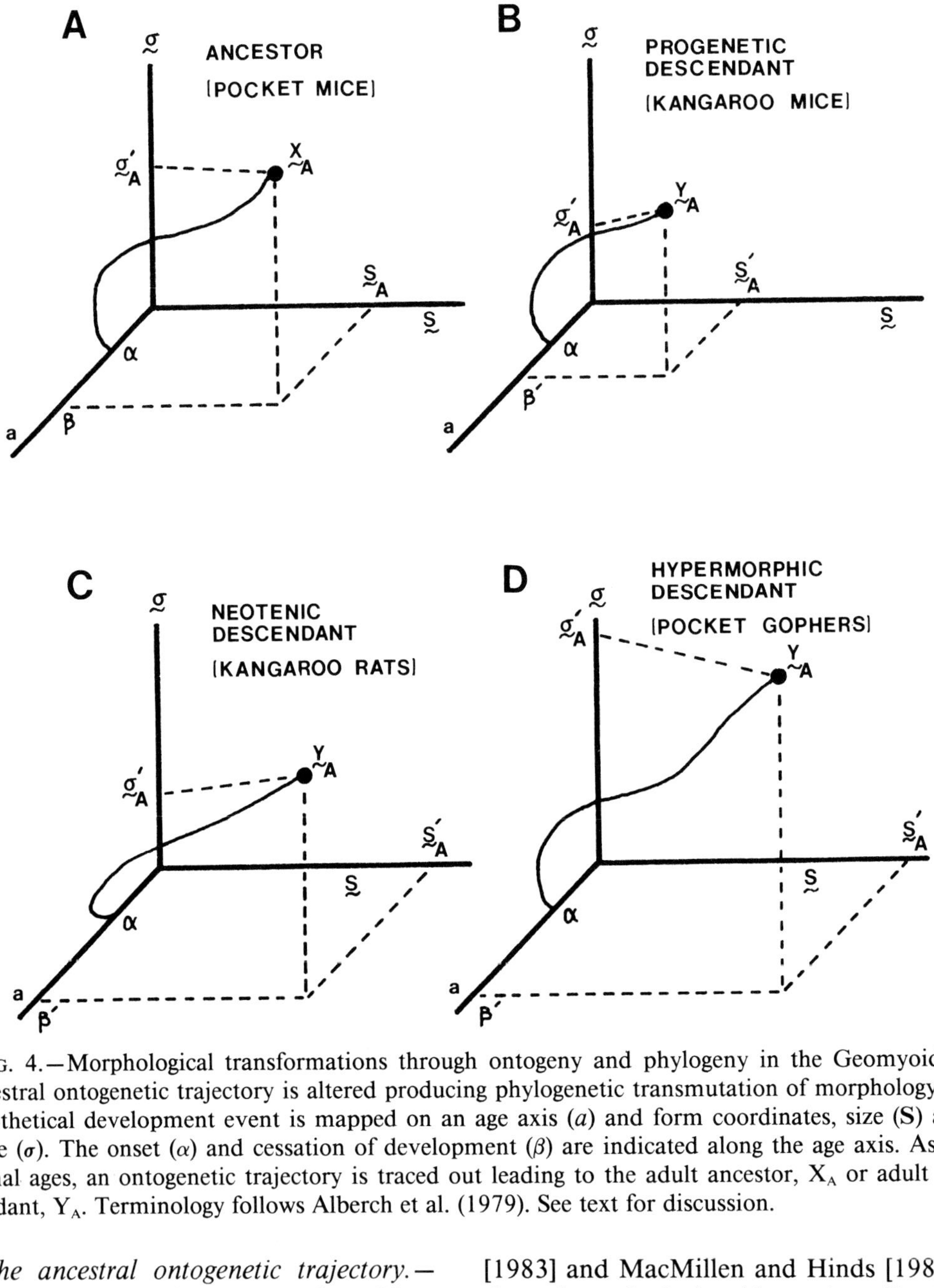

FIG. 4.—Morphological transformations through ontogeny and phylogeny in the Geomyoidea. Ancestral ontogenetic trajectory is altered producing phylogenetic transmutation of morphology. A hypothetical development event is mapped on an age axis (a) and form coordinates, size (S) and shape (σ). The onset (α) and cessation of development (β) are indicated along the age axis. As an animal ages, an ontogenetic trajectory is traced out leading to the adult ancestor, X_A or adult descendant, Y_A. Terminology follows Alberch et al. (1979). See text for discussion.

The ancestral ontogenetic trajectory.—Pocket mice, including *Perognathus, Chaetodipus, Liomys* and *Heteromys,* exhibit a generalized rodent *Bauplan* (Eisenberg, 1981:90) and probably represent a reasonable approximation of the ancestral geomyoid condition. It is difficult to specify the exact size of the ancestral *Bauplan,* but it was probably moderately small (and not moderately large as suggested by MacMillen [1983] and MacMillen and Hinds [1983]); the paleontological evidence (e.g., Reeder, 1956; Wahlert, this volume; Wood, 1935) shows that the earliest geomyoids (including *Heliscomys, Proheteromys* and *Mookomys*) varied from small rodents approximately the size of *Perognathus longimembris* (8 g; Jones, 1985), to medium-sized forms such as *Chaetodipus californicus* (23 g; Jones, 1985) and *Liomys* (43.8 g; Burt and Grossenheider, 1952; Jones, 1985; MacMillen and Hinds, 1983). An animal with this ancestral morphotype is likely to have conserved the developmental patterns of the geomyoid ancestor. As the animal ages from conception, its size and shape will change, following the ancestral ontogenetic trajectory. In Figure 4, I use the formalism proposed by Alberch et al. (1979) to analyze the morphological transformations through ontogeny and phylogeny in the Geomyoidea. This method describes ontogeny using a three-axis system: *a*, the age axis that details the timing of differentiation events; **S**, the size axis; and σ, the shape axis (Fig. 4). The ontogenetic-trajectory method is used here to represent the overall ontogeny of an individual and to illustrate the general effects of possible heterochronic changes. It is an idealized representation which describes a myriad of hypothetical structures and organs on a single pair of size and shape coordinates (for discussion see Alberch, 1980; Alberch et al., 1979). Perturbations (δ) of the "control parameters," including the onset of growth (α), cessation of development (β), size growth rate (k_S) and rate of change in shape (k_σ), deform the ancestral ontogenetic trajectory and lead to phylogenetic transmutations (Fig. 4). Much of the morphological diversification in the Geomyoidea is explicable from this ontogenetic perspective and the morphological novelties that are produced are predictable end products of heterochronic perturbations.

Kangaroo mice: the progenetic descendant.—Kangaroo mice, in comparison with other geomyoids, show retention of juvenile morphology and are very small in body size. It is hypothesized, then, that kangaroo mice represent a paedomorphic version of a somewhat larger (though still moderately small) geomyoid ancestor (see also Hafner and Hafner, 1983, 1988). Importantly, Shea (1983, 1988) and McKinney (1988) note that two different heterochronic events may lead to a small-sized paedomorph: time progenesis (= time hypomorphosis) and rate progenesis (= rate hypomorphosis). Time progenesis is a process in which ontogeny is truncated because the time required for reproductive maturation is abbreviated. This involves a negative perturbation ($-\delta\beta$) in the ancestral ontogenetic trajectory (Fig. 4B) and results in a small, rapidly maturing paedomorph. Rate progenesis, however, involves a reduction in size growth rate ($-\delta k_S$) and leads not only to a small descendant, but one that is also a paedomorph; this is true because change in size is inextricably associated with change in shape and few traits change isometrically with size. It is important to distinguish between these two kinds of progenesis, because these heterochronic mechanisms, though producing similar results, suggest different selective environments and adaptive scenarios. As noted by Shea (1988), the ecological correlates of the diminutive paedomorph would be expected to be different if the heterochronic mechanism involved truncated growth time (time progenesis) or reduced growth rate (rate progenesis). For time progenesis, selection may act mainly for reduced growth duration in an environment where early maturation and increased reproductive output are advantageous. Alternatively, if rate progenesis is involved, selection may act principally on reduced growth rate and smaller size as a means of reducing interspecific competition by exploiting an alternative food resource (e.g., insects).

Both Gould (1977) and McKinney (1986) observe that the key to understanding the immediate significance of heterochrony lies in the theory of *r* and *K* selection (life-history strategies). Gould (1977:293) predicts that "*progenesis will be associated with r strategies and neoteny with K strategies*" (italics his). Interestingly, *Microdipodops* inhabits an obvious *r*-selected environment (ephemeral sand-dune habitats in the Great Basin Desert) and, while the data are scanty, it seems to possess the attributes of an *r* strategist when compared to most other geomyoids. In addition to showing small body size, kangaroo mice seem to have somewhat reduced longevity (Egoscue et al., 1970) and subsist on insects (an ephemeral resource base) as well as seeds (Hall, 1941); they do not, however, seem to show larger litter sizes as was previously reported (cf., Hafner and Hafner, 1988; see Table 2). It should also be kept in mind that the genus is autochthonous in the Great Basin Desert (Hafner, 1978, 1981) and through its evolutionary history has faced the rigors of cold (high elevation), desert habitats that are characterized by a shortened growing season.

Given the complete lack of information on the growth and development of *Microdipodops,* it is impossible to identify which of the two mechanisms of progenesis may be involved (for discussion see McKinney, 1988). Indeed, future workers must gather the kinds of data for kangaroo mice that will allow for the distinction between these heterochronic processes (as well as a consideration of other possibilities including post-displacement or even neoteny) that may be responsible for paedomorphosis here. However, the small body size and obvious paedomorphosis suggest that either form of progenesis is the favored heterochronic hypothesis for *Microdipodops.* If time progenesis is involved in the evolution of kangaroo mice, then they may represent a parallel to New World callitrichid monkeys where precocious maturation as a life-history strategy may have been the principal object of selection (see Shea, 1988). Alternatively, selection favoring small body size and dietary specialization on insects may be the underlying effectual aspect of evolution involving rate progenesis; a parallel here may be the Old World talapoin monkeys (see Shea, 1988). In either case, it seems that the juvenilized morphology of kangaroo mice (a necessary byproduct of progenesis) may be entirely incidental.

The kangaroo mouse, as a progenetic descendant, harbors several morphological characters not seen in the kangaroo rat (a neotenic descendant, see beyond). Wood (1935:112) was first to notice that the enamel of the incisors of adult *Microdipodops* is white to very pale yellow in color and he suggested that this is a juvenile characteristic (enamel tends to darken with age). In contrast, adults of most other species of geomyoids usually have dark, yellow-colored enamel on their incisors. Kangaroo mice, despite having a relatively long tail, also possess the fewest caudal vertebrae (24) of all heteromyids (see Table 2). According to Hatt (1932:644) "*Microdipodops* then, has achieved a fairly long tail through lengthening of the units, while *Dipodomys* has its long tail, at least in part, by virtue of units added." Inasmuch as ossification occurs craniocaudally in heteromyids (Van De Graaff, 1973), this reduction in the number of vertebral elements may simply be due to the truncation of the size/shape pattern of the ancestral ontogeny that results from either time or rate progenesis (early offset signal of development or reduced rate of overall weight growth prevents further ossification of tail vertebrae). As discussed above, kangaroo mice have relatively large heads because they have short precaudal lengths (Fig. 3). However, kangaroo mice do have greatly inflated auditory bullae. It seems that the extremely enlarged bulla is a result of the unique development of an additional bullar lobe. Wood (1935:112, 242) determined that the "swelling" of the anterior lobe is due to the anterior migration of the temporalis muscle which carries the delicate squamosal up on the dorsum of the skull. Wood (1935:242) described this modification as "the most striking action of muscle on bone detected within the family . . . ." Lastly, the cheek teeth of kangaroo mice seem not to have fully developed roots when compared to those of pocket mice; the cheek teeth of *Microdipodops* possess molar roots that are irregularly present, greatly reduced in size and appear late in ontogeny (for discussion see Hall, 1941; Merriam, 1891; Wood, 1935).

Kangaroo rats: the neotenic descendant.—A retardation in the rate of change in shape of the ancestral ontogenetic trajectory ($-\delta k_{\sigma}$) will lead to the production of a neotenic descendant. The juvenilization of kangaroo rats may have occurred in this manner (Hafner and Hafner, 1983, 1988; see Fig. 4C). Neoteny occurs often in nature and may result from direct selection for juvenile features and/or larger body sizes in *K*-selected environmental regimes (Gould, 1977). Slow development may produce a descendant of the ancestral adult size but one retaining juvenile features. This decrement in the shape growth rate, coupled with a delay in β (Fig. 4C) is sufficient to explain the observed variety of medium- to large-sized kangaroo rat species. *Dipodomys,* in comparison to other geomyoids, possesses many of the classical features associated with the neotenic syndrome: moderate to long life span (Egoscue et al., 1970); slow development (Butterworth, 1961; Chew and Butterworth, 1959; Eisenberg and Isaac, 1963; Fleming, 1977; Hayden and Gambino, 1966; Lackey, 1967); long gestation period (Eisenberg and Isaac, 1963; Fleming, 1977); enlarged brain (Hafner and Hafner, 1984); and generally small litters (Butterworth, 1960; Eisenberg and Isaac, 1963; Fleming, 1977; Hall, 1946). This associated complex of characters (see Table 2) suggests that the kangaroo rats are near the *K* end of the *r*-*K* spectrum of life-history strategies. That some species of kangaroo rats inhabit desert regions does not contradict this hypothesis. Actually, several of the more strictly desert-dwelling species (e.g., *D. microps, D. spectabilis* and *D. deserti*) subsist almost entirely on leaves, flower heads and seeds from perennial species (a largely nonfluctuating resource base as compared with seeds of annuals and insects) and, strictly speaking, many other kangaroo rat species are not found in the desert at all (e.g., *D. californicus, D. heermanni, D. stephensi, D. venustus*). Gould's (1977) hypothesis that neoteny may result from direct selection for larger body size and/or juvenile features seems worthy of serious consideration here; Kotler (1984) observed that larger kangaroo rats were less vulnerable to predation by owls and canids and better able to forage in open habitats than smaller species of rodents (see also above discussion and Kotler, 1985).

Several morphological features unique to *Dipodomys* are germane to this discussion of neoteny: the reduction and/or loss of the hallux, the generally opaeodont (open-rooted teeth) condition of the molars, and the high number of caudal vertebrae (kangaroo rats, with generally 31 caudal elements, have the highest number of caudal vertebrae in the family; Hatt, 1932; Table 2). According to the retardation model, features that appear late in the ancestral ontogeny would be expected to be absent or reduced in size (and retain the juvenile shape) in the neotenic descendant. Most interestingly, there is evidence that both the distal phalanges (Van De Graaff, 1973) and the roots of molars (Zakrzewski, 1981) are features that appear late in ontogeny. Conversely, the high number of caudal vertebrae (Hatt, 1932) suggests that, although terminal ossification occurs relatively late (but probably earlier than the above two features), the retention and prolongation of fetal growth tendencies allow for the continued progressive development of caudal vertebrae.

Pocket gophers: the hypermorphic descendant.—It has been hypothesized that the process of hypermorphosis may be largely responsible for the gross shape and size of pocket gophers (Hafner and Hafner, 1983, 1988). Hypermorphosis leads not only to a descendant whose shape transcends that of the ancestral form (technically, peramorphosis), but one that is also large in size. Hypermorphosis can be achieved by two separate processes: time hypermorphosis and rate hypermorphosis (for discussion see Shea, 1983, 1988; McKinney, 1988). Time hypermorphosis is a process wherein there is a positive perturbation in the signal for cessation of growth ($+\delta\beta$) in the ancestral ontogeny (Fig. 4D) that produces a hypermorphic individual (also termed a peramorph; Alberch et al., 1979); that is, lengthened period of growth allows for the extension of the ancestral allometric trajectory. In contrast, rate hypermorphosis involves an increase in size growth rate ($+\delta k_S$) and also produces a large, peramorphic descendant. As noted by Shea (1988), increased rate of overall weight growth will result in extension of the size/shape pattern of the ancestor, but without changing the ancestral allometry or maturation time. Most of the general morphological attributes of pocket gophers seem to support this hypothesis of hypermorphosis. The adult pocket gopher possesses a heavily ossified skeleton, the cranial elements of which show a high degree of fusion; these are classical attributes of later mammalian ontogeny. Further, pocket gophers are large-sized geomyoids.

While peramorphosis may be produced by several other processes (e.g., acceleration and predisplacement; see Alberch et al., 1979), the marked similarities in size and shape between very young pocket gophers and adult pocket mice (Hafner and Hafner, 1988) suggest that hypermorphosis is involved in the morphological evolution of pocket gophers. This heterochronic perturbation allows pocket gophers to "go beyond" or extend the ancestral ontogenetic trajectory (Fig. 4D). Hence, juvenile levels of morphological differentiation in the pocket gophers may be adult features of the ancestor (recapitulation). In the developing pocket gopher, the "pocket mouse" morphotype is attained at the same size of the adult ancestor, but the pocket gopher is still a juvenile at this size. It is for this reason that I favor hypermorphosis (as opposed to recapitulation by acceleration) as the process leading to the peramorphosis in this case.

Gould (1977) notes that there is usually a link between *K*-selective regimes and hypermorphosis. In accord with Gould's observation, it seems that pocket gophers inhabit a relatively nonfluctuating environment (the subterranean niche), subsist on a stable food (primarily roots and tubers), are large in body size (relative to heteromyids), exhibit strong intra- and interspecific competitive abilities (e.g., Miller, 1964), and may be long lived (Howard and Childs, 1959). However, recent information available for *Thomomys*, the smallest of the pocket gopher genera, shows high annual population turnover and suggests that it may be a more *r*-selected strategist (Daly and Patton, 1986). Clearly, much more data are needed (particularly for the larger genera such as *Orthogeomys* and *Zygogeomys*) before any conclusions can be made on the possible relationship between hypermorphosis and life-history strategies among the pocket gophers.

Conclusions on ontogeny and phylogeny.—It has been hypothesized that heterochronic disturbances during development have been involved in macroevolution (sensu Simpson, 1944, 1953) in the Geomyoidea. To clarify a description of the observed phenomena, I have incorporated the "ontogenetic-trajectory" formalism of Alberch et al. (1979). Figure 5 summarizes these views concerning morphological transformations through ontogeny and phylogeny in geomyoid rodents by focusing on one structural complex: the cranium. Deformations in the ancestral (mouse-like) ontogenetic trajectory (Fig. 4) results in recapitulation (through hypermorphosis) and reverse recapitulation (through progenesis or neoteny). Adult descendants that are particularly large in size (e.g., the pocket gopher, *Orthogeomys grandis,* or the kangaroo rat, *D. deserti*; Fig. 5), but show the same shape as smaller relatives, may result from proportioncd giantism.

One of the principal aims of this review is to emphasize the possible role of heterochrony (mutations that cause altered developmental programs) in morphological evolution of heteromyid (and geomyid) rodents. The focus here on heterochrony, of course, is intended not to mitigate the role of natural selection in macroevolution. Clearly, the omnipresent force of selection serves ultimately in judging the success or failure of any evolutionary novelty. But by emphasizing heterochrony, I want to encourage a more eclectic view of morphological evolution in these rodents; specifically, a view that does not ignore the recent realization that evolutionary changes in size and shape arise from evolutionary change in developmental programs (see Atchley, 1987).

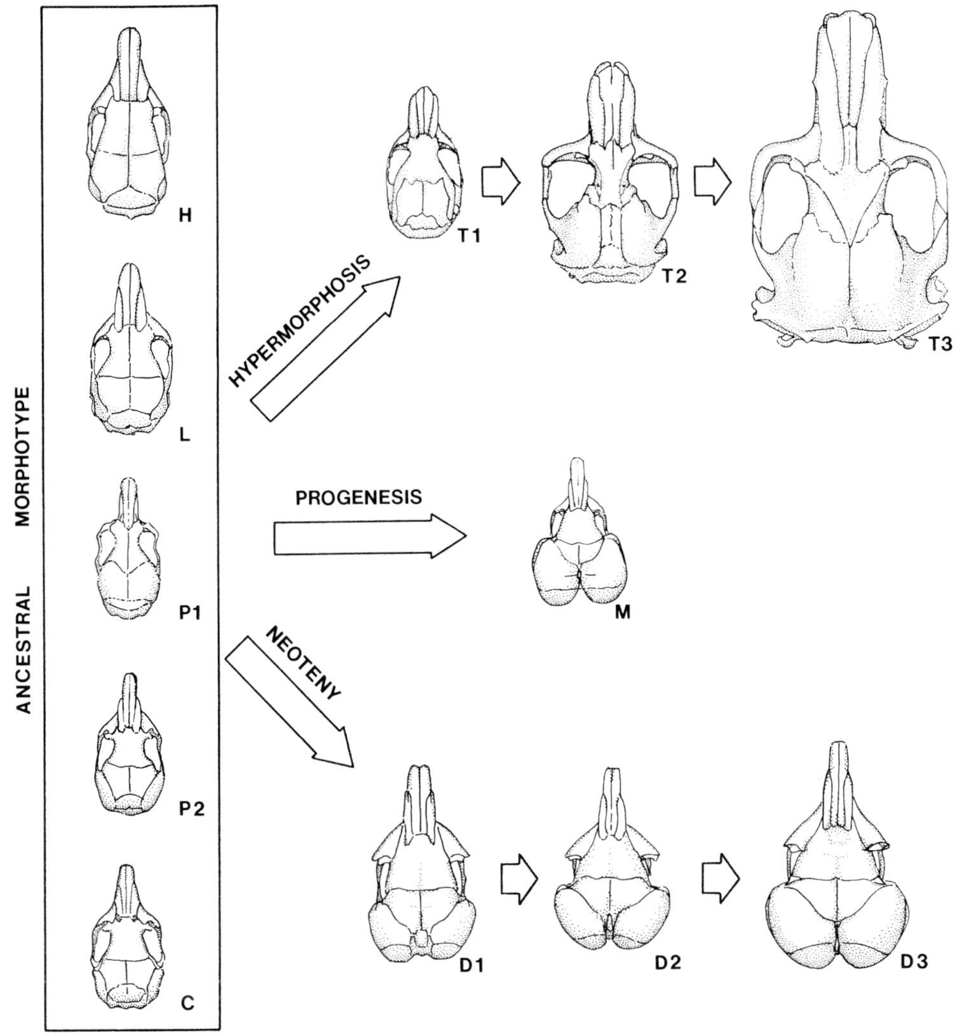

FIG. 5.—The evolution of geomyoid cranial differentiation based on the hypothesized heterochronic disturbances during development. Five genera illustrate the presumed ancestral ("mouse-like") morphotype: *Heteromys desmarestianus* (H), *Liomys irrortaus* (L), *Peromyscus truei* (P1), *Perognathus parvus* (P2), *Chaetodipus penicillatus* (C). Other crania are as follows: young *Thomomys monticola* (T1), *Thomomys umbrinus* (T2), *Orthogeomys grandis* (T3), *Microdipodops megacephalus* (M), *Dipodomys compactus* (D1), *Dipodomys merriami* (D2), and *Dipodomys deserti* (D3). All crania are drawn from adult individuals except for T1. Note the similarity between the young pocket gopher (T1) and the ancestral morphotype. Figures modified from Hall (1946, 1981), Setzer (1949), and Genoways (1973).

The hypotheses presented here (see also Hafner and Hafner, 1988) are based mainly on patterns of morphological differentiation seen in adult specimens. These hypotheses predict that the ontogenies of the derived novel forms, such as kangaroo mice and kangaroo rats, were altered in a specific fashion relative to the ancestral condition. It should be pointed out, however, that the epigenetic interactions affecting these morphological transformations are much more complex than the simple heterochronic changes that are outlined. The intent here is to suggest that a substantial portion of the morphology of each derived form represents an intercorrelated set of traits that may be produced by a single heterochronic perturbation. As such, I hypothesize that the kangaroo mouse is *mainly* progenetic, the kangaroo rat is *mainly* neotenic, and the pocket gopher is *mainly* hypermorphic. One must keep in mind that morphogenesis is a complex process and examples of "pure" types of heterochronic changes unaccompanied by other basic heterochronic events are probably very rare (for discussion see Alberch et al., 1979; Fink, 1982; Gould, 1977; McKinney, 1984). It follows then that not everything about these derived creatures would be expected to be consistent with a strict (global) interpretation of the heterochrony hypotheses that are presented. Indeed, not all traits are going to show the same heterochronies inasmuch as trait dissociation (mosaic evolution) is very common (for discussion see McKinney, 1988). For example, the short tail of pocket gophers seems to be truncated (only 16–18 caudal vertebrae [Hill, 1937]; cf. Table 2), but an obvious prediction from a global interpretation of the above hypothesis is that the peramorph should have a high number of caudal vertebrae (and, thus, a long tail). Also, the kangaroo mouse has greatly enlarged auditory bullae (perhaps the result of acceleration), but the animal is hypothesized to be mainly progenetic. Additionally, the large auditory bullae and elongated tail of kangaroo rats may be peramorphic, yet the form is hypothesized to be mainly neotenic. A single heterochronic perturbation may, but need not, cause rather global morphological changes when its effects are amplified by a myriad of pleiotropic and epigenetic effects that occur during development. More commonly, morphological patterns are produced by a combination of heterochronic events (Alberch et al., 1979), and individual traits evolve largely as a subunit of a developmentally integrated character complex that is, in turn, governed by various hierarchical developmental processes. Thus, it seems that dissimilar kinds of heterochronic changes, acting independently, may affect different developmentally integrated character complexes.

Future workers interested in morphological evolution in these rodents should seek to gather the kinds of data that can be brought to bear on these hypotheses of heterochrony. These hypotheses are testable (see Hafner and Hafner, 1988), and I hope that they will encourage others to investigate the embryological and postnatal ontogeny of geomyoid rodents, as was done recently in neotomine-peromyscine rodents (Creighton and Strauss, 1986). Before one is able to understand fully the morphological differentiation in these rodents, much more complete data are needed on growth and development, longevity, age at maturation, litter sizes, and food habits, among many other crucial natural history parameters. Despite the popularity of heteromyids, there is a surprising dearth of basic descriptive data on the postnatal growth for most species. Accordingly, I encourage future workers to perform these basic descriptive studies, as well as to perform manipulative embryological studies (e.g., DuBrul and Laskin, 1961) and studies of developmental integration and character correlation. In this regard, Zelditch's (1987) approach using confirmatory factor analysis in the study of developmental integration and Atchley's (1987) developmental quantitative genetics model stressing genetic variance-covariance structure seem to provide promising directions for future research.

In summary, I propose that much of the striking morphological modifications in the heteromyids may have resulted from heterochronic shifts in ontogeny. As such, the evolution of each trait might not be attributed solely to natural selection but, subsequent to the heterochronic event(s), each trait may have been modified by selection for its present utility. The novel features may have arisen in a rather serendipitous manner (nonadaptations), but later were available for cooptation in descendants by further selection regimes; the long hind feet, the large head, and the delicate, gracile body (among other novelties), then, might best be considered as exaptations rather than adaptations (see Gould and Vrba, 1982). Considering the examples of paedomorphosis and hypermorphosis within the Geomyoidea, it is reasonable to postulate that natural selection acted on heterochronic changes in ontogeny to effect adaptive *Baupläne* (and, perhaps, associated life-history features) rather than generating each morphological feature independently through orthoselection. Future work exploring the evolution of ontogenies may prove to be a profitable way of understanding macroevolution in this unique group.

Current Utility Versus the Evolution of Novel Features

It is important in the study of macroevolution to consider both the functional significance (current utility) and the evolution (historical genesis) of a novel character; the two must not be automatically considered as one in the same, nor should these ideas be taken to be directly competitive. The adaptive significance of the novel morphologies seen in the Heteromyidae have received much attention from several generations of biologists. The emphasis on the current utility of the novelties has been so overwhelming that few workers have attempted to explore the evolution of the novelties. In fact, most prior workers have assumed (usually implicitly) that they are explaining evolution when they explain present use; thus, they have failed to see the distinction between the two concepts. Moreover, it should also be clear that the questioning (or even refutation) of any hypothesized functional explanation does not necessarily lend support to any particular model for the evolution of the feature.

Adaptive significance of evolutionary novelties.—The adaptationist program has sought to demonstrate the adaptedness of novelties seen in the Heteromyidae by focusing on the functional significance of individual traits (e.g., the enlarged auditory bullae, the long hind feet, long tail, big eyes). Mayr (1983) stated repeatedly that a more holistic approach to the study of adaptation is appropriate only when all specific analyses of individual traits fail to determine an adaptive significance. I question this view because it is inappropriate, in principle, to dissect the phenotype into individual characters and concentrate on the adaptive significance of a trait (see Dobzhansky, 1956; Gould and Lewontin, 1979); the whole organism is always much more than the sum of its parts. However, I agree that the adaptationist program is a profitable method of scientific research, because of its great heuristic value (see Mayr, 1983). It is for this reason that the adaptationist question, "What is the function of a given trait?", is important in the biology of the Heteromyidae. Actually, there is no reason why the more atomistic and more holistic approaches cannot be pursued simultaneously. The research programs are not mutually exclusive and there is much promise for reciprocal illumination. The adaptationist approach in the Heteromyidae, however, has not yet produced convincing functional explanations for the various novel features. Accordingly, it is now appropriate to consider seriously more holistic approaches to explain the adaptiveness of the evolutionary novelties. In so doing, it might be profitable to investigate the adaptive significance of the *Bauplan* in its entirety (including life-history features) as opposed to individual morphological traits.

Distinction between selective and developmental constraints.—If we are to understand the evolution of the novel body plans in the Heteromyidae, we must find ways of distinguishing between developmental and selective constraints. Unfortunately, given our present ignorance of the mechanisms of developmental regulation and our inadequate understanding of the role of development in evolution, it may be exceedingly difficult to resolve this issue (for review, see Maynard Smith et al., 1985). Clearly, one of the principal tasks awaiting future generations of biologists is that of identifying the relative contribution of developmental and selective constraints in shaping macroevolutionary patterns.

Maynard Smith et al. (1985) suggest four possible ways to distinguish selective and developmental constraints in evolution: 1) testing quantitative predictions about the adaptive basis of certain traits; 2) direct measurement of the strength of selection; 3) direct assessment of heritable variation; and, 4) comparative analysis of variation (e.g., allometry). However, Maynard Smith et al. (1985) observed that none of these methods provides a foolproof means of distinguishing between developmental and selective constraints. They also noted that it will often be impossible to identify the constraints that are responsible for producing evolutionary patterns.

We will probably never be able to determine, unambiguously, the mechanism(s) responsible for the evolution of the novel forms because the production of a novelty, itself, is a unique event. Our best hope would be to identify what mechanism or combination of mechanisms is most likely responsible for the evolution of a particular trait or suite of characters. In the Heteromyidae, with obvious instances of paedomorphosis and peramorphosis, it seems clear that developmental constraints are involved (to some unknown extent) in effecting macroevolutionary patterns. As mentioned elsewhere (Hafner and Hafner, 1984), it is more parsimonious to suppose that the suite of juvenilized features of *Microdipodops* and *Dipodomys* evolved together through developmental heterochrony, rather than to suppose that each trait was shaped independently through natural selection. Clearly, we are in need of both descriptive and manipulative embryological studies to determine if, indeed, the suite of paedomorphic features of *Microdipodops* and *Dipodomys* are linked as has been hypothesized. In the interim, though, we should not discount the possible influence that heterochronic changes in ontogeny may have had in the macroevolutionary diversification of the Heteromyidae.

Summary

The rodent superfamily Geomyoidea is a monophyletic lineage that is autochthonous in North America. Extant geomyoids are divisible into two groups: 1) the Heteromyidae, whose members display an adaptive array of bipedal and scansorial forms; and, 2) the Geomyidae, all members of which are fossorial. The Heteromyidae, with its morphologically heterogeneous forms, provides an excellent opportunity to investigate major evolutionary divergence. This study treats three aspects of macroevolution in the Heteromyidae: 1) evolutionary relationships among the taxa; 2) adaptive significance of the novel morphologies; and, 3) mechanism(s) responsible for the diversification of the evolutionary novelties.

The extant heteromyids comprise three principal lineages that diverged during the Oligocene: 1) subfamily Heteromyinae (*Liomys* and *Heteromys*); 2) subfamily Perognathinae (*Chaetodipus* and *Perognathus*); and, 3) subfamily Dipodomyinae (*Dipodomys* and *Microdipodops*). The heteromyine and perognathine rodents are generalized in morphology. In contrast, kangaroo mice, *Microdipodops,* and kangaroo rats, *Dipodomys,* are morphologically derived; these forms represent evolutionary novelties.

There are two ways to explain the evolutionary novelties seen in the Heteromyidae: 1) functional explanations that focus on the adaptedness and present use of a novel feature; and, 2) causal (mechanistic) explanations for the evolution of the novel morphology. Most functional (adaptationist) explanations pertaining to kangaroo mice and kangaroo rats focus on the hypothesized "anti-predator morphology". These adaptationist arguments are reviewed and evaluated and it is concluded that there is much need for a clearer understanding of the functional significance of the novel morphologies. In addition, a model is presented to describe how heterochronic changes in ontogeny may explain the evolution of the morphological novelties seen in the superfamily. Thus, the initial evolution of each novel trait might not be attributed to the action of natural selection but, subsequent to its appearance, each nascent trait may have been modified by selection for its present utility.

Acknowledgments

This contribution has benefitted from many discussions with M. S. Hafner and I am most grateful to him for his invaluable advice and criticisms throughout the course of this work. Appreciation is extended to D. J. Hafner, M. S. Hafner, J. L. Patton, V. L. Roth and V. Sarich for their helpful comments on earlier drafts of this material. M. D. Carleton, A. L. Gardner, and R. W. Thorington, Jr. provided critical discussion during the early stages of this project. I thank S. Warschaw for clerical assistance. Figures 2 and 5 were drawn by M. E. Pereyra. This research was supported, in part, by a research grant from the National Science Foundation (BSR-8600644).

Literature Cited

ALBERCH, P. 1980. Ontogenesis and morphological diversification. American Zoologist, 20:653–667.

ALBERCH, P., S. J. GOULD, G. F. OSTER, AND D. B. WAKE. 1979. Size and shape in ontogeny and phylogeny. Paleobiology, 5:296–317.

ATCHLEY, W. R. 1987. Developmental quantitative genetics and the evolution of ontogenies. Evolution, 41:316–330.

AXELROD, D. I. 1950. Evolution of desert vegetation in western North America. Carnegie Institute of Washington, Publishers, 590:217–306.

———. 1958. Evolution of the Madro-Tertiary geoflora. Botanical Review, 24:433–509.

———. 1976. History of the coniferous forest, California and Nevada. University of California Publications in Botany, 70:1–62.

BARTHOLOMEW, G. A., JR., AND H. H. CASWELL, JR. 1951. Locomotion in kangaroo rats and its adaptive significance. Journal of Mammalogy, 32:155–169.

BARTHOLOMEW, G. A., JR., AND G. R. CARY. 1954. Locomotion in pocket mice. Journal of Mammalogy, 35:386–392.

BIEWENER, A., R. MCN. ALEXANDER, AND N. C. HEGLUND. 1981. Elastic energy storage in the hopping of kangaroo rats (*Dipodomys spectabilis*). Journal of Zoology, London, 195:369–383.

BOCK, W. J. 1979. The synthetic explanation of macroevolutionary change—a reductionist approach. Bulletin of the Carnegie Museum of Natural History, 13:20–69.

BROOKFIELD, J. F. Y. 1982. Adaptation and functional explanation in biology. Evolutionary Theory, 5:281–290.

BURT, W. H., AND R. P. GROSSENHEIDER. 1952. A field guide to the mammals. Houghton Mifflin Co., Boston, 289 pp.

BUTTERWORTH, B. B. 1960. A comparative study of sexual behavior and reproduction in the kangaroo rats, *Dipodomys deserti* Stephens and *Dipodomys merriami* Mearns. Ph.D. dissert., University of Southern California, 184 pp.

———. 1961. A comparative study of growth and development of the kangaroo rats, *Dipodomys deserti* Stephens and *Dipodomys merriami* Mearns. Growth, 25:127–139.

CHEW, R. M., AND B. B. BUTTERWORTH. 1959. Growth and development of Merriam's kangaroo rat, *Dipodomys merriami*. Growth, 23:75–95.

CRAIGHEAD, J. J., AND F. C. CRAIGHEAD, JR. 1969. Hawks, owls and wildlife. Dover Publications, Inc., New York, 443 pp.

CREIGHTON, G. K., AND R. E. STRAUSS. 1986. Comparative patterns of growth and development in cricetine rodents and the evolution of ontogeny. Evolution, 40:94–106.

DALY, J. C., AND J. L. PATTON. 1986. Growth, reproduction, and sexual dimorphism in *Thomomys bottae* pocket gophers. Journal of Mammalogy, 67: 256–265.

DOBZHANSKY, TH. 1937. Genetics and the origin of species. Columbia University Press, New York, 364 pp.

———. 1956. What is an adaptive trait? American Naturalist, 90:337–347.

DUBRUL, E. L., AND D. M. LASKIN. 1961. Preadaptive potentialities of the mammalian skull: an experiment in growth and form. American Journal of Anatomy, 109:117–132.

EGOSCUE, H. J., J. G. BITTMENN, AND J. A. PETROVICH. 1970. Some fecundity and longevity records for captive small mammals. Journal of Mammalogy, 51: 622–623.

EISENBERG, J. F. 1963. Behavior of heteromyid rodents. University of California Publications in Zoology, 69:1–100.

———. 1975. The behavior patterns of desert rodents. Pp. 189–224, *in* Rodents in desert environments (I. Prakash and P. K. Ghosh, eds.). Dr. W. Junk b.v., Publishers, The Hague, 624 pp.

———. 1981. The mammalian radiations: an analysis of trends in evolution, adaptation, and behavior. The University of Chicago Press, Chicago, 610 pp.

EISENBERG, J. F., AND D. E. ISAAC. 1963. The reproduction of heteromyid rodents in captivity. Journal of Mammalogy, 44:61–66.

FINK, W. L. 1982. The conceptual relationship between ontogeny and phylogeny. Paleobiology, 8:254–264.

FLEISCHER, G. 1978. Evolutionary principles of the mammalian middle ear. Adv. Anat. Embryol. Cell Biol., 55:1–70.

FLEMING, T. H. 1977. Growth and development of two species of tropical heteromyid rodents. American Midland Naturalist, 98:109–123.

FLINT, R. J. 1971. Glacial and Quaternary geology. John Wiley & Sons, Inc., New York, 892 pp.

GENOWAYS, H. H. 1973. Systematics and evolutionary relationships of spiny pocket mice, genus *Liomys*. Special Publication, The Museum, Texas Tech University, 5:1–368.

GOLDSCHMIDT, R. 1940. The material basis of evolution. Yale University Press, New Haven, 436 pp.

GOULD, S. J. 1977. Ontogeny and phylogeny. Harvard University Press, Cambridge, Massachusetts, 501 pp.

GOULD, S. J., AND R. C. LEWONTIN. 1979. The spandrels of San Marco and the Panglossian paradigm: a critique of the adaptationist programme. Proceedings of the Royal Society of London, B205:581–598.

GOULD, S. J., AND E. S. VRBA. 1982. Exaptation—a missing term in the science of form. Paleobiology, 8:4–15.

GRINNELL, J. 1922. A geographical study of the kangaroo rats of California. University of California Publications in Zoology, 24:1–124.

GUNDERSON, H. L. 1976. Mammalogy. McGraw-Hill Book Company, New York, 483 pp.

HAFNER, D. J., J. C. HAFNER, AND M. S. HAFNER. 1979. Systematic status of kangaroo mice, genus *Microdipodops*: morphometric, chromosomal, and protein analyses. Journal of Mammalogy, 60:1–10.

HAFNER, J. C. 1978. Evolutionary relationships of kangaroo mice, genus *Microdipodops*. Journal of Mammalogy, 59:354–366.

———. 1981. Evolution, systematics, and historical biography of kangaroo mice, genus *Microdipodops*. Ph.D. dissert., University of California, Berkeley, 269 pp.

HAFNER, J. C., AND M. S. HAFNER. 1983. Evolutionary relationships of heteromyid rodents. Great Basin Naturalist Memoirs, 7:3–29.

———. 1988. Heterochrony in rodents. Pp. 217–235, *in* Heterochrony in evolution (M. L. McKinney, ed.). Plenum Press, New York, 348 pp.

HAFNER, M. S. 1982. A biochemical investigation of geomyoid systematics (Mammalia: Rodentia). Zeitschrift für zool. Systematik und Evolutionsforschung, 20:118–130.

HAFNER, M. S., AND J. C. HAFNER. 1984. Brain size, adaptation and heterochrony in geomyoid rodents. Evolution, 38:1088–1098.

HALL, E. R. 1941. Revision of the rodent genus *Microdipodops*. Field Museum of Natural History, Zoology Series, 27:233–277.

———. 1946. Mammals of Nevada. University of California Press, Berkeley, 710 pp.

———. 1981. The mammals of North America. Second ed. John Wiley & Sons, New York, 1:1–600 + *90*, 2:601–1181 + *90*.

HALL, E. R., AND J. M. LINSDALE. 1929. Notes on the life history of the kangaroo mouse (*Microdipodops*). Journal of Mammalogy, 10:298–305.

HATT, R. T. 1932. The vertebral columns of ricochetal rodents. Bulletin of the American Museum of Natural History, 63:599–738.

HAYDEN, P., AND J. J. GAMBINO. 1966. Growth and development of the little pocket mouse, *Perognathus longimembris*. Growth, 30:187–197.

HEFFNER, R. S., AND H. E. HEFFNER. 1985. Hearing in mammals: the Least Weasel. Journal of Mammalogy, 66:745–755.

HILL, J. E. 1937. Morphology of the pocket gopher mammalian genus *Thomomys*. University of California Publications in Zoology, 42:81–172.

HOWARD, W. E., AND H. E. CHILDS, JR. 1959. Ecology of pocket gophers with emphasis on *Thomomys bottae mewa*. Hilgardia, 29:277–358.

HOWELL, A. B. 1923. Abnormal hairy growths upon the tails of the Heteromyidae. Journal of Mammalogy, 4:56–58.

———. 1932. The saltatorial rodent *Dipodomys*: the functional and comparative anatomy of its muscular and osseous systems. Proceedings of the American Academy of Arts and Sciences, 67:377–536.

———. 1944. Speed in animals, their specialization for running and leaping. The University of Chicago Press, Chicago, 270 pp.

JOHNSON, W. E., AND R. K. SELANDER. 1971. Protein variation and systematics in kangaroo rats (genus *Dipodomys*). Systematic Zoology, 20:377–405.

JONES, W. T. 1985. Body size and life-history variables in heteromyids. Journal of Mammalogy, 66: 128–132.

KOTLER, B. P. 1984. Risk of predation and the structure of desert rodent communities. Ecology, 65:689–701.

———. 1985. Owl predation on desert rodents which differ in morphology and behavior. Journal of Mammalogy, 66:824–828.

LACKEY, J. A. 1967. Growth and development of *Dipodomys stephensi*. Journal of Mammalogy, 48: 624–632.

LAY D. M. 1972. The anatomy, physiology, functional significance, and evolution of specialized hearing organs of gerbilline rodents. Journal of Morphology, 148:41–120.

———. 1993. Anatomy of the heteromyid ear. Pp. 270–290, *in* Biology of the family Heteromyidae (H. H. Genoways and J. H. Brown, eds.). Special Publication, American Society of Mammalogists, 10:1–719.

LINDSAY, E. H. 1972. Small mammal fossils from the Barstow Formation, California. University of California Publications in Geological Science, 93:1–104.

LONGLAND, W. S. 1983. Prey selection by great horned owls (*Bubo virginianus*) in central Nevada. M.S. thesis, University of Nevada, Reno, 95 pp.

LONGLAND, W. S., AND S. H. JENKINS. 1987. Sex and age affect vulnerability of desert rodents to owl predation. Journal of Mammalogy, 68:746–754.

LØVTRUP, S. 1981*a*. Introduction to evolutionary epigenetics. Pp. 139–144, *in* Evolution today (G. G. E. Scudder and J. L. Reveal, eds.). Hunt Institute for Botanical Documentation, Carnegie-Mellon University, Pittsburgh, 486 pp.

———. 1981*b*. The epigenetic utilization of the genomic message. Pp. 145–161, *in* Evolution today (G. G. E. Scudder and J. L. Reveal, eds.). Hunt Institute for Botanical Documentation, Carnegie-Mellon University, Pittsburgh, 486 pp.

MacMillen, R. E. 1983. Adaptive physiology of heteromyid rodents. Great Basin Naturalist Memoirs, 7:65–76.

MacMillen, R. E., and D. S. Hinds. 1983. Water regulatory efficiency in heteromyid rodents: a model and its application. Ecology, 64:152–164.

Mares, M. A. 1983. Desert rodent adaptation and community structure. Great Basin Naturalist Memoirs, 7:30–43.

———. 1992. Heteromyids and their ecological counterparts: a pandesertic view of rodent ecology and evolution. Pp. 652–714, *in* Biology of the family Heteromyidae (H. H. Genoways and J. H. Brown, eds.). Special Publication, American Society of Mammalogists, 10:1–719.

Maynard Smith, J., et al. 1985. Developmental constraints and evolution. Quarterly Review of Biology, 60:265–287.

Mayr, E. 1963. Animal species and evolution. Harvard University Press, Cambridge, Massachusetts, 797 pp.

———. 1983. How to carry out the adaptationist program? American Naturalist, 12:324–334.

McKinney, M. L. 1984. Allometry and heterochrony in an Eocene echinoid lineage: morphological change as a by-product of size selection. Paleobiology, 10: 407–419.

———. 1986. Ecological causation of heterochrony: a test and implications for evolutionary theory. Paleobiology, 12:282–289.

———. 1988. Classifying heterochrony: allometry, size, and time. Pp. 17–34, *in* Heterochrony in evolution (M. L. McKinney, ed.). Plenum Press, New York, 348 pp.

Merriam, C. H. 1891. Description of a new genus and species of dwarf kangaroo rat from Nevada (*Microdipodops megacephalus*). North American Fauna, 5:115–117.

Miller, R. S. 1964. Ecology and distribution of pocket gophers (Geomyidae) in Colorado. Ecology, 45: 251–272.

Patton, J. L. 1967*a*. Chromosomal studies of certain pocket mice, genus *Perognathus* (Rodentia: Heteromyidae). Journal of Mammalogy, 48:27–37.

———. 1967*b*. Chromosomes and evolutionary trends in the pocket mouse subgenus *Perognathus* (Rodentia: Heteromyidae). The Southwestern Naturalist, 12:429–438.

Patton, J. L., S. W. Sherwood, and S. Y. Yang. 1981. Biochemical systematics and chaetodipine pocket mice, genus *Perognathus*. Journal of Mammalogy, 62:477–492.

Price, M. V. 1978. Seed dispersion preferences of coexisting desert rodent species. Journal of Mammalogy, 59:624–626.

Price, M. V., and J. H. Brown. 1983. Patterns of morphology and resources use in North American desert rodent communities. Great Basin Naturalist Memoirs, 7:117–134.

Pye, A. 1965. The auditory apparatus of the Heteromyidae (Rodentia, Sciuromorpha). Journal of Anatomy, London, 99:161–174.

Rachootin, S. P., and K. S. Thomson. 1981. Epigenetics, paleontology, and evolution. Pp. 181–193, *in* Evolution today (G. G. E. Scudder and J. L. Reveal, eds.). Hunt Institute for Botanical Documentation, Carnegie-Mellon University, Pittsburgh, 486 pp.

Reeder, W. G. 1956. A review of the Tertiary rodents of the family Heteromyidae. Ph.D. dissert., The University of Michigan, Ann Arbor, 618 pp.

Reichman, O. J., and D. Oberstein. 1977. Selection of seed distribution types by *Dipodomys merriami* and *Perognathus amplus*. Ecology, 58:636–643.

Rensch, B. 1959. Evolution above the species level. Columbia University Press, New York, 419 pp.

Rogers, D. W., and D. J. Schmidly. 1982. Systematics of spiny pocket mice (genus *Heteromys*) of the *desmarestianus* species group from Mexico and northern Central America. Journal of Mammalogy, 63:375–386.

Sarich, V. M. 1985. Rodent macromolecular systematics. Pp. 423–452, *in* Evolutionary relationships among rodents: a multidisciplinary analysis (W. P. Luckett and J.-L. Hartenberger, eds.). Plenum Press, New York, 721 pp.

Seton, E. T. 1928. Lives of game animals. Vol. IV. Rodents, etc. Doubleday, Doran and Co., Inc., Garden City, New York, 949 pp.

Setzer, H. W. 1949. Subspeciation in the kangaroo rat, *Dipodomys ordii*. University of Kansas Publications, Museum of Natural History, 1:473–573.

Shea, B. T. 1983. Allometry and heterochrony in the African apes. American Journal of Physical Anthropology, 62:275–289.

———. 1988. Heterochrony in primates. Pp. 237–266, *in* Heterochrony in evolution (M. L. McKinney, ed.). Plenum Press, New York, 348 pp.

Simpson, G. G. 1944. Tempo and mode in evolution. Columbia University Press, New York, 237 pp.

———. 1953. The major features of evolution. Columbia University Press, New York, 434 pp.

Stanley, S. M. 1979. Macroevolution: process and product. W. H. Freeman Co., San Francisco, 332 pp.

Stebbins, W. C. 1983. The acoustic sense of animals. Harvard University Press, Cambridge, Massachusetts, 168 pp.

Thompson, S. D., R. E. MacMillen, E. M. Burke, and C. R. Taylor. 1980. The energetic cost of bipedal hopping in small mammals. Nature, 287: 223–224.

Thorington, R. W., Jr. 1966. The biology of rodent tails: a study of form and function. Arctic Aeromedical Lab., Aerospace Med. D. V., Air Force Sys. Comm., Fort Wainwright, Alaska (AAL-TR-65-8), 137 pp.

Van De Graaff, K. 1973. Comparative developmental osteology in three species of desert rodents, *Peromyscus eremicus, Perognathus intermedius,* and *Dipodomys merriami*. Journal of Mammalogy, 54: 729–741.

Vaughan, T. A. 1978. Mammalogy. Saunders College Publishing, Philadelphia, Pennsylvania, 522 pp.

Waddington, C. H. 1957. The strategy of the genes. George Allen & Unwin Ltd., London, 262 pp.

———. 1962. New patterns in genetics and development. Columbia University Press, New York, 271 pp.

———. 1975. The evolution of an evolutionist. Cornell University Press, Ithaca, New York, 323 pp.

Wahlert, J. H. 1992. The fossil record. Pp. 1–37, *in* Biology of the family Heteromyidae (H. H. Genoways and J. H. Brown, eds.). Special Publication, American Society of Mammalogists, 10:1–719.

Webster, D. B. 1962. A function of the enlarged middle-ear cavities of the kangaroo rat, *Dipodomys*. Physiological Zoology, 35:248–255.

Webster, D. B., and M. Webster. 1971. Adaptive value of hearing and vision in kangaroo rat predator avoidance. Brain, Behavior and Evolution, 4:310–322.

———. 1972. Kangaroo rat auditory thresholds before and after middle ear reduction. Brain, Behavior and Evolution, 5:41–53.

———. 1975. Auditory systems of Heteromyidae: functional morphology and evolution of the middle ear. Journal of Morphology, 146:343–376.

———. 1980. Morphological adaptations of the ear in the rodent family Heteromyidae. American Zoologist, 20:247–254.

Wenner, A. M., and P. H. Wells. 1990. Anatomy of a controversy: the question of a "language" among bees. Columbia University Press, New York, 399 pp.

Williams, D. F. 1978. Karyological affinities of the species groups of silky pocket mice (Rodentia: Heteromyidae). Journal of Mammalogy, 59:599–612.

Willson, M. F. 1984. Vertebrate natural history. CBS College Publ., Saunders College Publ., New York, 621 pp.

Wood, A. E. 1935. Evolution and relationships of the heteromyid rodents with new forms from the Tertiary of western Northern America. Annals of the Carnegie Museum, 24:73–262.

Zakrzewski, R. J. 1981. Kangaroo rats from the Borchers Local Fauna, Blancan, Meade County, Kansas. Transactions of the Kansas Academy of Science, 84:78–88.

Zelditch, M. L. 1987. Evaluating models of developmental integration in the laboratory rat using confirmatory factor analysis. Systematic Zoology, 36:368–380.

BIOGEOGRAPHY

DAVID J. SCHMIDLY, KENNETH T. WILKINS,
AND JAMES N. DERR

Introduction

The purpose of this chapter is to describe and explain the distribution of heteromyid rodents and to assess the importance of historical events in determining the present and past distributions of species and genera in the family. Our work concerns both ecological biogeography—the distributions of species in relation to their life conditions and dispersal potential—and historical biogeography—the distributions of species and higher taxa as a result of their past history and the paleogeographical evolution of the region in which they occur. Although these two approaches are not mutually exclusive, we find it convenient to review them separately.

The history of heteromyids spans at least 35 million years from the early Oligocene to the present (Gawne, 1975; Savage and Russell, 1983). For all but the last 3 million years, when *Heteromys* (or its ancestors) presumably entered northern South America, the group evolved entirely on the North American continent. The former geographic distribution of the family includes areas far east of those occupied by modern heteromyids. T. H. Patton (1969) reported heteromyid material from the late Oligocene (Whitneyan) Interstate 75 locality in north-central Florida. The genus *Proheteromys* likewise ranged eastward to Florida where it is known from Miocene deposits, notably Thomas Farm (Black, 1963).

Paramount among a myriad of factors that influence heteromyid distribution is climate, with both direct effects and indirect effects on animals via its impact on vegetation and habitat distributions. Efforts to understand the historical biogeography of any particular group of animals, therefore, must include not only a knowledge of the fossil record but also evidence of climatic and floral changes and of geological events. In situations where only a meager fossil record exists, climatic, vegetational, and geological evidence become even more important.

Many authors have discussed factors that influence the distribution, abundance, and diversity of extant heteromyids. These include (1) physical factors such as moisture, soils, and temperature (Grinnell, 1914, 1922; Huey, 1951; Munger et al., 1983; Reynolds, 1958; Rosenzweig and Winakur, 1969); (2) geographical barriers such as rivers, streams, and mountain ranges (Brown, 1973, 1975; Brown and Lieberman, 1973; Durrant, 1952; Hall, 1946); and (3) biotic factors such as competition (Bowers and Brown, 1982; Brown, 1973; Grinnell, 1914; Mares and Williams, 1977; Rosenzweig and Sterner, 1970), predation and parasites

(Munger et al., 1983; Rosenzweig, 1973; Thompson, 1982), and vegetation (Beatley, 1967; Brown, 1973; Dice and Blossom, 1937; Hafner, 1977; Munger et al., 1983).

There is much circumstantial evidence that the coexistence of many heteromyids is influenced by interspecific competition for limited resources (Bowers and Brown, 1982; Brown, 1973, 1975; Brown and Lieberman, 1973; Rosenzweig and Sterner, 1970; Price, 1978). These resources are ultimately determined by synergistic interactions of the biotic and abiotic factors discussed previously. Bowers and Brown (1982) proposed that significant differences in body size among sympatric seed eating rodents prevents or limits competition. Non-random patterns of sympatric occurrence among species of similar size is suggestive that competition may limit distribution through mechanisms of competitive exclusion. A number of authors have discussed the apparent correlation between parapatric geographical ranges of heteromyid rodents and body size (Bowers and Brown, 1982; Munger et al., 1983; Brown, 1973, 1975; Hallett, 1982; Wondolleck, 1978; Lemen and Freeman, 1983; Price and Brown, 1983; Stamp and Ohmart, 1978; Rosenzweig and Winakur, 1969), but interpretations are subject to alternative explanations and often there is no direct evidence of competitive interactions occurring on the boundaries of species ranges (Brown and Gibson, 1983; Price and Brown, 1983).

Heteromyids exhibit considerable variation in external appearance and size. Adult head and body length ranges from 50 to 180 mm, and tail length from 45 to 215 mm. We ranked all 57 heteromyid species by body size (head an body length) into five discrete categories ranging from very small to very large (Table 1). These body size rankings were then compared with intra and intergeneric distributions of all heteromyid taxa in order to further explore the influence of size and competition on their distribution.

Throughout this chapter we have used representative species distribution maps depicted in Hall (1981) except where noted otherwise. We employed J. M. Crowley's (1967) classification of the major ecoregions of North America to describe these distributions. In Crowley's scheme North America is divided into a hierarchical series of domains, divisions, and provinces. Provinces represent broad vegetative regions having a uniform regional climate and the same type of zonal soil throughout. A division is a group of provinces (when more than one is present) having definite vegetational affinities and falling within the same regional climate; usually the zonal soils also are related. A domain is a more inclusive category that includes a group of divisions characterized by loosely related climates. The four major ecological domains of North America are the Polar, Humid Temperate, Dry, and Humid Tropical domains (Fig. 1).

Ecological Biogeography

Distribution of the Family

The present distribution of the Heteromyidae, with respect to provinces within the four ecological domains of North America, is depicted in Figure 2, and a list of species that occur within each of the provinces is provided in Table 2. Heteromyids occur in all but the Polar Domain, which encompasses the Arctic tundra, subarctic parkland, and boreal forests of northern Canada and Alaska. Representatives of the family occur as far north as British Columbia and Saskatchewan, in the Humid Temperate Domain. In the east, heteromyids appear to be limited by the Mississippi River with the most easterly extension of their range in Minnesota, Iowa, eastern Kansas, Oklahoma, and central Louisiana. To the south they occur throughout the Humid Tropical Domain in Central America and the northern coast of South America in Colombia, Ecuador, and Venezuela.

The family is especially common in the Dry (37 species) and the western edge of the Humid Temperate (29 species) domains, with substantially fewer species (14) in the

FIG. 1.—The four major ecoregion domains of North America (Crowley, 1967).

Humid Tropical Domain. Heteromyids occur in all but four of the 15 provinces of the Humid Temperate Domain, but species numbers are especially abundant in the California Chaparral (18 species), California Grassland (9), and Prairie Brushland (7) provinces. They occur in each of the 15 provinces of the Dry Domain with highest diversity in the American Desert (19), Mexican Highlands Shrub Steppe (14), Colorado Plateau (13), Chihuahuan Desert (13), Baja California (12), Great Plains Short-Grass

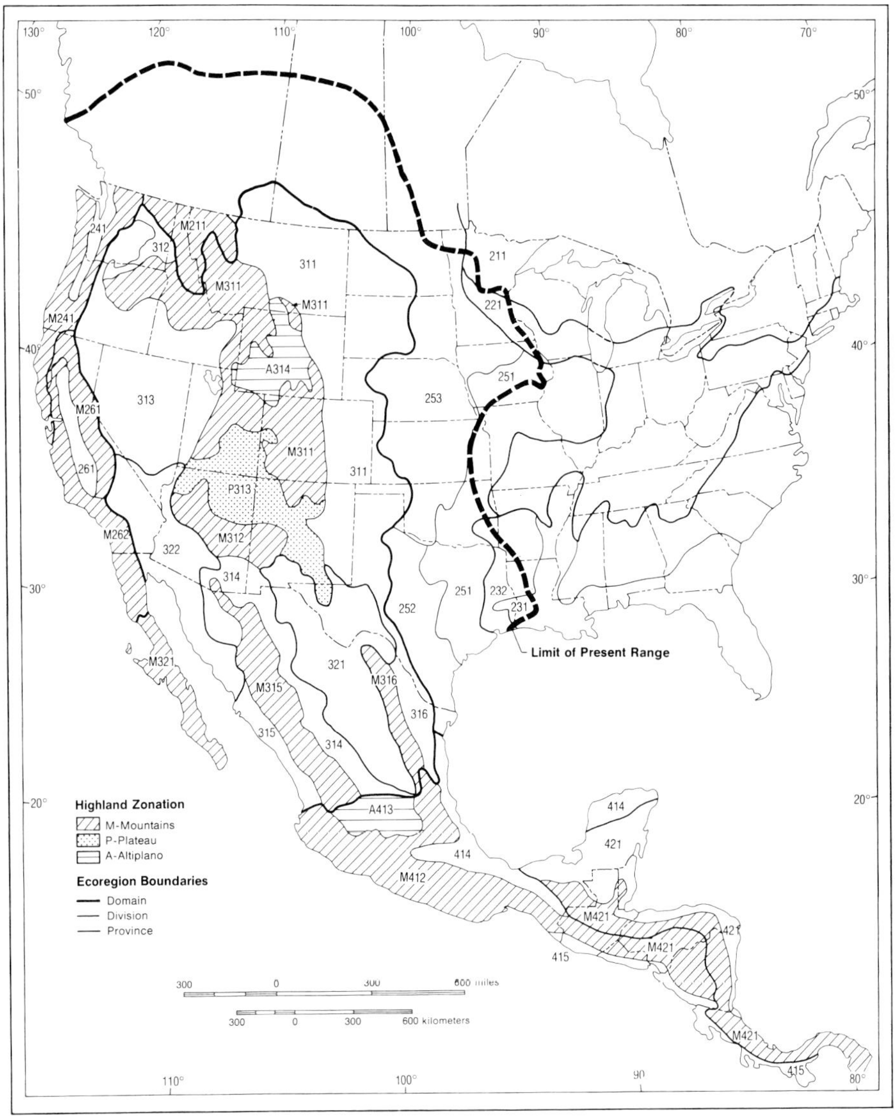

FIG. 2.—The present distribution of the family Heteromyidae with respect to the domains, divisions and provinces of North America (Crowley, 1967). Refer to Table 2 for a key to Crowley's original numbering system for each ecoregion and a list of heteromyid species that occur in each province.

Prairie (11), and Intermountain Sagebrush (11) provinces. The 14 heteromyids characteristic of the Humid Tropical Domain occupy six of the 10 provinces in that region with most of the diversity in the provinces of the Sierra Madre del Sur (9) and the Central American Ranges (6).

Distribution of Genera and Species

Liomys

Spiny pocket mice are distributed from Sonora, Mexico, and southern Texas south into Central America, and they are primarily restricted to the arid and semi-arid hab-

TABLE 1.—*Individual species grouped by body size. Species that overlap into two or more body size groupings were placed in the category that contained the greatest amount of variation found in that species. Ranges include: very small (50–66 mm), small (67–88 mm), medium (89–106 mm), large (107–140 mm), and very large (>141 mm).*

1) Very small		
P. amplus		
P. flavus		
P. longimembris		
2) Small		
P. fasciatus	*C. arenarius*	*M. megacephalus*
P. flavescens	*C. artus*	*M. pallidus*
P. inornatus	*C. fallax*	
P. parvus	*C. intermedius*	
	C. lineatus	
	C. nelsoni	
	C. penicillatus	
	C. pernix	
	C. spinatus	
3) Medium		
D. insularis	*P. alticola*	*C. baileyi*
D. margaritae		*C. californicus*
D. merriami		*C. formosus*
D. nitratoides		*C. goldmani*
D. ordii		*C. hispidus*
D. phillipsii		
D. compactus		
4) Large		
D. agilis	*H. australis*	*L. adspersus*
D. californicus	*H. gaumeri*	*L. irroratus*
D. elator		*L. pictus*
D. elephantinus		*L. salvini*
D. gravipes		*L. spectabilis*
D. heermanni		
D. microps		
D. nelsoni		
D. panamintinus		
D. spectabilis		
D. stephensi		
D. venustus		
5) Very large		
D. deserti	*H. anomalus*	
D. ingens	*H. desmarestianus*	
	H. goldmani	
	H. nelsoni	
	H. oresterus	

itats of the Dry and Humid Tropical domains. High moisture and extreme aridity are important limiting factors for this genus (Genoways, 1973). No species occurs where there is less than 250 mm of rainfall per year, and in more mesic areas *Liomys* seems to be displaced by species of *Heteromys.* Spiny pocket mice occur in 11 ecoregion provinces with their greatest diversity in the Central America Ranges and Pacific Savan-

Table 2.—*Major ecoregions within the range of the family Heteromyidae with a listing of individual species occurring within each region. The numbering system is after the original classification of Crowley (1967) and refers to the provinces in Fig. 2.*

200 Humid Temperate Domain	
Humid Warm-Summer Continental Regime Highlands Division	
M211—Columbian Forest Province	
P. parvus	
Humid Hot-Summer Continental Division	
221—Eastern Deciduous Forest Province	
P. flavescens	
Humid Subtropical Division	
231—Outer Coastal Plain Forest Province	
C. hispidus	
232—Southeastern Mixed Forest Province	
C. hispidus	
Humid Maritime Regime Highlands Division	
M241—Pacific Forest Province	
C. californicus	*D. heermanni*
D. californicus	*D. venustus*
Subhumid Prairie Division	
251—Prairie Parkland Province	
C. hispidus	
D. elator	
P. flavescens	
252—Prairie Brushland Province	
C. hispidus	*L. irroratus*
D. compactus	*P. flavescens*
D. elator	*P. flavus*
D. ordii	
253—Tall Grass Prairie Province	
C. hispidus	*P. flavus*
C. fasciatus	*D. ordii*
P. flavescens	
261—California Grassland Province	
C. californicus	*D. panamintinus*
D. heermanni	*P. inornatus*
D. ingens	*P. longimembris*
D. nitratoides	*P. parvus*
Mediterranean Regime Highland Division	
M261—Sierran Forest Province	
C. californicus	*D. panamintinus*
D. californicus	*P. longimembris*
D. microps	*P. parvus*
M262—California Chaparral Province	
C. arenarius	*D. gravipes*
C. baileyi	*D. heermanni*
C. formosus	*D. merriami*
C. penicillatus	*D. nitratoides*
C. californicus	*D. stephensi*
C. fallax	*D. venustus*
C. spinatus	*P. alticola*
D. agilis	*P. inornatus*
D. californicus	*P. longimembris*
D. deserti	
D. elephantinus	
300 Dry Domain	
Semiarid Steppe Division	
311—Great Plains Short-grass Prairie Province	
C. hispidus	*P. amplus*
C. formosus	*P. fasciatus*
D. merriami	*P. flavescens*
D. ordii	*P. flavus*
D. spectabilis	*P. parvus*
312—Palouse Grassland Province	
D. ordii	
P. parvus	
313—Intermountain Sagebrush Province	
C. formosus	*M. megacephalus*
D. deserti	*M. pallidus*
D. merriami	*P. longimembris*
D. microps	*P. parvus*
D. ordii	
D. panamintinus	
314—Mexican Highland Shrub Steppe Province	
C. baileyi	*D. phillipsii*
C. goldmani	*D. spectabilis*
C. hispidus	*P. amplus*
C. intermedius	*P. flavescens*
C. nelsoni	*P. flavus*
C. penicillatus	*L. irroratus*
D. merriami	*L. pictus*
D. ordii	
315—Sinaloa Coast Province	
C. artus	*C. pernix*
C. goldmani	*L. pictus*
316—Rio Grande Shrub Steppe Province	
C. hispidus	*L. irroratus*
D. merriami	*P. flavus*
D. ordii	*P. hispidus*
Semiarid Steppe Regime Highlands Division	
M311—Rocky Mountain Forest Province	
D. ordii	*P. flavus*
P. fasciatus	*P. parvus*
P. flavescens	
M312—Upper Gila Mountain Forest Province	
D. merriami	*P. flavescens*
D. ordii	*P. flavus*
D. spectabilis	*P. intermedius*
P. amplus	
M315—Sierra Madre Occidental Province	
C. baileyi	*D. merriami*
C. goldmani	*D. ordii*
C. penicillatus	*D. spectabilis*

TABLE 2.—*Continued.*

M316—Sierra Madre Oriental Province	
C. hispidus	*D. ordii*
C. nelsoni	*L. irroratus*
D. merriami	*P. flavus*
D. nelsoni	
P313—Colorado Plateau Province	
C. formosus	*D. spectabilis*
C. hispidus	*P. amplus*
C. intermedius	*P. fasciatus*
C. nelsoni	*P. flavus*
C. penicillatus	*P. longimembris*
D. merriami	*P. parvus*
D. microps	
D. ordii	
A314—Wyoming Basin Province	
C. baileyi	*P. fasciatus*
C. hispidus	*P. flavus*
D. ordii	*P. parvus*
P. amplus	
Arid Desert Division	
321—Chihuahuan Desert Province	
C. hispidus	*D. ordii*
C. intermedius	*D. phillipsii*
C. lineatus	*D. spectabilis*
C. nelsoni	*L. irroratus*
C. penicillatus	*P. flavescens*
D. merriami	*P. flavus*
D. nelsoni	
322—American Desert Province	
C. artus	*D. merriami*
C. baileyi	*D. microps*
C. fallax	*D. spectabilis*
C. formosus	*L. pictus*
C. goldmani	*P. alticola*
C. intermedius	*P. amplus*
C. penicillatus	*P. longimembris*
C. pernix	*P. flavus*
C. spinatus	
D. agilis	
D. deserti	
Arid Desert Regime Highlands Division	
M321—Baja California Province	
C. arenarius	*D. deserti*
C. baileyi	*D. insularis*
C. fallax	*D. margaritae*
C. formosus	*D. merriami*
C. spinatus	
D. agilis	
400 Humid Tropical Domain	
Tropical Savanna Division	
414—Campeche-Yucatan Savanna Province	
C. hispidus	*L. irroratus*
H. desmarestianus	*L. pictus*
H. gaumeri	

TABLE 2.—*Continued.*

415—Pacific Savanna Woodland Province	
H. desmarestianus	*L. adspersus*
H. nelsoni	*L. salvini*
Tropical Savanna Regime Highlands	
M412—Sierra Madre del Sur Province	
D. ordii	*L. irroratus*
D. phillipsii	*L. pictus*
H. desmarestianus	*L. salvini*
H. goldmani	
A413—Central Mexican Province	
C. hispidus	*L. irroratus*
D. phillipsii	*P. flavus*
Tropical Rainforest Division	
421—Caribbean Coast Rainforest Province	
H. gaumeri	
L. salvini	
Tropical Rainforest Regime Highlands Division	
M421—Central American Ranges Province	
H. anomalus	*H. oresterus*
H. australis	*L. adspersus*
H. desmarestianus	*L. salvini*

na Woodlands of Central America. All five species occur in the Central American Ranges which extend from Jalisco southward through Central America and into the extreme northern coast of South America. This large and varied area is characterized by high mountainous regions separated by valleys of wet tropical jungle.

The genus is comprised of two monotypic and three polytypic species with a total of 16 subspecies. The two monotypic species both have extremely limited ranges. *L. adspersus* is restricted to central Panama (Fig. 3), and *L. spectabilis* is known only from a few localities in southeastern Jalisco, Mexico (Fig. 4).

The three wide ranging polytypic species have geographically complementary ranges. *L. irroratus* occurs form southern Texas and Chihuahua, Mexico, throughout the Mexican Plateau to northern Veracruz and then southward to the mountainous regions of Oaxaca (Fig. 4). The distribution of *L. pictus* is continuous along the Pacific coast from Sonora, Mexico, southward to Guatemala, and the species also occurs along the Gulf coast of Veracruz (Fig. 4). The ranges of

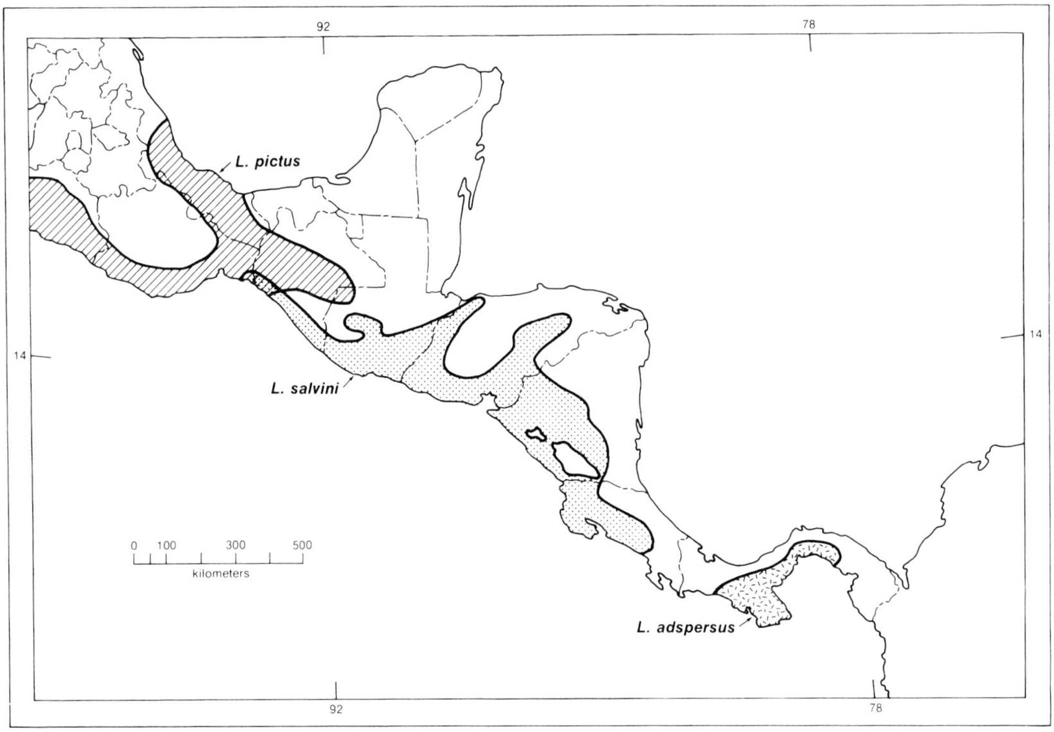

FIG. 3.—Distributions of three species of *Liomys* in Central America.

these two species marginally overlap in western Mexico, but they are characterized by different habitat preferences. *L. pictus* prefers moist habitats compared to the more xeric habitat requirements of *L. irroratus* (Genoways, 1973). *L. salvini* occurs along the Pacific coast from southern Oaxaca southward through Chiapas, Guatemala, El Salvador, Honduras, Nicaragua and into Costa Rica (Fig. 3). This species, which is found in the Central American Ranges and the Pacific Savanna Woodland provinces, seems to prefer the drier tropical lowland forests along the slopes of the mountain ranges.

All species of *Liomys* are large in body size, compared to other heteromyids, and there is some indication that competition among closely related species with similar habitat requirements may account for the unusual distribution patterns of some taxa. For example, *L. salvini* and *L. pictus* seem to prefer xeric lowland forest throughout their ranges, and Genoways (1973) reports that the northern range of *salvini* appears to be restricted by the presence of *pictus* in the central valley of Chiapas. Very little is known about the ecological requirements of these two species in areas of sympatry, but character displacement has been reported between males of the two species in southern Oaxaca and northwestern Chiapas (Genoways, 1973:314).

Heteromys

This genus is distributed exclusively in the Humid Tropical Domain of Middle America and northern South America (Fig. 5). These rodents inhabit lowland rain forest, tropical cloud forest, and limited areas in lower montane dry forest (Stuart, 1966). Representative species are found in three

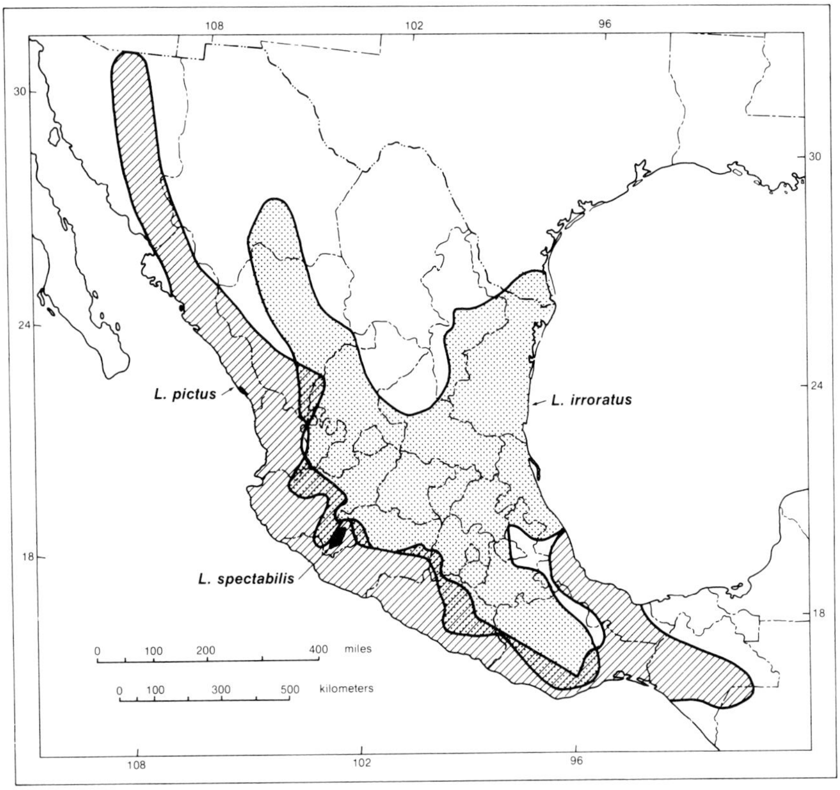

FIG. 4.—Distributions of three species of *Liomys*.

ecoregion provinces, the Campeche-Yucatan Savanna, Central American Ranges, and Pacific Savanna Woodland. Members of this genus are unique among heteromyids in their preference for mesic habitats. This ecological preference possibly results from resource partitioning with *Liomys* for suitable habitat that may have forced *Heteromys* out of the more xeric areas and into the remaining higher and wetter regions (Genoways, 1973).

Seven species are currently recognized in the genus, three of which are polytypic (*H. australis, H. anomalus,* and *H. desmarestianus*) and four are monotypic (*H. gaumeri, H. goldmani, H. nelsoni,* and *H. oresterus*). All of the species have restricted ranges except for *H. desmarestianus* which is continuously distributed from southern Veracruz and eastern Oaxaca southward through Guatemala, Honduras, El Salvador, Nicaragua, Costa Rica, Panama, and possibly into Colombia. The distribution of this species in South America is unknown. *H. anomalus* is the only heteromyid found exclusively in South America where it is restricted to Colombia, Venezuela, and the Caribbean Island of Trinidad (Goldman, 1911; Rogers, 1986). *H. australis* is limited to extreme southern Panama, the Pacific coast of Colombia, and the northern coast of Ecuador (Rogers, 1986). Of the monotypic species, *H. gaumeri* occurs in the Mexican states of Campeche, Chiapas, Quintana Roo, and Yucatan; *H. goldmani* (considered a subspecies of *H. desmarestianus* by Rogers, 1986) is found in southwestern Chiapas and western Guatemala; *H. nelsoni* is known only from restricted areas in southwestern Chiapas and western Guatemala; and *H.*

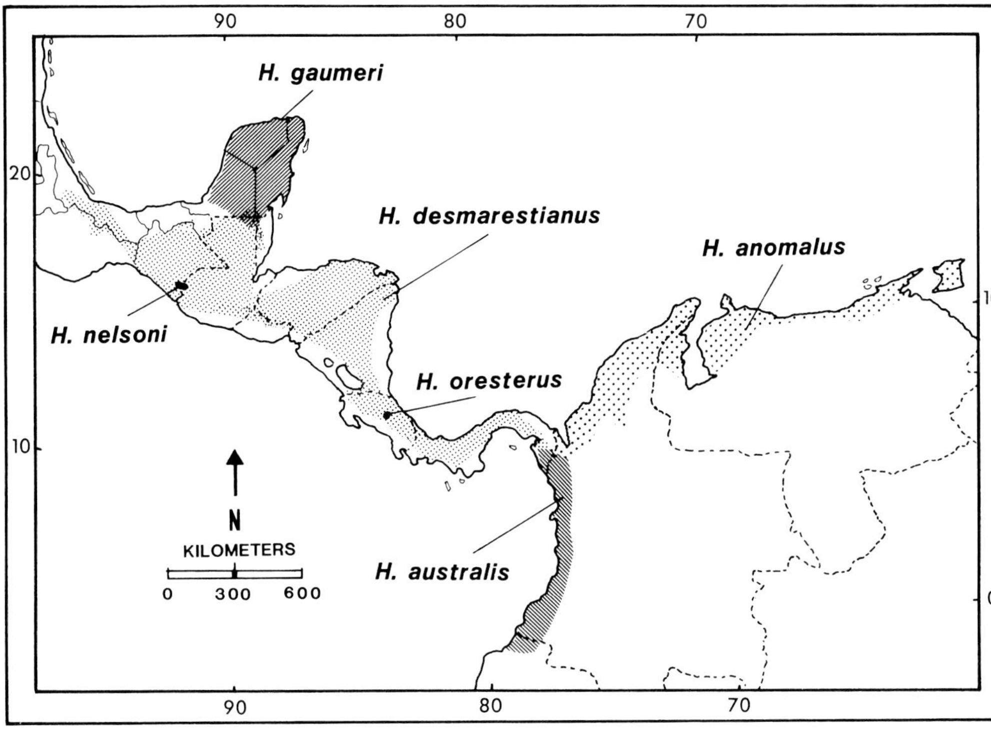

FIG. 5.—Distributions of the species in the genus *Heteromys*.

oresterus occurs in central Costa Rica. *H. gaumeri* and *H. oresterus* are large in body size, whereas the remaining five species in the genus are in the very large body size category.

Microdipodops

Kangaroo mice are restricted to extremely arid, sandy habitats in the Great Basin Desert in the Dry Domain of western North America. This genus contains only two species (*M. megacephalus* and *M. pallidus*) and occupies the smallest geographical range of all the heteromyid genera. Kangaroo mice are found in only one ecoregion province, the Intermountain Sagebrush (Fig. 6).

Both species are in the small body size category (Table 1), and only one other small heteromyid, *Perognathus parvus,* occurs sympatrically with them. *P. parvus* is distributed altitudinally from 3,900 (1,189 m) to 10,000 feet (3,048 m) and occurs in a wide variety of soil and vegetational habitats. In contrast, the two species of *Microdipodops* seem to be restricted to specific ecological areas.

M. megacephalus has the most extensive distribution of the two species, occurring in California, Nevada, Utah, and Oregon with isolated populations along the border of central California, Nevada, north-central Utah and southern Idaho. The range of *M. pallidus,* which is almost entirely confined to central Nevada except for a sandy region in western California, is broadly overlapped by that of *M. megacephalus. M. megacephalus* is less specialized in its adaption to desert environments and inhabits a much broader range of habitat types (Hafner, 1978). The former species is almost always found in gravely soils or gravel mixed with coarse sand and appears to avoid open sandy areas, whereas the latter is restricted to sand dune areas with fine, loose, wind-blown

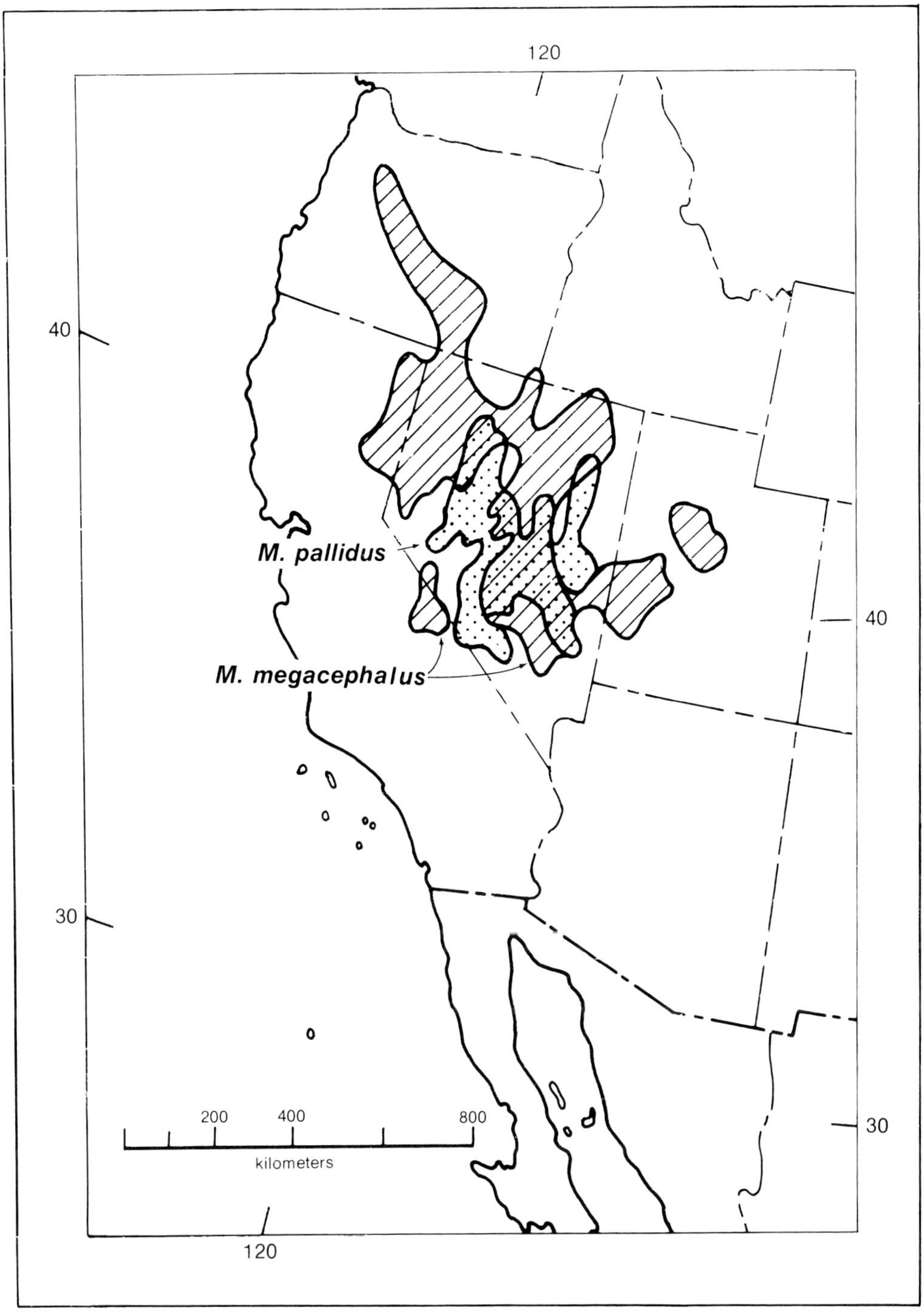

FIG. 6.—Distributions of the two species of *Microdipodops*.

sand. Hall (1946) reported this ecological separation at 11 of 12 places where the two species were taken together, and he suggested that edaphic factors were the most important parameters limiting their distribution.

Perognathus

Silky pocket mice occur as far north as British Columbia and Saskatchewan, throughout most of the central and western United States, and southward to Puebla, Mexico. Members of this genus occur farther north and east than other heteromyid taxa, but in central Mexico they are excluded from the Humid Tropical Domain. The genus is known from each of the ecoregion domains inhabited by the family, but the greatest diversity and abundance is in the deserts and grasslands of the Dry Domain. All eight species occur in the Dry Domain, six are found in the Humid Temperate Domain, and only one in the Humid Tropical Domain.

Although silky pocket mice have broad tolerances for various soil and vegetation types, they are usually associated with sandy or friable soils and sparse vegetation. Altitudinally, they occur in habitats ranging from lower elevations in the western deserts up to 10,000 feet (3,048 m) along the slopes of the Rocky Mountains (Hall, 1946).

The parapatric nature of the distributions depicted in Figs. 7 and 8 suggests that intrageneric competition among species of similar size may be a major factor influencing the distributions of these rodents (Bowers and Brown, 1982). Both very small and small species of *Perognathus* are distributed with little overlap, and in many cases they are parapatric over extended areas.

The distributions of two of the smallest heteromyid species (*P. flavus* and *P. longimembris*) are complementary to one another (Fig. 7). *P. flavus* has the largest geographic range in the genus occurring from Wyoming and South Dakota southward through the southwestern United States and into the Humid Tropical Domain of central Mexico. *P. longimembris* is primarily distributed in the arid and semi-arid regions of the Intermountain Sagebrush and the American Desert provinces, although it also occurs in the California Grassland and Sierran Forest provinces of the Humid Temperate Domain. A third very small species, *P. amplus,* is comprised of disjunct populations in Arizona, Nevada, and Sonora, Mexico, that are interspersed between the ranges of the other two species (Hoffmeister, 1986). The ranges of *P. flavus, P. amplus,* and *P. longimembris* are parapatric over a broad area in southern Arizona.

The distributions of four other small *Perognathus* (*P. fasciatus, P. flavescens, P. inornatus,* and *P. parvus*) also show a pattern of distributional complementarity with minimal overlap (Fig. 8). The major portion of the range of *P. fasciatus* is in the Great Plains Short Grass Prairie Province, but this species also is known from the Rocky Mountain Forest, Wyoming Basin, and Colorado Plateau provinces and it extends northward into the Tall Grass Prairie Province in Saskatchewan. *P. flavescens,* which occupies the extreme eastern extent of the range of the family of Heteromyidae (Christiansen and Sanz, 1978), occurs in 10 ecoregion provinces representing habitats ranging from deciduous forests, plains, and brushlands in the north to mountains, plateaus, and deserts in the southwest. The range of this species encompasses most of the Great Plains Short Grass Prairie and the Tall Grass Prairie provinces. *P. inornatus* is distributed primarily in the California Grassland Province, although it also occurs in the Central California Chaparral Province. The Great Basin Pocket Mouse, *P. parvus,* occurs from the Colorado Plateau Province in northern Arizona northward through the Intermountain Sagebrush Province in Utah, Nevada, Montana, California,

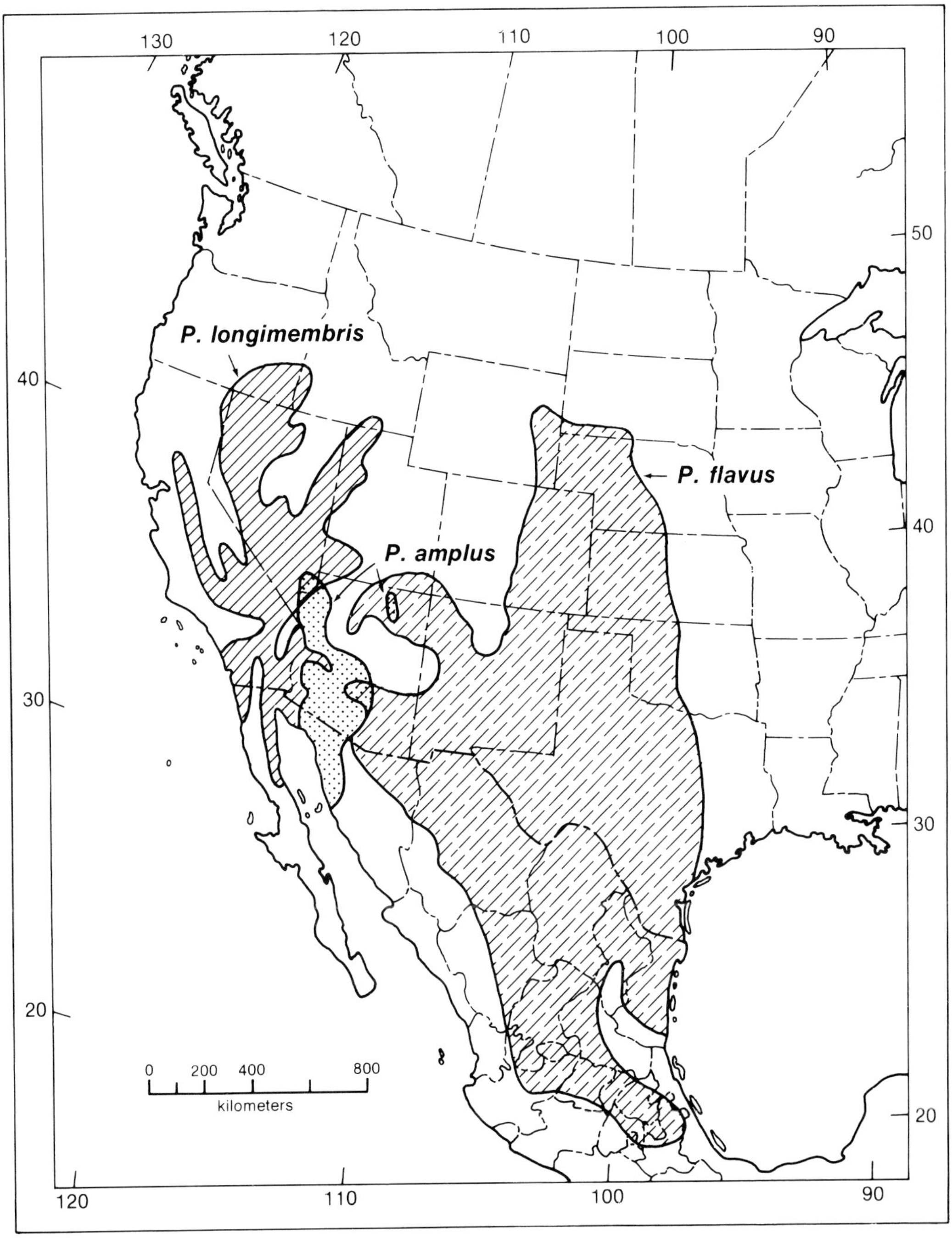

FIG. 7.—Distributions of the three smallest species of *Perognathus*.

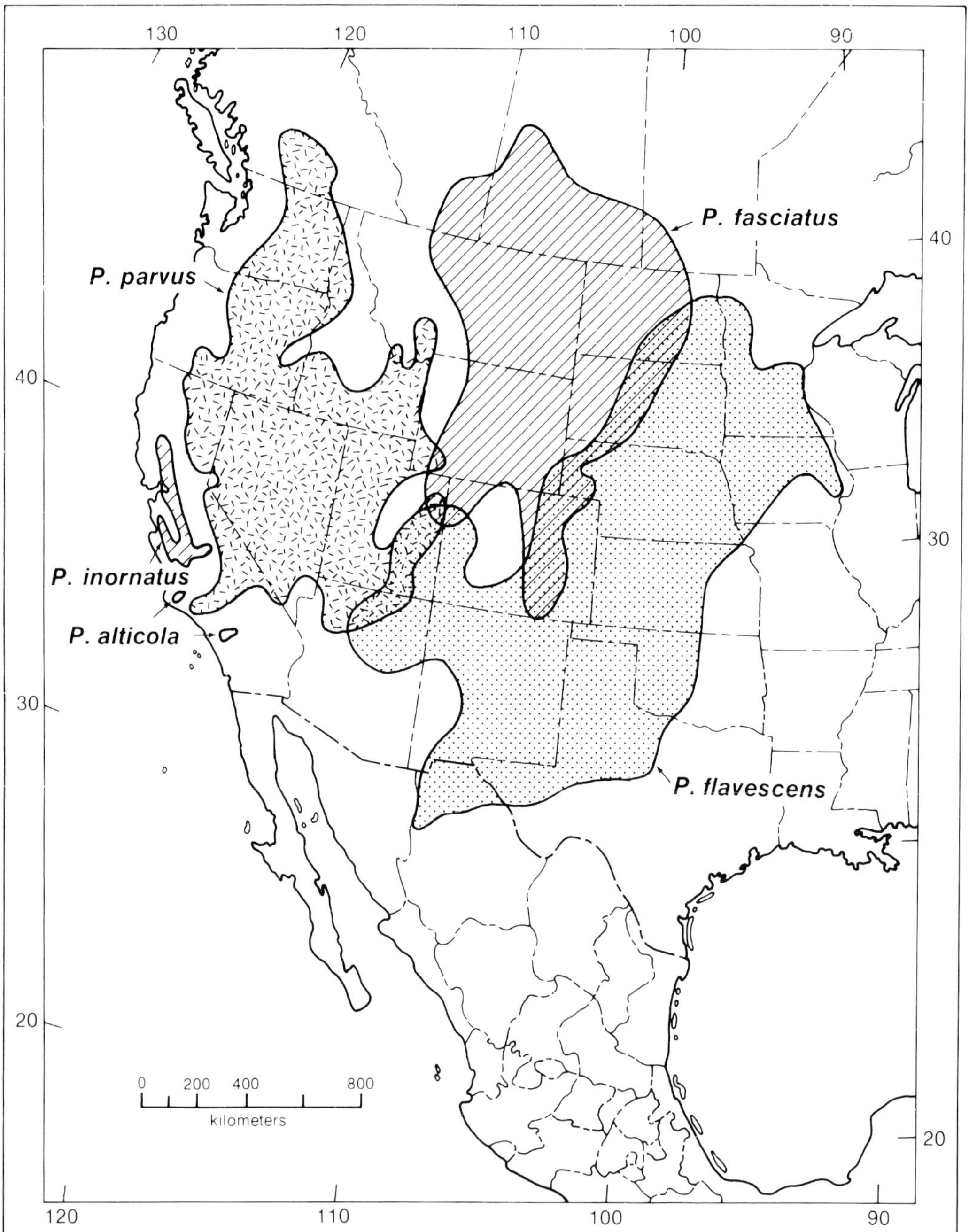

FIG. 8.—Distributions of the small body size species of *Perognathus* and *P. alticola,* the largest species in the genus.

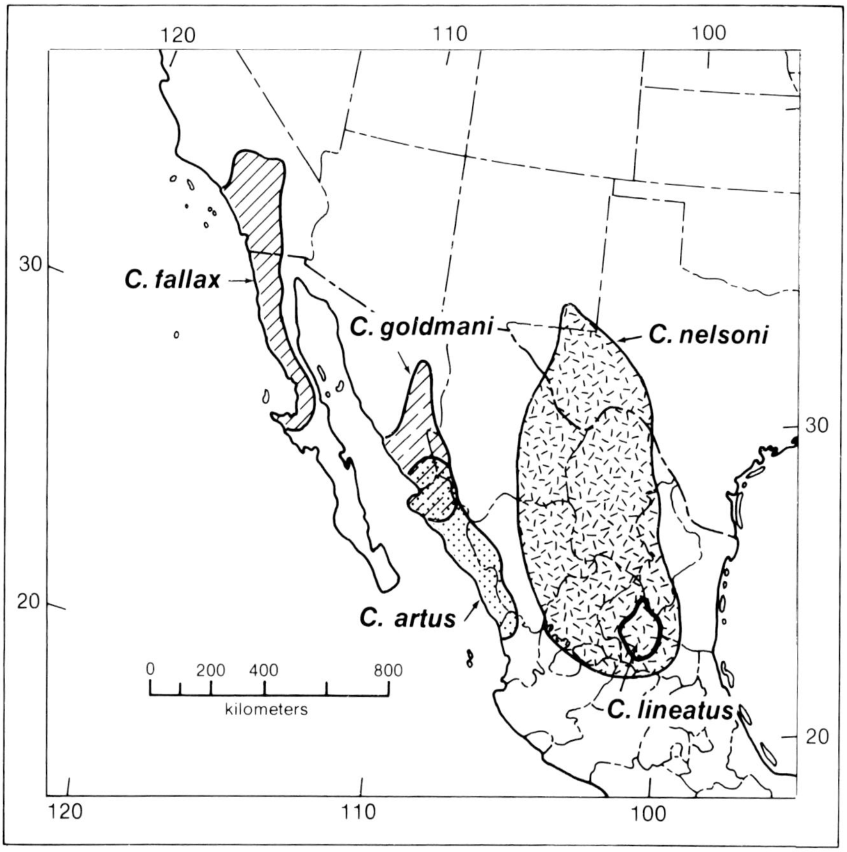

FIG. 9.—Distributions of four species of *Chaetodipus.*

Oregon, Washington, and British Columbia. *P. parvus* is found in eight ecoregion provinces, although the Intermountain Sagebrush Province comprises most of its range.

P. alticola is the largest species in the genus and the only one of medium body size. This species is known from two isolated populations in the California Chaparral and American Desert provinces of south-central California (Fig. 8).

Chaetodipus

Chaetodipus occupies more xeric habitat than the other heteromyid genera. Species occur in at least one and in some cases two of the major desert regions of the southwestern United States and northern Mexico. The range of the genus extends from northern Utah, Nevada, and North Dakota to the southern extent of the Mexican Plateau.

Chaetodipus is the second most speciose genus (*Dipodomys* is first) in the family with a total of 14 species, including three monotypic and 11 polytypic forms comprised of 83 subspecies. Species of *Chaetodipus* are generally larger in adult body size than *Perognathus* and smaller than most *Dipodomys* species. Nine species are small in body size, and five are medium in size (Table 1).

The greatest diversity (8 of 14 species) occurs within the Dry Domain of southern California, Arizona, and northern Baja California. This high species diversity seems to coincide with habitat complexity (Rosen-

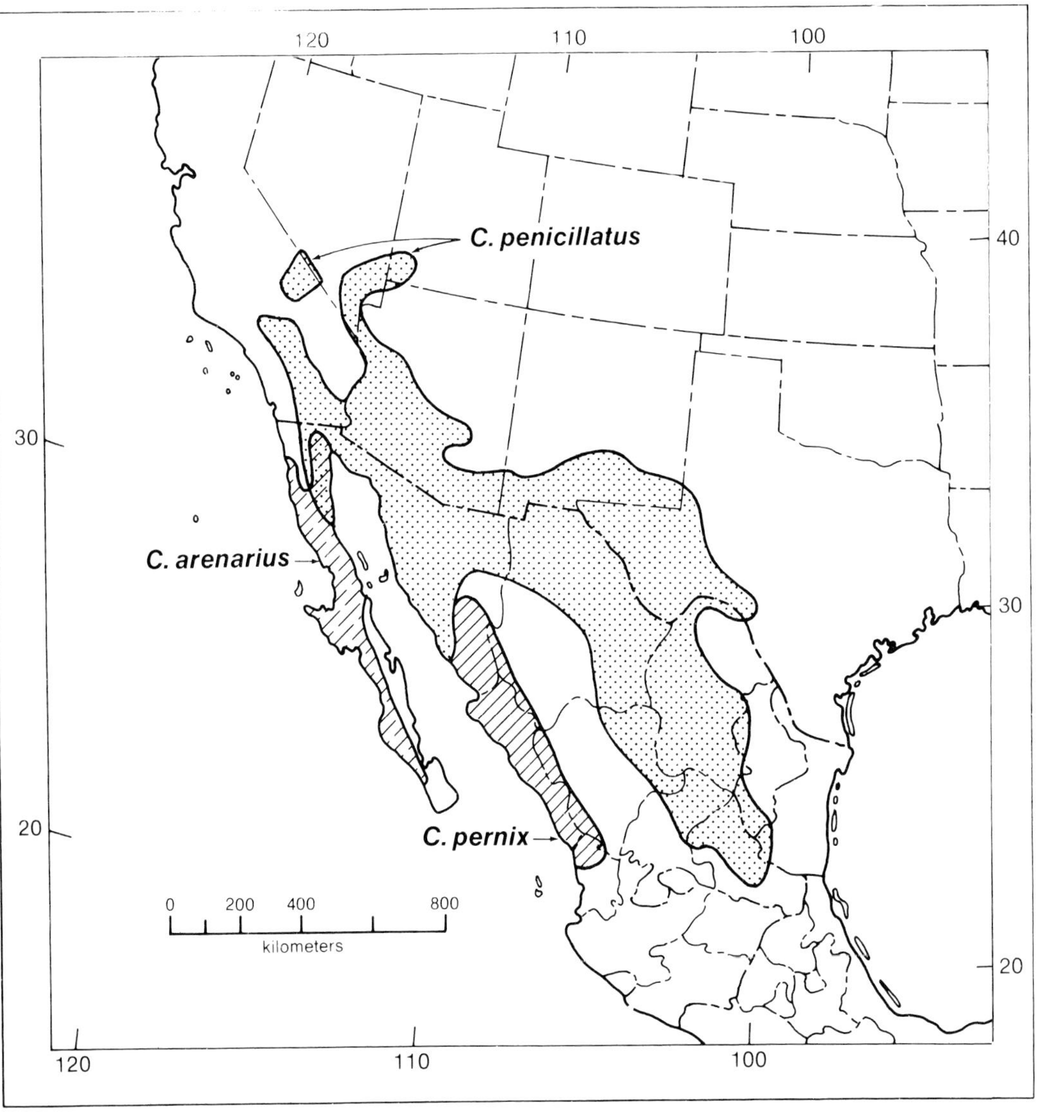

FIG. 10.—Distributions of three species of *Chaetodipus*.

zweig and Winakur, 1969) and annual precipitation (Brown, 1973, 1975; Brown et al., 1979; Hafner, 1977), which are two of the factors that strongly influence seed production and vegetation.

Every *Chaetodipus* species shares at least some of its range with that of at least one other member of the genus, which makes this the only heteromyid genus that does not have any totally allopatric forms. The three monotypic species have restricted distributions compared to the polytypic forms. *C. lineatus* is confined to San Luis Potosi and Zacatecas, Mexico, in the Chihuahuan Desert Province, *C. artus* occurs along the coast of the Gulf of California in the American Desert and Sinaloa Coast provinces, and *C. goldmani* occurs in the Mexican states of Chihuahua, Sinaloa, and Sonora in the American Desert as well as the Mexican Highland Shrub Steppe and Sinaloa Coast provinces (Fig. 9).

The polytypic forms include seven species in the small body size category and four

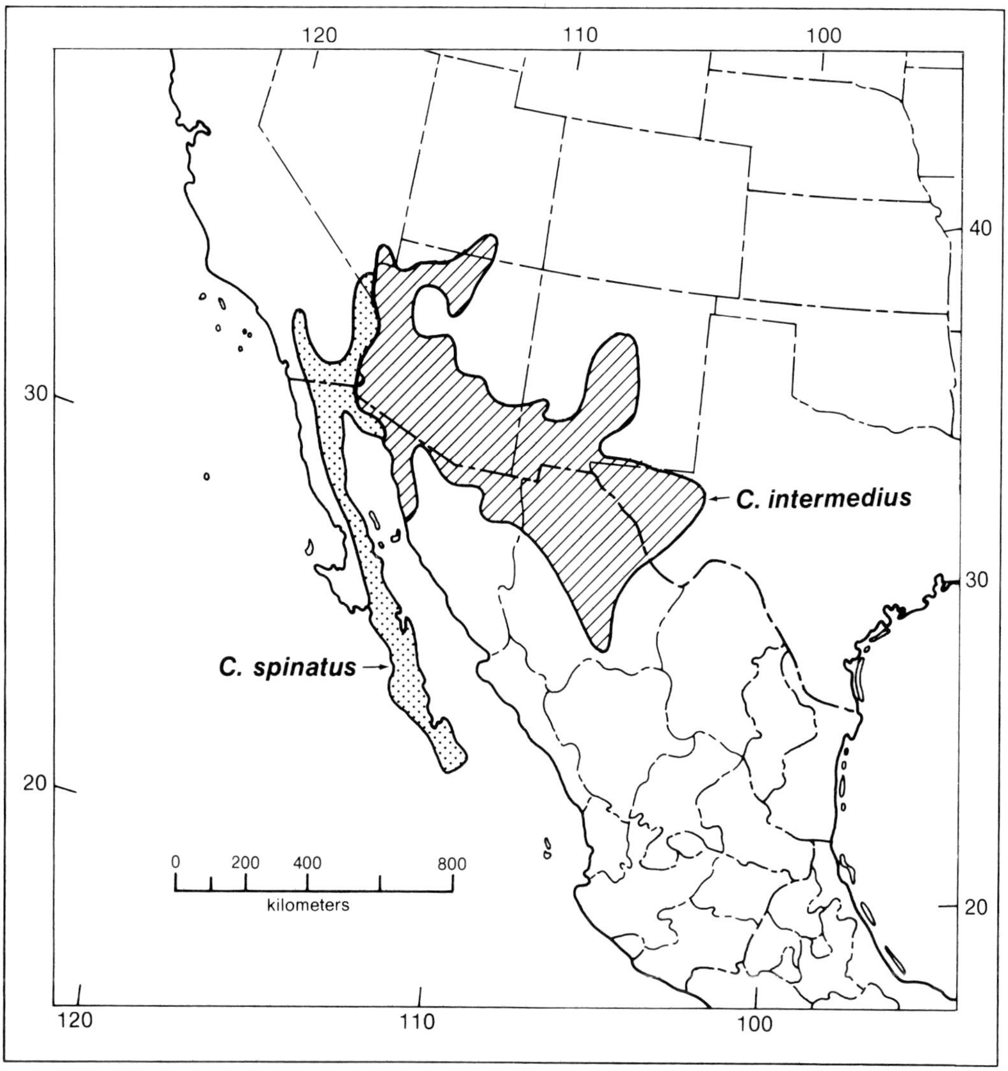

FIG. 11.—Distributions of *Chaetodipus intermedius* and *C. spinatus.*

medium-sized species. The small bodied forms include three species (*C. arenarius, C. fallax,* and *C. spinatus*) with distributions confined to the Baja California/southern California region. *C. fallax* occurs in the California grassland, American Desert, and Baja California provinces of southern California and northern Baja (Fig. 9). *C. arenarius* is restricted to the southern extension of the California grassland and Baja California provinces on the Baja Peninsula (Fig. 10). The range of *C. spinatus* (Fig. 11) overlaps that of *C. arenarius* (Fig. 10) throughout most of the Baja Peninsula, but the former occurs farther north into southern California.

C. intermedius and *C. penicillatus* are the two most widely distributed species in the small size category. Both have similar distributions that encompass southern California, Arizona, New Mexico, northern Sonora, Chihuahua, and Trans-Pecos Texas (Figs. 10, 11). The other two polytypic species in the small size category (*C. nelsoni* and *C. pernix*) both have relatively restricted distributions. *C. nelsoni* is confined

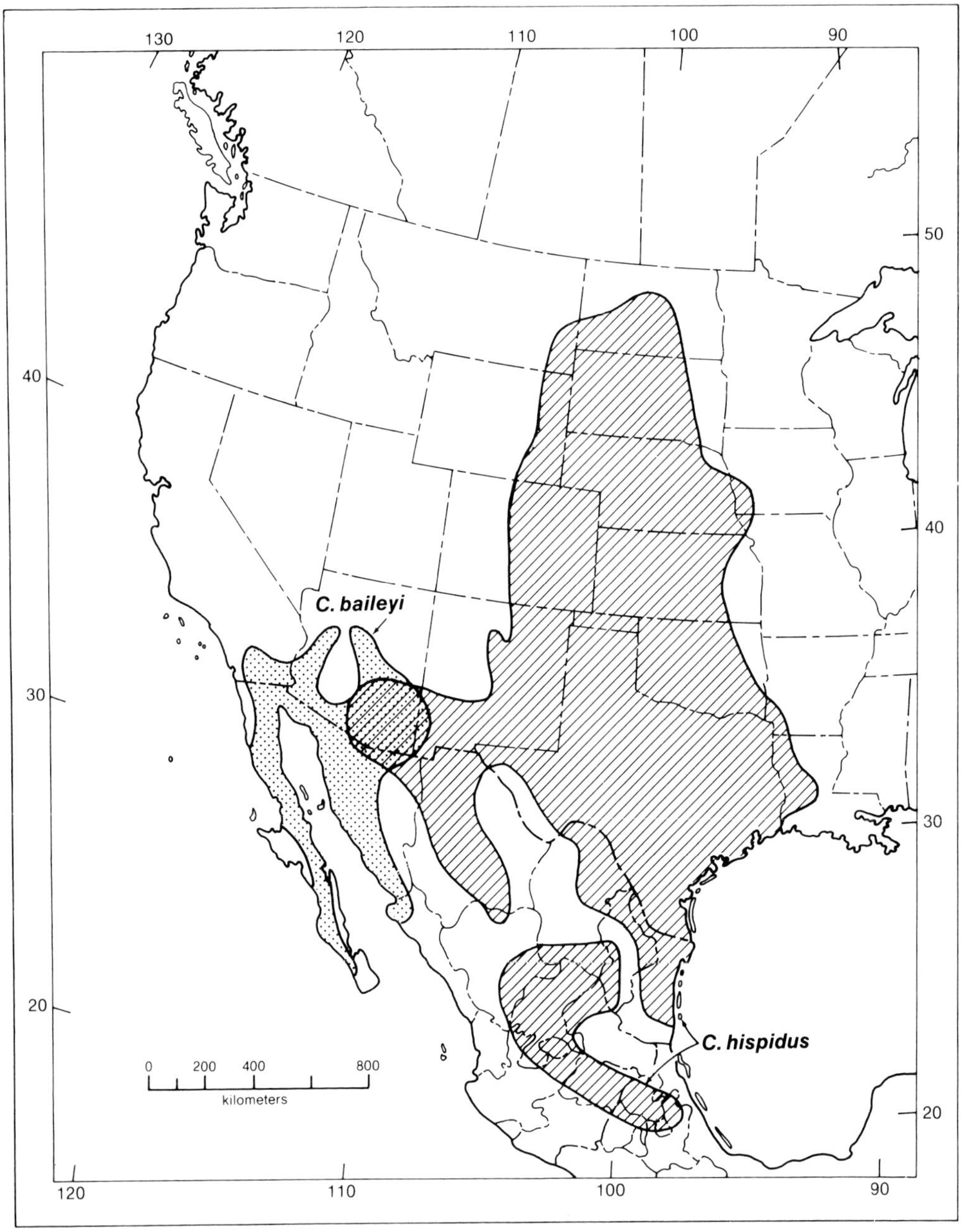

FIG. 12.—Distributions of *Chaetodipus baileyi* and *C. hispidus.*

mainly to the Chihuahuan Desert Province, occurring from southern New Mexico south through Trans-Pecos Texas and into Jalisco (Fig. 9). *C. pernix* (Fig. 10) has a distribution nearly identical to that of *C. artus* (Fig. 9), occurring in the Sinaloa Coast Province along the Pacific Coast of Mexico and barely entering the American Desert Province in southern Sonora.

The four medium-sized, polytypic species (*C. baileyi, C. hisidus, C. californicus* and *C. formosus*) typically have broader distributions than the small-sized polytypic forms. *C. baileyi* occurs throughout most of the Baja peninsula, southwest California, southern Arizona, western New Mexico and then south through Sonora into northern Sinaloa (Fig. 12). *C. hispidus* is distributed in two disjunct populations that encompass 14 ecoregion provinces (Fig. 12). It ranges farther north (North Dakota) and east (Louisiana) than any other species in the genus. *C. californicus* occurs in eastern and central California and south into Baja California (Fig. 13). This species is primarily distributed in the California Grassland Province, although it also occurs in the California Chaparral and American Desert provinces. *C. formosus* is restricted to arid regions in western California, Nevada, Utah, Arizona, and Baja California (Fig. 13). The majority of its range is in the Intermountain Sagebrush and American Desert provinces, but it also occurs in the Colorado Plateau, California Chaparral, and Baja California provinces. Although substantial distributional overlap is evident among the ranges of most chaetodipine species, there are several examples where species of similar size exhibit parapatric or nearly parapatric distributions. These have been depicted by representing the ranges of similar sized species with complementary distributions in Figs. 10–13 (three small species in Fig. 10; two small species in Fig. 11; two medium-sized species in Fig. 12, and two medium-sized species in Fig. 13).

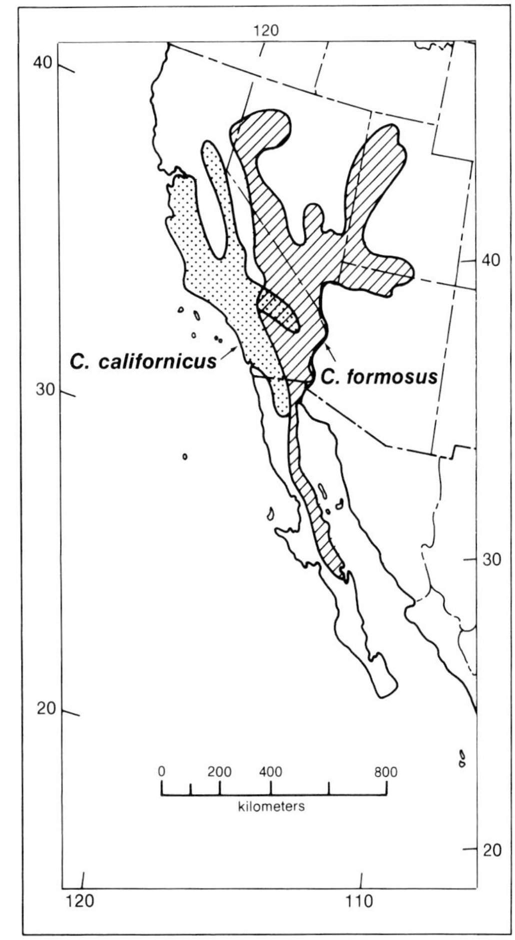

FIG. 13.—Distributions of *Chaetodipus californicus* and *C. formosus.*

Dipodomys

Dipodomys is the most speciose genus in the family with 21 species and 114 recognized subspecies. Twelve of the 21 species occur in California. Thirteen species are polytypic; the remaining eight, including two insular forms, are monotypic. In comparison to other heteromyids, most kangaroo rats are large in body size (12 species). However, the group also includes seven medium-sized and two very large species (Table 1).

The geographic range of kangaroo rats extends from Saskatchewan and Alberta, Canada, south to Oaxaca, Mexico, and from the Pacific Coast eastward through the Great Plains of the Dakotas, Kansas, Nebraska, and Oklahoma. Within this broad and diverse area, these animals occur primarily in the Humid Temperate Domain (15 species) and the Dry Domain (13 species) and only rarely in the Humid Tropical Domain (1 species). Kangaroo rats are found in 24 ecoregion provinces, but they are restricted mainly to sandy soils in arid and semi-arid areas within these regions.

The seven kangaroo rats of medium size (*D. insularis, D. margaritae, D. merriami, D. nitratoides, D. ordii, D. phillipsii* and *D.*

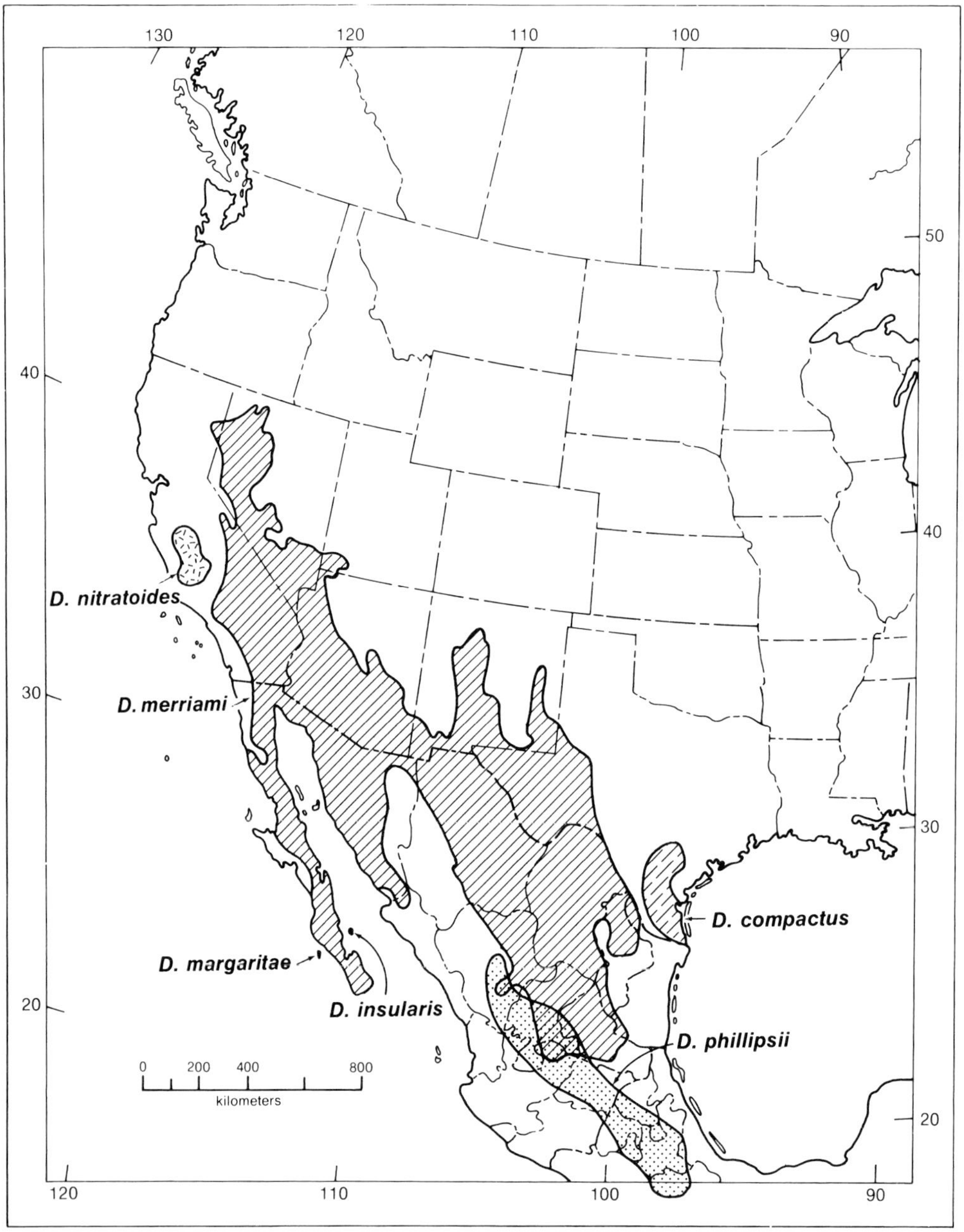

FIG. 14.—Distributions of three mainland and two insular species of *Dipodomys*.

compactus) include two insular forms and five mainland species (Figs. 14, 15). The two insular taxa are known from islands off Baja California, *D. insularis* from San José Island in the Gulf of California and *D. margaritae* from Santa Margarita Island off the Pacific coast of the Baja peninsula (Fig. 14).

The most widely distributed kangaroo rats are *D. merriami* and *D. ordii*. *D. merriami* ranges from northern Nevada throughout most of the southwestern United States and Baja California, east to Trans-Pecos Texas, and south to Aguascalientes, Mexico (Fig. 14). This species is found in 12 ecoregion

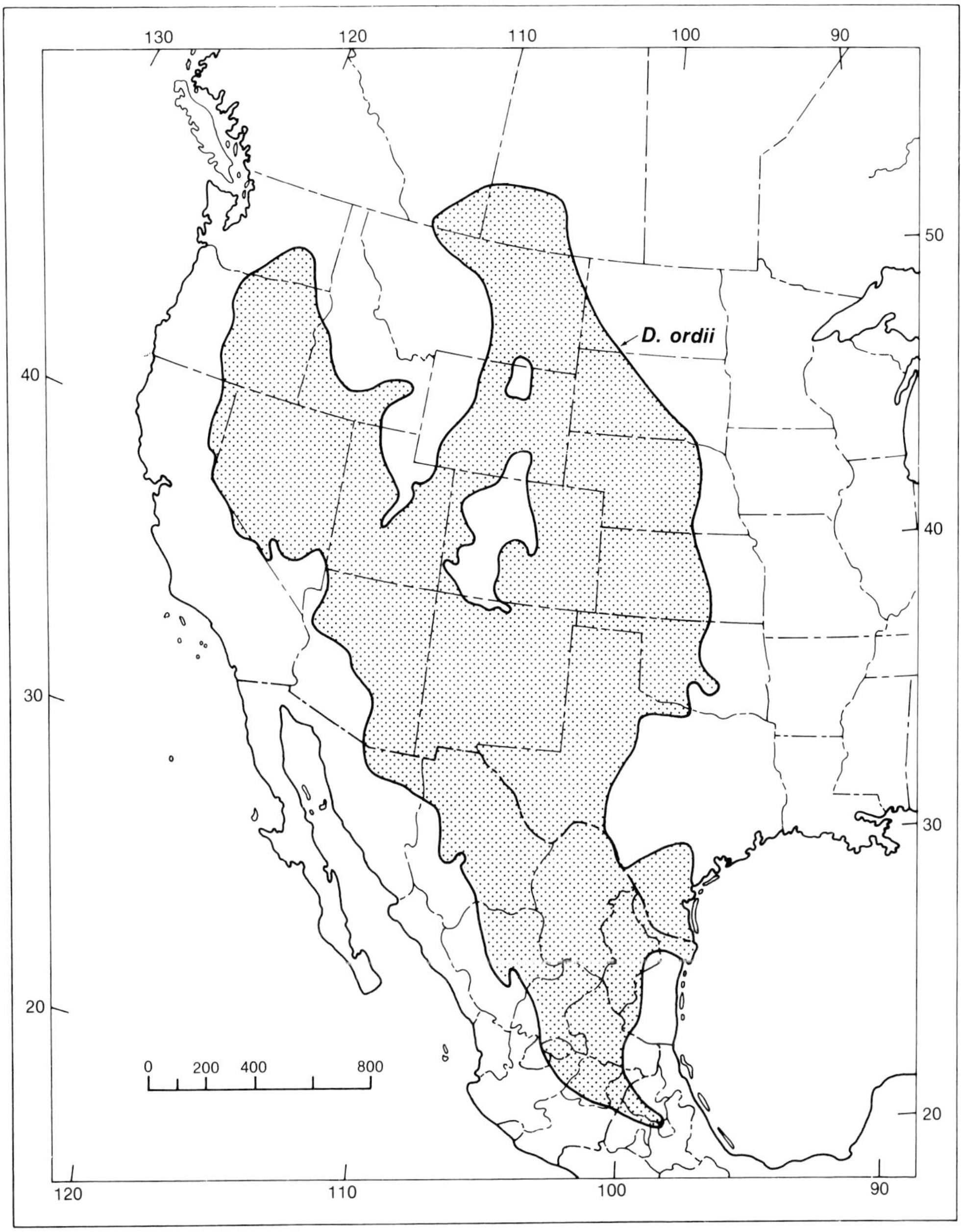

FIG. 15.—Distribution of *Dipodomys ordii.*

provinces within the Dry Domain and one province in the Humid Temperate Domain. *D. merriami* apparently prefers open environments (Bartholomew and Caswell, 1951; Wondolleck, 1978), although Wondolleck (1978) suggests it does not significantly favor any single microhabitat. Many factors seem to influence its distribution. Brown (1973) and Brown and Lieberman (1973) suggest that the Sierra Nevada Mountains form an effective dispersal barrier to *D. merriami* as well as a number of other heteromyids in the Great Basin Desert area. Additionally, Dawson (1955) has shown that

D. merriami has little ability to regulate its body temperature which may explain why the northern extent of its range coincides approximately with the 30°F (−1.1°C) isotherm for average January temperatures (Reynolds, 1958). Huey (1951) suggested that substrate is another factor influencing the distribution of *D. merriami* below 4,500 feet (1,370 m) in western North America. The species often is excluded from areas with very rocky ground cover, heavy clay, and densely packed soils because of difficulty in digging burrow systems (Hall, 1946; Huey, 1951; Reynolds, 1958).

D. ordii has the broadest distribution of any heteromyid, ranging from Saskatchewan and Alberta, south throughout most of the central and western United States, and through the Mexican Tableland to Hidalgo (Fig. 15). The species occurs in all three major Domains of North America that contain heteromyids and in 15 ecoregion provinces. Like many other kangaroo rats, this species is usually associated with sand dune areas and rarely occurs on hard or gravelly soils. Where *D. ordii* occurs sympatrically with *D. merriami,* the latter typically inhabits gravelly or hard soils and the former occupies sandy areas (Davis, 1966).

Three of the medium-sized kangaroo rats have restricted distributions in widely separated regions of North America. *D. nitratoides* occupies the arid grasslands and scrub plains of the California Chaparral and California Grassland provinces in the southern San Joaquin Valley of central California (Fig. 14). The Tehachapi Mountains in central California act as a physical barrier between *D. nitratoides* and *D. merriami* (Friesen, 1979). *D. phillipsii,* which is restricted to Mexico from Durango south to Oaxaca, occurs in two provinces of the Dry Domain (Mexican Highland Shrub Steppe and the Chihuahuan Desert) and two provinces of the Humid Tropical Domain (Sierra Madre Del Sur and the Central Mexican). *D. phillipsii* is the only kangaroo rat that occurs in the Humid Tropical Domain. *D. compactus* (Fig. 14) is restricted to the Prairie Brushland Province of south Texas and the barrier islands off Texas and Mexico (Baumgardner and Schmidly, 1981, 1985).

There are 12 species of large body size in this genus (Table 1). Five of these (*D. agilis, D. californicus, D. elephantinus, D. microps,* and *D. venustus*) are found in desert and chaparral habitats in the western United States and Baja California, and their geographic ranges are almost completely nonoverlapping (Fig. 16). *D. agilis* ranges from south-central California south throughout much of Baja California in the California Chaparral and the Baja California provinces and to a lesser extent in the American Desert Province. *D. californicus* is restricted to northern California and southern Oregon in the California Chaparral, California Grassland, Pacific Forest, and Sierra Forest provinces. *D. elephantinus* and *D. venustus* both have restricted ranges in central California. *D. venustus* occurs from San Francisco south along the Pacific coast for approximately 300 miles (500 km) in the Pacific Forest and California Chaparral provinces. *D. elephantinus* is found only in San Benito and Monterey counties in the California Chaparral Province. *D. venusuts* and *D. elephantinus* may occur in parapatry on the Monterey Peninsula (Friesen, 1979; Grinnell, 1922). *D. microps* occurs mainly in the Intermountain Sagebrush Province of southern Oregon, western Idaho, Nevada, western Utah, and south-central California as well as portions of the Colorado Plateau and American Desert provinces.

The range of *D. nelsoni* encompasses the Chihuahuan Desert Province in Chihuahua, Coahuila, Nuevo Leon, and San Luis Potosi, and shares an extensive border with *D. spectabilis* along both its northern and southern boundaries (Fig. 17). The latter species is known primarily from the American Desert and Chihuahuan Desert provinces of Arizona, Sonora, New Mexico, Texas, and Chihuahua, with a disjunct population in Central Mexico.

The five remaining kangaroo rats in the large body size category each have restricted

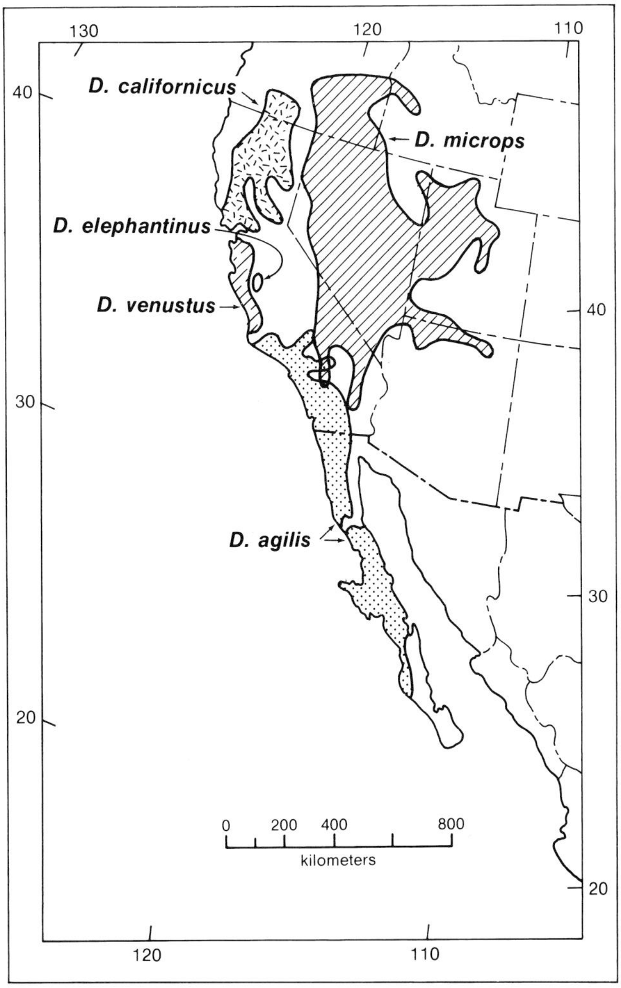

FIG. 16.—Distributions of four large species of *Dipodomys*.

ranges and are found in desert and arid grassland habitats in the southwestern United States and northern Mexico (Fig. 17). *D. elator* is limited to north Texas and southwestern Oklahoma in the Prairie Parkland and Prairie Brushland provinces. The preference of this species for habitats characterized by clay soils with sparse vegetation is unusual among kangaroo rats (Dalquest and Collier, 1964), and this could be an important factor accounting for its restricted distribution. *D. gravipes* occurs only along the western coast of Baja California in the California Chaparral Province. *D. panamintinus* occurs along the Sierra Nevada and San Gabriel Mountain ranges of western Nevada and south through central California with a disjunct population on the California-southern Nevada border. To the south, the Tehachapi Mountains form an effective barrier between this species and *D. heermanni* (Friesen, 1979). *D. heermanni* occurs along the Pacific coast and in the San Joaquin Valley from the San Francisco area south to Santa Barbara County in the California Grassland and California Chaparral provinces. *D. stephensi* is restricted to the California Chaparral Province of southern California. This species is sympatric over part of its range with three other species of

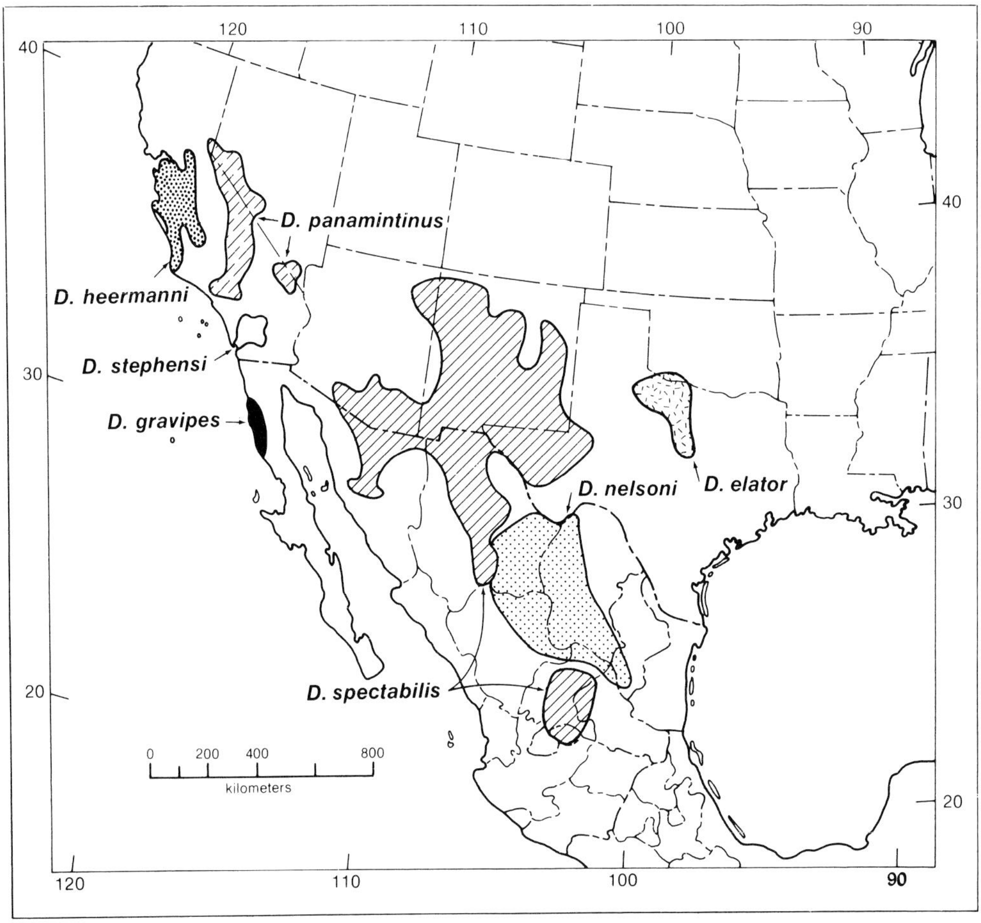

FIG. 17.—Distributions of seven large species of *Dipodomys.*

Dipodomys, namely the medium-sized *D. merriami,* the large *D. agilis,* and the very large *D. deserti.*

The two largest kangaroo rats (*D. deserti* and *D. ingens*) also occur in the southwest (Fig. 18). *D. deserti* ranges from northern Nevada into southern California and Arizona to Sonora and Baja California where it occurs in the Intermountain Sagebrush and American Desert provinces and to a lesser degree in the California Chaparral Province. In Nevada this species is found in low, hot areas along the eastern base of the Sierra Nevada, and it is always associated with wind-drifted sand not less than 20 inches (50.8 cm) deep (Hall, 1946). In California, it is found extensively in the Colorado and Mohave deserts through Death Valley and Owens Valley at an altitudinal range of −200 (−61 m) to 3,900 feet (1,189 m) (Grinnell, 1922). The ranges of six large kangaroo rats (*D. heermanni, D. californicus, D. agilis, D. spectabilis, D. gravipes* and *D. stephensi*) border that of *D. deserti.* However, other large kangaroo rats (*D. panamintinus* and *D. microps*) are widely sympatric with this species. *D. ingens,* which is the second largest species in the genus, is restricted to a narrow semi-arid strip along the southwestern border of the San Joaquin Valley in the California Grassland Province of central California. The altitudinal range of this species is from 500–2,500 feet (152–762 m) (Grinnell, 1922).

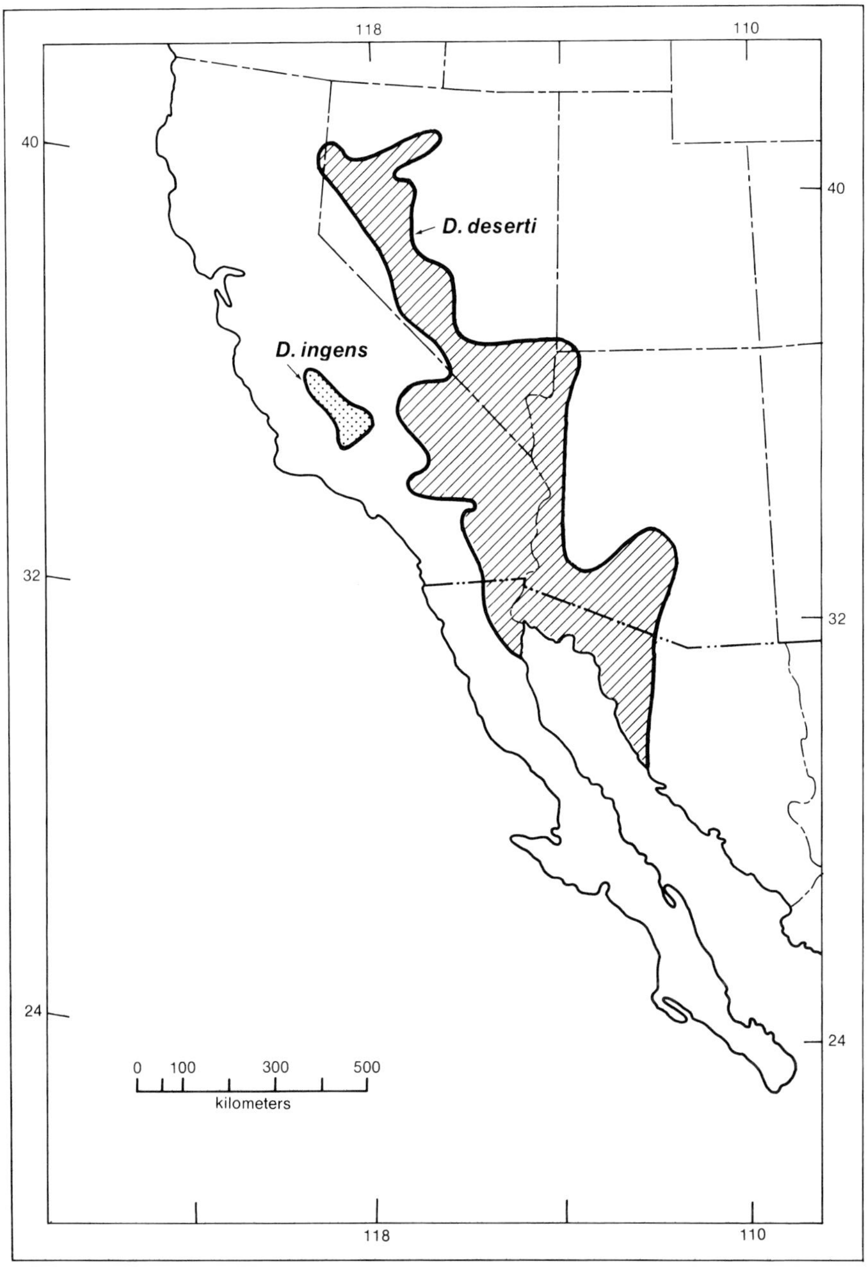

FIG. 18.—Distributions of the very large body size *Dipodomys*.

Summary of Ecological Zoogeography

The family Heteromyidae is distributed from southern Canada throughout the United States west of the Mississippi River, through Mexico and Central America and into northern South America. The greatest diversity in the family is in the desert, grassland, and chaparral habitats of the Dry Domain and Humid Temperate Domain in the southwestern United States and northern

Mexico. Within the broad and diverse areas inhabited by these rodents, no single ecological factor seems to have an overriding effect in defining the present distributions of members of this family but a few generalizations are evident. Edaphic factors such as substrate structure and soil moisture affect vegetative composition and density which, in turn, influence the habitat selection and local distribution of many heteromyids. Additionally, there is much circumstantial evidence that competition limits the geographic ranges of several species. There are numerous examples of ecologically similar, closely related species that occupy adjacent but non-overlapping geographic ranges. Examples of this sort are especially common in the genera *Chaetodipus, Dipodomys* and *Perognathus.*

The two heteromyid genera that comprise the subfamily Heteromyinae demonstrate differing degrees of affinity for neotropical habitats. *Liomys* occupies dry to arid habitats, whereas *Heteromys* tends to replace *Liomys* where drier habitats give way to moister cloud and rain forest.

The subfamily Dipodomyinae which contains only one genus, *Dipodomys,* the most speciose genus in the family. This genus is distributed throughout grasslands, deserts and chaparral from southern Canada to central Mexico. Kangaroo rats are found in many different ecological areas, but they are primarily restricted to sandy soils in arid and semi-arid habitats.

The subfamily Perognathinae, which includes the genera *Chaetodipus, Microdipodops,* and *Perognathus,* inhabits diverse ecological and geographic areas. *Chaetodipus* is a xeric adapted genus with its greatest diversity in the deserts of southern California. The ranges of all members of this genus extend into at least one of the major desert regions of North America. Kangaroo mice (*Microdipodops*), occupy the smallest geographic range of all the heteromyid genera, are restricted to the Great Basin Desert of western North America. Members of the genus *Perognathus* have broad tolerances for various soil and vegetation types and are distributed from southern Canada throughout most of the western and central United States and southward into Puebla, Mexico. Although common in both grasslands and deserts, these rodents are usually associated with sandy soils and sparse vegetation.

Ecological parameters are undoubtedly important in defining extant heteromyid distributions, but historical factors discussed in the next section must not be overlooked in evaluating the factors that have shaped the present distributions of these taxa.

Historical Biogeography

Historical aspects of heteromyid biogeography are treated herein as occurring in five theaters, (1) the northern neotropical, (2) the Great Plains, (3) the Great Basin, (4) the southwestern deserts and (5) central California. These are regions of high species richness in modern heteromyid communities, and they may well represent centers of heteromyid evolution. We acknowledge that evolution in one theater is not necessarily independent of that in other theaters. Additionally, each of these theaters is characterized by different geological histories and by different habitat types which have responded differently to changes in climate and sea level during the late Tertiary and Quaternary Periods. The effects of geological events and of changes in climatic and sea level on heteromyid distributions are presented below separately for each of these theaters. Scenarios of historical biogeography are developed for at least one species or species group within each theater.

Neotropical.—The two heteromyid genera that comprise the subfamily Heteromyinae demonstrate differing degrees of affinities for neotropical habitats. *Liomys* occupies dry to arid habitats, whereas *Het-*

eromys tends to replace *Liomys* where drier habitats give way to moister cloud and rain forests.

The fossil record of vertebrates in Central America is generally poor, and that of *Heteromys* is non-existent (Kurtén and Anderson, 1980; Savage and Russell, 1983). The history of *Heteromys* in South America is undoubtedly shorter than in Central America. The North and South American continents were separated from each other from the breakup of Pangaea during the early to middle Mesozoic (about 140 million years before present, mybp) until the Pliocene reunification of these continents. Rejoining of North and South America was via geological closure of the Bolivar Trough, a waterway that connected the Caribbean Sea with the Pacific Ocean, about 3 mybp (Webb and Marshall, 1982). This closure joined Panama with Colombia to provide a continuous mesic land connection between the two long-separated continents. The ensuing two-way traffic of flora and fauna between north and south constituted the Great America Interchange (Webb, 1977, 1978, 1985). The Heteromyidae, represented by *Heteromys* (or ancestors), was one of 16 mammalian families native to North America that successfully invaded South America beginning as early as 3 mypb (Simpson, 1980). Solbrig (1976) noted that rainforest predominated in northern South America throughout the Tertiary. It seems, therefore, that habitat suitable for *Heteromys* was probably available in South America well before access was provided by the Panamanian land bridge.

A more refined understanding of climatic and vegetational change over the last 40,000 years is available for tropical Mexico (Toledo, 1982). Extrapolation of this understanding into Central and northern South America may be useful in developing a biogeographic scenario for the genus *Heteromys* from the late Pleistocene (Wisconsinan glaciation) to present.

Toledo (1982, largely from Heine, 1973) summarized the major climatic changes in Mexico from 40,000 ybp to present. Seven major periods are evident: from 40,000 to 25,000 ybp, the climate in tropical Mexico was cold and wet. The next 5,000 years (25,000–20,000 ybp) were decidedly warm and wet. The latest Wisconsinan glaciation (20,000–12,000 ybp) was markedly cold and dry. The next 3,000 years (until 9,000 ybp) involved abrupt oscillations from cold and extremely wet to hot and dry and finally back to cold and wet. From 9,000 to 2,000 ybp a hot and dry climate prevailed. The present climatic regime was then achieved after a period of slightly lower temperature and slightly higher rainfall. Toledo (1982) concluded that temperatures during the coldest phases of the Pleistocene were at least 5 degrees Celsius lower than present.

Corresponding to these climatic shifts were changes in vegetational distribution. Toledo (1982:98) concluded that, over the last 40,000 years, the now dominant tropical rain forests of southern Mexico were disrupted, reduced in extent, and displaced to lower altitudes and/or latitudes. The particular habitats that replaced the tropical rain forests depended on the prevailing climatic regime as follows: during cold-dry times (20,000–12,000 ybp), pine and oak forests descended from higher altitudes to dominate the lowlands. Likewise, cloud forests, and probably oak forests, replaced tropical rain forests during cold-wet times which prevailed from 40,000 to 25,000 ybp, around 12,000 ybp, around 9,000 ybp, and within the last two millennia. During the warm-dry conditions around 11,000 ybp, from 9,000 to 2,000 ybp, and at present, tropical rain forests yielded to a variety of habitats including tropical deciduous and semi-deciduous forests, thorn forests, and savannas.

Even during the most extreme climatic fluctuations, tropical rain forests were not extirpated. Toledo (1982:104) identified eight refugia wherein tropical rain forests and their associated fauna persisted (Fig.

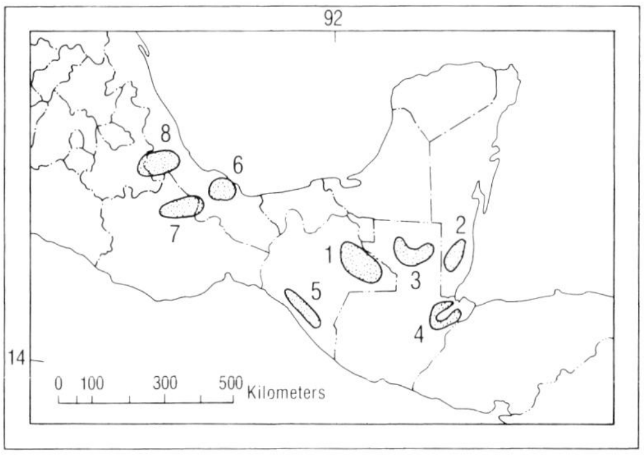

FIG. 19.—Primary (1–5) and secondary (6–8) Wisconsin refuges for tropical rain forest species in the northern portions of tropical Central America (after Toledo, 1982). Refuges in Chiapas include Region Lacandona (1) and Region del Soconusco (5). Refuge (2) is in the southernmost Maya Mountains of Belize. Two Guatemalan refuges are in the regions around Lago Peten-Itza (3) and Lago Izabel (4). The secondary refuges are Los Tuxtlas (6), Sierra de Juarez (7), and Cordoba (8).

19). His primary refugia were safe havens where both temperature and precipitation were unsuitable for rain forests; in secondary refugia, rain forests were protected from lowering of temperature or of precipitation, but not both. During this 40,000 year interval the Mexican (and presumably other neotropical) tropical rain forests probably experienced their least geographic extent between 20,000 to 12,000 ybp when climate was coldest and driest. It seems reasonable, therefore, that the rain forest-dwelling *Heteromys* probably were restricted to Toledo's five primary refugia (Fig. 19) at this time. Present subspecies (or even species) patterns could well be the result of divergence during this latest Wisconsinan isolation.

Glacial periods are times of low sea level because much ocean water is sequestered into glaciers. Hence, during glacial times, areas which were previously separated by shallow sea waters may become connected by emergent land. Such a scenario is useful in explaining the migration of *H. anomalus* from the Venezuelan mainland (if indeed this was the source) to the island of Trinidad, some 30 or fewer km distant. The approximately 100 m drop in sea level during the Wisconsinan glacial was sufficient to connect Trinidad with the Peninsula de Paria of Venezuela. Measured depths in the waters on the mainland sides of Trinidad are 16 m and 64 m (Brittanica, 1982). However, one would expect more or less continuous tropical rain forest habitat along the dispersal corridor to be necessary if colonization were to be achieved by dispersal. As noted above, glacial periods were times of tropical rain forest contraction such that *Heteromys* habitat was not continuous nor widespread along coastal lowlands. Fortunately for this scenario, numerous studies (summarized by De Granville, 1982) have identified tropical rain forest refugia occupying both the Peninsula de Paria and Trinidad during this glacial interval (Haffer, 1969; Prance, 1973; Brown, 1976). The presence of tropical rain forests coupled with the emergent land connection suggests that *H. anomalus* could have reached Trinidad via overland routes during the Wisconsinan or one (or more) of the previous glacials of the Quaternary (Eschelman and Morgan, 1985). Trinidad's location near the mouth of the Rio Orinoco suggests that dispersal on floating mats of vegetation from more interior forests is a viable alternative dispersal mechanism. Only dated fossil materials will shed light on the probable time that *Heteromys* invaded Trinidad and perhaps other Antillean islands.

As noted above, species of *Liomys* demonstrate affinities for various semi-arid to arid habitats of the northern Neotropics and the southern Nearctic (Genoways, 1973). From Toledo's (1982) discussion of late Quaternary Mexican vegetation, it is apparent that habitats (tropical deciduous and semi-deciduous forests, thorn forests, savannas) suitable to *Liomys* experienced at least three periods of expansion during the warm-dry intervals of the last 40,000 years: around 11,000 ybp, from 9,000 to 2,000 ybp, and at present (within the last few hundred years). Southward expansion of the range of *Liomys* probably coincided with contraction of suitable *Heteromys* habitat (tropical rain forests). Genoways (1973) suggested that the early evolution of *Liomys* centered on the southern Mexican Plateau in conjunction with the Madro-Tertiary Geoflora.

Quaternary evolution of *Liomys* may be viewed in three parts that correspond with the taxonomic recognition of species groups (Genoways, 1973). Evolution of *L. irroratus* from ancestral stock probably occurred in situ on the Mexican Plateau. Probably during the early Pleistocene, ancestral *Liomys* ranged more or less continuously along the Pacific coastline from about Nayarit, Mexico, into Costa Rica, a region occupied by the Madro-Tertiary Geoflora (see fig. 66B in Genoways, 1973). Subsequent evolution of the *salvini-adspersus* group and the *pictus-spectabilis* group probably took place to the south and north, respectively, of the Isthmus of Tehuantepec of Mexico. During glacial intervals, semi-arid environments suitable for *Liomys* were more extensive in the lowland of this isthmus. Conversely, moister conditions of inter-glacial intervals caused interruption of the semi-arid corridor and apparently led to isolation of the *salvini-adspersus* and *pictus-spectabilis* species groups. During these same (or later) glacial periods, representatives of the *salvini-adspersus* group dispersed farther southward into the savannas of the Pacific coast of Panama. Later isolation into *L. adspersus* of the Panamanian savannas and *L. salvini* of more northern Central America occurred during an ensuing inter-glacial when rising sea levels and the return of warmer and moister conditions interrupted gene flow between these populations.

Simultaneously, to the north of the Isthmus of Tehuantepec, the *pictus-spectabilis* group diverged into *L. spectabilis* and *L. pictus.* Apparently, *L. spectabilis* stock became geographically isolated within the interior drainages of Jalisco for sufficient time to become genetically isolated from *L. pictus* before *pictus* reinvaded the range of *spectabilis.* The situation and conditions under which this speciation event occurred have not yet been established, but it likely involved climatic fluctuations and regional vulcanism during the Wisconsinan (Genoways, 1973).

Great Plains.—In the Great Plains, particularly in the more northern regions, the heteromyid fauna experienced the direct effects of glaciation. At its maximum extent around 19,000 ybp, the Laurentide ice sheet of the late Wisconsinan covered much of the northern Great Plains to a level traced approximately by the Missouri and Ohio rivers (Brown and Gibson, 1983). Glaciated areas of the Great Plains now occupied by various heteromyids include all or parts of North and South Dakota, Montana, Minnesota, Iowa, Alberta, Saskatchewan, and Manitoba. Glaciers thereby usurped vast regions which formerly supported habitats suitable to various heteromyids during earlier interglacials. Boreal habitats (coniferous forests, tundra) preceded the glacial fronts into regions of lower latitudes and altitudes whose climates were moderated in part by the influence of nearby glaciers. Simultaneously, the steppe and savanna habitats were displaced southward. Hoffmann and Jones (1970) observed that, during the late Wisconsinan, steppe or savanna areas probably were restricted east of the Rocky Mountains in eastern Colorado and western Kansas. Modern Great Plains heteromyids affected by such shifts of steppe included *P. fasciatus, P. flavescens,* and *C. hispidus.*

Glaciers retreated with the post-glacial warming of climate, thereby permitting lower latitude floras and faunas to (re)invade higher latitudes and altitudes. One source of post-glacial invasion was the Great Basin from which *P. parvus* penetrated a short distance into Wyoming (Hoffmann and Jones, 1970). During the Wisconsinan, a barrier of boreal forest evidently existed between the Great Basin and the Wyoming Basin, thereby isolating the Great Basin from the northern Great Plains. Glacial retreat (re)established this corridor and facilitated the dispersal of *P. parvus* in areas of the Great Plains where sagebrush (*Artemesia tridentata*), a major constituent of Great Basin habitats, has also entered the northern plains. The eastward advance of *P. parvus* was checked at the Green River, perhaps due to competition with a congener, *P. fasciatus* (Hoffmann and Jones, 1970). Two other heteromyid species, *P. flavus* and *Dipodomys ordii,* (re)entered the Great Plains as another component of this post-glacial readjustment of ranges. These invaders from the southwest, which have affinities with the Chihuahuan and Sonoran deserts, attain their current northeastern distributional limits in the western Great Plains (Hoffmann and Jones, 1970).

Even within the last 10,000 to 12,000 years of post-Wisconsinan time, continued range adjustments are known for at least two Great Plains heteromyids. For example, *C. hispidus* occurred in the late Wisconsinan–early Holocene Crankshaft Cave fauna in east-central Missouri, a locality about 400 km east of the easternmost modern range for this species (Parmalee et al., 1969). Additionally, *P. flavus* apparently occurred farther east during the "Great Drought" of a generation ago than it does at present (Hoffmann and Jones, 1970). This scenario of expanding and contracting ranges of heteromyids in correspondence with alternating glacial advance and retreat undoubtedly was replayed numerous times during the Pleistocene and perhaps during even earlier times.

Williams (1978) proposed a scenario to account for the Quaternary evolution of the three Great Plains pocket mice of the *Perognathus fasciatus* species group. *P. fasciatus,* which is the most northerly distributed of this species group, is adapted to cooler, moister conditions. The other species of *fasciatus* group, *P. flavescens,* exists in warmer, drier climates to the south and east. *P. f. flavescens* occupies a range in the middle to southern Great Plains, whereas *P. f. apache* occurs in intermontane regions farther west. These taxa probably derived from an ancestral population that, during an interglacial period, ranged widely across the Great Plains and on intermontane plateaus. Glacial advance during the Wisconsinan or perhaps an earlier cycle obliterated northern portions of the range and eventually fragmented the ancestral population into two elements restricted to the southern plains and the Chihuahuan Plateau, respectively. An axis of mountains and plateaus coursing from eastern Chihuahua through Trans-Pecos Texas and into the southern Rocky Mountains was the obstacle that probably prevented genetic interchange. During the Wisconsinan glacial, *P. fasciatus* occupied the open yellow pine-sagebrush parkland that covered much of the southern plains (Wells, 1970); this species occupies such habitat in the northern plains today. The intermontane component to the west took refuge in drier situations which probably included the semi-arid grassland and pinyon-juniper woodland habitats in which several *P. flavescens* subspecies currently reside. Tracking vegetational changes, the two isolates later moved northward as the glacier retreated. A southern component of the western element apparently evolved in situ into extant races of *P. flavescens* by becoming adapted to increasingly warmer and drier conditions. *P. flavescens* then entered the southern plains as the Trans-Pecos montane barrier became ineffective. Interaction between the newly arrived *P. flavescens* and the resident *P. fasciatus* probably encouraged northward displacement of the latter. The other component of the western intermontane population evolved into various *P. flavescens* subspecies which followed suitable environmental conditions northward during this post-glacial. The opportunity for (probably) initial contact of western *P. flavescens* subspecies (*P. f. caryi* of Hall, 1981) and *P. fasciatus* has only recently occurred in the Uinta Basin of Utah (Williams, 1978).

Great Basin.—The metaphor of "upland islands" surrounded by intervening "lowland seas" has been employed in studies of evolution, biogeography, and natural history of many species on many continents, including studies addressing montane isolation of mammals in the Great Basin (Brown, 1978; Grayson, 1982). Almost invariably, such studies have focused upon species that occupy the uplands and ignored those of the lowlands. The heteromyids of the Great Basin generally are lowland occupants (see Brown, 1978, for lists of boreal species of mammals). In broad terms, the Quaternary range expansions and contractions of the lowland heteromyids may be thought of as the inverse of range shifts experienced by upland resident taxa.

Recent evaluations of plant macrofossils from packrat middens, pollen from intermontane lake cores, and vertebrate fossils have led to a relatively refined history of late Quaternary environments in the Great Basin, which was summarized by Thompson and Mead (1982), and is excerpted here. During the late Wisconsinan, montane plant communities occurred at much lower elevations than at present. Subalpine conifers (limber pine, *Pinus flexilis*; bristlecone pine, *P. longaeva*; Engelmann spruce, *Picea engelmanni*; common juniper, *Juniperus communis*) grew on rock outcrops and shallow soils of intermontane valleys. However, such woodlands apparently did not occupy the deeper alluvial and lacustrine sediments which, instead, supported steppe, meadow, and shrub communities much as today. Such a mosaic of habitats provided not only corridors (although somewhat discontinuous) connecting uplands but, or more importance to heteromyids, regional refugia for lowland species, even during the greatest extent of the Laurentide glacier. Meltwaters from the alpine glaciers accumulated in many of the intermontane basins to form pluvial lakes, several of which covered immense land areas (Lake Lahontan in Nevada; Lake Bonneville, whose much reduced modern remnant is Great Salt Lake). Warming and drying of the Holocene climate caused many changes. All pluvial lakes shrank in size and some disappeared altogether. Alpine communities were displaced to higher altitudes and latitudes. Their modern elevational and geographic ranges were achieved by about 8,000 ybp. Northward shifts of alpine communities were on the order of 5 to 6 degrees latitude (Wells, 1983). Shrub and steppe communities dispersed into most of the lowlands and, to a limited extent, upslope.

Undoubtedly, heteromyid populations tracked such glacial-interglacial contraction-expansion cycles of suitable lowland habitats. Detailed case histories of biogeography have been elucidated for few, if any, Great Basin heteromyids, despite the availability of considerable germane information regarding Quaternary vegetation, hydrology and geology of the region (Mehringer, 1977; Thompson and Mead, 1982; Wells, 1983).

The shifting distributions of kangaroo mice, however, offer a modern example that may be instructive in understanding longer-term biogeography of Great Basin heteromyids (Hafner, 1985). All subspecies of *Microdipodops pallidus* inhabit areas of fine, loose aeolian sands (Hall, 1941). Because the distribution of such sands is dynamic, populations of *M. pallidus* presumably are likewise mobile. Hafner (1985) recently described a new subspecies of kangaroo mice, *M. megacephalus atrirelictus,* from an isolated location in southern Idaho. This population resides in limited sands of an arroyo bottom in an area surrounded by otherwise unsuitable gravelly, hardpan soils. This disjunct population probably represents the relict of a formerly more widespread taxon whose requisite sandy substrates have been lost to aeolian and water erosion. Hence, habitats shifting in location due to physical factors within an interglacial (such as the current one) may well have been the most important agent in shaping the phylogenetic relationships among kangaroo mice.

Southwestern deserts.—The southwestern deserts are quite young, on the order of 2 million or fewer years (Axelrod, 1958; Webb, 1978). The Sonoran and Mohave deserts appeared during the Pleistocene, and an expansive Chihuahuan desert developed only as recently as the middle Pleistocene. By at least the late Pleistocene (early Rancholabrean), arid conditions prevailed over at least part of the Mexican Plateau (Mooser and Dalquest, 1975). These modern deserts probably derived from savannas as aridity increased (Webb, 1978). Heteromyid fossils (*Proheteromys*) thought to lie near the origin of Perognathinae are known from the Miocene (Homingforaian) Zia sand of New Mexico (Gawne, 1975).

Mammalogists traditionally have considered the southwestern deserts as the center of origin of the subfamily Dipodomyinae because that region currently supports the greatest diversity of kangaroo rats (Setzer, 1949; Lidicker, 1960). Indeed, some unspecialized, incipient kangaroo rats (*Cupidinimus* of southern California and Nebraska) inhabited the southwest as early as the upper Barstovian, middle Miocene, some 12 million years before these deserts evolved (Voorhies, 1975). Yet, *Eodipodomys celtiservator,* the earliest heteromyid known to possess several features (greatly reduced forelimbs and elongated hind limbs; markedly inflated auditory bullae; well developed dentine tracts on cheek teeth) associated with heteromyid adaptation to desert or steppe existence was found in the late Clarendonian, later Miocene (about 10 mybp) Ash Hollow formation of Nebraska (Voorhies, 1975). Although *Cupidinimus* is biostratigraphically younger than *Eodipodomys* by about 3 to 5 million years, Voorhies argued that the latter is closer to the ancestry of dipodomyines. If so, then saltatorial heteromyids probably first acquired their "desert adaptations" in the eastern Great Plains, a region then predominated by moist, lowland savanna, rather than in the desert southwest. Although early evolution of kangaroo rats may have occurred far east of their present center of distribution, these ancestors probably occupied the localized "deserts" which took the form of streamside accumulations of bare sand (Voorhies, 1975; see also Roth, 1984).

A more refined analysis of heteromyid biogeography may be feasible for the latest glacial-interglacial cycle due to the greater availability of vertebrate fossils and of botanical, climatic, and geological information for the last approximately 22,000 years. The southwestern deserts, although 1 to 2 million years old, have been dynamic systems whose geographic extents have varied tremendously. Wisconsinan vegetation often showed vast elevational and latitudinal displacements due largely to changes in temperature and precipitation regimes (Van Devender and Spaulding, 1979). For example, during the late Wisconsinan through its transition into the early Holocene (about 20,000–8,000 ybp), evergreen coniferous woodlands (primarily pinyon-juniper) extended downslope and tropicad to occupy most of the southwest (Wells, 1979). Wells and Hunziker (1976) noted that most of the Sonoran and probably all of the Chihuahuan and Mohave deserts then supported evergreen coniferous woodlands. Some of these woodlands might have extended to sea level in the northern Sonoran Desert and close to sea level in Death Valley. Additionally, this pinyon-juniper forest had not yet penetrated the central Great Basin from the adjacent Mohave Desert region (Wells, 1979).

During the glacial interval of woodland expansion, the more xeric and open habitats inhabited by modern heteromyids contracted considerably. Van Devender and Spaulding (1979) postulated that desert scrub communities of the Sonoran Desert persisted at elevations below about 300 to 400 m along the Colorado River. Wells and Hunziker (1976) developed a more severe climatic scenario in which desert refugia (probably occupied by arid-adapted heteromyids) occurred farther south in Mexico (central and southern Baja California; southern Sonora, Hidalgo, Puebla) and Guatemala (see fig. 1 in Wells, 1979).

The warmer, drier climate of the Holocene initiated a northward and upward retreat of evergreen woodland from lower elevations and lower latitudes which were then (re)invaded by desert-adapted communities (Van Devender and Spaulding, 1979). Creosote bush-white bursage communities dispersed into the Mohave and Sonoran lowlands. The formerly more expansive Joshua tree became restricted to higher elevations at the northern edges of the Mohave Desert. A grassland community occupying the Chihuahuan Desert during the middle Holocene was followed by the succulent desert scrub communities of today.

Very few detailed cases of historical biogeography have been elucidated for heteromyid species of southwestern deserts. One such study (Patton, 1969) addressed the late Quaternary history of *Perognathus goldmani,* a species which now occupies the thorn scrub and short-tree forests of the coastal plain of contiguous parts of Sonora, Sinaloa, and Chihuahua. Ranges of the six chromosomal races of *C. goldmani* exhibit contiguous allopatry with contact occurring between adjacent races along major rivers. Patton (1969, his fig. 12) developed a historical scenario which implicated rivers, a competitive sibling species (*C. artus*), and historical shifts in vegetation distribution as evolutionary agents resulting in the current race distributions. During pre-pluvial time (pre-Wisconsin glacial), the ranges of *C. goldmani* and *C. artus* overlapped only slightly. *C. artus* prefers the more mesic riparian and fully-developed short-tree forests in areas of contact with *C. goldmani.* With the cooling of climate, rich tropical deciduous forests now located near the base of the Sierra Madre and inhabited by *C. artus* expanded westward along the valleys of the Rios Yaqui and Fuerte. This encroachment of mesic habitat locally excluded *C. goldmani* and eventually split the distribution into disjunct northern and southern populations that evolved in isolation. These isolated populations probably dispersed as far along the riparian corridors as conditions permitted with continued evolution producing additional chromosomal races. Post-Wisconsinan re-expansion of *C. goldmani* into its former range tracked that of suitable xeric habitats as mesic habitats retreated eastward and achieved the present pattern of six chromosomal races.

Central valley of California.—California's central valley and the surrounding uplands host a heteromyid fauna whose great richness probably resulted, at least in part, from geological events of the late Tertiary and Quaternary Periods (Friesen, 1979; Martin and Schmidly, 1982; Smith, 1979). Friesen's (1979) exhaustive evaluation of the biogeography of kangaroo rats of California and adjacent regions represents a detailed case history. He identified numerous factors influencing the distributional history of *Dipodomys* and related forms including geomorphology, tectonic instability, sea level changes, and shifting distributions of suitable vegetational associations and soils in response to climatic changes during the Tertiary and Quaternary periods.

Fundamental to the historical biogeography of *Dipodomys* in this region is that the various California mountain ranges have been (and are) dynamic features whose elevations directly result from tectonic events. These mountain ranges have experienced repeated cycles of uplift and depression. Hence, some that now serve as dispersal barriers to kangaroo rats had much lower elevations (or did not even exist) as little as 1 or 2 mybp. Friesen (1979) identified the Tehachapi Mountains of southern California whose uplift occurred within the last 700,000 years as a crucial element in the later phases of *Dipodomys* biogeography.

The Tehachapi Mountains figure importantly in one of six corridors by which Friesen (1979) proposed that kangaroo rats might have crossed the Cascade-Sierra-Peninsular axis from the Great Basin and Mohave Desert into the Central Valley. Four valleys that now nearly cross the Tehachapi Mountains and that probably were drainages prior to the Quaternary uplift constitute the Tehachapi Valley System Corridor. Three other corridors traverse the southern end of the north-south montane axis; however, none of these avenues (Walker Pass-Kern River, Grapevine Canyon, and Cuyama River–Lockwood Valley) were usable since their middle Pleistocene uplift due to absence of suitable *Dipodomys* habitats. The Trans-Sierra Corridor of the central to south-central portion of the axis apparently allowed access through two passes (Tioga Pass and Walker Pass) which probably mark location of waterways that crossed the axis before the Pleistocene uplift of the Sierra Nevadas. Two major drainages, the Pit and Klamath rivers, now (and in the past) cross the northern portion of the axis. This, the Modoc-Cascade Corridor, operates even now to connect faunas of the Great Basin and Central Valley as evidenced by the presence of *Dipodomys californicus* in suitable habitats in these river valleys.

Changes in Tertiary and Quaternary sea levels also affected kangaroo rat distributions (Friesen, 1979). Until the middle Pleistocene, the San Joaquin Valley was open to the Pacific Ocean by at least one channel (Vallecitos channel along the modern Priest Valley). Such embayments filled with seawater during higher sea levels of interglacial times. Probably since the Eocene, mountains fringing the San Joaquin Valley were at least high enough to form an archipelago during ocean transgressions; the islands thus formed could have been occupied by isolated populations of kangaroo rats or their ancestors. During the Pliocene and Pleistocene, mountains surrounding the embayment (Tehachapi, Transverse, Peninsular, and Coastal Ranges and Sierra Nevadas) attained their more elevated positions such that the San Joaquin Valley was ringed by continuous uplands. During higher sea levels, kangaroo rats were displaced from valley floors into these uplands which probably served as dispersal routes from the Great Basin. With lower sea levels of glacial intervals, kangaroo rats of the uplands probably descended to colonize sediments of valley floors.

The Salton Trough provides another example of the impact of sea level changes on kangaroo rat distributions. The Gulf of California represents the southern portion of this trough which continues northward into southern California where it is occupied by the Salton Sea. As recently as the middle Pleistocene, the Gulf extended far northward into the Imperial Valley. Hence, much of the Salton Trough region was intermittently unsuitable for habitation by kangaroo rats.

Changes in sea level and precipitation patterns can also affect distributions of soils, a factor important in determining suitability of habitats for kangaroo rats. For example, typical habitat for *Dipodomys deserti* includes large tracts of sand dunes, a habitat not widespread in any California deserts. One source of this sand is the shorelines of ancient basin lakes, bodies of water whose extent correlated with sea level and precipitation. The vast alluvial fans of the many washes found throughout the Basin Ranges Province comprise another source of sands suitable to *D. deserti.* Fringing these sandy washes are areas of coarser sands occupied by other kangaroo rat species such as *D. merriami* and *D. ordii.* Changes in precipitation patterns affected the distribution of these sandy habitats, and, thereby, surely influenced the biogeography of these kangaroo rats.

Friesen's (1979) studies of geomorphology, sea level changes, climatic patterns and temporal shifts in vegetation and soils led him to propose several models explaining the relationships of *D. panamintinus, D. heermanni,* and *D. californicus* and their distributions in the Tehachapi Mountains. A series of these dispersal models centers on an ancestral population of kangaroo rats that arrived in an area (or evolved there) after the isolating barrier (Tehachapi Mountains) had become established. Friesen's (1979) complicated Pluvian Model incorporates altitudinal and latitudinal shifts of vegetational communities during at least four glacial-interglacial cycles.

Finally, in Friesen's (1979) Vicariant Model, one generalized species of kangaroo rat, derived from ancestral *heermanni* stock, spread through California during the late Pliocene and early Pleistocene. Entry into the Central Valley was probably through a combination of the corridors described above. Subsequently, this widespread population was vicariantly subdivided as barriers developed due to orogeny and the evolution and geographic shifting of vegetational associations to which various populations of kangaroo rats became adapted. The coincidence of distribution of subspecies of *D. agilis* with specific crustal blocks in Anderson's (1971) model of the San Andreas fault zone supports a vicariant model.

Summary of Historical Biogeography

This historical component of the evolution of heteromyids may be viewed as having occurred in five broad regions: northern portions of the Neotropics, the Great Plains, the Great Basin, the southwestern deserts, and central California. Distributional changes resulted from both physical and biotic factors whose relative importance varies according to these five regions.

Geological features and events are perhaps the fundamental consideration affecting distribution. This is particularly significant for *Heteromys*, a genus occurring on two continents. Entry of *Heteromys* into South America occurred only after North America and South America joined via the uplift of the Bolivar Trough about 3 mybp. On a lesser yet still meaningful scale, distributions of *Dipodomys* of California and adjacent regions have been affected greatly by dynamic geomorphology. Tectonic movements of crustal plates during the late Tertiary and Quaternary periods thrust up montane barriers to vicariantly dissect ranges of formerly widespread species. In other situations, tectonic events apparently felled barriers to open dispersal corridors for various *Dipodomys* species.

Sea level change is another physical phenomenon that has drastically affected distributions of heteromyids, especially species occurring in low elevation, coastal regions. Invasion of Trinidad and perhaps other islands off the northern shore of South America by *Heteromys* probably was facilitated by lowered sea levels of glacial intervals. Cycles of isolation and reunification of heteromyid populations along the Pacific and Gulf of Mexico coasts probably corresponded to the rises and falls of sea level during glacial and interglacial intervals, respectively.

A third physical factor, glaciation, directly impacted heteromyids of the northern Great Plains. Portions of the ranges of *Perognathus fasciatus* and *P. flavescens* (and perhaps others) were obliterated as the ice sheet swept southward.

Not all heteromyids experienced such direct impact on their distributional histories by physical factors as did those noted above. The distributional shifts seen for other species during the late Tertiary and Quaternary seemed to be responses more to biological factors (especially habitat shifts) which, in turn, correspond with physical factors such as climatic shifts and the associated glacial-interglacial cycles. The cooler glacial intervals were times (1) when Great Basin heteromyids were restricted to refugia in intermontane valley areas not occupied by lakes or unsuitable montane vegetation; (2) when southwestern desert species experienced range restrictions due to southward latitudinal and downward altitudinal displacement of boreal and montane habitats; (3) when ranges of some Great Plains species shifted southward into areas that are beyond current southern distribution limits; and (4) when tropical rain forest forms (*Heteromys*) became more isolated and species adapted to more arid conditions (*Liomys*) expanded their ranges in neotropical areas. Warmer interglacial times saw, in general, the reversal of distributional shifts noted for glacial periods.

As must be apparent from the preceding synthesis, efforts at detailing the biogeographic histories of heteromyids have been attempted for only a few species. Elucidation of these histories requires consultation of many types of information—geological, climatic, palynological, ecological, and paleontological. Requisite information is available for examination of biogeography of certain additional species and merely awaits assembly. For other species, biogeographic scenarios await collection of much more data.

Acknowledgments

We would like to thank Juan Carlos Morales for editorial assistance and George D. Baumgardner for his critique of the manuscript and numerous suggestions. William B. Davis and James R. Dixon kindly provided access to their professional libraries and Duke S. Rogers furnished the distribution maps of the genus *Heteromys*. Portions of this study were funded by the Texas Agricultural Experiment Station.

Literature Cited

ANDERSON, D. L. 1971. The San Andreas fault. Scientific American, 225:53–67.

AXELROD, D. I. 1958. Evolution of the Madro-Tertiary geoflora. Botanical Review, 24:433–509.

BARTHOLOMEW, G. A., JR., AND H. H. CASWELL, JR. 1951. Locomotion in kangaroo rats and its adaptive significance. Journal of Mammalogy, 32:155–169.

BAUMGARDNER, G. D., AND D. J. SCHMIDLY. 1981. Systematics of the southern races of two species of kangaroo rats (*Dipodomys compactus* and *D. ordii*). Occasional Papers, The Museum, Texas Tech University, 3:1–27.

———. 1985. Microhabitat relationships of kangaroo rats (*Dipodomys compactus* and *D. ordii*) in southern Texas. The Southwestern Naturalist, 30:315–318.

BEATLEY, J. C. 1967. Survival of winter animals in the Mojave desert. Ecology, 48:745–750.

BLACK, C. C. 1963. Miocene rodents from the Thomas Farm local fauna. Bulletin of the Museum of Comparative Zoology, Harvard University Press, Cambridge, Massachusetts, 128:483–501.

BOWERS, M. A., AND J. H. BROWN. 1982. Body size and coexistence in desert rodents: chance or community structure? Ecology, 62:391–400.

BRITANNICA ATLAS. 1982. Encyclopedia Britannica, Inc., Chicago, 320 pp.

BROWN, J. H. 1973. Species diversity of seed-eating desert rodents in sand dune habitats. Ecology, 54: 775–787.

———. 1975. Geographical ecology of desert rodents. Pp. 315–341, *in* Ecology and evolution of communities (M. L. Cody and J. M. Diamond, eds.). The Belknap Press of Harvard University Press, Cambridge, Massachusetts, 545 pp.

———. 1978. The theory of insular biogeography and the distribution of boreal birds and mammals. Great Basin Naturalist Memoirs, 2:209–227.

BROWN, J. H., AND A. C. GIBSON. 1983. Biogeography. C. V. Mosby Company, St. Louis, 643 pp.

BROWN, J. H., AND G. A. LIEBERMAN. 1973. Resource utilization and coexistence of seed-eating desert rodents in sand dune habitats. Ecology, 54:788–797.

BROWN, J. H., D. W. DAVIDSON, AND O. J. REICHMAN. 1979. An experimental study of competition between seed-eating desert rodents and ants. American Zoologist, 19:1129–1143.

BROWN, K. S., JR. 1976. Centros de evolucão, refugios Quaternarios, e conservacão de patrimonios genéticos na região neotropical: Padrões de diferenciacão em Ithomiinae (Lepidoptera: Nymphalidae). Acta Amazônica, 7:75–137.

CHRISTIANSEN, J. L., AND R. C. SANZ. 1978. A notable range extension for the plains pocket mouse *Perognathus flavescens perniger* Osgood, in Iowa. Proceedings of the Iowa Academy of Science, 85:97–98.

CROWLEY, J. M. 1967. Biogeography. Canadian Geographer, 11:312–326.

DALQUEST, W. W., AND G. COLLIER. 1964. Notes on *Dipodomys elator,* a rare kangaroo rat. The Southwestern Naturalist, 9:146–150.

DAVIS, W. B. 1966. The mammals of Texas. Texas Parks and Wildlife Department Publication, Austin, 267 pp.

DAWSON, W. R. 1955. The relation of oxygen consumption to temperature in-desert rodents. Journal of Mammalogy, 36:543–553.

DE GRANVILLE, J. 1982. Rain forest and xeric flora refuges in French Guiana. Pp. 159–181, *in* Biological diversification in the tropics (G. T. Prance, ed.). Proceedings of the Fifth International Symposium, Association of Tropical Biology, Columbia University Press, New York, 714 pp.

DICE, L. R., AND P. M. BLOSSOM. 1937. Studies of mammalian ecology in southwestern North America with special attention to the colors of desert mammals. Carnegie Institute Publication, Washington, 485 pp.

DURRANT, S. D. 1952. Mammals of Utah: taxonomy and distribution. University of Kansas Publication, Museum of Natural History, Lawrence, 549 pp.

ESCHELMAN, R. E., AND G. S. MORGAN. 1985. Tobagan Recent mammals, fossil vertebrates and their zoogeographical implications. National Geography Research Report, 21:137–144.

FRIESEN, R. D. 1979. Aspects of systematics and biogeography of *Dipodomys* (Rodentia: Heteromyidae) in the Tehachapi Mountain Region of California. Ph.D. dissert., University of Southern California, 621 pp.

GAWNE, C. E. 1975. Rodents from the Zia Sand Miocene of Mexico. American Museum Novitates, 2586: 1–25.

GENOWAYS, H. H. 1973. Systematics and evolutionary relationships of spiny pocket mice, genus *Liomys*. Special Publications, The Museum, Texas Tech University, 5:1368.

GRAYSON, D. K. 1982. Toward a history of Great Basin mammals during the past 15,000 years. Pp. 82–101, *in* Man and environment in the Great Basin (D. B. Madsen and J. F. O'Connell, eds.). Society of American Archaeology Papers, Number 2, 242 pp.

GRINNELL, J. 1914. An account of the mammals and birds of the lower Colorado Valley. California Publication in Zoology, 12:51–294.

———. 1922. A geographical study of the kangaroo rats of California. University of California Publication in Zoology, 24:1–174.

GOLDMAN, E. A. 1911. Revision of the spiny pocket mice (genera *Heteromys* and *Liomys*). North American Fauna, 34:1–70.

HAFFER, J. 1969. Speciation in Amazonian forest birds. Science, 165:131–137.

HAFNER, J. C. 1978. Evolutionary relationships of kangaroo mice genus *Microdipodops*. Journal of Mammalogy, 59:354–365.

———. 1985. New kangaroo mice, genus *Microdipodops* (Rodentia: Heteromyidae), from Idaho and Nevada. Proceedings of the Biological Society of Washington, 98:1–9.

HAFNER, M. S. 1977. Density and diversity in Mojave Desert rodent and shrub communities. Journal of Animal Ecology, 46:925–938.

HALL, E. R. 1941. Revision of the rodent genus *Microdipodops*. Field Museum of Natural History, Zoological Service, 27:233–277.

———. 1946. Mammals of Nevada. University of California Publication, Berkeley, 710 pp.

———. 1981. The mammals of North America. Second ed. John Wiley and Sons, New York, 1:1–600 + *90*, 2:601–1181 + *90*.

HALLETT, J. G. 1982. Habitat selection and the community matrix of a desert small-mammal fauna. Ecology, 63:1400–1410.

HEINE, K. 1973. Variaciones más importantes del clima durante los últimos 40,000 años en México. Comunicaciones, 7:51–56.

HOFFMEISTER, D. F. 1986. Mammals of Arizona. University of Arizona Press, Tucson, 602 pp.

HOFFMANN, R. S., AND J. K. JONES, JR. 1970. Influence of late-glacial and post-glacial events on the distribution of recent mammals on the northern Great Plains. Pp. 355–394, *in* Pleistocene and Recent environments of the central Great Plains, (W. Dort and J. K. Jones, Jr., eds.). University of Kansas, Geology Department, Special Publication, Number 3, 433 pp.

HUEY, L. M. 1951. The kangaroo rats (*Dipodomys*) of Baja California, Mexico. Transactions of the San Diego Society of Natural History, 11:205–256.

KURTÉN, B., AND E. ANDERSON. 1980. Pleistocene mammals of North America. Columbia University Press, New York, 442 pp.

LEMEN, C., AND P. W. FREEMAN. 1983. Quantification of competition among coexisting heteromyids in the southwest. The Southwestern Naturalist, 28:41–46.

LIDICKER, W. Z., JR. 1960. An analysis of intraspecific variation in the kangaroo rat *Dipodomys merriami*. University of California Publication in Zoology, 67: 125–218.

MARES, M. A., AND D. F. WILLIAMS. 1977. Experimental support for food particle size resource allocation in heteromyid rodents. Ecology, 58:454–457.

MARTIN, C. O., AND D. J. SCHMIDLY. 1982. Taxonomic review of the pallid bat, *Antrozous pallidus* (Le Conte). Special Publication, The Museum, Texas Tech University, Number 18, 48 pp.

MEHRINGER, P. J., JR. 1977. Great Basin late Quaternary environments and chronology. Pp. 113–167, *in* Models and Great Basin prehistory, a symposium (D. D. Fowler ed.). University of Nevada Desert Research Institute, Publication in Social Sciences, 213 pp.

MOOSER, O., AND W. W. DALQUEST. 1975. Pleistocene mammals from Aguascalientes, central Mexico. Journal of Mammalogy, 56:781–820.

MUNGER, J. C., M. A. BOWERS, AND W. T. JONES. 1983. Desert rodent populations: Factors affecting abundance, distribution and genetic structure. Pp. 91–116, *in* Biology of desert rodents (O. J. Reichman and J. H. Brown, eds.). Great Basin Naturalist Memoirs, Brigham Young University, Provo, Utah, Number 7.

PARMALEE, P. W., R. D. OESCH, AND J. E. GUILDAY. 1969. Pleistocene and Recent vertebrate faunas from Crankshaft Cave, Missouri. Ill. State Museum Report Number 14, 37 pp.

PATTON, J. L. 1969. Chromosome evolution in the pocket mouse *Perognathus goldmani*. Evolution, 23: 645–662.

PATTON, T. H. 1969. An Oligocene land vertebrate fauna from Florida. Journal of Paleontology, 43:543–546.

PRANCE, G. T. 1973. Phytogeographic support for the theory of Pleistocene forest refuges in the Amazon basin, based on evidence from distribution patterns in Caryoxcaraceae, Chrysobalanaceae, Dichapetalaceae and Lecythidaceae. Acta Amazônica, 3:5–28.

PRICE, M. V. 1978. The role of microhabitat in structuring desert rodent communities. Ecology, 59:910–921.

PRICE, M. V., AND J. H. BROWN. 1983. Patterns of morphology and resource use in North American desert rodent communities. Pp. 117–134, *in* Biology of desert rodents (O. J. Reichman and J. H. Brown, eds.). Great Basin Naturalist Memoirs, Brigham Young University, Provo, Utah, Number 7.

REYNOLDS, H. G. 1958. Relation of Merriam's kangaroo rats to range vegetation in southern Arizona. Ecology, 31:456–463.

ROGERS, D. S. 1986. Evolutionary relationships within the subfamily Heteromyinae (Genera *Heteromys* and *Liomys*). Ph.D. dissert., University of California, Berkeley.

ROSENZWEIG, M. L. 1973. Habitat selection experiments with a pair of coexisting heteromyid rodent species. Ecology, 54:111–117.

ROSENZWEIG, M. L., AND J. WINAKUR. 1969. Population ecology of desert rodent communities: habitats and environmental complexity. Ecology, 50:558–572.

ROSENZWEIG, M. L., AND P. STERNER. 1970. Population ecology of desert rodent communities: body size and seed husking as bases for heteromyid coexistence. Ecology, 51:217–224.

ROTH, E. L. 1984. Evolutionary significance of adaptations in heteromyid rodents in Baja California. Pp. 119–138, *in* Festschrift for Walter W. Dalquest in honor of his sixty-sixth birthday (N. V. Horner, ed.). Department of Biology, Midwestern State University, 163 pp.

SAVAGE, D. E., AND D. E. RUSSELL. 1983. Mammalian paleofaunas of the world. Addison-Wesley Publications, Reading, Massachusetts, 432 pp.

SETZER, H. W. 1949. Subspeciation in the kangaroo rat, *Dipodomys ordii*. University of Kansas Publication, Museum of Natural History, 1:473–573.

SIMPSON, G. G. 1980. Splendid isolation. The curious history of South American mammals. Yale University Press, New Haven, Connecticut, 266 pp.

SMITH, M. F. 1979. Geographic variation in genic and morphological characters in *Peromyscus californicus*. Journal of Mammalogy, 60:705–722.

SOLBRIG, O. T. 1976. The origin and floristic affinities of the South American temperate desert and semidesert regions. Pp. 7–51, *in* Evolution of desert biota (D. S. Gondall, ed.). University of Texas Press, Austin.

STAMP, N. E., AND R. D. OHMART. 1978. Resource utilization by desert rodents in the lower Sonoran Desert. Ecology, 59:700–707.

STUART, L. C. 1966. The environment of the Central American cold-blooded vertebrate fauna. Copeia, 684–699.

THOMPSON, S. D. 1982. Microhabitat utilization and foraging behavior of bipedal and quadrupedal heteromyid rodents. Ecology, 63:1303–1312.

THOMPSON, R. S., AND J. I. MEAD. 1982. Late Quaternary environments and biogeography in the Great Basin. Quaternary Research, 17:39–55.

TOLEDO, V. M. 1982. Pleistocene changes of vegetation in tropical Mexico. Pp. 93–111, *in* Biological diversification in the tropics (G. T. Prance, ed.). Proceedings of the Fifth International Symposium, Association for Tropical Biology, Columbia University, New York, 714 pp.

VAN DEVENDER, T. R., AND W. G. SPAULDING. 1979. Development of vegetation and climate in the southwestern United States. Science, 204:701–710.

VOORHIES, M. R. 1975. A new genus and species of fossil kangaroo rat and its burrow. Journal of Mammalogy, 56:160–176.

WEBB, S. D. 1977. A history of savanna vertebrates in the New World. Part I: North America. Annual Review of Ecology and Systematics, 8:355–380.

———. 1978. A history of savanna vertebrates in the New World. Part II: South America and the great interchange. Annual Review of Ecology and Systematics, 9:393–426.

———. 1985. Late Cenozoic mammal dispersals between the Americas. Pp. 357–386, *in* The great American biotic interchange (F. G. Stehli and S. D. Webb, eds.). Plenum Press, New York, 532 pp.

WEBB, S. D., AND L. G. MARSHALL. 1982. Historical biogeography of Recent South American land mammals. Pp. 39–52, *in* Mammalian biology in South America (M. A. Mares and H. H. Genoways, eds.). Special Publication No. 6, Pymatuning Laboratory of Ecology, University of Pittsburgh, 539 pp.

WEBB, S. D., AND L. G. MARSHALL. 1982. Historical biogeography of Recent South American land mammals. Pp. 39–52, *in* Mammalian biology in South America (M. A. Mares and H. H. Genoways, eds.). Special Publication No. 6, Pymatuning Laboratory of Ecology, University of Pittsburgh. 539 pp.

WELLS, P. V. 1970. Vegetation of the Southwest between 14,000 and 9,000 years ago: the macrofossil record. American Quaternary Association, Abstracts of First Meeting, 1:147–148.

———. 1979. An equable glaciopluvial in the West: pleniglacial evidence of increased precipitation on a gradient from the Great Basin to the Sonoran and Chihuahuan Deserts. Quaternary Research, 12:311–325.

———. 1983. Paleobiogeography of montane islands in the Great Basin since the last glaciopluvial. Ecological Monographs, 3:341–382.

WELLS, P. V., AND J. H. HUNZIKER. 1976. Origin of the creosote bush (*Larrea*) deserts of southwestern North America. Annals of the Missouri Botanical Garden, 63:843–861.

WILLIAMS, D. F. 1978. Systematics and ecogeographic variation of the Apache Pocket Mouse (Rodentia: Heteromyidae). Bulletin of the Carnegie Museum of Natural History, 10:1–57.

WONDOLLECK, J. T. 1978. Forage-area separation and overlap in heteromyid rodents. Journal of Mammalogy, 59:510–518.

THE EVOLUTIONARY MORPHOLOGY OF HETEROMYIDS

P. BRYLSKI

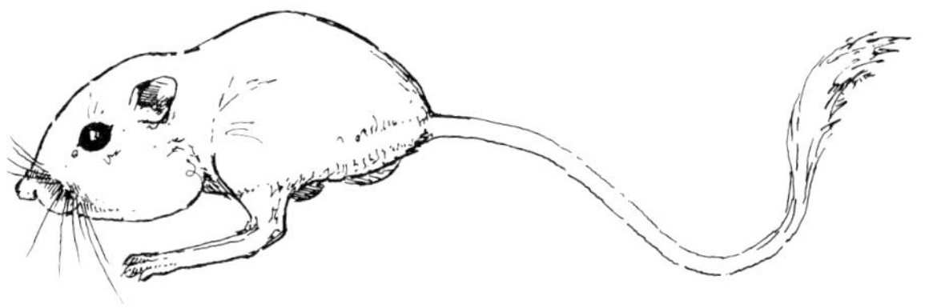

Introduction

Heteromyid rodents are a modest and morphologically diverse radiation indigenous to western North America (Wood, 1935). Since the first monograph on heteromyids over a century ago (Coues, 1875), their ecology, functional morphology, systematics, and ontophyletics have been studied extensively, making them useful models in the study of evolutionary morphology. The purpose of this chapter is to summarize some of this literature, and my own research on heteromyid anatomy, ontogeny, and evolution. In particular, emphasis is placed on discussing the phylogenetic and functional differences in the hard and soft anatomy of higher taxa (genus and subfamily) and interpreting these data in the light of modern theories of phenotypic evolution. This review is intended as a selective update of what is known and what remains to be known about the evolutionary morphology of heteromyids. Because the literature is too extensive for a complete review here, I concentrate on organ systems for which variation within the family is known. The descriptive details from the early monographs (Grinnell, 1922; Hatt, 1932; Howell, 1932; Wood, 1935) are included when they are important for a balanced view of heteromyid anatomy or if recent studies offer corrections or new insights.

External Anatomy

The external features considered are those of the integument, including the external ear, the pelage, and some of its subdermal specializations. The pattern of integumental differences among heteromyids is paralleled in various other hard and soft anatomical characters: heteromyines are typically generalized, and probably primitive, one or both perognathine genera are derived in a minority of characters, and dipodomyines are highly derived. First, however, I provide a synopsis of the differences in body weight among the genera.

Body Size

Body weight among heteromyids ranges from about eight grams in *P. merriami* to 180 grams in *D. ingens.* Figure 1 shows the ranges in body weight for 23 heteromyid

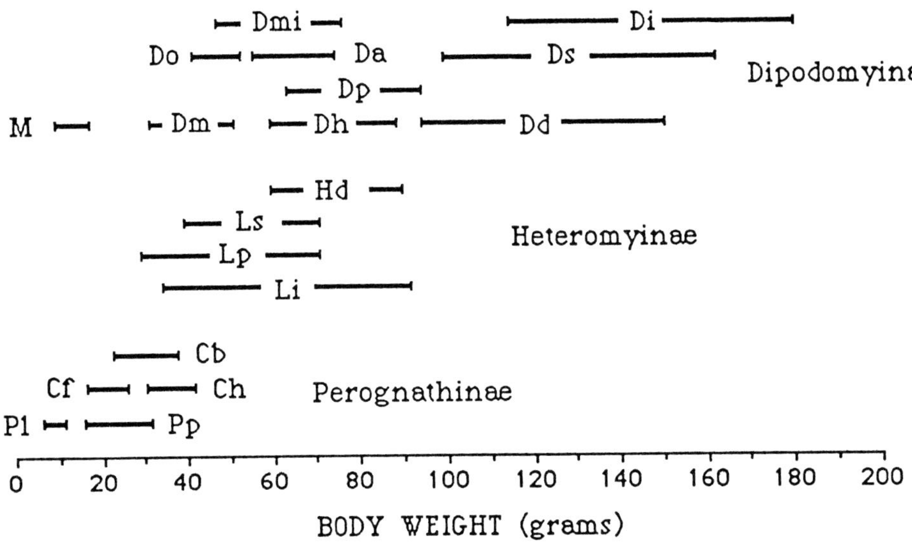

FIG. 1.—Approximate ranges in body weight for 22 species of heteromyid rodents (data from Hall, 1946; Genoways, 1973; Jones, 1985, and Brylski, unpubl. data). Taxonomic abbreviations: Perognathines: Cb, *Chaetodipus baileyi*; Cf, *C. formosus*; Ch, *C. hispidus*; Pa, *Perognathus amplus*; Pl, *P. longimembris*; Pp, *P. parvus*; Heteromyines: Hd, *Heteromys desmarestianus*; Li, *Liomys irroratus*; Lp, *L. pictus*; Ls, *L. salvini*; Dipodomyines: Da, *Dipodomys agilis*; Dd, *D. deserti*; Dh, *D. heermani*; Di, *D. ingens*; Dm, *D. merriami*; Dmi, *D. microps*; Do, *D. ordii*; Dp, *D. panamintinus*; Ds, *D. spectabilis*; Dv, *D. venustus*; M, *Microdipodops megacephalus* and *M. pallidus*.

species and provides a general picture of the body weight differences among the genera and subfamilies. The Dipodomyinae show the greatest intra-subfamilial range in weight: even the range within *Dipodomys* exceeds that of the Perognathinae and the Heteromyinae combined. The weight range of perognathines overlaps slightly with that of heteromyines and overlaps with dipodomyines only because of the small size of *Microdipodops*.

Ecologists interested in the evolution of community structure of desert rodents would benefit from a phylogenetic perspective on the origin of body size differences in the extant heteromyid lineages. This may soon be possible, as two ingredients essential to such an approach are currently being investigated: the cladistic relationships among fossil and recent taxa (see Wahlert, 1993) and the allometric relationships between various skeletal elements and body weight (presented later in this paper).

Pelage

The texture and color of the pelage varies widely among heteromyids. Pelage color ranges, with considerable overlap, from dark in heteromyines, intermediate in *Chaetodipus* and *P. parvus*, to light in dipodomyines and *Perognathus* (excluding *P. parvus*). Spectrophotometric comparisons are available only for *Heteromys* and *Liomys* (Genoways, 1973). A buff-colored lateral line is typically present, but variably developed, in all heteromyids. Dipodomyines are distinguished from other heteromyids in their derived pelage markings, which include white spots rostral and caudal to the eyes and caudal to the ears, and a white flank stripe across the lateral thigh. In *Dipodomys* the flank stripe is continuous from the base of the tail to the white ventral pelage, whereas in *Microdipodops* it ends at the anterior border of the thigh. Distantly related rodents that occupy arid habitats have similar markings,

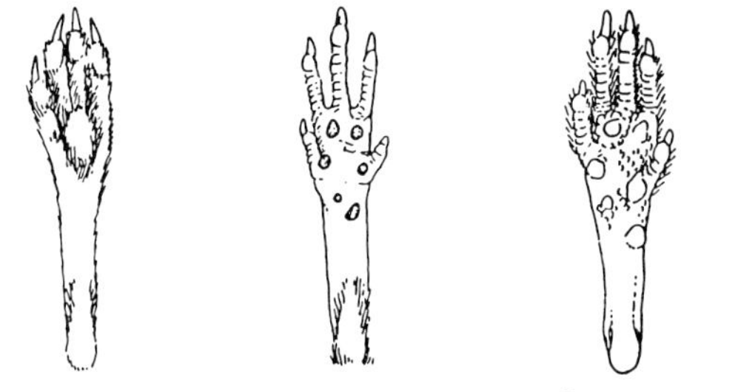

FIG. 2.—The plantar surface of the pes in three heteromyids, *D. merriami* (MVZ 171301), *L. salvini* (MVZ 171900), and *P. longimembris* (MVZ 171899). Not drawn to scale (from Brylski, 1985).

which has been interpreted to mean that they are convergent adaptations for concealment (Howell, 1932). The distinctive, often species-specific, arietiform facial marks found in *Dipodomys* (see Grinnell, 1922) is produced by a dark line at the bases of the mystacial whiskers superimposed on an unpigmented cheek region. The former character is primitive among heteromyids and the latter is derived. Some *Dipodomys* (*D. merriami, D. deserti, D. heermani,* and *D. spectabilis*) have a white-tipped tail, which may serve as a flag to distract pursuing predators or as a signal in intra-specific communication (Eisenberg, 1963).

Heteromyid hair is similar to that of other mammals, and therefore is primitive, in having imbricate cuticular scales and compound medullary patterns (Homan and Genoways, 1978). The underfur, when present, is short and curly (dipodomyines lack underfur). The dorsal pelage hairs are short in *Chaetodipus* ($\bar{X}$ = from 6.5 to 9.8 mm, 8 species) and *Perognathus* ($\bar{X}$ = from 5.4 to 8.6 mm, 5 species) and long in *Heteromys* ($\bar{X}$ = from 9.4 to 12.2 mm, 6 species), *Liomys* ($\bar{X}$ = from 9.9 to 12.3 mm, 3 species), *Microdipodops* ($\bar{X}$ = from 12.7 to 13.4 mm, 2 species), and *Dipodomys* ($\bar{X}$ = from 11.1 to 19.1 mm, 12 species). The guard hairs have a trough along their dorsal surface in heteromyines, *Chaetodipus,* and *P. amplus* and lack a trough in the remaining *Perognathus* and in dipodomyines. Hair morphology in heteromyids is species specific, but apparently is of little systematic utility above the species level (Homan and Genoways, 1978).

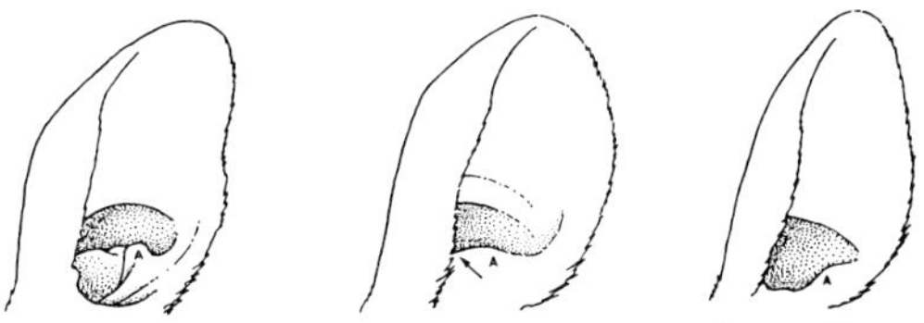

FIG. 3.—The external ear of *C. formosus* (PVB 654), *D. merriami* (MVZ 171301), and *P. longimembris* (MVZ 171899). A, antitragus; arrow points to attachment anti-tragus found in dipodomyines, showing absence of intertragic notch. Not drawn to scale (from Brylski, 1985).

Observations on the molt in *Perognathus* and *Liomys* can be found in Speth (1969) and Genoways (1973), respectively.

Hindfeet

The plantar tubercles on the hindfeet of heteromyines and perognathines are replaced in dipodomyines by a single foot pad located centrally at the base of the toes (Fig. 2). This derived feature of dipodomyines probably cushions the overlying foot bones during locomotion. The soles of the hindfeet are either naked or lightly haired in heteromyines and perognathines and are densely haired in dipodomyines. In heteromyines, the lateral hairs of the hindtoes are sparse and point anteriorly. In perognathines they project anteriorly and laterally, resulting in a "fringe-toed" effect, which may enhance locomotor efficiency on sandy substrates. Dipodomyines are more extensively fringe-toed owing to the greater density of hair on their hindsole.

External Ear

The antitragus of the pinna is well developed and primitive in *Heteromys, Lio-*

mys, Chaetodipus, and *P. parvus,* somewhat reduced and derived in other *Perognathus* (Fig. 3), and greatly reduced in *Microdipodops.* The antitragus is moderately developed in *Dipodomys,* but is derived in its attachment medial to the tragus, which has resulted in the loss of the intertragic notch (Fig. 3). It is not known whether there is any adaptive significance to any of these differences.

Secretory Cells of the Integument

Quay (1953, 1954, 1965*a*, 1965*b*, 1965*c*, 1966, 1972) examined the integument of heteromyids in a series of papers on the comparative anatomy of rodents from xeric and mesic habitats. Sebaceous glands, which function in pelage maintenance, and which may also reduce water loss, are larger and more active in desert heteromyids than in heteromyines (Quay, 1965*a*). Well-developed sebaceous glands are found in the oral lip and angle of all heteromyids (Quay, 1965*c*).

Specialized sebaceous glands are found in perognathines and in *Dipodomys.* Perognathines have a specialized glandular area in the ventral side of the tail, roughly one-third the distance from its base, which in other heteromyids is incipiently glandular. Hypertrophy of this caudal gland is usually greater in male pocket mice than in females (the reverse is true for *C. hispidus*). Its functional significance appears to be in communication, perhaps for territorial marking. The "perineal drag" behavior found in perognathines (Eisenberg, 1963) apparently serves to distribute exudate of the caudal gland on the ground.

Dipodomys is derived among heteromyids in having an enlarged sebaceous gland in the skin of the mid-back, between the shoulders. According the Howell (1932), grossly similar glands are found in some ground squirrels (*Callospermophilus* sp.) and rock hyraxes (*Hyrax* sp.). Superficially, the gland is a warty thickening of the skin that lacks hair follicles. Histologically, it is an aggregation of enlarged, holocrine sebaceous cells that secrete a granular lipid-rich substance. Since normal sebaceous glands are abundantly distributed throughout the dermis, and hypertrophied glands are found only in the mid-back region, the enlarged dorsal gland probably functions in olfactory communication rather than in pelage maintenance (Quay, 1954; Westerhaus, 1983; see also Randall, 1981). There is also marked seasonality in the size and secretory activity of the dorsal gland which corresponds to the breeding season in some species (*D. merriami* and *D. agilis,* and *D. deserti*), but not in others (*D. heermani* and *D. ordii*). The size and volume of secretion of the gland is generally greater in males (except in *D. ordii*) (Quay, 1953).

Cheek Pouch Development

Geomyoids are unique among rodents in having furred pouches that open externally (Fig. 4 shows a neonatal *D. elephantinus*), which contrast with the internal pouches of certain cricetids and sciurids that open into the buccal cavity. A comparison of external pouch development in the pocket gopher (*Thomomys bottae*) and several kangaroo rats (*D. merriami, D. panamintinus,* and *D. elephantinus*) and internal pouch development in the least chipmunk (*Eutamias minimus*), a sciurid, and the Syrian hamster (*Mesocricetus auratus*), a cricetid, shows that they all arise early in development by an evagination of the buccal epithlium (Brylski and Hall, 1988*a*, 1988*b*; Hardy et al., 1986). A schematic summary of pouch development in geomyoids is shown in Figure 5. Externalization of the internal pouches in geomyoids results from a seemingly simple change early in pouch development. Although ontogeny does not always parallel phylogeny, the similarities in the early development of internal and external pouches has been interpreted to mean that the external pouch was derived during phylogeny from an internal pouch (Brylski and Hall, 1988*a*, 1988*b*). External pouches may have

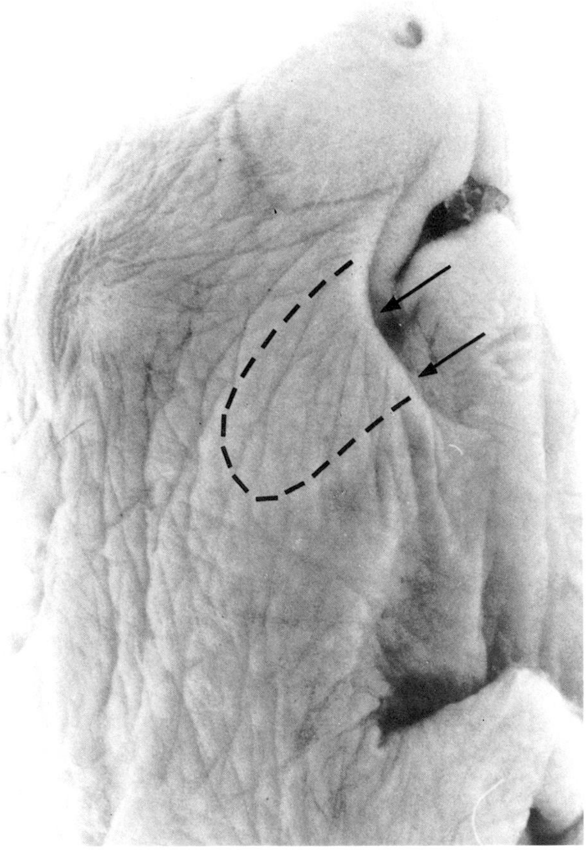

FIG. 4.—Neonatal (about 30 days of age) *D. elephantinus* showing the external cheek pouch. Arrows point to presumptive anterior opening and dashed line shows extent of pouch development. The position of the pouch is the same here as in adults. At this stage of development, the lumen epithelia are fused and keratinized facial epithelium covers the external opening.

been favored over internal pouches by natural selection because they conserve body water that would otherwise be lost when seeds are passed through the mouth and later cached (Long, 1976).

Muscles

Cheek Pouch Muscles

The cheek pouches are controlled by the facial, trapezius, and buccinator muscles; the number of their divisions differs between interpretations (Hill, 1937; Howell, 1932; Ryan, 1986). The pouch retractor muscle, which withdraws the everted pouch after grooming and emptying, originates from the last two thoracic vertebrae and inserts onto the caudal and dorsal pouch margins. Hill (1935) demonstrated experimentally that the caudal slip of the retractor is derived from the trapezius muscle and the cervical slip is from the facial muscle. This dual origin of the geomyoid retractor differs from the retractor in sciurids and cricetids, in which the retractor is entirely facial and trapezial, respectively (Ryan, 1986). Ontogenetically, the facial retractor in geomyoids is more prominent (and morphologically similar to the facial retractor of the least chipmunk, *Eutamias minimus*) early in development

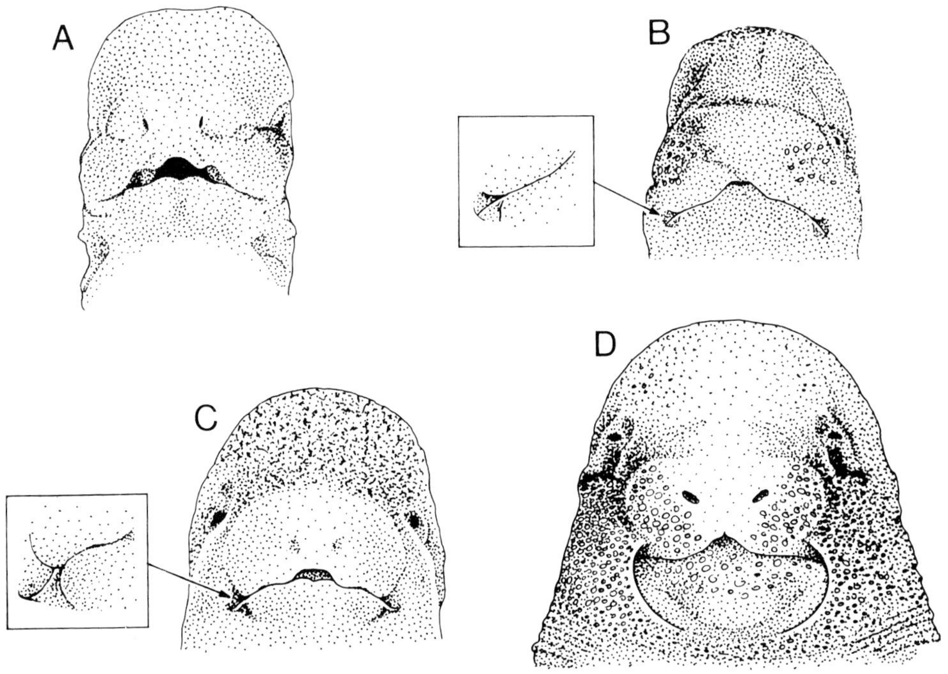

FIG. 5.—Four stages in the development of the geomyoid external pouch (the developmental stages represented range from Theiler (1972) stage 19 to 25 (11.5 to 17 days in the house mouse, *M. domesticus*)): (A) Evagination of the buccal epithelium commences, but is not yet evident externally. (×15); (B) The buccal evagination appears externally as a lateral extension of the mouth (see inset) (×12); (C) Externalization of the pouch coincides with the development of the snout—the inset shows that the corner of the mouth develops medial to the evagination and the evagination is still orientated laterally (see also Fig. 5) (×12); (D) The morphogenesis of the pouch is now complete (×6) (from Brylski and Hall, 1988*a*).

than the trapezial retractor (Brylski and Hall, 1988*a*). Smaller muscles are diffusely spread over the lateral and medial walls of the pouch, including slips of the facial muscle (the sphincter colli primitivus of Howell, 1932, = levator and protractor muscles of Hill, 1937) and eight slips of the buccinator muscle (Ryan, 1986). One of the buccinator slips, the orbicularis sacculi, exerts sphincter-like control of the lateral border of the cheek pouch opening and is unique to geomyoids.

Masseter Muscles and Mastication

The morphology of the masseter muscles and infraorbital foramen and their interrelations is one basis for the current higher-level systematics of rodents. Heteromyids and geomyids are sciuromorphous in two respects: the anterior lateral masseter muscle originates from the rostrum and adjacent zygomatic plate, and fibers of the anterior medial masseter muscle do not pass through the infraorbital foramen. Sciuromorphy appears to be a derived condition (Luckett and Hartenberger, 1985), but whether it means that geomyoids and sciuroids are monophyletic (Simpson, 1945) or convergent (Wood, 1965) is unresolved (Fahlbusch, 1985).

The masticatory muscles include the masseters (superficial, medial, and deep), temporalis (anterior and posterior), pterygoids (internal and external), and digastrics. Their roles during mastication have been

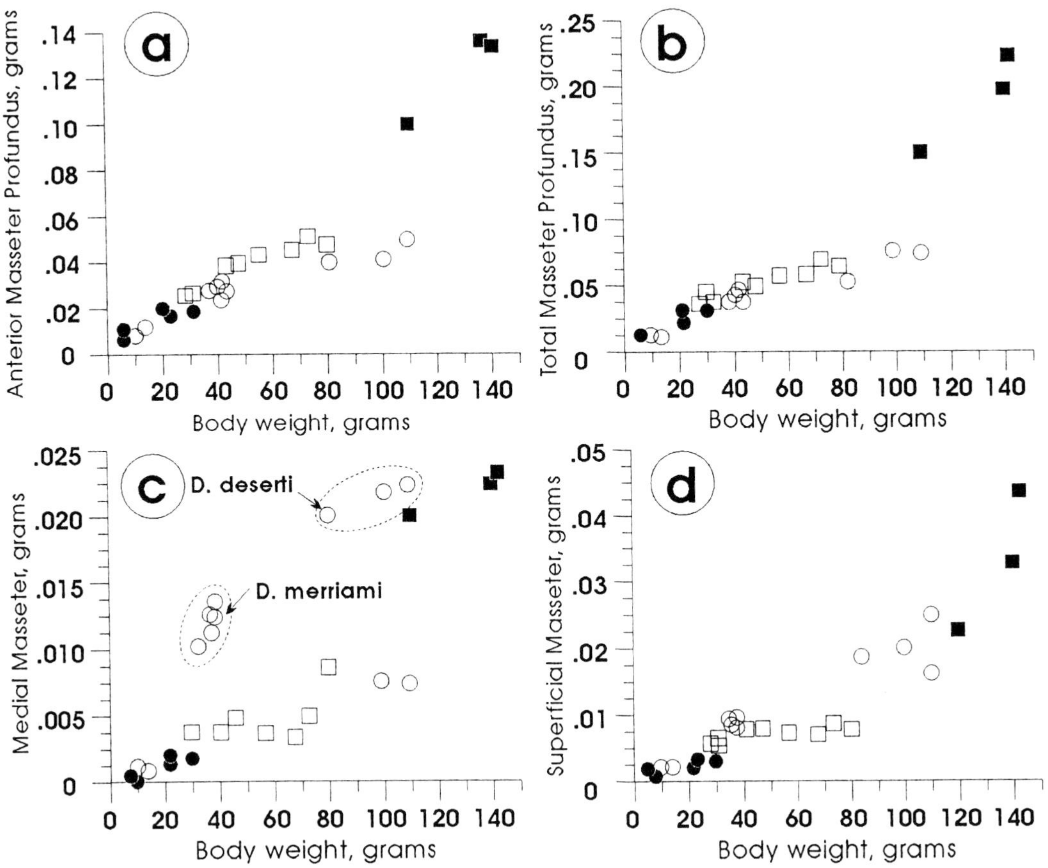

FIG. 6.—Bivariate plots of body weight and dry weight of four masseter muscles for individuals of 8 species of geomyoid rodents. A, Anterior profundus; B, Total profundus (anterior and posterior combined); C, Medial; D, Superficial. Open circles; the dipodomyines *M. megacephalus* (approx. 10 grams body weight), *D. merriami* (40 grams), and *D. deserti* (80–115 grams). Closed circles: the perognathines *P. longimembris* (8 grams) and *C. formosus* (20–30 grams). Open squares: the heteromyines *L. salvini* (30–45 grams) and *H. desmarestianus* (approx. 60–80 grams). Closed squares: the geomyine *T. bottae* (from Brylski, 1990).

studied in the rat (*Rattus norvegicus*; Hiiemae, 1971) and the hamster (*Mesocricetus auratus*; Gorniak, 1977; see also Gorniak, 1985 for a review of mammalian mastication). The masseters elevate the jaw and, with the assistance of the pterygoids, move it anteriorly during the power stroke. The jaw is lowered and retracted by the digastric muscles and elevated and retracted by the temporalis muscles. A comparison of the dry weights of four masseter muscles in eight geomyoids measured by Brylski (unpubl.) is shown in Figure 6. The masseter muscles in heteromyids vary in a size-dependent manner, with several exceptions: *D. merriami* and *D. deserti* have a more massive medial masseter muscle (which inserts onto an expanded maxillary arch) and the superficial masseter muscle appears to be enlarged in *D. deserti* (Fig. 6D).

Mastication is unstudied in heteromyids, but is propalinal (during the power stroke the lower cheek teeth move anteriorly against the uppers) and bilateral (both molar rows occlude simultaneously) in the closely related geomyids (Wilkins and Woods, 1985). The lack of transverse movement of the jaw during mastication in geomyids contrasts with the slight to moderate lateral movement reported for myomorph rodents (Gorniak, 1977; Hiiemae, 1971; Weijs, 1975), and the pronounced transverse jaw movement in hystricomorphs (Byrd, 1981). The masticatory patterns of geomyids are correlated with two features of their dental morphology: the enamel bands of the cheek teeth are perpendicular to the sagittal plane and the molar rows occlude simultaneously. Heteromyids are similar to geomyids in these respects and their mastication is probably also propalinal and bilateral. Patterns of mastication may differ among heteromyids, as reflected by 1) the larger maximum gape in heteromyines afforded by the uninflated auditory bullae (Nikolai and Bramble, 1983), which correlates with the large seeds known to be important in their diets (Fleming, 1974), 2) the restriction of the temporal muscle by the auditory bullae in perognathines and dipodomyines, 3) a slight movement of the insertion of the lateral masseter muscle to the diastema rostral to p4 in dipodomyines and *Perognathus,* and 4) the loss of enamel from the sides of the cheek teeth in *Dipodomys.*

Hindlimb Muscles and Locomotion

Heteromyines, perognathines (and most other rodents) locomote by quadrupedal saltation, a leaping mode of locomotion in which both feet of a pair strike the ground simultaneously (Gambaryan, 1974). Dipodomyines are bipedal saltators: they move upright by a series of rebounds with the hindfeet landing in unison and the forelimbs held close to the body (Hatt, 1932). Comparisons of heteromyid locomotion have been made in the field using tracks (Eisenberg, 1963; Pinkham, 1972) and in the laboratory using cinematography (Bartholomew and Caswell, 1951; Bartholomew and Carey, 1954; Biewiener et al., 1981; Nikolai and Bramble, 1983; Pinkham, 1972).

At slow speeds, heteromyines and perognathines move by alternate movement of the legs. Acceleration to moderate speeds leads to the typical ricochetal saltation where the forefeet act as shock absorbers and the hindlimbs provide the propulsive thrust. At moderate speeds, the hindfeet land before the forefeet leave the ground (Bartholomew and Carey, 1954; Pinkham, 1972) and at higher speeds, the hindlegs overreach the point of contact of the forelegs.

At slow speeds (e.g., when exploring objects or foraging) *Dipodomys* walks either quadrupedally like other heteromyids, or bipedally, whereby the hindlimbs move alternately and each plantar hindsole is placed flat on the ground (Bartholomew and Caswell, 1951). During the bipedal hop, only the toes (and presumably the foot pad) contact the ground and they are spread wide at the time of maximum thrust. Cinematographic analyses of hopping kangaroo rats revealed that movement is smallest at the hip joint and greatest at the knee, ankle, and metatarso-phalangeal joints (Biewiener et al., 1981). Therefore, the propulsive force for bipedal saltation comes mainly from the quadriceps, hamstring, gastrocnemius, and plantaris muscles. During maximum hindlimb and hip flexion, the toes are beneath the tip of the nose. During the swing phase, the hindlegs are extended downward and backward and the toes strike the ground at a point below the middle of the body. The fast and slow bipedal hops differ only with respect to the strength of the hindlimb propulsion. When alarmed, *Dipodomys* switches to an erratic bipedal hop with sudden and unpredictable changes in direction, which is thought to aid in escaping predators (Bartholomew and Carey, 1954; Pinkham, 1976).

There are conflicting accounts over the frequency of quadrupedal versus bipedal locomotion in *Microdipodops* (O'Farrell and Blaustein, 1974*a,* 1974*b*), although it is clearly more quadrupedal than *Dipodomys.* The reason for *Microdipodops* greater reliance on quadrupedality may be related to their small body size and their primitively elongate humerus. The relationships of body proportion and body mass to the quadrupedal/bipedal transition speed are discussed by Nikolai and Bramble (1983).

Three measures of locomotor performance: maximum velocity, jump height, and jump distance, vary directly with the mass of the propulsive musculature and the distance over which its force acts (the distance between the flexed and extended hindlimb; Emerson, 1985). Three adaptations for running and jumping have been described in dipodomyines: 1) larger quadriceps, hamstrings and ankle extensors relative to their body mass (Alexander et al., 1981; Berman, 1985), 2) elongate hindlimbs and pes (see section on body proportions), and 3) integumetary modifications of the plantar surface of the pes such as the foot pad and excessive hirsuteness.

There are two alternative, but not mutually exclusive, hypotheses on the selective advantages of bipedalism: 1) as an adaptation to escape predation through its effect on running and leaping performance (Hatt, 1932), and 2) as a solution to an optimal foraging problem through its effect on the energetic costs of locomotion (Riechman, 1981).

The anatomical basis of the energetic hypothesis is the elasticity of tendons and muscles, which allow energy to be stored in one stride and released in another (Alexander and Benet-Clark, 1980). The energetic effect of elastic energy release could be a decline, perhaps even a plateau, in the rate of increase of oxygen consumption over some interval of speed. Support for the energetic hypothesis has been sought in biomechanical, physiological, and ecological studies, without success (but see below). Biomechanical analyses of locomotion in *Dipodomys* indicate that little or no energy is stored in the tendons of the hindlimb extensor muscles, although the muscles themselves may store some energy (Biewiener et al., 1981). Tendon elasticity appears to be more important in large animals (>3 kg) than in small ones (Alexander et al., 1981). Physiological studies have shown that *Dipodomys* trained to run on a treadmill do not show an oxygen consumption plateau (Thompson, 1985; Thompson et al., 1980) like the one reported for the bipedal Australian murid, *Notomys cervinus* (Dawson, 1976). There may be an energy savings for bipedal heteromyids (compared to quadrupedal rodents generally) moving at high speeds (>7 km/h), but not at the speeds *Dipodomys* commonly use to move between foraging sites under natural conditions (modal speeds = from 3.0 to 3.5 km/h and from 6.0 to 6.5 km/h for *D. merriami* and *D. deserti,* respectively; Thompson, 1985).

Fedak and Seeherman's (1979) comparison of locomotor energetics in 66 species of reptiles, birds, and mammals showed that there is no consistent difference between the cost of locomotion for bipeds and quadrupeds of any size. However, these authors showed there is considerable size-independent variation in the cost of locomotion, and thereby emphasized the importance of making comparisons in a phylogenetic context. The question: "Do the locomotor energetics of comparably sized bipedal and quadrupedal heteromyids differ significantly?", has not been addressed (Taylor et al., 1970; Thompson, 1985; Thompson et al., 1980). The locomotor energetics of quadrupedal heteromyids, cricetids, and murids, may not differ significantly, but the number of confounding phylogenetic differences is potentially highest for distantly related taxa.

The hypothesis that bipedalism aids in escaping predation is supported by biomechanical and comparative morphological studies, but is difficult to test in nature. It is probably more accurate to recognize hindlimb elongation and its associated muscle changes as the principal adaptations for escaping predation, with bipedalism resulting from a decline in the ratio of forelimb length to hindlimb length (Hatt, 1932). Figure 7 shows the relationship between maximum running speed and body mass in some perognathine and dipodomyine rodents, based on treadmill studies by Djawdan and Garland (1988) and field observations on

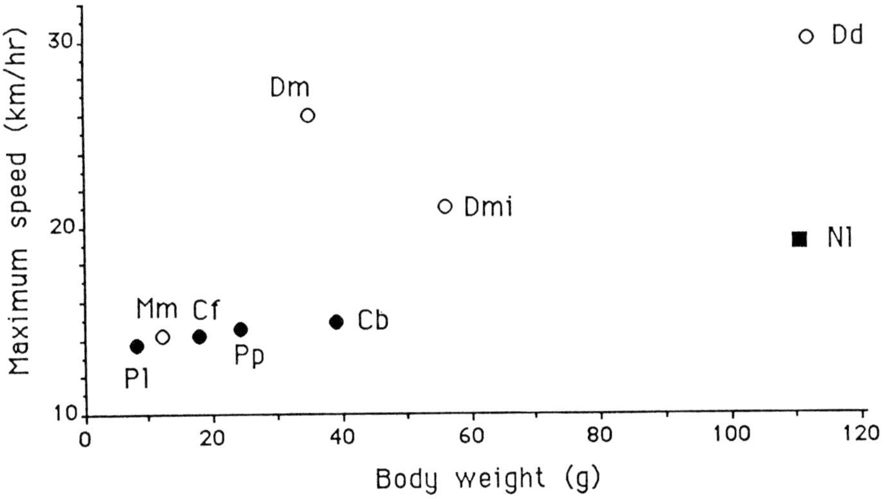

FIG. 7.—Relationship between maximum speed and body mass in perognathines (closed circles), dipodomyines (open circles) and the cricetid *Neotoma lepida*. The data are from treadmill studies (Djawdan and Garland, 1985) except data for *D. microps*, which is based on field studies (Kenagy, 1973). Taxonomic abbreviations: Cb, *Chaetodipus baileyi*; Cf, *C. formosus*; Pl, *Perognathus longimembris*; Pp, *P. parvus*; Dd, *D. deserti*; Dm, *D. merriami*; Dmi, *D. microps*; Mm, *M. megacephalus*; Nl, *Neotoma lepida*.

D. microps by Kenagy (1973). *Dipodomys* are considerably faster than the perognathines and cricetids studied (Djawdan and Garland, 1988). The lower value for *D. microps* may reflect the differences in performance between the laboratory and field (e.g., due to substrate differences), as this value was obtained by timing animals released from live-traps (Kenagy, 1973). Data on heteromyines are needed to test the hypothesis that large dipodomyines (e.g., *D. deserti*) are faster than large heteromyines of comparable size (e.g., *H. desmarestianus*). The fact that *M. megacephalus* is not significantly faster than perognathines of comparable body mass means that we do not know how this species benefits from its elongated hindlimbs. Kotler (1985) observed that bipedal heteromyids were underrepresented as owl prey items compared to quadrupedal heteromyids and cricetids, and argued that *Dipodomys* and *Microdipodops* may be better than quadrupedal forms at escaping predation because of differences in the auditory system rather than the locomotory. Preliminary data on stride length (the distance from footfall to footfall) in heteromyids indicates that stride length increases with body size, as expected, and that bipedal heteromyids have larger strides than quadrupeds of comparable body mass (Djawdan, pers. comm.). These data on maximum running speed and stride lengths in *Microdipodops* are consistent with Reichman's energetic efficiency argument for the evolution of bipedality.

Several authors have argued that a "ricochetal" (=saltatorial of Gambaryan, 1974) mode of locomotion evolved convergently in various living and extinct geomyoids (Wood, 1935; Munthe, 1981), apparently in reference to bipedal saltatorial locomotion. Quadrupedal saltation is the primitive locomotory mode among rodents (Gambaryan, 1974) and among geomyoids it is found in living heteromyines and perognathines. A comparison of the body proportions of recent and fossil taxa by Brylski (in press) indicates that the hindlimbs of the late Miocene heteromyids *Eodipodomys celtiservator* (Voorhies, 1975) were elongated, although to a lesser degree than in extant *Dipodomys*. Contrary to previous opinion, the early Miocene *Diprionomys agrarius* and the late Miocene *Cupidinimus nebraskensis* did not have elongated hindlimbs. The late Miocene *Schizodontomys harkseni* (Munthe, 1981) is the largest known geomorph and for this reason it is difficult to compare its limb proportions with smaller, living geomorphs. Nonetheless, it is likely that this species also possessed primitive limb proportions and locomotory mode. Since *E. celtiservator* is classified in the Dipodomyinae on other morphologic criteria, there is no evidence from fossil forms that supports the hypothesis that bipedality evolved more than once in geomyoids. The only support for the convergence hypothesis among living heteromyids (*Dipodomys* and *Microdipodops*) is tentative at best and is based on the current inability to identify a utility of bipedality that is shared by both taxa.

Internal Organs

The internal organs of heteromyids, with few exceptions, have attracted little attention. The few systems described below have been studied from either a systematic perspective (carotid circulation, placenta, and accessory reproductive glands) or from a functional perspective (brain and kidney). Some important work that is not summarized here is that on the ultrastructural and histochemical attributes of the choroid plexus of the brain and pineal gland (Karasek et al., 1982; Quay, 1960*a*) and pancreas (Quay, 1960*b*).

Carotid Circulation

The carotid circulation of the head was described for perognathines and dipodomyines by Bugge (1971) and later was examined in one or more species of the six extant genera of heteromyids by Brylski (1990). Figures 8 and 9 show the carotid arteries of *H. desmarestianus* and *D. merriami*. The principal difference in the carotid circulation of living heteromyids is whether the pterygopalatine artery of the braincase (the blackened artery in the dorsal skull views of Figs. 8 and 9) originates as a stapedial branch (in perognathines and dipodomyines) or as an internal maxillary branch (in geomyines and heteromyines). The stapedial artery originates from the internal carotid artery, enters the middle ear by the stapedial foramen, passes through the stapes, and enters the braincase through the alisphenoid canal, where it is renamed the pterygopalatine artery. The internal maxillary artery originates from the external carotid and enters the braincase directly through the alisphenoid canal. A study of this difference in developmental series of *T. bottae* (a geomyine) and *D. merriami* revealed that the stapedial artery is present in fetuses and juveniles of *T. bottae*, after which it is lost. Embryos of *L. salvini* also possess the stapedial artery. Brylski (1990) hypothesized that the primitive condition for geomyoids was the presence of both stapedial and internal maxillary arteries, that the stapedial artery was lost in geomyines and heteromyines because of the constraint on its size posed by the stapes, and that the stapedial artery was retained in perognathines and dipodomyines because enlargement of the stapes accompanied bullar inflation in these taxa and lifted the constraint on the size of the stapedial artery. The causative factor may have been either enlargement of the masseter muscles in geomyids and another non-neural cranial tissue (as yet unidentified) of heteromyines, or to general body mass enlargement in both taxa. For example, *H. desmarestianus* and *L. salvini* have enlarged olfactory lobes of the brain (Hafner and Hafner, 1984) which may reflect the enlargement of one or more olfactory tissues fed by the pterygopalatine artery.

Brain

Heteromyids provide a unique opportunity to study the functional significance of body size-independent differences in brain size. Much of the variation in brain size

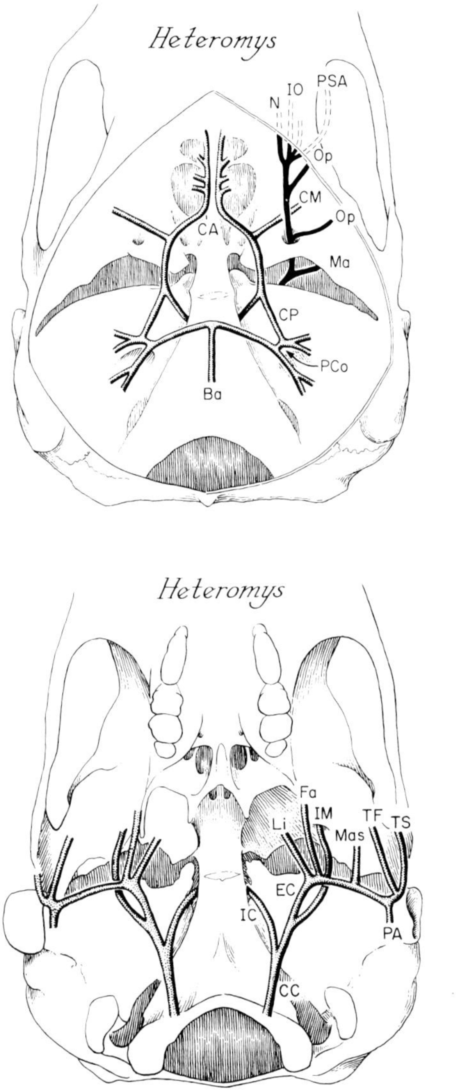

FIG. 8.—The major carotid arteries of *Heteromys desmarestianus.* Top: dorsal view of cranium with top of skull and brain removed; Bottom: ventral view with superficial neck muscles removed. Legend: Ba, basilar artery, CA, ant. cerebral; CC, common carotid; CM, medial cerebral; CP, posterior cerebral; EC, external carotid; IC, internal carotid; IM, internal maxilary; IO, infraoribital; Fa, facial; Li, lingual; M, mandibular (=inferior alveolar); Mas, masseteric; N, nasal; Op, ophthalmic; PA, posterior auricular; PCo, posterior communicating; PSA, posterior superior alveolar; Pt, pterygoid; TF, transverse facial; TS, superficial temporal (from Brylski, 1990).

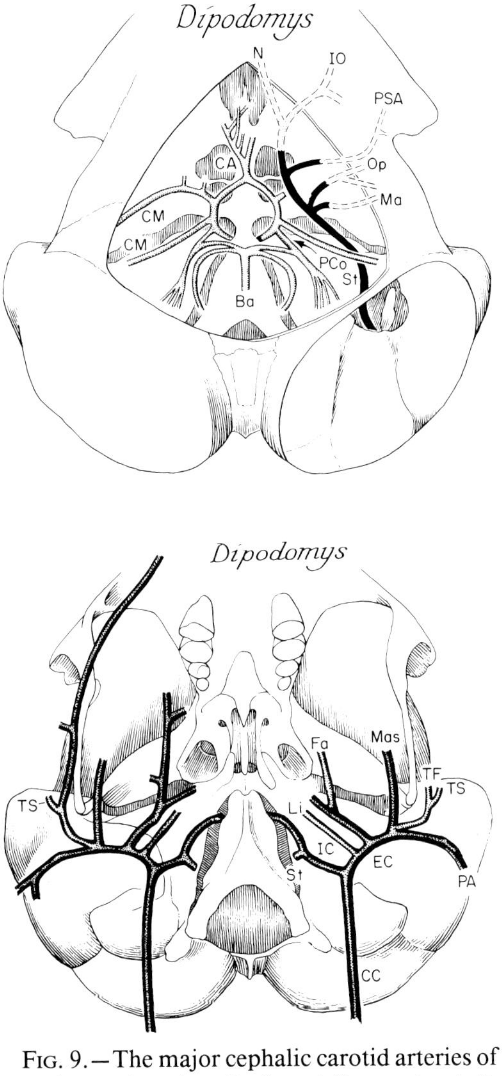

FIG. 9.—The major cephalic carotid arteries of *Dipodomys,* based on *D. deserti, D. merriami, D. ordii,* and *D. panamintinus.* Top: dorsal view of arteries of the cranium (with brain removed); Bottom: ventral view with superficial neck muscles removed. St, stapedial. See Figure 8 for legend of remaining abbreviations (from Brylski, 1990).

within the three heteromyid subfamilies, and between perognathines and heteromyines, is body size-dependent (Hafner and Hafner, 1984). Dipodomyines, however, have larger brains relative to their body weight and, in particular, have enlarged cerebral cortices. In addition, there are some smaller interspecific and intergeneric differences (after body size differences are taken into account): *Perognathus* (excluding *P. parvus*) has small cerebral hemispheres, *Microdipodops* has a small cerebellum, and heteromyines have large olfactory lobes. What is the functional significance of these differences? In general, larger brains may be functionally superior to smaller brains in storing and analyzing spatial or sensory information, in controlling motor output, or in mediating more complex behaviors (Bullock, 1984; Dressler, 1979). Dipodomyines have not been compared with other heteromyids in most of these respects, although preliminary results indicate that *Dipodomys merriami* possesses a slightly more acute spatial memory of seed caches than does *Chaetodipus intermedius* (Rebar, 1986).

Perognathines and dipodomyines are ecologically and behaviorally similar, so it appears that either brain enlargement in dipodomyines is unrelated to these functions or small differences in function are accompanied by a much greater difference in brain size. Hafner and Hafner (1984) argued that the small differences in brain size among heteromyids may be adaptive, but that the enlarged brains of dipodomyines were an indirect and adaptively neutral result of either a K-selected life history strategy or selection on another character (e.g., their large skulls, elongated hindlimbs, or large eyes). However, it is unlikely that the brain shares a growth regulating system with one or several skeletal elements. The pleiotropy hypothesis of cerebral enlargement should instead be pursued with respect to other hypertrophied brain parts, such as those belonging to the central auditory system (Webster et al., 1968). Compared to perognathines and heteromyines, dipodomyines have many derived morphological features, some of which may require neurological specializations. These functional relationships are largely unstudied, and for this reason it is premature to conclude that the enlarged brains of dipodomyines are not adaptive.

Kidney

The kidneys of desert heteromyids are thought to be highly adapted by virtue of their exceptional water-resorption ability. The renal papilla, part of the water collecting system of the kidney, is elongated in distantly related desert rodents worldwide (Sperber, 1944), including some heteromyids (perognathines) but not others (*Dipodomys*; Altschuler et al., 1979). Several parameters of kidney structure, namely total size (Altschuler et al., 1979) and thickness of the renal medulla (Lawlor and Geluso, 1986), are strongly body size-dependent among perognathines and dipodomyines. Figure 10 shows the relationships between kidney weight and thickness of the inner and outer medullae in several heteromyids and cricetids, based on the data in Altschuler et al. (1979). Relative thickness of the medullae, which contain the Henle's loops and collecting ducts, ostensibly reflects their degree of specialization for concentrating urine. Both outer and inner medullae are thicker relative to kidney weight in *C. penicillatus* than in other heteromyids and in cricetids; *D. merriami* shows enlargement of the inner medulla. It appears that enlargement of the medullae in heteromyids is not accompanied by changes in kidney weight (Altschuler et al., 1979). Ultrastructural studies indicate that the kidney of *C. penicillatus* does not contain novel types of medullary epithelia (Nagle et al., 1981). As there is considerable variation in water regulatory efficiency among perognathines (MacMillen and Hinds, 1983), the kidney morphology of *C. penicillatus* may ultimately prove to be more meaningful ecologically than phylogenetically (and therefore may not be typical of the genus). How these differences in kidney morphology relate to kidney function is uncertain. MacMillen and Hinds (1983) argued that a complex interplay of factors affects water regulation in heteromyids, including ambient temperature, diet, body mass, and phylogeny and that some level of

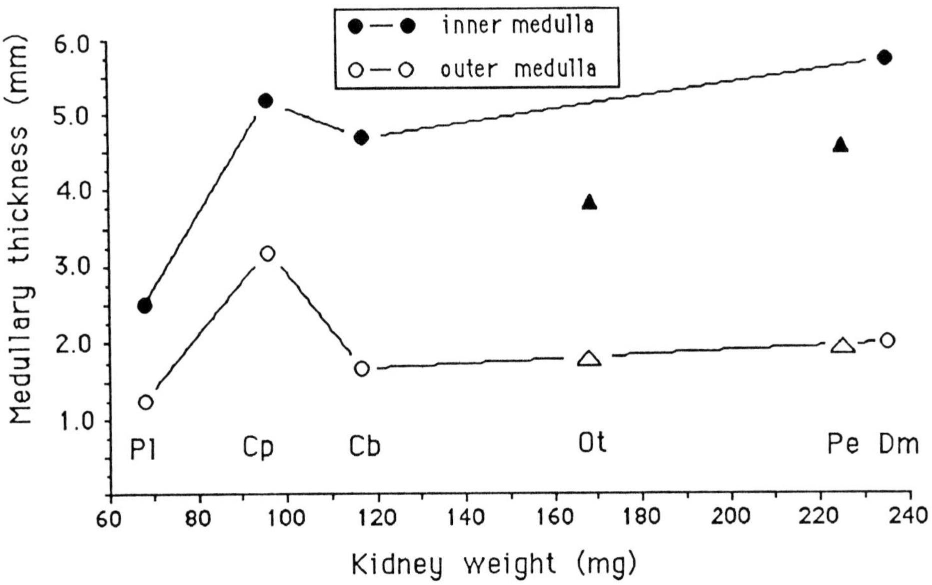

FIG. 10.—Relationships between thickness of the inner and outer medullae and kidney mass in six perognathine, dipodomyine, and cricetid rodents, based on data from Altschuler et al. (1979). The lines connect the data for heteromyids. Heteromyids: Pl, *P. longimembris*; Cp, *C. penicillatus*; Cb, *C. baileyi*; Dm, *D. merriami*; cricetids: Ot, *Onychomys torridus*; Pe, *Peromyscus eremicus*.

water independence is associated with granivory and was shared by early heteromyids that occupied semiarid to subtropical savannah habitats.

Accessory Reproductive Anatomy

Contrary to the findings of Hafner (1982), I found the ventromedial prostates to be present in *Microdipodops* and present, but reduced, in *H. desmarestianus* (Brylski, unpubl. data). The seminal vesicles of heteromyines, dipodomyines, and *Perognathus* are primitively elongate and tubular, whereas those of *Chaetodipus* are derived, being short and rounded (Hafner and Hafner, 1983).

Placentation

Placental anatomy is more derived in heteromyids than in other sciuromorphs and is more primitive than in myomorphs. A complete review of these data can be found in Luckett (1985).

Skeletal Anatomy

Cranial Anatomy

Conspicuous features of heteromyid cranial anatomy are the auditory bullae, which are uninflated in heteromyines, moderately inflated in perognathines, and greatly inflated in dipodomyines (see Lay, 1993). Bullar inflation in perognathines and dipodomyines occurs mostly antero-dorsally, and results in the following suite of craniomorphological changes: 1) the interparietals are reduced in perognathines and absent, or nearly so, in dipodomyines; 2) the shape of the squamosal bone is greatly altered and the parietals change from rectangular in heteromyines to triangular in dipodomyines; 3) the external auditory meatus points laterally and slightly posteriorly in dipodomyines, apparently as a result of inflation of the hypotympanic bulla between the meatus and the temporo-mandibular joint; 4) the posterior part of the temporal fossa, where the temporal muscle attaches to the braincase wall, is lost in *Dipodomys*; and 5) the

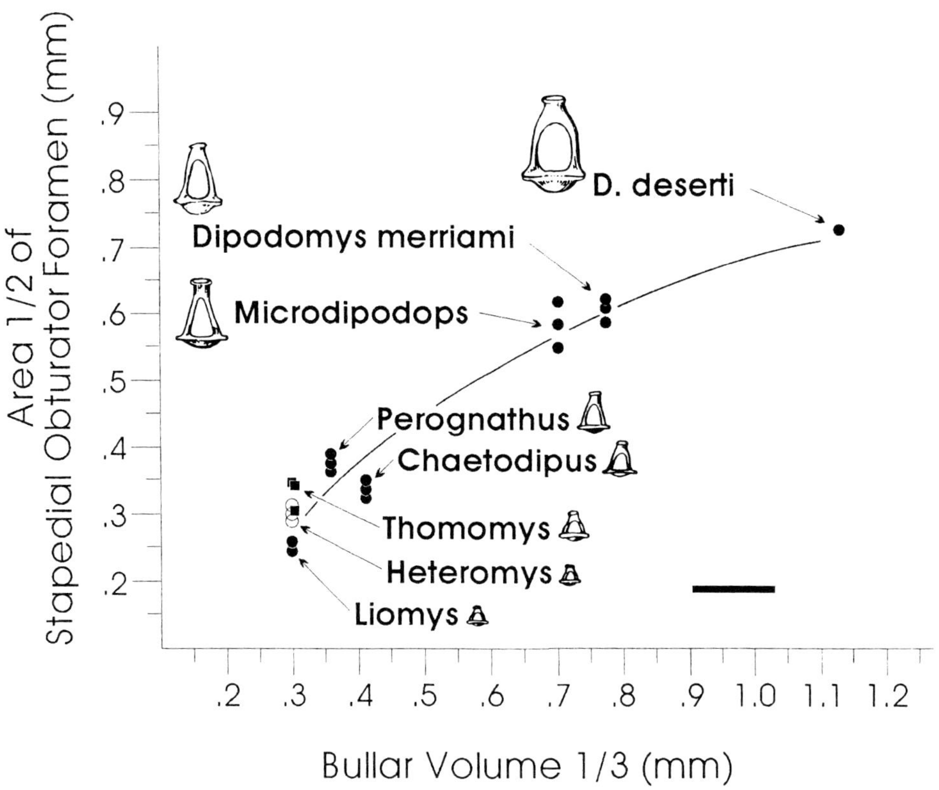

FIG. 11.—Bivariate plot of the (area of the stapedial obturator foramen)$^{1/2}$ and (bullar volume)$^{1/3}$ in six geomyoids. Scale bar for stapes = 2 mm (from Brylski, 1990).

angular processes of the mandible are everted as a solution to (or more likely, an epigenetic consequence of) the restriction on maximum gape caused by bullar inflation (Nikolai and Bramble, 1983).

The temporal region is a functional complex that includes the bony middle and inner ears, the auditory ossicles, and various soft tissues. Each of these shows phylogenetic variation among heteromyids. Cochlear length, measured from the fenestra cochleae to its apex, is short in perognathines and heteromyines (from 2.11 to 2.17 mm) and elongate in dipodomyines (from 3.17 to 4.0 mm). A medial constriction in the cochlea of dipodomyines separates the rostral and caudal modiolar sections, which may reflect their histological specialization. The stapes is unusually small in heteromyines (approximately 0.57 and 0.50 mm from the footplate to the stapes-incus articulation in *H. desmarestianus* and *L. irroratus*, respectively), larger in perognathines (0.73 and 0.83 mm in *C. formosus* and *P. longimembris*, respectively) and larger yet in dipodomyines (1.4, 1.4, and 1.83 mm in *M. megacephalus*, *D. merriami* and *D. deserti*, respectively). The area of the stapedial obturator foramen (the foramen enclosed by the stapedial crura) and bullar volume covary in a positive curvilinear fashion among geomyoids (Fig. 11): the obturator foramen is smallest in geomyines and heteromyines, which have uninflated bullae, intermediate in size in perognathines, which have moderately inflated bullae, and largest in dipodomyines, which have greatly inflated bullae. The footplate of the stapes is flattened in heteromyines and bullate in dipodomyines. Finally, dipodomyines are unique among heteromyids in having a stapedius muscle. Whether this muscle of the middle ear in

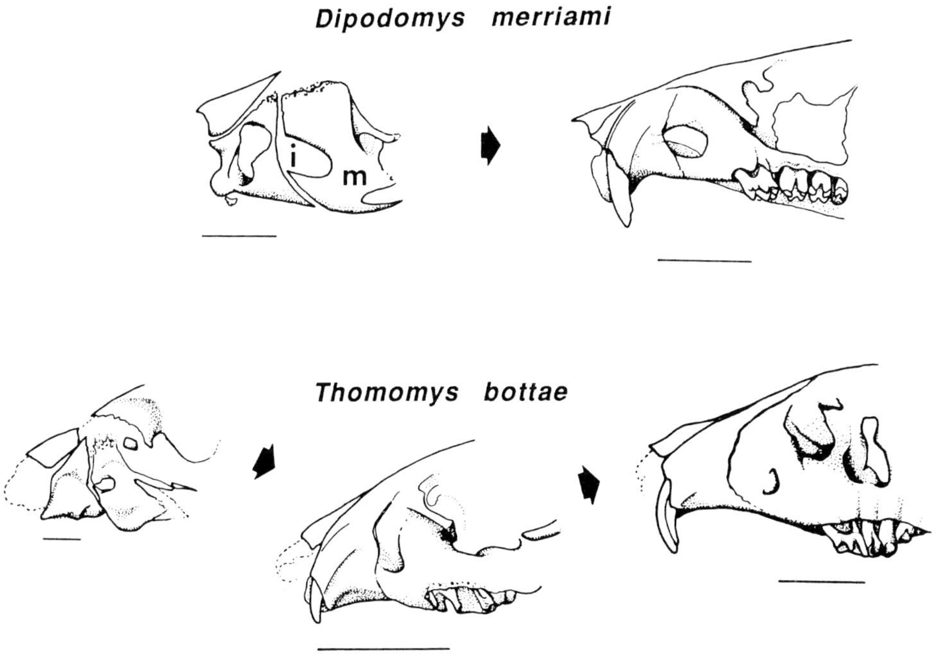

FIG. 12.—Stages in the development of the infraorbital region of the skull in the heteromyid *D. merriami* (above) and in the geomyid *T. bottae* (bottom). In *D. merriami,* the foramen (i) arises in the maxillary bone (m) of the prenatal individuals at the premaxillary suture (left), where it remains at juvenile (right) and adult stages (not shown). In *T. bottae,* the infraorbital foramen arises in the maxillary bone at the premaxillary suture as in *D. merriami,* but bone growth at its cranial margin (left) results in its gradual caudal movement (middle and bottom). Drawn from specimens that had been fixed, cleared, and stained for bone and cartilage. Scale bars = 1 mm for fetuses (top, bottom left) and 5 mm for others (from Brylski, unpubl.).

mammals functions in preventing the inner ear from overstimulation by excessive sound or in enhancing the frequency tuning of the inner ear is unresolved (Fleischer, 1978). The cochlear differences (including cochlear histology and size of the oval window; Webster and Webster, 1975), probably act in concert with the inflated bullae to heighten auditory sensitivity (Webster and Webster, 1984); the influence of the differences in stapedial height (see section on carotid circulation) on inner ear function has not been addressed in heteromyids.

The nasal bones are more tubular and elongated in *Dipodomys, Microdipodops* and *Perognathus* than in heteromyines; where *Chaetodipus* lies on this spectrum is uncertain. In *Dipodomys,* nasal elongation is thought to increase the amount of water resorbed from exhaled air (Schmidt-Nielsen et al., 1970). Several pairs of scroll-shaped maxilloturbinal bones are found in the nasal passages of heteromyids along the medial surface of each nasal bone. One of these ducts, the nasolacrimal, is continuous caudally with the lacrimal duct. In many rodents, the nasolacrimal duct carries a pheromone from the Harderian gland (medial to the eye) to the nose, where it is spread over the face and body with the aid of saliva (Theissen and Yahr, 1977).

The base of the external ear is supported by a cuff of elastic cartilage composed of one or more elements. In all heteromyids, a tubular element extends laterally from the ventral margin of the external auditory meatus. This tubular element is ossified in heteromyines and cartilaginous in perognathines and dipodomyines. The ossified condition is also found in the hystricomorphs *Chinchilla* sp., *Cavia* sp., and *Ctenodactylus* sp. (Allen, 1904). A second, crescent-shaped element lies flatly on the bulla rostral to the meatus, which is cartilaginous in perognathines, ossified in dipodomyines, and apparently absent in heteromyines.

Cranial Foramina

Cranial foramina mark the passage of nerves and vessels into and out of the skull. Their complete description in heteromyids and hypotheses on their phylogenetic significance are found in Wahlert (1985*a*, 1985*b*). The rostral region of heteromyids contains the infraorbital foramen, which is unique among rodents in two respects: 1) it is located anteriorly and pierces the rostrum rather than the zygomatic plate as it does in other rodents, and 2) it opens anteriorly with the premax-maxillary suture. Whether this rostral perforation has any functional significance is unknown. The rostral location of the infraorbital foramen has been argued to be an apomorphy of the Heteromyidae (Wahlert, 1985*a*). In Heteromyids, and in the geomyid *T. bottae,* the infraorbital foramen arises as a stirrup-shaped perforation on the rostral margin of the maxillary bone (Brylski, unpubl.). In heteromyids, the shape and position of the foramen do not differ between fetuses and adults, whereas in *T. bottae* the rostro-dorsal and rostro-ventral margins of the foramen grow and meet shortly after the appearance of the foramen (Fig. 12). Growth in this region continues, resulting in the formation of a bony wall medial to the foramen, producing the imperforate condition. The position of the infraorbital foramen appears to be displaced posteriorly in geomyids compared to heteromyids. However, comparison of the position of the posterior margin of the foramen relative to the maxillary-premaxillary suture in developmental series of *T. bottae, D. merriami, P. longimembris,* and *C. formosus* indicates that these species share a common ontogenetic trajectory in the position of the foramen. The developmental histories of both geomyids and heteromyids are unique among the rodents examined by me (including the sciuromorph *Eutamias minimus,* the myomorphs *Mus domesticus, Rattus rattus,* and *Meriones unguiculatus,* and the caviomorph *Cavia porcella*), obscuring the ontogenetic relationship of the geomorph infraorbital foramen to the other rodent outgroup taxa examined.

The orbit contains three main foramina of interest: the optic, ethmoid, and sphenopalatine. The optic foramen, which is surrounded by the orbitosphenoid bone and transmits the optic nerve and vessels, is small (~1.0 mm) in heteromyines and enlarged in perognathines and dipodomyines. The ethmoid foramen is found near the border of the orbitosphenoid and frontal bones in heteromyines. Perognathines and dipodomyines are derived in having an area of non-ossification where the ethmoid foramen is normally found. This area of non-ossification results from the failure of the orbitosphenoid to ossify dorsally and the failure of the frontal to ossify ventrally (Brylski, unpubl. data), the developmental and functional significance of which is unknown. The sphenopalatine foramen is located at the base of the orbit medial to the alveolar capsules and conducts vessels and a nerve of the same name. Wahlert (1985*a*) noted that the variation in its position was of uncertain phylogenetic significance. The ventral skull region contains the pterygoid and parapterygoid fossae, foramen ovale, and the stapedial foramen. Heteromyids (and geomyids) are unique in having parapterygoid fossae, the paired depressions between the meso- and parapterygoid plates. In geomyines, they house salivary glands and, in *Dipodomys,* fibers of the internal pterygoid muscle originate there. Two primitive features of sciuroids are the presence of a discrete foramen ovale which transmits the mandibular branch of the trigeminal nerve from the cranium, and separate buccinator and masticatory foramina. Alternative derived states are found in my-

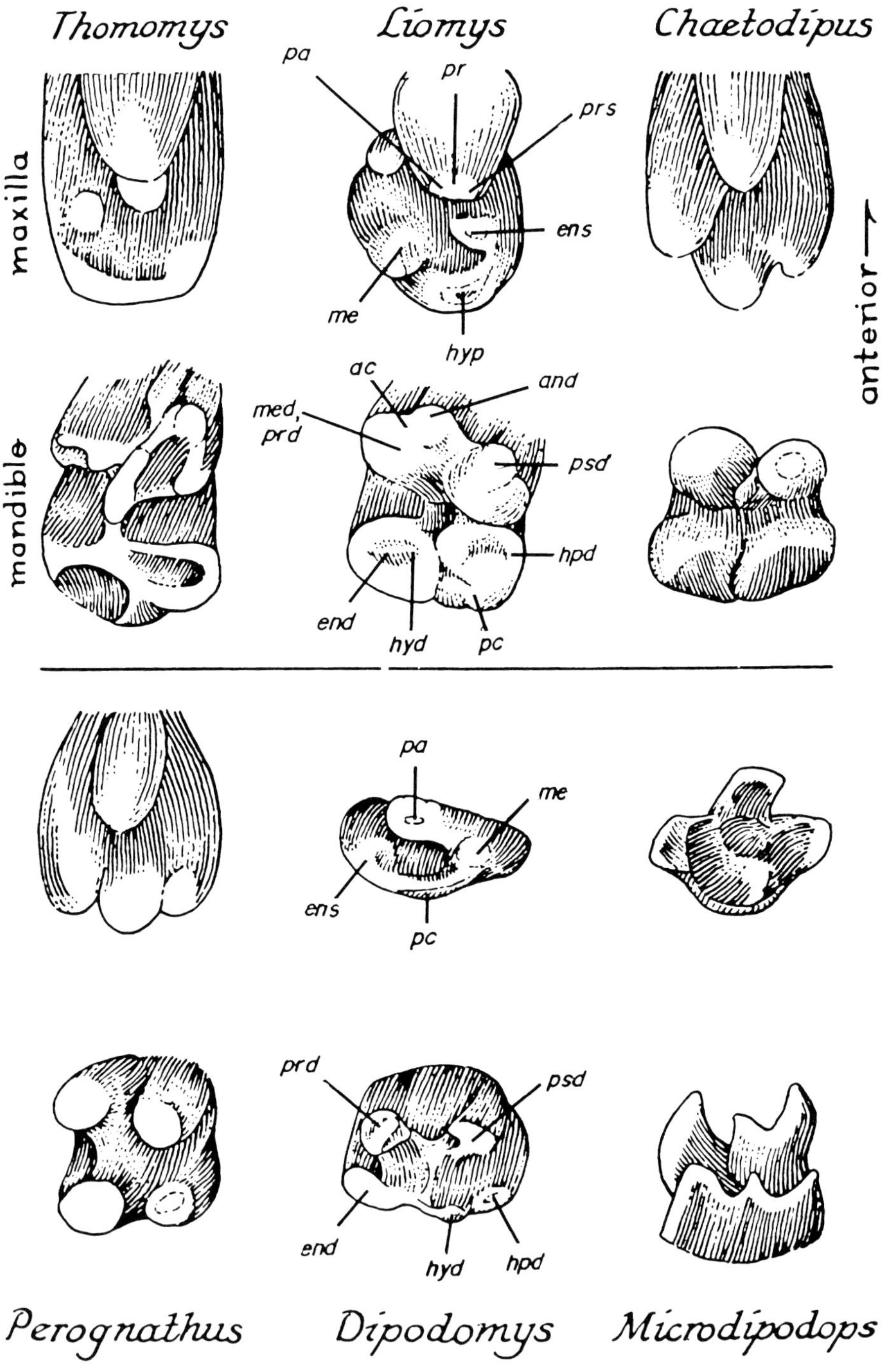

FIG. 13.—The upper (maxillary) and lower (mandibular) P4 of juvenile or subadult geomyoids (not drawn to scale). The cusp terminology follows Rensberger (1971), but the identifications and homologies of the hypoconid, paracone, and parastyle are tentative. (*T. bottae*: MVZ 23764, RP4; *L. salvini*: MVZ 104800, rp4; *C. californicus*: MVZ 5271, RP4; *P. longimembris*: MVZ 47351, RP4, MVZ 46348, rp4; *D. merriami*: MVZ 92779 LP^4, rp4; *M. megacephalus* MVZ 38807, RP4, MVZ 38668, rp4) (from Brylski, 1985). Legend: ac, anterior cigulum and anteroconid; asd, anterostylid;

omorphs and geomyoids: in myomorphs, the buccinator and masticatory foramina are absent; in geomyoids, the foramen ovale has moved posteriorly to become part of the post-alar fissure.

Teeth

The dental formula for all heteromyids is 1013/1013. The upper incisors are grooved in perognathines and dipodomyines and smooth in heteromyines. The cross sectional orientation of the enamel prisms in the incisors of heteromyids is uniserial, which is a derived pattern shared by myomorphs and other scuiromorphs (Koenigswald, 1985). All heteromyids have sexticuspidate and bilophodont molars. Dental variation among the heteromyid genera arises from differences in cuspularity, distribution and depth of enamel, and height of the crown. The molars of heteromyids consist of an anterior hypolophid and a posterior metalophid. There are three principal differences in the molars of heteromyids: 1) with slight to moderate wear, the individual cusps are obliterated in heteromyines and dipodomyines, but remain distinct in most perognathines (particularly *Chaetodipus*); 2) *Dipodomys* has hypsodont, open rooted molars, whereas Heteromyines show a slight tendency toward hypsodonty, but are not appreciably different from perognathines, which retain the primitive brachyodont condition; and 3) the molars of heteromyines and perognathines have enamel bands that surround the lophs and lophids, whereas in *Dipodomys* the enamel (after occlusal wear) is restricted to the anterior and posterior faces of the molars.

The premolars are morphologically more divergent among heteromyid genera than are the molars (Fig. 13). The morphology of P4 is similarly simple in heteromyines and perognathines and more derived in dipodomyines, particularly in *Dipodomys.* The morphology of P4 varies in the following respects: 1) the protoloph is large and three-cusped in heteromyines and small and single-cusped in perognathines and dipodomyines; 2) the hypostyle of the metaloph is positioned anteriorly in heteromyines, resulting in the J-shaped metaloph characteristic of this subfamily; and 3) the metacone has moved anteriorly in *Dipodomys,* resulting in its labial union with the paracone.

The morphology of p4 is simple in perognathines, complex in heteromyines, and intermediate in dipodomyines. The particular ways in which p4 differs are: 1) an anterior cingulum bearing a protostylid and one or more anterostylids is present in heteromyines and absent in other heteromyids—*Heteromys* has four to seven cingular cusps and *Liomys* has one or two; 2) the hypolophid of heteromyines bears an anteroconid as an anterior spur of the protoconid; 3) the metalophid and hypolophid are each three-cusped in heteromyines and dipodomyines and two-cusped in perognathines; and 4) the posterior cingulum is prominent in heteromyines, moderately well-developed in dipodomyines, and poorly developed to absent in perognathines.

There are several unresolved problems with respect to dental evolution in heteromyids (see also Wahlert, 1993). The first is the dental differences between the microsympatric and ecologically similar perognathines and dipodomyines. Both taxa are largely granivorous and seasonally forage on plant parts and insects, which elicits the logical question: What is the functional and evolutionary significance of their dental differences? It is unlikely that their food habits

←
end, entoconid; ens, entostyle; hpd, hypostylid; hyp, hypocone; hyd, hypoconid; hys, hypostyle; me, metacone; med, metaconid; pa, paracone; pc, posterior cingulum; pr, protocone; prd, protoconid; prs, protostyle; psd, protostylid.

are more different than has been recognized, although other differences such as locomotory mode (Rensberger, 1975), life span, or body size may have resulted in different selective pressures on dental function (and therefore structure). Dipodomyines do not appear to be more long-lived than perognathines (summarized by Jones, 1985) but, with the exception of *Microdipodops* (which lacks several derived dental characters found in *Dipodomys*) they are larger than perognathines.

A second problem is whether the simple p4 morphology found in perognathines is primitive to the complex patterns in heteromyines or vice versa. If the perognathine condition is primitive, then the derived cusp morphologies of heteromyines and geomyids argue strongly for the monophyly of these latter two taxa (and hence paraphyly of the currently recognized Heteromyidae) (Rensberger, 1971). On the other hand, if a complex p4 morphology is primitive, as argued by Wahlert (1978), then the hypothesis of heteromyid monophyly is more parsimonious than the paraphyly hypothesis (see also Wahlert, 1993). In either case, because perognathines and dipodomyines are monophyletic relative to heteromyines (based on other morphological characters), the complex dental features of dipodomyines and heteromyines are independently derived.

Postcranial Anatomy

Detailed descriptions of the postcranial anatomy of heteromyids can be found in Hatt (1932), Howell (1932), and Wood (1935). Lehmann (1963) and Nikolai and Bramble (1983) discuss the functional significance of some of these differences.

Cervical vertebrae.—Dipodomys is derived among heteromyids in that the second (axis) and third cervical vertebrae are fused dorsally. In *D. merriami,* fusion occurs between the cartilaginous neural processes late in juvenile life (30 to 40 days post-partum; Brylski, unpubl. data). The occasional fusion of cervicals three, four, and five in older individuals of *Microdipodops* and *Dipodomys* apparently occurs between previously ossified bones, and may result from their unique head posture and flexion of the neck (Hatt, 1932).

Scapulae.—The scapulae of all heteromyids have a pronounced subscapular fossa, which is correlated in fossorial rodents with the development of a large teres major muscle (Lehmann, 1963). The fact that heteromyines are like other heteromyids (as well as geomyids) in this respect has been interpreted to mean that living heteromyines may be less fossorial than were their ancestors (Nikolai and Bramble, 1983). In heteromyines and *Chaetodipus,* the supraspinous and infraspinous fossae of the lateral scapular surface are roughly equal in size. In *Dipodomys,* the supraspinous fossa is much reduced. This character is obscured in *Perognathus* and *Microdipodops* by the small overall size of the scapula. The supraspinous muscle is a shoulder joint extensor, and is not enlarged in rodents that are otherwise specialized for digging (Lehmann, 1963). The supraspinous muscle may be more important during locomotion in quadrupedal than in bipedal heteromyids. In *Rattus norvegicus,* experimental excision of the supraspinous muscle early in development results in an underdeveloped supraspinous fossa (Wolffsen, 1950) similar in appearance to that of *Dipodomys.* Thus, it is possible that reduction of this fossa in *Dipodomys* is related to their bipedal locomotion. Comparative studies on the size and function of the supraspinous muscle in heteromyids are needed to test this hypothesis.

Bacula.—Apart from considerable differences in length, the bacula is morphologically similar among heteromyids, being bulbous proximally and decurved distally. The bacula of heteromyines and *C. hispidus* are keeled distally. The baculum of *C. hispidus* also has two round knobs ventrolaterally, giving the tip a trifid appearance.

Examination of bacular and body length data for living heteromyids (Best and Schnell, 1974; Burt, 1936; Genoways, 1973) reveals that bacular length is: 1) short in the small bodied *Perognathus* (from 4.3 to 7.1 mm) and *M. megacephalus* (6.0 mm); 2) elongate in the larger bodied *Liomys* (from 7.9 to 9.5 mm), *Heteromys* (8 mm in *H. oresterus*), and *Dipodomys* (from 8.1 to 13.4 mm); and 3) elongate in *Chaetodipus* (from 9.0 to 14.8 mm), despite their smaller body size. Among heteromyids, only *D. spectabilis* has a larger baculum than *C. hipidus* (17.4 and 14.8 mm, respectively).

Body Proportions

The body proportions of some heteromyids differ in a size-independent fashion and are correlated with differences in their locomotory mode. This was first shown by Howell (1932) and Wood (1935) using various body proportion indices. Recent studies have addressed the problem of comparing body proportions in different sized animals by fitting the power function $y = ax^b$ to skeletal and body measurements, where y and x are the dependent and independent variables, respectively, b is the slope of the regression line, and a is its y-intercept (at x = 0). Isometry (slope = 1.0) assumes geometric similarity between linear variables and the cube root of body weight. Relationships which have slopes significantly greater and less than 1.0 show positive or negative allometry, respectively. Brylski (unpubl.) examined the interspecific allometry of various body proportions in 16 heteromyid species belonging to the six living genera. Four representative patterns from this survey are shown in Figure 14. Heteromyids are largely or entirely conservative in their basilar skull length and radius length to body weight, per length to tibia length, and hindlimb (=femur + tibia) length to naso-occipital skull length (e.g., Fig. 14A). This last relationship is conservative despite changes in hindlimb length in dipodomyines because both hindlimb and skull length are elongated, the latter by virtue of middle ear inflation posterior to the occiput.

The body proportions of heteromyines and perognathines are allometrically similar in most respects (their taxa lie along the same regression line), which is interesting considering the habitats they occupy range from deserts to tropical forests. The proximate explanation for this conservatism lies in their similar modes of locomotion (quadrupedal saltation). Perognathines and dipodomyines are monophyletic (Hafner, 1982; Wahlert, 1985*a*), and although the evolution of these two clades are linked to the Tertiary expansion of arid habitats in western North America (Reeder, 1956), no changes in body proportion underlie the heteromyine versus perognathine plus dipodomyine dichotomy.

Compared to perognathines and heteromyines, dipodomyines have a larger brain (Hafner and Hafner, 1984), and a more elongated skull (via bullar inflation), hindlimb, and pes (including metatarsus; Brylski, unpubl.; Fig. 14B). Both the femur and tibia contribute to hindlimb elongation but the contribution from the latter is greater. The hand of dipodomyines is reduced in size (Nikolai and Bramble, 1983). *Dipodomys* has a shortened humerus, which may be an adaptation for bipedal locomotion that permits the forelimbs to be tucked close to the neck during locomotion and may aid in food manipulation (Nikolai and Bramble, 1983).

Tail length is a special case of skeletal elongation in heteromyids (Fig. 14D). *Dipodomys* have more elongate tails than do heteromyines. However, perognathines occupy a position that is basal to both the short and long-tailed taxa. These allometric patterns make it difficult to determine whether perognathines are allometrically more similar to heteromyines or to *Dipodomys*. This issue is important because if elongated tails are not dissociable from body size, it cannot be presumed that they are adaptations for bipedal locomotion in *Dipodomys*. Such an interpretation is consistent with the presence of a short tail in the diminutive and bipedal *Microdipodops*. Since *Microdipodops* is bipedal, but not significantly faster than comparably sized perognathines (Djawdan and Garland, 1988), it is more likely that the elongated tail in *Dipodomys* is adaptive for maintaining balance during high speed locomotion.

Comparative Skeletal Development

The Ontogeny of Body Proportion Differences

The developmental bases of morphological diversification can be studied at various levels related to growth and differentiation. Figure 15 summarizes in schematic form three patterns of relative growth which underlie the evolutionary changes in body proportions among geomyids (Brylski, unpubl.). Each figure shows a growth relationship of two body parts Y and X (X may also be body weight) in taxa *a* and *b*. Provided that the time scale is the same for any pair of growing body parts shown in Figure 15, heterochrony results from changes in the relative rate of growth (Fig. 15-left), the onset of growth of one body part relative to another (Fig. 15-center), and an extension or truncation of a common growth trajectory (Fig. 15-right). If the regression slopes of Figure 15-right do not differ significantly from 1.0 (isometry), then the shapes of *a* and *b* are geometrically similar (there is no shape change). Significant deviations from isometry describe shape change both in ontogeny (along each line) and phylogeny (between taxa *a* and *b*).

The pattern shown in Figure 15-left underlies many morphological differences between heteromyids and geomyids. Humerus reduction in *D. merriami* and *D. deserti* arises by a truncation in development, illustrated in Figure 15-right, with the ancestral ontogeny depicted for taxon *b* and the derived *Dipodomys* ontogeny as taxon *a*. Elongation of the hindlimb and pes in *D. merriami* and *D. deserti* also arises late in development by the pattern depicted in Figure 15-right, this time with taxon *a* as primitive and taxon *b* (*Dipodomys*) as derived. In *D. merriami*, pes elongation commences earlier than hindlimb elongation. A comparison of growth of the epitympanic and antral parts of the middle ear in *T. bottae*, *C. formosus*, *P. longimembris*, and *D. merriami* suggests a somewhat more complex evolutionary history. If the greatly inflated bulla of dipodomyines were derived from a moderately inflated bulla (like that found in perognathines), and if this were derived from an uninflated bulla (like that found in heteromyines), then two heterochronic events are indicated by the relative growth data: 1) an acceleration in the rate of bullar growth (e.g., Fig. 15-left; taxon *b* is primitive), resulting in a moderately inflated bulla; and 2) a second acceleration accompanied by an extension in its duration to produce the greatly inflated condition.

The impressive array of derived morphological characters in *Dipodomys* has led to the question: Might these features be coordinately controlled during their development and evolution? Along the same lines, Hafner and Hafner (1983) hypothesized that the suite of derived morphological characters in *Dipodomys* arose from a heterochronic change in one or few developmental control parameters. Hypotheses of heterochrony are sometimes used to explain the relatively rapid emergence of complex morphologies and the absence of intermediate forms. However, little or nothing is known about the rates of morphological evolution in *Dipodomys* and a number of fossils display either moderately inflated bullae (Wood, 1935) or derived limb proportions (Voorhies, 1975; Wood, 1935). Testing the hypothesis that the elongated limbs, enlarged brains, and inflated middle ears of *Dipodomys* arose by a change in a common growth regulating mechanism requires a quantitative genetic analysis of these traits, a barrier to which is the difficulty of breed-

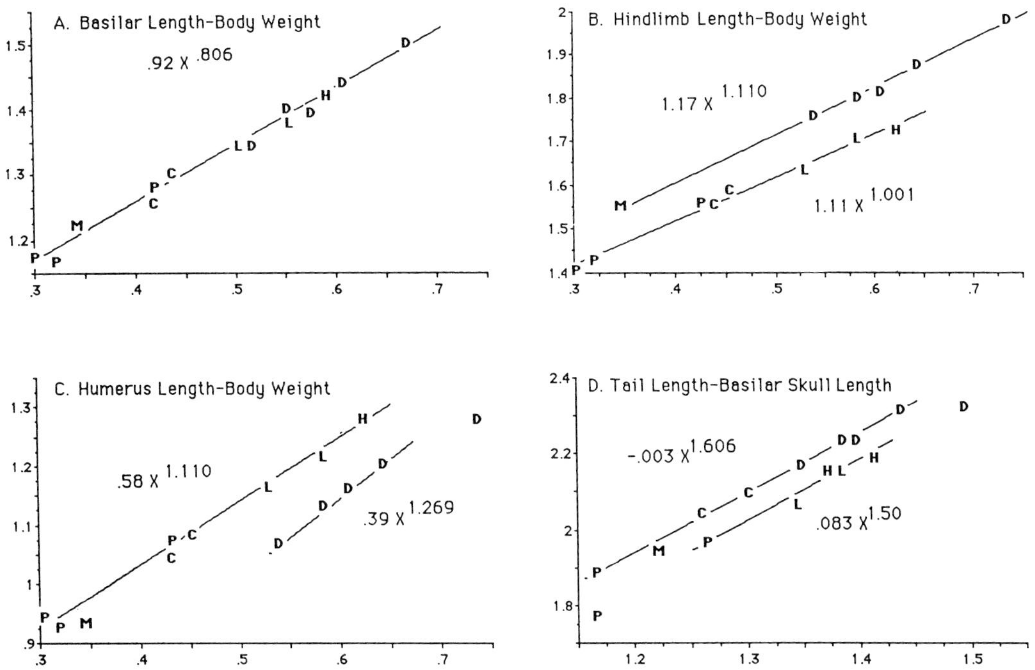

FIG. 14.—Four allometric patterns in the body proportions of 14 heteromyid species based on log-transformed means of adult measurements. The exponent of the power function is the slope of the nearest line. The x-axis scale is log (body $\text{weight}^{1/3}$) in *A–C* and log (basilar skull length) for *D*. *A*, Conservatism: basilar skull length; *B*, Elongation (in dipodomyines): hindlimb length (=femur + tibia lengths); *C*, Reduction (in *Dipodomys*): humerus length; *D*, Tail length: The two regression lines shown represent one possible interpretation. An alternative is that *P. longimembris* and *M. megacephalus* are allometrically related to *P. parvus* and heteromyines. Both assume that *P. merriami* and *D. spectabilis* are outliers (from Brylski, 1985). Taxonomic abbreviations, with congeners listed in ascending size along X-axis: C: *Chaetodipus formosus, C. baileyi*; D: *Dipodomys merriami, D. agilis, D. panamintinus, D. deserti, D. spectabilis*; M: *M. megacephalus*; P: *Perognathus merriami, P. longimembris, P. parvus*; H: *Heteromys desmarestianus*; Hg, *H. gaumeri* (only in *D*); Li, *L. irroratus*; Ls, *L. salvini.*

ing captive heteromyids. In the absence of such data, we can turn to interspecific comparisons because correlated evolutionary changes among body parts are a necessary, but not sufficient, condition of the hypothesis that they share a common growth regulating mechanism. Interspecific comparisons reveal that the list of correlated skeletal features does not include the auditory bullae and tail. This is because all perognathines, the sister group to dipodomyines, possess moderately inflated bullae and some have elongated tails, yet all lack elongated hindlimbs. Thus, phylogenetic changes in the auditory bullae and tail are independent of one another and of phylogenetic changes in the femur, tibia, and pes. The femur, tibia, and pes are correlated characters in the sense that they are all either primitive or derived in different heteromyid species. A longitudinal comparison of the origin of body proportion differences in heteromyids revealed that elongation of the pes in *D. merriami* and *D. deserti* begins earlier in development than elongation of the femur and tibia (Brylski, unpubl.). It is conceivable that pleiotropic changes in the cellular properties of developing limb buds are expressed at different times during growth (see Riska, 1986 for further discussion of this problem).

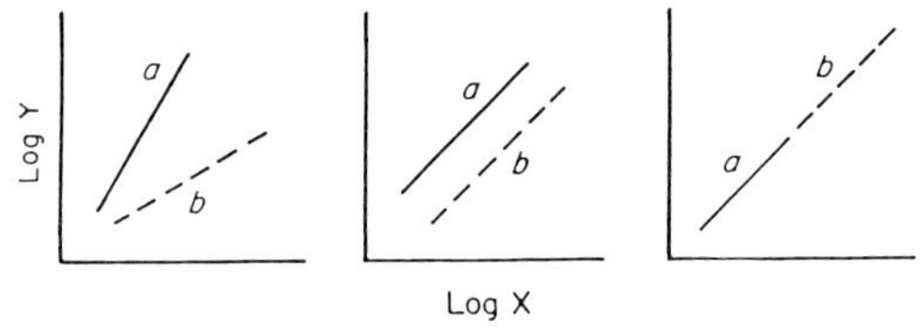

FIG. 15.—Schematic representation of three possible patterns of relative growth in two species (*a* and *b*) for two body parts (Y versus X), or one body part (Y) relative to body weight (X). Left: the regression slopes differ significantly. This pattern describes hypertrophy of the auditory bulla in perognathines and *Dipodomys* (taxon *a* is derived; *Dipodomys* also displays pattern on the right, compared to perognathines). Center: the slopes are identical but the y-intercepts differ significantly. This pattern was not encountered in the early development of heteromyids. Right: the slopes and y-intercepts are identical, but the trajectory for taxon *b* recapitulates that of taxon *a*. This pattern describes elongation of the hindlimb, pes, and tail (as taxon *b*) and humerus reduction (as taxon *a*) in *Dipodomys* (from Brylski, 1985).

Skeletal Ossification

Heterochrony is often studied at the level of developmental events rather than growth (Alberch et al., 1979). Brylski (1985) found that the changes in body proportion among heteromyids are not correlated with changes in the sequence and/or timing of skeletal ossification.

The heteromyids studied by Brylski (1985; *C. formosus, P. longimembris, D. merriami* and *D. deserti*) are conservative with respect to the sequence of skeletal ossification. The first bones to ossify are the membranous skull bones, the proximal and middle phalanges, and the diaphyses of the limbs, scapula, and clavicle. These are followed by the pelvic bones, ear ossicles, cochlea, tympanic ring, and certain distal limb bones (the capitulum of the humerus, metacarpals, metatarsals, calcaneus, and astragalus). The last bones to ossify are the epiphyses of the fibula and calcaneus, the femoral trochanters, the patella, and the lateral fabella of the knee (a sesamoid bone).

There is wide, size-related variation in the timing of ossification. Postcranial skeletal ossification in the diminutive *P. longimembris* is substantially advanced in time in comparison to *D. merriami,* which is more advanced than the larger and morphologically similar *D. deserti.* The timing of ossification in the temporal region shows a size continuum based on the size of the bulla rather than body size.

Are Dipodomyines Paedomorphic or Peramorphic?

Paedomorphosis and peramorphosis are morphological expressions of changes in the timing of development (Alberch et al., 1979). Paedomorphic adults retain juvenile features of their ancestors, whereas peramorphic adults (excluding proportional giants and dwarfs) have features not found in any ancestral stage. Hafner and Hafner (1983) and Hafner (1993) argue that dipodomyines are paedomorphic in two respects: 1) their large brains and skulls and elongate hindlimbs are juvenile-like; and 2) *Dipodomys* is K-selected, which Gould (1977) correlates with paedomorphosis by neoteny (*Microdipodops* is apparently secondarily derived in being r-selected, and therefore paedomorphic by progenesis).

This dual argument illustrates two integrated aspects of heterochrony: its ecological significance in relation to life history evolution and its morphological expression via allometric growth. For example, a K-selected increase in body size can be accomplished by extending the ancestral growth trajectory. Whether new shapes evolve depends on the allometric growth relationships among body parts that are inherited or evolved. For example, in *Homo sapiens* paedomorphosis occurs by a delay in maturation and is accompanied by the retention of certain morphological features found in juvenile gorillas and chimpanzees (Gould, 1977). However, paedomorphosis in *H. sapiens* is not global (organism-wide), as evidenced by the elongated trunk region and enlarged brains, which are peramorphic (Gould, 1977). Similarly, one derived feature in *Dipodomys,* the shortened humerus, is paedomorphic, whereas others, the middle ear, hindlimbs, and brain, are peramorphic.

The argument that *Dipodomys* is K-selected (Hafner and Hafner, 1983) implies that they evolved from small bodied (perognathine-like) r-selected ancestors. Although early heteromyids apparently were small bodied (Wood, 1935), this hypothesis on the polarity of body size in the phylogenetically much younger dipodomyines remains to be tested. Moreover, heteromyines (which lack specialized body proportions) and *Dipodomys* of comparable body mass do not differ importantly in their age of sexual maturation (Fleming, 1974), and one measure of somatic development (body growth) appears to be accelerated in *Dipodomys* (Fleming, 1977; retarded somatic development is typical of paedomorphs; Gould, 1977). Thus, the idea that the life-history attributes of *Dipodomys* are paedomorphic may also be incorrect.

Conclusions

Most of the morphological variation among heteromyids is found between the genera of different subfamilies: among these, perognathines and heteromyines are the least derived and dipodomyines are by far the most derived. More specifically, heteromyines are recognized by six derived characters, perognathines by two, and dipodomyines by 15 (Brylski, 1985). Why do the Perognathinae and Dipodomyinae, which are sympatric sister taxa, show such different amounts of morphological evolution? A potentially important clue is their microhabitat differences: perognathines typically forage beneath and around vegetation, whereas dipodomyines forage in the open areas between shrubs where predation pressure is presumably greater (Price and Brown, 1983). The highly derived morphology of dipodomyines may relate to niche specialization and the consequence of increased predation on taxa that occupy the most open habitat (Price and Brown, 1983), provided that this microhabitat difference has an historical basis, i.e., if it also characterized the primitive ancestors of dipodomyines and their contemporaneous perognathines. That the derived morphology of dipodomyines is adaptive in this way is supported by the observation that the specialized characters of the temporal bone and hindlimbs, which together account for 11 of the 15 derived characters shared by dipodomyines (Brylski, 1985), function in the detection and avoidance of predators, respectively.

There are also several alternative interpretations of the morphological gap between perognathines and dipodomyines. The first of these focuses on the suite of derived features and argues that some derived traits may be pleiotropic consequences of selection on others (Cheverud et al., 1983), in which case no functional explanation is required for their origin (in contrast to their current utility; Hafner and Hafner, 1983). The interspecific and ontogenetic allometry data indicate that, of the suite of derived features in dipodomyines, only the hindlimb elements are correlated characters, and whether these are pleiotropic remains to be demonstrated. The second interpretation focuses on morphological stasis as an outcome of the tendency of organisms to compensate environmental and genetic perturbations without changing their morphologies (Wake et al., 1983). This controversial hypothesis (see Lande, 1986 for an alternative view) is potentially applicable to perognathines. However, it remains to be demonstrated, by a quantitative assessment of morphological change in geological time in *Chaetodipus, Perognathus* and their ancestors, that the morphologically primitive perognathines indeed display stasis.

The anatomical data reviewed in this paper provide specific observations and questions on heteromyid evolution. For example, *Microdipodops* probably evolved from a bipedal species similar to *Dipodomys* prior to the evolution in *Dipodomys* of a reduced humerus, dorsal gland and several dental specializations. Thus, *Dipodomys* is not appreciably more derived than its sister taxon, *Microdipodops* (cf. Berman, 1985; Hafner and Hafner, 1983). Whether the complex dental anatomy of heteromyines is primitive or derived (Rensberger, 1971; Wahlert, 1985*a*, 1985*b*) awaits clarification from paleontologists. A number of anatomical features (e.g., body size, guard hairs, seminal vesicle shape, ear pinna) distinguish *Chaetodipus* from *Perognathus,* excluding *P. parvus.* Whether the smallest silky pocket mice (*Perognathus,* excluding *P. parvus*) represent a monophyletic clade awaits clarification from sytematists, as does the relationship of *P. parvus* to other *Perognathus* and *Chaetodipus.* The ways in which perognathines and dipodomyines are derived relative to heteromyines might be expected to exemplify adaptations to xeric versus mesic habitats, respectively. But perognathines and dipodomyines share relatively few derived traits, and only two of these, the elongated nasals (which might not be shared by *Chaetodipus*) and a water conserving kidney are candidates as adaptations to desert life. The absence of a clear physiological difference among heteromyids that is independent of body size suggests that some level of water independence was shared by early heteromyids that occupied semiarid to subtropical scrub savannah.

Three general questions have been pursued by students of heteromyid evolutionary morphology: 1) What are the major events in the phylogenetic history of heteromyids? The literature on heteromyid anatomy is extensive, but the events in their morphological evolution are not completely understood. A greater understanding of the anatomy of heteromyines is needed and a more conscientous effort should be made to present anatomical data on heteromyids in a phylogenetic context. The latter suggestion requires that morphological variation be assessed in the entire family, and in outgroup taxa as a basis for polarizing these characters. 2) How have these differences arisen during evolution? Ontogeny is the most direct means available to us for answering this question. Its validity requires that there be parallels between ontogeny and phylogeny. The question of how the patterns of morphological differences arose has been addressed with respect to body proportions, skeletal ossification, and external cheek pouches. The utility of a developmental perspective is evident from this meager start, which might profitably be applied to other aspects of heteromyid anatomy and to other mammalian clades. 3) How have these changes enhanced the fit between the organism and its environment? Some progress has been made on the experimental and functional morphology of heteromyids (Biewiener et al., 1981; Nikolai and Bramble, 1983; Thompson, 1985; Webster and Webster, 1984), but there are many unanswered questions and relatively few workers. These efforts would benefit from greater attention to phylogenetic hypotheses (e.g., comparing dipodomyines with heteromyines rather than, or in addition to, cricetids and sciurids).

Acknowledgments

I am grateful to W. Z. Lidicker, J. L. Patton, and K. Padian for their comments and discussions on the ideas contained in this paper. A. E. Wood, M. McLaren, W. Quay, and an anonymous reviewer offered helpful comments on the manuscript, although all errors remain my own. Gene Christman, Karen Christiansen, and Katrina Rice prepared some of the figures. Financial support during the preparation of this paper was provided by the Killam Foundation of Dalhousie University.

Literature Cited

Alberch, P., S. J. Gould, G. F. Oster, and D. B. Wake. 1979. Size and shape in ontogeny and phylogeny. Paleobiology, 5:296–317.

Alexander, R. McN., and H. C. Benet-Clark. 1980. Storage and elastic strain energy in muscle and other tissues. Nature, 265:114–117.

ALEXANDER, R. NCN., A. S. JAYES, G. M. O. MALOIY, AND E. M. WATHUTA. 1981. Allometry of the leg muscles of mammals. Journal of Zoology, 194:539–552.

ALLEN, J. A. 1904. The external ear bone in certain rodents. Bulletin of the American Museum of Natural History, 20:135–138.

ALTSCHULER, R. B., E. J. NAGLE, S. L. LINDSTEDT BRAUN, AND P. H. KRUTZSCH. 1979. Morphological study of the desert heteromyid kidney with emphasis on the genus *Perognathus*. Anatomical Record, 194:461–468.

BARTHOLOMEW, G. A., AND G. R. CAREY. 1954. Locomotion in *Perognathus*. Journal of Mammalogy, 35:386–392.

BARTHOLOMEW, G. A., AND H. H. CASEWELL. 1951. Locomotion in kangaroo rats and its adaptive significance. Journal of Mammalogy, 32:155–169.

BERMAN, S. L. 1985. Convergent evolution in the hindlimb of bipedal rodents. Zeitschrift zoologie Systematik v. Evolutionforshung, 23:59–77.

BEST, T. L., AND G. D. SCHNELL. 1974. Bacular variation in kangaroo rats (genus *Dipodomys*). American Midland Naturalist, 91:257–270.

BIEWIENER, A., R. MCNEIL-ALEXANDER, AND N. C. HEGLUND. 1981. Elastic energy storage in the hopping of kangaroo rats (*Dipodomys spectabilis*). Journal of Zoology, 195:369–383.

BRYLSKI, P. 1985. Morphological evolution and comparative development of geomyoid rodents. Ph.D. dissert., University of California, Berkeley, 196 pp.

———. 1990. The development and evolution of the carotid circulation of geomyoid rodents in relation to their craniomorphology. Journal of Morphology, 204:33–46.

———. In press. The identification of primitive and derived limb proportions and locomotory modes in fossil geomyoid rodents. Journal of Paleontology.

BRYLSKI, P., AND B. K. HALL. 1988*a*. The ontogeny of a macroevolutionary phenotype: the external cheek pouches of geomyoid rodents. Evolution, 42:391–395.

———. 1988*b*. Epithelial behaviors and threshold effects in the development and evolution of internal and external cheek pouches in rodents. Zeitschrift zoologie Systematik v. Evolutionforshung, 26:144–154.

BUGGE, J. 1971. The cephalic arterial system in sciuromorphs, with special reference to the systematic classification of rodents. Acta Anatomica, 80:336–361.

BULLOCK, T. H. 1984. The future of comparative neurology. American Zoologist, 24:693–700.

BURT, W. H. 1936. A study of the baculum in the genera *Perognathus* and *Dipodomys*. Journal of Mammalogy, 17:145–156.

BYRD, K. E. 1981. Mandibular movements and muscle activity during mastication in the guinea pig (*Cavia porcellus*). Journal of Morphology, 170:147–169.

CHEVERUD, J. M., J. J. RUTLEDGE, AND W. R. ATCHLEY. 1983. Quantitative genetics of development: genetic correlations among age-specific trait values and the evolution of ontogeny. Evolution, 37:895–905.

COUES, E. 1875. A critical review of the North American Saccomyidae. Proceedings of the Academy of Natural Sciences, Philadelphia, 272–327.

DAWSON, T. J. 1976. Energetic cost of locomotion in Australian jumping mice. Nature, 259:305–307.

DJAWDAN, M., AND T. GARLAND. 1985. Maximal running speeds of bipedal and quadrupedal rodents: relevance for coexistence? American Zoologist, 25: 8A.

———. 1988. Maximal running speeds of bipedal and quadrupedal rodents. Journal of Mammalogy, 69:765–772.

DRESSLER, J. B. 1979. An anatomical study of the brains of *Dipodomys* (Mammalia: Rodentia: Heteromyidae). Anatomischer Anzeizer, 145:359–368.

EISENBERG, J. F. 1963. The behavior of heteromyid rodents. University of California Publications in Zoology, 69:1–114.

———. 1981. The mammalian radiations. University of Chicago Press, Chicago, 610 pp.

EMERSON, S. B. 1985. Jumping and leaping. Pp. 58–72, *in* Functional vertebrate morphology (M. Hildebrand, D. M. Bramble, K. F. Liem, and D. B. Wake, eds.). Belknap Press, Cambridge, 430 pp.

FAHLBUSCH, V. 1985. Origin and evolutionary relationships among geomyoids. Pp. 617–629, *in* Evolutionary relationships of rodents (W. P. Luckett and J. L. Hartenberger, eds.). NATO ASI series A, vol. 92, 721 pp.

FEDAK, M. A. AND H. J. SEEHERMAN. 1979. Reappraisal of energetics of locomotion shows identical cost in bipeds and quadrupeds including ostrich and horse. Nature, 282:713–716.

FLEISCHER, G. 1978. Evolutionary principles of the mammalian middle ear. Advances in Anatomy, Embryology & Cell Biology, 55:1–70.

FLEMING, T. H. 1974. The population ecology of two species of Costa Rican heteromyid rodent. Ecology, 55:493–510.

———. 1977. Growth and development of two species of tropical heteromyid rodents. American Midland Naturalist, 98:109–123.

GAMBARYAN, P. P. 1974. How mammals run. Halsted Press, New York, 367 pp.

GENOWAYS, H. H. 1973. Systematics and evolutionary relationships of spiny pocket mice, genus *Liomys*. Special Publications, The Museum, Texas Tech University, 5:1–368.

GORNIAK, G. C. 1977. Feeding in the golden hamster, *Mesocricetus auratus*. Journal of Morphology, 154: 427–458.

———. 1985. Trends in the actions of mammalian masticatory muscles. American Zoologist, 25:331–338.

GOULD, S. J. 1977. Ontogeny and phylogeny. Belknap Press, Harvard, 501 pp.

GRINNELL, J. 1922. A geographical study of the kangaroo rats of California. University of California Publications in Zoology, 24:1–124.

HAFNER, J. C. 1993. Macroevolutionary diversification of heteromyid rodents: heterochrony and adaptation in phylogeny. Pp. 291–318, *in* biology of the Heteromyidae (H. H. Genoways and J. H. Brown, eds.). Special Publication, American Society of Mammalogists, 10:1–719.

HAFNER, M. S. 1982. A biochemical investigation of geomyoid systematics (Mammalia: Rodentia). Zeitschrift zoologie Systematik v. Evolutionforshung, 20: 118-130.

Hafner, M. S., and J. C. Hafner. 1983. Evolutionary relationships of heteromyid rodents. Pp. 3–29, *in* The biology of desert rodents (O. J. Reichman and J. H. Brown, eds.). Great Basin Naturalist Memoirs, No. 7, 134 pp.

———. 1984. Brain size, adaptation, and heterochrony in geomyoid rodents. Evolution, 38:1088–1098.

Hall, B. K. 1984. Developmental processes underlying heterochrony as an evolutionary mechanism. Canadian Journal of Zoology, 62:1–7.

Hall, E. R. 1946. The mammals of Nevada. University of California Press, Berkeley, CA.

Hardy, M. L., O. E. Vrablic, H. A. Covant, and S. V. Kandarakar. 1986. The development of the Syrian hamster cheek pouch. Anatomical Record, 214:273–282.

Hatt, R. T. 1932. The vertebral column of ricochetal rodents. Bulletin American Museum of Natural History, 63:599–738.

Hiiemae, K. 1971. The structure and function of the jaw muscles in the rat (*Rattus norvegicus* L.). Zoological Journal of the Linnaean Society, 50:111–132.

Hill, J. E. 1935. The retractor muscle of the pouch in the Geomyoidea. Science, 81:160.

———. 1937. Morphology of the pocket gopher mammalian genus *Thomomys*. University of California Publications in Zoology, 42:81–172.

Homan, J. A., and H. H. Genoways. 1978. An analysis of hair structure and its phylogenetic implications among heteromyid rodents. Journal of Mammalogy, 59:740–760.

Howell, A. B. 1932. The saltatorial rodent *Dipodomys*: the functional and comparative anatomy of its muscular and osseous systems. Proceedings of the American Academy of Arts and Sciences, 67:377–536.

Jones, W. K. 1985. Body size and life history in heteromyids. Journal of Mammalogy, 66:128–132.

Karasek, M., E. C. Hurlbert, J. T. Hansen, and R. J. Reiter. 1982. Ultrastructure of pinealocytes of the kangaroo rat (*Dipodomys ordi*). Cell Tissue Research, 226:167–175.

Kenagy, G. J. 1973. Daily and seasonal patterns of activity and energetics in a heteromyid rodent community. Ecology, 54:1201–1219.

Koenigswald, W. von. 1985. Evolutionary trends in the enamel of rodent incisors. Pp. 403–422, *in* Evolutionary relationships of rodents (W. P. Luckett and J. L. Hartenberger, eds.). NATO ASI series A, vol. 92, 721 pp.

Kotler, B. P. 1985. Owl predation on desert rodents which differ in morphology and behavior. Journal of Mammalogy, 66:824–828.

Lande, R. 1986. The dynamics of peak shifts and the pattern of morphological evolution. Paleobiology, 12:343–354.

Lawlor, R. M., and K. N. Geluso. 1986. Renal structure and body size in heteromyid rodents. Journal of Mammalogy, 67:367–372.

Lay, D. M. 1993. Anatomy of the heteromyid ear. Pp. 270–290, *in* Biology of the Heteromyidae (H. H. Genoways and J. H. Brown, eds.). Special Publication, American Society of Mammalogists, 10:1–719.

Lehmann, W. H. 1963. The forelimb architecture of some fossorial rodents. Journal of Morphology, 113: 59–76.

Long, C. A. 1976. Evolution of mammalian cheek pouches and a possibly discontinuous origin of a higher taxon (Geomyoidea). American Naturalist, 110:1093–1111.

Luckett, W. P. 1985. Superordinal and intraordinal affinities of rodents: developmental evidence from the dentition and placentation. Pp. 227–276, *in* Evolutionary relationships of rodents (W. P. Luckett and J. L. Hartenberger, eds.). NATO ASI series A, vol. 92, 721 pp.

Luckett, W. P., and J. L. Hartenberger. 1985. Evolutionary relationships among rodents: comments and conclusions. Pp. 685–712, *in* Evolutionary relationships of rodents (W. P. Luckett and J. L. Hartenberger, eds.). NATO ASI series A, vol. 92, 721 pp.

MacMillen, R. E., and D. S. Hinds. 1983. Water regulatory efficiency in heteromyid rodents: a model and its applications. Ecology, 64:152–164.

Munthe, K. 1981. Skeletal morphology and function of the Miocene rodent *Schizodontomys harkseni.* PaleoBios, 35:1–33.

Nagle, R. B., E. M. Altschuler, D. C. Dobyan, S. Dong, and R. E. Bulger. 1981. The ultrastructure of the thin limbs of Henle in kidneys of the desert heteromyid (*Perognathus penicillatus*). American Journal of Anatomy, 161:33–47.

Nikolai, J. C., and D. M. Bramble. 1983. Functional morphology of heteromyids. Pp. 44–64, *in* Biology of desert rodents (O. J. Reichman and J. H. Brown, eds.). Great Basin Naturalist Memoirs, No. 7, 134 pp.

O'Farrell, M., and A. Blaustein. 1974*a*. *Microdipodops megacephalus.* Mammalian Species, 46:1–3.

———. 1974*b*. *Microdipodops pallidus.* Mammalian Species, 47:1–2.

Pinkham, C. F. A. 1972. The evolutionary significance of locomotory patterns in the Mexican spiny pocket mouse, *Liomys irroratus.* Journal of Mammalogy, 54:742–746.

———. 1976. A biomechanical analysis of leaping in the kangaroo rat and the Mexican spiny mouse. Proceedings of the Utah Academy of Science, Arts, and Letters, 53:1–15.

Price, M. V., and J. H. Brown. 1983. Patterns of morphology and resource use in North American desert rodent communities. Pp. 117–134, *in* Biology of desert rodents (O. J. Reichman and J. H. Brown, eds.). Great Basin Naturalist Memoirs, No. 7, 134 pp.

Quay, W. B. 1953. Seasonal and sexual differences in the dorsal skin gland of the kangaroo rat (*Dipodomys*). Journal of Mammalogy, 34:1–14.

———. 1954. The dorsal holocrine skin gland of the kangaroo rat (*Dipodomys*). Anatomical Record, 119: 161–176.

———. 1960*a*. Experimental and comparative studies of succinic dehydrogenase activity in mammalian choroid plexuses, ependyma, and pineal organ. Physiological Zoology, 33:206–212.

———. 1960*b*. The pancreatic islets of desert rodents. American Midland Naturalist, 64:342–348.

———. 1965*a*. Integumentary modifications of North American rodents. Pp. 59–74, *in* Biology of the skin

and hair growth (A. G. Lyne and B. F. Short, eds.). Angus & Robertson, Sydney.
———. 1965*b*. Variation and taxonomic significance in the sebaceous caudal glands of pocket mice (Rodentia: Heteromyidae). Southwestern Naturalist, 10: 282–287.
———. 1965*c*. Comparative survey of the sebaceous and subordiferous glands of the oral lips and angle in rodents. Journal of Mammalogy, 46:23–37.
———. 1966. Stimulation of the palmar sweat glands of the kangaroo rat (*Dipodomys heermani*) with acetylcholine. American Zoologist, 6:351.
———. 1972. Integument and the environment: glandular composition, function, and evolution. American Zoologist, 12:95–108.
RANDALL, J. A. 1981. Olfactory communication at sand bathing loci by sympatric species of kangaroo rats. Journal of Mammalogy, 62:12–19.
REBAR, C. 1986. A comparison of learning abilities in *Dipodomys merriami* and *Perognathus intermedius.* Abstracts of the American Society of Mammalogists 66th Annual Meeting, Madison, Wisconsin.
REEDER, W. G. 1956. A review of the Tertiary rodents of the family Heteromyidae. Ph.D. dissert., University of Michigan, Ann Arbor, 618 pp.
REICHMAN, O. J. 1981. Factors influencing foraging in desert rodents. Pp. 195–213, *in* Foraging behavior (A. C. Kamil and T. D. Sargent, eds.). Garland Publ., New York, 534 pp.
RENSBERGER, J. M. 1971. Entoptychine pocket gophers (Mammalia, Geomyoidea) of the early Miocene John Day Formation, Oregon. University of California Publications in Geological Sciences, 90: 1–209.
———. 1975. Function in the cheek tooth evolution of some hypsodont geomyoid rodents. Journal of Paleontology, 49:10–22.
RISKA, B. 1987. Some models for development, growth, and morphometric correlation. Evolution, 40:1303–1311.
RYAN, J. M. 1986. Comparative morphology and evolution of cheek pouches in rodents. Journal of Morphology, 190:27–41.
SCHMIDT-NIELSEN, K., F. R. HAINSWORTH, AND D. E. MURRISH. 1970. Counter-current heat exchange in the respiratory passages: effect on water and heat balance. Respiratory Physiology, 9:263–276.
SIMPSON, G. G. 1945. The principles of classification and a classification of mammals. Bulletin American Museum of Natural History, 85:1–350.
SPERBER, I. 1944. Studies on the mammalian kidney. Zoologiska Bidragfran Uppsala, 22:249–431.
SPETH, R. L. 1969. Patterns and sequences of molts in the Great Basin pocket mouse, *Perognathus parvus.* Journal of Mammalogy, 50:284–290.
TAYLOR, C. R., K. SCHMIDT-NIELSEN, AND J. L. RAAB. 1970. Scaling of energetic cost of running to body size in mammals. American Journal of Physiology, 219:1104–1107.
THEILER, K. 1972. The house mouse: development and normal stages from fertilization to four weeks of age. Springer-Verlag, New York.
THEISSEN, D., AND P. YAHR. 1977. The gerbil in behavioral investigations. University of Texas Press, Austin, 224 pp.
THOMPSON, S. D. 1985. Bipedal hopping and seed dispersion selection by heteromyid rodents: the role of locomotor energetics. Ecology, 66:220–229.
THOMPSON, S. D., R. E. MACMILLEN, E. M. BURKE, AND C. R. TAYLOR. 1980. The energetic cost of bipedal hopping in small mammals. Nature, 287: 223-224.
VOORHIES, M. R. 1975. A new genus and species of fossil kangaroo rat and its burrow. Journal of Mammalogy, 56:160–176.
WAHLERT, J. H. 1978. Cranial foramina and relationships of the Eomyoidea (Rodentia, Geomorpha). Skull and upper teeth of *Kansasimys*. American Museum of Novitates, 2645:1–16.
———. 1985*a*. Skull morphology and relationships of geomyoid rodents. American Museum Novitates, 2812:1–20.
———. 1985*b*. Cranial foramina of rodents. Pp. 311–332, *in* Evolutionary relationships of rodents (W. P. Luckett and J. L. Hartenberger, eds.). NATO ASI series A, vol. 92, 721 pp.
———. 1993. The fossil record. Pp. 1–37, *in* Biology of the Heteromyidae (H. H. Genoways and J. H. Brown, eds.). Special Publication, American Society of Mammalogists, 10:1–719.
WAKE, D. B., G. ROTH, AND M. H. WAKE. 1983. On the problem of stasis in organismal evolution. Journal of Theoretical Biology, 101:211–224.
WEBSTER, D. B., R. F. ACKERMAN, AND G. C. LONGA. 1968. Central auditory system of the kangaroo rat. Journal of Comparative Neurology, 133:477–494.
WEBSTER, D. B., AND M. WEBSTER. 1975. Auditory systems of Heteromyidae: functional morphology and evolution in the middle ear. Journal of Morphology, 146:343–376.
———. 1984. The specialized auditory system of kangaroo rats. Contributions in Sensory Physiology, 8:161–196.
WEIJS, W. A. 1975. Mandibular movements of the albino rat during feeding. Journal of Morphology, 145:107–124.
WESTERHAUS, M. D. 1983. A histological comparison of the dorsal and generalized holocrine glands in the kangaroo rat *Dipodomys ordii*. Ohio Journal of Science, 253–255.
WILKINS, K. T., AND C. A. WOODS. 1985. Modes of mastication in pocket gophers. Journal of Mammalogy, 64:636–641.
WOLFFSEN, D. M. 1950. Scapula shape and muscle function with special reference to the vertebral border. American Journal of Physiological Anthropology, 8:331–341.
WOOD, A. E. 1935. Evolution and relationships of the heteromyid rodents. Annals of the Carnegie Museum, 24:73–263.
———. 1965. Grades and clades among rodents. Evolution, 19:115–130.

PARASITES

JOHN O. WHITAKER, JR., WILLIAM J. WRENN, AND ROBERT E. LEWIS

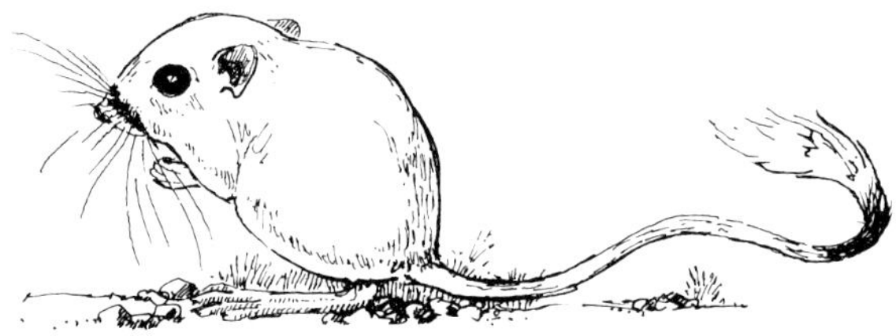

Introduction

Why should mammalogists, specifically those working with heteromyid rodents, be interested in parasites? A mammalogist is a biologist who studies mammals, presumably all aspects of mammalian biology. However, in practice none of us can study all aspects; most of us tend to specialize. The last few years have seen a burgeoning of mammalogists interested in genetics or population biology. However, parasitology is an area into which few mammalogists have ventured. This is unfortunate, because on the mammals we study there is a whole community of smaller organisms waiting to be examined. Parasites can cause diseases in mammals and in mammalogists. Research on the taxonomy of parasites can sometimes help us unravel some of the knotty taxonomic problems among their hosts. Indeed, many of the "parasites" are not parasites at all and we should be interested in what sort of relationships they have with their hosts. First, however, let us examine this community of organisms which we find on and in our hosts and which we often refer to as "parasites and other associates."

On and in mammals we find many organisms which we term ecto- and endo-parasites, respectively. Parasites, by definition, exist at the expense of (i.e., cause harm to) their host. Their host, of course, is a part of their environment; therefore, causation of harm to their host means harm to their environment. However, as we begin to study the organisms in this community, we immediately come to realize that "parasites" is not a very good collective term. The community of organisms on and in the animal consists of many true parasites, but also includes many species which cause little or no harm to the host. Some may actually be beneficial. How should we refer to an organism which does not feed, but simply clings to the hair and uses the host for a ride? One which parasitizes a parasite? One which is a predator upon parasites? Because these sorts of organisms exist, we have come to refer to members of this community as "parasites and other associates," a rather cumbersome term, but one which indicates that not all of the organisms in the community may actually be parasites. Perhaps the term "ectodytes" could be used instead of "ectoparasites and other associates." A similar term, "endodytes," could be used to replace "endoparasites and other associates." These terms remove the implication of particular life style and will be used in this chapter.

These species function as a community, as in any other habitat, and can be studied in the same way other communities are studied, but it is a much simpler community than most. This means it may be easier to attain an understanding of the interrelationships among the species. For this reason a parasite community might prove a valuable model for experimental ecological and evolutionary studies. There are no producers in the community. The energy, of course, ultimately comes from green plants through the host or through the organisms the host feeds upon. Likewise, there are very few predators in this community. Most, but not all, of the inhabitants of this community are parasitic or phoretic. This raises the interesting question of how population control is achieved in the absence of predators. Are there decomposers in this community, or do the dead organisms simply drop out of the fur and decompose in the host's environment?

The heteromyid biologist need only use a dissecting microscope to look through the fur of the host to begin finding ectodytes. This should then raise certain questions. What is the relationship of these organisms to the host? Do they suck blood, or other body juices? Do they feed on the dead or live skin of the host? Do they cause any harm to the host? Do they cause disease to the host or to the biologist? Do other related heteromyids have these same or related forms? Where did they come from, i.e., did they evolve along with the host? Did they invade from another host in the same environment? Do they occur on other hosts in the same area? If an ectodyte species evolved with the heteromyid in its respective niche, might it then help us to determine relationships of the hosts? Is there any chance that the parasites might have influenced the evolution, the behavior (dust baths, grooming), or the physiology of the hosts?

Why not let the parasitologist study the parasites? That is easy. The parasitologist does not have the expertise to collect, identify or otherwise work with the hosts. Besides, the parasitologist usually works with one or a few groups of species, not with the entire ectodyte and endodyte community. It is up to the heteromyid biologists to provide much of the information about the biology of ectodytes and endodytes, and to date they have not taken advantage of research opportunities. With a few rare exceptions, the mammalogist, at most, will brush off a few of the larger parasites and put them in a vial. Anybody who has collected very many heteromyids at all has had at their fingertips species unknown to science, if only they had realized and known what to do about it.

This chapter will attempt to provide partial answers to a few of the questions posed above, and may suggest ways to arrive at answers to others. It will also suggest new questions, some stated, and additional ones that surely will occur to the reader. The first part of this chapter will indicate methods for collecting and studying the organisms that live on and in heteromyid rodents. Information will also be presented about the life histories of the parasites and the parasite communities of the heteromyids. The discussion will then attempt to draw together this information and indicate some ways that it can be used by both mammalogists and parasitologists.

We hope this chapter will whet the appetites of the heteromyid biologists and perhaps entice some of them to venture into a whole new realm, one full of new species to be found, new discoveries to be made, and new interrelationships to be determined.

Methods

This section describes some of the methods that can be used in studying ectodytes and endodytes. Much of the information provided in the introductory paragraphs of the review of *Peromyscus* (Whitaker, 1968) applies to heteromyids and other mammalian hosts as well. Additional comments concerning that material follow. We feel strongly that the technique of using a dissecting microscope to examine the fur is the best single overall approach. It allows one to see the parasites in place, and thus to learn something of their location and biology. One can make estimates of the various numbers of organisms on the host by counting or by more indirect methods when the parasites are too numerous to count.

To use this method, simply put the animal under a good zoom dissecting microscope and examine the skin, hair, ears, and any areas you think worth examining. Pick off and put in 70% alcohol any organisms found. Notes should be taken on the numbers of various organisms found, using the best field identification available, such as: 12 fleas, 2 lice, 2 myobiid mites from nape, ±50 listrophorid mites (of an estimated 2,000 seen), and 14 small white mites. These data allow one to link up final identifications later with numbers of various types on each animal. We use McBee cards, one for each individual host examined. Each card contains all data for each host, including standard data plus information on habitat, internal and external parasites, food habits and reproduction. These cards allow rapid sorting and summary of information in different ways (specics, sex, age, locality, habitat, etc.), although now of course we often computerize the data for ease of sorting and analysis.

The ectodytes are then put in Nesbitt's solution for two to four days, and finally mounted on slides in Hoyer's solution. This method is especially valuable for mites, which are our main interest, but we use it for other groups also because of ease and convenience in processing large amounts of material. Tiny mites, especially those from hidden biotopes (see later discussion) can be taken immediately from the host and placed on a slide in Hoyer's solution. A washing technique was briefly mentioned (Whitaker, 1968, p. 256), but parasites may be found by visual examination that are not often dislodged by the washing technique. We do, however, often follow up our visual examination with washing, both to get better estimates of numbers and to obtain additional specimens. Overall, I believe this dual search is probably the best approach for ectodyte studies. As an addendum to the section on preservation of mites (p. 259) we now ring slides with Euparal or with fingernail polish rather than asphaltum, and 5 ml of HCl (not HCE) are used in Hoyer's solution. An outline of the washing technique along with other information on preservation and preparation of parasites is outlined below.

Washing technique.—The washing technique is especially useful for mammals too large for adequate direct examination with a dissecting microscope, but we also often use it as a secondary collecting method for smaller mammals. This method can be used in getting estimates of abundance. Items needed for this technique are a Buchner funnel, rubber stopper, filter paper, filtering flask, aspirator, non-collapsible hose, Alconox (or other detergent), and a container with a lid. Insert the Buchner funnel and rubber stopper into the filtering flask and attach the aspirator to a faucet (Fig. 1). Connect the flask and aspirator with the tight fitting non-collapsible hose. Place a piece of filter paper in the funnel. Place the host animal in the container and fill the container ¾ full of water (container should be of appropriate size, i.e., pint for mouse or shrew, quart for chipmunk, gallon for squirrel, etc.). Add a small amount of detergent to the container. A pinch per pint (about 0.1 gram) is usually sufficient. Too much detergent will make too many suds and clog the filter paper. Attach the lid and shake vigorously for 30 seconds to a minute. Turn the faucet on full and pour some of the liquid into the funnel. A vacuum is created which draws the water through, leaving parasites on the paper. Turn the water off, examine the filter paper under a dissecting microscope, and remove the parasites with a dissecting needle. The above procedure can be repeated until parasites no longer appear on the paper. Too much detergent may be remedied by diluting the solution with more water, using a larger container if necessary.

Making counts of ectoparasites.—As mentioned above, we normally examine the fur of the host using a dissecting microscope, often followed by washing, especially for larger organisms. For the second technique direct count is used, and for the first, direct count plus indirect estimation of numbers of those organisms occurring in large numbers.

However, for greater precision in making counts, dissolving techniques have been used which include digesting the skin and hair of the host in KOH, and screening the resulting fluid for ectoparasites. We have tried this method, and like Henry and McKeever (1971) rejected it. Ectoparasites containing chiton can be recovered and counted using this method. However, the method is difficult and time consuming, and many of the specimens are in poor condition for identification and completely inadequate for description. Rather than using this method, Henry and McKeever offered further modifications to the washing technique, and also provided some quantitative assessment of the washing versus dissolving techniques.

More recently, DeLoach (1985) described a new approach to counting ectoparasites using an electronic particle counter, or "Coulter counter." We have not tried this method but it would seem to have excellent possibilities.

Internal parasites.—Our examination for internal parasites begins with external examination of the lungs, liver, and bladder. If external signs of parasites are seen we investigate further. Most of our collections of internal parasites are from the digestive tract. Examination of the stomach occurs during routine stomach analysis. The intestinal contents are examined by squeezing out the material from the intestines using two pairs of forceps, and/or cutting open the intestines with a pair of scissors, depending on the size of the animal and whether or not the parasites can all be removed by the first technique. The larger internal parasites can easily be counted.

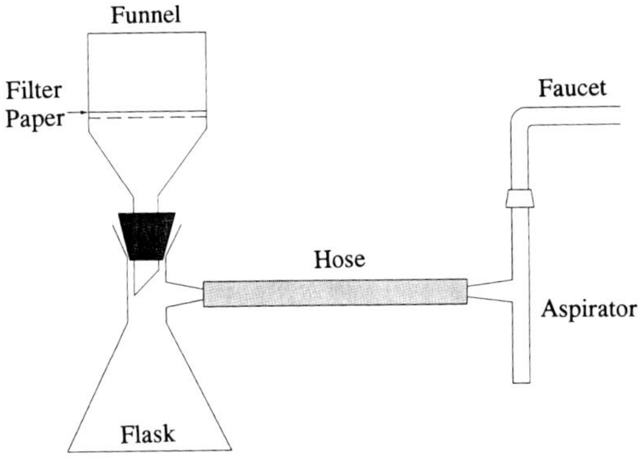

Fig. 1. Apparatus used in washing technique for collection of ectoparasites.

Many different techniques are available for processing internal parasites, thus it is best to check with specialists with whom you will work for specific techniques, but I have previously given some general information on handling trematodes and nematodes (Whitaker, 1968). In the absence of specific information one can generally preserve internal parasites in FAA (Formal-acetic acid: 5 cc formalin, 5 cc acetic acid, and 90 cc ethyl alcohol), with at least acceptable results.

Mites of hidden biotopes.—Specialized techniques are needed for finding parasitic mites of "hidden" or "cryptic" biotopes, such as of the eye sockets, ear canals, nasal passages, Meibomian or other gland areas, hair follicles, and in or under the skin. Nasal chiggers have been found in heteromyids (*Otorhinophila intrasola* and *O. sinaloa* from *Liomys* and *Chaetodipus*), whereas other closely related species, *O. parvisola* and *O. sola* (also primarily of *Chaetodipus*) are found in the ears. Also some hair follicle mites have been found in heteromyids.

To examine these areas, the eyes can be partially pulled out and the sockets and back sides of the eyes can be examined. Ear canals can be cleaned out with cotton swabs, nasal passages can be cleansed with a syringe and/or opened with scissors and examined with a dissecting microscope. Meibomian or other glands can be squeezed out with watchmakers forceps, and the surface of the skin can be examined for irritations or pustules. Further application of these sorts of techniques on heteromyids should turn up numerous additional species.

Identification of Heteromyid Parasites

Because of the great number of parasites, final identifications must generally be made by specialists. However, one cannot generally send a large collection to a parasitologist and expect him or her to make identifications. Since most parasitologists specialize on certain taxonomic groups, it is much more satisfactory to make preliminary identifications and then to send the specialists well mounted specimens of material in the appropriate taxa.

I do not have the experience necessary to provide keys to the internal parasites of heteromyids. However, for the ectodytes I have provided simple keys to the genera. With some experience, the heteromyid biologist should be able to use these keys to identify the common genera. In some groups, especially the fleas and laelapid mites, only the more important genera have been included.

For aid in determining structures, see Wenzel and Tipton (1966) and Brennan and Goff (1977*b*) for chiggers, and for other groups see other sources referred to in the text.

Keys to Important Genera of Ectodytes of Heteromyids

1. With four pairs of legs 2
– With three pairs of legs 4
2. Hypostome with retrorse teeth modified for piercing, adults large in size (usually at least 2 mm Ticks (adults and nymphs)
– Usually smaller and hypostome without retrorse teeth .... Mites
3. Body flattened laterally .... Fleas
– Body not flattened laterally 4
4. Mouthparts protruding noticeably forward beyond anterior edge of head; body with gnathosoma and opisthosoma, not segmented 5
– Mouthparts entirely confined within head capsule, body segmented and with head, thorax and abdomen .... Sucking lice, *Fahrenholzia*
5. Mouthparts a hypostome with rows of large teeth .... Ticks (larvae)
– Mouthparts of knifelike chelicerae, lacking prominent teeth .... Chiggers (Trombiculidae)

Ticks

1. Scutum (dorsal hard plate) absent in all stages .... Soft ticks (Argasidae) ... *Ornithodorus*
– Scutum present in all stages .... Hard ticks (Ixodidae) ... 2
2. Anal groove forming an arc in front of anus .... *Ixodes*
– Anal groove entirely behind anus or indistinct .... *Dermacentor*

Mites

1. With peritreme and lateral stigma dorsal to and at level between coxae III and IV .... Laelapidae ... 5
– No peritreme or lateral stigma 2
2. First pair of legs (not just claws) or maxillae highly modified for clasping hairs 3
– Neither legs nor claws highly modified for clasping hairs 4
3. Mites dorsoventrally flattened; first pair of legs highly modified for clasping hairs .... Myobiidae, *Radfordia bachai*
– Mites elongate fusiform; maxillae highly modified for clasping hairs .... Listrophoridae, *Geomylichus*
4. Lacking mouthparts, but with a posterior clasping organ for clasping hairs often present .... Glycyphagidae
– Mouthparts present; no clasping organ; sternal plate with triple coglike structure .... Ameroseiidae, *Sertitympanum*
5. One pair of setae on epigynial plate ... 6

- More than one pair of setae on epigynial plate 8
6. Epigynial plate with 3 or 4 pairs of setae 7
- With more than 6 pairs of setae; Leg II with heavy spines and spurs, especially on tarsus .... *Ischyropoda*
7. Epigynial plate with 3 pairs of setae .... *Steptolaelaps*
- Epigynial plate with 4 pairs of setae .... *Laelaps*
8. Peritremes not produced posterior to stigmata; large hairy mites generally over 1 mm long .... *Gigantolaelaps*
- Peritremes produced beyond stigmata, usually under 1 mm 9
9. Chelae slender with tips straight and pointed, triangular ventral spines on some of coxae .... *Echinonyssus*
- Chelae more robust, bent and with prominent teeth 10
10. Posterior setae of coxae III spinelike .... *Eubrachylaelaps*
- Posterior setae of coxae III not spinelike 11
11. Sternal plate about as wide as long, or longer .... *Hypoaspis*
- Sternal plate wider than long .... *Androlaelaps*
12. With well developed ventral clasping organ, mites living in fur .... *Dermacarus*
- Clasping organ small and at posterior of body or vestigial, mites living under skin or in hair follicles 13
13. Clasping organ vestigial, mites living under skin of *Dipodomys merriami* .... *Dipodomydectes americanus*
- Clasping organ small but evident, situated at very posterior of body rather than ventrally, mites living in hair follicles *Mediolabidophorus, Metalabidophorus* 14
14. Posterior clasper of clasping organ protruding beyond end of body .... *Dipodomyopus tuttlei*
- Posterior clasper of clasping organ not protruding beyond end of body 15
15. Claspers tiny, no enlarged setae on legs IV, in follicles of *Heteromys desmarestianus* ... *Medialabidophorus neotropicalis*
- Claspers larger, enlarged setae on legs IV .... *Metalabidophorus*

Fleas (Siphonaptera)

1. With genal ctenidium (comb) 2
- Lacking genal ctenidium 3
2. Genal ctenidium with 3 teeth .. *Carteretta*
- Genal ctenidium with 2 teeth ... *Meringis*
3. Pronotal comb present 5
- Pronotal comb absent 4
4. Eye absent 7
- Eye well developed and pigmented .... *Polygenis*
5. Abdominal terga with one row of setae .... *Euhoplopsyllus*
- Abdominal terga with more than one transverse rows of setae 6
6. Femur I with single setae on outside surface .... *Orchopeas*
- Femur I with several setae on outside surface .... *Oropsylla*
7. Fleas very small and almost without setae .... *Anomiopsyllus*
- Fleas larger and with setae .... *Wenzella*

Chiggers

(See Goff et al., 1982, Journal of Medical Entomology 19:221–238 for terminology.)

1. Leg segmentation 6-6-6 or 7-6-6 2
- Leg segmentation 7-7-7 10
2. Leg segmentation 6-6-6 3
- Leg segmentation 7-6-6 6
3. Scutum without nasus .... *Sasacarus*
- Scutum with nasus 4
4. Spiracles and tracheae absent .... *Comatacarus*
- Spiracles and tracheae present 5
5. Palpal tarsus with 5B ... *Xenodontacarus*
- Palpal tarsus with 7B ...... *Odontacarus*
6. With dorsal plates posterior to scutum *Polylopadium*
- Without dorsal plates posterior to scutum 7
7. PL setae on scutum .... *Cordiseta*
- PL setae off scutum 8
8. With 2–3 pairs of laterosternal setae between coxae II and III .... *Pseudoschoengastia*
- Without laterosternal setae 9
9. Genu I with 2 genualae .... *Anomalaspis*
- Genu I with 3 genualae .... *Vanidicus*
10. Sensilla flagelliform or only slightly thickened or expanded 11
- Sensilla expanded 26
11. Palpal tarsus with 4B 12
- Palpal tarsus with more than 4B 13
12. Coxa III multisetose, genuala II and gen-

uala III present, subterminala I and parasubterminala I present ...... *Crotonasis*
- Coxa III unisetose, genuala II and genuala III absent, subterminala I and parasubterminala I present or absent .... *Euschoengastiodes*

13. Palpal tarsus with 5B 14
- Palpal tarsus with more than 5B 15

14. Scutum pentagonal with acute posterior margin; 3 genualae I; tibiala III present; parasubterminala I absent .... *Boshkerria punctata*
- Scutum subrectangular; usually 1 genuala I; tibiala III absent; parasubterminala I usually present ...... *Otorhinophila*

15. Palpal tarsus with 6B .... *Microtrombicula perplexa*
Palpal tarsus with 7B ...... *Hyponeocula*
- Palpal tarsus with 7BS 16

16. Palpal claw with one prong ..... *Crotiscus*
- Palpal claw with more than one prong 17

17. Mastitarsala III usually present 18
- Mastitarsala III absent 24

18. Coxa III multisetose; genuala III and mastitarsala III absent .... *Miyatrombicula scottae*
- Coxa III unisetose; genuala III and mastitarsala III present 19

19. Eyes 2/2, on scutum .... *Peltoculus*
- Eyes 2/2, not on scutum 20

20. With 7 scutal setae; more than one pair humeral setae .... *Hoffmannina*
- With 5 scutal setae; one pair humeral setae 21

21. Parasubterminala I absent; tarsal claws with onychotriches .... *Hexidionis*
- Parasubterminala I present; tarsal claws without onychotriches 22

22. Palpal claw bifurcate .... *Eutrombicula*
- Palpal claw trifurcate 23

23. Scutum rectangular, with sinuous posterior margin .... *Parasecia*
- Scutum pentagonal, with acute or rounded posterior margin .... *Neotrombicula*

24. Genu I with 3 genualae .... *Parasecia*
- Genu I with 2 genualae 25

25. Scutum rectangular; palpal femoral, genual and tibial setae nude; galeal seta branched .... *Leptotrombidium*
- Without the above combination of characters .... *Trombicula*

26. Palpal tarsus with 4B 27
- Palpal tarsus with 5B or 5BS 30
- Palpal tarsus with 7B .... *Euschoengastia*

27. Subterminala I and parasubterminala I absent 28
- Subterminala I and parasubterminala I present 29

28. Without genuala II and genuala III ... . *Kayella*
- With genuala II and genuala III .... *Dermadelema*

29. Scutum subpentagonal with apex anterior; tarsus I with subterminal nude setae .... *Ectonyx*
- Scutum subrectangular with sinuous anterior and posterior margins; tarsus I without subterminal nude setae .... *Quadraseta*

30. Palpal tarsus with 5B .... *Kymocta*
- Palpal tarsus with 5BS 31

31. PL setae off scutum .... *Ascoschoengastia dyscrita*
- PL setae on scutum .... *Colicus*

Parasites of Heteromyids

In the listings below, records have generally been included when both host and parasite were identified to species, although there are selected exceptions. Some parasites identified only to genus are included when no species within that genus has been identified, or if a species of parasite was not recorded from heteromyids except on an unidentified heteromyid host.

Names of species have been updated in the present text without comment concerning synonymy, as the purpose here is to provide an indication of parasites as concisely as possible. One can usually determine synonymies by linking nomenclature used in this paper with that used in the original papers.

Some of the major summary references used during this work were Doran (1954*a*, 1954*b*, 1955*a*, 1955*b*) which contain parasite listings and host listings for protozoans and helminths, and Whitaker and Wilson (1974) which gives similar listings for parasitic mites other than chiggers.

Viruses

Five different viruses have been reported to infect heteromyid rodents, either by isolation or by the presence of antibodies. All are arboviruses (viruses which multiply in arthropod vectors as well as higher animals) and are classified in three different families. The Buttonwillow virus (Bunyaviridae) was first isolated from a leporid in 1961 and has since been isolated mainly from leporids and "punkies" or biting midges, *Culicoides variipennis* (Ceratopogonidae) in the southwestern United States (Hardy et al., 1972, 1974). *Culicoides* appears to be the vector, as the virus was transmitted experimentally between individuals of *Sylvilagus auduboni* via *Culicoides,* and viral multiplication occurred in that host. Two species of ticks, *Dermacentor parumapertus* and *Ornithodorus parkeri,* were tested but were found unsuitable as vectors for this virus. Hardy et al. (1972) experimentally infected one of six individuals of *Dipodomys nitratoides* tested with Buttonwillow virus; it developed HI antibodies to this virus.

The western equine encephalomyelitis virus (WEE, Togaviridae) was first isolated from a dying horse in California in July 1930. Thc case fatality rate in horses that year was 50% and seven major epidemics have occurred since then. The virus multiplies in the mosquito, *Culex tarsalis,* and a wide variety of birds are considered to be the major reservoir of the virus. Hardy et al. (1984) found antibodies for WEE in *D. nitratoides.*

The Powassan, Modoc and St. Louis encephalitis viruses are members of the recently established family Flaviviridae. The Powassan virus was first isolated from a fatal case of human encephalitis in Ontario in 1958 (McLean and Donohue, 1959). The virus is maintained in mammals, primarily squirrels and marmots, and is transmitted by ticks, *Ixodes marxi* and *I. cookei* in the eastern states, and *I. spinipalpus* in the western states. Hardy et al. (1974) found antibodies for Powassan virus in *Chaetodipus californicus, D. heermani,* and *D. nitratoides.*

St. Louis encephalitis was first recognized as a distinct virus following an outbreak in humans in Paris, Illinois, in 1932. It has been responsible for periodic outbreaks throughout the United States (Seymour and Yuill, 1981) with severe outbreaks in the midwest and Texas. The recognized vector is the mosquito, *Culex tarsalis,* but *C. pipiens* was associated with the original outbreak and other mosquitoes may be involved. There is serologic evidence (antibodies) of SLE in *Dipodomys nitratoides, D. heermani,* and *C. californicus* (Hardy et al., 1974).

Normally arthropod borne viruses cycle between their invertebrate and vertebrate maintenance hosts with no disease produced in either. Infection of humans occurs frequently for most of them but severe disease is rare. In this respect man is usually a dead-end host and does not play a role in the virus's normal cycle. (There are notable exceptions to this generalization, such as Dengue and Yellow Fever.) Usually when humans become infected the disease is not clinical. In one area where WEE outbreaks occur periodically 11% of the adults in the area had antibodies against the virus indicating that infection had occurred during their lifetime.

Viruses known from heteromyids are listed below. The numbers in parentheses refer to the corresponding entries in the Literature Cited section.

Dipodomys heermani
  Modoc virus (162)
  Powassan virus (162)
  St. Louis encephalitis (162, 184)
Dipodomys nitratoides
  Buttonwillow virus (161)
  Modoc virus (162)
  Powassan virus (162)
  St. Louis encephalitis (162)
  Western equine encephalitis (162)

Chaetodipus californicus
Powassan virus (162)
St. Louis encephalitis (162)

Rickettsia

Rickettsia are small, non-motile bacteria or bacterial-like organisms. They are spherical or rodlike and occur singly, in pairs or in strands. They have a cell wall, cell membrane and central bodies presumed to be nuclei. They multiply by binary fission. Like viruses they are obligate, intracellular parasites. Most can survive only briefly outside animal cells, although *Coxiella burnetii* is particularly resistant to heat and dryness as it passes between cells. Infection is generally by bites of invertebrates (lice, fleas, mites and ticks), except in *C. burnetii,* which is transmitted to man by infected dust or droplets. Rickettsia can pass from mammal to arthropod or vice versa during the bite of the arthropod, and between arthropods by ingestion of infected ova.

Two rickettsia have been reported from heteromyids, *Coxiella burnetii* that causes Q fever, and *Rickettsia rickettsii* that causes Rocky Mountain Spotted Fever. Both are human diseases and both have tick vectors. *Coxiella burnetii* is normally carried by *Dermacentor andersoni* and *R. rickettsii* has been found in a number of ticks. Untreated spotted fever often leads to fatalities, but Q fever is less severe. It is usually acute but self limited and non-fatal. It may become chronic (Bell, 1981). The causal organism, *C. burnetii,* is usually transmitted to man by inhaling dusts or aerosols, particularly around cattle barns. *Coxiella burnetii* infects man and a vast range of wild host species. It was first found in the United States in Utah by Stoenner et al. (1959) in *Dipodomys microps, D. ordii* and from a tick, *Dermacentor parumapertus,* that often parasitizes these hosts. *D. microps* was considered a probable reservoir species since it retained the rickettsia for several months. Vest et al. (1965) reported *C. burnetii* positive individuals of *Chaetodipus formosus* (7 of 481 individuals), *Perognathus longimembris* (20 of 389), *P. parvus* (36 of 494), *Dipodomys microps* (225 of 2,443), and *D. ordii* (105 of 214). Evidence of viable *C. burnetii* was found in these species except *P. parvus* (1 of 481, 3 of 389, 0 of 494, 40 of 2,443, and 6 of 2,154 respectively). In addition, *C. burnetii* was recovered from the following heteromyid parasites: *Dermacentor parumapertus* from *D. ordii, Ixodes kingi* from *D. ordii* and *D. microps,* and *Meringis* sp., probably *parkeri* from *D. ordii.*

Rickettsia rickettsii, which causes Spotted Fever, has also been found in several heteromyids. Lane et al. (1981) reported that *Dipodomys californicus* had a relatively high percentage (25%) of seropositive individuals for Spotted Fever Group (SFG) rickettsia. Also, Vest et al. (1965) reported antibodies in *D. microps, D. ordii, M. megacephalus, C. formosus, P. longimembris,* and *P. parvus.* However, Coultrip et al. (1973) found none of 44 pocket mice, *Chaetodipus californicus,* or 30 kangaroo rats *Dipodomys heermani* to be seropositive for Rocky Mountain Spotted Fever.

Rickettsia known from heteromyids are:

Dipodomys californicus
Rickettsia rickettsii (219)
Dipodomys microps
Coxiella burnetii (288, 316)
Rickettsia rickettsii (antibodics) (316)
Dipodomys ordii
Coxiella burnetii (288, 316)
Rickettsia rickettsii (antibodies) (316)
Chaetodipus formosus
Coxiella burnetii (316)
Microdipodops megacephalus
Rickettsia rickettsii (antibodies) (316)
Perognathus longimembris
Coxiella burnetii (316)
Rickettsia rickettsii (antibodies) (316)
Perognathus parvus
Coxiella burnetii (antibodies) (316)
Rickettsia rickettsii (antibodies) (316)

Spirochetes

Spirochetes, like rickettsia, are bacteria or bacteria-like organisms, but quite different from all other bacteria. They are motile and all pathogenic spirochetes are in the family Treponemataceae. The family includes three genera, including *Borrelia,* the genus reported from heteromyids. Members of this family are very long, spiraled, filamentous cells. Their movement is by a rotating motion so that they appear to bore through their environment like a corkscrew. Most spirochetes have to be viewed with darkfield illumination, because their transverse diameter is below the resolving power of an ordinary light microscope. Spirochetes divide by binary fission. The genus *Borrelia* includes mostly the relapsing fever spirochetes. They are larger and more loosely coiled than other Treponemataceae. Relapsing fever is sporadic in the western U.S. It is thought to be tick or sometimes louse transmitted. *Borrelia*-like spirochetes were found in the blood of one of eight individuals of *Chaetodipus hispidus* examined from Texas (Eads and Hightower, 1952*b*), but not enough is known of these to determine their effects on the host.

Spirochetes known from heteromyids are:

Chaetodipus hispidus
Borrelia-like spirochaetes (91)

Bacteria

Bacteria were originally classed as plants because they have a cell wall, they are able to synthesize complex protein molecules from simpler materials, and they reproduce in a manner similar to plants. They are very small, much smaller than protozoans. They have a diameter of about one micron. They are the smallest organisms that have all the necessities for growth and replication using simple nutrients. Bacteria are simpler than the cells of higher organisms. They are termed prokaryotic because they lack the organized nucleus of higher (eukaryotic) cells. Bacteria and blue-green algae are the lower or prokaryotic Protista. The algae, the slime molds, the fungi, and the protozoans comprise the higher, eukaryotic, Protista. Bacteria usually reproduce asexually by binary fission, but most forms have some capacity for genetic exchange and recombination between individuals. They also produce endospores which are resistant to desiccation and can remain dormant through unfavorable ecological conditions, the same as seeds do for higher plants.

Members of four different genera of bacteria have been recorded from heteromyids; two of them are major pathogens of humans, *Yersinia pestis,* causing plague, and *Francisella tularensis,* causing tularemia. The genus *Grahamella* appears to be nonpathogenic, and many of the species are not well distinguished. *Haemobartonella* sometimes shows varying degrees of pathogenicity.

Yersinia pestis (formerly *Pasteurella pestis*)

Hubbard (1947) stated (p. 423): "To what extent Kangaroo Rats are infected with plague is not known . . . but during April, 1939, 10 miles west of Las Cruces, Dona Ana County, New Mexico, these animals [Kangaroo Rats] were found plague positive." Holdenreid and Quan (1956) inoculated wild rodents with *Y. pestis. Perognathus flavus* from New Mexico was more susceptible to *Y. pestis* than white mice, whereas *Dipodomys spectabilis* and *D. ordii* were the least susceptible to experimental inoculation of the species tested. Holdenreid and Morlan (1955) found no *Y. pestis* infested fleas on six individuals of *Dipodomys ordii* or one of *D. spectabilis* from New Mexico. Coultrip et al. (1973) tested 28 kangaroo rats (*D. heermani*) and 42 pocket mice (*Chaetodipus californicus*) for *Y. pestis* by serology. All tests were negative.

Francisella tularensis

Tularemia (plaguelike disease of rodents, rabbit fever, rabbit disease) is primarily a disease of rodents and rabbits but is known from a number of mammalian species (Bell and Reilly, 1981). It is caused by *Francisella tularensis* and is transmitted by ectoparasites, or by contact with infected animals or infected material. Among heteromyids it has been detected in Utah by isolation or agglutination in *Perognathus parvus, C. formosus, Dipodomys ordii,* and *D. microps* (Thorp et al., 1965; Vest et al., 1965). Thorp et al. (1965) isolated *P. tularensis* also from the tick, *Ixodes kingi* from *Dipodomys ordii.*

Other bacteria

Wood (1952) reported the blood parasites, *Grahamella* sp. from *Chaetodipus californicus* and *Haemobartonella* sp. from *Perognathus inornatus.* A number of species of both *Grahamella* and also of *Haemobartonella* have been described from 9 varieties of mammal hosts.

Bacteria reported from heteromyids are:

Dipodomys microps
  Francisella tularense (298, 316 [antibodies])
Dipodomys ordii
  Pasteurella pestis (173, 298, 316)
  Francisella tularense (316)
Dipodomys spectabilis
  Pasteurella pestis (173)
Chaetodipus californicus
  Grahamella sp. (333)
Chaetodipus formosus
  Francisella tularense (20)
Perognathus flavus
  Pasteurella pestis (173)
Perognathus inornatus
  Haemobartonella sp. (333)
Perognathus parvus
  Francisella tularense (298, 316)

Fungi

Fungi are unicellular (yeasts) or multicellular filamentous colonial (molds and mushrooms) eukaryotic Protista with cell walls. They are usually non-motile. They reproduce by asexual or sexual spores, and many types produce thick-walled resting spores unlike those of higher plants. They do not possess chlorophyll, but absorb their nutrients from the environment. Fungi may be saprophytes, symbionts, parasites or hyperparasites. Dispersal is by spores, and infection is often by ingestion or inhalation.

Two species of fungi have been found in heteromyids. One, *Coccidioides immitis,* causes the respiratory infection Coccidiosus in man and other animals; the other, *Haplosporangium parvum,* causes pulmonary mycosis in rodents.

Coccidioides immitis is prevalent in the southwestern deserts of the United States (Emmons, 1942). It causes a primary respiratory infection (coccidioidomycosis) in man and several species of wild mammals. Emmons and Ashburn (1942) isolated this fungus from the lungs of 19 of 124 (15%) pocket mice, *Perognathus baileyi, P. penicillatus,* and *P. intermedius,* and from 5 of 29 (17%) kangaroo rats, *Dipodomys merriami.* Transmission is mainly by inhalation of spores from dust or soil (Davis, 1981). The disease can range from asymptomatic to fatal, but is usually a benign infection with lesions occurring most often in the lungs, although secondary lesions may be found elsewhere in the body.

Fungi reported from heteromyids are:

Dipodomys merriami
  Coccidioides immitis (11, 103, 104)
  Haplosporangium parvum (104)
Chaetodipus baileyi
  Coccidioides immitis (11, 103, 104)
  Haplosporangium parvum (104)
Chaetodipus intermedius
  Coccidioides immitis (103)
Chaetodipus penicillatus
  Coccidioides immitis (11, 103, 104)
  Haplosporangium parvum (104)

Protozoa

Protozoans are single-celled organisms consisting of cytoplasm surrounded by a cell membrane and usually containing one or more nuclei, thus more or less resembling the single cell of a multicellular organism. They differ from unicellular fungi and bacteria by the absence of a cell wall, from spirochetes by the presence of a well defined nucleus, and from the rickettsia and viruses by their much larger size.

Protozoans have diverse means of movement, morphology, and lifestyles. They generally reproduce asexually by simple cell division, but most have a simple sexual reproductive stage. There are four classes of protozoans, of which three have been found as parasites of heteromyids: Class Sarcodina, the amoeboid protozoans which move by pseudopodia; Class Mastigophora, the flagellates, which move by flagellae; and Class Sporozoa, which have no locomotor organs and are characterized by spore production. All sporozoans live inside of cells as parasites.

Class Sporozoa

Several protozoans have been reported from heteromyids. Most are sporozoans, including three genera: *Eimeria* (13 species reported), *Isospora* sp. and *Besnoitia. Eimeria* and *Isospora* are both intestinal protozoans which develop without an intermediate host.

Eimerians are very common coccidian protozoans of the intestinal tract. Numerous species of *Eimeria* and *Isospora* have been described from many different hosts. They can be pathogenic, but most do not overly harm their hosts. Thirteen species of *Eimeria* have been described from heteromyids, four from *Perognathus* and *Chaetodipus* (*E. merriami, penicillati, perognathi,* and *reedi*), two from *Liomys* (*E. liomysis* and *picta*), and seven from *Dipodomys* (*E. balphae, chihuahuaensis, chobotari, dipodomysis, mohavensis, scholtysecki,* and *utahensis*).

Sometimes more than one species of *Eimeria* may occur in the same host. For example, Short et al. (1980) described *Eimeria chihuahuaensis* from *D. ordii* and found that three of the individuals examined harbored three species at one time, two each had *E. chobotari* and *E. dipodomysis,* besides *E. chihuahuaensis,* while one had *E. chobotari* and *E. balphae.* One *D. merriami* harbored both *E. chobotari* and *E. chihuahuaensis.*

The work of Levine et al. (1957*b*) indicates that the *Eimeria* of heteromyids might offer fertile field for further study. They mention particularly that there are different strains of *E. arizonensis* from *Peromyscus truei* on different sides of the Colorado River. A study of *Eimeria* in the various genera, species, subspecies, and populations of heteromyid hosts in relation to breadth and completeness of isolating mechanisms should be of considerable interest.

Besnoitiosis is a protozoan disease that causes spherical pearly white cysts in various body tissues. Chobotar et al. (1970) found cysts of *Besnoitia jellisoni* in three of 88 Ord's kangaroo rats examined, but none in seven *Dipodomys microps.* Two of the infected rats were sick, whereas one was in good condition when captured, but died shortly thereafter. It had a heavy infection of cysts.

Class Mastigophora, The Flagellates

The second largest group of protozoans in heteromyids are the flagellates, with four genera having been reported, *Giardia, Leishmania, Tritrichomonas* and *Trypanosoma. Leishmania* can cause disease in man and rodents, and some species of *Giardia* and *Trypanosoma* cause problems in other hosts; however, not enough is known about them to determine if they are detrimental to heteromyids.

Leishmania is a protozoan transmitted by sandflies (Phlebotomidae). Species of *Leishmania* may cause lesions (leishmaniasis) of the skin and mucoid membranes of mammals. *Leishmania mexicana* has been isolated from cutaneous lesions in the tails of six of 58 individuals of *Heteromys desmarestianus* from Belize (Lainson and Strangways-Dixon, 1964*a*, 1964*b*), and *Leishmania braziliensis panamensis* was found in the same host in Costa Rica (Zolodon et al., 1977). Disney (1964) reported an *H. desmarestianus* from Belize with heavily infested ears, but the spleen, liver, and lungs were also involved. Esquivel et al. (1967) reported *Trypanosoma zeledoni* from *Liomys salvini* from Costa Rica.

Hegner (1926) described *Endamoeba dipodomysi* from the kangaroo rat, *Dipodomys spectabilis*, as the single member of the Sarcodina reported from heteromyids.

The flagellate protozoan, *Tritrichomonas muris*, was reported by Doran (1953a) from fecal smears of several species of heteromyids from California, including *Dipodomys agilis, D. panamintinus, D. nitratoides, Chaetodipus formosus, Perognathus longimembris,* and *parvus*. Doran (1953b) reported *Giardia duodenalis* var *perognathi* from *D. agilis* and *D. merriami* and *Endamoeba dipodomysi* from *D. agilis, heermani, merriami* and *D. panamintinus*, also from fecal smears. The infection rate of the latter ran from about 40 to 50%. Herman (1943) found a single trophozoite of *Giardia* in a blood smear from *D. heermani*. There was no evidence of contamination, but it was similar to *Giardia* from intestines of kangaroo rats.

Protozoans reported from heteromyids are:

Dipodomys agilis
- *Eimeria balphae* (290)
- *Eimeria chobotari* (290)
- *Eimeria scholtysecki* (290)
- *Eimeria utahensis* (290)
- *Endamoeba dipodomysi* (75)
- *Giardia duodenalis* var. *perognathi* (75)
- *Isospora* sp. (290)
- *Tritrichomonas muris* (74)

Dipodomys gravipes
- *Eimeria scholtysecki* (290)

Dipodomys heermani
- *Endamoeba dipodomysi* (75)
- *Giardia* sp. (166)

Dipodomys merriami
- *Eimeria balphae* (290)
- *Eimeria chihuahuaensis* (282, 290)
- *Eimeria chobotari* (109, 282, 290)
- *Eimeria dipodomysi* (290)
- *Eimeria merriami* (290)
- *Eimeria utahensis* (290)
- *Endamoeba dipodomysi* (75)
- *Giardia duodenalis* var. *perognathi* (75)

Dipodomys microps
- *Eimeria utahensis* (108)
- *Eimeria chobotari* (109)

Dipodomys nitratoides
- *Tritrichomonas muris* (74)

Dipodomys ordii
- *Besnoitia jellisoni* (61)
- *Eimeria balphae* (106, 282, 290)
- *Eimeria chobotari* (282, 290)
- *Eimeria dipodomysis* (282, 290)
- *Eimeria scholtysecki* (107)
- *Eimeria utahensis* (108, 105 [life cycle described in experimentally infected animals])

Dipodomys panamintinus
- *Eimeria mohavensis* (76, 81)
- *Eimeria scholtysecki* (290)
- *Endamoeba dipodomysis* (75)
- *Tritrichomonas muris* (74)

Dipodomys phillipsii
- *Eimeria dipodomysis* (222)

Dipodomys spectabilis
- *Eimeria balphae* (290)
- *Eimeria scholtysecki* (290)
- *Endamoeba dipodomysi* (164)

Dipodomys venustus
- *Tritrichomonas muris* (206)

Heteromys anomalus
- *Leishmania braziliensis* (307)
- *Leishmania mexicanus* (299)

Heteromys desmarestianus
- *Leishmania braziliensis panamensis* (342)
- *Leishmania mexicana* (217, 218, 342)

Liomys irroratus
- *Eimeria liomysis* (222)

Liomys pictus
- *Eimeria liomysis* (189, 222)
- *Eimeria picta* (222)

Liomys salvini
- *Trypanosoma zeledoni* (112)

Chaetodipus formosus
Eimeria reedi (109)
Tritrichomonas muris (74)
Chaetodipus intermedius
Eimeria perognathi (223)
Chaetodipus penicillatus
Eimeria penicillati (189)
Perognathus flavus
Eimeria penicillati (189)
Perognathus longimembris
Tritrichomonas muris (74)
Perognathus parvus
Tritrichomonas muris (74)

Acanthocephala: Spiny Headed Worms

Acanthocephalans occur as adults only in the intestines of vertebrate animals, but apparently there are no records of these from heteromyids.

Trematoda: Trematodes or Flukes

Trematodes have a complicated life cycle, but all require a snail as secondary host. There apparently are no records of trematodes in heteromyids. Perhaps this is because of the paucity of snails in the arid areas inhabited by all genera except *Heteromys,* but it may be related to the predominantly vegetarian diet.

Cestoda: Cestodes or Tapeworms

Most adult cestodes are parasitic in the digestive tract of vertebrates. They usually consist of a series of segments, or proglottids. There is no digestive system; nutrients are absorbed through the body wall. There is usually a vertebrate or arthropod intermediate host in which the larvae live. Larvae are variable, but often are bladderlike and are called bladderworms. Infection by cestodes is generally by feeding on infected intermediate hosts.

The main cestodes of kangaroo rats are quite clearly species of *Catenotaenia,* especially *C. linsdalei.* Most of the cestode records from heteromyids are from kangaroo rats, with a few from pocket mice.

Bienek and Klikoff (1974) demonstrated that *Dipodomys merriami* feeds on arthropods, since two of the cestode genera found, *Mathevotaenia* and *Catenotaenia,* and also two nematodes (*Rictularia* sp. and *Mastophorus numidica*) all require insect intermediate hosts. Additionally, they found insect parts in stomachs and intestines. They further state that *Dipodomys microps* is herbivorous throughout the year (Kenagy, 1972; Warnock and Grundmann, 1963) and harbors only *Trichuris dipodomis,* which requires no intermediate host.

There are several records of cestodes from heteromyids, as follows:

Dipodomyinae

Dipodomys deserti
Catenotaenia linsdalei (208)
Dipodomys heermani
Catenotaenia californica (82, 318)
Catenotaenia linsdalei (317, 318)
Hymenolepis citelli (283)
Dipodomys ingens
Hymenolepis citelli (283)
Dipodomys merriami
Andrya sp. (21)
Catenotaenia linsdalei (208, 318)
Mathevotaenia deserti (254, 318)
Schizorchodes dipodomi (21)
Dipodomys microps
Catenotaenia linsdalei (208)
Dipodomys nitratoides
Hymenolepis citelli (283)
Dipodomys ordii
Raillietina retractilis (154)
Dipodomys panamintinus
Catenotaenia californica (82, 318)
Catenotaenia linsdalei (318)
Mathevotaenia deserti (254)
Dipodomys phillippsi
Cysticercis sp. (221)
Dipodomys spectabilis
Catenotaenia linsdalei (157)
Dipodomys venustus
Catenotaenia linsdalei (317, 318)

Perognathinae

Chaetodipus californicus
Catenotaenia linsdalei (317, 338)
Chaetodipus formosus
Catenotaenia linsdalei (154)
Perognathus longimembris
Mathevotaenia deserti (254)

Nematoda: Nematodes or Roundworms

Parasitic nematodes generally have a single host and essentially a direct life cycle. Infection is usually by ingestion of a free living stage, often the egg, although free living larvae in some species actively burrow into the tissue of the host.

Fifteen species of nematodes have been reported from heteromyid rodents, the majority from *Dipodomys.* Seven species—*Gongylonema dipodomyis, Heteromoxyuris deserti, Protospirura dipodomis, Rictularia dipodomis, Trichuris dipodomis, Trichuris perognathi,* and *Wellcomia perognathi*—appear to be heteromyid nematodes; only four of these are known from more than two hosts. One species of *Trichuris, T. dipodomis,* occurs in kangaroo rats and a second *T. perognathi,* in pocket mice. In *Dipodomys ordii* nematodes have a division of habitat, with *Heteromoxyuris deserti* in the caecum and *Protospirura muris* in the stomach (Garner et al., 1976).

Dipodomyinae

Dipodomys deserti
Heteromoxyurus deserti (208)
Protospirura dipodomis (319)
Rictularia dipodomis (208)
Dipodomys heermani
Heteromoxyurus deserti (274)
Rictularia dipodomis (274)
Dipodomys merriami
Abbreviata sp. (208)
Gongylonema dipodomyis (208, 215)
Gongylonema neoplasticum (274)
Heteromoxyurus deserti (208, 274)
Mastophorus numidica (23)
Protospirura dipodomis (208, 274)
Protospirura numidica (23)
Rictularia dipodomis (208, 274, 301)
Trichuris minuta (159)
Wellcomia longejector (159)
Dipodomys microps
Capillaria americana (23)
Heligmosomum sp. (156)
Rictularia dipodomis (208)
Trichuris dipodomis (23, 153)
Dipodomys ordii
Capillaria americana (155)
Heligmosomum sp. (153)
Heteromoxyuris deserti (139)
Protospirura muris (139)
Protospirura numidica (153, 155)
Trichuris dipodomis (153, 272, 273)
Trichuris minuta (159)
Dipodomys panamintinus
Gongylonema dipodomydis (339)
Gongylonema neoplasticum (274)
Heteromoxyuris deserti (274)
Protospirura dipodomis (274)
Rictularia dipodomis (274)

Heteromyinae

Heteromys desmarestianus
"Hookworms" (216)
Heteromys sp.
Vexillata petteri (84)
Liomys pictus
Longistriata vexillata (55)
Trichuris sp. (55)

Perognathinae

Chaetodipus baileyi
Protospirura dipodomis (274)
Chaetodipus californicus
Trichuris perognathi (59, 319)
Chaetodipis formosus
Trichuris sp. (possibly *T. perognathi*) (153)

Chaetodipus intermedius
Wellcomia perognathi (214)
Chaetodipus penicillatus
Protospirura anodon (159, 274)
Protospirura tetradon (159, 274, 312)
Trichuris perognathi (59, 60)
Wellcomia sp. (10)
Microfilaria in blood (333)
Perognathus longimembris
Protospirura dipodomis (274)

Acari: Mites (Excluding chiggers)

The mites constitute a large and diverse group, and a large percentage of the heteromyid parasites are mites. Ticks and chiggers are specialized mites also but are treated separately. Parasitic mites typically have egg, larva, protonymph, deutonymph, and adult stages. The larva is often inactive and does not feed, whereas the deutonymph is often an active feeding stage. A good example of the amount of variation one finds is that the glycyphagid one most often finds on North American mammals consists of the deutonymph, which in this case is a nonfeeding form, phoretic rather than parasitic. To confuse the issue yet further, most of the glycyphagids described from heteromyids live in hair follicles, where they do not actively feed; they do not even have mouthparts developed, but it is possible that they absorb nutrients from the host. Some mites, the listrophorids, for example, are ovoviviparous: the egg and sometimes the larval stages are bypassed and the females produce living young.

A few further comments follow on some of the families of mites found on heteromyids.

Cheyletidae.—Most cheyletid mites found in the fur of mammals we suspect to be accidental there; they are said to be predatory on mites in the nest of the hosts. Some species are parasitic.

Dermanyssidae.—Dermanyssid mites are usually parasites of birds.

Glycyphagidae.—Adult glycyphagid mites are found in mammal nests where they probably are fungi feeders. The deutonymph or hypopial stage is usually phoretic in the fur or parasitic in the hair of mammal hosts.

Laelapidae.—Most of the larger mites found in the fur of rodent hosts are laelapid mites, with the genera *Androlaelaps, Echinonyssus, Ischyropoda* and *Steptolaelaps* often being common on heteromyids.

Listrophoridae.—Listrophorids are tiny tubular mites with legs modified for crawling up and down hairs. They feed by dipping their mouthparts into hair follicles and are often very abundant.

Macronyssidae.—Mites of this family are often found on bats and birds. One species, *Ornithonyssus bacoti,* is often found on rodents, including heteromyids.

Myobiidae.—Myobiids are small, light colored, dorso-ventrally flattened mites that lay eggs and live their entire life cycle in the fur of the host, where all of the stages may be found attached to the hairs.

A few non-parasitic families have been reported but, except for mites of the genus *Sertitympanum* discussed below, will not be commented upon further.

Important mites of heteromyids are the laelapids, *Ischyropoda* (all three species of the genus), the genus *Steptolaelaps* (two species, on *Liomys* and *Heteromys*), and *Hypoaspis leviculus*; the myobiid *Radfordia bachai*; and a number of species of the listrophorid genus *Geomylichus.*

A rather large number of species is known in the genus *Echinonyssus,* of which one is the heteromyid host group containing *E. hilli, E. perognathi, E. incomptis,* and *E. triacanthus.* It is of interest that another host group, also of four species, is the geomyid host group, and that these two groups apparently are closely related. *Echinonyssus hilli* is known mostly from *Perognathus flavus, longimembris,* and *parvus* from Utah, Nevada, Idaho, and California. *Echinonyssus perognathi* is known mostly from *Chaetodipus formosus* and *hispidus* from Nevada and Kansas. *E. incomptis* and *E. triacanthus* are known primarily from *Dipodomys merriami, ordii,* and *microps* from Utah, Nevada, and Texas. Thus, this group shows some degree of specialization on heteromyids.

Species of *Ischyropoda* commonly are taken at the base of the tail of the host, but are found elsewhere as well. *Ischyropoda armatus* may be common on *Onychomys* also (Eads et al., 1952).

Several literature records are listed as *Kleemania.* These mites probably are free living, but are reported often enough to be included here as associates of heteromyids. However, mites listed as *Kleemania* are surely *Sertitympanum* as recently described by Elsen and Whitaker (1985), and they are listed as *Sertitympanum* in this paper.

The mite genus *Geomylichus* (Listrophoridae) is particularly interesting. Information on it was summarized by Fain et al. (1978). Seven species were previously known, six additional ones were described in that paper, and additional species are being described. These mites are fusiform, hair-clasping mites, so small as to be overlooked by the casual collector using fur brushing techniques.

Of the 15 species of *Geomylichus,* eight (plus more undescribed ones) are from North American heteromyids, two are from geomyids, four are from cricetids (*G. klebergi* from *Sigmodon, G. neacomys* from *Neacomys, G. nectomys* on *Nectomys,* and *G. mexicanus* from *Teanopus*), and one is from a lagomorph, *Sylvilagus floridanus.*

Geomylichus can be described as widespread and abundant on North American heteromyid and geomyid rodents. However, there is only a sprinkling of species on North and South American cricetids. *Geomylichus klebergi* is known only from the type series from *Sigmodon hispidus* from near Kingsville, Kleberg County, Texas. A larva from the same host from Venezuela (Fain et al., 1978) is presumed to be the same species. *Geomylichus nectomys* is known from four specimens from *Nectomys* sp. from Costa Rica and *G. neacomys* is known from two specimens from *Neacomys tenuipes* from Columbia. *Geomylichus sylvilagus* is known only from a holotype nymph and paratype larva from *Sylvilagus floridanus* from Yucatan, Mexico. Thus, none of these species is known to be widespread or abundant. They may form a natural group since all of these four species have the striated membranes of coxae I and II serrated. No other species of *Geomylichus* has this character.

A cladistic approach to the genus *Geomylichus* of heteromyids has been attempted, although more data are needed. However, it appears that the species from heteromyids and geomyids are quite closely related, and that the cricetid-lagomorph branch might be separate. It would be exceedingly interesting to examine southern heteromyids, *Heteromys* and *Liomys,* first to see if they have *Geomylichus,* and if they do to see if they have typical heteromyid forms or if they have serrated membranes on coxa I or II (or an intermediate type). We suspect that southern heteromyids would have the intermediate or cricetid type *Geomylichus,* and that these have given rise to the cricetid-lagomorph branch of the genus.

Geomylichus of North American heteromyids is of further interest as it is in this host group that *Geomylichus* has undergone its greatest radiation. The one species described from *Liomys, G. postscutatus* (Fain, 1986), may support this view, as the male of this species is described as having the internal region of Coxae I–II with very thick striations. However, I have not seen the full description of this species.

Much evolution has occurred in the heteromyid *Geomylichus* mites, with nine species presently known. Six of these are each known from only one host, numbers 3, 4, and 6–9 below. *Geomylichus dipodomius* has been found on four species of *Dipodomys* and *G. texanus* is primarily on *D. ordii,* but has also been found on *D. merriami* and *Perognathus penicillatus. Geomylichus postscutatus* has also been found on two genera, *Liomys irroratus* and *Dipodomys* sp.

That they are quite host specific is demonstrated by the fact that some have been found on different hosts at the same locality, *G. perognathi* on *Perognathus parvus, Geomylichus microdipodops* on *Microdipodops megacephalus,* and *G. texanus* on *D. ordii* at Van Horn, Texas. At Simpson Springs, Juab Co., Utah, *Geomylichus formosus* was taken on *Perognathus formosus, G. dipodomius* (or a new species) on *D. microps, G. texanus* on *D. ordii,* and *G. perognathi* on *Perognathus parvus.* The latter three species and hosts were also taken below Hickison Summit, Lander Co., Nevada.

Undoubtedly examination of additional heteromyid host species and different localities will yield additional new species.

Information on geographic and host distribution of *Geomylichus* from heteromyids is summarized below.

1. *Geomylichus brevispinosus*
   P. penicillatus: Texas
2. *Geomylichus californicus*
   D. californicus: California
   D. elephantinus: California
   D. heermani: California
   D. venustus: California
3. *Geomylichus deserti*
   Dipodomys deserti: Nevada
4. *Geomylichus dipodomius*
   D. elator: Texas
   D. merriami: Nevada
   D. microps: New Mexico
   D. phillipsi: Mexico
   D. spectabilis: Utah
5. *Geomylichus formosus*
   Perognathus formosus: Utah
6. *Geomylichus inaequalis*
   C. hispidus: Texas
7. *Geomylichus microdipodops*
   Microdipodops megacephalus: Nevada
6. *Geomylichus multistriatus*
   D. merriami: Texas
   D. nitratoides: California
   M. megacephalus: Nevada
   C. penicillatus: Arizona
9. *Geomylichus perognathi*
   Perognathus parvus: Nevada
10. *Geomylichus postscutatus*
   Liomys irroratus: Mexico
   Dipodomys sp.: Nebraska
11. *Geomylichus texanus*
   D. elephantinus: California
   D. merriami: New Mexico
   D. ordii: Texas, Nevada
   D. venustus: California
   P. penicillatus: Ncvada, Arizona

A few species of glycyphagid hypopi have been reported from heteromyids. Hypopi are phoretic, i.e., are not true parasites; they simply use the host for transportation. Hypopi are non-feeding immature mites that use the host presumably as a means of distribution to other hosts, or more specifically to the nests of other hosts. Presumably they transform to their adult stage in the nest of the host. Hypopi of different species may be very similar, yet the adults may be very different. Thus it becomes necessary to culture hypopi in the laboratory for comparison with other adults, and to learn of other life stages. Most of the hypopi live in the fur of their mammal hosts, for example, *Dermacarus ornatus.* Others, including six from heteromyids, those in the genera *Dipodomyopus, Mediolabidophorus, Metalabidophorus,* and *Neolabidophorus,* live in hair follicles. One species, *Dipodomydectes americanus,* lives under the skin of its host, *Dipodomys merriami.* It is possible that those in hair follicles and under the skin absorb nutrients through their body wall, and thus would be more correctly classed as parasites. Those species of the follicles and subdermal areas show character reductions, thus appear to have evolved from the hair clasping species (see discussion for more details).

Mites other than chiggers recorded from Heteromyidae:

Dipodomyinae

Dipodomys californicus
  Listrophoridae
    Geomylichus californicus (125)
Dipodomys deserti
  Laelapidae
    Echinonyssus triacanthus (167)
    Ischyropoda armatus (198)

Listrophoridae
Geomylichus deserti (123)
Dipodomys elator
Laelapidae
Androlaelaps fahrenholzi (297)
Echinonyssus incomptis (297)
Listrophoridae
Geomylichus dipodomius (125, 297)
Dipodomys elephantinus
Listrophoridae
Geomylichus californicus (125)
G. texanus (125)
Dipodomys heermani
Listrophoridae
Geomylichus californicus (125)
Dipodomys merriami
Glycyphagidae
Dipodomydectes americanus (under the skin) (121)
Laelapidae
Androlaelaps casalis (7)
A. fahrenholzi (7, 144)
Brevisterna utahensis (7)
Echinonyssus incomptis (7, 144)
E. triacanthus (7, 144, 167, 190)
Ischyropoda armatus (7, 144, 198)
Listrophoridae
Geomylichus dipodomius (144)
G. multistriatus (125)
G. texanus (124)
Myobiidae
Radfordia bachai (261)
Amerosiidae
Sertitympanum sp. (144)
Dipodomys microps
Cheyletidae
Cheyletis linsdalei (100)
Laelapidae
Androlaelaps fahrenholzi (7, 100, 144, 199)
Echinonyssus hilli (144)
E. incomptis (7, 100, 144)
E. triacanthus (7, 100, 144, 167, 199)
E. utahensis (100)
Hypoaspis leviculus (144)
Ischyropoda armatus (7, 100, 144, 199)
Listrophoridae
Geomylichus dipodomius (100, 144)
Myobiidae
Radfordia bachai (100, 183)
Amerosiidae
Sertitympanum sp. (144)
Dipodomys nitratoides
Listrophoridae
Geomylichus multistriatus (125)
Dipodomys ordii
Cheyletidae
Cheyletus linsdalei (100)
Laelapidae
Androlaelaps fahrenholzi (6, 7, 100, 199, 246, 291, 276)
Brevisterna morlani (292)
Echinonyssus hilli (199, 246)
E. incomptis (6, 7, 90, 94, 100, 167, 246, 276)
E. longichelae (6)
E. neotomae (7)
E. triacanthus (6, 7, 94, 100, 167, 199, 246)
E. utahensis (6, 7, 100, 167, 246)
Eubrachylaelaps crowei (6)
E. debilis (6)
**Haemogamasus ambulans* (6, 7)
H. reidi (246)
H. onychomydis (246)
Hypoaspis leviculus (6, 7)
Ischyropoda armatus (6, 7, 94, 98, 100, 199, 246)
I. furmani (3, 6, 7, 200, 246)
Laelaps kochi (7)
Listrophoridae
Geomylichus dipodomius (100, 247)
G. texanus (124, 246)
Macrochelidae
Macrocheles sp. (246)
Macronyssidae
**Ornithonyssus bacoti* (6, 7, 171)
Myobiidae
Radfordia bachai (6, 100, 183)
Uropodidae
Uropoda sp. (94)
Amerosiidae
Proctolaelaps sp. (246)
Sertitympanum contiguum (99)

S. exarmatum (99)
Sertitympanum sp. (6, 246)
Cyrtolaelapidae
Euryparasitus sp. (246)
Dipodomys phillippsii
Listrophoridae
Geomylichus dipodomius (124)
Dipodomys spectabilis
Laelapidae
Androlaelaps fahrenholzi (94)
Echinonyssus incomptis (293)
Eubrachylaelaps crowei (94)
Ischyropoda armatus (94)
Listrophoridae
Geomylichus dipodomius (269)
Dipodomys venustus
Listrophoridae
Geomylichus californicus (125)
G. texanus (125)
Dipodomys sp.
Glycyphagidae
Dipodomyopus tuttlei (hair follicles) (120)
Listrophoridae
Geomylichus postscutatus (117)

Heteromyinae

Heteromys anomalus
Glycyphagidae
Dermacarus ornatus (114)
Metalabidophorus heteromys (243)
Laelapidae
Androlaelaps fahrenholzi (135, 138)
A. projecta (135)
A. rotunda (135)
Echinolaelaps boultoni (138)
Echinonyssus keenani (169)
***E. parvisoma* (169)
E. proctolaelaps (169)
***E. venezuelensis* (169)
Eubrachylaelaps rotundus (138)
Gigantolaelaps goyanensis (135)
G. inca (135)
G. wolffsohni (138)
***Laelaps dearmasi* (135)
L. ovata (135)
Mysolaelaps parvispinosus (135, 138)
***Steptolaelaps heteromys* (131, 134, 135, 138)
Macronyssidae
Acanthonyssus proechimys (279)
Heteromys australis
Laelapidae
Androlaelaps fahrenholzi (303)
Steptolaelaps heteromys (303)
Heteromys desmarestianus
Glycyphagidae
Mediolabidophorus neotropicalis (120)
Laelapidae
Androlaelaps fahrenholzi (303)
Echinonyssus heteromydis (294)
E. lunatus (294)
E. microchelae (294)
E. minutus (294)
E. panamensis (142, 294)
Eubrachylaelaps jamesoni (303)
Steptolaelaps heteromys (142, 303)
Tur uniscutatus (303)
Speleognathidae
Paraspeleognathopsis cricetidarum (62)
Heteromys gaumeri
Glycyphagidae
Dermacarus ornatus (265)
Neolabidophorus yucatanensis (265 [follicles])
Laelapidae
Androlaelaps fahrenholzi (142)
Eubrachylaelaps jamesoni (142)
Hypoaspis lubrica (142)
Steptolaelaps heteromys (142)
Heteromys sp.
Laelapidae
Steptolaelaps heteromys (134, 135)
Liomys adspersus
Laelapidae
Androlaelaps fahrenholzi (303)
Echinonyssus microchelae (294)
Steptolaelaps heteromys (303)
Macronyssidae
Ornithonyssus bacoti (341)

Liomys irroratus
Glycyphagidae
Dermacarus liomys (119)
Metalabidophorus liomys (243)
Laelapidae
Androlaelaps fahrenholzi (95)
Echinonyssus liomys (168)
E. neotomae (95)
E. perognathi (168)
Steptolaelaps liomydis (95, 134, 151)
Macronyssidae
Ornithonyssus bacoti (95)
O. sylviarum (95)
Listrophoridae
Geomylichus postscutatus (117, 124)
Liomys pictus
Laelapidae
Hypoaspis leviculus (142)
Steptolaelaps heteromys (142)
S. liomydis (134, 142)
Liomys salvini
Laelapidae
Androlaelaps fenilis (142)
Echinonyssus brevicalcar (169)
E. galindoi (142)
Eubrachylaelaps(?) *circularis* (142)
Hypoaspis lubrica (142)
Steptolaelaps heteromys (142)
Listrophoridae
Listrophorus sp. (142)
Cheyletidae
Eucheyletia n. sp. (142)

Perognathinae

Chaetodipus californicus
Laelapidae
Androlaelaps fahrenholzi (291)
Eubrachylaelaps hollisteri (133)
Ischyropoda armatus (198)
Chaetodipus fallax
Laelapidae
Androlaelaps frontalis (132)
Chaetodipus formosus
Laelapidae
Androlaelaps fahrenholzi (2, 7)
Brevisterna utahensis (7)
Echinonyssus affinis (7)
E. perognathi (2)
E. triacanthus (7)
Hypoaspis leviculus (2, 7)
Ischyropoda armatus (2, 7)
Amerosiidae
Sertitympanum sp. (2)
Chaetodipus hispidus
Laelapidae
Androlaelaps fahrenholzi (98, 188, 291, 297)
A. grandiculata (94, 188)
Echinonyssus neotomae (89)
E. perognathi (190, 297)
Hypoaspis leviculus (86)
Listrophoridae
Geomylichus inaequalis (124, 297)
Listrophorus klebergi (247)
Macronyssidae
Ornithonyssus bacoti (188)
Chaetodipus penicillatus
Glycyphagidae
Neolabidophorus verrucosus (under the skin?) (133)
Laelapidae
Ischyropoda spiniger (198)
Listrophoridae
Geomylichus brevispinosus (124)
G. multistriatus (125)
G. texanus (124)
Chaetodipus spinatus
Laelapidae
Ischyropoda spiniger (198, 278)
Microdipodops megacephalus
Laelapidae
Ischyropoda furmani (3, 199, 200)
Listrophoridae
Geomylichus multistriatus (125)
Perognathus fasciatus
Laelapidae
Androlaelaps fahrenholzi (143)
Listrophoridae
Geomylichus perognathi (122)
Perognathus flavescens
Laelapidae
Hypoaspis leviculus (7)
Perognathus flavus
Laelapidae
Androlaelaps grandiculatus (86, 94)

Echinonyssus hilli (7, 293)
Ischyropoda armatus (7)
Perognathus inornatus
Laelapidae
Ischyropoda armatus (198)
Perognathus longimembris
Laelapidae
Androlaelaps fahrenholzi (2, 7)
Echinonyssus hilli (7)
E. incomptis (7, 167)
E. triacanthus (7)
E. utahensis (7)
Eubrachylaelaps circularis (7)
Hypoaspis leviculus (2)
Ischyropoda armatus (2, 7)
I. furmani (200)
Ameroseiidae
Sertitympanum sp. (2)
Perognathus parvus
Dermanyssidae
Dermanyssus gallinae (6)
Laelapidae
Androlaelaps casalis (6)
A. fahrenholzi (2, 6, 7, 160, 246)
Echinonyssus hilli (2, 6, 7, 167, 190, 246)
E. incomptis (6, 7)
E. neotomae (7)
E. triacanthus (6, 167)
E. utahensis (6)
Eubrachylaelaps circularis (7)
E. debilis (6)
E. hollisteri (7)
Haemogamasus ambulans (7)
H. onychomydis (246)
Hypoaspis leviculus (7, 199)
H. lubrica (6)
Ischyropoda armatus (6, 7, 199, 246)
I. furmani (6)
Listrophoridae
Geomylichus perognathi (122, 246)
Amerosiidae
Sertitympanum exarmatum (99)
Sertitympanum sp. (6, 246)
Ascidae
Proctolaelaps sp. (246)
Macrochelidae
Macrocheles sp. (160)
Macronyssidae
Ornithonyssus bacoti (7)
Pygmephoridae
Bakerdania sp. (246)
Perognathus xanthonotus
Laelapidae
Ischyropoda armatus (198)

Acari: Trombiculidae: Chiggers

Chiggers are the parasitic larvae of mites of the family Trombiculidae. Adults and nymphal chiggers are not well known but on occasion they are found on the surface of, or in, the soil. Females lay their eggs in soil or other material such as decaying logs near or in the nests or runs of the hosts. The eggs develop into deutova from which the parasitic six-legged larvae, or chiggers, emerge. Most larvae feed on vertebrates. The chelicerae are used to abrate or pierce the skin of the host; at the same time, the larva introduces saliva into the wound. The saliva apparently contains enzymes that break down the host tissue and this predigested material is imbibed by the chigger. A tube-like structure, the stylostome, usually forms in the host's epidermis and predigested host tissue and saliva, respectively, are alternately pumped into and out of the chigger. Larvae do not normally feed on blood. Chiggers range in color from white to yellow, orange or red. Larvae often attach in the external auditory meatus, on the pinnae, and other areas of the host, including the head, genital region and nasal passages. The engorging larvae may remain in place several days on the host, then detach, drop off and complete the life cycle in the substrate. The postlarvae are free-living and predaceous upon eggs and immature forms of small arthropods. Many of the chiggers are parasitic on mammals and, in the Western Hemisphere, a large number of species are known to parasitize heteromyids.

Chiggers have by far the greatest number of species of any of the parasites considered here as infesting heteromyids, and also heteromyids are about as heavily parasitized by chiggers as any group of mammals. Listed here are 137 species. It is not known why chiggers are so abundant on heteromyids.

Loomis (1971*b*) discussed the desert habitat and *Perognathus* as a host for chiggers. He indicated that 81 species in 18 genera of chiggers were known from *Perognathus* (including *Chaetodipus*), and also provided additional information on certain genera. All 11 species of *Euschoengastoides* have been found on *Perognathus,* and Perognathus makes up about 40% of the host records (Loomis, 1971*a*). Similar figures for *Hyponeocula* are seven of 10 species on *Perognathus,* 33% of all hosts (data on this genus are summarized by Tanigoshi and Loomis, 1974). All seven species of *Otorhinophila* infest *Perognathus,* and members of *Perognathus* comprised 59% of the host records (Wrenn and Loomis, 1967). These three genera are essentially *Perognathus* chiggers, making up 44% of the host records (573 of 1,306). Other chigger genera described more recently from *Perognathus* are *Hexidionis* and *Dermadelema.* These five genera of chiggers and *Perognathus* have similar ranges, centering in the three major North American desert areas, the Mohave, Sonoran, and Chihuahuan. Loomis (1971*b*) further discussed these areas, all of which have hot summers. The Mohave is primarily in southeastern California and southern Nevada; it is the most northern of these three deserts and has winter precipitation. The Sonoran (southeastern California and southern Arizona, upper three fourths of Baja California, and western Sonora) has precipitation in summer and winter in the east, mostly in winter in the west. The Chihuahuan (southern New Mexico and northern central Mexico) has summer rain. The lowest, hottest, and driest is the Sonoran desert and it has the greatest diversity of vegetation, hosts, and chiggers. The Mohave desert is higher in elevation with much yucca, creosote bush, and burrow bush and has fewer chigger species.

Heteromyids thus are a major component of these desert faunas and of the chigger host community. A single pocket mouse may host hundreds of chiggers of one to several species. They may occur anywhere on the body, but are especially abundant on or in the ears, muzzle, cheek pouches, and genitalia.

Chiggers are thought to be habitat specific rather than host specific (Loomis, 1956). That is, they occur in the proper habitat and/or location, and thus are apt to occur on the particular host. However, other hosts that happen to occur in the proper place at the right time may be as likely to become infested as the "intended host."

It is difficult, at best, to indicate chiggers "often occurring" on heteromyids because of the small amount of information available on many of the chiggers and hosts and also because of the great amount of variation between papers in the amount of information given on frequency and abundance of the chiggers. However, it does seem valuable in a paper such as this to attempt some indication, even though primitive, of chiggers often found on heteromyids. Thus, a great deal of judgment has been exercised in marking with asterisks those species which have been reported several times or in abundance on heteromyids (Table 1).

Many of the most important chiggers of heteromyids are in the genera discussed above, *Euschoengastoides, Hyponeocula, Otorhinophila, Hexidionis,* and *Dermadelema.* However, there are a number of other genera for which a large proportion of the presently known records are from heteromyids, and these are listed with asterisks in Table 2.

Wrenn and Loomis (1967) examined about 5,000 small mammals from North America, including the nasal passages of more than 3,500. Numerous chiggers were found, including four closely related species from heteromyids for which the authors proposed a new genus, *Otorhinophila.* Two of the species were from the ears and two were from the nasal passages.

Intranasal species were found by cutting and lifting up the nasal bones, by flushing out the nasal passages with water, or by pulling out the nasal mucosa for examination as the skin was separated from the skull. Wrenn and Loomis (1967) also described rearing techniques for chiggers as follows: "Well-engorged larvae were kept alive for rearing. Usually a single larva was placed in a small culture vial nearly filled with a hardened mixture of activated charcoal and plaster of Paris (Wharton, 1946). Upon emergence of a nymph, the larval pelt was searched for and if found, was mounted for identification. After 24 hours the nymph was preserved in 75 percent ethyl alcohol. The preserved nymphs were usually cleared in warm lactophenol for approximately 24 hours before mounting. Both larvae and nymphs were mounted in polyvinyl alcohol lactophenol. The nymphs were fed freshly laid collembolan eggs and upon emergence of adults, nymphal pelts were recovered and prepared in the same way as larval pelts."

Chiggers (Trombiculidae)

Dipodomyinae

Dipodomys agilis
- *Dermadelema furmani* (266)
- *Euschoengastia ambocalis* (335)
- *Euschoengastia criceticola* (237)
- *Euschoengastia heteromyicola* (336)
- *Euschoengastia obscura* (336)
- *Hyponeocula arenicola* (296)
- *Odontacarus linsdalei* (234)

Dipodomys deserti
- *Dermadelema furmani* (266)
- *Euschoengastoides arizonae* (230)
- *Hexidionis deserti* (238)
- *Hyponeocula deserticola* (296)

Dipodomys elator
- *Euschoengastia decipiens* (297)

Dipodomys heermani
- *Dermadelema mojavense* (266)
- *Hyponeocula arenicola* (296)
- *Hyponeocula montanensis* (150)
- *Neoschoengastia americana* (150)

Dipodomys merriami
- *Dermadelema furmani* (266)
- *Dermadelema lynnae* (266)
- *Dermadelema mojavense* (266)
- *Dermadelema sleeperi* (233, 266)
- *Euschoengastia ambocalis* (335)
- *Euschoengastia decipiens* (38, 144, 336)
- *Euschoengastia hardyorum* (337)
- *Euschoengastia heteromyicola* (336)
- *Euschoengastia marginalis* (337)
- *Euschoengastia numerosa* (336)
- *Euschoengastia obscura* (336)
- *Euschoengastia simulans* (336)
- *Euschoengastoides arizonae* (230, 233)
- *Euschoengastoides hoplai* (233)
- *Euschoengastoides imperfectus* (230)
- *Euschoengastoides loomisi* (233, 241)
- *Euschoengastoides neotomae* (230)
- *Euschoengastoides webbi* (230)
- *Eutrombicula alfreddugesi* (233)
- *Hexidionis allredi* (233, 241)
- *Hexidionis deserti* (238)
- *Hexidionis doremi* (38, 238)
- *Hexiodinis harveyi* (233, 237)
- *Hexidionis jessiemae* (144)
- *Hyponeocula arenicola* (7, 144, 233, 296)
- *Hyponeocula fovea* (296)
- *Hyponeocula imitator* (296)
- *Leptotrombidium panamense* (233, 235)
- *Odontacarus linsdalei* (144)
- *Otorhinophila baccusi* (233, 240)
- *Otorhinophila desertorum* (240)
- *Otorhinophila parvisola* (240, 334)
- *Otorhinophila xerophila* (240)
- *Parasecia gurneyi* (233)

Dipodomys microps
- *Dermadelema furmani* (266)
- *Dermadelema sleeperi* (266)
- *Euschoengastia criceticola* (7, 100)
- *Euschoengastia decipiens* (7, 144, 336)
- *Euschoengastia hardyorum* (337)
- *Euschoengastia heteromyicola* (336)
- *Hexidionis jessiemae* (144)
- *Hyponeocula arenicola* (38, 144, 227, 296)
- *Hyponeocula fovea* (296)
- *Kayella lacerta* (144)
- *Odontacarus linsdalei* (38, 144)

Dipodomys nelsoni
- *Hyponeocula arenicola* (296)

Dipodomys nitratoides
Dermadelema mojavense (266)
Dipodomys ordii
Comatacarus americanus (6)
Euschoengastia cordiremus (6)
Euschoengastia criceticola (38, 100)
Euschoengastia decipiens (6, 7, 38)
Euschoengastoides arizonae (230, 241)
Euschoengastoides loomisi (228, 241)
Euschoengastoides tumidus (230)
Eutrombicula alfreddugesi (228, 332)
Eutrombicula batatas (228, 332)
Eutrombicula belkini (7)
Hexidionis allredi (241)
Hexidionis breviseta (241)
Hexidionis doremi (6, 7)
Hexidionis harveyi (233, 241)
Hyponeocula arenicola (6, 7, 38, 227, 228, 241, 296)
Hyponeocula fovea (296)
Hyponeocula montanensis (228, 296)
Odontacarus linsdalei (6, 7, 38, 44)
Odontacarus micheneri (6)
Otorhinophila baccusi (241)
Otorhinophila parvisola (240)
Parasecia gurneyi (228)
Pseudoschoengastia farneri (228)
Pseudoschoengastia hungerfordi (225, 228)
Trombicula bakeri (6, 7)
Xenodontacarus plumosus (228)
Dipodomys panamintinus
Dermadelema furmani (266)
Dermadelema mojavense (266)
Dermadelema sleeperi (266)
Euschoengastia decipiens (336)
Euschoengastia heteromyicola (336)
Euschoengastia marginalis (337)
Euschoengastia simulans (336)
Hyponeocula arenicola (296)
Hyponeocula fovea (296)
Hyponeocula imitator (296)
Otorhinophila xerophila (240)
Dipodomys spectabilis
Euschoengastoides arizonae (230)
Euschoengastoides imperfectus (230)
Euschoengastoides neotomae (230)
Hexidionis harveyi (237)
Hyponeocula arenicola (296)
Dipodomys stephensi
Dermadelema furmani (266)
Euschoengastia heteromyicola (336)
Euschoengastia obscura (336)
Dipodomys venustus
Euschoengastia criceticola (44, 150)
Euschoengastia radfordi (44, 150, 336)
Euschoengastia romola (44, 150)
Neotrombicula californica (44, 150)
Odontacarus hirsutus (44)

Heteromyinae

Heteromys anomalus
Anomalaspis ambiguus (27, 36, 51)
Boshkerria punctata (51)
Crotiscus danae (43)
Crotiscus desdentatus (28, 51)
Eutrombicula alfreddugesi (46)
Eutrombicula goeldii (46, 50, 51)
Kymocta faitkeni (33, 54)
Kymocta zulia (52)
Odontacarus comosus (275)
Odontacarus tubercularis (27, 51, 275)
Parasecia manueli (46, 51)
Polylopadium chaetolecanium (48)
Pseudoschoengastia bulbifera (51)
Quadraseta flochi (46, 51)
Quadraseta mirandae (147)
Vanidicus chalepus (37)
Vanidicus jojosti (37)
Heteromys australis
Eutrombicula goeldii (53)
Peltoculus almae (34)
Polylopadium tertium (34)
Pseudoschoengastia bulbifera (53)
Pseudoschoengastia zona (53)
Quadraseta trapidoi (34)
Trombicula dunni (34, 53)
Trombicula keenani (53)
Heteromys desmarestianus
Ascoschoengastia dyscrita (53)
Crotiscus desdendatus (53, 142)
Euschoengastia belgicae (53)
Eutrombicula alfreddugesi (53, 142)
Eutrombicula goeldii (53)
Hoffmannina handleyi (47, 53, 142)
Kymocta teratarsalis (53, 340)

Leptotrombidium hamaxiaia (142)
Pseudoschoengastia bulbifera (53, 142)
Pseudoschoengastia disparunguis (146)
Pseudoschoengastia finitima (53, 141)
Sasacarus furmani (53, 142)
Sasacarus vercammeni (40)
Trombicula dicrura (47, 53)
Trombicula dunni (53)
Trombicula keenani (53)
Vanidicus tricosa (53)

Heteromys gaumeri
Cordiseta mexicana (229)
Ectonyx fusicornis (229)
Leptotrombidium panamense (229)
Odontacarus tubercularis (229)
Parasecia gurneyi (229)
Pseudoschoengastia brennani (229)
Pseudoschoengastia extrinseca (229)
Pseudoschoengastia scitula (229)

Heteromys sp.
Parasecia aitkeni (51)

Liomys adspersus
Ascoschoengastia dyscrita (47, 53)
Colicus liomys (47, 53)
Crotonasis fissa (53)
Eutrombicula goeldii (53)
Leptotrombidium panamense (53)
Odontacarus tubercularis (53)
Polylopadium kramisi (47, 53)
Pseudoschoengastia bulbifera (30, 53)
Pseudoschoengastia zona (30, 53, 141)
Vanidicus tricosus (47, 53)

Liomys irroratus
Ectonyx fusicornis (29, 142)
Euschoengastia bigenuala (95, 235)
Euschoengastoides gagarini (30)
Euschoengastoides loomisi (95, 235)
Eutrombicula alfreddugesi (95)
Hexidionis allredi (95)
Hexidionis jessiemae (32)
Kayella lacerta (95, 235)
Leptotrombidium panamense (95, 235)
Odontacarus tubercularis (95, 235)
Parasecia universitatis (35, 170)
Pseudoschoengastia audyi (40, 95)
Pseudoschoengastia farneri (95, 235)
Pseudoschoengastia hoffmannae (142)
Pseudoschoengastia hungerfordi (142)
Trombicula bakeri (142)
Xenodontacarus plumosus (95, 235)

Liomys pictus
Ectonyx fusicornis (142)
Euschoengastoides arizonae (142, 230)
Euschoengastoides expansellus (142, 230)
Euschoengastoides gagarini (142)
Euschoengastoides tumidus (142, 230)
Eutrombicula alfreddugesi (142)
Hexidionis allredi (142)
Leptotrombidium panamense (142)
Otorhinophila intrasola (142, 334)
Otorhinophila sinaloae (142, 334)
Pseudoschoengastia aberrans (142)
Pseudoschoengastia audyi (40)
Pseudoschoengastia guatemalensis (40)
Pseudoschoengastia hoffmannae (30, 142)
Pseudoschoengastia hungerfordi (142)
Pseudoschoengastia scitula (45)
Sasacarus whartoni (142)

Liomys salvini
Ascoschoengastia dyscrita (142)
Cordiseta mexicana (142)
Eutrombicula alfreddugesi (142)
Leptotrombidium panamense (142)
Microtrombicula perplexa (323)
Pseudoschoengastia costaricensis (141)
Pseudoschoengastia guanacastensis (141)
Pseudoschoengastia hoguei (141)
Trombicula dunni (142)

Liomys sp.
Hoffmannina haramotoi (41)

Perognathinae

Chaetodipus arenarius
Otorhinophila parvisola (240)

Chaetodipus artus
Euschoengastoides annectens (230)
Euschoengastoides arizonae (230)
Euschoengastoides expansellus (230)
Euschoengastoides tumidus (230)
Hexidionis allredi (242)
Hyponeocula rugosa (296)
Otorhinophila intrasola (240, 334)
Otorhinophila sinaloae (334)

Chaetodipus baileyi
Dermadelema furmani (266)
Euschoengastoides arizonae (230)
Euschoengastoides imperfectus (230)
Euschoengastoides neotomae (230)
Euschoengastoides tanigoshii (230)
Euschoengastoides webbi (230)
Hexidionis navojoae (242)
Otorhinophila intrasola (334)
Otorhinophila parvisola (240, 334)
Chaetodipus californicus
Comatacarus stewarti (150)
Dermadelema furmani (44, 150, 266)
Euschoengastia ambocalis (335)
Euschoengastia criceticola (44, 237)
Euschoengastia enemi (44)
Euschoengastia heteromyicola (336)
Euschoengastia marginalis (337)
Euschoengastia multisetosa (239)
Euschoengastia nihi (44)
Euschoengastia pomerantzi (44)
Euschoengastia radfordi (44, 336)
Euschoengastia romola (44)
Euschoengastia simulans (336)
Euschoengastoides imperfectus (44)
Euschoengastoides neotomae (230)
Eutrombicula belkini (44, 149, 150, 315, 332)
Kayella lacerta (44, 150)
Miyatrombicula scottae (44)
Neotrombicula californica (44)
Neotrombicula dinehartae (44, 150)
Neotrombicula jewetti (44)
Odontacarus hirsutus (44, 113)
Odontacarus linsdalei (44)
Xenodontacarus brevicalcar (44, 236)
Chaetodipus fallax
Dermadelema furmani (266)
Dermadelema mojavense (266)
Euschoengastia ambocalis (335)
Euschoengastia criceticola (237)
Euschoengastia marginalis (337)
Euschoengastia multisetosa (239)
Euschoengastia obscura (336)
Euschoengastia simulans (336)
Euschoengastoides neotomae (230)
Euschoengastoides webbi (230)
Hexidionis deserti (238)
Hyponeocula fovea (296)
Otorhinophila desertorum (240)
Otorhinophila parvisola (240)
Otorhinophila xerophila (240)
Chaetodipus formosus
Dermadelema furmani (266)
Dermadelema mojavense (266)
Dermadelema sleeperi (266)
Euschoengastia criceticola (2, 7)
Euschoengastia decipiens (2, 38)
Euschoengastia hardyorum (337)
Euschoengastia heteromyicola (336)
Euschoengastia obesa (38)
Euschoengastia obscura (336)
Euschoengastoides arizonae (230)
Euschoengastoides imperfectus (230)
Euschoengastoides neotomae (230)
Euschoengastoides opimus (230)
Hexidionis deserti (238)
Hexidionis jessiemae (2)
Hyponeocula arenicola (2, 38, 296)
Hyponeocula deserticola (296)
Hyponeocula fovea (296)
Hyponeocula imitator (296)
Kayella utahensis (7)
Odontacarus linsdalei (2, 38, 44)
Otorhinophila desertorum (240)
Otorhinophila parvisola (240, 334)
Otorhinophila sola (240)
Chaetodipus goldmani
Euschoengastoides annectens (230)
Euschoengastoides arizonae (230)
Euschoengastoides expansellus (230)
Euschoengastoides tumidus (230)
Hexidionis allredi (242)
Hyponeocula deserticola (296)
Hyponeocula rugosa (296)
Otorhinophila intrasola (240, 334)
Chaetodipus hispidus
Euschoengastia cynomicola (228)
Euschoengastia decipiens (297)
Euschoengastia trigenuala (228)
Euschoengastoides arizonae (230)
Euschoengastoides hoplai (230)
Euschoengastoides loomisi (228, 235)
Eutrombicula alfreddugesi (228, 332)
Eutrombicula batatas (228)
Hyponeocula arenicola (227, 228, 296)
Hyponeocula montanensis (228, 296, 297)

Perognathus longimembris
- *Dermadelema furmani* (266)
- *Dermadelema lynnae* (266)
- *Dermadelema mojavense* (266)
- *Dermadelema sleeperi* (266)
- *Euschoengastia decipiens* (2, 38)
- *Euschoengastia heteromyicola* (336)
- *Euschoengastia obscura* (336)
- *Euschoengastia stephensi* (239)
- *Euschoengastoides imperfectus* (230)
- *Eutrombicula belkini* (2)
- *Hexidionis deserti* (238)
- *Hexidionis doremi* (38)
- *Hyponeocula arenicola* (2, 7, 38, 296)
- *Hyponeocula fovea* (296)
- *Hyponeocula imitator* (296)
- *Odontacarus linsdalei* (2)
- *Otorhinophila desertorum* (240)
- *Otorhinophila xerophila* (240)

Perognathus parvus
- *Euschoengastia cordiremus* (7)
- *Euschoengastia criceticola* (2)
- *Euschoengastia decipiens* (6, 7, 9, 38, 336)
- *Euschoengastia fasolla* (2)
- *Euschoengastia oregonensis* (7)
- *Euschoengastoides hoplai* (7)
- *Euschoengastoides loomisi* (2)
- *Hexidionis doremi* (6, 238)
- *Hyponeocula arenicola* (7, 38, 227, 296)
- *Leptotrombidium panamense* (38)
- *Neotrombicula harperi* (7)
- *Odontacarus linsdalei* (2, 38, 44)
- *Odontacarus micheneri* (38)

Ticks (Acarina: Ixodoidea)

Ticks are specialized mites that can be distinguished by their very large size and mouthparts consisting of a large, toothed proboscis.

Ticks fall into two families: the soft ticks (Argasidae) and the hard ticks (Ixodidae). Most of the heteromyid ticks are in the Ixodidae. The life cycle of a tick consists of four stages, egg, larva, nymph, and adult.

Soft ticks (Argasidae).—In soft ticks, males and females are similar, both lacking a hard dorsal scutum. Soft ticks have no pads or pulvilli on the tarsi, no spurs on the coxae and the spiracles are behind the third pairs of coxae. Soft ticks occur in the nesting or roosting areas, and emerge at intervals to feed on their host. They take a number of blood meals. Soft ticks are common on birds and bats but are much less common on rodents.

Hard ticks (Ixodidae).—In males of hard ticks the hard scutum covers the entire dorsal surface, whereas in females it covers only a portion of the anterior surface; thus adults of the two sexes generally look quite different. Hard ticks have pulvilli on the tarsi, the spiracles are behind the fourth coxae, and the coxae usually have spurs. In most ixodid ticks, larvae and nymphs take only one blood meal each, then drop off the host. Often the larval and nymphal stages are on one host, rodents for example, whereas the adults may seek out larger hosts such as ungulates. This may be an adaptation for providing a home and adequate nourishment for the enlarged females. These females may remain attached for extended periods and might not be tolerated by smaller host species. For example, the larvae of *Dermacentor parumapertus* commonly infest heteromyids as well as other rodents, whereas the nymphs attach to rodents or rabbits but the adults are found almost entirely on rabbits (Gastfriend, 1955). In *Dermacentor andersoni* the eggs hatch in about 30–35 days.

Adult and nymphal ticks are easy to recognize, but larval ticks can be confused with large adult mites. Larval ticks, however, have only three pairs of legs, whereas adult mites have four pairs. Larval mites also have three pairs of legs but are generally much too small to be confused with ticks of any stage.

Ticks often are collected by digging them out of the flesh, thereby leaving tissue on the proboscis. The proboscis is important in classification and identification of ticks and must be cleaned. This may be done with hydroxide solutions, but long ago, purely by accident, JOW learned a better way. On one occasion he neglected to immediately add preservative to the vial in which he had placed a tick. Within a half hour, the tick had extricated itself from the host tissue. One now simply removes ticks from the host, including skin tissue, and puts them in an empty vial for 30 minutes or so before preserving them. They generally clean themselves.

The regularly occurring ticks of heteromyids appear to be *Ornithodorus talaje, Dermacentor parumapertus, Ixodes eadsi, I. jellisoni,* and *I. venezuelensis.*

Gastfriend (1955) found that larval populations of *D. parumapertus* thought to be mainly on *Lepus californicus* appeared insufficient to maintain the high nymphal and adult populations found on that species. He found large numbers of *D. parumapertus* on *D. microps, D. ordii, M. megacephalus, C. formosus, P. longimembris* and *P. parvus.*

Examination of the data on ticks from *Heteromys* and *Liomys* points out an area in need of further research. Note the number of cases where ticks are identified only to genus. As Jones et al. (1972) point out, a great deal of information on life history of the South American heteromyid ticks is needed, and presumably would allow us to make many more identifications.

Ticks are important because many of them are vectors for various diseases. The Rocky Mountain Spotted Fever rickettsia has been found in *Dermacentor parumapertus,* an important heteromyid tick, as well as in other tick species that have been found on heteromyids.

Kierans and Clifford (1978) have presented a recent monograph on the genus *Ixodes.*

Ticks reported from heteromyids are listed below:

Argasidae—Soft Ticks

Dipodomys microps
  Ornithodorus parkeri (192)
Dipodomys sp.
  Ornithodorus parkeri (137)
  Ornithodorus talaje (212)
Liomys irroratus
  Ornithodorus talaje (95)
Chaetodipus formosus
  Ornithodorus sparnus (192)
Chaetodipus hispidus
  Ornithodorus talaje (188)

Ixodidae—Hard Ticks

Dipodomys agilis
  Dermacentor parumapterus (137)
  Ixodes jellisoni (137)
Dipodomys californicus
  Dermacentor occidentalis (219)
Dipodomys elator
  Ixodes sp. (297)
Dipodomys deserti
  Ixodes pacificus (137)
Dipodomys elephantinus
  Ixodes jellisoni (137)
Dipodomys heermani
  Dermacentor parumapterus (137)
  Ixodes sp. (66)
Dipodomys merriami
  Dermacentor parumapterus (63, 137)
  Ixodes spinipalpis (8)
Dipodomys microps
  Ixodes kingi (8, 100, 192)
  Dermacentor parumapterus (100, 140, 192)
Dipodomys nitratoides
  Ixodes jellisoni (137)
Dipodomys ordii
  Dermacentor andersoni (4, 192, 263)
  Dermacentor parumapterus (25, 63, 94, 100, 137, 140, 192)
  Haemaphysalis leporis-palustris (93, 94)
  Ixodes kingi (4, 8, 24, 100, 192)
Dipodomys panamintinus
  Dermacentor parumapterus (25, 137)
Dipodomys spectabilis
  Dermacentor parumapterus (94)
Dipodomys venustus
  Dermacentor occidentalis (137)

Heteromys anomalus
Amblyomma ovale (142, 196)
Ixodes venezuelensis (196, 209, 320)
Heteromys gaumeri
Amblyomma sp. (142)
Ixodes sp. (142)
Heteromys sp.
Amblyomma sp. (127)
Dermacentor sp. (127)
Liomys adspersus
Amblyomma sp. (127)
Liomys irroratus
Ixodes eadsi (95, 210)
Liomys pictus
Ixodes sinaloa (204, 211)
Liomys salvini
Amblyomma sp. (142)
Ixodes eadsi (142)
Ixodes sinaloa (204)
Chaetodipus californicus
Dermacentor occidentalis (137)
Ixodes jellisoni (64, 137)
Ixodes kingi (64)
Ixodes pacificus (137)
Chaetodipus formosus
Dermacentor parumapterus (140, 192)
Ixodes jellisoni (192)
Ixodes kingi (192)
Ixodes pacificus (192)
Ixodes spinipalpis (8, 137)
Microdipodops megacephalus
Dermacentor parumapterus (140, 192)
Ixodes kingi (192)
Perognathus fasciatus
Dermacentor andersoni (143)
Perognathus flavus
Dermacentor variabilis (25)
Perognathus longimembris
Dermacentor parumapterus (25, 137, 140, 192)
Ixodes kingi (8, 192)
Ixodes sculptus (8)
Perognathus parvus
Dermacentor andersoni (150, 192)
Dermacentor parumapterus (63, 137, 140, 192)
Haemaphysalis leporis-palustris (4)
Ixodes kingi (4, 8, 192)
Ixodes pacificus (192)
Perognathus sp.
Amblyomma americanum (270)

Anoplura: Sucking Lice

Sucking lice live their entire life cycle on the host and feed on blood. Transmission occurs during direct contact of the hosts. Eggs are glued to the hairs and hatch into larvae which go through several molts before transforming to the adults, although all stages are ecologically and morphologically quite similar. Anoplura occur only on mammals.

Fahrenholzia is the sucking louse of heteromyids, and all 12 species of *Fahrenholzia* are heteromyid lice (Ferris, 1922, 1951; Johnson, 1962; McDaniel, 1968). There are a number of records of other genera on heteromyids, most likely accidental and/or contaminants, but Johnson (1972) found enough cross infestation of lice from *Oryzomys* and *Zygodontomys* on *Heteromys* that she thought this may be of relatively common occurrence. *Fahrenholzia boleni* was described from *Perognathus flavus* (*merriami*) and *F. pinnata* was recorded from *C. penicillatus,* but otherwise all *Fahrenholzia* presently recorded from *Dipodomys* and *Perognathus* are *F. pinnata,* whereas none of those recorded from *Chaetodipus, Heteromys* or *Liomys* are of this species.

Louse infestations may be heavy and may exhibit different infestation rates between the host sexes. Beer et al. (1959) found 87.6% of male *D. merriami* (2,164 lice on 121 individuals) and 71.0% of females (707 on 93 rats) infested with lice, *Fahrenholzia pinnata,* a significantly different infestation rate between sexes ($\chi^2 = 8.201$, 3 *df*).

Sucking lice are exceedingly host specific, so much so that they can be used to help in classification of host species. More than once, a misidentification of a mammal has been discovered by examining a louse or other host-specific form.

Information on *Fahrenholzia* was summarized by McDaniel (1968):

Dipodomys deserti
Fahrenholzia pinnata (102, 128)
Dipodomys elator
Fahrenholzia pinnata (57, 297)
Dipodomys heermani
Fahrenholzia pinnata (128)
Dipodomys merriami
Fahrenholzia pinnata (18, 102, 128, 257)
Enderleinellus longiceps (18)
Dipodomys microps
Fahrenholzia pinnata (recorded with *D. ordii*) (100)
Dipodomys ordii
Fahrenholzia pinnata (6, 100, 128, 160, 253, 257, 276)
Hoplopleura arboricola (6)
Hoplopleura hesperomydis (6)
Neohaematopinus neotomae (257)
Neohaematopinus sp. (6)
Polyplax auricularis (6)
Dipodomys philliipsii
Fahrenholzia pinnata (128)
Dipodomys spectabilis
Fahrenholzia pinnata (257)
Heteromys anomalus
Fahrenholzia schwartzi (194, 195, 327)
Hoplopleura mesoryzomydis (195)
H. multilobata (195)
Heteromys desmarestianus
Fahrenholzia fairchildi (194, 324)
Fahrenholzia ferrisi (101)
Fahrenholzia hertigi (194, 324)
Heteromys gaumeri
Fahrenholzia ferrisi (102, 142)
Heteromys goldmani
Fahrenholzia ferrisi (128, 142, 194, 327)
Heteromys sp. (Guatemala)
Fahrenholzia ferrisi (194)
Liomys adspersus
Fahrenholzia fairchildi (194, 324)
Liomys irroratus
Fahrenholzia ehrlichi (95, 102, 128, 194)
Fahrenholzia microcephala (253)
Fahrenholzia texana (95, 128, 194, 289)
Liomys pictus
Fahrenholzia microcephala (102, 128, 142, 194)
Liomys salvini
Fahrenholzia fairchildi (101, 194)
Chaetodipus baileyi
Fahrenholzia reducta (102)
Chaetodipus californicus
Fahrenholzia tribulosa (128)
Chaetodipus formosus
Fahrenholzia reducta (128)
Chaetodipus hispidus
Fahrenholzia microcephala (253)
Fahrenholzia tribulosa (98)
Fahrenholzia zacatecae (128, 297)
Chaetodipus penicillatus
Fahrenholzia pinnata (257)
Microdipodops megacephalus
Fahrenholzia pinnata (128)
Perognathus flavus
Fahrenholzia pinnata (18, 257)
Fahrenholzia boleni (248)
Perognathus parvus
Fahrenholzia pinnata (6, 128, 160)
Hoplopleura arboricola (6)
Hoplopleura erratica (6)
Hoplopleura hesperomydis (6)
Neohaematopinus laeviusculus (6)
Neohaematopinus pacificus (6)
Polyplax auricularis (6)

Mallophaga: Biting Lice

Mallophagans, like Anoplurans, live their entire life cycle on the host. They have chewing rather than piercing mouthparts and feed on dead skin and hair and detritus on the skin. Their life cycle is similar to that of the Anoplura. Eggs are glued to the hairs, hatch into larvae, then go through several stages to the adult. Mallophagans are primarily on birds, but a few species occur on mammals.

There is one report of unidentified mallophagans from *Dipodomys ordii* and *Perognathus parvus* from Idaho (Allred, 1970). Although 17 mallophagans were found on *D. ordii* and five on *P. parvus,* these would still appear to be accidental occurrences. In

North America, mammal mallophagans occur only on deer, carnivores, porcupines, and pocket gophers and in Central America on marsupials. Since there are other parasite ties between pocket gophers and heteromyids, such as mites of the genera *Geomylichus* and *Echinonyssus,* it would be interesting to know if these specimens were pocket gopher lice, *Geomydoecus.*

Lepidoptera

Cornett (1980) found four adult moths on *Dipodomys merriami* from 4 km W Desert Hot Springs, Riverside Co., California. They were 8–9 mm long, with 3–4 mm protruding from the animal's hair; they withdrew their anterior portions from beneath the rodent's fur and fluttered away. Cornett interpreted this as a case of possible parasitism by lepidopterans. There are some well developed lepidopteran parasites in association with sloths, the sloth moths, *Cryptoses choloepi* (Marshall, 1981; Waage and Montgomery, 1976). It might be more likely that this was a phoretic association, as Davis et al. (1986) found three species of moths phoretic on heteromyids from Costa Rica. Two were described as new species, and the third is probably new but was represented by inadequate material for description. Two, *Amydria selvae* and *Ptilopsaltis* sp., were found on Spiny Pocket Mice, *Heteromys desmarestianus,* whereas *Ptilopsaltis santarosae* was described from *Liomys salvini.* Ten moths, *A. selvae,* were collected and 5 more seen but lost, from 58 individuals of *H. desmarestianus* taken in July 1984. Eleven moths, *P. santarosae,* were collected, and others were seen on the backs of *Liomys salvini.* Males and larvae have not yet been found for any of the three species; all females were collected from the backs of the mice. Sixteen species are known in the genus *Amydria,* but information on feeding is available for only two. Larvae of *Amydria effrentella* were found feeding on dried plant material in mountain beaver burrows, whereas *A. arizonella* is common in numerous southwestern bat caves where the larvae burrow in bat guano. Davis et al. thus suggested that the larvae of *A. selvae* might be scavengers in nests of *H. desmarestianus.* Davis et al. know of five other species, as yet undescribed, but the one other species of *Ptilopsaltis* described to date, *P. synchorista,* was from Trinidad. Its larvae were feeding on seeds disgorged by a Guacharo bird on the floor of a cave, suggesting that larvae of *P. sandarosae* might also feed on plant materials in nests of the mouse. Davis et al. are working further on this interesting situation, but we should watch for lepidopterans in association with heteromyids, either larvae or adults in the nests or burrows or adults in the fur.

Lepidopterans from Heteromyids

Dipodomys merriami
  Unidentified lepidopteran (63)
Heteromys desmarestianus
  Amydria selvae (67)
  Ptilopsaltis sp. (67)
Liomys salvini
  Ptilopsaltis santarosae (67)

Diptera: Cuterebridae

Liomys irroratus
  Cuterebra fontinella (262)
Chaetodipus hispidus
  Cuterebra sp. (145)

Parker and Chaney (1979) caught 66 spiny pocket mice, *Liomys irroratus,* and 65 white-footed mice, *Peromyscus leucopus,* near Brownsville, Texas, infested with *Cuterebra.* White-footed mice are a normal host for *Cuterebra fontinella* and 23% of them harbored this parasite in November and 17 in December. In the spiny pocket mice, similar percentages were 36 and 3%. Myiasis appeared similar in both hosts, and some individuals of each host were parasitized by more than one larva. Goertz (1966) found only a single individual of *Cuterebra* sp. in 34 mice, *Perognathus hispidus,* in Oklahoma, whereas *Neotoma* and *Peromyscus* were more heavily infested.

Siphonaptera: Fleas

Fleas are wingless, laterally compressed insects which as adults are obligate ectoparasites of warm-blooded vertebrates. They are found in the nest or on the host where they feed on blood. Their eggs are deposited in the nests or nesting material of the host. The larvae develop in the nesting material, feeding on detritus and other organic material, often including feces of the adults, and finally metamorphose there in a cocoon spun from salivary secretions. Adults have combs of spines which help them keep from falling or being brushed out of the fur.

Fleas exhibit much less host specificity than lice, listrophorid, or myobiid mites, yet some are fairly host specific. Host selection in fleas seems to be determined by ecological conditions in the nest or burrows of the host rather than on the host itself. One often finds fleas on host species that are characteristic of other hosts. This is the case with the heteromyid data. We have included records here for 82 species of fleas, but most are fleas characteristic of other hosts. For example, some are fleas of *Peromyscus* (*Aetheca wagneri, Orchopeas leucopus, Malaraeus* sp.); *Neotoma* (*Anomiopsyllus* sp., *Orchopeas sexdentatus*); lagomorphs (*Cediopsylla inaequalis,* most *Hoplopsyllus* and *Euhoplopsyllus* sp.); and *Spermophilus* (*Oropsylla montanus, Euhoplopsyllus anomalus, Oropsylla bacchi, Oropsylla* sp.).

Several species of fleas do seem to show regular association with heteromyids, particularly species of *Meringis* on *Dipodomys* and *Perognathus,* and *Oropsylla aridus* subspecies on *Dipodomys.* Relatively few fleas have been reported from *Chaetodipus,* but *Carterella carteri* and a few *Meringis* appear to be the most important.

Data presented by Holdenried and Morlan (1967) show a degree of host specificity among fleas. In Santa Fe Co., New Mexico, 137 pocket mice (*Perognathus flavus*) yielded 184 *Meringis parkeri.* In infested Ord's kangaroo rats (*Dipodomys ordii*) 260 hosts yielded 364 *Meringis nidi,* and 191 hosts yielded 267 *Meringis parkeri.* In *Dipodomys spectabilis* 792 hosts yielded 6,244 individuals of *Meringis rectus* and 462 hosts yielded 3,184 individuals of *M. nidi.*

Most of the fleas from *Heteromys* and *Liomys* are of the genus *Polygenis.* However, this relationship is weak. The species of *Polygenis* are not primarily fleas of heteromyids. *Polygenis* species and some other fleas build to such levels that they occur on most mammals in an area. In Panama *Polygenis dunni* had its maximum numbers on *Liomys adspersus* (Tipton and Mendez, 1966); however, in Venezuela (Tipton and Machado-Allison, 1972), *P. dunni* reached its greatest abundance on *Sigmodon* and occurred in much lower numbers on *Heteromys anomalus.* Most individuals of *P. peronis* from Venezuela were from *Heteromys.* Other species of *Polygenis* were abundant on *Didelphis, Oryzomys,* and *Proechimys.*

Casebeer (1965) reported two species of fleas from three heteromyids; from 189 *Dipodomys merriami,* he reported 11 *Oropsylla aridis* ssp. and 129 *Meringis dipodomys*; from 13 *Chaetodipus fallax* he collected a single *M. dipodomys*; and from 62 *Perognathus longimembris* only two specimens of *M. dipodomys.* On *D. merriami,* he found pronounced seasonal differences in flea infestation in both flea species, with all of the fleas being found between October and March (113 rodent captures and recaptures at this time versus 76 captures and recaptures from April through September). It was not clear whether there was a decline in flea population in summer or simply a change in flea behavior. These changes occurred in conjunction with high temperatures on the desert and perhaps fleas seek out cooler conditions at that time. However, it seems more likely that fleas were more abundant in nests in winter than in summer, paralleling the situation on the hosts.

Eads (1960) found *Meringis vitabilis* on *Dipodomys merriami* and *Meringis agilis* on *Perognathus,* at the same locality, and Eads (1978) found *Meringis facilis* primarily on *D. ordii,* and *M. disparalis* primarily on *D. merriami.* Much more information is needed but there appears to be some host specificity among the fleas of the genus *Meringis,* at least on *Dipodomys.*

Siphonaptera from Heteromyidae

Dipodomyinae

Dipodomys agilis
  Meringis cummingi (181)
Dipodomys californicus
  Meringis cummingi (185, 191)
Dipodomys deserti
  Aetheca wagneri (66)
  Foxella i. ignota (185)
  Malaraeus telchinus (66, 185)
  Meringis cummingi (66)
  Meringis deserti (181)
  Meringis dipodomys (17, 181)
  Oropsylla aridis ssp. (13, 66)
Dipodomys elator
  Meringis agilis (297)
  Meringis arachis (163)
Dipodomys heermani
  Eumolpianus e. eumolpi (286)
  Meringis cummingi (181, 286)
  Meringis parkeri (286)
  Opisodasys keeni (286)
  Oropsylla aridis (181)
  Rhadinopsylla sectilis ssp. (176)
Dipodomys merriami
  Anomiopsyllus amphibolus (15)
  Anomiopsyllus novomexicanensis (15, 152)
  Cediopsylla i. inaequalis (285)
  Echidnophaga gallinaceae (1, 152, 285)
  Euhoplopsyllus glacialis ssp. (152)
  Foxella ignota ssp. (185)
  Malaraeus euphorbia (13)
  Malaraeus sinomus (13)
  Malaraeus telchinus (185)
  Meringis altipectin (181, 191, 310)
  Meringis arachis (68)
  Meringis bilsingi (152)
  Meringis deserti (12)
  Meringis dipodomys (12, 13, 17, 58, 152, 181, 191, 285)
  Meringis disparalis (88)
  Meringis nidi (152)
  Meringis parkeri (17, 191)
  Meringis rectus (152)
  Meringis vitabilis (87)
  Oropsylla aridus ssp. (13, 58, 152, 191, 285, 312)
  Oropsylla bacchi setosis (13)
  Phalacropsylla paradisea (191)
Dipodomys microps
  Aetheca wagneri (160, 185)
  Hoplopsyllus anomalus (17)
  Meringis cummingi (186)
  Meringis dipodomys (97, 181, 185, 285)
  Meringis hubbardi (17, 185)
  Meringis parkeri (185, 285, 302)
  Meringis shannoni (185)
  Oropsylla bacchi ssp. (185)
Dipodomys nitratoides
  Meringis californicus (181)
Dipodomys ordii
  Aetheca wagneri (1, 5, 176, 285)
  Anomiopsyllus amphibolus (285)
  Anomiopsyllus novomexicanensis (15, 152)
  Anomioposyllus nudatus (15)
  Callistopsyllus terinus (306)
  Catallagia decipiens (5)
  Echinophaga gallinacea (152)
  Epitedia stanfordi (285)
  Epitedia w. wenmanni (5)
  Eumolpianus eumolpi (5, 267)
  Foxella ignota ssp. (5)
  Malaraeus telchinus (5)
  Megabothris quirini (160)
  Megarthroglossus divisus (285)
  Meringis altipectin (181, 191)
  Meringis bilsingi (152, 181)
  Meringis dipodomys (17, 152, 181, 185, 285, 331)
  Meringis disparalis (88)
  Meringis facilis (88)

Meringis hubbardi (5, 160, 181, 185)
Meringis nidi (152, 172, 181, 256, 331)
Meringis parkeri (5, 17, 97, 158, 160, 171, 172, 176, 181, 185, 220, 256, 285, 302)
Meringis rectus (152, 256)
Orchopeas leucopus (276)
Orchopeas sexdentatus ssp. (285)
Oropsylla aridis (13, 152, 273, 312)
Oropsylla bacchi ssp. (285)
Oropsylla aridis campestris(??) (271)
Oropsylla fota (152)
Oropsylla francisi ssp. (5)
Oropsyla bacchi gladiolis (302)
Oropsylla hirsutus (1, 285)
Oropsylla labis (5)
Oropsylla m. montana (1, 285)
Oropsylla t. tuberculata (295)
Pleochaetis exilis (5, 68, 197)
Rhadinopsylla sectilis ssp. (5)

Dipodomys panamintinus
Meringis parkeri (191)
Oropsylla aridis ssp. (312)
Rhadinopsylla sectilis ssp. (176)

Dipodomys spectabilis
Anomiopsyllus novomexicanensis (152)
Anomiopsyllus nudatus (15)
Atyphloceras m. multidenticulatum (152)
Echidnophaga gallinacea (152)
Euhoplopsyllus glacialis ssp. (152)
Megarthroglossus bisetis (152, 306)
Meringis altipectin (181, 310)
Meringis arachis (181, 191, 310)
Meringis bilsingi (152)
Meringis dipodomys (152)
Meringis jamesoni (256)
Meringis nidi (152, 172, 181, 191, 256, 331)
Meringis parkeri (256)
Meringis rectus (152, 171, 172, 181, 191, 256)
Oropsylla aridus ssp. (152)
Oropsylla fota (152)
Polygenis gwyni (152)
Pulex stimulans (152)

Dipodomys venustus
Anomiopsylla congruens (224)
Atheropsylla bakeri (287)
Carterella carteri (224)
Meringis cummingi (185, 224)
Oropsylla aridis ssp. (312)
Oropsylla m. montana (224)
Peromyscopsylla hesperomys spp. (224)

Heteromyinae

Heteromys anomalus
Adoratopsylla dilecta (305)
Ctenocephalides f. felis (305)
Neotyphloceras rosenbergi (305)
Polygenis bohlsi ssp. (193, 311)
Polygenis dunni (193, 305)
Polygenis peronis (131, 193, 305, 311)
Polygenis roberti (193, 311)

Heteromys australis
Polygenis bohlsi ssp. (251)
Polygenis klagesi ssp. (304)

Heteromys desmarestianus
Kohlsia traubi (304)
Pluseatis dolens ssp. (304)
Polygenis roberti ssp. (304)
Wenzella obscura (308)
Wenzella yunkeri (304)

Liomys adspersus
Ctenocephalides felis (304)
Polygenis dunni (304)
Polygenis klagesi ssp. (304)

Liomys irroratus
Euhoplopsyllus glacialis ssp. (142)
Polygenis gwyni (85, 88, 142)
Polygenis martinezbaezi (142)

Liomys pictus
Jellisonia wisemani (142)
Polygenis gwyni (142)
Polygenis martinezbaezi (142)
Polygenis vazquezi (187)
Polygenis vulcanius (142)

Liomys salvini
Polygenis vulcanius (142)

Perognathinae

Chaetodipus californicus
Anomiopsyllus falsicalifornicus congruens (224)

Anomiopsyllus f. falsicalifornicus (15)
Atyphloceras m. multidenticulatum (152)
Carteretta carteri (66, 181, 191, 224)
Echidnophaga gallinacea (224)
Hoplopsyllus anomalus (224)
Malaraeus telchinus (224)
Meringis cummingi (66, 185)
Orchopeas howardi ssp. (185)
Orchopeas sexdentatus ssp. (224)
Oropsylla m. montana (224)
Rhadinopsylla sectilis ssp. (224)
Chaetodipus fallax
Meringis dipodomys (58)
Chaetodipus formosus
Carteretta carteri (13, 97, 181, 191)
Meringis dipodomys (96)
Chaetodipus hispidus
Euhoplopsyllus g. affinis (188)
Orchopeas leucopus (163, 188)
Oropsylla fota (188)
Chaetodipus nelsoni
Meringis agilis (87)
Chaetodipus penicillatus
Carteretta carteri (13)
Meringis agilis (87)
Microdipodops megacephalus
Meringis hubbardi (185)
Meringis parkeri (96)
Perognathus fasciatus
Megabothris lucifer (143)
Meringis jamesoni (143)
Oropsylla bruneri (143)
Perognathus flavescens
Meringis hubbardi (285)
Meringis parkeri (181)
Perognathus flavus
Meringis agilis (87)
Meringis arachis (163)
Meringis facilis (88)
Meringis jamesoni (158, 172, 181)
Meringis dipodomys (152)
Meringis rectus (152)
Perognathus longimembris
Meringis dipodomys (58)
Meringis hubbardi (17, 185)
Meringis parkeri (96)
Rhadinopsylla sectilis ssp. (176)
Perognathus parvus
Aetheca wagneri (5, 160, 260)
Eumolpianus e. eumolpi (5, 286)
Foxella ignota (285)
Malaraeus sinomus (285, 302)
Malaraeus telchinus (5)
Megabothris quirini (160)
Meringis cummingi (181)
Meringis hubbardi (5, 17, 96, 97, 160, 181, 285)
Meringis parkeri (5, 176, 180, 181, 286)
Meringis shannoni (176, 181, 191, 260)
Oropsylla tuberculata (160)
Orcpsylla a. acamantis (285)

The following is a taxonomic list of parasites known from heteromyid hosts. Species often occurring on heteromyids are marked with asterisks.

Virus

Buttonwillow virus
D. nitratoides
Modoc virus
D. heermani, nitratoides
Powassan virus
D. heermani, nitratoides
C. californicus
St. Louis encephalitis
D. heermani, nitratoides
C. californicus
Western equine encephalitis
D. nitratoides

Rickettsia

***Coxiella burnetii* (Derrick, 1939)
D. californicus, microps, ordii
C. formosus
P. longimembris, parvus
***Rickettsia rickettsii* (Wolbach, 1919)
D. californicus, microps, ordii
C. formosus

M. megacephalus
P. longimembris, parvus

Spirochetes

Borrelia-like spirochetes
C. hispidus

Bacteria

***Francisella tularense* (McCoy and Chapin, 1912)
D. microps, ordii
P. formosus, parvus
Grahamella sp.
C. californicus
Haemobartonella sp.
P. inornatus
Pasteurella pestis (Lehmann and Neumann, 1896)
P. flavus
D. ordii, spectabilis

Fungi

***Coccidioides immitis* Rixford and Gilchrist, 1896
D. merriami
C. baileyi, intermedius
P. penicillatus
Haplosporangium parvum
D. merriami
C. baileyi, penicillatus

Protozoa

Besnoitia jellisoni
D. ordii
***Eimeria balphae* Ernst, Chobotar and Anderson, 1967
D. agilis, merriami, ordii, spectabilis
***Eimeria chihuahuaensis* Short, Mayberry and Bristol, 1980
D. merriami
***Eimeria chobotari* Ernst, Oaks and Sampson, 1970
D. agilis, merriami, microps, ordii
***Eimeria dipodomysis* Levine, Ivens and Kruidenier, 1957
D. ordii, phillipsii
***Eimeria liomysis* Levine, Ivens and Kruidenier, 1957
L. irroratus, pictus
***Eimeria merriami* Stout and Duszynski, 1983
C. intermedius
***Eimeria mohavensis* Doran and Jahn, 1952
D. panamintinus
***Eimeria penicillati* Ivens, Kruidenier and Levine, 1958
P. penicillatus, flavus
***Eimeria perognathi* Levine, Ivens and Kruidenier, 1957*b*
C. intermedius
***Eimeria picta* Levine, Ivens and Kruidenier, 1957
L. pictus
***Eimeria reedi* Ernst, Oaks and Sampson, 1970
C. formosus
***Eimeria scholtysecki* Ernst, Frydendall and Hammond, 1967
D. agilis, gravipes, ordii, panamintinus, spectabilis
***Eimeria utahensis* Ernst, Hammond and Chobotar, 1968
D. agilis, merriami, microps, ordii
***Endamoeba dipodomysi* Hegner, 1926
D. agilis, heermani, merriami, panamintinus, spectabilis
Giardia duodenalis var. *perognathi* Filice, 1952
D. agilis, merriami
Isospora sp.
D. agilis
Leishmania brasiliensis Vianna, 1911
H. desmarestianus
Leishmania mexicana Biagi, 1953
H. desmarestianus
***Tritrichomonas muris* (Grassi, 1879)
D. nitratoides, venustus

C. formosus
P. longimembris, parvus
Trypanosoma zeledoni Esquivel, Zuniga, Alfaro and Kotcher, 1967
L. salvini

Cestodes

Catenotaenia californica Dowell, 1953
D. heermani, panamintinus
***Catenotaenia linsdalei* McIntosh, 1941
D. deserti, heermani, merriami, microps, panamintinus, spectabilis, venustus
C. californicus, formosus
Hymenolepis citelli (McLeod, 1933)
D. heermani, ingens, nitratoides
***Mathevotaenia deserti* (Millemann, 1955)
D. merriami, panamintinus
P. longimembris
Raillietina retractilis Stiles, 1925
D. ordii
Schizorchodes dipodomi Bienek and Grundmann, 1973
D. merriami

Nematodes

Capillaria americana Read, 1949
D. microps, ordii
***Gongylonema dipodomyis* Kruidenier and Peebles, 1958
D. merriami, panamintinus
Gongylonema neoplasticum (Fibiger and Ditlevsen, 1914)
D. merriami, panamintinus
***Heteromoxyurus deserti* (Read and Millemann, 1953)
D. deserti, heermani, merriami, ordii, panamintinus
Longistriatus vexillata (Hall, 1916)
Liomys pictus
Protospirura anodon Hannum, 1943
C. penicillatus
***Protospirura dipodomis* (Read and Millemann, 1953)
D. deserti, merriami, panamintinus
C. baileyi
P. longimembris
Protospirura muris (Gmelin, 1790)
D. ordii
Protospirura numidica Seurat, 1914
D. merriami, ordii
Protospirura tetradon Hannum, 1943
C. penicillatus
***Rictularia dipodomis* Tiner, 1938
D. deserti, heermani, merriami, microps, panamintinus
***Trichuris dipodomis* Read, 1956*a*
D. microps, ordii
Trichuris minuta Rudulphi, 1819
D. merriami, ordii
***Trichuris perognathi* Chandler, 1945
C. californicus, formosus(?), *penicillatus*
Vexillata petteri Durette-Desset, 1970
Heteromys sp.
Wellcomia longejector Hannum, 1943
D. merriami
P. longimembris
***Welcomia perognathi* Kruidenier and Mehra, 1959
C. intermedius

Mites (other than chiggers) from Heteromyids

Cheyletidae
Cheyletus linsdalei Baker, 1949
D. microps, ordii
Eucheyletia n. sp.
L. salvini
Dermanyssidae
Dermanyssus gallinae (DeGeer, 1778)
P. parvus
Glycyphagidae
***Dermacarus liomys* Fain and Ide, 1978
L. irroratus
***Dermacarus ornatus* Fain, 1967
H. anomalus, gaumeri
***Dipodomydectes americanus* Fain and Lukoschus, 1978*b*
D. merriami

***Dipodomyopus tuttlei* Fain and Lukoschus, 1978*a*
Dipodomys sp.
***Mediolabidophorus neotropicalis* Fain and Lukoschus, 1978*a*
H. desmarestianus
***Metalabidophorus heteromys* Lukoschus, Janssen, Duijghuijsen and Fain, 1977
H. anomalus
***Metalabidophorus liomys* Lukoschus, Janssen, Duijghuijsen and Fain, 1977
L. irroratus
***Neolabidophorus verrucosus* Fain and Lukoschus, 1978
C. penicillatus
***Neolabidophorus yucatanensis* Pence and Genoways, 1974
H. gaumeri

Laelapidae
Androlaelaps casalis (Berlese, 1887)
D. merriami
P. parvus
Androlaelaps fahrenholzi (Berlese, 1911)
D. elator, merriami, microps, ordii, spectabilis
H. anomalus, australis, desmarestianus, gaumeri
L. adspersus, irroratus
C. californicus, formosus, hispidus
P. fasciatus, longimembris, parvus
Androlaelaps fenilis
L. salvini
Androlaelaps frontalis (Banks, 1910)
C. fallax
***Androlaelaps grandiculatus* Eads, 1951
D. merriami
C. formosus
Androlaelaps projecta Furman, 1972
H. anomalus
Androlaelaps rotundus Furman, 1972
H. anomalus
Brevisterna morlani Strandtmann and Allred, 1956
D. ordii
Brevisterna utahensis (Ewing, 1933)
D. merriami
C. formosus
Echinolaelaps boultoni (Furman and Tipton, 1961)
H. anomalus
Echinonyssus affinis Jameson, 1950
C. formosus
***Echinonyssus brevicalcar* (Herrin and Yunker, 1975)
L. salvini
Echinonyssus galindoi (Strandtmann and Yunker, 1966)
L. salvini
***Echinonyssus heteromydis* (Strandtmann and Yunker, 1966)
H. desmarestianus
***Echinonyssus hilli* (Jameson, 1950)
D. microps, ordii
P. flavus, longimembris, parvus
***Echinonyssus incomptis* (Eads and Hightower, 1952)
D. elator, merriami, microps, ordii, spectabilis
P. longimembris, parvus
Echinonyssus keenani (Strandtmann and Yunker, 1966)
H. anomalus
***Echinonyssus liomys* (Herrin and Yunker, 1973)
L. irroratus
Echinonyssus longichelae (Allred and Beck, 1966)
D. ordii
***Echinonyssus lunatus* (Strandtmann and Yunker, 1966)
H. desmarestianus
***Echinonyssus microchelae* (Strandtmann and Yunker, 1966)
H. desmarestianus
L. adspersus
***Echinonyssus minutus* (Strandtmann and Yunker,. 1966)
H. desmarestianus
***Echinonyssus neotomae* (Eads and Hightower, 1951)
D. ordii
L. irroratus
C. hispidus
P. parvus

**Echinonyssus panamensis* (Strandtmann and Yunker, 1966)
H. desmarestianus
**Echinonyssus parvisoma* (Herrin and Yunker, 1975)
H. anomalus
**Echinonyssus perognathi* (Herrin, 1970)
C. formosus, hispidus
H. anomalus
**Echinonyssus proctolatus* (Herrin and Yunker, 1975)
H. anomalus
**Echinonyssus triacanthus* (Jameson, 1950)
D. deserti, merriami, microps, ordii
C. formosus
P. longimembris, parvus
Echinonyssus utahensis (Allred and Beck, 1966)
D. microps, ordii
P. longimembris, parvus
**Echinonyssus venezuelensis* (Herrin and Yunker, 1975)
H. anomalus
Eubrachylaelaps circularis (Ewing, 1933)
L. salvini
P. longimembris, parvus
Eubrachylaelaps crowei Jameson, 1947
D. ordii, spectabilis
Eubrachylaelaps debilis Jameson, 1950
D. ordii
P. parvus
Eubrachylaelaps hollisteri (Ewing, 1925)
C. californicus
P. parvus
Eubrachylaelaps jamesoni Furman, 1955
H. desmarestianus, gaumeri
Eubrachylaelaps rotundus Fonseca, 1936
H. anomalus
Gigantolaelaps goyaensis Fonseca, 1939
H. anomalus
Gigantolaelaps inca Fonseca, 1960
H. anomalus
Gigantolaelaps wolffsohni (Oudemans, 1910)
H. anomalus
Haemogamasus ambulans (Thorell, 1872)
D. ordii
P. parvus
Haemogamasus onychomydis (Ewing, 1933)
D. ordii
P. parvus
Haemogamasus reidi Ewing, 1925
D. ordii
**Hypoaspis leviculus* (Eads, 1951)
D. microps, ordii
H. gaumeri
L. pictus
C. formosus, hispidus
P. flavescens, longimembris, parvus
Hypoaspis lubrica Voigts and Oudemans, 1904
L. salvini
P. parvus
**Ischyropoda armatus* Keegan, 1951
D. deserti, merriami, microps, ordii, spectabilis
C. formosus
M. megacephalus
P. flavus, inornatus, longimembris, parvus, xanthonotus
**Ischyropoda furmani* Keegan, 1956
D. ordii
P. longimembris, parvus
**Ischyropoda spiniger* Keegan, 1951
C. penicillatus, spinatus
**Laelaps dearmasi* Furman and Tipton, 1961
H. anomalus
Laelaps kochi Oudemans, 1936
D. ordii
Laelaps ovata Furman, 1972
H. anomalus
Mysolaelaps parvispinosus Fonseca, 1936
H. anomalus
**Steptolaelaps heteromys* (Fox, 1947)
H. anomalus, australis,

desmarestianus, gaumeri
L. adspersus, pictus, salvini
***Steptolaelaps liomydis* (Grant, 1917)
L. irroratus, pictus
Tur uniscutatus (Turk, 1946)
H. desmarestianus
Listrophoridae
***Geomylichus brevispinosus* Fain, Whitaker, Schwann and Lukoschus, 1978
C. penicillatus
***Geomylichus californicus* Fain, Whitaker, and Thomas, 1988
D. californicus, elephantinus, heermani, venustus
***Geomylichus deserti* Fain and Whitaker, 1987
D. deserti
***Geomylichus dipodomius* (Radford, 1953)
D. elator, merriami, microps, ordii, phillippsii, spectabilis
***Geomylichus formosus* Fain and Whitaker, 1987
C. formosus
***Geomylichus inaequalis* Fain, Whitaker, Schwann and Lukoschus, 1978
C. hispidus
Geomylichus klebergi (McDaniel, 1965)
C. hispidus
***Geomylichus microdipodops* Fain and Whitaker, 1987
M. megacephalus
***Geomylichus multistriatus* Fain, Whitaker, and Thomas, 1988
C. penicillatus
D. merriami, nitratoides
M. megacephalus
***Geomylichus perognathi* Fain and Whitaker, 1980
P. fasciatus, parvus
***Geomylichus postscutatus* Fain, 1976
Dipodomys sp.
***Geomylichus texanus* Fain, Whitaker, Schwann and Lukoschus, 1978
D. elephantinus, merriami, ordii, venustus
C. penicillatus
***Geomylichus utahensis* Fain and Whitaker, 1987
D. microps
Macronyssidae
Acanthonyssus proechimys Yunker and Saunders, 1973
H. anomalus
Ornithonyssus bacoti (Hirst, 1913)
D. ordii
L. adspersus, irroratus
C. hispidus
P. parvus
Ornithonyssus sylviarum (Canestrini and Fanzago, 1877)
L. irroratus
Myobiidae
***Radfordia bachai* Howell and Elzinga, 1962
D. merriami, microps, ordii
Speleognathidae
Paraspeleognathopsis cricetidarum Clark, 1967
H. desmarestianus

Miscellaneous Mite Associates, Not Parasitic

Ameroseiidae
Sertitympanum contiguum Elsen and Whitaker, 1985
D. ordii
Sertitympanum exarmatum Elsen and Whitaker, 1985
D. ordii
P. parvus
Sertitympanum sp.
D. merriami, microps, ordii
C. formosus
P. longimembris, parvus
Ascidae
Proctolaelaps sp.
D. ordii
P. parvus
Cyrtolaelapidae
Euryparasitus sp.
D. ordii

Macrochelidae

Macrocheles sp.

D. ordii

P. parvus

Pygmephoridae

Bakerdania sp.

P. parvus

Uropodidae

D. ordii

Chiggers

***Anomalaspis ambiguus* Brennan, 1952

H. anomalus

***Ascoschoengastia dyscrita* Brennan and Jones, 1961

H. desmarestianus

L. adspersus, salvini

***Boshkerria punctata* (Boshell and Kerr, 1942)

H. anomalus

***Colicus liomys* (Brennan and Jones, 1961)

L. adspersus

Comatacarus americanus Ewing, 1942

D. ordii

Comatacarus stewarti (Gould, 1956)

Chaetodipus californicus

Cordiseta mexicana (Hoffman, 1954)

H. gaumeri

L. salvini

Crotiscus danae Brennan and Goff, 1978

H. anomalus

***Crotiscus desdentatus* (Boshell and Kerr, 1942)

H. anomalus, desmarestianus

***Crotonasis fissa* Brennan and Yunker, 1966

L. adspersus

***Dermadelema furmani* (Gould, 1956)

D. agilis, deserti, merriami, microps, panamintinus, stephensi

C. baileyi, californicus, fallax, formosus, penicillatus, spinatus

P. longimembris

***Dermadelema lynnae* Pomeroy and Loomis, 1984

D. merriami

P. longimembris

***Dermadelema mojavense* Pomeroy and Loomis, 1984

D. heermani, merriami, nitratoides, panamintinus

C. fallax, formosus

P. longimembris

***Deremadelema sleeperi* Pomeroy and Loomis, 1984

D. merriami, microps, panamintinus

C. formosus

P. longimembris

***Ectonyx fusicornis* Brennan, 1960

H. gaumeri

L. irroratus

Euschoengastia ambocalis Wrenn and Loomis, 1973

D. agilis, merriami

C. californicus, fallax

***Euschoengastia belgicae* Brennan and Yunker, 1966

H. desmarestianus

Euschoengastia bigenuala Farrell, 1956

L. irroratus

Euschoengastia cordiremus Brennan, 1948

D. ordii

P. parvus

***Euschoengastia criceticola* Brennan, 1948

D. agilis, microps, ordii, venustus

C. californicus, fallax, formosus

P. parvus

Euschoengastia cynomyicola Crossley and Loomis, 1954

C. hispidus

***Euschoengastia decipiens* Gould, 1956

D. elator, merriami, microps, ordii, panamintinus

C. formosus, hispidus

P. longimembris, parvus

Euschoengastia enemi Brennan and Jones, 1954

C. californicus

Euschoengastia fasolla Brennan and Beck, 1955

P. parvus

***Euschoengastia hardyorum* Wrenn and Somerby, 1974
D. merriami, microps
C. formosus
***Euschoengastia heteromyicola* Wrenn and Loomis, 1974
D. agilis, merriami, microps, panamintinus, stephensi
C. californicus, formosus
P. longimembris
***Euschoengastia marginalis* Wrenn and Somerby, 1974
D. merriami, panamintinus
C. californicus, fallax
Euschoengastia multisetosa Loomis and Somerby, 1966
C. californicus, fallax
Euschoengastia nihi Brennan and Jones, 1954
C. californicus
Euschoengastia numerosa Wrenn and Loomis, 1974
D. merriami
***Euschoengastia obesa* Brennan and Beck, 1955
C. formosus
***Euschoengastia obscura* Wrenn and Loomis, 1974
D. agilis, merriami, stephensi
C. fallax, formosus
P. longimembris
Euschoengastia oregonensis (Ewing, 1929)
P. parvus
Euschoengastia pomerantzi Brennan and Jones, 1954
C. californicus
***Euschoengastia radfordi* Brennan and Jones, 1954
D. venustus
C. californicus
Euschoengastia romola Brennan and Jones, 1954
D. venustus
C. californicus
Euschoengastia simulans Wrenn and Loomis, 1974
D. merriami, panamintinus
C. californicus, fallax, formosus
Euschoengastia stephensi Loomis and Somerby, 1966
P. longimembris
Euschoengastia trigenuala Farrell, 1956
C. hispidus
***Euschoengastoides annectens* Loomis, 1971
C. artus, goldmani, pernix
***Euschoengastoides arizonae* Loomis, 1971
D. deserti, merriami, ordii, spectabilis
L. pictus
C. artus, baileyi, formosus, goldmani, hispidus, intermedius, nelsoni, penicillatus, pernix, spinatus
P. amplus, flavus
***Euschoengastoides expansellus* Loomis, 1971
L. pictus
C. artus, goldmani, pernix
***Euschoengastoides gagarini* (Brennan, 1962)
L. irroratus, pictus
***Euschoengastoides hoplai* (Loomis, 1954)
D. merriami
C. hispidus, nelsoni, penicillatus
P. flavescens, flavus, parvus
***Euschoengastoides imperfectus* (Brennan and Jones, 1954)
D. merriami, spectabilis
C. baileyi, californicus, formosus, intermedius, penicillatus, spinatus
P. flavus, longimembris
Euschoengastoides loomisi Crossley and Lipovsky, 1954
D. merriami, ordii
L. irroratus
C. hispidus, nelsoni, penicillatus
P. flavus, parvus
***Euschoengastoides neotomae* Loomis, 1971
D. merriami, spectabilis
C. baileyi, californicus, fallax, formosus, nelsoni, penicillatus
P. flavescens

**Euschoengastoides opimus* Loomis, 1971
C. formosus
**Euschoengastoides ryckmani* Brennan, 1966
C. penicillatus
**Euschoengastoides tanigoshii* Loomis, 1971
C. baileyi, intermedius, penicillatus
**Euschoengastoides tumidus* Loomis, 1971
D. ordii
L. irroratus
C. artus, goldmani, intermedius, penicillatus, pernix
**Euschoengastoides webbi* Loomis, 1971
D. merriami
C. baileyi, fallax, penicillatus, spinatus
Eutrombicula alfreddugesi (Oudemans, 1910)
D. merriami, ordii
H. anomalus, desmarestianus
L. irroratus, pictus, salvini
C. hispidus
P. flavescens
Eutrombicula batatas Linnaeus, 1758
D. ordii
C. hispidus
**Eutrombicula belkini* Gould, 1950
D. ordii
C. californicus
P. longimembris
**Eutrombicula goeldii* (Oudemans, 1910)
H. anomalus, australis, desmarestianus
L. adspersus
**Hexidionis allredi* (Brennan and Beck, 1956)
D. merriami, ordii
L. irroratus, pictus
C. artus, goldmani, nelsoni, penicillatus
P. flavus
Hexidionis breviseta (Loomis and Crossley, 1963)
D. ordii
P. flavus
**Hexidionis deserti* Loomis and Lucas, 1970
D. deserti, merriami
C. fallax, formosus
P. longimembris
**Hexidionis doremi* (Brennan and Beck, 1956)
D. merriami, ordii
P. longimembris, parvus
**Hexidionis harveyi* Loomis and Lucas, 1969
D. merriami, ordii, spectabilis
C. penicillatus
P. flavus
Hexidionis jessiemae (Gould, 1956)
D. merriami, microps
L. irroratus
C. formosus
**Hexidionis navojoae* Lucas and Loomis, 1968
C. baileyi, penicillatus, pernix
**Hoffmannina handleyi* Brennan and Jones, 1961
H. desmarestianus
**Hoffmannina haramotoi* Brennan and Goff, 1977
Liomys sp.
**Hyponeocula arenicola* (Loomis, 1954)
D. agilis, heermani, merriami, microps, nelsoni, ordii, panamintinus, spectabilis
C. formosus, hispidus, intermedius, penicillatus
P. fasciatus, flavus, longimembris, parvus
**Hyponeocula deserticola* Tanigoshi and Loomis, 1974
D. deserti
C. formosus, goldmani, penicillatus, spinatus
**Hyponeocula fovea* Tanigoshi and Loomis, 1974
D. merriami, microps, ordii, panamintinus
C. fallax, formosus
P. longimembris
Hyponeocula imitator Tanigoshi and Loomis, 1974

D. merriami, panamintinus
C. formosus, penicillatus
P. longimembris
Hyponeocula montanensis (Brennan, 1946)
D. heermani, ordii
C. hispidus
P. flavescens, flavus
***Hyponeocula rugosa* Tanigoshi and Loomis, 1974
C. artus, goldmani, penicillatus, pernix
Kayella lacerta (Brennan, 1948)
D. microps
L. irroratus
C. californicus, nelsoni, penicillatus
P. flavus
Kayella utahensis (Brennan and Beck, 1956)
C. formosus
***Kymocta faitkeni* Brennan, 1968a
H. anomalus
***Kymocta teratarsalis* (Yunker and Brennan, 1962)
H. desmarestianus
***Kymocta zulia* Brennan and van Bronswijk, 1973
H. anomalus
Leptotrombidium hamaxiaia (Brennan and Dalmat, 1960)
H. desmarestianus
***Leptotrombidium panamense* (Ewing, 1925)
D. merriami
H. gaumeri
L. adspersus, irroratus, pictus, salvini
C. nelsoni, penicillatus, pernix
P. flavus, parvus
***Microtrombicula perplexa* Webb and Loomis, 1971
L. salvini
Miyatrombicula scottae (Brennan, 1952)
C. californicus
Neoschoengastia americana (Hirst, 1921)
D. heermani
Neotrombicula californica (Ewing, 1942)
D. venustus
C. californicus
Neotrombicula dinehartae Brennan and Wharton, 1950
C. californicus
Neotrombicula harperi (Ewing, 1928)
P. parvus
Neotrombicula jewetti Brennan and Wharton, 1950
C. californicus
Odontacarus comosus Reed and Brennan, 1975
H. anomalus
Odontacarus dentatus (Ewing, 1925)
C. hispidus
Odontacarus hirsutus (Ewing, 1931)
D. venustus
C. californicus
Odontacarus linsdalei (Brennan and Jones, 1954)
D. agilis, merriami, microps, ordii
C. californicus, formosus
P. longimembris, parvus
***Odontacarus micheneri* (Greenberg, 1952)
D. ordii
C. hispidus
P. parvus
***Odontacarus tubercularis* (Brennan, 1952)
H. anomalus, gaumeri
L. adspersus, irroratus
***Otorhinophila baccusi* Loomis and Wrenn, 1973
D. merriami, ordii
C. hispidus, nelsoni, penicillatus
P. flavus
***Otorhinophila desertorum* Loomis and Wrenn, 1973
D. merriami
C. fallax, formosus, penicillatus
P. longimembris
***Otorhinophila intrasola* Wrenn and Loomis, 1967
L. pictus
C. artus, baileyi, goldmani, intermedius, penicillatus, pernix

**Otorhinophila parvisola* Wrenn and Loomis, 1967
D. merriami, ordii
C. arenarius, baileyi, fallax, formosus, intermedius, penicillatus, spinatus
**Otorhinophila sinaloae* Wrenn and Loomis, 1967
L. pictus
C. artus, pernix
**Otorhinophila sola* (Gould, 1956)
C. formosus, penicillatus
**Otorhinophila xerophila* Loomis and Wrenn, 1973
D. merriami, panamintinus
C. fallax, penicillatus
P. longimembris
Parasecia aitkeni (Brennan and Jones, 1960)
Heteromys sp.
Parasecia gurneyi (Ewing, 1937)
D. merriami, ordii
H. gaumeri
C. hispidus
P. flavus
**Parasecia manueli* (Brennan and Jones, 1960)
H. anomalus
Parasecia universitatis (Hoffman, 1963)
L. irroratus
**Peltoculus almae* (Brennan, 1968b)
H. australis
**Polylopadium chaetolecanium* Brennan and Reed, 1972
H. anomalus
**Polylopadium kramisi* Brennan and Jones, 1961
L. adspersus
**Polylopadium tertium* (Brennan, 1968*b*
H. australis
Pseudoschoengastia aberrans Brennan and Jones, 1959
L. pictus
**Pseudoschoengastia audyi* Brennan and Jones, 1959
L. irroratus, pictus
Pseudoschoengastia bisetosa Loomis, 1976
C. spinatus
Pseudoschoengastia brennani Hoffmann, 1960
H. gaumeri
**Pseudoschoengastia bulbifera* Brennan, 1960
H. anomalus, australis, desmarestianus
L. adspersus
Pseudoschoengastia costaricensis Geest and Loomis, 1968
L. salvini
**Pseudoschoengastia disparunguis* Goff, 1982
H. desmarestianus
Pseudoschoengastia extrinseca Brennan, 1960
H. gaumeri
Pseudoschoengastia farneri Lipovsky, 1951
D. ordii
L. irroratus
C. hispidus
P. flavus
**Pseudoschoengastia finitima* Brennan and Yunker, 1966
H. desmarestianus
Pseudoschoengastia guanacastensis Geest and Loomis, 1968
L. salvini
Pseudoschoengastia guatemalensis Brennan, 1952
L. pictus
**Pseudoschoengastia hoffmannae* Brennan, 1960
L. irroratus, pictus
Pseudoschoengastia hoguei Geest and Loomis, 1968
L. salvini
**Pseudoschoengastia hungerfordi* Lipovsky, 1951
D. ordii
L. irroratus, pictus
C. hispidus, nelsoni
P. flavus
**Pseudoschoengastia scitula* Brennan and Jones, 1959
H. gaumeri
L. pictus
**Pseudoschoengastia zona* Brennan, 1960
H. australis
L. adspersus
**Quadraseta flochi* (Brennan and Jones, 1960)

H. anomalus
**Quadraseta mirandae* Goff and Brennan, 1977
H. anomalus
**Quadraseta trapidoi* Brennan, 1968b
H. australis
Sasacarus furmani (Hoffmann, 1954)
H. desmarestianus
Sasacarus vercammeni (Brennan and Dalmat, 1960)
H. desmarestianus
Sasacarus whartoni (Hoffmann, 1951)
L. picta
Trombicula bakeri Ewing, 1946
D. ordii
L. irroratus
**Trombicula dicrura* Brennan and Jones, 1961
H. desmarestianus
Trombicula dunni Ewing, 1931
H. australis, desmarestianus
L. salvini
Trombicula keenani Brennan and Jones, 1961
H. australis, desmarestianus
**Vanidicus chalepus* Brennan, 1973
H. anomalus
**Vanidicus jojosti* Brennan, 1973
H. anomalus
Vanidicus tricosus Brennan and Jones, 1961
H. desmarestianus
L. adspersus
Xenodontacarus brevicalcar (Brennan and Jones, 1954)
C. californicus
Xenodontacarus plumosus (Greenberg, 1951)
D. ordii
L. irroratus

Ticks

Argasidae

Ornithodorus parkeri Cooley, 1936
Dipodomys sp.
Dipodomys microps
Ornithodorus sparnus Kohls and Clifford, 1963
Chaetodipus formosus
**Ornithodorus talaje* (Guerin-Meneville, 1849)
Dipodomys sp.
L. irroratus
C. hispidus

Ixodidae

Amblyomma americanum (Linne, 1758)
Perognathus sp.
Amblyomma ovale Koch, 1844
H. anomalus
Dermacentor andersoni Stiles, 1908
D. ordii
P. fasciatus, parvus
Dermacentor occidentalis Marx, 1892
D. californicus, venustus
C. californicus
**Dermacentor parumapertus* Neumann, 1901
D. agilis, heermani, merriami, microps, ordii, panamintinus, spectabilis
C. formosus
M. megacephalus
P. longimembris, parvus
Dermacentor variabilis (Say, 1821)
P. flavus
Haemaphysalis leporis-palustris (Packard, 1869)
D. ordii
P. parvus
**Ixodes eadsi* Kohls and Clifford, 1964
L. irroratus, salvini
**Ixodes jellisoni* Cooley and Kohls, 1938
D. agilis, elephantinus, nitratoides
C. californicus
Ixodes kingi Bishopp, 1911
D. microps, ordii
C. californicus
M. megacephalus
P. longimembris, parvus
Ixodes pacificus Cooley and Kohls, 1943
D. deserti
C. californicus
P. parvus
Ixodes sculptus Neumann, 1904
P. longimembris

Ixodes sinaloa Kohls and Clifford, 1966
L. pictus, salvini
Ixodes spinipalpis Hadwen and Nuttall, 1916
D. merriami
C. formosus
***Ixodes venezuelensis* Kohls, 1953
H. anomalus

Anoplura
(Only Fahrenholzia *Included)*

Fahrenholzia boleni McDaniel, 1968
P. flavus
Fahrenholzia ehrlichi Johnson, 1962
L. irroratus
Fahrenholzia fairchildi Johnson, 1962
H. desmarestianus
L. adspersus, salvini
Fahrenholzia ferrisi Werneck, 1952
H. desmarestianus, gaumeri
C. goldmani
Fahrenholzia hertigi Johnson, 1962
H. desmarestianus
Fahrenholzia microcephala Ferris, 1922
L. pictus
Fahrenholzia pinnata Kellogg and Ferris, 1915
D. deserti, elator, heermani, merriami, microps, ordii, phillipsii, spectabilis
C. penicillatus
P. flavus, parvus
Fahrenholzia reducta Ferris, 1933
C. baileyi, formosus
Fahrenholzia schwartzi Werneck, 1952
H. anomalus
Fahrenholzia texana Stojanovich and Pratt, 1961
L. irroratus
Fahrenholzia tribulosa Ferris, 1922
C. californicus, hispidus
Fahrenholzia zacatecae Ferris, 1922
C. hispidus

Diptera

Cuterebra fontinella Clark, 1927
L. irroratus

Fleas

Adoratopsylla dilecta Jordan, 1938
H. anomalus
Aetheca wagneri (Baker, 1904)
D. deserti, microps, ordii
P. parvus
Amaradix euphorbi (Rothschild, 1905)
D. merriami
Anomiopsyllus amphibolus Wagner, 1936
D. merriami, ordii
Anomiopsylla falsicalifornicus congruens Stewart, 1940
C. californicus
Anomiopsyllus f. falsicalifornicus C. Fox, 1929
C. californicus
Anomiopsyllus novomexicanensis Williams and Hoff, 1951
D. merriami, ordii
Anomiopsyllus n. nudatus (Baker, 1898)
D. ordii, spectabilis
Atyphloceras multidentata C. Fox, 1909
D. spectabilis
Callistopsyllus t. terinus (Rothschild, 1905)
D. ordii
***Carteretta carteri* Fox, 1927
C. californicus, formosus, penicillatus
Catallagia decipiens Rothschild, 1915
D. ordii
Cediopsylla i. inaequalis (Baker, 1895)
D. merriami
Ctenocephalides f. felis (Bouche, 1835)
H. anomalus
L. adspersus
Echidnophaga gallinacea (Westwood, 1875)
D. merriami, ordii, spectabilis
Epitedia stanfordi Traub, 1944
D. ordii
Epitedia w. wenmanni (Rothschild, 1904)
D. ordii
Euhoplopsyllus g. glacialis (Taschenberg, 1880)
D. merriami
L. irroratus
Euhoplopsyllus glacialis affinis (Baker, 1904)

C. hispidus
Eumolpianus e. eumolpi (Rothschild, 1905)
D. heermani, merriami, ordii
P. parvus
Foxella i. ignota (Baker, 1895)
D. deserti, merriami, ordii
P. parvus
Hoplopsyllus anomalus (Baker, 1904)
D. microps, spectabilis
Jellisonia wisemani Eads, 1951
L. pictus
Kohlsia traubi Tipton and Mendez, 1961
H. desmarestianus
Malaraeus sinomus (Jordan, 1925)
D. merriami
P. parvus
Malaraeus telchinus (Rothschild, 1905)
D. deserti, merriami, ordii, spectabilis
P. parvus
Megabothris lucifer (Rothschild, 1905)
P. fasciatus
Megabothris quirini (Rothschild, 1905)
D. ordii
P. parvus
Megarthroglossus bisetis Jordan and Rothschild, 1915
D. spectabilis
Megarthroglossus divisus Baker, 1895
D. ordii
***Meringis agilis* Eads, 1960
D. elator
C. nelsoni, penicillatus
P. flavus
***Meringis altipectin* Traub and Hoff, 1951
D. merriami, ordii, spectabilis
***Meringis arachis* (Jordan, 1929)
D. elator, merriami, spectabilis
P. flavus
Meringis bilsingi Eads and Menzies, 1949
D. merriami, ordii
Meringis californicus Augustson, 1953
D. nitratoides
***Meringis cummingi* (C. Fox, 1926)
D. agilis, californicus, deserti, heermani, microps, venustus
C. californicus
P. parvus
Meringis deserti Augustson, 1953
D. deserti, merriami
***Meringis dipodomys* Kohls, 1938
D. deserti, merriami, microps, ordii
C. fallax, formosus
P. flavus, longimembris
***Meringis disparalis* Eads, 1979
D. merriami, ordii
***Meringis facilis* Eads, 1979
D. ordii
P. flavus
***Meringis hubbardi* Kohls, 1938
D. microps, ordii
M. megacephalus
P. flavescens, longimembris, parvus
***Meringis jamesoni* Hubbard, 1943
D. spectabilis
P. fasciatus, flavus
***Meringis nidi* Williams and Hoff, 1951
D. merriami, ordii, spectabilis
***Meringis parkeri* Jordan, 1937
D. heermani, merriami, microps, ordii, panamintinus, spectabilis
M. megacephalus
P. flavescens, longimembris, parvus
***Meringis rectus* Morlan, 1953
D. merriami, ordii, spectabilis
P. flavus
***Meringis shannoni* (Jordan, 1929)
D. microps
P. parvus
***Meringis vitabilis* Eads, 1960
D. merriami
Neotyphloceras rosenbergi (Rothschild, 1904)
H. anomalus
Opisodasys keeni (Baker, 1896)
D. heermani
Orchopeas h. howardi (Baker, 1895)
C. californicus
Orchopeas leucopus (Baker, 1904)
D. ordii
C. hispidus
Orchopeas sexdentatus ssp. (Baker, 1905)
D. ordii, spectabilis
Oropsylla (Thrassis) acamantis acamantis (Rothschild, 1905)
P. parvus

**Oropsylla* (*Thrassis*) *aridis aridis* (Prince, 1944)
D. deserti, heermani, merriami, ordii, panamintinus, venustus
Oropsylla (*Thrassis*) *bacchi* ssp. (Rothschild, 1905)
D. microps, ordii
Oropsylla (*Opisocrostis*) *bruneri* (Baker, 1895)
P. fasciatus
Oropsylla (*Thrassis*) *aridis campestris* Prince, 1944
D. ordii
Oropsylla (*Thrassis*) *fota* (Jordan, 1925)
D. ordii
C. hispidus
Oropsylla (*Thrassis*) *francisi* ssp. (C. Fox, 1927)
D. ordii
Oropsylla (*Thrassis*) *bacchi gladiolis* (Jordan, 1925)
D. ordii
Oropsylla (*Opisocrostis*) *hirsuta* (Baker, 1895)
D. ordii
Oropsylla (*Opisocrostis*) *labis* (Jordan and Rothschild, 1922)
D. ordii
Oropsylla (*Diamanus*) *m. montana* (Baker, 1895)
D. ordii, spectabilis
Oropsylla (*Thrassis*) *bacchi setosis* (Prince, 1944)
D. merriami
Oropsylla (*Opisocrostis*) *t. tuberculata* (Baker, 1904)
D. ordii
P. parvus
Phalacropsylla paradisea Rothschild, 1915
D. merriami
Pleochaetis exilis (Jordan, 1937)
D. ordii
Pluseatis d. dolens (Jordan and Rothschild, 1914)
H. desmarestianus
Polygenis b. bohlsi (Wagner, 1901)
H. anomalus, australis
**Polygenis dunni* (Jordan and Rothschild, 1922)
H. anomalus
L. adspersus
Polygenis gwyni (Fox, 1914)
L. irroratus, pictus
Polygenis klagesi ssp. (Rothschild, 1904)
H. australis
L. adspersus
**Polygenis martinezbaezi* Vargas, 1951
L. irroratus, pictus
**Polygenis peronis* (Jordan and Rothschild, 1923)
H. anomalus
Polygenis roberti ssp. (Rothschild, 1905)
H. anomalus, desmarestianus
Polygenis vazquezi Vargas, 1951
L. pictus
**Polygenis vulcanius* Smit, 1958
L. pictus, salvini
Pulex stimulans Baker, 1895
D. merriami, spectabilis
Rhadinopsylla multidenticulata Morlan and Prince, 1954
D. merriami, spectabilis
Rhadinopsylla sectilis ssp. (Jordan and Rothschild, 1923)
D. heermani, ordii, panamintinus, spectabilis
P. longimembris
**Wenzella obscura* Traub, 1953
H. desmarestianus
**Wenzella yunkeri* Tipton and Mendez, 1966
H. desmarestianus

Discussion

The Parasite Community As an Ecosystem

The organisms living on or in a host form a community, and the interrelations within this community can be studied just as any other community can. Although there are many similarities between this community and a community we might find in a field or woods, there are some major differences. First, it is a much simpler community than many. It has far fewer individuals, far fewer species. As in cave systems, primary producers, the green plants that supply the energy for the whole ecosystem, are absent. The entire system is therefore dependent on an outside source of energy. This energy is supplied secondarily by the host.

The host is thus both the habitat and the food resource for the system. In parasite communities, since both producers and primary consumers (herbivores) are absent, the mammalian host itself is either a primary or secondary consumer (feeds on other organisms), and the primary organisms in the parasite community are at least secondary or tertiary consumers. Another major difference is the organization of "food webs." In most ecosystems the webs consist of "chains" of up to five or so "links," each link being a species that feeds on the link below it. Parasite ecosystems usually have fewer links, and the links are usually different in nature than in free-living communities. In free living communities, the first link usually consists of herbivores (primary consumers) and in turn, feeding on them are secondary consumers, tertiary consumers, and so on, with larger predators such as weasels, cats, hawks, and owls generally forming the top link. However, in parasite communities most species are carnivores but few can be classed as predators. Cheyletid mites are often said to be predators, but they are found on relatively few host species or individuals (Whitaker and Wilson, 1974). Not much information is available as yet, but probably more common than predators in this community are organisms that are parasitic (hyperparasites) or phoretic on parasites. A good example of these would be histiostomatid mites found on fleas (see Fain and Beaucournu, 1973). We suspect that additional searching will turn up many more parasites or phoretic associates of mammalian parasites than are now known.

Thus the parasite community is a community much reduced, not only in numbers of individuals and species, but also in ecosystem components and trophic structure. Presumably the decomposer levels would operate, but most decomposition probably occurs away from the host after the dead parasites fall off, are brushed off, or are excreted. Holmes and Price (1986) discuss parasite communities and other topics of interest here, such as effects of parasites on host communities. They further point out that there are many alternative life styles in parasites, leading to numerous variations in the processes regulating their abundance, distribution, community organization, and evolution.

Parasite Populations—Origin, Growth and Regulation

Population ecology attempts to explain the numbers and distributions of organisms in populations, but there has been relatively little emphasis on this in parasitology; rather the emphasis has been to describe the parasites and to indicate their relative abundance and distribution, but with relatively little emphasis on explanation. Most of this work remains to be done.

Esch et al. (1975) pointed out that parasites differ from free living populations in the way they form populations and subpopulations. Free living populations usually live in patches of suitable habitat separated in space. Parasites often occupy moving patchcs (the various hosts occupied) with various degrees of isolation between. These authors proposed the terms infrapopulation for all parasite individuals of a single species within a host, and suprapopulation for all individuals of a parasite species in all hosts of an ecosystem.

Some interesting questions are raised about parasite populations in relation to free-living communities. How do new endodyte and ectodyte populations get started, and what regulates their numbers? How does their evolution and distribution relate to that of their host (coevolution)? With essentially no predators one might think that populations would simply continue to increase, and eventually fill up and decimate their habitat. However, this usually does not occur. In looking at parasites of wild animals, one usually finds that most host individuals have relatively few parasites and only a few individuals have large numbers. Is this because the latter individuals are in poor health, and thus are more susceptible to parasites, or is this simply chance?

Parasites of wild species often do not seem to cause much harm to their hosts, probably for two reasons. First, they are usually not in large numbers on individual hosts, and second, host and parasite often have tended to evolve in such a way as to cause increasingly less adverse effect on each other. It would appear that parasites are more likely to cause harm when they invade a new host species, or a host individual already in poor health. The "perfect parasite" may be one that has evolved a commensual or even a mutualistic relationship with its host during a long history of association.

Under normal conditions, we suspect that infestation of new host individuals by parasites that spend their entire life on the host usually takes place when the host individuals are young; probably parasites are acquired from the parents before the young leave the nest areas. However, this tendency would not be of advantage in species spending only a portion of their life on the host. Invasion then could occur at any time. Since the parasite populations of the parents are generally small, relatively few individuals are passed to the offspring, giving rise to a small parasite population there. In addition to host individuals getting relatively small numbers of invading parasites, most parasites have relatively low reproductive rates on their hosts, so that parasite populations increase relatively slowly. If these ideas are correct, then older hosts should have more parasites than young ones. This is a testable hypothesis, and we have noticed that this sort of distribution does prevail in beavers. Beavers have many species and many individuals of beaver mites, *Schizocarpus*. However, older beavers have far more mites than do younger ones, both in terms of species and of individuals.

Different parasites have very different strategies. Most free living organisms and many parasites invade a new area, and then their populations increase through reproduction, but this is not always the case. Helminths, for example, increase within a mammalian host only through immigration (recruitment). In situations such as this, the numbers of eggs produced are often enormous as a strategy to increase the probability that some will ultimately infect new host individuals.

Esch et al. (1977) discussed typical r-selection (emphasis on reproduction, i.e., the production of large numbers of relatively cheap offspring, such as occurs in helminths) and K-selection (emphasis on production of only a few high-quality offspring). There has been much emphasis in parasitology on r-selection, in view of the large numbers of eggs produced in order to ensure dispersal of helminths. However, there are numerous examples, such as many mites and lice, of the other extreme: parasites that live their entire life cycle on one host and produce few offspring with relatively little energy wasted. As one would expect, given the diversity of parasite types, r, K and a whole range of strategies in between occur in parasites.

Population control comes about in several ways. Recruitment into populations is slow, either through invasion or reproduction. Thus there are usually small numbers of parasites per host of any one species. Also, there are few predators. Holmes et al. (1977) point out that most parasite population work has been done at the "infrapopulation" level, but that population regulation occurs at the suprapopulation level. These authors discuss several factors that can regulate parasite populations; these include immune responses of the host, crowding effects, and competition. In general, the higher the numbers of parasites that invade, the lower the survival, maturation rate, and reproductive rates. Holmes et al. also point out that parasite population control can be complex and that control in one host can lead to limitation in others. They also present models of such controls.

Population regulation can come about or be enhanced through immune reactions, interrupted development (see Schad, 1977), and the crowding effect thought to be the result of either infraspecific competition or an immune reaction. There is a great deal of information on population regulation in the book, *Regulation of Parasite Populations*, edited by Esch and Nickol (1977).

It was suggested earlier that hosts in poor condition might acquire large numbers of parasites or become diseased. If so, these individuals might be culled from the population, removing their genetic material and that of the parasites with them. This would be a constant selective factor against large parasite populations, and indeed, Croften (1971) has indicated that there may be a regulation of parasites by death of heavily infested hosts.

In addition to having low numbers of dispersers and slow reproductive rates, there is another important factor also. Unlike free-living populations, the habitat of the parasites abruptly disappears at frequent intervals, whenever an individual host dies. This automatically leads to death of thc parasites on that individual, bringing about immediate and frequent control of infrapopulations each time a host dies.

Price (1980; but see Holmes, 1979) presents evidence that competition has not been involved in determination of parasite community structure. Rather, most studies of parasite communities suggest non-interactive niche occupation. This, of course, implies the existence of numerous unfilled niches which could be utilized by invading parasites. Competition would presumably become more important as greater numbers of niches are filled.

To summarize these ideas then, dispersal occurs with low numbers of individuals invading host animals, often the young, followed by slow population growth, and finally as the population becomes larger, the individual host dies, eliminating that subpopulation. All of these factors would tend to keep populations low. However, the origin, growth and regulation of parasite populations is complex and variable, relating to the interrelations of the many different lifestyles of both parasites and hosts.

Coevolution: Fahrenholz's Rule and Resource Tracking

Timm (1983) discussed Fahrenholz's rule in relation to resource tracking as it applies to distribution of parasites on the host. Fahrenholz's rule states that in groups of permanent parasites, the taxonomic classification of the parasite usually corresponds directly to the taxonomic relationships of the host, implying that the parasite and host evolve together. Timm (1983) found Fahrenholz's rule to apply in pocket gophers, *Geomys*, and was able to use it in drawing conclusions about phylogenetic relationships in that genus. Timm also pointed out that Wenzel et al. (1966) and Machado-Allison (1967), using parasite relationships, suggested that the family Desmodontidae (Vampire Bats) had affinities with the family Phyllostomatidae, and that the Chilonycterinae (subfamily of Phyllostomatidae) should be elevated to family status. Subsequent studies confirmed this, placed the vampires as a subfamily of Phyllostomatidae (Forman et al., 1968), and created the family Mormoopidae for the former Chilonycterinae (Smith, 1972). Similarly Holland (1958, 1963) proposed two hypotheses concerning *Spermophilus parryii* based on their fleas: 1) a close affinity between the New World and Old World arctic ground squirrels, and 2) a distinct arctic-subarctic division in the New World forms. Nadler and Hoffmann (1977) subsequently concluded that northern and southern New World squirrels are more similar to those of Siberia, than either was to each other.

Resource tracking is a "contrasting model of host-parasite coevolution," where the parasite tracks a "particular and independently distributed resource on the host," and we thus "expect noncongruent host-parasite relationships" (Kethley and Johnston, 1975, p. 232). This model predicts no parallel relationships of parasite and host taxonomy, because "parasites are tracking a resource, such as a particular type of skin or hair." This process is predominant in bird lice, where, because the vagility of the birds gives the lice good powers of dispersal, the end result of several species of unrelated lice per bird species is best explained by resource tracking.

Timm examined "contradictions between Fahrenholz's rule and resource tracking" (Timm, 1979) and saw them as central to our understanding of parasite evolution. He concluded that "coevolutionary relationships can be greatly affected by the dispersal of parasites from one host species to another." This questions the assumption underlying Fahrenholz's rule: that there is no gene flow of parasites between unrelated hosts, versus the virtually unrestricted flow assumed by the resource tracking model.

Where do heteromyids fit into the ideas expressed in the paragraphs above? Every situation and every parasite group is different. On heteromyids it would appear that both of these factors come into play, with neither being dominant.

As one would expect from their lifestyles, heteromyid parasites fall into two groupings with regard to speciation. In general rickettsia, bacteria, fungi, cestodes, chiggers, ticks and fleas have remained rather generalized in relation to their hosts; i.e., each species tends to occur in several species and genera of hosts, and there is little evidence of convergent evolution.

The protozoans, glycyphagid mites and listrophorid mites, on the other hand, have tended to radiate into isolates, with species often tending to coevolve and speciate congruently with their hosts. In addition there has sometimes been secondary host switching, resulting in several rather than just one parasite per host. This former tendency (Fahrenholz's rule) has tended to increase the number of species per genus, but the latter tendency (resource tracking) has tended to break down the taxonomic parallelism between host and parasites.

The cestodes, nematodes, ticks and chiggers exemplify r selection, whereas the listrophorids and lice exhibit K-selection. We suspect much of the population regulation in heteromyid parasites is through low reproductive or recruitment rates, plus periodic death of host individuals. Fahrenholz's rule seems to be illustrated to some degree in the genera *Geomylichus* and *Eimeria,* although this is complicated by secondary host switching. Mites of the genus *Echinonyssus* and lice of the genus *Heteromys* and many of the chiggers and fleas are more influenced by resource tracking and less specific in their distribution among host species.

Comparison of Parasite Populations with Island Populations

Because the host represents an isolated, difficult to colonize patch of habitat, it is tempting to compare parasite populations with island populations. There are similarities, but there are also differences. Islands are separated from each other in space. Dispersal to islands or between islands occurs across water, or, in the case of "habitat islands," across physical barriers of unfavorable habitat. The distance of the island from the colonizing source is the major barrier to dispersal, although the size of the island, prevailing winds and currents, the habitat of the island, and undoubtedly other factors play a role in the amount and type of colonization. In the case of parasite populations on different hosts, the hosts often occur sympatrically, rather than being separated in space. Factors hindering dispersal thus are not the amount of space involved, but rather how closely related ecologically are the different species between which dispersal theoretically might occur. For example, parasites of burrowing mammals would be far more likely to appear on (disperse to) other burrowing species which use the same burrows than to mammals that live in tree holes or caves. The relative similarity of the niches in which the host species lives might be the best factor to introduce in place of distance or breadth of a barrier if one were to develop a model for parasite relationships following island theory. Holmes and Price (1986) further discuss parasites in relation to the theory of island biogeography.

The Heteromyid Parasite Community

Heteromyid rodents have a well developed parasite community, quite distinct from that of other mammals, including other rodents. This indicates that there has been much isolation during the development of this assemblage and, in turn, of their heteromyid hosts.

The main heteromyid parasite community consists of chiggers (numerous species), listrophorids (several species of *Geomylichus*), *Ischyropoda* (three species), one genus of sucking louse (*Fahrenholzia*), and some tick and flea species. Heteromyid ectodyte and endodyte data are summarized by heteromyid species (Table 1) and by important genera (Table 2). This parasite community description does not fit any other host taxon. The reason why heteromyids have this particular assemblage of parasites is not specifically known. In general terms, the life style of the heteromyids probably favors parasitism by these parasites, and the relative distinctiveness of this community reflects the fact that the hosts are relatively isolated taxonomically and ecologically from other kinds of mammals. Communities most similar to this one would likely be found on other desert species or groups ecologically similar to heteromyids. Some comparisons with other communities will be made later.

Some parasites have developed primarily as heteromyid specialists, whereas others are common on other hosts. A parasite community generally is made up of host-specific forms, non-host-specific but regular associates, occasional associates, and accidentals.

The main heteromyid protozoans are Coccidia, primarily *Eimeria,* a number of which are found only in heteromyids. Eimerians, however, have been described from numerous hosts, with most known from but one or few host species. There are no *Eimeria* shared among heteromyid genera, although one occurs in five species of *Dipodomys* (*E. scholtysecki*), three in four species of *Dipodomys* (*E. balphae, chobotari, utahensis*), one in two species of *Liomys* (*L. liomysis*), and one in two species of *Perognathus* (*E. penicillati*). Six of the species currently are known from only one host. Little has been written about natural groups within the coccidians, or on the significance of distribution and variation of species within the host groups.

It would be interesting to compare the coccidian species within and among the various genera of heteromyids, to see how the taxonomic relationships of the parasite groupings compare to the mammalian relationships. This statement also applies to other parasite groups which have a series of parasite species in different heteromyid hosts, particularly *Fahrenholzia* (Anoplura), *Geomylichus* (Acari: Listrophoridae), and many of the chiggers.

The main cestode parasitizing heteromyids is *Catenotaenia linsdalei.* It has been found in seven species of *Dipodomys* and two of *Chaetodipus.* No cestodes have been

TABLE 1.—*Numbers of species of parasites and other associates reported from heteromyids.*

	Virus	Rickett-sia and spiro-chetes	Bacteria	Fungi	Proto-zoa	Ces-todes	Nema-todes	Mites	Chig-gers	Ticks	Lice	Fleas	Diptera and Lepi-doptera	Totals
Dipodomys														
agilis					8				7	2		1		18
californicus		1						1		1		1		4
compactus														0
deserti						1	3	3	4	1	1	7		20
elator								3	1	1	1	2		8
elephantinus								2		1				3
gravipes					1									1
heermani	3				2	3	2	1	4	2	1	6		24
ingens						1								1
insularis														0
margaritae														0
merriami				2	8	4	10	11	35	2	2	22	1	96
microps		2	1		2	1	4	11	11	3	1	8		44
nelsoni									1					1
nitratoides	5				1	1		1	1	1		1		11
ordii		2	2		6	1	7	29	27	4	6	37		121
panamintinus					4	3	5		11	1		3		27
phillipsii					1	1		1			1			4
spectabilis			1		3	1		5	5	1	1	18		35
stephensi									3					3
venustus					1	1			5	1		7		15
Heteromys														
anomalus					2			19	17	2	3	7		50
australis								2	8			2		12
desmarestianus					2		1	11	17		3	5	2	41
gaumeri								6	8	2	1			17
goldmani											1			1
nelsoni														0
oresterus														0
Liomys														
adspersus								4	10	1	1	3		19
irroratus					1			10	17	2	3	3	1	37

TABLE 1.—*Continued.*

	Virus	Rickett-sia and spiro-chetes	Bacteria	Fungi	Proto-zoa	Ces-todes	Nema-todes	Mites	Chig-gers	Ticks	Lice	Fleas	Diptera and Lepi-doptera	Totals
pictus					2		2	3	17	1	1	5		31
salvini					1			8	9	3	1	1	1	24
spectabilis														0
Chaetodipus														
arenarius									1					1
artus									8					8
baileyi				2			1		9		1			13
californicus	2		1			1	1	3	24	4	1	12		49
fallax								1	15			1		17
formosus		1	1		2	1	1	8	24	5	1	2		46
goldmani									8					8
hispidus		1						8	16	1	3	3	1	33
intermedius				1	1		1		7					10
lineatus														0
nelsoni									10			1		11
penicillatus				2	1		5	4	26		1	2		41
pernix									9					9
spinatus								1	7					8
Microdipodops														
megacephalus		1						1		2	1	2		7
pallidus														0
Perognathus														
alticola														0
amplus									1					1
fasciatus								2	1	1		3		7
flavescens								1	4			2		7
flavus			1		1			3	15	1	2	6		29
inornatus			1					1						2
longimembris		2			1	1	1	10	18	3		4		40
parvus		2	1		1			24	13	5	7	12		65
xanthonotus								1						1

TABLE 2.—*Major parasites of genera of heteromyids, as compared with* Microtus *and* Peromyscus. (Microdipodops *is omitted because there are so few records).*

	Dipodomys	*Heteromys*	*Liomys*	*Chaetodipus*	*Perognathus*	*Microtus*	*Peromyscus*
Rickettsia						*	*
Coxiella	1			1	1		
Rickettsia	1			1	1		
Bacteria						*	*
Francisella	1				1		
Fungi						*	*
Coccidioides	1			1	1		
Protozoa						*	*
Eimeria	7		2	3			
Endamoeba	1						
Leishmania		1					
Tritrichomonas	1			1	1		
Acanthocephala							
Moniliformis						3	
Cestoda							
Catenotaenia	2			1			
Mathevotaenia	1				1		
Andrya						10	
Hymenolepis						8	4
Paranoplocephala						9	
Taenia						9	
Nematoda							
Gongylonema	2						
Heteromoxyurus	1						
Protospirura	3			3	1		
Rictularia	1						
Trichuris	2			1		4	
Welcommia	1			1	1		
Gongylonema							4
Heligmosomoides						5	
Heligmosomum						5	
Nematospiroides						5	
Protospirura							4
Syphacia						7	4
Trematoda	***					***	***
Mites							
Ameroseiidae							
Sertitympanum	3			1	2		
Dermanyssus							
Dermanyssus							6
Glycyphagidae							
Glycyphagus	1	1	1			8	
Dipodomydectes							
Dipodomyopus	1						
Mediolabidophorus		1					
Metalabidophorus		1	1				
Neolabidophorus		1		1			

TABLE 2.—*Continued.*

	Dipodomys	*Heteromys*	*Liomys*	*Chaetodipus*	*Perognathus*	*Microtus*	*Peromyscus*
Laelapidae							
Androlaelaps	3	3	2	3	2	13	11
Echinonyssus	6	10	5	4	5	8	7
Eubrachylaelaps	1	2	1	1	3		10
Hypoaspis	1	1	2	1	2		
Ischyropoda	2						
Laelaps	1	2				14	
Steptolaelaps		1	2				
Brevisterna							11
Eulaelaps						8	5
Haemogamasus						13	10
Listrophoridae							
Geomylichus	5			5	1	8	
Macronyssidae							
Ornithonyssus							10
Myobiidae							
Radfordia	1					7	
Myocoptidae							
Myocoptes						5	
Trichoecius						3	
Psorergatidae							
Psorergates						4	
Chiggers							
Anomalaspis		1					
Ascoschoengastia		1	1				
Boshkerria		1					
Colicus			1				
Crotiscus		2	1				
Crotonasis			1				
Dermadelema	4			3	4		
Ectonyx		1	1				
Euschoengastia	12	1	1	17	9	7	11
Euschoengastoides	7		5	12	5		
Eutrombicula	3	2	2	3	2		
Hexidionis	6		2	5	5		
Hoffmannina		1	1				
Hyponeocula	5			6	4		
Kymocta	3						
Leptotrombidium	1	2	1	1	1		
Microtrombicula			1				6
Odontacarus	3	2	1	3	2		4
Otorhinophila	4		2	7	3		
Parasecia	1	3	1	1			
Peltoculus		1					
Polylopadium		2	1				
Pseudoschoengastia	2	7	12	3	2		6
Quadraseta	3						
Trombicula	1	3	2	1			6
Vanidicus		2	1				
Miyatrombicula							4
Neotrombicula						11	6

TABLE 2.—*Continued.*

	Dipodomys	*Heteromys*	*Liomys*	*Chaetodipus*	*Perognathus*	*Microtus*	*Peromyscus*
Ticks							
Argasidae							
Ornithodorus	2		1	2			
Ixodidae							
Dermacentor	3			2	3	7	6
Ixodes	4	1	2	4	3	13	11
Anoplura							
Fahrenholzia	1	4	4	5	2		
Hoplopleura						9	14
Polyplax						10	10
Beetles							
Leptinus						4	
Flies							
Cuterebra						5	9
Siphonaptera							
Carteretta				1			4
Euhoplopsyllus	2		1	1			
Malaraeus	3				2	9	9
Meringis	16			3	10		5
Oropsylla	10			1	3		7
Polygenis		5	5				
Rhadinopsylla	3				1	8	4
Wenzella		2					
Anomiopsyllus							6
Atyphleroceras						9	8
Callistopsyllus							5
Catallagia						11	5
Ctenopthalmus						5	8
Delotelis						5	
Echidnophaga							4
Epitedia						13	5
Hoplopsyllus							7
Hystrichopsylla						12	
Megabothris						13	
Megarthroglossus							7
Monopsyllus						9	8
Opisodasys							5
Orchopeas						7	9
Peromyscopsylla						13	9
Pleochaetis							6
Stenistomera							4
Stenoponia						5	5
Subtotals	42	30	31	31	27	42	45
(excluding Rickettsia, Bacteria, Fungi and Protozoa)							
Totals	49	31	32	36	31	—	—

reported from either *Liomys* or *Heteromys.*

As with cestodes, most nematode records are from *Dipodomys,* although several nematodes have been reported from pocket mice, and one from *Liomys.* The more widespread and abundant nematodes of heteromyids are *Protospirura dipodomis, Heteromoxyurus deserti, Trichurus perognathi,* and *T. dipodomis,* which have been reported from five, five, three, and two species of heteromyids, respectively. One species, *Protospirura dipodomis,* has been found in three different genera of heteromyids, *Dipodomys, Chaetodipus,* and *Perognathus.*

Among mites, *Androlaelaps fahrenholzi* has been found on 17 species of heteromyids, including all five genera. This is logical as this mite occurs on far more hosts, including many other rodents, than any other mite in North America (Whitaker and Wilson, 1974). It has attained its success by remaining relatively unspecialized in morphology, habits, and habitat. It is the "*Peromyscus maniculatus* of the mite world."

Species of *Echinonyssus* on heteromyids in North America consist of *E. hilli, incomptis, perognathi,* and *triacanthus* (Herrin, 1970), from *Dipodomys, Chaetodipus,* and *Perognathus.* These species, and *Echinonyssus liomys,* described later by Herrin and Yunker (1973) from *Liomys irroratus,* form a natural group. Other species of this genus described from *Heteromys* and *Liomys* from Central America are *E. brevicalcar* (known from 4 specimens), *E. heteromydis* (numerous), *E. lunatus* (7), *E. microchelae* (23), *E. minutus* (7), *E. panamensis* (4), *E. parvisoma* (27), *E. proctolatus* (5), and *E. venezuelensis* (numerous). These have not been taken often, but tentatively I have classed them as heteromyid specialists.

Taxonomic comparison of the Central and North American heteromyid genus *Echinonyssus* would be of interest, but has not been made. The North American heteromyid *Echinonyssus* have a postero-marginal spur on coxa II, and two setae on tarsus II (avi, pvi) are normal rather than being stout and clawlike. The Central American species do not follow this pattern. One species, *E. proctolatus,* appears to have that combination of characters. Apparently the only other one that has the posteromarginal spur on coxa II is *E. lunatus.* It would be interesting to determine whether these relationships reflect biogeographic or host-specific affinities of the species in this genus.

Allred and Beck (1966) reported mites of the genus *Kleemania* and separated this genus by its sternal plate pattern resembling three asymmetrical cog-wheels (see their fig. 262). Other records of *Kleemania* are those of Allred (1962, 1963), Goates (1963), and Allred and Goates (1964*a*) from Nevada. The numbers of individuals per host were not given, but host records were as given below.

Of the 33 host records summarized by Allred and Beck, 18 (54.5%) were on heteromyids. On pocket mice from Nevada (Allred, 1963), the mite found in the greatest

Heteromyid Hosts		Other Hosts	
Dipodomys ordii	7	*Onychomys leucogaster*	2
D. merriami	1	*O. torridus*	1
D. microps	3	*Peromyscus crinitus*	3
Perognathus formosus	2	*P. maniculatus*	6
P. longimembris	3	*P. truei*	1
P. parvus	2	*Eutamias minimus*	1
		Spermophilus lateralis	1
Total, heteromyid records	18	Total, other records	15

abundance other than chiggers (Trombiculidae) was *Kleemania,* although only 48 individuals were taken on 156 individuals of *C. formosus,* seven on 473 of *P. longimembris* and none on 23 of *P. parvus.* Relatively little is known of their ecology. Allred and Beck (1966) believed that they were probably predaceous. Elsen and Whitaker (1985) found that no described *Kleemania* species other than *K. plumosus* exhibits a coglike structure, and in that species it is incomplete. For this reason, these workers established a new genus, *Sertitympanum,* with three new species for the peculiar mites with the coglike structure. Two of the species, *S. exarmatum* and *S. contiguum,* were described partly or entirely from heteromyids. The species figured by Allred and Beck keys to *S. contiguum.* It will be interesting to determine the host and geographic distribution and the number of species of *Sertitympanum,* and how closely associated to heteromyids these mites really are.

Another group of mites that could prove fascinating in terms of relationships of hosts are the glycyphagid mites. These mites are presumably non-feeding phoretic forms called hypopi. Nine species of hypopi have been described from heteromyids so far, and they show very interesting modifications and adaptations. Two species of typical hair-clasping glycyphagid hypopi have been found, *Dermacarus ornatus* from *Heteromys gaumeri* and *H. anomalus,* and *D. liomys* from *Liomys irroratus.* These two species, like most glycyphagid hypopi, attach to individual hairs by their large ventral, hair-clasping organs. Their mouthparts are vestigial. They presumably drop off in the nests of the hosts where transformation to the adult takes place, followed by breeding. This is the typical mammal-inhabiting glycyphagid life style.

The remaining seven species live in hair follicles or under the skin. These species have evolved from the hair-inhabiting form as indicated by various reductions of the clasping organs, the claspers themselves, claws, and setae. The seven are in five different genera with varying degrees of specialization for life in follicles or under the skin. *Metalabidophorus* has reduced and posteriorly placed claspers, whereas *Mediolabidophorus* has the claspers reduced even more. These genera are on *Heteromys* and *Liomys. Dipodomyopus,* of *Dipodomys,* has the setation and claws reduced and small claspers. *Dipodomydectus* has the claspers vestigial, claws 3 and 4 absent, and vestigial setation. The species of the genus *Neolabidophorus* have vestigial claspers and setation, and claws 3 and 4 absent. I suspect that close examination will turn up additional species of these interesting mites on heteromyids, and that we may glean something of relationships of the mites and their hosts from them.

The largest group of heteromyid parasites is the chiggers, family Trombiculidae, with 137 species known from heteromyids and 78 of them listed here as regularly occurring. Several of the genera are almost entirely restricted to heteromyids.

At least five species of chiggers occur in nasal passages of heteromyids, *Otorhinophila intrasola, O. sinaloae, Kymocta zulia, K. faitkeni,* and *Crotonasis fissa. Crotonasis fissa* was described from the nasal mucosa of *Liomys adspersus* (Brennan and Yunker, 1966). All but one of 516 larvae and 7 nymphs of *O. intrasola* were from the anterior portion of heteromyid nasal passages (the one was from nasal passages of *Peromyscus eremicus*). At the type locality of *O. intrasola* in Sonora, Mexico, there were 1-34 chiggers per host. The maximum number was 50 chiggers in one *L. pictus* from Sinaloa. These larvae were not attached. They moved freely in the nasal mucosa. *Otorhinophila sinaloae* also occurred in the anterior portions of the nasal passages. Eleven of 38 *P. pernix* contained 1–30 individuals, and two *L. pictus* contained 4 and 26 larvae. The larvae of *O. sola* and *O. parvisola* attach to an external site, usually deep within the ears and on a variety of small rodents. Brennan and Van Bronswijk (1973) found 37 individuals of *K. zulia* in *H. anomalus* from

TABLE 3.—*Distribution of* Fahrenholzia *among species of heteromyids.*

	Heteromys				*Liomys*			
	anomalus	*desmarestianus*	*gaumeri*	*goldmani*	*adspersus*	*irroratus*	*pictus*	*salvini*
schwartzi	×							
fairchildi		×			×			×
ferrisi		×	×	×				
hertigi		×						
ehrlichi						×		
microcephala						×	×	
texana						×		

	Chaetodipus					*Perognathus*			
	baileyi	*californicus*	*formosus*	*hispidus*	*penicillatus*	*flavus*	*parvus*	*Microdipodops*	*Dipodomys*
reducta	×		×						
tirbulosa		×		×					
zacatecae				×					
pinnata					×	×	×	×	×
boleni						×			

Venezuela, and Brennan (1968*a*) found 17 individuals of *K. faitkeni* in *H. anomalus* from Trinidad.

Ryckman and Roos (1955) found an individual of *Perognathus pernix* to have 5 wartlike growths on the rump, each containing numerous chiggers, *Leptotrombidium panamense.* The growths consisted of a "polypoid structure covered with benign stratified squamous epithelium which in several places showed papillary projections and elsewhere keratin filled crypts. The stroma was dense connective tissue infiltrated with lymphocytes."

Several ticks are regularly occurring inhabitants of heteromyids, particularly *Dermacentor parumaperetus,* which have their larvae on heteromyids and their adults on leporids. This species is one of the vectors for Rocky Mountain Spotted Fever and is known from 7 species of *Dipodomys,* one of *Chaetodipus,* one of *Microdipodops,* and two of *Perognathus.* Much more work is needed on ticks of heteromyids, especially of *Liomys* and *Heteromys.*

Twelve species of sucking lice, all in the genus *Fahrenholzia,* are regular parasites of heteromyids. Present data on heteromyid *Fahrenholzia* are summarized in Table 3.

Dipodomys and *Microdipodops* have one species of *Fahrenholzia,* two are known from *Perognathus,* and four have been reported from *Heteromys, Liomys* and *Chaetodipus.* The greater number of species of lice from *Heteromys* and *Liomys* might be related to greater diversity in the tropics, but this does not explain greater diversity in *Chaetodipus.* It would be interesting to compare these species and their hosts cladistically.

The louse of *Dipodomys* and *Microdipodops* is *Fahrenholzia pinnata.* This louse may be important on *Perognathus* also, although *F. boleni* McDaniel was described from *P. flavus.* This species has also been taken on *Chaetodipus. Fahrenholzia fairchildi* is known from three species of two genera, *Heteromys desmarestianus, Liomys adspersus* and *L. salvini.* This relationship is geographic, with all three occurring in southern Central America, and *H. desmarestianus* forming a link between the two species of *Liomys,* which are allopatric. *Heteromys desmarestianus* and *Chaetodipus hispidus* are the most wide-ranging species of their genera, and correspondingly, are the single species in their respective genera with records of three species of *Fahrenholzia*. In *Perognathus*, the only species with two *Fahrenholzia* species is *P. flavus*, again the species with the largest geographic range. It is not known why *Dipodomys* has essentially one species of *Fahrenholzia*, whereas adaptive radiation has apparently occurred in the lice of *Heteromys*, *Liomys* and *Chaetodipus*. This could indicate that the species of these genera have experienced greater degrees of isolation from each other than have the species of *Dipodomys*. However, the situation in the listrophorid mite genus *Geomylichus* with multiple species in *Dipodomys* would indicate the opposite. It will be interesting to see if *Geomylichus* occurs on *Heteromys* and *Liomys* species, and if so to determine its relationships on these genera.

A number of flea species have been found on heteromyids, but only the genus *Meringis* on *Dipodomys* appears to represent a specific association.

One parasite group lacking or reduced in heteromyids is the trematodes. We have postulated that trematodes might be lacking because of the dry environment in which their hosts live, although *Heteromys* is found in mesic forests. All trematodes require a snail intermediate host. Two groups of mites, the glycyphagids and myobiids, seem reduced on heteromyids, although both are present. Several glycyphagids are known from heteromyids but most are highly modified follicular mites. There are two typical hypopi known, but only from *Heteromys* and *Liomys*. Many other rodents have large numbers of hypopi (non-feeding phoretic forms) in the fur. In view of the highly modified forms, presumably evolved from the hypopi of the fur, one might expect less specialized forms of these mites in the fur. *Radfordia bachai* is a host specific myobiid on heteromyids, but it appears to be of infrequent occurrence, and other species in the genus occur more commonly on some other small mammals. Since *R. bachai* is a host specific form, one might expect it to be more successful in terms of numbers of individuals per host. We have no hypothesis as to why either of these forms are not more common on heteromyids.

Parasite Studies

Parasite studies consist of two stages. The descriptive first stage consists of finding, preparing and identifying the parasites, characterizing the parasite populations in terms of abundance, percentage of hosts infested, and determining which are regular associates and which are accidentals.

The synthetic second stage consists of determining the life histories and dispersal mechanisms of the parasites, effects on the host, and the type of relationships existing between the various components of the community.

For the heteromyid community the first stage is presently occurring rapidly. We can judge this by examining the rate at which new species are being described. (Table 4 summarizes the original dates of description of the heteromyid parasites. Many of these were originally described directly from heteromyids, but others were only later reported from heteromyids.) Two of the species, a chigger and a tick, were described in 1758 by Linnaeus, but a total of only four of the species had been described by 1800, and 27 by 1900. Between 20 and 24 of the species were described in each of the first five decades of this century, followed by 84 in the 1950's, 71 in the 1960's, 61 in the 1970's, and 15 through 1987. Thus there has been a tapering off in the 1970's and 1980's, but there are obviously many more species to be reported and to be described. The best studied host species give us the best idea of how many parasites there should be per host. There are 121 species of endodytes and ectodytes known from *Dipodomys ordii*, 50 from *Heteromys anomalus*, 49 from *Chaetodipus californicus*, and 40 from *Perognathus longimembris*, the species with the greatest number of known community members in each of their respective host genera. Only seven species have been reported from *Microdipodops megacephalus,* and none from *M. pallidus.*

We can probably do better than using maximum numbers of forms in the better studied species. We can probably predict numbers of regularly occurring species that should be found on species within heteromyid genera by constructing a "typical" or "composite" species list for each genus from the information on host records (for example, by looking at better studied hosts for each parasite group; if internal parasites have been studied in one species, chiggers in a second, and other mites in a third, we can use the combined data to construct the composite community for the genus, or perhaps even for a related genus). I have used this approach with the data in Table 1, excluding a few obvious accidentals. My estimates of the numbers of parasites that should occur in the various groups in the various host genera are given in Table 5.

These estimates would undoubtedly increase as further data are collected. One could go further and modify this table if enough data were available by applying geographic, habitat, size of host, or other variables through modelling techniques, once good estimates of the effects of these factors were known.

We can attempt to apply this information in a specific case, *Dipodomys ordii,* for which 116 species of parasites (other than some obviously accidental lice) have been reported. Estimates for this genus were 139 species. Thus we would expect about 23 additional parasites to be found even on *D. ordii,* the best studied heteromyid host. No viruses, fungi and only one cestode have been reported from *D. ordii*; we expect about five, two and three of these, respectively, to occur there. Presumably nobody has adequately looked for these parasites on this host. In addition, we expect more nematodes and chiggers to be found. A prediction from this would be that more examination will probably turn up approximately the numbers of species in Table 5 for the various host species once adequate examination has occurred, with perhaps somewhat lesser total numbers in less widespread species (see later discussion). On less well studied species we will expect many more parasite species to be found and some new species will yet be described.

TABLE 4.—*Rate of description of parasites found on heteromyids.*

Years included	No. new species described
1758	2
1760–69	
1770–79	1
1780–89	
1790–99	1
1800–09	
1810–19	1
1820–29	1
1830–39	1
1840–49	2
1850–59	
1860–69	1
1870–79	4
1880–89	2
1890–99	11
1900–09	21
1910–19	20
1920–29	22
1930–39	22
1940–49	24
1950–59	84
1960–69	71
1970–79	61
1980–89	15

Numbers of New Species Still to be Described

Can we now make use of these data to predict the numbers of new species still to be described? This could be done by examining variation within parasite groups, but across species, i.e., essentially by examining host specificity. Species of parasites occurring in several host species within a genus will likely be found in other species yet to be examined, whereas, at the other

TABLE 5.—*Estimated numbers of ectodyte and endodyte species on host species in the various heteromyid genera.*

	Dipodomys	*Heteromys*	*Liomys*	*Chaetodipus*	*Perognathus*	*Microdipodops*
Virus	5	2*	2*	2	2*	2*
Rickettsia	2	2*	2*	1	2	1
Bacteria	2	2*	2*	1	1	2*
Fungi	2	2*	2*	2	2*	2
Protozoa	8	2	2	2	1	1*
Cestoda	4	1*	1*	1	1	1*
Nematoda	10	2*	2	5	1	3*
Mites**	29	19	10	8	24	23*
Chiggers	35	17	17	26	18	18*
Ticks	4	2	3	5	5	2
Lice	1***	3	3	3	2***	1
Fleas	37	7	5	12	12	2
Total	139	61	51	68	71	58

* None (or almost none) recorded in this genus; estimate based on other genera.
** Mites excluding chiggers.
*** More were recorded but appear to be obvious accidentals and were omitted to avoid bloating the totals.

extreme, if each host within a genus has a different parasite species, one would expect additional new species to be found. Using this sort of an approach we would expect to find additional new species of *Eimeria*, especially in *Chaetodipus*; cestodes and nematodes, especially in *Liomys* and *Heteromys*; *Protospirura*, especially in *Chaetodipus*; *Geomylichus*, in all genera except *Microdipodops*, but several new species in that genus, particularly of *Eimeria*. Additional new glycyphagid species are likely to be found as well as new mites from the various hidden or cryptic biotopes.

As indicated above, the descriptive first stage is nowhere near complete, even for the best studied species, but especially as indicated by the numbers of heteromyids for which few parasites are known: there are less than 15 species per host known for 34 of the 58 (59%) of the heteromyid species (Table 1) considered here (below).

Even though the first stage is not complete, progress can occur on the second analytical and synthetic stage. We can study and compare the life histories of the forms we do know about. This has progressed to varying degrees, as has the study of effects of different kinds of parasites on the hosts. However, most of the contributions are yet to be made. We can deduce some information by examining the life histories of related species, and this, of course, is how we have gotten much of the information in the introductory material for the various parasite groups.

What Determines the Numbers of Parasites, Both Individuals and Species, on the Host?

This question is of particular interest with respect to the occurrence of parasites on free-living hosts in the field. On Indiana mammals, for example, burrowing species (moles, shrews, pocket gophers, pine voles) and hole-nesting species (many of the squirrels) are the forms which have the higher numbers of ectoparasites (J. O. Whitaker, unpubl.). The average body size of the host plays relatively little role, except in the smallest size classes (less than 20 g for Indiana mammals). However, there is little information of this sort for heteromyids.

What might we predict about the invasion of additional parasites onto heteromyid or other hosts? Price (1980) discusses this further, and suggests that numerous unfilled niches occur in host species, probably because of the difficulty of utilizing them. Parasites have evolved from free-living forms. In order to become permanently parasitic, species must accomplish all of the following. They must first invade the host, a chance occurrence. Second, they must be able to live in what may be an exceedingly harsh environment, especially in the case of internal parasites, which must cope with the internal physiology and immune defenses of the host. This is probably most likely accomplished in the case where the organism was preadapted in the sense that its previous environment was similar enough that it could survive (e.g., it may have inhabited a related host or it may have lived in rotting organic matter). Third, the would-be parasite must be able to reproduce in its new environment. Fourth, it must be able to disperse to other individuals of the host population if it is to become parasitic in nature. It is probably in part because of the difficulty, or more precisely, the unlikelihood of all of these necessities being fulfilled, that many niches for additional parasites remain unoccupied.

Heteromyids could be grouped by differing behavioral and ecological characteristics, and the parasite communities could be compared to determine if the ecological differences cause or otherwise relate to differences in the parasite communities. If so, such correlations might allow the development of hypotheses concerning why certain parasites are on certain hosts. Such hypotheses could then be further tested under field and/or laboratory conditions.

The data are pretty scanty on most heteromyid parasite communities, but one can attempt to determine effects of body size of heteromyid host and also of heteromyid host geographic range on ecto- and endodyte load, as they affect species diversity.

Placed in approximate order by decreasing body weight are some heteromyids as follows (weights from Bowers and Brown, 1982).

	Weight (grams)	Parasites Reported
Dipodomys spectabilis	120.1	35
D. deserti	102.7	20
D. panamintinus	80.4	27
D. microps	58.5	44
D. ordii	48.9	121
D. merriami	42.3	96
Chaetodipus hispidus	29.5	33
C. formosus	21.0	46
C. californicus	20.5	49
C. penicillatus	12.1	41
Perognathus parvus	16.9	65
C. fallax	13.1	17
P. flavus	7.2	29
P. longimembris	7.2	40

This approach is of interest but cannot be carried very far until more data are available. There is no obvious relationship. Any trends here appear to be masked by effects from differential collecting effort (e.g., the most parasites have been reported from the widely distributed and intensively studied species *D. ordii* and *D. merriami,* see Table 6 for data for *D. ordii*).

Area of Geographic Range of Host

To determine if there was a relationship between the area of the geographic range of the host and the number of parasites, the hosts in which at least 15 parasites were reported of *Dipodomys, Chaetodipus* and *Perognathus* were examined to determine the numbers of regularly occurring species for each. These data were then arranged in approximate order from those with the largest to those with the smallest geographic range, with the total and number of regularly occurring species indicated for each.

	Number of Parasites
Dipodomys ordii	121
D. merriami	96
D. spectabilis	35
D. microps	44

D. heermani	24
D. deserti	20
D. agilis	18
D. panamintinus	27
D. venustus	15
Perognathus flavus	29
P. parvus	65
P. longimembris	40
Chaetodipus hispidus	33
C. penicillatus	41
C. formosus	46
C. californicus	49
C. fallax	17

TABLE 6.—*Parasites from* Dipodomys ordii.

	Expected	Recorded
Viruses	5	0
Rickettsia	2	2
Bacteria	2	2
Fungi	2	0
Protozoa	8	6
Cestoda	4	1
Nematoda	10	7
Mites	29	29
Chiggers	35	27
Ticks	4	4
Lice	1	1*
Fleas	37	37
	139	116

* More lice were recorded but they appear to be obvious accidentals and were excluded.

There is a general decrease in numbers as one moves down the list for *Dipodomys,* with only *D. microps* and *D. panamintinus* being out of order. *Perognathus flavus* is out of order for that genus. For *Chaetodipus,* the trend seems reversed, with the first four species increasing in number.

This approach also is of interest; obviously many more data are needed. Differential collecting effort, rather than any interesting ecological or evolutionary relationships, may account for this trend.

Effects of Habitat on Heteromyid Parasite Populations

The question of effects of habitat as they relate to parasite community composition is of interest. Since the host is essentially the habitat of the parasite, one might expect there to be little or no effect of host habitat on the parasite population. Differing host habitats would seem to be of more importance for parasite species which spend part of their time off the host, such as ticks, chiggers or probably some of the mites, than in forms such as lice and listrophorid mites, which spend their entire life cycle on the host.

Allred (1963) studied ectoparasites of three species of heteromyids in six different habitats, and did find some differences in infestation rates within species but between habitats, as well as differences between host species. It would be interesting to examine these sorts of differences more fully in an attempt to determine causal factors.

Heteromyids as a Habitat

As mentioned previously, the host is the habitat for its parasites. All mammals have many factors in common, thus providing a similarity of available habitats (whether or not they are exploited) on all species. In addition, various hosts have their own ecological relationships and also morphological characteristics that affect the kinds and numbers of parasites involved. Heteromyids differ from many mammals in that most occur in deserts and all have external cheek pouches. This means that the parasites involved must be adapted in such a way as to avoid desiccation, at least during the life history stages that must survive outside the host. It also means that there is an additional area, the cheek pouches, that can serve as habitat for parasites.

Many parasites live their entire life in or on the host and thus have no particular problems with desiccation, especially since desert heteromyids themselves avoid desiccation and heat by living in humid burrows and being active at night. Most of the internal parasites live inside the host except for the spore or egg stage which is generally very hardy. Some of the nematodes and cestodes do occur inside an intermediate host, but again, desiccation is avoided there. Many of the ectoparasites are mites, and many of these, notably the listrophorids (*Geomylichus*), myobiids (*Radfordia bachai*), and *Fahrenholzia,* spend their entire life cycle on the host. Fleas are protected in having their eggs, larvae and pupae in the nests of the hosts. It is the chiggers (Trombiculidae), ticks (Ixodoidea), and some of the mites that need adaptations against desiccation, even though they occur in crevices, burrows and nest cavities, since the humidity is low even there. Loomis (1971) discusses some of these adaptations in chigger larvae, such as expanded body setae thought to reduce desiccation, smaller body and legs, less permeable cuticle and reduction in porous setae.

There is relatively little information on habitation of the cheek pouches of heteromyids. We believe lice, *Fahrenholzia,* are often found there. Attachment sites for chiggers of the genus *Hyponeocula* are in the cheek pouch or in a pocket or pustule on the venter (Tanigoshi and Loomis, 1974). Perhaps other chiggers are there also. We are uncertain how often systematic searches have been made for ectoparasites on the insides of cheek pouches, but this would be an interesting place to look. On the other hand, this may be too harsh an environment for many parasites because of the constant movement of food through the pouches.

Other parasites of heteromyids known from hidden biotopes are *Dipodomydectes americanus* and *Neolabidophorus verrucosus* (Glycyphagidae) from under the skin of *D. merriami,* and *Dipodomyopus tuttlei, Metalabidophorus heteromys, M. neotropicalis, M. liomys,* and *Neolabidophorus yucatanensis* from hair follicles. Also, the laelapid mites, genus *Ischyropoda,* often occur in groups at the base of the tail on their hosts.

Heteromyid Parasite Community Relationships

An attempt was made to determine overall relationships between parasite communities of different heteromyids. Similarity of parasite communities could be because of: a) taxonomic relationships of the hosts; b) ecological relationships of the hosts; c) geographic relationships of the hosts; or d) simply chance, with the most abundant parasites most often occurring on the hosts more often collected.

A table was formulated which included the parasites of those heteromyids which harbored 15 or more parasites. This table was much too large to include here, but can be obtained from the author by request.

Hosts included were *Dipodomys* (8 species), *Heteromys* (3), *Liomys* (4), *Chaetodipus* (5), and *Perognathus* (3). Many parasites were included which were found on only one host, which would seem to be indicative either of sampling artifacts or of specialization on that host, either evolutionarily or ecologically or both. The numbers of parasites occurring on only one host are indicated below.

	Species
Dipodomys merriami	3
D. ordii	1
D. panamintinus	1
Liomys pictus	2
L. irroratus	5
L. adspersus	3
L. salvini	1
Chaetodipus formosus	3
C. penicillatus	2
C. hispidus	2
Heteromys anomalus	18
H. desmarestianus	13
Perognathus flavus	1
P. parvus	1

After the parasites occurring on only one host were excluded, the following numbers of parasites were included: Protozoans (8), Cestodes (2), Nematodes (6), mites other than chiggers (17), chiggers (43), ticks (4), lice (4) and fleas (18). Thus data from 23 heteromyid hosts and 102 parasites were used the same as one might use morphological similarity to determine parasite community relationships.

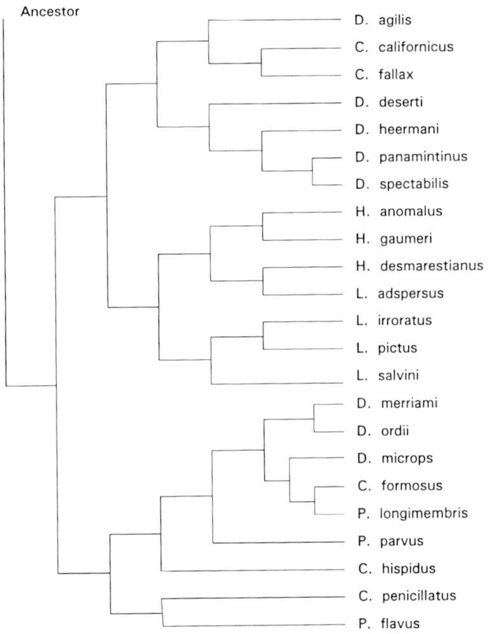

Fig. 2. Dendrogram showing relationships of heteromyid species as indicated by parasite community relationships.

The PAUP computer program was used to compare all pairs of host species and a hypothetical ancestor, using a zero character state for every character as the outgroup, and then to make a cladogram indicating those relationships (Fig. 2). The assumption was that if the cladogram from the parasite data matched the evolutionary tree of the hosts, then relationships being shown would be taxonomic. If the cladograms did not match, then one could examine the relationships to determine if they are ecological, geographical, or random, or perhaps that we are overlooking the causal factors.

There are three main stems in this dendrogram. The Heteromyinae all clustered in the middle stem, and that stem clustered roughly into two by genera, *Heteromys* and *Liomys,* except that *L. adspersus* grouped with the three *Heteromys* species rather than with the three *Liomys* species. In addition, it is on the same stem with *H. desmarestianus.* This probably was because *Liomys* is a more northern genus, centered in Mexico, whereas *Heteromys* is a more southern genus, ranging mostly south of Mexico. However, *L. adspersus* is the most southern member of its genus, occurring in Panama. In addition, *H. desmarestianus* is the only member of its genus included in the dendrogram which occurs in Panama.

The other two stems are more mixed taxonomically, and harder to explain. The upper stem includes five species of *Dipodomys* and two of *Chaetodipus,* whereas the lower includes 3 species each of *Dipodomys, Chaetodipus* and *Perognathus,* with members of a genus not even clustering together.

There does seem to be a fairly strong geographic effect, with the upper stem having species pretty much restricted to California and Baja, although *D. spectabilis* occurs further to the east. Most of the species in the lower stem occur further to the east and north and most have large ranges. *Perognathus longimembris* fits the least well into this pattern. It occurs in California and Baja, but also through most of Nevada.

Individual stems were examined. One contained *D. agilis, C. californicus* and *C. fallax.* These all occur in California and Baja. *Dipodomys ordii* and *D. merriami* occur together on one stem. These are both widespread and their ranges overlap rather broadly. Representatives of three different genera, *D. microps, C. formosus* and *P. longimembris* cluster together in one stem, but they occur essentially in the same area centering in Nevada. *Liomys irroratus, L. pictus* and *L. salvini* range through much of Central America, with *L. pictus* just to the west of *L. irroratus* and *L. salvini* to the south.

It seems clear that there are both taxonomic and geographical implications here; thus the method can be used to compare populations but one must then attempt to determine independently the meaning of the relationships.

A second approach is to develop a phenogram using characteristics of various host specific parasites. Since the parasites are host specific, they should have evolved from ancestral stocks on the host ancestor, and the evolutionary trees of the parasite and host should theoretically match. If they do not match, this could be viewed as evidence that perhaps the host or parasites are incorrectly classified, the parasites have switched to distantly related hosts, or of course the reconstruction of the cladograms could have been flawed. This approach would be particularly interesting for any of the following species groups of heteromyids: *Eimeria*, Glycyphagidae, *Echinonyssus*, *Geomylichus*, and *Fahrenholzia*.

Comparisons of Heteromyid Parasite Community with that of Other Host Species

It is difficult to make very extensive comparisons between the parasite community of heteromyids and those of other groups at this stage in our knowledge, because we have far too little summarized information for other groups.

In the deserts of southwestern North America one could compare the heteromyid community to the parasites of *Neotoma*, *Onychomys*, or *Spermophilus*. However, a much more appropriate and exceedingly interesting comparison would be to that of the Dipodidae, if the information was available. The dipodids are the Old World ecological counterparts of the heteromyids, and like the heteromyids contain a group of species and genera varying in size and niche. By examining the ecological differences of the mammals, and applying a knowledge of the parasite life histories, one might be able to develop and test hypotheses relating to reasons why certain groups of parasites are or are not present.

We do have summarized data for *Peromyscus* (Whitaker, 1968) and for *Microtus* (Timm, 1985). Information on the main parasite groupings in these genera is presented along with those for heteromyids (Table 2).

Included in this table are data from five genera of heteromyids (*Microdipodops* is omitted because there are too few data to make meaningful comparisons). The number of genera at the bottom of the table are summed for all except the Rickettsia through the Protozoa, since these groups are not covered for *Microtus* and *Peromyscus*.

The total numbers of genera are roughly similar in *Dipodomys*, *Microtus* and *Peromyscus* (42, 42 and 45), somewhat less in the other four genera of heteromyids, perhaps because they are less well studied, but perhaps because the observed differences are real. If so, the differences could be due to habitat differences, smaller body size, differences in ecology of the hosts, or perhaps to other factors.

It is of interest that the tropical species of *Liomys* and *Heteromys* have fewer genera, since tropical communities often have more taxa than temperate communities. We do not know if anyone has looked at latitudinal gradients in parasite diversity. The data in Table 2 have been further summarized for ease in examination (Table 7).

Some of the main points evident in this table are that *Microtus* has the following groups not represented in *Peromyscus* or heteromyids, Acanthocephala, myocoptids, psorergatids, and beetles; it has more cestodes, four, than the other groups; it has far fewer chigger genera, only two. It shares with *Peromyscus* having a fly, *Cuterebra*, whereas the heteromyids do not. It also shares with *Peromyscus* the trait of having far fewer chigger genera and far more flea genera than any of the heteromyids. *Peromyscus* had one dermanyssid and a macronyssid, whereas neither heteromyids nor *Microtus* did, and it lacked glycyphagids and listrophorids, which both heteromyids and *Microtus* had.

The only group that heteromyids had and that both *Peromyscus* and *Microtus* lacked was the ameroseiid mites; groups that heteromyids lacked but that were present in either *Peromyscus* or *Microtus* were mites of the families Dermanyssidae, Macronyssidae, Myocoptidae, Psorergatidae, and beetles and flies. In addition, heteromyids

TABLE 7.—*Number of genera in various groups parasitic on heteromyid genera, as compared with* Microtus *and* Peromyscus.

	Dipodomys	*Heteromys*	*Liomys*	*Chaetodipus*	*Perognathus*	*Microtus*	*Peromyscus*
Rickettsia	2			2	1	*	*
Bacteria	1				1	*	*
Fungi	1			1	1	*	*
Protozoa	3	1	1	2	1	*	*
Acanthocephala						1	
Cestoda	2			1	1	4	1
Nematoda	6			3	2	5	3
Trematoda							
Mites							
Ameroseiidae	1			1	1		
Dermanyssidae							1
Glycyphagidae	2	4	2	1		1	
Laelapidae	6	6	5	4	4	5	6
Listrophoridae	1			1	1	1	
Macronyssidae							1
Myobiidae	1					1	
Myocoptidae						2	
Psorergatidae						1	
Chiggers							
Trombiculidae	14	16	19	12	11	2	7
Ticks	3	1	2	3	2	2	2
Lice	1	1	1	1	1	2	2
Beetles						1	
Flies						1	1
Fleas	5	2	2	4	4	13	21
Subtotal	42	30	31	31	27	42	45

had far more chiggers, and far fewer flea genera than either *Peromyscus* or *Microtus.*

All three groups lacked prominent trematode genera. However, both *Microtus* and *Peromyscus* had some trematodes present, but none have been reported in heteromyids, probably because of a lack of snails in much of their environment.

Nematodes, laelapid mites, ticks and lice were all roughly similar in terms of numbers of genera in heteromyids, *Microtus* and *Peromyscus.*

Table 2 was used to look for relationships between heteromyid genera, and between heteromyids and the other genera. This relationship was summarized (Fig. 3) using a numerical taxonomy approach (PAUP) with only the regularly occurring genera (those listed in Table 2). The resulting dendrogram had three main stems, corresponding with present classification, i.e., with *Dipodomys, Chaetodipus* and *Perognathus* on one stem, *Heteromys* and *Liomys* on another, and with *Peromyscus* and *Microtus* clustering on the third stem, more distant from either of the other 2. This pattern corresponds quite well with hypothesized relationships among the heteromyid genera, and between heteromyids and other rodents.

There are many species of parasites essentially restricted to heteromyids (other than accidental or sporadic occurrence on other hosts), and there are also whole genera that can be classed essentially as heteromyid genera—for instance, *Fahrenholzia* is the only louse genus of heteromyids, and is found only there. The same is true or nearly true for the following genera of chiggers: *Dermalema, Hexidionis, Hyponeocula, Euschoengastoides,* and *Otorhinophila.* The listrophorid genus *Geomylichus* is characteristic of heteromyids and geomyids, as is one subgroup of laelapid mites of the genus *Echinonyssus.* These latter two generic connections suggest that parasites of heteromyids might follow phylogenetic relationships and show more similarity to the geomyid ectoparasite community than to any other.

Information on the ectodyte community of geomyids is given below in comparison with that of heteromyids.

Important Ectodytes

Geomyids	Heteromyids
Mites	
Androlaelaps	*Androlaelaps*
Echinonyssus	*Echinonyssus*
Haemogamasus	*Haemogamasus*
Geomylichus	*Geomylichus*
	Sertitympanum
	Eubrachylaelaps
	Hypoaspis
	Ischyropoda
	Laelaps
	Radfordia
Fleas	
Regular	
Foxella	
Dactylopsylla	
Occasional	
Orchopeas	
Diamanus	
Oropsylla	*Oropsylla*
Monopsylla	
Catallagia	
Epitedia	
Micropsylla	
Corypsylla	
	Euhoplopsyllus
	Malaraeus
	Meringis
	Rhadinopsylla

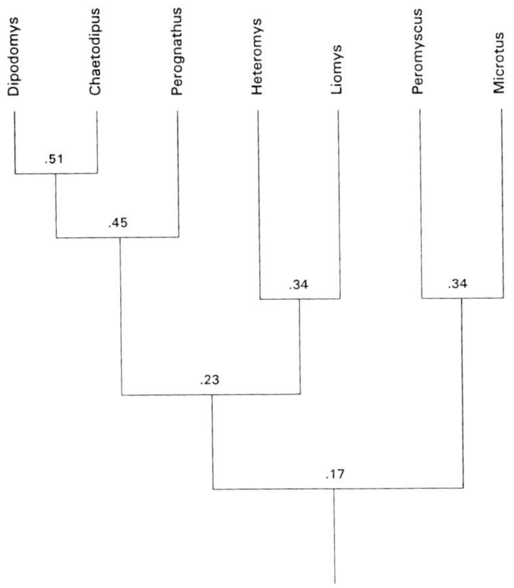

Fig. 3. Dendrogram showing parasite community relationships at generic level between heteromyids, *Microtus* and *Peromyscus.*

Lice	
Mallophaga:	Anoplura:
Geomydoecius	*Fahrenholzia*

The major similarities between the heteromyid and geomyid ectodytes are primarily in the genera *Androlaelaps, Echinonyssus,* and *Geomylichus,* but in addition the genera *Oropsylla* and *Haemogamasus* are on both. Of particular interest in the big genus *Echinonyssus,* is that the only two groups which have a posterior marginal spur on coxa II are the geomyid and heteromyid host groups.

Some of the major ectodyte groups of three heteromyid genera and the geomyids are given in Table 8. Three of the genera (*Androlaelaps, Echinonyssus* and *Geomylichus*) are found on all three of the heteromyid genera and on the geomyids. Groups found on all three of the heteromyids, however, were *Sertitympanum,* glycyphagids, *Hypoaspis* [although there is one record of *H. miles* from *Thomomys talpoides* and one of *H. lubrica* from *T. bottae* (Allred and Beck, 1966)], and *Ischyropoda* (there are records of *I. armatus* on *T. bottae* by Keegan, 1951,

TABLE 8.—*Major ectodyte community of* Dipodomys, Chaetodipus, Perognathus, *and geomyids.*

	Dipodomys	*Chaetodipus*	*Perognathus*	*Geomyids*
Sertitympanum	×	×	×	
Glycyphagids	×	×	×	
Androlaelaps	×	×	×	×
Echinonyssus	×	×	×	×
Eubrachylaelaps	×	×	×	
Hypoaspis	×	×	×	
Ischyropoda	×	×	×	
Haemogamasus	×		×	×
Geomylichus	×	×	×	×

and Miller and Ward, 1960). We do not know whether these relations are meaningful or are simply chance. We suspect the former, but if so, we do not know the meaning of the relationship. The fleas and lice show no relationship.

Parasites? Or Invertebrates Associated with Heteromyids?

This chapter is titled "Parasites"; however, it might better be titled "Invertebrate Associates of Heteromyids," since some of the species involved do not really have detrimental effects on their hosts. For example, the glycyphagid mite, *Dermacarus ornatus,* is found in the fur of *Heteromys anomalus.* The stage that inhabits the host is a hypopus or immature phoretic stage. It simply uses the host as a means of transportation. It probably drops off in the nest material of the host where it completes its life cycle. The hypopial stage does not feed; in fact the mouthparts are vestigial. Likewise mites of the genus *Sertitympanum* (Ameroseiidae) are quite common on heteromyids and other North American rodents. This could be a similar relationship, since many ameroseiids are phoretic.

Among species that more closely fit the definition of parasites, there is a great deal of variation in the degree of "harm," and one could argue in some cases that although they do use some nutrients, that actual damage to the host is minimal. This would appear true of many of the microorganisms such as *Eimeria* or *Isospora* and some of the nematodes which live naturally in the intestinal tract, as long as they are not too abundant, and do not cause or transmit diseases. Determination of the amount of nutrients usurped by the various parasites and the testing of the above hypothesis could be a fascinating topic for further work. It is also worth exploring the possibility that some "parasites" might actually benefit their hosts, perhaps by synthesizing certain nutrients that their hosts cannot produce.

On the other hand, there is little doubt of the harm and possible discomfort to the hosts caused by the ticks, lice, chiggers, and some of the other mites. However, even in these forms that naturally occur with the hosts, there is undoubtedly a degree of mutual adjustment on the part of parasite and host that minimizes harm to the host. This makes evolutionary sense, because the host is the habitat of the parasite. It is disadvantageous for the parasite to harm or cause discomfort to its host to the point that the host decreases in abundance, attempts to remove the parasite, or otherwise is adversely affected to the point that its survival and reproduction are reduced. We hypothesize that the more host-specific forms generally would cause less harm to their hosts than less host specific forms. There should be a selection among parasites for those forms which can exist in the greatest abundance yet cause the least harm and discomfort to their host.

Although the basic tendency is for an accommodation type evolution where the parasite causes little harm to the host, it does not always work this way. Parasitic forms can sometimes cause disease and even death to an individual host. This probably occurs most commonly either when the host organism is in a poor state of health, or when a parasite invades a species other than the normal host. We should not overlook this aspect of parasitism; one could write a whole book on parasitic diseases. [See Davis and Anderson's (1971) Parasitic Diseases of Wild Animals.] Parasitism as a host population control mechanism has been overemphasized, but this and modification of individual host behavior may sometimes occur.

Life Strategies

Ectodytes and endodytes, like other organisms, have adopted differing life history strategies. Some of these strategies are listed below:

a) Host-specific forms.—Some forms are very host specific. These include lice, especially those of the genus *Fahrenholzia* of heteromyids, and myobiid and listrophorid mites. Highly host-specific forms such as *Fahrenholzia* or *Geomylichus* (Listrophoridae) remain with their host throughout their life cycle. They reproduce on the host, and all life stages are found there. They have a low number of offspring and they evolve adaptations to better live on their hosts. As suggested above, they may have less tendency to damage their hosts than other less host-specific forms.

b) Habitat-specific forms.—Chiggers tend to be habitat specific; i.e., they live in a rather specific habitat and thus are likely to be picked up by the particular hosts living in that habitat and appear to be host specific. However, if another host does enter the habitat, chiggers will probably parasitize it as well. Since they may utilize any of the hosts in the area, they remain rather generalized in their characters. They tend to have higher numbers of offspring and have not evolved adaptations strongly fitting them for life on one particular host.

c) Non-host-specific forms.—Non-host-specific forms remain unspecialized rather than evolving to fit the form and ecology of any one particular host species. They have larger numbers of young and usually only one life stage occurs on any one host. *Androlaelaps fahrenholzi,* a large laelapid mite, is not host specific at all. It is known from well over a hundred different North American mammalian host species. Many of the fleas, likewise, are not very host specific, although they may be most often associated with certain hosts.

d) Phoretic forms.—Phoretic forms are those which use the host mainly as a means of transportation. Glycyphagid mites fall into this category. They usually drop off in the nest where they may change to the adult state and reproduce.

e) Accidentals.—All of the above are of regular occurrence on their respective hosts. However, if many hosts of a species are examined, one will find some species which are only of sporadic occurrence. They are inhabitants of other hosts. These are not really a regular part of the community, but are termed accidentals.

What Impact do Parasites Have on their Hosts?

We have indicated various effects of the different kinds of parasites in the group accounts, although little is specifically known of this in heteromyids. Price (1980) gives some detail on this in several broad categories, but the effects are many and varied since the numbers of both parasites and hosts are so large. Parasite infestations may influence or lead to behavior modifications such as grooming and preening, group defense, nest design, and sociality. Many diseases or other disorders are caused by parasites, and man has greatly increased these possibilities, by moving domestic and wild animals about, and by otherwise causing environmental disturbance. Davis and Anderson (1971) summarized much information on parasitic diseases of wild animals. Although there is often a general accommodation between parasite and host populations, it does not always hold, and parasite impact on host populations at times may be heavy and varied. However, little information is available on this topic in heteromyids and we will not explore it here.

Speciation and Subspeciation in Parasites

The idea of a biological subspecies concept has long been supported (Whitaker, 1970). One would use two criteria for separation of subspecies, (a) the presence of a primary isolating mechanism, i.e., any factor which prevents gene flow between two populations (most often geographical) and (b) evidence that evolution is proceeding, i.e., morphological or other evidence of differentiation.

In free-living forms these conditions are most commonly met when a species becomes divided geographically (i.e., by a geographical primary isolating mechanism) into separate populations that cannot interbreed. Once isolated, speciation can commence, but in the early stages (before secondary isolating mechanisms are established) the two forms would freely interbreed if reunited. Thus it is at this stage, when primary isolating mechanisms are in place but secondary isolating mechanisms have not yet been established (although they may be developing), that we would establish subspecies. Evolution can then go either way depending upon events, the two subspecies can merge if the primary isolating mechanisms break down, or if they remain in place long enough, speciation can occur.

Parasitologists in the past have often described new species on the basis of rather minor variations between closely related parasites on different host species. It might be better to use the subspecific category in this situation more often. The differing hosts would act as the primary isolating mechanisms, thus allowing morphological variation to evolve. The morphological differences themselves would serve as evidence of lack of gene flow. There are some parasites, such as the laelapid mite, *Androlaelaps fahrenholzi,* which, although considered as a single species, may consist of a species complex with different morphological variants on different hosts. We suspect that some of these too could best be treated as subspecies.

We have given much thought to sympatric speciation and can envision some ways it could occur, most simply through an immediate and sufficiently large genetic change that it can act as both primary and secondary isolating mechanism at the same time. However, this should not happen very often under normal circumstances. Other than through this sort of "macromutation," it is difficult to envision the formation and maintenance of primary isolating mechanisms which would separate members of the same species at the same location into separate groups. If they are members of the same species in the same area they will presumably occupy the same habitat and behave in the same way, thus should freely interbreed and lose their separate identities if any had been established.

However, with parasites another factor has been inserted. The habitat of parasites is their host. Especially the more host-specific species remain permanently with their host. Thus, if a parasite can establish residence on a new host in the same area, a primary isolating mechanism has been established, and speciation can then proceed as surely as if an allopatric situation existed.

This would appear to be the way that speciation might often occur in parasites. Examples from heteromyids might be the genera *Fahrenholzia, Geomylichus* and *Eimeria,* where it appears that we have closely related host-specific forms in different species with some of them remaining sympatric geographically, sometimes inhabiting different genera of hosts.

The more traditional means of viewing closely related host specific species in different hosts is that the species, both hosts and parasites, evolved and speciated together from common ancestors.

The mechanism discussed above gives another possibility, with the various host species thus acting as islands, and as in Darwin's Finches, there can be migration, sub-

TABLE 9.—*Distribution of species of* Eimeria *among species of heteromyids.*

	Dipodomys							
	ordii	*merriami*	*micropus*	*spect-abilis*	*pana-mintinus*	*agilis*	*phillipsii*	*gravipes*
Eimeria								
balphae	×	×		×		×		
chihuahuaensis		×						
chobotari	×	×	×			×		
dipodomysis	×						×	
mohavensis					×			
scholtysecki	×			×	×	×		×
utahensis	×	×	×			×		

	Liomys	
	pictus	*irroratus*
liomysis	×	×
picta	×	

	Chaetodipus			
	intermedius	*penicillatus*	*formosus*	*flavus*
merriami	×			
penicillati		×		×
perognathi	×			
reedi			×	

speciation, speciation, then reinvasion of a host, all in the same area. This may well be what has happened at least in part in *Fahrenholzia, Geomylichus* and *Eimeria* in heteromyids. For example, several species of *Eimeria* occur in heteromyids as in Table 9.

We have no way of knowing which of the species of *Eimeria* might be primitive, and there is no species of *Eimeria* shared between genera. However, let us consider the species in *Dipodomys,* seven in all. Two of them, *E. chihuahuaensis* of *D. merriami,* and *E. mohavensis* of *D. panamintinus* are found in only one species thus presumably evolved there. (Of course one of these might turn up in other species not yet examined, a factor affecting all of this discussion.)

There are five additional species of *Eimeria.* If we assume that each evolved in a separate host species (not necessarily the case), then we have the following potential hosts left, *D. ordii, microps, spectabilis, agilis, phillippsi* and *gravipes.* One of them, *E. dipodomysis,* is in two species, *D. ordii* and *D. phillipsi.* These two species share the southern part of the *ordii* range but *phillipsi* has a limited range. A likely possibility here is that *E. dipodomysis* evolved in *D. phillippsi* and later invaded *D. ordii,* or vice versa.

This leaves us with four species of *Eimeria* (*balphae, chobotari, scholtysecki* and *utahensis*), any one of which could have evolved in *D. ordii* or *D. agilis,* but the ranges of these two at least now are not sympatric. However, three of these *Eimeria* occur also in *D. merriami* which overlaps the ranges of both, whereas *E. scholtysecki* might have passed from one to the other through *D. spectabilis* and *D. panamintinus.* Either *E. chobotari* or *utahensis* might have evolved in *D. microps,* and either *balphae* or *scholtysecki* might have evolved in *D. spectabilis,*

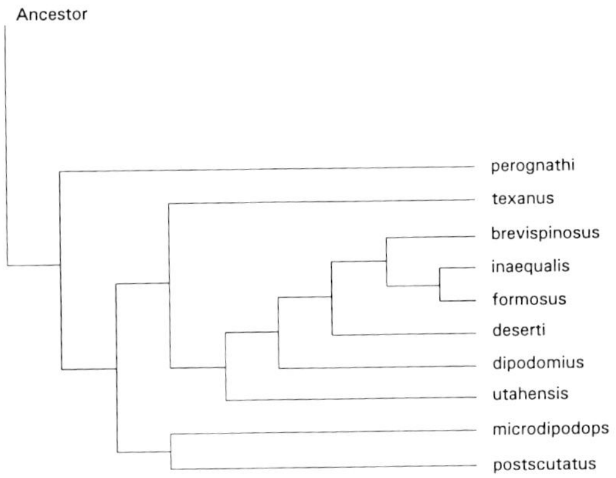

Fig. 4. Phenogram showing *Geomylicus* relationship using 23 female and 22 male characters, coded into discrete categories.

provided our initial assumption that one *Eimeria* evolved per host. However, it might not be that simple. The "island hopping" process might be in effect, yet one *Eimeria* could move from one host to another, undergo speciation there, then later move back to the first and again undergo speciation, creating three species in two hosts. Much more information on relationships of both hosts and parasites is needed before we can go very far in this direction. However, we have produced a phenogram using the heteromyid species used by *Geomylichus* (Fig. 4). It did not fit the heteromyid evolutionary tree very well. This may be because the species of *Geomylichus* moved between host species and even genera where the hosts were sympatric, then underwent speciation there, with the differential hosts acting as primary isolating mechanisms.

The situation is simpler in *Liomys* if we assume one species per host. *Eimeria picta* must have evolved in *L. pictus,* and *E. liomysis* in *L. irroratus,* but then the latter parasite invaded *L. pictus* after speciation had been completed. In *Chaetodipus,* more information is needed before much can be said. There are of course additional possibilities, especially involving reinvasion. Examination of additional hosts will undoubtedly throw additional light on the situation. We suspect that this is at least the general way in which evolution proceeded in this group.

Many authors have tended to think of parasites as overspecialized evolutionary dead ends. We agree with Price (1980) who expressed a different view in the introduction to his book, *Evolutionary Biology of Parasites.* He argued that "no group of organisms on this earth can surpass the parasites in their potential for continued adaptive radiation." The above discussion of possible evolutionary pathways in heteromyid parasite evolution certainly represents a dynamic evolutionary scenario.

Numbers of Parasites Known per Host

There is no information at all on parasites for nine of the 58 heteromyid species (Table 1): *Dipodomys compactus, insularis, margaritae*; *Heteromys nelsoni, oresterus*; *Liomys spectabilis*; *Chaetodipus lineatus*; *Microdipodops pallidus*; and *Perognathus alticola.*

There are only one to three (one unless otherwise indicated) parasites recorded for another twelve heteromyids: *D. californicus* (3), *D. elator* (2), *elephantinus, gravipes, ingens, nelsoni, stephensi* (3); *Heteromys goldmani*; *C. arenarius*; *Perognathus amplus, inornatus* (2), and *xanthonotus*; and minimal data (15 or less) for another 14 species, *Dipodomys nitratoides* (10), *phillipsii* (4), *venustus* (15); *H. australis* (12); *C. artus* (8), *baileyi* (13), *goldmani* (8), *intermedius* (10), *nelsoni* (11), *pernix* (9), *spinatus* (8), *Microdipodops megacephalus* (7), *Perognathus fasciatus* (7), and *flavescens* (7).

Thus we have no more than minimal information on the parasites and other associates of the majority of species of heteromyids with 15 records or less for 35 of the 58 (60%) heteromyid species.

There are only 11 species of the 58 (19%) for which there is anything even approaching adequate information as follows: *Dipodomys merriami* (96 species), *microps* (44), *ordii* (121); *Heteromys anomalus* (50),

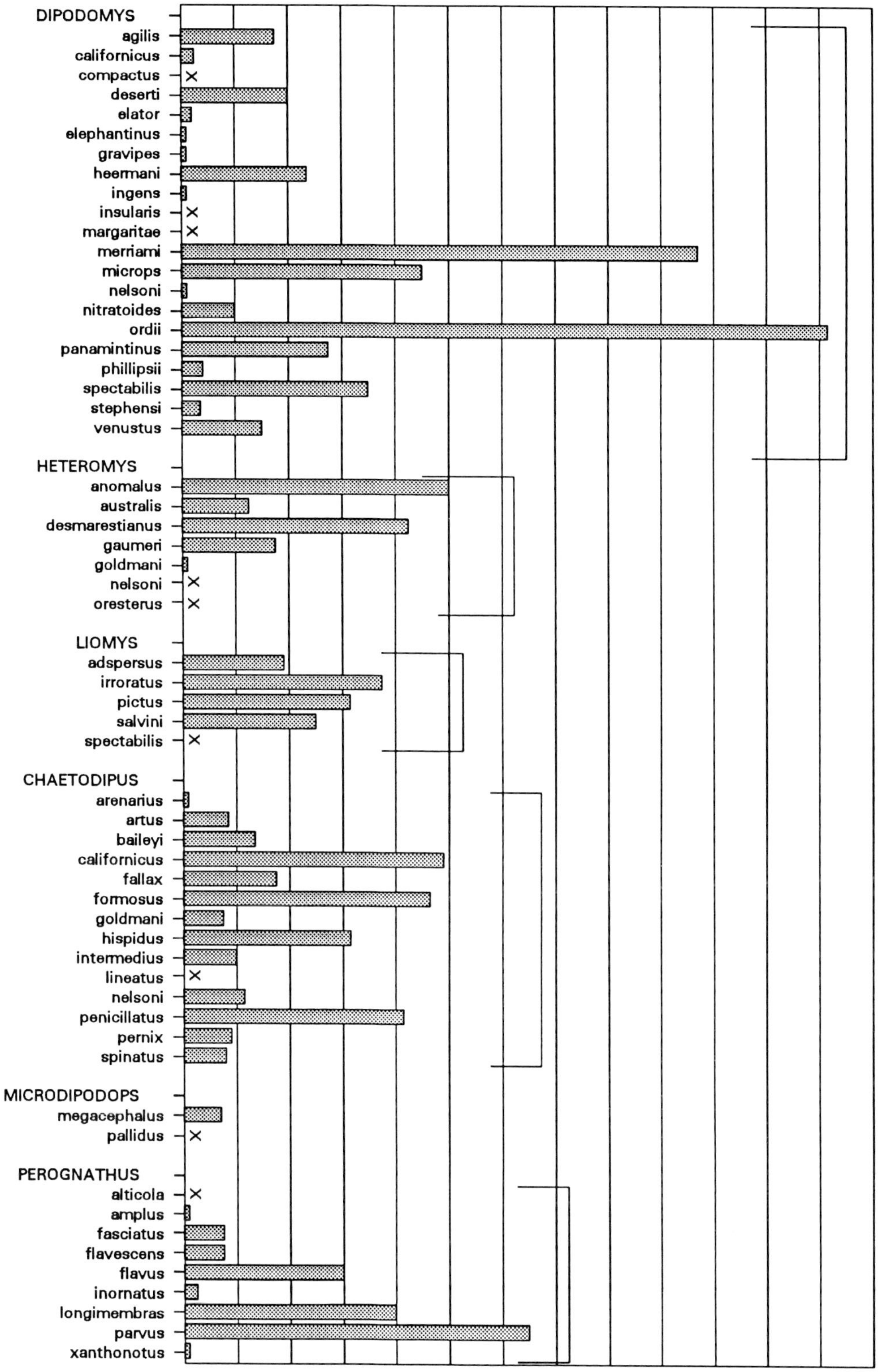

Fig. 5. Numbers of species of parasites reported from heteromyid hosts. Solid lines over genera indicate hypothesized numbers of parasites present. An × indicates that no parasites have been found on a particular host. Each horizontal line indicates ten species.

desmarestianus (41); *Liomys irroratus* (37); *Chaetodipus californicus* (49), *formosus* (47), *penicillatus* (41); *Perognathus longimembris* (40), *parvus* (65). It is primarily on these that we can begin to develop some tentative ecological generalizations.

The number of parasites per species is shown graphically in Fig. 5. The obvious reason so few species are known from most of the species is lack of collecting effort.

Areas for Further Study

(1) There is *no* parasite information on nine of the 58 heteromyid species, and only minimal information on over 60% of them. A thorough investigation of the internal and/or external parasite communities of any of these would be fruitful.

(2) There is little information on mites of hidden biotopes of heteromyids. A few nasal chiggers and follicular hypopi are known, but a systematic examination of hidden biotopes (eye sockets, ear canals, nasal passages, follicles, cheek pouches, meibomian glands, skin) should prove interesting. So few of these are known, and their occurrence on other rodents is so sporadic, that it would be difficult to predict what sorts of mites might be found. Except for some of the nasal chiggers and a few follicular mites, most of those found would undoubtedly be new species.

(3) Only females of the ameroseiid mites, *Sertitympanum,* have been found. It would be interesting to collect males and other life stages. Also, the relationship of *Sertitympanum* to heteromyids is unknown. What is the ecology of these species? Are they parasitic, predaceous, phoretic, or just there by accident? What is their life cycle?

(4) Hypopi of adults of *Dermacarus ornatus* and perhaps other glycyphagids could be kept in petri dishes and observed for other life stages. Their life cycles could be determined and the various stages could then be described.

(5) Much more information is desirable on the comparative ecological conditions and life history strategies that enable chiggers and many of the other parasites to exploit heteromyids.

(6) Much more information on ecological relationships between host and parasite would be desirable for nearly all of the parasite species. Where and when do they reproduce? How do they disperse? How do they manage to get on or in the heteromyids? How do they interrelate with the host? Do they feed on blood or other body juices, dead skin or detritus, or any other part of the host? What regulates their population size? Do they cause any harm to the host? Are they of any advantage to the host? What is their relationship to other organisms of the host? Are they on the host throughout its range, in certain habitats, and at restricted times of the year?

(7) Much more information is desirable on adaptations which enable those ectoparasites of desert heteromyids (and other desert forms) which have a free living stage to avoid desiccation.

(8) Much more information is needed on the listrophorid mites (genus *Geomylichus*) of heteromyids. To date, 13 species have been described, and it is suspected that there are more.

(9) Comparison of parasites between genera, species, subspecies, and populations of hosts in relation to taxonomic relationships and to breadth and completeness of isolating mechanisms would be fascinating for *Eimeria* (see discussion under Protozoa) as well as for many other parasite groups such as *Geomylichus, Echinonyssus,* and *Fahrenholzia.*

(10) Transfer experiments to determine degrees of host specificity between species and genera of heteromyids and perhaps even with cricetid rodents would be of interest, especially in *Eimeria, Geomylichus, Echinonyssus* and *Fahrenholzia.*

(11) Attempts at crossing closely related and less closely related species of *Geomylichus* should indicate the relative strength of the various secondary isolating mecha-

nisms and should be fascinating for the evolutionary biologist.

(12) Individuals with large numbers of parasites might be compared to those with small numbers to see if their overall condition is poor (healthwise) and to determine the relative amount of "harm" (energy loss) by those in "poor" versus "good" condition.

(13) We have postulated population control of parasites through low reproductive capacity coupled with periodic population reduction through death of the host individuals. It should be instructive to attempt to obtain adequate numerical data on reproduction and survival of parasite species to see if they are consistent with these ideas.

(14) Yet to be done on parasite-host energy flow are modelling studies such as that of Klekowski and Uchmanski (1980) on the non-parasitic mite, *Rhizoglyphus echinopus*. This type of study on some parasites would be a big step towards understanding how parasites affect the energy budget of their hosts and ultimately assessing the extent to which they harm the host.

Acknowledgments

Robert M. Timm and James H. Brown have read the entire manuscript and it has been greatly improved as a result of their suggestions. Robert F. Bozarth has made a similar contribution to the sections on microorganisms. Walter M. Shriner ran the PAUP computer programs. Chris Gummer made the drawing of the washing apparatus for collecting parasites. I also thank Laura Bakken who not only typed, but also greatly improved, the paper in numerous ways. I greatly appreciate the efforts of these individuals.

Literature Cited

1. Allred, D. M. 1952. Plague important fleas and mammals in Utah and the western United States. The Great Basin Naturalist, 12:67–75.
2. ———. 1963. Mites from pocket mice at the Nevada test site (Acarina). Proceedings of the Entomological Society of Washington, 65:231–233.
3. ———. 1965. Clarification of type data for *Ischyropoda furmani* Keegan. The Journal of Parasitology, 51:604.
4. ———. 1968*a*. Ticks of the National Reactor Testing Station. Brigham Young University Science Bulletin: Biological Series, 10(1):1–29.
5. ———. 1968*b*. Fleas of the National Reactor Testing Station. The Great Basin Naturalist, 28: 73–87.
6. ———. 1970. Mites and lice of the National Reactor Testing Station. Brigham Young University Science Bulletin: Biological Series, 12(1): 1–17.
7. Allred, D. M., and D E. Beck. 1966. Mites of Utah mammals. Brigham Young University Science Bulletin: Biological Series, 8(1):1–123.
8. Allred, D. M., D E. Beck, and L. D. White. 1960. Ticks of the genus *Ixodes* in Utah. Brigham Young University Science Bulletin: Biological Series, 1(4):1–42.
9. Allred, D. M., and M. A. Goates. 1964. Mites from mammals at the Nevada test site. The Great Basin Naturalist, 24:71–73.
10. Arnold, L. W. 1942. Notes on the life history of the sand pocket mouse. Journal of Mammalogy, 23:339–341.
11. Ashburn, L. L., and C. W. Emmons. 1942. The pathology of spontaneous coccidioidal granuloma in the lungs of wild rodents. American Journal of Pathology, 18:753–755.
12. Augustson, G. F. 1955. Records of fleas (Siphonaptera) from the Pacific Southwest. Bulletin of the Southern California Academy of Science, 54:36–39.
13. Augustson, G. F., and F. E. Durham. 1961. Records of fleas (Siphonaptera) from northwestern Arizona. Bulletin of the Southern California Academy of Science, 60:100–105.
14. Bacon, M., C. H. Drake, and N. G. Miller. 1959. Ticks (Acarina: Ixodoidea) on rabbits and rodents of eastern and central Washington. The Journal of Parasitology, 45:281–286.
15. Barnes, A. M., V. J. Tipton, and J. A. Wildie. 1977. The subfamily Anomiopsyllinae (Hystrichopsyllidae: Siphonaptera). I. A revision of the genus *Anomiopsyllus* Baker. The Great Basin Naturalist, 37:138–206.
16. Beck, D E. 1955. Some unusual distributional records of ticks in Utah. The Journal of Parasitology, 41:198–201.
17. Beck D E., and D. M. Allred. 1966. Siphonaptera (fleas) of the Nevada test site. Brigham Young University Science Bulletin: Biological Series, 7(2):1–27.
18. Beer, J. R., E. F. Cook, and R. G. Schwab. 1959. The ectoparasites of some mammals from the Chiricahua Mountains, Arizona. The Journal of Parasitology, 45:605–613.
19. Bell, J. F. 1981. Q Fever. Pp. 388–397, *in* Infectious diseases of wild mammals (J. W. Davis, L. H. Karstad, and D. O. Trainer, eds.). Second ed. Iowa State University Press, Ames, 446 pp.
20. Bell, J. F., and J. R. Reilly. 1981. Tularemia. Pp. 213–231, *in* Infectious diseases of wild mammals (J. W. Davis, L. H. Karstad, and D. O.

Trainer, eds.). Second ed. Iowa State University Press, Ames, 446 pp.
21. BIENEK, G. K., AND A. W. GRUNDMANN. 1973. A new tapeworm, *Schizorchodes dipodomi* gen. et sp. n. (Cestoda: Anoplocephalidae), from the Merriam kangaroo rat, *Dipodomys merriami vulcani*. Proceedings of the Helminthological Society of Washington, 40:192–195.
22. ———. 1974. *Catenotaenia utahensis* sp. n. (Cestoda: Catenotaeniidae) from the Merriam kangaroo rat, *Dipodomys merriami vulcani* in Utah. Proceedings of the Helminthological Society of Washington, 41:134–139.
23. BIENEK, G. K., AND L. G. KLIKOFF. 1974. Parasitological evidence of arthropods as food for *Dipodomys merriami vulcani*. The American Midland Naturalist, 91:251–253.
24. BISHOPP, F. C. 1911. Some new North American Ixodidae with notes on other species. Proceedings of the Biological Society of Washington, 24:197–208.
25. BISHOPP, F. C., AND H. L. TREMBLEY. 1945. Distribution and hosts of certain North American ticks. The Journal of Parasitology, 31:1–54.
26. BOWERS, M. A., AND J. H. BROWN. 1982. Body size and coexistence in desert rodents: chance or community structure? Ecology, 63:391–400.
27. BRENNAN, J. M. 1952. Two new Venezuelan chiggers (Acarina: Trombiculidae). The Journal of Parasitology, 38:143–146.
28. ———. 1957. Revival of *Crotiscus* Ewing, 1944, and descriptions of four new species of the genus from Peruvian rodents (Acarina: Trombiculidae). The Journal of Parasitology, 43:673–680.
29. ———. 1960*a*. *Ectonyx*, a new neotropical genus of chiggers (Acarina, Trombiculidae). Acarologia, 2:88–91.
30. ———. 1960*b*. Eight new species of *Pseudoschoengastia* from Mexico and Panama with a revised key to the species (Acarina: Trombiculidae). Acarologia, 2:480–492.
31. ———. 1962. Four new chiggers from Mexico. The Journal of Parasitology, 48:618–620.
32. ———. 1965. Two new species and other records of chiggers from Texas (Acarina: Trombiculidae). Acarologia, 7:79–83.
33. ———. 1968*a*. Endoparasitic chiggers: IV. Revision of *Kymocta* Yunker and Brennan, 1962 (Acarina: Trombiculidae), with description of a new species. The Journal of Parasitology, 54:613–615.
34. ———. 1968*b*. A collection of chiggers (Acarina: Trombiculidae) from rodents in southwestern Colombia. The Journal of Parasitology, 54:679–685.
35. ———. 1969. Three new species of subgenus *Parasecia* Loomis (genus *Fonsecia*) from northeastern Brazil and a key to the included species (Acarina: Trombiculidae). The Journal of Parasitology, 55:662–666.
36. ———. 1971. *Anomalaspis ambiguus* Brennan, 1952 redescribed and the questional association of South American species with *Pseudoschoengastia* (Acarina: Trombiculidae). Journal of Medical Entomology, 8:602–604.
37. ———. 1973. Revival of *Vanidicus* Brennan and Jones, 1961, descriptions of two new species of the genus from Venezuela, and notes on related taxa (Acarina, Trombiculidae). Journal of Medical Entomology, 10:107–109.
38. BRENNAN, J. M., AND D E. BECK. 1955. The chiggers of Utah. The Great Basin Naturalist, 15: 1–26.
40. BRENNAN, J. M., AND H. T. DALMAT. 1960. Chiggers of Guatemala (Acarina: Trombiculidae). Annals of the Entomological Society of America, 53:183–191.
41. BRENNAN, J. M., AND M. L. GOFF. 1977*a*. The neotropical genus *Hoffmannina*: four new species and other records from Mexico, Panama and Venezuela (Acarina: Trombiculidae). The Journal of Parasitology, 63:908–914.
42. BRENNAN, J. M., AND M. L. GOFF. 1977*b*. Keys to the genera of chiggers of the western hemisphere (Acarina: Trombiculidae). The Journal of Parasitology, 63:554–566.
43. BRENNAN, J. M., AND M. L. GOFF. 1978*a*. Two new species of *Crotiscus* (Acarina: Trombiculidae) from northern South America. Journal of Medical Entomology, 14:569–572.
44. BRENNAN, J. M., AND E. K. JONES. 1954. A report on the chiggers (Acarina: Trombiculidae) of the Frances Simes Hastings Natural History Reservation, Monterey County, California. Wasmann Journal of Biology, 12:155–194.
45. ———. 1959. *Pseudoschoengastia* and four new neotropical species of the genus (Acarina: Trombiculidae). The Journal of Parasitology, 45:421–429.
46. ———. 1960. Chiggers of Trinidad, B.W.I. (Acarina: Trombiculidae). Acarologia, 2:493–540.
47. ———. 1961. New genera and species of chiggers from Panama (Acarina: Trombiculidae). The Journal of Parasitology, 47:105–124.
48. BRENNAN, J. M., AND J. T. REED. 1972. Two new Venezuelan chiggers of the genus *Polylopadium* (Acarina: Trombiculidae). Journal of Medical Entomology, 9:461–463.
49. ———. 1973. More new genera and species of chiggers (Acarina: Trombiculidae) from Venezuela. The Journal of Parasitology, 59:706–710.
50. ———. 1974. The genus *Eutrombicula* in Venezuela (Acarina: Trombiculidae). The Journal of Parasitology, 60:699–711.
51. ———. 1975. A list of Venezuela chiggers, particularly of small mammalian hosts (Acarina: Trombiculidae). Brigham Young University Science Bulletin: Biological Series, 20(1):45–75.
52. BRENNAN, J. M., AND J. E. M. H. VAN BRONSWIJK. 1973. Endoparasitic chiggers. VII. A redescription of *Kymocta* Yunker and Brennan, 1962, a new species from Venezuela, and other records (Acarina: Trombiculidae). Journal of Medical Entomology, 10:449–451.
53. BRENNAN, J. M., AND C. E. YUNKER. 1966. The chiggers of Panama (Acarina: Trombiculidae). Pp. 221–266, *in* Ectoparasites of Panama (R. L. Wenzel and V. J. Tipton, eds.). Field Museum of Natural History, Chicago, 861 pp.
54. BRENNAN, J. M., AND C. E. YUNKER. 1969. En-

doparasitic chiggers: V. New genera, species and records from Venezuela and Brazil (Acarina: Trombiculidae). Journal of Medical Entomology, 6:299–304.

55. Caballero y C., E. 1959. Estudios helmintologicos de la region oncocercosa de Mexico y de la Republica de Guatemala. Nematoda. 10a parte. Anales de la Escuela Nacional de Ciencias Biologicas (Mexico), 9:61–76.

56. Carter, C. H., and H. H. Genoways. 1978. Liomys salvini. Mammalian Species, 84:1–5.

57. Carter, D. C., W. D. Webster, J. K. Jones, Jr., C. Jones, and R. D. Suttkus. 1985. Dipodomys elator. Mammalian Species, 232:1–3.

58. Casebeer, R. S. 1965. Seasonal fluctuations in ectoparasite populations of desert rodents. I. Siphonaptera. Bulletin of the Southern California Academy of Science, 64:75–88.

59. Chandler, A. C. 1945. *Trichuris* sp. from California rodents. The Journal of Parasitology, 31: 284–287.

60. ———. 1946. *Trichuris peromysci* n. sp. from *Peromyscus californicus,* and further notes on *T. perognathi* Chandler, 1945. The Journal of Parasitology, 32:208.

61. Chobotar, B., L. C. Anderson, J. V. Ernst, and D. M. Hammond. 1970. Pathogenicity of *Besnoitia jellisoni* in naturally infected kangaroo rats (*Dipodomys ordii*) in northwestern Utah. The Journal of Parasitology, 56:192–193.

62. Clark, G. M. 1967. New Speleognathinae from Central and South American mammals (Acarina, Trombidiformes). Proceedings of the Helminthological Society of Washington, 34:240–243.

63. Coffey, M. C. 1954. A study of some Rocky Mountain spotted fever vectors and their hosts in Utah. The Great Basin Naturalist, 14:31–37.

64. Cooley, R. A., and G. M. Kohls. 1945. The genus *Ixodes* in North America. National Institutes of Health Bulletin, 184:1–246.

65. Cornett, J. W. 1980. A possible parasitic lepidopteran. The Journal of Parasitology, 66:149.

66. Coultrip, R. L., R. W. Emmons, L. J. Legters, J. D. Marshall, Jr., and K. F. Murray. 1973. Survey for the arthropod vectors and mammalian hosts of Rocky Mountain Spotted Fever and plague at Fort Ord, California. Journal of Medical Entomology, 10:303–309.

67. Crofton, H. D. 1971. A model of host-parasite relationships. Parasitology, 63:343–364.

68. da Costa Lima, A., and C. R. Hathaway. 1946. Pulgas: Bibliografia, catalogo e animais por elas sugados. Monografias do Instituto Oswaldo Cruz Rio de Janeiro, 4:1–522.

69. Davis, D. R., D. H. Clayton, D. H. Janzen, and A. P. Brook. 1986. Neotropical Tineidae, II: Biological notes and descriptions of two new moths phoretic on spiny pocket mice in Costa Rica (Lepidoptera: Tineoidea). Proceedings of the Entomological Society of Washington, 88:98–109.

70. Davis, J. W. 1981. Coccidiosus, Pp. 361–365, *in* Infectious diseases of wild mammals (J. W. Davis, L. H. Karstad, and D. O. Trainer, eds.). Second ed. Iowa State University Press, Ames, 446 pp.

71. Davis, J. W., and R. C. Anderson. 1971. Parasitic diseases of wild mammals. Iowa State University Press, Ames, 364 pp.

72. DeLoach, J. R. 1985. Precision electronic counting of *Psoroptes ovis* (Acari: Psoroptidae). Journal of Medical Entomology, 22:18–21.

73. Disney, R. H. L. 1964. Visceral involvement with dermal leishmaniasis in a wild-caught rodent in British Honduras. Transactions of the Royal Society of Tropical Medicine and Hygiene, 58:581.

74. Doran, D. J. 1953*a*. The occurrence of *Tritrichomonas muris* (Grassi) in heteromyid and cricetid rodents of California. The Journal of Parasitology, 39:569.

75. ———. 1953*b*. A note on *Giardia duodenalis* var. *perognathi* Filice and *Endamoeba* (=*Entamoeba*) *dipodomysi* Hegner in California rodents. The Journal of Parasitology, 39:668.

76. ———. 1953*c*. Coccidiosus in the kangaroo rats of California. University of California Publications in Zoology, 59(3):31–60.

77. ———. 1954*a*. A catalogue of the Protozoa and helminths of North American rodents. I. Protozoa and Acanthocephala. The American Midland Naturalist, 52:118–128.

78. ———. 1954*b*. A catalogue of the Protozoa and helminths of North American rodents. II. Cestoda. The American Midland Naturalist, 52:469–480.

79. ———. 1955*a*. A catalogue of the Protozoa and helminths of North American rodents. III. Nematoda. The American Midland Naturalist, 53:162–175.

80. ———. 1955*b*. A catalogue of the Protozoa and helminths of North American rodents. IV. Trematoda. The American Midland Naturalist, 53:446–454.

81. Doran, D. J., and T. L. Jahn. 1952. Preliminary observations on *Eimeria mohavensis* n. sp. from the kangaroo rat *Dipodomys panamintinus mohavensis* (Grinnell). Transactions of the American Microscopical Society, 71:93–101.

82. Dowell, A. M. 1953. *Catenotaenia californica,* sp. nov., a cestode of the kangaroo rat, *Dipodomys panamintinus mohavensis.* The American Midland Naturalist, 49:738–742.

83. Dowler, R. C., and H. H. Genoways. 1978. *Liomys irroratus.* Mammalian Species, 82:1–6.

84. Durette-Desset, M.-C. 1970. Description de *Vexillata petteri* n. sp. nematode heligmosome parasite d'un rongeur nearctique. Annales Parasitologie Humaine Comparee, 45(3):289–293.

85. Eads, R. B. 1950. The fleas of Texas. Texas State Health Department, Austin, 85 pp.

86. ———. 1951. New mites of the genus *Androlaelaps* Berlese. The Journal of Parasitology, 37: 212–216.

87. ———. 1960. Two new fleas (Hystrichopsyllidae) from the kangaroo rat and pocket mouse. The Journal of Parasitology, 46:213–218.

88. ———. 1978. Two new species of fleas of the genus *Meringis* (Siphonaptera: Hystrichopsyllidae). The Great Basin Naturalist, 38:447–455.

89. Eads, R. B., and B. G. Hightower. 1951. A

new *Neoichoronyssus* from the pack rat, *Neotoma micropus* (Acarina, Laelaptidae). Proceedings of the Entomological Society of Washington, 53:295–298.

90. ———. 1952*a*. A new mite of the genus *Neoichoronyssus* (Acarina, Laelaptidae). Proceedings of the Entomological Society of Washington, 54: 32–35.

91. ———. 1952*b*. Blood parasites of Southwest Texas rodents. The Journal of Parasitology, 38: 89–90.

92. Eads, R. B., and G. C. Menzies. 1949. A preliminary list of the Siphonaptera of Texas. Texas Journal of Science, 1:33–39.

93. Eads, R. B., G. C. Menzies, and B. G. Hightower. 1956. The ticks of Texas, with notes on their medical significance. Texas Journal of Science, 8:7–24.

94. Eads, R. B., G. C. Menzies, and V. I. Miles. 1952. Acarina taken during West Texas plague studies. Proceedings of the Entomological Society of Washington, 54:250–253.

95. Eads, R. B., H. A. Trevino, and E. G. Campos. 1965. Ectoparasites of the spiny pocket mouse, *Liomys irroratus texensis* Merriam. The Southwestern Naturalist, 10:17–21.

96. Egoscue, H. J. 1966. New and additional host-flea associations and distributional records of fleas from Utah. The Great Basin Naturalist, 26:71–75.

97. ———. 1976. Flea exchange between deer mice and some associated small mammals in western Utah. The Great Basin Naturalist, 36:475–480.

98. Ellis, L. L., Jr. 1955. A survey of the ectoparasites of certain mammals in Oklahoma. Ecology, 36:12–18.

99. Elsen, P., and J. O. Whitaker, Jr. 1985. *Sertitympanum,* a new genus of Ameroseiidae (Acarina, Mesostigmata) taken from rodents in the United States: including descriptions of three new species in the genus. Acarologia, 26:117–122.

100. Elzinga, R. J. 1964. The importance of the Berlese technique in studying ectoparasite populations upon rodent hosts. Journal of the Kansas Entomological Society, 37:52–56.

101. Emerson, K. C. 1971*a*. Records of a collection of Mallophaga and Anoplura from Nicaraguan mammals. Journal of the Kansas Entomological Society, 44:332–334.

102. ———. 1971*b*. New records of Anoplura from Mexico. Journal of the Kansas Entomological Society, 44:374–377.

103. Emmons, C. W. 1942. Isolation of *Coccidioides* from soil and rodents. Public Health Report, 57: 109–111.

104. Emmons, C. W., and L. L. Ashburn. 1942. The isolation of *Hoplosporangium parvum* n. sp. and *Coccidioides immitis* from wild rodents. Public Health Report, 57:1715–1727.

105. Ernst, J. V., and B. Chobotar. 1978. The endogenous stages of *Eimeria utahensis* (Protozoa: Eimeriidae) in the kangaroo rat, *Dipodomys ordii.* The Journal of Parasitology, 64:27–34.

106. Ernst, J. V., B. Chobotar, and L. C. Anderson. 1967. *Eimeria balphae* n. sp. from the Ord kangaroo rat *Dipodomys ordii.* Journal of Protozoology, 14:547–548.

107. Ernst, J. V., M. J. Frydendall, and D. M. Hammond. 1967. *Eimeria scholtysecki* n. sp. from the Ord's kangaroo rat *Dipodomys ordii.* Journal of Protozoology, 14:181–182.

108. Ernst, J. V., D. M. Hammond, and B. Chobotar. 1968. *Eimeria utahensis* sp. n. from kangaroo rats (*Dipodomys ordii* and *D. microps*) in northwestern Utah. Journal of Protozoology, 15:430–432.

109. Ernst, J. V., E. C. Oaks, and J. R. Sampson. 1970. *Eimeria reedi* sp. n. and *E. chobotari* sp. n. (Protozoa: Eimeriidae) from heteromyid rodents. Journal of Protozoology, 17:453–455.

110. Esch, G. W., J. W. Gibbons, and J. E. Bourque. 1975. An analysis of the relationship between stress and parasitism. The American Midland Naturalist, 93:339–353.

111. Esch, G. W., T. C. Hazen, and J. M. Aho. 1977. Parasitism and r- and K selection. Pp. 9–62, *in* Regulation of parasite populations (G. W. Esch and B. B. Nickol, eds.). Academic Press, New York, 253 pp.

112. Esquivel, R. R., J. A. Zuniga, M. Alfaro, and E. Kotcher. 1967. Trypanosomes of Costa Rican feral mammals. The Journal of Parasitology, 53:951–955.

113. Ewing, H. E. 1931. A catalogue of the Trombiculinae, or chigger mites, of the New World with new genera and species and a key to the genera. Proceedings of the United States National Museum, 80(8):1–19.

114. Fain, A. 1967. Nouveaux hypopes vivant en association phoretique sur des rongeurs et des marsupiaux (Acarina: Glycyphagidae). Acarologia, 9:415–434.

115. ———. 1973. Les listrophorides d'Amerique Neotropicale (Acarina: Sarcoptiformes). I. Familles Listrophoridae et Chirodiscidae. Bulletin de l'Institute Royal des Sciences Naturelles de Belgique, 49(6):1–149.

116. ———. 1973. Diagnoses d'Acariens nouveaux (Listrophoroidea et Myobiidae). Revue de Zoologie et de Botanique Africaines, 87:330–332.

117. ———. 1976. Nouveaux acariens parasites de la superfamille Litrophoroidea (Astigmates). Acta Zoologica et Pathological Antverpiensia, 64:37–67.

118. Fain, A., and J. C. Beaucournu. 1973. Description de trois nouveaux hypopes d'Anoetidae phoretiques sur les pes de mammiferes. Acarologia, 15:514–520.

119. Fain, A., and G. S. Ide. 1978. A new hypopus, *Dermacarus liomys* sp. n. (Acari: Glycyphagidae) from a Mexican rodent, *Liomys irroratus.* International Journal of Acarology, 4:233–234.

120. Fain, A., and F. S. Lukoschus. 1978*a*. New endofollicular or subcutaneous hypopi from mammals (Acarina: Astigmata). Acarologia, 19: 484–494.

121. ———. 1978*b*. *Dipodomydectes americanus* gen. et sp. n. (Acari: Hypoderidae) from the kangaroo rat. The Journal of Parasitology, 64:137–138.

122. Fain, A., and J. O. Whitaker, Jr. 1980. *Geo-*

mylichus perognathi sp. n. (Acari: Listrophoridae) from *Perognathus* spp. in the United States. The Journal of Parasitology, 66:839–840.
123. ———. 1987. New observations on the genus *Geomylichus* Fain, 1970 (Acari, Listrophoridae) with description of four new species and a new subgenus. International Journal of Acarology, 13: 15–28.
124. Fain, A., J. O. Whitaker, Jr., T. G. Schwan, and F. S. Lukoschus. 1978. Notes on the genus *Geomylichus* Fain, 1970 (Astigmata: Listrophoridae) and descriptions of six new species. International Journal of Acarology, 4:101–114.
125. Fain, A., J. O. Whitaker, Jr., and H. H. Thomas. 1988. Two new species of *Geomylichus* Fain, 1970 (Acari, Listrophoridae) from California, USA. International Journal of Acarology, 14:121–125.
126. Fairchild, G. B. 1943. An annotated list of the bloodsucking insects, ticks and mites known from Panama. American Journal of Tropical Medicine, 23:569–571.
127. Fairchild, G. B., G. M. Kohls, and V. J. Tipton. 1966. The ticks of Panama (Acarina: Ixodoidea). Pp. 167–219, *in* Ectoparasites of Panama (R. L. Wenzel and V. J. Tipton, eds.). Field Museum of Natural History, Chicago, 861 pp.
128. Ferris, G. F. 1922. Contributions toward a monograph of the sucking lice. Part II. Stanford University Publications, University Series Biological Sciences, 2:135–178.
129. ———. 1951. The sucking lice. Memoirs of the Pacific Coast Entomological Society, 1:1–320.
130. Forman, G. L., R. J. Baker and J. D. Gerber. 1968. Comments on the systematic status of vampire bats (family Desmodontidae). Systematic Zoology, 17:417–425.
131. Fox, I. 1947. Notes on ectoparasites from Venezuela (Siphonaptera and Acarina). Zoologica: Scientific Contributions of the New York Zoological Society, 32:117–119.
132. Furman, D. P. 1954. A new species of *Androlaelaps* from *Perognathus* in southern California. Pan-Pacific Entomologist, 30:119–124.
133. ———. 1955*a*. Revision of the genus *Eubrachylaelaps* (Acarina: Laelaptidae) with the descriptions of two new species from Mexico. Annals of the Entomological Society of America, 48: 51–59.
134. ———. 1955*b*. *Steptolaelaps* (Acarina: Laelaptidae) a new genus of mites parasitic on neotropical rodents. The Journal of Parasitology, 41:519–525.
135. ———. 1972*a*. Laelapid mites (Laelapidae: Laelapinae) of Venezuela. Brigham Young University Science Bulletin: Biological Series, 17(3): 1–58.
136. ———. 1972*b*. New species of *Laelaps* (Acarina: Laelapidae) from Venezuela. Journal of Medical Entomology, 9:35–46.
137. Furman, D. P., and E. C. Loomis. 1984. The ticks of California (Acari: Ixodida). Bulletin of the California Insect Survey, 25:1–239.
138. Furman, D. P., and V. J. Tipton. 1961. Acaros parasitos laelaptinae (Acarina: Laelaptinae) de Venezuela. Memoria de la Sociedad de Ciencias Naturales La Salle, 21:166–212.
139. Garner, H. W., L. W. Richardson, and L. A. Felts. 1976. Alimentary helminths of *Dipodomys ordii*: effects on the host population. The Southwestern Naturalist, 21:327–334.
140. Gastfriend, A. 1955. New host records for the immature stages of the tick *Dermacentor parumapertus*. The Journal of Parasitology, 41:63–65.
141. Geest, J. C., and R. B. Loomis 1968. Chiggers of the genus *Pseudoschoengastia* (Acarina: Trombiculidae) from Costa Rica. Contributions in Science, Natural History Museum of Los Angeles County, 150:1–49.
142. Genoways, H. H. 1973. Systematics and evolutionary relationships of spiny pocket mice, genus *Liomys*. Special Publications, The Museum, Texas Tech University, 5:1–368.
143. Genoways, H. H., and J. K. Jones, Jr. 1972. Mammals from southwestern North Dakota. Occasional Papers, The Museum, Texas Tech University, 6:1–36.
144. Goates, M. A. 1963. Mites on kangaroo rats at the Nevada Test Site. Brigham Young University Science Bulletin: Biological Series, 3(11): 1–12.
145. Goertz, J. W. 1966. Incidence of warbles in some Oklahoma rodents. The American Midland Naturalist, 75:242–245.
146. Goff, M. L. 1982. A new species of *Pseudoschoengastia* (Acari: Trombiculidae) from a forest spiny pocket mouse (Rodentia: Heteromyidae) taken in Panama. Journal of Medical Entomology, 19:12–14.
147. Goff, M. L., and J. M. Brennan. 1977. A new monotypic genus of chiggers and four new species of *Quadraseta* from Venezuela (Acari: Trombiculidae). The Great Basin Naturalist, 37:501–509.
148. Goff, M. L., R. B. Loomis, W. C. Welbourn and W. J. Wrenn. 1982. A glossary of chigger terminology (Acari: Trombiculidae). Journal of Medical Entomology, 19:221–238.
149. Gould, D. J. 1950. A new species of chigger from California (Acarina: Trombiculidae). Wasmann Journal of Biology, 8:367–370.
150. ———. 1956. The larval trombiculid mites of California (Acarina: Trombiculidae). University of California Publications in Entomology, 11:1–115.
151. Grant, C. D. 1947. North American mites of the genus *Laelaps* (Arachnida: Acarina: Parasitidae). Microentomology, 12:2–21.
152. Graves, G. N., W. C. Bennett, J. R. Wheeler, B. E. Miller, and D. L. Forcum. 1974. Sylvatic plague studies in Southeast New Mexico I. Flea-host relationships from six years' study. Journal of Medical Entomology, 11:488–498.
153. Grundmann, A. W. 1957. Nematode parasites of mammals of the Great Salt Lake Desert of Utah. The Journal of Parasitology, 43:105–112.
154. ———. 1958. Cestodes of mammals from the Great Salt Lake Desert region of Utah. The Journal of Parasitology, 44:425–429.
155. Grundmann, A., and J. C. Frandsen. 1960. *Capillaria bonnevillei* n. sp. (Nematoda: Trichu-

ridae) from the Ord kangaroo rat in Utah. Proceedings of the Helminthological Society of Washington, 27:206–208.

156. Grundmann, A., and R. G. Warnock. 1964. A population study of eight parasitic nematode species from mammals of the Bonneville Basin of Utah. Proceedings of the Utah Academy of Science, Arts and Letters, 41:228–233.

157. Guay, J., and C. M. Senger. 1962. The occurrence of *Catenotaenia* sp. in *Dipodomys spectabilis* in Arizona. The Journal of Parasitology, 48:451.

158. Haas, G. E., R. P. Martin, M. Swickard, and B. E. Miller. 1973. Siphonaptera-mammal relationships in northcentral New Mexico. Journal of Medical Entomology, 10:281–289.

159. Hannum, C. A. 1943. Nematode parasites of Arizona vertebrates. Publications of the University of Washington, Thesis Series, 7:229–231.

160. Hansen, C. G. 1964. Ectoparasites of mammals from Oregon. The Great Basin Naturalist, 24:75–81.

161. Hardy, J. L., R. N. Lyness, and W. A. Rush. 1972. Experimental vector and wildlife host ranges of buttonwillow virus in Kern County, California. American Journal of Tropical Medicine and Hygiene, 21:100–109.

162. Hardy, J. L., W. C. Reeves, R. P. Scrivani, and D. R. Roberts. 1974. Wild mammals as hosts of group A and group B arboviruses in Kern County, California: A five year serologic and virologic survey. American Journal of Tropical Medicine and Hygiene, 23:1165–1177.

163. Hedeen, R. A. 1953. Ectoparasites occurring on mammals in the vicinity of Fort Hood, Texas. Texas Journal of Science, 5:125–129.

164. Hegner, R. W. 1926. *Giardia beckeri* n. sp. from the ground squirrel and *Endamoeba dipodomysi* from the kangaroo rat. The Journal of Parasitology, 12:203–206.

165. Henry, L. G., and S. McKeever. 1971. A modification of the washing technique for quantitative evaluation of the ectoparasite load of small mammals. Journal of Medical Entomology, 8:504–505.

166. Herman, C. M. 1943. *Giardia* in the blood of a kangaroo rat. The Journal of Parasitology, 29: 423.

167. Herrin, C. S. 1970. A systematic revision of the genus *Hirstionyssus* (Acari: Mesostigmata) of the Nearctic region. Journal of Medical Entomology, 7:391–437.

168. Herrin, C. S., and C. E. Yunker. 1973. A new species of *Hirstionyssus* from *Liomys irroratus* in Texas (Acari: Mesostigmata). Journal of Medical Entomology, 10:269–273.

169. ———. 1975. Systematics of neotropical *Hirstionyssus* mites with special emphasis on Venezuela (Acarina: Mesostigmata). Brigham Young University Science Bulletin: Biological Series, 20(2):93–127.

170. Hoffmann, A. 1963. Contribuciones al conocimiento de los trombiculidos mexicanos (Acarina: Trombiculidae) 10 a parte. Anales de la Escuela Nacional de Ciencias Biologicas (Mexico), 12:101–109.

171. Holdenried, R., and H. B. Morlan. 1955. Plague-infected fleas from northern New Mexico wild rodents. Journal of Infectious Diseases, 96: 133–137.

172. ———. 1956. A field study of wild mammals and fleas of Santa Fe County, New Mexico. The American Midland Naturalist, 55:369–381.

173. Holdenried, R., and S. F. Quan. 1956. Susceptibility of New Mexico rodents to experimental plague. Public Health Reports, 71:979–984.

174. Holland, G. P. 1958. Distribution patterns of northern fleas (Siphonaptera). Proceedings of the Tenth International Congress on Entomology, 1:645–658.

175. ———. 1963. Faunal affinities of the fleas (Siphonaptera) of Alaska with an annotated list of species. Pp. 45–63, *in* Pacific Basin Biogeography (J. L. Gressitt, ed.). Tenth Pacific Science Congress, Bishop Museum, Honolulu, 563 pp.

176. ———. 1985. The fleas of Canada, Alaska and Greenland (Siphonaptera). Memoirs of the Entomological Society of Canada, 130. 1–631.

177. Holmes, J. C. 1979. Parasite populations and host community structure. Pp. 27–46, *in* Host-parasite interfaces (B. B. Nickol, ed.). Academic Press, New York.

178. Holmes, J. C., R. P. Hobbs, and T. S. Leong. 1977. Populations in perspective: community organization and regulation of parasite populations. Pp. 209–245, *in* Regulation of parasite populations (G. W. Esch, ed.). Academic Press, New York.

179. Holmes, J. C., and P. W. Price. 1986. Communities of parasites. Pp. 187–213, *in* Community ecology: pattern and process (J. Kikkawa and D. J. Anderson, eds.). Blackwell Scientific Publications, Melbourne.

180. Hopkins, D. 1985. Host associations of Siphonaptera from central Oregon. Northwest Science, 59:108–114.

181. Hopkins, G. H. E., and M. Rothschild. 1962. An illustrated catalogue of the Rothschild collection of fleas (Siphonaptera) in the British Museum (Natural History). III. Hystrichopsyllidae (Acedestiinae, Anomiopsyllinae, Hystrichopsyllinae, Neopsyllinae, Rhadinopsyllinae and Stenoponiinae). British Museum (Natural History) Publications, 410:1–560.

182. ———. 1966. An illustrated catalogue of the Rothschild collection of fleas (Siphonaptera) in the British Museum (Natural History). IV. Hystrichopsyllidae (Ctenophthalminae, Dinopsyllinae, Doratopsyllinae and Listropsyllinae). British Museum (Natural History) Publications, 652: 1–549.

183. Howell, J. F., and R. J. Elzinga. 1962. A new *Radfordia* (Acarina: Myobiidae) from the kangaroo rat and a key to the known species. Annals of the Entomological Society of America, 55:547–555.

184. Howitt, B. F., and V. Van Herrick. 1942. Relationship of the St. Louis and the western

equine encephalitis viruses to fowl and mammals in California. Journal of Infectious Diseases, 71: 179–191.
185. HUBBARD, C. A. 1947. Fleas of western North America. Their relation to the public health. Iowa State University Press, Ames. 533 pp.
186. ———. 1949. Fleas of the state of Nevada. Bulletin of the Southern California Academy of Science, 48:115–128.
187. ———. 1958. Mexican jungle and desert fleas with three new descriptions. Entomological News, 69:161–166.
188. IRONS, J. V., R. B. EADS, C. W. JOHNSON, O. L. WALKER, AND M. A. NORRIS. 1952. Southwest Texas Q fever studies. The Journal of Parasitology, 38:1–5.
189. IVENS, V., F. J. KRUIDENIER, AND N. D. LEVINE. 1958. Further studies of *Eimeria* (Protozoa: Eimeriidae) from Mexican rodents. Transactions of the Illinois Academy of Science, 51:53–57.
190. JAMESON, E. W., JR. 1950. Notes on mites of the genus *Neoichoronyssus*, with the description of a new subgenus and three new species of the subgenus *Hirstionyssus* (Acarina, Laelaptidae, Macronyssinae). Proceedings of the Entomological Society of Washington, 52:161–172.
191. JELLISON, W. L., AND C. M. SENGER. 1976. Fleas of western North America except Montana in the Rocky Mountain Laboratory collection. Pp. 55–136, *in* Papers in honor of Jerry Flora (H. C. Taylor, Jr., and J. Clark, eds.). Western Washington State College, Bellingham, Washington, 281 pp.
192. JOHNSON, D. E. 1966. Ticks of Dugway Proving Ground and vicinity and their host associations. Proceedings of the Utah Academy of Sciences, Arts and Letters, 43:49–66.
193. JOHNSON, P. T. 1957. A classification of the Siphonaptera of South America with descriptions of new species. Memoirs of the Entomological Society of Washington, 5:1–299.
194. ———. 1962. The species of *Fahrenholzia* Kellogg and Ferris from spiny pocket mice (Anoplura: Hoplopleuridae). Annals of the Entomological Society of America, 55:415–428.
195. ———. 1972. Sucking lice of Venezuelan rodents, with remarks on related species (Anoplura). Brigham Young University Science Bulletin: Biological Series, 17(5):1–62.
196. JONES, E. K., C. M. CLIFFORD, J. E. KEIRANS, AND G. M. KOHLS. 1972. The ticks of Venezuela (Acarina: Ixodoidea) with a key to the species of *Amblyomma* in the western hemisphere. Brigham Young University Science Bulletin: Biological Series, 17(4):1–40.
197. JORDAN, K. 1938. On some nearctic Siphonaptera. Novitates Zoologicae, 41:119–124.
198. KEEGAN, H. L. 1951. The mites of the subfamily Haemogamasinae (Acari: Laelaptidae). Proceedings of the United States National Museum, 101:203–268.
199. ———. 1953. Collections of parasitic mites from Utah. The Great Basin Naturalist, 13:35–42.
200. ———. 1956. *Ischyropoda furmani* n. sp., a new ectoparasitic mite from Utah. The Journal of Parasitology, 42:311–315.
201. KENAGY, G. J. 1972. Saltbush leaves: excision of hypersaline tissue by a kangaroo rat. Science, 178(4065):1094–1096.
202. KENNEDY, C. R. 1977. The regulation of fish parasite populations. Pp. 63–109, *in* Regulation of parasite populations (G. W. Esch and B. B. Nickol, eds.). Academic Press, New York, 253 pp.
203. KIERANS, J. E., AND C. M. CLIFFORD. 1978. The genus *Ixodes* in the United States: A scanning electron microscope study and key to the adults. Journal of Medical Entomology Supplement, 2:1–149.
204. KIERANS, J. E., AND E. K. JONES. 1972. Description of the immature stages of *Ixodes* (*I.*) *sinaloa* Kohls and Clifford (Acarina: Ixodidae), from rodents in Mexico and Nicaragua. Acarologia, 13:471–475.
205. KING, S. R., AND B. B. BABERO. 1974. Helminths of kangaroo rats (*Dipodomys* spp.) in Nevada with reports of other worm parasites from these hosts. Proceedings of the Helminthological Society of Washington, 41:241–248.
206. KIRBY, H., AND B. HONIGBERG. 1949. Flagellates of the caecum of ground squirrels. University of California Publications in Zoology, 53: 315–366.
207. KLEKOWSKI, R. Z. AND J. UCHMANSKI. 1980. 5. Simulation model of energy flow through a *Rhizoglyphus echinopus* (F. et R.) individual during its development. Polish Ecological Studies, 6:61–80.
208. KOHLS, G. M. 1937. Hosts of the immature stages of the Pacific coast tick *Dermacentor occidentalis* Neum. (Ixodidae). Public Health Reports, 52(16):490–496.
209. ———. 1953. *Ixodes venezuelensis*, a new species of tick from Venezuela, with notes on *Ixodes minor* Neumann, 1902 (Acarina: Ixodidae). The Journal of Parasitology, 39:300–303.
210. KOHLS, G. M., AND C. M. CLIFFORD. 1964. *Ixodes eadsi*, a new species of tick from rodents in southern Texas (Acarina: Ixodidae). The Journal of Parasitology, 50:466–470.
211. ———. 1966. Three new species of *Ixodes* from Mexico and description of the male of *I. auritulus auritulus* Neumann, *I. conepati* Cooley and Kohls, and *I. lasallei* Mendez and Ortiz (Acarina: Ixodidae). The Journal of Parasitology, 52:810–820.
212. KOHLS, G. M., AND R. A. COOLEY. 1936. Notes on the occurrence and host relationships of the tick *Ornithodoros talaje* in Arizona. United States Public Health Reports, 51:512–513.
213. KREIER, J. P., AND M. RISTIC. 1968. Haemobartonellosis, eperxthrozoonosis, grahamellosis, and ehrlichiosis. Pp. 387–472, *in* Infectious blood diseases of man and animals. Vol. II. The pathogens, the infections, and the consequences (D. Weinman and M. Ristic, eds.). Academic Press, New York, 576 pp.
214. KRUIDENIER, F. J., AND K. N. MEHRA. 1959. *Wellcomia perognathi* n. sp. (Nematoda, Oxyuri-

dae) from pocket mice, *Perognathus intermedius* (Merriam, 1889) of the Grand Canyon, Arizona. Transactions of the Illinois Academy of Science, 51(3&4):20–25.

215. KRUIDENIER, F. J., AND C. R. PEEBLES. 1958. Gongylonema of rodents: *G. neoplasticum* (redefinition): *G. dipodomysis* n. sp. and *G. peromysci* n. sp. Transactions of the American Microscopical Society, 67:307–315.
216. KUNS, M. L., AND R. E. TASHIAN. 1954. Notes on mammals from northern Chiapas. The Journal of Mammalogy, 35:100–103.
217. LAINSON, R., AND J. STRANGWAYS-DIXON. 1964*a*. *Leishmania mexicana*: epidemiology of dermal leishmaniasis in British Honduras. Transactions of the Royal Society of Tropical Medicine and Hygiene, 58:3.
218. ———. 1964*b*. The epidemiology of dermal leishmaniasis in British Honduras. Part II. Reservoir hosts of *Leishmania mexicana* among the forest rodents. Transactions of the Royal Society of Tropical Medicine and Hygiene, 58:136–153.
219. LANE, R. S., R. W. EMMONS, D. V. DONDERO, AND B. C. NELSON. 1981. Ecology of tick-borne agents in California 1. Spotted fever group rickettsiae. American Journal of Tropical Medicine and Hygiene, 30:239–252.
220. LARSON, O. R., M. G. MCKENNA, AND N. R. FELLOWS. 1985. North Dakota fleas. IX. Siphonapterans of mammals in southwestern North Dakota. The Prairie Naturalist, 17:241–247.
221. LEIPER, R. T. 1936. Deaths caused by animal parasites. *In* Report on the deaths occurring in the society's gardens during the year 1935 (A. M. Hammerton, ed.). Proceedings of the Zoological Society of London, 1936:682–684.
222. LEVINE, N. D., V. IVENS, AND F. J. KRUIDENIER. 1957*a*. New species of *Eimeria* (Protozoa: Eimeriidae) from Mexican rodents. Transactions of the Illinois Academy of Science, 50:291–298.
223. ———. 1957*b*. New species of *Eimeria* from Arizona rodents. Journal of Protozoology, 4:80–88.
224. LINSDALE, J. M., AND B. S. DAVIS. 1956. Taxonomic appraisal and occurrence of fleas at the Hastings Reservation in central California. University of California Publications in Zoology, 54: 293–370.
225. LIPOVSKY, L. J. 1951. A new genus of Walchiinae (Acarina: Trombiculidae). Journal of the Kansas Entomological Society, 24:95–102.
226. LIPOVSKY, L. J., D. A. CROSSLEY, JR., AND R. B. LOOMIS. 1955. A new genus of chigger mites (Acarina, Trombiculidae). Journal of the Kansas Entomological Society, 28:136–141.
227. LOOMIS, R. B. 1954. A new subgenus and six new species of chigger mites (genus *Trombicula*) from the central United States. University of Kansas Science Bulletin, 36:919–941.
228. ———. 1956. The chigger mites of Kansas (Acarina, Trombiculidae). University of Kansas Science Bulletin, 37:1195–1443.
229. ———. 1969. Chiggers (Acarina, Trombiculidae) from vertebrates of the Yucatan Peninsula, Mexico. Miscellaneous Publications of the Museum of Natural History, University of Kansas, 50:1–22.
230. ———. 1971*a*. The genus *Euschoengastoides* (Acarina: Trombiculidae) from North America. The Journal of Parasitology, 57:689–707.
231. ———. 1971*b*. Chiggers (family Trombiculidae) of the North American deserts. Proceedings of the 3rd International Congress of Acarology, 1971:435–440.
232. ———. 1976. Two new species of the genus *Pseudoschoengastia* (Acarina, Trombiculidae) from Mexico. Bulletin of the Southern California Academy of Science, 75:278–281.
233. LOOMIS, R. B., J. T. BACCUS, AND W. J. WRENN. 1972. The chiggers (Acarina, Trombiculidae) of Big Bend National Park, Texas. National Park Service Report. Mimeo.
234. LOOMIS, R. B., AND M. BUNNELL. 1962. New species of chigger, genus *Euschoengastia* (Acarina, Trombiculidae), with notes on other species of chiggers from the Santa Ana Mountains, California. Bulletin of the Southern California Academy of Science, 61:177–184.
235. LOOMIS, R. B., AND D. A. CROSSLEY, JR. 1963. New species and new records of chiggers (Acarina: Trombiculinae) from Texas. Acarologia, 5:371–383.
236. LOOMIS, R. B., AND M. L. GOFF. 1973. A new genus and two new species of North American leeuwenhoekiine chiggers (Acarina, Trombiculidae). Journal of Medical Entomology, 10:113–117.
237. LOOMIS, R. B., AND J. L. LUCAS. 1969. A species of *Hexidionis* (Acarina: Trombiculidae) from kangaroo rats (genus *Dipodomys*) of western North America. Bulletin of the Southern California Academy of Science, 68:225–228.
238. ———. 1970. A new subgenus and two new species of *Hexidionis* (Acarina, Trombiculidae) from North America. Bulletin of the Southern California Academy of Science, 69:52–59.
239. LOOMIS, R. B., AND R. E. SOMERBY. 1966. New species and new records of *Euschoengastia* (Acarina, Trombiculidae) from western Mexico. Bulletin of the Southern California Academy of Science, 65:211–224.
240. LOOMIS, R. B., AND W. J. WRENN. 1973*a*. The North American genus *Otorhinophila* Wrenn and Loomis (Acarina: Trombiculidae), with the descriptions of three new species. The Journal of Parasitology, 59:175–182.
241. ———. 1973*b*. Additional chiggers (Acarina, Trombiculidae) from vertebrate hosts of Big Bend National Park and vicinity, Brewster County, Texas. National Park Service Report; Mimeo.
242. LUCAS, J. L., AND R. B. LOOMIS. 1968. The genus *Hexidionis* ((Acarina, Trombiculidae) with the description of a new species from western Mexico. Bulletin of the Southern California Academy of Science, 67:233–239.
243. LUKOSCHUS, F. S., G. H. S. JANSSEN DUIJGHUIJSEN, AND A. FAIN. 1977. Three new species of genus *Metalabidophorus* Fain 1967 (Glycyphagidae:

Sarcoptiformes) from American rodents. International Journal of Acarology, 3:9–18.
244. MACHADO-ALLISON, C. E. 1967. The systematic position of the bats *Desmodus* and *Chilonycteris*, based on host-parasite relationships (Mammalia; Chiroptera). Proceedings of the Biological Society of Washington, 80:223–226.
245. MARSHALL, A. G. 1981. The ecology of ectoparasitic insects. Academic Press, London. 459 pp.
246. MASER, C., AND J. O. WHITAKER, JR. 1980. Mites found in the fur of a small sample of heteromyid rodents from Oregon. Northwest Science, 54:279–280.
247. MCDANIEL, B. 1965. The subfamily Listrophorinae Gunther with a description of a new species of the genus *Listrophorus* Pagenstecher from Texas (Acarina: Listrophoridae). Acarologia, 7:704–712.
248. ———. 1968. A new species of *Fahrenholzia* (Anoplura: Hoplopleuridae) from Texas, with a key to the species of the genus and a list of their known hosts. Annals of the Entomological Society of America, 61:402–405.
249. MCGHEE, M. E., AND H. H. GENOWAYS. 1978. *Liomys pictus*. Mammalian Species, 83:1–5.
250. MCLEAN, D. M. AND W. L. DONOHUE. 1959. Powassan virus: isolation of virus from a fatal case of encephalitis. Canadian Medical Association Journal, 80:708–712.
251. MENDEZ, E. 1977. Mammalian-Siphonapteran associations, the environment and biogeography of mammals of southwestern Colombia. Quaestiones Entomologicae, 13:91–182.
253. MENZIES, G. C., R. B. EADS, AND B. G. HIGHTOWER. 1951. List of Anoplura from Texas. Proceedings of the Entomological Society of Washington, 53:150–152.
254. MILLEMANN, R. 1955. Studies on the life history and biology of *Oochorista deserti* n. sp. (Cestoda: Linstowiidae) from desert rodents. The Journal of Parasitology, 41:424–440.
255. MILLER, R. S., AND R. A. WARD. 1960. Ectoparasites of pocket gophers from Colorado. The American Midland Naturalist, 64:382–391.
256. MORLAN, H. B. 1955. Mammal fleas of Santa Fe County, New Mexico. Texas Reports on Biology and Medicine, 13:93–125.
257. MORLAN, H. B., AND C. C. HOFF. 1957. Notes on some Anoplura from New Mexico and Mexico. The Journal of Parasitology, 43:347–351.
258. MUMFORD, R. E., AND J. O. WHITAKER, JR. 1982. Mammals of Indiana. Indiana University Press, Bloomington, 537 pp.
259. NADLER, C. F., AND R. S. HOFFMANN. 1977. Patterns of evolution and migration in the arctic ground squirrel, *Spermophilus parryii* (Richardson). Canadian Journal of Entomology, 55:748–758.
260. O'FARRELL, T. P. 1975. Small mammals, their parasites and pathologic lesions on the Arid Lands Ecology Reserve, Benton County, Washington. The American Midland Naturalist, 93:377–387.
261. PARAN, T. P. 1966. A new fur mite, *Lavoimyobia hughesi* n. g., n. sp. (Acarina: Myobiidae) from a North American rodent. Journal of Medical Entomology, 3:172–178.
262. PARKER, O. S., AND A. H. CHANEY. 1979. *Liomys irroratus* (Rodentia: Heteromyidae), a new host for *Cuterebra fontinella* (Diptera: Cuterebridae). Journal of Medical Entomology, 15:573.
263. PARKER, R. R., AND R. W. WELLS. 1917. Some facts of importance concerning the Rocky Mountain spotted fever tick (*Dermacentor venustus* Banks) in eastern Montana. Second Biennial Report, Montana State Board of Entomology, 1915-1916:45–56.
264. PECK, S. B. 1982. A review of the ectoparasitic *Leptinus* beetles of North America (Coleoptera: Leptinidae). Canadian Journal of Zoology, 60: 1517–1527.
265. PENCE, D. B., AND H. H. GENOWAYS. 1974. *Neolabidophorus yucatanensis* gen. et sp. n. and a new record for *Dermacarus ornatus* Fain, 1967 (Acarina: Glycyphagidae) from *Heteromys gaumeri* Allen and Chapman, 1897, Gaumer's spiny pocket mouse (Rodentia: Heteromyidae). The Journal of Parasitology, 60:712–715.
266. POMEROY, L. V., AND R. B. LOOMIS. 1984. A new genus of trombiculine chiggers (Acari: Trombiculidae) from western North America. Journal of Medical Entomology, 21:268–273.
267. POORBAUGH, J. H., AND H. T. GIER. 1961. Fleas (Siphonaptera) of small mammals in Kansas. Journal of the Kansas Entomological Society, 34: 198–204.
268. PRICE, P. W. 1980. Evolutionary biology of parasites. Monogr. in Population Biology. Princeton University Press, Princeton, NJ, 237 pp.
269. RADFORD, C. D. 1953. Four new species of 'harvest mite' or 'chigger' and a new fur-mite (Acarina: Trombiculidae and Listrophoridae). Parasitology, 43:210–214.
270. RANDOLPH, N. M., AND R. B. EADS. 1946. An ectoparasitic survey of mammals from Lavaca County, Texas. Annals of the Entomological Society of America, 39:597–601.
271. RAPP, W. F., JR., AND D. B. GATES. 1957. A distributional check-list of the fleas of Nebraska. Journal of the Kansas Entomological Society, 30: 50–53.
272. READ, C. P. 1956*a*. *Trichuris dipodomis* n. sp. from Ord's kangaroo rats. Proceedings of the Helminthological Society of Washington, 23:119.
273. ———. 1956*b*. *Trichuris* from kangaroo rats. Proceedings of the Helminthological Society of Washington, 23:2.
274. READ, C. P., AND R. E. MILLEMANN. 1953. Helminth parasites in kangaroo rats. University of California Publications in Zoology, 59(3):61–80.
275. REED, J. T., AND J. M. BRENNAN. 1975. The subfamily Leeuwenhoekinae in the neotropics (Acarina: Trombiculidae). Brigham Young University Science Bulletin: Biological Series, 20(1): 1–42.
276. REISEN, W. K., AND T. L. BEST. 1973. An ectoparasite survey of *Dipodomys ordii* (Rodentia: Heteromyidae) from central Oklahoma. Annals

of the Entomological Society of America, 66:599–601.
277. Ryckman, R. E., and J. C. Roos. 1955. Chiggers associated with squamous papillomata on a *Perognathus* mouse. The Journal of Parasitology, 41:639.
278. Ryser, F. A. 1964. Spiny pocket mouse in Nevada. The Journal of Mammalogy, 45:301–302.
279. Saunders, R. C. 1975. Venezuela Macronyssidae (Acarina: Mesostigmata). Brigham Young University Science Bulletin: Biological Series, 20(2)2:73–90.
280. Schad, G. A. 1977. The role of arrested development in the regulation of nematode populations. Pp. 111–167, *in* Regulation of parasite populations (G. W. Esch and B. B. Nickol, eds.). Academic Press, New York, 253 pp.
281. Seymour, C., and T. M. Yuill. 1981. Arboviruses. Pp. 54–86, *in* Infectious diseases of wild mammals (J. W. Davis, L. H. Karstad, and D. O. Trainer, eds.). Second ed. Iowa State University Press, Ames, 446 pp.
282. Short, J. A., L. F. Mayberry, and J. R. Bristol. 1980. *Eimeria chihuahuaensis* sp. n. and other coccidia from *Dipodomys* spp. in El Paso County, Texas. Journal of Protozoology, 27:361–364.
283. Simpson, R. L., and W. Harmon. 1968. Occurrence of *Hymenolepis citelli* McLeod, 1933 in the rodent genus *Dipodomys* Gray, 1841. The Journal of Parasitology, 54:769.
284. Smith, J. D. 1972. Systematics of the chiropteran family Mormoopidae. Miscellaneous Publications, Museum of Natural History, University of Kansas, 56:1–132.
285. Stark, H. E. 1958. The Siphonaptera of Utah. Their taxonomy, distribution, host relations, and medical importance. United States Department of Health, Education, and Welfare, Public Health Service. Atlanta, Georgia, 239 pp.
286. Stark, H. E., and A. R. Kinney. 1969. Abundance of rodents and fleas as related to plague in Lava Beds National Monument, California. Journal of Medical Entomology, 6:287–294.
287. Stewart, M. A. 1940. New Siphonaptera from California. Pan-Pacific Entomologist, 16:17–28.
288. Stoenner, H. G., R. Holdenried, D. Lackman, and J. S. Osburn, Jr. 1959. The occurrence of *Coxiella burneti, Brucella* and other pathogens among fauna of the Great Salt Lake Desert in Utah. American Journal of Tropical Medicine and Hygiene, 8:590–596.
289. Stojanovich, C. J., and H. D. Pratt. 1961. *Fahrenholzia texana,* new species, with a key to the United States species of *Fahrenholzia* (Anoplura: Hoplopleuridae). Annals of the Entomological Society of America, 54:693–696.
290. Stout, C. A., and D. W. Duszynski. 1983. Coccidia from kangaroo rats (*Dipodomys* spp.) in the western United States, Baja California, and northern Mexico with descriptions of *Eimeria merriami* sp. n. and *Isospora* sp. The Journal of Parasitology, 69:209–214.
291. Strandtmann, R. W. 1949. The blood-sucking mites of the genus *Haemolaelaps* (Acarina: Laelaptidae) in the United States. The Journal of Parasitology, 35:325–352.
292. Strandtmann, R. W., and D. M. Allred. 1956. Mites of the genus *Brevisterna* Keegan, 1949 (Acarina—Haemogamasidae). Journal of the Kansas Entomological Society, 29:113–132.
293. Strandtmann, R. W., and H. B. Morlan. 1953. A new species of *Hirstionyssus* and a key to the known species of the world. Texas Reports on Biology and Medicine, 11:627–637.
294. Strandtmann, R. W., and C. E. Yunker. 1966. The genus *Hirstionyssus* Fonseca in Panama (Acarina: Dermanyssidae). Pp. 105–124, *in* Ectoparasites of Panama (R. L. Wenzel and V. J. Tipton, eds.). Field Museum of Natural History, Chicago, 861 pp.
295. Svihla, R. D. 1941. A list of the fleas of Washington. University of Washington Publications in Biology, 12:9–20.
296. Tanigoshi, L. K., and R. B. Loomis. 1974. The genus *Hyponeocula* (Acarina, Trombiculidae) of western North America. Melanderia, 17:1–27.
297. Thomas, H. H., J. O. Whitaker, Jr., and Troy L. Best. 1990. Ectoparasites of *Dipodomys elator* from North Central Texas with some data from sympatric *Chaetodipus hispidus* and *Perognathus flavus.* The Southwestern Naturalist, 34: 111–114.
298. Thorpe, B. D., R. W. Sidwell, D. E. Johnson, K. L. Smart, and D. D. Parker. 1965. Tularemia in the wildlife and livestock of the Great Salt Lake Desert Region, 1951 through 1964. American Journal of Tropical Medicine and Hygiene, 14:622–637.
299. Tikasingh, E. S. 1975. La Leishmaniasis enzootica de los roedores en Trinidad, Indias Occidentales. Boletin de la Oficina Sanitaria Panamerica, 29:312–324.
300. Timm, R. H. 1983. Fahrenholz's rule and resource tracking: A study of host parasite coevolution. Pp. 225–265, *in* Coevolution (H. M. Nitecki, ed.). University of Chicago Press, Chicago, 392 pp.
301. Tiner, J. D. 1948. *Rictularia dipodomis* n. sp. (Nematoda: Thelazidae) from the kangaroo rat *Dipodomys* sp. The Journal of Parasitology, 34: 332–335.
302. Tipton, V. J. 1950. New distributional records for Utah Siphonaptera. The Great Basin Naturalist, 10:62–65.
303. Tipton, V. J., R. M. Altman, and C. M. Keenan. 1966. Mites of the subfamily Laelaptinae in Panama (Acarina: Laelaptidae). Pp. 23–82, *in* Ectoparasites of Panama (R. L. Wenzel and V. J. Tipton, eds.). Field Mus. Nat. Hist., Chicago, 861 pp.
304. Tipton, V. J., and E. Mendez. 1966. The fleas (Siphonaptera) of Panama. Pp. 289–385, *in* Ectoparasites of Panama (R. L. Wenzel and V. J. Tipton, eds.). Field Museum of Natural Hist., Chicago, 861 pp.
305. Tipton, V. J., and C. E. Machado-Allison. 1972. Fleas of Venezuela. Brigham Young University Science Bulletin: Biological Series, 17(6): 1–115.
306. Tipton, V. J., H. E. Stark, and J. A. Wildie. 1979. Anomiopsyllinae (Siphonaptera: Hystrichopsyllidae), II. The genera *Callistopsyllus,*

Conorhinopsylla, Megarthroglossus, and *Stenistomera.* The Great Basin Naturalist, 39:351–418.
307. Torrealba, J. W., J. C. Gomez-Nunez, and G. Ulloa. 1972. Isolation of *Leishmania braziliensis* by intraperitoneal inoculation of blood from a reservoir host into hamsters. Transactions of the Royal Society of Tropical Medicine and Hygiene, 66:361.
308. Traub, R. 1953. *Wenzella obscura,* a new genus and new species of flea from Guatemala (Siphonaptera). Journal of the Washington Academy of Science, 43:77–85.
309. ———. 1962. Some considerations of mites and ticks as vectors of human disease. Pp. 123–134, *in* Biological transmission of disease agents (K. Maramorosch, ed.). Academic Press, New York. 192 pp.
310. Traub, R., and C. C. Hoff. 1951. Records and descriptions of fleas from New Mexico (Siphonaptera). American Museum Novitates, 1530:1–23.
311. Traub, R., and P. T. Johnson. 1952. Fleas collected during a plague survey in Venezuela. Bulletin of the Pan-American Sanitary Bureau, 32(2):111–135.
312. Traub, R., M. Rothschild, and J. F. Haddow. 1983. The Rothschild collection of fleas. The Ceratophyllidae: key to the genera and host relationships, with notes on their evolution, zoogeography and medical importance. Cambridge University Press/Academic Press (Cambridge and London), 288 pp.
313. Traub, R., and H. Starcke. 1980. Some adaptive modifications in fleas. Pp. 33–67, *in* Fleas (R. Traub and H. Starcke, eds.). Proceedings of the International Conference on Fleas, Ashton Wold/Peterborough/UK/21–25 June 1977. A. A. Balkema, Publishers, Rotterdam, 420 pp.
314. Traub, R., and H. Starcke. 1980. The zoogeography and evolution of some fleas, lice and mammals. Pp. 93–172, *in* Fleas (R. Traub and H. Starcke, eds.). Proceedings of the International Conference on Fleas, Ashton Wold/Peterborough/UK/21–25 June 1977. A. A. Balkema, Publishers, Rotterdam, 420 pp.
315. Vercammen-Grandjean, P. H., and J. R. Audy. 1965. Revision of the genus *Eutrombicula* Ewing 1983 (Acarina: Trombiculidae). Acarologia, 7:280–294.
316. Vest, E. D., D. L. Lundgren, D. D. Parker, D. E. Johnson, E. L. Morse, J. B. Bushman, R. W. Sidwell, and B. D. Thorpe. 1965. Results of a five year survey for certain enzootic diseases in the fauna of western Utah. American Journal of Tropical Medicine and Hygiene, 14:124–135.
317. Voge, M. 1948. New rodent hosts for *Catenotaenia linsdalei,* with an additional description of this cestode. Transactions of the American Microscopical Society, 67:266–267.
318. ———. 1955. A list of cestode parasites from California mammals. The American Midland Naturalist, 54:413–417.
319. ———. 1956. A list of nematode parasites from California mammals. The American Midland Naturalist, 56:423–429.
320. Vogelsang, E. G., and J. A. Travassos Dias. 1953. Nueva contribucion al estudio de la fauna ixodologica en Venezuela. Revista de Medicina Veterinaria y Parasitologia Caracas, 12:63–89.
321. Waage, J. K., and G. G. Montgomery. 1976. *Cryptoses choloepi*: a coprophagous moth that lives on a sloth. Science, 193:157–158.
322. Warnock, R. G., and A. W. Grundmann. 1963. Food habits of some mammals of the Bonneville Basin as determined by the contents of the stomach and intestine. Proceedings of the Utah Academy of Sciences, Arts and Letters, 40:66–73.
323. Webb, J. P., Jr., and R. B. Loomis. 1971. Trombiculid mites of the genus *Microtrombicula* (Acarina) from Costa Rica. Contributions in Science, Natural History Museum of Los Angeles County, 207:1–15.
324. Wenzel, R. L., and P. T. Johnson. 1966. Checklist of the sucking lice of Panama (Anoplura). Pp. 273–279, *in* Ectoparasites of Panama (R. L. Wenzel and V. J. Tipton, eds.). Field Museum of Natural History, Chicago, 861 pp.
325. Wenzel, R. L., and V. J. Tipton, editors. 1966. Ectoparasites of Panama. Field Museum of Natural History, Chicago, 861 pp.
326. Wenzel, R. L., V. J. Tipton, and A. Kiewlicz. 1966. The streblid batflies of Panama (Diptera: Calypterae: Streblidae). Pp. 405–675, *in* Ectoparasites of Panama (R. L. Wenzel and V. J. Tipton, eds.). Field Museum of Natural History, Chicago, 861 pp.
327. Werneck, F. L. 1952. Contribuicao ao conhecimento dos Anopluros. I. Revista Brasileira de Biologia, 12(1):69–78.
328. Wharton, G. W. 1946. Observations on *Ascoschoengastia indica* (Hirst, 1915) (Acarina: Trombiculidae). Ecological Monographs, 16:151–184.
329. Whitaker, J. O., Jr. 1968. Parasites. Pp. 254–311, *in* Biology of *Peromyscus* (Rodentia) (J. A. King, ed.). Special Publications, American Society of Mammalogists, 2:1–593.
330. Whitaker, J. O. Jr., and N. Wilson. 1974. Host and distribution lists of mites (Acari), parasitic and phoretic, in the hair of wild mammals of North America, north of Mexico. The American Midland Naturalist, 91:1–67.
331. Williams, L. A., and C. C. Hoff. 1951. Fleas from the upper Sonoran Zone near Albuquerque, New Mexico. Proceedings of the United States National Museum, 101:305–313.
332. Wolfenbarger, K. A. 1952. Systematic and biological studies on North American chiggers of the genus *Trombicula,* subgenus *Eutrombicula* (Acarina, Trombiculidae). Annals of the Entomological Society of America, 45:645–677.
333. Wood, S. F. 1952. Mammal blood parasite records from southwestern United States and Mexico. The Journal of Parasitology, 38:85–86.
334. Wrenn, W. J., and R. B. Loomis. 1967. *Otorhinophila,* a new genus of chiggers (Acarina: Trombiculidae) from western North America. Acarologia, 9:152–178.
335. ———. 1973. A new species of *Euschoengastia* (Acarina: Trombiculidae) from western North

America, and the status of *E. californica* (Ewing). Journal of Medical Entomology, 10:97–100.
336. ———. 1974. The *Euschoengastia radfordi* species complex (Acarina: Trombiculidae) from western North America, with descriptions of five new species. Annals of the Entomological Society of America, 67:241–256.
337. WRENN, W. J., AND R. E. SOMERBY. 1974. Descriptions of two new species of *Euschoengastia* (Acarina: Trombiculidae) from California and Baja California Norte, Mexico. Bulletin of the Southern California Academy of Science, 73:5–8.
338. YAMAGUTI, S. 1959. Systema Helminthum. Vol. II. The cestodes of vertebrates. Interscience, New York, 860 pp.
339. ———. 1961. Systema Helminthum. Vol. III. The nematodes of mammals. Interscience, New York (2 vols.), 1261 p.
340. YUNKER, C. E., AND J. M. BRENNAN. 1962. Endoparasitic chiggers: II. Rediscovery of *Doloisia synoti* Oudemans, 1910, with descriptions of a new subgenus and two new species (Acarina: Trombiculidae). Acarologia, 4:570–576.
341. YUNKER, C. E., AND F. J. RADOVSKY. 1966. The dermanyssid mites of Panama (Acarina: Dermanyssidae). Pp. 83–103, *in* Ectoparasites of Panama (R. L. Wenzel and V. J. Tipton, eds.). Field Museum of Natural History, Chicago, 861 pp.
342. ZELEDON, R., B. MCPHERSON, AND C. PONCE. 1977. Isolation of *Leishmania braziliensis* from a wild rodent in Costa Rica. American Journal of Tropical Medicine and Hygiene, 26:1044–1045.

ONTOGENY

JOHN F. EISENBERG

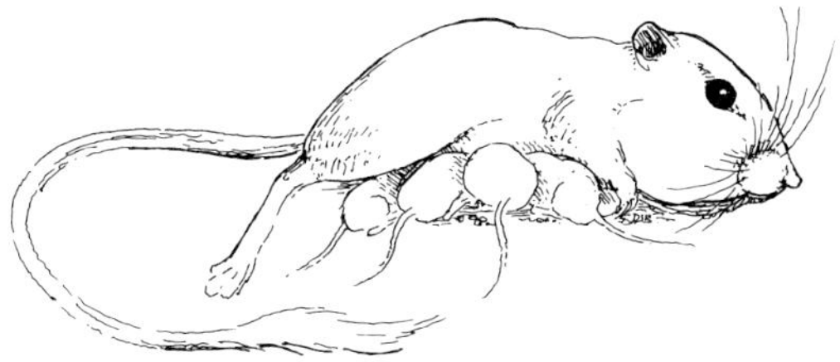

Introduction

In the broadest sense ontogeny is the study of behavioral development, growth, and associated morphological changes. One can begin the study of ontogeny at the moment of the first division of the fertilized egg and thus include embryology and the behavior of embryos within the context of ontogenetic development. In this study I will mainly consider growth and development from birth until the attainment of adulthood.

Physical maturation as determined by growth and the attainment of adult proportions allows one to analyze behavioral development from the standpoint of different stages. An animal can only behave in patterns determined by its skeletal, neural, and muscular systems. Since during growth these systems are in a dynamic state, the attainment of behavioral capacity is often constrained until a critical stage has been passed in physical development (Grand, 1983). Development involves change and thus the terminology employed in the description of behavioral development must have a dynamic aspect. By and large behavior patterns can be divided into those involved in the maintenance activity of the individual and those that are interactive or require the participation of another organism. The ontogeny of simple maintenance behaviors has been explored in depth with laboratory rodents but the analysis of the ontogeny of socialization and interaction systems has lagged (Fentress, 1983).

The development of behavior and growth in heteromyid rodents is a challenging area for future research. Not a great deal of work has been done mainly because most of the species are rather difficult to breed in captivity. Most forms are seasonally polyestrous and aside from the mating association between the male and the female, adults behave aggressively toward one another and a breeding program has to be conducted with great care. Fights to the death are not uncommon even between opposite-sexed adults when the female is unreceptive (Eisenberg and Isaac, 1963). Although we have developmental data on approximately 17 species, only 9 species have been studied in sufficient detail to give a full account of development from conception to weaning (see Table 1).

Prenatal Development and Gestation

Intrauterine development has been analyzed for two species: *Dipodomys merriami* and *Perognathus intermedius*. Van de Graaf

TABLE 1.—*Reproductive and developmental data for selected heteromyids.*

Taxon	Adult weight (g) ♂	Adult weight (g) ♀	Gestation (days)	Litter size	Days of age: Incisors appear	Days of age: Ears open	Days of age: Eyes open	Days of age: Weaning	Days of age: Sexual maturity	Neonatal weight (g)	$\bar{X}$ litter mass (g)	References
Heteromyinae												
Heteromys desmarestianus	83	62	27	3.1	3	14	19	28		3.0	9.3	Fleming, 1977
Liomys pictus	52	41	25	3.5	1–3	14	19	24–28	90	2.5	8.75	Eisenberg, 1963
Liomys salvini	51	39	27	3.8	3	10	14	26		2.5	9.5	Fleming, 1977
Perognathinae												
Perognathus longimembris		8.3	25	3.3	5	13	14	14–18	42	1.3	4.29	Hayden and Gambino, 1966
P. parvus		19	23	5.3			15			1.5	8	O'Farrell et al., 1975; Eisenberg, unpublished
P. flavus		10	≥26	4	5	14	15					Eisenberg and Isaac, 1963
Chaetodipus penicillatus		20	≥26	4	~9	≥14	14					Eisenberg and Isaac, 1963
C. californicus		29	25	3	10	12	15	22–24		1.5	4.5	Eisenberg, 1963
Micropipodops pallidus		13		3.9						1	3.9	Eisenberg, 1963
Dipodomyinae												
Dipodomys deserti		145	30.5	2.5	9+	12	16	15–25		5.5	13.7	Butterworth, 1961
D. merriami		40	28.7	2.8	2+	9	12	17–22	60	3.3	8.3	Butterworth, 1961
D. heermani		60		3.2	11–12	16	14–15	17–25	58	3.7	11.8	Tappe, 1941
D. stephensi		67		2.7	8	10	14	18–20		4.4	11.9	Lackey, 1967
D. nitratoides		42	32	2.3	5	8	10	21–24	84	3.3	7.6	Eisenberg, 1963
D. panamintinus		78	29	3.5		13	17	27–29		4.5	15.8	Eisenberg, 1963
D. ordii		52	29	3	?		?	?	83			Day et al., 1956
D. spectabilis		123	≤27	2.4	?		?	?		7.8	18.7	Bailey, 1931

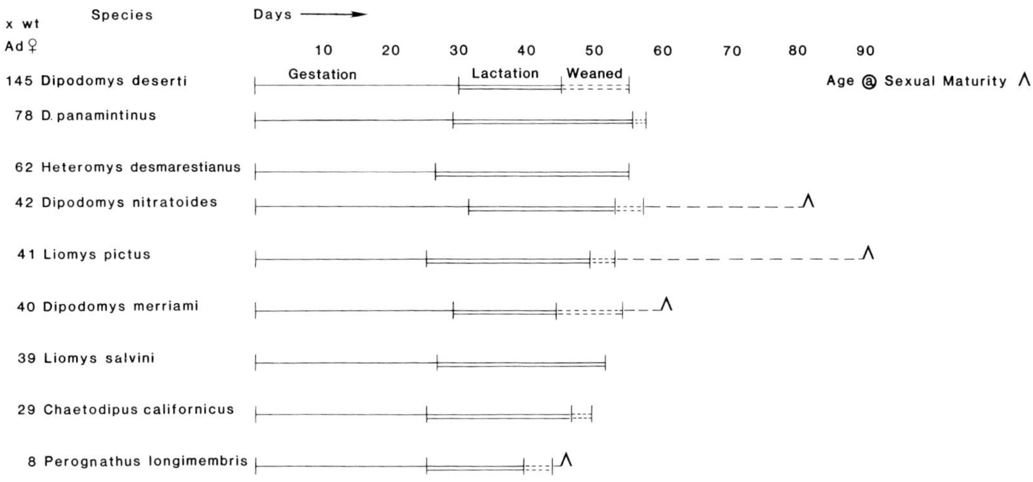

FIG. 1.—Developmental schedule from conception to sexual maturity for nine species of heteromyid rodents. Gestation periods for *Dipodomys* appear to be longer than those recorded for other genera. Data taken from Table 1.

(1973) concentrated on bone development and noted that both species displayed ossification that proceeded initially from the cranium and then caudally. In *P. intermedius* ossification was slower than in *D. merriami.* Pronounced allometric bone development was noted in the mastoid region of the skull and in the tibia and metatarsals of *D. merriami.* These data are consistent with observed hypertrophy of the mastoid bullae and hind limbs shown by adult *Dipodomys.*

Prenatal development in heteromyid rodents tends to be somewhat longer than comparable-sized sigmodontine rodents. Furthermore, there appears to be some correlation between gestation length and taxonomic affinity. The Heteromyinae and Perognathinae have gestations in the range of 23 to 27 days. The kangaroo rats (Dipodomyinae) have gestation lengths from approximately 29 to 32 days. For their body size, heteromyid rodents have modest litter sizes averaging from 2.4 to 5.3. Litter mass at birth ranges from 4.3 to 18.7 grams. This may be as high as 50% of the adult female's mean weight in a small species such as *Perognathus longimembris,* or as low as 10% of mean adult female weight in *Dipodomys deserti* (Table 1).

The young of all species are born naked except for the tactile hairs. The toes are somewhat separated at birth, the pouches are represented as small grooves on either side of the mouth. The incisors have not erupted at the time of birth and the eyes, as well as the auditory meatus, are closed. The condition at birth is not very different from the newborn of mice of the genus *Peromyscus.* However, in *Peromyscus* the toes are usually fused. Although the young heteromyid rodent may be somewhat more precocious than a comparable-sized cricetine, by any criteria they are altricial (Eisenberg, 1963).

Birth, Maternal Behavior, and Lactation

Maternal behavior patterns are basically the same for all species that have been studied. The female generally shows heightened activity on the day preceding birth. Much of this activity is directed toward nest building and caching seeds within the burrow system. The female stretches frequently prior to the delivery of the neonate. The neonate is pulled from the birth canal using the incisors and forepaws. The female licks blood and mucus from the neonate and consumes the placenta. Females crouch over the newborn, usually arching the back. The young are permitted to nurse and she adopts this same posture even when not nursing during the time that the young are unable to thermoregulate (Eisenberg, 1963).

Throughout the early parental care period the female grooms and removes the neonate's feces and urine by licking the anal and genital regions. In handling the young the female employs several fundamental patterns: pulling under, patting, and pushing. The latter two patterns are identical to pushing and patting materials within the nest. A pup that has wandered out of the nest chamber is retrieved by the female. She picks it up by a fold of skin with her incisors and transports it back to the nest. The female may initially seize the skin anywhere on the neonate's body but generally shifts to carry it by the skin of the back. While studying *Liomys pictus,* on two occasions I noticed a female transport young by placing them in her cheek pouches and moving them from one portion of the tunnel system to the next (Eisenberg, 1963).

The lactation period of heteromyid rodent females varies from 15 to some 29 days. The shorter lactation time is characteristic of the smaller species. Weaning is defined differently by different authors. In some cases, it is considered as the last day nursing was observed; in other cases it is where there is a noticeable transition from milk dependency to consuming adult foods. Clearly, weaning is a process and does not occur abruptly. In Figure 1, I have attempted to portray for 9 species the developmental cycle from conception to weaning. The larger species complete the rearing cycle within 53 to 60 days. The smaller species may accomplish rearing from 45 to 50 days.

Growth

The most exhaustive analysis of growth in heteromyid rodents concerns two species: *Dipodomys merriami* and *Dipodomys deserti* (Butterworth, 1961; Chew and Butterworth, 1959). However, growth data have also been assembled for *Perognathus longimembris* (Hayden and Gambino, 1966), *Perognathus californicus* (Eisenberg, 1963), *Dipodomys heermani* (Tappe, 1941), *Dipodomys merriami* (Reynolds, 1960), *Dipodomys nitratoides* (Eisenberg, 1963), *Dipodomys stephensi* (Lackey, 1967), *Liomys pictus* (Eisenberg, 1963), *Liomys salvini* (Fleming, 1977), and *Heteromys desmarestianus* (Fleming, 1977). The recent publication by Zulinger et al. (1984) discusses the problems of fitting sigmoidal equations to mammalian growth curves. They chose the Gompertz model to fit a sample of 331 species and calculated growth rate constants (K). Their constants ranged from a low of 0.0265 for *D. heermani* to a high of 0.099 for *H. desmarestianus.*

Butterworth (1961) compared growth in *D. deserti* and *D. merriami*; he plotted his measurements on a semilogarithmic scale against age on an arithmetic scale. Linear segments of such a plot indicated periods when growth increments were constant percentages of the previous sizes. From these linear sections he calculated instantaneous growth rates. His comparison showed that the smaller *D. merriami* had a more rapid growth rate from birth until approximately 10 days of age than that shown by *D. deserti.* The growth rate from 10 to 23 days of age for *D. merriami* slowed to approach a value achieved by *D. deserti* at approximately 12 to 29 days of age.

His analyses of linear measurements are especially interesting because they indicate the very different rates of growth when different segments of the body are compared. Since kangaroo rats are bipedal, and the relative size of the hind foot is not terribly exaggerated at birth, it is not surprising that the growth rate of the hind foot is quite rapid from birth until approximately 24 days of age. The necessity of attaining a long hind foot at an early age is directly related to the fact that the young animals begin to explore and locomote outside the burrow at ap-

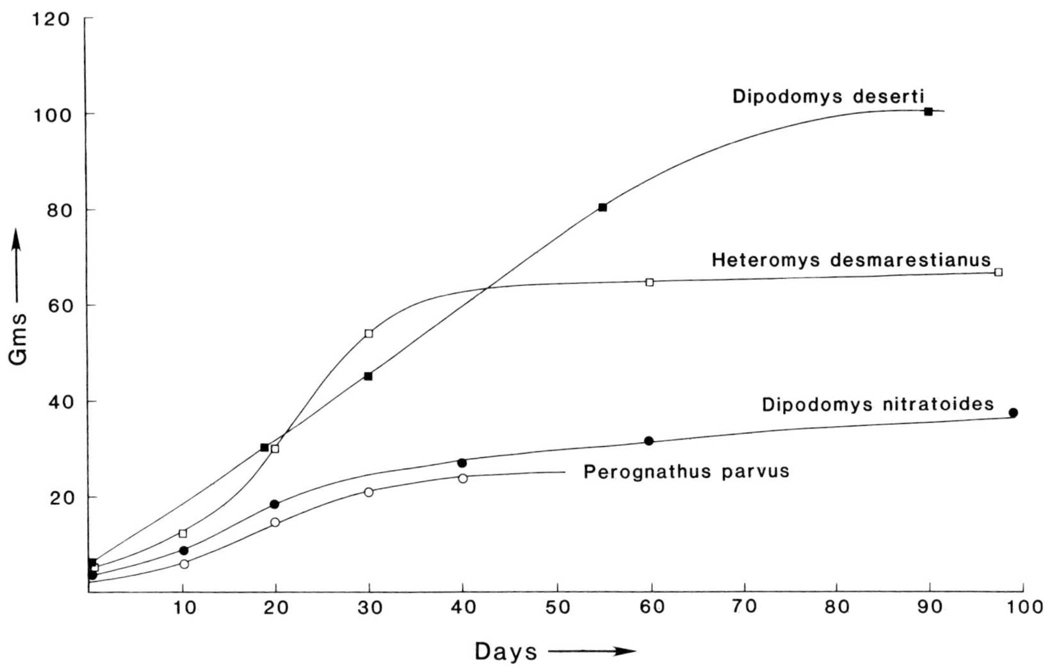

FIG. 2.—Growth curves for four species of heteromyid rodents. Average values for several litters are plotted. Data for *D. deserti* from Butterworth (1961); *H. desmarestianus,* Fleming (1977); *Dipodomys nitratoides,* Eisenberg (1963); and *P. parvus,* Eisenberg, unpublished.

proximately one month of age. In this case the relative growth of the body parts determines the form of locomotion that the young animal can display. Similar results were noted in the growth studies by Lackey (1967) with *D. stephensi* and Tappe (1941) with *D. heermani.* Growth curves for selected species of heteromyids are portrayed in Figure 2.

Physical Development and Molts

At birth the neonate has several sets of tactile hairs which may be important in orientation to litter mates and the mother. The eyes and auditory meatus are closed. In general the meatus opens before the eyes (Table 1). Since lactation requires that young seek out and find the nipple I assume that tactile, olfactory, and gustatory cues may be involved. The young respond to temperature changes since cooling causes a repetitive vocalization (to be discussed below). The incisors do not erupt until several days after birth. The lower incisors appear first (see Table 1). At birth the pouches are represented by small grooves on either side of the mouth. They gradually deepen but do not become conspicuous until after the second week. In all species during the first week of life the tail is often coiled and exhibits some tendency to coil if stroked with a probe (Eisenberg, 1963).

The young are born naked except for the tactile hairs. The skin is usually a light pink that darkens dorsally as the dorsal pelage begins to appear. The venter becomes haired after the dorsum and the first coat is shed shortly after the body is covered. As the dorsal hairs grow, the final effect is lighter and grayer than the first appearance of the hairs. This pelage is softer in texture than that of the adult. The final molt results in a pattern and texture approximating the adult coat. Table 2 presents the molting sequence for three species of heteromyids. As will be

TABLE 2.—*Molt and pelage in the Heteromyidae. (Numbers refer to days unless otherwise indicated.)*

	First coat					
	Dorsum		Ventrum			
Species	Begins	Completed	Begins	Completed	Subadult pelage	Molt to adult pelage
Liomys pictus	9–10 tan	16 brown	13 white	19 white	22–27 gray-brown	6–9 wk; gradual molt to adult brown begins
Perognathus californicus	9 gray	16 brown	16 white	20 white	20–25 gray-brown	8–10 wk; molt to dark brown and spines
Dipodomys nitratoides	6 dark brown	7 dark brown	9 white	10 white	13–18 light-brown	ca. 12 wk; gradual molt to dark brown begins

noted for *Liomys pictus,* dorsal natal pelage is complete at 16 days and the ventral at 19 days; while for *Perognathus californicus* the age is similar being 16 and 20 days, respectively. *Dipodomys nitratoides* attains its pelage at an earlier age, the dorsum being completed at 7 days and the venter at 10. The subadult pelage begins to appear at 18 days in *D. nitratoides,* but not until 25 and 27 days for *Perognathus californicus* and *Liomys pictus.* The molt to the adult pelage is attained at 9 weeks in *Liomys,* approximately the same in *Perognathus californicus,* but not until 12 weeks of age in *D. nitratoides.* The attainment of the natal pelage at an early age by *D. nitratoides* is characteristic of species of the genus *Dipodomys* and perhaps this precociality can be related to the longer gestation manifested by the dipodomyine rodents.

The Ontogeny of Communication

The young of heteromyid rodents produce sounds of four types: the first sound is a faint peeping that accompanies grooming by the female; the second is the tic sound accompanying sucking movements when the animal is removed from the teat; the third is a low scratchy whine when the animal is cold or isolated from the litter and the female; and the fourth is a sharp squeak produced when the young are roughly handled (Eisenberg, 1963). Recordings were made of the scratchy whines for *L. pictus, P. californicus,* and *D. nitratoides.* The scratchy whine of the neonate is similar to the adult growl. The scratchy whine is composed of two syllables, a long syllable ranging from 0.05 to 0.25 seconds and a short syllable that may be as brief as 0.01 seconds in duration. When this latter syllable is repeated several times it may be called a cackle sound. The syllables are harmonic systems with a fundamental of around 500 cycles per second. However, the harmonic structure may be blurred or lost when syllables are of very short duration. The results from analyzing the long syllable are portrayed in Table 3. The frequencies with the greatest energy lie between 1,000 and 3,200 Hz. Durations are similar ranging from 0.07 to 0.12 seconds.

The pitch of the scratchy whine seems to be higher in *Perognathus californicus* and *Liomys pictus* than in *Dipodomys nitratoides. Dipodomys* almost always emphasizes the second harmonic and several higher ones, whereas *Liomys* and *Perognathus* emphasize the third, fourth, and other higher harmonics. It will be appreciated that the so-called abandoned cry of *Peromyscus* (Eisenberg, 1968) is higher in pitch than the scratchy whine of the heteromyids. When the sounds of scratchy whines are played back to a female it causes great activity and random searching movements. Clearly, the scratchy whine functions as an abandoned cry and allows the mother to locate a young displaced from the nest (Eisenberg, 1963; LaVick, 1982). As the animals mature all vocalizations increase in amplitude so that

TABLE 3.—*Physical description of the long syllable of the scratchy whine for 3 heteromyid species.*

Species	Age in days	Number of harmonics recorded	Lowest frequency (c)	Frequencies with the greatest energy (c)	Harmonics emphasized	Duration (sec)
Dipodomys nitratoides	4	. . .[a]	ca. 1,000	1,500	. . .[a]	0.07–0.12 (avg. 1)
Dipodomys nitratoides	6	ca.6	250–800	1,000–3,200	II, V, VI; II, III, IV	0.07–0.12 (avg. 11)
Dipodomys nitratoides	7	ca.6	800–1,000	1,600–2,000	II	0.08–0.12 (avg. 9)
Perognathus californicus	28	ca. 7	250–500	1,700–2,800	III, IV, VII, VIII	0.05–0.07
Liomys pictus	1	5–7	500	1,800–3,500	III, IV, V	0.07–0.09

[a] Harmonic structure blurred.

the scratchy whine approximates the adult growl. Table 4 summarizes the vocalizations of the neonate and their transformation to adult sounds.

In all species by the time the young are three weeks of age they exhibit competence in vision and hearing. Affiliative behaviors among litter mates while in the nest include grooming one another. Clearly at this stage tactile and olfactory input are extremely important. Active grooming among siblings appears at 16 days of age in *Perognathus californicus,* 14 days in *Dipodomys nitratoides,* 24 days in *Dipodomys panamintinus,* and 20 days in *Liomys pictus* (Eisenberg, 1963). Communication among adults involves all of the major sensory systems (Eisenberg, 1963). Olfactory communication is important in intraspecific interaction, and the coordination of sexual behavior (Martin, 1977; Randall, 1983, 1985). Sandbathing is important not only in dressing the pelage but also in chemical communication (Eisenberg, 1963; Randall, 1981). Auditory communication need not involve only vocalizations since foot drumming is an important communication system in some species of *Dipodomys* (Eisenberg, 1963; Randall, 1984).

Behavioral Development to Weaning

Development of Individual Behavior Patterns

At birth the young animal can breathe and vocalize weakly. It tries to suck and in so doing mouths the belly fur of the female. The neonate can move its forelimbs forward and backward but the hind limbs seem to

TABLE 4.—*Summary of vocalizations in the young Heteromyidae and their transformation to adult sounds.*

Sound in neonate	Circumstances	Sound in adult
Tic	Accompanies sucking movements	(Disappears at end of sucking period)
Squeal	Accompanies "painful" stimuli	Pain squeal
Peeping	Accompanies grooming by the female (comfort sound)	(Disappears)
Scratchy whine	Accompanies cooling and foreign odors (?)	Growl
Repeated scratchy whine (abandoned cry)	Stimulates the female to begin searching	(Disappears)

TABLE 5.—*Schedule of first appearance of various behavior patterns. (Numbers refer to age in days.)*

Species	Locomotion: Diagonal limb pattern	Locomotion: Quadrupedal hop	Feeding on solids	Stretch	Scratch	Shake	Dig	Complete wash
Liomys pictus	13–19	17–21	13–14	[a]	16–17	[a]	20	15–17
Perognathus californicus	11–16	18	13	[a]	14	[a]	[a]	11–12
Dipodomys nitratoides	7–10	11–16	9	9	15	9	10	13

[a] Uncertain of first occurrence.

be weaker. They move more or less with a gross movement of the whole body. The torso can be extended and flexed and the head raised and lowered.

When the animal moves its forepaws it may brush its nose with them or push them against the female's venter while nursing (treading). It can crawl using its forelimbs and slightly moving its hind legs as the torso is wriggled from side to side. If the neonate is placed on a table top, it attempts to crawl and switches its head from side to side (seeking). At the end of each lateral swing the young animal presses its mouth against the substrate (nosing and head bobbing). As it begins to become cold it utters the scratchy whine which repeated synchronously is the "cry of the abandoned young."

Coordination develops in an anterior to posterior direction. The forelimbs predominate in the early phases of locomotion. Gradually the hind limbs and forelimbs are synchronized into the crossed extension walk. As the young animal gathers strength the venter can be held off the ground during a bout of running or walking. The transition to quadrupedal and (in *Dipodomys*) bipedal saltation follows as the young animal gathers strength and ability to coordinate. Most of the behavior patterns seem to develop gradually but the scratch reflex appears suddenly and often occurs at first without being directed against any one part of the body. The young animal simply begins to move the hind leg in a rapid, undirected, up and down manner. Digging develops gradually from the alternate movements of the forepaws. Washing develops by a combination of wiping the nose and licking the forepaws. Shaking seems to appear spontaneously. Elimination is stimulated for the first few days by the mother's licking. The female eats the feces and licks up the urine. But as the young begin to move about the burrow they soon begin to urinate at the female's elimination spot and the female's attentions wane. A developmental schedule for the first appearance of several behavior patterns in heteromyids is included in Table 5.

The young *Liomys* remain blind for almost three weeks, but they have already begun to nibble solid food and use the place of elimination. The young animal usually encounters the spot of urination, sniffs, licks, and then turns and urinates. This suggests that the smell of urine stimulates active excretion by the young. The significance of the behavior in adult forms of chemi-communication is evident. It appears that many adult mammals use urine for communication, and the smell of urine may serve to trigger a reciprocal response (Eisenberg, 1963).

The abandoned cry is given by young *D. nitratoides* and *L. pictus* up until a few days after their eyes open. Feeding and gnawing begin gradually after the emergence of the incisors. As the young continue to grow and develop their ability to dig and locomote they try to follow the female on her nightly forays. The female usually plugs the burrow as she leaves in order to prevent this, and if a pup should slip out it is promptly returned to the nest by the female. As the young mature and are weaned the female's retrieving response wanes and they begin to leave the burrow and forage at night also.

Sandbathing seems to develop spontaneously, but because rubbing the side and stretching have been practiced by the young animal before it begins to actually sandbath it is difficult to establish absolute times of onset.

Although foraging above ground does not usually occur until weaning, the young manipulate and move food within the burrow. Cheek pouches are used for transporting food at the time of weaning and Lackey (1967) noted this behavior at 20 to 21 days of age in *D. stephensi.*

Development of Social Behavior Patterns

Social behavior begins with litter mate and maternal-neonate interaction and it is difficult to assign precise beginnings to any one set of patterns. Initially the neonate sucks, noses, treads, and suckles and the interaction is primarily directed at the mother. The young, when unattended by the mother, tend to huddle in a pile, push and lick one another, and also crawl over and under. They begin to groom one another while they are washing themselves. The young animal readily shifts from licking its own to licking a sibling's fur. In a similar manner the young will begin to groom the mother's fur and investigate its own anogenital region and that of its associates. Pushing onc another develops into the warding behavior seen in adult interactions, while crawling over and under involves many positions assumed in mounting and other sexual behaviors when they attain adulthood.

When siblings are removed from the nest for weighing and measurement and placed in a holding container they tend to crawl until contacting one another and then huddle. Genuine orientation movements such as pausing, touching noses, and sniffing at each other, however, occur at a later date. Oriented activity to one another in litter mates who were separated by several inches was observed first at 16 days in *D. nitratoides,* and 19 days in *L. pictus.*

Behavioral Development Post Weaning

The weaning period varies among heteromyid rodents depending on their size. *Perognathus longimembris* weans at a rather early age, while the larger kangaroo rats wean at later ages (see Table 1). Weaning is a gradual process taking up to 4 days. As the young animals mature and about the time of weaning they begin to locomote outside of the burrow. At first the excursions are brief and the distance traversed not far from the burrow mouth. In the captive studies reported on by Eisenberg (1963), leaving the burrow commenced at 18 days for *D. nitratoides* and 29 days for *Liomys pictus.* These two species represent the extremes. *Perognathus californicus* began to leave the burrow between 22 and 25 days of age, and *Dipodomys panamintinus* at 28 days of age. For all species the expansion of activity outside the burrow by the young increases rapidly so that longer excursions are being made at the age of about one month. Estimated ages of one month to five weeks correspond well with field trapping data when juveniles are taken (Eisenberg, 1963).

In the captive situations attempts at mounting behavior among siblings may be shown at 39 days of age in *Perognathus californicus,* 35 days of age in *Dipodomys nitratoides,* and 56 days of age in *D. panamintinus.* Some of these "sexual" behaviors may be considered pseudosexual because actual maturity may not be achieved until considerably later. Onset of sexual activity with conceptions does not occur in *D. nitratoides* until they are approximately 50 days of age. *D. merriami,* however, can ex-

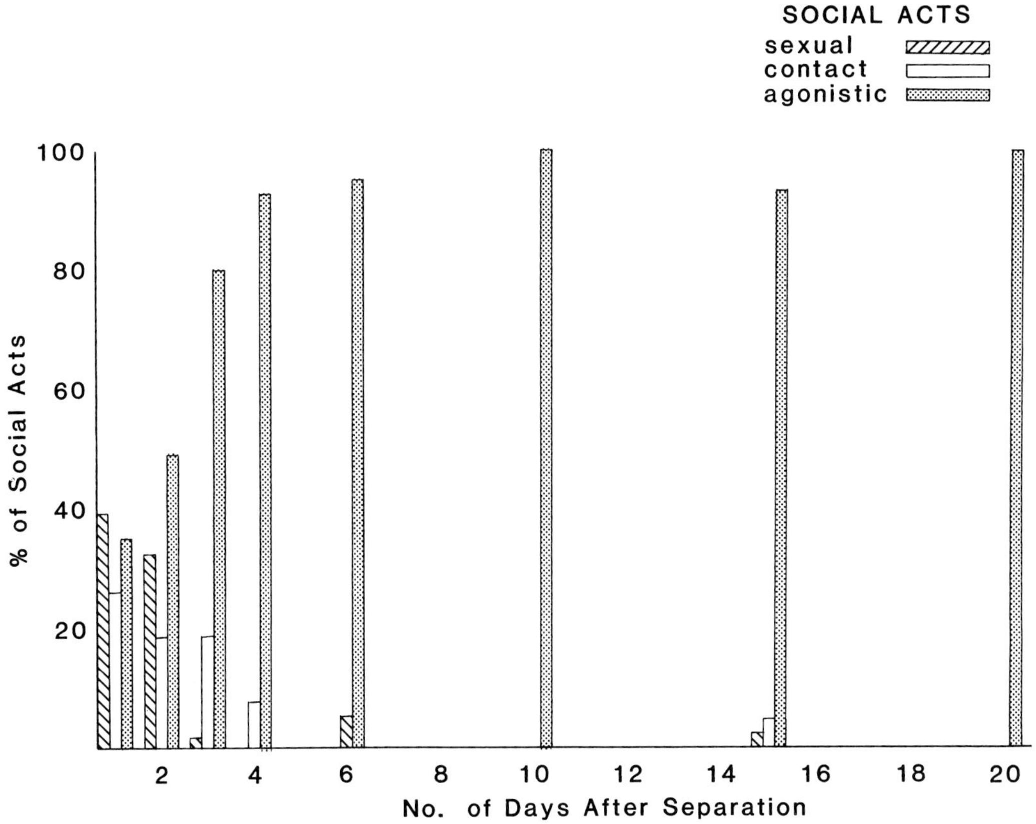

FIG. 3.—Onset of agonistic behavior following separation of litter mates. Agonsitic behavior increases with increasing time following separation. Data from Eisenberg, unpublished. See text.

hibit estrus as young as 32 days of age, and the small *Perognathus longimembris* in the laboratory has been recorded as exhibiting estrus at 21 days of age. Clearly in some great basin species of heteromyids that experience severe winters the onset of reproductive activity may not occur until the year following the animal's birth, for example, *Perognathus formosus* (French et al., 1974) and *Perognathus parvus* (O'Farrell et al., 1975). Some of the very large species of kangaroo rats similarly may not exhibit reproductive activity until the year following their birth (*Dipodomys spectabilis*; Jones, 1982).

Although litter mates born in captivity may often be caged together compatibly prior to and even beyond the onset of sexual maturity, usually aggressive tendencies will eventually manifest themselves. The litter mates of *P. parvus* can be maintained together until eight weeks of age but rapidly become aggressive toward one another if the litter mates are separated and then reintroduced. Figure 3 portrays the results of an isolation experiment utilizing four sibling *P. parvus* young. It can be easily seen that within 48 hours the young no longer respond to each other with affiliative behaviors but increasingly agonistic elements begin to appear until, after a week of separation, the young behave toward one another as if they were complete strangers.

Dispersal of young from the natal burrow and their establishment in a burrow system of their own is poorly understood. There is some field evidence to indicate that at certain seasons of the year not all burrows in a given area will be occupied. In one population study 42 out of 121 *D. spectabilis* mounds were unoccupied (Schroder and Geluso, 1975). There is evidence that an individual may utilize more than one burrow. Chapman and Packard (1974) found that male *Dipodomys merriami* use 6 to 7 burrows on the average, and females use 5. These data suggest that under some conditions old burrows may be occupied by dispersing young. Jones (1982) reports and documents several interesting cases of young remaining in the natal mound and in some cases "inheriting it" from their mother in *D. spectabilis.* Best (1972) notes that *D. spectabilis* may under some circumstances construct their own burrows. No doubt patterns of residency and dispersal vary widely from species to species depending on the onset of breeding after weaning. Some species clearly are geared to breed in the year of their birth, e.g., *Perognathus longimembris, Liomys pictus,* and *Dipodomys merriami.* In these early breeding species the attainment of adult display patterns occurs prior to or at the onset of their first reproductive activity.

Discussion and Comparisons

Jones (1985) concluded from an analysis of both field and laboratory data that there was no general relationship between body size and life history variables in heteromyids. I concur that the heteromyids show only a small amount of variation in gestation length and weaning time. However, reference to Figure 1 will indicate for those species where we have good data that the total developmental time from conception to weaning does indeed demonstrate that the smaller forms attain above ground competency earlier than the larger species. Further, my original conclusions (Eisenberg, 1963) that, given equivalent adult body mass, gestation length in the Heteromyinae and Perognathinae is shorter than in the Dipodomyinae appears to be valid. Thus, phylogenetic affinity and size do play a role in some of the variation we can detect when different species of heteromyid develomental schedules are compared.

Behavioral development in the young follows an anterior to posterior gradient. From a ground plan of basic reflexes the integrated patterns are gradually developed. The bipedal locomotion in *Dipodomys* is preceded by a pattern of diagonal limb coordination followed by the quadrupedal ricochet. The relative growth rates of the different body segments set limits at different stages of development defining what type of locomotive behavior and postures can be assumed. Most of the fundamental patterns seem to develop without imitative learning.

Literature Cited

Bailey, V. 1931. Mammals of New Mexico. United States Department of Agriculture, Division of Biological Survey. North American Fauna, 53:1–412.

Best, T. 1972. Mound development by a pioneer population of banner tailed kangaroo rats *Dipodomys spectabilis baileyi* Goldman in eastern New Mexico. American Midland Naturalist, 87:201–206.

Butterworth, B. B. 1961. A comparative study of growth and development of the kangaroo rats, *Dipodomys deserti* Stephens and *Dipodomys merriami* Mearns. Growth, 25:127–139.

Chapman, B., and R. L. Packard. 1974. An ecological study of Merriam's pocket mouse in southern Texas. Southwestern Naturalist, 19:281–291.

Chew, R. M., and B. B. Butterworth. 1959. Growth and development of Merriam's kangaroo rat, *Dipodomys merriami.* Growth, 23:75–95.

Day, B. H., H. J. Egosque, and A. Woodbury. 1956. Ord's kangaroo rat in captivity. Science, 124:485–486.

Eisenberg, J. F. 1963. The behavior of heteromyid rodents. University of California Publications in Zoology, 69:1–100.

———. 1968. Behavior patterns. Pp. 451–495, *in* Biology of *Peromyscus* (Rodentia) (J. A. King, ed.). American Society of Mammalogists, Special Publication No. 2, Allen Press, Lawrence, Kansas, 593 pp.

Eisenberg, J. F., and D. E. Isaac. 1963. The reproduction of heteromyid rodents in captivity. Journal of Mammalogy, 44:61–67.

Fentress, J. 1983. A view of ontogeny. Pp. 24–64, *in* Advances in the study of mammalian behavior (J. F. Eisenberg and D. G. Kleinman, eds.). American Society of Mammalogists, Special Publication No. 7, Allen Press, Lawrence, Kansas, 753 pp.

Fleming, T. H. 1977. Growth and development of two species of tropical heteromyid rodents. American Midland Naturalist, 98:109–123.

FRENCH, N. R., B. G. MAZA, H. O. HILL, A. P. ASCHWANDEN, AND H. W. KAAZ. 1974. A population study of irradiated desert rodents. Ecological Monographs, 44:45–72.

GRAND, T. I. 1983. The anatomy of growth and its relation to locomotor capacity in *Macaca*. Pp. 5–23, *in* Advances in the study of mammalian behavior (J. F. Eisenberg and D. G. Kleinman, eds.). American Society of Mammalogists, Special Publication No. 7, Allen Press, Lawrence, Kansas, 753 pp.

HAYDEN, P., AND J. J. GAMBINO. 1966. Growth and development of the little pocket mouse *Perognathus longimembris*. Growth, 30:187–197.

JONES, W. T. 1982. Natal nondispersal in kangaroo rats. Ph.D. dissert., Purdue University, West Lafayette, Indiana, 185 pp.

———. 1985. Body size and life history variables in heteromyids. Journal of Mammalogy, 66:128–132.

LACKEY, J. A. 1967. Growth and development of *Dipidomys stephensi*. Journal of Mammalogy, 48: 624–632.

LAVICK, J. P. 1982. Maternal response to neonate vocalizations in Ord's kangaroo rat (*Diplomys ordii*). Southwestern Naturalist, 27:122–123.

MARTIN, R. E. 1977. Species preference of allopatric and sympatric populations of silky pocket mice, genus *Perognathus* (Rodentia: Heteromyidae). American Midland Naturalist, 198:124–136.

O'FARRELL, T. P., R. J. OLSON, R. O. GILBERT, AND J. D. HEDLUND. 1975. A population of Great Basin pocket mice, *Perognathus parvis* in the shrub-steppe of south central Washington. Ecological Monographs, 45:1–28.

RANDALL, J. A. 1981. Olfactory communication at sandbathing loci by sympatric species of kangaroo rats. Journal of Mammalogy, 62:12–19.

———. 1983. Olfactory communication in kangaroo rats (*D. merriami*). *In* Chemical signals in vertebrates III (D. Muller-Schwarze and R. M. Silverstein, eds.). Plenum Press, New York.

———. 1984. Territorial defense and advertisement by footdrumming in bannertail kangaroo rats (*Dipodomys spectabilis*) at high and low population densities. Behavioral Ecology and Sociobiology, 16:11–20.

———. 1985. Role of urine in coordinating reproduction in a desert rodent (*Dipodomys merriami*). Physiological Behaviour, 34:199–203.

REYNOLDS, H. G. 1960. Life history notes on Merriam's kangaroo rat in southern Arizona. Journal of Mammalogy, 41:48–58.

SCHRODER, G. D., AND K. N. GELUSO. 1975. Spatial distribution of *Dipodomys spectabilis* mounds. Journal of Mammalogy, 56:363–368.

TAPPE, D. T. 1941. Natural history of the Tulare kangaroo rat. Journal of Mammalogy, 22:117–148.

VAN DE GRAAF, K. 1973. Comparative developmental osteology in three species of desert rodents *Peromyscus eremicus, Perognathus intermedius* and *Dipodomys merriami*. Journal of Mammalogy, 54:729–741.

ZULINGER, E., R. E. RICKLEFS, K. H. REDFORD, AND G. M. MACE. 1984. Fitting sigmoidal equations to mammalian growth curves. Journal of Mammalogy, 65:607–636.

THE PROXIMAL COLON OF HETEROMYID RODENTS: POSSIBLE MORPHOPHYSIOLOGICAL CORRELATES TO ENHANCED WATER CONSERVATION

G. LAWRENCE FORMAN AND CARLETON J. PHILLIPS

Introduction

Many species of rodents of the family Heteromyidae have remarkable physiological capabilities, as discussed in several chapters in this book. In particular, numerous species are adapted to life in the deserts of western North America. In order to survive in harsh environments, a positive water balance must be maintained even in the absence of any available free water. Numerous studies have documented physiological and behavioral features that serve to maximize water conservation in these animals. Metabolically-produced water, nocturnal activity, fossorial inactivity, excretion of hyperosmotic urine, storage of food in a moist burrow, and reduced evaporative and fecal water loss are among the best-studied of these features. It is the prospective role of the proximal colon in water balance that is the subject of this chapter.

The importance of the proximal colon in water absorption in rodents is well-established (see Lange and Staaland, 1970, 1971; Steggerda, 1935). Reduced water content in the feces of desert dwelling rodents, when compared with laboratory *Rattus,* has been reported by several investigators (Nasledova and Pechurikina, 1979; Schmidt-Nielsen, 1964). MacMillen and Lee (1967) have presented values for percent fecal water content in several Australian desert rodents (Muridae), which are comparable to those reported by Schmidt-Nielsen (1964) for *Dipodomys merriami.* Additionally, data presented in several studies of water balance in heteromyids suggest that reduced fecal water loss might be particularly important to survival during water deprivation. In a study of the effects of diet on water balance in desert-dwelling *Peromyscus maniculatus* (Cricetidae) and *Perognathus parvus* (Heteromyidae), Withers (1982) reported that *P. maniculatus* showed higher rates of fecal water loss than did *P. parvus* when water-deprived on a diet of millet seed. *Peromyscus maniculatus* also had a somewhat lower metabolic water production. The capability of retaining water is wide-spread in the heteromyids; Christian et al. (1978) reported that *Liomys pictus* and *L. irroratus,* which are adapted to semiarid or even somewhat mesic environments, are able to produce extremely dry feces. Indeed, feces produced by *Liomys* are lower in water content than are those of desert-dwelling heteromyids for which data are available for comparison.

Regardless of the likely importance of the proximal colon to our understanding of the unusual physiological capabilities of heteromyids, until very recently no data on its morphology were available in the scientific literature. However, in a recent paper (Forman and Phillips, 1988) we described the general histology of the mucosal glands in the proximal colon in several genera of North American heteromyid and cricetid rodents, including one species each of *Liomys* and *Chaetodipus,* and two species of *Dipodomys.* We reported two general forms of asymmetry in the mucosa of the proximal colon, one in the heteromyids and another one in an arid land cricetid, *Peromyscus crinitus.* Colonic asymmetry in heteromyids was by far the most striking and in this chapter we report: a) additional details concerning the microanatomy of this asymmetry; b) the existence of variation in the histological pattern; and c) the presence of colonic asymmetry in an additional genus, *Heteromys.* We also provide new data on the distribution of a regulatory neuropeptide, methionine-enkephalin (met-ENK), and a neurotransmitter, serotonin (5-HT). Information about the innervation of the colon ultimately might be significant to our understanding of the role of the proximal colon in the ecological physiology of heteromyids.

Methods

Portions of the proximal colon (a 3 cm portion of the colon, beginning at the junction with the caecum) were extracted from live-trapped specimens of *Liomys, Dipodomys, Chaetodipus,* and *Perognathus.* These "fresh" portions were immersed in fixative [either in acetic acid (5%)-formalin (20%)-alcohol (75%), or in 10% non-buffered formalin]. An entire lower digestive tract of *Heteromys* was removed from a field-collected museum specimen whose abdominal cavity had been opened prior to submergence of the whole animal in 10% non-buffered formalin, with subsequent storage in 70% ethanol. We selected the proximal colon for study because of its primary role in water absorption in mammals and avoided the caecum and lower colon (including rectum), which may have their own distinctive histology. Tissue samples were embedded in paraffin, sectioned at 8–10 μm, and sections were mounted on slides so that longitudinal and transverse sections could be examined simultaneously. Sections were stained with 1) gallocyanin and eosin Y (general oversight stains), 2) Milligan trichrome stain (after Humason, 1979) for muscle and collagen, and 3) combined periodic acid-Schiff (PAS) reaction with Alcian Blue 8GX 300 at pH 2.5 (after Lillie and Fullmer, 1976) for mucopolysaccharides and glycoproteins.

Met-ENK- and 5-HT-like immunoreactivity in the nerve fibers and nerve cell bodies in the proximal colon and adjacent mesentery was demonstrated by means of the peroxidase-anti-peroxidase (PAP) technique of Sternberger et al. (1970). Specificity of each antiserum was tested by either preabsorption control or by substitution of normal serum for primary antiserum. The methodology was as follows: 1) paraffin sections were incubated in 3% H_2O_2, placed in 0.05 M Tris buffer (pH 7.6), and treated with normal swine serum; 2) they were incubated in primary antiserum to met-ENK, or 5-HT for 1 hr at 37°C and rinsed in 0.05 M Tris buffer; 3) swine anti-rabbit serum was used as the second antibody and H_2O_2 and amino-ethylcarbazole (DAKO Corp.) were used as the substrate.

The primary antisera, their sources, and the dilutions at which they were used are as follows: 1) anti-methionine-enkephalin, MILAB, Malano, Sweden, batch R-810705-1; 1:640; 2) anti-serotonin, Immuno Nuclear Corp., lot 8431011, 1:100.

This PAP method results in the deposition of a red precipitate over antigenic sites. Thus, one can microscopically visualize the possible locations of particular molecules in situ. It should be mentioned, however, that immunoreactivity does not, in itself, prove the presence of the molecule in question and, therefore, we refer to the stained reaction sites as "ENK-like" or "serotonin-like" immunoreactivity.

The specimens examined, their collecting localities, and their disposition are as follows: 1) spiny pocket mice, *Liomys pictus* KU 120585, 120593 (Jalisco, Mexico), *Heteromys desmarestianus,* KU 84370 (Costa Rica), University of Kansas Museum of Natural History; 2) pocket mice, *Chaetodipus arenarius,* MSB 42718, 42719, 42720, *C. baileyi,* MSB 42686, 42688, 42690, 42691, *C. californicus,* MSB 43126, *C. fallax,* MSB 42862, *C. penicillatis,* MSB 42674, 42710, *C. spinatus,* MSB 42695; *Perognathus flavus,* MSB 42531, *P. formosus,* MSB 42805; 3) kangaroo rats, *Dipodomys agilis,* MSB 42872, *D. gravipes,* MSB 42995, 42996, *D. merriami,* MSB 42548, *D. spectabilis,* MSB 42535 (all from southern California, southwestern New Mexico, Baja California, or Sonora, Mexico), Museum of Southwestern Biology, University of New Mexico (MSB). All of the specimens used herein were collected between mid-June and late July.

Results

Histology

Mucosa

Because the histology of the mucosal portion of the proximal colon is interspecifically variable in the heteromyids, we have arranged the following descriptions by genera, beginning with *Dipodomys.*

Dipodomys. —Approximately three-fourths of the circumference of the proximal colon of the San Quintin kangaroo rat, *Dipodomys gravipes,* is lined with simple tubular glands containing abundant goblet cells mixed with columnar epithelial cells (Fig. 1). Villi are lacking and cellular organization within the tubular gland units resembles that described for small mammals such as laboratory rats and mice. In sharp contrast, the mucosa on the side of the gut tube adjacent to its suspensory mesentery (dorsal) is highly modified, which results in an asymmetrical mucosa.

Two mucosal histomorphs can be distinguished on the mesenteric side. The most commonly observed is shown in Figure 2. The gland units are slightly taller (190–230 μm) than are the contra-mesenteric ones (120–170 μm), are nearly devoid of goblet cells, and possess a thick absorptive-type epithelium within the upper one-half of the gland unit. The basal one-half of the gland unit is narrow; the upper portions are funnel-like and open into broad glandular "pits" at the luminal surface. As a result of the shape of gland units, each intestinal fold, which is made up of many gland units, appears to be constricted at the base and thus presents a broadly fan-shaped outline in cross section. A few of the gland units seen in this region differ in having broad, bulbular bases with large lumina (Fig. 2).

The second type of modified mucosa in *D. gravipes* is illustrated in Figure 3. The transition to this mucosa is abrupt. These mesenteric glands are four to five times longer (up to 740 μm) than the contra-mesenteric ones. Shallow, but broad, "pits" are present at the luminal terminus of each gland and this gives the surface epithelium a wrinkled appearance not found on the other side of the colon. A small number of apparently mature goblet cells (Fig. 3) are found within and below these shallow pits.

In the agile kangaroo rat, *Dipodomys agilis* (Fig. 4), the mucosa on the mesenteric side of the proximal colon is similar to the fan-shaped epithelium described above for *D. gravipes* (Fig. 2), although regions of the mucosa have gland units that histologically are somewhat intermediate between the histomorphs described above. Goblet cells are extremely sparse within the mesenteric mu-

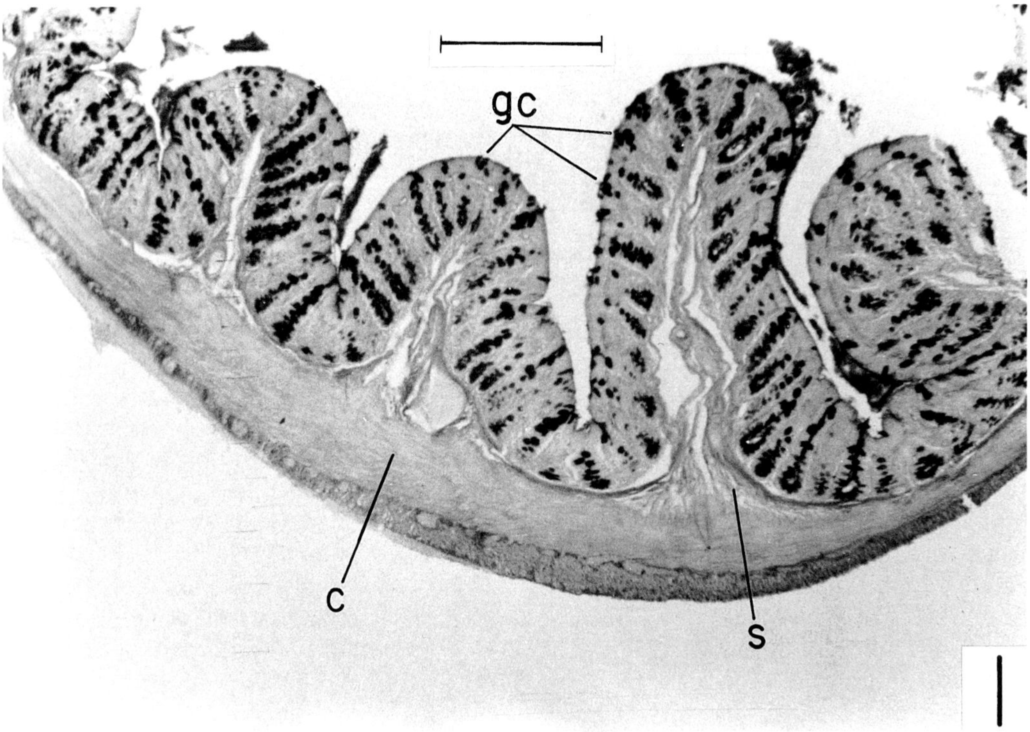

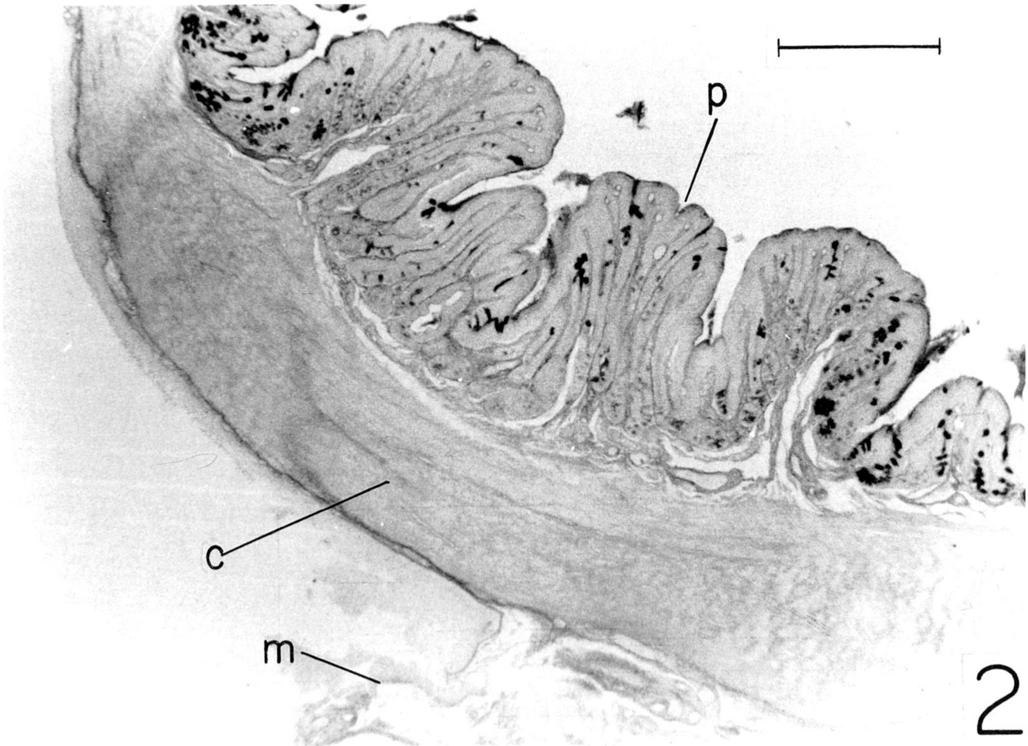

FIG. 1.—Transverse section of the contra-mesenteric side of the proximal colon of the San Quintin kangaroo rat, *Dipodomys gravipes* (MSB 42995); c = circular muscle, gc = goblet cells, s = submucosa.

cosa and hardly ever are observed in the surface epithelium of these glands (Fig. 4).

In Merriam's kangaroo rat, *Dipodomys merriami* (Fig. 5) and in the banner-tailed kangaroo rat, *D. spectabilis,* a third type of modified mesenteric mucosa is present. This mesenteric epithelium is typically arranged on several folds (3 to 5), all of which are covered with tubular glands. Glandular "pits" on the luminal surface of the tubules are broader and deeper than are those in the previous two kangaroo rats. As a result, the surface of mesenteric epithelium is composed of crevices separated by digitiform processes formed by surface columnar cells along with small amounts of underlying lamina propria (Fig. 5). Large vascular channels, probably lymphatic vessels, are found within the loose submucosa both within and between these folds. An unusual feature of these glands is that goblet cells are present in the basal-most portion of the gland unit, but either are absent or unrecognizable within the upper portion and surface of the glands unit (Fig. 5).

Chaetodipus *and* Perognathus.—The proximal colon in these two genera is like that of *Dipodomys* in that it is 1) asymmetrical when viewed in transverse sections, and 2) possesses more than one histomorphological arrangement in the modified "mesenteric" portion of the gut wall. In *Chaetodipus baileyi* (Bailey's pocket mouse), the mesenteric mucosa is unusual in that the modified glands often form a single, enlarged, bulbous intestinal fold (Fig. 6). This fold is rounded at its apex, constricted at its base, and due to its large size resembles a "typhlosole." Its associated glands are three to four times the length of others in the proximal colon and are similar in form to some of those seen in *Dipodomys gravipes* (Fig. 3). Beneath these glands there is a large assemblage of lymphatic vessels with abundant associated collagenous tissue (Fig. 6). Nowhere else in our transverse sections of the colon of this species is there such an extensive complement of large lymphatic vessels.

In *C. arenarius* (little desert pocket mouse), the glands on the mesenteric side of the bowel are elongated and goblet cells are relatively more abundant than they are in the other heteroymids. In the spiny pocket mouse, *C. spinatus* (Fig. 7), the modified, mesenteric mucosa occupies nearly 50% of the circumference of the bowel; the glands are nearly indistinguishable from those of several other species of *Chaetodipus* (e.g., *C. californicus, C. fallax, C. penicillatus*).

With regard to the contra-mesenteric side of the proximal colon, typical colonic mucosa is present in all species of *Chaetodipus* and *Perognathus* examined. However, plicae (folds) are absent in *C. arenarius, C. fallax* (San Diego pocket mouse), *C. californicus,* and *Perognathus flavus* (silky pocket mouse). The remaining species have plicae throughout the circumference of the bowel.

Chaetodipus californicus and *C. penicillatus* (desert pocket mouse) are similar to *C. spinatus* in distribution and morphology of mesenteric mucosae, whereas *Perognathus formosus* (long-tailed pocket mouse) is most like *C. baileyi. C. fallax* appears to have the least amount of modified mucosa, about 20% of the colonic circumference. Shallow luminal "pits" are present in the mesenteric glands of all *Chaetodipus* spp. and *Perognathus* spp.; however, they are relatively small in *C. fallax*. Deep, broad glan-

←

Note the abundant goblet cells and smooth surface of the mucosa where it contacts the intestinal lumen. Scale = 0.25 mm.

FIG. 2.—Transverse section of the proximal colon (mesenteric side) of *D. gravipes.* Note the rarity of goblet cells, greater thickness of the surface epithelium, and greater length of the gland units in comparison with those of the histomorph shown in Figure 1; m = mesentery; p = glandular pit. Scale = 0.25 mm.

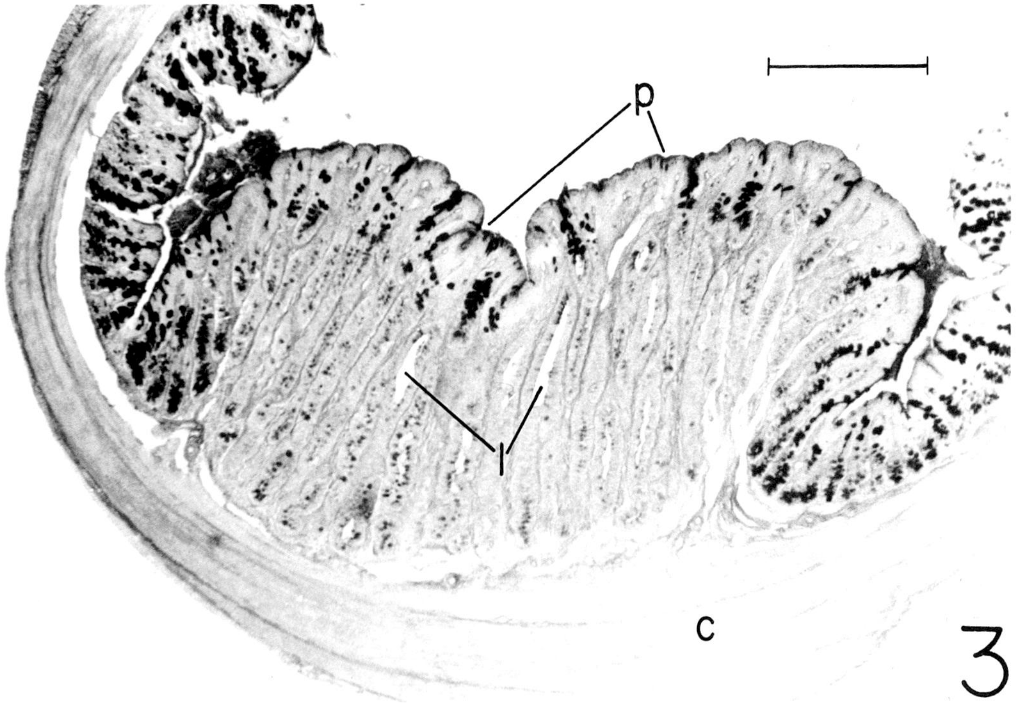

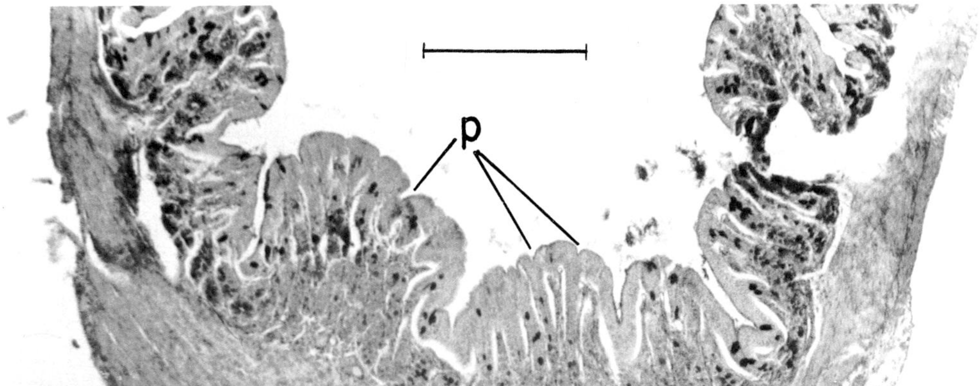

FIG. 3.—Transverse section of the colon (mesenteric side) of *D. gravipes*. Note the extremely elongate mesenteric glands, thickened circular muscle layer, and large glandular lumina; c = circular muscle, l = gland lumen; p = glandular pits. Scale = 0.25 mm.

FIG. 4.—Proximal colon (mesenteric side) of the agile kangaroo rat, *Dipodomys agilis* (MSB 42872); c = circular muscle; m = mesentery; p = glandular pits; s = submucosa. Scale = 0.25 mm.

dular pits, as found in *D. spectabilis* and *D. merriami,* are not found in *Perognathus* or *Chaetodipus.*

Liomys.—In *Liomys pictus* (painted spiny pocket mouse), there is a region of the mucosa that is unusual in its histological appearance. This region occupies 10–20% of the circumference of the colon and is positioned directly over the dorsal mesentery attachment. The epithelium is formed into mucosal folds that are only slightly different (in cross-section) from other folds within the proximal colon. The unusual coiled "mesenteric" gland units are approximately one-third taller than others within the mucosa, have extremely narrow, shallow luminal pits (barely distinguishable in most sections), and surface epithelial cells that are approximately twice the height of those in glands found within the contra-mesenteric surface of the colon (Forman and Phillips, 1988). Goblet cells within the base of the mesenteric glands are PAS positive, but show a reduced alcian blue staining when compared with goblet cells elsewhere in the proximal colon which react positively to both PAS and alcian blue.

Collagenous tissue was sparse beneath the modified mucosa, and vascular elements and ganglionic masses generally are equally abundant within all plicae.

The positioning and morphology of the modified epithelium, to some extent, is individually variable; sometimes the coiled glands are positioned within a slight "depression" in the muscularis externa. These gland units are similar to those described previously for *Dipodomys gravipes.* However in *Liomys,* the associated mucosa is only about one-third thicker than the adjacent unmodified mucosa.

Heteromys.—The proximal-most two centimeters of the colon of Desmarest's spiny pocket mouse, *Heteromys desmarestianus,* contain an extremely large typhlosole-like fold (Fig. 8), which gives way to more typical colonic mucosa beyond this point. This fold contains two cores of submucosal tissue, which suggests it might have formed as a result of fusion of two smaller folds. At its greatest diameter this fold is 1.3 mm wide and extends 1.5 mm into the lumen. In cross sections of the proximal colon, this fold decreases in diameter from a broad base to a narrower midregion, then enlarges to a broad, bulbous apex at its lumenal surface. Goblet cells are nearly absent within the relatively thick apical columnar epithelium covering the fold. However, a large central zone of the fold (Fig. 8) has elongated, convoluted colonic glands which do contain goblet cells throughout the gland unit. Glands on the sides of this large fold are shorter, broader, and similar in structure to those found elsewhere in the proximal colon.

A very unusual histological feature is seen in the subepithelial vasculature within this large fold. Glomerular-like masses of convoluted capillaries are present between glands, immediately below the surface columnar cells at the apex of the fold (Fig. 9). Occasionally veins extend to the luminal surface of the fold and form "varicosities" (Fig. 10).

Muscularis and Lamina Propria

The circular layer of the muscularis externa is thickened immediately below the mucosal folds (plicae) of the large bowel in all heteromyids. The amount of thickening varies, but appears to be loosely proportional to the depth of the other portions of the submucosa. In particular, it is proportional to the loose lamina, which is composed principally of collagen and vascular elements, and to the mucosa itself. Thickening always is greatest in the proximity of the mesenteric attachment to the gut. At this position the circular musculature is expanded tremendously into a bulbous mass (Figs. 2, 4, 5, 6, 8) with an internal complexity not generally observed elsewhere in the circular layer. This complexity often takes the form of large bundles of fibers arranged in laminated sheets or bundles (Fig. 3). On the mesenteric side of the proximal colon the outer longitudinal layer of the

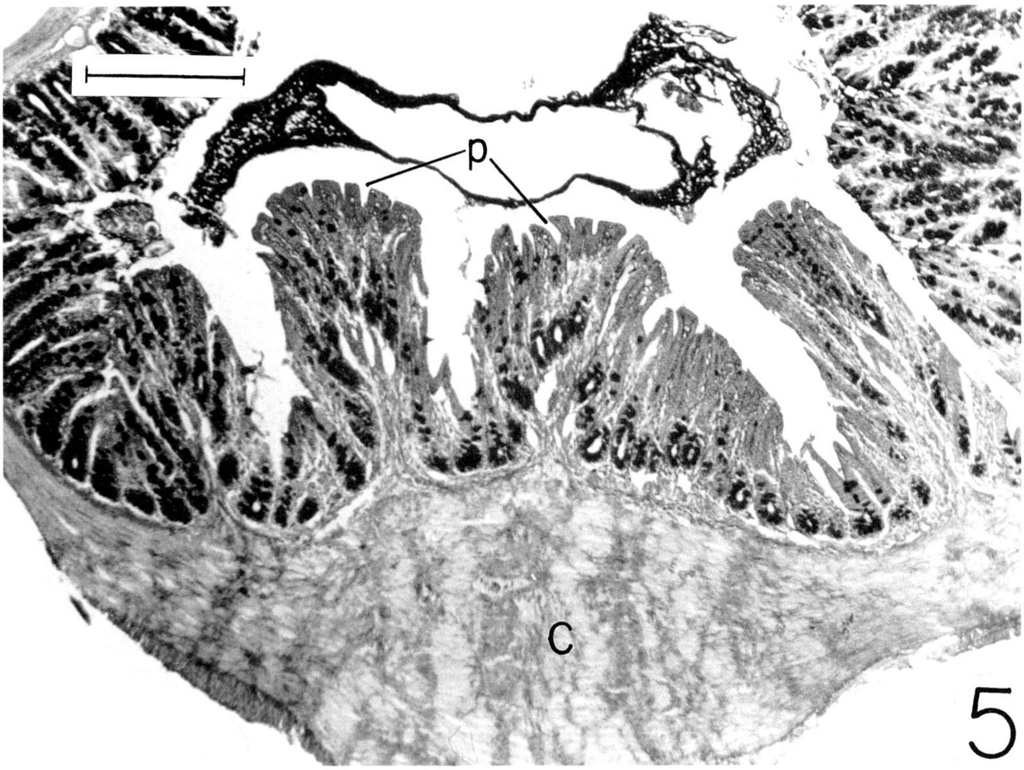

5

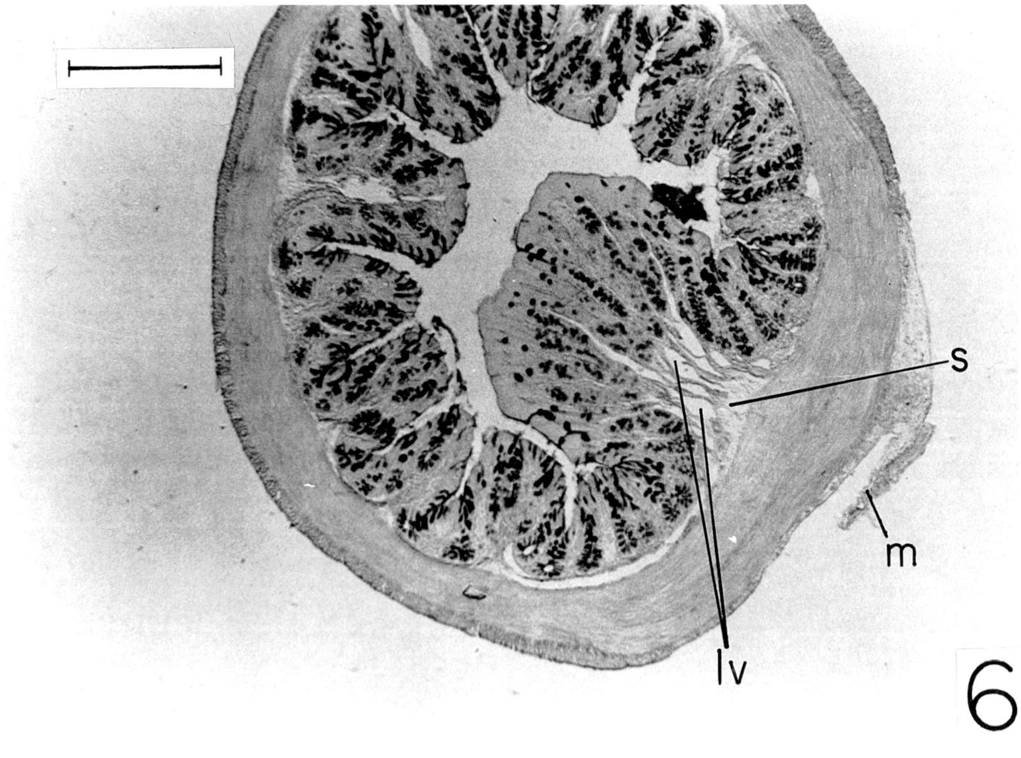

6

muscularis is reduced to a vestige and sometimes appears to be absent.

The muscularis mucosae, which underly the glands, are present in the proximal colon of heteromyids. However, on the mesenteric side these layers generally are reduced and possibly absent. The presence of even a very thin layer of muscle cannot always be demonstrated, even with Milligan trichrome, which specifically stains muscle. In such regions the loose connective tissue appears to be greatly expanded.

Innervation

The proximal colon of heteromyids is largely innervated by the enteric nervous system, which is intrinsic to the gastrointestinal tract of mammals. Each of the components of the enteric nervous system to some degree can be visualized with general histological techniques and, as expected, each is present in all of the heteromyids examined to date. The submucosal (Meissner's) plexus consists of ganglia of nerve cell bodies located circumferentially in the dense connective tissue of the submucosa. Generally, these ganglia are situated at the base of the plicae as well as within the lamina propria between plicae. Associated bundles of nerve fibers (which often are difficult or impossible to visualize with general histology) are found adjacent to many of the blood vessels and also overlying the basal surface of gland units. The myenteric (Auerbach's) plexus is located between the circular and longitudinal muscle layers; it consists of ganglia of nerve cell bodies and nerve bundles that extend into the smooth muscle. The nerve fibers and nerve ganglia in the mesentery are part of the inferior mesenteric plexus and are extrinsic to the gut.

In *Chaetodipus baileyi,* the submucous plexus typically has from 3–7 nerve cell bodies. The myenteric plexus, on the other hand, usually has relatively small ganglia and only 2–4 nerve cell bodies typically are seen in histological profiles. Furthermore, cross-sections of the proximal colon suggest that these ganglia tend to be concentrated asymmetrically toward the contra-mesenteric side of the colon.

Antisera to met-ENK and serotonin (5-HT) demonstrate the presence of enkephalin-like and 5-HT-like immunoreactivity in the enteric nerves of *C. baileyi.* With anti-ENK, the immunoreactivity is virtually absent in the myenteric plexus, but is intense in the large ganglia of the submucous plexus, especially on the mesenteric side of the colon. Some immunoreactive fine nerve fibers and bundles of nerves also are associated with these ganglia. The strongest ENK-like immunoreactivity is seen in extrinsic ganglia associated with the mesenteric blood supply. The antiserum to serotonin shows scattered immunoreactivity in nerve cell bodies of the submucous plexus; the only reactive nerve cell bodies in the myenteric plexus are those situated adjacent to the mesenteric blood supply. As with ENK, strong immunoreactivity to serotonin is seen in ganglia situated outside of the proximal colon, in association with the mesentery.

In *Dipodomys spectabilis,* submucous

←

FIG. 5.—Transverse view, proximal colon of Merriam's kangaroo rat, *Dipodomys merriami* (MSB 42548). Note the absence or lack of goblet cells within the upper one-half of each mesenteric gland unit (central portion of the photograph); c = thickened circular muscle; p = glandular pits. Scale = 0.25 mm.

FIG. 6.—Transverse section, proximal colon of Bailey's pocket mouse, *Chaetodipus baileyi* (MSB 42688). Note the absence of goblet cells at the luminal surface of the typhlosole-like fold, and the proximity of the mesentery (m) to this fold; lv = lymphatic vessels; s = submucosa. Scale = 0.25 mm.

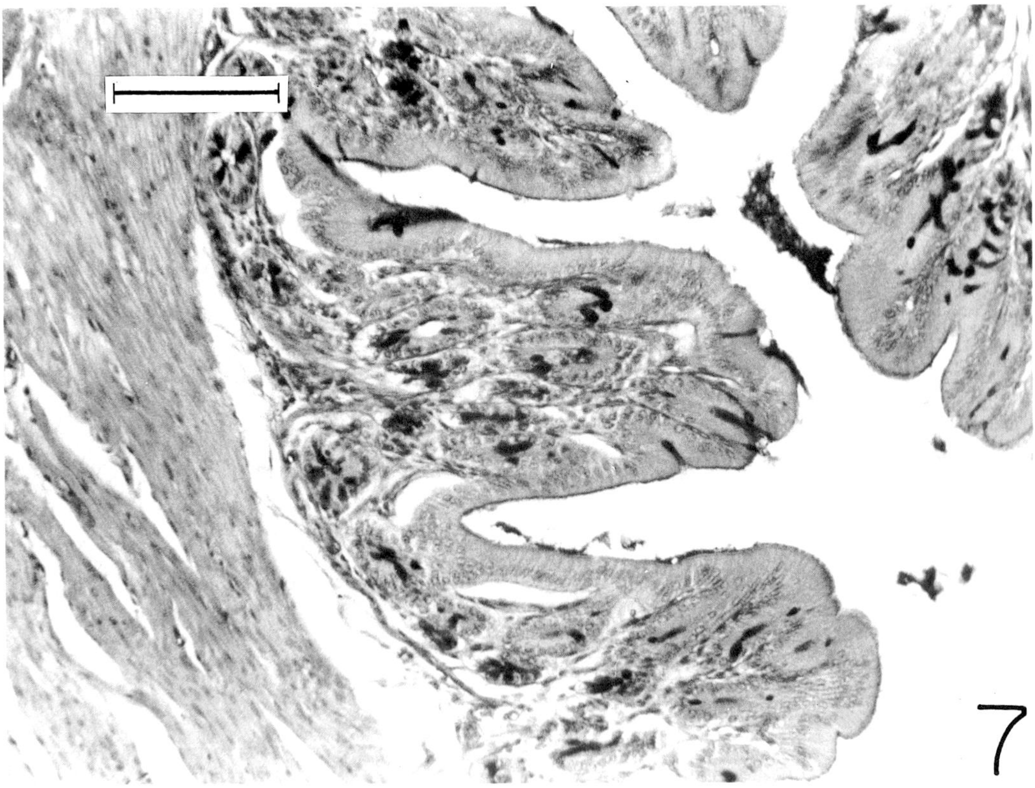
7

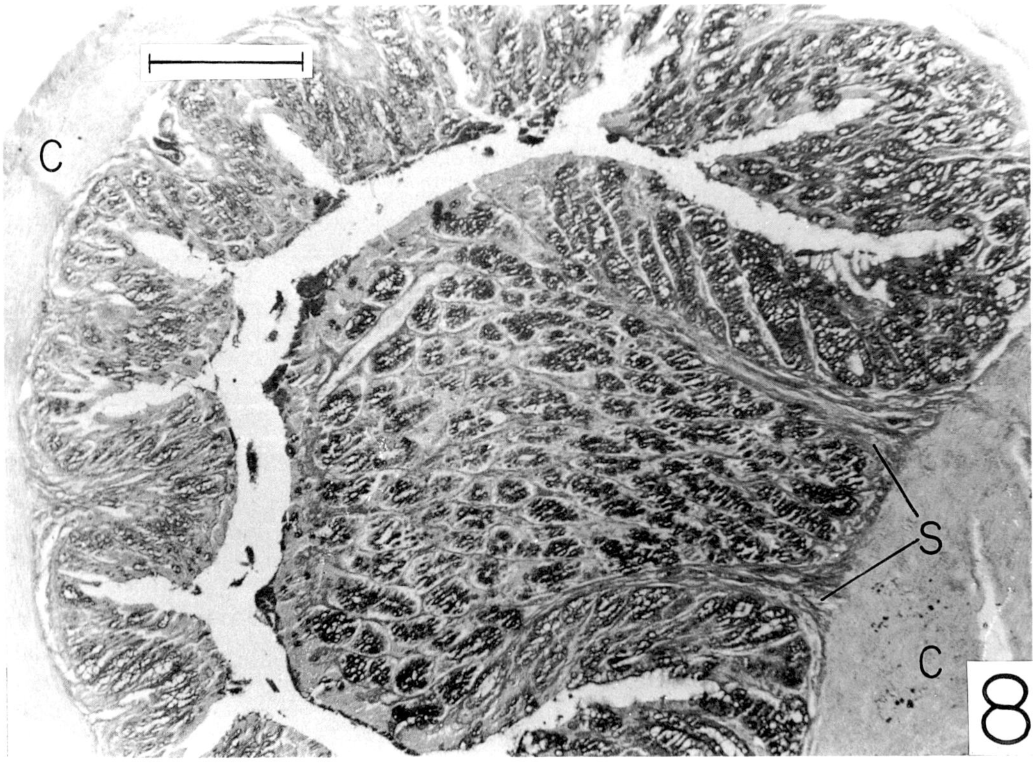
C
S
C
8

ganglia of 3–5 neurons are found at the base of each plica circulares. The myenteric plexus differs from that of *Chaetodipus baileyi* in that the ganglia are larger and more abundant on the mesenteric side of the proximal colon (10–11 nerve cells) than on the contramesenteric side (3–5 cells). Met-ENK like immunoreactivity is scarse in the proximal colon of *D. spectabilis,* but clusters of 1–3 immunoreactive submucous nerve cells are found throughout the circumference of the colon. Immunoreactive nerve cell bodies are very uncommon in the myenteric plexus and the ones that are immunoreactive are individual cell bodies directly adjacent to the mesenteric blood supply. As in *C. baileyi,* strongly immunoreactive nerve cell bodies are found within the loose connective tissue of the mesentery. The distribution of endogenous 5-HT, as judged by immunoreactivity, parallels that of ENK and thus is similar to that in *C. baileyi.*

Discussion

The participation of the lower colon in homeostasis in mammals includes activities related to processing or absorption of numerous molecules and ions. These include, but are not restricted to, the modification, utilization or uptake of amino acids, urea, ammonia, sugars, volatile fatty acids, sodium, potassium, bicarbonate, chloride, and water. It thus is to be expected that the colon might vary, both structurally and functionally, in ways that may correlate with ecological factors such as water availability and diet. Heteromyid rodents represent an excellent model for use in testing these relationships because they a) are distributed in habitats which range from being extremely mesic to extremely xeric, and b) partition food resources in order to maximize resource allocation in areas where food and water are in short supply, and species diversity and competition for resources are extensive.

Beyond information for laboratory species (*Rattus, Mus, Mesocricetus*), little comparative data are available about the histology of the colon of rodents. In studies of gastrointestinal structure in small mammals, accounts of colon morphology generally are brief, and rarely include mention of the microstructure of the large bowel posterior to the caecum. It thus follows that data on lower bowel morphology of heteromyids are extremely meager. In a comprehensive review of lower intestinal morphology in rodents, Gorgas (1967) provided general descriptions of the gross anatomy of the caecum and colon in species representing 23 families of sciuromorph (including the Heteromyidae), caviomorph, and hystricomorph rodents. Gorgas included measurements for lengths of the small intestine, large intestine, and caecum for one species each of *Dipodomys, Heteromys,* and *Perognathus.* However, no information on morphology of the bowel was included for heteromyids. Nevertheless, some information was included on gross morphology of the colon in several caviomorph and hystricomorph rodents. This information suggests to us that there are possible parallels with our microanatomical data for heteromyids. For instance, Gorgas reported the presence of a "typholosole-like system" within the mesenteric wall of the colon in a caviomorph, *Dolichotis patagonum* (Caviidae). This unusual structure contained one or

←

FIG. 7.—"Mesenteric" mucosa within the proximal colon of the spiny pocket mouse, *Chaetodipus spinatus* (MSB 42695). Scale = 0.10 mm.

FIG. 8.—Giant typhlosolar mucosal fold (transverse section) within the proximal-most colon of Desmarest's spiny pocket mouse, *Heteromys desmarestianus* (KU 84370); c = circular muscle; s = submucosa. Scale = 0.25 mm.

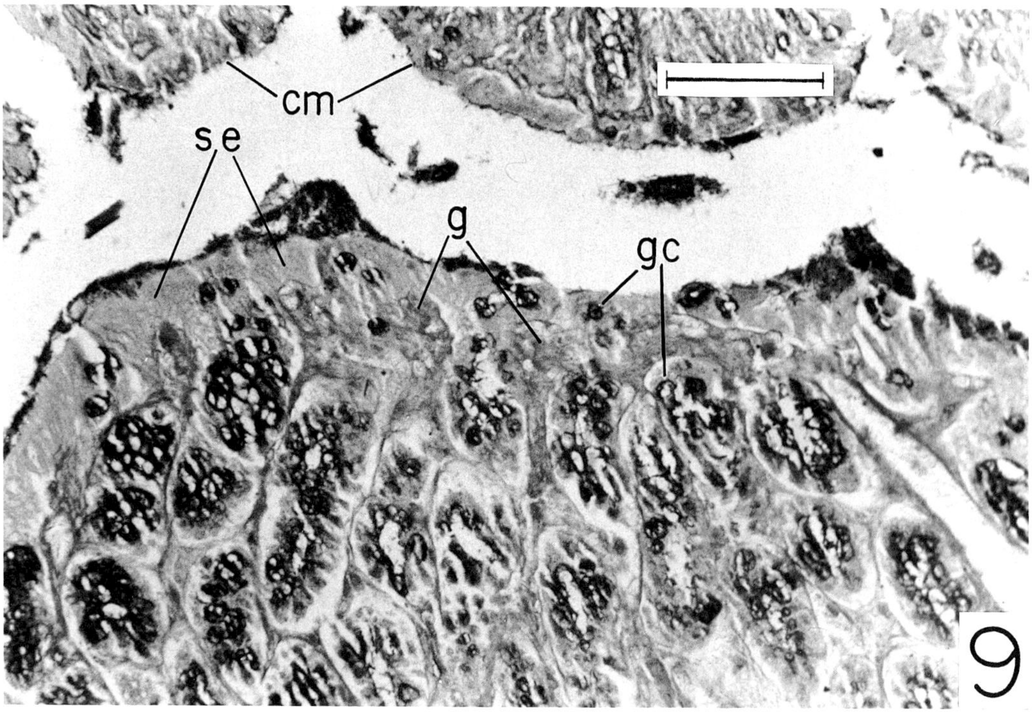

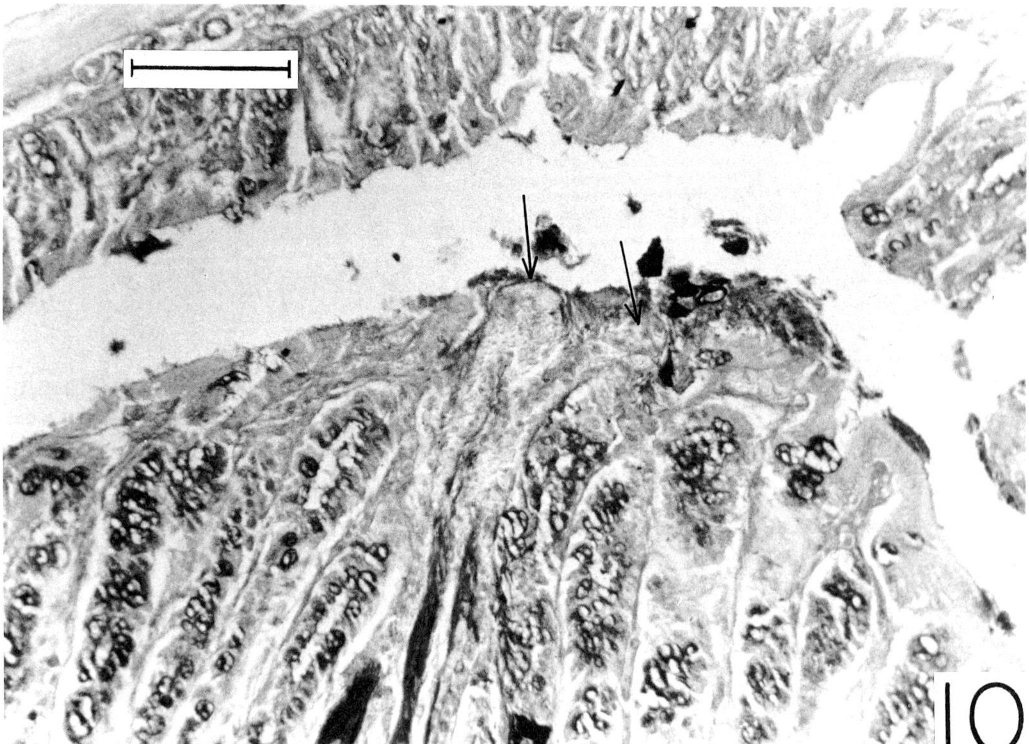

FIG. 9.—Apex of the typhlosolar fold in *Heteromys desmarestianus*; cm = contra-mesenteric glands; g = glomerulus; gc = goblet cell; se = surface epithelium. Scale = 0.10 mm.

FIG. 10.—Apex of the typhlosolar fold in *H. desmarestianus* showing surface vascularities (arrows) within the apical epithelium. Scale = 0.10 mm.

more large veins. There was no mention of any histological asymmetry in the colon of this species. Although no histological data were provided for other species of rodents, surface views of colonic mucosa (Gorgas' figures) showed similar, if less well developed, mesenteric "ridges" in several other species of the Caviidae, as well as in some genera of the Capromyidae and Chinchillidae. Likewise, the colon in several genera from three hystricomorph families (Hystricidae, Thryonomidae, and Bathyergidae) has some form of mucosal specialization in the proximity of the mesentery.

Our work (Forman and Phillips, 1988) represents the only investigations of which we are aware of colonic specialization in heteromyids. The Heteromyidae are the only rodents for which these morphological specializations may be widespread, if not characteristic. Unfortunately, no corroborative physiological data are available to test their possible functional significance. Therefore, our conclusions regarding their possible role or roles in water conservation or nutrient recovery are tentative.

Regarding water conservation or enhanced water absorption, the "mesenteric" mucosa is particularly interesting. The importance of reduction in goblet cell complements may be twofold. First, absence of goblet cells at the surface of the bowel suggests that absorptive-type columnar cells are present in their place, thus increasing absorptive surface area. Second, in areas where goblet cells are reduced or absent there will be less mucus, including its aqueous component, added to the fecal mass. Broad glandular pits at the luminal surface, particularly evident in *Dipodomys merriami* and *D. spectabilis,* could serve as "sinks" for the collection and absorption of water, or as means of increasing absorptive surface area by way of the epithelial projections which would lie between them (Forman and Phillips, 1988).

It is possible that interspecific variation in the extent to which the mesenteric side of the colon is morphologically specialized is related in some way to availability of preformed water, availability of substrates for production of metabolic water, or other physiological parameters important to the lives of heteromyids. For example, *Dipodomys agilis* and *D. gravipes* both occur in the coastal scrub of northwestern Baja California. However, in the areas where our specimens were collected, *D. agilis* is more frequently found in habitats with greater elevation and denser vegetation than is *D. gravipes* (Best and Lackey, 1985). Interestingly, Carpenter (1966) reported that the water dependent *D. agilis* does not survive indefinitely without preformed water, and survives with difficulty on dry diets when compared with *D. merriami.* Our data reveal that of the four *Dipodomys* spp. that we examined, the relatively water dependent *D. agilis* had the least specialized proximal colon.

Colonic asymmetry was least distinctive in *Liomys pictus,* an animal adapted to relatively mesic conditions. In *Liomys,* slight colonic asymmetry likely represents retention of a primitive or generalized state in an animal that is, overall, the most generalized of the living heteromyids (Genoways, 1973). Even though colonic asymmetry is less well developed in *Liomys,* than in kangaroo rats or desert pocket mice, it likely contributes to maximization of water retention in a genus that is an efficient water regulator (Christian et al., 1978).

In *Heteromys desmarestianus,* an even more mesic-adapted form, the unusual typhlosolar fold was found only in the proximal-most two centimeters of the post-caecal colon, and contained abundant goblet cells. This observation, in concert with data for *Liomys,* is consistent with our hypothesis that colonic asymmetry, overall, is least well developed in heteromyids occupying the least xeric environments. However, it is important, and somewhat perplexing, to note that MacMillen and Hinds (1983) reported that overall water regulatory efficiency in *H. desmarestianus* was greater than that of *Liomys,* and was comparable to that of some *Perognathus* spp. and *Dipodomys* spp. tested. No data are available for fecal water content for *Heteromys,* so it is not known to what degree the lower bowel may contribute to the high water regulatory efficiency in this animal. *Heteromys* represents a unique lineage of heteromyids that shifted to mesic habitats, in contrast to others in the family (Genoways, 1973). Colonic asymmetry in *Heteromys,* however limited in distribution or specialized in form, may represent retention of a primitive characteristic present in early heteromyid stock prior to their diversification and radiation into drier habitats.

Entire caecal-colon units were not available for study, except for *Liomys* and *Heteromys.* Therefore, we cannot comment on the lengthwise distribution of asymmetrical mucosa in the lower bowels (caecum-colon) of any but these genera. Our histological data and interpretation must be considered preliminary to additional investigations which should include histological examinations of the caecum and caecal-colon junctions in heteromyids.

The elaborate, thickened circular muscle in the proximal colon of heteromyids could function directly or indirectly to influence metabolic processes within the lumen of the colon, or enhance water absorption into the tissues of the colon, or into those of the caecum. This muscularis could serve to a) delay distal progression of scybala, thus augmenting water absorption by the colon, b) assist in propulsion of especially dry scybala which are difficult to transport due to little surface lubricant from the modified mucosa, c) "squeeze" from the feces liquid which is then collected by the closely applied mesenteric glands or, d) return scybala to the caecum for further concentration or absorptive action.

The anatomy of innervation of the proximal colon in heteromyids appears to be conservative, at least insofar as can be determined from the methods used by us. On the other hand, the importance of the innervation to function seems obvious and raises the question of how control and regulation by the nervous system relates to the interspecific variations in the epithelial glands and musculature as discussed in the foregoing paragraphs.

Over the past decade it has become clear that the neurotransmitter 5-hydroxytryptamine (5-HT) is abundant in neurons intrinsic to the gut (Gershon, 1981; Mawe et al., 1986). Additionally, many regulatory peptides also have been isolated from gut tissue. A large percentage of these now are known to occur in nerve fibers and nerve cell bodies as well as in endocrine cells. Some of these peptides eventually might be regarded as neurotransmitters, but presently are still categorized as "neuropeptides" or "regulatory peptides." The discovery of these peptides, and the gradual accumulation of physiological data about their modes of action and their function, has begun to revolutionize our understanding of the gastrointestinal tract of mammals (Miller, 1984).

Most of the comparative data that are available about the distribution of various neuropeptides and 5-hydroxytryptamine (5-HT) have been derived from investigations of laboratory mammals (Keast et al., 1985). Because different laboratory species generally have quite different evolutionary histories, no clear patterns have emerged that can be correlated with ecology, diet, or taxonomy. The only exceptions to this rule involve the distribution of the peptide hormone gastrin in the pylorus of the stomach in bats (Mennone et al., 1986), and the distribution of a variety of neurotransmitters in the retinas of ecologically divergent Neotropical bats (Studholme et al., 1987).

Because of the complete absence of any data about the innervation of the digestive tract in heteromyids, we used antisera to 5-HT and met-ENK as a means of demonstrating their possible presence and their distributional pattern in the proximal colon, using *Chaetodipus baileyi* and *Dipodomys spectabilis* as models.

The action of 5-HT is not fully understood, but apparently it serves as a transmitter of the interneurons, and thus acts indirectly by affecting a variety of intrinsic gut neurons. Depending on circumstances, it can indirectly cause either contraction or relaxation of the gut musculature (Gershon, 1981). Several other reported actions of serotonin are particularly interesting in terms of investigating the proximal colon of heteromyids. Physiological data suggest that 5-HT can: a) stimulate production and release of colonic mucus; b) promote net water and electrolyte secretion in the small intestine; and c) influence mesenteric blood flow (Ormsbee and Fondacaro, 1985).

Although 5-HT-like immunoreactivity was found in nerve cells in both *Chaetodipus* and *Dipodomys,* the strongest reactivity in our tissue samples, by far, was in nerve ganglia physically associated with the mesenteric veins and arteries. Only very few of the nerve cell bodies within the myenteric and submucous ganglia were immunoreactive. The fact that we found strong immunoreactivity in some cells but not in others within single histological sections suggests that our data might reflect the real distribution of 5-HT in these species. However, it must be noted that by using field-fixed specimens we were limited to localizing endogenous 5-HT; consequently, we cannot be certain of the significance of negative results. If 5-HT-containing neurons are indeed rare in the submucous plexus of the proximal colon, then our results would be reasonable in terms of the known roles of 5-HT. For example, our histological data show that at least some areas of the colon have reduced numbers of mucus-secreting goblet cells and therefore we would not necessarily expect to find abundant indications of the presence of an amine that has been associated with stimulation of mucus or promotion of water loss across an epithelium. The presence of 5-HT-like immunoreactivity in numerous neurons in the mesentery is interesting, but once again the role of 5-HT in affecting blood flow is not well enough understood to allow us to relate its presence to a single effect. Nevertheless, some physiological studies have shown that, in vivo, serotonin relaxes resistance vessels in the mesenteric microvasculature and thus promotes blood flow (Ormsbee and Fondacaro, 1985).

We selected the opiate-active peptide met-enkephalin for study along with 5-HT because: met-ENK a) has been demonstrated by both immunohistochemistry and HPLC to be present in enteric nerves and nerve cell bodies in mammals (Keast et al., 1985; Konturek, 1981; Schultzberg et al., 1980); b) has been associated with decreased gut motility, which indirectly results in increased water retention by the colon (Bueno and Fioramonti, 1987; Konturek, 1981); and c) recently has been linked to aspects of mucosal ion transport (Miller, 1984). Inhibition of both stimulated and resting release of acetylcholine (ACh) by cholinergic axon terminals is one means by which met-enkephalin acts and although its exact mode of action is unknown, it possibly hyperpolarizes neurons (Jodal and Lundgren, 1983).

The presence of ENK-like immunoreactivity in the submucous plexus in *Chaetodipus* and *Dipodomys* is noteworthy because enkephalin has not been reported previously from the submucous nerve cell bodies of any studied species, including laboratory rats (*Rattus norvegicus*) and guinea pigs (*Cavia porcellus*) (Keast et al., 1985; Schultzberg et al., 1980). Additionally, it is known to have an inhibitory effect on the neurons of the submucous plexus, as it does on those of the myenteric plexus (North and Williams, 1976; Polak et al., 1983). However, pharmacological data available from in vitro studies suggest that a) enkephalin does affect ion transport across the epithelium by causing a decrease in short circuit current (Isc), and b) that enkephalin has a strong antisecretory effect on the mucosa (Miller, 1984). Because ion transport obviously is a significant factor in water absorption, enkephalin-containing neurons in the submucous plexus of the proximal colon in heteromyids might be species-specific and might play a role in mediating this process.

The myenteric plexus in mammals typically includes ENK-containing neurons (Keast et al., 1985; Schultzberg et al., 1980). The scarcity of ENK-like immunoreactivity in the myenteric plexus in *Chaetodipus* and *Dipodomys* is thus unusual and somewhat surprising. Before offering an interpretation of this finding, several factors need to be considered. First, the absence of immunoreactivity does not necessarily prove the absence of a neuropeptide. Antigenic sites can be obscured in several ways so caution must be exercised in interpreting negative data. Second, the possibility exists that expression of particular neuropeptides is seasonably variable. Although this has not been demonstrated in mammals, Gesser and Larsson (1985) have shown that enkephalin-like immunoreactivity in invertebrates can be expressed by certain, specific neurons only during certain seasons. The heteromyid rodents that we examined immunohistochemically were collected during mid-summer so the possibility that neuropeptides vary seasonally in heteromyids remains an open issue. Finally, the fact that ENK-like immunoreactivity in particular histological sections was strong in neurons other than those in the myenteric plexus suggests that our data are not artifactual.

The near absence of ENK-like immunoreactivity in the myenteric plexus in *Chaetodipus* and *Dipodomys* is exactly the opposite of what one might expect, given the role of enkephalin in retarding motility (Bueno and Fioramonti, 1987; Grider and Makhlouf, 1987; Jodal and Lundgren, 1983). However, a physiological study of digesta passage through the large intestine of the laboratory rabbit provides a possible answer. Pickard and Stevens (1972) found that in rabbits it is mechanical separation of water from digesta, rather than decreased motility, that makes more water available for absorption by the colon. Either mechanism would result in the dry feces for which the heteromyids are well-known. Enkephalin inhibits the cholinergic nerves, whose normal action is to stimulate strong contractions by the muscularis externa (Jodal and Lundgren, 1983; North and Williams, 1976). Thus, in heteromyids it appears possible that a near absence of met-enkephalin in the myenteric plexus could be an adaptive feature of the intrinsic nervous system of the proximal colon. It seems that such a feature would facilitate muscular contraction stimulated by cholinergic nerves. Our hypothesis is given added support by the extraordinarily thick muscularis externa in heteromyids. If our interpretation of the met-ENK data is correct, then one might hypothesize that in heteromyids water is made available for absorption by mechanical separation rather than by prolonged retention of digesta.

Currently it is thought that ENK-containing nerve fibers in the inferior mesenteric ganglion of the guinea pig come from the lumbar splancnic nerves (Dalsgaard et al., 1983). The presence of strong met-ENK-like immunoreactivity in nerve cell bodies, but not in nerve fibers, of the inferior mesenteric ganglion is yet another interesting feature of the heteromyids because this differs from guinea pigs and because histologically these neurons are associated with the mesenteric blood supply and are especially prominent in *Chaetodipus.* In one study enkephalin was shown, in vitro, to cause hypotension (Rhee and Tyler, 1984) and according to Konturek (1981) opiate-active peptides relax intestinal arterioles and precapillary sphincters, thus increasing blood flow. In another study, of human beings, met-ENK caused a slight dilatation in some of the mesenteric arteries and veins that were tested (Tornebrandt et al., 1987), but the effects were not as pronounced as expected from the previous reports. Overall, enkephalin and serotonin both appear to have at least some effect on mesenteric blood flow. This is noteworthy because we found strong ENK- and 5-HT-like immunoreactivity in mesenteric neurons in both *Chaetodipus* and *Dipodomys.* Precise control of blood flow, volume and rate, to and from the proximal colon probably would be especially significant to water conservation and, possibly, to successful use of certain diets. The relationships among mesenteric innervation, ENK and 5-HT, and mesenteric blood flow to and from the proximal colon remain to be elucidated but are probably an important part of the story in heteromyids.

The eco-physiological significance of our immunohistochemical data on enkephalin and serotonin partly lies with the fact that innervation of the proximal colon might determine the ability of a species to exploit particular nutrient resources and thus subdivide habitats. Species differences in the localization and presence or absence of neuropeptides and neurotransmitters are virtually unstudied within mammalian orders, but their elucidation should be a valuable addition to our understanding of how animals interface with their environment.

In summary, we think that the unusual histology of the proximal colon in heteromyid rodents most likely relates in some way to the considerable capacity for water retention exhibited by these rodents. The production of exceptionally dry feces is one consequence of this histomorph. The innervation of the proximal colon, particularly in terms of the distribution and presence or absence of particular neuropeptides and neurotransmitters, probably is significant to our understanding of the role of this structure-function complex in the evolution and physiological ecology of heteromyids. A great deal remains to be learned about the proximal colon and the remainder of the digestive tract, which is virtually unstudied. Future investigations in which histology, immunohistochemistry, and physiology are integrated, and hypotheses tested through experimentation, should prove to be of considerable interest to students of a variety of disciplines in biology.

Acknowledgments

Financial support for this investigation came from a grant from Rockford College (GLF), Hofstra University (HCLAS grant to CJP), the Research Corporation (Grant C-1855 to CJP), and the National Science Foundation (grant CD801-8653 to CJP and CD801-8971 to Dr. G. W. Grimes). Various colleagues supplied the specimens that we used in this investigation, but in particular we are appreciative of the efforts of Drs. Donald Duszinski and Terry Yates, and Dr. David Reduker of the University of New Mexico. Dr. Robert S. Hoffmann, formerly of the University of Kansas Museum of Natural History, kindly supplied the specimens of *Liomys* and *Heteromys*. Support for field collection of material was through NSF grant DEB-8004685 to TLY. Albert Mennone and David Principe, Department of Biology, Hofstra University, assisted us with immunohistochemistry and microphotography.

Literature Cited

Best, T. L., and J. A. Lackey. 1985. Dipodomys gravipes. Mammalian Species, No. 236, pp. 1–4.

Bueno, L., and J. Fioramonti. 1987. Enkephalins, other endogenous opioids and colonic motility in dog and man. Gastroenterology and Clinical Biology, 11:69B–76B.

Carpenter, R. E. 1966. A comparison of thermoregulation and water metabolism in the kangaroo rats *Dipodomys agilis* and *Dipodomys merriami*. University of California Publications in Zoology, 78: 1–36.

Christian, D., et al. 1978. Comparative water balance in two species of *Liomys* (Rodentia: Heteromyidae). Comparative Biochemistry and Physiology, 61(1):589–594.

Dalsgaard, C.-J., et al. 1983. Origin of peptide-containing fibers in the inferior mesenteric ganglion of the guinea pig: immunohistochemical studies with antisera to substance P, enkephalin, vasoactive intestinal polypeptid, cholecystokinin and bombesin. Neurosciences, 9:191–211.

Forman, G. L., and C. J. Phillips. 1988. Unusual mucosal epithelium in the colon of heteromyid and cricetid rodents from an arid environment. Journal of Mammalogy, 69:144–149.

Genoways, H. H. 1973. Systematics and evolutionary relationships of spiny pocket mice, genus *Liomys*. Special Publication Number 5, The Museum, Texas Tech University, Lubbock, 368 pp.

Gershon, M. D. 1981. The enteric nervous system. Annual Review of Neurosciences, 4:227–272.

Gesser, B. P., and L.-I. Larsson. 1985. Changes from enkephalin-like to gastrin/cholecystokinin-like immunoreactivity in snail neurons. Journal of Neuroscience, 5:1412–1417.

Gorgas, M. 1967. Vergleichend-anatomische untersuchungen am magan-darm-kanal der Sciuromorpha, Hystricomorpha und Caviomorpha (Rodentia). Zeitschrift für Wissenschaftliche Zoologie, 175:237–404.

GRIDER, J. R., AND G. M. MAKHLOUF. 1987. Role of opioid neurons in the regulation of intestinal peristalsis. American Journal of Physiology, 253 (Gastrointestinal and Liver Physiology 16):G226–G231.

HUMASON, G. L. 1979. Animal tissue techniques, 4th ed. W. H. Freeman and Co., San Francisco, 661 pp.

JODAL, M., AND O. LUNDGREN. 1983. Nerves of the colon. Pp. 187–209, *in* Colon: structure and function (L. Bustos-Fernandez, ed.) Plenum Medical Book Co., New York, 308 pp.

KEAST, J. R., J. B. FURNESS, AND M. COSTA. 1985. Distribution of certain peptide-containing nerve fibers and endocrine cells in the gastrointestinal mucosa in five mammalian species. Journal of Comparative Neurology, 236:403–422.

KONTUREK, S. J. 1981. Action of enkephalins on the digestive system. *In* Gut hormones (S. R. Bloom and J. M. Polak, eds.), 2nd ed. Churchill Livingstone, London, 605 pp.

LANGE, R., AND H. STAALAND. 1970. Adaptations of the caecum-colon structure of rodents. Comparative Biochemistry and Physiology, 35:905–919.

———. 1971. On the mechanisms of water absorption in the rat colon. Comparative Biochemistry and Physiology, 40A:823–831.

LILLIE, R. D., AND H. M. FULLMER. 1976. Histopathologic technique and practical histochemistry, 4th ed. McGraw-Hill, New York, 942 pp.

MACMILLEN, R. E., AND A. L. LEE. 1967. Australian desert mice: independence of exogenous water. Science, 158:383–385.

MACMILLEN, R. E., AND D. S. HINDS. 1983. Water regulatory efficiency in heteromyid rodents: a model and its application. Ecology, 64:152–164.

MAWE, G. M., T. A. BRANCHEK, AND M. D. GERSHON. 1986. Peripheral neural serotonin receptors: identification and characterization with specific antagonists and agonists. Proceedings of the National Academy of Sciences USA, 83:1–5.

MENNONE, A., C. J. PHILLIPS, AND D. E. PUMO. 1986. Evolutionary significance of interspecific differences in gastrin-like immunoreactivity in the pylorus of phyllostomid bats. Journal of Mammalogy, 67:373–384.

MILLER, R. J. 1984. The mucosa as a target tissue for gut neuropeptides. Trends in Neuroscience, 7:2–3.

NASLEDOVA, N. L., AND N. I. PECHURIKINA. 1979. Importance of the gastrointestinal tract of desert rodents in osmoregulation. Journal of Evolution, Biochemistry and Physiology, 15:152–156 (translation from Russian).

NORTH, R. A., AND J. T. WILLIAMS. 1976. Enkephalin inhibits firing of myenteric neurons. Nature, 264: 461.

ORMSBEE, H. S. III, AND J. D. FONDACARO. 1985. Action of serotonin on the gastrointestinal tract. Proceedings of the Society of Experimental Biology and Medicine, 178:333–338.

PICKARD, D. W., AND C. E. STEVENS. 1972. Digesta flow through the rabbit large intestine. American Journal of Physiology, 222:1161–1166.

POLAK, J. M., A. E. BISHOP, AND S. R. BLOOM. 1983. The regulatory peptides of the colon. Pp. 167–186, *in* Colon: structure and function (L. Bustos-Fernandez, ed.). Plenum Med. Book Co., New York, 308 pp.

RHEE, H. M., AND L. TYLER. 1984. Potential role of MET-enkephalin in short term regulation of blood pressure in anesthetized rabbits. Clinical Experiments: Theory and Practice, A6:2063–2067.

SCHMIDT-NIELSEN, K. 1964. Desert animals: physiological problems of heat and water. Oxford University Press, London, 277 pp.

SCHULTZBERG, M., ET AL. 1980. Distribution of peptide- and catecholamine-containing neurons in the gastrointestinal tract of the rat and guinea pig: immunohistochemical studies with antisera to substance P, vasoactive intestinal polypeptide, enkephalins, somatostatin, gastrin/cholecystokinin, neurotensin and dopamine-hydroxylase. Neuroscience, 5:689–744.

STEGGERDA, F. R. 1935. Observations on the variation of water content of fecal material along the colon. American Journal of Physiology, 112:559–564.

STERNBERGER, L. A., P. H. HARDY, JR., J. J. CUCALIS, AND H. G. MEYER. 1970. The unlabeled antibody enzyme method of immunohistochemistry: preparation and properties of soluble antigen-antibody complex (horseradish peroxidase–anti-horseradish peroxidase) and its use in identification of spirochetes. Journal of Histochemistry and Cytochemistry, 18:315–333.

STUDHOLME, K. M., S. YAZULLA, AND C. J. PHILLIPS. 1987. Interspecific comparisons of immunohistochemical localization of retinal neurotransmitters in four species of bats. Brain, Behavior and Evolution, 30:160–173.

TORNEBRANDT, K., A. NOBIN, AND CH. OWMAN. 1987. Contractile and dilatory action of neuropeptides on isolated human mesenteric blood vessels. Peptides, 8:251–256.

WITHERS, P. C. 1982. Effect of diet and assimilation efficiency on water balance for two desert rodents. Journal of Arid Environments, 5:375–384.

PHYSIOLOGICAL ECOLOGY OF THE HETEROMYIDAE: ECONOMICS OF ENERGY AND WATER UTILIZATION

ALAN R. FRENCH

Introduction

The family Heteromyidae is distributed from southern Canada to the northern part of South America in habitats that span a wide range of both water and energy availability. Spiny pocket mice in the genus *Heteromys* are restricted to rainforests in the tropics where the climate is continually warm and wet, but the rest of the family are found in regions that have distinct dry seasons during which plant productivity is reduced and seeds are relied upon as the primary source of energy and water. These latter environments range in aridity from seasonally-dry tropical forests inhabited by species of *Liomys* to the driest and hottest deserts of North America where *Dipodomys, Chaetodipus,* and *Perognathus* are abundant. Superimposed on this variation in aridity is a marked north-south gradient in the magnitude of seasonal temperature changes. Unlike their tropical relatives, many northernmost species not only must face hot summers, but also endure long, cold winters when thermoregulatory costs are high and food is scarce.

Physiological adaptations are generally most elaborate in animals living at environmental extremes, and therefore heteromyids, particularly desert-dwellers, are of great interest. The high summer temperatures, unpredictable precipitation, and ephemeral primary productivity of their arid habitats require adaptations for frugal use of available energy and water. This chapter describes the patterns of energy and water balance among the Heteromyidae, and attempts to correlate them with factors such as body size, phylogenetic association, and habitat type.

Energy Use during Normeothermy

Energy utilization in mammals that faithfully regulate their body temperatures at high and relatively constant levels can be conveniently divided into four areas: energy for (1) basal metabolism, (2) thermoregulation, (3) work or activity, and (4) growth. There is a large body of data on basal metabolism of heteromyids, somewhat less on the costs of thermoregulation, and comparatively little on the costs of activity and growth, including reproduction. Information on growth and reproduction is contained in chapter X of this volume; the other topics are covered here.

Basal Metabolic Rate

Since the initial measurements on kangaroo rats by Dawson (1955), heteromyids have been known for their low rates of basal metabolism compared to other eutherian mammals (Table 1). Such low basal rates are common to members of the family inhabiting arid or semiarid environments (McNab, 1979), with the greatest reductions found in desert species (Brower, 1970*a*; Hinds and MacMillen, 1985). Among heteromyids, only *Heteromys anomalus* and *H. desmarestianus* from tropical rainforests have been shown to have rates of basal metabolism that exceed those predicted for rodents of their size (Hinds and MacMillen, 1985; McNab, 1979). Therefore, the magnitude by which basal metabolism is reduced within the family is correlated in a general sense with the aridity and maximum temperature of the habitat in which the animals are found.

Although much speculation has been made concerning the adaptiveness of different levels of basal metabolism, a comparatively low rate of metabolism is clearly tied to a low rate of energy use. For desert heteromyids, this energy savings can be substantial if the animals spend much time at thermal neutral temperatures. The reduction in basal metabolism of desert species has been calculated by Brower (1970*a*) to be 25% and by Hinds and MacMillen (1985) to be 31%, but both of these are conservative estimates because their animals were not fasting during the measurements. At least two conditions must occur in order for energy economy to have been a major force in the selection for this metabolic trait: 1) the animals must live in environments that provide them regular access to thermal-neutral temperatures, and 2) the value of saving energy must outweigh any costs associated with a reduction in the rate of fundamental metabolic processes. Heteromyids certainly meet the first condition. Not only are high environmental temperatures seasonally available to most species, but there is a general correlation between their magnitude and the extent of the reduction in basal metabolism (Hinds and MacMillen, 1985). Furthermore, the few species that have been studied (*P. longimembris, D. merriami,* and *D. microps*) rest at temperatures near the lower end of thermal neutrality whenever possible (Kenagy, 1973*a*). Thus, not only do these species have low basal rates, but they metabolize at those levels much of the time. The second condition is more speculative. The value of reduced energy use is a function of the quantity and predictability of resources in the environment. Most arid-zone species are primarily granivorous, and seed availability fluctuates greatly. Virtually all heteromyids store food during periods of abundance, and it is likely that frugal consumption is frequently important for survival until the next mast crop is set. The high number of species that reduce their body temperatures during episodes of daily torpor or seasonal hibernation (see below) reinforces the concept that these animals often are energy stressed. Unfortunately, little information is available on the costs of having a low rate of basal metabolism. Basal metabolism may be tied in some way to other aspects of the animals' biology, such as maximum aerobic capacity, growth rate, or reproductive output. If so, parallel reductions in such parameters might also be present in many heteromyids, but there are few data to evaluate this possibility.

Low rates of basal metabolism might be evolutionarily linked to a reduction of evaporative water loss in arid regions (McNab, 1979) and to the prevention of overheating in hot climates (MacMillen and Lee, 1970; McNab, 1966). The former is unlikely to have been of any selective advantage because reductions in respiratory water loss associated with a reduced metabolism are matched by equal reductions in metabolic water production (see the following section on water balance). The latter suggestion is a distinct possibility. The dissipation of endogenously produced heat may become critical at high temperatures, especially in small species that can ill afford to evaporate water. A reduction in basal metabolism could thus allow the exploitation of habitats where high temperatures are unavoidable. However, even in deserts, moderately-high temperatures are usually avoidable by behavioral means. From a heat dissipating point of view, a low basal rate of metabolism might be most important to individuals, such as dispersing juveniles, that do not have extensive burrow systems to provide refuge from high soil temperatures, but it could also have general value in minimizing thermal restrictions on the timing of activity.

Basal rates of metabolism have been reported to rise in the winter and fall in the summer in at least one species, *C. intermedius* (Bradley et al., 1975). These changes suggest that the balance between the costs and benefits of a particular minimum level metabolism also change seasonally. The environmental factors that stimulate such metabolic changes are problematical. Captive *D. ordii* have been shown to alter resting (presumably non-fasting) metabolism at 29°C in response to changes in photoperiod even though similar metabolic differences are not evident among individuals captured at different seasons of the year (Gettinger et al., 1986).

Thermoregulation

There is no suggestion that heteromyids differ significantly from other small rodents with which they are sympatric in their capacity to regulate body temperature in cool environments. Brower (1970*a*) found that the body temperatures of eight species in the genera *Perognathus, Chaetodipus,* and *Dipodomys* declined no more than 2°C as ambient temperature was lowered from 30°C to 10°C. Greater variations have been reported in a few cases (e.g., *P. longimembris,* Chew et al., 1967; *C. intermedius,* Bradley et al., 1975), but always in species capable of torpor. Few measurements have been made in below-freezing conditions. Hoover and his coworkers (Hoover et al., 1977) found that *C. intermedius* but not *C. penicillatus* could survive a 2 hr exposure to −5°C. Similarly, Kenagy (1973*a*) found that at −25°C the deep body temperatures of *D. microps, D. merriami,* and *P. longimembris* fell at rates of 0.1°C/min, 0.3°C/min, and 0.9°C/min, and their tails froze during exposures of 1 hr, 40 min, and 20 min respectively. These animals have been observed active on the surface in temperatures of −19°C (kangaroo rats) and −10°C (pocket mouse), and presumably they could forage in colder environments if their exposure was limited to a few minutes at a time (Kenagy, 1973*a*).

The ability to produce heat changes dramatically on a seasonal basis in many small mammals, and it is likely that similar seasonal changes influence the thermoregulatory abilities of most heteromyids from temperate regions. Seasonal changes in non-shivering thermogenic capacity have been directly measured in at least one kangaroo rat, *Dipodomys ordii* (Gettinger et al., 1986). Field-caught animals had the capacity to produce twice as much heat in the middle of winter as they did during the warmest months of the year, and such changes could be reproduced (actually magnified) in the laboratory by changing photoperiod (Gettinger and Ralph, 1985) and, more importantly, ambient temperature (Gettinger et al., 1986). Such acclimation allows resting animals to thermoregulate successfully in cold winter temperatures without having to maintain the heat producing machinery, presumably brown fat, year-round.

The energetic costs of thermoregulation in the cold appear to be somewhat less for heteromyids from arid regions than for

TABLE 1.—*Factors related to energy consumption in heteromyid rodents.*

Species	Mass (g)	T_{lc}[a] (°C)	BMR[b] ($cc\ O_2/g\ hr$)	C_{min}[c] ($cc\ O_2/g\ hr\ °C$)	Critical T_b[d] (°C)	Citation[e]
P. longimembris	8	30.7	1.76 (81%)			Hinds and MacMillen, 1985
	8	32.2	1.70 (78%)	0.273 (77%)		Brower, 1970*a*
	9	30.5	1.07 (51%)	0.277 (83%)	<5	Chew et al., 1967
P. flavus	9	26.6	2.09 (100%)	0.286 (85%)		Hinds and MacMillen, 1985
					5	Wolff and Bateman, 1978
P. parvus	15			0.115 (44%)		Brower, 1970*a*
	16				2	Meehan, 1977
P. inornatus	16				<4	French, A. R. (unpublished data)
P. amplus	13			0.283 (100%)		Chew et al., 1963
	17				7.5	Hayden and Lindberg, 1970
M. megacephalus	11			0.229 (75%)		Hinds and MacMillen, 1985
M. pallidus	12			0.185 (64%)		Brown and Bartholomew, 1969
	15	30	1.3 (71%)		<5	Bartholomew and MacMillen, 1961
C. intermedius	13			0.242 (87%)		Chew et al., 1963
	15	30	1.3 (71%)	0.22 (85%)	7	Bradley et al., 1975
	15	32.9	1.27 (69%)			Brower, 1970*a*
C. penicillatus	16	31.7	1.38 (77%)	0.202 (80%)		Brower, 1970*a*
	19				13–15	Grubbs, 1974
	26				12–16	Hayden and Lindberg, 1970
C. formosus	17			0.198 (81%)		Brower, 1970*a*
	17			0.216 (89%)		Mullen and Chew, 1973
	21				11–12	Hayden and Lindberg, 1970
C. fallax	18			0.187 (79%)		Brower, 1970*a*
	20	24.7	1.37 (81%)	0.161 (72%)		Hinds and MacMillen, 1985
	20				11	Nishimoto, 1980
	26				12–13	Hayden and Lindberg, 1970
C. californicus	22	32.5	0.97 (59%)	0.19 (89%)	15	Tucker, 1965
C. baileyi	29	27.6	1.19 (78%)	0.131 (70%)		Hinds and MacMillen, 1985
	31			0.185 (103%)		Chew et al., 1963
	42				13–14	Hayden and Lindberg, 1970
C. hispidus	31	31.0	1.25 (83%)	0.196 (109%)		Brower, 1970*a*
	32	24.6	1.43 (96%)	0.147 (83%)		Hinds and MacMillen, 1985
	39	30.5	1.25 (88%)	0.20 (125%)	12	Wang and Hudson, 1970
	48		1.28 (96%)	0.16 (111%)		Morrison and Ryser, 1962

TABLE 1.—*Continued.*

Species	Mass (g)	T_{lc}[a] (°C)	BMR[b] (cc O_2/g hr)	C_{min}[c] (cc O_2/g hr °C)	Critical T_b[d] (°C)	Citation[e]
D. merriami	35	29.8	1.18 (80%)	0.136 (80%)		Brower, 1970*a*
	35		1.2 (82%)			Dawson, 1955
	36	27.9	1.26 (87%)	0.129 (77%)		Hinds and MacMillen, 1985
	37			0.109 (66%)		Brower, 1970*a*
	38	26	1.13 (79%)	0.108 (68%)	<16	Carpenter, 1966
					15	Yousef and Dill, 1971
D. ordii	47	25.9	1.37 (102%)	0.12 (82%)		Hinds and MacMillen, 1985
	56	28.5	1.05 (82%)	0.091 (68%)		Brower, 1970*a*
D. microps	57	26	1.17 (92%)	0.11 (83%)	<19	Breyen et al., 1973
D. agilis	47	28.3	1.07 (80%)	0.134 (92%)		Brower, 1970*a*
	61	26	1.05 (84%)	0.073 (55%)	nt	Carpenter, 1966
D. panamintinus	55	28.8	1.08 (83%)	0.108 (80%)		Brower, 1970*a*
	57		1.2 (94%)		<18	Dawson, 1955
	64	25.2	1.16 (94%)	0.080 (64%)		Hinds and MacMillen, 1985
	77			0.083 (73%)		Scelza and Knoll, 1975
D. deserti	103	29.5	0.75 (69%)	0.065 (66%)		Brower, 1970*a*
	105	26.5	0.90 (83%)	0.072 (74%)		Hinds and MacMillen, 1985
	106	28.5	0.87 (81%)	0.081 (84%)		McNab, 1979
L. irroratus	45	28.4	1.34 (99%)	0.164 (110%)		Hinds and MacMillen, 1985
	48	31	1.12 (84%)	0.176 (125%)		Hudson and Rummel, 1966
L. salvini	43	30.0	1.31 (95%)	0.211 (138%)		Hinds and MacMillen, 1985
	44	31	1.07 (78%)	0.186 (126%)		Hudson and Rummel, 1966
					nt	Fleming, 1977
H. anomalus	69	26	1.45 (120%)	0.153 (127%)		McNab, 1979
H. desmarestianus	76	27.9	1.31 (111%)	0.147 (128%)		Hinds and MacMillen, 1985
					nt	Fleming, 1977

[a] Lower limit of thermal neutrality.
[b] Basal metabolic rate; percentage of value expected on the basis of mass ($3.8M^{-0.27}$, Brody, 1945).
[c] Minimal thermal conductance; percentage of value expected on the basis of mass ($1.02M^{-0.505}$, Herreid and Kessel, 1967).
[d] Body temperature below which torpid individuals could not spontaneously arouse; nt = no torpor was observed during periods of starvation. Data from Hayden and Lindberg (1970) were obtained during hypoxia.
[e] Data from Brower (1970*a*), Hinds and MacMillen (1985), and Scelza and Knoll (1975) were taken from animals that were not postabsorptive.

mammals in general. Below-average measurements of minimal thermal conductance have been frequently reported for members of the family (Table 1), and summaries of such data have been made previously by Brower (1970*a*) and McNab (1979). Recently, Hinds and MacMillen (1985) found that the average minimal thermal conductance of 10 arid-adapted species within the genera *Perognathus, Chaetodipus, Microdipodops,* and *Dipodomys* was 26% lower than that predicted for similar-sized mammals. This agrees well with the 24% reduction estimated by Brower (1970*a*). Because body temperatures are similar, this difference in conductance suggests that, at any cool ambient temperature, arid-adapted heteromyids use only about 75% of the energy that is consumed by an average mammal of similar size. Desert habitats are also seasonally cold. The most tropical species that have been measured (*L. irroratus, L. salvini, H. desmarestianus,* and *H. anomalus*) rarely encounter very cold temperatures and all have minimal thermal conductances equal to or above those predicted for mammals of their size (Table 1).

The fact that heteromyids from dry, temperate regions are comparatively well insulated suggests that energy conservation in the cold is of greater importance than the ability to dissipate metabolic heat at high ambient temperatures. Most heteromyids are forced to thermoregulate in cool temperatures, both above and below ground, throughout many months of the year when food supplies are unpredictable. At these times, energy is saved not only physiologically by having a low minimal thermal conductance, but also behaviorally by selecting to rest at the warmest available temperatures (French, 1976; Kenagy, 1973*a*) and by constructing winter nests. Such behavior is not energetically inconsequential. Kenagy (1973*a*) calculated that *D. microps* and *D. merriami* reduced their resting metabolism at 5°C by 21% and 14%, respectively, when they occupied nests collected from the field. On the other hand, heat dissipation is critical at very high temperatures and during exercise. Wunder (1974) found that *D. ordii* normally regulate their body temperature 1.5–3.0°C above resting levels during sustained (10–15 min) activity, but are unable to limit this temperature rise when running in moderately-warm environments (20 m/min at 30°C or 30 m/min at 25°C). Presumably heat dissipation is rarely a problem, even in deserts, because the animals 1) are relatively-inactive underground during the day, 2) have low basal rates of heat production when at rest, and 3) probably never run continuously for 15 minutes or more, and therefore would seldom experience problems with hyperthermia on warm nights. Despite this, it would not be surprising if heteromyids were shown to change thermal conductance on a seasonal basis. Bradley and his coworkers (1975) found no differences in minimal conductance between *C. intermedius* captured in the winter and spring, but they did not make measurements on animals acclimatized to the hottest times of the year. This phenomenon has not been investigated systematically.

Activity Metabolism

There is no reason to believe that heteromyids differ from other rodents either in their metabolic capacities or in the energetic costs of various activities. Data are limited, however. Maximum aerobic metabolism has been determined for *D. ordii* running on a treadmill, and that value falls within the 95% confidence intervals of the allometric relationship derived by Lechner (1978) for small mammals (MacMillen, 1983). The costs of locomotion have been calculated for a few species, with particular emphasis placed on comparing bipedal and quadrupedal movement. It is widely felt that bipedality (and the erratic movements associated with ricochetal bounding) allowed the forelimbs of kangaroo rats and mice to become specialized for sifting seeds from fine soils, and aided their ability to evade predators in open habitats (Bartholomew and Caswell, 1951). In addition, small bipedal mammals may be able to move faster and more economically than quadrupedal species (Dawson, 1976), and this concept has been incorporated into models of heteromyid foraging strategy and resource partitioning (Reichman, 1981). Metabolic expenditure does appear to be independent of speed in rapidly-hopping kangaroos (Dawson and Taylor, 1973), but this is not the case for small bipeds including kangaroo rats (Thompson et al., 1980), which have different elastic properties of the hind limbs (Biewener et al., 1981). Thompson and his coworkers found that both *D. merriami* and *D. deserti* exhibit a linear relationship between oxygen consumption and speed even when they are hopping bipedally, and their costs of locomotion do not differ from those of similar-sized quadrupedal mammals. MacMillen (1983) did observe an uncoupling of oxygen consumption and speed in bipedal *D. ordii,* but he concluded in part from measurements of blood lactate that anaerobic metabolism continued to rise after aerobic capacity was reached. Thus, it appears that the advantages of bipedal locomotion to *Dipodomys* and *Microdipodops* are other than energetic.

Energetics of Facultative Heterothermy

Many heteromyids can vary the level at which body temperature is regulated and, as a result, alter the energy needed for thermoregulation. Slight reductions in temperature and energy use occur during sleep in most species, although traditionally these are considered to be fluctuations within the normeothermic range. More profound reductions are referred to as either shallow or deep torpor, depending upon whether body temperature is above or below some arbitrary level, usually around 15°C.

Physiological Characteristics

Sleep and deep torpor at a body temperature near freezing probably represent the thermal extremes of a continuum of physiological performance and energy conservation (Walker and Berger, 1980). It has been shown in the kangaroo rats *D. ingens, D. panamintinus,* and *D. heermanni* that the hypothalamic regulator of body temperature is reset downward and has a lower thermosensitivity during slow-wave sleep compared to wakefulness (Glotzbach and Heller, 1976). This supports the hypothesis that a primary function of slow-wave sleep is energy conservation (Berger, 1975). Torpor appears to be an extension of sleep. *Perognathus longimembris* enter torpor through slow-wave sleep, and remain in that neurophysiological state 96% of the time they are in shallow torpor (Harris et al., 1984). Furthermore, studies on ground squirrels have shown that changes in the CNS regulator of body temperature between wakefulness and deep torpor are greater than, but qualitatively similar to, those that occur during entrance into slow-wave sleep (Heller et al., 1978). There is no reason to suspect that heteromyids capable of deep torpor differ from this pattern.

Like sleep, sequential episodes of daily torpor reoccur with a strong circadian periodicity. Heteromyids exposed to a light-dark cycle usually rewarm (arouse) at about the same time each day, and variations in the duration of torpor are attained predominantly by changing the time at which torpor is initiated (Brower, 1970*b*; French, 1977*a*; Tucker, 1966). For example, as sequential one-day torpors increased in duration in *P. longimembris,* the time of entry was advanced three times as much as the time of arousal was delayed (French, 1977*a*). Because arousal times have a fairly precise phase relationship with dusk, the animals become euthermic in anticipation of the time to start foraging.

Torpors lasting longer than 24 hours apparently occur when one or more time windows for arousal are skipped (Brower and Cade, 1971; French, 1977*a*). Like in all hibernators, this process is profoundly affected by body temperature. For example, torpors last for a maximum of only two days in *P. longimembris* kept in an 18°C environment, but they can last for as long as five days if the animals are kept at 8°C (French, 1977*a*). Metabolic rates during torpor also show an approximate 2.5 fold change in magnitude after a 10°C change in body temperature, suggesting that some metabolic factor may be influencing this aspect of the timing mechanism.

Variations in the Use of Torpor

There is a broad spectrum of capabilities for torpor within the family. Torpor has not been observed in the most tropical genera, *Heteromys* and *Liomys* (Fleming, 1977; MacMillen, 1983), and it appears to be poorly developed in *Dipodomys* (Table 1). Shallow torpor has been documented only during starvation in *D. merriami* (Carpenter, 1966; Dawson, 1955; Yousef and Dill, 1971), *D. panamintinus* (Dawson, 1955); *D. microps* (Breyen et al., 1973) and *D. deserti* (MacMillen, 1983), and its ecological significance is questionable (Kenagy, 1973*a*; MacMillen, 1983). On the other hand, all the species of *Microdipodops, Chaetodipus,* and *Perognathus* that have been studied enter torpor readily, and this capacity is probably universal within these genera. Two basic strategies are evident. Some species apparently forage year-round and presumably utilize brief episodes of torpor only during short-term energetic emergencies. Many of these such as *C. californicus* (Tucker, 1965), *C. fallax* (Nishimoto, 1980), *C. hispidus* (Wang and Hudson, 1970), and *C. baileyi* (Hayden and Lindberg, 1970) live in relatively mild habitats and only can tolerate body temperatures down to 10–15°C (Table 1) for less than 24 hours at a time. At the opposite extreme are true hibernators, mainly in the genus *Perognathus,* that use torpor to make their hoarded food supplies last during a period of seasonal dormancy. Those that inhabit seasonally cold environments, such as *P. parvus,* can remain torpid continuously for up to 8 days with body temperatures as low as 2°C (Meehan, 1977). There are a few species whose behavior appears to be intermediate between these two extremes. Both *M. pallidus* and *P. flavus* regularly undergo multi-day torpors at low body temperatures (Table 1) even though some individuals may forage nightly all winter (Brown and Bartholomew, 1969; Wolff and Bateman, 1978). The utilization of torpor in these species along with the complex changes in behavior and thermoregulation associated with seasonal dormancy are discussed later in more detail.

The capacity for torpor appears to represent an evolutionary compromise between the energetic benefits and the ecological costs of low body temperatures. Any drop in body temperature followed immediately by arousal requires less energy than continuous euthermy (Tucker, 1965). The lower the body temperature and the longer the duration of torpor, the greater the energetic savings. However, sensory function and mobility also decrease as body temperature declines. Torpid heteromyids are much less aware of their environment and more susceptible to the theft of their food stores and to predation than are euthermic individuals. It follows that the diverse physiological capabilities seen in the family probably reflect differences in environmental productivity, competition and foraging success, energetic costs of foraging, and risks of predation.

The balance between the costs and benefits of torpor is reflected in the response of individuals to changes in their energetic stresses. Such a torpor response was shown first in *C. californicus* (Tucker, 1966) and later documented for *M. pallidus* (Brown and Bartholomew, 1969), *P. flavus* (Wolff and Bateman, 1978), and *C. fallax* (Nishimoto, 1980). These mice do not become torpid if they have enough food to continuously maintain high body temperatures. However, on reduced rations they are able to match the time they spend in torpor to their daily energetic deficit such that they maintain body weight under a wide variety of conditions that would otherwise be lethal. This ability changes seasonally in *M. pallidus.* Kangaroo mice captured in winter used torpor in excess of the minimum necessary for short-term survival, and as a result they accumulated seeds even under the most restrictive daily rations (Brown and Bartholomew, 1969). In contrast, kangaroo mice captured in late spring and summer were unable to survive any of the restrictive regimes to which Brown and Bartholomew subjected winter mice (French, 1989*a*). These summer mice were similar to kangaroo rats in that they only entered torpor after they had lost weight and were near death from starvation.

Seasonal Dormancy

The majority of the heteromyids that become dormant on a seasonal basis are members of the genus *Perognathus.* Winter inactivity has been reported in populations of *P. parvus* (O'Farrell et al., 1975; Scheffer, 1938), *P. longimembris* (Chew and Butterworth, 1964; Grinnell and Swarth, 1913; Kenagy, 1973*a*), *P. fasciatus* (Criddle, 1915), *P. amplus* (Reichman and Van De Graaff, 1973), *P. flavus merriami* (Chapman and Packard, 1974), and *P. flavescens* (Jones, 1964). In addition, I have found that captive *P. inornatus* remain underground continuously for many months when kept in artificial burrow systems (unpublished observations). The only species within the genus for which data are unavailable is *P. alticola,* but it would be surprising if these mice did not also undergo seasonal dormancy given that they are found at high elevations in mountains of southern California where the winter climate can be severe. Convincing documentation of seasonal dormancy in the other genera of heteromyids is rare, but does include *C. formosus* (French et al., 1966; Kenagy and Bartholomew, 1985), *C. penicillatus* (Arnold, 1942; Reynolds and Haskell, 1949), and *M. megacephalus* (O'Farrell, 1974; John H. Harris, unpublished data).

The suggestion that seasonal dormancy is a characteristic common to all members of the genus *Perognathus* is likely to be controversial because numerous investigators have trapped many of these species year-round. Certainly some individuals of an otherwise hibernating population, perhaps those that do not accumulate a sufficient store of food (see below), may remain active all winter. However, it is also difficult to determine the occurrence of seasonal inactivity from trapping records because the timing of dormancy can vary among individuals within a population. The best documentation of this is the long-term field study of O'Farrell and his colleagues on the Great Basin pocket mouse, *P. parvus* (O'Farrell et al., 1975). They found that individual mice were active above-ground for only two or three months each year, but the active seasons of all members of the population were not synchronous. Yearling males were the first to emerge from dormancy in the early spring, followed in turn by older males and then females of all ages. Young of the year usually left their natal burrows for the first time after their fathers and only shortly before their mothers resumed dormancy. In years with high rainfall and plant productivity, adult females had an average of two litters and these offspring also bred. Mice born in the last litters of the year from females born earlier that season comprised the trappable population of late fall and early winter. Thus, although adults spent 9–10 months continuously underground, the species was captured in a minimum of 9 months in some years and in all

12 months in others. These data point out that 1) if dormancy is timed by environmental events, then sex and age-class differences exist in the time cues used, 2) dormancy, at least for adults, spans both hot and cold times of the year, thereby merging estivation with hibernation, and 3) it is not justifiable to assume that members of a population do not hibernate simply because unmarked individuals are trapped year-round.

Timing.—There appears to be a strong endogenous component to the timing of dormancy. When the pocket mice *P. longimembris, P. inornatus,* and *P. parvus* were kept in complex artificial burrow systems, their behavior alternated between periods of activity when they regularly foraged on the surface at night and periods of dormancy during which they remained continuously underground (French, 1977*b,* and unpublished observations). In most animals, this cycle of behavior persisted throughout the several years they were kept under constant environmental conditions of temperature (5, 8, or 18°C), photoperiod (LD 12:12, or constant darkness), and above-ground food availability. The durations of the active and dormant phases of the cycle were quite variable, and thus in nature the timing of these changes in behavior must be greatly influenced by changes in environmental conditions.

These *Perognathus* may cease surface activity as soon as reproduction is terminated and a minimal cache of seeds is accumulated. It is highly unlikely that dormancy is initiated directly in response to environmental events such as changes in temperature or food availability. This is suggested by the observations that the disappearance of mice in nature is asynchronous (O'Farrell et al., 1975), and that captive mice will stop foraging even though food is always available on the surface and the temperature is constant (French, 1977*b*). However, the quality and quantity of food may indirectly determine the time pocket mice cease surface activity by influencing the extent of the breeding season or the foraging success of the animals. *Perognathus* undergo gonadal regression prior to entering dormancy, and in those years when reproductive activity is prolonged, the initiation of dormancy is delayed (Kenagy, 1973*a*; O'Farrell et al., 1975). I have found that high levels of testosterone, induced by silastic capsule implants, act to postpone dormancy indefinitely in captive *P. longimembris* regardless of the temperature or the availability of food (unpublished observations). However, reproductively-quiescent *P. longimembris* will not cease nightly surface activity if they are not allowed to accumulate an underground food cache. I have found that the active "season" of these mice can be prolonged to over a year by providing them only a few seeds each day. The need for a minimal food cache apparently is responsible for the observation that in years of normal food availability, individual *P. parvus* remain active for an average of only 60 days, but in unproductive years they do not breed and spend an average of 90 days foraging for the scarce seeds (O'Farrell et al., 1975). The minimal cache size necessary for dormancy is not known for any species, and it probably varies among sexes and age-classes when those segments of a population have different durations of dormancy.

Some incidents of winter activity in *Perognathus* may be related to environmental conditions that either promoted reproductive development in animals born late in the summer or precluded individuals from caching enough seeds to permit entrance into dormancy. For example, James Brown (pers. comm.) has found that *P. flavus* usually, but not always, become dormant. In some years he has trapped individuals in midwinter that were in reproductive condition. Wolff and Bateman (1978) also found members of this species active all winter, but they were unable to locate more than a few days supply of food upon excavation of their burrow systems. A similar situation may exist for *Microdipodops. M. palladus* has been collected throughout the winter in some years (Brown and Bartholomew, 1969), but at the same location in other winters I have failed to trap a single individual (French, 1989*a*). *M. megacephalus* is known to become dormant on a seasonal basis (O'Farrell, 1974), but the status of *M. palladus* awaits further clarification.

In contrast to the start of dormancy, the date surface activity is resumed in the spring is remarkably constant from year to year and appears to be synchronized with changes in soil temperature (French, 1977*b*; O'Farrell et al., 1975). *Perognathus longimembris* select the warmest available environment (below about 30°C) at all times of the year (French, 1976). In the summer, the soil temperature is highest near the surface, but in the winter, the gradient is reversed and the mice build their winter nests in the deepest parts of their burrow systems. Winter nests have been located up to 193 cm deep in *P. parvus* (Scheffer, 1938) and over 200 cm deep in *P. fasciatus* (Criddle, 1915). In the spring and autumn there is a slow overturn of the existing gradients, and the spring emergence of *P. longimembris* is correlated with the time of vertical uniformity of soil temperatures (French, 1977*b*). In the early spring these mice rest in warm soil within a couple of centimeters of the surface (Kenagy, 1973*a*), and I hypothesize that exposure to such high temperatures may trigger the termination of dormancy. Curiously, prolonged warm spells in midwinter greatly heat the surface of the ground in the desert, but mice do not emerge. If the mice avoid moving up through progressively-cooler soils, they might not experience the diurnal warming of the surface soil until the temperature gradient of the deeper layers is eliminated in the spring. Such isolation would be enhanced if the mice seal off the passages to their hibernacula with soil, as they do when kept in artificial burrows in the laboratory. The mice may possibly bc refractory to such thermal cues until late in their dormant seasons. However, I was able to stimulate emergence in 7 out of 11 captive *P. longimembris* after only two and one-half months of dormancy by heating their entire burrow systems (so that they were guaranteed to experience the temperature change) from 5°C to either 12 or 23°C for no more than four days (unpublished observations). Because the environmental conditions in the laboratory have been far different from those experienced by animals in nature, the resolution of this problem awaits further study.

Like most hibernating species, male pocket mice emerge a few weeks prior to females (Kenagy, 1973*a*; O'Farrell et al., 1975). If emergence is triggered by a warming of the soil, then males might be more sensitive to thermal changes or they might perceive those changes sooner than females. Some differences between males and females have been noted in captivity. In *P. inornatus,* the time spent at high body temperatures during the spring phase of hibernation is greater in males than females (see below). In addition, male *P. longimembris* spend less time underground than females in the absence of environmental cues. The partial results of a study of mine now in progress suggest that these mice exhibit two basic patterns of behavior when kept in artificial burrows at 5°C. Once dormancy has been initiated, some individuals remain continuously underground until they have consumed nearly all of their stored seeds, whereas others spontaneously resume surface activity after a certain period of time has elapsed. Females are more likely to have open-ended dormancy seasons than males. In this study, 50% (12 of 24) of the females remained underground until they nearly ran out of food in contrast to only 11.1% (2 of 18) of the males. The length of time these mice remained dormant was a function of the amount of food they originally stored and the amount of time they spent in torpor. Some individuals still in the experiment have remained continuously underground for almost three years and, because their seed caches are still plentiful, I expect their already long dormant intervals to be greatly extended. There were also obvious differences between the males and females that spontaneously resumed surface activity before their food supply was depleted. Of the males in this group, 50% had emerged by the 200th day of dormancy and 75% were active by the 218th day. In contrast, 50% of the females had not emerged until 322 days of dormancy had elapsed, and 75% of them were not active until the 355th day. How these sexual differences in physiology and behavior translate into differences in the timing of emergence in nature remain to be elucidated.

Patterns of temperature regulation. — Pocket mice alternate between brief episodes of torpor and euthermia throughout dormancy and, like other hibernators, progressively change the durations of these different thermoregulatory states. I have found that in *P. inornatus* torpors are short in the autumn phase of hibernation, become long and relatively-constant in duration throughout the winter phase, and then shorten once again in the spring (Fig. 1). In contrast, euthermic episodes shorten in midwinter and then lengthen again in the spring. These changes are presumably endogenous because the mice were given food caches at the start of the experiment and then kept under constant environmental conditions with minimal disturbances until their energy supplies were nearly exhausted. A few individuals remained in the experiment long enough to begin a second seasonal cycle, but at no time did any animal cease hibernation altogether.

The amount of time spent in torpor during the hibernation season appears to be a function of the animal's initial food supply and its sex. *Perognathus inornatus* given 400 g of seeds at the start of dormancy had shorter episodes of torpor, both in the winter and spring phases of hibernation, than those individuals that started with 200 g or less (French, 1989*b*; Fig. 1). The match between time spent at low body temperatures and energy supplies was imperfect, but it does indicate that there are advantages to being euthermic even during the time of year when the animals are continuously underground. Furthermore, the springtime increase in the frequency and duration of euthermic episodes suggests that the value of high body temperatures increases at the time when emergence from dormancy is likely. Euthermic mice should be better than torpid ones at assessing environmental conditions, especially near the surface, and thus an increase in the time spent at high body temperatures should increase the accuracy of the timing of emergence. The fact that males, which emerge first, spend more time euthermic during the spring phase of hibernation than females is consistent with this observation (French, 1989*b*).

Water Balance

The dependency of heteromyid rodents on dietary water is in general inversely correlated with the aridity of the habitat in which they live. For example, *H. desmarestianus* from wet tropical forests can survive for only a few days without a source of free water (Fleming, 1977), whereas many desert forms, such as *D. merriami* (Schmidt-Nielsen, 1964), are well known for their ability to live indefinitely on a diet consisting exclusively of air dried seeds. Although variations among species are conspicuous, the family as a whole has attained the reputation of being adapted for water conservation because the majority of members live in areas that are dry at least on a seasonal basis.

Obviously the degree to which any mammal must rely on dietary water (either drinking-water or free water in the food) depends upon the balance between the amount of water produced by oxidation of hydrogen in its food (metabolic water) and the amount of water lost from the body in urine, milk, feces, and evaporation from the skin and respiratory surfaces. Heteromyids do not produce an exceptionally-large amount of metabolic water; if anything, their rates of metabolism, and hence water production, are slightly lower than other mammals of a

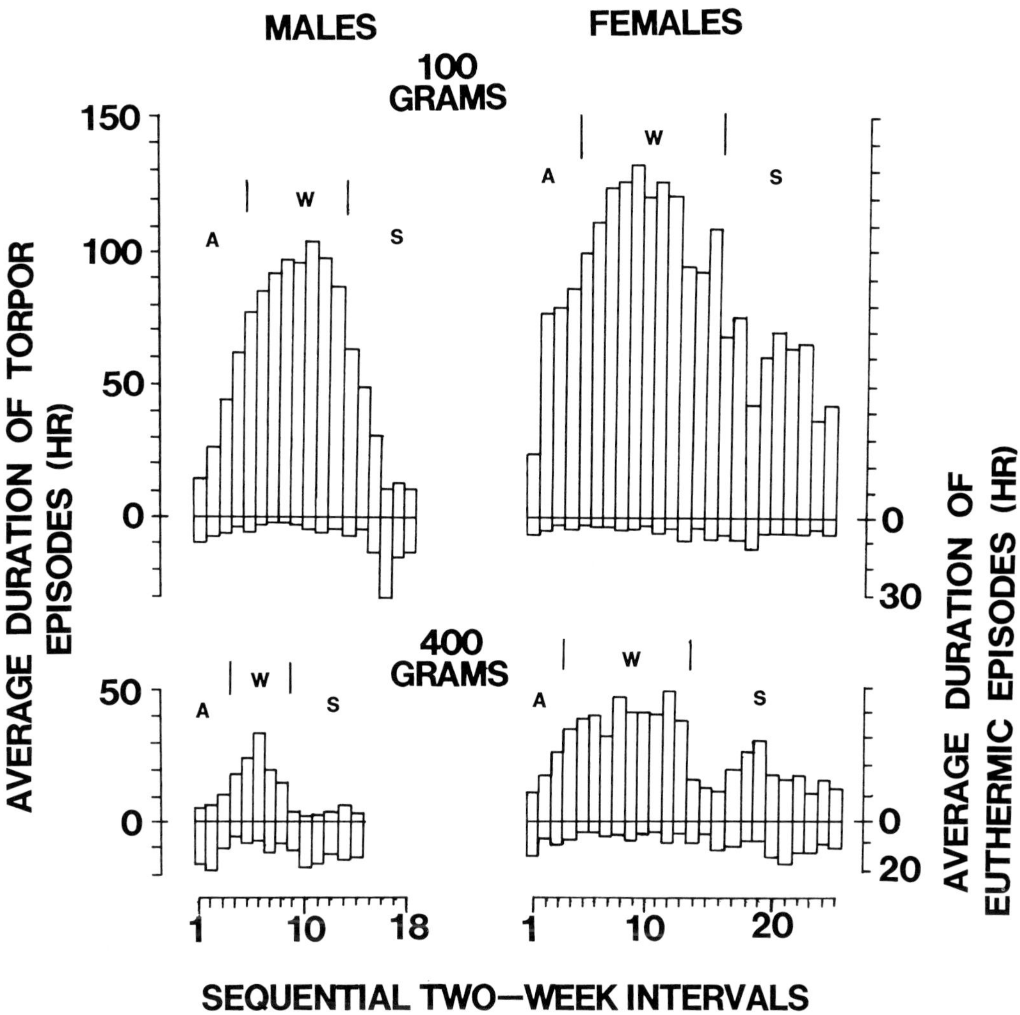

FIG. 1.—Patterns of thermoregulation during hibernation at 5°C in representative male and female *Perognathus inornatus*. Times of entrance into and arousal from torpor were determined from records of thermocouples placed in each animal's nest as described by French (1977*a*). Two-week averages of the durations of episodes of torpor (above the base line) and euthermia (below the base line) are plotted sequentially for each animal. Mice were given either 100 g (top) or 400 g (bottom) of mixed seeds at the start of the experiment, and were monitored until this was completely consumed. Autumn (A), winter (W), and spring (S) phases of hibernation were arbitrarily determined from changes in torpor duration.

similar size (Brower, 1970*a*; Hinds and MacMillen, 1985). Those species able to survive on a dry diet do so because they have enhanced physiological and behavioral abilities to reduce losses when water stressed. However, quantifying those abilities and making comparisons among species is not a trivial task. Rates of water loss vary in response to factors such as the animal's size, its state of hydration, its activity, ambient temperature, and environmental humidity. Ideally, information on the most rigorous set of conditions under which an animal can remain in water balance should be used as a basis of comparison among species. Unfortunately, such data are

TABLE 2.—*Evaporative water loss of heteromyids exposed to air of low humidity at cold temperatures below thermal neutrality and at 35°C.*

Species	Mass (g)	EWL_{cold} (mg H_2O/ g hr)	$EWL_{35°C}$ (mg H_2O/ g hr)	Citation
P. longimembris	8	2.079	2.484	Hinds and MacMillen, 1985
P. flavus	9	3.025	3.025	Hinds and MacMillen, 1985
M. megacephalus	11	2.017	2.809	Hinds and MacMillen, 1985
C. intermedius	15	2.4*		Bradley et al., 1975
C. fallax	20	1.904	2.735	Hinds and MacMillen, 1985
C. baileyi	29	1.417	1.417	Hinds and MacMillen, 1985
C. hispidus	32	1.930	1.930	Hinds and MacMillen, 1985
D. merriami	36	1.698	2.527	Hinds and MacMillen, 1985
		1.08		Church, 1969
		1.25*		Carpenter, 1966
		1.21*		Schmidt-Nielsen and Schmidt-Nielsen, 1950*a*
D. ordii	47	1.326	2.052	Hinds and MacMillen, 1985
D. microps	57	1.01*		Breyen et al., 1973
D. agilis	61	1.1*		Carpenter, 1966
D. panamintinus	64	0.939	1.592	Hinds and MacMillen, 1985
D. venustus	85	0.91		Church, 1969
D. spectabilis	100	0.80*		Schmidt-Nielsen and Schmidt-Nielsen, 1950*a*
D. deserti	105	0.998	1.592	Hinds and MacMillen, 1985
L. pictus	40	1.96		Christian et al., 1978
L. irroratus	45	1.970	2.804	Hinds and MacMillen, 1985
	48	2.2*		Hudson and Rummel, 1966
	68	1.17		Christian et al., 1978
L. salvini	43	2.384	2.916	Hinds and MacMillen, 1985
	44	1.75*		Hudson and Rummel, 1966
H. desmarestianus	76	1.512		Hinds and MacMillen, 1985

* Calculated from data presented in a different format.

not available. Water loss has been measured in only a small percentage of the species in the family, and most studies have been concerned with the effects of only one or two of the above factors. Furthermore, experimental conditions are rarely the same in studies conducted in different laboratories, potentially masking subtle differences among species. Nevertheless, some general patterns have emerged recently thanks mainly to the work of MacMillen and Hinds who incorporated several species from different genera and habitats in their studies (Hinds and MacMillen, 1985; MacMillen and Hinds, 1983). These patterns will be enumerated in the following discussion as each of the avenues of water flux are examined separately.

Evaporative Water Loss

Heteromyids in general, and those from desert habitats in particular, appear to have low rates of evaporative water loss compared to most other mammals (Table 2). This long-standing conclusion (Hart, 1971) has been reinforced recently by Hinds and MacMillen (1985) who found that data from most of the heteromyids they studied fell below values predicted by the allometric relationship for eutherians derived by Chew (1965). A direct comparison of these two data sets (Table 2 and Fig. 2) reveals two additional points. First, the heteromyids from relatively mesic habitats (*Liomys, Heteromys*) lose more water by evaporation than similar-sized heteromyids from arid

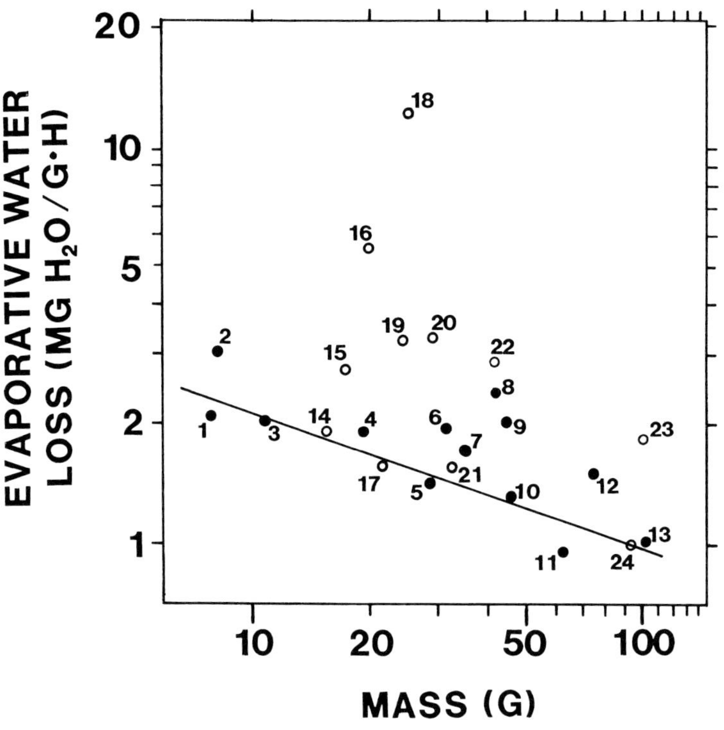

FIG. 2.—Evaporative water loss in heteromyid rodents compared to other similar-sized mammals. Data for heteromyids (solid symbols) are from Hinds and MacMillen (1985) and data for other mammals (open symbols) are those used by Chew (1965) to construct his allometric relationship between EWL and mass for eutherians in general. All data are from animals measured when quiet in dry air and at cool temperatures where evaporation was not used to augment heat loss. The regression of EWL against mass for xeric-adapted heteromyids (*Liomys* and *Heteromys* were excluded) is represented by the equation $5.267g^{-0.368}$ and the solid line. Species are : 1 *P. longimembris*; 2 *P. flavus*; 3 *M. megacephalus*; 4 *C. fallax*; 5 *C. baileyi*; 6 *C. hispidus*; 7 *D. merriami*; 8 *L. salvini*; 9 *L. irroratus*; 10 *D. ordii*; 11 *D. panamintinus*; 12 *H. desmarestianus*; 13 *D. deserti*; 14 *Antrozous pallidus*; 15 *Peromyscus maniculatus*; 16 *Peromyscus leucopus noveboracensis*; 17 *Peromyscus crinitus*; 18 *Blarina brevicauda*; 19 *Mus musculus*, albino; 20 *Peromyscus leucopus tornillo*; 21 *Mus musculus*, feral; 22 *Microtus ochrogaster*; 23 *Rattus norvegicus*, albino; 24 *Mesocricetus auratus*.

and semiarid locations, and second, the adaptations of desert heteromyids are not unique. Other small mammals that often frequent dry habitats, such as *Antrozous pallidus, Peromyscus crinitus, Mus musculus,* and *Mesocricetus auratus,* have evaporative water losses equivalent to those of desert heteromyids even though they do not have equivalent reputations for water conservation.

Evaporation has been minimized in arid-adapted heteromyids by a reduction in both respiratory and cutaneous water losses. Evaporation from the respiratory tract is a major avenue of water loss, comprising nearly 70% of the total in kangaroo rats when they are excreting highly concentrated urine (Schmidt-Nielsen, 1964). Nevertheless. these rodents are extremely efficient compared to large mammals such as humans. Respiratory losses are a function of both the difference in water content between the inspired and expired air, and the volume of air that is passed across the respiratory surfaces. Expired air is always saturated, and thus its water content is directly related to its temperature. The nasal passages of kangaroo rats and presumably other heteromyids serve as counter-current heat exchangers in which the saturated warm air leaving the lungs is cooled to temperatures near to, or even slightly lower than, that of inspired air (Jackson and Schmidt-Nielsen, 1964). Depending on the ambient temperature and humidity, approximately ⅔ to ¾ of the water vapor added to the inspired air is recovered in the nasal passages during exhalation (Collins et al., 1971). However, such adaptations may be common among rodents. Even the laboratory rat, whose total water losses greatly exceed those of kangaroo rats, is just as efficient at cooling exhaled air (Schmidt-Nielsen and Schmidt-Nielsen, 1952). Therefore, the water lost per ml of air breathed is not unusually low in desert heteromyids. However, respiratory water loss is also directly related to the magnitude of air moved through the respiratory system, which in turn is a function of the animal's metabolic rate (assuming that the extraction efficiency for oxygen is constant). As discussed in the previous section, the metabolic rates of arid-adapted heteromyids are approximately 25% less than those of average mammals of a similar size (Brower, 1970*a*; Hinds and MacMillen, 1985), suggesting that respiratory water loss in this group of rodents has been reduced by a similar amount.

Cutaneous evaporation appears to have been reduced in arid-adapted species even more than respiratory evaporation. Chew and Dammann (1961) calculated that 84% of the water lost by evaporation in *D. merriami* came from the respiratory tract and only 16% came from the skin, confirming the Schmidt-Nielsens' (1950*a*) previous estimate that cutaneous losses were relatively insignificant in kangaroo rats. This situation is in marked contrast to that found in other rodents (*Peromyscus maniculatus sonoriensis,* Chew, 1955; laboratory rats, Stupfel and Geloso, 1959; Tennent, 1945) where diffusion across the skin amounts to approximately half of the total evaporative loss. Similar high percentages of cutaneous water loss also occur in mammals belonging to other orders (Chew, 1965). Thus, even though respiratory water losses have been reduced, they make up a greater percentage of the total evaporation in *D. merriami,* and presumably other arid-adapted heteromyids, because evaporation from the skin has been reduced to an even greater extent.

Evaporative water loss can be greatly influenced by the conditions under which it is measured, thereby complicating interspecific comparisons. Evaporation declines as animals become dehydrated. This was shown for *D. venustus,* a species from mesic habitats that was unable to maintain weight and presumably fluid volume when kept on a dry diet (Church, 1969). The influence of changes in environmental temperature and humidity can be more profound. Schmidt-Nielsen and Schmidt-Nielsen (1951) showed that kangaroo rats at 25°C double their evaporative losses when moved from a saturated to a dry environment. Such a move increases the vapor pressure deficit, thereby increasing the evaporation from both the skin and the pulmonary surfaces. Temperature effects are more complex. A reduction in ambient temperature below thermal neutrality causes metabolism, and hence the volume of air passed through the respiratory system, to increase. This would increase evaporative water loss except for the fact that the vapor pressure deficit and the temperature of exhaled air simultaneously decline. Apparently these effects cancel one another and evaporative water loss remains relatively constant over a wide range of ambient temperatures. Hinds and MacMillen (1985) found this to be true between 5–35°C for some 5–25°C for all of the heteromyids they studied. Only at higher temperatures did evaporation increase to augment heat dissipation. Thus, species measured at different temperatures but similar humidities can be compared, as in Figure 2 and Table 2, with reasonable confidence.

TABLE 3.—*Mean maximum urine concentrations. Species are grouped on the basis of the length of time they can survive without an external source of free water under conditions of low humidity and moderate temperatures.*

Species	Mass (g)	UC_{max}[a] (mOsm/l)	Citation
Independent of exogenous water			
C. penicillatus	19	4,684	Grubbs, 1974
C. fallax	20	4,617	Christopher, 1975
D. merriami	36	4,800	Carpenter, 1966
		4,675	Kenagy, 1973*b*
		5,020	Christopher, 1975
Somewhat dependent on exogenous water			
D. agilis	61	4,100	Carpenter, 1966
L. pictus	40	3,849	Christian et al., 1978
L. irroratus	48	3,580	Hudson and Rummel, 1966
	68	3,820	Christian et al., 1978
L. salvini	43	4,000	Hudson and Rummel, 1966
Strongly dependent on exogenous water			
D. microps	57	2,827	Kenagy, 1973*b*

[a] Urine was determined to be maximally concentrated if it was obtained from animals that were dehydrated.

Urinary Water Loss

Limited data on heteromyids suggest that their urine concentrating ability is related to the aridity of their habitat and the degree to which they are dependent on exogenous water within their food (Table 3). Some xeric-adapted species are capable of producing among the most concentrated urine known, and thus are able to excrete wastes with a minimum of water loss. Urine concentrations as high as 7,500 and 6,048 mOsm/kg water have been reported for individual *C. penicillatus* (Altschuler et al., 1979) and *D. merriami* (Schmidt-Nielsen et al., 1948) respectively; values that are equaled or exceeded by only a very few desert rodents from Africa and Australia (Buffenstein and Jarvis, 1985; MacMillen and Lee, 1967). However such very high concentrations are not commonplace, even for those two species. There is always a great deal of variation among individuals kept under the same conditions, and maximum concentrations average between 4,600 and 5,000 mOsm/l for those species as well as for *C. fallax*, which is also independent of exogenous water under most environmental situations (Table 3). In comparison, *Liomys* spp. and *D. agilis* from arid but non-desert habitats have somewhat lesser abilities to concentrate urine (around 3,600 to 4,100 mOsm/l), and *D. microps*, which consumes succulent plants, can concentrate its urine to only 2,800 mOsm/l when dehydrated (Table 3). Unfortunately, most other species have not been similarly stressed to the point of desiccation, and therefore appropriate data for further comparisons within the family are lacking.

The average values for maximum urine concentration in desert heteromyids are certainly above those for most mammals, but they are matched by those from other "water-independent" rodents that have evolved in parallel elsewhere. Similar average maximums have been reported in dipodids from Asia and North Africa (Schmidt-Nielsen, 1964), and murids from Australia (MacMillen and Lee, 1967) and southern Africa (Buffenstein et al., 1985). In virtually all cases, the high osmotic concentrations were due to extraordinarily high concentrations of urea. Nevertheless, up to 47% of the nitrogen in the urine of the African murids was not dissolved, but instead was eliminated as a crystalline precipitate of allantoin (Buffenstein et al., 1985). Similar crystals have been observed in the urine of *D. merriami,* and were especially prevalent in water stressed animals on high protein diets (Carpenter, 1966). However, this phenomenon has not been seen in other heteromyids, or confirmed in *D. merriami,* despite numerous investigations of urinary water losses.

Maximum concentrating ability is often inferred indirectly from kidney morphology, particularly the prominence of the renal medulla. The nephron operates on a countercurrent principle and maximum urine concentration is related to the length of the loops of Henle and collecting ducts that traverse the medulla. In small mammals, the inner medulla is organized into a single papilla (Sperber, 1944), and in desert forms it may extend beyond the margins of the renal capsule well into the ureter. This extension is quite pronounced in some of the smaller heteromyids within the genera *Perognathus* and *Chaetodipus* (Altschuler et al., 1979), suggesting that they have quite powerful kidneys. However, these heteromyids are not unique. Equally spectacular elaborations are found in certain other small desert rodents (MacMillen and Lee, 1969), shrews (Lindstedt, 1980), and bats (Geluso, 1978; Vogel and Vogel, 1972).

Anatomical evidence also suggests that desert heteromyids perfuse a greater proportion of their renal blood flow through juxtamedullary nephrons. and have a greater density of vasa recta than rodents with lower urine concentrating capacities. Although many midcortical nephrons may be endowed with long loops of Henle, there is a clear tendency for juxtamedullary corpuscles to give rise to the loops that extend farthest into the papilla (Jamison and Kriz, 1982). In the species *P. amplus, C. baileyi, C. penicillatus,* and *D. merriami,* the glomeruli of the juxtamedullary nephrons are about 1.5–1.6 times greater in diameter than those of the superficial and midcortical nephrons (Altschuler et al., 1979). Furthermore, approximately 20% of all glomeruli are of the large juxtamedullary type in these heteromyids compared to 11% and 14% in the sympatric murids *Peromyscus eremicus* and *Onychomys torridus,* respectively. Altschuler and coworkers calculated that the ratio of the glomerular blood volume in juxtamedullary nephrons to that in superficial nephrons was 1.6 in *C. penicillatus* and 1.3 in *D. merriami,* compared to only 0.71 and 0.52 in *Peromyscus* and *Onychomys.* In addition, desert rodents including *C. penicillatus, D. spectabilis,* and *D. merriami,* have an above average density of vasa recta bundles in the outer medulla (Munkacsi and Palkovits, 1977). The vasa recta acts as a countercurrent diffusion exchanger thereby helping to maintain the concentration gradient developed in the medullary interstitium, and their density is correlated with urine concentrating ability. Thus it appears that arid-adapted heteromyids have extended medullae that include relatively long loops of Henle from juxtamedullary nephrons, have relatively more of these nephrons, perfuse a relatively higher proportion of their renal blood flow through these nephrons, and surround these nephrons with a greater density of blood vessels than species that produce less concentrated urine.

Medullary prominence was first quantified by Sperber (1944), who introduced the concept of relative medullary thickness (RMT) as an index of morphological adaptation. The RMT is the length of the medulla from the cortico-medullary boundary to the area cribrosa, divided by the cube root of the product of the maximum length, width, and thickness of the kidney. Sperber (1944) found that the RMT's of mammals living in arid areas were greater than those of mammals from more mesic environments, and Blake (1977) found this to be true even for members of the same species. The relationship between RMT and maximum concentrating capacity of the kidney was first quantified by Schmidt-Nielsen and O'Dell (1961), and a regression equation for their data, based on a diversity of species

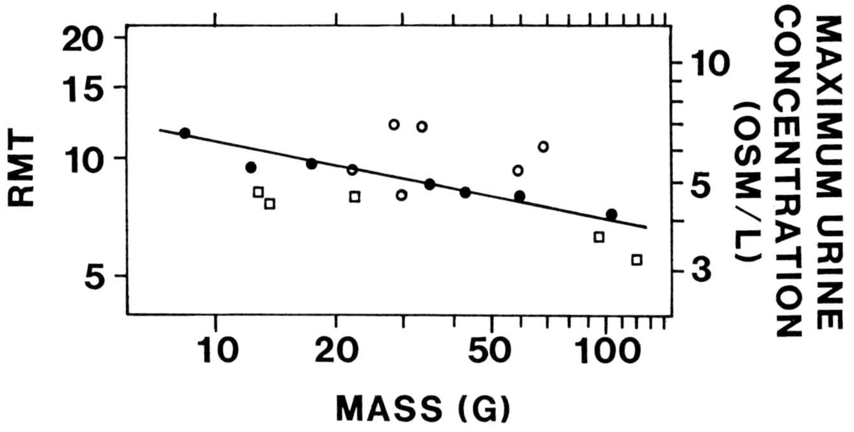

FIG. 3.—Relative Medullary Thickness and from it a derivation of maximal urine concentration as a function of body mass for desert-dwelling heteromyids and other similar-sized mammals. Heteromyids (solid circles), in order of ascending mass, are: *P. longimembris, M. pallidus, C. formosus, D. merriami, D. ordii, D. agilis,* and *D. deserti*. RMT for *D. agilis* is from Sperber (1944); average values for the other heteromyids are from Lawler and Geluso (1986) and their regression against mass is represented by the solid line and the equation RMT = $16.22g^{-0.181}$. Other points are RMT data from similar-sized non-heteromyids used by Greegor (1975) to construct his allometries for xeric (open circles) and mesic (open squares) mammals. Conversion of RMT to maximum urine concentrations was based on the equation mOsm/l = 580RMT − 39 (Lindstedt, 1980).

from differing habitats, was presented by Lindstedt (1980). RMT was also found to be approximately proportional to $\text{mass}^{-0.09}$ when differences in habitat were factored out (Greegor, 1975), suggesting that, in general, small species have more powerful kidneys than large species.

The kidney morphology of desert heteromyids is about what one would predict for arid-adapted mammals of their size. The RMT values reported for a series of six desert heteromyids (Lawler and Geluso, 1986) are generally greater than those of similar-sized mammals from mesic or aquatic habitats, but are matched or exceeded by those of several species from xeric environments elsewhere (Fig. 3). The negative correlation with body size has led Lawler and Geluso to conclude that small pocket mice are capable of producing more concentrated urine than larger kangaroo rats even though they are from the same habitat. The average maximum urine concentrations for these species have been estimated (Fig. 3), and values range from 4,143 mOsm/l in *D. deserti* to 6,660 mOsm/l in *P. longimembris*. Direct measurements of maximum urine concentrations in these species are too few to substantiate these predictions. The mean maximum concentration of 4,100 mOsm/l recorded for *D. agilis* (Carpenter, 1966) is lower than the prediction of 4,600 mOsm/l based on a RMT measurement of 8.0 for this species (Sperber, 1944). However the average of 5,020 mOsm/l found in *D. merriami* by Christopher (1975) is quite close to the 4,950 mOsm/l predicted in Figure 3.

Like concentrating ability, RMT varies greatly among individuals of the same species. The range in RMT reported for *D. merriami* by Lawler and Geluso (1986) would translate into maximum urine concentration values ranging from approximately 3,700 to 5,750 mOsm/l. The higher value is close to the 6,040 mOsm/l maximum estimated by Schmidt-Nielsen and coworkers (1948), but is well above the average maximum. This intraspecific variability demands that caution be used when interpreting the precision of the correlations between kidney morphology determined from one set of animals and the maximum urine concentration ever reported for those species in the literature (e.g., Blake, 1977; Brownfield and Wunder, 1976; Greegor, 1975; Munkacsi and Palkovits, 1977; Schmidt-Nielsen and O'Dell, 1961). Some of this variability may be the consequence of differences in water availability during juvenile development. Although data are lacking for heteromyids, many rodents are born with immature kidneys, and in some, renal development may not be complete until after weaning. Water stress during this time has been shown to cause medullary hypertrophy in *Mus musculus* (Blount and Blount, 1968), as well as in the xeric-adapted species *Notomys alexis* from Australia (Hewitt, 1981) and *Aethomys namaquensis* and *Tatera leucogaster* from Africa (Buffenstein and Jarvis, 1985). These morphological changes result in an elevated capacity to concentrate urine which, in the case of the African animals, was 1.5 fold. Thus, developmental plasticity can account for variations in maximum urine concentrating capacity equivalent to those estimated for *D. merriami*.

The accuracy of using RMT as universal predictor of maximum urine concentration has been questioned (Brownfield and Wunder, 1976), and the significance of size-related differences in RMT among similar species from the same habitat has never been tested directly. The denominator in the RMT index is a measure of kidney size. Calder (1984) predicted that kidney size should scale in parallel with metabolic rate (proportional to body $mass^{0.75}$) because, in well hydrated mammals, the excretion of nitrogenous compounds (Brody, 1945), glomerular filtration rate (Edwards, 1975), and urine volume (Edwards, 1975) all do so. However, kidneys scale with a larger mass exponent (0.85, Brody, 1945) than the metabolic functions they provide, leading Calder (1984) to suggest that there is a systematic increase with body size in the amount of supporting tissue surrounding the nephrons. If this proves to be the case, then there would be a systematic bias in the RMT index of almost exactly the same magnitude ($mass^{0.75}/mass^{0.85} = mass^{-0.1}$) as the decline in RMT with increasing body size found by Greegor (1975) and Blake (1977). Perhaps the size-related differences in RMT among desert heteromyids have little physiological consequences, and all species have, on average, nearly the same ability to concentrate urine. This possibility needs to be tested.

Other renal indexes have proven to be better predictors of maximum concentrating ability than RMT. Brownfield and Wunder (1976) found that relative medullary area (medullary area/cortical area) was better for comparing mammals of broad taxonomic diversity. Such measurements account for differences in the relative numbers of short and long looped nephrons which can affect renal performance. Geluso (1978) found that in bats, where the relative populations of the nephron types were similar, the ratio of the thickness of the inner medulla to the thickness of the cortex (IM/C) was the best predictor of maximum concentrating ability. Presumably heteromyid rodents are also morphologically similar in this regard. Values for the IM/C index derived for the six species of desert heteromyids studied by Lawler and Geluso (1986) lie within the range reported for bats that had maximum urine concentrations of approximately 2,800–4,000 mOsm/l. Such maximum concentrations, however, appear to be too low for desert heteromyids.

Fecal Water Loss

Fecal water loss is a function of the water content and amount of fecal matter produced, both of which appear to be low for heteromyid rodents. Schmidt-Nielsen and Schmidt-Nielsen (1951) calculated that the feces of *D. merriami* were over 2.5 times as dry as those of white rats (834 vs 2,246 mg of water per g of dry fecal matter). Additional water appears to be reclaimed from the feces when animals are water stressed. This has been demonstrated in *P. parvus* which reduced the water content of their feces from 667 to 562.5 mg water/g dry mass (calculated from Withers, 1982), and also in *L. pictus* and *L. irroratus* which produced equally-dry (540 and 575 mg water/g dry mass, respectively) feces upon dehydration (Christian et al., 1978). Furthermore, heteromyids on seed diets often assimilate over 90% of the food that they ingest (French, 1976; Johnson and Groepper, 1970; Schreiber, 1979; Soholt, 1973; Withers, 1982), thereby minimizing the amount of undigested material and associated water that is passed from the body. Schmidt-Nielsen and Schmidt-Nielsen (1951) calculated that their kangaroo rats produced only half as much fecal matter and, as a result, had only one-fifth the fecal water loss of white rats when both ate the same amount of food.

The colons of heteromyids are morphologically distinct from those of most other rodents (Forman and Phillips, this volume). The mucosa just distal to the caecum is asymmetrical when viewed in cross section. On the mesenteric side of this region the epithelial cells are thickened and appear to be specialized for absorption, secretory goblet cells are few in number, and the venous and lymphatic drainages are prominent. This peculiar morphology may be related to the animals' enhanced ability to resorb water from the fecal material. Likewise, the habit of coprophagy known to occur in heteromyids (Howell and Gersh, 1935; Kenagy and Hoyt, 1980) may also contribute to their high assimilation efficiencies and low rates of fecal water loss.

Lactational Water Loss

The milk of *D. merriami* is highly concentrated, averaging 50.4% water (Kooyman, 1963). The only other mammals known to have such concentrated milk are seals and whales. Presumably other heteromyids are similarly adapted for minimizing water loss during lactation, although data are unavailable for them as well as other desert rodents. In addition to producing highly concentrated milk, it is possible that heteromyids are able to reclaim a significant fraction of its water. It has been shown in desert mammals as diverse as rodents, canids, and kangaroos, that mothers consume the dilute urine and feces of their young, thereby regaining approximately one-third of the water secreted as milk (Baverstock and Green, 1975).

Body Size and Environmental Limits to Water Balance

Water balance is strongly influenced by ambient temperature. Metabolic water production (MWP), the major avenue of water input in most heteromyids (Schmidt-Nielsen and Schmidt-Nielsen, 1951), is directly related to metabolic rate and therefore increases as ambient temperature declines below thermal neutrality. However as mentioned above, the major avenue of water output, evaporative water loss (EWL), is independent of temperature over this range. This means that there will be some temperature below which MWP exceeds EWL by a sufficient amount to allow for water lost in excretory products and still render the animal independent of exogenous water. That "break even" temperature is a function of the relative humidity, but not the activity of the animal. Relative humidity alters EWL but not MWP, whereas changes in the level of activity change EWL and MWP by equal amounts (Raab and Schmidt-Nielsen, 1972). The concept of thermal dependency was first developed by Bartholomew (1972) for granivorous birds, and later was applied to heteromyid rodents by MacMillen and his coworkers (MacMillen, 1972; MacMillen and Christopher, 1975; MacMillen and Hinds, 1983).

The influence of temperature on water balance is reflected in urine concentration. Mammals modify urinary water output via

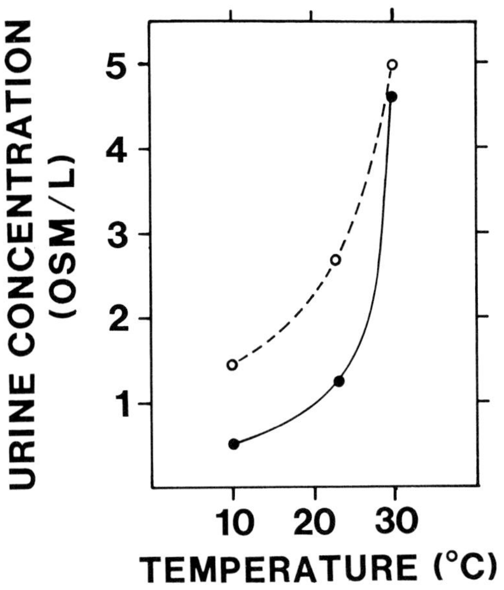

FIG. 4.—The relationship between average urine concentration and ambient temperature in 37 g *D. merriami* (open circles) and 21 g *C. fallax* (closed circles) fed hulled barley. Data are from Christopher (1975).

antidiuretic hormone so that water losses are matched to water gains. Because (MWP − EWL) declines as ambient temperature increases, urine volume in heteromyids without exogenous water also declines and therefore urine concentration increases (Fig. 4). This figure also shows that there is a wide range of ambient temperatures in which the animals are in favorable water balance, and there is a great potential for behaviorally adjusting water balance by selecting different thermal locations for rest during the day or different times (and temperatures) for activity at night.

In general, small heteromyids should be in more favorable water balance than large ones when exposed to similar environmental conditions. Mass-specific rates of MWP and EWL both decline with increasing body size but, at temperatures between 5–25°C, the rate of change is greater for MWP (Fig. 5). Thus (MWP − EWL) declines not only with increasing ambient temperature but also with increasing body size. This concept was formulated by MacMillen and Hinds (1983), and supported by the data on urine concentrations that they compiled. Those data suggest that species of *Dipodomys* typically produce urine that is twice as concentrated as that produced by small species of *Perognathus* or *Chaetodipus* kept under the same laboratory conditions (also see Fig. 4). The regressions of MWP and EWL with mass (Fig. 5) suggest that, on average, medium to large heteromyids probably would be unable to maintain water balance at 25°C if the humidity were low and they ate air-dried seeds. The Schmidt-Nielsens (1951) found this to be true for *D. merriami* when the relative humidity was 10% or below. At moderate relative humidities of 30–60%, the corresponding reduction in EWL and increase in preformed water in the food make it likely that many species could survive indefinitely without having to drink free water.

The regression lines in Figure 5 represent data from 13 species from all 6 genera in the family, and therefore conceal potentially interesting exceptions to the general trend. MacMillen and Hinds (1983) quantified the thermal dependence of each species by calculating the ambient temperature at which MWP = EWL. This index was then plotted against body mass, but the expected negative allometry was weak. In fact, their calculated index was independent of mass throughout the size range represented by the genera *Chaetodipus, Dipodomys, Liomys,* and *Heteromys. Microdipodops megacephalus* and one of the two species of *Perognathus* studied (*P. longimembris,* not *P. flavus*) had indices that were more than two standard deviations higher than average. Because these two species were among the smallest, a slight negative allometry existed for the family as a whole. I suggest that this relatively poor correlation might be the result of the way the thermal dependency index was calculated. MacMillen and Hinds calculated the best fit exponential equation relating the MWP:EWL ratios to temperature over the range of 5–35°C for each spe-

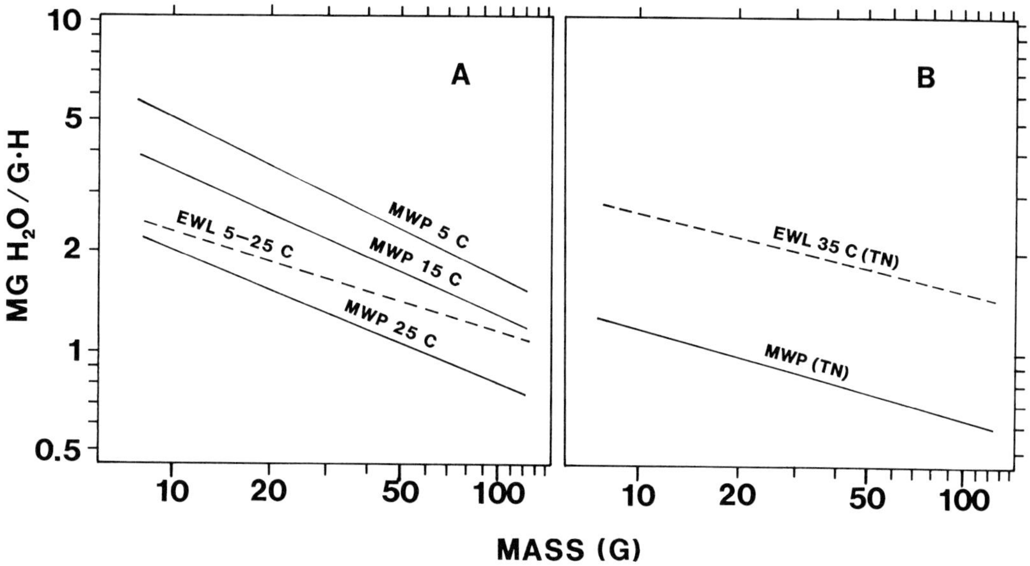

FIG. 5.—The scaling of metabolic water production and evaporative water loss as a function of body mass in heteromyid rodents at moderate (A) and hot (B) temperatures. Equations for the regression lines are: $MWP_{5°C} = 15.46g^{-0.48}$, $MWP_{15°C} = 9.179g^{-0.421}$, $MWP_{25°C} = 5.153g^{-0.41}$ (Table 4, MacMillen and Hinds, 1983); $MWP_{TN} = 2.292g^{-0.284}$, $EWL_{5-25°C} = 4.51g^{-0.295}$, and $EWL_{35°C} = 4.656g^{-0.244}$ (equations 5, 9, 14 in Hinds and MacMillen, 1985). EWL at 35°C includes only species that are in thermal neutrality at that temperature. Since metabolism does not change throughout the thermal neutral zone, MWP was calculated for those animals by multiplying the equation for basal metabolism (#5, Hinds and MacMillen, 1985) by 0.62 mg water/ml O_2 (MacMillen and Hinds, 1983).

cies, and then determined from that equation the temperature at which MWP/EWL = 1. However the MWP:EWL ratio theoretically should not decline exponentially with increasing temperature. As temperature rises from 5 to near 25°C, MWP declines linearly and EWL remains constant (the ratio declines linearly). Then as air temperature rises throughout the thermal neutral zone, MWP does not continue to decline but instead remains constant reflecting basal rates of metabolism, whereas EWL is no longer a constant but increases curvilinearly in order to dissipate heat (the ratio declines curvilinearly). Macmillen and Hinds' estimations of the temperatures of equality could have been distorted by their use of single exponential equations to describe their data, and if that error was not equal among species, an analysis of the role of body size would be compromised.

Despite the potential imprecision in the calculation of the MWP:EWL index, the allometries in Figure 5A strongly suggest a negative relationship between body size and water stress. What this means ecologically is open to question. MacMillen and Hinds (1983) proposed that heteromyids were size matched to the aridity of their microhabitats in historical times, with the smallest species occupying the most xeric regions. The fact that such a correlation does not exist today was explained by recent and rapid climatic changes that resulted in faunal mixing soon after the rain-shadow deserts of the southwest were created. This line of reasoning leads to the implication that large kangaroo rats are able to survive alongside small pocket mice in the most xeric of deserts because they have either lower nonevaporative water losses or greater access to exogenous water. Although urine concentrations at low temperatures and humidities are indeed inversely proportional to size

(MacMillen and Hinds, 1983; Fig. 4), this just means that large species are forced to eliminate less excess water than small species. If coexistence occurs under conditions that require small species to produce maximally concentrated urine, then larger heteromyids might be expected to have more powerful kidneys than their smaller relatives. However, the limited data on maximum urine concentrations do not suggest an increase in concentrating power of the kidney with increases in body size. In fact, measurements of renal morphology have been interpreted to reflect just the opposite trend (Lawler and Geluso, 1986). MacMillen and Hinds (1983) suggest that the larger species might gain additional metabolic water by utilizing seeds that have a high carbohydrate content, whereas small heteromyids that are less water stressed could specialize in seeds that are high in lipids and proteins and therefore yield, per kcal, slightly less oxidative water. Kangaroo rats do select diets that have the effect of enhancing metabolic water production (Frank, 1988), but there are no data to support the speculation that such food preferences are size-dependent.

When temperatures rise to near 25°C and the humidity is low, large species might be expected to retreat to their humid burrows more frequently than would small species. However, such size related behavior would not be expected to occur at temperatures that fall within the zone of thermal neutrality. MacMillen and Hinds (1983) report that at 35°C, as at cooler temperatures, the decline in MWP with mass ($g^{-0.319}$) is more steep than that of EWL ($g^{-0.218}$). However, these regressions include data both from species that are in thermal neutrality at 35°C, as well as species (*Liomys*) that are more severely stressed by 35°C because it is above their upper critical temperature (Hinds and MacMillen, 1985). When the analysis is restricted to animals that are in thermal neutrality, the difference in the slopes of the allometries of MWP and EWL is much less and of questionable significance (Fig. 5B). If the difference between MWP and EWL at high temperatures is not a function of body size, then size-related differences in habitat, food preferences, or behavior might not be expected. All species would be stressed equally and forced to retire to humid microenvironments.

What are the greatest challenges to water balance faced by different heteromyids that might act through natural selection to maintain the animals' maximum capabilities for urine concentration? Even in the most extreme deserts, nocturnal and fossorial habits allow heteromyids to avoid the heat and dryness of summer days. Kenagy (1973*a*) found that the desert species *P. longimembris, D. merriami,* and *D. microps* spend the majority of their time resting underground at temperatures as close as possible to, but not much above, the lower end of thermal neutrality even though cooler burrow microhabitats were always available. This undoubtedly is a behavioral adaptation to minimize energy use, and it probably does not tax the animals' urine concentrating ability because humidity in plugged burrows is usually near saturation (Schmidt-Nielsen and Schmidt-Nielsen, 1950*b*). For example, *D. merriami* eating seeds in saturated air at 25°C must produce a relatively-dilute urine in order to eliminate excess water (Schmidt-Nielsen, 1964). Perhaps this preference for high temperatures is made possible by the possession of a powerful kidney, but it does not appear to have been the selective force for the maintenance of the upper limits of concentrating capabilities. Above ground activity, even though at night, appears to be occasionally more stressful. Kenagy (1973*a*) found that the above three species were active in air temperatures as high as 30°C when the relative humidity was only 15%. Such conditions are likely to result in a net loss of water and the production of maximally-concentrated urine (Fig. 4). Assuming similar MWP:EWL ratios (Fig. 5B), those individuals with the most powerful kidneys could theoretically remain exposed to desiccating conditions the longest before being forced to retire to a humid burrow for rehydration. However, there is no obvious reason why such an extension of activity time, and by inference urine concentrating ability, should be related to body size. Probably the most stressful time for a desert rodent is when it is displaced from its burrow and forced to locate or construct a new home. Juvenile dispersal often occurs during the hottest and driest parts of the summer, and this event might be critical not only from an ecological but also a physiological point of view.

Certainly the selection pressures for water conservation vary among habitats and this is reflected in the physiological attributes of the different members of the family. *Heteromys desmarestianus* from wet tropical forests lost weight rapidly and died within two to four days when kept on a dry diet without water (Fleming, 1977). This species is a weak burrower (Fleming and Brown, 1975), has a relatively high rate of evaporative water loss (Fig. 2), and a smaller renal papilla compared to other members of the family (Fleming, 1977). Several species of *Liomys* also have a tropical distribution, but they usually live in areas that have distinct dry seasons when water conservation should be important. *L. salvini* from dry tropical forests near Managua, Nicaragua were independent of drinking water when kept at 25°C, in variable conditions of humidity, and on a mixed birdseed diet (Hudson and Rummel, 1966). Their body weight declined by 17% during the first week of water restriction, but it then stabilized and presumably the animals could have survived well beyond the 24-day experimental period. However these conditions must be near the most rigorous tolerated because the mice continuously lost weight when the diet was changed to sunflower seeds that have a slightly higher protein content. It appears that both *L. irroratus* from arid grasslands and *L. pictus* from lowland deciduous forests also can be independent of exogenous water if temperatures and humidities are moderate (Christian et al., 1978), but *L. irroratus* from a more mesic riparian habitat have a relatively high rate of evaporative water loss and can survive without water only if allowed to burrow (Hudson and Rummel, 1966).

Similar variations exist within the genus *Dipodomys*. Both *D. agilis* (Carpenter, 1966) and *D. venustus* (Church, 1969) from relatively mesic environments need free water or succulent food to survive in "typical" laboratory conditions under which *D. merriami*, and presumably many other desert species, are water independent. Other differences can be related to the microhabitats that the animals exploit. *D. microps* is a desert species, but many populations are phytophagous year round. These populations have learned to shave off the salty epithelial layer of *Atriplex* leaves with their chisel-shaped lower incisors and then consume the more nutritious and succulent interior. They have the lowest mean maximum urine concentrations (2,827 mOsm/l) reported for the genus and cannot maintain their weight under moderate laboratory conditions on a dry seed diet (Kenagy, 1973*b*). In contrast, other populations of this species live in non-*Atriplex* zones of the desert where shrubs can be completely deciduous and seeds are the sole food source. At least at certain times of the year, these individuals can survive without water in the laboratory, although this capability may require some acclimation. Leaf shaving appears to be a learned behavior in both groups, but there is a distinct difference in kidney size, suggestive of modifications in physiological performance (Csuti, 1979). More subtle microhabitat differences also occur among the desert-dwelling species. When on a natural diet, *D. spectabilis* may produce an alkaline urine which suggests that a considerable portion of its diet consists of succulents that have a high content of salts of organic acids (Schmidt-Nielsen et al., 1948). Although this species can survive for many weeks on a dry diet, it eats succulent food in captivity more readily than other heteromyids with which it is sympatric. *D. spectabilis* also fails to show the hyperdevelopment of the ADH neurosecretory cells that are present in other xeric-adapted species including *D. merriami* (Hatton et al., 1972), again suggesting that it has a higher reliance on exogenous water. Undoubtedly, similar differences must exist among species within the genera *Perognathus* and *Chaetodipus,* but little information is available. The only thing that can be said is that none of those species is known to require succulent food or drinking water when kept under conditions of moderate temperature and humidity.

Summary

The family Heteromyidae has attracted a great deal of attention because many members have been able to successfully colonize harsh deserts by utilizing some of the most extreme physiological adaptations for water and energy conservation known in mammals. Most of these desert species do not need to drink or eat succulent food. They can rely exclusively on the water produced during oxidative metabolism because their rates of water loss are so low. Evaporation, especially from the skin, is well below that of average mammals, and some species are capable of producing among the driest feces and most concentrated milks and urines known. Similar adaptations are found in other small mammals, but the water independence that results from the combination of all these is matched by only a few rodents that have evolved in parallel in deserts elsewhere in the world. Heteromyids from deserts and other arid regions with ephemeral plant productivity also tend to be frugal with their energy resources. The rates of metabolism of desert species are about a third less than those of average mammals when at rest, and further metabolic reductions are possible by lowering body temperature in times of food scarcity. Most species forage year round and use brief episodes of shallow torpor during short-term energetic emergencies, although some, mainly in the genus *Perognathus,* are hibernators that employ comparatively long and deep torpors on a regular basis during a season of dormancy. Such facultative heterothermy is not unique to heteromyids. However, some species are able to modify the time they spend in torpor to match their anticipated energetic deficits; a process that requires a remarkable integration of complex factors related to their food stores, foraging success, and environmental conditions.

The magnitude of energy and water turnover is related to the harshness of the environment. Heteromyids from semiarid habitats have somewhat higher rates of metabolism and water loss than do desert species, and the most tropical members of the family deviate little or not at all from the patterns documented for average mammals. However, much work remains to be done. Many avenues of water loss have been measured only infrequently, some species have never been studied, and broad, comparative investigations are rare. In particular, the role of body size on water balance and urine concentrating ability needs to be clarified.

Literature Cited

Altschuler, E. M., R. B. Nagle, E. J. Braun, S. L. Lindstedt, and P. H. Krutzsch. 1979. Morphological study of the desert heteromyid kidney with emphasis on the genus *Perognathus.* Anatomical Record, 194:461–468.

Arnold, L. W. 1942. Notes on the life history of the sand pocket mouse. Journal of Mammalogy, 23:339–341.

Bartholomew, G. A. 1972. The water economy of seed-eating birds that survive without drinking. Proceedings of the International Ornithological Congress, 15:237–254.

Bartholomew, G. A., and H. H. Caswell. 1951. Locomotion in kangaroo rats and its adaptive significance. Journal of Mammalogy, 32:155–169.

Bartholomew, G. A., and R. E. MacMillen. 1961. Oxygen consumption, estivation, and hibernation in the kangaroo mouse, *Microdipodops pallidus.* Physiological Zoology, 34:179–198.

Baverstock, P., and B. Green. 1975. Water recycling in lactation. Science, 187:657–658.

Berger, R. J. 1975. Bioenergetic functions of sleep and activity rhythms and their possible relevance to aging. Federation Proceedings, 34:97–102.

Biewener, A., R. M. Alexander, and N. C. Heglund.

1981. Elastic storage in the hopping of kangaroo rats (*Dipodomys spectabilis*). Journal of Zoology, London, 195:369–383.

Blake, B. H. 1977. The effects of kidney structure and the annual cycle on water requirements in golden-mantled ground squirrels and chipmunks. Comparative Biochemistry and Physiology, 58A:413–419.

Blount, R. F., and I. H. Blount. 1968. Adaptive changes in size of renal papilla with altered function. Texas Reports on Biology and Medicine, 26:473–484.

Bradley, W. G., M. K. Yousef, and I. M. Scott. 1975. Physiological studies on the rock pocket mouse, *Perognathus intermedius*. Comparative Biochemistry and Physiology, 50A:331–337.

Breyen, L. J., W. G. Bradley, and M. K. Yousef. 1973. Physiological and ecological studies on the chisel-toothed kangaroo rat, *Dipodomys microps*. Comparative Biochemistry and Physiology, 44A: 543–555.

Brody, S. 1945. Bioenergetics and growth. Hafner, New York, 1,023 pp.

Brower, J. E. 1970*a*. Metabolic and thermal adaptations of heteromyid rodents to the desert. Ph.D. dissert., Syracuse University, 218 pp.

———. 1970*b*. The relationship of food and temperature to torpidity in pocket mice. Pennsylvania Academy of Science, 44:160–171.

Brower, J. E., and T. J. Cade. 1971. Bicircadian torpor in pocket mice. BioScience, 21:181–182.

Brown, J. H., and G. A. Bartholomew. 1969. Periodicity and energetics of torpor in the kangaroo mouse, *Microdipodops pallidus*. Ecology, 50:705–709.

Brownfield, M. S., and B. A. Wunder. 1976. Relative medullary area: a new structural index for estimating urinary concentrating capacity of mammals. Comparative Biochemistry and Physiology, 55A:69–75.

Buffenstein, R., W. E. Campbell, and J. U. M. Jarvis. 1985. Identification of crystalline allantoin in the urine of African Cricetidae (Rodentia) and its role in their water economy. Journal of Comparative Physiology B, 155:493–499.

Buffenstein, R., and J. U. M. Jarvis. 1985. The effect of water stress on growth and renal performance of juvenile Namib rodents. Journal of Arid Environments, 9:232–236.

Calder, W. A., III. 1984. Size, function, and life history. Harvard University Press, Cambridge, 431 pp.

Carpenter, R. E. 1966. A comparison of thermoregulation and water metabolism in the kangaroo rats *Dipodomys agilis* and *Dipodomys merriami*. University of California Publications in Zoology, 78: 1–36.

Chapman, B. R., and R. L. Packard. 1974. An ecological study of Merriam's pocket mouse in southeastern Texas. Southwestern Naturalist, 19:281–291.

Chew, R. M. 1955. The skin and respiratory water losses of *Peromyscus maniculatus sonoriensis*. Ecology, 36:463–467.

———. 1965. Water metabolism of mammals. Pp. 43–178, *in* Physiological mammalogy, Vol. 2 (W. V. Mayer and R. G. Van Gelder, eds.). Academic Press, New York, 326 pp.

Chew, R. M., and B. B. Butterworth. 1964. Ecology of rodents in Indian Cove (Mojave Desert), Joshua Tree National Monument, California. Journal of Mammalogy, 45:203–225.

Chew, R. M., and A. E. Dammann. 1961. Evaporative water loss of small vertebrates, as measured with an infrared analyzer. Science, 133:384–385.

Chew, R. M., R. G. Lindberg, and P. Hayden. 1963. Metabolic characteristics of pocket mice (*Perognathus*) especially the little pocket mouse (*P. longimembris*). Pp. 5–59, *in* Investigation of *Perognathus* as an experimental organism for research in space biology. Northrup Space Laboratories, Hawthorne, California.

———. 1967. Temperature regulation in the little pocket mouse, *Perognathus longimembris*. Comparative Biochemistry and Physiology, 21:487–505.

Christian, D. P., J. O. Matson, and S. G. Rosenberg. 1978. Comparative water balance in two species of *Liomys* (Rodentia: Heteromyidae). Comparative Biochemistry and Physiology, 61A:589–594.

Christopher, E. A. 1975. The ecology of non-captive desert rodents. Ph.D. dissert, University of California, Irvine.

Church, R. L. 1969. Evaporative water loss and gross effects of water privation in the kangaroo rat, *Dipodomys venustus*. Journal of Mammalogy, 50:514–523.

Collins, J. C., T. C. Pilkington, and K. Schmidt-Nielsen. 1971. A model of respiratory heat transfer in a small mammal. Biophysical Journal, 11:886–914.

Criddle, S. 1915. The banded pocket mouse, *Perognathus fasciatus* Wied. The Ottawa Naturalist, 28:130–134.

Csuti, B. A. 1979. Patterns of adaptation and variation in the Great Basin kangaroo rat (*Dipodomys microps*). University of California Publications in Zoology, 111:1–69.

Dawson, T. J. 1976. Energetic cost of locomotion in Australian hopping mice. Nature, 259:305–307.

Dawson, T. J., and C. R. Taylor. 1973. Energetic cost of locomotion in kangaroos. Nature, 246:313–314.

Dawson, W. R. 1955. The relation of oxygen consumption to temperature in desert rodents. Journal of Mammalogy, 36:543–553.

Edwards, N. A. 1975. Scaling of renal functions in mammals. Comparative Biochemistry and Physiology, 52A:63–66.

Fleming, T. H. 1977. Response of two species of tropical heteromyid rodents to reduced food and water availability. Journal of Mammalogy, 58:102–106.

Fleming, T. H., and G. J. Brown. 1975. An experimental analysis of seed hoarding and burrowing behavior in two species of Costa Rican heteromyid rodents. Journal of Mammalogy, 56:301–315.

Forman, G. L., and C. J. Phillips. 1993. The proximal colon of heteromyid rodents: possible morphophysiological correlates to enhanced water conservation. Pp. 491–508, *in* Biology of the Heteromyidae (H. H. Genoways and J. H. Brown, eds.), Special Publication, American Society of Mammalogy, 10: 1–719.

Frank, C. L. 1988. Diet selection by a heteromyid rodent: role of net metabolic water production. Ecology, 69:1943–1951.

French, A. R. 1976. Selection of high temperatures for hibernation by the pocket mouse, *Perognathus longimembris*: ecological advantages and energetic consequences. Ecology, 57:185–191.

———. 1977*a*. Periodicity of recurrent hypothermia during hibernation in the pocket mouse, *Perognathus longimembris*. Journal of Comparative Physiology, 115:87–100.

———. 1977*b*. Circannual rhythmicity and entrainment of surface activity in the hibernator, *Perognathus longimembris*. Journal of Mammalogy, 58: 37–43.

———. 1989*a*. Seasonal variation in use of torpor by pallid kangaroo mice, *Microdipodops pallidus*. Journal of Mammalogy, 70:839–842.

———. 1989*b*. The impact of variations in energy availability on the time spent torpid during the hibernation season. Pp. 129–136, *in* Living in the cold II (A. Malan and B. Canguilhem, eds.). John Libby, London, 525 pp.

French, N. R., B. G. Maza, and A. P. Aschwanden. 1966. Periodicity of desert rodent activity. Science, 154:1194–1195.

Geluso, K. N. 1978. Urine concentrating ability and renal structure of insectivorous bats. Journal of Mammalogy, 59:312–323.

Gettinger, R. D., P. Arnold, B. A. Wunder, and C. L. Ralph. 1986. Seasonal effects of temperature and photoperiod on thermogenesis and body mass of the kangaroo rat (*Dipodomys ordii*). Pp. 505–509, *in* Living in the cold (H. C. Heller, X. J. Musacchia, and L. C. H. Wang, eds.). Elsevier, New York, 587 pp.

Gettinger, R. D., and C. L. Ralph. 1985. Thermoregulatory responses to photoperiod by kangaroo rats (*Dipodomys ordii*): influence of night lighting on nonshivering thermogenesis and resting metabolism. Journal of Experimental Zoology, 234:335–340.

Glotzbach, S. F., and H. C. Heller. 1976. Central nervous regulation of body temperature during sleep. Science, 194:537–539.

Greegor, D. H., Jr. 1975. Renal capabilities of an Argentine desert armadillo. Journal of Mammalogy, 56:626–632.

Grinnell, J., and H. S. Swarth. 1913. An account of the birds and mammals of the San Jacinto area of Southern California. University of California Publications in Zoology, 10:197–406.

Grubbs, D. E. 1974. Desert rodents and desert agriculture: the ecological and physiological relationships of a desert rodent community to native and modified environments. Ph.D. dissert., University of California, Irvine.

Harris, D. V., J. M. Walker, and R. J. Berger. 1984. A continuum of slow-wave sleep and shallow torpor in the pocket mouse *Perognathus longimembris*. Physiological Zoology, 57:428–434.

Hart, J. S. 1971. Rodents. Pp. 1–49, *in* Comparative physiology of thermoregulation, Vol. 2 (G. C. Whittow, ed.). Academic Press, New York, 410 pp.

Hatton, G. I., J. I. Johnson, and C. Z. Malatesta. 1972. Supraoptic nuclei of rodents adapted for mesic and xeric environments: numbers of cells, multiple nucleoli, and their distributions. Journal of Comparative Neurology, 145:43–60.

Hayden, P., and R. G. Lindberg. 1970. Hypoxia-induced torpor in pocket mice (genus: *Perognathus*). Comparative Biochemistry and Physiology, 33:167–179.

Heller, H. C., J. M. Walker, G. L. Florant, S. F. Glotzbach, and R. J. Berger. 1978. Sleep and hibernation: electrophysiological and thermoregulatory homologies. Pp. 225–265, *in* Strategies in cold: natural torpidity and thermogenesis (L. C. H. Wang, and J. W. Hudson, eds.). Academic Press, New York, 715 pp.

Herreid, C. F., III, and B. Kessell. 1967. Thermal conductance in birds and mammals. Comparative Biochemistry and Physiology, 21:405–414.

Hewitt, S. 1981. Plasticity of renal function in the Australian desert rodent *Notomys alexis*. Comparative Biochemistry and Physiology, 69A:297–304.

Hinds, D. S., and R. E. MacMillen. 1985. Scaling of energy metabolism and evaporative water loss in heteromyid rodents. Physiological Zoology, 58:282–298.

Hoover, K. D., W. G. Whitford, and P. Flavill. 1977. Factors influencing the distributions of two species of *Perognathus*. Ecology, 58:877–884.

Howell, A. B., and I. Gersh. 1935. Conservation of water by the rodent *Dipodomys*. Journal of Mammalogy, 16:1–9.

Hudson, J. W., and J. A. Rummel. 1966. Water metabolism and temperature regulation of the primitive heteromyids, *Liomys salvani* (sic) and *Liomys irroratus*. Ecology, 47:345–354.

Jackson, D. C., and K. Schmidt-Nielsen. 1964. Countercurrent heat exchange in the respiratory passages. Proceedings of the National Academy of Sciences of the United States of America, 51:1192–1197.

Jamison, R. L., and W. Kriz. 1982. Urinary concentrating mechanism: structure and function. Oxford University Press, New York, 340 pp.

Johnson, D. R., and K. L. Groepper. 1970. Bioenergetics of north plains rodents. American Midland Naturalist, 84:537–548.

Jones, J. K. 1964. Distribution and taxonomy of mammals of Nebraska. University of Kansas Publications, Museum of Natural History, 16:1–356.

Kenagy, G. J. 1973*a*. Daily and seasonal patterns of activity and energetics in a heteromyid rodent community. Ecology, 54:1201–1219.

———. 1973*b*. Adaptations for leaf eating in the Great Basin kangaroo rat, *Dipodomys microps*. Oecologia (Berlin), 12:383–412.

Kenagy, G. J., and G. A. Bartholomew. 1985. Seasonal reproductive patterns in five coexisting California desert rodent species. Ecological Monographs, 55:371–397.

Kenagy, G. J., and D. F. Hoyt. 1980. Reingestion of feces in rodents and its daily rhythmicity. Oecologia (Berlin), 44:403–409.

Kooyman, G. L. 1963. Milk analysis of the kangaroo rat, *Dipodomys merriami*. Science, 147:1467–1468.

Lawler, R. M., and K. N. Geluso. 1986. Renal structure and body size in heteromyid rodents. Journal of Mammalogy, 67:367–372.

Lechner, A. J. 1978. The scaling of maximal oxygen consumption and pulmonary dimensions in small mammals. Respiratory Physiology, 34:29–44.

Lindstedt, S. L. 1980. Energetics and water economy of the smallest desert mammal. Physiological Zoology, 53:82–97.

MacMillen, R. E. 1972. Water economy of nocturnal desert rodents. Symposia of the Zoological Society of London, 31:147–174.

———. 1983. Adaptive physiology of heteromyid rodents. Great Basin Naturalist Memoirs, 7:65–76.

MacMillen, R. E., and E. A. Christopher. 1975. The water relations of two populations of noncaptive desert rodents. Pp. 117–137, *in* Environmental physiology of desert organisms (N. F. Hadley, ed.). Dowden, Hutchinson, and Ross, Stroudsburg, Pennsylvania, 283 pp.

MacMillen, R. E., and D. S. Hinds. 1983. Water regulatory efficiency in heteromyid rodents: a model and its application. Ecology, 64:152–164.

MacMillen, R. E., and A. K. Lee. 1967. Australian desert mice: independence of exogenous water. Science, 158:383–385.

———. 1969. Water metabolism of Australian hopping mice. Comparative Biochemistry and Physiology, 28:493–514.

———. 1970. Energy metabolism and pulmocutaneous water loss of Australian hopping mice. Comparative Biochemistry and Physiology, 35:355–369.

McNab, B. K. 1966. The metabolism of fossorial rodents: a study of convergence. Ecology, 47:712–733.

———. 1979. Climatic adaptation in the energetics of heteromyid rodents. Comparative Biochemistry and Physiology, 62A:813–820.

Meehan, T. E. 1977. The occurrence, energetic significance and initiation of spontaneous torpor in the Great Basin pocket mouse, *Perognathus parvus*. Ph.D. dissert., University of California, Irvine, 111 pp.

Morrison, P., and F. A. Ryser. 1962. Hypothermic behavior in the hispid pocket mouse. Journal of Mammalogy, 43:529–533.

Mullen, R. K., and R. M. Chew. 1973. Estimating the energy metabolism of free-living *Perognathus formosus*: a comparison of direct and indirect methods. Ecology, 54:633–637.

Munkacsi, I., and M. Palkovits. 1977. Measurements on the kidneys and vasa recta of various mammals in relation to urine concentrating capacity. Acta Anatomica, 98:456–468.

Nishimoto, J. T. 1980. Characteristics and energetics of torpor in the San Diego pocket mouse, *Perognathus fallax*. M.A. thesis, University of California, Riverside, 58 pp.

O'Farrell, M. J. 1974. Seasonal activity patterns of rodents in a sagebrush community. Journal of Mammalogy, 55:809–823.

O'Farrell, T. P., R. J. Olson, R. O. Gilbert, and J. D. Hedlund. 1975. A population of Great Basin pocket mice, *Perognathus parvus*, in the shrub-steppe of south-central Washington. Ecological Monographs, 45:1–28.

Raab, J. L., and K. Schmidt-Nielsen. 1972. Effect of running on water balance of the kangaroo rat. American Journal of Physiology, 222:1230–1235.

Reichman, O. J. 1981. Factors influencing foraging patterns in desert rodents. Pp. 195–213, *in* Foraging behavior: ecological, ethological, and psychological approaches (A. C. Kamil, and T. D. Sargent, eds.). Garland-STPM Press, New York, 534 pp.

Reichman, O. J., and K. M. Van De Graaff. 1973. Seasonal activity and reproductive patterns of five species of Sonoran desert rodents. American Midland Naturalist, 90:118–126.

Reynolds, H. G., and H. S. Haskell. 1949. Life history notes on Price and Bailey pocket mice of southern Arizona. Journal of Mammalogy, 30:150–156.

Scelza, J., and J. Knoll. 1975. The rate of oxygen consumption, carbon dioxide production, and evaporative water loss in *Dipodomys panamintinus*. Comparative Biochemistry and Physiology, 52A: 339–341.

Scheffer, T. H. 1938. Pocket mice of Washington and Oregon in relation to agriculture. United States Department of Agriculture Technical Bulletin, 608: 1–15.

Schmidt-Nielsen, B., and R. O'Dell. 1961. Structure and concentrating mechanism in the mammalian kidney. American Journal of Physiology, 200: 1119–1124.

Schmidt-Nielsen, B., and K. Schmidt-Nielsen. 1950*a*. Pulmonary water loss in desert rodents. American Journal of Physiology, 162:31–36.

———. 1950*b*. Evaporative water loss in desert rodents in their natural habitat. Ecology, 31:75–85.

———. 1951. A complete account of the water metabolism in kangaroo rats and an experimental verification. Journal of Cellular and Comparative Physiology, 38:165–182.

Schmidt-Nielsen, K. 1964. Desert animals: physiological problems of heat and water. Clarendon Press, London, 277 pp.

Schmidt-Nielsen, K., and B. Schmidt-Nielsen. 1952. Water metabolism of desert mammals. Physiological Reviews, 32:135–166.

Schmidt-Nielsen, K., B. Schmidt-Nielsen, and A. Brokaw. 1948. Urea excretion in desert rodents exposed to high protein diets. Journal of Cellular and Comparative Physiology, 32:361–380.

Schreiber, R. K. 1979. Coefficients of digestibility and caloric diet of rodents in the northern Great Basin desert. Journal of Mammalogy, 60:416–420.

Soholt, L. F. 1973. Consumption of primary production by a population of kangaroo rats (*Dipodomys merriami*) in the Mojave desert. Ecological Monographs, 43:357–376.

Sperber, I. 1944. Studies on the mammalian kidney. Zoologiska Bidrag Fran Uppsala, 22:249–430.

Stupfel, M., and J. P. Geloso. 1959. Étude des pertes d'eau par les voies cutanée et respiratoire du rat blanc non anesthésié, placé dans une atmosphère sèche normal et dans une atmosphère sèche renfermant 10% de gaz carbonique. Comptes Rendus des Seances de la Societe de Biologie et de ses Filiales, 248:740–742.

Tennent, D. M. 1945. A study of the water losses through the skin in the rat. American Journal of Physiology, 145:436–440.

Thompson, S. D., R. E. MacMillen, E. M. Burke, and C. R. Taylor. 1980. The energetic cost of bipedal hopping in small mammals. Nature, 287: 223–224.

Tucker, V. A. 1965. Oxygen consumption, thermal conductance, and torpor in the California pocket mouse *Perognathus californicus.* Journal of Cellular and Comparative Physiology, 65:393–403.

———. 1966. Diurnal torpor and its relation to food consumption and weight changes in the California pocket mouse *Perognathus californicus.* Ecology, 47: 245–252.

Vogel, V., and W. Vogel. 1972. Uber das Konzentrationsvermogen der Nieren zweier Fledermausarten (*Rhinopoma hardwickei* und *Rhinolophus ferrum equinum*) mit unterschiedlich langer Nierenpapille. Zeitschrift fur vergleichende Physiologie, 76:358–371.

Walker, J. M., and R. J. Berger. 1980. Sleep as an adaptation for energy conservation functionally related to hibernation and shallow torpor. Progress in Brain Research, 53:255–278.

Wang. L. C. H., and J. W. Hudson. 1970. Some physiological aspects of temperature regulation in the normothermic and torpid hispid pocket mouse, *Perognathus hispidus.* Comparative Biochemistry and Physiology, 32:275–293.

Withers, P. C. 1982. Effect of diet and assimilation efficiency on water balance for two desert rodents. Journal of Arid Environments, 5:375–384.

Wolff, J. O., and G. C. Bateman. 1978. Effects of food availability and ambient temperature on torpor cycles of *Perognathus flavus* (Heteromyidae). Journal of Mammalogy, 59:707–716.

Wunder, B. A. 1974. The effect of activity on body temperature of Ord's kangaroo rat (*Dipodomys ordii*). Physiological Zoology, 47:29–36.

Yousef, M. K., and D. B. Dill. 1971. Daily cycles of hibernation in the kangaroo rat, *Dipodomys merriami.* Comparative Biochemistry and Physiology, 33:821–836.

ECOLOGICAL ASPECTS OF HETEROMYID FORAGING

O. J. REICHMAN AND M. V. PRICE

Introduction

In this chapter we will view heteromyid foraging behavior from an ecological perspective. We begin with a description of the diets of heteromyids and an overview of their methods of securing food. Subsequently, we consider in sequence the decisions individuals must make while foraging: when and where to forage (activity patterns, microhabitat and patch choice), what seeds to harvest from those available (diet choice), and what to do with the harvested food (immediate consumption, caching, and cache management). For each decision level we review what is known about the range of behaviors within the Heteromyidae and consider what factors might be responsible for the observed patterns of variation. We conclude the chapter with a summary, suggestions for future research, and a brief discussion of how of contemporary foraging theory provides a valuable tool for further study of heteromyid foraging ecology.

Much more is known about foraging of desert-dwelling genera (*Dipodomys, Microdipodops, Perognathus,* and *Chaetodipus*) than of tropical and subtropical forms (*Liomys, Heteromys*). Even within desert genera, only a few populations of a few species have been investigated thoroughly. Hence, it must be kept in mind that our attempts to draw conclusions about heteromyids in general are based on a limited sample of taxa and locations.

Several reviews of heteromyid behavior have been published. In the interest of brevity, we may treat a subject only in passing and refer the reader to these reviews. Eisenberg's (1963) survey of heteromyid behavior and his chapter in this volume deal with behaviors we will not cover. Several chapters in the current volume (e.g., those by Brylski, Kenagy, French, Brown and Harney) relate to our topic, and volume seven of the Great Basin Naturalist Memoirs (1983) is devoted to desert rodents.

Overview of Heteromyid Foraging Ecology

Diets

Heteromyids show little variation in dentition compared to other desert-dwelling rodent families; all have a total of 20 teeth and simple lophodont cheek teeth with the enamel restricted to plates on the anterior and posterior surfaces. The lower jaw is not robust, jaw musculature is relatively weak, and the inflation of the auditory bullae restricts jaw gape, especially in kangaroo rats and kangaroo mice (Nikolai and Bramble, 1983).

This uniformity of structure would suggest uniformity in basic diet, and the tooth and jaw structure suggest a granivorous diet. Indeed, with only one exception, desert heteromyids are primarily granivorous, although their diets can reflect seasonal availability of insects and green vegetation (Alcoze and Zimmerman, 1971; Best and Hoditschek, 1982; Bradley and Mauer, 1971; Brand, 1975; Chapman, 1972; Chew and Chew, 1970; Csuti, 1979; Dunham, 1968; Flake, 1973; Harris, 1986; Hawbecker, 1940; Holdenreid, 1957; Kritzman, 1974; LaTourette, 1971; Lemen, 1978; M'Closkey, 1980; Meserve, 1976; Monson, 1943; Monson and Kessler, 1940; O'Connell, 1979; Pulliam and Brand, 1975; Reichman, 1975, 1978; Reynolds, 1958, 1960; Reynolds and Haskell, 1949; Schreiber, 1978; Shaw, 1934; Smigel and Rosenzweig, 1974; Smith, 1942; Soholt, 1973, 1977; Stamp and Ohmart, 1978; Tappe, 1941; Vorhies and Taylor, 1922). In keeping with their small gape, the seeds used by desert heteromyids tend to be small (less than 3 mm in length and less than 25 mg) and primarily from grasses or forbs (Brown et al., 1979; Reichman, 1975, 1976, 1978; Reynolds, 1958).

The diets of *Heteromys* and *Liomys* are poorly known. Fleming (1970, 1974) noted that these tropical and subtropical genera are also primarily granivorous, harvesting a variety of fruits, nuts, and seeds from several tree and shrub species and storing the propagules in underground burrows. Janzen (1982, 1986) found that *Liomys salvini* would harvest guanacaste (*Enterolobium cyclocarpum*) seeds from fallen fruit, forest litter, and piles of horse dung. The seeds available to *Liomys* and *Heteromys* tend to be much larger than those available in North American deserts, and more appear to contain toxins, which are relatively uncommon in desert plant seeds (Janzen, pers. comm.).

Insects are seasonally abundant in the diets of heteromyids and are especially valuable because they are high in protein and contain enough preformed water to make up any deficit incurred from digesting and excreting the metabolic wastes of a high protein item (Schmidt-Nielsen, 1964). Insects can make up as much as 50% of a kangaroo rat's diet over the short term (Reichman, 1975, 1978) and 78% of the summer diet of *Microdipodops megacephalus* (Harris, 1986). Flake (1973) found that "animal matter" made up 10% of the annual diet of *D. ordii* although the diet was composed of as much as 35% adult coleoptera and 27% lepidoptera larvae in some seasons.

Although fungi have not been found in the diet of heteromyids, it is reasonable to presume that fungi would be ingested if available, especially in tropical areas where they are common. Vorhies and Taylor (1922) reported fungi stored in the caches of *D. spectabilis* and Reichman et al. (1985) found the spores of over two dozen species of molds associated with the cheek pouches and caches of this kangaroo rat species. These molds were common in caches and any that might be ingested would probably be incidental to consuming seeds.

Heteromyid diets have been investigated by analysis of stomach and fecal contents under a microscope, and by inspection of cheek pouch contents. The first two techniques express diets in terms of relative frequencies of masticated fragments of dietary items (Brand, 1975; Hansson, 1970), while the third expresses diets as relative weights or numbers of whole items. The techniques used can influence results because estimates of relative abundance based on number or mass may not equal those based on surface area if the various dietary items are different sizes. This point should be considered when comparing, for example, stomach and cheek pouch contents.

We will discuss factors which impinge on diet choice in a later section.

TABLE 1.—*Movement patterns while foraging for four desert heteromyid species. Values are based on direct observation of free-foraging animals. Data compiled from Thompson (1985) and Bowers (1982).*

Data source	Rodent species	Body mass (g)	Stop duration (s)	Between stop: Velocity (m/s)	Between stop: Distance (m)
Thompson (1985)	*Dipodomys deserti*	120	59	1.74	7
	Dipodomys merriami	37	18	0.91	5
	Perognathus longimembris	7	10	0.49	3
Bowers (1982)	*Dipodomys merriami*	37	—	1.01	9
	Perognathus formosus	20	—	0.27	4
	Perognathus longimembris	7	—	0.16	3

Foraging Behavior and Methods

Limb structure and locomotion.—Heteromyids share a number of morphological traits that affect locomotion and foraging methods. All heteromyids have hind feet that are somewhat elongated, in comparison to other North American desert-dwelling rodent families, which provide good leaping abilities; and all heteromyids use bounding bipedal or quadrupedal (Bartholomew and Cary, 1954; Bartholomew and Caswell, 1951) gaits that are very different from the running gait used by many cricetid rodents. Plantar pads are reduced or absent on the hind feet, which restricts climbing abilities, especially in the bipedal forms, and the tail has only rudimentary prehensile capability in a few quadrupedal species. Although no systematic functional comparison has been made of heteromyid forelimbs, they appear uniform in structure. All have some modification for scratch-digging, such as a relatively short humerus, pronounced deltoid process, and large scapular area for attachment of the teres major muscle (Heinz, 1983*a*; Howell, 1932). All have prominent medial and lateral epicondyles, suggesting that forelimb pronation and supination are important motions, and the manus of desert species has well-developed claws, a large palmar pad, and a semiopposable reduced first digit (Nikolai and Bramble, 1983).

Hence, it is no surprise that heteromyids share a basic mode of foraging. All emerge at night from a burrow or nest and spend most of their time moving about the surface, alternating periods of locomotion with stops, during which the animals may extract seeds from the soil with the forefeet while balancing on their hind feet.

While heteromyids spend a majority of their time foraging on the ground, scattered reports exist of heteromyids climbing in vegetation while foraging. Kenagy (1972) observed *Dipodomys microps,* a specialized leaf-eater, regularly climbing in *Atriplex* bushes. Lemen and Freeman (1985, 1986) noted that *Perognathus flavescens, Chaetodipus hispidus,* and even *Dipodomys ordii* occasionally climb to collect seed heads of low grasses and forbs, and Daly (pers. comm.) has occasionally observed similar behavior in *Dipodomys merriami* and *Chaetodipus penicillatus.* Reichman (pers. obs.) has seen *D. spectabilis* awkwardly climbing into, and falling out of, the lower branches of *Ephedra* bushes while harvesting buds, and insect galls associated with *Larrea* stems have been found in the cheek pouches of *D. merriami* (Reichman, 1978). The types of galls harvested do not fall off the bushes, so the kangaroo rats probably climbed to obtain them.

Although few systematic field studies of arboreal foraging have been attempted, all indications are that this type of foraging is rare, at least in desert dwelling species. Lemen and Freeman (1986) observed only 17 vertical excursions when they followed hun-

TABLE 2.—*Arboreal activity of heteromyid rodents. Values are the number of captures per 100 trap-nights in live traps placed on platforms at various heights above the ground; N = total number of captures. Values for Studies 1 and 2 are taken from Rosenzweig et al. (1975); those for Study 3 are from Price (unpubl. data). See text for details.*

	Study 1				Study 2			Study 3			
	Height (cm)				Height (cm)			Height (cm)			
Species	0	25	60	N	0	45	N	0	60	150	N
Chaetodipus penicillatus	4.7	5.3	0.0	19	7.2	4.8	31	10.0	1.2	0	28
Chaetodipus baileyi	—	—	—	—	—	—	—	4.8	0	0	12
Chaetodipus hispidus	—	—	—	—	2.0	0.5	5	—	—	—	—
Perognathus amplus	—	—	—	—	—	—	—	10.4	0	0	26
Dipodomys merriami	18.4	5.3	1.1	47	17.0	0	26	13.6	0	0	34
Dipodomys spectabilis	—	—	—	—	3.9	0	6	—	—	—	—
Cricetid species	4.2	16.8	6.8	53	28.6	13.3	84	—	—	—	—

dreds of meters of horizontal trails left by mice dusted with fluorescent powder. Rosenzweig et al. (1975) performed two experiments to investigate arboreal activity during the summer rainy season in Chihuahuan desert scrub, one ("study 1" in Table 1) where traps at three heights were open simultaneously, and one ("study 2," Table 1) in which the heights were sampled sequentially. They found (Table 1) that heteromyid species were rarely captured in traps set above ground. Quadrupedal species were captured more frequently above ground than bipeds (but not as often as cricetid rodent species). Price (unpubl.) performed a similar experiment in southcentral Arizona during early spring when trees and shrubs were not fruiting. She established 50 trapping stations at 15 m intervals and at each station placed one trap in an open area, one under a shrub canopy, one on a platform 60 cm up in a small shrub, and one 1.5 m in a *Cercidium* or *Prosopis* tree. Trap stations were separated by at least 5 m horizontal distance, and all were opened and baited simultaneously. Again, few captures were recorded from elevated traps, and quadrupedal species climbed somewhat more than bipedal species (Table 1, "study 3").

Major differences among heteromyids in locomotion patterns involve gait and characteristics of horizontal movement rather than climbing frequency. In the laboratory, tropical heteromyids (*Liomys* and *Heteromys*) climb more readily than do bipedal forms and as readily as desert-dwelling quadrupeds (Eisenberg, 1963), but this does not necessarily mean that their foraging is primarily arboreal (Fleming and Brown, 1975). The apparent rarity of climbing has lead investigators to concentrate their analyses on heteromyids foraging for seeds within reach of the ground.

Species differ in the proportion of time spent in rapid travel, distances moved between stops, and velocities used during rapid running (Bowers, 1982; Thompson, 1982*a*, 1985; Table 2). Gaits vary among genera; bipedal *Dipodomys* and *Microdipodops* put weight on all four limbs only during very slow movement and hop exclusively on the hind feet when running rapidly (Bartholomew and Caswell, 1951). The other four genera are quadrupedal, although they may use erratic bipedal leaps to escape attack (Bartholomew and Cary, 1954; Pinkham, 1973).

Methods of seed harvest.—Heteromyids use two basic methods to harvest seeds. Some species, primarily kangaroo rats, have been reported collecting most of their seeds directly from plants, either by clipping fruiting stalks and extracting seeds from the felled seed heads (Hawbecker, 1944; Vorhies and Taylor, 1922) or by plucking seeds directly from fruits that are within 15–20 cm of the ground (Reynolds, 1958; Daly, pers. comm.). Other species appear to collect seeds from the soil surface or sieve them from the soil. The relative contribution of these two collection modes is not known. Species such as *D. spectabilis* or *D. ingens,* which make large belowground caches, could conceivably harvest a season's supply of food directly from plants during the short fruiting season, but other species must rely to some extent on seeds gleaned from the soil. The efficiency with which heteromyids can collect buried seeds (see below) argues for the general importance of this foraging method.

Sensory modes.—Heteromyids appear to locate seeds by olfaction, but very little research has been done on the constraints this places on foraging. Reichman (1981) presents a simple model based on the assumption that the odor concentration in soil decreases as the square of the distance from the odor source, and that total odor concentration at the soil surface is the sum of all concentrations from nearby seeds. Heteromyids could use the pattern of odors at the soil surface to assess characteristics of buried seeds. One critical prediction of the model is that in some circumstances (when seeds are relatively deep compared to their horizontal separation beneath the soil), odors should be more concentrated between seed clumps than directly over either clump, thus misleading a forager using olfaction about the exact location of seed clumps. Indeed, in a laboratory study Reichman and Oberstein (1977; see also Reichman, 1981) noted that under such confounding circumstances, *D. merriami* individuals often dug halfway between buried seed piles. *Perognathus amplus* exhibited a different pattern of digging which suggested that they responded to changes in the odor gradient rather than simply to the strongest odors present in the experimental arena (Reichman and Oberstein, 1977). How the use of olfaction by seed consumers affects the evolution of seed characters has not been explored (Price and Jenkins, 1986).

Seed extraction methods.—Although olfaction is used to locate seeds, touch and taste are probably involved in recognizing seeds and separating them from all the mineral and other organic particles encountered while foraging (perhaps because with dorsally-located eyes heteromyids cannot see under their head to direct forelimb movement; Lawhon and Hafner, 1981; Nikolai and Bramble, 1983). High-speed films, taken from below, of heteromyids gathering loose millet seeds indicate a stereotyped sequence of movements (Fig. 1; Nikolai and Bramble, 1983; Price, unpubl.). The rodents balance on their hind feet and position the nose over a seed before reaching forward with both forefeet, opening the hands, and rotating them so the palms face downward. Seeds are clutched at the end of each limb extension; the limbs are then retracted and rotated so that the palms face up and backwards towards the cheek pouch openings. Digits of each hand are inserted briefly into

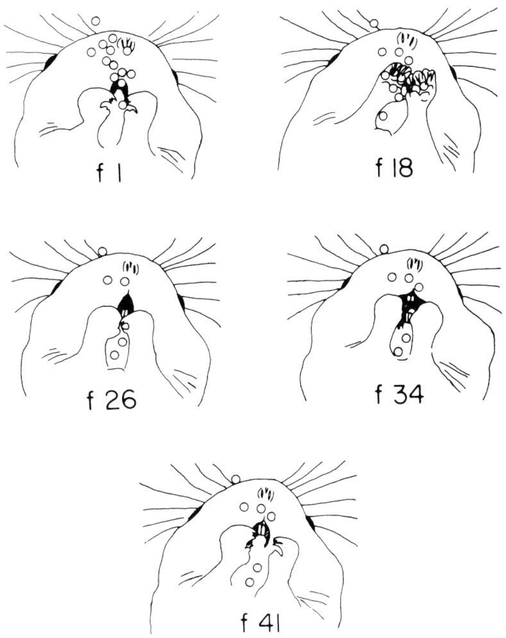

FIG. 1.—Seed pouching behavior by *Dipodomys deserti.* The drawings are tracings of representative frames of slow motion film (200 fps) and illustrate one complete pouching cycle (see text for details; from Nikolai and Bramble, 1983).

the ipsilateral pouch and removed after the seeds are released. As seeds are inserted into the pouch the mouth is drawn slightly open, putting tension on the pouch opening and preventing seeds from falling out. Occasionally, seeds are pinched between the lips or lips and incisors before being inserted into the cheek pouches or being eaten, perhaps to ascertain the suitability of the seeds for harvesting (Nikolai and Bramble, 1983; Price, unpubl.). Pouching movements under these conditions are rapid and vary with body size from a low of 10 forelimb cycles per second for 120 gm *D. deserti* to 16 cycles per second for 10 gm *P. amplus* and *P. longimembris* (Price and Heinz, 1984; see also Nikolai and Bramble, 1983). Rates are probably lower when seeds are being harvested from the soil.

When extracting seeds from the soil, heteromyids may also use nose position to orient forelimb motions, but little in known about this behavior. While digging, soil is loosened with front-to-back movements of the forefeet. As soil accumulates it is periodically kicked from under the belly with rearward movements of the hind limbs.

Exactly how heteromyids extract seeds from the soil matrix remains unknown. That the rodents are efficient is indisputable; although young heteromyids often place inappropriate food items in their cheek pouches (e.g., sticks, stones, feces), adults rarely do so (Reichman, 1975). Lawhon and Hafner (1981) and Price and Heinz (1984) observed very few non-food items in the pouches when heteromyids were allowed to extract seeds from a matrix of particles similar to the seeds themselves. Not only are heteromyids accurate, they are fast. Price and Heinz (1984) measured maximum harvest rates of 11 seeds/sec for *D. deserti* foraging for millet at a density of four seeds/cm^2 in patches of fine soil particles. Harvest rates varied with seed density, heteromyid body size, and soil characteristics. We will return to this in the context of microhabitat and patch choice (see below).

Use of cheek pouches.—One other morphological feature of heteromyids has a profound effect on what they do while gathering food items—the external fur-lined cheek pouches that are characteristic of all members of the family. The cheek pouches allow the act of collecting seeds to be separated from ingesting because they facilitate the transfer of a large number of seeds back to the burrow. Thus, the heteromyids can incur the costs associated with processing and eating seeds in their burrows rather than the inhospitable above-ground environment (Reichman, 1977). Indeed, heteromyids rarely appear to eat seeds while foraging above ground and often do not husk seeds before pouching them. Husking probability appears variable among species and individuals, however, perhaps because the value of taking time to pack cheek pouches with high density, husked material varies with such factors as body size, gait, and competitor and predator densities (Lemen, 1978; Morton et al., 1980; Nikolai and Bramble, 1983; Rosenzweig and Sterner, 1970).

Cheek pouch volume scales with body size for smaller heteromyids (*Perognathus* and *Chaetodipus* up to about 30 g), but this relationship does not hold for larger species (Morton et al., 1980). Nikolai and Bramble (1983) found that a more accurate relationship exists across all body sizes between head mass (rather than whole body mass) and cheek pouch volume; the authors suggested that the requirement for balance and stability has affected the structure of heteromyid cheek pouches and the ways they can be used.

Cheek pouches may influence the characteristics of heteromyid foraging bouts. For example, energetic costs of carrying a load of seeds may make it beneficial for individuals to ignore seeds when leaving the burrow and collect them only on the homeward leg of their foraging trip. Maximum seed load could be constrained by the torque exerted on the head during sudden acceleration or deceleration (rather than cheek pouch volume), and animals should load their pouches symmetrically to eliminate torsional rolling of the head (Nikolai and Bramble, 1983). These possibilities have not been investigated.

Because cheek pouches allow heteromyids to separate seed harvest and consumption in time, the rodents do not become "satiated" readily when they encounter a rich seed patch, but will return repeatedly until it is depleted (Janzen, 1982; Lockard and Lockard, 1971; Price, pers. obs.).

Caching.—All heteromyids appear to take advantage of the fact that dormant seeds can be stored for long periods of time by caching them in excess of immediate requirements, either in numerous small surface caches (scatterhoards) placed throughout their home range, or in large larder hoards in the burrows (see Smith and Reichman, 1984, and citations therein). The type and extent of caching varies among heteromyid species (Eisenberg, 1963; Fleming and Brown, 1975; Lawhon and Hafner, 1981), but detailed information on caching behavior is only now beginning to be collected (Reichman et al., 1985).

Interactions During Foraging

Some aspects of heteromyid foraging behavior may represent strategies for dealing with other individuals. Interactions between individual heteromyids can be mediated through resource acquisition or through aggression. Reichman (pers. obs.) has observed aggression in the field between kangaroo rats and pocket mice, and between two species of kangaroo rats. Eisenberg (1963), Blaustein and Risser (1974, 1976), Hoover et al. (1977), and Vorhies and Taylor (1922) have observed aggression in laboratory settings, and Congdon (1974) and Frye (1983) detected dominance of large species over smaller species in field populations. Of course, aggressions between individuals could pertain to a number of resources (such as burrows or mates; Kenagy, 1976; Randall, 1984), although interactions between species are likely to relate to food. Some authors have suggested that large heteromyids should be dominant over smaller species, and Bowers et al. (1987) have interpreted asymmetrical size-dependent competitive effects as an outcome of direct aggression between species. Changes in seed use may actually be mediated through shifts in microhabitat use brought on by aggression from the larger species in the preferred microhabitat (see subsequent microhabitat use section). Interestingly, potential competitors other than heteromyids (e.g., *Neotoma, Onychomys* may also be affected; Bowers et al., 1987). As Price (1983*b*) noted, however, asymmetrical size-dependent competitive effects could occur even if interactions were solely exploitative if size influences the efficiency with which preferred resources can be sequestered (Wilson, 1975).

Vorhies and Taylor (1922) noticed attempts by various species of heteromyids to make excavations in the mounds of banner-tailed kangaroo rats, presumably to steal seed stores. Reichman (pers. obs.) twice observed *D. merriami* chasing larger *D. spectabilis* away from the former species' burrow, and Daly (pers. comm.) has noticed *Chaetodipus penicillatus* regularly entering burrows of larger *D. merriami.* Clark and Comanor (1973) discovered kangaroo rats excavating ant mounds and pilfering the seeds stored belowground.

Influence of Torpor on Foraging

The physiology of torpor and the selection pressures that might promote its use by heteromyids are discussed by French in this volume and MacMillen (1983). Several features of the use of torpor, however, impinge directly on foraging choices and deserve consideration in that context. Only the smallest of the pocket mice are known to regularly rely on torpor as a winter energy-saving strategy. Whether or not pocket mice use torpor depends on environmental conditions (primarily temperature; French, 1976), and on the quantity of food available to them (Reichman and Van De Graaff, 1973). If resources are low, or unavailable, pocket mice are more likely to enter into torpor. If seeds are available, but distributed so as to make them costly to harvest, pocket mice may also become torpid (Reichman and Brown, 1979). Furthermore, the animals appear to respond to the rate at which resources become available rather than just the absolute amount. Brown and Bartholomew (1969) found that kangaroo mice went into torpor when given a set amount of seed at the beginning of an experimental trial. If, however, the same amount of seeds was spread over several days, the mice would stay active. This interesting behavior may allow the rodents to stay active as long as there is a reasonable expectation of resource availability, but to minimize their costs by quickly dropping into torpor when the rate of resource income falls below a profitable level. It is probably not profitable for larger species to enter and recover regularly from torpor. Rather, several of the large kangaroo rat species are known to cache large quantities of seeds for use during the winter when seeds are less available.

Factors Affecting Heteromyid Foraging Decisions

The Sequence of Foraging Decisions

At any point in time, a heteromyid can initiate alternative behaviors, the set of options being dependent on the animal's current behavioral state. For example, if the animal is in its burrow, it can either leave or remain in it. Once outside, the options expand; the animal can initiate foraging, mate search, exploration, territory maintenance, or return to the burrow. Each point in time can be considered a behavioral branch point of "decision node," and the animal can be thought of as having made a choice when it changes behavioral states. Terms such as "decision" and "choice" do not imply conscious choice, but instead the act of following one behavioral option rather than another.

The sequence of decisions involved in foraging is as follows. An animal in its burrow must first "decide" when to leave and initiate a foraging bout, and for how long to forage. Once outside the burrow, it must choose where within its home range to begin searching for food. Once searching, it must decide which microhabitats to search, and once a seed patch is encountered, the forager must decide whether to harvest the seed or continue searching. Once a seed has been harvested, the animal must decide whether to discard it, eat it, or put it in the cheek pouches for transport back to the burrow. Seeds carried back to the burrow can be eaten or cached, and if they are cached decisions must be made about how to store them.

In the following sections we will use an optimal behavior perspective to identify what factors might affect decisions about activity time, home range use, microhabitat choice, diet choice, and seed caching. By this, we simply mean that we will focus on how behavioral choice affects an animal's fitness—its ability to survive and reproduce. Such a focus is useful in understanding factors likely to have molded the evolution of heteromyid foraging behavior and in developing predictive models about their behavior. We will suggest additional ways in which such an approach can be applied to heteromyid foraging in the discussion.

When to Forage: Patterns of Activity

Cost-benefit models.—Rosenzweig (1974) developed a cost-benefit model for aboveground activity in which he proposed that the fitness benefits and costs of emerging from a burrow vary through time. According to this approach, benefits of aboveground activity are measured by rates at which resources such as food or mates are accumulated, and by the fitness gain that accrues per unit of resource obtained. Benefits should vary with reproductive condition, abundance of competitors, and seasonal and year-to-year patterns in food availability. Costs of aboveground activity are measured by rates at which resources such as energy or water are expended, by probabilities of being eaten by a predator or infected by a disease or parasite, and by the fitness that is lost per unit of expenditure. Both benefits and costs will vary with environmental conditions such as temperature, humidity, and ambient light levels, and with abundances of predators, competitors, or parasites.

The net value of aboveground activity is equal to the difference (not the ratio) between benefits and costs of activity, and animals are expected to spend time above ground only when this difference is greater than the net value of inactivity. The utility of this model is that it can be used to predict how activity should vary if a particular factor is important to fitness; if the activity pattern does not change in the predicted direction, then either that factor is unimportant or its influence is balanced by other factors. Rosenzweig (1974) used a cost-benefit model to analyze moonlight avoidance by *Dipodomys spectabilis* (Lockard and Owings, 1974), but the model can easily be applied to other factors. We will rely on the basic cost-benefit conceptual framework in what follows.

Daily timing of foraging. —Predators, competitors (as mediated through aggression or resource acquisition), and environmental conditions all may affect the fitness advantages and disadvantages of foraging diurnally or nocturnally. Heteromyids forage almost exclusively at night. Although crepuscular activity is routinely observed, reports of diurnal activity are extremely rare. Lockard (1978) suggested that *Dipodomys spectabilis* extends its activity into daytime during summer when food availability is low enough to require foraging periods longer than the duration of darkness. His observations have not been confirmed, however, and it is possible that diurnal squirrels, rather than kangaroo rats, were responsible for the daytime visitation to his automatic seed dispensers. The prevalence of nocturnal activity suggests that diurnally-active heteromyids would have low fitness and this leads us to ask why nocturnal activity is advantageous.

Potential predators and competitors of heteromyids are active both during daylight and night so it is not clear that nocturnal activity would allow even partial escape from predation or competition. It is clear, however, that the physiological costs of foraging vary on a daily cycle because water loss and energy expenditure are functions of radiant energy flux, air temperature, and humidity, which show pronounced diurnal fluctuations in virtually all terrestrial habitats. For homeotherms in dry climates, rates of water loss generally decrease with decreasing ambient temperature and metabolic rates are lowest at thermal neutrality (about 30°C for heteromyids with no net radiant energy exchange; Hinds and MacMillen, 1985). Periods of minimal energy and water loss occur sometime during the night for much of the year in the warm environments heteromyids presumably experienced during much of their evolutionary history.

While physiological characteristics can explain why heteromyids are nocturnal during a majority of the year, some other factor must prevent them from foraging diurnally in cool seasons when temperatures may be too cool for efficient nocturnal foraging. It is possible that heteromyids have poor daylight vision and that predation prevents them from becoming diurnally active in the winter, but this possibility has not been investigated.

The few detailed studies of nightly activity patterns (Behrends, 1984; Braun, 1985; Kenagy, 1973; Lockard, 1978; O'Farrell, 1974; Schroder, 1979) suggest that heteromyids exhibit a peak of activity just after dusk and frequently another near dawn.

These periods would usually be associated with low heat and water loss, respectively. Kenagy (1973) noted that the decline in activity after the initial peak was most pronounced in the smallest species, which would suffer the greatest relative heat loss at low temperatures. To our knowledge, however, only two studies (Behrends, 1984; O'Farrell, 1974) have systematically compared warm and cool season activity patterns to determine whether variation in the temporal temperature gradient modifies the onset or cessation of evening activity. Behrends (1984) noted a tendency for *D. merriami* to be more evenly active throughout warm nights than cool nights, but the trend was not statistically significant, and O'Farrell (1974) did not observe consistent seasonal patterns.

By initiating foraging early in the evening, heteromyids could increase their chances of obtaining seeds that became available during the preceding day through seed production or redistribution by daytime winds. Seed renewal rates appear to be very slow, however. At a Chihuahuan Desert site in Arizona, Waser (1988) observed extremely slow influx of seeds into artificial depressions 3 cm in diameter and 2 cm deep, even during seasons of peak seed production. Mean rates of influx per depression were 0.8 seeds/day in fall, vs. 0.1–0.2 seeds/day during spring and summer. Influx to sterile soil patches lacking depressions was essentially zero. Reichman (1984) found similarly low average influx rates into artificial depressions (0.4–0.8 seeds/day) at a Sonoran Desert site, although one rain event did trigger substantial flooding and seed deposition. While it could be that low net influx was due to high gross inflow followed by high removal rates by rodents (the depressions were available to rodents), this seems unlikely. Hence, it is unlikely that the early evening peak in heteromyid activity is a response to newly-available seeds, but we need better estimates of seed renewal rates to be secure in this conclusion.

Deviations from the bimodal nocturnal activity patterns have been correlated with timing of moonrise and moonset (presumably, a response to moonlight-related predation risk; Behrends, 1984; Braun, 1985; Lockard, 1978; Lockard and Owings, 1974; O'Farrell, 1974; but see Schroder, 1979), and with rainfall or cloud cover (Behrends, 1984; Braun, 1985; Kenagy, 1973).

These observations suggest that heteromyids have adopted a bimodal nocturnal activity pattern to minimize their physiological costs during a majority of the year, and adjust the timing of their above-ground activity in relation to predation-related costs.

Total foraging time and seasonal distribution of foraging.—A heteromyid must decide not only when to forage, but also whether to forage at all and if so, for how long. Such decisions should be dictated by seasonal variation in the benefits realized from foraging and seasonal and diurnal variation in physiological and predator-related cost of foraging relative to other activities (e.g., Rosenzweig, 1974). Animals should forage only as long as net benefits of foraging outweigh net benefits of other activities; in general, the "window" of time during the day when this occurs, and therefore, the expected time spent foraging, should vary seasonally.

A major difficulty with evaluating these expectations is that it is difficult to distinguish foraging from other aboveground activities in these nocturnal rodents, so that most available information presents only total aboveground activity. Aboveground activity of heteromyids does seem to change seasonally, however. Time spent above ground is high when the immediate reproductive benefits of foraging are high: female *D. merriami* are more active when pregnant or lactating, times when energetic demands are high and anticipated reproductive gains from foraging are immediate (Behrends, 1984; Behrends et al., 1986*a*). Females are also more active during estrus, and males are more active when in reproductive condition, but presumably here the greater activity involves mate search in addition to foraging time (Behrends et al., 1986*a*). Lockard and Owings (1974; Lockard, 1978; but see Schroder, 1979) reported that in fall and winter, presumably periods of abundant food and long nights in which to gather it, *D. spectabilis* were inactive when the moon was up (possibly a time of high predation risk), but were forced to forage during moonlight in the lean times of late spring or summer when seed stores were likely depleted and nights were short. These observations of seasonal moonlight avoidance may be artifacts of estimating activity from rates of visitation to automatic seed dispensers, for direct observations of radio- or light-tagged individuals have not revealed any pronounced seasonal differences in the effect of moonlight on aboveground activity level (Behrends, 1984; Braun, 1985; Schroder, 1979).

Total activity time seems to be small when costs of activity are high. Many heteromyids appear to be less active in bright moonlight (Kaufman and Kaufman, 1982; Kotler, 1984*a*; Lockard and Owings, 1974; O'Farrell, 1974; Price et al., 1984; but see Braun, 1985; Chew and Butterworth, 1964; Schroder, 1979). Larger heteromyids may reduce activity in summer, when warm nighttime temperatures would promote relatively high rates of water loss and when spring seed production has been depleted (Braun, 1985; Lockard, 1978; Reichman and Van De Graaff, 1973). Small-bodied pocket mice and kangaroo mice (French, 1976; Kenagy, 1973; Reichman and Van De Graaff, 1973), which by virtue of their size have relatively high heat conductance, can probably tolerate warm summer temperatures better than larger heteromyids, but at the cost of having to remain inactive below ground in winter when cold temperatures increase rates of heat loss. Interestingly, in light of the fact that in the laboratory torpor can be forestalled by supplying seeds continuously rather than in a pulse (Brown and Bartholomew, 1969), winter pocket mouse activity is higher in years of exceptionally good fall seed production than in years of poor production (Brown, pers. comm.; Kenagy 1973).

Regardless of this variation, elapsed time from first exit to last entry into a burrow is often less than two hours (Braun, 1985; Kenagy, 1973; Langford, 1983; Schroder, 1979), and not all of this time is spent outside the burrow. Most reported exceptions to this are anecdotal; Shaw (1934) described *D. ingens* spending all night to transfer "cured" surface caches to the burrow, and we have observed (pers. obs.) animals returning repeatedly for hours to a rich bait station that was continually replenished. Chew and Chew (1970), however, reported *D. merriami* to be active for up to six hours each night and Behrends' (1984) systematic investigation of radio-tagged *D. merriami* revealed that elapsed time from first exit to last entry was rarely less than two hours, although the total time spent outside the burrow often was less than two hours.

The general impression gained from these studies is that heteromyids spend, on an annual basis, close to the minimum time above ground necessary to find required food and mates, as expected of animals whose burrow environments are more benign than those they face outside the burrow.

Where to Forage

Once an individual has started foraging, it must decide where within the home range to begin looking for food, and then what patches within that area to harvest. Being small and nocturnal, heteromyids are difficult to observe in the field, and most information on use of the home range or of microhabitats or patches within regions of the home range is based on live trapping or visitation to artificial bait stations. A few recent studies have used miniature radio transmitters or light emitters, or direct observation, to analyze movements in more detail (Behrends, 1984; Behrends et al., 1986*a*, 1986*b*; Jones, 1982; Langford, 1983;

Randall, 1984; Schroder, 1979; Thompson, 1982*a*).

Home range.—Cost-benefit analysis can be applied to the use of space as well as time (e.g., Price, 1983*b*; Rosenzweig, 1981). Imagine that the home range consists of areas that differ in quality, either intrinsically or because of the actions of competitors. Under these conditions, home range use is unlikely to resemble a single-peaked bivariate normal distribution that is usually envisioned, but should instead consist of multiple activity peaks that decrease in size irregularly with distance from the home range center (Don and Rennolls, 1983; Getty, 1981). Detailed analysis of the movements of individuals indicates this pattern of home range use for *D. spectabilis* (Schroder 1979), *D. merriami* (Behrends et al., 1986*b*; Daly, pers. comm.), and *D. ingens* (Braun, 1985; Fig. 2), but to date no attempt has been made to understand why some areas are heavily used and others are not.

What total area should animals use? Most general theories of home range size (e.g., Covich, 1976; Harestad and Bunnell, 1979; Jenkins, 1981; McNab, 1963; Schoener, 1968) are based on the notion that animals move only as far as they must to obtain food requirements. This minimum distance should be a function of body size-related energetic requirements and the abundance of suitable prey. Indeed, home range size generally appears to increase with body size in mammals (Harestad and Bunnell, 1979; McNab, 1963), and increases faster for carnivores than herbivores (Jenkins, 1981). In contrast to these patterns, heteromyid home range size does not seem to vary with body size (Braun, 1985; Jenkins, 1981; Maza et al., 1973; O'Farrell, 1978), perhaps because of the confounding influence of variation in habitat productivity (large kangaroo rats, such as *D. spectabilis* or *D. ingens,* may inhabit regions that are more productive than the average for the region). Lacking good estimates of the amount of food actually available to heteromyids per unit area, we cannot say whether energetic models of home range size are reasonable. Certainly, the home ranges of heteromyids seem very small for inhabitants of deserts, which are reputed to be unproductive. Measurements vary from a minimum of about 0.05 ha for two 120-g kangaroo rats (Braun, 1985; Schroder, 1979) to a maximum of about 2.6 ha for a 35 g kangaroo rat (Behrends et al., 1986*b*; Schroder, 1987; York, 1949).

For animals such as heteromyids that exploit a slowly renewing resource, foraging of one individual substantially changes food availability for others (Reichman, 1979). This situation should inhibit sociality and home range overlap (Waser, 1981). Indeed, heteromyids are solitary and the burrows of some species tend to be overdispersed (Schroder and Geluso, 1975). Home range overlap is often high, however, especially between different-sex individuals of the same species (Behrends et al., 1986*b*; O'Farrell, 1978, 1980; Schroder, 1979). Apparently high home range overlap may be misleading, because core areas around day burrows do not overlap (Behrends et al., 1986*b*), and neighboring individuals utilize distinct portions of the region of overlap. Effective space use overlap can be much less than home range overlap might suggest (Daly, in litt.).

Microhabitat Use

Uneven home range use in heteromyids presumably occurs because the home range consists of a mosaic of patches that vary in quality (Getty, 1981). In the desert, heterogeneous topography and perennial vegetation cause patchiness not only in the three-dimensional structure of the habitat, which could affect risk of predation and rates of radiant heat loss (Dice, 1945, 1947; Kotler, 1984*b*; Lowe and Hinds, 1971), but also in characteristics of surface soil and soil seed pools (Bagnold, 1954; Price and Reichman, 1987; Price and Waser, 1985; Reichman, 1984; Shreve and Wiggins, 1964), which could influence seed harvest rates (Price,

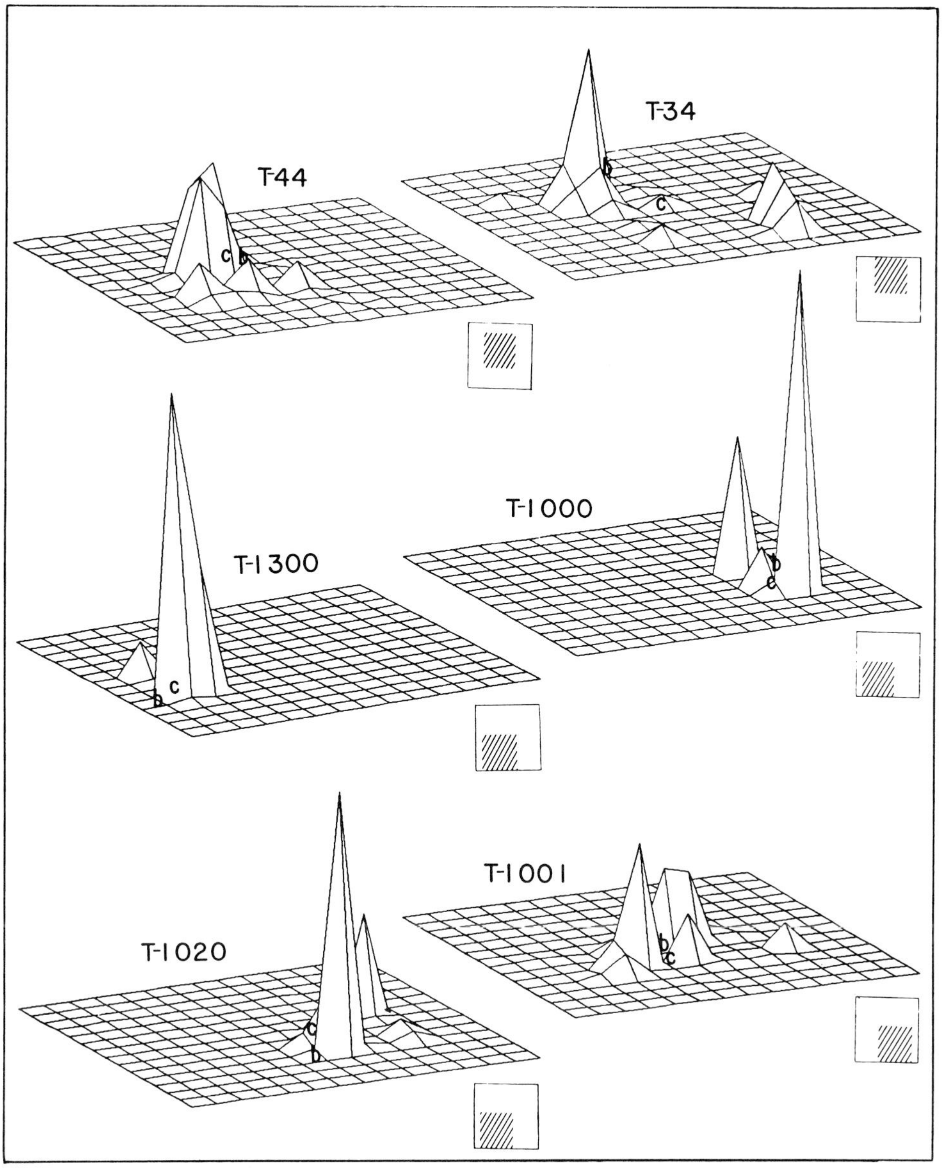

FIG. 2.—Utilization distributions of individual *Dipodomys ingens* based on independent observations. The X and Y axes indicate location on a grid and the Z axis presents the frequency with which each portion of the home range was used. Each grid unit is 5 m on a side; b = location of home burrow and c = location of arithmetic mean center of activity. Shaded areas in small squares indicate location of the grid within a 1.0 ha study area (from Braun, 1985).

1983*b*; Price and Heinz, 1984). Areas that differ in physical structure therefore should differ also in rates with which animals gain and lose fitness while foraging in them, and animals should prefer areas in which they have the highest net rates of fitness gain. We shall follow Pulliam (1976) and use the term "habitat" for large structurally-distinct areas and "microhabitat" for areas so small that many are traversed during the normal course of foraging. Within microhabitats, heterogeneity in soil or seed densities can further define seed patches on a scale of a few centimeters (Price and Reichman, 1987).

Microhabitat use can be measured in a number of ways, not all of which are strictly comparable. Two commonly-used methods quantify microhabitat use in terms of relative capture frequency in predefined microhabitats (or relative occurrence in the diet of marked bait placed in predefined microhabitats), or the multivariate distribution of environmental characteristics at points-of-capture (Price and Kramer, 1984). These methods typically use bait, and so measure relative response of individuals to food placed in different places. Baitless methods that quantify microhabitat use in terms of relative track frequency or deposition of fluorescent pigment by dusted animals (e.g., Harris, 1984; Kotler, 1985*a*; Lemen and Freeman, 1986) or proportion of foraging time in the microhabitat (Schroder, 1987; Thompson, 1982*a*) measure microhabitat use under some unknown distribution of resources. Hence, different methods may not provide identical estimates of proportional microhabitat use, even if applied to the same community (*contra* Thompson, 1982*a*). Two studies that compared microhabitat use as measured by trapping and a baitless method (Kotler, 1985*a*; Lemen and Freeman, 1986) found no significant differences between the techniques, but more such comparisons will be necessary before a robust generalization can be made.

Despite these methodological difficulties, most evidence indicates that heteromyid species exhibit nonrandom microhabitat associations, and that specific associations are correlated in a qualitative way with overall body shape. Detailed analyses of tropical forms are lacking, but they also show large-scale habitat associations (Bonoff and Janzen, 1980; Fleming, 1970), as do temperate forms (e.g., Grinnell, 1932; Hoover et al., 1977; Reynolds, 1958; Rosenzweig and Winakur, 1969), and may also discriminate on a smaller spatial scale. Among North American arid-zone species, bipedal forms (*Dipodomys, Microdipodops*) utilize open spaces more than do quadrupedal forms (*Chaetodipus, Perognathus*), which prefer more structurally complex areas under shrub or tree canopies, in areas of abundant grasses or forbs, or in the vicinity of rock crevices (Brown, 1973, 1975; Brown and Lieberman, 1973; Harris, 1984; Heinz, 1983*b*; Kotler, 1984*b*, 1985*b*; Lemen and Freeman, 1986; Lemen and Rosenzweig, 1978; M'Closkey, 1978; Meserve, 1976; O'Connell, 1979; O'Farrell, 1980; Price, 1977, 1978*b*, 1984, 1986; Price and Brown, 1983; Price and Kramer, 1984; Price and Waser, 1984; Rosenzweig, 1973; Rosenzweig and Winakur, 1969; Rowland and Turner, 1964; Thompson, 1982*a*; Wondolleck, 1978). To some extent, each species' local density is determined by the local availability of preferred microhabitats (Kotler, 1984*b*; Price, 1978*b*; Price and Waser, 1984; Thompson, 1982*b*).

Why heteromyids prefer some microhabitats over others, and why morphologically-distinct species have different preferences is not known. Because the enlarged auditory bullae and leaping ability of kangaroo rats enable them to avoid predators in open spaces in the laboratory (Webster and Webster, 1971; also see Lay, 1974), and because owls appear to be less capable of capturing prey in structurally complex areas (Dice, 1947), some authors have emphasized the probable importance of predation risk in restricting quadrupedal heteromyids to shrub microhabitats and in allowing bipedal heteromyids to exploit the more "risky" open spaces (Kotler, 1984*b*, 1985*b*; Kotler et al., 1988; Price, 1984; Price and Brown, 1983; Rosenzweig, 1973; Thompson, 1982*a*, 1982*b*). Alternatively, because shrub and open microhabitats contain different types of soil and seed resources, and because these variables affect rates of seed harvest, other authors have emphasized the probable importance of foraging economics for patterns of microhabitat use (Price, 1984, 1986; Price and Brown, 1983; Price and Waser, 1985; Reichman, 1981).

To date, there is very little direct evidence

regarding the relative importance of these alternatives, because few studies have set out to determine whether shifts in microhabitat use accompany known changes in the relative economic value or riskiness of microhabitats. The primary difficulty is knowing exactly how an experimental manipulation affects risk or economic value (Price, 1986; Price and Brown, 1983).

The role of foraging economics in microhabitat selection.—The importance of foraging economics is suggested by studies that relate inter-microhabitat variation in characteristics of soil and seed pools to seed harvest rates and patch choice in the laboratory and field. Reichman (1979, 1984; Reichman and Oberstein, 1977; see also Mott and McComb, 1974; Nelson and Chew, 1977) noted that soil samples taken from under shrub canopies in the Sonoran Desert contain on average more seeds than samples from open spaces between shrubs, but that the coefficient of variation in seed number per sample from open spaces was large. This suggests that microhabitats differ not only in average seed abundance, but also in the degree of seed clumping. Subsequent soil analyses (Price and Brown, 1983; Price and Reichman, 1987; Price and Waser, 1985; Reichman, 1984) have reinforced and extended these findings: not only do microhabitats differ in seed numbers and spatial distributions, but they also differ in the species of seed they contain and hence probably differ in distributions of seed sizes, morphologies, and chemical characteristics. The texture and density of the soil in which seeds are embedded also varies among microhabitats (Price and Reichman, 1987; Price and Waser, 1985): soil under shrubs and trees is coarser in texture and lighter in density than soil in open spaces between perennial plants. Furthermore, soils vary in their propensity to form crusts (primarily in relation to their silt content), which may obscure underlying seeds (although we know of no data relating such conditions to specific microhabitats). Such inter-microhabitat differences could promote microhabitat preferences if soil density and texture, and/or seed density or type, affect heteromyid foraging efficiency.

These factors do indeed affect at least one component of foraging efficiency—the rate with which seeds are extracted from the soil and placed into cheek pouches. Price and Heinz (1984; Podolsky and Price, 1990) found that the rate with which a variety of heteromyid species harvested millet seeds in the laboratory increased with seed density, decreased as soil particle diameter increased to equal that of millet, and increased slightly with soil density. Although the qualitative effect of these variables was the same for all heteromyid species tested, the quantitative effect was not: a given change in seed density or soil conditions caused much larger changes in harvest rate for larger species (Price, 1983*b*; Price and Heinz, 1984). Size and gait also appear to affect rates of travel between stops (Table 2, Garland, 1983*a*, 1983*b*; Thompson, 1985). Price (1983*b*) incorporated these allometric variables into an optimality model of seed patch choice, which predicted that size alone could promote interspecific differences in patch choice. Whether each species actually prefers the most profitable patches remains to be determined, but observations that heteromyids do differ in the degree to which they specialize on profitable seed clumps, as Price's model would predict (Bowers, 1982; Hutto, 1978; Price, 1978*a*; Reichman, 1981; Reichman and Oberstein, 1977; Thompson, 1982*a*; Trombulak and Kenagy, 1980; but see Behrends, 1986; Frye and Rosenzweig, 1980), suggest that energetics may well be important in causing observed differences among heteromyid species in foraging behavior.

The role of inter-microhabitat differences in seed species composition is not clear; for such differences by themselves to account for microhabitat preference, quadrupeds and bipeds on average must prefer different kinds of seeds. This does not appear to be the case: Price (1983*a*, and unpubl.) has found that eight heteromyid species spanning three genera and an order of magnitude in body size show strong concordance in preferences for eight commercially available seed species, and for ground wheat particles ranging from 0.3 to 24 mg in weight. It is possible, however, that differences among microhabitats in seed size distributions could interact with soil texture differences to affect microhabitat profitability (Price and Podolsky, 1989).

Because soil texture and density affect seed harvest rates, we would expect heteromyids to discriminate among different soils. The importance of soil to heteromyid distributions has long been recognized by naturalists (e.g., Hardy, 1945), but only recently have systematic studies of soil preferences and digging abilities been initiated. Price and Waser (1985; also see Price and Longland, 1989) presented four heteromyid species with eight artificial seed patches in the laboratory. All four species harvested significantly more millet seeds from some patch types than from others, patterns of preference were correlated with body size and morphology, and preferred substrates were qualitatively similar to soils contained in preferred microhabitats in nature. Furthermore, the authors found that microhabitat use by *Dipodomys merriami* could be shifted in the field by shifting the microhabitats into which preferred artificial seed patches were placed, and that the animals did not respond to the location of patches of nonpreferred substrate. These observations strongly suggest that microhabitat use to some extent reflects the spatial location of preferred combinations of seeds and soils.

It is not yet known whether substrate preferences occur because substrates affect the energetic costs of digging for buried seeds or because of some other effect. For example, soil characteristics may affect the intensity of odor gradients established around buried seeds which in turn would affect their detectability. The latter possibility has yet to be investigated. Heinz's (1983*a*) observation that variation in forelimb structure within the genus *Dipodomys* is correlated with edaphic associations certainly suggests that digging energetics may be involved, but it is not conclusive in the absence of physiological studies of digging. Hoover et al. (1977) attributed differences in the substrate affinities of *Chaetodipus penicillatus* and *C. intermedius* to differences in their abilities to maintain water balance under the regimes of temperature and humidity characteristic of burrows located in the rocky vs. sandy soils.

The role of predation risk in microhabitat selection.—The role of predation risk in microhabitat choice is even less well-documented than is the role of foraging economics (Price, 1986; Price and Brown, 1983). Predators are significantly more difficult to manipulate than resources, and observation of predation events in the field is very rare. To date, evidence for the importance of risk comes primarily from observations that microhabitat use shifts in response to natural or experimental changes in moonlight intensity (Bowers, 1988; Kotler, 1984*b*, 1985*a*; Kotler et al., 1988; Price, 1986; Price and Brown, 1983; Price et al., 1984, and references therein). The link between moonlight intensity and risk is as yet tenuous, coming from observations that owls have greater success in capturing dead prey as light intensity increases (Dice, 1947), and that they require less time to capture live *Peromyscus maniculatus* at high ambient light levels (Clarke, 1983). In the latter experiment, it is not certain that lower capture time under the experimental conditions would translate into higher capture probabilities in nature; the probability a mouse escaped when attacked did not change with light intensity, and in fact there was some evidence that mice were more efficient in detecting predators at high light levels because owls missed more often on the initial strike (Clarke, 1983). This suggests that both prey escape ability and predator attack accuracy may be improved by light, a result consistent with Webster and Webster's (1971) finding that hearing-impaired kangaroo rats released back into the field disappeared primarily duing a new moon (see also Price and Brown, 1983:122). Given conflicting effects of moonlight, it is not clear whether the balance is struck in favor of prey or predator. Of course, even if prey capture probabilities are lower in the light, reduced mortality may be bought at the cost of higher vigilance. In this case, the cost of predation would be economic—the loss in time that could be spent foraging—rather than in terms of mortality. Unambiguous interpretation of the significance of the heteromyid response to light is simply not possible until controlled experiments are performed to measure how microhabitat and light level affect relative escape abilities and time budgets of quadrupedal and bipedal heteromyids.

Results from flight cage experiments are beginning to quantify effects of various factors on predation risk for heteromyids. Kotler et al. (1988) found that adding 15% shrub cover to one of a pair of enclosures decreased both the number of individuals captured by owls over a 5-night period and the capture rate per gram of seeds harvested from seed trays in the bush enclosure compared to the open enclosure. Capture rates per unit of foraging activity were also higher for all species under high light intensities, and larger species appeared to be more at risk than smaller ones. A bipedal species (*D. merriami*) suffered slightly higher capture rates per unit estimated foraging activity than a quadrupedal species (*C. baileyi*) of the same size. From these data, it is difficult to derive precise estimates of the risk an individual incurs per unit of time spent foraging, because samples sizes are small (less than 3 captures per night with two owls foraging for 40 mice) and information is lacking on attack rates of owls and rodent activity times in the various microhabitats. Brown et al. (1987) noted that activity (measured by grams of seed removed from trays) was indeed variable and differed among rodent species, microhabitats, light levels, and owl treatments.

Another flight cage study used direct observation of attack and capture rates (Longland, 1989) to get detailed estimates of vulnerability under conditions similar to those of Kotler et al. (1988) and Brown et al. (1987). Results of 459 hours of observation confirm that rodents tend to spend more time above ground in enclosures that contain shrubs. Analysis of 95 captures from 761 attacks by owls indicate the probability that a rodent is attacked is higher in the open (approximately 0.20 attacks/min of foraging) than under a shrub (approximately 0.05 attacks/min foraging), and the probability it is captured when attacked is higher if it is in the open (approximately 0.17) than if it is under a shrub (0.07). There were no pronounced differences between rodent species in overall microhabitat-specific vulnerability, although species did differ in the probability of attack (*Microdipodops megacephalus* was the most vulnerable) and the probability of capture given an attack (bipeds were less vulnerable than quadrupeds of the same size). Light intensity had an insignificant effect on components of risk, although for 3 of 4 species capture probability was slightly higher under bright conditions. These results suggest that predation probably does not play a major role in promoting interspecific differences in microhabitat affinity, because all species incur similar risks in one microhabitat relative to another.

The role of competitors in microhabitat selection.—The densities of other rodent species can have a significant effect on microhabitat use by heteromyids. Wondolleck (1978) observed that pocket mice increased use of open spaces when kangaroo rats were removed from unenclosed populations. Price (1978*a*) introduced four heteromyid species to enclosures singly and in pairs, and found that all species decreased their use of competitor-preferred microhabitats and specialized more heavily on their own preferred microhabitat when a competitor species was present. M'Closkey (1981) noted that as heteromyid densities increased naturally on an unenclosed study area, microhabitat use became less specialized and interspecific overlaps increased. In a geographic comparison of sand-dune habitats Larsen (1986) observed microhabitat specialization decrease as species numbers decreased. Bowers et al. (1986) noted that exclusion of a large heteromyid species, *Dipodomys spectabilis,* resulted in microhabitat use shifts by 7 of the 9 remaining species. Rebar and Conley (1983) observed that *D. ordii* increased use of preferred open microhabitats when an insectivorous cricetid species, *Onychomys leucogaster,* was present in experimental enclosures. They interpreted the shift as a response to interspecific competition or predation, but an equally likely interpretation is that the kangaroo rats were responding to a decrease in conspecific densities in the two-species treatment (total rodent density, rather than conspecific density, was kept constant across treatments), rather than to the other species. Under interspecific competition, species should diverge in microhabitat use, rather than converging as happened in the Rebar and Conley study. Finally, Lemen and Freeman (1987) observed that *Perognathus longimembris* increased use of open microhabitats when *Dipodomys merriami* (but not *D. microps,* a foliovore) were removed from unenclosed populations. The larger species did not respond to removal of the smaller pocket mouse species.

It is not clear whether competitive shifts in microhabitat use are the result of behavioral interactions ("interference competition") or of changes in food abundance and distribution caused by the foraging activities of competitors ("exploitative competition"). The latter seems a more likely mechanism; it is difficult to imagine that a heteromyid species could behaviorally exclude other species from a microhabitat, because the "resource" consists of small, widely dispersed patches that would be difficult to defend. On the other hand, if direct encounters are frequent enough or expensive enough, the effective value of a microhabitat to a subordinate species could be reduced by the possibility of encountering a dominant individual in it, and subordinate species should avoid the competitor's preferred microhabitats.

Distinguishing the mechanism of competition-induced microhabitat shifts is not easy, but the symmetry of competitive responses and the rapidity with which they occur may provide clues. Direct behavioral interactions would cause a competitive hierarchy with larger species dominating smaller ones (Blaustein and Risser, 1974, 1976; Eisenberg, 1963; Trombulak and Kenagy, 1980). In such cases, large competitors should have strong effects on smaller ones, but small competitors would have weak effects, if any, on larger species. Indirect, exploitative, interactions would also produce asymmetrical size-dependent effects (cf. Price, 1983*b*) if large size increases speed or effectiveness of resource acquisition, but small species should also affect larger ones to some degree. The rate of response to reciprocal removals should also be different for interference and exploitative interactions. Because removal of a dominant competitor instantaneously changes the relative cost of utilizing the competitor-preferred microhabitats, shifts in microhabitat by the remaining competitor should be swift if interference is operating. Conversely, if exploitative interactions predominate, the rate of response by the remaining species should be determined by the rate with which unused resources accumulate after the removal of the competitor.

As yet, very little is known either about the symmetry of heteromyid competitive interactions in the field or about rates of resource removal or of responses when species are removed. Inferences about the importance of aggression have come from observations that large species tend to exclude smaller ones from the vicinity of the burrow (Frye, 1983), or that the removal of large species yields an increase in the density (Brown and Munger, 1985; Lemen and Freeman, 1983; Munger and Brown, 1981) or changes in the microhabitat use (Bowers et al., 1987; Lemen and Freeman, 1987;

Wondolleck, 1978) of smaller species. Unfortunately, in few of these cases was the effect of the smaller species on the larger quantified. Lemen and Freeman (1983) did find that whereas the removal of kangaroo rats resulted in a significant short-term increase of pocket mice, removal of pocket mice did not increase kangaroo rat density. On the other hand, Price (1978*b*) observed that both kangaroo rats and pocket mice shifted microhabitat use when in each other's presence, and that the shifts were apparent soon after the introduction of kangaroo rats. This reciprocal response is suggestive of exploitative interactions, but the rapidity of the response seems to imply interference. We have a very poor understanding of the dynamics of seed renewal and consumption, however, and so cannot yet predict how soon detectable effects of exploitative interactions should be apparent. Clearly, the issue cannot be resolved until more data are gathered.

Food Choice

After a heteromyid has decided where to forage, it must decide which food items to harvest. This type of choice has been treated in the literature of optimal foraging (e.g., Krebs and Davies, 1984), which assumes that individuals select food items that yield the highest net return on energy invested to harvest and process them, which in turn is dictated by the morphological and physiological characteristics of the forager (Reichman, 1977).

A number of factors have contributed to the evolution of heteromyids as seed specialists, and to the choices made by individuals as to what specific seeds to consume and cache. Heteromyids probably evolved in deserts (Hafner and Hafner, 1983), subsequently moving into tropical areas, taking their granivore habit with them. A number of desert organisms consume seeds (Brown et al., 1979; Reichman et al., 1979), suggesting that they are a valuable dietary item in deserts. Seeds can be very abundant in desert soils (Nelson and Chew, 1977; Reichman, 1984), and persist for long periods during which no further seed production occurs (Tevis, 1958). Other resources, such as insects and green vegetation, are more ephemeral. Furthermore, because seeds have evolved anatomical and physiological strategies to remain viable in the soil for long periods, presumably to survive until conditions are appropriate for germination, they are ideal high-energy storage units for caching.

Although seeds are a persistent and profitable food resource, they are expensive to extract from the soil. Thus, a single seed may not be worth collecting unless it is very large or contains a rare, essential nutrient. When individual seeds form aggregations, the clumps may become profitable for extraction (Reichman, 1981). In the following sections we will consider how intrinsic characteristics of seeds (size, chemical composition) and extrinsic features of the environment (spatial distribution, competitors and predators) affect seed choice by heteromyids.

Effects of Intrinsic Seed Characteristics

Seed size.—Early investigations dealing with the allocation of seed resources between heteromyid species suggested that there might be a positive correlation between body size of heteromyids and the average size of seeds used (Brown, 1975; Brown and Lieberman, 1973; Mares and Williams, 1977). Subsequent investigations have not confirmed this pattern (Lemen, 1978; Price, 1983*a*). In fact, constraints imposed by auditory bullae on jaw gape (Nikolai and Bramble, 1983) may cause the largest heteromyids to be unable to process extremely large seeds. Laboratory and field studies (e.g., Price, 1983*a*; Stamp and Ohmart, 1978) tend to support the idea that large kangaroo rats select smaller seeds than do smaller pocket mouse species, perhaps in part due to the fact that small seeds tend to end up in high density clumps (Reichman, 1984).

Energy.—Seeds are generally high in calories, providing a discrete source of energy (Chew and Chew, 1970; Reichman, 1976). Seed size has an overwhelming effect on the amount of energy obtained per seed consumed. If the costs of husking and consuming a seed do not increase faster with seed size than total caloric content, heteromyids should prefer large over small seeds. This appears to be the case for seeds smaller than about 10 mg. In the field, heteromyids avoid seeds smaller than about 0.3 mg (Brown et al., 1979; Reichman, 1975, 1978; Reynolds, 1958), and in the laboratory consumption increases with seed size up to about 5 mg (Price, 1983*a*). Seed consumption goes down, however, at larger sizes, perhaps because handling time begins to increase rapidly with size at this point (cf. Rosenzweig and Sterner, 1970; Rosenzweig et al., 1975)

Reichman (1977) found that some heteromyids gather seeds which are significantly higher in energy content than those randomly available in the soil, and choose to ingest, from those gathered, energetically superior seeds (as noted earlier, a problem may occur when comparing seed fragments in stomach contents and whole seeds from cheek pouches).

Water balance.—Desert organisms are likely to have adapted to periodic water stress and many of the physiological studies of heteromyids have taken advantage of this (French, this volume; Hinds and MacMillen, 1985; MacMillen, 1983; MacMillen and Hinds, 1983). Heteromyid kidneys are extremely efficient and the rodents employ a variety of strategies to minimize evaporative water loss (Schmidt-Nielsen, 1964). In addition to anatomical and physiological adaptations, desert heteromyids rely on their diet to provide significant metabolic water. Dietary choices relating to metabolic water and preformed water may be especially important, as heteromyids rarely have water available to drink.

Schmidt-Nielsen et al. (1948) found that the chemical composition of the diet influenced water balance in desert heteromyids. Individuals fed on high protein mesquite seeds quickly developed water balance problems and switched to eating the green, succulent, mesquite pods rather than the seeds. Protein content, per se, may not be critical, however, if it is packaged with enough free water to offset the costs of protein digestion and excretion, as would be the case with ingested insects.

The basic carbohydrates, fats, and proteins that comprise seeds provide different metabolic water returns. While oxidation of fats produces more grams of water per gram of seed oxidized (1.07) than oxidation of carbohydrates (0.56) or proteins (0.39), carbohydrates provide the greatest return per kcal produced because lipids provide relatively more calories than water per gram than do carbohydrates. When the evaporative water loss associated with respiration required to oxidize these components is taken into consideration, however, the net amount of water gained changes significantly. For example, for *D. spectabilis,* carbohydrates provide the greatest return of metabolic water for each gram consumed (approximately 0.084 g), while lipids yield a slight deficit in water for each gram consumed, and protein metabolism and excretion generates considerable water loss (Frank, 1987*a*). Schmidt-Nielsen (1975) indicates that the few other heteromyids that have been tested for oxidative water loss yielded similar results.

It is reasonable to assume that heteromyids take water into consideration when choosing a diet. Lockard and Lockard (1971) and Price (1983*a*) found that heteromyids in general prefer high-carbohydrate seeds over oil- or protein-rich seeds and Kelrick et al. (1986) extended this to several desert granivore groups. Soholt (1977) found that *Dipodomys merriami* consumed enough food to yield 4.8 grams of water/d (from preformed and metabolic sources). When females were lactating, however, they took in 10.7 grams of water/d, providing an average of 2.1 g/d to each nursling. In addition, there was a high correlation between the amount of water taken in and the litter size of the individual (r^2 = 0.81; Soholt, 1977). Some Australian desert mammals, including rodents, are known to ingest the urine of their nurslings, thereby reclaiming up to one third of the water they lose during lactation (Baverstock and Green, 1975). This has not been investigated in heteromyids, although one might suspect this type of behavior to occur.

Kenagy (1972) found that *D. microps* is able to specialize on the vegetation of *Atriplex* by stripping off the salt-laden epidermis of the leaves and ingesting the remainder. Interestingly, Csuti (1979) found that individual *D. microps* which were trapped in areas without *Atriplex* quickly developed the epidermis-stripping behavior when given saltbush leaves to eat, whereas sympatric *D. ordii* did not. This suggests a genetic component to the propensity for developing this element of foraging behavior.

Some heteromyids exhibit a preference for slightly moldy seeds over very moldy and non-moldy seeds (Rebar and Reichman, 1983; Reichman and Rebar, 1985). This may be related to nutritional advantages of slightly moldy seeds (Reichman and Rebar, 1985), but recent work indicates that the levels of moldiness may serve as a conditioned clue to the moisture content of seeds (Frank, 1987*b*). Furthermore, some species of heteromyids appear to be able to distinguish between seeds that differ in moisture content by only 2%, always preferring those with the highest moisture content (Frank, 1987*a*). A number of heteromyid species cache seeds (Smith and Reichman, 1984), and in the high humidity of their burrows (Kay and Whitford, 1978; Reichman et al., 1985) the seeds absorb significant quantities of free water, perhaps tipping the water balance equation in favor of their preference (Morton and MacMillen, 1982).

Nutrition and toxins.—Very little is known about the nutritional requirements of heteromyids. Lockard and Lockard (1971) found preferences for specific types of seeds related to their nutritional content, especially the ratios of carbohydrates, lipids, and proteins, and Henderson (in litt.) found a positive relationship between seed selectivity by *D. ordii* and nitrogen content. One species, *Chaetodipus baileyi,* ingests the seeds of jojoba (*Simmondsia chinensis*) and apparently can metabolize the waxes it contains (Sherbrooke, 1976). M'Closkey (1983), however, did not find *C. baileyi* associated with the jojoba plant, suggesting that this pocket mouse species is not an obligate specialist on jojoba seeds.

Kenagy and Hoyt (1980) found some heteromyids to be coprophagic. The rodents tended to ingest the first fecal pellets produced after a feeding bout (those produced within 4 hours of feeding). These pellets were visually distinct from those produced later, and were found to be low in inorganic ions and relatively high in moisture and nitrogen. Coprophagy has been shown to increase nutrient extraction efficiency in other herbivores (Schmidt-Nielsen, 1975).

Green vegetation presumably provides a variety of nutrients and is frequently incorporated into the diet when seasonally available. In addition, greenery may stimulate reproduction directly (through stimulatory hormones; Beatley, 1969; Berger et al., 1981; Sanders et al., 1981), or indirectly (as a cue to future seed availability; Reichman and Van De Graaff, 1975; Van De Graaff and Balda, 1973).

Effects of Extrinsic Seed Characteristics

Availability and distribution.—Food availability obviously dictates the opportunity for diet preferences to be manifested, for animals can specialize on dietary items only if they are abundant enough to meet dietary requirements. Many heteromyid species occur in unpredictable environments across a broad geographic range and in a broad variety of habitats which support distinct plants and seeds. For example, *D.*

ordii ranges across 3,500 km of latitude from low desert to cold sagebrush deserts and arid grasslands. Because of this, heteromyids (indeed, all organisms) must be opportunistic. We might ask, however, to what extent do the sorts of food preferences we have just discussed actually affect diets in the field?

Most evidence suggests that heteromyid diets reflect what is available. Diets of members of single populations vary substantially between seasons and years in overall proportions of seeds, green vegetation, and insects (Reichman, 1975, 1978), essentially in concert with the temporal pattern of abundances of these general food categories.

The species of seeds taken also reflect patterns of availability. Overall use of forb and grass seeds changes seasonally in response to the timing of seed production (Brown et al., 1979; Reichman, 1975; Reichman and Van De Graaff, 1973), although the use of certain seeds may be abruptly curtailed by the seasonal appearance of a more preferred species. For example, McAdoo et al. (1983) noted that heteromyids switched from an almost pure diet of nonpreferred *Salsola paulsenii* (Russian Thistle) to a pure diet of highly preferred *Oryzopsis hymenoides* (Indian Rice Grass) when the latter became abundant. Rodents might also modify seed choice in response to environmental conditions or physiological requirements. For example, heteromyids might need to eat more high-carbohydrate seeds in the summer than in winter to satisfy metabolic water requirements. In the laboratory lactating females select a diet which increases metabolic water production (Soholt, 1977), but Alcoze and Zimmerman (1973), Hendersen (in litt.), and Reichman (1975) found no differences in the diets of males and females in the field, even during times when the latter were pregnant and lactating.

Reichman (1975) found a close match between the seeds available in the soil at points of capture and seeds ingested by four sympatric Sonoran Desert heteromyid species. The four species had quite similar diets overall, and interspecific differences were associated with habitat differences. *Dipodomys merriami* and *P. amplus,* which were sympatric on sandy flats, had very similar diets and their diets were distinct from those of *Chaetodipus baileyi* and *C. penicillatus,* which inhabited adjacent rocky hillsides and streambeds (Reichman, 1975). M'Closkey (1980) and O'Connell (1979) obtained similar results, and Hendersen (in litt.) has also noted a remarkable correlation between what is available and what is consumed.

It is reasonable to ask whether heteromyid populations are limited by seed availability. Because different species of heteromyids appear to specialize on specific seed distributions or microhabitats, specific subsets of seed availability may limit populations. For example, Brown and Munger (1985) found that when seeds of several sizes (barley seeds ground to different sizes) were added to plots, the density of *D. spectabilis* increased at the expense of other kangaroo rats, while smaller rodents were not affected. Furthermore, the average body size of the banner-tailed kangaroo rats also increased (although no increase in reproductive output was detected). While there may always be seeds in the soil, the costs associated with extracting them must vary between locations and seasons, which, in turn, must be partially responsible for the range in foraging behaviors exhibited by heteromyids.

Conversely, granivores, including heteromyids, have probably had a significant effect on the evolution of seed characteristics which govern their availability. Pulliam and Brand (1975) describe the relationship between several taxa of desert granivores and the types of seeds produced seasonally in the Sonoran Desert. Mares and Rosenzweig (1978) go even further by suggesting that in the absence of countering selection by two granivore taxa (e.g., ants and mammals), seeds could escape predation by moving toward the morphology that was least acceptable to a single consumer taxon. Hay and Fuller (1981) also discuss means by which seeds can reduce their losses to grani-

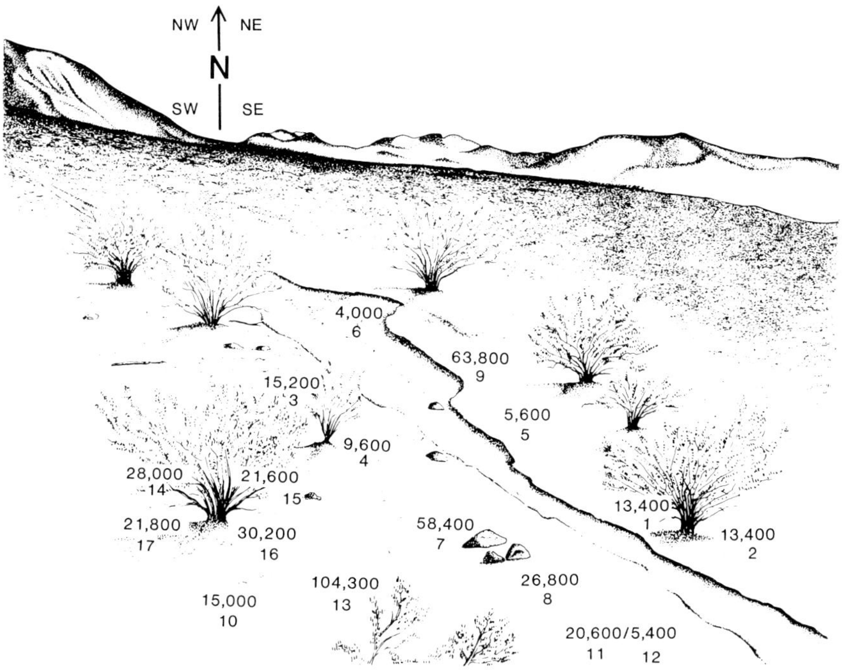

FIG. 3.—Average seed densities per m^2 in seventeen microhabitats in the Sonoran Desert. The values are extrapolated from numerous smaller soil samples or small seed traps as though they were for an entire square meter, but do not imply that such densities actually occur over an entire square meter (from Reichman, 1984). The microhabitats are 1—NW of *Larea* bush, 2—SE of *Larrea* bush, 3—NW of *Ambrosia* bush, 4—SE of *Ambrosia* bush, 5—area between bushes, 6—normally dry wash, 7—NW of small obstruction, 8—SE of obstruction, 9—natural depression, 10—area of extensive rodent digging, 11—upper 2 mm of soil, 12—upper 2 cm of soil, 13—seed trap in open area, 14—trap NW of *Larrea* bush, 15—trap NE of *Larrea* bush, 16—trap SE of *Larrea* bush, 17—trap SW of *Larrea* bush.

vores by evolving physical traits that promote their distribution into microhabitats which are unacceptable to rodents.

Extrinsic seed characteristics such as clumping and microhabitat distribution appear to be very important for heteromyid foraging. As noted in an earlier section, larger bipedal species preferentially harvest seeds from high density patches while quadrupedal pocket mice include low density patches. Applying economic principles to foraging, Price's (1983*b*) model of patch choice predicts that body size should generally affect the optimal degree of specialization on dense, profitable seed patches. One attempt to detect specializations on patch densities in the field (Frye and Rosenzweig, 1980) failed, perhaps because a single millet seed on the soil surface may be perceived as a profitable "patch" relative to what is normally available (cf. Bowers et al., 1987). Movement patterns of free-ranging animals (Table 2, Bowers, 1982; Thompson, 1982*a,* 1985) suggest that larger species are utilizing fewer of the available seed patches than are smaller species. Several species of heteromyids have been observed to gather seeds while they are still clumped on the parent plants (Reichman, 1983; Schroder, 1979).

The pronounced microhabitat affinities exhibited by heteromyids may, in part, be due to the distribution of seeds within various subsets of the habitat. Microhabitat

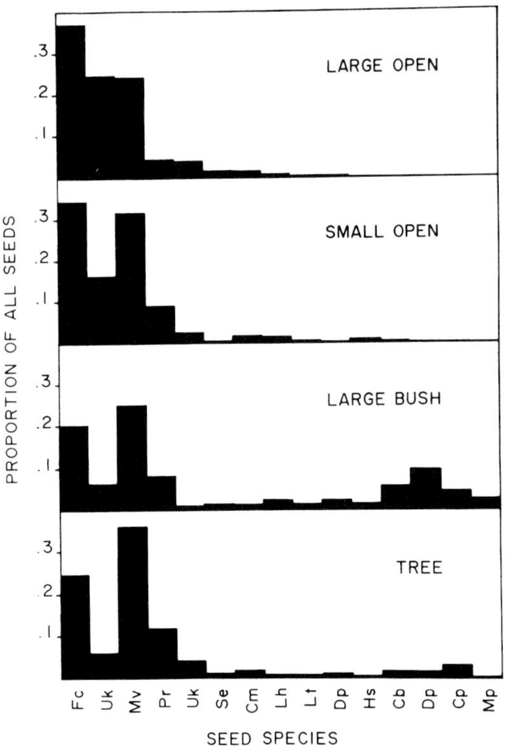

FIG. 4.—Relative abundances of 15 species of seeds in four microhabitats at a Sonoran Desert site. Histograms represent the proportional representation of the most common plants species in the seed pools of each microhabitat (number of seeds of species *i*/total number of seeds), based on total sample sizes of 9,203 seeds for Large Open, 12,773 for Small Open, 13,826 for Large Bush, and 11,912 for Tree microhabitats. Fc = *Filago californica,* Uk = Unknown, Mv = *Mollugo verticillata,* Pr = *Pectocarya recurvata,* Se = *Spermolepis echinata,* Cm = *Cryptantha micrantha,* Lh = *Lotus humistratus,* Lt = *L. tomentellus,* Dp = *Descurainia pinnata,* Cp = *Cryptantha pterocarpa,* Mp = *Muhlenbergia porteri.* Data are taken from Price and Reichman (1987), Appendix I.

specialization may occur for other reasons (e.g., predator avoidance; see above), and the microspatial distribution of seeds will influence what subset of all seeds produced are encountered. Price and Reichman (1987), Price and Waser (1985), Reichman (1984), and Reichman and Oberstein (1977) found distinct differences in seed densities (Fig. 3) and species composition (Fig. 4). Hence heteromyid species that forage in specific microhabitats should have different seeds available to them. Such differences in seed availability may explain differences between sympatric species in the composition of diets (cf. Reichman, 1975).

Availability and distribution of seeds are affected by the foraging effectiveness of closely and distantly related taxa of granivores (Brown et al., 1979). Brown and Davidson (1977) noted reciprocal increases when granivorous rodents or ants were removed from experimental plots and suspected that this was the result of alterations in seed availability left by the experimentally extracted taxon. Brown et al. (1979) and Reichman (1979) noted actual changes in seed densities and distributions as a result of ant and/or rodent removal on the same plots used by Brown and Davidson (1977). If seeds disperse through a series of distributions, from the parent plants, to relatively scattered, to relatively clumped, the infringement of any taxon along the distribution patch should affect all subsequent distributions, and the taxa that exploit them.

Predation.—Microhabitats, and hence certain types of seed distributions, also differ in predation risk, which may affect seed selectivity. If so, the foragers could be expected to be choosier in risky than in safe microhabitats. Hay and Fuller (1981) tested this prediction by placing artificial seed patches in presumably risky open spaces and under the protective canopy of shrubs. The difference in removal of preferred and nonpreferred seeds was significantly greater in the open areas than under the shrubs. The authors interpreted this pattern of selectivity as a response to the presumed increase in predator risk in the open. This interpretation is confounded, however, by the greater overall seed removal (caused by heavy rodent activity in general) under the shrubs; if the rodents removed preferred seeds first, then high seed removal rates under shrubs necessarily mean apparent selectivity will be lower. It would be useful to repeat this experiment with controls for overall removal rates.

Caching Behavior

Food caching occurs in many mammalian taxa (Smith and Reichman, 1984). It is a behavior, like torpor, that allows cachers to cope with periods of stress such as environmental extremes (e.g., winter) or low resource availability. Caching is an integral part of the cycle of entering torpor and arousing to feed exhibited by many small species of heteromyids, and for larger species, cached food may be the sole means of escaping seasonal extremes. Caching also removes food from the environment, and hence from potential competitors. Several important considerations impinge on the relationship between caching and feeding. One is that the act of gathering food and the consumption of food are separated in time and space. In addition, a caching animal is foraging for the future, essentially increasing its current demand (it requires enough food to satisfy its current requirements plus whatever amount it must secure for future use), perhaps requiring it to gather food to store that it might not choose to ingest were it only foraging for its current needs. Finally, cached food can be lost to thieves, competitors, and microbes (spoilage) before the harvest costs can be recouped (Smith and Reichman, 1984).

As is true for all caching species, heteromyids must decide whether to cache their harvested food, and what type of cache should be employed. Two features associated with heteromyids may have promoted the evolution of elaborate caching behaviors. All heteromyids have cheek pouches and those that have been studied carry food items back to their burrows. Caching behavior appears to be at least 10 million years old, as Voorhies (1974) has reported fossil evidence of that age showing pocket mice associated with seed caches. It is not clear how many species actually cache for extended periods, but the possession of cheek pouches certainly can promote seed gathering and transport back to burrows, whether the seeds are then eaten or stored. The use of seeds as a primary resource is also an advantage, as seeds have evolved traits for longevity and resistance to spoilage which promote their own survival in soil for years.

It appears that many heteromyid species cache in the field (Reichman et al., 1986) or in the lab (M. Hafner, in litt.). Although few reports exist of substantial caches from the burrows of pocket mice (Kenagy, 1972), they probably cache seeds to ingest during brief periods of arousal from torpor. Several of the large species of kangaroo rats are known to make extensive larderhoards (Hawbecker, 1940; Holdenreid, 1957; Monson, 1943; Monson and Kessler, 1940; Reichman et al., 1985; Tappe, 1941; Vorhies and Taylor, 1922). The extent of caching behavior in small kangaroo rats is poorly known, but they may use scatterhoards (Daly, in litt.). Even less is known about the caching behavior of *Liomys* and *Heteromys* (Fleming and Brown, 1975; Matson and Christian, 1977; Vandermeer, 1979).

Once heteromyids have chosen to cache resources, they must decide between two basic approaches (Smith and Reichman, 1984). In one, food is gathered and brought to a central location where it can be defended from theft. The second approach is to scatter hoard the resource at a density that is not high enough to promote the evolution of a pilfering specialist, but is sufficiently more dense than the background resource distribution to be profitable for the cacher to retrieve, especially given the advantage of some form of memory for the location of caches (Smith and Reichman, 1984; Stapanian and Smith, 1978, 1984). One could speculate that the smallest and largest species should larder hoard seeds, as the smallest species have narrow burrows that are unaccessible by larger forms, and the largest species can defend their caches against all thieves (Hafner, pers. comm.). The caches of the intermediate forms, such as the smallest kangaroo rats, would be susceptible to pilfering by smaller species, and larger species could sequester stored food by force. Therefore, the intermediate-sized species might be expected to scatter-hoard, thereby reducing the chance of losing all their seeds to one thief.

Generally, the literature supports this pattern of cache use. As predicted, the largest kangaroo rat species primarily larder hoard (see Reichman et al., 1985, and citations therein). What little in known about pocket mice suggests that they may also form larderhoards (Kenagy, 1972). Some smaller kangaroo rats appear to make dozens or even hundreds of small, shallow caches of seeds around their burrow openings (Daly, in litt.; Reynolds, 1958, 1960).

The most detailed information about caches is available for a few of the large kangaroo rats, perhaps because they make the largest and most spectacular caches (see Reichman et al., 1985 and the citations therein). These heteromyids cache in monospecific piles, apparently sorting harvested seeds into discrete caches. The sizes of the caches can be substantial, with some containing several liters and hundreds of thousands of seeds (Reichman et al., 1985). The banner-tailed kangaroo rat is also known to cache large quantities of grass seed heads cut into 3 cm lengths (Schroder, 1979).

The caches of *D. spectabilis* tend to be in large mounds that require up to two years to construct (Best, 1972), and which tend to be inherited between generations (Jones, 1984; Jones, 1993). In one study of *D. spectabilis,* caches tended to be approximately half way between the center of a mound and the edge, on the NW side, and concentrated in a narrow band at about 30 cm deep or at depths below 50 cm (Reichman et al., 1985). The authors attempted to identify a pattern to the sequence of cache use (e.g., whether the first material cached was the first used), but no pattern was obvious. Although the details of how long food is cached are unknown, comparisons between seeds in caches and the time of seed drop indicate that seeds may remain cached for several months (pers. obs.).

It is easy to imagine that the diet of an animal eating from its stored food would be different from that of an animal eating what it gathered each day, even if the array and quality of food available were the same. Optimal foraging theory yields hypotheses that suggest a forager should choose to ingest those food items which maximize its net return on the effort required to secure the items; in doing so, a forager would extract the highest quality items from the environment before a competitor can secure them. This same behavior should be true for a caching species when it is gathering resources for its cache. Once the cache is secured it should be relatively safe from theft. Thus one would predict that the diet of a forager eating items as they are gathered would exhibit low diversity as the animal consumed items with the highest net reward for the effort and ignored other items that are available but lie outside the acceptance criteria. An individual eating from a cache, however, could afford to eat a little of all items cached (perhaps maintaining a more balanced diet and avoiding being left with only poorer quality food at the end of the cache-use period) as there would be little pressure from competitors and thieves. This would be an appropriate strategy for any forager, but is probably a strategy not available to an individual continually competing for resources above ground.

To test the hypothesis that a caching species would have a more diverse diet than a non-caching species feeding on the same items, Reichman and Fay (1983) compared the diets of a pocket mouse (*Chaetodipus intermedius*) and a deer mouse (*Peromyscus maniculatus*). The results revealed that the caching pocket mice maintained a diverse diet during the period of cache use and each day ate some of each of five food items presented during the experiment. The deer mice, conversely, exhibited a relatively low diet diversity, and depleted the two most favored dietary items in the first four days of the 10-day experimental trial.

Recent investigations have identified at least 30 species of fungi associated with the cheek pouches of some heteromyids and the caches of *D. spectabilis* (Reichman et al., 1985). The burrow environments are ideal for the growth of fungi and the production of numerous beneficial and harmful products, from nutritional enhancements to toxins and carcinogens. Details of the degree of mold growth and production of chemicals are poorly known, but it is clear that there is the potential for significant post-harvest alteration of seeds by fungal action and for the fungi themselves to affect the quality of a diet garnered from stored seeds.

The nature of the growth of the fungi in the burrows is such that moderate levels of moldiness could generate beneficial byproducts, making this level of mold growth more advantageous than non-moldy seeds, and less toxic than very moldy seeds. Preference experiments for seeds of varying levels of moldiness revealed that, indeed, *Chaetodipus intermedius* and *D. spectabilis* preferred seeds exhibiting intermediate degrees of fungal infestation over sterile, non-moldy seeds and very moldy seeds (Rebar and Reichman, 1983; Reichman and Rebar, 1985). There has been no analysis of this type of preference in the field, and recent, more detailed experiments analyzing preferences for moldy seeds indicate that the preferences may be based on moisture content of the seed, with molds serving as a cue to moisture levels (Frank, 1987*b*).

If there are advantages to intermediate levels of moldiness in cached seeds, and if heteromyids do exhibit preferences for these levels, it is reasonable to predict that caches might be managed by heteromyids to promote appropriate degrees of fungal growth. Laboratory experiments have shown that *D. spectabilis* tends to move sterile seeds to high humidity locations. The same individuals move seeds molded to preferred levels to humidities below which the fungi show slowed growth rates and no mycotoxin production (Reichman et al., 1986). The results from the above experiments can be interpreted as an attempt by the rodents to promote beneficial mold growth on sterile seeds and to inhibit further growth on seeds that have reached a preferred level of mold growth.

An interesting feature of the caching behavior of heteromyids has come from the above analyses of the caches. It has been noted in the field and laboratory that seeds in the presence of rodents do not germinate, while those kept in similar conditions without rodent supervision show extensive germination (Reichman et al., 1985). The mechanism for this inhibition is not known, but preliminary investigations have begun.

Discussion

In this review of heteromyid foraging ecology we first outlined the general foraging options that heteromyids adopt and the morphological, physiological, and environmental features that appear to be involved in defining the broad options. We noted, for example, that heteromyids possess skeletal features that seem to facilitate locomotion and predator avoidance on the ground while reducing climbing ability. We also noted that seeds represent the mainstay of the heteromyid diet and that their tooth and jaw structure are well suited to such a diet. Furthermore, heteromyids possess external cheek pouches that facilitate the transport of particulate food items such as seeds, and hence allow the rodents to buffer themselves against fluctuating food availability by caching seeds when they are abundant.

Environmental conditions interact with morphology and physiology as further determinants of heteromyid foraging options. The characteristic ephemeral life history of many desert plants makes available a reservoir of dormant seeds, and heteromyids possess the requisite locomotory, olfactory, and digging abilities to exploit this resource. Generally, harsh daytime temperatures and low moisture conditions probably favor nocturnal activity during much of the year, and this has molded organ systems involved in water balance, such as kidneys, as well as mechanisms of temperature regulation. These modifications in turn dictate which diets, and which activity periods, allow heteromyids to maintain water and energy balance.

While heteromyids are somewhat restricted to a set of foraging behaviors compatible with their nocturnal, burrowing, seed-eating life-style, the individual animal still faces a series of foraging decisions; when and where to forage, which food items to harvest upon encounter, and how to manage harvested food. As a means of understanding these decisions, we considered, where possible, how the fitness of the individual might be affected by alternative behavioral options. This approach is efficacious because if one can specify how the behavior of an individual affects its ability to survive and reproduce then one can develop predictive models on how animals should behave. The basic premise of this "optimality" approach to behavior is that those individuals which behave in a way that, on average, allows them to produce more successful offspring than individuals which adopt other behaviors will be favored by natural selection and eventually the entire population will exhibit the "optimal" behavior.

Clearly, successful organisms must accommodate a variety of selection pressures, some of which will conflict with one another. A fundamental premise of the optimality approach is that morphological and physiological characteristics of an animal define how well it can perform fitness-related tasks—that is, the net gain in fitness it would realize when engaged in a particular activity under a particular set of environmental conditions. The net fitness gain in turn defines the value to the animal of a particular behavioral option relative to others. Hence, if we can specify what behavioral options are available to an animal, and what the relative fitness value of each is, we should be able to predict what the animal would do.

Two components of fitness—reproductive success and survival probability—should be affected by foraging behavior. It is well known that female reproductive efforts and sizes of heteromyid populations are tied to the patterns of seed production (see review in Munger et al., 1983). This suggests that food availability limits individual reproductive success. If so, then the amount of food an individual obtains is very likely tied to its fitness.

The acquisition of food, however, also incurs risks. Environmental conditions are generally harsher above ground than in burrows, and the probability of being captured by a predator should be directly related to the amount of time spent out of the burrow. Risk may not be zero in the burrow—some snakes and mammals can enter burrows—but in all probability the suite of predators outside the burrows make risks associated with foraging very high. Hence, the time spent foraging should directly affect survival probabilities, and individuals which restrict their foraging to less risky times or locations, or reduce their total foraging time, should have higher fitness than less careful individuals.

These considerations make it feasible to develop optimality models to analyze heteromyid foraging behavior. The difficulty of doing so is obvious: to specify the fitness return for all behavioral options is an overwhelmingly difficult task. Hence, we may never be able to predict behavior precisely except, perhaps, in the most simple laboratory settings. The optimality approach is nonetheless extremely useful as a strategy for developing precise and testable hypotheses about determinants of behavior (Krebs and Davies, 1984; Stephens and Krebs, 1986). We urge that it be applied to the study of heteromyid behavior, not only because it is an efficient research tool, but also because any success in developing predictive models for heteromyids would be of great interest to behavioral ecologists in general, and not just to heteromyid specialists.

Heteromyids possess a number of characteristics that make them excellent subjects for sophisticated behavioral analyses. They are abundant and relatively easy to catch, they forage willingly in the presence of investigators, and they are amenable to manipulation in the field and laboratory. For these reasons, heteromyids, especially desert species, are among the ecologically best known terrestrial organisms. Nevertheless, much remains to be learned about the factors that impinge on the broad patterns of behavior that have been uncovered. We suggest that the following areas of investigation would lead to a greater understanding of the evolutionary and ecological forces that have molded the foraging behavior of heteromyids in particular, and animals in general.

Analysis of geographic variation within a single species similar to Csuti's (1979) excellent study of *D. microps* should indicate major features of the environment that affect the evolution of heteromyids. Comparisons of tropical and desert heteromyids would also be useful in this regard. For example, tropical forms have to deal with relatively widely dispersed, large seeds, which may be quite toxic while desert forms are probably more water-stressed, and must garner many small seeds to meet their requirements. These environmental difference may have promoted divergence between tropical and desert forms in physiological tolerances, foraging methods, morphology, and patterns of seed selection.

Information on spatial and temporal activity patterns and time budgets, detailing the behaviors of individuals over extended periods of time, is important. Long term studies of marked individuals similar to those of Behrends (1984), Behrends et al. (1986*a*, 1986*b*), Jones (1984 and this volume), and Randall (1984) should yield insight into the social structure of populations. If these types of analyses were tied to data on seed densities and distributions, and the actions of predators, the results would be even more revealing.

Although we have a good idea of broad patterns of microhabitat use and morphology in desert heteromyid communities, much less is known about the proximate and ultimate causes of these patterns (Price, 1986). Detailed mechanistic analyses of relationships between morphology, predator avoidance, and foraging efficiency under a variety of conditions are required to tease apart the effects of predation costs and foraging economics on microhabitat use.

There is especially sound experimental field evidence that coexisting heteromyids compete (cf. Brown and Harney, this volume), but less information is available about interactions with more distantly related taxa (Brown et al., 1979; Rebar and Conley, 1983) or about the mechanisms of any of the interspecific interactions. It is difficult to design experiments that unambiguously reveal the importance of direct aggression relative to exploitation under field conditions. Perhaps the degree to which competitive effects between different-sized species are symmetrical will provide a clue (Price, 1986).

The relative importance of intrinsic and extrinsic seed factors for diet choice is not known. The potential for seed types to provide nutritional and metabolic requirements is known, but these features have not been tied to actual diets in the field. Even less is known about the presence of toxins in desert seeds and their possible effects on diet choice.

The consequences of cheek pouches and cache use on foraging decisions are not known for heteromyids. Investigations of the forces generating scatter- or larder-hoards would be pertinent, and almost nothing is known about the schedule of use of cached seeds. It is possible that economic models of inventory control would be helpful in this regard.

We anticipate that research on heteromyid foraging will continue to reveal important features about ecological characteristics and interactions within this taxon, and between heteromyids and other taxa. Systematic studies of heteromyid foraging ecology should ultimately allow a thorough understanding of the forces that have caused the family to diversify in body size, mode of locomotion, and spatial foraging patterns while remaining remarkably uniform in basic diet and foraging methods.

Acknowledgments

We thank James H. Brown for stimulating our interest in heteromyid ecology with his seminal work on this group and with his personal enthusiasm for the subject. E. L. Cockrum and H. R. Pulliam introduced MVP to the heteromyid system and theoretical evolutionary ecology, respectively. Mark Hafner provided important insight into heteromyid caching behavior. We thank N. M. Waser for critical input and M. L. Daly for critiquing a draft of the manuscript; thorough reviews by Burt Kotler and an anonymous reviewer also strengthened the manuscript. Support from several National Science Foundation grants to one or both authors provided an opportunity to investigate heteromyids and MVP further acknowledges the financial support of the University of California Academic Senate. This manuscript was originally submitted on December 23, 1986. Minor changes were last incorporated in March, 1989.

Literature Cited

Alcoze, T., and E. Zimmerman. 1971. Food habits and dietary overlap of two heteromyid rodents from the mesquite plains of Texas. Journal of Mammalogy, 54:900–908.

Bagnold, R. A. 1954. The physics of blown sand and desert dunes. Methuen and Co., Ltd., London, 345 pp.

Bartholomew, G. R., and G. A. Cary. 1954. Locomotion in pocket mice. Journal of Mammalogy, 35:386–392.

Bartholomew, G. A., and H. Caswell. 1951. Locomotion in kangaroo rats and its adaptive significance. Journal of Mammalogy, 32:155–169.

Baverstock, P., and B. Green. 1975. Water recycling in lactation. Science, 187:657–658.

Behrends, P. 1984. Spatiotemporal activity patterns of Merriam kangaroo rats (*Dipodomys merriami*). Ph.D. dissertation, McMaster University, Hamilton, Ontario, 324 pp.

———. 1986. Utilization of seed dispersions by two sympatric kangaroo rat species. Southwestern Naturalist, 31:548–551.

Behrends, P., M. Daly, and M. Wilson. 1986*a*. Aboveground activity of Merriam's kangaroo rat (*Dipodomys merriami*) in relation to sex and reproduction. Behaviour, 96:210–226.

———. 1986*b*. Range use patterns and spatial relationships of Merriams's kangaroo rat (*Dipodomys merriami*). Behaviour, 96:187–209.

Beatley, J. 1969. Dependence of desert rodents on winter annuals and precipitation. Ecology, 50:721–724.

Berger, P., N. Negus, E. Sanders, and P. Gardner. 1981. Chemical triggering of reproduction in *Microtus montanus*. Science, 214:69–70.

Best, T. 1972. Mound development by a pioneer population of banner-tailed kangaroo rats, *Dipodomys spectabilis baileyi* Goldman in eastern New Mexico. American Midland Naturalist, 87:201–206.

Best, T., and B. Hoditschek. 1982. Analysis of cheek pouch contents of Ord's Kangaroo rat (*Dipodomys ordii*). Southwestern Naturalist, 27:117–124.

Blaustein, A., and A. Risser. 1974. Dominance relationships of the dark kangaroo mouse (*Microdipodops megacephalus*) and the little pocket mouse (*Perognathus longimembris*) in captivity. Great Basin Naturalist, 34:312–316.

———. 1976. Interspecific interactions between three species of kangaroo rats (*Dipodomys*). Animal Behaviour, 24:381–385.

Bonoff, M. B., and D. H. Janzen. 1980. Small terrestrial rodents in eleven habitats in Santa Rosa National Park, Costa Rica. Brenesia, 17:163–174.

Bowers, M. A. 1982. Foraging behavior in heteromyid rodents: field evidence for resource partitioning. Journal of Mammalogy, 63:361–367.

———. 1988. Seed removal experiments on desert rodents: the microhabitat by moonlight effect. Journal of Mammalogy, 69:201–204.

Bowers, M. A., D. B. Thompson, and J. H. Brown. 1987. The spatial organization of a desert rodent community: food addition and species removal. Oecologia, 72:77–82.

Bradley, W., and R. Mauer. 1971. Reproduction and food habits of Merriam's Kangaroo rat, *Dipodomys merriami*. Journal of Mammalogy, 52:497–507.

Brand, M. R. 1975. Determination of rodent diet using a microtechnique of fecal analysis. M.S. thesis, University of Arizona, Tucson, 87 pp.

Braun, S. 1985. Home range and activity patterns of the giant kangaroo rat, *Dipodomys ingens*. Journal Mammalogy, 66:1–12.

Brown, J. H. 1973. Species diversity of seed-eating rodents in sand dune habitats. Ecology, 54:775–787.

———. 1975. Geographical ecology of desert rodents. Pp. 315–341, *in* Ecology and evolution of communities (M. L. Cody and J. Diamond, eds.). Belknap Press, Cambridge, Massachusetts, USA, 545 pp.

Brown, J. H., and G. A. Bartholomew. 1969. Periodicity and energetics of torpor in the Kangaroo mouse, *Microdipodops pallidus*. Ecology, 50:705–709.

BROWN, J. H., AND D. W. DAVIDSON. 1977. Competition between seed-eating rodents and ants in desert ecosystems. Science, 196:880–882.

BROWN, J. H., D. W. DAVIDSON, AND O. J. REICHMAN. 1979. An experimental study of competition between seed-eating desert rodents and ants. American Zoologist, 19:1129–1143.

BROWN, J. H., AND B. A. HARNEY. 1993. Population and community ecology of heteromyid rodents in temperate habitats. Pp. 618–651, *in* Biology of the Heteromyidae (H. H. Genoways and J. H. Brown, eds.). Special Publication, American Society of Mammalogists, 10:1–719.

BROWN, J. H., AND G. LIEBERMAN. 1973. Resource utilization and coexistence of seed-eating desert rodents in sand dune habitats Ecology, 54:788–797.

BROWN, J. H., AND J. C. MUNGER. 1985. Experimental manipulation of a desert rodent community: food addition and species removal. Ecology, 66:1545–1563.

BROWN, J. H., O. J. REICHMAN, AND D. W. DAVIDSON. 1979. Granivory in desert ecosystems. Annual Review of Ecology and Systematics, 10:201–227.

BROWN, J. S., B. P. KOTLER, R. J. SMITH, AND W. O. WIRTZ II. 1987. The effects of owl predation on the foraging behavior of heteromyid rodents. Oecologia, 76:408–415.

CHAPMAN, B. R. 1972. Food habits of Loring's Kangaroo rat, *Dipodomys elator*. Journal of Mammalogy, 53:877–880.

CHEW, R. M., AND B. B. BUTTERWORTH. 1964. Ecology of rodents in Indian Cove (Mojave Desert), Joshua Tree Monument California. Journal of Mammalogy, 45:203–225.

CHEW, R. M., AND A. CHEW. 1970. Energy relationships of the mammals of a desert shrub (*Larrea tridentata*) community. Ecological Monographs, 40:1–21.

CLARK, W., AND P. COMANOR. 1973. The use of western harvester ant, *Pogonomyrmex occidentalis* (Cresson), seed stores by heteromyid rodents. Occasional Papers Biological Society of Nevada, 34:1–6.

CLARKE, J. A. 1983. Moonlight's influence on predator–prey interactions between short-eared owls (*Asio flammeus*) and deermice (*Peromyscus maniculatus*). Behavioral Ecology and Sociobiology, 13:205–209.

CONGDON, J. 1974. Effect of habitat quality on distributions of three sympatric desert rodents. Journal of Mammalogy, 55:659–662.

COVICH, A. P. 1976. Analyzing shapes of foraging areas: some ecological and economic theories. Annual Review of Ecology and Systematics, 7:235–257.

CSUTI, B. 1979. Patterns of adaptation and variation in the Great Basin Kangaroo rat (*Dipodomys microps*). University of California Publications in Zoology, 111:1–69.

DICE, L. R. 1945. Minimum intensities of illumination under which owls can find dead prey by sight. American Naturalist, 79:385–416.

———. 1947. Effectiveness of selection by owls of deer-mice (*Peromyscus maniculatus*) which contrast in color with their background. Contributions to the Laboratory of Vertebrate Biology, University of Michigan, 34:1–20.

DON, B. A. C., AND K. RENNOLLS. 1983. A home range model incorporating biological attraction points. Journal of Animal Ecology, 52:69–81.

DUNHAM, M. K. 1968. A comparative food habit study of two species of kangaroo rats, *Dipodomys ordii* and *Dipodomys merriami*. M.S. thesis, University of New Mexico, Albuquerque, 61 pp.

EISENBERG, J. F. 1963. The behavior of heteromyid rodents. University of California Publications in Zoology, 69:1–100.

FLAKE, L. 1973. Food habits of four species of rodents on a short-grass prairie in Colorado. Journal of Mammalogy, 54:636–647.

FLEMING, T. 1970. Notes on the rodent faunas of two Panamanian forests. Journal of Mammalogy, 51:473–490.

———. 1974. The population ecology of two species of Costa Rican heteromyid rodents. Ecology, 55:493–510.

FLEMING, T., AND G. BROWN. 1975. An experimental analysis of seed hoarding and burrowing behavior in two species of Costa Rican heteromyid rodents. Journal of Mammalogy, 56:301–315.

FRANK, C. 1987*a*. The influence of moisture content on heteromyid rodent seed preferences. Journal of Mammalogy, 69:358–362.

———. 1987*b*. The effects of moldiness level on seed preferences of *Dipodomys spectabilis*. Journal of Mammalogy, 69:353–357.

FRENCH, A. 1976. Selection of high temperatures for hibernation by the pocket mouse *Perognathus longimembris*: ecological advantages and energetic consequences. Ecology, 57:185–191.

———. 1993. Physiological ecology of the Heteromyidae: economics of energy and water utilization. Pp. 509–538, *in* Biology of the Heteromyidae (H. H. Genoways and J. H. Brown, eds.). Special Publication, American Society of Mammalogists, 10:1–719.

FRYE, R. 1983. Experimental field evidence of interspecific aggression between two species of kangaroo rat (*Dipodomys*). Oecologia, 59:74–78.

FRYE, R., AND M. L. ROSENZWEIG. 1980. Clump size selection: a field test with two species of *Dipodomys*. Oecologia, 47:323–327.

GARLAND, T. 1983*a*. Scaling the ecological costs of transport to body mass in terrestrial mammals. American Naturalist, 121:571–587.

———. 1983*b*. The relations between maximal running speed and body mass in terrestrial mammals. Journal of Zoology, London, 199:157–170.

GETTY, T. 1981. Analysis of central-place space-use patterns: the elastic disc revisited. Ecology, 62:907–914.

GRINNELL, J. 1932. Habitat relations of the giant kangaroo rat. Journal of Mammalogy, 13:305–320.

HAFNER, J. C., AND M. S. HAFNER. 1983. Evolutionary relationships of heteromyid rodents. Pp. 3–29, *in* Biology of desert rodents (O. J. Reichman and J. H. Brown, eds.). Great Basin Naturalist Memoirs, No. 7, 134 pp.

HANSSON, L. 1970. Methods of morphological diet microanalysis in rodents. Oikos, 21:255–267.

HARDY, R. 1945. The influence of types of soil upon the local distribution of some mammals in southeastern Utah. Ecological Monographs, 15:71–108.

HARESTAD, A. S., AND F. L. BUNNELL. 1979. Home

range and body weight—a reevaluation. Ecology, 60: 389–402.
HARRIS, J. 1984. Experimental analysis of desert rodent foraging ecology. Ecology, 65:1578–1584.
———. 1986. Microhabitat segregation in two desert rodent species: the relation of prey availability to diet. Oecologia, 68:417–421.
HAWBECKER, A. 1940. The burrowing and feeding habits of *Dipodomys venustus*. Journal of Mammalogy, 22:388–396.
———. 1944. The giant kangaroo rat and sheep forage. Journal of Wildlife Management, 8:161–165.
HAY, M., AND P. FULLER. 1981. Seed escape from heteromyid rodents: the importance of microhabitat and seed preference. Ecology, 62:1395–1399.
HEINZ, K. M. 1983*a*. Scratch-digging adaptations in kangaroo rats. American Zoologist, 23:890.
———. 1983*b*. Species replacement along elevational gradients. M.S. thesis, University of California, Riverside.
HINDS, D. S., AND R. MACMILLEN. 1985. Scaling of energy metabolism and evaporative water loss in heteromyid rodents. Physiological Zoologist, 58:282–298.
HOLDENREID, R. 1957. Natural history of the banner-tailed kangaroo rat in New Mexico. Journal of Mammalogy, 38:330–350.
HOOVER, K., W. WHITFORD, AND P. FLAVILL. 1977. Factors influencing the distribution of two species of *Perognathus*. Ecology, 58:877–884.
HOWELL, A. B. 1932. The saltatorial rodent *Dipodomys*: the functional and comparative anatomy of its muscular and osseous systems. Proceedings of the American Academy of Arts and Sciences, 67:377–536.
HUTTO, R. 1978. A mechanism for resource allocation among sympatric heteromyid rodent species. Oecologia, 33:115–126.
JANZEN, D. 1982. Removal of seeds from horse dung by tropical rodents: influence of habitat and amount of dung. Ecology, 63:1887–1900.
———. 1986. Mice, big mammals, and seeds: it matters who defecates what where. Pp. 251–271, *in* Frugivores and seed dispersal (A. Estrada and T. H. Fleming, eds.). Dr. W. Junk Publishers, Dordrecht, 392 pp.
JENKINS, S. H. 1981. Common patterns in home range–body size relationships of birds and mammals. American Naturalist, 118:126–128.
JONES, W. T. 1982. Natal nondispersal in kangaroo rats. Ph.D. dissertation, Purdue University, West Lafayette, Indiana, 185 pp.
———. 1984. Natal philopatry in bannertailed kangaroo rats. Behavioral Ecology and Sociobiology, 15: 151–155.
KAUFMAN, D. W., AND G. KAUFMAN. 1982. Effect of moonlight on activity and microhabitat use by Ord's Kangaroo rat (*Dipodomys ordii*). Journal of Mammalogy, 63:309–312.
KAY, F., AND W. WHITFORD. 1978. The burrow environment of the banner-tailed kangaroo rat, *Dipodomys spectabilis*, in south-central New Mexico. American Midland Naturalist, 99:270–279.
KELRICK, M. I., A. MACMAHON, R. R. PARMENTER, AND D. V. SISSON. 1986. Native seed preferences of shrub-steppe rodents, birds, and ants: the relationship of seed attributes and seed used. Oecologia, 68:327:337.
KENAGY, G. J. 1972. Saltbush leaves: excision of hypersaline tissue by a kangaroo rat. Science, 178:1094–1096.
———. 1973. Daily and seasonal patterns of activity and energetics in a heteromyid rodent community. Ecology, 54:1201–1219.
———. 1976. Field observations of male fighting, drumming, and copulation in the Great Basin Kangaroo rat, *Dipodomys microps*. Journal of Mammalogy. 57:781–785.
KENAGY, G. J., AND D. HOYT. 1980. Reingestion of feces in rodents and its daily rhythmicity. Oecologia, 44:403–409.
KOTLER, B. 1984*a*. Effects of illumination on the rate of resource harvesting in a community of desert rodents. American Midland Naturalist, 11:383–389.
———. 1984*b*. Risk of predation and the structure of desert rodent communities. Ecology, 65:689–701.
———. 1985*a*. Microhabitat utilization in desert rodents: a comparison of two methods of measurement. Journal of Mammalogy, 66:374–378.
———. 1985*b*. Owl predation on desert rodents which differ in morphology and behavior. Journal of Mammalogy, 66:824–828.
KOTLER, B. P., J. S. BROWN, R. J. SMITH, AND W. O. WIRTZ II. 1988. The effects of morphology and body size on rates of owl predation on desert rodents. Oikos, 53:145–152.
KREBS, J. R., AND N. B. DAVIES. 1984. Behavioral ecology: an evolutionary approach. Sinauer Associates, Sunderland, Massachusetts, 493 pp.
KRITZMAN, E. B. 1974. Ecological relationships of *Peromyscus maniculatus* and *Perognathus parvus* in eastern Washington. Journal of Mammalogy, 55:172–188.
LANGFORD, A. 1983. Pattern of nocturnal activity of male *Dipodomys ordii* (Heteromyidae). Southwestern Naturalist, 28:341–346.
LATOURRETTE, J. E., J. A. YOUNG, AND R. A. EVANS. 1971. Seed dispersal in relation to rodent activities in seral big sagebrush communities. Journal of Range Management, 24:118–120.
LARSEN, E. 1986. Competitive release in microhabitat use among coexisting desert rodents: a natural experiment. Oecologia, 69:231–237.
LAWHON, D., AND M. HAFNER. 1981. Tactile discriminatory ability and foraging strategies in kangaroo rats and pocket mice (Rodentia: Heteromyidae). Oecologia, 50:303–309.
LAY, D. M. 1974. Differential predation of gerbils (*Meriones*) by the little owl, *Athena brahma*. Journal of Mammalogy, 55:608–614.
LEMEN, C. 1978. Seed size selection in heteromyids: a second look. Oecologia, 35:13–19.
LEMEN, C., AND P. W. FREEMAN. 1983. Quantification of competition among coexisting heteromyids in the southwest. Southwestern Naturalist, 28:41–46.
———. 1985. Tracking animals with fluorescent pigments: a new technique. Journal of Mammalogy, 66: 134–136.
———. 1986. Habitat selection and movement patterns in sandhill rodents. Prairie Naturalist, 18:129–141.
———. 1987. Competition for food and space in a

heteromyid community in the Great Basin Desert. Great Basin Naturalist, 46:1–6.

LEMEN, C., AND M. L. ROSENZWEIG. 1978. Microhabitat selection in two species of heteromyid rodents. Oecologia, 33:127–135.

LOCKARD, R. 1978. Seasonal change in the activity pattern of *Dipodomys spectabilis*. Journal of Mammalogy, 59:563–568.

LOCKARD, R., AND J. LOCKARD. 1971. Seed preference and buried seed retrieval of *Dipodomys deserti*. Journal of Mammalogy, 52:219–221.

LOCKARD, R., AND D. OWINGS. 1974. Seasonal variation in moonlight avoidance by bannertail kangaroo rats. Journal of Mammalogy, 55:189–193.

LONGLAND, W. S. 1989. Risk of predation and microhabitat choice by heteromyid rodents. Ph.D. dissertation, University of California, Riverside, 187 pp.

LOWE, C., AND D. S. HINDS. 1971. Effects of paloverde (*Cercidium*) trees on the radiation flux at ground level in the Sonoran Desert in winter. Ecology, 52: 916–922.

MACMILLEN, R. 1983. Adaptive physiology of heteromyid rodents. Great Basin Naturalist Memoirs, 7:65–76.

MACMILLEN, R., AND D. S. HINDS. 1983. Water regulatory efficiency in heteromyid rodents: a model and its application. Ecology, 64:152–164.

MARES, M. A., AND M. L. ROSENZWEIG. 1978. Granivory in North and South American deserts: rodents, birds and ants. Ecology, 59:235–241.

MARES, M. A., AND D. WILLIAMS. 1977. Experimental support for food particle size resource allocation in heteromyid rodents. Ecology, 58:1186–1190.

MATSON, J., AND D. CHRISTIAN. 1977. A laboratory study of seed caching in two species of *Liomys* (Heteromyidae). Journal of Mammalogy, 58:670–671.

MAZA, B. G., N. R. FRENCH, AND A. P. ASCHWANDEN. 1973. Home range dynamics in a population of heteromyid rodents. Journal of Mammalogy, 54:405–425.

MCADOO, J. K., C. C. EVANS, B. A. ROUNDY, J. A. YOUNG, AND R. A. EVANS. 1983. Influences of heteromyid rodents on *Oryzopsis hymenoides* germination. Journal of Range Management, 36:61–64.

M'CLOSKEY, R. T. 1978. Niche separation and assembly in four species of Sonoran Desert rodents. American Naturalist, 112:683–694.

———. 1980. Spatial patterns in types of seeds collected by four species of heteromyid rodents. Ecology, 61:488–489.

———. 1981. Microhabitat use in coexisting desert rodents: the role of population density. Oecologia, 50:310–315.

———. 1983. *Perognathus baileyi* and jojoba (*Simmondsia chinensis*): a test of their association. Journal of Mammalogy, 64:499–501.

MCNAB, B. K. 1963. Bioenergetics and the determination of home range size. American Naturalist, 97:133–140.

MESERVE, P. L. 1976. Habitat and resource utilization by rodents of a California coastal sage scrub community. Journal of Animal Ecology, 45:647–666.

MONSON, G. 1943. Food habits of bannertail kangaroo rats in Arizona. Journal of Wildlife Management, 7:98–102.

MONSON, G., AND W. KESSLER. 1940. Life history notes on the bannertail kangaroo rats, Merriam's kangaroo rat and the white-throated woodrat in Southern Arizona and New Mexico. Journal of Wildlife Management, 4:37–43.

MORTON, S., D. S. HINDS, AND R. E. MACMILLEN. 1980. Cheek pouch capacity in heteromyid rodents. Oecologia, 46:143–146.

———. 1982. Seeds as a source of preformed water for desert-dwelling granivores. Journal of Arid Environments, 5:61–67.

MOTT, J. J., AND A. J. MCCOMB. 1974. Patterns in annual vegetation and soil microrelief in an arid region of western Australia. Journal of Ecology, 62: 115–126.

MUNGER, J. C., M. A. BOWERS, AND W. T. JONES. 1983. Desert rodent populations: factors affecting abundance, distribution, and genetic structure. Great Basin Naturalist Memoirs, 7:91–116.

MUNGER, J. C., AND J. H. BROWN. 1981. Competition in desert rodents: an experiment with semi-permeable exclosures. Science, 211:510–512.

NELSON, J., AND R. CHEW. 1977. Factors affecting seed reserves in the soil of a Mojave Desert ecosystem, Rock Valley, Nye County, Nevada. American Midland Naturalist, 97:300–320.

NIKOLAI, J., AND D. BRAMBLE. 1983. Morphology, structure, and function in desert heteromyid rodents. Pp. 44–63, *in* Biology of desert rodents (O. J. Reichman and J. H. Brown, eds.). Great Basin Naturalist Memoirs, No. 7, 134 pp.

O'CONNELL, M. A. 1979. Coexistence of two species of kangaroo rats (genus *Dipodomys*) in the Guadeloupe Mountains National Park, Texas. Pp. 349–371, *in* Biological investigation in the Guadeloupe Mountains National Park, Texas (H. H. Genoways and R. J. Baker, eds.). National Park Service, Washington, D.C.

O'FARRELL, M. J. 1974. Seasonal activity patterns of rodents in a sagebrush community. Journal of Mammalogy, 55:809–823.

———. 1978. Home range dynamics of rodents in a sagebrush community. Journal of Mammalogy, 58: 657–668.

———. 1980. Spatial relationships of rodents in a sagebrush community. Journal of Mammalogy, 61: 589–605.

PINKHAM, C. F. A. 1973. The evolutionary significance of locomotory patterns in the Mexican pocket mouse, *Liomys irroratus*. Journal of Mammalogy, 54:742–746.

PODOLSKY, R. H., AND M. V. PRICE. 1990. Patch use by *Dipodomys deserti* (Rodentia: Heteromyidae): profitability, preference, and depletion dynamics. Oecologia, 83:83–90.

PRICE, M. W. 1977. The validity of live trapping as a measure of foraging activity of heteromyid rodents. Journal of Mammalogy, 58:107–110.

———. 1978*a*. Seed dispersion preferences of coexisting desert rodent species. Journal of Mammalogy, 59:624–626.

———. 1978*b*. The role of microhabitat in structuring desert rodent communities. Ecology, 59:910–921.

———. 1983*a*. Laboratory studies of seed size and species selection by heteromyid rodents. Oecologia, 60:259–263.

———. 1983*b*. Ecological consequences of body size: a model of patch choice in desert rodents. Oecologia, 59:384–392.
———. 1984. Microhabitat use in rodent communities: predator avoidance or foraging economics? Netherlands Journal of Zoology, 34:63–80.
———. 1986. Structure of desert rodent communities: a critical review of questions and approaches. American Zoologist, 26:39–49.
PRICE, M. W., AND J. H. BROWN. 1983. Patterns of morphology and resource use in North American desert rodent communities. Pp. 117–134, *in* Biology of desert rodents (O. J. Reichman and J. H. Brown, eds.). Great Basin Naturalist Memoirs No. 7, 134 pp.
PRICE, M. V., AND K. HEINZ. 1984. Effects of body size, seed density, and soil characteristics on rates of seed harvest by heteromyid rodents. Oecologia, 61: 420–425.
PRICE, M. V., AND S. H. JENKINS. 1986. Rodents as seed consumers and dispersers. Pp. 191–235, *in* Seed dispersal (D. R. Murray, ed.). Academic Press (Australia), North Ryde, Australia.
PRICE, M. V., AND K. A. KRAMER. 1984. On measuring microhabitat affinities with special reference to small mammals. Oikos, 42:349–354.
PRICE, M. V., AND W. S. LONGLAND. 1989. Use of artificial seed patches by heteromyid rodents. Journal of Mammalogy, 70:316–322.
PRICE, M. V., AND R. H. PODOLSKY. 1989. Mechanisms of seed harvest by heteromyid rodents: soil texture effects on harvest rate and seed size selection. Oecologia, 81:267–273.
PRICE, M. V., AND O. J. REICHMAN. 1987. Distributions of seeds in Sonoran Desert soils: implications for heteromyid rodent foraging. Ecology, 68: 1797–1811.
PRICE, M. V., AND N. M. WASER. 1984. On the relative abundance of species: post-fire changes in a coastal sage scrub rodent community. Ecology, 65: 1161–1169.
———. 1985. Microhabitat use by heteromyid rodents: effects of artificial seed patches. Ecology, 66: 211–219.
PRICE, M. V., N. M. WASER, AND T. A. BASS. 1984. Effects of moonlight on microhabitat use by desert rodents. Journal of Mammalogy, 65:353–356.
PULLIAM, H. R. 1976. The principle of optimal behavior and the theory of communities. Pp. 311–322, *in* Perspectives in ethology, Vol. 2 (P. P. G. Bateson and P. H. Klopfer, eds.). Plenum Press, New York, 340 pp.
PULLIAM, R., AND M. BRAND. 1975. The production and utilization of seeds in plains grassland of southeastern Arizona. Ecology, 56:1158–1166.
RANDALL, J. A. 1984. Territorial defense and advertisement by foot drumming in bannertail kangaroo rats (*Dipodomys spectabilis*) at high and low population densities. Behavioral Ecology and Sociobiology, 16:11–20.
REBAR, C., AND W. CONLEY. 1983. Interactions in microhabitat use between *Dipodomys ordii* and *Onychomys leucogaster*. Ecology, 64:984–988.
REBAR, C., AND O. J. REICHMAN. 1983. Ingestion of moldy seeds by heteromyid rodents. Journal of Mammalogy, 64:713–715.
REICHMAN, O. J. 1975. Relationships of desert rodent diets to available resources. Journal of Mammalogy, 56:731–751.
———. 1976. Relationships between dimensions, weights, volume and calories of some Sonoran Desert seeds. Southwestern Naturalist, 20:573–575.
———. 1977. Optimization of diets through food preferences by heteromyid rodents. Ecology, 58:454–457.
———. 1978. Ecological aspects of the diets of Sonoran Desert rodents. Museum of Northern Arizona Research Reports No. 15, 96 pp.
———. 1979. Desert granivore foraging and its impact on seed densities and distributions. Ecology, 60: 1085–1092.
———. 1981. Factors influencing foraging in desert rodents. Pp. 195–213, *in* Foraging behavior: ecological, ethological, and psychological approaches (A. Kamil and T. Sargent, eds.). Garland-STMP Press, New York, 534 pp.
———. 1983. Behavior of heteromyid rodents. Great Basin Naturalist Memoirs, 7:77–90.
———. 1984. Spatial and temporal variation in seed distributions in desert soils. Journal of Biogeography, 11:1–11.
REICHMAN, O. J., AND J. H. BROWN. 1979. The use of torpor by *Perognathus amplus* in relation to resource distribution. Journal of Mammalogy, 60:550–555.
REICHMAN, O. J., A. FATTAEY, AND K. FATTAEY. 1986. Management of sterile and mouldy seeds by a desert rodent. Animal Behaviour, 34:221–225.
REICHMAN, O. J., AND P. FAY. 1983. Comparison of the diets of a caching and non-caching rodent. American Naturalist, 122:576–581.
REICHMAN, O. J., AND D. OBERSTEIN. 1977. Selection of seed distribution types by *Dipodomys merriami* and *Perognathus amplus*. Ecology, 58:636–643.
REICHMAN, O. J., I. PRAKASH, AND V. ROIG. 1979. Food selection and consumption. Pp. 681–716, *in* Arid-land ecosystems: structure, functioning, and management, Vol. 1 (R. A. Perry and D. W. Goodall, eds.). Cambridge University Press, 881 pp.
REICHMAN, O. J., AND C. REBAR. 1985. Seed preferences by desert rodents based on levels of mouldiness. Animal Behaviour, 33:726–729.
REICHMAN, O. J., AND K. VAN DE GRAAFF. 1973. Seasonal reproductive and activity patterns of five species of Sonoran Desert rodents. American Midland Naturalist, 90:118–126.
———. 1975. Influence of green vegetation on desert rodent reproduction. Journal of Mammalogy, 53: 503–506.
REICHMAN, O. J., D. WICKLOW, AND C. REBAR. 1985. Ecological and mycological characteristics of caches in the mounds of *Dipodomys spectabilis*. Journal of Mammalogy, 66:643–651.
REYNOLDS, H. G. 1958. Ecology of Merriam's kangaroo rat on the grazing lands of southern Arizona. Ecological Monographs, 28:111–127.
———. 1960. Life history notes on Merriam's kangaroo rat in southern Arizona. Journal of Mammalogy, 41:48–58.
REYNOLDS, H. G., AND H. S. HASKELL. 1949. Life history notes on Price and Bailey pocket mice of southern Arizona. Journal of Mammalogy, 30:150–156.
ROSENZWEIG, M. L. 1973. Habitat selection experi-

ments with a pair of coexisting heteromyid species. Ecology, 54:111–117.
———. 1974. On the optimal above ground activity of bannertail kangaroo rats. Journal of Mammalogy, 55:193–199.
———. 1981. A theory of habitat selection. Ecology, 62:327–335.
ROSENZWEIG, M. L., B. SMIGEL, AND A. KRAFT. 1975. Patterns of food, space, and diversity. Pp. 241–268, *in* Rodents in desert environments (I. Prakash and P. K. Ghosh, eds.). Dr. Junk b. v., Publishers, The Hague, The Netherlands, 624 pp.
ROSENZWEIG, M. L., AND P. STERNER. 1970. Population ecology of desert rodent communities: body size and seed husking as a basis for heteromyid coexistence. Ecology, 51:217–224.
ROSENZWEIG, M. L., AND J. WINAKUR. 1969. Population ecology of desert rodent communities: habitats and environmental complexity. Ecology, 50:558–571.
ROWLAND, R. H., AND F. B. TURNER. 1964. Correlation of the local distribution of *Dipodomys microps* and *Dipodomys merriami* and the annual grass *Bromus rubens*. Southwestern Naturalist, 9:56–61.
SANDERS, E., P. GARDNER, P. BERGER, AND N. NEGUS. 1981. 6-methoxybenzoxazolinone: a plant derivative that stimulates reproduction in *Microtus montanus*. Science, 214:67–69.
SCHMIDT-NIELSEN, K. 1964. Desert animals: physiological problems of heat and water. Clarendon Press, Oxford, England, 277 pp.
———. 1975. Animal physiology. Cambridge University Press, Oxford, England, 699 pp.
SCHMIDT-NIELSEN, K., B. SCHMIDT-NIELSEN, AND A. BROKAW. 1948. Urea excretion in desert rodents exposed to high protein diets. Journal of Cell Physiology, 32:361–380.
SCHOENER, T. W. 1968. Sizes of feeding territories among birds. Ecology, 49:123–141.
SCHREIBER, R. K. 1978. Bioenergetics of the Great Basin pocket mouse, *Perognathus parvus*. Acta Theriologica, 23:469–487.
SCHRODER, G. 1979. Foraging behavior and home range utilization of the bannertail kangaroo rat (*Dipodomys spectabilis*). Ecology, 60:657–665.
———. 1987. Mechanisms for coexistence among three species of *Dipodomys*: habitat selection and an alternative. Ecology, 68:1071–1073.
SCHRODER, G., AND K. GELUSO. 1975. Spatial distribution of *Dipodomys spectabilis* mounds. Journal of Mammalogy, 56:363–368.
SHAW, W. T. 1934. The ability of the giant kangaroo rat as a harvester and storer of seeds. Journal of Mammalogy, 15:275–286.
SHERBROOKE, W. C. 1976. Differential acceptance of toxic jojoba seed (*Simmondsia chinensis*) by four Sonoran Desert heteromyid rodents. Ecology, 57: 596–602.
SHREVE, F., AND I. L. WIGGINS. 1964. Vegetation and flora of the Sonoran Desert. Vol. 1. Stanford University Press, Stanford, California.
SMIGEL, B., AND M. L. ROSENZWEIG. 1974. Seed selection in *Dipodomys merriami* and *Perognathus penicillatus*. Ecology, 55:329–339.
SMITH, C. 1942. The fall food of brushfield pocketmice. Journal of Mammalogy, 23:337–339.
SMITH, C. C., AND O. J. REICHMAN. 1984. The evolution of food caching behavior by birds and mammals. Annual Review of Ecology and Systematics, 15:329–351.
SOHOLT, L. 1973. Consumption of primary production by a population of kangaroo rats (*Dipodomys merriami*) in the Mojave Desert. Ecological Monographs, 43:357–376.
———. 1977. Consumption of herbaceous vegetation and water during reproduction and development of Merriam's kangaroo rat, *Dipodomys merriami*. American Midland Naturalist, 98:445–457.
STAMP, N., AND R. OHMART. 1978. Resource utilization by desert rodents in the Lower Sonoran Desert. Ecology, 55:329–339.
STAPANIAN, M. A., AND C. C. SMITH. 1978. A model for seed scatterhoarding: coevolution of fox squirrels and black walnuts. Ecology, 59:884–896.
———. 1984. Density dependent survival of scatterhoarded nuts: an experimental approach. Ecology, 65:1387–1396.
STEPHENS, D. W., AND J. R. KREBS. 1986. Foraging theory. Princeton University Press, Princeton, New Jersey, 252 pp.
TAPPE, D. T. 1941. Natural history of the Tulare kangaroo rat. Journal of Mammalogy, 22:117–148.
TEVIS, L. 1958. Germination and growth of ephemerals induced by sprinkling a sandy desert. Ecology, 39:681–688.
THOMPSON, S. D. 1982*a*. Microhabitat utilization and foraging behavior of bipedal and quadrupedal heteromyid rodents. Ecology, 63:1303–1312.
———. 1982*b*. Structure and species composition of desert heteromyid species assemblages: effects of a simple habitat manipulation. Ecology, 63:1313–1321.
———. 1985. Bipedal hopping and seed-dispersion selection by heteromyid rodents: the role of locomotion energetics. Ecology, 66:220–229.
TROMBULAK, S. C., AND G. J. KENAGY. 1980. Effects of seed distribution and competitors on seed harvesting efficiency in heteromyid rodents. Oecologia, 44:342–346.
VAN DE GRAAFF, K., AND R. P. BALDA. 1973. Importance of green vegetation for reproduction in the kangaroo rat, *Dipodomys merriami*. Journal of Mammalogy, 54:509–512.
VANDERMEER, J. H. 1979. Hoarding behavior of captive *Heteromys desmarestianus* (Rodentia) of fruits of *Welfia georgi*, a rainforest dominant plant in Costa Rica. Brenesia, 16:107–116.
VOORHIES, M. R. 1974. Fossil pocket mouse burrows in Nebraksa. American Midland Naturalist, 91:429–498.
VORHIES, C. T., AND W. P. TAYLOR. 1922. Life history of the kangaroo rat *Dipodomys spectabilis spectabilis* Merriam. Bulletin of the United States Department of Agriculture, 109:1–40.
WASER, P. M. 1981. Sociality or territorial defense? The influence resource renewal. Behavioral Ecology and Sociobiology, 8:231–237.
———. 1988. Resources, philopatry, and social interactions among mammals. Pp. 109–130, *in* Ecology of social behavior (N. Slobodchikoff, ed.). Academic Press, San Diego, California, 429 pp.
WEBSTER, D. B., AND M. WEBSTER. 1971. Adaptive value of hearing and vision in kangaroo rat predator avoidance. Brain, Behaviour and Evolution, 4:310–322.

WILSON, D. S. 1975. The adequacy of body size as a niche difference. American Naturalist, 109:769–784.
WONDOLLECK, J. T. 1978. Forage-area separation and overlap in heteromyid rodents. Journal of Mammalogy, 59:510–518.
YORK, C. L. 1949. Notes on home ranges and population density of two species of heteromyid rodents in southwestern Texas. Texas Journal of Science, 1:42–46.

THE SOCIAL SYSTEMS OF HETEROMYID RODENTS

W. Thomas Jones

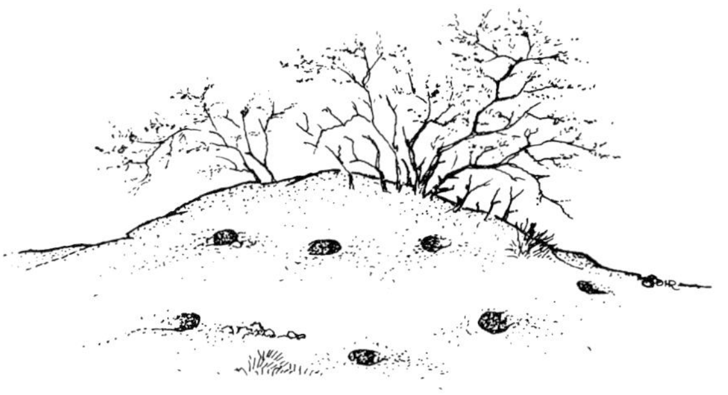

Introduction

Although heteromyid rodents have yielded much information on community ecology and physiology, until recently few studies have directly addressed aspects of social behavior in this family. Yet the characteristics that make heteromyid species well suited to studies of community ecology and interspecific competition are advantageous for investigations of social behavior and intraspecific competition as well. They are readily trapped, resources can be defined and measured for many species, and most species occupy open habitats where direct observation is possible.

In this chapter I review what is currently known of heteromyid social behavior. The first section describes some of the behavior patterns important in social interactions and the forms of communication. Following this, for each of the six heteromyid genera I review three aspects of social organization. First, what are the forms of their spacing systems? Are most species territorial? Do groups ever occur? Second, what are the forms of their mating systems, that is, are they monogamous, polygynous, or promiscuous? Third, what are the patterns of dispersal? Is dispersal biased by sex, and if so in what direction? Is natal dispersal delayed in any species? Finally, Bekoff et al. (1981), Armitage (1981) and Eisenberg (1981) have noted correspondences between life history variables and the degree of sociality in mammals. I describe the results of similar analyses for heteromyids that characterize broad trends in variation in sociality in this family.

Much of the information summarized in this review only indirectly answers questions about the nature of heteromyid social systems. The best evidence of territoriality is direct observation of defense of burrows or space in the field. However, one can infer territorial behavior from aggression in laboratory encounters, regular dispersion of home ranges, or the absence of home range overlap (Table 1). A description of mating systems ideally would include a comparison of variation in annual and lifetime reproductive success of males and females, but one can infer monogamy from observations of bisexual pairs sharing dens or home ranges. Indications of polygyny or promiscuity include larger home ranges or other evidence of greater mobility in males than in females, or possibly male-biased size dimorphism. As Ralls (1977) noted, polygyny in mammals is not always characterized by discrete harem groups of a single male and several females, but instead it may take a more dispersed form in which each male's range encompasses those of several females. Promiscuity may occur when each male's home range overlaps several females' and each female home range is overlapped by several males. The existence of copulatory plugs is an indication that several males compete for access to each female and that a mating system is potentially promiscuous. The desired dispersal data include the timing of dispersal and distributions of natal to breeding site distances for juveniles of each sex and breeding site to breeding site distances for adults. In the absence of this information, the proportion of each sex recaptured in mark–recapture studies provides some indication of the direction and degree of any sex bias in dispersal.

Behavior Patterns and Communication

Behavior Patterns

Most of our knowledge of basic behavior patterns in heteromyids comes from Eisenberg's (1963*a*) careful study of 13 heteromyid species. In laboratory encounters of two conspecifics, contact was generally preceded by a slow approach in which the animals assumed elongate postures with eyes partially closed and ears folded. Initial contacts then usually took two basic forms: naso–nasal contact and naso–anal contact. In naso–nasal contact one animal may place its head under the forepaws of the other in a submissive gesture with its eyes half-closed and its ears back. In naso–anal contact one animal may crawl under the other or there may be mutual perineal examination so that the pair assumes a circling movement. Through these patterns the areas investigated are usually glandular and moist, and thus they are potential sources of information during an encounter.

Agonistic encounters often involve rushing by one animal. The rushing animal may strike its opponent with its forepaws and body, and if the opponent flees, a chase may occur with the rushing animal attempting to bite the retreating animal's rump. If an attacked animal does not flee, there may be locked fighting in which the two animals grip each other with all four limbs, chew at each other's fur, and scratch and kick with their hind legs. Locked fights are brief and usually followed by chasing. Other agonistic behaviors include, but are not limited to: sandkicking, in which one animal kicks sand in another's face; the attack leap, in which one animal leaps into the air and lands on another with its hind limbs; the escape leap, wild erratic jumps to avoid a rush; jockeying, in which the animals turn and hop back and forth with both individuals in an upright posture (*Dipodomys* and *Microdipodops* only); warding, an upright posture with outstretched forepaws to ward off an attacker; and sparring, in which two individuals in warding posture jump back and forth to strike one another with their forepaws. Locked-fighting is more common in the quadrupedal genera, while warding and sparring are more common in the bipedal *Dipodomys* and *Microdipodops.* In withdrawal or submission the animal assumes a rounded hunched posture, it moves slowly, its eyes may be half-closed, and its ears may be folded back.

Sexual behavior patterns typically begin with driving, in which the male follows the female and sniffs the substrate behind her. The female often marks (perineal drag) and sandbathes while the male is driving. During driving the female suddenly assumes lordosis, a posture in which she freezes and raises her hindquarters allowing the male to mount. As the male begins to mount he grooms the female and often pats her rump with his forepaws. In mounting he grips the female's flanks with his forelimbs, and in

TABLE 1.—*Summary of spacing and dispersal patterns in heteromyids.*

Species	Evidence of territoriality	Home range overlap	Home range size	Dispersal pattern
H. anomalus		extensive		
L. salvini	home ranges regularly dispersed			
L. adspersus	home ranges randomly dispersed		distance between captures greater in males	
C. fallax		intersexual, but not intrasexual		
C. formosus		intersexual, intrasexual	male > female	no sex bias
P. longimembris			distance between captures greater in males	no sex bias
P. parvus			male > female	
M. megacephalus		large overlap around small territories	male > female	
D. merriami	defense of a core area observed	male/female; male/male > female/female	male = female	male bias in long movement
D. ordi		male/male > female/female	male = female	
D. agilis		male/female; male/male > female/female	male = female	
D. microps		male = female	no sex bias	
D. heermani	observations of burrow defense			female bias in long movement
D. panamintinus		male/female; male/male > female/female	male = female	
D. spectabilis	observations of burrow defense	no female/female	male > female	female bias in long movement
D. ingens		male/female; no female/female	male = female	

Dipodomys he grasps the female's fur on her back with his incisors. Probably several intromissions must precede an intromission with ejaculation. In some cases an estrous female may initiate sexual contact by rushing and striking the male with her forepaws and then assuming lordosis. Engstrom and Dowler (1981) reported field observations of mating in *Dipodomys ordii* that included many of the behaviors that Eisenberg (1963*a*) described. Naso–anal contact was followed by running and circling by the male and some sparring. Three unsuccessful copulation attempts were followed by 30 sec and 3 min copulations. Allan (1944) released two male and one female *D. ordii* in his garage and saw the female pursue one of the males. Again naso–anal contact and sparring were involved. After 13 unsuccessful copulation attempts, there were two copulations each exceeding 2 min. Similar behavior was reported for *Dipodomys merriami* and *Dipodomys spectabilis,* although in the latter species ejaculation may occur with a single intromission (Randall, 1987*a*, 1991*b*).

Eisenberg (1963*a*) stated that heteromyids in general are not "contact animals" (p. 48). Most encounters do not involve physical contact, and even in male-female encounters, ". . . mutual avoidance and agonistic behaviors predominate unless the female is in estrus" (p. 49). Later laboratory observations supported these findings (Eisenberg, 1967). Bisexual heteromyid pairs nested together only when the female was in estrus, in contrast to other rodents in which the male and female often nested together through rearing. Laboratory-raised litters of *Liomys pictus* broke up because of aggression among litter mates and their mothers (Eisenberg 1963*a*). The same was true of *Perognathus californicus*. However, "*Dipodomys* litters did not show extreme avoidance and aggression by the female or aggressive interaction among littermates. . . . There seems to be a social flexibility among some species of *Dipodomys* that is not present in the other (heteromyid) genera studied" (Eisenberg, 1963*a*, p. 61).

Communication

Eisenberg (1963*a*) observed visual, olfactory, and auditory communication among heteromyids in the laboratory. He noted poor eyesight in *Heteromys*, but he found that heteromyids did respond to the visual stimulus of a conspecific. Ward and Randall (in press) demonstrated that *D. spectabilis* responded more strongly to a visual stimulus (a stuffed conspecific) than to an auditory stimulus (playbacks of foot-drumming).

Olfactory communication appears to be important. Besides the olfactory information that is exchanged in naso–nasal and naso–anal contact, olfactory information is also exchanged at sites of sandbathing, perineal dragging (marking), and urination. Urination sites in a burrow and perineal dragging sites are investigated by intruders. In laboratory encounters there was a tendency to investigate and sandbathe in sites where another animal had sandbathed (Eisenberg, 1963*a*). Sandbathing is a stereotypical behavior in heteromyids that may have originated from a combination of stretching and the perineal drag (Eisenberg, 1963*b*). Typically, the animal first ". . . digs rapidly with its forepaws in the substrate. It then lowers its cheeks to the sand and extends its body while sliding forward on its side. Alternate extension and flexion of the body results in a series of side-rubs. In other cases, the extension and flexion may be performed with the ventrum pressed against the sand: in this instance the movement is termed a ventral-rub. In general an animal will perform isolated side-rubs or ventral-rubs, but anextended sequence of sandbathing usually includes both acts" (Eisenberg, 1963*b*, p. 17).

There are some data on what information is exchanged through olfactory communication in *Dipodomys*, and this varies among species. Randall (1981) determined that *Dipodomys merriami* and *Dipodomys microps* were attracted to sandbathing loci of conspecifics rather than to those of other species, which might indicate that these kangaroo rats can identify species from sandbathing deposits. In *D. merriami* female reproductive status is not communicated to males through sandbathing deposits (Lepri and Randall, 1983). Randall (1985) also ruled out scent gland secretions and urine as indicators of female reproductive condition in *D. merriami*. She suggested that this information may be contained in vaginal secretions and that this would explain at least one role of the naso–anal contact prior to mating observed by Eisenberg (1963*a*). As in *D. merriami*, *D. spectabilis* males do not learn female reproductive condition from sandbathing deposits (Randall, 1987*b*), although experiments by Laine and Griswold (1976) suggested an ability to recognize sex from sandbathing deposits. But in contrast to *D. merriami*, *D. spectabilis* males can distinguish urine of estrous and non-estrous females (Randall, 1986). Sandbathing in *D. spectabilis* and *D. merriami* probably functions as a territorial scent-marking behavior—it is independent of gonadal control and reproductive condition and sandbathing rates are independent of sex and age (Randall, 1987*b*, 1991*a*, 1991*b*). Both sexes can distinguish the sandbathing deposits of familiar vs. unfamiliar conspecifics.

Auditory communication includes both vocal and nonvocal means. Gibbs (1955) reported hearing a high-pitched "peeee" sound lasting about a second in *D. spectabilis*. He witnessed exchanges of the sound between different individuals in the wild, and he actually saw one individual emitting the sound. Eisenberg (1963*a*) reported hearing a "repeated low-intensity, high-pitched squeal of around two seconds in length" (p. 39) from a female *Dipodomys panamintinus*. The function of these calls (and whether they are homologous) is unknown, but Eisenberg stated that ". . . the female generally uttered this cry while digging at the corner of the divider (in her enclosure) when the male was resting directly opposite her" (p. 39). Eisenberg also described a "scratchy growl" given during nest defense or harassment by a conspecific. When an animal is attacked the growl may become a squeal of higher pitch and intensity. A low grunt may be given when an animal is startled or escaping a conspecific.

Another means of auditory communication is foot-drumming with the hind feet. It is performed with one foot repeatedly striking the substrate. It usually occurs when the animal is aroused, such as during aggression (Eisenberg, 1963*a*). Randall (1984, 1989*b*) studied foot-drumming in *D. spectabilis*. Of 6,949 footrolls ($n = 39$ animals), 20.6% were in response to neighbors' footrolls, 16.5% were given during challenges of burrow ownership, and no stimulus was evident for 62.9% of footrolls. Highest rates of foot-drumming occurred during challenges of burrow ownership. From playback experiments Randall determined that *D. spectabilis* spent more time near the speaker in response to footrolls of strangers than to those of neighbors and more time near the speaker in response to neighbors and strangers than to themselves. She suggested that *D. spectabilis* may be able to recognize neighbor-stranger classes and even specific individuals from footrolls (Randall, 1989*a*). Ward and Randall (in press) argued that *D. spectabilis* foot-drumming may be a long distance territorial display with a function similar to visual displays and singing in birds. Randall and Stevens (1987) found that *D. spectabilis* foot-drum at high rates in response to snakes but not in response to owls. They concluded that foot-drumming also functions as an individual alarm signal that probably informs a snake it has been detected and thus may cause it to leave.

Spacing Patterns, Mating Systems, and Dispersal

Heteromys

Eisenberg (1963*a*) suggested that *Heteromys* species are generally "somewhat (more) social" than other heteromyids and that they are characterized by greater intraspecific tolerance. In laboratory encounters Eisenberg (1963*a*) documented less agression in *Heteromys anomalus* than in other heteromyids. Rood and Test (1968) reported extensive home range overlap in this species. They suggested that territorial behavior is absent or poorly developed. Wagner (1961) reported field observations of *Heteromys goldmani* living in pairs during the reproductive season. He also described groups of six or more individuals sharing a burrow during seasons of reproductive inactivity. However, he did not determine whether such groups were comprised of related or unrelated individuals. The most detailed study of *Heteromys* social structure is that of Fleming (1974*a*, 1974*b*) for *Heteromys desmarestianus*. He showed that, like *H. anomalus*, *H. desmarestianus* are more tolerant

TABLE 2.—*Life history data for 30 heteromyids in order of increasing adult size.*

	Adult weight (gm)	Neonate weight (gm)	Litter size	Gestation time (days)	Age at maturity (days)	Maximum lifespan (years)	Age at eye opening (days)	Age at weaning (days)	Age adult weight reached (days)	G + AEO	RE	|RI|	|SRE|	MW:FW
Perognathus														
P. flavus	8		3.8	26.0		2.5	15.5			41.5				
P. longimembris	8	0.9	4.0	23.0		5.0	14.5	23.0	55	37.5	3.6	0.10	0.5	1.0
P. fasciatus	9		5.0	28.0										
P. flavescens	10		4.1											
P. amplus	12		3.3											
P. parvus	19	1.5	5.3	23.5	330	4.0					8.0	0.10	0.4	1.1
Microdipodops														
M. megacephalus	12		4.5											1.0
M. pallidus	13	1.0	3.5								3.5	0.10	0.3	
Chaetodipus														
C. intermedius	15		4.5											
C. nelsoni	16		2.9											
C. penicillatus	18		3.8	26.0			14.0			40.0				1.1
C. formosus	19	1.0	5.6		365	4.0					5.6	0.10	0.3	1.1
C. fallax	20		4.0	25.0										
C. californicus	23	1.5	3.8	25.0			15.0	23.0		40.0	5.7	0.10	0.2	1.1
C. baileyi	28		4.0											1.2
C. hispidus	39		5.5											
Liomys														
L. pictus	44	2.5	3.9	25.0			19.5	26.0		44.5	9.8	0.10	0.2	
L. salvini	45	1.9	3.9	27.0	105	3.0	14.0	24.5	105	41.0	7.4	0.04	0.2	1.3
L. adspersus			3.6		90	1.8								
Heteromys														
H. desmarestianus	73	3.0	3.3	27.0	237	2.5	19.0	24.5	180	46.0	9.9	0.04	0.1	1.4
Dipodomys														
D. nitratoides	34	3.3	2.3	32.0			12.0	22.5		44.0	7.6	0.10	0.2	
D. merriami	40	3.2	2.5	30.0	60	2.0	14.6	20.3	203	44.6	8.0	0.10	0.2	1.1
D. ordii	52		3.0	29.0										1.0

TABLE 2.—*Continued.*

	Adult weight (gm)	Neonate weight (gm)	Litter size	Gestation time (days)	Age at maturity (days)	Maximum lifespan (years)	Age at eye opening (days)	Age at weaning (days)	Age adult weight reached (days)	G + AEO	RE	|RI|	|SRE|	MW:FW
D. microps	59	4.0	2.3	31.0							9.2	0.10	0.2	1.1
D. heermanni	60	3.7	3.2			2.4	15.0	23.0			11.8	0.10	0.2	1.2
D. stephensi	66	4.4	2.7				14.0	20.0			11.9	0.10	0.2	
D. panamintinus	69	4.5	3.5	29.0			17.5	28.0		46.5	15.8	0.10	0.2	
D. deserti	100		3.6	31.0										
D. spectabilis	123	7.8	2.4	27.0	300	6.0	14.0		365	41.0	18.7	0.10	0.2	1.1
D. ingens	154		5.5											1.0
CV	89	63.0	25.0	10.0	61	34.0	14.0	11.0	68	7.0	47.0	23.00	42.0	10.0

References utilized in Table 2: Alcorn, 1941; Arnold, 1942; Asdell, 1964; Bailey, 1931; Banfield, 1974; Bradley and Mauer, 1971; Breyen et al., 1973; Butterworth, 1961*b*; Chew, 1958; Chew and Butterworth, 1959, 1964; Conley et al., 1977; Culbertson, 1946; Dale, 1939; Daly et al., 1984; Davis, 1966; Day et al., 1956; Doran, 1952; Eisenberg, 1963*a*, 1981; Eisenberg and Isaac, 1963; Fitch, 1948; Fleming, 1971, 1974*a*, 1977; French et al., 1967; French et al., 1974; Grinnell, 1932; Grinnell et al., 1930; Hall and Kelson, 1959; Hayden and Gambino, 1966; Hayden et al., 1966; Holdenreid, 1957; Johnston, 1956; Jones, 1982, unpublished data; Kenagy, 1973; Kenagy and Bartholomew, 1981, 1985; Lackey, 1967; Lemen 1978; Lidicker, 1960; McCulloch and Inglis, 1961; M'Closkey, 1978; Miller and Stebbins, 1964; O'Farrell, 1978; O'Farrell et al., 1975; Reichman and Brown, 1979; Reichman and van de Graaff, 1973; Reynolds, 1960; Reynolds and Haskell, 1949; Rosenzweig and Sterner, 1970; Scelza and Knoll, 1980; Scheffer, 1938; Schmidly, 1971, 1977; Smith and Jorgensen, 1975; Tappe, 1941; Trombulak and Kenagy, 1980; Tucker, 1966; Vorhies and Taylor, 1922; Wagner, 1961; Wilson et al., 1985.

of conspecifics in laboratory encounters than are other heteromyids. In his field studies Fleming (1974*a*) did not report groups like those in *H. goldmani,* but he stated that "... there is no strong evidence ... that repulsion plays an important part in determining the spatial distributions of individuals ..." and that this suggests that "... mutually exclusive home ranges do not exist in this species" (Fleming, 1974*a,* p. 555).

The details of the mating system of *H. desmarestianus* are not clear. It is among the most strongly dimorphic species within the Heteromyidae, with males larger than females (Table 2). However, Fleming (1974*a*) reported that size was not a good indicator of dominance in males and that there were no significant correlations between male weight and the number of adjacent females. Female weight was uncorrelated with dominance but positively correlated with home range size.

Liomys

The tendencies toward intraspecific social tolerance present in *Heteromys* species do not characterize the other tropical genus, *Liomys.* Both Eisenberg (1963*a*) and Fleming (1974*a*) demonstrated greater aggression in *Liomys* species than in *Heteromys* species in laboratory encounters. Wagner (1961) reported that, although both sexes of *Liomys pictus* may share dens during nonbreeding seasons, reproductively active females live alone and maintain distinct home ranges. Fleming (1974*a*) found that adult *L. salvini* home ranges tended toward a regular dispersion. In *L. adspersus* "... home ranges were oriented randomly with respect to members of the same or opposite sex" (Fleming, 1971, p. 53).

Available evidence on the mating system of *L. salvini* suggests that they are promiscuous or effectively polygynous. They are more strongly dimorphic than most other heteromyids, with males larger than females (Table 2). Fleming (1974*a*) showed that male size was a good predictor of dominance in the lab. His field data suggested that larger males tend to have more female neighbors than smaller males. Survivorship was higher in larger males. Intrasexual aggression was significantly higher in males than in females in laboratory encounters. In *L. adspersus* the average and maximum distances between captures were significantly greater for males than for females. Males, but not females, made significantly longer movements during the breeding season than at other times of year (Fleming, 1971).

Chaetodipus

Like *Liomys, Chaetodipus* species for the most part appear to have a social system in which all adults live alone in separate home ranges. *Chaetodipus californicus* was among the most aggressive of six heteromyid species tested in laboratory encounters by Eisenberg (1963*a*). Field data thus far show no evidence of social groupings in any *Chaetodipus* species. MacMillen (1964) reported very little intrasexual home range overlap in *Chaetodipus fallax,* though there was intersexual overlap. Maza et al. (1973) reported both intrasexual and intersexual overlap in male and female *Chaetodipus formosus.*

Information on mating systems exists for one *Chaetodipus* species, *C. formosus.* In this case greater mobility in males than in females (Table 1) suggests polygyny or promiscuity. Males' home ranges were larger than females' in one study (Maza et al., 1973). In another study mean distance between farthest capture points was greater in males than in females (Allred and Beck, 1963). Maza et al. (1973) reported that male *C. formosus* make more long distance excursions than do females during periods of sexual activity (170 excursions for 474 males, 13 for 371 females). Shifts in home range were made primarily by males in the reproductive season.

Limited data on dispersal exist for *C. for-*

mosus. French et al. (1968) found that 25% to 30% of the individuals in their study made dispersal movements greater than 500 ft. In contrast, Maza et al. (1973) found that only 5% of *C. formosus* shifted home range during their study. Allred and Beck (1963) recapture 62% of 45 marked males and 54% of 33 females ($P = 0.5$, G-test of independence for sex differences).

Perognathus

What information is available on *Perognathus* species suggests that their social systems are probably not very different from those of *Chaetodipus.* Along with *C. californicus, Perognathus inornatus* was among the most aggressive heteromyids studied in laboratory encounters by Eisenberg (1963*a*). Males appear to be more mobile than females in two species (Table 1); males had larger home ranges than females in *Perognathus parvus* (O'Farrell et al., 1975), and the mean distance between farthest capture points was greater in males than in females in *Perognathus longimembris* (Allred and Beck, 1963). In their mark–recapture study, Allred and Beck (1963) recaptured 25% of 214 male and 29% of 165 female *P. longimembris* ($P = 0.4$, G-test for sex differences).

Microdipodops

Little information is available on the social systems of the two species of *Microdipodops.* Eisenberg (1963*a*) found that *Microdipodops pallidus* was nearly as aggressive as *Liomys, Chaetodipus, Perognathus,* and *Dipodomys* species in intraspecific laboratory encounters and markedly more aggressive than *H. anomalus.* Observations of male-female pairs in the lab revealed no tendency to nest together, except during estrus, as in other heteromyids studied (Eisenberg, 1967).

O'Farrell and Blaustein (1974) reported larger home range size for males than for females in *M. megacephalus.* Individuals maintained small territories around burrows, but also used a larger area that considerably overlapped neighbors' home ranges. Home range size changed seasonally, but there was a high degree of fidelity to a site.

Dipodomys

Information on social behavior and social structure is most extensive for this genus (Table 1). The evidence indicates that *Dipodomys* species are generally characterized by the same solitary spacing pattern that occurs in *Liomys, Chaetodipus,* and *Perognathus* species, but with some interesting variations. Adults probably defend territories, or at least burrows, in all *Dipodomys* species. Behavioral evidence of territoriality, that is, direct observations of individuals defending burrows or territories, exists for *Dipodomys heermanni* (Tappe, 1941) and for *D. spectabilis* (Randall, 1984). (Both were observed defending burrows.) *D. spectabilis* concentrate their activity near their burrows, which are contained within large mounds. Using radiotelemetry, Schroder (1979) found that they spent 78% of their time within 6 m of the mound. (Home ranges averaged 23 m to 30 m in diameter depending on the method of calculation.) Another radiotracking study generally confirmed thus, but it also found that *D. spectabilis* spent significantly more time away from their mounds in late summer and autumn, when seed caches are accumulated, then in other seasons (Jones, 1982). Four studies have documented a regular dispersion of active burrows in *D. spectabilis* (Jones, 1984; Randall, 1984, 1989*b*, 1991*a*, 1991*b*; Schroder, 1979; Schroder and Geluso, 1975). In *D. merriami,* Behrends et al. (1986*a*) reported aggressive encounters between neighbors, but they also reported nonaggressive encounters. This study and that of Randall (1989*c*) suggest that both *D. spectabilis* and *D. merriami* may be intolerant of conspecifics around their core areas and burrows, but relatively more tolerant of familiar neighboring individuals in more peripheral and less frequently used areas of their ranges. These observations parallel observations by Eisenberg (1963*a*) of less aggression and more social flexibility in *Dipodomys* than in other heteromyids.

Some general patterns emerge from present data on spacing behavior in *Dipodomys*. Females maintain exclusive home ranges in *Dipodomys agilis* during the breeding season (MacMillen, 1964), in *D. spectabilis* (Schroder, 1979), in *Dipodomys ingens* (Braun, 1985), and in most populations of *D. merriami* (Jones, 1982; O'Farrell, 1980; Reynolds, 1958, 1960; but see Chew and Butterworth, 1964). Male-male home range overlap seems to be more common than female-female overlap in most species (*D. agilis,* MacMillen, 1964; *D. merriami,* Chew and Butterworth, 1964; O'Farrell, 1980; *D. ordii,* O'Farrell, 1980; *D. panamintinus,* O'Farrell, 1980; but not in *D. ingens,* Braun, 1985), and male home ranges usually overlap female ranges (*D. agilis,* MacMillen, 1964; *D. merriami,* Behrends et al., 1986*a*; Chew and Butterworth, 1964; O'Farrell, 1980; Reynolds, 1958; *D. panamintinus,* O'Farrell, 1980; *D. ingens,* Braun, 1985).

Several exceptions to the usual solitary existence of adult *Dipodomys* have been reported in *D. spectabilis.* Three of 44 occupied burrows excavated by Monson and Kessler (1940) and four of 53 occupied burrows excavated by Monson (1943) contained two adults. The relationships of these individuals to each other were unknown. However, Jones (1984) found an adult female and her son occupying the same burrow system for 18 months, at least six months after the son reached sexual maturity.

As with the other heteromyid genera, our understanding of *Dipodomys* mating systems is still sketchy, but they appear to be promiscuous or polygynous. Copulatory plugs form in *Dipodomys deserti* (Butterworth, 1961*a*), in *D. merriami* (pers. obs.), and in *D. spectabilis* (Vorhies and Taylor, 1922; pers. obs.). Kenagy (1976) observed two adult male *Dipodomys microps* grappling at the burrow of an adult female, after which one male ran off and the other copulated with the female. Randall (1987*a*) saw a single *D. merriami* female mate with three different males in one night. Randall (1991*b*) observed aggression among males at the mound of an estrous female *D. spectabilis.* Although these observations suggest intense mate competition among males, in some *Dipodomys* species males are not clearly more mobile than females. Male home range size is approximately equal to that for females in some cases (*D. agilis,* MacMillen, 1964; *D. ordii,* O'Farrell, 1978; *D. panamintinus,* O'Farrell, 1978; *D. ingens,* Braun, 1985; and possibly *D. microps,* Maza et al., 1973). In *D. merriami* male home range size is reported to be greater than that for females in some studies (Allred and Beck, 1963; Maza et al., 1973; O'Farrell, 1980; Soholt, 1973), but not in others (Chew and Butterworth, 1964; Reynolds, 1958, 1960). The most detailed study of *D. merriami* home ranges (Behrends et al., 1986*a*, 1986*b*) found no significant sexual dimorphism in home range size, but males moved significantly farther than females between successive radio-fixes during breeding seasons. Female choice may play some role in *D. merriami.* Behrends et al. (1986*b*) suggested that increased movement by estrous females might indicate that they are actively choosing mates. Finally, in *D. spectabilis* males do appear to be more mobile than females. Schroder (1979) found that male home ranges were twice the size of female ranges in this species. Randall (1984) reported that most visits by *D. spectabilis* to neighbors' burrows were made by adult males to adult females' burrows (67% of 65 visits by adults). Furthermore, she found that visits occurred more frequently in spring, when *D. spectabilis* reproduce, than at other times of year.

Good data on dispersal patterns exist for three *Dipodomys* species, and limited data are available for a fourth. In all four cases dispersal movements are restricted; most individuals are never trapped far from their birthplaces or points of first capture. In three cases dispersal patterns do not conform to the usual male-biased pattern that characterizes most mammals (Greenwood, 1980). In *D. merriami* 24 of 27 males and 34 of 36 females remained through reproductive maturity within 100 m of the centers of their juvenile home ranges (Jones, 1982; home ranges were approximately 50 m in diameter). Zeng and Brown (1987) reported dispersal data from a seven year study of *D. merriami*. Sixty-six percent of 140 males and 76% of 124 females that survived at least four months moved less than 50 m from first to last capture, and 84% of all individuals moved less than 100 m during their lifetimes. Their data suggest that males predominate among individuals moving beyond 100 m, and that movement by adults is common. Reynolds (1960) reported that immature *D. merriami* "... tended to take up (residence in) areas not occupied by adults. Also, more home ranges of immature animals were located within the home ranges of adult females than of adult males" (p. 57).

In *D. microps* Allred and Beck (1963) recaptured 58% of 315 males and 53% of 238 females tagged ($P = 0.23$, G-test for sex differences). The distance between farthest capture points averaged 249 feet for males and 225 feet for females, and 79% of males and 87% of females ranged less than 400 feet.

Fitch (1948) recorded dispersal movements for 652 male and 618 female *D. heermanni* in a study based on over 10,000 trapping records. There was no significant sex difference in the proportion that moved less than 300 feet (85% of males and 83.4% of females, $P = 0.42$, G-test; home ranges were 100 to 400 feet in diameter). However, there was a female bias in long distance dispersal. Nineteen males and 41 females moved more than 900 feet ($P < 0.001$, G-test), and 12 of 15 rats moving more than 1,600 feet were females ($P = 0.01$, G-test).

Dipodomys spectabilis' dispersal pattern is almost identical to that in *D. heermanni*. Jones (1987) found that 80% of 96 males and 77% of 99 females remained within 50 m (the approximate average inter-burrow distance among adults) of their natal sites or points of first capture (no significant sex bias, $P = 0.66$, G-test), but again there were more females than males among long distance dispersers. Only one male moved more than 200 m, while 6 females did so ($P = 0.046$, G-test). Adult dispersal in *D. spectabilis* is more restricted than in juveniles (Jones, 1987). Seventy percent of 70 males and 61% of 72 females remained in the same burrows throughout their adult lives ($P = 0.26$, G-test), and 89% of males and 92% of females remained within 50 m of the burrows they occupied at the beginning of their first breeding season ($P = 0.54$, G-test). Other data, however, suggest that in other populations of *D. spectabilis* individuals may disperse longer distances as both juveniles and adults.

Body Size, Life History Traits, and Sociality in Heteromyids

Eisenberg (1963*a*) discussed the relationships between life history parameters, such as gestation time and litter size, and sociality in heteromyids. He suggested that "... some Dipodomyinae have departed from the short-gestation–larger-litter reproductive pattern, having lengthened gestation and reduced litter ..." and that this trend "... reaches its greatest development in *D. spectabilis*" (p. 81). He observed that the "... Perognathinae are geared to a faster reproductive turnover, and the opportunity for a more permanent social organization is absent. On the other hand, those *Dipodomys* with slower rates of reproduction may have a very well-integrated social system even though the adults dwell separately" (p. 85). These observations anticipate Eisenberg's (1981) own study of body size and life history traits in mammals, as well as the analyses of body size, life history traits, and sociality that have been done in two other mammalian classes, Canidae and Sciuridae. Bekoff et al. (1981) found that large canids are more social than their smaller relatives, and that this trend is related to several other trends: offspring of larger species achieve independence and breed later than offspring of smaller species; larger females give birth to absolutely larger young, but relative to their own size they allocate fewer resources to bringing a pup to term. Armitage (1981) found similar trends in the Sciuridae; larger species tend to be more social, to disperse at a later age, and to breed for the first time later than smaller species do.

If life history tactics, rates of development, body size, and behavior patterns are closely correlated in heteromyids, as they appear to be in canids and sciurids, then trends in life history variation can provide clues to understanding the overall patterns of social organization in heteromyids. A quantitative analysis of the relationships between life history parameters and degree of sociality in heteromyids is not yet possible because there are not enough data on heteromyid social systems to provide an index of sociality, such as Bekoff et al. (1981) and Armitage (1981) used in their studies. However, there is 20-fold variation in body size among the heteromyids, and there are enough data on life history parameters that we can at least ask whether the relationships follow the same trends as in canids and sciurids. If so, then we might also expect similar relationships between body size, life history variables, and sociality.

Data on life history parameters are available for 30 of the approximately 57 heteromyid species. I compiled data for nine life history parameters: adult weight, neonate weight, litter size, gestation time, age at maturity for females in wild populations, maximum lifespan in wild populations, age at eye opening, age at weaning, and age at which adult weight is reached (Table 2). These are the variables included in Bekoff et al. (1981) and Armitage (1981) for which data are available for heteromyids. Following Bekoff et al. (1981) I calculated four composite parameters, three of which provide partial measurements of parental expenditure on reproduction. Reproductive index (RI) is neonate weight divided by adult weight, and it estimates the effort needed to bring one offspring to term relative to adult size. Reproductive effort (RE) is the product of neonate weight and litter size, and it should be related to the amount of resources devoted to bringing a litter to term. Specific reproductive effort (SRE) is RE divided by adult weight, and it assesses the amount of resources devoted to reproduction relative to adult weight. The ratio of male weight to female weight (MW:FW) is a measure of the degree of sexual dimorphism. I calculated a fifth composite parameter, the sum of gestation time and age at eye opening, as a measure of maturation rate. I attempted to trace all data to primary references, but when I could not obtain primary references, or when secondary references gave values different from primary references, I included secondary references. I attempted to use all published estimates of each parameter, and I derived single values for each species as described in Jones (1985). The natural log of each variable was regressed against the natural log of each of the other variables (Sokal and Rohlf, 1981), except in cases where one was used to derive the other (e.g., neonate weight and RI). For all regressions I used two-tailed t-tests to evaluate the significance of correlations.

In general, patterns of variation among these life history parameters do not appear to be as closely related in heteromyids as in canids or sciurids. Only 19 of the 79 (24%) possible cells are filled in heteromyids (Table 3), which is relatively few compared to corresponding data for canids (40 of 55 cells, 73%, Bekoff et al., 1981) and sciurids (64

TABLE 3.—*Coefficients of determination for significant regressions of heteromyid life history variables. AWT = adult weight; NW = neonate weight; LS = litter size; G = gestation period; AM = age at maturity; ML = maximum lifespan; AEO = age at eye opening; AW = age at weaning; AAW = age adult weight reached. Other symbols are defined in the text.*

	AWT	NW	LS	G	AM	ML	AEO	AW	AAW	G + AEO	RE	|RI|	|SRE|
NW	0.86***												
LS	—	0.55**											
G	0.29*	0.46*	0.46**										
AM	—	—	—	—									
ML	—	0.52*	—	0.86***	0.66*								
AEO	—	—	—	—	—	—							
AW	—	—	—	—	—	—	—						
AAW	0.83*	0.98***	0.81*	—	—	—	—	—					
G + AEO	0.37*	—	—		—	0.76*		—	—				
RE	0.86***			—	—	—	—	—	0.92*	—			
|RI|			—	—	—	—	—	—	—	—			
|SRE|				—	—	0.64*	—	—	—	0.53*			
MW:FW	—	—	—	—	—	—	—	—	—	—	—	0.77***	0.61**

* $P < 0.05$; ** $P < 0.01$; *** $P < 0.001$.

of 135 cells, 47%, Armitage, 1981). There is no evidence that reproductive maturity is delayed in larger heteromyids, a trend that occurs in both sciurids and canids. Nor do the smaller *Chaetodipus* and *Perognathus* species generally have faster reproductive turnover than the larger *Dipodomys* species. Life span and age at maturity in *Chaetodipus* and *Perognathus* are similar to those in *Dipodomys,* and so there appears to be as much opportunity for permanent well-integrated social systems in *Chaetodipus* and *Perognathus* as in *Dipodomys.*

Demographic schedules in heteromyids are more similar than one would expect from the wide variation in body size and reproductive traits (Kenagy and Bartholomew, 1985), hence the lack of correlations between many life history parameters in this family. For instance, the absence of significant correlations between body size and age at maturity and lifespan may be due partially to the ability of some of the smaller *Chaetodipus* and *Perognathus* species to enter torpor, thereby delaying maturity or prolonging their lives. Three of the four *Chaetodipus* and *Perognathus* species for which lifespan data are available are known to enter torpor (see Jones, 1985). *D. merriami* also has adaptations that result in prolonged lifespan and slow maturation rates. Zeng and Brown (1987) found high rates of adult survival and a flexible reproductive strategy. A few small, widely-spaced litters per year are timed to favorable resource availability. Adults reproduce without trade-offs in survivorship. *D. merriami* is the second smallest kangaroo rat, but it has a suite of life history traits that are often associated with larger species (see also Behrends et al., 1986*a*). The flexible life history strategy of *D. merriami* may be an adaptation to unpredictable variation in the availability of food and other resources in deserts (Zeng and Brown, 1987). In both this species and *D. spectabilis* (Waser and Jones, 1991), individuals appear to adjust the timing of breeding and the magnitude of reproductive effort to environmental conditions so as to minimize the cost of reproduction. Kenagy and Bartholomew (1985) examined reproductive life histories of four desert heteromyids. These species all have different reproductive traits, but all four species are characterized by long lifespans and stable population densities. Long lifespans may be essential in unpredictable environments in which the chances of complete reproductive failure in any one year are high (Kenagy and Bartholomew, 1985). Many of the desert heteromyids are subject to unpredictable variation in environmental conditions, and many of these species are relatively long-lived.

Some of the trends found in canids and sciurids are repeated in heteromyids. Larger species produce larger neonates. Larger species tend to devote fewer resources to reproduction relative to their own body size; the slope is less than one in the log-log plot of litter mass against adult mass (Jones, 1985).

Bekoff et al. (1981) and Armitage (1981) concluded that one of the most important life history correlates of sociality was age at independence or dispersal. They argued that delayed dispersal, which was most pronounced in larger species, was one of the primary factors leading to increased sociality in canids and sciurids. Good data on age of dispersal exist for only one heteromyid species, but the data are intriguing. In *D. spectabilis,* the second largest heteromyid, juveniles delay dispersal from natal burrows for up to six months after weaning (Jones, 1984). During this period mothers may share burrows with as many as five offspring, and juvenile females from earlier litters occasionally reproduce in these circumstances (P. Waser, pers. comm.).

Even after temporary family groups break up in *D. spectabilis,* offspring frequently settle in the natal burrow or in one of their mother's subsidiary burrows within the maternal home range. Some of these incidences of philopatry involve opportunistic philopatry in which the offspring inherits its mother's mound or territory when the mother dies. However, the frequency of philopatry is too high to attribute it all to opportunism. Waser (1988) estimated that of 196 juveniles in 115 families that survived to the age of one year, 63 should have been opportunistically philopatric based on adult mortality rates. In reality, 156 of these 196 juveniles remained within a home range diameter of their natal mounds. Waser attributes the extra philopatary to three causes: mothers sharing mounds with offspring, mothers donating secondary mounds, and mothers abdicating their primary mounds and moving elsewhere. All three phenomena involve philopatry by parental consent rather than opportunistic philopatry.

Questions and Prospects for Future Research

The social systems of heteromyids range from a solitary dispersed pattern to incipient sociality. In *Chaetodipus, Perognathus,* most smaller *Dipodomys,* and *Liomys,* all adults appear to live separately. There is no evidence of social groupings among adults in any of these species with the single exception of *L. pictus,* and even in this species adult females live alone during breeding seasons. Home range overlap between females is generally slight. Male home ranges usually overlap female home ranges.

There are two departures from this basic solitary plan in heteromyids. In *Heteromys,* laboratory and field evidence indicates less intraspecific agression than in other heteromyid genera. Groups of six or more cohabiting individuals were reported for one species. However, little more than this is known of *Heteromys* social behavior. Most importantly, we do not know the kin relationships of individuals in the groups reported.

The second departure from the basic solitary system occurs in large *Dipodomys* species, where it has been best documented in *D. spectabilis.* Although adults of this species virtually always live separately and defend their burrows against other adults, delayed dispersal by juveniles leads to the formation of temporary family groups. These family groups do not usually persist in *D. spectabilis,* but they may be similar to the evolutionary precursors of social groups of related adult females that characterize gregarious sciurids. Thus *D. spectabilis* may be viewed as a species with rudimentary sociality.

The absence of social groups in most heteromyids does not mean there are no well-integrated social systems—even many smaller species have long lifespans that might permit the development of long-term relationships between neighbors. We know something of the extent to which heteromyids are aware of their social environment. There is evidence that *D. merriami* and *D. microps* can recognize species from sandbathing deposits and that *D. spectabilis* can distinguish sex from sandbathing deposits. Eisenberg's (1963*b*, 1967) observations indicate that information is exchanged also at urination and marking sites. The means by which males recognize estrous females is known only for *D. spectabilis,* although sandbathing and urination sites and scent gland secretions have been ruled out in *D. merriami.* Perhaps actual contact of the sort described by Eisenberg is necessary. We need to know the extent to which heteromyids can recognize individuals among their neighboring conspecifics. *D. spectabilis* may be able to distinguish neighboring individuals from sandbathing deposits and foot-drumming patterns (Randall, 1987*b*, 1989*a*). In *D. spectabilis,* because juveniles tend to disperse short distances or to settle within maternal home ranges, neighboring adults are sometimes close relatives (Jones, 1984). The question that arises now is whether behavior and degree of social tolerance between related neighbors differ from that between unrelated neighbors in *D. spectabilis.*

Armitage (1981) suggested that delayed dispersal in sciurids may be a means of continuing parental investment beyond weaning. He argued that offspring cannot achieve maturity and independence in one growing season, so they are "allowed" to remain for additional years within parental home ranges and thus retain access to parental resources. A similar explanation may apply to delayed dispersal in *D. spectabilis.* This species inhabits complex burrows contained in large mounds (Vorhies and Taylor, 1922). New mounds are rarely built, and instead existing mounds are passed on through generations (Holdenried, 1957; Jones, 1984). Inside these mounds, *D. spectabilis* accumulate large seed caches (Monson, 1943; Monson and Kessler, 1940; Vorhies and Taylor, 1922). The limited availability of suitable vacant mounds and the relatively low availability of seeds in spring when juveniles are born and weaned suggest that both mounds and seed caches may be essential resources that are not readily available to dispersing juveniles (Jones, 1984). By allowing delayed dispersal, mothers provide their offspring with access to these resources until the offspring can acquire their own. Survival costs associated with dispersal are well documented in *D. spectabilis.* Juveniles that eventually acquire maternal mounds and seed caches are significantly more likely to reach reproductive age than those that move elsewhere (Jones, 1986). When juveniles were experimentally forced to disperse at early ages, they suffered higher mortality than philopatric controls, primarily as a result of predation (Waser, 1988). Some data suggest that increased exposure to predation in forced dispersers may be a consequence of increased time spent foraging above ground.

Interestingly, *D. heermanni,* a medium-size species, also builds fairly elaborate burrow systems and caches seed inside them (Dale, 1939; Tappe, 1941). Fitch (1948, p. 31) found that *D. heermanni* "... young are not apt to shift to new areas at the time they become able to forage independently ..." and that young animals marked when they were very small "... did not shift home range for periods of several weeks; in many instances they stayed in the same location, where they grew to adult size." If limited availability of burrows and the need for access to maternal caches are the primary causes of delayed dispersal by juveniles in these species, then we might expect to find no delayed dispersal in a species such as *D. merriami,* which has little caching and abundant burrows (Jones, 1982; Monson and Kessler, 1940). No groups of females with weaned young have been reported in *D. merriami,* but more work is certainly necessary before we can determine whether natal dispersal occurs at an earlier age in *D. merriami* than in *D. spectabilis.* Tendencies to move among adults apparently do differ between these two species. Most *D. spectabilis* never move from their original home range during their adult lives (Jones, 1987), but shifts in home range are common in adult *D. merriami* (Zeng and Brown, 1987).

There are two aspects of natal philopatry and incipient gregariousness in *D. spectabilis,* and possibly in *D. heermanni,* that may be important in the initial development of gregarious social systems in mammals. First, offspring delay departures from natal burrows so that groups of close relatives temporarily reside in the same burrow. Second, after offspring reach independence, they often settle permanently within natal home ranges or in immediately adjacent areas. Neighboring adults are sometimes close relatives. Either of these phenomena could be a precursor of social groups in gregarious species. As species evolve toward gregariousness, periods of common residence in the natal burrow may be lengthened into adulthood, or neighboring adult relatives may develop behaviors characteristic of gregarious species (e.g., alarm calls, cooperative defense of territories) so that loose aggregations of close relatives gradually become complex closed social units.

Another noteworthy aspect of *Dipodomys* dispersal is the possible female bias in long-distance movements in *D. spectabilis* and *D. heermanni.* Its occurrence here may be due to more stringent burrow requirements for females than for males. Perhaps females require larger and more elaborate burrows that can hold large seed caches because they must accommodate litters for prolonged periods. The importance of the burrow for seed caches is apparent from work by Reichman et al. (1985) that indicates that the burrow environment is important in maintaining proper temperature and humidity levels for seed caches, thereby controlling mold growth on the seeds. When young females leave their natal burrows, some of them may move long distances before finding a suitable vacancy. Consistent with this hypothesis is the absence of a female bias in long-distance moves in *D. merriami.* The availability of suitable burrows is probably not a limiting factor for females in *D. merriami* as it is for female *D. spectabilis* or *D. heermanni*; Monson and Kessler (1940) reported an average of 4.11 burrows per individual in the population they studied.

A complete explanation of dispersal patterns in *D. spectabilis, D. heermanni,* and other heteromyids will require more knowledge of mating systems in these species; explanations for dispersal patterns in mammals usually invoke aspects of mate competition (e.g., Greenwood, 1980) or inbreeding avoidance (e.g., Packer, 1979; Pusey, 1980). At this point, however, we know little about the mating system of any heteromyid. Males apparently compete for access to multiple females in *Dipodomys* and probably in most other genera as well. The role of female-choice in heteromyid mating systems is unknown. Both Eisenberg (1963*a*) and Allan (1944) reported that females may initiate sexual contacts, and Behrends et al. (1986*b*) noted increased activity by estrous females in *D. merriami.* Most heteromyids are probably promiscuous, but the extent of sex differences in variation in annual or lifetime reproductive success or in the intensity of mate competition are not known for any species. Greater mate competition among males than among females in mammals is thought to be associated with male-biased dispersal (Clutton-Brock and Harvey, 1976; Dobson, 1982; Greenwood, 1980). If mate competition is greater in males than in females in *Dipodomys,* as we would expect if they are polygynous or promiscuous, then why is male-biased dispersal absent in most *Dipodomys* species for which data are available? Again, a partial answer may be that competition among females for resources such as burrows or food may be as intense as competition among males for mates.

A related problem involves patterns of size dimorphism in heteromyids. Ralls (1977) observed that although monogamy is a good predictor of monomorphism and extreme polygyny a good predictor of extreme dimorphism, there are many non-monogamous species, especially in Rodentia, that show little sexual dimorphism. The reason for this last pattern may be that size is often as important to females in competition for burrows and food as it is to males in competition for mates. In heteromyids, size dimorphism is generally slight in the four desert-dwelling genera, *Dipodomys, Chaetodipus, Perognathus,* and *Microdipodops,* but it is probably more pronounced in the two tropical genera, *Heteromys* and *Liomys.* Is this difference related to a difference in the intensity of competition among females for resources? The evidence for intraspecific tolerance in *Heteromys* suggests this possibility at least in this genus, but the absence of similar tendencies in *Liomys* is not consistent with this hypothesis. Behrends et al. (1986*a*) noted that the absence of strong size dimorphism in kangaroo rats is paralleled by an absence of sexual dimorphism in home range size and in survivorship schedules in most *Dipodomys* species.

Finally, the highly viscous dispersal patterns in both sexes of some *Dipodomys* species raise questions about inbreeding. If most members of both sexes settle permanently near their birth sites in these species, then what is the extent of inbreeding and does it have deleterious consequences? Perhaps *Dipodomys* populations can tolerate fairly high levels of inbreeding with little or no inbreeding depression. On the other hand, demographic schedules may be such that close relatives rarely co-occur as adults, and in cases where they do so, avoidance may be the rule. These are not mutually exclusive hypotheses, and both may have some validity for kangaroo rats.

A picture of heteromyid social organization is emerging from field and laboratory studies of behavior patterns, communication, dispersal, and life histories. Furthermore, these studies are yielding information on the causes of variation in social organization in this family. Most heteromyids are characterized by solitary dispersed social organizations. Mutual avoidance and agonistic behavior predominate even in male-female encounters unless the female is in estrus. In general, home range overlap between females is slight. Male-male and male-female overlap is more extensive. In some species individuals defend small core areas around their burrows, but they are more tolerant of conspecifics in peripheral areas of their home ranges.

Exceptions to the predominant solitary plan occur in *Heteromys* species and in some of the larger *Dipodomys* species. In *D. spectabilis,* and perhaps in *D. heermanni,* offspring delay dispersal from natal burrows for several months after weaning. Delayed dispersal results in the formation of temporary family groups resembling the social units of gregarious rodents. The cause of delayed dispersal in these species may be limited availability of resources—burrows and food caches—outside natal home ranges. Philopatric offspring retain access to maternal burrows and resources until opportunities for settlement arise elsewhere. In *D. spectabilis,* many offspring of both sexes settle permanently within maternal home ranges. Consequently, neighboring adults may sometimes be close relatives. *D. spectabilis* may be able to distinguish neighboring individuals from one another from sandbathing deposits and foot-drumming sounds, but whether their behavior toward relatives differs from that toward non-relatives is unknown.

Although in some mammalian taxa life history parameters are correlated with each other and with degree of sociality, these relationships are not all repeated in heteromyids. A correspondence between life history traits and sociality in heteromyids is probably confounded by adaptations to variable and unpredictable environments. Despite wide variation in body size and in reproductive traits, many heteromyid species are characterized by similarly long life spans and stable population densities. Long life span may be important when the chances of reproductive failure in any one year are high, as is the case with many desert heteromyids. Longevity and slow maturation rates appear to be characteristics of some species with solitary dispersed social organizations as well as species with rudimentary sociality. Hence the degree of sociality does not seem to be as closely related to life history strategies in heteromyids as it is in other mammalian taxa.

Acknowledgments

Jim Brown, John Eisenberg, Theodore Fleming, Lauraine Hawkins, Jan Randall, Jim Reichman, and Peter Waser read drafts of this chapter, and I thank them for their suggestions.

Literature Cited

Alcorn, J. R. 1941. Counts of embryos in Nevadan kangaroo rats (genus *Dipodomys*). Journal of Mammalogy, 22:88–89.

Allan, P. F. 1944. Mating behavior of *Dipodomys ordii richardsoni.* Journal of Mammalogy, 25:403–404.

Allred, F. M., and D. E. Beck. 1963. Range of movement and dispersal of some rodents at the Nevada Atomic Test Site. Journal of Mammalogy, 44: 190–200.

Armitage, K. B. 1981. Sociality as a life-history tactic of ground squirrels. Oecologia, 48:36–49.

Arnold, L. W. 1942. Notes on the life history of the sand pocket mouse. Journal of Mammalogy, 23:339–341.

Asdell, S. A. 1964. Patterns of mammalian reproduction. Cornell University Press, Ithaca, New York, 670 pp.

Bailey, V. 1931. Mammals of New Mexico. North American Fauna, 53, 412 pp.

BANFIELD, A. W. F. 1974. The mammals of Canada. University of Toronto Press, Toronto, 438 pp.

BEHRENDS, P. R., M. DALY, AND M. I. WILSON. 1986*a*. Range use patterns and spatial relationships of Merriam's kangaroo rats (*Dipodomys merriami*). Behaviour, 96:187–209.

———. 1986*b*. Aboveground activity of Merriam's kangaroo rat (*Dipodomys merriami*) in relation to sex and reproduction. Behaviour, 96:210–226.

BEKOFF, M., J. DIAMOND, AND J. B. MITTON. 1981. Life history patterns and sociality in canids: body size, reproduction, and behavior. Oecologia, 50:386–390.

BRADLEY, W. G., AND R. A. MAUER. 1971. Reproduction and food habits of Merriam's kangaroo rat, *Dipodomys merriami*. Journal of Mammalogy, 52: 497–507.

BRAUN, S. E. 1985. Home range and activity patterns of the giant kangaroo rat, *Dipodomys ingens*. Journal of Mammalogy, 66:1–12.

BREYEN, L. J., W. G. BRADLEY, AND M. K. YOUSEF. 1973. Physiological and ecological studies on the chisel-toothed kangaroo rat, *Dipodomys microps*. Comparative Biochemistry and Physiology, 44A: 543–555.

BUTTERWORTH, B. B. 1961*a*. The breeding of *Dipodomys deserti* in the laboratory. Journal of Mammalogy, 42:413–414.

———. 1961*b*. A comparative study of growth and development of the kangaroo rats, *Dipodomys deserti* Stephens and *Dipodomys merriami* Mearns. Growth, 25:127–138.

CHEW, R. M. 1958. Reproduction by *Dipodomys merriami* in captivity. Journal of Mammalogy, 39: 597–598.

CHEW, R. M., AND B. B. BUTTERWORTH. 1959. Growth and development of Merriam's kangaroo rat, *Dipodomys merriami*. Growth, 23:75–95.

———. 1964. Ecology of rodents in Indian Cove (Mojave Desert), Joshua Tree National Monument, California. Journal of Mammalogy, 45:203–225.

CLUTTON-BROCK, T. H., AND P. H. HARVEY. 1976. Evolutionary rules and primate societies. Pp. 195–237, *in* Growing points in ethology (P. P. G. Bateson and R. A. Hinde, eds.). Cambridge University Press, Cambridge, 548 pp.

CONLEY, W., J. D. NICHOLS, AND A. R. TIPTON. 1977. Reproductive strategies in desert rodents. Transactions of the Symposium on the Biological Resources of the Chihuahuan Desert Region, United States Department of Interior, National Park Service Transactions and Proceedings Series, No. 3.

CULBERTSON, A. E. 1946. Observations on the natural history of the Fresno kangaroo rat. Journal of Mammalogy, 27:189–203.

DALE, F. H. 1939. Variability and environmental responses of the kangaroo rat, *Dipodomys heermanni saxatilis*. American Midland Naturalist, 22:703–731.

DALY, M., M. I. WILSON, AND P. BEHRENDS. 1984. Breeding of captive kangaroo rats, *Dipodomys merriami* and *D. microps*. Journal of Mammalogy, 65: 338–341.

DAVIS, W. B. 1966. The mammals of Texas. Texas Parks and Wildlife Department, Austin, Texas, 267 pp.

DAY, B. N., H. J. EGOSCUE, AND A. M. WOODBURY. 1956. Ord kangaroo rat in captivity. Science, 124: 485–486.

DOBSON, F. S. 1982. Competition for mates and predominant juvenile male dispersal in mammals. Animal Behavior, 30:1183–1192.

DORAN, D. J. 1952. Observations on the young of the Merriam kangaroo rat. Journal of Mammalogy, 33:494–495.

EISENBERG, J. F. 1963*a*. The behavior of heteromyid rodents. University of California Publications in Zoology, 69:1–100.

———. 1963*b*. A comparative study of sand-bathing behavior in heteromyid rodents. Behaviour, 22:16–23.

———. 1967. A comparative study of rodent ethology with emphasis on evolution of social behavior. I. Proceedings of the United States National Museum, No. 3597, vol. 122, pp. 1–51.

———. 1981. The mammalian radiations. University of Chicago Press, Chicago, 610 pp.

EISENBERG, J. F., AND D. E. ISAAC. 1963. The reproduction of heteromyid rodents in captivity. Journal of Mammalogy, 44:61–67.

ENGSTROM, M. D., AND R. C. DOWLER. 1981. Field observations of mating behavior in *Dipdomys ordi*. Journal of Mammalogy, 62:384–386.

FITCH, H. S. 1948. Habits and economic relationships of the Tulare kangaroo rat. Journal of Mammalogy, 29:5–35.

FLEMING, T. H. 1971. Population ecology of three species of neotropical rodents. Miscellaneous Publications of the Museum of Zoology, University of Michigan, 143:1–77.

———. 1974*a*. Social organization in two species of Costa Rican heteromyid rodents. Journal of Mammalogy, 55:543–561.

———. 1974*b*. The population ecology of two species of Costa Rican heteromyid rodents. Ecology, 55:493–510.

———. 1977. Growth and development of two species of tropical heteromyid rodents. American Midland Naturalist, 98:109–123.

FRENCH, N. R., B. G. MAZA, AND A. P. ASCHWANDEN. 1967. Life spans of *Dipodomys* and *Perognathus* in the Mojave desert. Journal of Mammalogy, 48:537–548.

FRENCH, N. R., T. Y. TAGAMI, AND P. HAYDEN. 1968. Dispersal in a population of desert rodents. Journal of Mammalogy, 49:272–280.

FRENCH, N. R., B. G. MAZA, H. O. HILL, A. P. ASCHWANDEN, AND H. W. KAAZ. 1974. A population study of irradiated desert rodents. Ecological Monographs, 44:45–72.

GIBBS, R. H., JR. 1955. Vocal sound produced by the kangaroo rat, *Dipodomys spectabilis*. Journal of Mammalogy, 36:463.

GREENWOOD, P. J. 1980. Mating systems, philopatry and dispersal in birds and mammals. Animal Behavior, 28:1140–1162.

GRINNELL, J. 1932. Habitat relations of the giant kangaroo rat. Journal of Mammalogy, 13:305–320.

GRINNELL, J., J. DIXON, AND J. M. LINSDALE. 1930. Vertebrate natural history of a section of northern California through the Lassen Peak region. University of California Publications in Zoology, 35:1–594.

HALL, E. R., AND K. R. KELSON. 1959. The mammals

of North America. Ronald Press, New York, 1083 pp.

HAYDEN, P., AND J. J. GAMBINO. 1966. Growth and development of the little pocket mouse, *Perognathus longimembris*. Growth, 30:187–197.

HAYDEN, P., J. J. GAMBINO, AND R. G. LINDBERG. 1966. Laboratory breeding of the little pocket mouse, *Perognathus longimembris*. Journal of Mammalogy, 47:412–423.

HOLDENRIED, R. 1957. Natural history of the bannertail kangaroo rat in New Mexico. Journal of Mammalogy, 38:330–350.

JOHNSTON, R. F. 1956. Breeding of the Ord kangaroo rat (*Dipodomys ordii*) in southern New Mexico. Southwestern Naturalist, 1:190–193.

JONES, W. T. 1982. Natal nondispersal in kangaroo rats. Unpublished Ph.D. dissertation, Purdue University, West Lafayette, Indiana, 185 pp.

———. 1984. Natal philopatry in bannertailed kangaroo rats. Behavioral Ecology and Sociobiology, 15: 151–155.

———. 1985. Body size and life history variables in heteromyids. Journal of Mammalogy, 66:128–132.

———. 1986. Survivorship in philopatric and dispersing kangaroo rats (*Dipodomys spectabilis*). Ecology, 67:202–207.

———. 1987. Dispersal patterns in kangaroo rats (*Dipodomys spectabilis*). Pp. 119–127, *in* Mammalian dispersal patterns: the effects of social structure on population genetics (D. Chepko-Sade and Z. Halpin, eds.). University of Chicago Press, Chicago.

KENAGY, G. J. 1973. Daily and seasonal patterns of activity and energetics in a heteromyid rodent community. Ecology, 54:1201–1219.

———. 1976. Field observations of male fighting, drumming, and copulation in the Great Basin kangaroo rat, *Dipodomys microps*. Journal of Mammalogy, 57:781–785.

KENAGY, G. J., AND G. A. BARTHOLOMEW. 1981. Effects of day length, temperature, and green food on testicular development in a desert pocket mouse, *Perognathus formosus*. Physiological Zoology, 54:62–73.

———. 1985. Seasonal reproductive patterns in five coexisting California desert rodent species. Ecological Monographs, 55:371–397.

LACKEY, J. A. 1967. Growth and development of *Dipodomys stephensi*. Journal of Mammalogy, 48: 624–632.

LAINE, H., AND J. G. GRISWOLD. 1976. Sandbathing in kangaroo rats (*Dipodomys spectabilis*). Journal of Mammalogy, 57:408–410.

LEMEN, C. A. 1978. Seed size selection in heteromyids: a second look. Oecologia, 35:13–19.

LEPRI, J. J., AND J. A. RANDALL. 1983. Hormonal regulation of sandbathing in male kangaroo rats (*Dipodomys merriami*). Behavioral Neural Biology, 37: 256–264.

LIDICKER, W. Z., JR. 1960. An analysis of intraspecific variation in the kangaroo rat *Dipdomys merriami*. University of California Publications in Zoology, 67: 125–218.

MACMILLEN, R. E. 1964. Population ecology, water relations, and social behavior of a southern California semidesert rodent fauna. University of California Publications in Zoology, 71:1–59.

MAZA, B. G., N. R. FRENCH, AND A. P. ASCHWANDEN. 1973. Home range dynamics in a population of heteromyid rodents. Journal of Mammalogy, 54:405–425.

MCCULLOCH, C. Y., AND J. M. INGLIS. 1961. Breeding periods of the Ord kangaroo rat. Journal of Mammalogy, 42:337–344.

M'CLOSKEY, R. T. 1978. Niche separation and assembly in four species of Sonoran Desert rodents. American Naturalist, 112:683–694.

MILLER, A. H., AND R. C. STEBBINS. 1964. The lives of desert animals in Joshua Tree National Monument. University of California Press, Berkeley, 452 pp.

MONSON, G. 1943. Food habits of the banner-tailed kangaroo rat in Arizona. Journal of Wildlife Management, 7:98–102.

MONSON, G., AND W. KESSLER. 1940. Life history notes on the banner-tailed kangaroo rat, Merriam's kangaroo rat, and the white-throated woodrat in Arizona and New Mexico. Journal of Wildlife Management, 4:37–43.

O'FARRELL, M. J. 1978. Home range dynamics of rodents in a sagebrush community. Journal of Mammalogy, 59:657–668.

———. 1980. Spatial relationships of rodents in a sagebrush community. Journal of Mammalogy, 61: 589–605.

O'FARRELL, M. J., AND A. R. BLAUSTEIN. 1974. Microdipodops megacephalus. Mammalian Species, No. 46, pp. 1–3.

O'FARRELL, T. P., R. J. OLSON, R. O. GILBERT, AND J. D. HEDLUND. 1975. A population of Great Basin pocket mice, *Perognathus parvus*, in the shrub-steppe of south-central Washington. Ecological Monographs, 45:1–28.

PACKER, C. 1979. Inter-troop transfer and inbreeding avoidance in *Papio anubis*. Animal Behavior, 27:1–36.

PUSEY, A. E. 1980. Inbreeding avoidance in chimpanzees. Animal Behavior, 28:543–552.

RALLS, K. 1977. Sexual dimorphism in mammals: avian models and unanswered questions. American Naturalist, 111:917–938.

RANDALL, J. A. 1981. Olfactory communication at sandbathing loci by sympatric species of kangaroo rats. Journal of Mammalogy, 62:12–19.

———. 1984. Territorial defense and advertisement by footdrumming in bannertail kangaroo rats (*Dipodomys spectabilis*) at high and low population densities. Behavioral Ecology and Sociobiology, 16:11–20.

———. 1985. Role of urine in coordinating reproduction in a desert rodent (*Dipodomys merriami*). Physiological Behavior, 34:199–203.

———. 1986. Preference for estrous female urine by male kangaroo rats (*Dipodomys spectabilis*). Journal of Mammalogy, 67:736–739.

———. 1987*a*. Field observations of male competition and mating in Merriam's and bannertail kangaroo rats. American Midland Naturalist, 117:211–213.

———. 1987*b*. Sandbathing as a territorial scent-mark in the bannertail kangaroo rat, *Dipodomys spectabilis*. Animal Behavior, 35:426–434.

———. 1989*a*. Individual foot drumming signatures in banner-tailed kangaroo rats, *Dipodomys spectabilis*. Animal Behavior, 38:620–630.

———. 1989*b*. Territorial-defense interactions with neighbors and strangers in banner-tailed kangaroo rats. Journal of Mammalogy, 70:308–315.

———. 1989*c*. Neighbor recognition in a solitary desert rodent (*Dipodomys merriami*). Ethology, 82:123–133.

———. 1991*a*. Sandbathing to establish familiarity in the Merriam's kangaroo rat, *Dipodomys merriami*. Animal Behavior, 41:267–275.

———. 1991*b*. Mating strategies of a nocturnal, desert rodent (*Dipodomys spectabilis*). Behavioral Ecology and Sociobiology, 28:215–220.

RANDALL, J. A., AND C. M. STEVENS. 1987. Footdrumming and other anti-predator responses in the bannertail kangaroo rat (*Dipodomys spectabilis*). Behavioral Ecology and Sociobiology, 20:187–194.

REICHMAN, O. J., AND J. H. BROWN. 1979. Use of torpor by *Perognathus amplus* in relation to resource distribution. Journal of Mammalogy, 60:550–555.

REICHMAN, O. J., AND K. M. VAN DE GRAAFF. 1973. Seasonal activity and reproductive patterns of five species of Sonoran Desert rodents. American Midland Naturalist, 90:118–126.

REICHMAN, O. J., D. T. WICKLOW, AND C. REBAR. 1985. Ecological and mycological characteristics of caches in the mounds of *Dipodomys spectabilis*. Journal of Mammalogy, 66:643–651.

REYNOLDS, H. G. 1958. The ecology of the Merriam kangaroo rat (*Dipodomys merriami* Mearns) on the grazing lands of southern Arizona. Ecological Monographs, 28:111–127.

———. 1960. Life history notes on Merriam's kangaroo rat in southern Arizona. Journal of Mammalogy, 41:48–58.

REYNOLDS, H. G., AND H. S. HASKELL. 1949. Life history notes on Price and Bailey pocket mice of southern Arizona. Journal of Mammalogy, 30:150–156.

ROOD, J. P., AND F. H. TEST. 1968. Ecology of the spiny rat, *Heteromys anomalus*, at Rancho Grande, Venezuela. American Midland Naturalist, 79:89–102.

ROSENZWEIG, M. L., AND P. W. STERNER. 1970. Population ecology of desert rodent communities: body size and seed-husking as bases for heteromyid coexistence. Ecology, 51:217–224.

SCELZA, J., AND J. KNOLL. 1980. Effects of acclimatization on body weight and oxygen consumption in *Dipodomys panamintinus*. Comparative Biochemistry and Physiology, 65A:77–84.

SCHEFFER, T. H. 1938. The pocket mice of Washington and Oregon in relation to agriculture. United States Department of Agriculture Technical Bulletin 608, 15 pp.

SCHMIDLY, D. J. 1971. Population variation in *Dipodomys ordii* from western Texas. Journal of Mammalogy, 52:108–120.

———. 1977. The mammals of Trans-Pecos Texas. Texas A&M University Press, College Station, Texas, 225 pp.

SCHRODER, G. D. 1979. Foraging behavior and home range utilization of the bannertail kangaroo rat (*Dipodomys spectabilis*). Ecology, 60:657–665.

SCHRODER, G. D., AND K. N. GELUSO. 1975. Spatiaal distribution of *Dipodomys spectabilis* mounds. Journal of Mammalogy, 56:363–368.

SMITH, H. D., AND C. D. JORGENSEN. 1975. Reproductive biology of North American desert rodents. Pp. 305–330, *in* Rodents in desert environments (I. Prakash and P. K. Ghosh, eds.). Dr. W. Junk b.v., Publishers, The Hague, The Netherlands, 624 pp.

SOHOLT, L. F. 1973. Consumption of primary production by a population of kangaroo rats (*Dipodomys merriami*) in the Mojave Desert. Ecological Monographs, 43:357–376.

SOKAL, R. R., AND F. J. ROHLF. 1981. Biometry. W. H. Freeman and Co., San Francisco, 859 pp.

TAPPE, D. T. 1941. Natural history of the Tulare kangaroo rat. Journal of Mammalogy, 22:117–148.

TROMBULAK, S. C., AND G. J. KENAGY. 1980. Effects of seed distribution and competitors on seed harvesting efficiency in heteromyid rodents. Oecologia, 44:342–346.

TUCKER, V. A. 1966. Diurnal torpor and its relation to food consumption and weight changes in the California pocket mouse, *Perognathus californicus*. Ecology, 47:245–252.

VORHIES, C. T., AND W. P. TAYLOR. 1922. Life history of the kangaroo rat, *Dipodomys spectabilis spectabilis*. United States Department of Agriculture Bulletin No. 1091:1–40.

WAGNER, H. O. 1961. Die Nagetiere einer Gebirgsabdachung in Sudmexiko und ihre Beziehungen zur Umwelt. Zoologisch Jahrbucher, 89:177–242.

WARD, D. W., AND J. A. RANDALL. In press. Territorial defense in the bannertail kangaroo rat (*Dipodomys spectabilis*): foot-drumming and visual threats. Behavioral Ecology and Sociobiology.

WASER, P. M. 1988. Resources, philopatry, and social interactions among mammals. Pp. 109–130, *in* The ecology of social behavior (C. N. Slobodchikoff, ed.). Academic Press, New York.

WASER, P. M., AND W. T. JONES. 1991. Survival and reproductive efforts in banner-tailed kangaroo rats. Ecology, 72:771–777.

WILSON, M., M. DALY, AND P. BEHRENDS. 1985. The estrous cycle of two species of kangaroo rats (*Dipodomys microps* and *D. merriami*). Journal of Mammalogy, 66:726–732.

ZENG, Z., AND J. H. BROWN. 1987. Population ecology of a desert rodent: *Dipodomys merriami* in the Chihuahuan desert. Ecology, 68:1328–1340.

ECOLOGY OF TROPICAL HETEROMYIDS

VICTOR SANCHEZ-CORDERO AND THEODORE H. FLEMING

Introduction

The family Heteromyidae is distributed over a wide geographic area that includes desert, grassland and subtropical and tropical forest habitats from North America to northern South America. Sixteen species of the genera *Heteromys* and *Liomys* occur in non-desert habitats. These closely related genera probably evolved from a common ancestor (Hafner and Hafner, 1983; Ryan, 1989) and usually occur in different tropical habitats. *Heteromys* is found in humid forests in montane or lowland regions from the Atlantic coast of Mexico to northern South America, while *Liomys* generally occurs in seasonally dry forests or thorn scrub along the Pacific lowlands of Mexico and Central America (Hall, 1981).

Ecological information about tropical heteromyids is scanty and only a few population studies have been conducted in forested habitats. Most of these studies have dealt with one species, *Heteromys desmarestianus,* and were conducted in tropical wet forests located at: Los Tuxtlas in Veracruz, México (Sánchez-Cordero, in press *a*); Monteverde, Costa Rica (Anderson, 1982); La Selva, Costa Rica (Fleming, 1974*a*); and Panama (Fleming, 1971). Population studies of several species of *Liomys* have been conducted in dry tropical or subtropical forests located at: Chamela in Jalisco, México (Pérez-Saldaña, 1978); La Pacifica, Costa Rica (Fleming, 1974*a*) and Panama (Fleming, 1971). Recently, several studies dealing with the foraging behavior of some *Heteromys* and *Liomys* species have been conducted, particularly in Costa Rica (Janzen, 1982*a,* 1982*b,* 1982*c,* 1982*d,* 1983*a,* 1986; Perry and Fleming, 1980). These studies provide an opportunity to detect and explore possible ecological patterns of tropical heteromyids based on within- and between-species comparisons.

The objectives of this study are to: 1) summarize available information on the ecology of tropical heteromyids; 2) compare and contrast long-term studies conducted in different tropical environments; and 3) pose some questions for future research. The chapter is organized into three broad sections. First, we compare the demography, reproduction and social organization of several species, and discuss various abiotic and biotic factors that may influence their ecology. Second, we examine heteromyid community structure and highlight some important gaps in our current information. Finally, we consider the role of heteromyids as seed predators and food in tropical ecosystems. We hope that this review will stimulate ecological research on tropical heteromyids.

Since several studies on tropical heteromyids have not yet been published, it is important to describe briefly the study sites (see Table 1).

Tropical Wet Forests

Los Tuxtlas.—The biological station Estación de Biología Tropical Los Tuxtlas is located in Veracruz, México (18°31′N, 95°W) at an elevation of 150 m. The vegetation is Tropical Wet Forest (sensu Holdridge et al., 1971). Detailed vegetational information can be found in Gómez-Pompa and Vázquez-Yañez (1976) and Ibarra (1985). The rainfall pattern is strongly seasonal with a mean annual total of 4,500 mm. The wet season begins in mid June and lasts through February. August, September and October are characterized by heavy rainfall. The dry season begins in March and extends through May. Seasonal variation in temperature is small with mean daily temperatures ranging from 29°C in the dry season to 17°C in the wet season. Three grids of 0.45 ha were established and trapped for four nights each month between August 1982 and May 1984, except in December 1982, July and October 1983, and February and April 1984 (Sánchez-Cordero, in press *a*).

Monteverde.—This site is located in the Tilarán mountain range of northern Costa Rica (10°N, 85°W) at an elevation of 1,380–1,700 m (Anderson, 1982). The vegetation is Lower Montane Tropical Wet Forest according to the Holdridge et al. (1971) classification. The average annual precipitation is 2,500 mm. Monteverde experiences a dry season from January through April, when less than 100 mm per month of rain falls. The wet season runs from May through November; September and October are usually the wettest months. Daily temperatures are relatively constant but vary seasonally from 14° to 28°C. Three grids of approximately one ha each were established and trapped for five nights each month between September 1978 and December 1980 (Anderson, 1982). *H. desmarestianus* was most abundant in Anderson's upper grid, and these data will be considered here.

La Selva.—This site is located at Finca La Selva in Costa Rica (10°30′N, 84°00′W) at an elevation of 100 m, in the Wet Tropical Forest Zone of Holdridge et al. (1971). Fleming (1974*a*) provides a detailed description of the vegetation. Rainfall is relatively high throughout the year, except in February and March when less than 100 mm of rain falls; average annual precipitation is 4,370 mm. Temperature is relatively constant and varies from a monthly average of 23° to 25°C. One grid, covering 4.8 ha was established and trapped for seven consecutive nights per month for a year between August 1970 and August 1971. The grid was retrapped in July 1972.

Panama (Panama-S).—This site is located on the Atlantic Coast (9°20′N, 79°57′W) at an elevation of 5 m three km west of Cristobal. This locality is a Tropical Wet Forest according to Holdridge et al. (1971). Mean annual temperature is about 25.3°C, and rainfall is seasonal with an eight-month wet season from May through December; total annual rainfall averages 3,250 mm. The dry season lasts from January through April which receive less than 100 mm of rain per month. A grid covering 5.06 ha was established and trapped for nine consecutive nights each month for one year between June 1966 and June 1967 (Fleming, 1971).

Tropical Dry Forests

Chamela.—The biological station at Chamela is located on the Pacific coast of Jalisco, México (19°30′N, 105°03′W). The vegetation is Dry Tropical Forest according to Holdridge et al. (1971). Lott et al. (1986)

TABLE 1.—*Information on the abiotic and biotic characteristics of the study sites.*

Study site	Location	Altitude (m)	Annual precipitation (mm) Range	Total	Dry season (<100 mm/mo)	Annual temperature (°C)	Vegetation	Trapping methods Grid (ha)	Trap nights/mo	Duration of study (mo)	Source
Tropical wet forest											
Los Tuxtlas	18°31′N, 95°W Veracruz, México	150	50–900	4,500	3 mos	17–19	Wet tropical forest	1.3	4	17	Sánchez-Cordero (in press *a*)
Monteverde	10°N, 85°W Tilarán, Costa Rica	1,700	30–500	2,500	3	14–28	Lower montane tropical wet forest	1.0	5	17	Anderson (1982)
La Selva	10°30′N, 84°00′W Finca La Selva, Costa Rica	100	180–520	4,370	0	23–25	Wet tropical forest	4.8	7	12	Fleming (1974*a*)
Panamá-S	9°20′N, 57°W Cristóbal, Panamá	5	20–350	325	6	25	Wet tropical forest	5.1	9	12	Fleming (1971)
Tropical dry forest											
Chamela	19°30′N, 105°03′W Jalisco, México	100	0–325	750	6	15–32	Dry tropical forest			12	Pérez-Saldaña (1978)
La Pacífica	10°28′N, 85°09′W La Pacífica, Costa Rica	45	20–350	1,560	5	27–29	Dry tropical forest	4.8	7	12	Fleming (1974*a*)
Panamá-R	8°51′N, 9°37′W Rodman, Panamá	50	0–120	175	4	25	Dry tropical forest	5.1	9	12	Fleming (1971)

provide a detailed description of the vegetation, rainfall pattern, and temperature of this area. Average annual precipitation is 748 mm, and the rainy season begins in July and ends in late November. A long dry season with less than 50 mm per month occurs from January through June. Mean monthly temperature ranges from 14.8° to 32°C. This study is based on the information obtained from 391 individuals of *Liomys pictus* that were collected and removed from Chamela during one year (Pérez-Saldaña, 1978).

La Pacifica.—This site is located at Finca La Pacifica in Costa Rica (10°28′N, 85°09′W) at an elevation of 45 m, in the Dry Tropical Forest zone of Holdridge et al. (1971). Fleming (1974*a*) provides a detailed description of the vegetation. Rainfall averages 1,562 mm annually and is seasonally distributed. The dry season begins in November and lasts until May. June, September and October are characterized by heavy rains. Temperature is relatively constant and varies from a monthly mean of 27° to 29°C. Trapping methods were the same as those at La Selva (Fleming, 1974*a*).

Panama (Panama-R).—This study site is located on the Pacific coast of Panama (8°51′N, 79°37′W) at an elevation of 50 m. The vegetation is characterized as Dry Tropical Forest (Holdridge et al., 1971). Fleming (1971) provides a detailed description of the vegetation. Rainfall is seasonal with the dry season running from January through April and the wet season, from May through December; total annual precipitation averages 1,750 mm. Temperature is constant throughout the year and averages 25.3°C. Trapping methods were the same as those at Panama-S (Fleming, 1971).

Population Ecology

Most population studies included in this chapter are based on standard mark-recapture techniques, but they differ in several important ways including the duration of the study, frequency of trapping, number and size of grids, etc., that necessarily limit an analytical comparison.

Population density of *H. desmarestianus* can vary significantly between sites and from year-to-year. At Los Tuxtlas, density fluctuated from two to 50 individuals/ha, being low during the rainy season in 1982, and high during the dry and rainy seasons in 1983 (Fig. 1). Immatures (juveniles and subadults) were abundant during the dry and early rainy seasons in 1983 and decreased in density during the late rainy seasons in 1982 and 1983 (Sánchez-Cordero, in press *a*). Population density at Monteverde varied from three to 25 individuals/ha. A decrease in density occurred during the last part of 1978 when rainfall declined, and increased during the rainy season in 1979. This pattern was not repeated in 1980, when population density remained relatively low during the rainy season (Fig. 1). Immatures were abundant during the rainy season of 1979 but represented a small proportion of the population during the rainy seasons of 1978 and 1980 (Anderson, 1982). Population density of *H. desmarestianus* at La Selva ranged from seven to 18 individuals/ha. An increase in density was observed during the rainy season in 1971, and density peaked during the dry season of 1972 (Fig. 1). Immatures were most abundant during the rainy season in 1971. As the rainy season progressed, adult individuals dominated the population structure (Fleming, 1974*a*). *H. desmarestianus* density at Panama-S ranged from zero to two individuals/ha; it increased gradually during the rainy season in 1966, and reached a maximum during the 1967 dry season (Fig. 1). Immatures were most numerous during the 1966–67 dry season (Fleming, 1970).

Densities of *H. desmarestianus* appeared to be related to both the seasonality and the year-to-year variation in total precipitation at these sites. Rainfall at Los Tuxtlas is highly seasonal and shows strong year-to-year variation. Precipitation at Monteverde is less seasonal than at Los Tuxtlas, but also shows year-to-year variation. Rainfall at La Selva

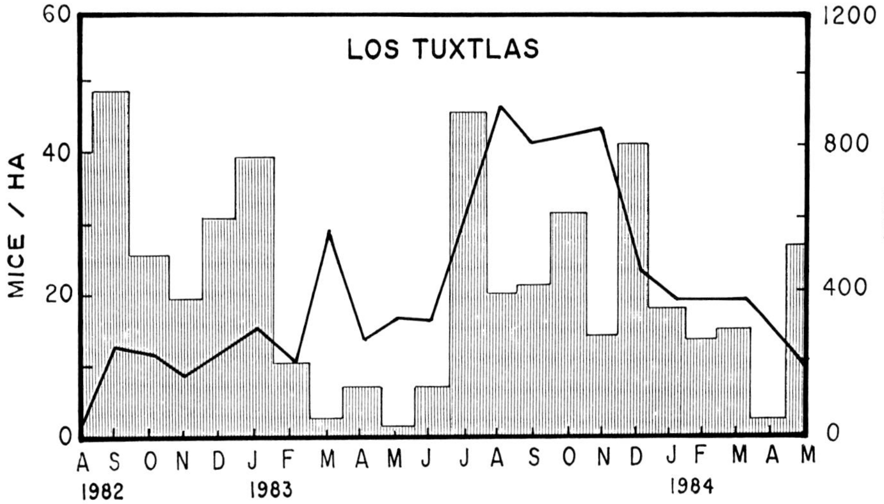

FIG. 1.—Population densities of *Heteromys* and *Liomys* at different locations in relation to rainfall. *H. desmarestianus* densities at Los Tuxtlas, Monteverde, La Selva and Panama-S are shown. Data from La Pacifica and Panama-R represent *L. salvini* and *L. adspersus*, respectively. All precipitation records correspond to years when the study was undertaken, except at La Selva and La Pacifica where monthly mean records for ten years were graphed. Lines represent adults, and shaded areas, precipitation records.

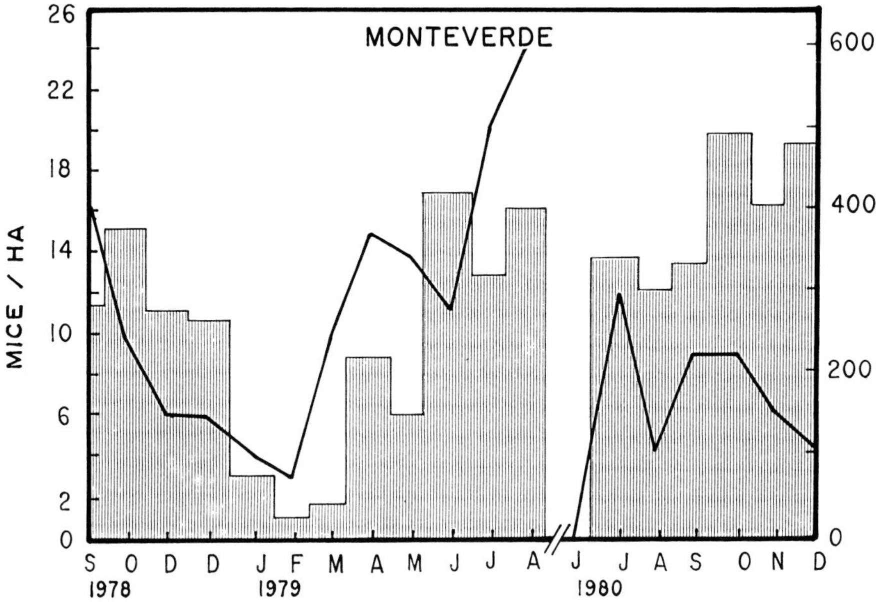

FIG. 1.—Continued.

is relatively non-seasonal compared with Los Tuxtlas and Monteverde, and Panama-S exhibited a seasonal pattern but had the lowest total precipitation. Correspondingly, *H. desmarestianus* density at Los Tuxtlas varied more than 20-fold, while at Monteverde variation was less than 10-fold. Population levels at La Selva showed a two-

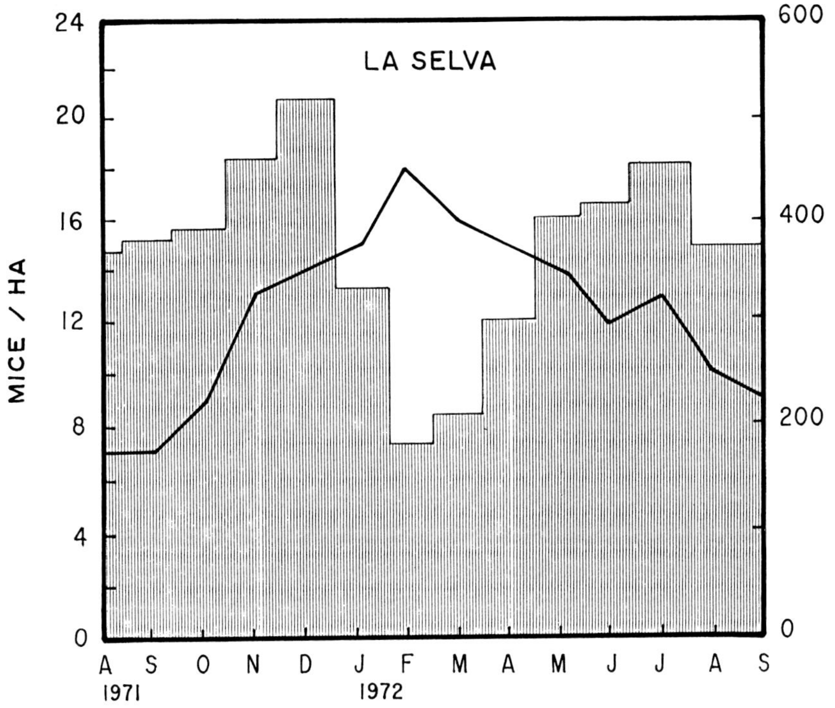

FIG. 1.—Continued.

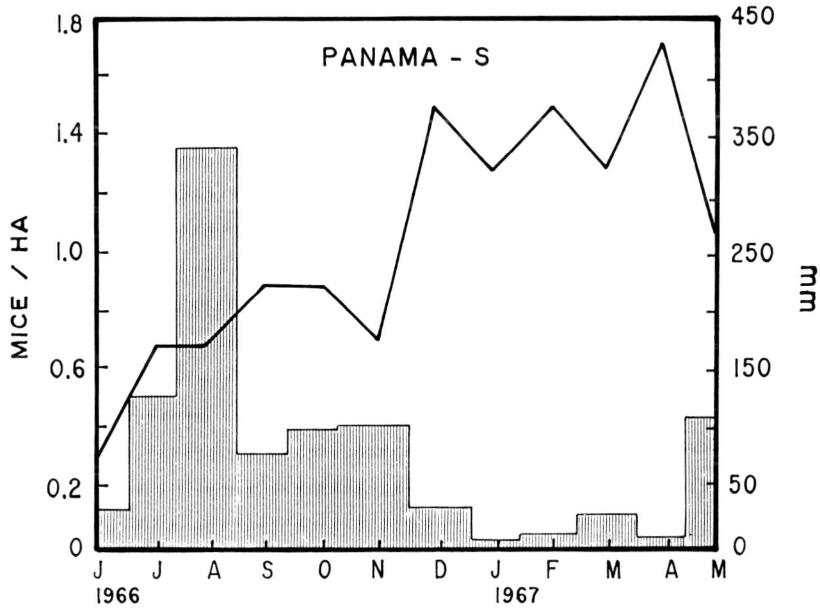

FIG. 1.—Continued.

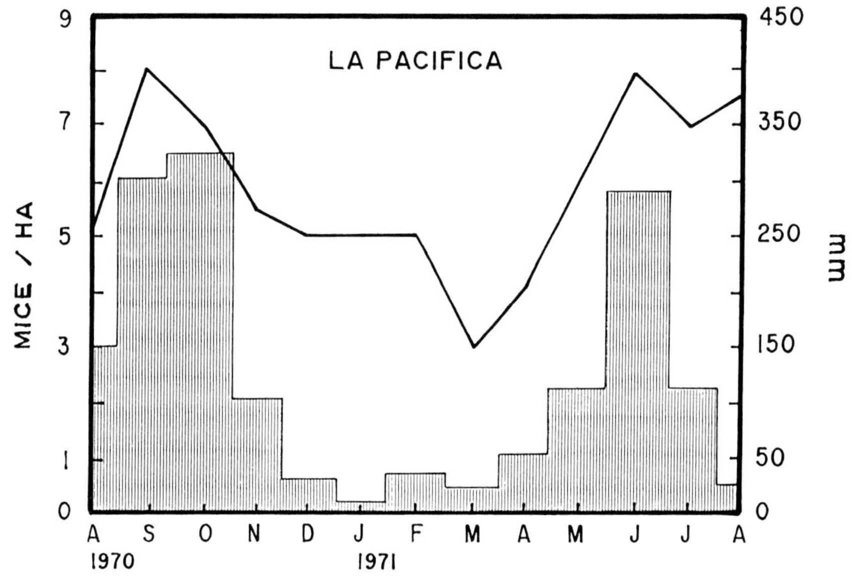

FIG. 1.—Continued.

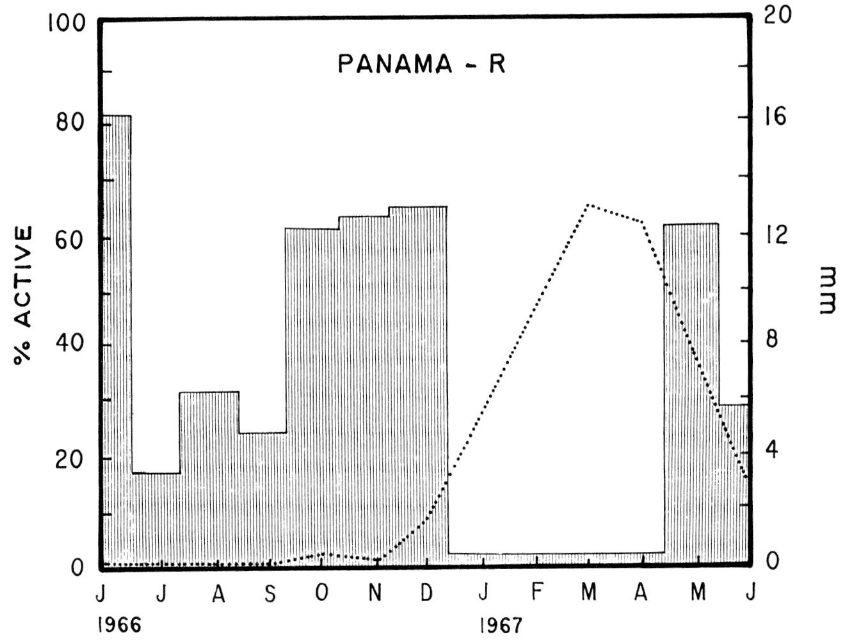

FIG. 1.—Continued.

fold variation, and at Panama-S density was the lowest of all locations (Table 2). Year-to-year variation in population density was evident at Los Tuxtlas and Monteverde. Fluctuations in density were greater at Los Tuxtlas in 1983 than in 1984, and at Monteverde in 1979 compared with 1980 (Fig. 1).

Major population increases in *H. desmarestianus* resulted from high recruitment of immature individuals during the rainy season, a period which usually corresponds to a fruitfall peak in most tropical wet forests (Los Tuxtlas: Alvarez, 1984; Monteverde: Anderson, 1982; Wheelwright, 1985; La Selva: Fleming, 1974*a*). This pattern ap-

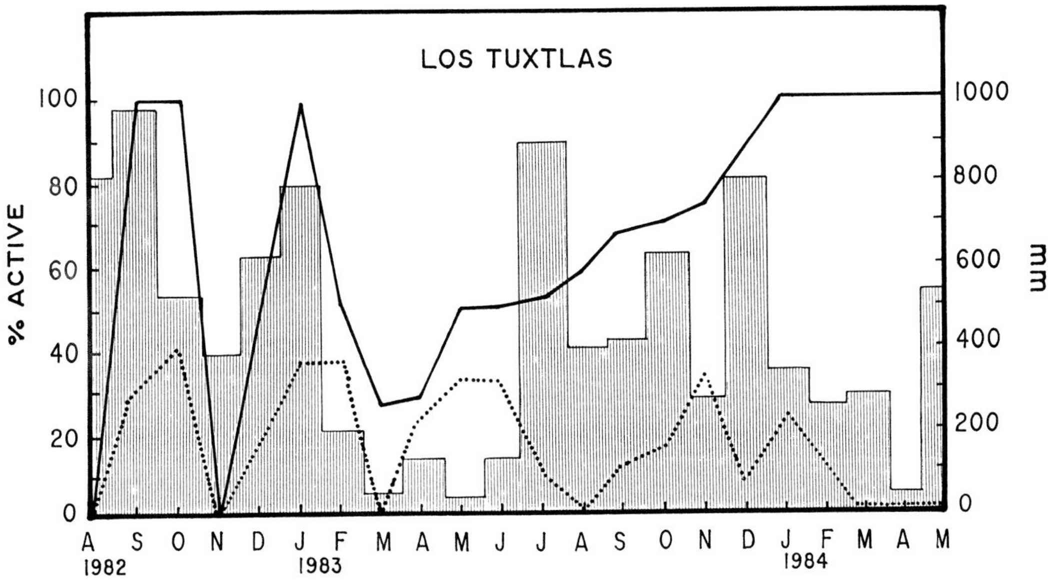

FIG. 2.—Reproduction in *Heteromys* and *Liomys* at different locations in relation to rainfall. *H. desmarestianus* was studied at Los Tuxtlas and La Selva-S. *Liomys pictus, L. salvini* and *L. adspersus* were studied at Chamela, La Pacifica and Panama-R, respectively. Breeding activity includes pregnancy or lactation in females and scrotal testes in males. Lines represent males; broken lines, females; and shaded areas, precipitation records.

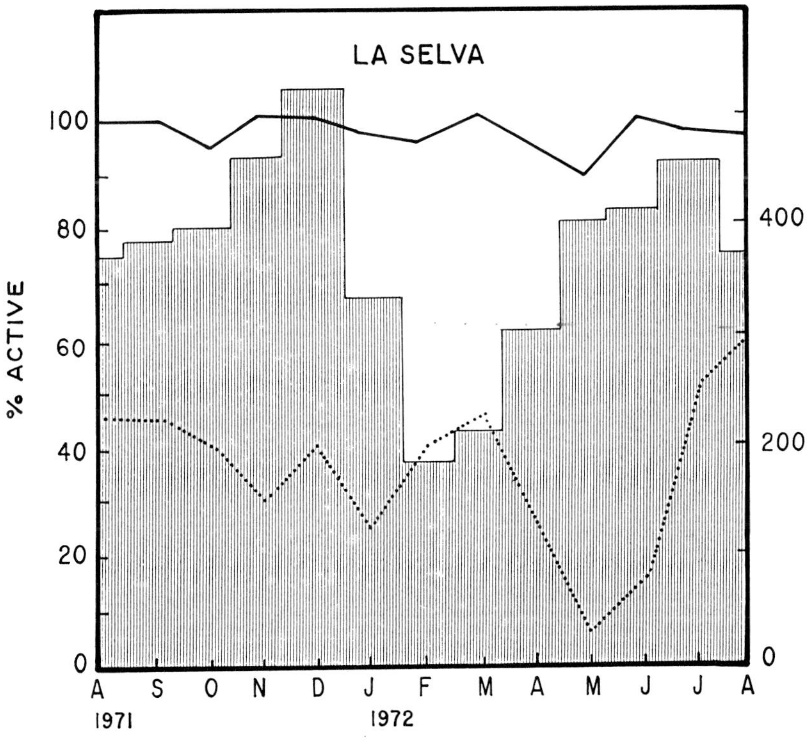

FIG. 2.—Continued.

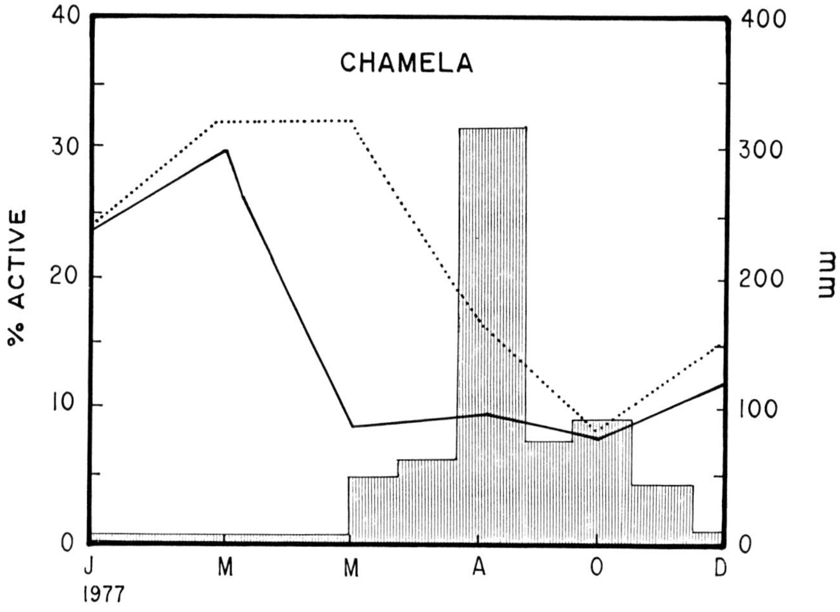

FIG. 2.—Continued.

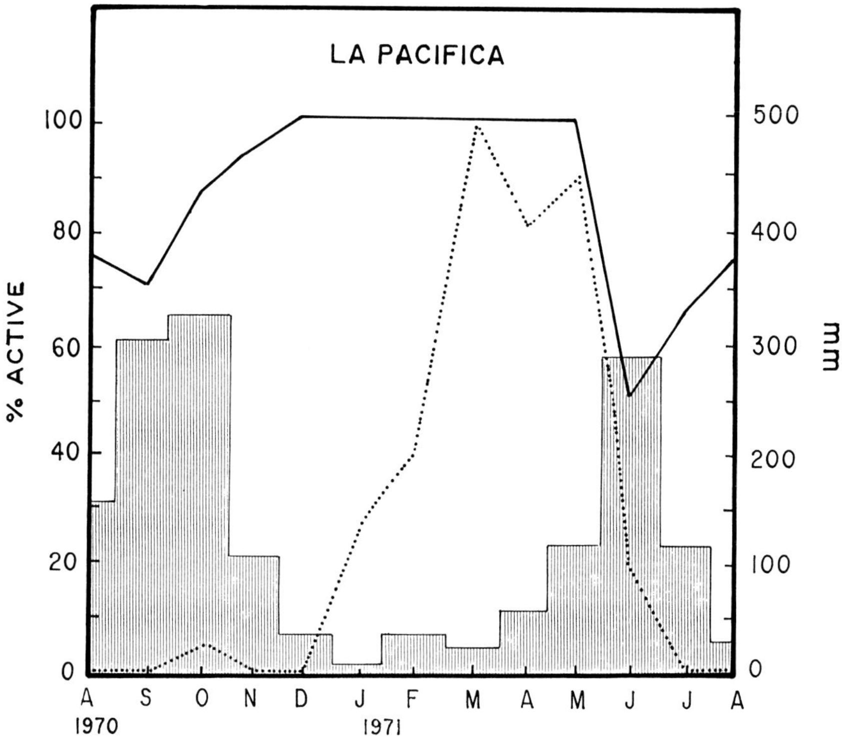

FIG. 2.—Continued.

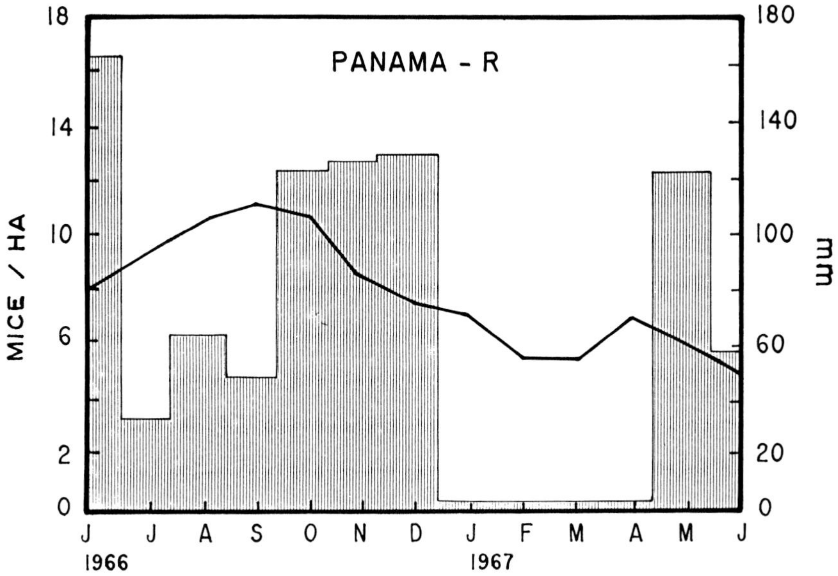

FIG. 2.—Continued.

peared not to hold at Panama-S where immatures were relatively numerous late in the rainy season when low levels of fruitfall were observed (Fleming, 1971: table 18). Survivorship differed among populations of *H. desmarestianus*. At La Selva, individuals lived longer and population annual survival was higher than at Los Tuxtlas (Table 2). In both locations, however, most individuals that were trapped regularly disappeared from the grid after one year (Fleming, 1974*a*; Sánchez-Cordero, in press *a*). No data on survivorship of adults or immatures are available at Monteverde or Panama-S for comparison.

Liomys densities fluctuated seasonally and appeared to be related to the rainfall pattern. Population density of *Liomys salvini* at La Pacifica ranged from four individuals/ha in the dry season to eight individuals/ha in the rainy season in 1971 (Fig. 1). A high proportion of adults was present in the dry season, and most immatures were captured in the rainy season (Fleming, 1974*a*). At Panama-R, *L. adspersus* density averaged 10 individuals/ha during the 1966 rainy season, declining to five individuals/ha during the dry season (Fig. 1). Recruitment of juveniles began in the dry season and peaked early in the rainy season of 1967 (Fleming, 1971).

L. salvini and *L. adspersus* densities increased early in the rainy season and then decreased as the rainy season progressed and during the dry season. No information is available for evaluating year-to-year demographic variation in tropical dry forests, except for the limited information from *L. salvini*. Fleming (1974*a*) observed that population levels appear to be stable between years at La Pacifica, but D. Janzen (pers. comm.) has found considerable year-to-year variation in *L. salvini* density at Santa Rosa National Park in Costa Rica. Certainly, more data on this topic are badly needed to determine the effects of temporal habitat variation on population dynamics in tropical heteromyids. Scanty data for *L. pictus* at Chamela indicated a density fluctuation of 5–30 individuals/ha. Reproductive data (see below) suggest that juvenile numbers peaked at the beginning of the rainy season in 1977.

Presumably, this recruitment would lead to an increase in population density.

In *L. salvini* and *L. adspersus*, high recruitment of juveniles at the beginning of the rainy season contributed significantly to increased population densities. At the end of the rainy season, and especially during the dry season, population structure was composed mostly of adult individuals. In both species, each year-class dominated the population only for one year, but some individuals were known to live almost two years (Fleming, 1971, 1974*a*; Table 2).

In general, *Heteromys* reached higher densities than *Liomys*, probably because of the greater productivity of tropical wet forests than dry forests. *Liomys* appeared to have a lower adult and immature survivorship compared to *Heteromys*, and their populations consisted of well-defined cohorts due to their greater reproductive seasonality (Table 2). Population densities in tropical heteromyids appear to vary considerably more than those of desert heteromyids, in which densities usually vary two-fold (Fleming, 1974*a*, table 9; but see Brown and Harney, this volume). However, both heteromyid groups show strong spatial and temporal variations in population density (see also Brown and Harney, this volume).

Reproduction

In general, tropical heteromyids reproduced both during the dry and rainy seasons. At Los Tuxtlas, peaks in percent lactating were observed at the end of the dry season in 1983 and in the 1982 and 1983 rainy seasons. Males with scrotal testes were observed during the rainy seasons of 1982 and 1983 (Sánchez-Cordero, in press *a*). At Monteverde, females were pregnant or lactating during the dry and rainy seasons in 1978–79 and rainy season in 1980 (Anderson, 1982). At La Selva, lactating females were observed in the dry and rainy seasons in 1971 and 1972, and males were found with scrotal testes throughout the study

TABLE 2.—*Comparative data on population ecology of several* Heteromys *and* Liomys *species at different locations in tropical wet and dry forests.*

Species	Locality	Population density (individuals/ha)			Survivorship		Reproduction					Source
		Range	Mean	*CV*	Adult longevity (mo)	Annual survival	Breeding season (mo)	% Breeding females	Litters/yr	Avg. litter	Annual product.	
Heteromys desmarestianus	Los Tuxtlas	2–50	21	0.63	24	0.20–0.26	8	0–40	2.1	3.0	6.3	Sánchez-Cordero (in press *a*)
H. desmarestianus	Monteverde	3–25	11	0.58	24		7		2.0			Anderson (1982: table 31)
H. desmarestianus	La Selva	7–18	12	0.27	27	0.21–0.31	10	5–60	2.9	3.1	9.0	Fleming (1974*a*)
H. desmarestianus	Panama-S	0–2	1	0.37			12					Fleming (1970)
Liomys pictus	Chamela	5–30					6					Pérez-Saldaña (1978) Sánchez-H, and G. Ceballos (pers. comm.)
L. salvini	La Pacifica	4–8	6	0.25	17	0.18	6	0–100	1.8	3.8	6.8	Fleming (1974*a*)
L. adspersus	Panama-R	5–11	8	0.24	21	0.28	6	0–70	1.4	4.0	5.6	Fleming (1971)

(Fleming, 1974*a*). At Panama-S, pregnant or lactating females were observed during the rainy seasons of 1966 and 1967 (Fleming, 1970) (Fig. 2).

The reproductive pattern in *H. desmarestianus* was somewhat different between locations. The breeding season appeared to be shorter in seasonal than non-seasonal wet forests, but more data are needed to confirm this pattern. The percentage of breeding females was slightly higher, and female annual productivity appeared to be higher, at La Selva than at Los Tuxtlas (Table 2). Perhaps La Selva provides more uniformly distributed and less-limited food resources due to a more regular precipitation pattern. In both locations, juveniles appeared to mature sexually in the reproductive season following their birth, at an age of approximately eight months (Fleming, 1974*a*; Sánchez-Cordero, in press *a*). Few data on reproduction of other species of *Heteromys* are available. *H. anomalus* is known to begin breeding early in the rainy season in Venezuela (Rood, 1963; Rood and Test, 1968).

Reproduction was far more seasonal in *Liomys* than in *Heteromys*. At Chamela, *L. pictus* females were observed lactating during most of the dry and early rainy seasons. Most males had inguinal testes during the dry season of 1977; their frequency decreased in the rainy season in 1977 (Pérez-Saldaña, 1978). Breeding in *L. salvini* was seasonal at La Pacifica, with female reproduction occurring in the dry and early rainy seasons in 1971. Males had inguinal testes most of the year (Fleming, 1974*a*). At Panama-R, female and male reproduction in *L. adspersus* peaked in the dry and early wet seasons in 1967 (Fleming, 1971) (Fig. 2).

Although breeding activity in *Liomys* was mostly restricted to the dry and early rainy seasons, there were some interspecific differences in *Liomys*' reproductive patterns. The percentage of breeding adults differed between locations. At Chamela, only 32% of *L. pictus* females in the population were lactating during the dry season, but pregnant females were found throughout most of the year (Pérez-Saldaña, 1978). Percent female *L. salvini* breeding reached 100% in the population during the dry season at La Pacifica; no evidence of reproduction was observed in the rainy season. Percent female *L. adspersus* breeding reached 70% in the population, but no reproduction occurred for almost seven months of the year (Fleming, 1971). Females of *L. salvini* and *L. adspersus* had similar annual productivities of one or two litters (see Table 2).

Heteromys and *Liomys* differed in their longevities and breeding frequencies. In *Heteromys*, a few individuals were long-lived and bred several times, but most individuals were relatively short-lived and reproduced only once (Fleming, 1974*a*; Sánchez-Cordero, in press *a*). In *Liomys*, annual turnover is almost complete and in the next season yearlings constitute nearly 100% of the breeding population (Fleming, 1971, 1974*a*). These patterns result in differences in the reproductive success of individuals. The reproductive output in *Heteromys* may be concentrated in a few long-lived individuals; breeding in *Liomys* may be nearly equal for individuals having similar (restricted) longevity (see Munger et al., 1983).

In general, tropical heteromyids showed longer breeding seasons than desert heteromyids, but there appeared to be no striking differences regarding average litter size or female annual productivity between both heteromyid groups (Fleming, 1974*a*: table 9). In contrast, reproductive characteristics between tropical heteromyids and cricetids, which commonly occur sympatrically, differ significantly. Cricetids have similar litter sizes, but female annual productivity is higher and annual survivorship shorter compared with heteromyids (Fleming, 1971: table 22).

Several hypotheses have been advanced to explain the demographic and reproductive patterns of heteromyids in desert habitats (Brown et al., 1979; Munger et al., 1983). Below, we consider these explanations for tropical heteromyids. We discuss the importance of various factors influencing demographic and reproductive characteristics given the present ecological knowledge of this group and present some suggestions for future research.

Resource Availability

Resources such as food and water play a major role in the population dynamics of desert heteromyids. In desert species, population fluctuations and reproduction are affected by the timing and amount of rainfall, which in turn is positively associated with seed and plant productivity, and thus food availability (Beatley, 1974; Brown, 1975; Brown et al., 1979; French et al., 1974; Price, 1978; Reichman and Van der Graaff, 1975; see also Conley et al., 1977 and Munger et al., 1983 for reviews). Do population fluctuations in tropical heteromyids result from spatial and temporal differences in habitat productivity? There is some evidence that resource availability is associated with rainfall pattern in tropical ecosystems. In tropical wet forests, the availability of potential heteromyid food items such as seeds, fruits and litter arthropods can vary seasonally, with peaks usually occurring in the rainy season (Los Tuxtlas: Alvarez, 1984; Monteverde: Anderson, 1982; La Selva: Fleming, 1974*a*; Panama-R: Fleming, 1971; Barro Colorado Island (BCI): Foster, 1982; see Levings and Windsor, 1982 for review on litter arthropod densities). It has been shown, however, that significant year-to-year variation in the production of fruits and seeds may occur in some tropical wet forests (e.g., Los Tuxtlas: Alvarez, 1984; Monteverde: Wheelwright, 1985; BCI: Foster, 1982).

Information pointing to a relationship between tropical dry forest productivity and precipitation is scanty. Fleming (1971, 1974*a*) demonstrated that peaks in fruitfall occurred at the end of the dry and early rainy seasons and that food levels decreased significantly later in the rainy season at La Pacifica and in Panama. At Chamela, Pérez-Saldaña (1978: fig. 18) reported a peak in seed species diversity during the dry season and a significant decrease during the rainy season in 1977. These results support the general idea that tropical mammals synchronize reproduction to produce offspring during food peaks (Anderson, 1982; Delany, 1972; Field, 1975; Fleming, 1970, 1971, 1974*a*; Glanz et al., 1982; Happold, 1977; Rood, 1963; Rood and Test, 1968; Russell, 1982; Taylor and Green, 1976).

Observed correlations between rodent demography and tropical food availability are mostly based on general measures of fruit or insect numbers. It is unlikely that such measures correspond directly to the food items actually consumed by heteromyids, since many tropical seeds are inedible or too big for them, and thus, these patterns must be interpreted with caution. Furthermore, the seasonality of fruitfall may be buffered for heteromyids if the seeds they harvest are durable enough to be stored for a later season. Some species of *Heteromys* and *Liomys* have shown a strong hoarding behavior in captivity (Fleming, 1974*b*). Cheek-pouch contents from trapped animals may also provide information on what these mice eat and when it is available. Long-term studies that monitor spatial and temporal variation in food resources such as fruits, seeds or litter arthropods known to be consumed by heteromyids are needed to determine the effect of resource variation on population dynamics of tropical heteromyids.

Predation

To our knowledge, no studies designed to determine the impact of predation on rodent demography have been conducted in tropical habitats. Much of the evidence documenting predation is indirect or anecdotal (Anderson, 1982; Fleming, 1971, 1974*a*;

Sánchez-Cordero, in press *a*, in press *b*). Potential mammalian predators in tropical forests include coyotes (*Canis latrans*), tayras (*Eira barbara*), skunks (*Spilogale putorius*), coatis (*Nasua narica*), raccoons (*Procyon lotor*), long-tailed weasels (*Mustela frenata*), opposums (*Philander opposum, Didelphis marsupialis* and *D. virginiana*) and ocelots (*Felis pardalis*); avian predators include owls (*Ciccaba virgata* and *Tyto alba*) and semiplumbeous hawks (*Leucopternis semiplumbea*). Other predators include the common boa (*Boa constrictor*) and other species of snakes (*Spilotes pullatus, Bothrops atrox* and *Lachesis muta*). Hooper and Carleton (1976) provide a more complete list of rodent predators in tropical habitats.

The diversity of predators suggests that predation may play an important role in the population dynamics of tropical heteromyids. Do demographic differences among *H. desmarestianus* populations result from differences in predation pressure? The diversity and abundance of predators can vary in time and space in tropical habitats. Perhaps differences in diversity and/or abundance of predators at Los Tuxtlas, Monteverde, La Selva and Panama-S account for geographic differences in the demography of *H. desmarestianus*. It would be interesting, for example, to determine if the observed differences in survivorships between these locations and between seasons of the year are the result of different predation rates.

Unfortunately, the most direct way to determine the effect of predation on heteromyids is to remove predators and measure the demographic responses of rodents. This is particularly difficult to do in tropical habitats due to the great diversity of predators. Therefore, most information regarding predator-prey interactions must rely on indirect evidence. One possible way to examine the importance of predation is to study rodent demography in contrasting sites where predator diversity and abundance are known to differ. Anderson (1982) made an initial effort along these lines with *H. desmarestianus*. He observed the frequency of trap disturbances, presumably produced by predators on his three grids. A significant difference in the frequency of trap disturbances was observed between grids. Interestingly, population density of *H. desmarestianus* at grids with high trap disturbances was lower, perhaps as a result of different predation rates (Anderson, 1982).

Parasitism

Parasitism has been recognized as another factor influencing the population dynamics of small mammals. Several studies have demonstrated that parasitic infections affect survivorship and fecundity of rodents (Errington, 1954; Timm and Cook, 1979). Little information is available on host-parasite relationships in tropical heteromyids. Anderson (1982) observed that *H. desmarestianus* was commonly infested by fleas, ticks and mites. Perhaps a higher parasite diversity might permit more frequent outbreaks of lethal parasites, leading to greater population fluctuations among tropical than desert heteromyids. Future research should quantitatively evaluate and determine the influence of parasitism on the demography and reproduction of tropical heteromyids. Munger et al. (1983) discuss approaches to this problem.

Competitive Interactions

Competitive interactions have been considered as a key factor affecting rodent densities in communities of desert heteromyids. Much indirect evidence suggests that strong competitive interactions occur between species, particularly for food resources (Abramsky, 1978; Brown, 1975; Brown and Harney, this volume, and references therein; Price and Brown, 1983). More recently, direct evidence through experimental removal of competing species resulted in an increased density in some (Munger and Brown, 1981), but not all (Schroeder and Rosenzweig, 1975), of the remaining species. This experimental approach provides strong evidence for competitive interactions between species of rodents, although it is unclear whether these interactions are ruled by exploitative or interference competition (Rosenzweig et al., 1975).

So far, research on tropical heteromyids has not addressed the question of the role of competitive interactions on population dynamics. However, indirect evidence suggests that competition may be an important factor affecting heteromyid densities in communities of tropical rodents. It is likely that other species, particularly those storing seeds, may be the most important mammal competitors of *Heteromys* and *Liomys*. These undoubtedly include peccaries (*Tayassu* and *Dicotyles*), dasyproctids (*Agouti* and *Dasyprocta*), squirrels (*Sciurus*), spiny rats (*Proechimys*), and rice rats (*Oryzomys*), among others (Kiltie, 1981; Smythe, 1970, 1986; Smythe et al., 1982). If competition is important, then forests with a high diversity and density of competitors should lead to a lower heteromyid density compared to forests where competitors are absent or occur at low densities. *H. desmarestianus* population patterns with and without these potential competitors suggest interesting trends. For example, at Los Tuxtlas, peccaries, agoutis and rice rats occur at extremely low densities (Estrada and Coates, 1986), and spiny rats are absent from this forest (Navarro, 1982; Sánchez-Cordero, in press *b*). At Monteverde and La Selva, the rodent fauna is more diverse than at Los Tuxtlas, but spiny rats are absent from these forests too (Anderson, 1982; Fleming, 1974*a*). In contrast, all above potential competitors occur sympatrically with *H. desmarestianus* at Panama-S. Population densities of *H. desmarestianus* were much higher at Los Tuxtlas than at Monteverde and La Selva; densities were lowest at Panama-R (Table 2).

Social Organization

Studies of the social organization of tropical heteromyids have involved the analysis of spatial distribution and home range dynamics under field conditions and behavioral tests staged in neutral arenas under laboratory conditions (Eisenberg, 1963; Fleming, 1971, 1974*a*, 1974*b*; Quintero and Sánchez-Cordero, 1989; Rood, 1963; Rood and Test, 1968).

At Los Tuxtlas, females moved farther between recaptures in the dry season than in the rainy season, but male movements were greater during the rainy season than in the dry season. Female home ranges were largest during the dry season, but male ranges were larger in the rainy season than in the dry season (Quintero and Sánchez-Cordero, 1989). At Monteverde, females showed greater movements than males, but both sexes had equal home range sizes (Anderson, 1982). At La Selva, *H. desmarestianus* males showed seasonal differences with more restricted movements occurring during the dry season, but females showed no seasonal differences. In general, males tended to move greater distances than females and had larger home ranges (Fleming, 1974*a*; see Table 3). No information on the seasonality of home range is available for other sites.

Distances moved by individuals and home range sizes showed several trends in *H. desmarestianus*. Seasonal differences in movements and home range size for both sexes appeared to be related to the reproductive condition of individuals. At Los Tuxtlas, lactating females showed smaller home ranges and shorter movements and appeared to be territorial compared to non-reproductive females (Quintero and Sánchez-Cordero, 1989). Perhaps lactating females establish territories in microhabitats with high food availability—e.g., "seed shadows"—to fulfill the high energetic demands imposed by reproduction (Millar, 1979; Quintero and Sánchez-Cordero, 1989). Compared to non-reproductive males and females, reproductive males with scrotal testes tended to move greater distances

TABLE 3.—*Seasonal movements and home range variations in several species of tropical heteromyid rodents.*

		Average distance between recaptures (m)				Home range (ha)			
		Females		Males		Females		Males	
Species	Study site	Dry	Wet	Dry	Wet	Dry	Wet	Dry	Wet
Heteromys desmarestianus	Los Tuxtlas	20.0	16.5	12.5	17.8	0.020	0.014	0.012	0.023
H. desmarestianus	Monteverde	20.0		19.0		0.15		0.15	
H. demarestianus	La Selva	24.8	26.5	28.1	39.4	1.17	1.23	1.51	1.60
Liomys salvini	La Pacifica	31.5	32.8	37.8	47.1	1.67	1.55	2.21	1.95

and had larger home ranges, perhaps to gain access to mates and food (Fleming, 1974*a*; Quintero and Sánchez-Cordero, 1989).

Field and laboratory data support the idea that *Heteromys* has evolved towards a "loose" type of social organization (Eisenberg, 1963). Extensive home range overlap occurred among individuals of *H. desmarestianus* at La Selva, but not at Los Tuxtlas, and laboratory behavioral tests demonstrated that *H. desmarestianus* (Fleming, 1974*b*) and *H. anomalous* (Rood, 1963; Rood and Test, 1968) were tolerant of conspecifics. A similar social behavior has been suggested for *H. goldmani* (Wagner, 1961).

Movement patterns in *L. salvini* indicated that males moved greater distances than females throughout the year at La Pacifica. Males showed greater movements in the breeding season, but no seasonal differences in female movements were observed. Males had larger home ranges than females (Fleming, 1974*a*). In *L. adspersus,* males moved greater distances between captures than females at Panama-R. There were seasonal differences in the movements of males, which moved greater distances in the dry season, when individuals were reproductively active, than in the wet season. Females showed no significant differences in recapture distances between seasons. There were no significant differences in home range size between females and males (Fleming, 1971; see Table 3).

Liomys movements and home range dynamics differ somewhat from the patterns observed in *Heteromys.* Seasonal differences were observed in males, but not in females. Like *Heteromys,* these differences appeared to be related to reproductive condition. Reproductive males move greater distances than females in *L. salvini* (Fleming, 1974*a*; McPherson, 1967) and *L. adspersus* (Fleming, 1971), and home range size differed between sexes in *L. salvini* but not in *L. adspersus* (Fleming, 1971, 1974*a*). Unlike *Heteromys,* female *Liomys* showed no seasonal differences in movements and home ranges. These results are surprising because food availability is far more seasonal in tropical dry forests than tropical wet forests (Fleming, 1974*a*).

Laboratory experiments demonstrated that *Liomys* was less tolerant of conspecifics than *Heteromys* (Eisenberg, 1963; Fleming, 1974*b*). Dominance, survivorship during the breeding season, and perhaps reproductive success appeared to be related to individual body size in *Liomys* (Fleming, 1974*b*). Home range overlap was less in *Liomys* than in *Heteromys,* suggesting a more pronounced aggressive behavior towards conspecifics (Eisenberg, 1963; Fleming, 1971, 1974*a*; Wagner, 1961).

Differences in the degree of environmental seasonality appear to affect the social organization of tropical heteromyids. Conspecific intolerance appears to increase as the environment becomes more seasonal (e.g., increased rainfall seasonality) and when food scarcity increases. Non-overlapping home ranges among individuals of *H. desmarestianus* were observed at Los Tuxtlas, a seasonal wet forest (Quintero and Sánchez-Cordero, 1989). In contrast, extensive home range overlap and conspecific tolerance in this species were observed at La Selva, an aseasonal wet forest (Fleming, 1974*a*, 1974*b*). Thus, we should expect more pronounced aggressive behavior in *Heteromys* species inhabiting drier and less productive wet forests than in *H. desmarestianus* at the study sites. Aggressive behavior should also increase in *Liomys* species inhabiting highly seasonal and less productive habitats, such as semiarid regions (e.g., *L. pictus* and *L. irroratus*). In dry tropical forests, for example, home ranges of *L. adspersus* overlapped extensively in Panama, whereas home ranges of *L. salvini* were non-overlapping seasonally at La Pacifica. Curiously, however, *L. adspersus* had larger home ranges than those of *L. salvini,* although the Panamanian habitat appeared to be more productive (Fleming, 1974*a*). In arid regions of Mexico, individuals of *L. pictus* and *L. irroratus* segregate by sex into different parts of their habitat (Wagner, 1961).

Community Structure

Tropical habitats are characterized by a low diversity of heteromyids. Rarely can one find two species of heteromyids occurring sympatrically, whereas six or more species of heteromyids can coexist in desert habitats. One explanation for the difference of species richness between tropical and desert habitats lies in the availability and diversity of seeds (Fleming, 1984). For a seed-eating rodent, desert habitats are far more productive than tropical habitats. For example, in North American deserts seeds of annual plants can reach a density of $91{,}000/m^2$. Further, several studies focusing on the structure of desert heteromyid communities have demonstrated that coexistence between species resulted from differential exploitation of resources (Brown, 1975; Brown and Harney, this volume, and references therein; Rosenzweig, 1973). Large species belonging to the genus *Dipodomys* consume large seeds, whereas small species of the genera *Perognathus* and *Microdipodops* consume small seeds. A characteristic feature of tropical heteromyids, however, is that neither *Heteromys* nor *Liomys* contain such small-sized species. Although tropical forests are rich in fruits and seeds, many of them are too large and hard, heavily-armored or contain poisonous chemical compounds that make them unavailable even for large-sized seed-eating heteromyids (Janzen, 1971, 1981, 1982*d*; Janzen and Higgins, 1979; Martínez-Gallardo and Sánchez-Cordero, in press). Nevertheless, species of *Heteromys* and *Liomys* eat fruits and seeds of *Enterolobium cyclocarpum* and *Nectandra ambigens,* which have poisonous compounds (Fleming, 1974*a*; Janzen, 1982*a*, 1982*b*; Martínez-Gallardo and Sánchez-Cordero, in press). Also, many seeds in tropical forests are rapidly dispersed or removed by competitors, reducing the profitability of seed sources to heteromyids (Janzen, 1983*a*, and references therein). It would be interesting to investigate experimentally which seeds are inedible and why, and which seeds eaten by heteromyids are rapidly dispersed by other animals in a particular forest.

Furthermore, *Heteromys* and *Liomys* appear to differ strongly in their physiological characteristics, which reduces their chances of geographic overlap. *Liomys* is more resistant to water and food deprivation than is *Heteromys* (Fleming, 1977). The distribution of *Liomys* corresponds to dry forests along the Pacific coastline and inland, whereas the distribution of *Heteromys* corresponds to wet forests along the Atlantic coast and mountain ranges.

Importance of Heteromyids in Tropical Habitats

Here we attempt to provide an overview of the importance of heteromyids in tropical habitats by emphasizing their role as predators of fruits and seeds and as prey for a large number of vertebrates.

A characteristic feature of the family Het-

TABLE 4.—*List of the genera of plants whose fruits and seeds are known to be consumed by* Heteromys *and* Liomys *at different locations in tropical wet and dry forests, respectively. Plant species are given when available.**

Heteromys	
Spondias sp.	Anacardiaceae
Cymbopetalum bailloni	Annonaceae
Pleuranthodendron mexicana	Flacourtiaceae
Nectandra ambigens	Lauraceae
Pithecellobium sp.	Leguminosae
Pterocarpus sp.	Leguminosae
Vatairea sp.	Leguminosae
Guarea glabra	Meliaceae
Brosimum alicastrum	Moraceae
Cecropia obtusifolia	Moraceae
Ficus sp.	Moraceae
Poulsenia armata	Moraceae
Pseudolmedia oxyphylaria	Moraceae
Astrocaryum mexicanum	Palmae
Chamaedora tepejilote	Palmae
Psychotria sp.	Rubiaceae
Sapindus sp.	Sapindaceae
Turpinia occidentalis	Staphyleaceae
Liomys	
Mangifera indica	Anacardiaceae
Spondias mombin	Anacardiaceae
Forsteronia sp.	Apocynaceae
Cordia elaeagnoides	Boraginaceae
Couepia polyandra	Chorysobalanaceae
Cochlospermum sp.	Cochlospermaceae
Combretum farinosum	Combretaceae
Ipomea arborescens	Convolvulaceae
Momordica charantia	Cucurbitaceae
Cnidoscolus tubulosus	Euphorbiaceae
Euphorbia cnidoscolus	Euphorbiaceae
Manihot colimensis	Euphorbiaceae
Lasiacis divaricata	Graminea
Acacia farnesiana	Leguminosae
Aploplanesia paniculata	Leguminosae
Bahuinia pauletta	Leguminosae
Cassia nutans	Leguminosae
Caesalpina sclerocarpa	Leguminosae
Entadopsis polystachya	Leguminosae
Enterolobium cyclocarpum	
Hymenaea sp.	Leguminosae
Lonchocarpus	Leguminosae
parviflorus	Leguminosae
Nissolia fruticosa	Leguminosae
Phaseolus adenanthus	Leguminosae
Phaseolus lunatus	Leguminosae
Phaseolus microcarpus	Leguminosae
Phaseolus speciosus	Leguminosae
Pithecellobium sp.	Leguminosae
Pityrocarpa constricta	Leguminosae
Sesbania sp.	Leguminosae
Tephrosia sp.	Malvaceae
Abutilon sp.	Malvaceae
Malvaviscus sp.	Malvaceae
Sida sp.	Malvaceae
Ficus sp.	Moraceae
Astrocaryum standleyanum	Palmae
Corozo oleifera	Palmae
Scheelea rostiata	Palmae
Recchia mexicana	Simaroubaceae
Cissus sp.	Vitaceae

* Sources: Fleming (1971, 1974*a*), Martínez-Gallardo and Sánchez-Cordero (in press), for *Heteromys*; and Janzen (1981); Pérez-Saldaña (1978) for *Liomys*.

eromyidae is that all species possess external cheek pouches that are used for gathering food items. This adaptation permits the efficient collection and transport of food in a short period of time (Reichman, 1983). Heteromyids appear to be important seed predators and have been shown experimentally to influence the composition and abundance of communities of annual plants in desert habitats (Brown et al., 1979). Tropical heteromyids are fruit and seed consumers, but their diets also include other plant material and arthropods. Table 4 lists species whose fruits and seeds are consumed by heteromyids. The seeds were either provided to caged animals or found in the cheek-pouches of live-trapped individuals (Fleming, 1971, 1974*a*; Janzen, 1981, 1982*a*, 1982*b*, 1982*d*, 1986; Pérez-Saldaña, 1978; Perry and Fleming, 1980; Vandermeer, 1979; Martínez-Gallardo and Sánchez-Cordero, in press).

Tropical heteromyids play a major role in the complex interactions of forest food webs (Janzen, 1983*b*). Recent studies on their foraging ecology have demonstrated that heteromyids consitute important post-dispersal seed predators of a large number of plant species (Janzen, 1982*a*, 1982*b*, 1982*c*, 1982*d*, 1983*a*, 1983*b*, 1986; Fleming, 1974*b*; Fleming and Brown, 1975; Perry and Fleming, 1980; Vandermeer, 1979). Some of these studies have focused on the interactions between seed shadows, large mammal herbivores as post-dispersal agents (horses, cattle, tapirs, etc.), and heteromyids as post-dispersal seed predators. Presumably, these interactions have had ecological, as well as evolutionary, significance in tropical habitats, since they are probably representative of interactions that occurred between the extinct Pleistocene megafauna and contemporary tropical rodent and plant species (Janzen and Martin, 1982).

Experimental studies have investigated the seed removal and consumption rates by heteromyids beneath parent trees, or in the dung of large mammals (Janzen, 1982*b*, 1982*c*, 1982*d*, 1986). Several trends have been observed regarding the foraging behavior of heteromyids, particularly in tropical dry forests. First, heteromyids can remove a high proportion of seeds from seed shadows. For example, *L. salvini* removes about 97% of an *Enterolobium cyclocarpum* seed crop containing as many as 108 seeds/m^2 in 46 nights (Janzen, 1982*a*). Moreover, *L. salvini* avidly removes seeds from dung piles and develops an attraction response to the dung (Janzen, 1982*c*). Second, the removal of seeds from dung depends on various factors like habitat where the dung is located, amount and type of dung, and seed density in the dung (Janzen, 1982*b*, 1982*d*, 1986). *L. salvini* harvested more seeds from a) dung piles located in the forest than in the grassland, b) dung piles with low volume than with high volume, c) horse dung rather than cattle dung, and d) seed-rich dung piles rather than seed-poor dung piles.

These studies are provocative and demonstrate complex interactions among seeds, seed dispersal agents, and seed predators that may have been established during the Pleistocene, or even earlier (Janzen and Martin, 1982). Undoubtedly, seed removal and hoarding behaviors by heteromyids now have, and probably long have had, important consequences for plant demography in tropical habitats.

Finally, the importance of heteromyids as food resources for predators has not been evaluated partially due to the difficulty in quantifying predation rates. Numerous vertebrate species are known or suspected to eat tropical heteromyids (see above). This gap in our understanding of tropical predator-prey interactions requires much further investigation.

Summary

Mice of the genera *Heteromys* and *Liomys* occur primarily in tropical wet forest and tropical dry forest habitats, respectively. We compared the demography, reproduction and behavior within and between species in contrasting habitats. Rainfall patterns, presumably through their effect on plant reproduction, influence the demographic, reproductive, and behavioral characteristics of tropical heteromyids. Population densities usually increase early in the rainy season due to the recuitment of immature individuals and decrease gradually as the rainy season progresses. *Heteromys* densities generally are higher than those of *Liomys*, probably because of the greater or more equitable productivity of tropical wet forests. In *Liomys*, reproduction is more seasonal than in *Heteromys*, but all species appear to synchronize reproduction so as to produce offspring at food peaks. *Heteromys desmarestianus* shows considerable demographic and reproductive variation between habitats and years. Differences in the demography and reproduction of *Heteromys* and *Liomys* may result from differences in the productivity of habitats, although the roles of predation, parasitism and competitive interactions need further investigation. Higher conspecific tolerance and overlapping home ranges suggest a "looser" type of social organization in *Heteromys* compared with the more asocial behavior of *Liomys*. Conspecific intolerance may increase with environmental seasonality and food scarcity. Tropical habitats are characterized by low heteromyid species richness. Differences in species richness between desert and tropical habitats presumably result from differences in the diversity and availability of seeds. In the tropics, some fruits and seeds are too large and hard, or contain poisonous compounds that make then unavailable to seed-eating rodents. Tropical heteromyids appear to play an important role as post-dispersal seed predators of a large number of plant species and as prey for a variety of vertebrate predators.

Acknowledgments

We thank B. Villa-Ramírez for his continuing support throughout the course of this study. Dr. J. H. Brown and two anonymous reviewers made constructive comments that greatly improved an earlier version of the manuscript. S. Anderson kindly provided unpublished data. M. Canela, M. Prieto, G. Magaña and R. Martínez-Gallardo helped in the preparation of the manuscript. G. Ibarra and S. Sinaca helped in the identification of seeds, and F. Villegas helped in the preparation of the figures. Financial and logistic support was provided by the Universidad Nacional Autónoma de México and Estación de Biología Tropical Los Tuxtlas to V.S-C.

Literature Cited

ABRAMSKY, Z. 1978. Small mammal community ecology: changes in species diversity in response to manipulated productivity. Oecologia, 34:113–123.

ALVAREZ, F. J. 1984. Dinámica de la hojarasca en una selva alta perennifolia: Los Tuxtlas, Veracruz. Tesis de Maestría. Facultad de Ciencias, UNAM, México, 213 pp.

ANDERSON, D. S. 1982. Comparative population ecology of *Peromyscus mexicanus* in a Costa Rican wet forest. Ph.D. dissertation, University of Southern California, Los Angeles.

BEATLEY, J. C. 1974. Phenological events and their environmental triggers in Mojave Desert ecosystems. Ecology, 55:856–863.

BROWN, J. H. 1975. Geographical ecology of desert rodents. Pp. 315–341, *in* Ecology and evolution of communities (M. L. Cody and J. M. Diamond, eds.). Belknap Press of Harvard University Press, Cambridge, Massachusetts. 545 pp.

BROWN, J. H., O. J. REICHMAN, AND D. W. DAVIDSON. 1979. Granivory in desert ecosystems. Annual Review of Ecology and Systematics, 10:201–227.

BROWN, J. H., AND B. HARNEY. 1993. Population and community ecology of heteromyid rodents in temperate habitats. Pp. 618–651, *in* Biology of the Heteromyidae (H. H. Genoways and J. H. Brown, eds.). Special Publication, American Society of Mammalogists, 10:1–719.

CONLEY, W., J. D. NICHOLS, AND A. R. TRIPTON. 1977. Reproductive strategies in desert rodents. Pp. 193–215, *in* Transactions of the symposium on the biological resources of the Chihuahuan Desert Region, United States and Mexico (R. A. Wauer and D. H. Riskind, eds.). Proceedings and Transactions of the National Park Service, 3:XXII + 1–658 pp.

DELANY, M. J. 1972. The ecology of small mammals in tropical Africa. Mammal Review, 2:1–42.

EISENBERG, J. F. 1963. The behavior of heteromyid rodents. University of California Publications in Zoology, 69:1–100.

ESTRADA, A., AND R. COATES. 1986. Manual de los mamíferos de la Estación de Biología Tropical Los Tuxtlas. Instituto de Biología, UNAM, Mexico. 151 pp.

ERRINGTON, P. L. 1954. The special responsiveness of minks to epizootics in muskrat populations. Ecological Monographs, 24:377–393.

FIELD, A. C. 1975. Seasonal changes in reproduction, diet and body composition of two equatorial rodents. East African Wildlife Journal, 13:221–236.

FLEMING, T. H. 1970. Notes on the rodent faunas of two Panamanian forests. Journal of Mammalogy, 51:473–490.

———. 1971. Population ecology of three species of neotropical rodents. Miscellaneous Publications Museum of Zoology, University of Michigan, 143: 1–47.

———. 1974*a*. The population of two species of Costa Rican heteromyid rodents. Ecology, 55:493–510.

———. 1974*b*. Social organization in two species of Costa Rican heteromyid rodents. Journal of Mammalogy, 55:543–561.

———. 1977. Response of two species of tropical heteromyid rodents to reduced food and water availability. Journal of Mammalogy, 58:102–106.

———. 1984. Scaly-tailed squirrels and pocket mice. Pp. 632–633, *in* Encyclopedia of mammals (D. MacDonald, ed.). Quinox Ltd., Oxford.

FLEMING, T. H., AND G. J. BROWN. 1975. An experimental analysis of seed hoarding and burrowing behavior in two species of Costa Rican heteromyid rodents. Journal of Mammalogy, 56:301–316.

FOSTER, R. B. 1982. Famine on Barro Colorado Island. Pp. 201–202, *in* The ecology of a tropical forest: seasonal rhythms and long-term changes (E. G. Leigh Jr., A. S. Rand, and D. M. Windsor, eds.). Smithsonian Institution Press, Washington, D.C., 454 pp.

FRENCH, N. R., B. G. MAZA, H. O. HILL, A. P. ASCHWANDEN, AND H. W. KAAZ. 1974. A population study of irradiated desert rodents. Ecological Monographs, 44:45–72.

GLANZ, W. E., R. W. THORINGTON, JR., J. GIACALONE-MADDEN, AND L. R. HEANEY. 1982. Seasonal food use and demographic trends in *Sciurus granatensis*. Pp. 239–252, *in* The ecology of a tropical forest: seasonal rhythms and long-term changes (E. G. Leigh Jr., A. S. Rand, and D. M. Windsor, eds.). Smithsonian Institution Press, Washington, D.C., 454 pp.

GOMEZ-POMPA, A., AND C. VAZQUEZ YAÑES. 1976. Estudios sobre sucesión secundaria en los trópicos cálido-húmedos: el ciclo de vida de especies naturales. Pp. 579–593, *in* Regeneración de selvas (A. Gómez-Pompa, C. Vázquez Yañes, S. del Amo, and A. Butanda, eds.). Instituto de Investigación sobre Recursos Bióticos (INIREB), Jalapa, Veracruz, México. 676 pp.

HAFNER, J. C., AND M. S. HAFNER. 1983. Evolutionary relationships of heteromyid rodents. Pp. 3–29, *in* Biology of desert rodents (O. J. Reichman and J. H. Brown, eds.). Great Basin Naturalist Memoirs No. 7, 134 pp.

HALL, E. R. 1981. The mammals of North America. Second edition. John Wiley & Sons, New York, 1: 1–600 + 90.

HAPPOLD, D. C. 1977. A study of small rodents in the tropical rainforest of Nigeria. Terr e Vie, 31:385–457.

HOLDRIDGE, L. R., W. C. GRENKE, W. H. HATHEWAY, T. LIANG, AND J. A. TOSI, JR. 1971. Forest environments in tropical life zones: a pilot study. Pergamon Press, New York, 747 pp.

HOOPER, E. T., AND M. D. CARLETON. 1976. Reproduction, growth and development in two contiguously allopatric rodent species, genus *Scotinomys*. Miscellaneous Publications Museum of Zoology, University of Michigan, 151:1–52

IBARRA, M. G. 1985. Estudios preliminares sobre la flora leñosa de la Estación de Biología Tropical Los Tuxtlas, Veracruz, México. Tesis Profesional Facultad dc Ciencias, UNAM, Mexico, 322 pp.

JANZEN, D. H. 1971. Seed predation by animals. Annual Review of Ecology and Systematics, 2:465–492.

———. 1981. Lectins and plant-herbivore interactions. Recent advances in phytochemistry, 15:241–258.

———. 1982*a*. Seed removal from fallen guanacaste fruits (*Enterolobium cyclocarpum*) by spiny pocket mice (*Liomys salvini*). Brenesia, 19/20:425–429.

———. 1982*b*. Removal of seeds from horse dung by tropical rodents: influence of habitat and amount of dung. Ecology, 63:1887–1900.

———. 1982*c*. Attraction of *Liomys* mice to horse dung and the extinction of this response. Animal Behaviour, 30:483–489.

———. 1982*d*. Fruit traits, and seed consumption by rodents, of *Crescentia alata* (Bignoniaceae) in Santa Rosa National Park, Costa Rica. American Journal of Botany, 69:1258–1268.

———. 1983*a*. Dispersal of seeds by vertebrate guts. Pp. 232–262, *in* Coevolution (D. J. Futuyma and M. Slatkin, eds.). Sinauer Associates, Sunderland, Massachusetts.

———. 1983*b*. Food webs: who eats what, why, how, and with what effects in a tropical forest? Pp. 167–182, *in* Tropical rain forest ecosystems. A. Structure and Function (F. B. Golley, ed.). Elsevier Scientific Publications Co., Amsterdam, The Netherlands. 410 pp.

———. 1986. Mice, big mammals, and seeds: it matters who defecates what where. Pp. 251–271, *in* Frugivores and seed dispersal (A. Estrada and T. H. Fleming, eds.). Dr. W. Junk Publishers, Dordrecht, Netherlands, 392 pp.

JANZEN, D. H., AND P. S. MARTIN. 1982. Neotropical anachronisms: the fruits the gomphoteres ate. Science, 215:19–27.

JANZEN, D. H., AND M. L. HIGGINS. 1979. How hard are *Enterolobium cyclocarpum* seeds? Brenesia, 16: 61–67.

KILTIE, R. A. 1981. Distribution of palm fruits on a rain forest floor: why white-lipped peccaries forage near objects. Biotropica, 13:141–145.

LEVINGS, S. C., AND D. M. WINDSOR. 1982. Seasonal and annual variation in litter arthropod populations. Pp. 355–388, *in* The ecology of a tropical forest: seasonal rhythms and long-term changes (E. Leigh, Jr., A. S. Rand, and D. M. Windsor, eds.). Smithsonian Institution Press, Washington, D.C., 454 pp.

LOTT, E. J., S. H. BULLOCK, AND J. A. SOLIS-MAGALLANES. 1986. Listado florístico de la Estación de Biología Chamela, México. Instituto de Biología, UNAM, Mexico, 47 pp.

MARTINEZ-GALLARDO R., AND V. SANCHEZ-CORDERO. In press. Dietary value of fruits and seeds to spiny pocket mice, *Heteromys desmarestianus* (Heteromyidae).

MCPHERSON, A. B. 1967. Population dynamics of *Liomys salvini* (Mammalia: Rodentia). M.S. thesis, Louisiana State University, Shreveport.

MILLAR, J. S. 1979. Energetics of lactation in *Peromyscus maniculatus*. Canadian Journal of Zoology, 57:1015–1019.

MUNGER, J. C., AND J. H. BROWN. 1981. Competition in desert rodents: an experiment with semipermeable exclosures. Science, 210:510–512.

MUNGER, J. C., M. A. BOWERS, AND W. T. JONES. 1983. Desert populations: factors affecting abundance, distribution and genetic structure. Pp. 91–116, *in* Biology of desert rodents (O. J. Reichman and J. H. Brown, eds.). Great Basin Naturalist Memoirs No. 7, 134 pp.

NAVARRO, D. 1982. Mamíferos de la Estación de Biología Tropical "Los Tuxtlas," Veracruz. Tesis Profesional Facultad de Ciencias, UNAM, México, 128 pp.

PEREZ-SALADAÑA, G. 1978. Observaciones sobre la variación morfológica, alimentación y reproducción de *Liomys p. pictus*, Rodentia: Heteromyidae. Tesis Profesional, Facultad de Ciencias, UNAM, México, 61 pp.

PERRY, A. E., AND T. H. FLEMING. 1980. Ant and rodent predation on small animal-dispersed seeds in a tropical dry forest. Brenesia, 17:11–22.

PRICE, M. V. 1978. The role of microhabitat in structuring desert rodent communities. Ecology, 59:910–921.

PRICE, M. V., AND J. H. BROWN. 1983. Patterns of morphology and resource use in North American desert rodent communities. Pp. 117–134, *in* Biology of desert rodents (O. J. Reichman and J. H. Brown, eds.). Great Basin Naturalist Memoirs. No. 7, 134 pp.

QUINTERO, G., AND V. SANCHEZ-CORDERO. 1989. Estudio del área de actividad de *Heteromys desmarestianus* (Rodentia: Heteromyidae) en una selva alta

J. Reichman and J. H. Brown, eds.). Great Basin Naturalist Memoirs. No. 7, 134 pp.

REICHMAN, O. J., AND K. M. VAN DE GRAAFF. 1975. Association between ingestion of green vegetation and desert rodent reproduction. Journal of Mammalogy, 56:503–506.

ROOD, J. P. 1963. Observations on the behavior of the spiny rat *Heteromys melanoleucus* in Venezuela. Mammalia, 27:186–192.

ROOD, J. P., AND F. H. TEST. 1968. Ecology of the spiny rat, *Heteromys anomalus* at Rancho Grande, Venezuela. American Midland Naturalist, 79:89–102.

ROSENZWEIG, M. L. 1973. Habitat selection experiments with a pair of coexisting heteromyid species. Ecology, 54:111–117.

ROSENZWEIG, M. L., B. SMIGEL, AND A. KRAFT. 1975. Patterns of food, space and diversity. Pp. 241–268, *in* Rodents in desert environments (I. Prakash and P. K. Ghosh, eds.). W. Junk, The Hague, The Netherlands, 624 pp.

RUSSELL, J. K. 1982. Timing of reproduction by Coatis (*Nasua narica*) in relation to fluctuations in food. Pp. 413–431, *in* The ecology of a tropical forest: seasonal rhythms and long-term changes (E. G. Leigh, Jr., A. S. Rand, and D. M. Windsor, eds.). Smithsonian Institution Press, Washington, D.C., 454 pp.

RYAN, J. M. 1989. Comparative myology and phylogenetic systematics of the Heteromyidae (Mammalia: Rodentia). Miscellaneous Publications Muperennifolia. Anales Instituto de Biología, UNAM, Zoología, 60:223–240.

REICHMAN, O. J. 1983. Behavior of desert heteromyids. Pp. 77–90, *in* Biology of desert rodents (O. seum of Zoology, University of Michigan No. 176: 1–103.

SANCHEZ-CORDERO, V. In press *a*. Estudio poblacional de la rata espinosa *Heteromys desmarestianus* en una selva húmeda en Veracruz, México. *In* Avances en el estudio de los mamíferos de México (G. Ceballos and R. Medellín, eds.). Universidad Nacional Autónoma de México. México, D. F. México.

———. In press *b*. Mamíferos de la selva húmeda de México. *In* Selvas húmedas de México (S. Guevara y S. Purata, eds.). Facultad de Ciencias, Universidad Nacional Autónoma de México. México, D. F. México.

SCHROEDER, G. D., AND M. L. ROSENZWEIG. 1975. Perturbation analysis of competition and overlap in habitat utilization between *Dipodomys ordii* and *Dipodomys merriami*. Oecologia, 19:9–28.

SMYTHE, N. 197C. The relationships between fruiting season and seed dispersal in a neotropical forest. American Naturalist, 104:25–35.

———. 1986. Competition and resource partitioning in the guild of neotropical terrestrial frugivorous mammals. Annual Review of Ecology and Systematics, 17:169–188.

SMYTHE, N., W. G. GLANZ, AND E. G. LEIGH JR. 1982. Population regulation in some terrestrial frugivores. Pp. 227–238, *in* The ecology of a tropical forest (E. G. Leigh Jr., A. S. Rand, and D. M. Windsor, eds.). Smithsonian Institution, Washington, D.C., 454 pp.

TAYLOR, K. D., AND M. G. GREEN. 1976. The influence of rainfall on diet and reproduction in four African rodent species. Journal of Zoology, 180:367–389.

TIMM, R. M., AND E. F. COOK. 1979. The effect of bot fly larvae on reproduction in white-footed mice, *Peromyscus leucopus*. American Midland Naturalist, 101:211–217.

VANDERMEER, J. H. 1979. Hoarding behavior of captive *Heteromys desmarestianus*, (Rodentia) on the fruits of *Welfia georgii*, a rainforest dominant palm in Costa Rica. Brenesia, 16:107–116.

WAGNER, H. O. 1961. Die Nagetiere einer Gebirgsabdachung in Súdmexiko und ihre Beziehungen zur Umwelt. Zoologische Jahrbücher, 89:177–242.

WHEELWRIGHT, N. T. 1985. Competition for dispersers, and the timing of flowering and fruiting in a guild of tropical trees. Oikos, 44:465–477.

POPULATION AND COMMUNITY ECOLOGY OF HETEROMYID RODENTS IN TEMPERATE HABITATS

JAMES H. BROWN AND BARBARA A. HARNEY

Introduction

The heteromyid rodents of temperate North America have been the subject of numerous studies in physiological, behavioral, population, and community ecology. In fact, along with a few other organisms, such as Caribbean *Anolis* lizards and Galapagos finches, the desert-dwelling heteromyids have become a model system to investigate population interactions and community organization (Brown, 1984*a*).

Studies of heteromyids have contributed importantly to recent advances in population and community ecology for several compelling reasons. These rodents maintain abundant populations and diverse communities in the physically severe, structurally simple, unpredictably variable, and unproductive arid habitats of western North America. They are easily studied by trapping, by biotelemetry, and through experimental manipulations. Through their seed consumption and other activities they have major impacts on the structure and function of desert ecosystems. If only heteromyids were diurnal and as easy to observe as lizards and birds, they would be the ideal terrestrial vertebrate for ecological studies.

The goal of this chapter is primarily to synthesize the extensive information on population and community ecology of the temperate heteromyids in the genera *Perognathus, Chaetodipus, Microdipodops,* and *Dipodomys.* This is a difficult task, not only because there have been so many studies on diverse aspects of heteromyid ecology, but also because of differences in opinion among investigators and apparently conflicting data. Although we cannot claim to be unbiased, we have tried to sort through this mass of information in order to identify the patterns and processes that seem to be well documented and those that seem to warrant further study. This chapter addresses only the temperate heteromyids, especially the well-studied populations and communities inhabiting desert and semiarid habitats. Sanchez and Fleming (this volume) summarize what is known about the ecology of the much less intensively studied *Heteromys* and *Liomys* of tropical America.

We have arranged the chapter hierarchically, dealing in turn with increasingly complex levels of ecological organization. We begin with the ecology of single species populations: limiting factors, life histories, and population dynamics. We then address community organization: the highly variable individualistic nature of species assemblages, the extrinsic environmental variables that influence the number and kinds of coexisting species, and the intrinsic interspecific interactions that also affect the composition of desert rodent communities. We conclude by considering the role of heteromyids in arid ecosystems.

Population Ecology

The Niche

The abundance and distribution of a population is determined primarily by the extent to which individual organisms can tolerate conditions and acquire essential resources. Those different environmental variables that limit abundance and distribution because they are necessary for survival and reproduction have been characterized as the axes of the multidimensional ecological niche (Hutchinson, 1958). These include both physical and biotic factors. Although in theory each variable could act independently to limit survival and reproduction, in reality the various niche dimensions usually interact in complex ways to affect populations, such that no single variable can be identified as "the" limiting factor (Pianka, 1983) (Fig. 1). This makes it difficult to identify and characterize the effects of particular limiting factors and to assess their relative importance.

Unfortunately, there are no studies of heteromyid populations that have systematically identified and evaluated all important niche parameters. Consequently, it is necessary to develop our understanding of the numbers, kinds, relative importance, and interactions among limiting factors by integrating the results of diverse studies that have focused on effects of particular environmental variables on many different populations. Although not ideal, this approach provides examples of the ways that different types of abiotic and biotic factors limit heteromyid populations. It also reveals the kinds of complex interactions among variables that can make it extremely difficult to elucidate mechanisms of population regulation, even when controlled experimental manipulations have been performed.

Nonrandom patterns of abundance and distribution ultimately can be attributed primarily to variation in the physical environment, but the effects of abiotic factors on a population can be either direct or mediated through other organisms. There is abundant evidence that the niches of heteromyids are delimited by both the direct and indirect influence of physical variables. For desert rodents, as for most other terrestrial organisms, the most important physical factors are climate and geology. Interestingly, although the climate of western North America is quite variable in both space and time, and although certain heteromyid species exhibit geographic distributions or local population fluctuations that are obviously related to climatic variation, it is difficult to demonstrate that populations are limited by the *direct* effects of climate. Thus, despite the diverse and spectacular adaptations shown by individuals of different populations and species of heteromyids to the extreme temperature and moisture regimes of desert regions (e.g., French, this volume), there is little evidence that survival and reproduction are limited by the direct effects of thermal stress or water deprivation (e.g., Bartholomew, 1958). Instead, most of the influence of climate appears to be indirect, mediated through other organisms, such as the plants that provide both food and shelter. Conversely, geologic features (e.g., mountain ranges and rivers) acting as dispersal barriers directly limit species distributions at large geographic scales (Schmidly et al., this volume).

However, soil texture is a physical factor that may act directly to limit species' distributions. Produced by the interaction of geology, climate, and organisms, soils affect

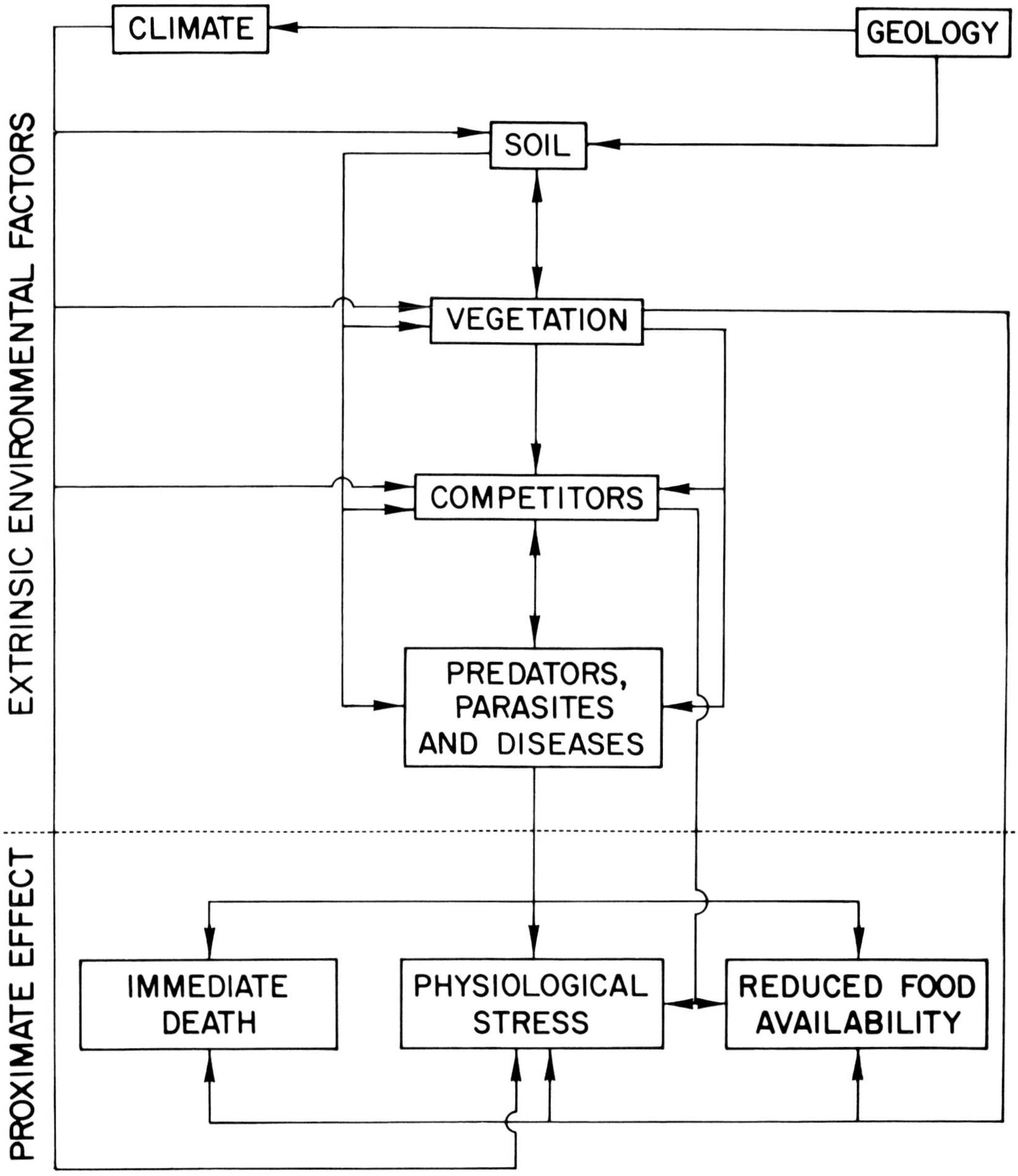

FIG. 1.—A flow chart showing some of the important interactions between niche variables that affect the distribution and abundance of heteromyid rodent populations. Note that the ultimate limiting factors are features of the physical environment, chiefly climate and geology, but that their effects are often mediated by biotic factors, such as vegetation, predators, and competitors.

rodents directly through their influence on burrowing, locomotion, and foraging. Ever since Grinnell (1914) described the restriction of *D. deserti* to sand dunes and other habitats with deep, unconsolidated, sandy soil, it has been apparent that the local abundances and distributions of many heteromyids are limited in part by soil texture. Although there have been many studies correlating occurrence of a particular heteromyid species with a particular soil structure (e.g., Baumgardner and Schmidly, 1985; Deynes, 1954; Hardy, 1945; Hoover et al., 1977; Roberts and Packard, 1973), most have been unable to distinguish unequivocally whether the species is limited directly by its ability to survive on that substrate or indirectly by interactions with other organisms that have even more specific soil requirements.

Yet, some of the patterns are so striking that it seems reasonable to assume that the effect is direct. Perhaps the best example is the complete absence of bipedal heteromyids (both *Dipodomys* and *Microdipodops*) from extremely rocky soils where quadrupedal heteromyids (especially *Chaetodipus*) and murids (*Peromyscus* and *Neotoma*) are often abundant (e.g., Rosenzweig and Winakur, 1969; Rosenzweig et al., 1975). The frequent repetition of this pattern in different localities suggests that it is not due to competitive exclusion of bipedal rodents by quadrupedal ones (these functional groups coexist on other substrates), to vegetation, or to some other indirect effect. Instead, it is probable that kangaroo rats and kangaroo mice experience sufficiently decreased foraging efficiency, impeded locomotion and burrowing capability, and enhanced predation risk (an indirect effect) to preclude their occupation of rocky substrates. Other more subtle associations between heteromyid species and substrate may reflect equally direct limitations, but are more difficult to document and deserve further study. In particular, soil and substrate can act to limit local species distributions and abundances, both indirectly by influencing the distribution and dispersion of seed resources and directly by affecting the mechanics of harvesting (Reichman and Price, this volume, and references therein).

The importance of vegetation structure in limiting species abundance and distribution is well documented. This is an interesting case; in one sense the physical effect is indirect and mediated by other organisms, because the physiognomy reflects the ultimate influence of climate and soil on the composition of plant communities. But in another sense, the effect is abiotic because it is the physical structure of the vegetation that often appears to be responsible for its effects on rodent populations. Thus, there are many studies correlating the abundance and distribution of heteromyid populations with some aspect of physiognomy (e.g., Hafner, 1977; M'Closkey, 1976). Rosenzweig clearly documented the association of bipedal species with open habitats and of quadrupeds with denser vegetation (Rosenzweig and Winakur, 1969; Rosenzweig et al., 1975), and performed "habitat tailoring" experiments (Rosenzweig, 1973) to test directly for the effects of vegetative cover. He removed woody vegetation from some habitat patches and used it to augment cover in other patches. Then he showed that, as predicted, the relative abundance of bipeds increased in the cleared areas and quadrupeds increased in the patches with supplemental cover. This elegant experiment demonstrated that it was the physical structure of the vegetation that ultimately affected the rodent populations, because the added brush was dead and thus unable to have a biotic effect such as providing food. Similarly, Price (1978*b*) demonstrated an increased number of captures of *D. merriami* at trapping stations cleared of vegetation and debris, indicating that this species may be limited locally by the abundance of its preferred microhabitat. Yet, these studies do not provide unequivocal evidence for the direct, proximal effect of vegetative structure, because the habitat manipulations could have affected the populations indirectly by influencing the rodents' ability to compete or to avoid predation rather than by changing their ability to make a living in the modified habitats independent of the effects of other species.

Many of the effects of abiotic factors are clearly indirect and mediated by the responses of other organisms. Perhaps the best example is the effect of variation in precipitation on population density. A positive correlation between precipitation and the abundance of heteromyid species is observed both over space, between structurally similar habitats that differ in mean annual rainfall (Brown, 1973, 1975; Hafner, 1977), and over time, within a local habitat in response to variation in rainfall among years (Beatley, 1969; Brown and Heske, 1990*a*; Petryszyn, 1982; Whitford, 1976). Primary productivity and seed production in arid environments are highly correlated with precipitation (Brown et al., 1979*b* and references therein; Hillel and Tadmore, 1962; Rosenzweig, 1968), suggesting that increases in precipitation enhance food availability. Thus, these studies also complement more direct experimental studies demonstrating that at least some populations of granivorous heteromyids are food limited, because they increase when supplemental seeds are provided (Abramsky, 1978; Brown and Munger, 1985).

In addition to soil texture, vegetative structure, and food availability, competition and predation also limit desert rodent populations (Munger et al., 1983). We consider these biotic interactions when we discuss community structure.

Abundance and distribution of heteromyid populations are limited by many abiotic and biotic factors that often interact in complex ways. The spatial and temporal variability of these limiting factors is often extreme in desert environments, enhancing the opportunity for coexistence of species that have different requirements. Interactions among limiting factors are also critical in determining community structure. Excessive overlap among species on one niche axis may be compensated by differentiation in other niche parameters. This niche complementarity promotes locally high species diversity, a characteristic feature of many desert rodent communities.

Long-term studies of population dynamics have the potential to contribute importantly to our understanding of the combinations of limiting factors that define the niche of each species population. Brown and Heske (1990*a*) used time series analysis to investigate the temporal changes in abundance and species composition of desert rodents on our experimental study site in the Chihuahuan Desert near Portal, Arizona. Although it was difficult to pick out the influence of particular environmental conditions (except for the wet winters that occurred in el Nino years and were correlated with increases in several species), it was apparent that there was much variation in the patterns of fluctuation among the 11 common species (five heteromyids and six murids). This suggests that, although these species all occur together in the same environment, they differ substantially in the kinds and combinations of environmental variables that limit local abundance.

Life History and Demography

The life history of heteromyids has been treated elsewhere (Jones, this volume). Rather than reiterate, we emphasize those aspects that are particularly relevant to population and community ecology.

Desert-dwelling heteromyids are well adapted to tolerate the wide and unpredictable fluctuations in the climate and productivity of arid habitats (Kenagy and Bartholomew, 1985). These adaptations in life history have emphasized survival of adults through stressful droughts rather than rapid recruitment of juveniles during favorable periods following sufficient rains. In this respect heteromyids appear to differ significantly from the murid rodents (e.g., *Peromyscus* and *Neotoma*) with which they usually coexist (Brown and Zeng, 1989; Conley et al., 1977; Whitford, 1976).

Compared to most other small rodents, heteromyids are remarkably long-lived. Individuals of both *Perognathus* and *Dipodomys* have both been recorded living 5 years in the field (French et al., 1967, 1974; Zeng and Brown, 1987*a*; Table 1). Concomitant with the emphasis on survival, reproductive efforts are modest. Litter size varies

TABLE 1.—*Comparative data on population ecology of eight species of granivorous desert rodents in a Chihuahuan Desert shrub habitat near Portal, Arizona. The 20-ha study area contained 24 experimental plots, each 0.25 ha in area (50 × 50 m). Different-sized holes in these plots permitted free access to different sizes of rodents. All species of rodents were excluded from four plots, the three species of* Dipodomys *were excluded from four plots, and* D. spectabilis *was excluded from two plots. Additional plots had ants removed or millet seeds added (see Brown and Munger, 1985 for details of experimental design and layout of plots). Two values are given for population density: mean density for all plots to which the species had access regardless of experimental treatment, and mean density for the two control plots only ( ). Data are from Brown and Zeng (1989) and Brown and Kurzius (1989).*

Species	Captures on all grids	Number of months out of 90 total that species was captured	Density/ha all plots (control plots)	Individual body weight (g)	Species biomass (g/ha)	Maximum longevity (mo)	Coeff. variation of density between years
Dipodomys spectabilis	1,940	89	5.30 (4.09)	123.4	504.8	50	0.31
Dipodomys ordii	773	87	1.82 (1.35)	48.1	64.9	35	0.60
Dipodomys merriami	3,489	90	7.93 (11.81)	42.8	505.1	43	0.31
Chaetodipus penicillatus	180	61	0.39 (0.56)	16.0	9.0	26	0.12
Perognathus flavus	456	75	0.93 (0.56)	7.0	3.9	35	0.73
Reithrodontomys megalotis	398	57	0.65 (0.56)	10.2	5.7	40	0.93
Peromyscus maniculatus	141	47	0.28 (0.19)	21.4	4.1	19	1.16
Peromyscus eremicus	256	72	0.43 (0.26)	21.2	5.5	19	0.76

from approximately 2 in some populations of *Dipodomys* to 4 or more in certain species of *Perognathus* and *Chaetodipus* (Conley et al., 1977; Jones, this volume). Gestation periods are moderate (20–30 days; Jones, this volume; Smith and Jorgensen, 1975), and growth and development are relatively slow so that weaning occurs within 2–3 weeks of birth (Eisenberg, this volume). Because of these constraints and the costs of allocating scarce resources to reproduction, the number of litters per year is low. Some populations of kangaroo rats and pocket mice may occasionally produce 3 litters in a season, but the usual number is 1 or 2 (Bradley and Mauer, 1971; Conley et al., 1977; Reynolds, 1960; Smith and Jorgensen, 1975). In contrast, some microtines may live at most 1 year, but are continuous breeders producing 3 to 8 young per litter and more than 6 litters per year, and females may become pregnant at 3 weeks of age (Keller, 1985). Although the seasonality of reproductive activity varies among heteromyid species and among populations of the same species, in many populations the number and timing of litters is quite flexible and depends on the suitability of immediate environmental conditions (e.g., Beatley, 1969, 1976; Kenagy and Bartholomew, 1985; Zeng and Brown, 1987*a*, 1987*b*). In extremely bad years there may be no reproduction at all (French et al., 1967; Kenagy and Bartholomew, 1985; Maza et al., 1973).

This type of life history has important demographic consequences. Populations are comprised primarily of adults, which use efficient foraging, seed storage, facultative torpor and dispersal, and predator-avoidance tactics to survive through stressful periods of drought and food shortage. Mortality rates are correspondingly low. Most studies have probably overestimated death rates by assuming that all those marked adults which disappeared from permanent study grids had died. In fact, there appears to be a high frequency of long-distance adult dispersal in many species (Brown and Zeng, 1989; Zeng and Brown, 1987*a*, 1987*b*). When this is accounted for, mortality rates are probably on the order of 25% per year, and perhaps even lower in some populations. Difficulty in distinguishing death from dispersal in mark-recapture studies complicates the assessment of survival probabilities as a function of age. Similarly, because of problems in determining the fates of litters, we know little about the reproductive contribution of females as a function of their age. However, the pattern of slow growth and development is apparently associated with delayed sexual maturation. For example, in *D. merriami* individuals of both sexes require 2–3 months after weaning to achieve the size at which the first signs of reproductive activity appear (Zeng and Brown, 1987*a*).

While sex ratios of most populations do not differ markedly from 1:1, those reported in several studies appear to favor males. In trying to account for these sex ratios, it will be important to distinguish between real differences owing to sex-biased birth and survival rates, and artifacts resulting from differential dispersal and sex-related differences in home range size.

Despite the popularity of heteromyids for ecological studies, much remains to be learned about the basic life history and demography of even the most abundant and best studied species. Most of the standard mark-recapture techniques for obtaining demographic data have been developed primarily for murid rodents in more mesic environments. Interpretation of these data is based on assumptions about trapability (that can vary for different species; Petryszyn, 1982), dispersal, survival, and fecundity that need to be evaluated rigorously for all small mammal populations. It is particularly important to question the validity of these methods and assumptions for the desert-dwelling heteromyids, which live in such different environments and have very different life histories from the murid rodents commonly studied by mammalian population ecologists.

Population Dynamics

Temporal and spatial variation in population density is largely a consequence of interactions between the environmental limiting factors and the life history attributes described above. The life history mediates the response to spatial and temporal variation in the niche variables. In general, the dynamics of heteromyid populations can be characterized quite simply. Local populations exhibit large, irregular fluctuations in response to a variable environment (Fig. 2), but the magnitude of these fluctuations is moderated by life history traits that promote survival of adults through unfavorable periods at the expense of rapid recruitment of juveniles during favorable times (Conley et al., 1977; Whitford, 1976).

Populations appear to be largely food limited, so population fluctuations are correlated with and driven by temporal variation in productivity. In arid habitats primary production is controlled primarily by precipitation (Hillel and Tadmore, 1962; Rosenzweig, 1968). The seeds, green vegetation, and insects that comprise heteromyid diets are produced in great abundance following periods of sufficient rainfall. But the relationship between precipitation, primary production, food availability, and population increase is complex. Production of seeds and other food depends on the timing, as well as the amount of precipitation (Beatley, 1974; Petryszyn, 1982), and rodent populations may attain their highest densities only after a succession of two or more exceptionally productive seasons.

Populations typically increase during the favorable periods of high food supply following sufficient precipitation and then de-

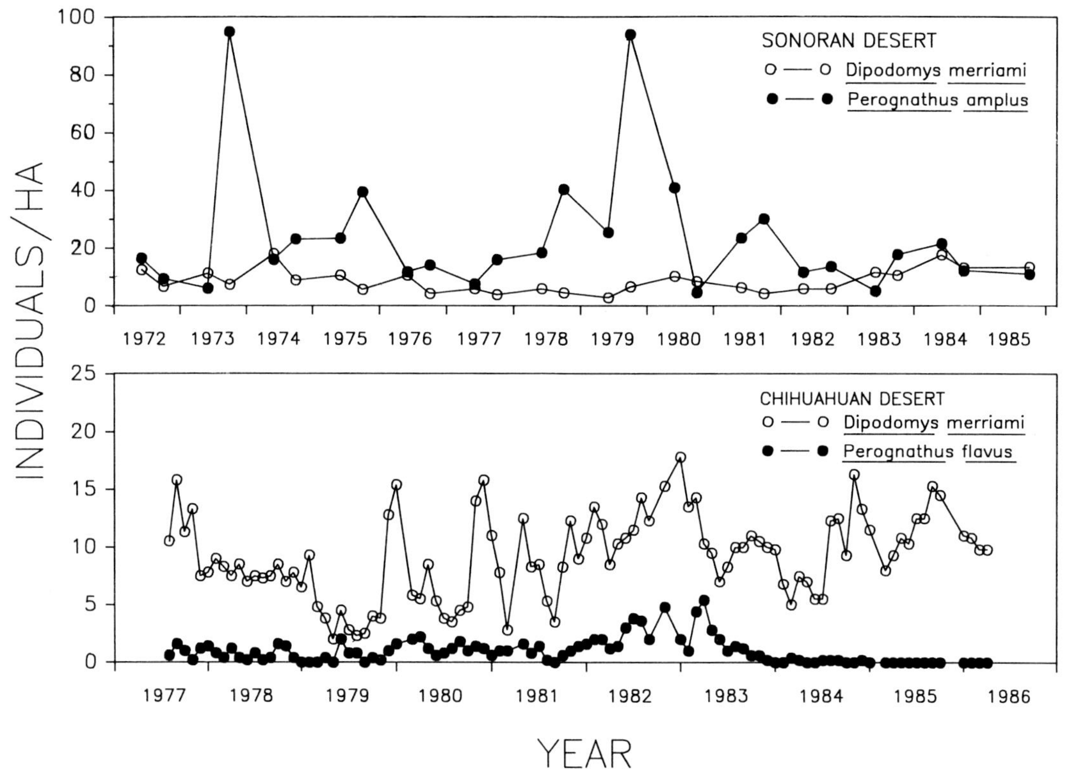

FIG. 2.—Patterns of long-term fluctuation of heteromyid populations at two sites approximately 200 km apart: above, *D. merriami* and *P. amplus* in the Sonoran Desert approximately 40 km northwest of Tucson, Arizona (data from Petryszyn, 1982, and unpubl.); below, *D. merriami* and *P. flavus* in the Chihuahuan Desert approximately 5 km east of Portal, Arizona (data from Zeng and Brown, 1987*a,* and Brown and co-workers, unpubl.). Note that *D. merriami* shows substantial fluctuations at one site but not the other, whereas the *Perognathus* populations are highly variable at both sites. Both species of *Perognathus* become torpid, accounting especially for the midwinter lows at the Chihuahuan Desert site.

cline during the subsequent droughts. The timing of reproduction appears to be cued, at least in part, by substances ingested along with new plant growth, so that the first litters are weaned at about the time that abundant seed crops are produced (Bradley and Mauer, 1971; Chew and Butterworth, 1964; Reichman and Van De Graaff, 1975; Reynolds, 1958; Van De Graaff and Balda, 1973). Despite facultative timing of reproduction to coincide with high food availability, the rate of population increase is limited by the relatively small size of litters and the long intervals between litters. Even during favorable times, adults appear to devote much of their energy to activities, such as seed collection and caching, that tend to promote their own survival at the expense of reproduction. The ability to utilize stored food, torpor (in all genera except *Dipodomys*), and effective predator avoidance strategies facilitates survival and tends to prevent high rates of population decrease during the unfavorable dry periods after most of the available food has been harvested.

Nevertheless, despite life histories that tend to dampen the rates of both population increase and decrease, fluctuations in population density can be very large and extremely variable. As shown in Figure 2, the populations of the same species can fluctuate to different degrees at different sites, and within the same site different species can show very different patterns of fluctuation (see also Brown and Heske, 1990*a*). In the most arid parts of the Mojave and Sonoran deserts several years may elapse between precipitation events of sufficient magnitude to significantly increase food supplies (Brown, 1973). Brown (unpubl. data) has estimated that the population of *D. deserti* inhabiting the Kelso Dunes of southern California varied at least 100-fold in density between 1968 and 1972, and the population almost became extinct by the end of a 3-year drought (see also Miller and Stebbins, 1964). Even in much more productive deserts, a succession of favorable seasons can permit populations to erupt to extremely high densities. For example, in the summer following the exceptionally wet spring of 1973, Brown (unpubl. data) observed local densities of *P. flavus* in the Animas Valley in the Chihuahuan Desert of southwestern New Mexico that attained an estimated 50–100 individuals/ha, at least 100 times their usual abundance.

Although we are confident that the above overview of heteromyid population dynamics is correct in its major features, admittedly it has been pieced together from rather fragmentary data. Unfortunately, there have been few intensive, long-term studies. Most of those that have been done (e.g., Brown and Heske, in press *a*; Brown and Zeng, 1989; Kenagy, 1973; Petryszyn, 1982; Whitford, 1976; Zeng and Brown, 1987*a*) have been performed in the more productive deserts where precipitation is not only greater, but more frequent and predictable. Of course, there is good reason for this. Populations are not only greater, but probably also more stable than those inhabiting the most arid regions. Given the variation in both amount and timing of precipitation, even these studies are not of sufficient duration to characterize accurately the relationship between precipitation, primary production, food supply, and population density. In addition to monitoring populations for much longer periods, it would be most desirable to initiate intensive, long-term studies of populations inhabiting extremely arid and unpredictable habitats (e.g., French et al., 1967; Maza et al., 1973).

Community Organization

Arid and semiarid habitats in western North America typically have from 1 to more than 15 species of rodents, of which more than half feed primarily on seeds (Fig. 3). Because they share requirements for a common, limited food resource, the granivorous species often have been designated as a "guild" and singled out for intensive community-level studies. As defined by most investigators, this guild contains all of the heteromyid species (except perhaps for some populations of *D. microps* that are primarily folivorous; see Kenagy, 1972) as well as murid rodents of the genera *Peromyscus, Reithrodontomys,* and *Baiomys.* In the discussion that follows, we shall include these other granivorous rodents, which comprise less than 50% of the individuals in most assemblages, in our discussion of community organization for two reasons. The first is purely practical. The authors of the original papers included both heteromyids and murids in their analyses and discussions, and it is difficult to extract and synthesize the data on heteromyids alone without losing much valuable information. The second reason is biological. There is evidence that some heteromyid and murid species are more similar in ecology or affect each other more than some coexisting heteromyid species (e.g., Rebar and Conley, 1983). Consequently, in order to study community organization in a realistic functional context, it is necessary to define assemblages on the basis of ecological rather than taxonomic similarity of the component species.

The Individualistic Nature of Species Assemblages

By definition, each species is unique. It differs in its tolerances and requirements

FIG. 3.—Photographs of habitats in the four subdivisions of the North American Desert that support substantial populations of several species of heteromyid and murid rodents: A, Great Basin Desert: a sand dune on the shore of a dry lake bed near Mina, Nevada, where *D. deserti, D. merriami, D. ordii, M. pallidus, P. longimembris, Reithrodontomys megalotis* and *Peromyscus maniculatus* occur; the rocky hillsides in the background are occupied by *C. formosus, Peromyscus crinitus* and *P. maniculatus*; B, Mojave Desert: a site in Joshua Tree National Monument, California, where *D. merriami, P. longimembris* and *Peromyscus eremicus* occur on the flats in the foreground and *C. fallax* and *Peromyscus crinitus* inhabit the rock outcrops in the background; C, Sonoran Desert: a locality in Organ Pipe Cactus National Monument, Arizona, where *D. spectabilis, D. merriami, C. baileyi,* and *C. penicillatus* occur; D, Chihuahuan Desert: a bajada (alluvial plain) near Portal, Arizona, where ten species of granivorous rodents have been captured and *D. spectabilis, D. ordii, D. merriami, C. penicillatus, P. flavus, Peromyscus eremicus, Peromyscus maniculatus,* and *Reithrodontomys megalotis* are relatively common.

from all other species, and as a result of these niche differences it has a unique pattern of abundance, distribution, and association with other species. Ecological communities are assemblages of species that live together within some specified habitat or region. The composition of communities is determined largely by the ability of each species individually to meet its own particular requirements. These include not only abiotic conditions that are necessary for existence, but also the resolution of interactions with other species that permit coexistence in the same environment. One consequence of unique niches is that species should be assembled into communities in a highly individualistic way, so that the occurrence of each species is largely, but *not entirely,* independent of the co-occurrence of other species (Brown and Heske, 1990*a*; Brown and Kurzius, 1989; Gleason, 1926).

The composition of granivorous desert rodent assemblages provides abundant evidence for the effect of unique species niches

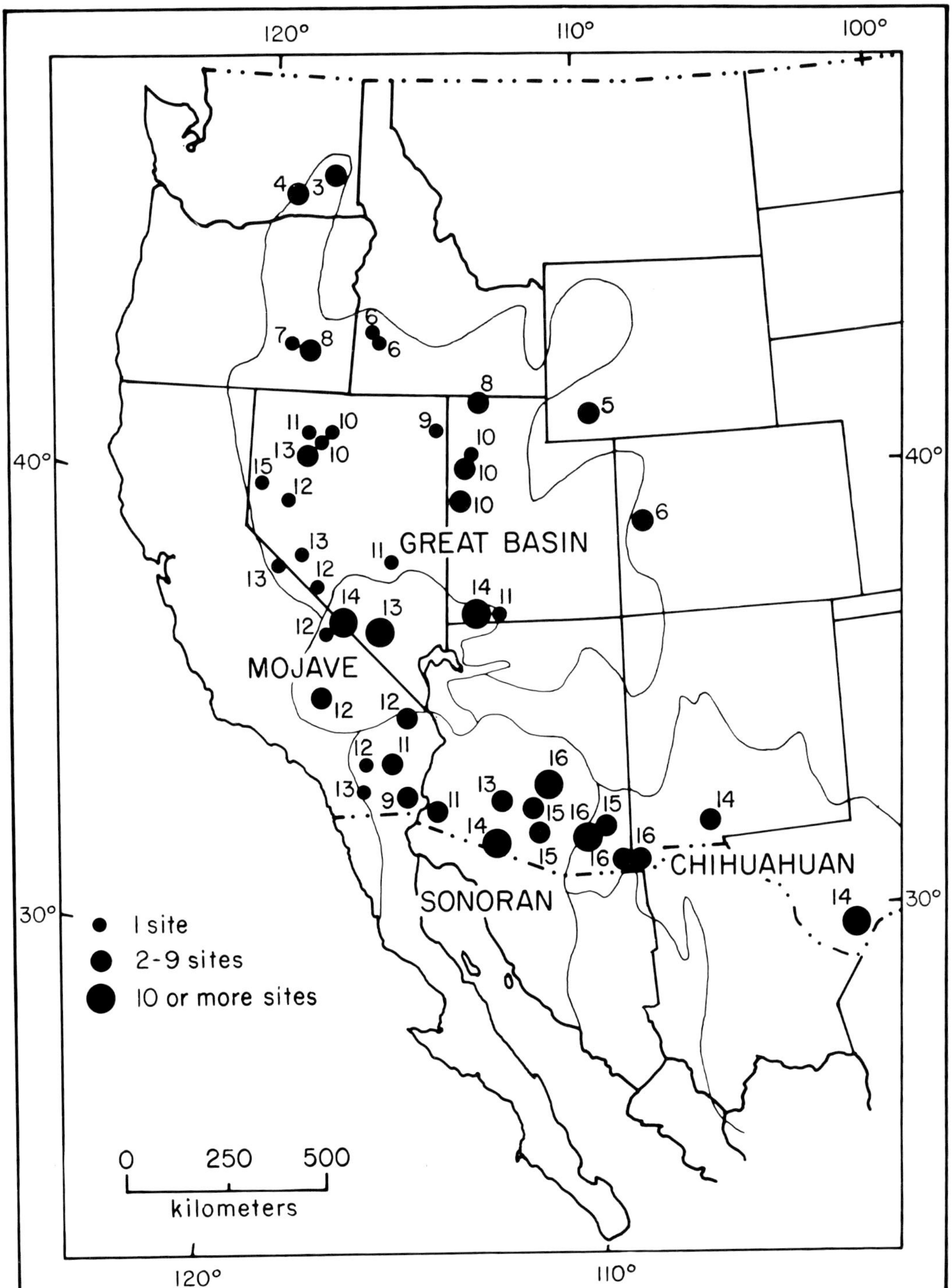

FIG. 4.—Map of the southwestern United States showing the number of granivorous rodent species (both heteromyid and murid) whose geographic ranges overlap selected sites (data compiled from Hall, 1981). These are the numbers of species in the regional species pool that are potentially available to occur at each site. Note the low species richness in the northern and eastern Great Basin Desert and in the arid regions to the east.

TABLE 2.—*Distribution and co-occurrence of granivorous rodent species at 202 sites surveyed in the southwestern United States.*

Species	Number of sites where found	Number of co-occurring species	Number of different combinations	Area of geog. range (sq. km × 10^2)	% sites in range occupied	Number of species in geog. range	% species with which it co-occurred
Perognathus							
flavus	30	17	21	2,237	45%	22	77%
longimembris	39	15	31	232	43	24	63
amplus	19	5	7	132	45	21	24
parvus	35	10	18	960	38	19	53
Chaetodipus							
penicillatus	45	16	28	883	46	26	62
intermedius	8	10	6	511	11	24	42
nelsoni	5	5	3	582	25	16	31
fallax	2	4	2	90	67	14	29
formosus	34	12	22	417	41	19	63
baileyi	18	8	13	359	26	21	38
hispidus	6	11	6	2,381	21	20	55
Microdipodops							
pallidus	5	9	5	32	100	15	60
megacephalus	3	10	3	237	27	15	67
Dipodomys							
microps	31	13	26	385	43	20	65
spectabilis	8	12	8	394	12	21	57
panamintinus	7	9	5	62	25	17	82
ordii	31	16	25	3,410	21	27	59
deserti	21	12	16	313	32	26	46
merriami	107	25	76	1,497	65	28	89
Reithrodontomys							
montanus	1	2	1	1,873	3	19	11
megalotis	33	21	28	4,819	16	28	75
fulvescens	5	10	4	2,491	8	20	50
Peromyscus							
eremicus	39	29	29	1,525	26	28	68
crinitus	26	18	18	1,055	26	22	64
maniculatus	85	59	59	13,758	42	28	82
boylei	2	2	2	2,406	2	28	21

on what we call community organization: the number, identity, abundance, and morphological, physiological, and behavioral attributes of coexisting species. Assemblages exhibit enormous variation in their composition in both space and time. Analysis of censuses of rodents in 202 local patches of desert shrub habitat dispersed throughout the southwestern United States revealed 137 different combinations of 29 granivorous rodent species (19 heteromyids and 10 murids, Table 2; Brown, 1987; Brown and Kurzius, 1987). Similarly, analysis of 90 monthly censuses over an 8-year period on our experimental site of 20 ha of Chihuahuan Desert shrub habitat revealed 37 different combinations of 10 granivorous species (6 heteromyids and 4 murids; Brown and Kurzius, 1989).

At one level, the interpretation of this variation is straightforward. It reflects the individualistic requirements of the species, and the open, nonequilibrium organization of communities. Each species tends to occur wherever and whenever its unique requirements are met, but its abundance and distribution are limited by some combination of biogeographic barriers to dispersal, local physical conditions, and interactions with coexisting species. Species vary greatly in their ecological and biogeographic amplitudes. Some are tolerant of a wide range of conditions, have large geographic ranges, and maintain high population densities in many habitats; over their entire geographic ranges these species also tend to coexist in local habitat patches with many other species in diverse combinations. For example, *D. merriami* occurred at 107 of the 202 sites analyzed by Brown and Kurzius (1987) and it was found together with a total of 25 other species in 76 different combinations of species; it occurred at 64% of the sites within its geographic range and coexisted with 89% of the species with which its geographic range overlapped (Table 2). *Dipodomys merriami* was also the most abundant species at our Chihuahuan Desert study site and it was the only one of the 8 common granivorous species that was captured in every monthly trapping period (Table 1). Other species, apparently because of their much narrower tolerances and requirements, have smaller geographic ranges and lower population densities. These species, such as *D. deserti, P. amplus,* and *P. flavus,* tend to occur in fewer local habitat patches and in a smaller proportion of the sites within their geographic ranges, and they coexist with fewer other species and combinations of species (Table 2; see also Brown, 1984*b*; Brown and Kurzius, 1989). These species also tend to be more ephemeral even in habitats where they normally occur. For example, at our Chihuahuan Desert study site *P. flavus* fluctuated greatly in abundance, although it was present in 75 of the first 100 monthly trapping periods (Fig. 2, Table 1; see also Brown and Heske, 1990*a*).

The individualistic nature of species niches in conjunction with the great spatial and temporal variation in the species composition of granivorous rodent assemblages might suggest that the communities have little or no real "structure"; they are essentially random collections of species. This is not the case. The following two subsections will demonstrate that community organization can be characterized in terms of predictable patterns in the number, identity, and biological attributes of those species which coexist locally. Furthermore there is increasing evidence, much of it from controlled, manipulative experiments conducted in the field, that these patterns are largely the result of deterministic processes of interaction of species with both their physical environment and with other organisms. In discussing these complex relationships, we shall first consider the patterns of species diversity and the processes that regulate the number of species that coexist on various spatial scales.

Regulation of Species Diversity: Effects of Climate, Soil, and Vegetation

Like the abundance and distribution of single-species populations, the size and composition of multi-species assemblages is determined ultimately by physical factors, principally aspects of climate and geology. These have both direct effects through their influence on dispersal, survival, and reproduction, and indirect effects through their influence on other species that provide food and shelter or act as competitors, predators, parasites, pathogens, or mutualists.

Although the Heteromyidae are confined to a limited region of the North American continent, the ecological range of the family is impressive. Temperate representatives are found from below sea level in Death Valley to near timberline in the Sierra Nevada and Rocky Mountains, from the subtropics of northern Mexico to the prairies of southern Canada. They inhabit desert, steppe, grassland, chaparral, woodland, and thorn forest. However, the greatest diversity of genera and species occurs in the arid habitats of the southwestern United States and northwestern Mexico (Schmidly et al., this volume). Climatic and geological factors ultimately limit species to restricted areas and habitats within this region, and severely curtail the diversity of forms that have been able to colonize the generally more mesic environments beyond the arid southwestern United States. The composition of assemblages of heteromyids inhabiting different regions and habitats reflect the extent to which local environments meet the individualistic requirements of particular species. But what determines how many and which of these species coexist within a local habitat?

Climate and geology affect community structure over a wide range of spatial and temporal scales. At one extreme, by influencing the geographic distributions of species they limit the composition of the regional pool of species from which local communities must be derived (Fig. 5). Such limits on the regional geographic pool can have important effects on the number as well as the identity of species that comprise local assemblages (Fig. 6). Biogeographic barriers of inhospitable climate, substrate, and vegetation have prevented certain species of desert rodents from colonizing regions where they otherwise could live. This was dramatically demonstrated by the successful introduction of a population of *D. ordii* from Oklahoma to a sand dune on the shore of Lake Erie in Ohio, approximately 1,000 km east of the geographic range of any heteromyid (Bole and Moulthrop, 1942). In southwestern North America the effects of biogeographic barriers are apparent in the reduced diversity of the assemblages in certain desert valleys isolated by mountain ranges (Brown 1973, 1987), on islands in the Gulf of California (especially the "oceanic" islands; Lawlor, 1983), and also in the distinctive composition of the highly endemic faunas of the Central Valley of California and Baja California (see Schmidly et al., this volume).

At the other extreme, climate and geology also influence which of the species in the regional geographic pool that could potentially occur in a local habitat actually live there. Some of these controls on species diversity and composition are relatively direct and straightforward consequences of the limiting niche variables of individual species or functional groups of species with similar requirements. For example, pocket mice range into colder regions than any other heteromyids. *Perognathus parvus* is found in the sagebrush steppe above 2,700 m elevation in the Wasatch Mountains of Utah (J. Cranford, pers. comm.) and *P. flavescens* inhabits the prairies of southern Saskatchewan and Alberta. These species are able to forage and reproduce during the brief summer season and then become torpid during the cold winter months when they cannot be active and maintain energetic balance. It is not hard to explain why kangaroo rats are absent from these environments (although they inhabit others that are only slightly less severe). All *Dipodomys* species are apparently unable to hibernate because they cannot tolerate prolonged hypothermia (Carpenter, 1966).

Within regions having a common species pool and similar climates, the composition of assemblages varies conspicuously with habitat structure, which is determined ultimately by local geology and climate and more proximally by soil and vegetation type. Thus, hillsides covered with bare rock or boulders tend to have communities comprised of 1 or 2 species of *Chaetodipus* and *Peromyscus*; bipedal *Dipodomys* and *Microdipodops* are conspicuously absent (e.g., Brown, 1975). Quadrupedal forms also dominate densely vegetated grass and shrub habitats (e.g., Rosenzweig and Winakur, 1969; Rosenzweig et al., 1975). The highest diversity, both in terms of number of species and variety of functional groups, is found in habitats with sandy soils and a mosaic of vegetation that includes both open spaces and dense shrub cover (e.g., Brown, 1975, 1984*b*; Rosenzweig and Winakur, 1969; Rosenzweig et al., 1975). Rosenzweig (1973) and Price (1978*b*) were able to change the distribution and abundance of individ-

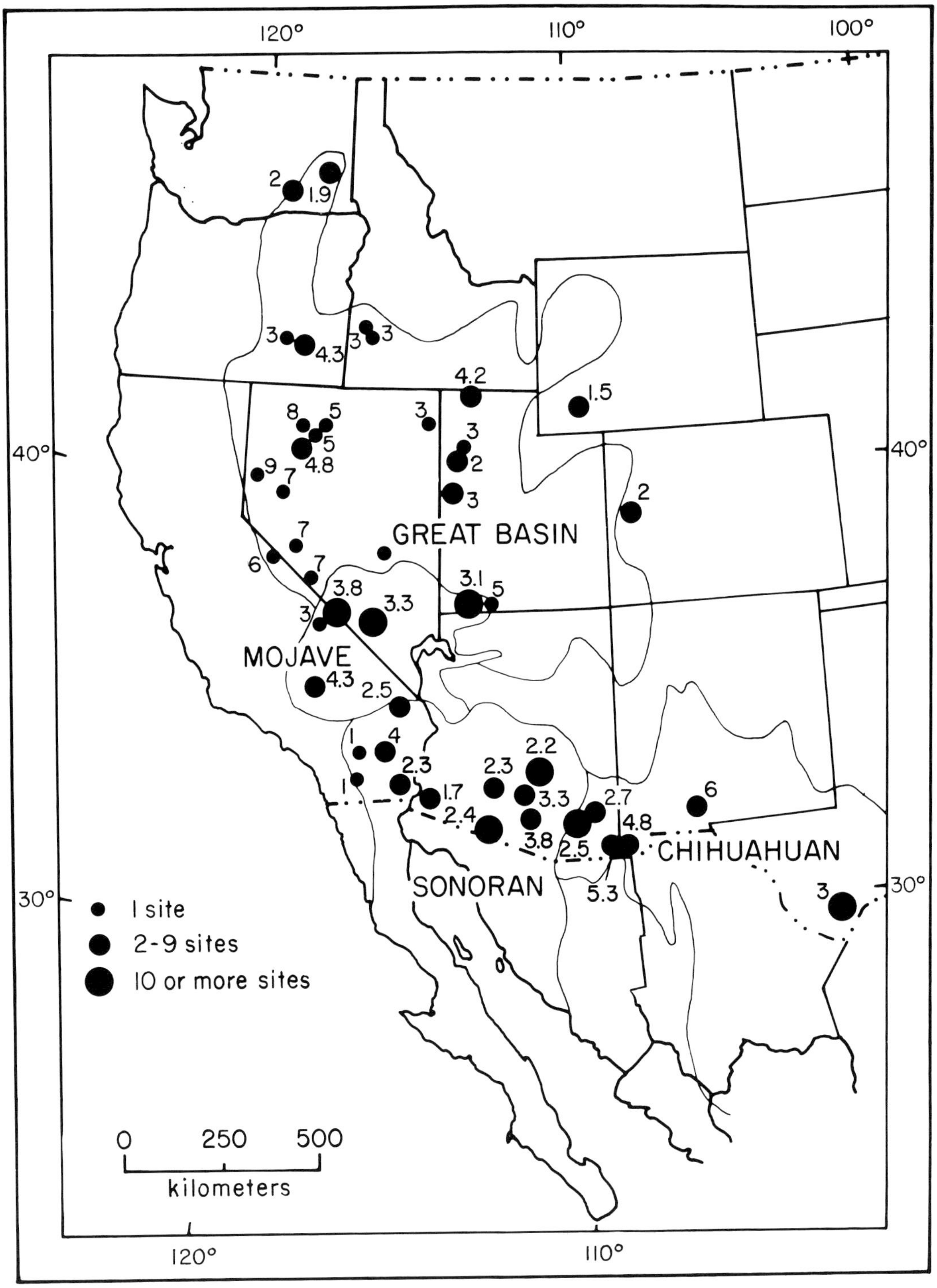

FIG. 5.—Map of the southwestern United States showing the number of granivorous rodent species inhabiting 202 sites, each containing a few hectares of relatively homogeneous desert shrub habitat (data compiled from the literature survey of Brown and Kurzius, 1987, and replotted from Brown, 1987). When several sites were too close together to plot individually, a larger symbol was used

ual species (see above) and hence the composition of communities in a predictable direction by experimentally "tailoring" the vegetation to change habitat structure. The presence of particular types of structural habitat may allow certain species or functional groups to persist in a locality by directly facilitating locomotion or foraging (by altering the distribution of resources), or by diminishing predation risk. Heterogeneous habitats support more diverse assemblages in part because they provide opportunities for effective microhabitat segregation among species in different functional groups. Changes in habitat structure may also affect competitive relationships and risk of predation, and thereby alter the allocation of resources among coexisting species (see beyond).

Other ultimate effects of climate and geology are equally apparent but not quite so easily understood in terms of the tolerances of individual species. Perhaps the best example is the effect of productivity on diversity. When habitat structure is held constant in geographic gradients of varying climate, the diversity of rodents within local communities increases with increasing precipitation (Brown, 1973, 1975). Since total rodent populations and biomass also vary directly with precipitation, the capacity of the environment to support rodents can be attributed to the indirect effect of precipitation on primary production (see above) and hence on food availability. Abramsky (1978) demonstrated this phenomenon experimentally. He added seeds to shortgrass prairie habitat and observed an increase in rodent species diversity due to the colonization and establishment of a population of *D. ordii.*

It is not obvious, however, why enhanced food resources should result in an increased number of species rather than increased populations of the same species. As shown in Figure 7, we can make a simple graphical model in which each species is limited by the availability of specific food resources (e.g., the concentration of seeds in a particular microhabitat or substrate necessary for efficient harvesting and predator avoidance). When productivity is low, only one or a few resource types will be sufficiently abundant to support a species, but an overall increase in productivity will raise the availability of other resource types above the thresholds where they too can support a species. Although increases in species diversity in response to spatial (Brown, 1973, 1975) and temporal (Abramsky, 1978; Congdon, 1974) variation in food availability are consistent with this model, alternative explanations exist. It should be possible to design seed addition experiments that would provide a good test.

Interspecific Competition and Community Structure

In addition to the extrinsic factors, the organization of communities depends upon intrinsic factors—interactions among the species that comprise the assemblage (Table 3). Since we are dealing here with a single trophic guild, the interactions among rodent species should be primarily competitive. Although we shall focus on the influence of interspecific competition on local community composition, we point out that it can also operate at larger spatial scales. For example, competitive exclusion may limit the

---

←
(numbers indicate the average species richness for all those sites). These are the numbers of species that actually coexist in small patches of relatively uniform habitat. Note the small size of local assemblages in the eastern Great Basin, Mojave, and western Sonoran deserts. This low diversity can be attributed to some combination of biogeographic barriers that have limited the size of the regional species pool (see Fig. 5) and local abiotic and biotic conditions that limit the number of species that can coexist.

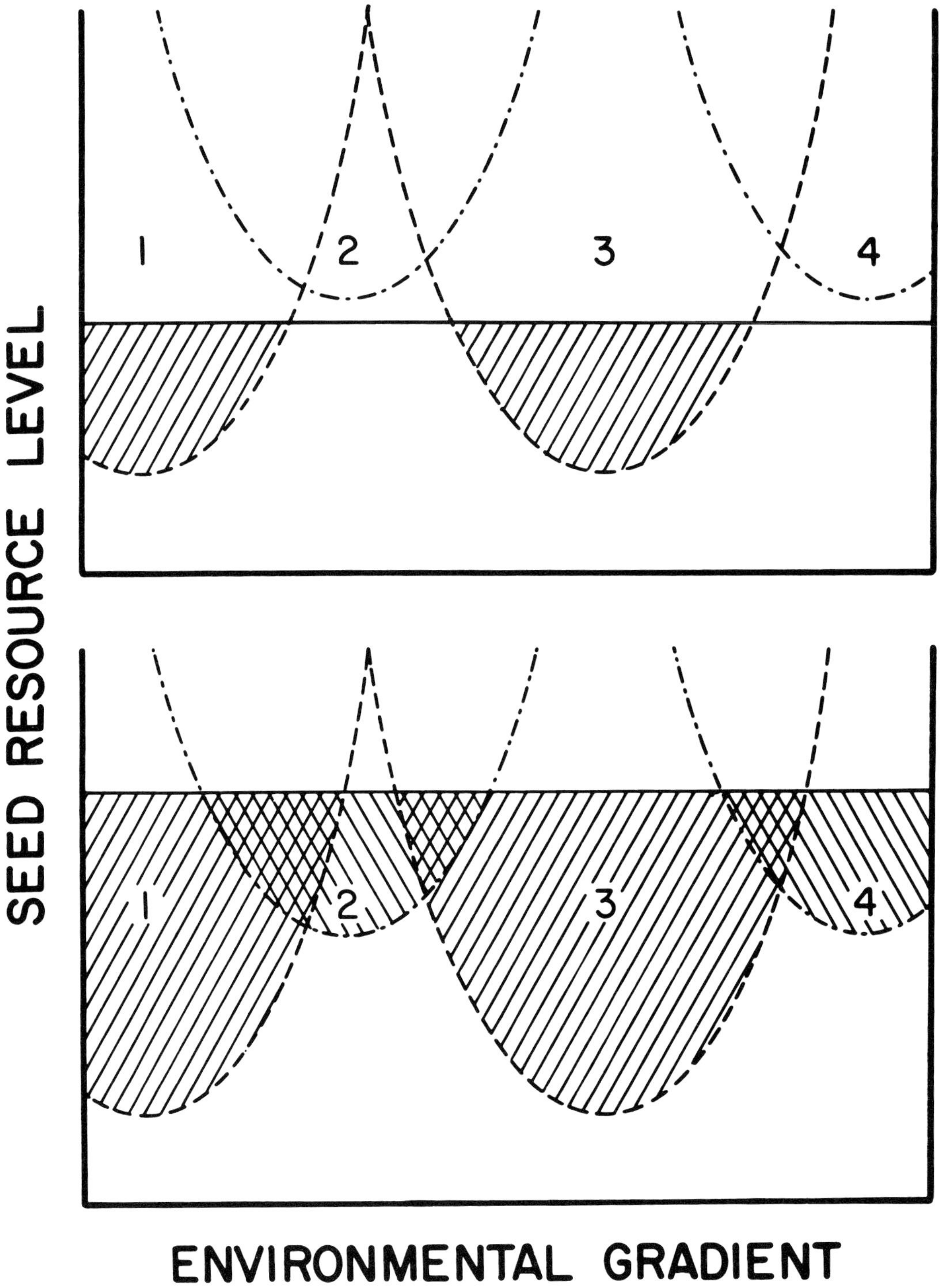

FIG. 6.—A conceptual model of how productivity can limit local species diversity. The environmental gradient can be a gradient either in characteristics of the seed resources themselves (e.g., seed size or chemical composition), or in the habitat (e.g., shrub cover or substrate particle size) that affects the profitability with which seeds can be harvested. The dashed curves depict the minimum level to

geographic ranges of species, and this in turn will affect the composition of the regional species pools from which local communities are assembled.

Desert rodents have long been one of the groups that have provided evidence for the importance of interspecific competition in population dynamics and community organization. This evidence is basically of two types. First, there are nonexperimental studies documenting nonrandom patterns in the kinds of species that coexist in local habitats. Additionally, there are experimental studies demonstrating increases in population density or shifts in microhabitat utilization of certain species in response to removal of others.

Ever since Grinnell and Orr's (1934) note about the "gauges" of mice, ecologists have been impressed by the differences in body size among coexisting species of desert rodents (Fig. 8; see also Bowers and Brown, 1982; Brown, 1973, 1975; Price and Brown, 1983; Rosenzweig and Sterner, 1970). More sophisticated statistical tests of null hypotheses, performed on increasingly larger data sets, have shown repeatedly that species which coexist in local habitats are highly nonrandom assemblages with respect to body size (Bowers and Brown, 1982; Brown, 1973; Hopf and Brown, 1986). Two patterns are evident: the ratios of body weights between adjacent species in the size hierarchy are more similar (even), and the minimum ratios between the two most similar species are larger than expected on the basis of chance (Hopf and Brown, 1986). These larger than random differences support the hypotheses that interspecific competition plays a major role in community organization of desert rodents and that differences in body size reduce competition and promote local coexistence.

This interpretation is supported by nonrandom patterns in attributes of coexisting species other than body size. For example, we analyzed the 61 local two-species communities in the data set compiled by Brown and Kurzius (1987) by determining the frequency of coexistence of pairs of species in three functional groups: bipedal heteromyids (*Dipodomys* and *Microdipodops*), quadrupedal heteromyids (*Perognathus* and *Chaetodipus*), and quadrupedal murids (*Peromyscus* and *Reithrodontomys*). Pairs of species from two different functional groups (bipedal and quadrupedal heteromyid, and quadrupedal heteromyid and murid) occur together more frequently than expected by chance, while pairs of quadrupedal heteromyids coexist less frequently than expected (Table 4). These patterns provide further evidence that species in the same functional group probably compete most intensely and may exclude each other from local patches of habitat. Interestingly, one combination of different functional groups, bipedal heteromyid and quadrupedal murid, also occurs together less frequently than expected at random. We hypothesize that this kind of two-species assemblage is usually unstable, because it is susceptible to invasion by a quadrupedal heteromyid that would convert it into a three-species community. Thus analysis by functional groups that are not defined on the basis of body size also supports the hypothesis that coexisting granivorous desert rodents are non-

---

←

which a species can harvest resources as a function of position on the gradient. The horizontal line indicates the level of productivity in the habitat. Above, when productivity is low, only two species can profitably harvest resources as indicated by the shaded portions of the utilization curves of species 1 and 3. Below, when productivity is higher, all four species can forage profitably and coexist. This model does not assume or preclude any changes in niche breadth, position, or overlap as the number of coexisting species changes. Somewhat similar models have been developed by Tilman (1982) to conceptualize species interactions in plant communities.

TABLE 3.—*Mechanisms of resource allocation or competition that promote coexistence in heteromyid communities. These are interrelated, and each may occur independently or as a consequence of other mechanisms (e.g., rodents may select habitats directly or indirectly as a result of resource dispersion). Several mechanisms may operate in a community and their relative importance will vary with locality and time. Relevant examples are listed under the primary mechanism investigated. Parentheses indicate locality or type of investigation.*

Mechanism	Examples
Habitat selection/restriction	Supporting
	Rosenzweig and Winakur, 1969 (Chihuahuan Desert)
	Rosenzweig, 1973 (Chihuahuan Desert)
	Brown and Lieberman, 1973 (Great Basin and Mojave deserts)
	Schroder and Rosenzweig, 1975 (Chihuahuan Desert)[a]
	Hoover et al., 1977 (Chihuahuan Desert)
	Price, 1978*b* (Sonoran Desert)
	Lemen and Rosenzweig, 1978 (Sonoran Desert)
	Wondolleck, 1978 (Sonoran Desert)
	Stamp and Ohmart, 1978 (Sonoran Desert)
	M'Closkey, 1976, 1978 (Sonoran Desert)
	Bowers, 1982 (Sonoran Desert)
	Larsen, 1986 (Great Basin and Mojave deserts)
	Conflicting
	Thompson, 1982*a* (Mojave Desert)
Neutral ground (non-equilibrium)	Supporting
	Schroder, 1987 (Chihuahuan Desert)[b]
Independent adaptation	Supporting
	Hallet, 1982 (in part; Chihuahuan Desert)[c]
Food partitioning/variable foraging efficiency	
Seed size	Supporting
	Brown and Lieberman, 1973 (Great Basin and Mojave deserts)
	Reichman, 1975 (Sonoran Desert)
	Mares and Williams, 1977 (Laboratory)
	M'Closkey, 1980 (Sonoran Desert)
	Conflicting
	Rosenzweig and Sterner, 1970 (Laboratory)
	Smigel and Rosenzweig, 1974 (Chihuahuan Desert)
	Lemen, 1978 (Laboratory)
	Stamp and Ohmart, 1978 (Sonoran Desert)
	Price, 1983 (Laboratory)
Seed distribution	Supporting
	Reichman and Oberstein, 1977 (Sonoran Desert)
	Price, 1978*a* (Laboratory)
	Hutto, 1978 (Laboratory)
	Trombulak and Kenagy, 1980 (Laboratory)
	Harris, 1984 (Great Basin Desert)
	Conflicting
	Frye and Rosenzweig, 1980 (Chihuahuan Desert)
	Thompson, 1985 (Laboratory and Mojave Desert)
Resource variability	Supporting
Environmental	J. S. Brown, 1988 (Sonoran Desert)
	Price and Waser, 1985 (Laboratory and Sonoran Desert)
Species-induced	J. S. Brown, 1988 (Sonoran Desert)

TABLE 3.—*Continued.*

Mechanism	Examples
Predator-mediated coexistence	Supporting Thompson, 1982*b* (Mojave Desert) Kotler, 1984*a* (Great Basin Desert) Price et al., 1984 (Sonoran Desert)
Aggressive interference	Supporting Congdon, 1974 (Mojave Desert) Blaustein and Risser, 1974, 1976 (Laboratory) Hoover et al., 1977 (Laboratory and Chihuahuan Desert) Trombulak and Kenagy, 1980 (Laboratory) Frye, 1983 (Chihuahuan Desert) Bowers et al., 1987 (Chihuahuan Desert)

[a] Species coexist without significant competition due to evolved differences in habitat preference resulting from competitive interactions in the past.
[b] Species coexist but compete intensely in areas of restricted sympatry where neither species has a competitive advantage.
[c] Species coexist because of differences in habitat preference that evolved independently of species interactions.

random assemblages, whose composition is determined at least in part by interspecific competition.

That desert rodents compete interspecifically has been tested directly in experiments in which certain species were removed and the densities and/or microhabitat utilization of the remaining species was monitored to detect the predicted changes. Two examples of substantial increases in the population densities of the small quadrupedal species in response to experimental exclusion of large bipedal kangaroo rats have been reported: one in the Chihuahuan Desert of southeastern Arizona where four of five species of small heteromyids and murids increased an average of approximately three times (Brown and Munger, 1985; Munger and Brown, 1981), and one in the Chihuahuan Desert of New Mexico where *Chaetodipus penicillatus* increased approximately 2.5 times (Freeman and Lemen, 1983). Both of these experiments documented competition between species of different size and functional group, but the most intense interaction would be expected between more similar species that do not usually coexist in the same local habitats.

It is surprising, then, that the one experiment in which the predicted increase in population density was not observed was the only one in which very similar species were manipulated—Schroder and Rosenzweig's (1975) reciprocal removal of *D. merriami* and *D. ordii.* These investigators attributed the lack of compensatory density changes in these kangaroo rat species to inflexible habitat selection. But, during most of the experiment they were unable to maintain significant reductions in populations of the species designated for removal. Schroder (1987), using data from high resolution radio-tracking, suggested that the coexistence of these two species in shrub-grassland habitat mosaic is only partially explained by habitat selection. Assuming that neither species has a competitive advantage in this mixed habitat, Schroder proposed that they may coexist as competitors in a nonequilibrial state due to continual immigration from nearby allopatric populations. Evaluation of this "neutral ground hypothesis" requires further experimental study.

Shifts in the microhabitats used by the remaining species in response to experimental addition or removal of other species also suggest that interspecific competition

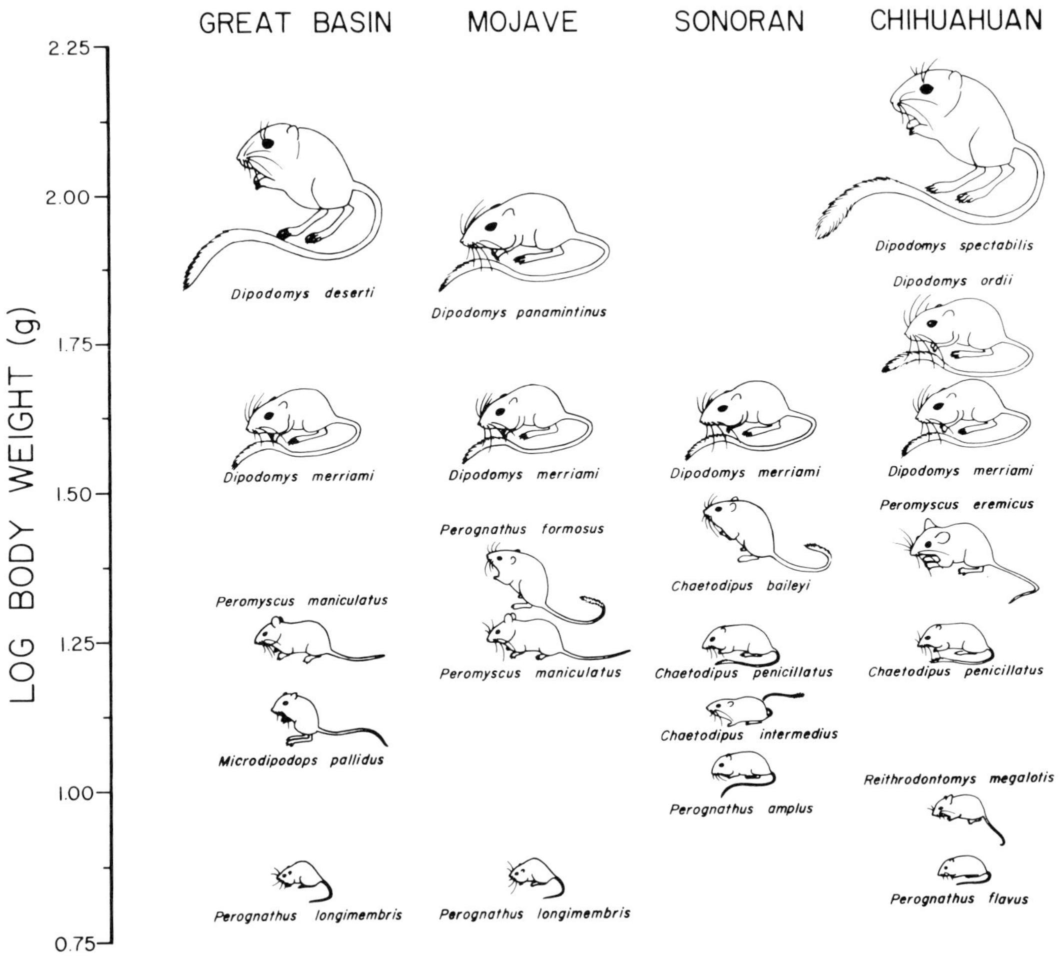

FIG. 7.—The species composition and morphologies of the abundant (>2.5% of total sample) granivorous rodents that coexist to form diverse assemblages at selected sites in the four subdivisions of the North American Desert: Great Basin Desert, a sand dune in Fish Lake Valley, Nevada (dune 7 in Brown, 1973); Mojave Desert, desert shrub habitat near Johannesburg, San Bernardino Co., California (site 5 from Hafner, 1977); Sonoran Desert, upper Silverbell bajada northwest of Tucson, Arizona (from Petryszyn, 1982); Chihuahuan Desert, the Cave Creek Bajada near Portal, Arizona (Rodeo B from Brown, 1975).

plays a major role in community organization. Working in the Sonoran Desert of southern Arizona, Price (1978*b*) observed that both *D. merriami* and three species of pocket mice (*P. amplus, C. penicillatus,* and *C. baileyi*) predictably altered their patterns of microhabitat use when the other species were removed or added to experimental enclosures. Larsen (1986) obtained similar results in a natural experiment in the Great Basin Deserts of Utah and Nevada. He showed that in otherwise comparable habitats, *D. ordii* and *D. merriami* showed a higher variance in microhabitat use when fewer other heteromyid species were present locally. This suggests that interspecific interactions could influence microhabitat use and the local distribution of heteromyid species. Analyzing the effects of experimental removal of kangaroo rats in the Chihua-

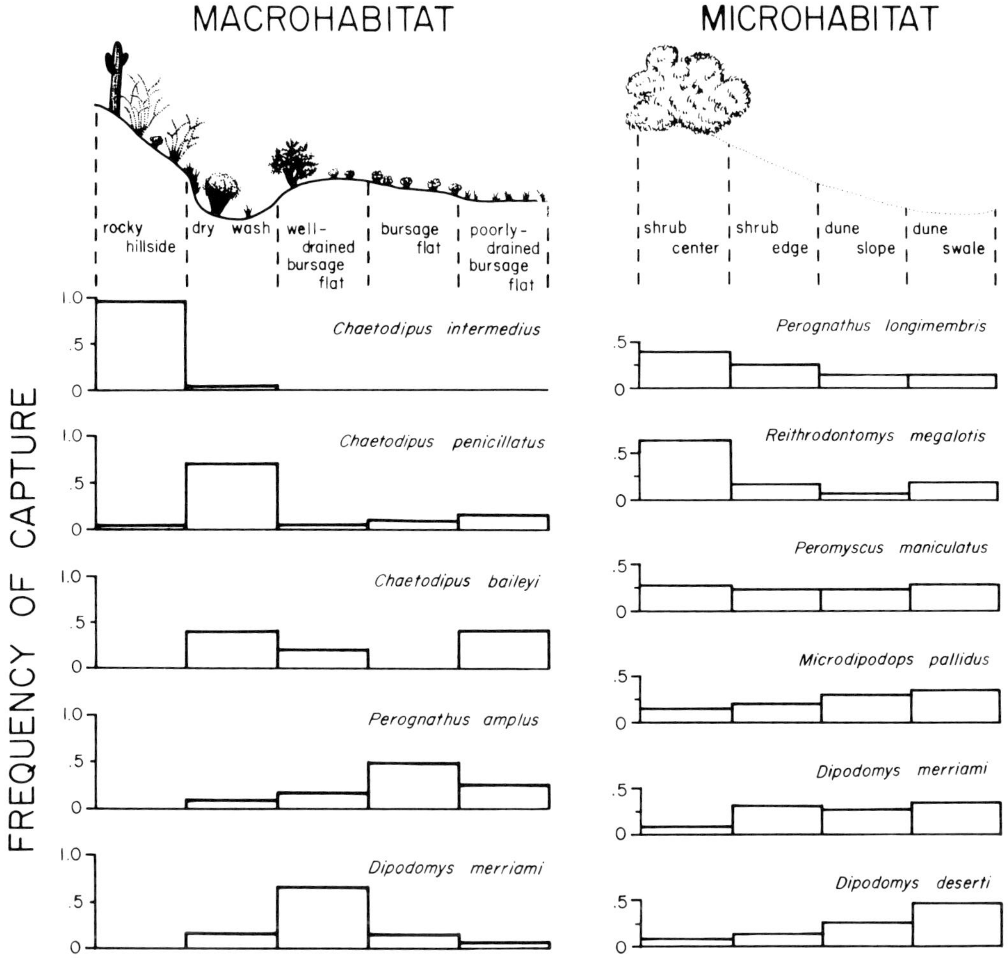

FIG. 8.—Examples of segregation of granivorous desert rodent species with respect to macro- and microhabitat. Left, relative abundance of five species in four distinctive types of macrohabitat in the Sonoran Desert, Organ Pipe Cactus National Monument, Arizona (unpubl. data from Petryszyn, J. S. Brown, and Harney). Right, frequency distribution of six species with respect to microhabitat features on sand dunes in the Great Basin and Mojave deserts (data from Brown and Lieberman, 1973). Note that macrohabitat types change on spatial scales of 10 to 10^3 m whereas microhabitats change on scales of less than 10 m.

huan Desert of southeastern Arizona, Bowers et al. (1987) showed that 7 of 9 rodent species significantly shifted their microhabitat use in response to exclusion of *D. spectabilis,* and both quadrupedal heteromyids *P. flavus* and *C. penicillatus* shifted to more open microhabitats in response to removal of all three kangaroo rats (*D. spectabilis, D. merriami,* and *D. ordii*). Taken together with the compensatory increases in density, these shifts in microhabitat utilization in response to missing species make a very strong case for the importance of interspecific competition in the resource utilization, population dynamics, and community organization of desert rodents.

But if rodents compete, what are the resources that are in short supply, and what are the mechanisms of interaction that allow some combinations of species but not others to coexist? There are many hypotheses, most are supported to varying degrees by data, but much remains to be done to achieve a comprehensive understanding. The two primary mechanisms of competition, resource exploitation and aggressive interference, have both been implicated in desert rodents. It is difficult to distinguish between these mechanisms, in part because rodents are nocturnal and secretive so that it is hard to observe their microhabitat use, foraging behavior, and aggressive interactions directly. As a result, there has been considerable debate about the relative merits of indirect methods, such as trapping and telemetry, and more direct techniques, such as observations of individuals in the laboratory, at feeding stations, or with attached lights. All of these methods have their problems, and none of them have successfully "turned nocturnal rodents into diurnal birds" (M. L. Rosenzweig, pers. comm.). Another problem with trying to distinguish between exploitation and interference is that the two mechanisms are not independent—species that coexist and overlap in utilization necessarily compete by exploitation, but they may also interact aggressively in ways that restrict access to resources.

Abundant evidence is available suggesting that desert rodents interact by both mechanisms. Aggressive interspecific interference has been documented by direct observation, both in the laboratory (Blaustein and Risser, 1974, 1976; Congdon, 1974; Eisenberg, 1963; Hoover et al., 1977; MacMillen, 1964) and at artificial feeding stations in the field (Congdon, 1974; Wondolleck, 1978). Shifts in use of space by smaller species in response to experimental exclusion of large kangaroo rats have been attributed to the absence of a behaviorally dominant competitor (Bowers et al., 1987; Frye, 1983; see also Rebar and Conley, 1983).

There is equally abundant evidence that coexisting species differ in their use of resources in ways that would tend to reduce the intensity of competition. This is not surprising since those species that occur together locally tend to differ in body size, mode of locomotion, and other behavioral and morphological traits that undoubtedly affect their exploitation of food and microhabitats (see above, and Price and Brown, 1983). Coexisting species often differ markedly in their use of both space and time. Differences in habitat use are well documented and generally agreed upon (Fig. 9; see also Brown, 1975; Brown and Lieberman, 1973; Congdon, 1974; Hallett, 1982; Hoffmeister and Goodpaster, 1954; Hoover et al., 1977; Kotler, 1985*a*; Lemen and Rosenzweig, 1978; M'Closkey, 1981; Price, 1978*b*; Rosenzweig, 1973; Rosenzweig and

TABLE 4.—*Observed and expected* ( ) frequencies of co-occurrence of rodents in different taxonomic and functional groups for 61 two-species assemblages. Rodent species are classified as follows: bipedal heteromyids (*Dipodomys *and* Microdipodops*), quadrupedal heteromyids (*Perognathus *and* Chaetodipus*), and quadrupedal murids (*Peromyscus *and* Reithrodontomys*). Observed frequencies of co-occurrence differ significantly from that expected by random association (chi-square, 4* d.f., *= 9.7, 0.025 <* P *< 0.050). Quadrupedal heteromyids occur together less often than expected by chance. Species in different taxonomic or functional groups coexist more often than expected, except in the case of bipedal heteromyids and quadrupedal murids. See text for possible explanation.*

		Coexistence of:		
		Bipedal heteromyid	Quadrupedal heteromyid	Quadrupedal murid
with:	Bipedal heteromyid	4 (4)	19 (13)	3 (9)
	Quadrupedal heteromyid		6 (12)	25 (17)
	Quadrupedal murid			5 (6)

* Expected frequencies were calculated by multiplying the number of sites by the product of the probabilities of independent species occurrence. Probabilities were estimated from 122 species-site occurrences at the 61 local sites: 30 for bipeds, 54 for quadrupeds (heteormyid), and 38 for murids.

Winakur, 1969; Rosenzweig et al., 1975; Schroder and Rosenzweig, 1975; Wondolleck, 1978), although the details may still be controversial (see Thompson, 1982*a*, 1987). Interspecific differences in the microhabitats where rodents forage are particularly important in effecting division of seed resources among locally coexisting species. Communities are also usually comprised of species that differ in seasonal activity patterns—some are active throughout the year, whereas others spend as much as several months in torpor or estivation (e.g., Brown, 1975; Brown and Munger, 1985; Brown and Zeng, 1989; J. S. Brown, 1988; MacMillen, 1964; O'Farrell, 1974; Petryszyn, 1982). It is not as obvious how these differences in timing of activity facilitate resource division and coexistence, but J. S. Brown (1988) has developed a theory of coexistence based on differential exploitation of varying resources and has tested this mechanism with experiments on desert rodents. Other possible mechanisms that might reduce exploitative competition and promote coexistence are much more controversial (Table 3). These include differential utilization of different sizes and/or spatial aggregations of seeds (e.g., Brown, 1975; Brown and Lieberman, 1973; Frye and Rosenzweig, 1980; Lemen, 1978; M'Closkey, 1980; Mares and Williams, 1977; Price, 1978*b*; Price and Heinz, 1984; Price and Waser, 1985; Reichman and Oberstein, 1977; Rosenzweig and Sterner, 1970; Smigel and Rosenzweig, 1974; Stamp and Ohmart, 1978; Trombulak and Kenagy, 1980).

Seeds of different plant species differ in physical, chemical, and nutritional properties (e.g., Kelrick and MacMahon, 1985). Rodents are known to be somewhat selective in their foraging, differentially harvesting seeds of certain sizes, chemical composition, water content, or other properties (e.g., Brown et al., 1986; Frank, 1988; Kelrick et al., 1986; Reichman, 1975, 1977; Sherbrooke, 1976). It is questionable, however, to what extent partitioning of seeds on the basis of these properties, with the possible exception of seed size (see above and Table 3), influences heteromyid community organization. Sherbrooke's (1976) study of the relationship between *C. baileyi* and the desert shrub *Simmondsia chinensis* (jojoba) and Kenagy's (1972) work on the specializations of *D. microps* for feeding on *Atriplex* leaves provide examples of apparently coevolved dependencies that may affect abundance, distribution, and community structure. Even in these cases, however, the data suggest that the relationship, while important, is by no means obligate. Most species are extremely opportunistic, and within some very broad dietary constraints (in the extent to which they are strictly granivorous) they utilize a wide variety of food resource species.

Predation, Parasitism, and Disease

In addition to competition, other biotic interactions can affect community structure directly, by limiting the distribution and abundance of species, and more indirectly, by affecting the availability of resources and the outcome of interspecific competition.

Unfortunately our knowledge of these interactions is meager (but see Munger et al., 1983). Most attention has been devoted to predation. It is well documented that a variety of vertebrate predators, including snakes, owls, hawks, and mammalian carnivores, prey upon heteromyids. Studies of the food habits of these predators suggest that their direct impact can be severe, at least in some cases (e.g., Egoscue, 1962; French et al., 1967; Kotler, 1985*b*). Some of the most conspicuous morphological and behavioral specializations of heteromyids have been interpreted as adaptations for avoiding predators (Mares, 1983, but see Hafner, this volume). These include the dorsally-placed eyes and background-matching coloration of all genera (Benson, 1933; Dice and Blossom, 1937) and the inflated auditory bullae and bipedal, saltatory locomotion of kangaroo rats and kangaroo

FIG. 9.—A typical mound of *D. spectabilis* in the Chihuahuan Desert near Portal, Arizona. This mound is approximately 2 m in diameter, 0.5 m above ground with burrows and chambers extending >0.5 m below the soil surface. Mounds frequently contain stored seeds and many commensal organisms. As major features of the landscape, these persist for many years and are utilized by many successive generations of banner-tailed kangaroo rats (see text and Jones, this volume).

mice (e.g., Bartholomew and Caswell, 1951; Eisenberg, 1975; Webster, 1962; Webster and Webster, 1971, 1975, 1980, 1984). Behavioral adaptations (e.g., moonlight avoidance: Kaufman and Kaufman, 1982; Kotler, 1984*a*; Lockard, 1978; Lockard and Owings, 1974; Price et al., 1984) also suggest the importance of predator evasion.

Much of the influence of predation on community organization is probably subtle and indirect. There is increasing evidence that actual and perceived risk of predation influences the foraging and behavior and microhabitat use of individual rodents (e.g., Bowers, 1988; Kotler 1984*a,* 1984*b*; Thompson, 1982*b*; see also references on moonlight avoidance, above). There is every reason to expect that these responses to predation should also affect allocation of resources and interspecific competitive relationships in the community.

Unfortunately, effects of predation risk on community organization are difficult to assess. There are few good "natural experiments," in which the intensity of predation or the occurrence of particular kinds of predators varies between regions but other variables that might affect community structure remain relatively constant. Because most kinds of predator have extensive geographic and habitat distributions, most desert rodent communities probably experience similar predation pressure. When the predators differ, many other environmental factors also vary, making it difficult to attribute differences in the rodent communities primarily or exclusively to predation. Nevertheless, some of the most promising cases have not been investigated thoroughly. For example, the islands in the Gulf of California have different combinations of both predators and rodent species. Also, the restriction of *Microdipodops* to the cold Great Basin Desert and the similar confinement of the small, bipedal, and apparently convergent dipodids to the cold Asian deserts and steppes may be related to the low abundance and diversity of pit vipers which locate their prey by thermal sensors.

In other kinds of organisms, experimental exclusion of particular predators has been shown to cause dramatic changes in prey community organization. There is no reason to expect that this would not also be true of desert rodents, but the appropriate experiments have not yet been done. There are obvious practical problems in effectively excluding mammalian, avian, or reptilian predators on a sufficient scale to assess realistically their impact on rodent communities. Nevertheless, we hope that someone will overcome these obstacles and perform the definitive experiments.

Parasites and diseases may also significantly affect the community organization of their hosts, and the dynamics of these interactions may resemble those of predator and prey. Unfortunately the parasitology of desert rodents is still in the descriptive stage. Although the geographic distribution and host affinities of parasites are becoming better understood (see Whitaker, this volume) little is known about the effect of these organisms on host individuals and populations, not to mention communities (but see Munger et al., 1983). There is even less information on viral, bacterial, and fungal diseases.

A major problem facing community ecologists is to evaluate the relative contributions of and interactions between the numerous mechanisms that are known or suspected to influence species diversity and composition. This is a formidable task, because none of the mechanisms are necessarily mutually exclusive. Not only is it probable that they interact with each other in complex ways, but as we suggested previously, the intrinsic competitive relationships among the rodent species themselves are likely to be influenced by interactions with components of the extrinsic environment, such as climatic and soil conditions, habitat structure, and predators. Several investigators have begun to explore the potentially complex interrelationships between microhabitat use, predation risk, foraging behavior, and interspecific competition (e.g., Hay and Fuller, 1981; Kotler, 1984*a*, 1984*b*, 1985*a*, 1985*b*; Price and Heinz, 1984; Price and Waser, 1985; Thompson, 1982*b*), but much remains to be done.

Much has been learned about community organization in desert rodents since Rosenzweig, Brown, and others began their work in the late 1960s and early 1970s. Inevitably, setbacks, errors, and controversies have developed about both fact and interpretation. Many issues remain to be resolved. Nevertheless, it is a testimony to the contributions of many investigators that the community ecology of desert rodents is probably as well known as that of any other group of organisms. This background of invaluable information should serve to make desert rodents even more attractive for future studies.

Impact of Rodents on Arid Ecosystems

Activities of rodents have major effects on the structure and dynamics of desert ecosystems. Some of these impacts can be attributed unambiguously to particular species of heteromyids. Others are clearly assignable to rodents, but effects of heteromyids cannot be or have not been separated from those of coexisting murids, sciurids, and geomyids. The impacts can be divided into four general categories: modification of the physical environment, consumption, production, and indirect effects. Although there is good reason to believe that all of these activities have major effects, the evidence for them is often scanty.

Through their digging, rodents move large quantities of soil and modify the physical environment. Soil movement, creation of disturbed patches of loose, bare soil, and mixing of organic and inorganic material is an important impact of rodents. In many habitats, it is accomplished mainly by pocket gophers, but geomyids are rare in many desert regions and much of the soil transport is done by heteromyids and sometimes by sciurids. The short- and long-term effects of these activities are only beginning to be appreciated.

Experiments in which we removed rodents from Chihuahuan Desert scrub habitat resulted in obvious changes (Brown and Heske, 1990*b*; Brown et al., 1986). Four to twelve years after three species of kangaroo rats (*D. spectabilis, D. merriami,* and *D. ordii*) were excluded the experimental plots exhibited much less disruption of the soil surface, higher densities of tall perennial and annual grasses, increased accumulation of litter, decreased foraging by granivorous birds, and differential colonization by rodents typical of grassland habitats. Although selective predation on large seeds (see below) may have contributed to some of these changes, it appears that physical soil disturbance as a result of foraging and burrowing activities was the most important mechanism. In this desert, kangaroo rats are a "keystone guild," and their long-term removal caused a conversion of desert shrub habitat to grassland (Brown and Heske, 1990*b*). Whether exclusion of *Dipodomys* or other heteromyid rodents would cause such dramatic changes in other desert sites or in other kinds of habitats remains to be seen.

Rodents also have important physical effects on other organisms by constructing burrows. The burrows and dens of all rodents probably serve as homes and refuges for other kinds of animals. The huge mounds constructed by the large kangaroo rats are particularly important. For example, the mounds of *D. spectabilis* are 2–3 m in diameter and approximately 0.5 m high; they have several entrances and a labyrinth of tunnels and chambers that may extend 0.5 m below ground level (Fig. 9). Mounds typically take several years to construct, but then they persist for decades, used and maintained by successive generations of banner-tailed kangaroo rats. These mounds constitute a unique, patchy microenvironment that contains distinctive communities of surprising diversity. In addition to providing physical refuges and favorable microclimates for a variety of reptiles and invertebrates (Kay and Whitford, 1978), the mounds also support a more obligate community of microbes and invertebrates that use the seeds stored by the kangaroo rats (Rebar and Reichman, 1983; Seastedt et al., 1986).

Granivorous rodents are seed predators that kill the juvenile stages of plants. As seed consumers they have major impacts on desert ecosystems. Rodents forage preferentially for seeds of certain species, sizes, and physical and chemical properties (Brown and Davidson, 1977; Brown et al., 1979*b*, 1986; Reichman, 1975). In particular, they selectively harvest large seeds and this has a number of important direct and indirect consequences. Rodents severely suppress populations of their preferred prey, so that large-seeded annual plant species increase, sometimes as much as 3 orders of magnitude, when rodents are experimentally removed (Table 5; Brown et al., 1986; Inouye et al., 1980). Because large-seeded plants compete asymmetrically with small-seeded species, which are fed upon by other specialized granivores and herbivores, removal of rodents also has major indirect effects on the abundances of both small-seeded plants and their consumers (Brown et al., 1986; Davidson et al., 1984). Since rodents transport seeds and store them in their caches (Reichman and Price, this volume), and some of these seeds ultimately germinate, rodents may also play significant roles as seed dispersers. There are also other indirect

TABLE 5.—*Effects of seed predation by granivorous desert rodents on plant community structure at three sites in southeastern Arizona. Plant species listed are large-seeded annuals that dominate the annual community when rodents are excluded (from Brown et al., 1986).*

Site and plant species	Plant family	Average seed mass (mg)	Relative increase (density on rodent removal/density on control)
Sonoran Desert (sandy bajada)			
Erodium cicutarium	Geraniaceae	1.62	9.2**
Erodium texanum	Geraniaceae	1.60	1.5*
Lotus humistratus	Fabaceae	1.50	3.6**
Sonoran Desert (rocky hillside)			
Astragalus nuttallianus	Fabaceae	1.36	157.3***
Lupinus sparsiflorus	Fabaceae	1.51	8.2***
Chihuahuan Desert (sandy bajada)			
Erodium cicutarium	Geraniaceae	1.62	7,936.4*
Erodium texanum	Geraniaceae	1.60	1,000.0***
Lesquerella gordonii	Brassicaceae	0.94	9.7*
Astragalus nuttallianus	Fabaceae	1.36	4.3***

* $P < 0.05$; ** $P < 0.01$; *** $P < 0.005$.

effects, because, as mentioned above, some seed stores are exploited by specialized animals and microbes. The full impact of rodent seed-harvesting activities is probably much greater than has been appreciated. Many of the effects were unexpected until rodents were experimentally removed, and the experiments have probably been of insufficient duration to show complete results, especially with respect to perennial plants and their associated biota.

As consumers of seeds, rodents also affect other granivores. Despite significant differences in diet and foraging behavior, rodents overlap substantially with harvester ants and seed-eating birds in the sizes and species of seeds that they collect and the times and places where they forage (Brown and Davidson, 1977; Brown et al., 1979*a,* 1979*b,* 1986). Since populations of all these granivores appear to be limited at least in part by food availability, and since all of these seed-eaters have been shown to reduce substantially stocks of available seeds (Brown et al., 1979*a,* 1979*b,* 1986; Reichman, 1979), there is good circumstantial evidence that these unrelated granivores should compete. Competition has been tested more directly by removing one taxon and assessing the numerical response of the remaining taxa. The results strongly support the competition hypothesis; ants in particular increased dramatically on plots from which rodents had been excluded (Brown and Davidson, 1977; Brown et al., 1979*a,* 1986; Davidson et al., 1984). Responses of rodents to ant removals and of birds to rodent and ant removals were less conclusive, but consistent with the competition hypothesis (see Brown and Davidson, 1986; Brown et al., 1979*a,* 1986; Galindo, 1986; Reichman, 1979). On balance the data strongly indicate that granivorous rodents compete interspecifically not only among themselves, but also with other taxa of seed eaters.

The evolutionary consequences of competition between rodents and other kinds of organisms are intriguing but poorly understood. Mares and Rosenzweig (1978) and Pulliam and Brand (1975) have used the similarities and differences in seed utilization by rodents, ants, and birds as a basis to speculate on the coevolution of seeds and their predators, suggesting that the morphologies of seeds and the species composition of seed plants in different habitats reflect both short-term (ecological time) competition and long-term (evolutionary time) mutualism between the different taxa of granivores. These ideas are promising but largely untested. The demonstration of plant-mediated indirect mutualistic interactions in ecological time between rodents and ants (Brown et al., 1986; Davidson et al., 1984) should encourage further study of such coevolutionary relationships.

Rodents are among the most important secondary producers in arid ecosystems in terms of their numbers, biomass, and diversity of species. Consequently rodents are largely responsible for supporting substantial populations of carnivores, including snakes, owls, and mammalian carnivores. None of these generalized predators feed exclusively on rodents, let alone heteromyids. Nevertheless rodents are quantitatively important in their diets and are probably also qualitatively important because their population dynamics make them particularly dependable prey. Because their seed caching behavior and life histories enable the desert rodents, especially some of the heteromyids (see above), to maintain stable populations despite fluctuations in climate and productivity, rodents provide a relatively predictable food resource. By maintaining populations of carnivores through periods of food scarcity, rodents may have significant indirect effects on population dynamics of other prey, such as lizards, lagomorphs, and some birds.

Summary

The heteromyid rodents that occur in the deserts and semiarid habitats of southwestern North America have been the subject of numerous ecological studies, especially during the last two decades. The resulting information has played a significant role in the development and testing of ecological theory.

Populations of heteromyids are limited by combinations of interacting biotic and abiotic factors, including climate, substrate, vegetation, food, competitors, and predators. The life histories of these rodents seem to emphasize adult survival during unfavorable dry periods at the expense of rapid recruitment of offspring when resource availability increases following seasonal rains. Compared to other small mammals, desert heteromyids have fewer and smaller litters but prolonged adult survival and perhaps more facultative reproductive seasons and activity patterns. Foraging and caching behavior promote efficient exploitation of seasonally variable and often unpredictable resources. Although populations tend to increase in the favorable periods following rains and to decline gradually during the long intervening droughts, the timing and amplitude of these fluctuations varies greatly among species and habitats.

Several species of heteromyids typically coexist with granivorous murid rodents to form surprisingly diverse communities in harsh, structurally simple, unproductive desert habitats. Because of the distinct requirements of different species, the number, identity, and relative abundance of the component species vary greatly in both time and space. The composition of these assemblages is not random. Instead, it is determined by interactions of the species with the physical environment, with other kinds of organisms, and with other rodent species. Extrinsic factors that affect community organization include climate, substrate, vegetation, productivity, and predators. Intrinsic competitive interactions among rodent species tend to prevent local coexistence of morphologically and ecologically similar species. Generally, those species that do coexist tend to differ in patterns of seasonal activity and microhabitat use.

Heteromyids also modify the desert environment. Their burrowing moves soil, creating refuges for other organisms including soil arthropods, molds, and fungi. Burrowing may also alter characteristics of the soil, vegetation, and topography. Collection, storage, and consumption of seeds by rodents have dramatic direct and indirect effects on the distribution and population dynamics of associated plant and animal species. Heteromyids not only are abundant and diverse inhabitants of desert regions, but they are also highly interactive, functionally important components of desert ecosystems.

For good reasons, heteromyid rodents have played a central role in the development of modern community ecology, and to a lesser extent population ecology as well. Rodents are abundant and diverse in the desert and semiarid habitats of southwestern North America. In terms of both biomass and abundance of individuals, representatives of the family Heteromyidae are usually the dominant members of these assemblages. These and other attributes have facilitated the intensive study of populations and communities of desert rodents by many investigators and through a variety of techniques, including laboratory experiments, comparative field observations, and controlled, manipulative field experiments. These studies have been rewarded with an increasing understanding of the population dynamics and community organization of these rodents and of the important roles that they play in arid ecosystems.

We hope that studies of desert rodent ecology will continue to play a major role in modern ecology. We hope that the greatest value of the information summarized in this chapter will be in stimulating new studies that will correct mistakes, expand knowledge, and increase the importance of studies of heteromyid rodents to the development and testing of ecological theory.

Acknowledgments

We thank the many researchers whose work provided the essential data for this synthesis, and especially all those colleagues and students who have contributed and criticized the ideas expressed here. We are indebted to M. Kurzius for technical advice and assistance, and to J. S. Brown and R. T. M'Closkey for careful reading of and critical comments on the manuscript.

Literature Cited

ABRAMSKY, Z. 1978. Small mammal community ecology: changes in species diversity in response to manipulated productivity. Oecologia, 34:113–123.

BARTHOLOMEW, G. A. 1958. The role of physiology in the distribution of terrestrial vertebrates. Pp. 81–95, *in* Zoogeography (C. I. Hubbs, ed.). American Association for the Advancement of Science, Publication No. 51, Washington, D.C., 500 pp.

BARTHOLOMEW, G. A., AND H. H. CASWELL. 1951. Locomotion in kangaroo rats and its adaptive significance. Journal of Mammalogy, 32:155–169.

BAUMBARDNER, G. D., AND D. J. SCHMIDLY. 1985. Microhabitat relationships of kangaroo rats (*Dipodomys compactus* and *D. ordii*) in southern Texas. Southwestern Naturalist, 30:315–317.

BEATLEY, J. C. 1969. Dependence of desert rodents on winter annuals and precipitation. Ecology, 50: 721–724.

———. 1974. Phenological events and their environmental triggers in Mohave Desert ecosystems. Ecology, 55:856–863.

———. 1976. Environments of kangaroo rats (*Dipodomys*) and effects of environmental change on populations in southern Nevada. Journal of Mammalogy, 57:67–93.

BENSON, S. B. 1933. Concealing coloration among some desert rodents of the southwest United States. University of California Publications in Zoology, 40: 1–70.

BLAUSTEIN, A. R., AND A. C. RISSER. 1974. Dominance relationships of the dark kangaroo mouse (*Microdipodops megacephalus*) and the little pocket mouse (*Perognathus longimembris*) in captivity. Great Basin Naturalist, 34:312–316.

———. 1976. Interspecific interactions between three sympatric species of kangaroo rats (*Dipodomys*). Animal Behaviour, 24:381–385.

BOLE, B. P., AND P. N. MOULTHROP. 1942. The Ohio recent mammal collection in the Cleveland Museum of Natural History. Scientific Publications of the Cleveland Museum of Natural History, 5:83–181.

BOWERS, M. A. 1982. Foraging behavior in heteromyid rodents: field evidence of resource partitioning. Journal of Mammalogy, 63:361–367.

———. 1988. Seed removal experiments on desert rodents: the microhabitat by moonlight effect. Journal of Mammalogy, 69:201–204.

BOWERS, M. A., AND J. H. BROWN. 1982. Body size and coexistence in desert rodents: chance or community structure. Ecology, 63:391–400.

BOWERS, M. A., D. B. THOMPSON, AND J. H. BROWN. 1987. Foraging and microhabitat use in desert rodents: the role of a dominant competitor. Oecologia, 72:77–82.

BRADLEY, W. G., AND R. A. MAUER. 1971. Reproduction and food habits of Merriam's kangaroo rat, *Dipodomys merriami*. Journal of Mammalogy, 52: 497–507.

BROWN, J. H. 1973. Species diversity of seed-eating desert rodents in sand dune habitats. Ecology, 54: 775–787.

———. 1975. Geographical ecology of desert rodents.

Pp. 315–341, *in* Ecology and evolution of communities (M. L. Cody and J. M. Diamond, eds.). Belknap Press, Cambridge, Massachusetts, 545 pp.

———. 1984*a*. Desert rodents: a model system. Acta Zoologica Fennica, 172:45–49.

———. 1984*b*. On the relationship between abundance and distribution of species. American Naturalist, 124:255–279.

———. 1987. Variation in desert rodent guilds: patterns, processes, and scales. Pp. 185–203, *in* Organization of communities: past present (J. H. R. Gee and P. S. Giller, eds.). Blackwell Scientific Publishers, Oxford, England, 576 pp.

Brown, J. H., and D. W. Davidson. 1977. Competition between seed-eating rodents and ants in desert ecosystems. Science, 196:880–882.

———. 1986. Reply to Galindo. Ecology, 67:1423–1424.

Brown, J. H., and E. J. Heske. 1990*a*. Temporal changes in a Chihuahuan Desert rodent community. Oikos, 59:290–302.

——— 1990*b*. Mediation of a desert-grassland transition by a keystone rodent guild. Science, 250:1705–1707.

Brown, J. H., and M. A. Kurzius. 1987. Composition of desert rodent faunas: combinations of coexisting species. Annales Zoologici Fennici 24:227–237.

Brown, J. H., and M. A. Kurzius. 1989. Spatial and temporal variation in guilds of North American granivorous desert rodents. Pp. 71–90, *in* Patterns in the structure of mammalian communities (D. W. Morris, Z. Abramsky, B. J. Fox, and M. R. Willig, eds.). Special Publications, The Museum, Texas Tech University, No. 28, 122 pp.

Brown, J. H., and G. A. Lieberman. 1973. Resource utilization and coexistence of seed-eating desert rodents in sand dune habitats. Ecology, 54:788–797.

Brown, J. H., and J. C. Munger. 1985. Experimental manipulation of a desert rodent community: food addition and species removal. Ecology, 66:1545–1563.

Brown, J. H., and Z. Zeng. 1989. Comparative population ecology of eleven species of rodents in the Chihuahuan Desert. Ecology. 70:1507–1525.

Brown, J. H., D. W. Davidson, and O. J. Reichman. 1979*a*. An experimental study of competition between seed-eating desert rodents and ants. American Zoologist, 19:1129–1143.

Brown, J. H., O. J. Reichman, and D. W. Davidson. 1979*b*. Granivory in desert ecosystems. Annual Review of Ecology and Systematics, 10:201–227.

Brown, J. H., D. W. Davidson, J. C. Munger, and R. S. Inouye. 1986. Experimental community ecology: the desert granivore system. Pp. 41–61, *in* Community ecology (J. Diamond and T. J. Case, eds.). Harper & Row, New York, 665 pp.

Brown, J. S. 1988. The role of resource variability in structuring desert rodent communities. Pp. 141–154, *in* Patterns in the structure of mammalian communities (D. W. Morris, Z. Abramsky, B. J. Fox, and M. R. Willig, eds.). Special Publications, The Museum, Texas Tech University, No. 28, 266 pp.

Carpenter, R. E. 1966. A comparison of thermoregulation and water metabolism in the kangaroo rats *Dipodomys agilis* and *Dipodomys merriami*. University of California Publications in Zoology, 78: 1–36.

Chew, R. M., and B. B. Butterworth. 1964. Ecology of rodents in Indian Cove (Mojave Desert), Joshua Tree National Monument, California. Journal of Mammalogy, 45:203–225.

Congdon, J. 1974. Effect of habitat quality on distributions of three sympatric species of desert rodents. Journal of Mammalogy, 55:659–662.

Conley, W., J. D. Nichols, and A. R. Tipton. 1977. Reproductive strategies in desert rodents. Pp. 193–214, *in* Transactions of the symposium on the biological resources of the Chihuahuan Desert region (R. H. Wauer and D. H. Riskind, eds.). United States Department of the Interior, National Park Service, Transactions & Proceedings Series, 3:1–658.

Davidson, D. W., R. S. Inouye, and J. H. Brown. 1984. Granivory in a desert ecosystem: experimental evidence for indirect facilitation of ants by rodents. Ecology, 65:1780–1786.

Deynes, A. 1954. Habitat restriction and the digging ability of certain pocket mice. Journal of Mammalogy, 35:453.

Dice, L. R., and P. M. Blossom. 1937. Studies of mammalian ecology in southwestern North America with special attention to the colors of desert mammals. Carnegie Institute of Washington, Publications, 485:1–129.

———. 1962. Ecology and life history of the kit fox in Tooele County, Utah. Ecology, 43:481–497.

Eisenberg, J. F. 1963. The behavior of heteromyid rodents. University of California Publications in Zoology, 69:1–114.

———. 1975. The behavior patterns of desert rodents. Pp. 189–224, *in* Rodents in desert environments (I. Prakash and P. Ghosh, eds.). W. Junk, The Hague, The Netherlands, 624 pp.

———. 1992. Ontogeny. Pp. 479–490, *in* Biology of the Heteromyidae (H. H. Genoways and J. H. Brown, eds.). Special Publication, American Society of Mammalogists, 10:1–719.

Frank, C. L. 1988. The influence of moisture content on seed selection by kangaroo rats. Journal of Mammalogy, 69:353–357.

Freeman, P. W., and C. Lemen. 1983. Quantification of competition among coexisting heteromyids in the southwest. Southwestern Naturalist, 28:41–46.

French, A. R. 1993. Physiological ecology of the Heteromyidae: economics of energy and water utilization. Pp. 509–538, *in* Biology of the Heteromyidae (H. H. Genoways and J. H. Brown, eds.). Special Publication, American Society of Mammalogists, 10: 1–719.

French, N. R., B. G. Maza, and A. P. Aschwanden. 1967. Life spans of *Dipodomys* and *Perognathus* in the Mohave Desert. Journal of Mammalogy, 48:537–548.

French, N. R., B. G. Maza, H. O. Hill, A. P. Aschwanden, and H. W. Kaaz. 1974. A population study of irradiated desert rodents. Ecological Monographs, 44:45–72.

Frye, R. J. 1983. Experimental field evidence of interspecific aggression between two species of kangaroo rat (*Dipodomys*). Oecologia, 59:74–78.

FRYE, R. J., AND M. L. ROSENZWEIG. 1980. Clump size selection: a field test with two species of *Dipodomys*. Oecologia, 47:323–327.

GALINDO, C. 1986. Do desert rodent populations increase when ants are removed? Ecology, 67:1422–1423.

GLEASON, H. A. 1926. The individualistic concept of the plant association. Bull. Torrey Bot. Club, 53:7–26.

GRINNELL, J. 1914. An account of the mammals and birds of the lower Colorado River Valley. University of California Publications in Zoology, 12:51–294.

GRINNELL, J., AND R. T. ORR. 1934. Systematic review of the Californicus group of the rodent genus *Peromyscus*. Journal of Mammalogy, 15:210–217.

HAFNER, J. C. 1993. Macroevolutionary diversification in heteromyid rodents: heterochrony and adaptation in phylogeny. Pp. 291–318, *in* Biology of the Heteromyidae (H. H. Genoways and J. H. Brown, eds.). Special Publication, American Society of Mammalogists, 10:1–719.

HAFNER, M. S. 1977. Density and diversity in Mojave Desert rodent and shrub communities. Journal of Animal Ecology, 46:925–938.

HALL, E. R. 1981. The mammals of North America. Second edition. John Wiley, New York, 1:526–589 and 2:605–776.

HALLETT, J. G. 1982. Habitat selection and the community matrix of a desert small-mammal fauna. Ecology, 63:1400–1410.

HARDY, R. 1945. The influence of types of soil upon local distribution of small mammals in southwestern Utah. Ecological Monographs, 15:71–108.

HARRIS, J. H. 1984. An experimental analysis of desert rodent foraging ecology. Ecology, 65:1579–1584.

HAY, M. E., AND P. J. FULLER. 1981. Seed escape from heteromyid rodents: the importance of microhabitat and seed preference. Ecology, 62:1395–1399.

HILLEL, D., AND N. TADMOR. 1962. Water regime and vegetation in the central Negev highlands of Israel. Ecology, 43:33–41.

HOFFMEISTER, D. F., AND W. W. GOODPASTER. 1954. The mammals of the Huachuca Mountains, Southeastern Arizona. Illinois Biological Monographs, 24:1–152.

HOOVER, K. D., W. G. WHITFORD, AND P. FLAVILL. 1977. Factors influencing the distributions of two species of *Perognathus*. Ecology, 58:877–884.

HOPF, F. A., AND J. H. BROWN. 1986. The bull's-eye method for testing randomness in ecological communities. Ecology, 67:1139–1155.

HUTCHINSON, G. E. 1958. Concluding remarks. Cold Spring Harbor Symposia on Quantitative Biology, 22:415–427.

HUTTO, R. L. 1978. A mechanism for resource allocation among sympatric heteromyid rodent species. Oecologia, 33:115–126.

INOUYE, R. S., G. S. BYERS, AND J. H. BROWN. 1980. Effects of predation and competition on survivorship, fecundity, and community structure of desert annuals. Ecology, 61:1344–1351.

JONES, W. T. 1993. Social systems of heteromyid rodents. Pp. 575–595 *in* Biology of the Heteromyidae (H. H. Genoways and J. H. Brown, eds.). Special Publication, American Society of Mammalogists, 10:1–719.

KAUFMAN, D. W., AND G. KAUFMAN. 1982. Effect of moonlight on activity and microhabitat use by Ord's kangaroo rat (*Dipodomys ordii*). Journal of Mammalogy, 63:309–312.

KAY, F., AND W. WHITFORD. 1978. The burrow environment of the banner-tailed kangaroo rat, *Dipodomys spectabilis*, in south-central New Mexico. American Midland Naturalist, 99:270–279.

KELLER, B. L. 1985. Reproductive patterns. Pp. 725–778, *in* Biology of the New World *Microtus* (R. H. Tamarin, ed.). American Society of Mammalogists, Special Publications No. 8, 893 pp.

KELRICK, M. I., AND J. A. MACMAHON. 1985. Nutritional and physical attributes of seeds of some common sagebrush-steppe plants: some implications for ecological theory and management. Journal of Range Management, 38:65–69.

KELRICK, M. I., J. A. MACMAHON, R. R. PARMENTER, AND D. V. SISSON. 1986. Native seed preferences of shrub-steppe rodents, birds and ants: the relationships of seed attributes and seed use. Oecologia, 68:327–337.

KENAGY, G. J. 1972. Saltbush leaves: excision of hypersaline tissue by a kangaroo rat. Science, 178:1094–1096.

———. 1973. Daily and seasonal patterns of activity and energetics in a heteromyid rodent community. Ecology, 54:1201–1219.

KENAGY, G. J., AND G. A. BARTHOLOMEW. 1985. Seasonal reproductive patterns in five coexisting California desert rodent species. Ecological Monographs, 55:371–397.

KOTLER, B. P. 1984*a*. Predation risk and the structure of desert rodent communities. Ecology, 65:689–701.

———. 1984*b*. Harvesting rates and predatory risk in desert rodents: a comparison of two communities on different continents. Journal of Mammalogy, 65:91–96.

———. 1985*a*. Microhabitat utilization in desert rodents: a comparison of two methods of measurement. Journal of Mammalogy, 66:374–378.

———. 1985*b*. Owl predation on desert rodents which differ in morphology and behavior. Journal of Mammalogy, 66:824–828.

LARSEN, E. C. 1986. Competitive release in microhabitat use among coexisting desert rodents: a natural experiment. Oecologia, 69:231–237.

LAWLOR, T. E. 1983. The mammals. Pp. 265–289 and 482–500, *in* Island biogeography in the Sea of Cortez (T. J. Case and M. L. Cody, eds.). University of California Press, Berkeley, 508 pp.

LEMEN, C. A. 1978. Seed size selection in heteromyids. Oecologia, 35:13–19.

LEMEN, C. A., AND M. L. ROSENZWEIG. 1978. Microhabitat selection in two species of heteromyid rodents. Oecologia, 33:127–135.

LOCKARD, R. B. 1978. Seasonal change in the activity pattern of *Dipodomys spectabilis*. Journal of Mammalogy, 59:563–568.

LOCKARD, R. B., AND D. H. OWINGS. 1974. Seasonal variation in moonlight avoidance by bannertail kangaroo rats. Journal of Mammalogy, 55:189–193.

M'Closkey, R. T. 1976. Community structure in sympatric rodents. Ecology, 57:728–739.
———. 1978. Niche separation and assembly in four species of Sonoran Desert rodents. American Naturalist, 112:683–694.
———. 1980. Spatial patterns in sizes of seeds collected by four species of heteromyid rodents. Ecology, 61:486–489.
———. 1981. Microhabitat use in coexisting desert rodents—the role of population density. Oecologia, 50:310–315.
MacMillen, R. E. 1964. Population ecology, water relations, and social behavior of a southern California semidesert rodent fauna. University of California Publications in Zoology, 71:1–66.
Mares, M. A. 1983. Desert rodent adaptation and community structure. Pp. 30–43, *in* Biology of desert rodents (O. J. Reichman and J. H. Brown, eds.). Great Basin Naturalist Memoirs No. 7, 134 pp.
Mares, M. A., and M. L. Rosenzweig. 1978. Granivory in North and South American deserts: rodents, birds and ants. Ecology, 59:235–241.
Mares, M. A., and D. F. Williams. 1977. Experimental support for food particle size resource allocation in heteromyid rodents. Ecology, 58:1186–1190.
Maza, B. G., N. R. French, and A. P. Aschwanden. 1973. Home range dynamics in a population of heteromyid rodents. Journal of Mammalogy, 54:405–425.
Miller, A. H., and R. C. Stebbins. 1964. The lives of desert animals in Joshua Tree National Monument. University of California Press, Berkeley, 452 pp.
Munger, J. C., and J. H. Brown. 1981. Competition in desert rodents: an experiment with semipermeable exclosures. Science, 211:510–512.
Munger, J. C., M. A. Bowers, and W. T. Jones. 1983. Desert rodent populations: factors affecting abundance, distribution, and genetic structure. Pp. 91–116, *in* Biology of desert rodents (O. J. Reichman and J. H. Brown, eds.). Great Basin Naturalist Memoirs No. 7, 134 pp.
O'Farrell, M. J. 1974. Seasonal activity patterns of rodents in a sagebrush community. Journal of Mammalogy, 55:809–823.
Petryszyn, Y. 1982. Population dynamics of nocturnal desert rodents: a nine-year study. Ph.D. dissert., University of Arizona, Tucson, 108 pp.
Pianka, E. R. 1983. Evolutionary ecology. Third edition. Harper and Row, New York, 416 pp.
Price, M. V. 1978*a*. Seed dispersion preferences of coexisting desert rodent species. Journal of Mammalogy, 59:624–626.
———. 1978*b*. The role of microhabitat in structuring desert rodent communities. Ecology, 59:910–921.
———. 1983. Laboratory studies of seed size and species selection by heteromyid rodents. Oecologia, 60:259–263.
Price, M. V., and J. H. Brown. 1983. Patterns of morphology and resource use in North American desert rodent communities. Pp. 117–134, *in* Biology of desert rodents (O. J. Reichman and J. H. Brown, eds.). Great Basin Naturalist Memoirs No. 7, 134 pp.
Price, M. V., and K. M. Heinz. 1984. Effects of body size, seed density, and soil characteristics on rates of seed harvest by heteromyid rodents. Oecologia, 61:420–425.
Price, M. V., and N. Waser. 1985. Microhabitat use by heteromyid rodents: effects of artificial patches. Ecology, 66:211–219.
Price, M. V., N. W. Waser, and T. A. Bass. 1984. Effects of moonlight on microhabitat use by desert rodents. Journal of Mammalogy, 65:353–356.
Pulliam, H. R., and M. R. Brand. 1975. The production and utilization of seeds in plains grassland of southeastern Arizona. Ecology, 56:557–566.
Rebar, C., and W. Conley. 1983. Interactions in microhabitat use between *Dipodomys ordii* and *Onychomys leucogaster*. Ecology, 64:984–988.
Rebar, C., and O. J. Reichman. 1983. Ingestion of moldy seeds by heteromyid rodents. Journal of Mammalogy, 64:713–715.
Reichman, O. J. 1975. Relation of desert rodent diets to available resources. Journal of Mammalogy, 56:731–751.
———. 1977. Optimization of diets through food preferences by heteromyid rodents. Ecology, 58:454–457.
———. 1979. Desert granivore foraging and its impact on seed densities and distributions. Ecology, 60:1085–1092.
Reichman, O. J., and D. Oberstein. 1977. Selection of seed distribution types by *Dipodomys merriami* and *Perognathus amplus*. Ecology, 58:636–643.
Reichman, O. J., and M. V. Price. 1993. Ecological aspects of heteromyid foraging. Pp. 539–574, *in* Biology of the Heteromyidae (H. H. Genoways and J. H. Brown, eds.). Special Publication, American Society of Mammalogists, 10:1–719.
Reichman, O. J., and K. M. Van De Graff. 1975. Association between ingestion of green vegetation and desert rodent reproduction. Journal of Mammalogy, 56:503–506.
Reynolds, H. G. 1958. Ecology of Merriam's kangaroo rat on the grazing lands of southern Arizona. Ecological Monographs, 28:111–127.
———. 1960. Life history notes on Merriam's kangaroo rat in southern Arizona. Journal of Mammalogy, 41:48–58.
Roberts, J. D., and R. L. Packard. 1973. Comments on movements, home range, and ecology of the Texas kangaroo rat *Dipodomys elator*. Journal of Mammalogy, 54:957–962.
Rosenzweig, M. L. 1968. Net primary productivity of terrestrial communities: prediction from climatological data. American Naturalist, 102:67–74.
———. 1973. Habitat selection experiments with a pair of coexisting heteromyid rodent species. Ecology, 54:111–117.
Rosenzweig, M. L., and P. W. Sterner. 1970. Population ecology of desert rodent communities: body size and seed husking as bases for heteromyid coexistence. Ecology, 51:217–224.
Rosenzweig, M. L., and J. Winakur. 1969. Population ecology of desert rodent communities: habi-

tats and environmental complexity. Ecology, 50:558–572.

ROSENZWEIG, M. L., B. SMIGEL, AND A. KRAFT. 1975. Patterns of food, space and diversity. Pp. 241–268, *in* Rodents in desert environments (I. Prakash and P. Gosh, eds.). W. Junk, The Hague, The Netherlands, 624 pp.

SANCHEZ-CORDERO, V., AND T. H. FLEMING. 1993. Ecology of tropical heteromyids. Pp. 596–617, *in* Biology of the Heteromyidae (H. H. Genoways and J. H. Brown, eds.). Special Publication, American Society of Mammalogists, 10:1–719.

SCHMIDLY, D. J., K. T. WILKINS, AND J. N. DERR. 1993. Biogeography. Pp. 319–356, *in* Biology of the Heteromyidae (H. H. Genoways and J. H. Brown, eds.). Special Publication, American Society of Mammalogists, 10:1–719.

SCHRODER, G. D. 1987. Mechanisms for coexistence among three species of *Dipodomys*: habitat selection and an alternative. Ecology, 68:1071–1083.

SCHRODER, G. D., AND M. L. ROSENZWEIG. 1975. Perturbation analysis of competition and overlap in habitat utilization between *Dipodomys ordii* and *Dipodomys merriami*. Oecologia, 19:9–28.

SEASTEDT, T. R., O. J. REICHMAN, AND T. C. TODD. 1986. Microarthropods and nematodes in kangaroo rat burrows. Southwestern Naturalist, 31:114–116.

SHERBROOKE, W. C. 1976. Differential acceptance of toxic jojoba seed (*Simmondsia chinensis*) by four Sonoran Desert heteromyid rodents. Ecology, 57: 596–602.

SMIGEL, B. W., AND M. L. ROSENZWEIG. 1974. Seed selection in *Dipodomys merriami* and *Perognathus penicillatus*. Ecology, 55:329–339.

SMITH, H. D., AND C. D. JORGENSEN. 1975. Reproductive biology of the North American desert rodents. Pp. 305–330, *in* Rodents in desert environments (I. Prakash and P. K. Ghosh, eds.). W. Junk, The Hague, The Netherlands, 624 pp.

STAMP, N. E., AND R. D. OHMART. 1978. Resource utilization by desert rodents in the Lower Sonoran Desert. Ecology, 59:700–707.

THOMPSON, S. D. 1982*a*. Microhabitat utilization and foraging behavior of bipedal and quadrupedal heteromyid rodents. Ecology, 63:1303–1312.

———. 1982*b*. Structure and species composition of desert heteromyid rodent species assemblages: effects of a simple habitat manipulation. Ecology, 63: 1313–1321.

———. 1985. Bipedal hopping and seed-dispersion selection by heteromyid rodents: the role of locomotion energetics. Ecology, 66:220–229.

———. 1987. Resource availability and microhabitat use by Merriam's kangaroo rats, *Dipodomys merriami*, in the Mojave Desert. Journal of Mammalogy, 68:256–265.

TILMAN, D. 1982. Resource competition and community structure. Princeton University Press, Princeton, New Jersey, 296 pp.

TROMBULAK, S. C., AND G. J. KENAGY. 1980. Effects of seed distribution and competitors on seed harvesting efficiency in heteromyid rodents. Oecologia, 44:342–346.

VAN DE GRAFF, K. M., AND R. P. BALDA. 1973. Importance of green vegetation for reproduction in the kangaroo rat *Dipodomys merriami merriami*. Journal of Mammalogy, 54:509–512.

WEBSTER, D. B. 1962. A function of the enlarged middle ear cavities of the kangaroo rat, *Dipodomys*. Physiological Zoology, 35:248–255.

WEBSTER, D. B., AND M. WEBSTER. 1971. Adaptive value of hearing and vision in kangaroo rat predator avoidance. Brain, Behavior, and Evolution, 4:310–322.

———. 1975. Auditory systems of Heteromyidae: functional morphology and evolution of the middle ear. Journal of Morphology, 146:343–376.

———. 1980. Morphological adaptations of the ear in the rodent family Heteromyidae. American Zoologist, 20:247–254.

———. 1984. The specialized auditory system of kangaroo rats. Pp. 161–196, *in* Contributions to sensory physiology (W. D. Neff, ed.). Academic Press, Orlando, Florida, 198 pp.

WHITAKER, J. O. JR., W. J. WRENN, AND R. F. LEWIS. 1993. Parasites. Pp. 386–478 *in* Biology of the Heteromyidae (H. H. Genoways and J. H. Brown, eds.). Special Publication, American Society of Mammalogists, 10:1–719.

WHITFORD, W. G. 1976. Temporal fluctuations in density and diversity of desert rodent populations. Journal of Mammalogy, 57:351–369.

WONDOLLECK, J. T. 1978. Forage-area separation and overlap in heteromyid rodents. Journal of Mammalogy, 59:510–518.

ZENG, Z., AND J. H. BROWN. 1987*a*. Life history of a desert rodent: a seven year study of *Dipodomys merriami* in the Chihuahuan Desert. Ecology, 68: 1328–1340.

ZENG, Z., AND J. H. BROWN. 1987*b*. A method for distinguishing dispersal from death in mark-recapture studies. Journal of Mammalogy, 68:656–665.

HETEROMYIDS AND THEIR ECOLOGICAL COUNTERPARTS: A PANDESERTIC VIEW OF RODENT ECOLOGY AND EVOLUTION

MICHAEL A. MARES

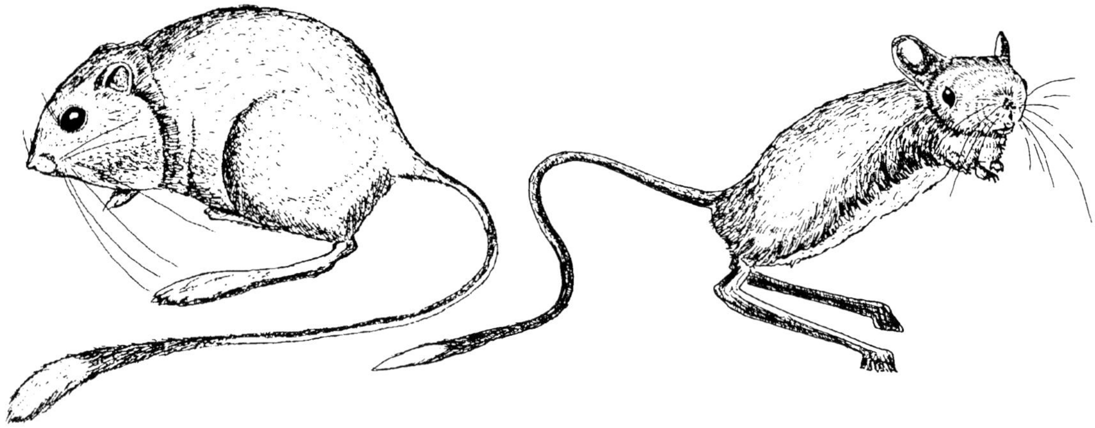

Dipodomys *Jaculus*

Introduction

The previous chapters in this volume have shown quite clearly that the heteromyid rodents constitute a remarkably successful family of small mammals. Although they are most diverse in some of the driest portions of the North American continent, they are also found over a broad array of habitats ranging from grasslands through cloud forests, seasonally dry tropical forests, and moist forests of Central and northern South America (Schmidly et al., this volume). Nevertheless, it is the highly desert-adapted species that primarily have interested biologists. The ability of the desert species to survive and prosper in what to man is an extremely hostile environment, their ease of trapability, and their proximity to universities and research institutes in the United States have stimulated investigations on anatomy, physiology, ecology, community structure, behavior, cell biology, systematics, and other biological subdisciplines (see Reichman and Brown, 1983 for review). Indeed, most North Americans tend to view heteromyids, particularly *Dipodomys,* as the paragon of desert adaptation. The bipedal, largely granivorous, semifossorial, water-independent rodents (*Dipodomys* and *Microdipodops,* Figs. 1, 2) and their granivorous quadrupedal relatives, *Perognathus* and *Chaetodipus* (Figs. 3, 4), have been the focus of many studies dealing with adaptation and coexistence. The complex strategies employed by these taxa for living in the desert will continue to yield exciting scientific information over the coming years, for many important questions concerning the biology of these mammals have yet to be answered. Their ability to compete successfully with confamilials, as well as with other competitors as taxonomically distinct as birds and ants (see Brown and Harney, this volume, and Brown et al., 1979, for reviews), coupled with their coevolutionary

FIG. 1.—*Dipodomys ordii,* a bipedal dipodomyine heteromyid. (Photo: J. F. Eisenberg)

FIG. 3.—*Perognathus inornatus,* a small, silky, quadrupedally-saltatorial perognathine. (Photo: J. F. Eisenberg)

interactions with predators (e.g., Kotler, 1984, 1985) and with plants (Reichman and Rebar, 1985; Reichman et al., 1986), make this assemblage of desert rodents an enticing experimental system—one likely to yield results of broad application to evolutionary ecology.

The desert areas of North America have the advantage of more than a century of research effort by thousands of biologists. The result of this labor is a growing appreciation of how complex the biotic interactions are that permit organisms to inhabit hot, arid regions that are subject to extensive droughts. Heteromyid rodents are found in all North American xeric areas, regardless of the climate and substrate; given a few requisite plant products (e.g., seeds) and friable soil for burrow placement, one or more heteromyid species will be present. Clearly, the heteromyid adaptive strategy is quite effective in allowing colonization of the North American deserts.

Precisely because these rodents are so successful at inhabiting areas where few other small mammals can survive, it might be expected that the suite of desert adaptations shown by members of the family Heteromyidae would be likely to appear in mammals living in other deserts of the world. Should there prove to be only a limited number of ways for a small mammal to exist in the rigorous desert environment (e.g., Mares, 1975*a,* 1976; and see Hafner, this volume), and should the adaptive suite of the heteromyids reflect some of the more successful characteristics for desert existence, then evolution in other deserts might have led to the development of species that were strongly convergent on the heteromyid type. Repeated examples of such convergent evolution would suggest that deserts, being climatically challenging habitats for small homeothermic mammals, exact similar adaptations from any species that successfully inhabit them. Moreover, given the wide distribution of deserts and the divergent phylogenetic histories of the small mammals that have had an opportunity to inhabit any particular desert, such convergent evolution

FIG. 2.—*Microdipodops pallidus,* a small bipedal dipodomyine. (Photo: J. F. Eisenberg)

FIG. 4.—*Chaetodipus californicus,* a spiny, quadrupedally-saltatorial perognathine. (Photo: J. F. Eisenberg)

would argue strongly for the genetic plasticity of small mammals. That similar adaptations occurring at all levels, from subcellular to organismal (Mares, 1983), could have developed from very distantly related organisms having little or no previous evolutionary exposure to deserts would give strong evidence as to the molding power of natural selection acting on the genetic variability inherent among the individuals comprising a population. An understanding of such convergent evolutionary processes would thus permit a much more general understanding of the evolutionary process itself. To understand a process is to be able to predict its outcome. This level of understanding of much of evolution remains to be developed.

Hafner (this volume) and Wahlert (this volume) have pointed out how heteromyids have evolved into three separate lineages: the Dipodomyinae, Perognathinae, and Heteromyinae. Two of these, the Dipodomyinae (bipedal) and Perognathinae (quadrupedal) are desert specialists. Members of both groups can occur in the driest habitats. In searching for possible ecological equivalents, there are a number of adaptations that can be examined. Elsewhere (Mares, 1983; see also Eisenberg, 1967, 1975), I have reviewed many of these and will not detail them here; however, a few comments are required to clarify the analysis that follows.

Dipodomys and *Microdipodops*, being bipedal, have many unusual traits that are easy to compare with the members of other faunas. Their long hind limbs, reduced forelegs, greatly inflated bullae, long tufted tail (in *Dipodomys*), ability to exist without free water, semifossorial nocturnal existence, and obligate granivory with seed hoarding (*Microdipodops* may not share seeds and has a more catholic diet) offer many avenues of comparison. The perognathines, on the other hand, being less derived (and thus more generalized) are less unusual than the dipodomyines. They are small obligate granivores that may hoard seeds and frequently experience torpor (e.g., MacMillen, 1983). Data on reproduction, behavior and general biology for *Dipodomys, Perognathus* and *Chaetodipus* have been summarized by Reichman and Brown (1983) and Eisenberg (1963, 1975). Given the pronounced differences between these two heteromyid subfamilies, it might be expected that species that are convergent with dipodomyines could be selected in a less ambiguous manner than those that might be convergent upon perognathines. This must be kept in mind in the comparisons that follow.

In this paper I will discuss the rodent faunas of the deserts of the world and compare the biology of the species inhabiting these deserts with the biology of the heteromyids. Is there a general niche type that characterizes the heteromyids and does one see this repeated in other deserts? What are the heteromyid equivalents in other deserts and does each desert support species that are similar to the North American heteromyids?

The World's Deserts

The biology, geology and geography of the world's deserts have been covered in depth in numerous important publications and do not require detailing here (e.g., Amiran and Wilson, 1973; Barbault and Halfetter, 1981; Bender, 1982; G. W. Brown, 1968, 1974; Cogger and Cameron, 1984; Evenari et al., 1971, 1985, 1986; Goodall, 1976; Goodin and Northington, 1985; Grenot, 1983; Hills, 1966; Jaeger, 1957; Mabry et al., 1977; McGinnies and Goldman, 1969; McGinnies et al., 1968; Nir, 1974; Orians and Solbrig, 1977; Petrov, 1976; Prakash and Ghosh, 1975; Schmidt-Nielsen, 1964; B. B. Simpson, 1977; Wauer and Riskind, 1977; West, 1983*a*). It is sufficient to be aware that deserts are caused by a number of factors (Logan, 1968), are widely distributed over the earth's surface (McGinnies et al., 1968) (Figs. 5, 11, 20, 23, 32, 36), and are frequently hot, sometimes cold, but always arid, except for brief periods of rainfall (Petrov, 1976). Because deserts occur on all continents, they have been subjected to colonization by diverse groups of mammals. I will briefly discuss the deserts of the world from the standpoint of their climate, phytogeography and small mammal faunas. This will provide a foundation for the general discussion that will follow.

In this review I include regions that, by virtue of their extreme aridity, are easily classified as true deserts. However, I also include areas that are more readily considered semideserts, because their extremes of heat and/or aridity are less pronounced. Aridity is a product of many interacting factors, including total precipitation, seasonality of rainfall, periodicity and extent of droughts, insolation, latitude, topographical factors, ground-water retention, and others. Arid and semiarid areas, therefore, occupy a climatological continuum. Some true deserts are hot, others cold. Some are subtropical, others temperate. Some semiarid areas occur within the tropics, while others are at higher latitudes (see Logan, 1968; Oberlander, 1979; and Petrov, 1976, for a general discussion of deserts, as well as citations below). Regardless of the actual climatological classification of an area, however, deserts or semideserts can be challenging habitats for small mammals. Both areas demand adaptations to extremes of aridity, temperature, and resource scarcity. Because it is frequently difficult to distinguish truly xeric habitats from semiarid habitats, and because such regions frequently grade imperceptibly from one to another, I include in this paper a review of all deserts and semideserts and their faunas. Moreover, I may occasionally refer to some regions as "deserts" when they may be considered semideserts by some authors (e.g., much of the North American "desert" system is considered a semidesert by many authors). I do this to facilitiate readability when discussing the faunas of regions that are either popularly known as deserts, or that include both arid and semiarid habitats within a broad geographic area.

North American Deserts and Semideserts

Much information on heteromyids in North American deserts (Figs. 5–10) has been presented in the previous chapters. Most habitats in the arid areas of North America support heteromyids, cricetids and sciurids. Soricids and geomyids are also occasional desert inhabitants, while leporids (*Sylvilagus* and *Lepus*) are larger rodentiform mammals that are ubiquitous in deserts. The great majority of localities will have their small mammal fauna comprised of two leporids, one sciurid, one to five cricetids, two to six heteromyids and an occasional soricid or geomyid. Diversity in most desert and semidesert sites can be quite high, with up to 14 species co-occurring (e.g., Bradley and Mauer, 1973; Chew and Chew, 1970; Hallett, 1982; Hoffmeister and Goodpaster, 1954). The high diversity reported for parts of the Sonoran Desert is paralleled by elevated diversity in the Mojave (Bradley and Deacon, 1971), Chihuahuan (e.g., Grenot, 1983; Packard, 1977; Schmidly, 1977) and Great Basin (O'Farrell, 1974) deserts. Although the Chihuahuan Desert has received less scientific attention than the other North

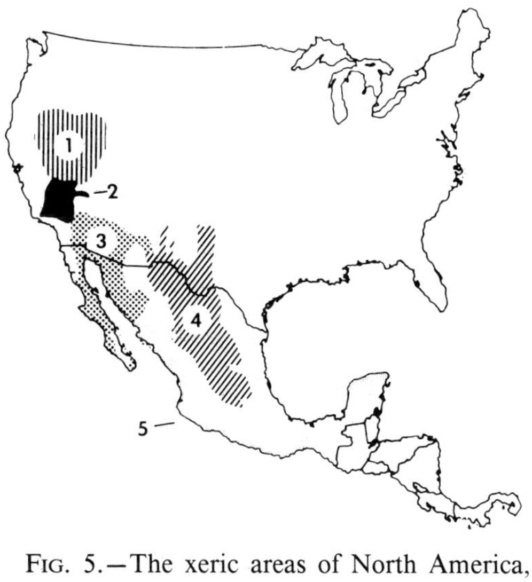

FIG. 5.—The xeric areas of North America, including: 1) Great Basin, 2) Mojave, 3) Sonoran, 4) Chihuahuan, 5) Sinaloan lowlands and other semiarid shrublands.

FIG. 6.—Typical Sonoran Desert habitat near Tucson, Arizona, with columnar cacti (*Carnegiea*), ocotillo (*Fouquieria splendens*), and assorted shrubs. (Photo: M. A. Mares)

American deserts, overall faunal patterns in all four deserts are clear (see MacMahon and Wagner, 1985; West, 1983*b,* 1983*c,* 1983*d,* 1983*e,* for general discussion of the North American desert system). Heteromyids are among the most abundant and ubiquitous species in each desert. In the most arid portions of the desert, usually only one or two species of small mammals, both heteromyids, can manage to exist (see Brown and Davidson, 1977; and Brown and Harney, this volume, for a discussion of patterns of coexistence).

The many adaptations of heteromyids to deserts have been discussed elsewhere (e.g., Berman, 1985; Brylski, this volume; Chew, 1965; Eisenberg, 1963, 1975; Hafner, this volume; Lawler and Geluso, 1986; MacMillen, 1972, 1983; Mares, 1983; McNab, 1979; Nikolai and Bramble, 1983; Reichman, 1983; Schmidt-Nielsen, 1964; Webster, 1962). Basically, almost all species are water-independent, seed-eating, nocturnal burrow dwellers. They hoard seeds, and several bipedal and quadrupedal species may coexist (Brown, 1973, 1975; Reichman and Price, this volume). Coexisting heteromyids differ in body size, such that each coexisting species is the sole occupant of a body size category (Brown, 1975; Simberloff and Boecklen, 1981). This stepped pattern of body sizes has been interpreted to reflect mechanisms permitting coexistence (Bowers and Brown, 1982; Brown, 1975; Brown and Harney, this volume), perhaps on the basis of differential food particle size utilization (e.g., Brown and Lieberman, 1973; Mares and Williams, 1977).

It is difficult to generalize about species coexistence in the deserts of North America, for the number of species can vary greatly from one site to the next. However, patterns of species co-occurrence have been reported for a number of localities within the North American desert system. A generic-level view of desert small mammal faunas in North America is given in Appendix 1. Note that heteromyids may occur in sandy areas, in wash areas, or on rocky hillsides. The North American deserts do not differ greatly from one another in the types of species that occur in each region. As Mares (1979, 1985) has discussed, given the paleohistory of the deserts and their faunas, it is to be expected that great niche equivalency be exhibited between these areas. In general, North American deserts have rather high species richness and relative abundance of rodents, with almost all lowland localities (as opposed to desert mountains) supporting both bipedal and quadrupedal heteromyids, as well as quadrupedal cricetines (see also Brown and Harney, this volume).

South American Deserts and Semideserts

South America contains numerous arid and semiarid areas (Fig. 11). Five of these

FIG. 7.—Mojave Desert scrubland near Goffs, California, with tall *Yucca* and various low scrubs (e.g., *Larrea, Encelia*). (Photo: M. A. Mares)

(the Monte, Patagonia, Puna, Atacama, and Peruvian) are deserts in the strict sense, while the Chaco and Caatinga are semiarid thorn scrub areas. In addition to these xeric areas, the continent also supports some semiarid scrubland in northernmost Venezuela and Colombia (Lahey, 1973), but these dry areas are of very limited extent and have not de-

FIG. 8.—Chihuahuan Desert habitat near Boquillas, Mexico, with the Rio Grande River in the background. Ground bromeliads (*Hechtia*) are visible in the foreground. (Photo: M. A. Mares)

FIG. 9.—Great Basin Desert habitat near Mono Lake, California. The dominant plants are chenopodiaceous shrubs (e.g., *Artemisia*). (Photo: M. A. Mares)

veloped autochthonous species of mammals; they will not be considered further in this report.

Atacama/Peruvian.—These two deserts will be considered together, as they are continuous along the west coast of South America west of the Andean mountain chain. The deserts extend from near 2°S latitude to near

FIG. 10.—Thorn scrub of the Sinaloan lowlands of Mexico, near Guamuchil, Sinaloa, with scattered columnar cacti. (Photo: M. A. Mares)

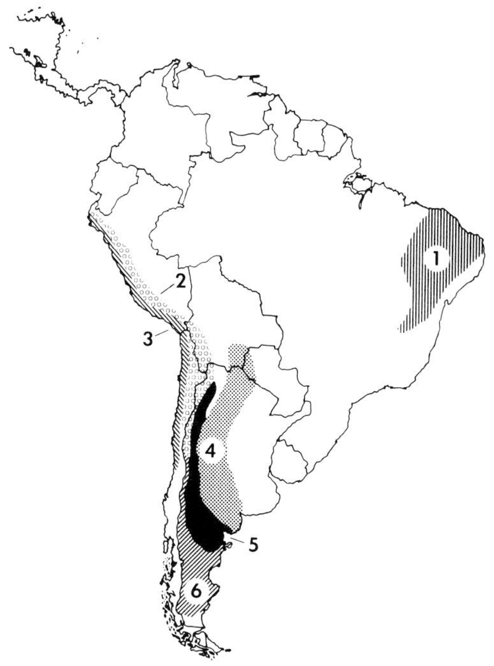

FIG. 11.—The principal arid and semiarid areas of South America: 1) Caatinga, 2) Puna, 3) Sechura-Atacama, 4) Chaco, 5) Monte, 6) Patagonia.

33°S latitude, a distance of more than 3,700 km. Various aspects of the Chilean and Peruvian deserts have recently been reviewed (Borgel, 1973; Rauh, 1985). Like most desert areas, there is a great variety of habitats in the region; aridity extends from sea level to about 2,500 m elevation along the western Andean slopes. The most notable and extensive lowland habitat, however, is the coastal desert, which contains broad areas supporting little or no vegetation, as well as areas where the principal moisture is received as fog (Fig. 12). This is primarily a desert where extremely cold offshore sea water cools the ocean wind currents such that air reaching the land is quite cool and forms fog. This surface layer of cool, moist air is of very low volume, perhaps extending upwards only a few hundred meters, where it encounters hot dry air, thus eliminating the possibility for rainfall (Logan, 1968). Hence fog is the primary source of moisture in these deserts, which are among the driest on earth (Borgel, 1973; Oberlander, 1979). In some localities, rainfall has never been recorded.

FIG. 12.—Looking eastward over a nearly abiotic portion of the Atacama Desert of Chile, southeast of Antofagasta. (Photo: P. Meserve)

Diversity of small mammals in the lowland arid and semiarid habitats is low, but abundance may be high (e.g., for Chile: Fischer, 1978; Fulk, 1975; Meserve and Glanz, 1978; for Peru: Pearson, 1982; Pearson and Ralph, 1978; Pine et al., 1979). In northern Peru, frequently only one or two species of small mammals occur in arid areas and both are cricetids, *Phyllotis gerbillus* and *P. amicus.* In some areas other cricetids, such as *P. darwini* or *Akodon mollis,* may co-occur with either *P. amicus* or *P. gerbillus.* Introduced *Mus musculus* are common in arid and semiarid portions of the coastal desert (e.g., Pearson, 1982; Pearson and Ralph, 1978; Péfaur et al., 1979). *P. gerbillus,* as might be surmised from the specific epithet, is a gerbil-like phyllotine that inhabits some of the more arid, sandy portions of the northern Peruvian desert, frequently being the only native rodent found over large areas. Externally, this small (16 g) rodent strongly resembles an old world gerbil; it is a light blond color dorsally with a white venter and sparsely-haired white tail. Koford (1968) examined an aspect of water balance in two individuals and his work strongly suggests that the species is well adapted to withstand dessication, although he believed that this ability was less pronounced than in other desert rodents such as heteromyids or dipodids. Unfortunately, the species has not been studied in detail and it is not known whether it is an obligate granivore, a seed hoarder, or undergoes torpor. Nevertheless, it is a highly desert-adapted rodent and more research may show that *P. gerbillus* is a perognathine analogue that inhabits the northern tip of this extensive desert region.

In Chile, Meserve and Glanz (1978) and Fulk (1975) have shown that small mammal diversity declines with decreasing rainfall. Only one to three species are found in the most arid localities supporting vegetation. In areas where vegetation is more abundant, one herbivorous caviomorph (*Octodon degus*) may coexist with several cricetids, including *Phyllotis darwini, Akodon olivaceous* and *A. longipilis.* Meserve and Glanz (1978) noted that *P. darwini* consumes many seeds and considered it a granivore (albeit not an obligate one). Meserve (1978) found that none of the cricetids he examined from the Chilean arid zone was independent of free water. Thus no heteromyid analogue inhabits the southern parts of this desert. An overview of the small mammals of these deserts is given in Appendix 1; a systematic listing of genera of small mammals inhabiting the world's deserts is presented in Appendix 2.

Monte/Patagonia.—These two deserts are also continuous and will be considered together, although Patagonia is more allied phytogeographically with the high altitude Puna Desert (Mares et al., 1985*a*). Nevertheless, the Monte and Patagonian deserts are not separated by barriers and are both low-elevational deserts, conducive to faunal mixing by their small mammals. Various characteristics of these deserts have been reviewed by Mares et al. (1985*a* and references therein; see also Soriano et al., 1983). This arid region extends about 3,000 km along western and southern Argentina (Fig. 11). In the northern section, the desert is limited to the lowland intermontane valleys lying east of the main Andean Cordillera and west of the pre-Andean and other mountain ranges. Rain in the north falls in summer; this gradually grades to a pattern of winter rainfall in the south, especially south of the Rio Negro, where Patagonia begins. The Monte Desert (Fig. 13) is among the most studied phytogeographic arid regions in South America (e.g., Mabry et al., 1977; Morello, 1958; Orians and Solbrig, 1977; B. B. Simpson, 1977). The faunal history of Argentina's arid areas has been reviewed by Mares (1985); part of the fossil story is pertinent to this report.

South America's fossil history is complex. Basically, the continent formed a huge island from the Late Cretaceous to the Late Pliocene, a span of some 90 million years (G. G. Simpson, 1980). The native fauna during the earliest stages of isolation included marsupials, xenarthrans, notoungulates, and various ungulate groups (e.g., G. G. Simpson, 1980). Rodents did not colonize the continent until the Oligocene, and only the caviomorphs are known from that period (e.g., Patterson and Pascual, 1972). Cricetids do not appear as fossils until the Late Pliocene (Marshall et al., 1982) and, although it has been argued that they may have colonized South America as early as the Miocene (Hershkovitz, 1972; Reig, 1981), there is strong evidence that they did not enter South America until the Central American land bridge was nearly completed (e.g., Baskin, 1978, 1986; Mares, 1985; G. G. Simpson, 1980). Whenever they entered, it seems clear that they did not colonize the Monte/Patagonian desert before the Late Pliocene.

In a series of papers, I have examined the degree of desert specialization of the small mammals of the Monte Desert (Blair et al., 1976; Mares, 1973, 1975*a,* 1975*b,* 1976, 1977*a,* 1977*b,* 1977*c,* 1977*d,* 1979, 1980, 1983; Mares and Hulse, 1977; Mares and Rosenzweig, 1978; Mares et al., 1977*a,* 1977*b,* 1985*a*; Orians et al., 1977; Solbrig et al., 1977; Williams and Mares, 1978). The Monte and Patagonian deserts share numerous taxa of small mammals (Appendix 1), particularly at the generic level. Basically, the fauna is composed of the old inhabitants (the caviomorph rodents, marsupials and xenarthrans) and the newer immigrants (cricetids). Most of the old inhabitants show pronounced bullar hypertrophy (Roig, 1969, 1972). Two small marsupials (*Marmosa pusilla* and *Lestodelphis halli*) and one small xenarthran (*Chlamyphorus truncatus*) are found in this xeric region. The small mammal niches filled by these species are very different from those of the heteromyids; all are small insectivores. Their physiology is poorly known (McNab, 1982), but the invertebrate food source does supply significant amounts of free water. A similar species of *Marmosa* (*M. elegans*) is known to hibernate (Roig, 1971); *M. pusilla* may also experience torpor (Mares, 1973). Unlike the insectivorous *Chlamyphorus,* other edentates of this region (*Chaetophractus, Zaedyus*) are larger-bodied insectivore/omnivores.

FIG. 13.—Aerial view of the Monte Desert near Andalgalá, Catamarca, Argentina. Visible are creosote bush (*Larrea*) flats and riparian desert woodland (*Prosopis, Acacia*) along the generally dry Rio Amanao. (Photo: M. A. Mares)

The caviomorph rodents of the Monte Desert, well-adapted to xeric existence (Bozinic and Contreras, 1990), also fill niches that are quite distinct from heteromyids. *Octomys mimax* is similar in many ways to woodrats (*Neotoma*), while *Ctenomys* are burrowing species remarkably similar to pocket gophers (Geomyidae). *Tympanoctomys* is also a burrowing browsing herbivore that inhabits halophytic plant communities near saline flats (Roig, 1962; Torres-Mura et al., 1989). The caviids (*Galea, Cavia, Microcavia*) are similar in many respects to the ground squirrels of North America, being diurnal burrowing cursorial/scansorial herbivores that range in size from 150–300 g (e.g., Rood, 1972). The physiology of the caviomorphs of Argentina has not yet been examined, but there are suggestions that they do have abilities to limit water requirements. Mares (1973), for example, found *Octomys mimax* inhabiting the driest portions of the Monte and foraging on cacti. The association of *Tympanoctomys barrerae* with halophytes also suggests a pronounced ability to process saline solutions. The small caviids are usually associated with green vegetation such as *Prosopis,* but may occur in very hot, dry scrublands that support little green vegetation in winter or during the dry season; they appear to require green vegetation to exist in the desert, however. A fairly large (about 10 kg) caviomorph, *Dolichotis patagonum* (the Patagonian "hare" or mara), inhabits the Monte/Patagonian desert. It is a cursorial grazing/browsing herbivore ecologically similar to *Lepus* of other deserts.

Only the cricetids of the Monte/Patagonian region are potential analogues of heteromyids, but research has shown that few species are even partial equivalents. In the northern Monte Desert, most cricetids avoid the desert, inhabiting mesic enclaves within the desert. Species included in this category are members of the genera *Akodon* and *Oryzomys.* The phyllotines, *Graomys griseoflavus* and *P. darwini,* are similar to *Peromyscus* in some ways (though larger). *G. griseoflavus* mainly inhabits *Prosopis* woodlands and forested gullies, while *P. darwini* is found on rocky cactus-covered hillsides (Mares, 1977*a*). No cricetid in this desert is bipedal, none has extensive bullar hypertrophy (although octodontids and ctenomyids have greatly inflated bullae), and none is an obligate granivore. *Calomys musculinus,* a species that inhabits more mesic localities in the northern Monte, *Prosopis* woodlands in the ecotonal areas between Monte and Chaco (where it co-occurs with *Akodon molinae,* Contreras and Rosi, 1980), and grass-scrublands in the southern Monte and in Patagonia, is water independent (Mares, 1977*d*). *Andalgalomys olrogi,* a cricetid apparently limited to the Bolsón de Pipanaco in the northern Monte (Catamarca Province), has not been examined physiologically; however, it possesses the very light blond coloration of "classic" desert rodents and shows some bullar hypertrophy (Williams and Mares, 1978).

Among living mammals, only members of the genus *Eligmodontia* (Fig. 14) show several adaptations that are reminiscent of heteromyids. Recent research on the genus (Mares and Braun, unpubl.) suggests that there are seven species of *Eligmodontia.* Four of these inhabit the Monte/Patagonian desert region. *Eligmodontia* sp. and *E. moreni* occur in the northern Monte, *E. typus* in the central and southern Monte and parts of Patagonia (Fig. 15), and *E. morgani* in the southernmost Monte and over much of Patagonia. Some ecological information is available on *E. typus* and *E. moreni.* These small cricetids have a light coloration (especially *E. moreni*), long hind legs (but are not bipedal), and inhabit arid scrublands. In the driest habitats *Eligmodontia* may be the only small mammal present. Unlike heteromyids, they forage on insects and leaves, as well as seeds; there is no evidence of seed hoarding in the genus. Physiologically, however, they are well adapted to xeric life (Mares, 1975*b*, 1977*b*; Mares et al., 1977*a*). At least a few of the animals examined by Mares (1975*b*) were water independent, al-

FIG. 14.—An Argentine gerbil mouse, *Eligmodontia moreni,* from near Andalgalá, Argentina. (Photo: M. A. Mares)

though most were not. Nevertheless, the animals had an extraordinary ability to process electrolytes; they were able to maintain weight on 1.6 M salt solution, or about three times the salt concentration of sea water. In many of the habitats frequented by *Eligmodontia,* halophytic plants are the dominant vegetation. Mares (1977*b*) suggested that the marked ability to process electrolytes might allow the animals to utilize these plants as a food or water source. Mares (1988) examined growth rates in *E. typus* born in the laboratory.

In the southern Monte and Patagonia, the faunal composition is not strikingly different from that of the northern Monte. *Oryzomys longicaudatus* becomes more common in scrublands as rainfall becomes more equitable and as grass cover increases. *Akodon xanthorinus* and *A. iniscatus* are also found in scrub/grass areas, while *A. longipilis* usually occurs in more mesic habitats. *Phyllotis darwini* and *Auliscomys micropus* frequent rocky areas in Patagonia, and *G. griseoflavus* is found in scrublands as far south as southern Patagonia. *Eligmodontia typus* and *E. morgani* are found in the Patagonian scrublands. Various *Ctenomys* species are the only small caviomorphs in arid areas. As in the northern Monte, no heteromyid equivalents are found in this region.

There is a curious twist to the story of heteromyid equivalency among small mammals of the arid areas of Argentina. Elsewhere (Mares 1975*a*, 1976, 1983), I have attributed the lack of ecological equivalency between the Monte and Sonoran deserts

FIG. 15.—Patagonian scrub desert in Rio Negro Province, Argentina. (Photo: M. A. Mares)

(strongly similar areas, physiognomically speaking) to the lack of time for pronounced desert adaptations to have evolved among the cricetids, which only encountered the Monte Desert, at the earliest, in the latest Pliocene (and more likely in the Early Pleistocene). However, such a scenario does not explain why a heteromyid type had not evolved among the caviomorph rodents, which were present since the desert's inception (e.g., Patterson and Pascual, 1972). Indeed, if a *Dipodomys*-type rodent is in fact the paragon of desert existence, one would assume strong selection pressures acting toward the development of an ecological equivalent to this morphoecological type in other deserts. Given the fact that cricetids were absent, which presumably lessened the competition for seeds and other resources, the desert-specialist niche-type seems to have been available in this desert.

In fact, there was a family of small to mid-sized bipedal mammals in the Monte and adjoining regions from the Oligocene through part of the Pleistocene. These animals were strongly similar to modern kangaroo rats (or Old World dipodids) and possessed a diastema, reduced forelimbs, elongated hindlimbs, a long tail, some bullar hypertrophy, and many other characteristics similar to dipodomyines (Fig. 16)—this was the marsupial family, Argyrolagidae (G. G. Simpson, 1970; Wolff, 1984). Their presence in the Monte strongly suggests that there was a group of highly convergent rodentiform marsupials that were at least loose heteromyid equivalents (Mares and Rosenzweig, 1978). They became extinct in the Pleistocene during the great vegetational changes of that period and have not been replaced—yet. Given enough time, *Eligmodontia,* which is somewhat intermediate in several heteromyid characteristics, could give rise to a "classic" desert rodent of South America.

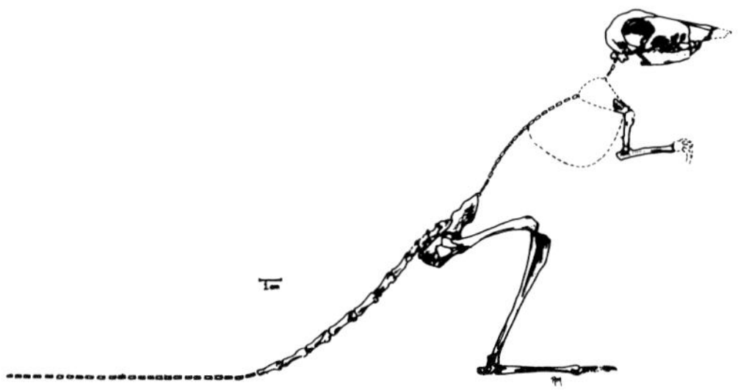

FIG. 16.—Depiction of the skeleton of an extinct, bipedal marsupial argyrolagid—a possible heteromyid analogue of the deserts of southern South America. (After Simpson, 1970.)

Puna.—The Puna is a high elevational desert (above 3,000 m) that extends for more than 3,000 km along the Andean and pre-Andean mountains from northern Peru to south-central Chile (Cabrera, 1976; Weber, 1969). It is a cold desert with great daily and seasonal temperature fluctuations. For example, Mares (1975*b*) noted that some localities may undergo daily temperature changes of up to 38°C. Although the Puna contains a number of habitats, from bunch grass areas to arid scrubland, the region generally supports only sparse vegetation except in localized mesic areas (Fig. 17). Many salt flats and salt lakes are scattered throughout the Puna.

Mammals of the Puna have been studied primarily in Peru (Pearson, 1951, 1957, 1959, 1982; Pearson and Ralph, 1978) and Chile (Greer, 1965; Mann, 1945; Osgood, 1943; Pine et al., 1979), with some information available for Argentina (Mares et al., 1981*a*, 1989; Ojeda and Mares, 1988; Roig, 1962). The small mammal fauna of the Puna is rich in both genera and species (e.g., Pearson, 1951, 1982; Appendix 1 and 2), although many of these are limited to the more mesic microhabitats. All of the caviomorphs and many of the cricetines are herbivores. Some cricetids (e.g., *Akodon*) may be primarily insectivorous, while others (e.g., *Phyllotis, Oryzomys, Eligmodontia*) are micro-omnivores, with a tendency toward granivory. There are two species of *Eligmodontia* in the Puna, *E. hirtipes* and *E. puerulus.* Both are larger than the lowland species, hairier and more heavy-bodied. The tails are densely furred and the animals appear less desert-specialized than

FIG. 17.—Puna bunchgrass habitat at 3,400 m elevation near Cachi, Salta Province, Argentina, with a group of domestic llamas (*Lama glama*). (Photo: M. A. Mares)

those of the lowlands, although their ecophysiology has not been examined.

Pearson (1982) has shown that the northern Puna is more species poor than the southern Puna within Peru. The Peruvian Puna is continuous with that of Bolivia, Chile and Argentina, and many faunal elements are shared. As in southern Peru, diversity in Chile and Argentina can be quite high. Also, most species are herbivores and no heteromyid analogues occur.

Chaco.—The Chaco (including the Espinal) is the semiarid thorn scrub of Argentina, Bolivia and Paraguay that covers an area of 1 million square kilometers (Fig. 18). Its small mammals have been examined in various reports (e.g., Mares et al., 1981*b*, 1989; Myers, 1982; Myers and Wetzel, 1979; Ojeda and Mares, 1988; Olrog and Lucero, 1981; Wetzel and Lovett, 1974). In the drier (western) portions of the Chaco (I here exclude the more mesic eastern portion from further consideration), the small mammal fauna is quite poor in any one locality, although overall species richness of the Chaco is high (Appendix 1, and see Redford et al., 1990). Generally, the fauna consists of a marsupial (*Marmosa,* or occasionally *Monodelphis*), a small edentate (*Chlamyphorus*), several larger edentates (*Chaetophractus, Euphractus, Cabassous, Tolypeutes*), one or two caviomorphs (*Galea, Ctenomys*), and several cricetids. The latter are either ground dwelling micro-omnivores/insectivores (*Bolomys, Akodon, Calomys*) or scansorial micro-omnivores (*Graomys, Andalgalomys*). Ground cover in extensive portions of the region is exceedingly sparse to non-existent during much of the year and the substrate varies from clay to sand; summer temperatures are high and the yearly drought may extend for 7.5 months. No heteromyid equivalents inhabit the area. Phytogeographically, this region strongly resembles the thorn scrub of Texas, New Mexico and Mexico, which supports several species of heteromyids.

Caatinga.—The Caatinga is an unusual semiarid area in northeastern Brazil (Fig. 19). Located between 3° and 16°S latitude and 35° and 45°W longitude, it extends over 650,000 square kilometers (Frota-Pessoa et

FIG. 18.—Xeric chaco thorn scrub habitat in extreme northeastern Salta Province, Argentina. Photo taken during the dry season, with leafless *Geoffroya, Cercidium,* and *Acacia* visible. (Photo: M. A. Mares)

al., 1971; Reis, 1976). The entire area undergoes periodic droughts that may extend up to a year or more during which vegetation cover becomes exceedingly sparse. Recent research in the Caatinga has begun to clarify the distribution and ecology of small mammals (e.g., Lacher, 1981; Lacher et al., 1982; Mares et al., 1981*b*, 1985*b*; Streilein,

FIG. 19.—Caatinga thorn scrub habitat in northeastern Brazil near Serra Talhada, Pernambuco. In the foreground is the common cactus, xique-xique (*Pilocereus*). (Photo: M. A. Mares)

1982*a*, 1982*b*, 1982*c*, 1982*d*, 1982*e*, 1982*f*; Willig and Mares, 1989). The region contains few endemics and those species that do persist in the area undergo marked population reductions during dry periods. Persistence of most species occurs because they inhabit mesic microhabitats limited to granitic outcroppings that form hills and mesa-like structures. Such areas support green vegetation during the driest periods and, being both extensive and widely distributed throughout the Caatinga, act as mesic refugia for most species.

Almost all species of Caatinga mammals are shared with the neighboring Cerrado grassland biome, a more mesic savanna woodland that occupies upland plateau areas of central Brazil (Mares et al., 1985*b*; Mares et al., 1989). As many as 10 species of small mammals are found in the Caatinga proper, but most are ground-dwelling or scansorial herbivores, micro-omnivores, or insectivores. Like the Chaco, no heteromyid equivalents inhabit the region.

Australia

Australia is essentially a desert continent, with 55% (or 4.2 million square kilometers) of the continental land area being desert or semidesert habitats (Beadle, 1981; Mabbutt, 1984; Williams and Calaby, 1985). The paleohistory of the Australian continent is fascinating. Once connected to the Antarctic land mass and thence to South America and the remainder of Gondwanaland (in the Cretaceous, 115 million years ago, Norton and Sclater, 1979), the continent underwent great climatic and vegetational changes as it moved via sea floor spreading to its present position (e.g., Raven and Axelrod, 1972; Specht, 1981*a*). Because it is now astride 30°S latitude, and because the continent is largely free of extensive mountainous areas to act as barriers to the prevailing westerlies, its deserts are of the subtropical type (Logan, 1968), with diverse substrates and habitats (e.g., Beard, 1984; Mabbutt, 1984; Williams, 1984); dunes make up 40% of Australia's arid zone.

Like South America, much of the evolutionary history of the continent is associated with an autochthonous fauna which developed in isolation and, also like South America, colonization by rodents occurred relatively recently. Murids immigrated into Australia during the Miocene, about five to ten million years ago (Watts and Aslin, 1981). Although they entered from regions that were largely tropical, they underwent a pronounced adaptive radiation after reaching Australia. Again, this is quite similar to what occurred in South America. Indeed, the Australian murids radiated into 14 genera and 53 species (Watts and Aslin, 1981), while in South America about 46 genera and 214 species of cricetids evolved (Honacki et al., 1982). An important difference in the history of rodents on the two continents is that the Australian murids arrived perhaps five million years earlier than did their South American counterparts. Moreover, because Australia is largely desertic (Figs. 20, 21), the immigrants encountered semiarid and arid habitats shortly after colonization. In South America, by contrast, immigrants entered at the northern tip of the continent across the Central American land bridge. Only after traversing thousands of kilome-

FIG. 20.—Map of Australian Desert.

FIG. 21.—Sandhill habitat near Sandringham Station, southwest Queensland, Australia; *Notomys alexis* and *Pseudomys hermannsburgensis* occur in this area. (Photo: R. E. MacMillen)

ters of forest, savannas or developing mountainous regions did they reach the southern deserts. This encounter with aridity probably occurred many millions of years later in South America than it did in Australia.

Morton and Baynes (1985) point out that the present-day fauna of Australian desert mammals is quite distinct from the prehistoric fauna, presumably due to the effects of human activities, including the introduction of exotic species (e.g., Low, 1984). Prehistoric species diversity of both marsupials and rodents in deserts was about 40% higher than what is found there today, although information on abundance in the past is unavailable. Nevertheless, the mammal fauna of Australia's deserts today is still quite diverse (73 species), with marsupials and rodents predominating (Archer, 1981; Lee et al., 1981; Main and Bakker, 1981), although relative abundance in any particular area is low (Morton, 1979; Appendix 1). Mammal insectivory is much more pronounced in Australian arid areas than in North America, while the reverse is true for granivory (Morton, 1979, 1985; Morton and Baynes, 1985; Watts, 1977). Many species of murids inhabit Australia's deserts, with six genera (*Rattus, Leporillus, Notomys, Zyzomys, Leggadina, Pseudomys*) frequenting dry areas and all but *Rattus* being primarily limited to xeric areas (Watts and Aslin, 1981). Like North American desert rodents, several Australian species show pronounced physiological adaptation to aridity (e.g., Baudinette, 1972; Baverstock and Watts, 1974, 1975; Baverstock et al., 1976, 1979; MacMillen and Lee, 1967, 1969; MacMillen et al., 1972; Purohit, 1974*a*, 1974*b*; Smith et al., 1972); several are water independent or produce highly concentrated urine. Some, notably species of the rodent genus *Notomys*, have also developed bipedal locomotion. Insectivorous marsupials of the genera *Sminthopsis* and *Antechinomys* are quadrupedally saltatorial (Ride, 1970), although some species may move bipedally (Nowak and Paradiso, 1983). There are several other small to mid-size bipedal marsupials in Australia's desert. These are primarily grazing and brows-

FIG. 22.—*Notomys cervinus*, a bipedal desert-adapted Australian murid. (Photo: R. E. MacMillen)

ing herbivores, and some insectivores (Appendix 1) and will not be considered further with reference to heteromyid similarities; however, the hopping mice, *Notomys*, offer a possible Australian analogue to heteromyids.

Notomys is a desert-adapted rodent genus containing about 10 species, five of which are probably still extant (Watts and Aslin, 1981). These mice are almost entirely granivorous and have many of the adaptations associated with bipedality: long hind limbs, reduced forelimbs and long tufted tail (Fig. 22). They inhabit the most arid parts of the Australian scrub desert and are highly adapted physiologically for desert life. There are some notable differences, however, between them and heteromyids (Watts and Aslin, 1981). Australian hopping mice differ anatomically from heteromyids in having very large pinnae; bullae are inflated, but not as much as in heteromyids. *Notomys* are primarily colonial, although the social structure of the groups has not yet been clarified; up to 12 individuals may inhabit a burrow. They apparently do not store seeds. Despite the many differences between the desert-dwelling *Notomys* and heteromyids, *Notomys* can be considered moderately convergent on the *Dipodomys* pattern.

Finally, another Australian murid is trophically similar to heteromyids. *Pseudomys hermannsburgensis* can be considered granivorous (Morton and Baynes, 1985), although it includes diverse vegetable material in its diet at times (Watts and Aslin, 1981). The diet of this tiny (12 g) desert mouse is thus roughly similar to *Perognathus*. However, like *Notomys*, it may also be colonial (Watts and Aslin, 1981). It is probably more similar to some North American cricetines than to perognathines.

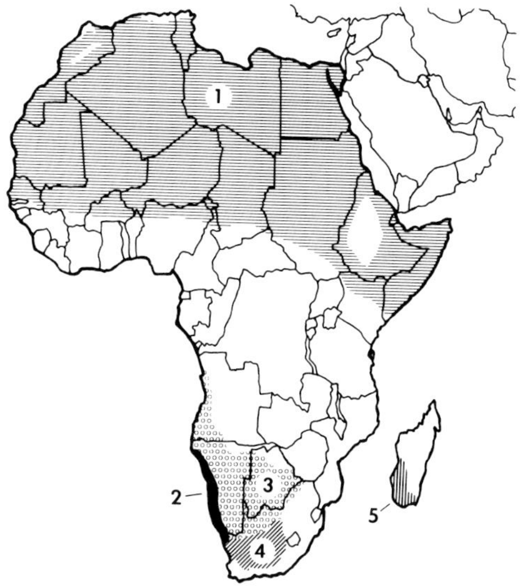

FIG. 23.—Map of the xeric areas of Africa and Madagascar: 1) Sahara, 2) Namib, 3) Kalahari, 4) Karroo, 5) Madagascar scrubland.

African Deserts and Semideserts

There are two major desert regions on the African continent: the massive Sahara, which covers much of the northern third of Africa (including the Sahel region) and the Namib/Karroo and Kalahari desert system at the southern tip of the African land mass. Each merits separate consideration in this report.

Sahara.—The Somali-Chalbi Desert is herein considered a coastal extension of the Sahara, although there are notable differences in the geology and climate of the two deserts. Nevertheless, the regions are continuous along the western coast of the Red Sea (see Meigs, 1953). The Sahara is the world's largest desert (Fig. 23), containing about nine million square kilometers of arid and semiarid habitat (McGinnies et al.,

FIG. 24.—Xeric scrubland of the Sahara Desert in Wadi Natroun, Egypt. (Photo: M. A. Mares)

1968), more than the rest of the world's deserts combined. It is among the hottest and driest of deserts and contains a great diversity of habitats, including extensive dune areas, mountain ranges, wadis and other areas of topographic relief and substrate diversity (Ayyad and Ghabbour, 1986; Grenot, 1974; Le Houérou, 1986; Monod, 1986). The Sahara is arid at its northern limits, extremely arid in the central portions, and semiarid at the southern limits. In its most arid portions, Grenot (1974:104) noted that "One can sometimes travel for 400–500 km without seeing anything but sand, gravel, or slabs of rock." In such areas the desert is ". . . devoid of a drop of water or a sprig of grass." It might be expected that in such an arid and extensive desert natural selection would have acted in such a manner that a high degree of desert adaptation would be favored among the regular inhabitants of the desert, and this is indeed the case.

The Sahara has a long history of aridity, although its present great expanse may be more recent than previously thought. Like subtropical deserts elsewhere, part of the reason that the Sahara is arid is its position astride the 30° latitudinal band, where descending dry air causes arid zones around the world if other climatic and geologic factors do not mitigate this aridity (Logan, 1968). Thus, some portions of the globe have probably always been drier than other portions, although not necessarily desertic (Axelrod, 1950, 1972). Originally, a portion of the Sahara formed one of two major Afro-Asian arid areas in the Early Tertiary (Walter et al., 1983), with a portion of the Gobi Desert being the second (below). However, much evidence supports the idea that the extent of the Saharan arid zone was much less pronounced in the last several thousand years and, indeed, that large portions of the desert had been more mesic in the Quaternary and subjected to pluvial cycles (e.g., McCauley et al., 1982; Le Houérou, 1986). Within the last 2,500 years, woodlands covered much of the North African Saharan region. Deforestation by humans very likely helped cause the massive desert that is seen today, especially when changes in the surface albedo are considered (Otterman, 1974). Nevertheless, the relatively frequent and

FIG. 25.—*Gerbillus gerbillus,* a common desert-adapted species of the Western Desert of Egypt. (Photo: M. A. Mares)

FIG. 26.—The gerbilline, *Dipodillus,* a quadrupedally-saltatorial seed-hoarding granivore of the Sahara Desert. (Photo: M. A. Mares)

complex climatic oscillations that have occurred throughout the Sahara over the last two to three million years may well have provided an excellent mechanism for species multiplication (e.g., Conrad, 1971; Grenot, 1974; Mares, 1979). As shown below, small mammals in the Sahara desert are quite diverse.

Because of its vast size, it is impractical in this report to examine the Sahara's fauna in great detail. There are numerous publications, however, that allow a reasonable assessment of the occupancy of the heteromyid niche in this desert (e.g., Cockrum, unpublished; Kingdon, 1974; Osborn and Helmy, 1980; Petter, 1961; Ranck, 1968; Yalden et al., 1976; see also Eisenberg, 1975; Mares, 1980).

The Sahara is rich in species of smaller mammals. For example, in arid areas of Egypt (Fig. 24) may be found hedgehogs (*Hemiechinus* spp. and *Paraechinus* spp.), shrews of the genus *Crocidura,* hares (*Lepus*), porcupines (*Hystrix*) and a large number of species (about 25) of small rodents (Appendix 1 for generic diversity, and Appendix 2 for systematics of genera), including members of the Gerbillinae (Cricetidae), the fossorial herbivorous Spalacidae, Muridae (excluding Cricetidae), Muscardinidae and Dipodidae (Osborn and Helmy, 1980). Many of these species, by virtue of their food habits or basic natural history are not potential equivalents of the heteromyids. Thus, the insectivorous hedgehogs and shrews, the large-bodied herbivores, *Lepus* and *Hystrix,* and the burrowing gopherlike *Spalax* can be excluded from further consideration. Similarly the omnivorous squirrellike dormouse, *Eliomys,* which inhabits rocky hillsides in Egypt can be excluded, as can the murid spiny mice, *Acomys,* which forage readily on seeds, but include much green matter and animal material in their diets (see also Müller and Van Aken, 1990). *Mus* also occurs in some arid and semiarid parts of this desert, but is not a heteromyid equivalent. The remaining species to be considered are the gerbillines which, in Egypt, include *Gerbillus, Dipodillus, Sekeetamys, Meriones, Pachyuromys, Psammomys,* and the dipodids *Allactaga* and *Jaculus.* Of these, *Sekeetamys* is a rock-inhabiting granivore/herbivore (somewhat woodratlike), *Psammomys* is a squirrellike herbivore-folivore from sandy areas supporting halophytic vegetation (Degen et al., 1988; Kam and Degen, 1988, 1989), and *Pachyuromys* is a small insectivore/herbivore inhabiting areas of desert pavement (Daly and Daly, 1973; Wassif and Soliman, 1979); none is comparable to a heteromyid.

All of the remaining genera are quite similar to heteromyids. *Gerbillus* (Fig. 25) and *Dipodillus* (Fig. 26) are strongly convergent on *Perognathus* and *Chaetodipus.* They are similar in body size, are obligate seed-hoarding granivores, and are quadrupedally saltatorial (e.g., Happold, 1968). They are also similar in various other behavioral and ecological aspects (Eisenberg, 1967, 1975). They are well-adapted physiologically to

FIG. 27.—The Mongolian jird, *Meriones unguiculatus,* illustrating the general body form of the granivorous burrowing members of this genus. (Photo: M. Andera)

aridity (e.g., Haim, 1984; Kirmiz, 1962 for *Jaculus,* but see Ghobrial and Nour, 1975; Schmidt-Nielsen, 1964). *Meriones, Jaculus* and *Allactaga* are all similar in some ways to *Dipodomys. Meriones* (Fig. 27) is a seed-hoarding burrowing granivore that is quadrupedal, while *Jaculus* (Fig. 28) is remarkably convergent in body form on *Dipodomys* (Berman, 1985; Happold, 1967; Wassif, 1960) and, while not a seed hoarder, is a pronounced granivore, although it cannot eat hard seeds (Happold, 1968). *Allactaga* is also bipedal (Fig. 29) and, except for its elongate pinnae, also externally resembles *Dipodomys.* Unlike *Dipodomys,* however, its diet is largely herbivorous, containing many bulbs and tubers (for diet and general biology of the genera considered see Eisenberg, 1975; Happold, 1975; Mares, 1980; Nowak and Paradiso, 1983; Osborn and Helmy, 1980; Petter, 1961; Ranck, 1968; Rosevear, 1969).

In Egypt, therefore, located in the eastern Sahara, there are a number of species that are convergent on the heteromyid type. Moreover, since Egypt is representative of much of the Sahara Desert, a major segment of the mammal fauna of this desert is ecologically and morphologically similar to the North American desert mammal fauna. Indeed, as other portions of the Sahara are included, the overall similarity of this desert to the North American desert system increases. For example, desert ground squirrels (Sciuridae: *Atlantoxerus, Xerus*) are common in some portions of the Sahara (e.g., Happold, 1975; Ranck, 1968), as are xeric-adapted squirrels in North America. *Lemniscomys,* a murid, is a quadrupedal herbivorous rodent of African deserts. Some

FIG. 28.—The bipedal desert-adapted dipodid, *Jaculus jaculus,* represents a genus that is strongly convergent morphologically (and in some species, ecologically) on North American *Dipodomys.* (Photo: M. Andera)

Saharan species, however, such as the herbivorous browsing members of the rodent family Ctenodactylidae (*Ctenodactylus, Massoutiera*) or the hyrax family Procaviidae (*Heterohyrax, Procavia*) have no equivalents in the North American deserts, although they do show many ecological similarities to other rodent and non-rodent rock-dwelling mammals found in other semiarid or arid regions (Gouat and Gouat, 1987; Grenot, 1973; Mares and Lacher, 1987). Other Saharan gerbils (*Tatera, Taterillus*) and the murid *Steatomys* are more like North American desert peromyscines or neotomyines than heteromyids in their food habits, although *Desmodilliscus braveri* from the Sudan may be another perognathine equivalent (cf. Ayyad and Ghabbour, 1986; Happold, 1975; Nowak and Paradiso, 1983; Rosevear, 1969).

Namib/Karroo-Kalahari.—This arid region (Fig. 23) is composed of three separate deserts (e.g., Meigs, 1953; Petrov, 1976; Werger, 1986). The Namib is a narrow, cool, coastal desert lying between 17° and 35°S latitude (Walter, 1986). Causative factors in the formation of this desert are similar to those of the Atacama/Peruvian desert system of South America—latitudinal placement, mountain ranges forming barriers to easterly winds, and cold ocean currents immediately offshore. The Namib is extremely arid (yearly precipitation ranges from 15 to 43 mm, Petrov, 1976) and contains extensive dune systems supporting little vegetation (Fig. 30) (Seely, 1978); fog and dew are important sources of water in the desert for both plants and animals. Other habitats support sparse scrub vegetation, low grass cover and, in more mesic seasonal river-

FIG. 29.—The bipedal dipodid, *Allactaga elater,* is typical of the members of this genus which possess elongated pinnae. Although similar to *Dipodomys* morphologically, they are more like geomyids trophically. (Photo: M. Andera)

beds, desert gallery forests of *Acacia* and other woody vegetation (Walter, 1986; Werger, 1978). The Namib is limited to that area lying between the coast and the quickly rising escarpment that forms the western limits of the Kalahari; thus the desert varies from only 80 to 150 km in width (Walter, 1986). Despite its relatively small areal extent and its pronounced aridity, however, this desert supports a rich mammal fauna (Bigalke, 1978; Coetzee, 1969; Rautenbach, 1978; Stuart, 1975) (Appendix 1).

The Karroo is adjacent to the southern Namib, lying south and east of that desert on the South African Plateau. It differs from the Namib in that precipitation is received as summer rain (Werger, 1978, 1986). Rainfall may total 250 mm in some sites and as little as 50 mm in others; elevation is about 900 m. Many plant species are shared with the Namib, although vegetation in the Karroo is much more luxuriant than in the arid Namib, with tall Euphorbiaceae and leguminous trees being common (Werger, 1978, 1986).

The Kalahari (including the Namaland Domain) is found to the east of the Namib and north of the Karroo. In its northern portion (the Namaland Domain) it includes scrublands of the escarpment zone (Werger, 1978), with substrate ranging from rocky to deeper soil. Grasses are common in large parts of this area. In the southern portion, the Kalahari includes extensive stable dunes (90% of this region is covered by red dunes, Werger, 1978) and clayey or silty soil areas (Fig. 31). Rain falls mainly in summer, while maximum vegetative growth occurs from

FIG. 30.—Sand dune area in the Namib Desert in Ludervitz District, Namibia (25°53′S, 16°07′E). (Photo: D. A. Schlitter)

late winter to early spring. Vegetation ranges from extensive bare areas (e.g., salt pans) to scrublands to areas supporting trees; vegetation of the inland sand desert is especially sparse.

The entire southern African region has probably been arid or semiarid for a long period of time. At least a portion of the aridity is due to the 30°S latitudinal zone of adiabatically heated dry, descending air masses (Logan, 1968), but cold offshore currents and geological factors also combine to cause aridity. Although there are numerous disjunctions of southern African botanical

FIG. 31.—A portion of the Kalahari Desert approximately 100 km NE of Mariental, Namibia. (Photo: D. P. Christian)

elements with plants found in the deserts of North Africa and the Middle East or Asia, the arid regions of southern Africa contain many endemics (Werger, 1978). When Africa separated from South America about 85 million years ago forming the southern Atlantic Ocean, the cold water currents that began to flow along the west coast of Africa led to the formation of the present-day Namib Desert with its extreme aridity, although aridity in the general region very likely extends back to the Cretaceous (Raven and Axelrod, 1974). Nevertheless, like other deserts, climatic fluctuations occurred throughout the later Tertiary and the Quaternary (e.g., Van Zinderen Bakker, 1975, 1978) and led to the complex desert flora that today characterizes the region.

Among the smaller mammals found in xeric habitats of the Namib are members of the insectivorous elephant shrew family Macroscelididae (*Macroscelides, Elephantulus*), nearly bipedal desert species. Other insectivores include shrews, *Crocidura,* and fossorial golden moles (Chrysochloridae: *Eremitalpa*). Like the Sahara, hyraxes (Procaviidae: *Procavia*) are present, as are lagomorphs (*Lepus, Pronolagus*) and porcupines (*Hystrix,* e.g., Haim et al., 1990). Unlike the Sahara, squirrels are absent from the Namib. In a manner similar to the North American and Saharan deserts, however, the Namib supports up to 16 species of smaller rodents, including gopherlike fossorial herbivores (Reichman and Jarvis, 1989) of the family Bathyergidae (*Bathyergus, Cryptomys*) and the squirrellike herbivorous rock specialist *Petromus* (Petromyidae), which appears to be much like *Kerodon* of the Brazilian Caatinga (Mares and Lacher, 1987). A rich assemblage of murids and cricetids includes arboreal herbivores (the *Acacia* mouse, *Thallomys*), ground-dwelling herbivores (*Rhabdomys, Lemniscomys, Malacothrix, Parotomys*), micro-omnivores such as *Mus, Praomys, Petromyscus* and *Steatomys,* and the gerbilline, *Tatera,* which was discussed above in reference to the Sahara and which is a broad herbivore/omnivore.

Of the remaining rodents, *Gerbillus* is similar to heteromyids and was discussed above (*Gerbillus* of southern Africa are also highly desert specialized, e.g., *G. pusillus,* Buffenstein and Jarvis, 1985). Another gerbilline, *Desmodillus,* is an endemic that forages on and hoards seeds, although it consumes significant amounts of insect matter as well (Nowak and Paradiso, 1983). Its habit of storing seeds in fairly complex burrows makes it similar in this regard to perognathines. The pouched rat, *Saccostomus,* a cricetid, is a kangaroo-rat sized granivore that stores large numbers of seeds in burrow chambers for use during the winter. While the species *S. campestris* strongly resembles North American *Onychomys* externally, it is strongly converged on heteromyids in its diet and seed-storing habit. Finally, the springhare (Pedetidae: *Pedetes*) is a bipedal herbivore of medium body size (3–4 kg) that is strongly convergent morphologically on bipedal heteromyids, particularly in aspects of its hind limb morphology (Berman, 1985). Nevertheless, ecologically the species is not comparable to heteromyids.

Thus in the Namib, the only bipedal (or nearly so) small mammals found in the desert are an herbivore (*Pedetes*) and insectivores (Macroscelididae). Several granivorous species (a cricetid and two gerbillines) are also seed hoarders and in this and other aspects of their biology are quite convergent on the heteromyids of North America.

The fauna of the Karroo is generally similar to that of the Namib, although it is richer in species (e.g., Rautenbach, 1978). It contains more insectivorous forms (elephant shrews, shrews, and golden moles), more fossorial herbivores (bathyergids), desert ground squirrels (*Xerus*), dormice (*Graphiurus*) and a wide variety of murids and cricetids (additional genera include *Dendromus, Mystromys, Aethomys* and *Zelotomys*). None is an ecological equivalent of a heteromyid. All species that could be considered convergent on heteromyids are in the same genera as the Namib species.

The Kalahari mammal fauna is basically similar to that of the Namib/Karroo Desert.

No species found in the Kalahari, but absent from the Namib/Karroo system, are ecological equivalents of heteromyids.

Several rodents found in the deserts of southern Africa have been studied ecologically and physiologically (Christian, 1978, 1979*a*, 1979*b*, 1980*a*, 1980*b*; Nel and Rautenbach, 1975; Taylor and Green, 1976; Withers et al., 1980). While not all of the species examined were independent of free water, all produced highly concentrated urine, especially *Desmodillus auricularis* and *Gerbillus paeba*, with the former being the most efficient at conserving water (*Rhabdomys pumilio*, an herbivore, was much less efficient at conserving water than the gerbillines). Thus it is clear that physiological similarities to heteromyids characterize those African species that are ecologically similar. Rodents from more mesic parts of the southern African scrublands appear to be less xeric adapted (Taylor and Green, 1976).

Madagascar

It might seem surprising that Madagascar, a large (600,000 square kilometers) island lying 400 km off the southeastern African coast (Fig. 23) would be included in a chapter dealing with deserts, but Madagascar contains an extensive semiarid zone having a pronounced dry season (Rauh, 1986). This semiarid zone is largely the result of rain shadow effects of a mountainous region that parallels the east coast of the island and causes most of the moisture-bearing southeast trade winds to deposit their moisture as orographic rainfall on the eastern slopes of the mountains. Although rainfall in the xeric parts of the island may vary from 350–900 mm, the eight-month droughts in the portion receiving the greatest amounts of precipitation magnify the effects of aridity. The sub-arid zone may receive up to 350 mm of rain per year, but the rainfall is unpredictably sporadic. Indeed, the year's total may be received in a single month (Rauh, 1986). Insolation during the droughts is intense and evapotranspiration is high; the vegetation of the dry zones is xerophytic. Both shrub and tree layers are present in some areas, although ground cover between shrubs may be sparse. As Rauh (1986) notes, in general physiognomy, the Madagascar scrub recalls the Caatinga or the Chaco of South America. In Madagascar, as in both of those South American habitats, succulence and thorniness are pronounced (Koechlin, 1972).

Only six orders of mammals occur in Madagascar (Eisenberg and Gould, 1970), and of these, the smaller-bodied species are found in the Insectivora and Rodentia (Heim de Balsac, 1972; Petter, 1972). Tenrecs of the genera *Echinops* and *Geogale* are found in the semiarid portion of Madagascar; both genera are monotypic and each species undergoes torpor (Eisenberg and Gould 1970).

Of the rodents, only one subfamily of cricetids is native to Madagascar, the Nesomyinae, containing seven genera and 10 species (Bigalke, 1972; Petter, 1972). Most genera and species are found in the more mesic forested portions of the island. One species, *Hypogeomys antimena* (the Malagasy Giant Rat), is a large-bodied, long-eared cursorial and hopping herbivore/frugivore

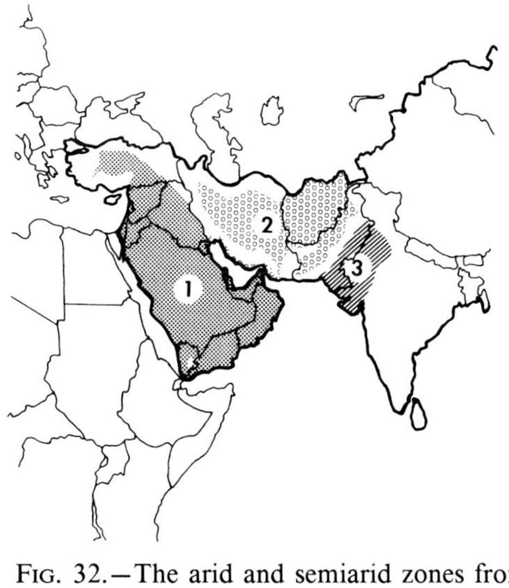

FIG. 32.—The arid and semiarid zones from the Middle East to India: 1) Arabian Desert, 2) Iranian Desert, 3) Thar Desert.

FIG. 33.—A portion of the arid Negev Desert of Israel in the Middle East Desert System. (Photo: M. A. Mares)

that may fill the role of a rabbit in the sandy coastal forests; the seventh genus, *Macrotarsomys,* contains two species, one of which, *M. bastardi,* occurs in dry habitats.

Petter (1972), Eisenberg (1975), and others (e.g., Bigalke, 1972) have noted that *Macrotarsomys* is similar in many ways to gerbils. Externally, it strongly resembles the gerbils of Africa and Asia, the Peruvian phyllotine (*Phyllotis gerbillus*), or the Argentine gerbil mice (*Eligmodontia*). It appears to be a broadly foraging herbivore, however, and is not similar trophically to those species of desert rodents to which it is similar in external aspect.

The Middle East-Pakistan Arid Belt

An extensive belt of aridity occurs along 30°N latitude and gives rise to the Arabian Desert (including the deserts of the Middle East and Arabian Peninsula) and the Iranian Desert (including xeric areas in Iran, Afghanistan and parts of Pakistan) (Fig. 32). For information on the climate and vegetation of this region see Abd El Rahman (1986), Evenari et al. (1971), Hassinger (1968), Logan (1968), McGinnies et al. (1968), Orshan (1986) and Petrov (1976). Tchernov (1975), Turnbull (1975) and Walter and Box (1983*d*) review the climatic and botanic history of this region and show that great changes in the distribution and degree of aridity have characterized the area since the early Tertiary.

Arabian Desert.—This desert includes large areas of shrub steppe vegetation, with vegetation of a more mesic nature occurring along ephemeral watercourses (Fig. 33). Palm oases are also found here (McGinnies, 1968), as are extensive sand deserts (Petrov, 1976). Indeed, Petrov (1976) noted that 95 percent of the Arabian Peninsula is composed of primitive desert soil with moving or semistabilized sand areas; the Rub' Al Khali of the Arabian Peninsula is the largest continuous expanse of sand in the world (650,000 square kilometers, Abd El Rahman, 1986). Rainfall occurs primarily in the winter-spring period and is quite sparse over most of the desert (<200 mm annually).

Information on the mammals of the Arabian Desert is available from a number of sources (Atallah, 1977, 1978; Atallah and Harrison, 1967; Gutterman et al., 1990;

FIG. 34.—The Iranian Desert (Dasht-e-Kavir) in the Kavir Protected Region, northcentral Iran. (Photo: M. A. Mares)

Haim and Tchernov, 1974; Harrison, 1964, 1968, 1972; Hatt, 1959; Lewis et al., 1967). Faunistically, the Arabian Desert does not differ greatly from the Sahara Desert to the west, or from the Iranian Desert to the east (Ranck, 1968). Ranck suggested that this entire xeric strip could be placed within a single zoogeographic subregion, the Saharo-Sindien Faunal Region. He also included the arid portions of southeastern Russia (the Kara Kum and Kyzyl Kum deserts and adjacent regions, i.e., the Turkestan Desert) within this faunal subregion. Ranck noted that only 11 genera of rodents occurred throughout this immense zone, a very sparse fauna given the size of the area involved. As additional support for his hypothesis, Ranck noted that fully 23 non-rodent species are endemic to this faunal zone. As shown below, however, the Turkestan Desert contains several genera of small mammals that do not occur in the Middle East-Pakistan arid zone.

As would be expected from the above discussion, small mammal ecology in the Arabian Desert is very like the Sahara. While several additional species of desert-adapted rodents are added (especially in the genera *Gerbillus* and *Meriones*), no new genera appear in this region. Thus the same generic equivalents to heteromyids that were suggested for the Sahara would be representative of this desert.

Iranian Desert.—The arid parts of Iran, Afghanistan and Pakistan form one of the world's smaller deserts. It is an area that receives winter precipitation and experiences very cold winters and summers that range from warm to hot (e.g., McGinnies, 1968). This desert includes the largest salt desert in the world, an area practically devoid of vegetation that is found in Iran. Outside of this essentially abiotic region, however, the Iranian Desert supports diverse but sparse shrub communities, with more complex vegetation in seepage areas or other mesic microhabitats (Breckle, 1983) (Fig. 34). The hottest portions of the Iranian Desert are among the hottest on earth, with summer air temperatures exceeding 54°C (McGinnies, 1968). Substrates vary from sand to gravel or rock; some portions have extensive moving dunes.

The mammals of this region have been examined by Brown (1980), DeBlase (1980), Gaisler (1975), Hassinger (1973), Lay

(1967), Mares (1980), Misonne (1959), Neronov (1976), Niethammer (1965), Roberts (1977) and Siddiqi (1961). As in the Sahara and Arabian deserts, hedgehogs and shrews are present (in fact, the same genera occur in xeric areas of both deserts). Unlike the western deserts, however, hyraxes are absent, as are several groups of rodents, such as ctenodactylids and spalacids. Sciurids (*Spermophilopsis*) occur over a small portion of this desert in northern Afghanistan and Iran. Hares and porcupines are present and shared with deserts found to the west, as are the gerbilline genera *Gerbillus, Meriones, Rhombomys* and *Tatera,* the dipodids *Jaculus* and *Allactaga,* and numerous murid and cricetid genera.

A few genera of rodents found in this desert are not common to the Arabian or North African deserts, however. The cricetid, *Calomyscus bailwardi* (the long-tailed hamster), is found on barren, rocky hillsides and feeds on seeds and other plant parts (Lay, 1967; Roberts, 1977), although Mares (1980) suggested that this species may also consume insects. The gray hamster, *Cricetulus migratorius,* is also a seed-gathering small rodent that inhabits upland rocky habitats in parts of this region (Hassinger, 1973); gray hamsters occasionally take animal matter (Harrison, 1972). In the Dipodidae, the dwarf jerboa, *Salpingotus michaelis,* is a tiny (head–body 44 mm) biped limited to sand dunes, gravel plains, and sandy flats in western Pakistan. Roberts (1977) found that this tufted-tailed desert mouse is colonial and inhabits areas where many burrows are placed. He suggested that the animals are granivorous and herbivorous, with an apparent dislike for insect matter. Another species, *S. crassicaudata* of the U.S.S.R., was reported to be mainly insectivorous (Naumov and Lobachev, 1975). Regardless of its dietary specialization, however, this diminutive dipodid is similar in many characteristics to heteromyids (see also Berman, 1979; Mares, 1980). Another dipodid found in the Iranian Desert is the lesser five-toed jerboa, *Alactagulus pumilio.* This is also a biped with a long tufted white-tipped tail. The animals store food in burrows and are herbivores that forage on roots, tubers, bulbs and green vegetation (Naumov and Lobachev, 1975; Nowak and Paradiso, 1983). Although not a granivore, this species is very similar morphologically to *Dipodomys* (Berman, 1979, 1985; Mares, 1980). Finally, in some xeric parts of the Iranian Desert, *Rattus* (=*Mus*) *gleadowi* are common. This murid (90 mm head–body length) is colored like a desert gerbil, being a light sandy color above and white below. It is primarily granivorous and is found in shifting dune areas of Pakistan (Roberts, 1977); it is probably independent of free water.

Thar Desert

The Thar Desert is a relatively small arid region located in northwestern India and eastern Pakistan between about 24° to 32°N latitude and 67° to 75°E longitude (e.g., Gupta, 1986; Mani, 1974*a*; Petrov, 1976) (Fig. 32). The Thar contains several major physiographic regions including a sand "sea" (the Thar), a zone of salt lakes (the Pat), an extensive plain (Ghaggar Plain) and the desert of the Rajasthan upland plateau (steppe desert) (Gupta, 1986). Vegetation is extremely sparse in sand dune areas and generally of a low scrub nature in the plain (Fig. 35). Rainfall occurs primarily (89%) in a summer (June–September) monsoon season, with the remaining precipitation coming in winter (Gupta, 1986).

There is a great deal of information available on mammals in India (e.g., Chakraborty, 1984; Ellerman, 1947; Ghosh, 1975; Kurup, 1986; Prakash, 1962, 1963, 1964, 1968, 1975*a,* 1975*b,* 1981), while information on Pakistani mammals is given in Roberts (1977). The mammal fauna of the Thar Desert is rather depauperate due to the geological history of the region.

India was originally a part of the Gondwanic supercontinent, breaking away from

FIG. 35.—Small dune areas in the Thar Desert near Jodhpur, Rajasthan, India. (Photo: M. A. Mares)

that land mass in the Early Cretaceous and colliding with the Asian land mass about 53 million years ago (e.g., Crawford, 1974; Kurup, 1986; Molnar and Tapponnier, 1975; Norton and Sclater, 1979; Sahni, 1984); the Himalayan Chain is a result of that tectonic event. Because of the isolation of the subcontinent during much of the Cenozoic, the present-day fauna of India includes elements from several biogeographic regions (Kurup, 1974); the arid western portion of India and adjacent Pakistan was colonized primarily by Ethiopian and Palaearctic elements (Mani, 1974*b*). The Thar Desert is not sharply distinguishable taxonomically (either botanically or zoologically) from those deserts located further to the west, although a few Oriental and southern Indian forms are found in the Thar (Mani, 1974*b*, 1974*c*). The relatively late arrival of mammals in the Thar Desert and the formidable barriers they had to cross in order to enter the arid region (e.g., the Kirthar Range, the Himalayas and Hindu Kush), coupled with the relatively recent occurrence of an extensive zone of aridity in the region of the present-day Thar Desert (Gupta, 1986; Prakash, 1974; Wadia, 1960, 1966), resulted in a fauna that is rather impoverished (Prakash, 1975*a*; Roberts, 1977). No bipedal species are present, and all desert-inhabiting genera are shared with the Iranian, Arabian and Sahara deserts to the west (Appendix 1). *Gerbillus* is the major genus containing granivorous, seed-hoarding species, while *Tatera*, *Mus* and *Rattus* are also common desert inhabitants, as is the porcupine (*Hystrix*). *Meriones hurrianae*, the Indian Desert gerbil, is an omnivorous species that is quite resistant to dessication but not water independent (Ghosh et al., 1962; see also Goyal, 1988) and takes much green plant matter, as well as invertebrate food items, in addition to many seeds; *Tatera* also consumes many seeds (Prakash, 1981), and has a pronounced ability to concentrate urine (Goyal, 1988). Among other small to midsize mammals in the Thar are *Lepus nigricollis*, hedgehogs (*Hemiechinus*), shrews (*Suncus*) and pangolins (Pholidota: *Manis*) (Prakash, 1963). Only *Gerbillus gleadowi* and *G. nanus* are good candidates for convergence with perognathines; no species is the equivalent of a dipodomyine.

Central Asian Deserts and Semideserts

Taken together, the deserts and semideserts of the U.S.S.R. and China cover an enormous land area extending from near 55° to about 38°N latitude and from near 40° to 120°E longitude (Meigs, 1953; see also Petrov, 1976; Walter and Box, 1983*a*, 1983*b*, 1983*c*, 1983*d*, 1983*e*; Walter et al., 1983) (Fig. 36). As might be expected, these deserts are composed of diverse landforms and substrates. The Taklamakan Desert is a huge sand desert (about 270,000 square kilometers) surrounded by mountain ranges; in some areas soils may be gravelly, rocky, or clayey. The Alashan Desert, forming the eastern limit of the Asian xeric region, consists of moving dune areas, scattered hills and, in the north, extensive plains with a sandy-pebbly surface. These plains are termed gobis and give the Gobi Desert its name (Fig. 37). The Gobi Desert is the name commonly applied to the Chinese-Mongolian desert region that includes the Alashan, Tsaidam, Gachoun Gobi, and the colder semidesert areas of Mongolia and China (e.g., Dzungaria) (see Meigs, 1953; Petrov, 1976).

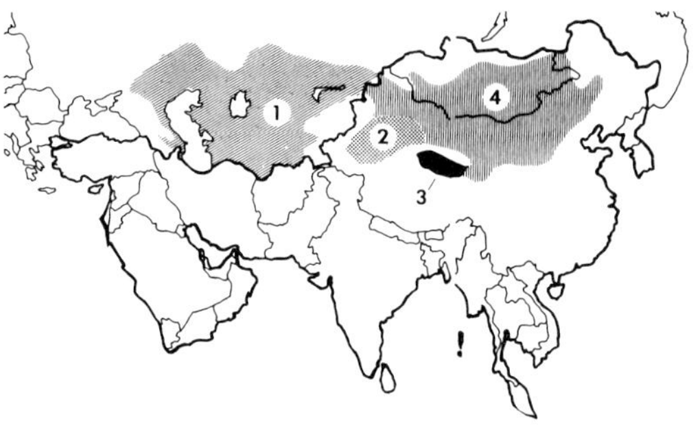

FIG. 36.—The arid and semiarid areas of Asia: 1) Turkestan Desert, 2) Takla Makan Desert, 3) Tsaidam Desert, 4) Gobi Desert.

Most areas in the Gobi Desert receive less than 200 mm annual precipitation; elevation varies, but over much of the zone it is between 600–1,800 m. Drainage from the region is poor, but little surface water occurs; ground water, however, is abundant. Vegetation over much of the Gobi and associated deserts is extremely sparse (McGinnies, 1968; Petrov, 1976; Richardson, 1966). Since the region being considered is so extensive, it might be expected that climate in the various deserts would vary greatly. Indeed, in the northern Gobi it is temperate, arid and generally cool. The Dzungaro-Kazakhstan area is less arid and less continental, with cool wet springs. Generally, the deserts of China are more biologically arid than those of the U.S.S.R., when potential soil moisture is included in the calculations of aridity (Oberlander, 1979). This fact is due to winter or spring precipitation occurring when evapotranspiration rates are lower in the deserts of the U.S.S.R.

In order to give a more detailed view of these deserts, I will first discuss the westernmost xeric regions lying principally within the U.S.S.R., followed by a more in-depth discussion of the deserts of China and the intervening arid areas of the high mountains located within and around this vast arid zone.

Turkestan Desert.—I here include most arid and semiarid areas west of about 75°E longitude within the Turkestan Desert. This encompasses the Caspian lowlands, a semidesert shrub steppe supporting many halophytes and receiving significant winter precipitation in the form of snow (Walter and Box, 1983*a*). General vegetative physiognomy is that of a sparsely vegetated low scrubland. Farther east is the desert and semidesert zone of Kazakhstan. Aridity increases from west to east across this region and vegetative patterns reflect this gradient as shrub steppe gives way to desert vegetation, with many grasses predominating in the former and halophytes common in the latter. South of Kazakhstan are the northern parts of the Irano-Turanian desert, which extends through the lower elevations eastward to the mountains of Middle Asia (Tien Shan and Pamiro-Alay ranges) and includes the Kyzylkum and Karakum deserts, both

FIG. 37.—The Gobi Desert between Turpan and Urumchi in Xinjiang (northwest China). Photo taken at near sea level looking toward the Bogda Feng (5,445 m) in the Tien Shan. (Photo: A. T. Smith)

of which are sand deserts. Winter rainfall predominates across this region, with heavy snowfall occurring occasionally; summers are hot. In some areas along flood plains, forests of *Populus* may appear, with trees reaching 7 m in height.

The Karakum Desert is one of the best-studied Old World deserts and includes about 350,000 square kilometers of xeric habitats (Kashkarov and Kurbatov, 1930; Walter and Box, 1983*e*). In many respects it is typical of the entire Turkestan Desert, for it contains a complex array of habitats, including extensive dune areas, riparian forests, scrub deserts, and saline flats. As noted above, Ranck (1968) suggested that there were few zoogeographic differences between the North African, Iranian and Turkestan deserts. Walter and Box (1983*e*) summarize information on the mammals of a portion of the Karakum (see also Stalmakova, 1954, 1955). Additional information is available in Naumov and Lobachev (1975).

The broad area encompassed by the Turkestan Desert supports a rather diverse fauna at the generic level, when the entire arid and semiarid region is considered (Appendix 1). There are a number of generic differences with the Middle East-Pakistan band of aridity, however, Ranck's (1968) comments notwithstanding. Among insectivores, *Hemiechinus* and *Crocidura* are shared with other Old World deserts, but the shrew, *Diplomesodon pulchellum,* is a monospecific taxon endemic to this desert. *Lepus, Spermophilopsis, Cricetulus, Meriones, Rhombomys* (see Dubyanskaya, 1989, and references therein for ecological information on this genus), *Jaculus, Allactaga, Salpingotus* and *Hystrix* are genera also shared with deserts to the south. Among those rodent genera found in the arid or semiarid portions of this desert are a number that are not shared with the more southern arid areas, however. *Lagurus* and *Microtus* are two genera of arvicolid grazing herbivores that occur in semiarid steppes. The desert dormouse family (Seleviniidae: *Selevinia betpakdalensis*) is monotypic and endemic to the Turkestan Desert. It is insectivorous and fills the role of small-bodied (24 g) insectivore.

The remaining genera that are common in this desert, but are not shared with deserts to the south, include a cricetid, *Phodopus,* and several dipodids. *Phodopus,* the cricetid dwarf hamster (Fig. 38), is a seed-hoarding granivore of small body size (50–102 mm head–body length, Nowak and Paradiso, 1983). Possible competitive interactions between species in this genus have been described (Hamann, 1987). The dipodids, *Dipus, Paradipus, Eremodipus, Stylodipus* and *Pygeretmus,* are all broadly foraging bipedal herbivores. *Cardiocranius* is a tiny (head–body length 50–75 mm) bipedal dipodid that is granivorous. I could find no information as to whether it stores seeds, but it accumulates large amounts of body fat in summer and hibernates (Naumov and Lobachev, 1975). It appears to be similar to *Microdipodops* in a number of ecological and morphological respects. Thus, this desert supports several genera and many species that are morphologically convergent on the dipodomyines of North America, but few species that are trophically similar. Moreover, most species that are trophically like dipodomyines are quadrupedal, with the exception of *Cardiocranius. Jaculus* and *Cardiocranius* may be considered loose equivalents of *Dipodomys* and *Microdipodops,* respectively, while *Meriones, Cricetulus* and *Phodopus* are broadly similar to perognathines.

Deserts of China and Mongolia.—Several desert areas, largely basins encircled by massive mountain ranges, together form the enormous cold desert of China and Mongolia. To the lowland basin deserts must be added the vast montane deserts, or xeric orobiomes, of the mountainous regions—high altitude, semiarid shrub steppe habitats. Like most extensive desert zones having pronounced topographic relief, the habitats of the China-Mongolia arid region vary from sand deserts to gravelly plateaus to high elevation rock deserts (e.g., Walter et al., 1983). This entire region is situated mainly between 75° and 120°E longitude.

The orobiomes of Asia extend to over 7,000 m elevation, with vegetation in the region being sparse, but varied (Walter and Box, 1983*f*). In some areas ephemerals predominate, with the perennial shrub *Artemisia* also being present. At higher elevations (1,600–1,800 m) grasses are common, while above this level *Juniperus* and cushion plants are found. This pattern occurs in the western limits of the Asian orobiomes. The Tien Shan ranges extend from western Asia eastward to the western limits of the Gobi Desert in middle Asia. Various forests, from broadleaf to evergreen, are found at lower to middle elevations, especially below 3,000 m. In some areas below 2,000 m elevation, a semidesert belt of aridity is common and supports chenopodiaceous shrubs. At the highest elevations, the Asian orobiomes support xeric cryophilic plants. Extensive information on xeric habitats of one of the Asian orobiomes, the Pamir, is available in Walter and Box (1983*g*). In the Pamir, desert and semidesert vegetation, largely consisting of sparse grass and shrubs, may extend above 5,000 m elevation. Small mammals in the xeric parts of the Asian orobiomes are not especially abundant and are similar to those found in the arid mountains of Iran, Afghanistan, and Pakistan (see earlier discussion). No good analogues of heteromyids inhabit the high deserts.

Walter et al. (1983) note that during the Early Tertiary the Sahara and Central Asia formed the major desert areas of Eurasia. With the disappearance of the Tethys Sea in the Late Tertiary, some floral (and probably faunal) mixing occurred between these two areas, but the eastern portion of the Central Asian desert, the Gobi, retained many of its endemic elements. Each of the major substrate deserts in the Gobi Desert (sand deserts, gravel deserts, salt plains, and rock-rubble deserts) supports very distinctive floral elements that are limited to that general region (Walter et al., 1983). The "true" Gobi Desert has its northern limits near 46°N latitude in the west and 43°N latitude in the east, with elevation varying from 700–1,300 m. Aridity is greater in the

FIG. 38.—The dwarf hamster, *Phodopus sungorus*, of the Turkestan, Chinese and Mongolian deserts, a small, quadrupedal, seed-hoarding obligate granivore. (Photo: M. Andera)

west, where mean yearly rainfall may be as low as 40 mm, with some years having no rain; in the east, yearly precipitation may reach 150 mm. Vegetation over much of the region is sparse grass-shrub steppe, with Chenopodiaceae, Tamaricaceae, Zygophyllaceae, Fabaceae, Polygonaceae, Convolvulaceae and Asteraceae being the predominant plant families. Numerous forbs and grasses are also present, including many summer ephemerals which produce important seed crops. Walter et al. (1983) present an extensive discussion of plant ecology of the Gobi Desert and its adjoining arid and semiarid areas. They note that in some of the Gobi's subunits, such as the Ala Shan and Tien Shan deserts, or in the Tarim Basin (Takla Makan Desert), isolated patches of trees (*Populus, Eleagnus, Haloxylon*) may appear. These may be quite large in surface depressions or other areas where soil moisture accumulates.

Faunistically, the Gobi Desert shares several elements with the Turkestan Desert (Appendix 1), although there are a number of mammal genera found in the Gobi that are endemic. Among genera (and species) shared with the western Asian deserts are the insectivores *Hemiechinus* and *Crocidura*, the leporid *Lepus*, the sciurid *Spermophilus*, the cricetids *Meriones, Rhombomys* and *Cricetulus*, the arvicolids *Microtus* and *Lagurus*, and the dipodids *Dipus, Stylodipus, Allactaga, Cardiocranius*, and *Salpingotus*. The cricetid dwarf hamster *Phodopus* is also shared with the Turkestan Desert and is a tiny rodent that is strongly granivorous (Nowak and Paradiso, 1983). Genera that are endemic to the Gobi and its associated deserts include the cricetid, *Brachiones,* a gerbil that has not been well studied, but may be either a granivore or an herbivore. The arvicolid mole vole, *Ellobius,* is found within arid and semiarid parts of the Gobi that support moist vegetation in localized areas such as the mesic vegetation found along riverbanks. The only dipodid that is endemic to the Gobi is the long-eared jerboa, *Euchoreutes naso,* that resembles *Allactaga* externally, but has much longer ears. Its biology is unknown.

Like the deserts located to the west in the U.S.S.R. and adjacent countries, the Gobi supports a rich small mammal fauna, but only a few species are ecological equivalents of the North American heteromyids. *Phodopus* and *Cricetulus* are seed-hoarding quadrupedal granivores well adapted to aridity (e.g., Schierwater and Klingel, 1985), while the various bipedal species inhabiting the Gobi Desert are probably either herbivorous or rather omnivorous.

General Discussion

The small mammals of the world's deserts and semideserts represent nine orders, 41 families, and 164 genera (Appendix 2). Each group has colonized these arid areas at different times and under differing geographic and climatological conditions. The deserts themselves have geological, climatological and floristic histories that often differ from one another. Yet there are several similarities evident among various deserts. Most xeric areas have been subjected to alternating periods of cool, wet times and warm, dry times. These climatic oscillations have had a number of effects on both their flora and fauna. In arid areas having broad areal extent and containing considerable topographic diversity within the xeric region itself, such as the deserts of western North America, the desert was broken up into arid refugia with intervening forest vegetation during mesic times (e.g., Van Devender and Spaulding, 1979). These refugia may have influenced both extinction rates of larger mammals and speciation rates of smaller mammals (e.g., Hubbard, 1974; Mares, 1979).

The Sahara (Le Houérou, 1986) and Eurasian deserts (Frenzel, 1968) show patterns broadly similar to that of North America, while the South American arid areas, being either long or narrow (Atacama, Monte) or lacking extensive topographic diversity (e.g., Chaco, Caatinga), do not support elevated generic diversity (e.g., Mares, 1975*a,* 1980; Mares et al., 1985*a,* 1985*b*). The Atacama-Peruvian system of Chile and Peru is a fairly old desert, but has been insulated from colonization by the enormous geographic barrier of the Andean Mountains. Its equivalent desert, the Namib, has been more readily colonized due to its proximity to adjacent arid and semiarid zones and a lack of effective geographic barriers. Consequently its overall diversity is much higher than that found along the arid western coast of South America.

Australia, which has been arid in some portions since the Miocene (Specht, 1981*b*), also fits this pattern if diversity of rodents is examined. When desert marsupials (largely insectivorous) are included in the species numbers found in desert areas (Morton, 1979; Morton and Baynes, 1985), species diversity is roughly comparable to other desert areas (although recent human-induced extinctions must be included in the totals). In general, however, species diversity of small mammals in xeric habitats of Australia is lower than what is found in the extensive Holarctic arid zones, but much higher than what is found in South American arid areas. Pianka (1972) has hypothesized that the very high species richness of lizards in Australia evolved through habitat disruption, with particular substrate/floral groupings becoming isolated for extended periods and promoting speciation (see also Kershaw, 1981). This may help account for the high small mammal diversity as well, although it must be remembered that the continent was colonized by rodents during several periods since the Early Pliocene

TABLE 1.—*Bipedal small mammals and their general food habits in the deserts and semideserts of the world.*

	Bipedal			
Desert	Granivore	Herbivore	Insectivore	Omnivore
North America	*Microdipodops, Dipodomys*	None	None	None
South America	argyrolagids*	None	None	None
Australia	*Notomys*	various marsupials	*Sminthopsis*	None
Sahara	*Jaculus*	*Allactaga*	*Elephantulus*	None
Southern Africa	None	*Pedetes*	*Elephantulus, Macroscelides*	None
Madagascar	None	None	None	None
Middle East-Thar	*Jaculus*	various dipodids	None	*Salpingotus*
Turkestan	*Cardiocranius*	various dipodids	None	*Pygeretmus, Salpingotus, Dipus*
Chinese/Mongolian	*Cardiocranius*	various dipodids	None	*Salpingotus, Dipus*

* Extinct.

(Archer, 1981; Watts and Aslin, 1981). This pulsated colonization could also have led to an increase in species numbers.

Species diversity and taxonomic makeup of many of the world's desert mammal faunas differ, but there are a number of similarities evident among the adaptive niche types of the various arid areas (Appendix 1). These are more succinctly summarized in Tables 1–4. Bipedality among small mammals, which has long been associated with sparse desert habitats, is indeed ubiquitous among deserts. When Pleistocene extinctions are included, no arid habitat in the world has lacked bipedal species. However, the popular belief that bipedality and granivory are generally associated is not tenable (Table 1). Only in North America, among all of the world's deserts, do we find a diverse group of bipedal granivores (*Microdipodops, Dipodomys*). From West Africa to the Middle East and in the deserts of Eurasia, a region encompassing about 22 million square kilometers, only about two species (in two genera) of bipedal granivores are found, and neither of these may be an obligate granivore. In Australia, one genus of bipedal granivore has evolved. Much more common in the world's deserts are bipedal herbivores, insectivores and omnivores. Indeed, when viewed globally, the most common bipeds in deserts by far are herbivorous species foraging on above- and belowground plant parts. Such adaptive types occur in all deserts except those found in the New World (and the extinct marsupial argyrolagids of the South American Monte Desert may have been somewhat to moderately herbivorous, G. G. Simpson, 1970). Clearly, ideas associating bipedality and granivory have resulted from the fact that *Dipodomys* and *Microdipodops,* bipedal granivores, are found in the world's most intensively studied desert. The assumption that patterns discerned in North American deserts can be extrapolated to other deserts is quite understandable, especially when other deserts support species that strongly resemble dipodomyines in external morphology. Nevertheless, investigations that formulate hypotheses dealing with patterns of foraging of bipeds on seeds should keep in mind the fact that only a small percentage of bipedal desert species forage on seeds; most desert granivores are quadrupedal species of relatively small body size (Table 2). Bipedality in sparse desert habitats may be more a response to predation pressure than to foraging strategy (e.g., Djawdan and Garland, 1988), although foraging in open microhabitats on widely spaced resources could make bipedality an evolutionary response

TABLE 2.—*Quadrupedal small mammals and their general food habits in the deserts and semideserts of the world.*

Desert	Smaller quadrupedal			
	Granivore	Herbivore	Insectivore	Omnivore
North America	perognathines	various rodents	*Notiosorex, Onychomys*	various rodents
South America	None	various rodents	*Marmosa, Lestodelphis*	various rodents
Australia	*Pseudomys*	various rodents	various marsupials	None
Sahara	gerbillines	*Psammomys*	*Crocidura*	various rodents
Southern Africa	gerbillines, *Saccostomus*	various rodents	*Crocidura*	various rodents
Madagascar	None	*Hypogeomys, Macrotarsomys*	*Geogale*	*Echinops*
Middle East-Thar	gerbillines, *Cricetulus*	*Rhombomys*	soricids, erinaceids	various rodents
Turkestan	*Meriones, Cricetulus, Phodopus*	various rodents	*Crocidura, Diplomesodon, Hemiechinus*	None
Chinese/Mongolian	*Meriones, Cricetulus, Phodopus*	various rodents	*Hemiechinus, Crocidura*	*Spermophilus*

to an interaction between spatial resource distribution, vegetation cover and predation pressures (cf., Berman, 1979; Brown and Harney, this volume; Eisenberg, 1975; Hafner, this volume; Harris, 1984, 1986; Howell, 1932; Kotler, 1984; Mares, 1980, 1983; Price, 1978; Price and Brown, 1983; Price and Reichman, 1987; Reichman and Price, this volume; Schröpfer and Klemer-Fringes, 1985).

Quadrupedal species of small mammals are much more common in the world's deserts than are bipedal ones (Table 2); both generic and species diversity of quadrupeds is high in all major trophic categories. Only South America and Madagascar lack quadrupedal granivores. Interestingly, although most desert granivorous mammals are quadrupedal, most quadrupedal desert mammals forage on foods other than seeds, with the most common food niche type being broadly foraging herbivores and omnivores. All of the world's major arid zones support small quadrupedal herbivores and insectivorous forms, and most deserts support a variety of small quadrupedal omnivores.

Various herbivorous genera from deserts throughout the world are listed in Table 3. Clearly, the leporid niche is one of the most ubiquitous found in deserts. Australia's cursorial herbivores happen to be bipedal marsupials, but their niche is broadly similar to that filled by leporids in most deserts, by caviids in South America, and possibly by a cricetid, *Hypogeomys,* in Madagascar. Saxicolous species and slow-moving herbivores such as *Hystrix* occur in some deserts, while fossorial species are found in deserts that lack bipedal root and tuber feeders. Thus most bipedal dipodids fill a trophic niche similar to geomyids of North America, ctenomyids of South America, spalacids of parts of North Africa, and bathyergids of southern Africa. These various phylogenetically unrelated fossorial groups provide a classic example of convergent evolution in morphology and general ecology. Curiously, the evolution of a "gopher" type of animal in the extensive deserts of Eurasia may have been prevented by the evolution and diversification of dipodids, bipedal rodents that forage heavily on underground plant parts. Only localized *Ellobius* are general ecological equivalents of gophers in Chinese deserts.

Several additional niche types occur with

TABLE 3.—*Various types of herbivores (excluding small quadrupedal and bipedal forms) found in the deserts and semideserts of the world.*

Desert	Other herbivore categories			
	Cursorial	Larger-bodied	Saxicolous	Fossorial
North America	leporids	*Erethizon*	None	geomyids
South America	*Dolichotis, Pediolagus*	*Lagostomus*	*Kerodon, Lagidium*	ctenomyids
Australia	various marsupials	None	None	None
Sahara	leporids	*Hystrix*	ctenodactylids, procaviids	*Spalax*
Southern Africa	leporids	None	procaviids, *Pronolagus, Petromus*	bathyergids
Madagascar	*Hypogeomys*	None	None	None
Middle East-Thar	leporids	*Hystrix*	None	None
Turkestan	leporids	*Hystrix*	None	None
Chinese/Mongolian	leporids	None	None	*Ellobius*

regularity in the deserts of the world (Table 4). One of the most common is what I termed the squirrel niche—terrestrial or scansorial medium-sized diurnal omnivores. In most deserts, sciurids fill this niche, but in South America caviid rodents seem to be loose equivalents of squirrels, whereas in parts of North Africa the glirid, *Eliomys,* is similar in many ways to squirrels. Australia and Madagascar lack such equivalents. Species that are broadly similar to woodrats are found in several deserts, as are fossorial insectivores and larger-bodied omnivores. These latter three niche types are of only sporadic occurrence, however.

In several deserts there are small mammals that appear to fill a role that is unique to that particular desert (Appendix 1). *Malacothrix typica,* for example, is a very small (7–13 g) southern African herbivorous rodent that feeds almost entirely on green vegetation. Generally, desert rodents of such small body size are either granivorous or omnivorous; the strictly herbivorous diet of this species is unusual and makes it unique among desert species. Similarly, *Rhombomys opimus,* the great gerbil of Eurasian deserts, is unique in that it is a diurnal herbivore that is also a food hoarder (up to 60 kg of food may be stored in a single burrow, Nowak and Paradiso, 1983); surface storage of plant food is also common in this species. Several of these "unusual" niche types (unusual because they are different from those found in North American deserts) are apparent in Appendix 1 (see footnotes).

In this very broad overview of small mammal ecology and evolution in deserts, it is clear that each type of small mammal occurring in a desert has had to be compared to some loosely-defined niche type. I have chosen to compare all genera (and occassionally species) to those ecological and morphological types found in North American deserts. I have done this for several reasons, but these reasons are germane to further discussion of patterns.

The North American deserts are by far the best-studied in the world. Many thousands of papers have been published on various aspects of the fauna that is found in this area. Because of this massive research literature, other deserts have often been viewed with a North American bias. North American deserts are rich in species, generally semiarid, rather than arid, and frequently support great numbers of both species and individuals in any one locality. Also, the diversity of niche types represented in these deserts is broad. By comparing these well-studied mammals to those of other deserts, I assumed it would be simpler to examine possible ecological equivalence. Moreover, by assigning small mammals

TABLE 4.—*Several other common niche types that are regularly found in the deserts and semideserts of the world.*

Desert	Other niche types			
	Woodrat niche	Squirrel niche	Fossorial insectivore	Larger omnivore
North America	*Neotoma*	sciurids	None	None
South America	*Abrocoma*, octodontids, *Thrichomys*, *Chinchillula*	caviids	*Chlamyphorus*	armadillos
Australia	various rodents	None	*Notoryctes*	various marsupials
Sahara	*Sekeetamys*	sciurids, *Eliomys*	None	None
Southern Africa	*Paratomys*	*Xerus*	*Eremitalpa*	None
Madagascar	None	None	None	*Echinops*
Middle East-Thar	*Sekeetamys*	*Spermophilus*	None	None
Turkestan	None	*Spermophilus*	None	None
Chinese/Mongolian	None	*Spermophilus*	None	None

from other deserts to the North American niche types, I surmised that it would make these lesser-known species more understandable to North American ecologists. Since this volume is dedicated to heteromyids, it seemed logical to place my review in a North American perspective. Because of the complexity of the adaptive types evidenced by small mammals in the deserts of the world, however, I have had to expand the discussion well beyond the taxon of main concern in this volume. Nevertheless, it is my hope that it will help to point out the uniqueness of the heteromyid type (particularly the dipodomyine bipedal obligate seed-hoarding granivores). Both bipedalty and granivory are commonly found in deserts, but not in the way they are associated in North American deserts. This is perhaps the most surprising result of this research. Species possessing these adaptations, so commonly accepted and well-studied in North American deserts, are actually quite unusual when viewed from a global perspective. Had I chosen small mammal niche types common to the southern African or Asian deserts, for example, *Dipodomys* and *Microdipodops* would have appeared almost as unique as *Malacothrix*.

Clearly, as we learn more about the mammals that inhabit the world's deserts, we may find that the loose assignation of niche types employed herein is no longer tenable. Groups that now appear similar may, in fact, be shown to be quite different. There is utility in attempting to examine the broad picture, however. In a very real sense, the groupings of mammals under a particular niche type are predictions of equivalency. If our understanding of adaptation is limited only to patterns found in a single desert, it could be argued that our level of understanding is not very profound. Indeed, it may be that our theories of nature can only really be tested for robustness by comparing them to other faunas that have evolved to solve similar problems in sometimes similar, but frequently different, ways. If a theory is really predictive, it should not be limited by geographical, or even phylogenetic, boundaries. By restricting the scope of our research to a rather unusual desert system (North America), we may find that the questions we are asking are trivial when their ability to predict patterns beyond the limits of the North American desert system is examined. Ecology is often not viewed as a science because it lacks a formal body of powerful theories and is largely concerned with unrepeatable experiments that ask very limited questions on a few species which in themselves are subject to innumerable uncontrolled variables of nature. The many theories that have been generated from studies of North American desert species have generally not proven robust even within the deserts of the United States (but see Brown and Heske, 1990; and arguments and citations in Rosenzweig, 1987, 1989). Their predictive abilities in other deserts remain untested. We should possess the humility to recognize much current research as possibly being weak in its ability to predict patterns of foraging behavior, coexistence, adaptation or faunal development among desert species. Much research to date (with the exception of the rich literature on physiology and, possibly, adaptive morphology) is quite rudimentary. If we do not test these hypotheses on other faunas, they may remain as a very limited scientific view of ecology in one of the world's most unusual deserts.

Acknowledgments

I thank J. K. Braun for assisting with all aspects of this research. Hugh Genoways refused to accept my withdrawal of this paper from the volume when the deadline for submission (1986) appeared impossible to meet. J. F. Eisenberg, R. E. MacMillen, D. A. Schlitter, P. L. Meserve, D. P. Christian, A. T. Smith, and M. Andera very kindly provided photographs of desert habitats and animals. J. K. Braun produced the maps and R. Macedo drew the argyrolagid and the *Dipodomys* and *Jaculus* that introduced the chapter. D. A. Schlitter provided some hard-to-find reprints. O. J. Reichman and J. F. Eisenberg provided very helpful reviews of the manuscript, although they are not necessarily in agreement with all of the views expressed. Finally, I thank those researchers who have left the comfortable and well-studied areas of the world and ventured into regions that cry out for mammalogical research. Without them, this paper would have been impossible.

Literature Cited

Abd El Rahman, A. A. 1986. The deserts of the Arabian Peninsula. Pp. 29–54, *in* Ecosystems of the world. 12B. Hot deserts and arid shrublands (M. Evenari, I. Noy-Meir, and D. W. Goodall, eds.). Elsevier, New York, 451 pp.

Amiran, D. H. K., and A. W. Wilson (eds.). 1973. Coastal deserts. University of Arizona Press, Tucson, 207 pp.

Archer, M. 1981. A review of the origins and radiations of Australian mammals. Pp. 1435–1488, *in* Ecological biogeography of Australia (A. Keast, ed.). Junk, b.v., The Hague, The Netherlands, 3:1435–2142.

Atallah, S.I. 1977. The mammals of the eastern Mediterranean region: their ecology, systematics and zoogeographical relationship. Part I. Saugetierkundliche Mitteilungen, 25:241–320.

———. 1978. The mammals of the eastern Mediterranean region: their ecology, systematics and zoogeographical relationship. Part II. Saugetierkundliche Mitteilungen, 26:1–50.

Atallah, S. I., and D. L. Harrison. 1967. New records of rodents, bats and insectivores from the Arabian Peninsula. Journal of Zoology (London), 153: 311–319.

Axelrod, D.I. 1950. Evolution of desert vegetation in western North America. Carnegie Institute Publications, 590:217–306.

———. 1972. Ocean-floor spreading in relation to ecosystematic problems. Pp. 15–70, *in* A symposium on ecosystematics (R. T. Allen and F. C. James, eds.). Occasional Paper No. 4, University of Arkansas Museum, Fayetteville, Arkansas, 235 pp.

Ayyad, M. A., and S. I. Ghabbour. 1986. Hot deserts of Egypt and the Sudan. Pp. 149–202, *in* Ecosystems of the world. 12B. Hot deserts and arid shrublands (M. Evenari, I. Noy-Meir, and D. W. Goodall, eds.). Elsevier, New York, 451 pp.

Barbault, R., and G. Halfetter (eds.). 1981. Ecology of the Chihuahuan Desert. Instituo de Ecología, Mexico, D.F., 167 pp.

Baskin, J. A. 1978. *Bensonomys, Calomys,* and the origin of the phyllotine group of Neotropical cricetines (Rodentia, Cricetidae). Journal of Mammalogy, 59:125–135.

———. 1986. The late Miocene radiation of neotropical sigmodontine rodents in North America. Contributions to Geology, University of Wyoming, 3:287–303.

Baudinette, R. V. 1972. The impact of social aggregation on the respiratory physiology of Australian hopping mice. Comparative Biochemistry and Physiology, A. Comparative Physiology, 41:35–38.

Baverstock, P. R., L. Spencer, and C. Pollard. 1976. Water balance of small lactating rodents. II. Concentration and composition of milk of females on *ad libitum* and restricted water intakes. Comparative Biochemistry and Physiology, A. Comparative Physiology, 53:47–52.

Baverstock, P. R., and C. H. S. Watts. 1974. Water balance and kidney function in some rodents native to Australia. Laboratory Animal Science, 24:743–751.

———. 1975. Water balance of small lactating rodents. I. *Ad libitum* water intakes and effects of water restriction on growth of young. Comparative Biochemistry and Physiology, A. Comparative Physiology, 50:819–825.

BAVERSTOCK, P. R., C. H. S. WATTS, AND L. SPENCER. 1979. Water-balance of small lactating rodents. V. The total water-balance picture of the mother-young unit. Comparative Biochemistry and Physiology, A. Comparative Physiology, 63:247–252.

BEADLE, N. C. W. 1981. The vegetation of the arid zone. Pp. 695–716, *in* Ecological biogeography of Australia (A. Keast, ed.). Junk, b.v., The Hague, The Netherlands, 1:1–805.

BEARD, J. S. 1984. The vegetation of the Australian arid zone. Pp. 113–119, *in* Arid Australia (H. G. Cogger and E. E. Cameron, eds.). Australian Museum, Sydney, 338 pp.

BENDER, G. L. (ED.). 1982. Reference handbook on the deserts of North America. Greenwood Press, Westport, Connecticut, 594 pp.

BERMAN, S. L. 1979. Convergent evolution in the hind limb of bipedal rodents. Ph.D. dissertation, University of Pittsburgh, Pittsburgh, Pennsylvania, 302 pp.

———. 1985. Convergent evolution in the hindlimb of bipedal rodents. Sonderdruck aus Zeitschrift für zool. Systematik und Evolutionsforschung, 23:59–77.

BIGALKE, R. C. 1972. The contemporary mammal fauna of Africa. Pp. 141–194, *in* Evolution, mammals, and southern continents (A. Keast, F. C. Erk, and B. Glass, eds.). State University of New York Press, Albany, 543 pp.

———. 1978. Mammals. Pp. 981–1048, *in* Temperate deserts and semi-deserts (N. E. West, ed.). Elsevier, New York, 522 pp.

BLAIR, W. F., A. C. HULSE, AND M. A. MARES. 1976. Origins and affinities of vertebrates of the North American Sonoran Desert and the Monte Desert of northwestern Argentina. Journal of Biogeography, 3:1–18.

BORGEL, O. R. 1973. The coastal desert of Chile. Pp. 111–114, *in* Coastal deserts (D. H. K. Amiran and A. W. Wilson, eds.). University of Arizona Press, Tucson, 207 pp.

BOWERS, M. A., AND J. H. BROWN. 1982. Body size and coexistence in desert rodents: chance or community structure? Ecology, 62:391–400.

BOZINIC, F., AND L. C. CONTRERAS. 1990. Basal rate of metabolism and temperature regulation of two desert herbivorous octodontid rodents: *Octomys mimax* and *Tympanoctomys barrerae*. Oecologia, 84: 567–570.

BRADLEY, W. G., AND J. E. DEACON. 1971. The ecology of small mammals at Saratoga Springs, Death Valley National Monument, California. Journal of the Arizona Academy of Science, 6:206–215.

BRADLEY, W. G., AND R. A. MAUER. 1973. Rodents of a creosote bush community in southern Nevada. The Southwestern Naturalist, 17:333–344.

BRECKLE, S.-W. 1983. Temperate deserts and semi-deserts of Afghanistan and Iran. Pp. 271–319, *in* Ecosystems of the world. 5. Temperate deserts and semi-deserts (N. E. West, ed.). Elsevier, New York, 522 pp.

BROWN, G. W., JR. (ED.). 1968. Desert biology. Academic Press, New York, 1:1–635.

———. 1974. Desert biology. Academic Press, New York, 2:1–601.

BROWN, J. H. 1973. Species diversity of seed-eating rodents in sand dune habitats. Ecology, 54:775–787.

———. 1975. Geographical ecology of desert rodents. Pp. 315–341, *in* Ecology and evolution of communities (M. L. Cody and J. M. Diamond, eds.). Harvard University Press, Cambridge, Massachusetts, 545 pp.

BROWN, J. H., AND D. W. DAVIDSON. 1977. Competition between seed-eating rodents and ants in desert ecosystems. Science, 196:880–882.

BROWN, J. H., AND B. A. HARNEY. 1993. Population and community ecology of heteromyid rodents in temperate habitats. Pp. 618–651, *in* Biology of the Heteromyidae (H. H. Genoways and J. H. Brown, eds.). Special Publication, American Society of Mammalogists, 1:1–719.

BROWN, J. H., AND E. J. HESKE. 1990. Control of a desert-grassland transition by a keystone rodent guild. Science, 250:1705–1707.

BROWN, J. H., AND G. A. LIEBERMAN. 1973. Resource utilization and coexistence of seed-eating rodents in sand dune habitats. Ecology, 54:788–797.

BROWN, J. H., O. J. REICHMAN, AND D. W. DAVIDSON. 1979. Granivory in desert ecosystems. Annual Review of Ecology and Systematics, 10:201–227.

BROWN, R. E. 1980. Rodents of the Kavir National Park, Iran. Mammalia, 44:89–96.

BRYLSKI, P. V. 1993. Anatomy. Pp. 370–377, *in* Biology of the Heteromyidae (H. H. Genoways and J. H. Brown, eds.). Special Publication, American Society of Mammalogists, 10:1–719.

BUFFENSTEIN, R., AND J. U. M. JARVIS. 1985. Thermoregulation and metabolism in the smallest African gerbil, *Gerbillus pusillus*. Journal of Zoology (London), 205:107–121.

CABRERA, A. L. 1976. La vegetación de la Puna Argentina. Revista de Investigaciones Agrícolas, 11: 317–412.

CHAKRABORTY, S. 1984. Research on mammals of the Indian subcontinent by the scientists of the Zoological Survey of India. Zoological Survey of India, Calcutta, 35 pp.

CHEW, R. M. 1965. Water metabolism of mammals. Pp. 43–178, *in* Physiological mammalogy (W. V. Mayer and R. G. Van Gelder, eds.). Academic Press, New York, 2:1–326.

CHEW, R. M., AND A. E. CHEW. 1970. Energy relationships of the mammals of a desert shrub. Ecological Monographs, 40:1–21.

CHRISTIAN, D. P. 1978. Effects of humidity and body size on evaporative water loss in three desert rodents. Comparative Biochemistry and Physiology, A. Comparative Physiology, 60:425–430.

———. 1979*a*. Comparative demography of three Namib Desert rodents: responses to the provision of supplementary water. Journal of Mammalogy, 60: 679–690.

———. 1979*b*. Physiological correlates of demographic patterns in three sympatric Namib Desert rodents. Physiological Zoology, 52:329–339.

———. 1980*a*. Patterns of recovery from low numbers in Namib Desert rodents. Acta Theriologica, 25:431–450.

———. 1980*b*. Vegetative cover, water resources, and

microdistributional patterns in a desert rodent community. Journal of Animal Ecology, 49:807–816.

COCKRUM, E. L. The mammals of Tunisia. Unpublished manuscript.

COETZEE, C. G. 1969. The distribution of mammals in the Namib Desert and adjoining inland escarpment. Scientific Papers of the Namib Desert Research Station, 40:23–36.

COGGER, H. G., AND E. E. CAMERON (EDS.). 1984. Arid Australia. Australian Museum, Sydney, 338 pp.

CONRAD, G. 1971. Synthese de l'evolution continentale post-hercynienne du Sahara algerien. Bulletin du Service de la carte Geologique de l'Algerie, 41:143–159.

CONTRERAS, J. R., AND M. I. ROSI. 1980. Comportamiento territorial y fidelidad al habitat en una población de roedores del centro de la provincia de Mendoza. Ecología, 5:17–29.

CRAWFORD, A. R. 1974. The Indus suture line, the Himalaya, Tibet and Gondwanaland. Geological Magazine, 111:369–383.

DALY, M., AND S. DALY. 1973. On the feeding ecology of *Psammomys obesus* (Rodentia, Gerbillidae) in the Wadi Saoura, Algeria. Mammalia, 37:545–561.

DEBLASE, A. F. 1980. The bats of Iran: systematics, distribution, ecology. Field Columbian Museum Publications, Zoology Series, 4:1–424.

DEGEN, A. A., M. KAM, AND D. JURGRAU. 1988. Energy requirements of fat sand rats (*Psammomys obesus*) and their efficiency of utilization of the saltbush *Atriplex halimus* for maintenance. Journal of Zoology (London), 215:443–452.

DJAWDAN, M., AND T. GARLAND, JR. 1988. Maximal running speeds of bipedal and quadrupedal rodents. Journal of Mammalogy, 69:765–772.

DUBYANSKAYA, L. D. 1989. Effect of weather conditions from the previous year on winter survival of greater gerbil. The Soviet Journal of Ecology, 20: 360–367. [Translated from Russian.]

EISENBERG, J. F. 1963. The behavior of heteromyid rodents. University of California Publications in Zoology, 69:1–100.

———. 1967. A comparative study in rodent ethology with emphasis on evolution of social behavior. I. Proceedings of the United States National Museum, 122:1–51.

———. 1975. The behavior patterns of desert rodents. Pp. 189–224, *in* Rodents in desert environments (I. Prakash and P. K. Ghosh, eds.). Junk, b.v., The Hague, The Netherlands, 624 pp.

EISENBERG, J. F., AND E. GOULD. 1970. The tenrecs: a study in mammalian behavior and evolution. Smithsonian Contributions to Zoology, 27:1–137.

ELLERMAN, J. R. 1947. A key to the Rodentia inhabiting India, Ceylon and Burma, based on collections in the British Museum. Journal of Mammalogy, 28: 249–278; 357–386.

EVENARI, M., I. NOY-MEIR, AND D. W. GOODALL (EDS.). 1985. Ecosystems of the world. 12A. Hot deserts and arid shrublands. Elsevier, New York, 365 pp.

———. (EDS.). 1986. Ecosystems of the world. 12B. Hot deserts and arid shrublands. Elsevier, New York, 451 pp.

EVENARI, M., L. SHANAN, AND N. TADMOR. 1971. The Negev. Harvard University Press, Cambridge, Massachusetts, 345 pp.

FISCHER, G. M. 1978. Los pequeños mamíferos de Chile. Gayana, 40:1–342.

FRENZEL, B. 1968. The Pleistocene vegetation of northern Eurasia. Science, 161:637–649.

FROTA-PESSOA, O., ET AL. 1971. Biologia Nordeste. I. Ecologia e taxonomia. Universidade do Federal de Pernambuco, Recife, Brazil.

FULK, G. W. 1975. Population ecology of rodents in the semiarid shrublands of Chile. Occasional Papers, The Museum, Texas Tech University, 33:1–40.

GAISLER, J. 1975. Comparative ecological notes on Afghan rodents. Pp. 59–74, *in* Rodents in desert environments (I. Prakash and P. K. Ghosh, eds.). Junk, b.v., The Hague, The Netherlands, 624 pp.

GHOBRIAL, L. I., AND T. A. NOUR. 1975. The physiological adaptations of desert rodents. Pp. 413–444, *in* Rodents in desert environments (I. Prakash and P. K. Ghosh, eds.). Junk, b.v., The Hague, The Netherlands, 624 pp.

GHOSH, P. K. 1975. Thermo-regulation and water economy in Indian desert rodents. Pp. 397–412, *in* Rodents in desert environments (I. Prakash and P. K. Ghosh, eds.). Junk, b.v., The Hague, The Netherlands, 624 pp.

GHOSH, P. K., K. G. PUROHIT, AND I. PRAKASH. 1962. Studies on the effects of prolonged water deprivation on the Indian Desert Gerbil, *Meriones hurrianae*. Pp. 301–306, *in* Environmental physiology and psychology in arid conditions. Proceedings of the Lucknow Symposium, UNESCO, Paris.

GOODALL, D. W. (ED.). 1976. Evolution of desert biota. University of Texas Press, Austin, 250 pp.

GOODIN, J. R., AND D. K. NORTHINGTON (EDS.). 1985. Plant resources of arid and semiarid lands. Academic Press, New York, 338 pp.

GOUAT, P., AND J. GOUAT. 1987. Le répertoire comportemental du goundi *Ctenodactylus gundi* (Rongeurs, Ctenodactylidae). I. Description. Mammalia, 51:3–26.

GOYAL, S. P. 1988. Field urine concentration in two India gerbils. Journal of Mammalogy, 69:418–421.

GREER, J. K. 1965. Mammals of Malleco Province, Chile. Publications of the Museum, Michigan State University, Biological Series, 3:51–151.

GRENOT, C. 1973. Sur la biologie d'un Rongeur heliophile du Sahara, le "Goundi" (Ctenodactylidae). Acta Tropica, 30:237–250.

GRENOT, C. J. 1974. Physical and vegetational aspects of the Sahara Desert. Pp. 103–164, *in* Desert biology (G. W. Brown, Jr., ed.). Academic Press, New York, 2:1–601.

———. 1983. Desierto Chihuahuense: fauna de Bolsón de Mapimi. Departamento de Zonas Aridas, Universidad Autónoma Chapingo, Mexico, 63 pp.

GUPTA, R. K. 1986. The Thar Desert. Pp. 55–99, *in* Ecosystems of the world. 12B. Hot deserts and arid shrublands (M. Evenari, I. Noy-Meir, and D. W. Goodall, eds.). Elsevier, New York, 451 pp.

GUTTERMAN, Y., T. GOLAN, AND M. GARSANI. 1990. Porcupine diggings as a unique ecological system in a desert environment. Oecologia, 85:122–127.

HAFNER, J. C. 1993. Macroevolutionary diversification in heteromyid rodents: heterochrony and adaptation in phylogeny. Pp. 291–318, *in* Biology of the Heteromyidae (H. H. Genoways and J. H. Brown, eds.). Special Publication, American Society of Mammalogists, 10:1–719.

HAIM, A. 1984. Adaptive variations in heat production within gerbils (genus *Gerbillus*) from different habitats. Oecologia, 61:49–52.

HAIM, A., AND E. TCHERNOV. 1974. The distribution of myomorph rodents in the Sinai Peninsula. Mammalia, 38:201–223.

HAIM, A., R. J. VAN AARDE, AND J. D. SKINNER. 1990. Metabolism and thermoregulation in the Cape porcupine, *Hystrix africaeaustralis*. Physiological Zoology, 63:795–802.

HALLETT, J. G. 1982. Habitat selection and the community matrix of a desert small-mammal fauna. Ecology, 63:1400–1410.

HAMANN, V. U. 1987. Zu Aktivität und Verhalten von drei Taxa der Zwerghamster der Gattung *Phodopus* Miller, 1910. Zeitschrift für Säugetierkunde, 52:65–76.

HAPPOLD, D. C. D. 1967. Biology of the jerboa, *Jaculus jaculus butleri* (Rodentia, Dipodidae), in the Sudan. Journal of Zoology (London), 151:257–275.

———. 1968. Observations on *Gerbillus pyramidum* (Gerbillinae, Rodentia) at Khartoum, Sudan. Mammalia, 32:44–53.

———. 1975. The ecology of rodents in the northern Sudan. Pp. 15–45, *in* Rodents in desert environments (I. Prakash and P. R. Ghosh, eds.). Junk, b.v., The Hague, The Netherlands, 624 pp.

HARRIS, J. H. 1984. An experimental analysis of desert rodent foraging ecology. Ecology, 65:1579–1584.

———. 1986. Microhabitat segregation in two desert rodent species: the relation of prey availability to diet. Oecologia, 68:417–421.

HARRISON, D. L. 1964. The mammals of Arabia. Insectivora, Chiroptera, Primates. Ernest Benn, London, Vol. 1, 1–192.

———. 1968. The mammals of Arabia. Carnivora, Hyracoidea, Artiodactyla. Ernest Benn, London, Vol. 2, 193–380.

———. 1972. The mammals of Arabia. Lagomorpha, Rodentia. Ernest Benn, London, Vol. 3, 381–670.

HASSINGER, J. D. 1968. Introduction to the mammal survey of the 1965 Street Expedition to Afghanistan. Fieldiana: Zoology, 55:1–81.

———. 1973. A survey of the mammals of Afghanistan. Fieldiana: Zoology, 60:1–195.

HATT, R. T. 1959. The mammals of Iraq. Miscellaneous Publications of the Museum of Zoology, University of Michigan, 106:1–113.

HEIM DE BALSAC, H. 1972. Insectivores. Pp. 629–660, *in* Biogeography and ecology in Madagascar (R. Battistina and G. Richard-Vindard, eds.). Junk, b.v., The Hague, The Netherlands, 765 pp.

HERSHKOVITZ, P. 1972. The Recent mammals of the Neotropical Region: a zoogeographic and ecological review. Pp. 311–431, *in* Evolution, mammals, and southern continents (A. Keast, F. C. Erk, and B. Glass, eds.). State University of New York Press, Albany, 543 pp.

HILLS, E. S. (ED.). 1966. Arid lands. A geographical appraisal. Methuen Co., London, 461 pp.

HOFFMEISTER, D., AND W. GOODPASTER. 1954. The mammals of the Huachuca Mountains, southeastern Arizona. Illinois Biological Monographs, 24:1–152.

HONACKI, J. H., R. E. KINMAN, AND J. W. KOEPPL (EDS.). 1982. Mammal species of the world: a taxonomic and geographic reference. Allen Press, Inc. and The Association of Systematics Collections, Lawrence, Kansas, 694 pp.

HOWELL, A. B. 1932. The saltatorial rodent *Dipodomys:* the functional and comparative anatomy of its muscular and osseous systems. Proceedings of the American Academy of Arts and Sciences, 67:377–536.

HUBBARD, J. P. 1974. Avian evolution in the aridlands of North America. Living Bird, 12:155–196.

JAEGER, E. C. 1957. The North American deserts. Stanford University Press, Stanford, California, 308 pp.

KAM, M., AND A. A. DEGEN. 1988. Water, electrolyte and nitrogen balances of fat sand rats (*Psammomys obesus*) when consuming the saltbush *Atriplex halimus*. Journal of Zoology (London), 215:453–462.

———. 1989. Efficiency of use of saltbush (*Atriplex halimus*) for growth by fat sand rats (*Psammomys obesus*). Journal of Mammalogy, 70:485–493.

KASHKAROV, D., AND V. KURBATOV. 1930. Preliminary ecological survey of the vertebrate fauna of the Central Kara-Kum Desert in West Turkestan. Ecology, 11:35–60.

KERSHAW, A. P. 1981. Quaternary vegetation and environments. Pp. 81–101, *in* Ecological biogeography of Australia (A. Keast, ed.). Junk, b.v., The Hague, The Netherlands, 805 pp.

KINGDON, J. 1974. East African mammals. Hares and rodents. University of Chicago Press, 2B:342–704.

KIRMIZ, J. P. 1962. Adaptations to desert environment: a study of the jerboa, rat and man. Butterworths, London, 168 pp.

KOECHLIN, J. 1972. Flora and vegetation of Madagascar. Pp. 145–190, *in* Biogeography and ecology in Madagascar (R. Battistina and G. Richard-Vindard, eds.). Junk, b.v., The Hague, The Netherlands, 765 pp.

KOFORD, C. B. 1968. Peruvian desert mice: water independence, competition, and breeding cycle near the Equator. Science, 160:552–553.

KOTLER, B. P. 1984. Risk of predation and the structure of desert rodent communities. Ecology, 65:789–802.

———. 1985. Owl predation on desert rodents which differ in morphology and behavior. Journal of Mammalogy, 66:824–828.

KURUP, G. U. 1974. Mammals of Assam and the mammal geography of India. Pp. 585–613, *in* Ecology and biogeography in India (M. S. Mani, ed.). Junk, b.v., The Hague, The Netherlands, 773 pp.

———. 1986. Mammal-zoogeography of India. Pp. 69–93, *in* Wildlife wealth of India (resources and management) (T. C. Majupuria, ed.). Tec Press Service, L. P. Bangkok, Thailand, 656 pp.

LACHER, T. E., JR. 1981. The comparative social behavior of *Kerodon rupestris* and *Galea spixii* in the xeric Caatinga of northeastern Brazil. Bulletin of Carnegie Museum of Natural History, 17:1–71.

LACHER, T. E., JR., M. R. WILLIG, AND M. A. MARES. 1982. Food preference as a function of resource abundance with multiple prey types: an experimental analysis of optimal foraging theory. American Naturalist, 120:297–316.

Lahey, J. F. 1973. On the origin of the dry climate in northern South America and the southern Caribbean. Pp. 75–90, *in* Coastal deserts (D. H. K. Amiran and A. W. Wilson, eds.). University of Arizona Press, Tucson, 207 pp.

Lawler, R. M., and K. N. Geluso. 1986. Renal structure and body size in heteromyid rodents. Journal of Mammalogy, 67:367–372.

Lay, D. M. 1967. A study of the mammals of Iran, resulting from the Street Expedition of 1962–63. Fieldiana: Zoology, 54:1–282.

Lee, A. K., P. R. Baverstock, and C. H. S. Watts. 1981. Rodents—the late invaders. Pp. 1521–1553, *in* Ecological biogeography of Australia (A. Keast, ed.). Junk, b.v., The Hague, The Netherlands, 3: 1435–2142.

Le Houérou, H. N. 1986. The desert and arid zones of northern Africa. Pp. 101–147, *in* Hot deserts and arid shrublands (M. Evenari, I. Noy-Meir, and D. W. Goodall, eds.). Elsevier, New York, 451 pp.

Lewis, R. E., J. H. Lewis, and S. A. Atallah. 1967. A review of Lebanese mammals, Lagomorpha and Rodentia. Journal of Zoology (London), 153:45–70.

Logan, R. F. 1968. Causes, climates, and distribution of deserts. Pp. 21–50, *in* Desert biology (G. W. Brown, Jr., ed.). Academic Press, New York, 1:1–635.

Low, W. A. 1984. Interactions of introduced mammals in the arid regions of Australia. Pp. 301–313, *in* Arid Australia (H. G. Cogger and E. E. Cameron, eds.). Australian Museum, Sydney, 338 pp.

Mabbutt, J. A. 1984. The desert physiographic setting and its ecological significance. Pp. 87–109, *in* Arid Australia (H. G. Cogger and E. E. Cameron, eds.). Australian Museum, Sydney, 338 pp.

Mabry, T. J., J. H. Hunziker, and D. R. DiFeo, Jr. (eds.). 1977. Creosote bush. Dowden, Hutchinson and Ross, Inc., Stroudsburg, Pennsylvania, 284 pp.

MacMahon, J. A., and F. H. Wagner. 1985. The Mohave, Sonoran and Chihuahuan deserts of North America. Pp. 105–202, *in* Ecosystems of the world. 12A. Hot deserts and arid shrublands (M. Evenari, I. Noy-Meir, and D. W. Goodall, eds.). Elsevier, New York, 365 pp.

MacMillen, R. E. 1972. Water economy of nocturnal desert rodents. Symposia of the Zoological Society of London, 31:147–174.

———. 1983. Adaptive physiology of heteromyid rodents. Pp. 65–76, *in* Biology of desert rodents (O. J. Reichman and J. H. Brown, eds.). Great Basin Naturalist Memoirs, 7:1–134.

MacMillen, R. E., and A. K. Lee. 1967. Australian Desert mice: independence of exogenous water. Science, 158:383–385.

———. 1969. Water metabolism of Australian hopping mice. Comparative Biochemistry and Physiology, A. Comparative Physiology, 28:493–514.

MacMillen, R. E., R. V. Baudinette, and A. K. Lee. 1972. Water economy and energy metabolism of the sandy inland mouse, *Leggadina hermannsburgensis.* Journal of Mammalogy, 53:529–539.

Main, A. R., and H. R. Bakker. 1981. Adaptation of macropod marsupials to aridity. Pp. 1489–1519, *in* Ecological biogeography of Australia (A. Keast, ed.). Junk, b.v., The Hague, The Netherlands, Vol. 3, pp. 1435–2142.

Mani, M. S. 1974*a*. The flora. Pp. 159–177, *in* Ecology and biogeography in India (M. S. Mani, ed.). Junk, b.v., The Hague, The Netherlands, 773 pp.

———. 1974*b*. Biogeography of the western borderlands. Pp. 682–697, *in* Ecology and biogeography in India (M. S. Mani, ed.). Junk, b.v., The Hague, The Netherlands, 773 pp.

———. 1974*c*. Biogeographical evolution in India. Pp. 698–724, *in* Ecology and biogeography in India (M. S. Mani, ed.). Junk, b.v., The Hague, The Netherlands, 773 pp.

Mann, G. 1945. Mamíferos de Tarapacá. Biologica, 2:23–98.

Mares, M. A. 1973. Climates, mammalian communities, and desert rodent adaptations: an investigation into evolutionary convergence. Ph.D. dissertation, University of Texas, Austin, 345 pp.

———. 1975*a*. South American mammal zoogeography: evidence from convergent evolution in desert rodents. Proceedings of the National Academy of Sciences, 72:1702–1706.

———. 1975*b*. Observations of Argentine desert rodent ecology, with emphasis on water relations of *Eligmodontia typus.* Pp. 155–175, *in* Rodents in desert environments (I. Prakash and P. R. Ghosh, eds.). Junk, b.v., The Hague, The Netherlands, 624 pp.

———. 1976. Convergent evolution in desert rodents: multivariate analysis and zoogeographic implications. Paleobiology, 2:39–63.

———. 1977*a*. Water balance and other ecological observations on three species of *Phyllotis* in northwestern Argentina. Journal of Mammalogy, 58:514–520.

———. 1977*b*. Water economy and salt balance in a South American desert rodent, *Eligmodontia typus.* Comparative Biochemistry and Physiology, A. Comparative Physiology, 56:325–332.

———. 1977*c*. Aspects of water balance of *Oryzomys longicaudatus* from northwest Argentina. Comparative Biochemistry and Physiology, A. Comparative Physiology, 57:237–238.

———. 1977*d*. Water independence in a non-desert South American rodent. Journal of Mammalogy, 58: 653–656.

———. 1979. Small mammals and creosote bush: patterns of richness. Pp. 57–94, *in* Larrea (E. C. Lopez, T. J. Mabry, and S. F. Tavizon, eds.). Centro de Investigacíon en Química Aplicada, Saltillo, Coahuila, México, 411 pp.

———. 1980. Convergent evolution among desert rodents: a global perspective. Bulletin of Carnegie Museum of Natural History, 16:1–51.

———. 1983. Desert rodent adaptation and community structure. Pp. 30–43, *in* Biology of desert rodents (O. J. Reichman and J. H. Brown, eds.). Great Basin Naturalist Memoirs, 7:1–134.

———. 1985. Mammal faunas of xeric habitats and the Great American Interchange. Pp. 489–520, *in* The Great American Interchange (F. Stehli and S. D. Webb, eds.). Plenum, New York, 532 pp.

———. 1988. Reproduction, growth, and development in Argentine gerbil mice, *Eligmodontia typus*. Journal of Mammalogy, 69:852–854.

MARES, M. A., AND A. C. HULSE. 1977. Patterns of some vertebrate communities in creosote bush deserts. Pp. 209–226, *in* Creosote bush: biology and chemistry of *Larrea* in New World deserts (T. J. Mabry, J. H. Hunziker, and D. R. DiFeo, Jr., eds.). Dowden, Hutchinson and Ross, Stroudsburg, Pennsylvania, 284 pp.

MARES, M. A., AND T. E. LACHER, JR. 1987. Ecological, morphological, and behavioral convergence in rock-dwelling mammals. Current Mammalogy, 1:307–345.

MARES, M. A., AND M. L. ROSENZWEIG. 1978. Granivory in North and South American deserts. Ecology, 59:235–241.

MARES, M. A., AND D. F. WILLIAMS. 1977. Experimental support for food particle size resource allocation in heteromyid rodents. Ecology, 58:454–457.

MARES, M. A., J. K. BRAUN, AND D. D. GETTINGER. 1989. Observations on the distribution and ecology of the mammals of the cerrado grasslands of central Brazil. Annals of Carnegie Museum, 58:1–60.

MARES, M. A., J. MORELLO, AND G. GOLDSTEIN. 1985*a*. The Monte Desert and other subtropical semiarid biomes of Argentina, with comments on their relation to North America arid areas. Pp. 203–237, *in* Ecosystems of the world. 12A. Hot deserts and arid shrublands (M. Evenari, I. Noy-Meir, and D. W. Goodall, eds.). Elsevier, New York, 365 pp.

MARES, M. A., R. A. OJEDA, AND R. M. BARQUEZ. 1989. Guide to the mammals of Salta Province, Argentina. University of Oklahoma Press, 303 pp.

MARES, M. A., R. A. OJEDA, AND M. P. KOSCO. 1981*a*. Observations on the distribution and ecology of the mammals of Salta Province, Argentina. Annals of Carnegie Museum, 50:151–206.

MARES, M. A., M. R. WILLIG, AND T. E. LACHER, JR. 1985*b*. The Brazilian Caatinga in South American zoogeography: tropical mammals in a dry region. Journal of Biogeography, 12:57–69.

MARES, M. A., M. R. WILLIG, K. E. STREILEIN, AND T. E. LACHER, JR. 1981*b*. The mammals of northeastern Brazil: a preliminary assessment. Annals of Carnegie Museum, 50:80–137.

MARES, M. A., F. ENDERS, J. KINGSOLVER, J. NEFF, AND B. B. SIMPSON. 1977*b*. *Prosopis* as a niche component. Pp. 123–149, *in* Mesquite: its biology in two desert ecosystems (B. B. Simpson, ed.). Dowden, Hutchinson and Ross, Stroudsburg, Pennsylvania, 250 pp.

MARES, M. A., ET AL. 1977*a*. The strategies and community patterns of desert animals. Pp. 107–163, *in* Convergent evolution in warm deserts (G. H. Orians and O. T. Solbrig, eds.). Dowden, Hutchinson and Ross, Stroudsburg, Pennsylvania, 333 pp.

MARSHALL, L. G., D. WEBB, J. J. SEPKOSKI, JR., AND D. M. RAUP. 1982. Mammalian evolution and the Great American Interchange. Science, 215:1351–1357.

MCCAULEY, J. F., ET AL. 1982. Substrate valleys and geoarcheology of the eastern Sahara revealed by shuttle radar. Science, 218:1004–1020.

MCGINNIES, W. G. 1968. Vegetation of desert environments. Pp. 379–566, *in* Deserts of the world (W. G. McGinnies, B. J. Goldman, and P. Paylore, eds.). University of Arizona Press, Tucson, 788 pp.

MCGINNIES, W. G., AND B. J. GOLDMAN (EDS.). 1969. Arid lands in perspective. University of Arizona Press, Tucson, 421 pp.

MCGINNIES, W. G., B. J. GOLDMAN, AND P. PAYLORE (EDS.). 1968. Deserts of the world. University of Arizona Press, Tucson, 788 pp.

MCNAB, B. K. 1979. Climatic adaptation in the energetics of heteromyid rodents. Comparative Biochemistry and Physiology, A. Comparative Physiology, 62:813–820.

———. 1982. The physiological ecology of South American mammals. Pp. 187–207, *in* Mammalian biology in South America (M. A. Mares and H. H. Genoways, eds.). Special Publications Series, Pymatuning Laboratory of Ecology, University of Pittsburgh, 6:1–539.

MEIGS, P. 1953. World distribution of arid and semiarid homoclimates. Pp. 203–209, *in* Reviews of research on arid zone hydrology. UNESCO, Paris. Arid Zone Programme, No. 1.

MESERVE, P. L. 1978. Water dependence in some Chilean arid zone rodents. Journal of Mammalogy, 59:217–219.

MESERVE, P. L., AND W. E. GLANZ. 1978. Geographical ecology of small mammals in the northern Chilean arid zone. Journal of Biogeography, 5:135–148.

MISONNE, X. 1959. Analyse zoogeographique des mammiferes de l'Iran. Memoires de l'Institut Royal des Sciences naturelles de Belgique, 59:1–157.

MOLNAR, P., AND P. TAPPONNIER. 1975. Cenozoic tectonics of Asia: effects of a continental collision. Science, 189:419–426.

MONOD, T. 1986. The Sahel zone north of the Equator. Pp. 203–243, *in* Ecosystems of the world. 12B. Hot deserts and arid shrublands (M. Evenari, I. Noy-Meir, and D. W. Goodall, eds.). Elsevier, New York, 451 pp.

MORELLO, J. 1958. La provincia fitogeográfica del Monte. Opera Lilloana, 2:1–155.

MORTON, S. R. 1979. Diversity of desert-dwelling mammals: a comparison of Australia and North America. Journal of Mammalogy, 60:253–264.

———. 1985. Granivory in arid regions: comparison of Australia with North and South America. Ecology, 66:1859–1866.

MORTON, S. R., AND A. BAYNES. 1985. Small mammal assemblages in arid Australia: a reappraisal. Australian Mammalogy, 8:159–169.

MÜLLER, E. F., AND R. VAN AKEN. 1990. Thermoregulation and evaporative water loss in spiny mice (*Acomys cahirinus* Desmarest, 1819). Zeitschrift für Säugetierkunde, 55:244–252.

MYERS, P. 1982. Origins and affinities of the mammal fauna of Paraguay. Pp. 85–93, *in* Mammalian biology in South America (M. A. Mares and H. H. Genoways, eds.). Special Publications Series, Pymatuning Laboratory of Ecology, University of Pittsburgh, 6:1–539.

MYERS, P., AND R. M. WETZEL. 1979. New records of mammals from Paraguay. Journal of Mammalogy, 60:638–641.

NAUMOV, N. P., AND V. S. LOBACHEV. 1975. Ecology of desert rodents of the U.S.S.R. (jerboas and gerbils). Pp. 465–598, *in* Rodents in desert environments (I. Prakash and P. R. Ghosh, eds.). Junk, b.v., The Hague, The Netherlands, 624 pp.

NEL, J. A. J., AND I. L. RAUTENBACH. 1975. Habitat use and community structure of rodents in the southern Kalahari. Mammalia, 39:9–29.

NERONOV, V. M. 1976. A zoogeographical analysis of the rodent fauna of Iran. Byulleten Moskovskogo Obshchestva Ispytatelei Prirody, Biologia, Zoology, 81:32–47 (in Russian, Translation 1168, Medical Zooloogy Department USNAMRU 3, Cairo, Egypt).

NIETHAMMER, J. 1965. Die Saugetiere Afghanistans (Teil II): Insectivora, Lagomorpha, Rodentia. Science (Faculty of Science, Kabul University), August 1965: 18–42.

NIKOLAI, J. C., AND D. M. BRAMBLE. 1983. Morphological structure and function in desert heteromyid rodents. Great Basin Naturalist Memoirs, 7:44–64.

NIR, D. 1974. The semi-arid world: man on the fringe of the desert. Longman, New York, 187 pp.

NORTON, I. O., AND J. G. SCLATER. 1979. A model for the evolution of the Indian Ocean and the breakup of Gondwanaland. Journal of Geophysical Research, 84:6803–6830.

NOWAK, R. M., AND J. L. PARADISO. 1983. Walker's mammals of the world. Fourth edition. The Johns Hopkins University Press, Baltimore, Maryland, 2:569–1362 + xi–xxv.

OBERLANDER, T. M. 1979. Characterization of arid climates according to combined water balance parameters. Journal of Arid Environments, 2:219–241.

O'FARRELL, M. J. 1974. Seasonal activity patterns of rodents in a sagebrush community. Journal of Mammalogy, 55:809–823.

OJEDA, R. A., AND M. A. MARES. 1988. A zoogeographic analysis of the mammals of Salta Province, Argentina: patterns of community assemblage in the Neotropics. Special Publications, The Museum, Texas Tech University, 27:1–66.

OLROG, C. C., AND M. M. LUCERO. 1981. Guía de los mamíferos argentinos. Ministerio de Cultura y Educación, Fundación Miguel Lillo, Tucumán, Argentina, 151 pp.

ORIANS, G. H., AND O. T. SOLBRIG (EDS.). 1977. Convergent evolution in warm deserts. Dowden, Hutchinson and Ross, Stroudsburg, Pennsylvania, 333 pp.

ORIANS, G. H., ET AL. 1977. Resource utilization systems. Pp. 164–224, *in* Convergent evolution in warm deserts (G. H. Orians and O. T. Solbrig, eds.). Dowden, Hutchinson and Ross, Stroudsburg, Pennsylvania, 333 pp.

ORSHAN, G. 1986. The deserts of the Middle East. Pp. 1–28, *in* Ecosystems of the world. 5. Temperate deserts and semi-deserts (N. E. West, ed.). Elsevier, New York, 522 pp.

OSBORN, D. J., AND I. HELMY. 1980. The contemporary land mammals of Egypt (including Sinai). Fieldiana: Zoology, 5:1–579.

OSGOOD, W. H. 1943. The mammals of Chile. Field Museum of Natural History, Zoology Series, 30:1–268.

OTTERMAN, J. 1974. Baring high-albedo soils by overgrazing: a hypothesized desertification mechanism. Science, 186:531–533.

PACKARD, R. L. 1977. Mammals of the southern Chihuahuan Desert: an inventory. Pp. 141–153, *in* Transactions of the symposium on the biological resources of the Chihuahuan Desert Region, United States and Mexico (R. H. Wauer and D. H. Riskind, eds.). U.S. Department of the Interior, National Park Service Transactions and Proceedings Series, 3:1–658.

PATTERSON, B., AND R. PASCUAL. 1972. The fossil mammal fauna of South America. Pp. 247–309, *in* Evolution, mammals, and southern continents (A. Keast, F. D. Erk, and B. Glass, eds.). State University of New York Press, Albany, 543 pp.

PEARSON, O. P. 1951. Mammals of the highlands of Peru. Bulletin of the Museum of Comparative Zoology, 106:117–174.

———. 1957. Additions to the mammalian fauna of Peru and notes on some other Peruvian mammals. Breviora, 73:1–7.

———. 1959. Biology of the subterranean rodents, *Ctenomys,* in Peru. Memorias del Museo de Historia Natural "Javier Prado," 9:1–56.

———. 1982. Distribución de pequeños mamíferos en el altiplano y los desiertos del Peru. Pp. 263–284, *in* Zoología tropical (P. Salinas, ed.). Actas del VIII Congreso Latinoamericano de Zoología, Mérida, Venezuela.

PEARSON, O. P., AND C. P. RALPH. 1978. The diversity and abundance of vertebrates along an altitudinal gradient in Peru. Memorias del Museo de Historia Natural "Javier Prado," 18:1–97.

PÉFAUR, J., J. L. YAÑEZ, AND F. JAKSIC. 1979. Biological and environmental aspects of a mouse outbreak in the semi-arid region of Chile. Mammalia, 43:313–322.

PETROV, M. P. 1976. Deserts of the world. John Wiley and Sons, New York, 447 pp.

PETTER, F. 1961. Répartition géographique et écologie des Rongeurs désertiques (du Sahara occidental a l'Iran oriental). Mammalia, 25:1–222.

———. 1972. The rodents of Madagascar: the seven genera of Malagasy rodents. Pp. 661–665, *in* Biogeography and ecology in Madagascar (R. Battistina and G. Richard-Vindard, eds.). Junk, b.v., The Hague, The Netherlands, 765 pp.

PIANKA, E. R. 1972. Zoogeography and speciation of Australian desert lizards: an ecological perspective. Copeia, 1972:127–145.

PINE, R. H., S. D. MILLER, AND M. L. SCHAMBERGER. 1979. Contributions to the mammalogy of Chile. Mammalia, 43:339–376.

PRAKASH, I. 1962. Ecology of gerbils of the Rajasthan Desert, India. Mammalia, 24:311–331.

———. 1963. Zoogeography and evolution of the mammalian fauna of Rajasthan Desert, India. Mammalia, 27:342–351.

———. 1964. Taxonomical and ecological account of the mammals of Rajasthan Desert. Annals of Arid Zone, 2:150–161.

———. 1968. Biology of the rodents of Rajasthan Desert. Proceedings of the Symposium on the Natural Resources of Rajasthan, University of Jodhpur, India: 337–352.

———. 1974. The ecology of vertebrates of the Indian Desert. Pp. 369–420, *in* Ecology and biogeography in India (M. S. Mani, ed.). Junk, b.v., The Hague, The Netherlands, 773 pp.

———. 1975*a*. Population ecology of rodents in the Rajasthan Desert, India. Pp. 75–186, *in* Rodents in desert environments (I. Prakash and P. R. Ghosh, eds.). Junk, b.v., The Hague, The Netherlands, 624 pp.

———. 1975*b*. The ecology and zoogeography of mammals. Pp. 448–467, *in* Environmental analysis of the Thar Desert (R. R. Gupta and I. Prakash, eds.). English Book Depot, Dehra Dun, India.

———. 1981. Ecology of the Indian desert gerbil, *Meriones hurrianae*. Central Arid Zone Research Institute, Cazri Monograph, 10:1–87.

PRAKASH, I., AND P. K. GHOSH (EDS.). 1975. Rodents in desert environments. Junk, b.v., The Hague, The Netherlands, 624 pp.

PRICE, M. V. 1978. The role of microhabitat in structuring desert rodent communities. Ecology, 59:910–921.

PRICE, M. V., AND J. H. BROWN. 1983. Patterns of morphology and resource use in North American desert rodent communities. Pp. 117–134, *in* Biology of desert rodents (O. J. Reichman and J. H. Brown, eds.). Great Basin Naturalist Memoirs, 7:1–134.

PRICE, M. V., AND O. J. REICHMAN. 1987. Distribution of seeds in Sonoran Desert soils: implications for heteromyid rodent foraging. Ecology, 68:1797–1811.

PUROHIT, K. G. 1974*a*. Observations on histomorphology of kidneys and urine osmolarities in some Australian mammals and their ecophysiological appraisal. Zoologishe Anzeizer, Jena, 193:221–227.

———. 1974*b*. Observations on size and relative medullary thickness in kidneys of some Australian mammals and their ecophysiological appraisal. Zeitschrift fur Angewandte Zoologie, 61:495–505.

RANCK, G. L. 1968. The rodents of Libya: taxonomy, ecology, and zoogeographical relationships. Bulletin of the United States National Museum, 275:1–264.

RAUH, W. 1985. The Peruvian–Chilean deserts. Pp. 239–267, *in* Ecosystems of the world. 12A. Hot deserts and arid shrublands (M. Evenari, I. Noy-Meir, and D. W. Goodall, eds.). Elsevier, New York, 365 pp.

———. 1986. The arid region of Madagascar. Pp. 361–377, *in* Ecosystems of the world. 12A. Hot deserts and arid shrublands (M. Evenari, I. Noy-Meir, and D. W. Goodall, eds.). Elsevier, New York, 451 pp.

RAUTENBACH, I. L. 1978. A numerical re-appraisal of the southern African biotic zones. Pp. 175–187, *in* Ecology and taxonomy of African small mammals (D. A. Schlitter, ed.). Bulletin of the Carnegie Museum of Natural History, 6:1–214.

RAVEN, P. H., AND D. I. AXELROD. 1972. Plate tectonics and Australian paleobiogeography. Science, 176:1379–1386.

———. 1974. Angiosperm biogeography and past continental movement. Annals of the Missouri Botanical Garden, 61:539–673.

REDFORD, K. H., A. TABER, AND J. A. SIMONETTI. 1990. There is more to biodiversity than the tropical rain forests. Conservation Biology, 4:328–330.

REICHMAN, O. J. 1983. Behavior of desert heteromyids. Pp. 77–90, *in* Biology of desert rodents (O. J. Reichman and J. H. Brown, eds.). Great Basin Naturalist Memoirs, 7:1–134.

REICHMAN, O. J., AND J. H. BROWN (EDS.). 1983. Biology of desert rodents. Great Basin Naturalist Memoirs No. 7, 134 pp.

REICHMAN, O. J., AND J. U. M. JARVIS. 1989. The influence of three sympatric species of fossorial mole-rats (Bathyergidae) on vegetation. Journal of Mammalogy, 70:763–771.

REICHMAN, O. J., AND M. V. PRICE. 1993. Ecological aspects of heteromyid foraging. Pp. 539–574, *in* Biology of the Heteromyidae (H. H. Genoways and J. H. Brown, eds.). Special Publication, American Society of Mammalogists, 10:1–719.

REICHMAN, O. J., AND C. REBAR. 1985. Seed preference by desert rodents based on levels of mouldiness. Animal Behaviour, 33:726–729.

REICHMAN, O. J., A. FATTAEY, AND K. FATTAEY. 1986. Management of sterile and mouldy seeds by a desert rodent. Animal Behaviour, 34:221–225.

REIG, O. A. 1981. Teoría del origen y desarollo de la fauna de mamíferos de América del Sur. Monografie Naturae, Publicadas por el Museo Municipal de Ciencias Naturales "Lorenzo Scaglia," Mar del Plata, Argentina, 1:1–162.

REIS, A. C. DE SOUZA. 1976. Clima da Caatinga. Anales de la Academia Brasileira de Ciencias, 48:325–335.

RICHARDSON, S. D. 1966. Forestry in communist China. Johns Hopkins Press, Baltimore, Maryland, 237 pp.

RIDE, W. D. L. 1970. A guide to the native mammals of Australia. Oxford University Press, Melbourne, 249 pp.

ROBERTS, T. J. 1977. The mammals of Pakistan. Ernest Benn, London, 361 pp.

ROIG, V. G. 1962. Aspectos biogeográficos y planteos ecológicos de la fauna de mamíferos de las zonas áridas de Mendoza. Revista de la Facultad de Ciencias, Universidad de Cuyo, Argentina, 9:59–81.

———. 1969. La hipertrofia de la bula timpánica en los mamíferos del desierto. Scientia, 104(689–690): 1–7.

———. 1971. La presencia de estados de hibernación en *Marmosa elegans*. Acta Zoológica Lilloana, 28: 5–12.

———. 1972. La hipertrofia de la bula timpánica y su significado adaptativo en los edentados de zonas áridas. Deserta, 2:87–97.

ROOD, J. P. 1972. Ecological and behavioural comparisons of three genera of Argentine cavies. Animal Behaviour Monographs, 5:1–82.

ROSENZWEIG, M. L. 1987. Community organization from the point of view of habitat selectors. Pp. 469–490, *in* Organization of communities past and present (J. H. R. Gee and P. S. Giller, eds.). Blackwell Scientific Publications, Oxford.

———. 1989. Habitat selection, community organization, and small mammal studies. Pp. 5–21, *in* Patterns in the structure of mammalian communities (D. W. Morris, Z. Abramsky, B. J. Fox, and M. R. Willig, eds.). Texas Tech University Press, Lubbock, 266 pp.

ROSEVEAR, D. R. 1969. The rodents of West Africa.

British Museum of Natural History Publications, 604 pp.
SAHNI, A. 1984. Cretaceous-Paleocene terrestrial faunas of India: lack of endemism during drifting of the Indian Plate. Science, 226:441–443.
SCHIERWATER, B., AND H. KLINGEL. 1985. Food digestibility and water requirements in the Djungarian hamster *Phodopus sungorus*. Zeitschrift für Säugetierkunde, 50:35–39.
SCHMIDLY, D. J. 1977. Factors governing the distribution of mammals in the Chihuahuan Desert Region. Pp. 163–192, *in* Transactions of the symposium on the biological resources of the Chihuahuan Desert Region, United States and Mexico (R. H. Wauer and D. H. Riskind, eds.). U.S. Department of the Interior, National Park Service Transactions and Proceedings Series, No. 3, 658 pp.
SCHMIDLY, D. J., K. T. WILKINS, AND J. N. DERR. 1993. Biogeography. Pp. 319–356, *in* Biology of the Heteromyidae (H. H. Genoways and J. H. Brown, eds.). Special Publication, American Society of Mammalogists, 10:1–719.
SCHMIDT-NIELSEN, K. 1964. Desert animals. Oxford University Press, Oxford, 277 pp.
SCHRÖPFER, R., B. KLEMER-FRINGES, AND E. NAUMER. 1985. Locomotion patterns and habitat utilisation of the two jerboas *Jaculus jaculus* and *Jaculus orientalis* (Rodentia, Dipodidae). Mammalia, 49:445–454.
SEELY, M. K. 1978. The Namib Dune Desert: an unusual ecosystem. Journal of Arid Environments, 1:117–128.
SIDDIQI, M. S. 1961. Checklist of mammals of Pakistan with particular reference to the mammalian collection in the British Museum (Natural History), London, Biologia, 7:93–225.
SIMBERLOFF, D., AND W. BOECKLEN. 1981. Santa Rosalia reconsidered: size ratios and competition. Evolution, 35:1206–1228.
SIMPSON, B. B. (ED.). 1977. Mesquite: its biology in two desert ecosystems. Dowden, Hutchinson and Ross, Stroudsburg, Pennsylvania, 250 pp.
SIMPSON, G. G. 1970. The Argyrolagidae, extinct South American marsupials. Bulletin of the Museum of Comparative Zoology, 139:1–86.
———. 1980. Splendid isolation. Yale University Press, New Haven, 266 pp.
SMITH, J. R., C. H. S. WATTS, AND E. G. CRICHTON. 1972. Reproduction in the Australian Desert rodents. Australian Mammalogy, 1:1–7.
SOLBRIG, O. T., ET AL. 1977. The biota: the dependent variable. Pp. 50–66, *in* Convergent evolution in warm deserts (G. H. Orians and O. T. Solbrig, eds.). Dowden, Hutchinson and Ross, Stroudsburg, Pennsylvania, 333 pp.
SORIANO, A., ET AL. 1983. Deserts and semi-deserts of Patagonia. Pp. 423–472, *in* Ecosystems of the world. 5. Temperate deserts and semi-deserts (N. E. West, ed.). Elsevier, New York, 522 pp.
SPECHT, R. L. 1981*a*. Major vegetation formations in Australia. Pp. 163–297, *in* Ecological biogeography of Australia (A. Keast, ed.). Junk, b.v., The Hague, The Netherlands, 1:1–805.
———. 1981*b*. Evolution of the Australian flora: some generalizations. Pp. 783–805, *in* Ecological biogeography of Australia (A. Keast, ed.). Junk, b.v., The Hague, The Netherlands, 1:1–805.
STALMAKOVA, V. A. 1954. Rodents of the Kara-Kum and their economic importance. Pustyni. S.S.S.R. Ikh Osvoyeniye, 2:756–782.
———. 1955. Mammals of the sand-desert natural reserve Repetek and the neighboring Kara-Kum Desert. Pp. 307–345, *in* Papers of the Sand-Desert Station Repetek (M. P. Petrov, ed.). Ashkhabad, 358 pp.
STREILEIN, R. E. 1982*a*. Ecology of small mammals in the semiarid Brazilian Caatinga. I. Climate and faunal composition. Annals of the Carnegie Museum, 51:79–107.
———. 1982*b*. Ecology of small mammals in the semiarid Brazilian Caatinga. II. Water relations. Annals of the Carnegie Museum, 51:109–126.
———. 1982*c*. The ecology of small mammals in the semiarid Brazilian Caatinga. III. Reproductive biology and population ecology. Annals of the Carnegie Museum, 51:251–269.
———. 1982*d*. The ecology of small mammals in the semiarid Brazilian Caatinga. IV. Habitat selection. Annals of the Carnegie Museum, 51:331–343.
———. 1982*e*. The ecology of small mammals in the semiarid Brazilian Caatinga. V. Agonistic behavior and overview. Annals of the Carnegie Museum, 51: 345–369.
———. 1982*f*. Behavior, ecology, and distribution of South American marsupials. Pp. 231–250, *in* Mammalian biology in South America (M. A. Mares and H. H. Genoways, eds.). Special Publications Series, Pymatuning Laboratory of Ecology, University of Pittsburgh, 6:1–539.
STUART, C. T. 1975. Preliminary notes on the mammals of the Namib Desert Park. Madoqua, Series II, 4:5–68.
TAYLOR, K. D., AND M. G. GREEN. 1976. The influence of rainfall on diet and reproduction in four African rodent species. Journal of Zoology (London), 180:367–389.
TCHERNOV, E. 1975. Rodent faunas and environmental changes in the Pleistocene of Israel. Pp. 331–362, *in* Rodents in desert environments (I. Prakash and P. K. Ghosh, eds.). Junk, b.v., The Hague, The Netherlands, 624 pp.
TORRES-MURA, J. C., M. L. LEMUS, AND L. C. CONTRERAS. 1989. Herbivorous specialization of the South American desert rodent *Tympanoctomys barrerae*. Journal of Mammalogy, 70:646–648.
TURNBULL, P. F. 1975. Prehistoric rodents of the Middle East. Pp. 363–378c, *in* Rodents in desert environments (I. Prakash and P. K. Ghosh, eds.). Junk, b.v., The Hague, The Netherlands, 624 pp.
VAN DEVENDER, T. R., AND W. G. SPAULDING. 1979. Development of vegetation and climate in the southwestern United States. Science, 204:701–710.
VAN ZINDEREN BAKKER, E. M. 1975. The origin and paleoenvironment of the Namib Desert zone. Journal of Biogeography, 2:65–73.
———. 1978. Quaternary vegetation changes in southern Africa. Pp. 131–143, *in* Biogeography and ecology of southern Africa (M. J. A. Werger and A. C. Van Bruggen, eds.). Junk, b.v., The Hague, The Netherlands, 1,439 pp.

WADIA, D. N. 1960. The post-glacial desiccation of Central Asia: evolution of the arid zone of Asia. National Institute of Science, India Monographs, 10: 1–25.

———. 1966. Geology of India. Third Edition. MacMillan, London.

WAHLERT, J. H. 1993. The fossil record. Pp. 1–37, *in* Biology of the Heteromyidae (H. H. Genoways and J. H. Brown, eds.). Special Publication, American Society of Mammalogists, 10:1–719.

WALTER, H. 1986. The Namib Desert. Pp. 245–282, *in* Ecosystems of the world. 12B. Hot deserts and arid shrublands (M. Evenari, I. Noy-Meir and D. W. Goodall, eds.). Elsevier, New York, 451 pp.

WALTER, H., AND E. O. BOX. 1983*a*. Overview of Eurasian continental deserts and semi-deserts. Pp. 3–7, *in* Ecosystems of the world. 5. Temperate deserts and semi-deserts (N. E. West, ed.). Elsevier, New York, 522 pp.

———. 1983*b*. Caspian lowland biome. Pp. 9–41, *in* Ecosystems of the world. 5. Temperate deserts and semi-deserts (N. E. West, ed.). Elsevier, New York, 522 pp.

———. 1983*c*. Semi-deserts and deserts of Central Kazakhstan. Pp. 43–78, *in* Ecosystems of the world. 5. Temperate deserts and semi-deserts (N. E. West, ed.). Elsevier, New York, 522 pp.

———. 1983*d*. Middle Asian deserts. Pp. 79–104, *in* Ecosystems of the world. 5. Temperate deserts and semi-deserts (N. E. West, ed.). Elsevier, New York, 522 pp.

———. 1983*e*. The Karakum Desert, an example of a well-studied Eu-biome. Pp. 105–159, *in* Ecosystems of the world. 5. Temperate deserts and semi-deserts (N. E. West, ed.). Elsevier, New York, 522 pp.

———. 1983*f*. The orobiomes of Middle Asia. Pp. 161–191, *in* Ecosystems of the world. 5. Temperate deserts and semi-deserts (N. E. West, ed.). Elsevier, New York, 522 pp.

———. 1983*g*. The Pamir—an ecologically well-studied high mountain desert biome. Pp. 237–269, *in* Ecosystems of the world. 5. Temperate deserts and semi-deserts (N. E. West, ed.). Elsevier, New York, 522 pp.

WALTER, H., E. O. BOX, AND W. HILBRIG. 1983. The deserts of Central Asia. Pp. 193–236, *in* Ecosystems of the world. 5. Temperate deserts and semi-deserts (N. E. West, ed.). Elsevier, New York, 522 pp.

WASSIF, R. 1960. Studies on the osteology of Egyptian jerboas. Bulletin of the Zoological Society of Egypt, 15:71–92.

WASSIF, K., AND S. SOLIMAN. 1979. The food of some wild rodents in the Western Desert of Egypt. Bulletin of the Zoological Society of Egypt, 29:43–51.

WATTS, C. H. 1977. The foods eaten by some Australian rodents (Muridae). Australian Wildlife Research, 4:151–157.

WATTS, C. H. S., AND H. J. ASLIN. 1981. The rodents of Australia. Angus and Robertson, Sydney, Australia, 321 pp.

WAUER, R. H., AND D. H. RISKIND (EDS.). 1977. Transactions of the symposium on the biological resources of the Chihuahuan Desert Region, United States and Mexico. U.S. Department of the Interior, National Park Service Transactions and Proceedings Series, 3:1–658.

WEBER, H. 1969. Zur naturlichen vegetations gliederung von Sudamerika. Pp. 475–518, *in* Biogeography and ecology in South America (E. J. Illies, H. Klinge, G. H. Schwabe, and H. Sioli, eds.). Junk, b.v., The Hague, The Netherlands.

WEBSTER, D. B. 1962. A function of the enlarged middle-ear cavities of the kangaroo rat, *Dipodomys*. Physiological Zoology, 35:248–255.

WERGER, M. J. A. 1978. The Karoo-Namib Region. Pp. 231–299, *in* Biogeography and ecology of southern Africa (M. J. A. Werger and A. C. Van Bruggen, eds.). Junk, b.v., The Hague, The Netherlands, 1,439 pp.

———. 1986. The Karoo and southern Kalahari. Pp. 283–359, *in* Ecosystems of the world. 12B. Hot deserts and arid shrublands (M. Evenari, I. Noy-Meir, and D. W. Goodall, eds.). Elsevier, New York, 451 pp.

WEST, N. E. (ED.). 1983*a*. Ecosystems of the world. 5. Temperate deserts and semi-deserts. Elsevier, New York, 522 pp.

———. 1983*b*. Overview of North American temperate deserts and semideserts. Pp. 321–330, *in* Ecosystems of the world. 5. Temperate deserts and semideserts (N. E. West, ed.). Elsevier, New York, 522 pp.

———. 1983*c*. Great Basin-Colorado Plateau sagebrush semi-desert. Pp. 331–349, *in* Ecosystems of the world. 5. Temperate deserts and semi-deserts (N. E. West, ed.). Elsevier, New York, 522 pp.

———. 1983*d*. Intermountain salt-desert shrubland. Pp. 375–397, *in* Ecosystems of the world. 5. Temperate deserts and semi-deserts (N. E. West, ed.). Elsevier, New York, 522 pp.

———. 1983*e*. Colorado Plateau-Mohavian blackbrush semi-desert. Pp. 399–411, *in* Ecosystems of the world. 5. Temperate deserts and semi-deserts (N. E. West, ed.). Elsevier, New York, 522 pp.

WETZEL, R. M., AND J. W. LOVETT. 1974. A collection of mammals from the Chaco of Paraguay. Occasional Papers, University of Connecticut, Biological Science Series, 2:203–216.

WILLIG, M. R., AND M. A. MARES. 1989. Mammals from the Caatinga: an updated list and summary of recent research. Revista Brasileira de Biologia, 49: 361-367.

WILLIAMS, D. F., AND M. A. MARES. 1978. A new genus and species of phyllotine rodent (Mammalia: Muridae) from northwestern Argentina. Annals of the Carnegie Museum, 47:192–221.

WILLIAMS, M. A. 1984. Cenozoic evolution of arid Australia. Pp. 59–78, *in* Arid Australia (H. G. Cogger and E. E. Cameron, eds.). Australian Museum, Sydney, 338 pp.

WILLIAMS, O. B., AND J. H. CALABY. 1985. The hot deserts of Australia. Pp. 269–312, *in* Ecosystems of the world. 12A. Hot deserts and arid shrublands (M. Evenari, I. Noy-Meir, and D. W. Goodall, eds.). Elsevier, New York, 365 pp.

WITHERS, P. C., W. R. SIEGFRIED, AND G. N. LOUW. 1980. Energetics and water relations of Namib Desert rodents. South African Journal of Zoology, 15: 131–137.

Wolff, R. G. 1984. A new Early Oligocene argyrolagid (Mammalia: Marsupialia) from Salla, Bolivia. Journal of Vertebrate Paleontology, 4:108–113.

Yalden, D. W., M. J. Largen, and D. Kock. 1976. Catalogue of the mammals of Ethiopia. Monitore Zoologico Italiano, N. S. Supplemento, 8:1–118.

APPENDIX 1.—*Genera of terrestrial small mammals of the arid and semiarid regions of the world. Included is information on diet, habitat, physiology, anatomy, general habits and North American niche equivalence (see footnotes for abbreviations utilized). Some genera (e.g.,* Holochilus, Sigmodon, Lestodelphis, Didelphis, Erethizon*) are generally found in mesic areas, but will occasionally occur in arid and semiarid habitats.*

	Diet[a]				Habitat[b]						Physiology[c]			Anatomy[d]					Habits[e]				North American niche types[f]								
Genus	G	H	I	O	S	R	Sc	Ri	F	H	WI	HS	H/E	BP	QS	QP	TT	BH	SS	B	N/D	S/C	U[g]	G	D	P	S	Y	N	I	F
North American arid and semiarid areas																															
O. Insectivora																															
F. Soricidae																															
Notiosorex			x		x		x		x							x					N/D	S								x	
O. Rodentia																															
F. Sciuridae																															
Ammospermophilus				x	x	x	x		x	x			x			x		x		x	D	S					x				
Eutamias		x				x	x		x	x			x			x				x	D	S					x				
Spermophilus				x	x	x	x		x				x			x		x		x	D	S					x				
F. Geomyidae																															
Pappogeomys		x			x		x		x							x		x		x	N/D	S									x
Thomomys		x			x		x		x							x		x		x	N/D	S									x
F. Heteromyidae																															
Chaetodipus	x				x	x	x	x	x	x	x		x		x			x		x	N	S				x					
Dipodomys	x				x		x		x		x			x			x	x		x	N	S			x						
Liomys	x				x		x		x						x			x		x	N	S				x					
Microdipodops	x				x		x		x		x		x	x				x		x	N	S			x						
Perognathus	x				x	x	x	x	x	x	x		x		x			x		x	N	S				x					
F. Cricetidae																															
Neotoma		x			x	x	x	x	x	x		x				x					N	S							x		
Onychomys			x		x		x		x							x					N	S								x	
Peromyscus				x	x	x	x	x	x	x						x					N	S						x			
Reithrodontomys				x	x		x	x	x							x					N	S						x			
Sigmodon		x			x		x	x	x							x					N	S		x							
F. Arvicolidae																															
Lagurus		x			x		x		x							x					N	S		x							
F. Erethizontidae																															
Erethizon		x			x	x	x	x	x	x						x					N	S		x							
O. Lagomorpha																															
F. Leporidae																															
Lepus		x			x		x	x	x							x					N/D	S		x							
Sylvilagus		x			x		x	x	x							x		x			N/D	S		x							
South American arid and semiarid areas																															
Atacama/Peruvian																															
O. Marsupialia																															
F. Didelphidae																															
Marmosa			x		x	x	x		x	x			?			x					N	S								x	

APPENDIX 1.—*Continued.*

	Diet[a]				Habitat[b]						Physiology[c]			Anatomy[d]					Habits[e]				North American niche types[f]								
Genus	G	H	I	O	S	R	Sc	Ri	F	H	WI	HS	H/E	BP	QS	QP	TT	BH	SS	B	N/D	S/C	U[g]	G	D	P	S	Y	N	I	F
O. Rodentia																															
F. Cricetidae																															
Akodon			x		x	x	x	x	x	x						x				x	N	S						x			
Oryzomys				x	x		x	x	x	x						x				x	N	S						x			
Phyllotis sp.				x	x	x	x	x	x	x						x				x	N	S						x			
Phyllotis (*gerbillus*)	?			?	x		x		x		x		?			x		x		x	N	S				?					
F. Octodontidae																															
Octodon		x			x	x	x		x	x		x				x		x		x	D	C					x		x		
Monte/Patagonia																															
O. Marsupialia																															
F. Argyrolagidae (+)	x	x			x		x		x					x			?	x							x						
F. Didelphidae																															
Didelphis				x			x	x	x							x					N	S	13								
Lestodelphis			x			x	x		x	x			?			x					N	S								x	
Marmosa			x		x	x	x	x	x	x			?			x					N	S								x	
O. Xenarthra																															
F. Dasypodidae																															
Chaetophractus			x	x	x		x	x	x							x		x		x	N/D	S	1								
Chlamyphorus			x		x		x		x				?			x		x		x	N/D	S								x	
Zaedyus			x	x	x		x	x	x							x		x		x	N/D	S	1								
O. Rodentia																															
F. Cricetidae																															
Akodon			x	x	x	x	x	x	x	x						x				x	N	S						x			
Andalgalomys				x	x		x	x	x		?					x		x		x	N	S				x		x			
Auliscomys				x		x	x		x	x						x				x	N	S						x			
Calomys				x	x		x	x	x		x					x				x	N	S						x			
Eligmodontia				x	x		x		x		x	x			x	x	x	x		x	N	S				x		x			
Euneomys		x		x		x		x		x						x				x	N	S						x			
Graomys				x	x	x	x	x	x	x						x		x		x	N	S						x			
Oryzomys				x	x	x	x	x	x	x						x				x	N	S						x			
Phyllotis				x		x	x			x						x				x	N	S						x			
Reithrodon		x			x		x		x							x				x	N	S		x							
F. Caviidae																															
Dolichotis		x			x		x	x	x			x				x		x			N/D	C		x							
Galea		x			x		x	x	x							x		x		x	D	C		x			x				
Microcavia		x			x		x	x	x							x		x		x	D	C		x			x				
F. Octodontidae																															
Octomys		x			x	x	x		x	x		x				x		x		x	N	S		x					x		
Tympanoctomys		x			x		x		x			x				x		x		x	N	?		x					x		
F. Ctenomyidae																															
Ctenomys		x			x		x	x	x							x		x		x	N/D	S		x							x

APPENDIX 1.—*Continued.*

Genus	Diet[a]				Habitat[b]						Physiology[c]			Anatomy[d]					Habits[e]				North American niche types[f]								
	G	H	I	O	S	R	Sc	Ri	F	H	WI	HS	H/E	BP	QS	QP	TT	BH	SS	B	N/D	S/C	U[g]	G	D	P	S	Y	N	I	F
Puna																															
O. Marsupialia																															
F. Didelphidae																															
Marmosa			x		x	x	x		x	x						x				x	N	S							x		
O. Xenarthra																															
F. Dasypodidae																															
Chaetophractus			x	x	x		x		x							x		x		x	N/D	S	1								
O. Rodentia																															
F. Cricetidae																															
Akodon				x	x	x	x		x	x						x				x	N/D	S						x			
Auliscomys				x		x	x			x						x				x	N	S						x			
Calomys				x	x		x		x							x				x	N	S						x			
Chinchillula		x				x	x		x	x						x				x	N	S							x		
Eligmodontia				x	x		x		x			x				x		x		x	N	S						x			
Neotomys				?	x		x	x	x							x				x	N/D	S	2	x							
Phyllotis				x		x	x			x						x				x	N	S						x			
Punomys		x			x	x	x		x							x				x	D	S	2	x							
F. Caviidae																															
Galea		x			x		x	x	x							x		x		x	D	C		x			x				
Microcavia		x			x		x	x	x							x		x		x	D	C		x			x				
F. Chinchillidae																															
Chinchilla		x				x	x			x						x		x		x	N	C		x					x		
Lagidium		x				x	x			x						x		x		x	D	C		x							
F. Octodontidae																															
Octodontomys		x										x				x		x		x	N	S		x					x		
F. Ctenomyidae																															
Ctenomys		x			x		x		x							x		x		x	N/D	S		x							x
F. Abrocomidae																															
Abrocoma		x				x	x			x						x		x		x	N	C		x					x		
Chaco																															
O. Marsupialia																															
F. Didelphidae																															
Didelphis				x			x	x	x							x					N	S	13								
Marmosa			x		x	x	x	x	x	x			?			x					N	S								x	
Monodelphis			x	x	x		x		x							x					N	S	3							x	
O. Xenarthra																															
F. Dasypodidae																															
Cabassous			x	x	x		x	x	x							x		x		x	N/D	S	1								
Chaetophractus			x	x	x		x	x	x							x		x		x	N/D	S	1								

APPENDIX 1.—*Continued.*

	Diet[a]				Habitat[b]						Physiology[c]			Anatomy[d]					Habits[e]				North American niche types[f]								
Genus	G	H	I	O	S	R	Sc	Ri	F	H	WI	HS	H/E	BP	QS	QP	TT	BH	SS	B	N/D	S/C	U[g]	G	D	P	S	Y	N	I	F
Chlamyphorus			x		x		x		x				?			x		x		x	N/D	S	1								
Dasypus			x	x	x		x	x	x							x				x	N/D	S	1								
Euphractus			x	x	x		x	x	x							x		x		x	N/D	S	1								
Tolypeutes			x	x	x		x		x							x		x		x	N/D	S	1								
O. Rodentia																															
F. Cricetidae																															
Akodon			x	x	x	x	x	x	x	x						x				x	N	S						x			
Andalgalomys				x	x		x	x	x							x				x	N	S						x			
Calomys				x	x		x	x	x							x				x	N	S						x			
Graomys				x	x	x	x	x	x	x						x		x		x	N	S						x			
Holochilus		x			x			x	x							x					N	?	4	x							
Oryzomys				x	x		x	x	x	x						x				x	N	S						x			
Pseudoryzomys				x	x			x	x							x					N	S						x			
F. Caviidae																															
Galea		x			x		x	x	x							x		x		x	D	C		x			x				
Microcavia		x			x		x	x	x							x		x		x	D	C		x			x				
Pediolagus		x			x		x	x	x							x		x		x	D	C		x							
F. Chinchillidae																															
Lagostomus		x			x		x	x	x							x		x		x	N	C	5	x							
F. Ctenomyidae																															
Ctenomys		x			x		x	x	x							x		x		x	N/D	S		x							x
O. Lagomorpha																															
F. Leporidae																															
Sylvilagus		x			x			x	x							x					D	S		x							
Caatinga																															
O. Marsupialia																															
F. Didephidae																															
Didelphis				x			x	x	x							x					N	S	13								
Marmosa			x		x	x	x	x	x	x			?			x					N	S								x	
Monodelphis			x	x		x	x		x	x						x					N	S	3								
O. Xenarthra																															
F. Dasypodidae																															
Dasypus			x	x	x		x	x	x							x				x	N/D	S	1								
Euphractus			x	x	x		x	x	x							x				x	N/D	S	1								
Tolypeutes			x	x	x		x	x	x							x				x	N/D	S	1								
O. Rodentia																															
F. Cricetidae																															
Bolomys				x	x		x	x	x							x				x	N	S						x			
Calomys				x	x		x	x	x							x				x	N	S						x			

APPENDIX 1.—*Continued.*

	Diet[a]				Habitat[b]						Physiology[c]			Anatomy[d]					Habits[e]				North American niche types[f]								
Genus	G	H	I	O	S	R	Sc	Ri	F	H	WI	HS	H/E	BP	QS	QP	TT	BH	SS	B	N/D	S/C	U[g]	G	D	P	S	Y	N	I	F
Oryzomys				x	x	x	x	x	x	x						x				x	N	S						x			
Wiedomys				x	x	x	x	x	x							x					N	S						x			
F. Caviidae																															
Galea		x			x		x	x	x							x					D	S		x			x				
Kerodon		x				x	x		x	x						x					D	C	6	x							
F. Echimyidae																															
Thrichomys		x				x	x		x	x						x					N/D			x					x		
O. Lagomorpha																															
F. Leporidae																															
Sylvilagus		x			x		x	x	x							x					N/D	S		x							
Australia																															
O. Monotremata																															
F. Tachyglossidae																															
Tachyglossus			x		x	x	x	x	x	x			x			x				x	N	S	1								
O. Marsupialia																															
F. Dasyuridae																															
Antechinomys			x		x		x		x						x		x				N	S								x	
Antechinus			x		x	x	x		x							x					N	S								x	
Dasycercus				x		x	x		x							x				x	N/D	S	3								
Dasyuroides				x	x		x		x							x	x			x	N	C								x	
Dasyurus			x		x		x		x							x					N	S	3								
Ningaui			x		x		x		x							x					N									x	
Phascogale			x		x		x		x							x	x				N									x	
Planigale			x		x	x	x	x	x							x					N									x	
Sminthopsis			x		x	x	x		x				x	x	x					x	N	C?								x	
F. Myrmecobiidae																															
Myrmecobius			x		x		x	x	x							x				x	D	S	1								
F. Notoryctidae																															
Notoryctes			x		x		x		x							x				x	N/D	S									x
F. Peramelidae																															
Chaeropus				x	x		x		x								x				N		7								
F. Thylacomyidae																															
Macrotis				x	x		x		x						x		x			x	N	S	3								
F. Macropodidae																															
Bettongia		x			x		x		x					x							N	C		x							
Caloprymnus		x			x		x		x					x							N	S		x							
Lagorchestes		x			x		x	x	x					x							N	S		x							
Lagostrophus		x			x		x	x	x					x							N	C		x							
Onychogalea		x			x		x	x	x					x							N	S		x							
Petrogale		x				x	x			x				x							N	C		x							

APPENDIX 1.—*Continued.*

	Diet[a]				Habitat[b]						Physiology[c]			Anatomy[d]					Habits[e]				North American niche types[f]								
Genus	G	H	I	O	S	R	Sc	Ri	F	H	WI	HS	H/E	BP	QS	QP	TT	BH	SS	B	N/D	S/C	U[g]	G	D	P	S	Y	N	I	F
O. Rodentia																															
F. Muridae																															
Leggadina		x			x		x		x		?	x				x				x	N	S						x			
Leporillus		x			x		x	x	x			x				x					N	C						x			
Notomys	x				x		x		x		x			x						x	N	C			x						
Pseudomys	x	x			x		x		x		x					x				x		C							x		
Rattus		x			x		x		x							x				x	N	S						x			
Zyzomys		x				x	x	x	x	x						x					N								x		
												Africa Sahara																			
O. Insectivora																															
F. Erinaceidae																															
Hemiechinus			x		x	x	x		x							x				x	N	S	1								
Paraechinus			x		x	x	x		x							x				x	N	S	1								
F. Soricidae																															
Crocidura			x		x	x	x	x	x	x						x					N/D	S								x	
O. Hyracoidea																															
F. Procaviidae																															
Heterohyrax		x				x	x			x						x					D	C	6	x							
Procavia		x				x	x			x						x					D	C	6	x							
O. Rodentia																															
F. Sciuridae																															
Atlantoxerus	x	x				x	x			x						x		x		x	D	S					x				
Xerus				x	x	x	x		x	x						x		x		x	D	C					x				
F. Cricetidae																															
Ammodillus	?				x		x		x									x								?					
Desmodilliscus	x				x		x		x						x			x	?	x	N					x					
Dipodillus	x				x	x	x	x	x	x	x				x			x	x	x	N	S				x					
Gerbillus	x				x		x		x		x				x			x	x	x	N	S				x					
Meriones	x				x	x	x	x	x	x	x					x		x	x	x	N	S			x	x					
Pachyuromys		x	x			x	x		x							x		x		x	N									x	
Psammomys		x			x		x		x			x				x		x		x	D	S		x				x		x	
Sekeetamys	x	x				x	x		x	x						x	x	x		x	N	S							x		
Tatera				x	x		x		x						x			x		x	N	C						x			
Taterillus				?	x		x		x							x		x		x	N		?								
F. Spalacidae																															
Spalax		x			x		x		x							x		x		x	N/D	S									x
F. Muridae																															
Acomys				x	x	x	x	x	x	x						x					N							x			

APPENDIX 1.—*Continued.*

Genus	Diet[a]				Habitat[b]						Physiology[c]			Anatomy[d]					Habits[e]				North American niche types[f]								
	G	H	I	O	S	R	Sc	Ri	F	H	WI	HS	H/E	BP	QS	QP	TT	BH	SS	B	N/D	S/C	U[g]	G	D	P	S	Y	N	I	F
Lemniscomys		x			x		x		x							x					N	S	2								
Mus				x	x		x		x							x				x	N	S						x			
F. Gliridae																															
Eliomys				x		x				x			H			x	x				N	C					x				
F. Dipodidae																															
Allactaga		x			x		x		x		x			x			x	x		x	N	S		x	x						
Jaculus	x	x			x		x		x		x?			x			x	x		x	N	S			x						
F. Hystricidae																															
Hystrix		x			x		x		x							x					N	S		x							
F. Ctenodactylidae																															
Ctenodactylus		x				x	x		x							x		x			D	C	6	x							
Massoutiera		x				x	x		x							x		x			D	C	6	x							
O. Lagomorpha																															
F. Leporidae																															
Lepus		x			x		x	x	x							x					D	S		x							
O. Macroscelidea																															
F. Macroscelididae																															
Elephantulus			x			x	x		x	x				x						x	D	C								x	
Namib/Karroo-Kalahari																															
O. Insectivora																															
F. Chrysochloridae																															
Eremitalpa			x		x				x							x				x	D	S								x	x
F. Soricidae																															
Crocidura			x		x	x	x	x	x	x						x					N/D	S								x	
O. Hyracoidea																															
F. Procaviidae																															
Procavia		x				x	x			x						x					D	C	6	x							
O. Rodentia																															
F. Sciuridae																															
Xerus				x	x		x	x	x							x		x		x	D	C					x				
F. Pedetidae																															
Pedetes		x			x		x		x					x			x	x		x	N	S		x							
F. Cricetidae																															
Dendromus				x	x		x	x	x							x				x	N	C						x			
Desmodillus	x				x		x	x			x					x		x	x	x	N	S				x					
Gerbillus	x				x		x		x		x				x			x	x	x	N	S				x					
Malacothrix		x			x			x	x							x				x	N	S	9								
Mystromys	x	x			x		x		x							x				x	N							x			
Parotomys		x			x		x		x							x		x		x	D	C		x					x		

APPENDIX 1.—*Continued.*

	Diet[a]				Habitat[b]						Physiology[c]			Anatomy[d]					Habits[e]				North American niche types[f]								
Genus	G	H	I	O	S	R	Sc	Ri	F	H	WI	HS	H/E	BP	QS	QP	TT	BH	SS	B	N/D	S/C	U[g]	G	D	P	S	Y	N	I	F
Petromyscus				x		x				x						x					N							x			
Saccostomus	x				x		x	x	x							x			x	x	N	S				x					
Steatomys				x	x		x		x				x			x				x	N							x			
Tatera				x	x		x		x						x			x		x	N	C						x			
F. Muridae																															
Aethomys		x				x	x									x				x	N	C						x			
Lemniscomys		x			x		x		x							x					N	S	2								
Mus				x	x	x	x			x						x				x	N	S						x			
Praomys				x	x		x	x								x				x	N	C						x			
Rhabdomys		x			x			x	x							x					D	S	2								
Thallomys		x			x		x	x	x							x					N	C	8								
Zelotomys	x		x	x	x		x		x							x					N	S								x	
F. Gliridae																															
Graphiurus				x		x				x			x			x					N	C					x				
F. Petromyidae																															
Petromus		x				x	x			x						x					D	S	6								
F. Bathyergidae																															
Bathyergus		x			x		x									x				x	N/D	C		x							x
Cryptomys		x			x		x									x				x	N/D			x							x
O. Lagomorpha																															
F. Leporidae																															
Lepus		x			x		x	x	x							x					D	S		x							
Pronolagus		x				x										x					D	?	6	x							
O. Macroscelidea																															
F. Macroscelididae																															
Elephantulus			x		x	x	x		x					x						x	D	S								x	
Macroscelides			x		x	x	x	x	x					x						x	N/D	S								x	
Madagascar																															
O. Insectivora																															
F. Tenrecidae																															
Echinops			x	x	x		x	x	x				E			x					N	S	1								
Geogale			x		x		x		x				E			x				x	N	S								x	
O. Rodentia																															
F. Cricetidae																															
Hypogeomys		x			x		x	x	x							x				x	N	S		x							
Macrotarsomys		x			x		x		x							x	x	x		x	N		10								
Middle East—Thar Desert arid zone																															
O. Insectivora																															
F. Erinaceidae																															

APPENDIX 1.—*Continued.*

Genus	Diet[a]				Habitat[b]						Physiology[c]			Anatomy[d]					Habits[e]				North American niche types[f]								
	G	H	I	O	S	R	Sc	Ri	F	H	WI	HS	H/E	BP	QS	QP	TT	BH	SS	B	N/D	S/C	U[g]	G	D	P	S	Y	N	I	F
Hemiechinus			x		x	x	x		x							x				x	N	S	1								
Paraechinus			x		x	x	x		x							x				x	N	S	1								
F. Soricidae																															
Crocidura			x		x	x	x	x	x	x						x					N/D	S								x	
Suncus[h]			x				x	x	x							x															
O. Pholidota																															
F. Manidae																															
Manis[h]			x		x		x		x							x				x		S	1								
O. Rodentia																															
F. Sciuridae																															
Spermophilopsis				x	x		x									x				x	D	C					x				
F. Cricetidae																															
Calomyscus[j]				x		x	x			x						x		x		x	N/D	S						x			
Cricetulus[i,j]	x			x		x				x			x			x		x	x	x	N/D	S				x		x			
Gerbillus	x				x		x		x		x				x			x	x	x	N	S				x					
Meriones	x			x[h]	x	x	x	x	x	x	x					x		x	x	x	N	S			x	x					
Rhombomys[h,j]		x			x		x		x	x						x		x		x	D	C	11								
Sekeetamys[i]	x	x				x	x		x	x						x	x	x		x	N	S							x		
Tatera				x	x		x		x						x			x		x	N	C						x			
F. Muridae																															
Acomys[i,j]				x	x	x	x	x	x	x						x					N							x			
Mus				x	x		x		x							x				x	N	S						x			
Rattus	x	x			x	x	x		x	x	x	x?				x				x	N	S				x		x			
F. Dipodidae																															
Alactagulus[j]		x			x		x		x				H	x			x	x		x	N	S	12								
Allactaga		x			x		x		x		x		H	x			x	x		x	N	S	12								
Jaculus	x	x			x		x		x		x		H	x			x	x		x	N	S			x						
Salpingotus[j]	?		?	x	x		x		x					x				x		x	N	C			x					x	
F. Hystricidae																															
Hystrix		x			x		x		x							x					N	S		x							
O. Lagomorpha																															
F. Leporidae																															
Lepus		x			x		x	x	x							x					D	S		x							
Turkestan Desert																															
O. Insectivora																															
F. Erinaceidae																															
Hemiechinus			x		x	x	x		x							x				x	N	S	1								
F. Soricidae																															
Crocidura			x		x	x	x	x	x	x						x					N/D	S								x	
Diplomesodon			x	x	x		x		x	x						x					N/D	S								x	

APPENDIX 1.—*Continued.*

Genus	Diet[a]				Habitat[b]						Physiology[c]			Anatomy[d]					Habits[e]				North American niche types[f]								
	G	H	I	O	S	R	Sc	Ri	F	H	WI	HS	H/E	BP	QS	QP	TT	BH	SS	B	N/D	S/C	U[g]	G	D	P	S	Y	N	I	F
O. Rodentia																															
F. Sciuridae																															
Spermophilopsis				x	x		x									x				x	D	C					x				
Spermophilus				x	x	x	x		x				x			x		x		x	D	S					x				
F. Cricetidae																															
Cricetulus	x			x		x				x			x			x		x	x	x	N/D	S				x		x			
Meriones	x				x	x	x	x	x	x	x					x		x		x	N	S		x	x						
Phodopus	x				x		x		x							x		x	x	x	N/D					x					
Rhombomys		x			x		x		x	x						x		x		x	D	C	11								
F. Arvicolidae																															
Lagurus		x			x		x		x	x						x				x	N/D		2, 11								
Microtus		x			x		x		x	x						x				x	N/D		2								
F. Seleviniidae																															
Selevinia			x		x		x		x				H			x				x	N	S								x	
F. Dipodidae																															
Allactaga		x			x		x		x		x		H	x			x	x		x	N	S	12								
Cardiocranius	x				x	x	x				x		H	x				x		x	N				x						
Dipus		x		x	x		x		x	x			H	x			x	x		x	N	S	12								
Eremodipus		x			x		x		x				H	x			x	x		x	N	C	12								
Jaculus		x			x		x		x		x		H	x			x	x		x	N		12								
Paradipus		x			x		x		x				H	x			x	x		x	N		12								
Pygeretmus		x	x		x		x		x				H	x				x		x	N		12								
Salpingotus	?		?	x	x		x		x					x				x		x	N	C			x					x	
Stylodipus		x			x		x		x				H	x				x		x	N		12								
F. Hystricidae																															
Hystrix		x			x		x		x							x					N	S		x							
O. Lagomorpha																															
F. Ochotonidae																															
Ochotona		x				x	x		x	x						x				x	D	C		x							
F. Leporidae																															
Lepus		x			x		x	x	x							x					D	S		x							
Deserts of China and Mongolia (excluding orobiomes)																															
O. Insectivora																															
F. Erinaceidae																															
Hemiechinus			x		x	x	x		x							x				x	N	S	1								
F. Soricidae																															
Crocidura			x		x	x	x	x	x	x						x					N/D	S								x	

APPENDIX 1.—*Continued.*

Genus	Diet[a]				Habitat[b]						Physiology[c]			Anatomy[d]					Habits[e]				North American niche types[f]								
	G	H	I	O	S	R	Sc	Ri	F	H	WI	HS	H/E	BP	QS	QP	TT	BH	SS	B	N/D	S/C	U[g]	G	D	P	S	Y	N	I	F
O. Rodentia																															
F. Sciuridae																															
Marmota		x				x				x			x			x				x	D	S		x							
Spermophilus				x	x	x	x		x				x			x		x		x	D	S					x				
F. Cricetidae																															
Brachiones	?	?					x		x							x				x	?	?		?		?			?		
Cricetulus	x			x		x				x			x			x		x	x	x	N/D	S				x		x			
Meriones	x				x	x	x	x	x	x	x					x		x		x	N	S		x	x						
Phodopus	x				x		x		x							x		x	x	x	N/D					x					
Rhombomys		x			x		x		x	x						x		x		x	D	C	11								
F. Arvicolidae																															
Ellobius		x					x	x	x							x				x	N/D	S									x
Lagurus		x			x		x		x	x						x				x	N/D		2, 11								
Microtus		x			x		x		x	x						x				x	N/D		2								
F. Dipodidae																															
Alactagulus		x			x		x		x		x		x	x			x	x		x	N	S	12								
Allactaga		x			x		x		x		x		x	x			x	x		x	N	S	12								
Cardiocranius	x				x	x	x				x		x	x				x		x	N				x						
Dipus		x		x	x		x		x	x			x	x			x	x		x	N	S	12								
Euchoreutes		?			x		x		x				x	x			x	x		x	?	?		?							
Salpingotus	?		?	x	x		x		x					x				x		x	N	C			x					x	
Stylodipus		x			x		x		x				x	x				x		x	N		12								
O. Lagomorpha																															
F. Ochotonidae																															
Ochotona		x				x	x		x	x						x				x	D	C		x							
F. Leporidae																															
Lepus		x			x		x	x	x							x					D	S		x							

[a] *Diet:* Granivore (G), Insectivore (I), Omnivore (O), Herbivore (H); [b] *Habitat:* Sandy or clayey (S), Rocky (R), Scrub (Sc), Riverine (Ri), Flats (F), Hills (H); [c] *Physiology:* Water independent (WI), Halophyte or succulent food (HS), Hibernation or estivation (H or E); [d] *Anatomy:* Bipedal (BP), quadrupedal saltation (QS), Quadrupedal (QP), Tufted tail (pronounced and frequently with a black and white pattern) (TT), Bullar hypertrophy (BH); [e] *Habits:* Seed storage (SS), Burrow existence (B), Nocturnal or diurnal (N or D), Solitary or colonial social organization (S or C); [f] *North American niche type equivalence:* Unique (U, see below), Grazing herbivore, frequently cursorial (G), Dipodomyine (D), Perognathine (P), Squirrel (S), Peromyscine (Y), Neotomyine (N), Insectivore (I), Fossorial herbivore (F); U[g] 1) armadillo—insectivorous or omnivorous burrower and equivalents; 2) microtine-like grazer; 3) small carnivorous species; 4) semiaquatic grazing herbivore; 5) prairie dog-like grazer; 6) saxicolous colonial herbivore; 7) cursorial omnivore; 8) arboreal folivore/herbivore; 9) tiny desert herbivore; 10) small fruit, seed, root and stem herbivore; 11) broadly herbivorous food hoarder; 12) bipedal broadly foraging herbivore; 13) broadly foraging omnivore; [h] Thar Desert species; [i] Arabian Desert; [j] Iranian Desert.

APPENDIX 2.—*Systematic listing and distribution in deserts and semideserts of the genera of small mammals considered in this report. Systematics generally follows Honacki et al. (1982).*

Order MONOTREMATA
  Family Tachyglossidae (Australia)
    Tachyglossus
Order MARSUPIALIA
  Family Argyrolagidae (†) (South American)
  Family Didelphidae
    Lestodelphis (Patagonia)
    Marmosa (South American)
    Monodelphis (Chaco, Caatinga)
    Didelphis (South American)
  Family Dasyuridae (Australia)
    Antechinus
    Antechinomys
    Dasyuroides
    Dasyurus
    Ningaui
    Phascogale
    Planigale
    Sminthopsis
  Family Myrmecobiidae (Australia)
    Myrmecobius
  Family Notoryctidae (Australia)
    Notoryctes
  Family Peramelidae (Australia)
    Chaeropus
  Family Thylacomyidae (Australia)
    Macrotis
  Family Macropodidae (Australia)
    Bettongia
    Caloprymnus
    Lagorchestes
    Lagostrophus
    Onychogalea
    Petrogale
Order XENARTHRA
  Family Dasypodidae
    Cabassous (Chaco)
    Chaetophractus (South American)
    Chlamyphorus (Monte, Chaco)
    Dasypus (Chaco, Caatinga)
    Euphractus (Chaco, Caatinga)
    Tolypeutes (Chaco, Caatinga)
    Zaedyus (Patagonia)
Order INSECTIVORA
  Family Tenrecidae (Madagascar)
    Echinops
    Geogale
  Family Chrysochloridae (Namib-Kalahari)
    Eremitalpa
  Family Erinaceidae
    Hemiechinus (Sahara, Middle East-Thar, Turkestan, Gobi)
    Paraechinus (Sahara, Middle East-Thar)

APPENDIX 2.—*Continued.*

  Family Soricidae
    Crocidura (Africa, Middle East-Thar, Turkestan, Gobi)
    Diplomesodon (Turkestan)
    Notiosorex (North American deserts)
    Suncus (Thar)
Order HYRACOIDEA
  Family Procaviidae
    Heterohyrax (Sahara)
    Procavia (Sahara, Namib/Karroo-Kalahari)
Order PHOLIDOTA
  Family Manidae (Thar)
    Manis
Order RODENTIA
  Family Sciuridae
    Ammospermophilus (North American)
    Atlantoxerus (Sahara)
    Eutamias (North American)
    Spermophilopsis (Middle East-Thar, Turkestan)
    Spermophilus (North American, Turkestan, Gobi)
    Xerus (African)
  Family Geomyidae (North American)
    Pappogeomys
    Thomomys
  Family Heteromyidae (North American)
    Chaetodipus
    Dipodomys
    Liomys
    Microdipodops
    Perognathus
  Family Pedetidae (Namib-Kalahari)
    Pedetes
  Family Cricetidae
    Akodon (South American)
    Ammodillus (Sahara)
    Andalgalomys (Monte)
    Auliscomys (Patagonia, Puna)
    Bolomys (Caatinga)
    Brachiones (Gobi)
    Calomys (South American)
    Calomyscus (Iranian)
    Chinchillula (Puna)
    Cricetulus (Middle East-Thar, Turkestan, Gobi)
    Dendromus (Namib-Kalahari)
    Desmodillus (Namib-Kalahari)
    Demodilliscus (Sahara)
    Dipodillus (Sahara)
    Eligmodontia (South American)
    Euneomys (Patagonia)
    Gerbillus (African, Middle East-Thar)
    Graomys (South American)
    Holochilus (Chaco)
    Hypogeomys (Madagascar)
    Macrotarsomys (Madagascar)
    Malacothrix (Namib-Kalahari)
    Meriones (African, Middle East-Thar, Turkestan, Gobi)

Appendix 2.—*Continued.*

Mystromys (Namib-Kalahari)
Neotoma (North American)
Neotomys (Puna)
Onychomys (North American)
Oryzomys (South American)
Pachyuromys (Sahara)
Parotomys (Namib-Kalahari)
Peromyscus (North American)
Petromyscus (Namib-Kalahari)
Phodopus (Turkestan, Gobi)
Phyllotis (South American)
Psammomys (Sahara)
Pseudoryzomys (Chaco)
Punomys (Puna)
Reithrodon (Patagonia)
Reithrodontomys (North American)
Rhombomys (Iranian, Thar, Turkestan, Gobi)
Saccostomus (Namib-Kalahari)
Sekeetamys (Namib-Kalahari)
Sigmodon (North American)
Steatomys (Namib-Kalahari)
Tatera (African, Middle East-Thar)
Taterillus (Sahara)
Wiedomys (Caatinga)
Family Spalacidae (Sahara)
Spalax
Family Arvicolidae
Ellobius (Gobi)
Lagurus (North American, Turkestan, Gobi)
Microtus (North American, Turkestan, Gobi)
Family Muridae
Acomys (Sahara, Middle East)
Aethomys (Namib-Kalahari)
Leggadina (Australia)
Lemniscomys (African)
Leporillus (Australia)
Mus (African, Middle East-Thar)
Notomys (Australia)
Praomys (Namib-Kalahari)
Pseudomys (Australia)
Rattus (Australia, Middle East-Thar)
Rhabdomys (Namib-Kalahari)
Thallomys (Namib-Kalahari)
Zelotomys (Namib-Kalahari)
Zyzomys (Australia)
Family Gliridae
Eliomys (Sahara)
Graphiurus (Namib-Kalahari)
Family Seleviniidae (Turkestan)
Selevinia
Family Dipodidae
Alactagulus (Iranian, Gobi)
Allactaga (Sahara, Middle East-Thar, Turkestan, Gobi)
Cardiocranius (Turkestan, Gobi)

Appendix 2.—*Continued.*

Dipus (Turkestan, Gobi)
Eremodipus (Turkestan)
Euchoreutes (Gobi)
Jaculus (Sahara, Middle East-Thar, Turkestan)
Paradipus (Turkestan)
Pygeretmus (Turkestan)
Salpingotus (Iranian, Turkestan, Gobi)
Stylodipus (Turkestan, Gobi)
Family Hystricidae (African, Middle East-Thar, Turkestan)
Hystrix
Family Erethizontidae (North American)
Erethizon
Family Caviidae
Dolichotis (Monte, Patagonia)
Galea (South American)
Kerodon (Caatinga)
Microcavia (South American)
Pediolagus (Chaco)
Family Chinchillidae
Chinchilla (Puna)
Lagidium (Puna)
Lagostomus (Chaco)
Family Octodontidae
Octodon (Chilean Atacama)
Octodontomys (Puna)
Octomys (Monte)
Tympanoctomys (Monte)
Family Ctenomyidae (South American)
Ctenomys
Family Abrocomidae (Puna)
Abrocoma
Family Echimyidae (Caatinga)
Thrichomys
Family Petromyidae (Namib-Kalahari)
Petromus
Family Bathyergidae (Namib-Kalahari)
Bathyergus
Cryptomys
Family Ctenodactylidae (Sahara)
Ctenodactylus
Massoutiera
Order LAGOMORPHA
Family Ochotonidae (Turkestan, China and Mongolia)
Ochotona
Family Leporidae
Lepus (North American, African, Middle East-Thar, Turkestan, Gobi)
Pronolagus (Namib-Kalahari)
Sylvilagus (North American, Chaco, Caatinga)
Order MACROSCELIDEA
Family Macroscelididae
Elephantulus (African)
Macroscelides (Namib-Kalahari)

INDEX *

* This index is primarily a guide to subject areas covered in the text. Topics and species briefly mentioned in the text or in abstracts, summaries, figures, references, or appendices may not be indexed. Also check CONTENTS for subject areas.